Lecture Notes in Mechanical Engineering

Lecture Notes in Mechanical Engineering (LNME) publishes the latest developments in Mechanical Engineering—quickly, informally and with high quality. Original research reported in proceedings and post-proceedings represents the core of LNME. Volumes published in LNME embrace all aspects, subfields and new challenges of mechanical engineering. Topics in the series include:

- Engineering Design
- Machinery and Machine Elements
- Mechanical Structures and Stress Analysis
- Automotive Engineering
- Engine Technology
- Aerospace Technology and Astronautics
- Nanotechnology and Microengineering
- Control, Robotics, Mechatronics
- MEMS
- Theoretical and Applied Mechanics
- Dynamical Systems, Control
- Fluid Mechanics
- Engineering Thermodynamics, Heat and Mass Transfer
- Manufacturing
- Precision Engineering, Instrumentation, Measurement
- Materials Engineering
- Tribology and Surface Technology

To submit a proposal or request further information, please contact the Springer Editor of your location:

China: Dr. Mengchu Huang at mengchu.huang@springer.com
India: Priya Vyas at priya.vyas@springer.com
Rest of Asia, Australia, New Zealand: Swati Meherishi at swati.meherishi@springer.com
All other countries: Dr. Leontina Di Cecco at Leontina.dicecco@springer.com

To submit a proposal for a monograph, please check our Springer Tracts in Mechanical Engineering at http://www.springer.com/series/11693 or contact Leontina.dicecco@springer.com

Indexed by SCOPUS. All books published in the series are submitted for consideration in Web of Science.

More information about this series at http://www.springer.com/series/11236

Banh Tien Long · Yun-Hae Kim ·
Kozo Ishizaki · Nguyen Duc Toan ·
Ivan A. Parinov · Ngoc Pi Vu
Editors

Proceedings of the 2nd Annual International Conference on Material, Machines and Methods for Sustainable Development (MMMS2020)

Springer

Editors
Banh Tien Long
Vietnam Association for Science Editing
Hanoi University of Science and Technology
Hanoi, Vietnam

Kozo Ishizaki
School of Mechanical Engineering
Nagaoka University of Technology
Nagaoka, Japan

Ivan A. Parinov
Vorovitch Research Institute of Mechanics
and Applied Mathematics
Southern Federal University
Rostov-on-Don, Russia

Yun-Hae Kim
Department of Mechanical Engineering
Korea Maritime and Ocean University
Busan, Korea (Republic of)

Nguyen Duc Toan
Hanoi University of Science and Technology
Hanoi, Vietnam

Ngoc Pi Vu
Faculty of Mechanical Engineering
Thai Nguyen University of Technology
Thai Nguyen, Vietnam

ISSN 2195-4356 ISSN 2195-4364 (electronic)
Lecture Notes in Mechanical Engineering
ISBN 978-3-030-69609-2 ISBN 978-3-030-69610-8 (eBook)
https://doi.org/10.1007/978-3-030-69610-8

This Springer imprint is published by the registered company Springer Nature Switzerland AG
The registered company address is: Gewerbestrasse 11, 6330 Cham, Switzerland

Preface

On behalf of the Organizing Committee, we are delighted to issue the volume of Lecture Notes in Mechanical Engineering which contains selected papers presented at the Second International Conference on Material, Machines and Methods for Sustainable Development (MMMS2020) from 12–15 November 2020 in Nha Trang, Vietnam. MMMS2020 has been focused on extending overall sustainable development engineering to engage more researchers from different areas and various fields since 2018. Following the new beginning in Nalod, Da Nang, Vietnam of 2018, this year's 2020 in Nha Trang would like to evolve into one of the leading academic conferences in the field of sustainable development engineering, underpinned by insightful lectures and presentations for even more attendees. MMMS2020 continues enabling researchers from all over the world to share their research and accomplishments as well as review the research currently being conducted by their peers.

During the conference, we were particularly pleased with the extended exchanges between the existing participants and Vietnamese researchers. The Vietnam Association of Science Editing (VASE) held joint workshops in collaboration with Hanoi University of Science and Technology (HUST), Nha Trang University (NTU), Phenikaa University (PKA), Thai Nguyen University (TNU), Kyungpook National University (KNU-Korea), Nagaoka University of Technology (NUT-Japan), National Kaohsiung University of Science and Technology (NKUST), VUSTA, SFU, Vietnam Federation of Mechanical Engineering Associations (VFMEA), and Council of Asian Science Editors (CASE). In addition, we could also promote networking between the participants through special sessions organized by VASE researchers specializing in "The method of improving writing papers publishing in the advanced international journals" presented by Prof. Hyusun Kim—President of Council of Asian Science Editors (CASE). Existing participants from VASE, various universities, institutions met and shared their intelligence and insights with newcomers, making the conference even more meaningful and fruitful.

The MMMS2020 Organizing Committee has received 315 papers of authors from many countries (Korea, Belgium, Taiwan, Vietnam, Russia, Malaysia, Thailand, etc.), and there were only 135 accepted full papers after blind peer review. The accepted rate to be published in the present book is about 42.8%.

The purpose of this volume of Lecture Notes in Mechanical Engineering is to explore and ensure an understanding of the critical aspects which contributed to sustainable development, especially materials, machines, and methods. From such understanding, the conference aims at an overall approach to assist policymakers, industries, and researchers at various levels to position the local technological development toward sustainable development in the global context, to assist decision making toward a greener approach especially for materials, machines, and methods.

We would like to express our gratitude to the members of the International Organizing and Academic Committees of the conference, for their hard work and advice which were helpful in maintaining the high level of the conference. We also thank Springer Nature Switzerland AG 2020 for publishing the selected papers in the Lecture Notes in Mechanical Engineering. We hope that all the papers will become a source of valuable information in the scientific work for academics, researchers, engineers, and students.

Best regards

Banh Tien Long
Yun-Hae Kim
Kozo Ishizaki
Nguyen Duc Toan
Ivan A. Parinov
Ngoc Pi Vu

Committees

International Organizing Committee

Banh Tien Long	Vietnam Association of Science Editing, Vietnam
Tran Quoc Khanh	Ministry of Science and Technology, Vietnam
Nguyen Van Phuc	Ministry of Education and Training, Vietnam
Dang Vu Minh	Vietnam Union of Science and Technology Associations, Vietnam
Do Canh Duong	General Department of Geology and Minerals of Vietnam, Vietnam
Dinh Van Phong	Hanoi University of Science and Technology, Vietnam
Do Huu Hao	Vietnam Federation of Mechanical Engineering Associations, Vietnam
Hoang Van Phong	Council for National Policy of Science and Technology, Vietnam
Kim Hyungsun	Council of Asian Science Editors, Korea
Kim Kyuman	Kyungpook National University, Korea
Kim Youngsuk	Kyungpook National University, Korea
Kozo Ishizaki	Nagaoka University of Technology, Japan
Parinov I. A.	Southern Federal University, Russia
Chang S.-H.	National Kaohsiung University of Science and Technology, Taiwan
Nghiem Vu Khai	Vietnam Union of Science and Technology Associations, Vietnam
Nobuhiko Azuma	Nagaoka University of Technology, Japan
Tran Van Nghia	Ministry of Science and Technology, Vietnam
Trang Si Trung	Nha Trang University, Vietnam
Vu Ngoc Pi	Thai Nguyen University, Vietnam
Pham Van Hung	Hanoi University of Science and Technology, Vietnam

Nguyen Dinh Tung	Research Institute of Agricultural Machinery, Vietnam
Tran Duc Quy	Hanoi University of Industry, Vietnam
Bui Trung Thanh	Hung Yen University of Technology and Education, Vietnam
Vu Ngoc Pi	Thai Nguyen University, Vietnam
Cao Hung Phi	Vinh Long University of Technical Education, Vietnam
Pham Xuan Duong	Vietnam Maritime University, Vietnam
Nguyen Quoc Hung	Vietnamese-German University, Vietnam
Huynh Trung Hai	Hanoi University of Science and Technology, Vietnam
Pham Van Hung	Hanoi University of Science and Technology, Vietnam
Luong Cong Nho	Vietnam Maritime University, Vietnam
Tran Van- Nghia	Ministry of Science and Technology, Vietnam
V. K. Gupta	PDPM-Indian Institute of Information Technology, Design and Manufacturing, Jabalpur, India
R. Hastijanti	University of 17 Agustus 1945 Surabaya, Indonesia
K. Hasyim	University of Darul Ulum, Jombang, Indonesia
H.-C. Huang	National Kaohsiung University of Science and Technology, Taiwan, ROC
M. A. Jani	University of 17 Agustus 1945 Surabaya, Indonesia
V. V. Kalinchuk	South Scientific Center of Russian Academy of Sciences, Russia
P. K. Kankar	Indian Institute of Technology Indore, India
C.-T. Lin	National Kaohsiung University of Science and Technology, Taiwan, ROC
M. B. Manuilov	Southern Federal University, Russia
A. V. Metelitsa	Southern Federal University, Russia
V. I. Minkin	Southern Federal University, Russia
E. L. Mukhanov	Southern Federal University, Russia
M. Musta'in	University of Darul Ulum, Jombang, Indonesia
T. Nakayama	Nagaoka University of Technology, Japan
M. Nanko	Nagaoka University of Technology, Japan
Nurmawati	University of 45 Surabaya, Indonesia
I. Prasetyawan	Lloyd Register Asia, Malaysia
M. A. Rahim	University of 17 Agustus 1945 Surabaya, Indonesia
Sajiyo	University of 17 Agustus 1945 Surabaya, Indonesia
T. Sekino	Osaka University, Japan

Local Organizing Committee

Tran Thi Hanh	Nha Trang University, Vietnam
Nguyen The Han	Nha Trang University, Vietnam
Dang Xuan Duong	Nha Trang University, Vietnam
Nguyen Van Tuong	Nha Trang University, Vietnam
Nguyen Thanh Tuan	Nha Trang University, Vietnam
Huynh Van Vu	Nha Trang University, Vietnam
Do Quang Thang	Nha Trang University, Vietnam
Tran Hung Tra	Nha Trang University, Vietnam
Duong Dinh Hao	Nha Trang University, Vietnam
Le Nguyen Anh Vu	Nha Trang University, Vietnam
Duong Tu Tien	Nha Trang University, Vietnam
Le Cong Lap	Nha Trang University, Vietnam
Le Chi Cong	Nha Trang University, Vietnam
Nguyen Ngoc-Chuong	Vietnam Federation of Mechanical Engineering Associations, Vietnam
Vu Ngoc Hai	Phenikaa University, Vietnam
Pham Huong Tra	Phenikaa University, Vietnam
Chu Anh My	Le Quy Don Technical University, Vietnam
Le Chi Hieu	Greenwich University, UK

Scientific Committee

Banh Tien Long	Hanoi University of Science and Technology, Vietnam
Bui Van Ga	Da Nang University, Vietnam
Le Anh Tuan	Hanoi University of Science and Technology, Vietnam
Tran Van Nam	Da Nang University, Vietnam
Nguyen Thanh Nam	Ho Chi Minh City University of Technology (HCMUT), Vietnam
Nguyen Hay	Nong Lam University, Vietnam
Chu Van Dat	Military Technical Institute, Vietnam
Nguyen Lac Hong	Le Quy Don University, Vietnam
Luong Cong Nho	Vietnam Maritime University, Vietnam
Trang Si Trung	Nha Trang University, Vietnam
Nguyen Duc Toan	Hanoi University of Science and Technology, Vietnam
Chinda Charoenphonphanich	King Mongkut's Institute of Technology Ladkrabang, Thailand
Dao Hong Bach	Hanoi University of Science and Technology, Vietnam
Tran Hung Tra	Nha Trang University, Vietnam
Dang Xuan Phuong	Nha Trang University, Vietnam

Bui Anh Hoa	Hanoi University of Science and Technology, Vietnam
Dinh Van Hai	Hanoi University of Science and Technology, Vietnam
Truong Hoanh Son	Hanoi University of Science and Technology, Vietnam
Vu Toan Thang	Hanoi University of Science and Technology, Vietnam
Hoang Hong Hai	Hanoi University of Science and Technology, Vietnam
Nguyen Dinh Tung	Research Institute of Agricultural Machinery, RIAM, Vietnam
Truong Viet Anh	Hanoi University of Science and Technology, Vietnam
Nguyen Phong Dien	Hanoi University of Science and Technology, Vietnam
Nguyen Ngoc Trung	Hanoi University of Science and Technology, Vietnam
Nguyen Huy Bich	Nong Lam University, Vietnam
Erik Bohez	Asian Institute of Technology, Thailand
Gyu Man Kim	Kyungpook National University, Korea
Jung-Il Jin	Korea University, Korea
Le Anh Tuan	Hanoi University of Science and Technology, Vietnam
Le Quang	Hanoi University of Science and Technology, Vietnam
Nobuhiko Azuma	Nagaoka University of Technology, Japan
Nobumasa Sekishita	Toyohashi University of Technology, Japan
Nghiem Trung Dung	Hanoi University of Science and Technology, Vietnam
Nguyen Thanh Nam	Ho Chi Minh City University of Technology (HCMUT), Vietnam
Nguyen Hay	Nong Lam University, Vietnam
Nguyen Dang Binh	Viet Bac University, Vietnam
Nguyen Hong Lanh	Military Technical Institute, Vietnam
Nguyen The Mich	Hanoi University of Science and Technology, Vietnam
Nguyen Thi Thu Hien	Hanoi University of Science and Technology, Vietnam
Nguyen Viet Dung	Hanoi University of Science and Technology, Vietnam
Pham Minh Tuan	Hanoi University of Science and Technology, Vietnam
Pham Van Hung	Hanoi University of Science and Technology, Vietnam

Contact

Email: mmms@hust.edu.vn
Website: mmms.hust.edu.vn
Facebook: facebook.com/mmms.conference

Contact Person

1. Dr. Hoang Van Quy
   Hai Phong University
   Email: Quyhv@dhhp.edu.vn
2. Dr. Mac Thi Thoa (Mobile: 0985288366)
   Hanoi University of Science and Technology, Vietnam
   Email: Thoa.MacThi@hust.edu.vn

Science Secretary

Dr. Pham Duc An (Mobile: 0971561418)
Hanoi University of Science and Technology, Vietnam
Email: An.PhamDuc@hust.edu.vn

Local Organizer

1. Assoc. Prof. Nguyen Thi Hong Minh
   Hanoi University of Science and Technology, Vietnam
   Email: minh.nguyenthihong@hust.edu.vn

2. Assoc. Prof. Nguyen Duc Toan
   Hanoi University of Science and Technology, Vietnam
   Email: Toan.NguyenDuc@hust.edu.vn

Contents

Advance Materials Toward Sustainability

Improved Fatigue Strength of a Dissimilar Metal T-Lap Joint Fabricated by Controlled Double-Pass Friction Stir Welding

Masakazu Okazaki[1(✉)], Hao Dinh Duong[1], Satoshi Hirano[2], and Tra Hung Tran[3]

[1] Nagaoka University of Technology, 1603-1 Kamitomioka-machi, Nagaoka-shi, Niigata 940-2188, Japan
okazaki@mech.nagaokaut.ac.jp
[2] Hitachi Research Lab, Hitachi 319-1292, Japan
[3] Department of Engineering Mechanics, Nha Trang University, 02 Nguyen Dinh Chieu Street, Nha Trang, Vietnam

Abstract. Dissimilar metal welding can answer the social requirements to achieve higher efficiency and performances of machines and components, where Friction Stir Welding (FSW) is expected as one of industrial solutions to fabricate them. However, the strength of FSWed T-lap joint, especially for the fatigue failures, has not been satisfactory so far, due to formation of some undesirable defects in the joint; e.g. tunnel, kissing bond, zigzag line, bonding line, oxide line defects, *etc.*, those directly degraded the mechanical properties of the joint. In this work the strength of the dissimilar metal FSWed T-lap joint between AA7075 and AA5083 were optimized by controlling the FSW conditions. The fatigue strength was also evaluated. Special attention were paid to minimize the defect size and the interface morphologies to improve the strength, together with some reasonable fracture mechanics backgrounds.

1 Introduction

Fusion welding technology has been widely applied for many various materials. Some methods are known as arc welding, gas welding, radiant energy welding, and resistance welding, *etc*. In these methods, the materials are heated to a melting temperature before they are diffused each other [1–5]. There are some cases where these are generally difficult to be applied for the light metal alloys. For example, the joint of titanium alloys produced conventional welding method frequently occur porosity, hot cracking, distortion, and high residual stress [1–5]. Specially, the solidification cracking of aluminum alloys of 7xxx and 2xxx series are more susceptible than that of other series. Friction Stir Welding (FSW) that was invented by The Welding Institute (TWI) in 1991 is one of solutions [1]. This technique uses frictional heat induced by rotating tool to weld materials. It means that the welding temperature during FSW process is lower than the melting temperature of base materials. The FSW is expected to be very effective to produce a large amount of various industries such as automotive, shipbuilding, high-speed train manufacturing, aerospace, and railway. Here the T-joints

B. T. Long et al. (Eds.): MMMS 2020, LNME, pp. 3–11, 2021.
https://doi.org/10.1007/978-3-030-69610-8_1

composed of stringer and skin plates is one of the common junctions. The FSW must be also useful to produce the joints between dissimilar metals with different physical and mechanical properties.

The main goal of this work is to investigate the interface morphology and fracture behavior of the FSWed dissimilar metal T-lap joints between AA7075 and AA5083 aluminum alloys, those make platform for the improvement of the mechanical properties of the joints. Special attention was paid to the FSW conditions to improve the strength associating with enough rupture ductility, and to provide some academic backgrounds on how and why the joint strength was affected by many FSW process parameters.

2 Classification of Defects in the T-Joint

The FSWed T-Lap joint was fabricated according to a process illustrated in Fig. 1. Many experimental variables were tried in this work, those are rotational speed, tool offset, welding speed and pin depth. More or less, some types of defects were often found in the joint, as classified in Fig. 1(b). Here, the interface locates outside stirred zone is referred to as kissing bond defects that were also mentioned in literatures [1–3]. The interface locates inside the SZ with hook geometry is referred to as a hook defects that are often observed in the FSWed lap-joint. The other interface located inside the SZ with oxide film is referred to as bonding line defects.

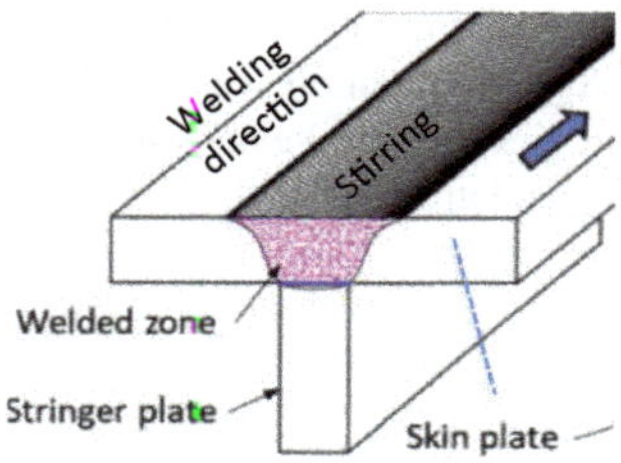

(a) T-lap joint by FSW

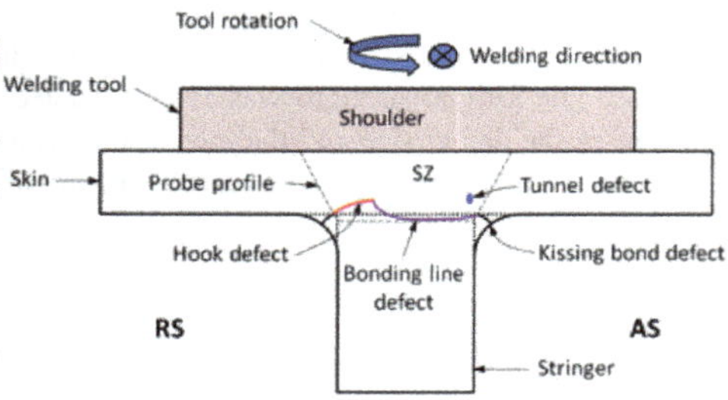

(b) Classification of typical defects

Fig. 1. Fabrication of T-lap joint by FSW

3 Preparation of T-Lap Joints

3.1 Fabrication by Single Pass FSW

In order to produce advanced panel structure with main T-joint configuration, two dissimilar metals were used in this work; aluminum alloys of 7075-T651 (denoted as AA7075) for the stringer and 5083-H116 (denoted as 5083) for the skin plate with the dimensions of 300 × 50 × 8.2 mm and 300 × 150 × 3.0 mm, respectively. The welding speed was arranged from 50 to 200 mm/min and the rotational speed was kept constant at 400 rpm.

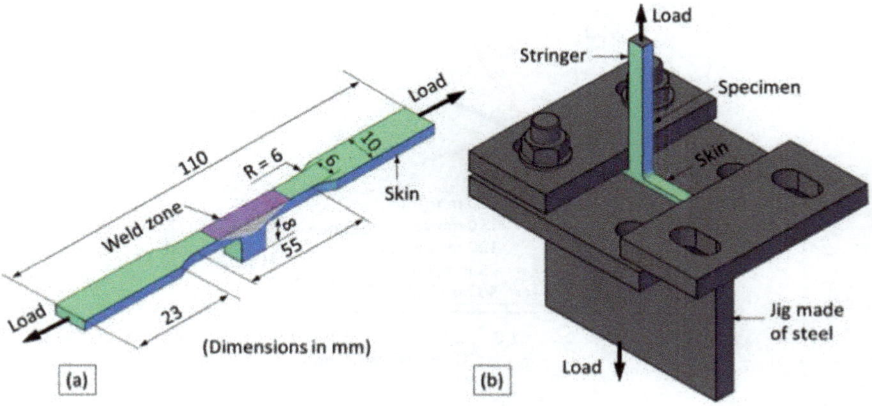

Fig. 2. Two kinds of tests to evaluate the FSW properties.

The performance of the joint was evaluated by two types of tests by extracting small samples from the T joint; one is the tensile test along the skin part (denoted as "skin test"), and the other is along the stringer part (denoted as "stringer test": see Fig. 2.

The effect of welding speed on the formation of four types of defects in T-lap joints is quantitatively summarized in Fig. 3 and Fig. 1(b), as a representative result showing the influence of FSW variables. Here, the defect size was directly measured based on oxide film on the cross-section of the specimens by means of the optical microscope. In addition, the difference in the bonding strength of defect interfaces would lead to the dissimilar fracture surface. The size of defects can be determined based on these differences by means of the SEM. Figure 3 showed that the welding speed significantly impacted the defect formations. The tunnel and bonding line defects were drastically

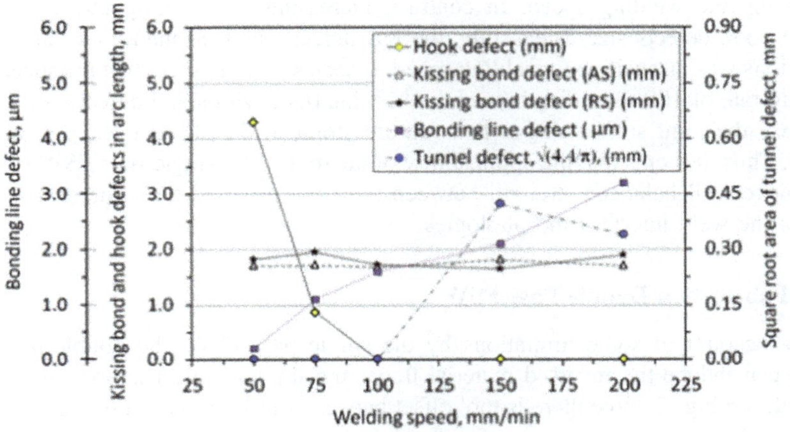

Fig. 3. Influence of welding speed on the formation of each type of defect

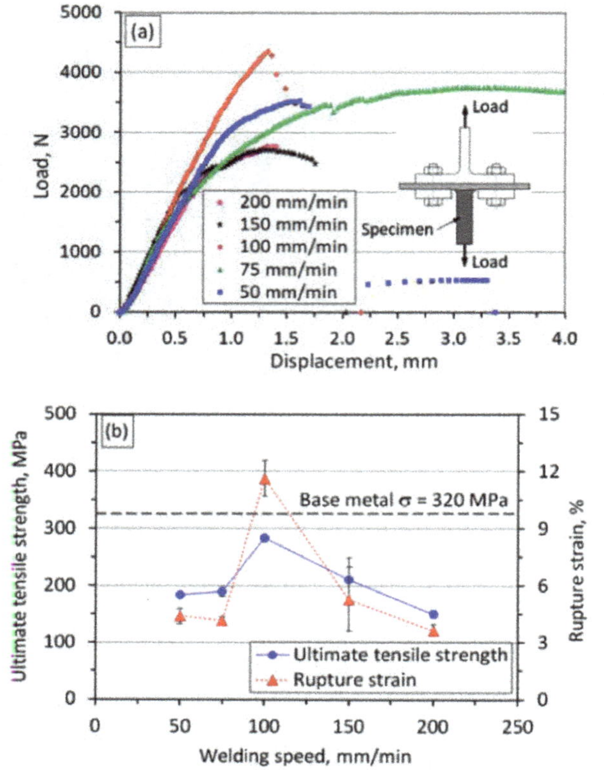

Fig. 4. Stringer test results. (a) Load-displacement curves and (b) maximum load and displacement at maximum load.

changed by controlling welding speed. These types of defects might be minimized by decreasing the welding speed. In contrast, increasing the welding rate is useful to reduce hook defects size. Meanshile, the KB defect which formed under all welding conditions was insensitive to welding speed. It seems that the specimen produced at the welding rate of 100 mm/min showed the best interface with small defect size. Even in this case the joint suffered from low rupture strain in the stringer test, as shown in Fig. 4. Thus, it seems that the applicability of the traditional single-pass FSW is limited to achieve well balanced strength between the skin tests and the stringer test and to control the weld interface morphologies.

3.2 Fabrication Double Pass FSW

Taking account of some limitations by the single pass FSW, the double-pass FSW which can induce the reversed material flows and the tool offset is applied as a new method; see Fig. 5. Note there is tool offset between the 1st pass and the 2nd pass, and the second pass has a role to minimize the defect size which have been nucleated by the 1st pass welding process. These effects were so effective not only in the skin tests but

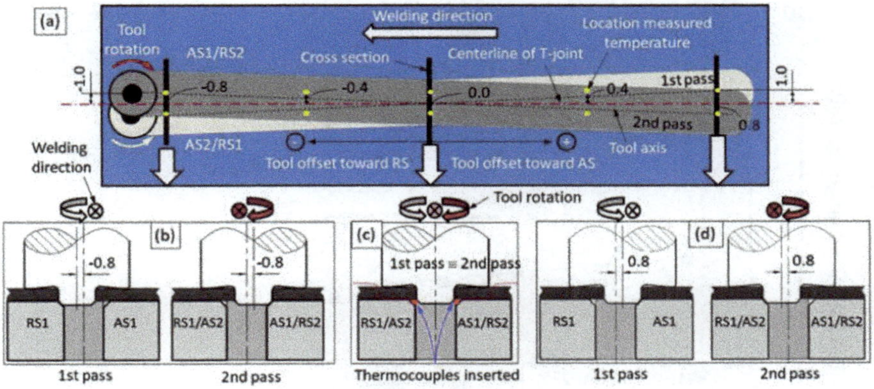

Fig. 5. Illustration of double FSW associating with tool offset.

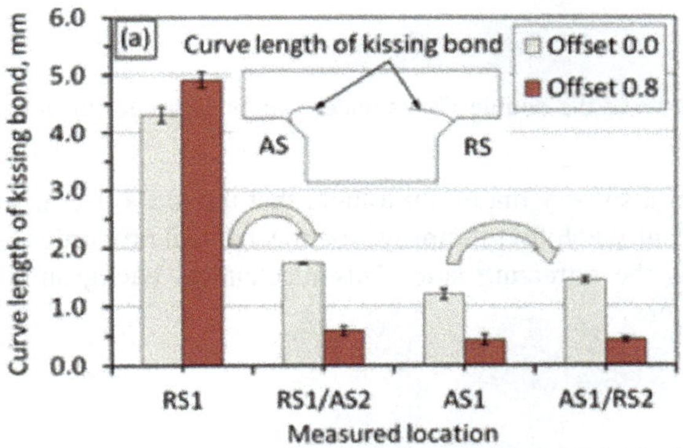

(a) Change in interface morphology

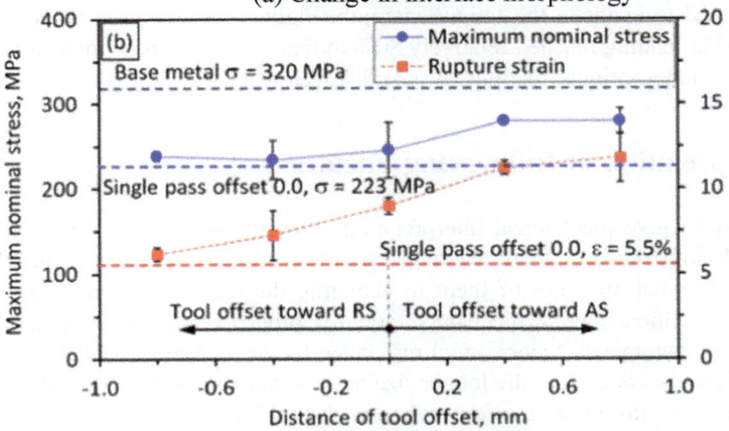

(b) Change of strength in the skin test

Fig. 6. Effectiveness of the double FSWs associating with tool offset.

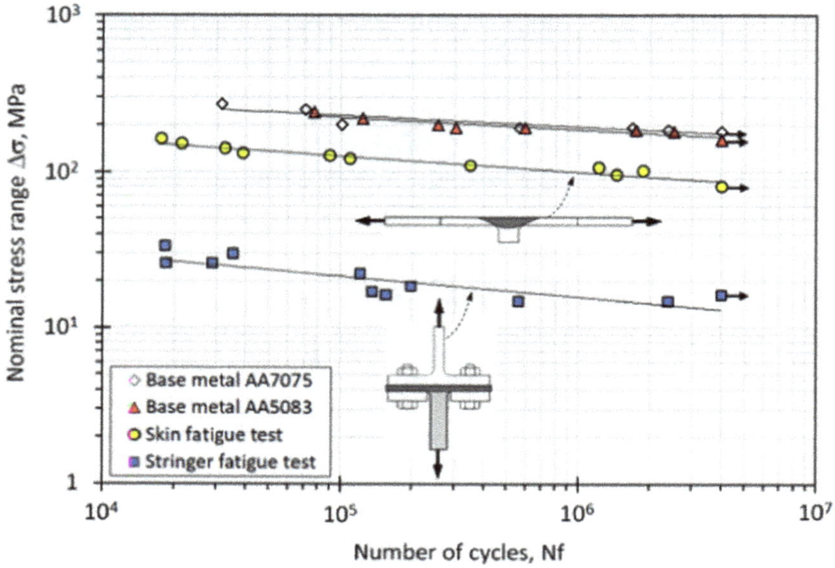

Fig. 7. Fatigue strengths of the double FSWs under skin and stringer loadings.

also in the stringer test. It is also very important to note that the tool offset towards the advancing side was significantly helpful to improve the mechanical properties as shown in Fig. 6 than that towards the retreating side. This mechanical background will be discussed in the next chapter.

3.3 Fatigue Strength

Following some successful results presented by the previous chapter, the FSW joint was fabricated to evaluate the fatigue strength. Figure 7 reveals, the fatigue strength under the skin loading reached relatively 80% to that of the two base metals. However, the strength under stringer loadings was still lower.

4 Interpretation of Influencing Factors

It is worthy to give mechanical interpretations why the strength of FSW joint were significantly influenced by many welding variables presented in the previous chapters, together with what are roles of them in changing the mechanical properties. In this section, a simplified fracture mechanics model is proposed from this aspect by introducing a new parameter, "geometrical resistance factor of defects (GRFD)".

As indicated in Sect. 3.1, the longer size defects were nucleated at the RS side with concave angle to the prior interface, whereas at the AS side they were with convex angle. The fracture predominantly began in the former site. As shown in Sect. 3.2, the double pass welding dispersed the defects into the AS and RT side, which made the

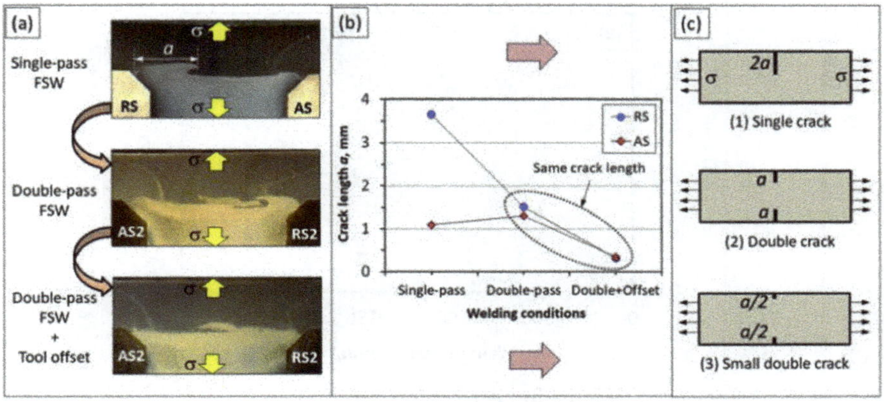

Fig. 8. A simplified model explaining the effects of various welding conditions on the strength of FSWed T-lap joints.

joint strength higher, compared with the single FSW. These behavior can be rationalized based on fracture mechanics as follows; see a simplified model in Fig. 8.

In fracture of structures, the stress intensity factor (SIF) play plays an essential role. Note in Fig. 8 that the magnitude of SIF of double edged crack is lower than that of single edge crack under the same defect size. More precisely, when the kinked edge crack make an inclined angle to loading axis (Fig. 9(a)), which was often seen in the single FSW T-joint at the corner, the SIF is given by [6, 7]

$$K_I = F \times Y_I \sigma \sqrt{\pi a} \quad Y_I = 1.12^2 (\sin\theta)^2 / \sqrt{\sin\theta} \tag{1}$$

$$K_{II} = F \times Y_{II} \sigma \sqrt{\pi a} \quad Y_{II} = \frac{1}{1.12^2} \sin\theta \cos\theta / \sqrt{4} \sin 2\theta \tag{2}$$

In this case the crack is under the mixed mode, for which an equivalent parameters, Y_{eq} is effective upon considering the fracture [8, 9].

$$Y_{eq} = \sqrt{(Y_I)^2 + (Y_{II})^2} \tag{3}$$

In this work, the parameter, $1/(F \times Y - {}_I\sqrt{a})$ and $1/(F \times Y - {}_{II}\sqrt{a})$ are called by "Geometrical resistance factor of defects (GRFD)". On the other hand, the SIF of the double edged crack which was formed under the double FSW process, the boundary correction factor F appearing in Eq. (4) is replaced by [7],

$$F = \left[1 + 0.122\left(cos\frac{\pi\xi}{2}\right)^4\right] \times \sqrt{\frac{2}{\pi\xi} tan\frac{\pi\xi}{2}} \tag{4}$$

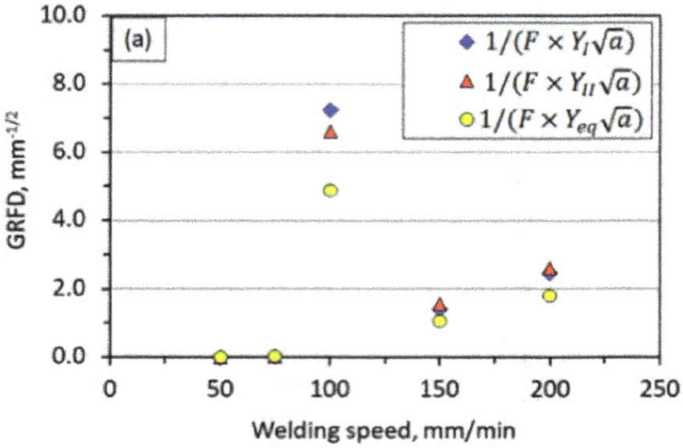

(a) Change in GRFD by welding speed.

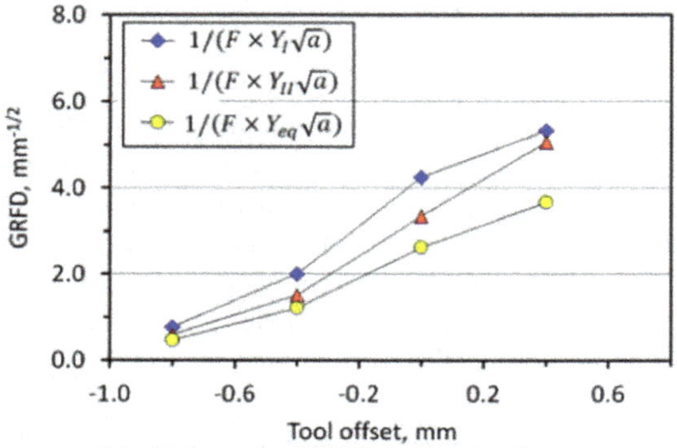

(b) Change in GRFD by tool offset.

Fig. 9. Change in GRFD by changing FSW conditions.

The effects of welding speed and probe length on the GRFD under skin test are shown in Fig. 9, respectively. Note that the GRFD showed the peak value at the welding speed of 100 mm/min, explaining well the experimental results that the strength showed maximum under this condition. The similar good agreement is also found in Fig. 9(b). Figure 8 directly explains the difference in the strength between the single and double FSW joints. It is also important to note that the change in FSW conditions changes the interface geometry resulting in the change in inclined angle of defect, $\boldsymbol{\theta}$ □□ Eqs. (1) and (2).□

5 Conclusion

The effects of FSW parameters; welding speed, tool offset, pin depth, and application of double FSW, on the mechanical properties were explored by the skin test, stringer test and the fatigue tests. It was found that the performance of dissimilar T-joint might be surely optimized by controlling these factors. Based on fracture mechanics a new concept has been proposed to make the analysis of the dissimilar FSWed T-lap joints, via introducing the "GRFD" parameter. This concept could provide reasonable explanations to the experimental works.

Acknowledgement. A part of this work is based on results obtained from a future pioneering project commissioned by the New Energy and Industrial Technology Development Organization (NEDO).

References

1. Khan, B.H.: A Framework for Web-Based Learning. Educational Technology Publications, Englewood Cliffs (2000)
2. Liaw, S.-S., Huang, H.-M., Chen, G.-D.: Surveying instructor and learner attitudes toward e-learning. Comput. Educ. **49**, 1066–1080 (2007)
3. Wang, Y.-S., Wang, H.-Y., Shee, D.Y.: Measuring e-learning systems success in an organizational context: scale development and validation. Comput. Hum. Behav. **23**, 1792–1808 (2007)
4. Chikh, A., Berkani, L.: Communities of practice of e-learning, an innovative learning space for e-learning actors. Procedia Soc. Behav. Sci. **2**, 5022–5027 (2010)
5. Sultan, N.: Cloud computing for education: a new dawn? Int. J. Inf. Manage. **30**, 109–116 (2010)
6. Arshad, J., Townend, P., Jie, X.: A novel intrusion severity analysis approach for Clouds. Future Gener. Comput. Syst. (2011). https://doi.org/10.1016/j.future.2011.08.009
7. Loganayagi, B., Sujatha, S.: Enhanced cloud security by combining virtualization and policy monitoring techniques. Procedia Eng. **30**, 654–661 (2012)

Machine Learning in Design of Optimization of MR Actuators: A New View of Solution

Xuan Phu Do(✉)

Department of Mechatronics and Sensor Systems Technology, Vietnamese-German University, Thu Dau Mot city, Binh Duong, Vietnam
phu.dx@vgu.edu.vn

Abstract. In this study, a new progress of optimization of smart structure is carried out. The application of machine learning through neural networks brings a new view in design. The data using for optimization is collected from the simulation results, the real values of the closed structure with the proposed model. The combination of these data and the outstanding property of the neural networks helps improve the obtained variables, and then performance of the proposed structure is also improved. The proposed structure is built based on cantilever beam model. Smart material is covered outside of the beam based on magnetorheological elastomer (MRE in short). This model is used for evaluate the performance of the neural networks in optimization. Results of the progress show that this method can use in case of no specialize software, and satisfy the initial requirements.

Keywords: Magneto-rheological elastomer (MRE) · Machine learning · MR actuator · Neural networks optimization · Valve optimization · Block-up phenomenon

1 Introduction

MR actuators are currently applied in many devices with high precision and large impact related force, torque, or moment. These devices can be seen in advanced automobile such as BMW, Mercedes, Lamborghini, Audi,…or in military devices such as mount of engine in ship. However, a main state in design of MR actuator is optimization. The conventional method for optimize is use ANSYS software with ADPL module/or ANSYS workbench. This method was shown in [1–15]: brake [1], mount [2–7], damper [5, 9, 10], clutch [8], seal [12], valve [11], skin [13], and control [14, 15]. The cost function of the optimization is designed following the maximal dynamic parameter such as force or torque. In this function, main variables of the system are included to improve performance of the process. However, the operation of the manufactured devices following results of the optimization may have vital problem related efficiency because the real impact is not support in the software. Recently, the era of machine learning in our life has open a new window for optimization. This trend is also effect to the optimization of MR actuators. The features studies for this development are expressed in [16–19]. Hence, this study presents a new view of optimization of MR devices using machine learning (neural networks, deep learning).

B. T. Long et al. (Eds.): MMMS 2020, LNME, pp. 12–19, 2021.
https://doi.org/10.1007/978-3-030-69610-8_2

2 Machine Learning for Optimization

Currently, there are two tools for optimization using machine learning. The first is neural networks, and the second is deep learning. The difference of these methods is the number of layers. The increase of layer helps to improve the performance/or precision of results.

2.1 Neural Networks

The conventional model of neural networks model includes three elements such as input layer, hidden layer, and output layer. The hidden layer is derived as many sub-layers which the number of layer belongs to the requirement standard of output. However, the time of calculation of the progress will increase if the sub-layers increases. The neural networks can be improved its performance by using two modified model such as neural fuzzy model and fuzzy neural model as shown in Fig. 1. These modification are new group of machine learning as fuzzy machine learning model. In fact, the calculation of the new group is also used studied models which has been developed, but the data for the group is bigger than the former data. The data can be called as big data.

2.2 Deep Learning

Deep learning model is developed based on neural networks model. The main layers are also similar neural networks including input layer, hidden layer, and output layer. The output layer includes many output variables, and the number of these variables is larger than the conventional neural networks. Main difference point of neural networks and deep learning is the chosen data for improve performance. The chosen data may be as a small group /or a big group. These groups are also effect to the function for every node of layer. Normally, the big data is used for deep learning, hence the function for node is softmax function, and the updated method is stochastic gradient descent. It is remarked that the batch gradient descent is one of choices for the model, but time of calculation is long.

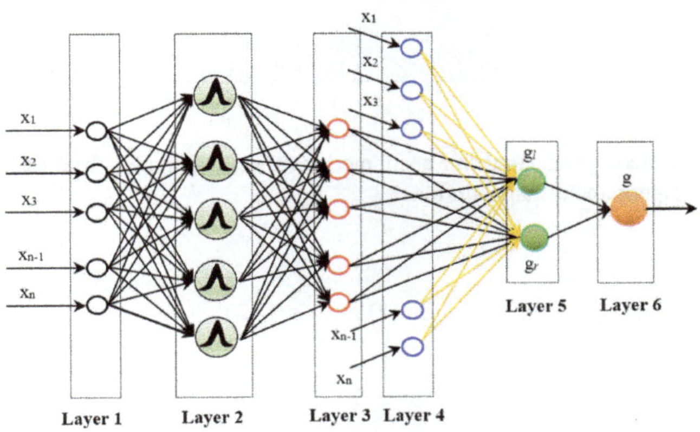

Fig. 1. Neural networks based on interval type 2 fuzzy model.

3 Application for a Magneto-Rheological Sensor

3.1 Magneto-Rheological Cantilever Beam Model

Model of magnetorheological cantilever beam is shown in Fig. 2. In Fig. 2a, the general view is shown which includes a cover of magnetorheological elastomer (MRE) as shown in Fig. 2b. The core is a magnetic beam which can be changed its stiffness by using a permanent magnetic as shown in Fig. 2c. This model is a basis structure for smart sensor using MR. The variation of magnetic field will change the stiffness of the beam, and then dynamic variables such as force, displacement, velocity, acceleration are also changed. Based on this variation, the properties are used for measure. The simplified model is shown in Fig. 3 with all parameters related the equation of two cases: large and small deflections.

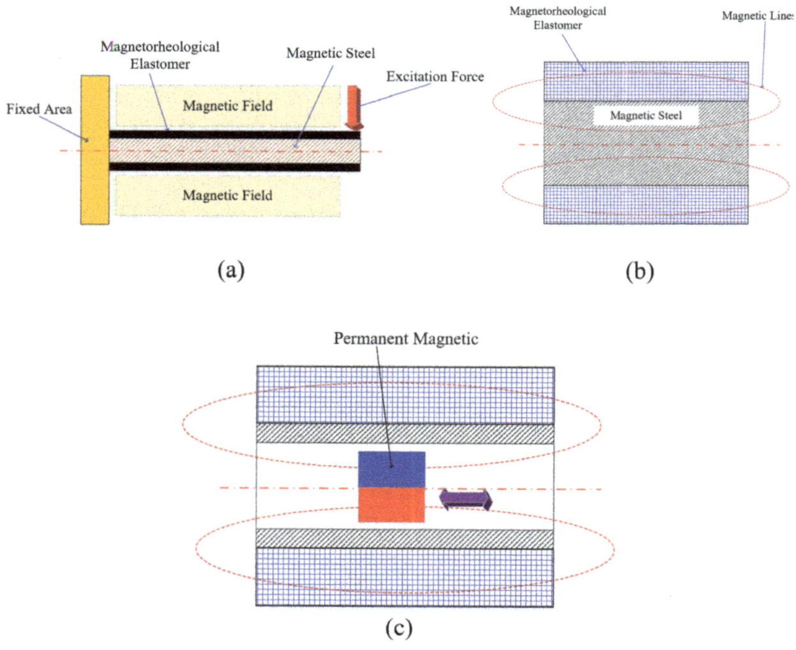

Fig. 2. Magneto-rheological cantilever beam model: (a) general model, (b) distribution of layers, (c) movement of permanent magnetic block.

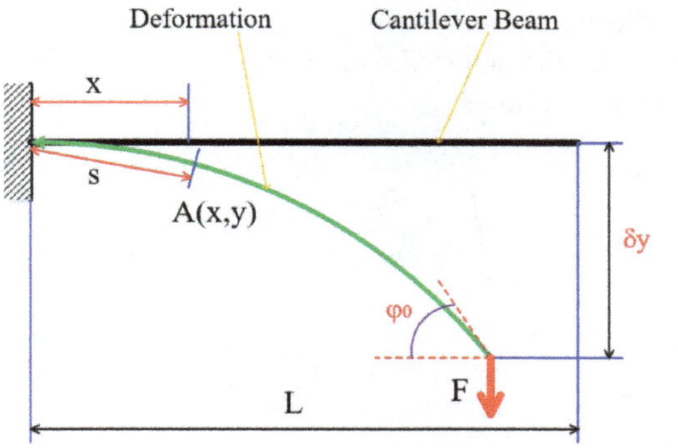

Fig. 3. Simplified magneto-rheological cantilever beam model

a. **Case 1.** Large deflection.
The Bernoulli-Euler bending moment curvature relationship for a uniform section rectangular beam is obtained as follows [20]:

$$EI\frac{d\varphi}{ds} = M \tag{1}$$

Where M and $\frac{d\varphi}{ds}$ are the bending moment and the curvature at any pint of the beam, I is the moment of inertia, E is the modulus of elasticity. The arc length s is determined as follows:

$$s = \sqrt{\frac{EI}{2F}}\int_0^{\varphi}\frac{d\varphi}{\sqrt{\sin\varphi_0 - \sin\varphi}} \tag{2}$$

The length L is determined following angle φ_0 as follows:

$$L = \sqrt{\frac{EI}{2F}}\int_0^{\varphi_0}\frac{d\varphi}{\sqrt{\sin\varphi_0 - \sin\varphi}} \tag{3}$$

The coordinates x and y are calculated as follows:

$$x = \sqrt{\frac{2EI}{F}}\left(\sqrt{\sin\varphi_0} - \sqrt{\sin\varphi_0 - \sin\varphi}\right) \tag{4}$$

$$y = \sqrt{\frac{EI}{2F}}\int_0^{\varphi}\frac{\sin\varphi d\varphi}{\sqrt{\sin\varphi_0 - \sin\varphi}} \tag{5}$$

b. **Case 2.** Small deflection.
If φ and φ_0 are small, the approximation $\sin\varphi \approx \varphi$ and $\sin\varphi_0 \approx \varphi_0$. Hence the Eqs. (2–5) is obtained as follows:

$$\int_0^{\varphi_0} \frac{d\varphi}{\sqrt{\varphi_0 - \varphi}} - 2\sqrt{\alpha} = 0 \tag{6}$$

$$\zeta = 1 - \sqrt{1 - \frac{\varphi}{\alpha}} \tag{7}$$

$$\mu = \frac{2}{3}\left(\alpha - (\varphi + 2\alpha)\sqrt{1 - \frac{\varphi}{\alpha}}\right) \tag{8}$$

Where $\alpha = \frac{FL^2}{2EI}, \zeta = \frac{x}{L}, \mu = \frac{y}{L}$. The magnetic field into the dynamic variables of the cantilever beam is determined. The magnetic dipole moment m of the entire magnetic seed is obtained as follows:

$$m = \iiint_{volume} MdV = MAL_s \tag{9}$$

Where M is the magnetization of permanent magnetic seed, A is cross-section of beam, L_s is length of the seed. The magnetic torque is written as:

$$T = m \times B = mB\sin(\gamma - \varphi) \tag{10}$$

Where γ denotes the initial angle between vector m and vector B. Relation between the torque and the force F is expressed as follows:

$$T = FL \tag{11}$$

Using the Eqs. (9–11), the magnetic force is defined as follows:

$$F = MAB\sin(\gamma - \varphi) \tag{12}$$

The result of the Eq. (12) is used for calculation the dynamic parameters of the system as shown in (2–8).

3.2 Data for Optimization

Data for optimization is collected from simulation based on the Eqs. (2–12). In this study, the data is designed based on MATLAB. In addition, a real factor 0.7 is used and embedded into the data for overcome the decrease of damping force of the MRE. In this optimization, the initial values of the system is $x \in [0, 0.001]m$, $y \in [0, 0.009]m$, L = 0.02 m, $B \in [0, 3]$Tesla, $A = 0.06, \mathrm{m}^2$, $\gamma = [20°, 35°]$, $\varphi = [0°, 60°]$.

3.3 Optimization Results

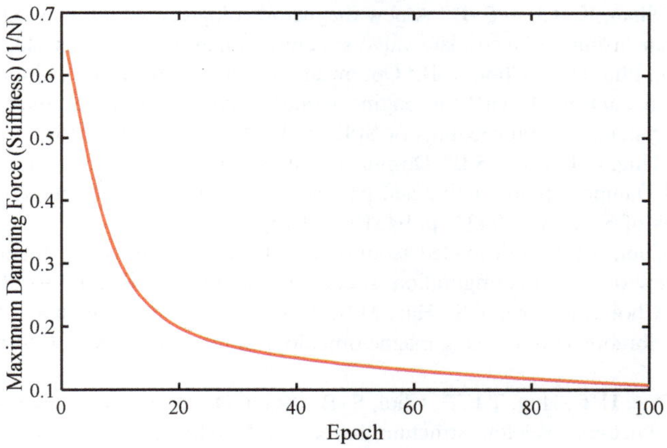

Fig. 4. Result of optimization.

Using this value, the normalization force is 0.1 (1/N). Hence the progress is setup with maximum epoch 200. After 100 epochs as shown in Fig. Fig. 4 , the value of force is 0. 11 (1/N). This value is also the optimized bound of force. Based on this value, the dynamic variables are $B = 1.43\,\mathrm{T}$, $A = 0.042\ \mathrm{m}^2$, $\varphi = 53°$. These results can be used for design of final cantilever beam of the proposed sensor.

4 Conclusion

In this study, a new progress of optimization of smart structure is carried out. The application of machine learning through neural networks brings a new view in design. The data using for optimization is collected from the simulation results, the real values of the closed structure with the proposed model. The combination of these data and the outstanding property of the neural networks helps improve the obtained variables, and then performance of the proposed structure is also improved. This method is effective in case of no specialized software such as ANSYS because of high cost.

Acknowledgement. This research is funded by Vietnam National Foundation for Science and Technology Development (NAFOSTED) under grant number 107.02-2020.13.

References

1. Phu, D.X., Hung, N.Q., Park, J.H., Choi, S.-B.: Design of a new MR brake mount system considering vertical and horizontal vibrations, active and passive smart structures and integrated systems. In: Proceedings of SPIE, vol. 9057, p. 90570O (2014)

2. Yang, S.-Y., Phu, D.X., Choi, S.-B.: Design of magneto-rheological mount for a cabin of heavy equipment vehicles, active and passive smart structures and integrated systems. In: Proceedings of SPIE, vol. 9799, 97992S (2016)
3. Phu, D.X., Shah, K., Choi, S.-B.: A new magnetorheological mount featured by changeable damping gaps using a moved-plate valve structure. Smart Mater. Struct. **23**, 125022 (2014)
4. Chung, J.U., Phu, D.X., Choi, S.-B.: Optimization of a new magnetorheological fluid mount for vibration control of start/stop engine mode, active and passive smart structures and integrated systems. In: Proceedings of SPIE, vol. 9431, 94311O (2015)
5. Phu, D.X., Ung, C.J., Choi, S.B.: Design and analysis of an innovative combined Magneto-rheological Damper-Mount, active and passive smart structures and integrated systems. In: Proceedings of SPIE, vol. 9431, p. 943130 (2015)
6. Do, X.P., Choi, S.B.: High loaded mounts for vibration control using magnetorheological fluids: review of design configuration, shock and vibration, Article ID 915859 (2015)
7. Phu, D.X., Choi, S.B., Lee, Y.S., Han, M.S.: Design of a new engine mount for vertical and horizontal vibration control using magnetorheological fluid. Smart Mater. Struct. **23**, 117001 (2014)
8. Do, X.P., Duc, H.T., Huy, T.L.T., Choi, S.-B., Yoon, D.: Design of a new inside multi-coils clutch for knee-exoskeleton structure based on helmholtz phenomenon and magneto-rheological fluid, sensors and smart structures technologies for civil, mechanical, and aerospace systems. In: Proceedings of SPIE, vol. 10598, p. 105983M (2018)
9. Do, X.P., Quoc, L.M.B., Choi, S.-B., Nguyen, Q.H., Cha, A.-R.: Design of a new exoskeleton based on the combination of two magnetorheological damper, sensors and smart structures technologies for civil, mechanical, and aerospace systems. In: Proceedings of SPIE, vol. 10598, p. 05983N (2018)
10. Do, X.P., Quoc, L.M.B., Kang, B.-H., Choi, S.-B.: Design of a new magneto-rheological damper featuring a hybrid type of piston for lower limb exoskeleton. In: Proceedings Volume 10970, Sensors and Smart Structures Technologies for Civil, Mechanical, and Aerospace Systems 2019, 109702O (2019)
11. Phu, D.X., Choi, S.-B.: Magnetorheological fluid based devices reported in 2013–2018: mini-review and comment on structural configurations. Front. Mater. **6**(19) (2019)
12. Phu, D.X., Long, M.B.Q., Kang, B.-H., Choi, S.-B.: Design of a new magneto-rheological pressure seal for rotary shaft. In: Proceedings Volume 10970, Sensors and Smart Structures Technologies for Civil, Mechanical, and Aerospace Systems 2019, 109702M (2019)
13. Do, X.P., Le, T.H.T., Kang, B.H., Choi, S.-B.: A new magneto-rheological skin for controlling pressure of haptic devices. In: Proceedings Volume 10970, Sensors and Smart Structures Technologies for Civil, Mechanical, and Aerospace Systems 2019, 109702N (2019)
14. Phu, D.X., Hung, N.Q., Choi, S.-B.: A novel adaptive controller featuring inversely fuzzified values with application to vibration control of magneto-rheological seat suspension system. J. Vibr. Control **24**(21), 5000–5018 (2018)
15. Do, X.P., Nguyen, Q.H., Choi, S.-B.: New hybrid optimal controller applied to a vibration control system subjected to severe disturbances. Mech. Syst. Signal Process. **124**, 408–423 (2019)
16. Han, Y., Hong, W., Faidley, L.E.: Field-stiffening effect of magneto-rheological elastomers. Int. J. Solids Struct. **50**, 2281–2288 (2013)
17. Shi, J., Song, J., Song, B., Lu, W.F.: Multi-objective optimization design through machine learning for drop-on-demand. Bioprint. Eng. **5**, 586–593 (2019)

18. Isaev, D., Semisalova, A., Alekhina, Y., Makarova, L., Perov, N.: Simulation of magnetodielectric effect in magnetorheological elastomers. Int. J. Mol. Sci. **20**, 1457 (2019)
19. Richiedei, D., Trevisani, A.: Optimization of the energy consumption through spring balancing of servo-actuated mechanisms. J. Mech. Des. **142**, 012301-1-10 (2020)
20. Belendez, T., Neipp, C., Belendez, A.: Large and small deflections of a cantilever beam. Eur. J. Phys. **23**, 371–379 (2002)

A New Design of Variable Stiffness Mechanism Based on Variation of Magnetic of Magnetorheological Elastomer

Xuan Phu Do[1(✉)] and Thinh Nguyen Duc[2]

[1] Department of Mechatronics and Sensor Systems Technology, Vietnamese-German University, Thu Dau Mot City, Binh Duong, Vietnam
phu.dx@vgu.edu.vn

[2] Electrical Engineering and Information Technology, Vietnamese-German University, Thu Dau Mot city, Binh Duong, Vietnam
thinh.nd@vgu.edu.vn

Abstract. In this study, a new design of variable stiffness mechanism based on magnetorheological elastomer is proposed. The variation of magnetic from the permanent magnet is formulated as main input for the proposed mechanism. Principle of the magnetic through the permanent magnet is applied for the design. The applied force for the mechanism is also find based on relation of vibration and the magnetic field of permanent magnet. In addition, the arrangement of MR elastomer blocks is suggested to find optimized shape for improve performance of the mechanism. The results of simulation show that the control of magnetic field with new calculation can improve and remain the stable of mechanism under severe vibration. This is also open a new direction for design of sensor in the future.

Keywords: Magneto-rheological elastomer (MRE) · Machine learning · MR actuator · Neural networks optimization · Valve optimization · Block-up phenomenon

1 Introduction

MR actuators are currently applied in many devices with high precision and large impact related force, torque, or moment. These devices can be seen in advanced automobile such as BMW, Mercedes, Lamborghini, Audi,... or in military devices such as mount of engine in ship. The structure was shown in [1–15]: brake [1], mount [2–7], damper [5, 9, 10], clutch [8], seal [12], valve [11], skin [13], and control [14, 15]. In these structure, MR fluid was used for design. This fluid has some disadvantages such as block phenomenon, dirty, high cost. Another type of MR material is MR elastomer. This material has overcome the disadvantage of MRF, and then has many application in design of brake, clutch, damper. Recently, this material uses for design of mechanism of sensor or robotic. In this study, a new design of mechanism with cover of MRE is presented. In addition, a new progress of optimization using machine learning is carried out.

B. T. Long et al. (Eds.): MMMS 2020, LNME, pp. 20–27, 2021.
https://doi.org/10.1007/978-3-030-69610-8_3

2 Proposed MRE Mechanism

In proposed design, a new cover of MRE is applied in structure of [16]. Vibration of the system can be controlled by two methods: moving the mount and magnetic field through stiffness of MRE. These methods improve performance of the system, and then can expanded the range of stiffness for obtaining high precision. The proposed structure is shown in Fig. 1. It is remarked that the magnetic field is from movement of permanent magnet which is placed inside the core of the cantilever beam.

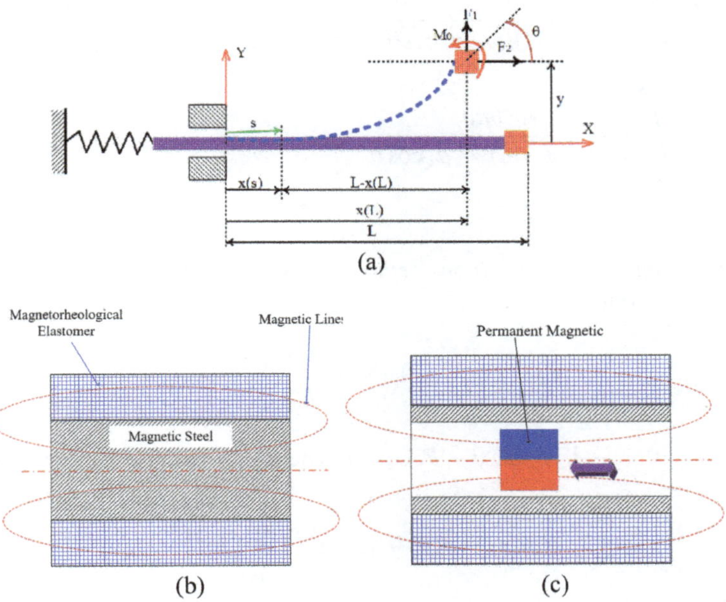

Fig. 1. Proposed structure with MRE: (a) general model, (b) distribution of MRE, (c) movement of permanent magnet

The governing equation of the system is defined following beam of Euler-Bernoulli as follows:

$$M(x, y) = F_y(x(L) - x) - F_x(y(L) - y) + M_0 \quad (1)$$

Where L is length of beam (m), F_x is horizontal force (N), F_y is vertical force (N), L_y^{max} is maximum deflection following the vertical direction (m), M_0 is excitation moment (N.m), δ_x is horizontal deflection (m), x and y are coordinates. In this analysis, angular deflection $d\theta$ of the beam is determined by arc dimension ds. The Eq. (1) is rewritten as follows:

$$EI\frac{d\theta}{ds} = F_y(x(L) - x) - F_y(y(L) - y) + M_0 \tag{2}$$

Where θ is angular deflection (rad), E is elasticity modulus N/m^2, I is inertial moment of cross-section m^4, y is vertical deflection (m), s is position and can be seen as a deflection function. Define $\alpha_x = \frac{F_x L^2}{EI}, \alpha_y = \frac{F_y L^2}{EI}$, $\beta_1 = \alpha_y \sin\theta_L - \alpha_x \cos\theta_L + \frac{1}{2}m^2, m = \frac{M_0 L}{EI}$. The dynamic equations of the system are determined as follows [16]:

$$\sqrt{2} = \int_{\theta(0)}^{\theta_L} \frac{d\theta}{\sqrt{\beta_1 - \alpha_y \sin(\theta) - \alpha_x \cos(\theta)}}, \frac{x(s)}{L} = \frac{1}{\sqrt{2}} \int_{\theta(0)}^{\theta(s)} \frac{\cos\theta d\theta}{\sqrt{\beta - \alpha_y \sin\theta - \alpha_x \cos\theta}} a,$$
$$\frac{y}{L} = \frac{1}{\sqrt{2}} \int_{\theta(0)}^{\theta(s)} \frac{\sin\theta d\theta}{\sqrt{\beta - \alpha_y \sin\theta - \alpha_x \cos\theta}} \tag{3}$$

Using the *Adomian Decomposition Method* (ADM) for approximation of variables of the structure as follows:

$$\theta(s) = \theta(0) + \dot{\theta}(L)s + \int_0^s \int_t^L N(z)dzdt \tag{4}$$

Where $N(\theta) = -\frac{1}{EI}\left(F_x \sin\theta - F_y \cos\theta\right)$. The Eq. (4) can be written as follows:

$$\theta(s) = \theta(0) + \gamma s + \ell^{-1}\left(\sum_{n=0}^{\infty} A_n\right) \tag{5}$$

Where $A_{n\text{-}}$ is series of ADM, $\gamma = \frac{M_0}{EI}$. Set $\theta(s) = \sum_{n=0}^{\infty} \theta_n(s)$. Hence the components of $\theta_n(s)$ can be calculated by using definitions as:

$$\begin{aligned} \theta_0 &= \theta(0) + \gamma s \\ \theta_n &= \ell^{-1}(A_{n-1}) \end{aligned} \tag{6}$$

Where, series ADM of $f(\theta) = -\frac{1}{EI}\left(F_x \sin(\theta) - F_y \cos(\theta)\right) = -\frac{1}{L^2}(\alpha_x \sin(\theta) - \alpha_y \cos(\theta))$ is given by:

$$
\begin{aligned}
A_0 &= f(\theta_0) = -\frac{1}{L^2}\left(\alpha_x \sin\theta_0 - \alpha_y \cos\theta_0\right) \\
A_1 &= \theta_1 \frac{d}{d\theta_0}[f(\theta_0)] \\
A_2 &= \theta_2 \frac{df(\theta_0)}{d\theta_0} + \frac{\theta_1^2}{2!}\frac{d^2 f(\theta_0)}{d\theta_0^2} \\
A_3 &= \theta_3 \frac{df(\theta_0)}{d\theta_0} + \theta_1\theta_2 \frac{d^2 f(\theta_0)}{d\theta_0^2} + \frac{\theta_0^3}{3!}\frac{d^3 f(\theta_0)}{d\theta_0^3} ac
\end{aligned} \tag{7}
$$

Substituting the Eqs. (4–7), elements of Eq. (7) are found as follows:

$$
\begin{aligned}
\theta_0(s) &= \gamma s \\
\theta_1(s) &= -\frac{1}{2L^2}\left[s(-2L+s)\left(\alpha_y \cos(\gamma s) - \alpha_x \sin(\gamma s)\right)\right] \\
\theta_2(s) &= -\frac{1}{4L^2}\left[s^2(-2L+s)^2\left(\alpha_y \cos(\gamma s) - \alpha_x \sin(\gamma s)\right)\left(\alpha_y \sin(\gamma s) + \alpha_x \cos(\gamma s)\right)\right] \\
\theta_3(s) &= -\frac{1}{32L^6}\left[-\alpha_y \cos(\gamma s) + \alpha_x \sin(\gamma s)\right]\left[\alpha_x^2 + \alpha_y^2 + 3\left(\alpha_x^2 - \alpha_y^2\right)\cos(2\gamma s) + 6\alpha_x\alpha_y \sin(2\gamma s)\right]
\end{aligned} \tag{8}
$$

Hence approximated results of $\theta(s)$ are expressed as:

$$
\theta(s) = \theta_0(s) + \theta_1(s) + \theta_2(s) + \theta_3(s) \tag{9}
$$

Consequently, positions of $x(s)$ and $y(s)$ is calculated as follows:

$$
x(s) = s - \frac{\left(\alpha_y + m\right)^2 s^3}{6L^2} + \frac{\left(\alpha_y + m\right)\left(\alpha_y + 2\alpha_x\alpha_y + 2m\alpha_x\right)s^4}{8L^3} \tag{10}
$$

$$
y(s) = \frac{1}{2L}\left(\alpha_y + m\right)s^2 - \frac{\left(\alpha_y + 2\alpha_x\alpha_y + 2m\alpha_x\right)s^3}{6L^2} \tag{11}
$$

Vibration of the system at the end beam $s = L$ is determined as follows:

$$
\Delta\theta(L) = \frac{1}{2}\alpha_y + m - \alpha_x\left(\alpha_y + m\right) \tag{12}
$$

$$
\Delta x(L) = L - \frac{1}{6}L\left(\alpha_y + m\right)^2 + \frac{1}{8}L\left(\alpha_y + m\right)\left(\alpha_y + 2\alpha_x\alpha_y + 2\alpha_x m\right) \tag{13}
$$

$$
\Delta y(L) = \frac{1}{2}L\left(\alpha_y + m\right) - \frac{1}{6}L\left(\alpha_y + 2\alpha_x\alpha_y + 2\alpha_x m\right) \tag{14}
$$

The magnetic field into the dynamic variables of the cantilever beam is determined. The magnetic dipole moment m of the entire magnetic seed is obtained as follows:

$$m = \iiint_{volume} MdV = MAL_s \tag{15}$$

Where M is the magnetization of permanent magnetic seed, A is cross-section of beam, L_s is length of the seed. The magnetic torque is written as:

$$T = m \times B = mB\sin(\gamma - \varphi) \tag{16}$$

Where γ denotes the initial angle between vector m and vector B. Relation between the torque and the force F is expressed as follows:

$$T = FL \tag{17}$$

Using the Eqs. (15–17), the magnetic force is defined as follows:

$$F = MAB\sin(\gamma - \varphi) \tag{18}$$

The result of the Eq. (18) is used for calculation the dynamic parameters of the system as shown in (1–14).

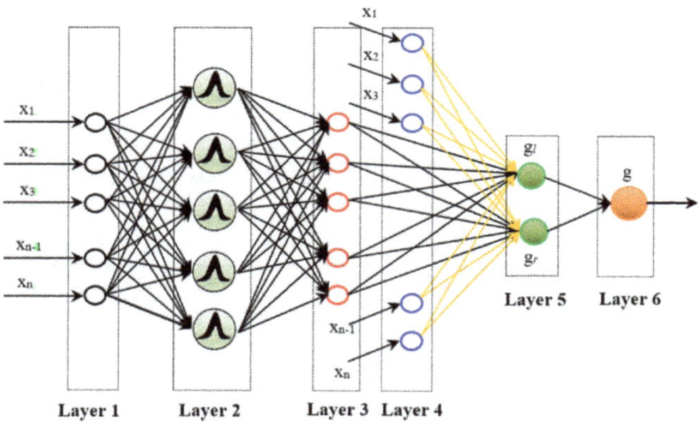

Fig. 2. Neural networks based on interval type 2 fuzzy model.

3 Application for a Magneto-Rheological Sensor

Currently, there are two tools for optimization using machine learning. The first is neural networks, and the second is deep learning. The difference of these methods is the number of layers. The increase of layer helps to improve the performance/or precision of results. In this study, model of neural networks is applied for optimization. The conventional model of neural networks model includes three elements such as input layer, hidden layer, and output layer. The hidden layer is derived as many sub-layers which the number of layer belongs to the requirement standard of output. However, the time of calculation of the progress will increase if the sub-layers increases. The neural networks can be improved its performance by using two modified model such as neural fuzzy model and fuzzy neural model as shown in Fig. 2. These modification are new group of machine learning as fuzzy machine learning model. In fact, the calculation of the new group is also used studied models which has been developed, but the data for the group is bigger than the former data. The data can be called as big data.

3.1 Data for Optimization

Data for optimization is collected from simulation based on the Eqs. (1−14). In this study, the data is designed based on MATLAB. In addition, a real factor 0.7 is used and embedded into the data for overcome the decrease of damping force of the MRE. In this optimization, the initial values of the system is $x \in [0, 0.001]m$, $y \in [0, 0.009]m$, L = 0.02 m, $B \in [0, 3]Tesla$, $A = 0.06\,\mathrm{m}^2$, $\gamma = [20°, 35°]$, $\varphi = [0°, 60°]$.

3.2 Optimization Results

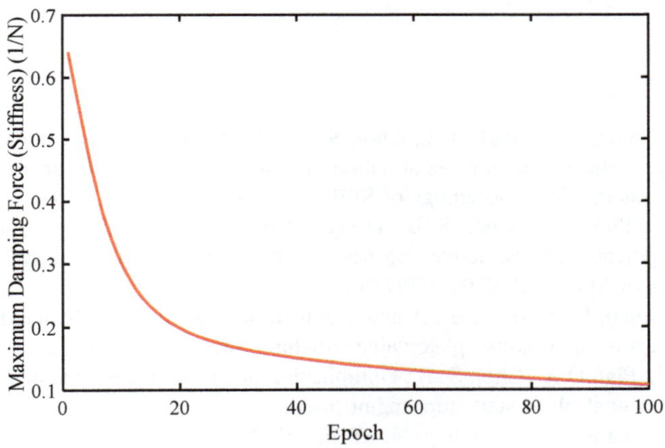

Fig. 3. Result of optimization.

The optimization cost function is designed following maximum damping force 10 N. Using this value, the normalization force is 0.1 (1/N). Hence the progress is setup with maximum epoch 200. After 100 epochs as shown in Fig. 3, the value of force is 0.11 (1/N). This value is also the optimized bound of force. Based on this value, the dynamic variables are $B = 1.5\,\mathrm{T}$, $A = 0.048\,\mathrm{m}^2$, $\varphi = 52°$. These results can be used for design of final cantilever beam of the proposed sensor.

4 Conclusion

In this study, a new structure and progress of optimization of smart structure is carried out. The structure is proposed based on variable stiffness mechanism with MRE. The layer of MRE is covered outside of the cantilever beam. This layer can be changed its stiffness by varying of magnetic field. The magnetic field is established by a block of permanent magnet. The vibration of the structure is controlled by variation of magnetic field, movement of mount. Besides the optimization of the proposed model is done by using machine learning (neural networks). The data using for optimization is collected from the simulation results, the real values of the closed structure with the proposed model. The combination of these data and the outstanding property of the neural networks helps improve the obtained variables, and then performance of the proposed structure is also improved. This method is effective in case of no specialized software such as ANSYS because of high cost.

Acknowledgement. This research is funded by Vietnam National Foundation for Science and Technology Development (NAFOSTED) under grant number 107.02-2020.13.

References

1. Phu, D.X., Hung, N.Q., Park, J.H., Choi, S.-B.: Design of a new MR brake mount system considering vertical and horizontal vibrations, active and passive smart structures and integrated systems. In: Proceedings of SPIE, vol. 9057, 90570O (2014)
2. Yang, S.-Y., Phu, D.X., Choi, S.-B.: Design of magneto-rheological mount for a cabin of heavy equipment vehicles, active and passive smart structures and integrated systems. In: Proceedings of SPIE, vol. 9799, 97992S (2016)
3. Phu, D.X., Shah, K., Choi, S.-B.: A new magnetorheological mount featured by changeable damping gaps using a moved-plate valve structure. Smart Mater. Struct. **23**, 125022 (2014)
4. Chung, J.U., Phu, D.X., Choi, S.-B.: Optimization of a new magnetorheological fluid mount for vibration control of start/stop engine mode, active and passive smart structures and integrated systems. In: Proceedings of SPIE, vol. 9431, 94311O (2015)
5. Phu, D.X., Ung, C.J., Choi, S.B.: Design and analysis of an innovative combined magneto-rheological damper-mount, active and passive smart structures and integrated systems. In: Proceedings of SPIE, vol. 9431, 943130 (2015)
6. Do, X.P., Choi, S.-B.: High loaded mounts for vibration control using magnetorheological fluids: review of design configuration. Shock Vibr., Article ID 915859 (2015)
7. Phu, D.X., Choi, S.B., Lee, Y.S., Han, M.S.: Design of a new engine mount for vertical and horizontal vibration control using magnetorheological fluid. Smart Mater. Struct. **23**, 117001 (2014)
8. Do, X.P., Duc, H.T., Huy, T.L.T., Choi, S.-B., Yoon, D.: Design of a new inside multi-coils clutch for knee-exoskeleton structure based on Helmholtz phenomenon and magneto-rheological fluid, sensors and smart structures technologies for civil, mechanical, and aerospace systems. In: Proceedings of SPIE, vol. 10598 105983M (2018)
9. Do, X.P., Quoc, L.M.B., Choi, S.-B., Nguyen, Q.H., Cha, A.-R.: Design of a new exoskeleton based on the combination of two magnetorheological damper, sensors and smart structures technologies for civil, mechanical, and aerospace systems. In: Proceedings of SPIE 10598 05983N (2018)

10. Do, X.P., Quoc, L.M.B., Kang, B.-H., Choi, S.-B.: Design of a new magneto-rheological damper featuring a hybrid type of piston for lower limb exoskeleton. In: Proceedings Volume 10970 Sensors and Smart Structures Technologies for Civil, Mechanical, and Aerospace Systems, 109702O (2019)
11. Phu, D.X., Choi, S.-B.: Magnetorheological fluid based devices reported in 2013–2018: mini-review and comment on structural configurations. Front. Mater. **6**(19) (2019)
12. Phu, D.X., Long, M.B.Q., Kang, B.-H., Choi, S.-B.: Design of a new magneto-rheological pressure seal for rotary shaft. In: Proceedings Volume 10970, Sensors and Smart Structures Technologies for Civil, Mechanical, and Aerospace Systems 2019, 109702M (2019)
13. Do, X.P., Le, T.H.T., Kang, B.H. Choi, S.-B.: A new magneto-rheological skin for controlling pressure of haptic devices. In: Proceedings Volume 10970, Sensors and Smart Structures Technologies for Civil, Mechanical, and Aerospace Systems 2019, 109702N (2019)
14. Phu, D.X., Hung, N.Q., Choi, S.-B.: A novel adaptive controller featuring inversely fuzzifield values with application to vibration control of magneto-rheological seat suspension system. J. Vibr. Control **24**(21), 5000–5018 (2018)
15. Do, X.P., Nguyen, Q.H., Choi, S.-B.: New hybrid optimal controller applied to a vibration control system subjected to severe disturbances. Mech. Syst. Signal Process. **124**, 408–423 (2019)
16. Phu, D.X., Hung, N.Q., Huy, T.D.: Analysis and summarization of a mechanism featuring variable stiffness. In: Proceedings of the International Conference on Advances in Computational Mechanics 2017. Lecture Notes in Mechanical Engineering, pp. 977–990 (2018)

A New Adaptive Hybrid PID Controller for Severe Disturbance of Bus Suspension Featuring Smart Damper

Xuan Phu Do(✉)

Department of Mechatronics and Sensor Systems Technology, Vietnamese-German University, Thu Dau Mot City, Binh Duong, Vietnam
phu.dx@vgu.edu.vn

Abstract. A new proposed control based on the modified Riccati-like equation is developed in this study. The interval type 2 fuzzy model is applied and embedded in the controller for control the uncertainty and also support for calculation progress of the proposed control. To design of the control, the first sliding surface based on the conventional model is established and its elements using for combine with the other elements of the Riccati-like equation. The robustness control is designed including three components of PID control and matrices of the traditional model of Riccati equation. Model of the proposed control is evaluated by applying to the bus suspension system. In the suspension, a magneto-rheological (MR in short) damper with its stiffness following the variation of applied current is used. The main input control is applied to change the value of the applied current. Belong to the magnitude of vibration, the current is also changed. The results of simulation of the proposed control are shown that the proposed control is better for control real system.

Keywords: PID control · H-infinity technique · Interval type 2 fuzzy control · Hybrid control · Vibration control · Magneto-rheological (MR) damper · Suspension control

1 Introduction

The development of machine learning brings many fields of application in control complex system. However, the classical control and its modern controls are remain as important roles in the system. The less memories are, the more application of modern control are. In this research, the modern controls are concentrated. Firstly, adaptive control has two kinds of style as indirect and direct controls. In addition, the main advantage of the adaptive control is its combination with other controls such as fuzzy model [1, 3, 5, 7, 9–14], PID [4–10], optimal control [2], sliding mode control [4, 6, 7, 9–14]. The method using PID [4–10] is continuously analyzed. The sliding surface of sliding mode control was used as a first solid function for design of adaptive control in [4]. Then the PID [4] was used as the robustness control, and its components were adapted through adaptation laws. However, the technique in [4] was follow the conventional design of adaptive control with non-modification of the robustness control to improve performance of the system. It is similar the technique which was shown in [5–7]. The technique [8] was

B. T. Long et al. (Eds.): MMMS 2020, LNME, pp. 28–36, 2021.
https://doi.org/10.1007/978-3-030-69610-8_4

basically similar the control [4–7]. The adaptation laws [8] used the matrices of the conventional Lyapunov equation and error vector. The property of the adaptation laws with the Lyapunov equation and PID was studied in [9]. The study [9] presented a new combination of the Riccati-like equation with PID control, which the sliding surface was main parameter of the adaptation laws. Another solution is use exponential function which was presented in [10]. The application of the modified Riccati-like equation was also studied in [11–15].

In this study, the model of interval type 2 fuzzy (IT2FNN) is used in the proposed controller. The featured studies of IT2FNN are presented in [17–19]. To increase the fast calculation, the granular clustering method [16] is used. The new proposed control is designed based on the PID control and the fuzzy model. The fuzzified parameter is embedded in the Riccati-like equation to robust the control. This is also presented in the adaptation laws.

2 Adaptive Fuzzy PID Control

2.1 Fuzzy Model

The model of interval type 2 fuzzy (IT2FNN in short) is applied for control the uncertainty of the proposed control. The rule base of IT2FNN can be expressed as follows [19].

$$R_f^j : \textbf{If } h_1 \text{ is } \mathbf{H}_{f1}^{\mathbf{j}} \text{ and} \ldots \text{and } h_n \text{ is } \mathbf{H}_{fn}^{\mathbf{j}} \textbf{ Then } g \text{ is } a_0^j + \sum_{i=1}^{n} a_i^j h_i \tag{1}$$

Where, $\mathbf{H}_{fi}^{\mathbf{j}}(i = 1, \ldots, n; j = 1, \ldots, m)$ are fuzzy sets, m is the number of rules, and a_i^j are interval sets. The calculation process of IT2FNN is clearly depicted in [18, 19]. The defuzzified output is then determined by

$$g_f = \frac{g_l + g_r}{2} = \frac{\boldsymbol{\theta}_l^T \boldsymbol{\xi}_l^f + \boldsymbol{\theta}_r^T \boldsymbol{\xi}_r^f}{2} \tag{2}$$

2.2 Adaptive Fuzzy PID Control

The nth-order nonlinear system is used to define a new proposed controller:

$$\dot{\mathbf{x}}_s = \mathbf{f}(\mathbf{x}_s) + \mathbf{g}(\mathbf{x}_s)u(t) + \mathbf{d}(t) \tag{3}$$

In Eq. (3), the functions $\mathbf{f}(\mathbf{x}_s) \in R^n$ and $\mathbf{g}(\mathbf{x}_s) \in R^n$ are two unknown non-linear functions, $u(t) \in R^1$ is the input control function, $\mathbf{d}(t) \in R^n$ is the disturbance, $|\mathbf{d}(t)| \le \delta\mathbf{d}$ where $\delta\mathbf{d} \in R^n$ is the upper boundary of $\mathbf{d}(t)$, and $\mathbf{x}_s = [x_1, x_2, \ldots, x_n] = \left[x_1, \dot{x}_1, \ldots, x_1^{(n-1)}\right]^T \in R^n$. The Eq. (3) can be described as follows.

$$\dot{\mathbf{x}}_s = \mathbf{f_0}(\mathbf{x}_s) + \mathbf{g_0}(\mathbf{x}_s)u(t) + \mathbf{D}(t) \tag{4}$$

Where, $\mathbf{D}(t) = \delta\mathbf{f} + \delta\mathbf{g}u(t) + \mathbf{d}(t)$ denotes the uncertain disturbances as $\mathbf{D} = [0, 0, \ldots, D_0]^T$, $\mathbf{f_0}(\mathbf{x}_s) = [x_2, \ldots, x_n, f_0]^T$, $\mathbf{g_0}(\mathbf{x}_s) = [0, \ldots, 0, g_0]^T$, $\delta\mathbf{f} = [0, 0, \ldots, \delta f_0]^T$, $\delta\mathbf{g} = [0, 0, \ldots, \delta g_0]^T$. Based on Eq. (2), the relationship between the Eq. (4) and IT2FNN is found as follows:

$$\mathbf{f_{00}}(\mathbf{x}_s) = \mathbf{f_0}(\mathbf{x}_s)g_f = \boldsymbol{\theta}_f\boldsymbol{\xi}_f \tag{5}$$

$$\mathbf{g_{00}}(\mathbf{x}_s) = \mathbf{g_0}(\mathbf{x}_s)g_f = \boldsymbol{\theta}_g\boldsymbol{\xi}_g \tag{6}$$

The sliding surface s_Δ is defined as follows.

$$s_\Delta = k_1x_1 + k_2x_2 + k_3x_3 + \ldots + k_nx_n = \sum_{i=1}^{n} k_ix_i \tag{7}$$

Where, $\mathbf{K} = [k_n, k_{n-11}, k_{n-2}, \ldots, k_1]$ is defined as the coefficient vector. The sliding surface (7) is determined using the state variables as follows:

$$x_n = -k_1x_1 - k_2x_2 - k_3x_3 - \ldots - k_{n-1}x_{n-1} + s_\Delta \tag{8}$$

A new vector $\tilde{\mathbf{x}}_s$ is defined by $\tilde{\mathbf{x}}_s = [x_1\, x_2\, x_3 \ldots x_{n-1}]^T$, and thus the Eq. (4) is rewritten as follows.

$$\dot{\tilde{\mathbf{x}}}_s = \mathbf{S}_1\tilde{\mathbf{x}}_s + \mathbf{S}_2^T s_\Delta \tag{9}$$

The tracking error is then defined as $e_\nabla(t) = x_s - x_{ds}$, where x_{ds} is the desired value. It is noted that x_{1s} is the same as x_s in the Eq. (9). The second (prescribed) sliding surface is defined as follows [12]:

$$\sigma_{s\nabla} = \dot{\varphi} + c_{s\nabla}\varphi \tag{10}$$

where, $c_{s\nabla} > 0$. Then, the derivative of Eq. (10) is obtained as follows.

$$\dot{\sigma}_{s\nabla} = \ddot{\varphi} + c_{s\nabla}\dot{\varphi} = M_{1\nabla} + M_{2\nabla} + M_{3\nabla}(f_0(x_s) + g_0(x_s)u(t) + D(t) - \ddot{x}_{ds}) + c_{s\nabla}\dot{\varphi} \tag{11}$$

The lumped uncertainty of system is then defined by

$$w_\nabla = M_{3\nabla}\tilde{\gamma}_{f\nabla}\xi_{f\nabla} + M_{3\nabla}\tilde{\gamma}_{g\nabla}\xi_{g\nabla}u_\nabla + M_{3\nabla}D(t) \tag{12}$$

where, $\gamma_{f\nabla} = f_0(x_s) - f_0^*(x_s)$, $\gamma_{g\nabla} = g_0(x_s) - g_0^*(x_s)$. Using Eq. (11) and Eq. (12), the derivative of Eq. (11) with respect to time is rewritten as follows:

$$\dot{\sigma}_{s\nabla} = M_{1\nabla} + M_{2\nabla} + M_{3\nabla}f_0^*(x_s) + M_{3\nabla}g_0^*(x_s)u(t) - M_{3\nabla}\ddot{x}_{ds} + c_{s\nabla}\dot{\varphi} + w_{\nabla} \quad (13)$$

The relationship between Eq. (13) and IT2FNN model is expressed as follows:

$$\dot{\sigma}_{s\nabla} = M_{1\nabla} + M_{2\nabla} + M_{3\nabla}\theta_f^*\xi_f + M_{3\nabla}\theta_g^*\xi_g u - M_{3\nabla}\ddot{x}_{ds} + c_{s\nabla}\dot{\varphi} + w_{\nabla} \quad (14)$$

Now, an equivalent control is determined from Eq. (14) based on the assumption $\dot{\sigma}_{s\nabla} \approx 0$:

$$u_1 = \frac{1}{M_{3\nabla}\hat{\theta}_g\xi_g}\left(-M_{1\nabla} - M_{2\nabla} - M_{3\nabla}\hat{\theta}_f\xi_f + M_{3\nabla}\ddot{x}_{ds} - c_{s\nabla}\dot{\varphi}\right) \quad (15)$$

To guarantee the robustness, a new robust control u_2 is proposed as follows:

$$u_2 = \frac{1}{M_{3\nabla}\hat{\theta}_g\xi_g}\begin{pmatrix} -\sum_{i=1}^{n-1} P_{(n-1)i}x_i - \frac{\sigma_{s\Delta}^2}{\beta} + \frac{1}{2}M_{3\nabla}\Gamma\xi_{fz}\tilde{\boldsymbol{x}}_s\boldsymbol{P}\boldsymbol{S}_2\boldsymbol{S}_2^T\boldsymbol{P}\tilde{\boldsymbol{x}}_s^T + K_P M_{3\nabla}\boldsymbol{S}_2^T\boldsymbol{E}\boldsymbol{E}^T\boldsymbol{P}\boldsymbol{B}_P \\ +K_I\int_0^\infty M_{3\nabla}\boldsymbol{S}_2^T\boldsymbol{E}\boldsymbol{E}^T\boldsymbol{P}\boldsymbol{B}_P dt + K_D M_{3\nabla}\boldsymbol{S}_2^T\dot{\boldsymbol{E}}\dot{\boldsymbol{E}}^T\boldsymbol{P}\boldsymbol{B}_P \end{pmatrix} \quad (16)$$

Where, $\mathbf{E} = [e_{\nabla 1}, e_{\nabla 2}, \ldots, e_{\nabla n}]$ is tracking error, x_d is the desired value. The value of $\mathbf{B}_P$ is determined as follows.

$$\dot{\mathbf{x}}_s = \mathbf{A}_P(x_s(t), t)\mathbf{x}_s(t) + \mathbf{B}_P(x_s(t), t)u(t) + \mathbf{D}(t) \quad (17)$$

It is remarked that $\mathbf{A}_P \in R^{n\times n}, \mathbf{B}_P \in R^{n\times m}, u(t) \in R^m (m = 1)$. The value Γ is the adaptive parameter where its boundary is given by $\Delta\Gamma = \{\Gamma \in R, \|\Gamma\| \le \varepsilon_\Gamma, \Gamma\xi_{fz} \le \rho\}$, and ε_Γ is constant boundary. The matrix $\mathbf{P} = \mathbf{P}^T \ge 0$ in which its result is a solution of a new fuzzified Riccati-like equation given by

$$\mathbf{P}\mathbf{S}_1 + \mathbf{S}_1^T\mathbf{P} + \mathbf{Q} - \sigma_{s\Delta}\Gamma\xi_{fz}\mathbf{P}\mathbf{S}_2\mathbf{S}_2^T\mathbf{P} + \rho\mathbf{P}\mathbf{S}_2\mathbf{S}_2^T\mathbf{P} - K_P\xi_{fz}\mathbf{S}_2^T\mathbf{E}\mathbf{E}^T\mathbf{P}\mathbf{B}_P - K_I\xi_{fz}\int \mathbf{S}_2^T\mathbf{E}\mathbf{E}^T\mathbf{P}\mathbf{B}_P dt$$
$$- K_D\xi_{fz}\mathbf{S}_2^T\dot{\mathbf{E}}\dot{\mathbf{E}}^T\mathbf{P}\mathbf{B}_P = 0 \quad (18)$$

The control u of the system is determined as follows.

$$u = u_1 + u_2 \quad (19)$$

To evaluate the stability and robustness of the system, Lyapunov function of the Eq. (3) is suggested as follows.

$$V = \frac{1}{2}s_\Delta^2 + \frac{1}{2}\tilde{\mathbf{x}}_s\mathbf{P}\tilde{\mathbf{x}}_s^T + \frac{1}{2\eta_1}\tilde{\gamma}_f^2 + \frac{1}{2\eta_2}\tilde{\gamma}_g^2 + \frac{1}{2\eta_3}\Gamma^2 + \frac{1}{2\eta_4}K_P^2 + \frac{1}{2\eta_5}K_I^2 + \frac{1}{2\eta_6}K_D^2 \quad (20)$$

Based on the evaluation of the Lyapunov equation, six adaptation laws are found as follows.

$$\dot{\tilde{\gamma}}_f = -\eta_1 M_{3\nabla}\sigma_{s\Delta}\xi_f \tag{21}$$

$$\dot{\tilde{\gamma}}_g = -\eta_2 M_{3\nabla}\sigma_{s\Delta}\xi_g u \tag{22}$$

$$\dot{\Gamma} = -\eta_3 M_{3\nabla}\sigma_{s\Delta}\xi_{fz}\tilde{\mathbf{x}}_s\mathbf{P}\mathbf{S_2}\mathbf{S_2^T}\mathbf{P}\tilde{\mathbf{x}}_s^T \tag{23}$$

$$\dot{K}_P = -\eta_4\left(M_{3\nabla}\sigma_{s\Delta}\mathbf{S}_2^T\mathbf{E}\mathbf{E}^T\mathbf{P}\mathbf{B}_P + \frac{1}{2}\xi_{fz}\tilde{\mathbf{x}}_s\mathbf{S}_2^T\mathbf{E}\mathbf{E}^T\mathbf{P}\mathbf{B}_P\tilde{\mathbf{x}}_s^T\right) \tag{24}$$

$$\dot{K}_I = -\eta_5\begin{pmatrix} M_{3\nabla}\sigma_{s\Delta}\int\mathbf{S}_2^T\mathbf{E}\mathbf{E}^T\mathbf{P}\mathbf{B}_P dt \\ +\frac{1}{2}\xi_{fz}\tilde{\mathbf{x}}_s\left[\int\mathbf{S}_2^T\mathbf{E}\mathbf{E}^T\mathbf{P}\mathbf{B}_P dt\right]\tilde{\mathbf{x}}_s^T\end{pmatrix} \tag{25}$$

$$\dot{K}_D = -\eta_6\left(M_{3\nabla}\sigma_{s\Delta}\mathbf{S}_2^T\dot{\mathbf{E}}\dot{\mathbf{E}}^T\mathbf{P}\mathbf{B}_P + \frac{1}{2}\xi_{fz}\tilde{\mathbf{x}}_s\left[\mathbf{S}_2\mathbf{S}_2^T\dot{\mathbf{E}}\dot{\mathbf{E}}^T\mathbf{P}\mathbf{B}_P\right]\tilde{\mathbf{x}}_s^T\right) \tag{26}$$

Proof: From the Eq. (20), the derivative is obtained as follows.

$$\begin{aligned}\dot{V} &= \sigma_{s\nabla}\dot{\sigma}_{s\nabla} + \frac{1}{2}\dot{\tilde{\mathbf{x}}}_s\mathbf{P}\tilde{\mathbf{x}}_s^T + \frac{1}{2}\tilde{\mathbf{x}}_s\mathbf{P}\dot{\tilde{\mathbf{x}}}_s^T + \frac{1}{\eta_1}\dot{\tilde{\gamma}}_f\tilde{\gamma}_f + \frac{1}{\eta_2}\dot{\tilde{\gamma}}_g\tilde{\gamma}_g + \frac{1}{\eta_3}\dot{\Gamma}\Gamma + \frac{1}{\eta_4}\dot{K}_P K_P + \frac{1}{\eta_5}\dot{K}_I K_I + \frac{1}{\eta_6}\dot{K}_D K_D \\ &= \sigma_{s\nabla}\begin{bmatrix} M_{3\nabla}\tilde{\gamma}_f\xi_f + M_{3\nabla}\tilde{\gamma}_g\xi_g u + w_\nabla - \sum_{i=1}^{n-1}P_{(n-1)i}x_i - \frac{\sigma_{s\Delta}^2}{\beta} + \frac{1}{2}M_{3\nabla}\Gamma\xi_{fz}\tilde{\mathbf{x}}_s\mathbf{P}\mathbf{S}_2\mathbf{S}_2^T\mathbf{P}\tilde{\mathbf{x}}_s^T \\ + K_P M_{3\nabla}\mathbf{S}_2^T\mathbf{E}\mathbf{E}^\mathbf{T}\mathbf{P}\mathbf{B}_\mathbf{P} + K_I\int_0^\infty M_{3\nabla}\mathbf{S}_2^T\mathbf{E}\mathbf{E}^\mathbf{T}\mathbf{P}\mathbf{B}_\mathbf{P}dt + K_D M_{3\nabla}\mathbf{S}_2^T\dot{\mathbf{E}}\dot{\mathbf{E}}^\mathbf{T}\mathbf{P}\mathbf{B}_\mathbf{P}\end{bmatrix} \\ &+ \left[\frac{1}{2}\tilde{\mathbf{x}}_s^T(\mathbf{P}\mathbf{S_1} + \mathbf{S_1^T}\mathbf{P})\tilde{\mathbf{x}}_s + \sigma_{s\nabla}\sum_{i=1}^{n-1}P_{(n-1)i}x_i\right] + \frac{1}{\eta_1}\dot{\tilde{\gamma}}_f\tilde{\gamma}_f + \frac{1}{\eta_2}\dot{\tilde{\gamma}}_g\tilde{\gamma}_g + \frac{1}{\eta_3}\dot{\Gamma}\Gamma + \frac{1}{\eta_4}\dot{K}_P K_P \\ &+ \frac{1}{\eta_5}\dot{K}_I K_I + \frac{1}{\eta_6}\dot{K}_D K_D\end{aligned} \tag{27}$$

The equivalence function of Eq. (16) is written by

$$\begin{aligned}\mathbf{P}\mathbf{S_1} + \mathbf{S_1^T}\mathbf{P} &= -\mathbf{Q} + \sigma_{s\nabla}\Gamma\xi_{fz}\mathbf{P}\mathbf{S_2}\mathbf{S_2^T}\mathbf{P} - \rho\mathbf{P}\mathbf{S_2}\mathbf{S_2^T}\mathbf{P} + K_P\xi_{fz}\mathbf{S}_2^T\mathbf{E}\mathbf{E}^T\mathbf{P}\mathbf{B}_P + K_I\xi_{fz}\int\mathbf{S}_2^T\mathbf{E}\mathbf{E}^T\mathbf{P}\mathbf{B}_P dt \\ &+ K_D\xi_{fz}\mathbf{S}_2^T\dot{\mathbf{E}}\dot{\mathbf{E}}^T\mathbf{P}\mathbf{B}_P\end{aligned} \tag{28}$$

From Eqs. (27) and (28), using the adaptation laws (20–25), Eq. (27) is rewritten by

$$\dot{V} \leq -\frac{1}{2}\tilde{\mathbf{x}}_s^T \mathbf{Q} \tilde{\mathbf{x}}_s + \beta w_{\nabla}^2 \tag{29}$$

Now the integrating of Eq. (29) from $t = 0$ to $t = T$ yields the following equation.

$$V(0) - V(T) + \beta \int_0^T w_{\nabla}^2 dt \geq \frac{1}{2} \int_0^T \tilde{\mathbf{x}}_s^T \mathbf{Q} \tilde{\mathbf{x}}_s dt \geq 0 \tag{30}$$

The proof of the adaptation laws is completed.

3 Application to Vibration Control

3.1 Vehicle Seat Suspension System

A model of vehicle MR seat suspension system is used for simulation of the proposed control. The governing equations of the MR seat suspension system are derived as follows [9].

$$\begin{cases} \dot{x}_{11} = \dot{x}_s = x_{22} \\ \dot{x}_{22} = f_{11}(x_{11}, x_{22}, x_{33}, x_{44}) + g_{11}(x_{11}, x_{22}, x_{33}, x_{44})u \\ \dot{x}_{33} = \dot{x}_1 = x_{44} \\ \dot{x}_{44} = f_{22}(x_{11}, x_{22}, x_{33}, x_{44}) \end{cases} \tag{31}$$

The voltage to be applied to MR damper v is found as follows.

$$v = \frac{F_{MR} - [c_a(x_{44} - x_{22}) + k_0(x_{33} - x_{11}) + \alpha_a \phi]}{c_b(x_{44} - x_{22}) + \alpha_b \phi} \tag{32}$$

The voltage v will be changed to the applied current, and then this current uses for the electric coil of the MR damper as input control $u = F_{MR}$.

3.2 Results of Simulation

In this simulation, an excitation signal as random bump road as shown in Fig. 1 is used. The sigma value for Gaussian function of the fuzzy model is chosen as 0.4. The constant values of the sliding surface $[k_1, k_2]$ are chosen as $[0.025, 1]$. The initial value Γ of the Riccati-like equation is chosen by 0.01. Besides, the matrix $\mathbf{Q}$ of the Riccati-like equation is chosen as $\mathbf{Q} = [-2\,0; 0\ -2]$, and the value β is 100. The constants $\eta_1, \eta_2, \eta_3, \eta_4, \eta_5, \eta_6$ of adaptation laws are chosen as 500 for all. The values of ε_p, ε_q, ε_Γ, ε_{KP}, ε_{KI}, ε_{KD} of the expanded adaptation laws are chosen by 10 for all. The values of δ_1, δ_2, δ_3, δ_4, δ_5, δ_6 are chosen by 0.05 for all. The observer [20] is used in this simulation.

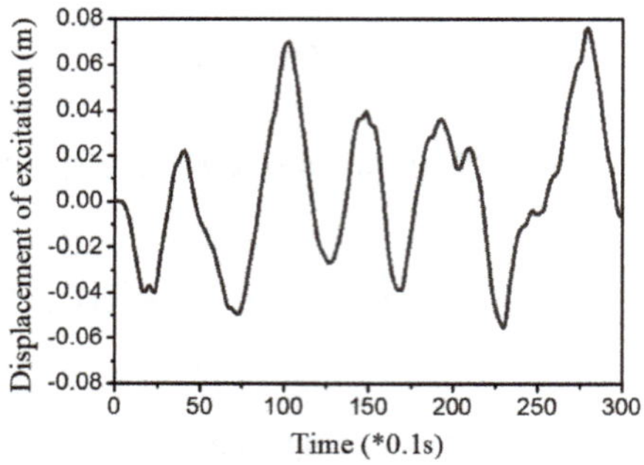

Fig. 1. Excitation signals: random bump road.

Results of simulation are shown in Fig. 2. The responses of displacement of both seat and driver (man) are good performance under severe disturbance as shown in Fig. 1. The magnitude of vibration is dramatically decrease, and the state of system is stable after using the control. From the above results, the proposed control has good properties in control vibration.

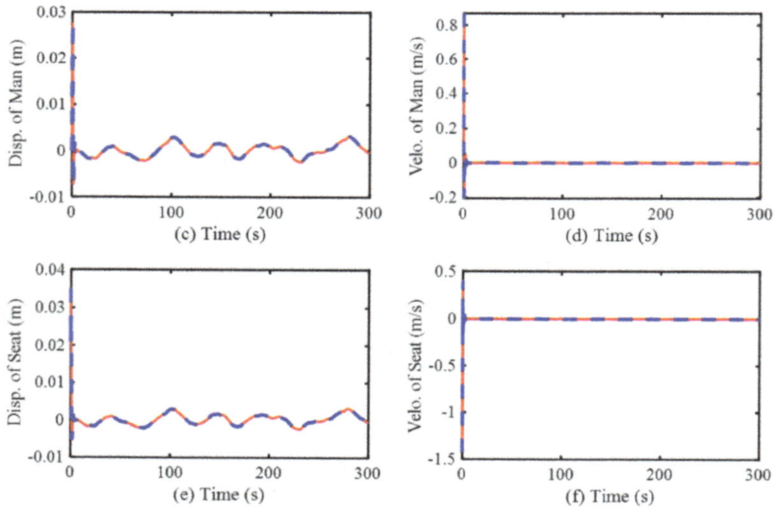

Fig. 2. Simulation result of displacement and velocity of seat and man (driver)

4 Conclusion

In this work, a new adaptive fuzzy PID controller based on the H-infinity control technique, sliding mode controller and PID controller was formulated and applied to vibration control of MR seat suspension system to demonstrate its effectiveness. The interval type 2 fuzzy and a new fuzzified Riccati-like equation were also used for combining three controllers. Parameters of the PID control was embedded in the Riccati-like equation. The objective of this approach is to improve performance of the system, especially in severe disturbances. The result of simulation showed that the proposed control is good and can be applied in a real system to obtain the best performance following the initial requirements.

Acknowledgement. This research is funded by Vietnam National Foundation for Science and Technology Development (NAFOSTED) under grant number 107.02-2020.13.

References

1. Bounemeur, A., Chemachema, M., Essounbouli, N.: Indirect adaptive fuzzy fault-tolerant tracking control for MIMO nonlinear systems with actuator and sensor failures. ISA Trans. **79**, 45–61 (2018)
2. Davanipour, M., Khayatian, A.R., Dehghani, M., Arefi, M.M.: A solution for enhancement of transient performance in nonlinear adaptive control: optimal adaptive reset based on barrier Lyapunov function. ISA Trans. **80**, 169–175 (2018)
3. Zhang, W., Hang, S.: Fuzzy adaptive control of nonlinear MIMO systems with unknown dead zone outputs. J. Franklin Inst. **355**, 5690–5720 (2018)
4. Moradi, M., Fekih, A.: Adaptive PID-sliding-mode fault-tolerant control approach for vehicle suspension systems subject to actuator faults. IEEE Trans. Veh. Technol. **63**(3), 1041–1054 (2014)
5. Cao, K., Gao, X., Lam, H.-K., Vasilakos, A.: H fuzzy PID control synthesis for Takagi-Sugeno fuzzy systems. IET Control Theory Appl. **10**(6), 607–616 (2016)
6. Chamsai, T., Jirawattana, P., Radpukdee, T.: Robust adaptive PID controller for a class of uncertain nonlinear systems: an application for speed tracking control of an SI engine. Math. Probl. Eng. **2015**, 510738 (2015)
7. Mahmoodabadi, M.J., Abedzadeh Maafi, R., Taherkhorsandi, M.: An optimal adaptive robust PID controller subject to fuzzy rules and sliding modes for MIMO uncertain chaotic systems. Appl. Soft Comput. **52**, 1191–1199 (2017)
8. GanapathySubramanian, R., Elumalai, V.K., Karuppusamy, S., Canchi, V.K.: Uniform ultimate bounded robust model reference adaptive PID control scheme for visual servoing. J. Franklin Inst. **354**(4), 1741–1758 (2017)
9. Phu, D.X., An, J.-H., Choi, S.-B.: A novel adaptive PID controller with application to vibration control of a semi-active vehicle seat suspension. Appl. Sci. **7**, 1055 (2017)
10. Van, M., Do, X.P., Mavrovouniotis, M.: Self-tuning fuzzy PID-nonsingular fast terminal sliding mode control for robust fault tolerant control of robot manipulators. ISA Trans. (2019). https://doi.org/10.1016/j.isatra.2019.06.017
11. Phu, D.X., Choi, S.-M., Choi, S.-B.: A new adaptive hybrid controller for vibration control of a vehicle seat suspension featuring MR damper. J. Vibr. Control **23**(20), 3392–3413 (2017)
12. Phu, D.X., Huy, T.D., Mien, V., Choi, S.-B.: A new composite adaptive controller featuring the neural network and prescribed sliding surface with application to vibration control. Mech. Syst. Signal Process. **107,** 409–428 (2018)
13. Do, X.P., Nguyen, Q.H., Choi, S.-B.: New hybrid optimal controller applied to a vibration control system subjected to severe disturbances. Mech. Syst. Signal Process. **124**, 408–423 (2019)

14. Do, X.P., Nguyen, Q.H., Choi, S.-B.: A novel adaptive controller featuring inversely fuzzified values with application to vibration control of magnetorheological seat suspension system. J. Vib. Control **24**(21), 5000–5018 (2018)
15. Do, X.P., Choi, S.-B.: A new adaptive fuzzy PID controller based on Riccati-like equation with application to vibration control of vehicle seat suspension. Appl. Sci. **9**(21), 4540 (2019)
16. Wang, X., Liu, X., Zhang, L.: A rapid fuzzy rule clustering method based on granular computing. Appl. Soft Comput. **24**, 534–542 (2014)
17. Mendel, J.M.: Computing derivatives in interval type-2 fuzzy logic systems IEEE Trans. Fuzzy Syst. **12**(1), 84–98 (2004)
18. Juang, C.-F., Chen, C.-Y.: Data-driven interval type-2 neural fuzzy system with high learning accuracy and improved model interpretability. IEEE Trans. Cybern. **43**(6), 1781–1795 (2013)
19. Liang, Q., Mendel, J.M.: Interval type-2 fuzzy logic systems: theory and design. IEEE Trans. Fuzzy Syst. **8**(5), 535–550 (2000)
20. Ciccarella, G., Dalla Mora, M., Germani, A.: A Luenberger-like observer for nonlinear systems. Int. J. Control **57**(3), 537–556 (1993)

Theoretical and Experimental Study of Vibration-Assisted Turning

Hoang Trung Kien[1(✉)], Dang Quang Khoa[1], Pham Huy Tuan[1], Mai Khoan[1], and Duong Van Ba[2]

[1] Faculty of Mechanical Engineering, Ho Chi Minh City University of Technology and Education, Ho Chi Minh City, Vietnam
kienhtr@hcmute.edu.vn

[2] Faculty of Mechanical Engineering Technology, Cao Thang Technical College, Ho Chi Minh City, Vietnam

Abstract. The development of the manufacturing industry requires an improvement in machining technology for cutting high strength and hard material such as stainless steel or alloys. There are many methods to improve the machining ability of the turning process such as vibration-assisted methods, ultrasonic-aided methods, hybrid-process methods and optimization methods, etc. Among these methods, vibration-assisted method is one of the most efficient methods in improving the machining ability and efficiency of the turning process. Many researches have proved that vibration helps reduce the cutting force of the cutting process therefore improving the tool life and machining accuracy. Moreover, the surface roughness was improved and the power consumption was reduced. The improvement is attributed to the intermittent cutting mechanism of the hybrid process. This paper describes an attempt to apply non-ultrasonic frequency vibration to the turning process to prove the feasibility of non-ultrasonic frequency vibration-assisted turning (VAT). The fundamental theory was analysed. The vibration-assisted turning experiments were conducted and the experimental results were analysed. It was observed that the cutting surface was improved with the increasing of the vibration frequency.

Keywords: Vibration-assisted turning (VAT) · Compliant mechanism · Surface roughness · Machining ability · Turning process

1 Introduction

There are many applications for ultrasonic vibration. Typical applications include non-destructive measuring equipment, cleaning and medical technology. Applied ultrasonic processes in manufacturing including ultrasonic welding or ultrasonic-assisted machining such as ultrasonic assisted EDM (discharge processes) are popular. Significant benefits can be archived when applying assisted vibration to the cutting processes. The reduction of temperature helps reduce cutting tool wear [1] Hard steel can be cut easier [2]. Furthermore, the difficult-to-cut materials such as glass can also be cut when they add vibrations into the cutting tool [3]. Cutting forces were significantly reduced when drilling deep holes on aluminum alloy silicon hypereutectic [4, 5].

B. T. Long et al. (Eds.): MMMS 2020, LNME, pp. 37–44, 2021.
https://doi.org/10.1007/978-3-030-69610-8_5

In material processing, the surface quality is one of the most specific requirements of customers. The surface roughness is a key factor to evaluate the surface quality of the machined parts. Many theoretical and empirical studies on surface roughness of machined products have been reported. In the case of turning process, there are many factors that affect the surface roughness such as cutting parameters, cutter material, tool geometry and its abrasion, material structure and motional characteristics between tool and workpiece, lubrication and cooling conditions, and system rigidity, etc. It is important that the cutting performance and surface roughness be taken into consideration to ensure the machining quality and reducing machining costs. Numerous studies have developed accurate prediction models for surface roughness and tool wear in traditional machining.

In this paper, a method to improve the turning process has been presented, the vibration-assisted turning. The purpose of this presented study is to apply axial support oscillations to the cutting tool of the lathe machining to explore the effects of vibration parameters such as amplitude and frequency on the surface roughness when machining stainless steel. The experiment results showed the significant effects of frequency and amplitude on the surface roughness.

2 Vibration Assisted Turning

2.1 Fundamental Theory

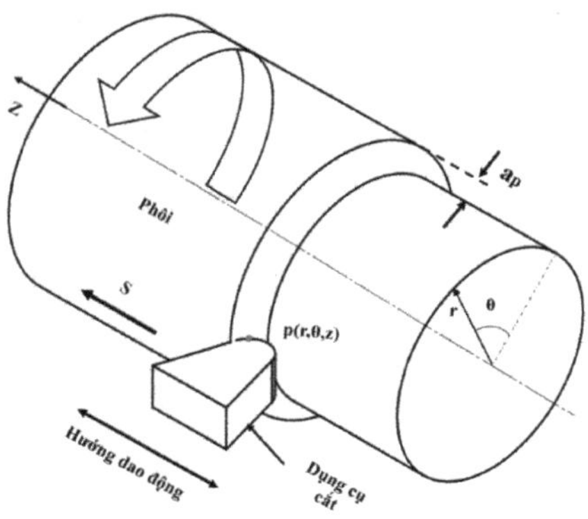

Fig. 1. Axial vibration-assisted turning

Figure 1 describes the diagram of the application of axial vibration in turning. A point in the cutting plane is chosen to describe the kinematic equation of the cutting tool. The positional and velocity of the cutting point at time t represented by z(t) and z′(t) respectively are given by:

$$z(t) = Asin(wt) + v_f.t \tag{1}$$

$$z'(t) = wA\cos(wt) + v_f \tag{2}$$

The motional and positional characteristic at point P are as below:

$$\begin{cases} r = r_p \\ \theta = w_n \cdot t = \frac{n.\pi}{30} \cdot t \\ z = Asin(2\pi Ft) + v_f \cdot t \end{cases} \tag{3}$$

where A is the amplitude of vibration (mm), F is the vibration frequency (Hz), vf is the speed of the cutting tool (mm/s), w = 2πF is the angular velocity (rad/s). rp is cutting plane radius (mm); ap is the cutting depth (mm); wn is the angular speed of turning spindle (rad/s); n is the spindle speed (rev/min) and s is the feed rate (mm/rev).

Time of turning an angle θ and cutting tool speed are given as:

$$t = \frac{30.\theta}{n.\pi}, v_f = \frac{n.s}{60} \tag{4}$$

Substitute Eq. (4) in to z of Eq. (3), we have:

$$z = Asin.\left(\frac{60.F}{n}\theta\right) + \frac{s}{2\pi}\theta \tag{5}$$

From Eq. (5), the angular frequency can be expressed as:

$$w_f = \frac{60F}{n} = K + \varepsilon \tag{6}$$

where K is integer and $0 \leq \varepsilon \leq 1$.

If ε = 0, the path of the cutting tool will be sequential and parallel. That means the original phase of both cuts are the same and there is no phase difference. If $\varepsilon \neq 0$, the path of the cutting tool will be intersected. That means there is a phase difference.

The phase difference is given as:

$$\varphi = 2\pi\varepsilon \tag{7}$$

Kinematic equation of cut N is:

$$\mathrm{z_N = Asin.(w_f\theta + (N-1)\varphi) + \frac{s}{2\pi}\theta + (N-1).s} \tag{8}$$

The relationship between cut N and N + 1 is:

$$z_{N+1} - z_N = A\left[sin.\left(w_f.\theta + (N)\varphi\right) - sin.\left(w_f\theta + (N-1)\varphi\right)\right] + s \tag{9}$$

Avoiding cutting tool continuously contact with machined surface, cut of N and N + 1 need to be intersected or min value of $z_{N+1} - z_N \leq 0$.

$$-2A + s \leq 0 \Rightarrow s/2A \leq 1 \quad (10)$$

Cutting tool contact continuously with machined surface, cut of N and N + 1 does not intersect, or $z_{N+1} - z_N > 0$.

$$-2A + s \leq 0 \Rightarrow s/2A \leq 1 \quad (11)$$

When s/2A $\leq$ 1, the cutting path can be intersected and the cutting tool and cutting plane can be separated according to cutting direction such as Fig. 2.

When s/2A > 1, the cutting path is not intersect, continuous cutting occurs.

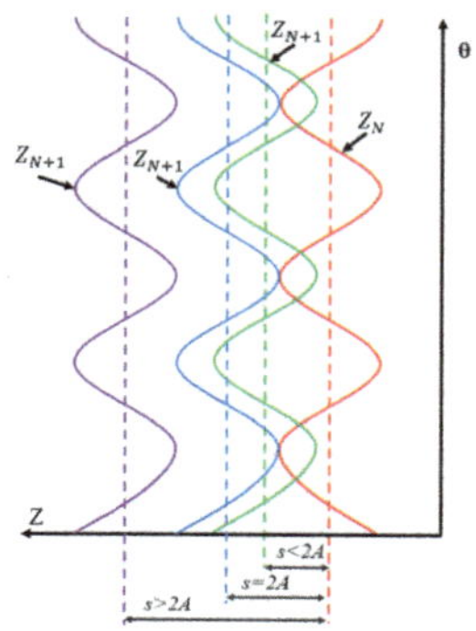

Fig. 2. Tool's cut line at s < 2A, s = 2A, s > 2A

Influence of Oscillation Frequency on Surface Roughness

The wavelength of the cut is calculated as in the formula below:

$$\lambda = \frac{v}{F} \quad (12)$$

If the frequency decreases, the linearity of the structure increases in the direction of the circumference. When the frequency is low, the surface structure of VAM shows no significant difference from the surface structure created in traditional machining. When the frequency is high enough, it will create a sine-pattern microstructure on the machining surface.

Influence of Phase Shift Angle in Case of s $\leq$ 2A

According to Eq. (9), coefficients such as amplitude A, feedrate s and phase deviation φ are important factors for discontinuous tool conditions. As shown in Fig. 3, although s $\leq$ 2A, the two cut marks do not intersect. Therefore discontinuous cutting has not yet happened.

Consider the relationship between N and N + 1 cut at two tangent point A and B:

$$\begin{cases} z_{N+1} = z_N \\ \dfrac{\partial z_{N+1}}{\partial \theta} = \dfrac{\partial z_N}{\partial \theta} \end{cases} \tag{13}$$

Substitute Eq. (8) and (9) in to Eq. (13) and solve we get the phase differences:

$$\begin{cases} \varphi_A = 2\arcsin\left(\frac{s}{2A}\right) \\ \varphi_B = 2\pi - 2\arcsin\left(\frac{s}{2A}\right) \end{cases} \tag{14}$$

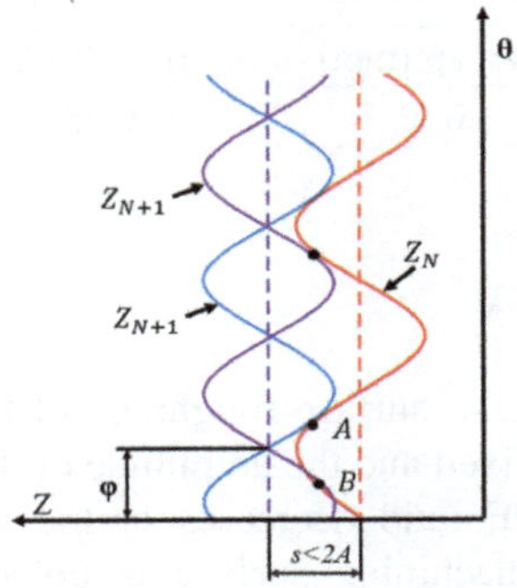

Fig. 3. Tangential cut lines

Thus, for a continuous cut to occur, the phase angle must be within:

$$\frac{\arcsin\left(\frac{s}{2A}\right)}{\pi} < \varepsilon < 1 - \frac{\arcsin\left(\frac{s}{2A}\right)}{\pi} \tag{15}$$

From Eq. (10) and (15), the conditions for discontinuous cutting occur when applying vibration to the turning process is given as:

$$\begin{cases} s \leq 2A \\ \frac{\arcsin\left(\frac{s}{2A}\right)}{\pi} < \varepsilon < 1 - \frac{\arcsin\left(\frac{s}{2A}\right)}{\pi} \end{cases} \tag{16}$$

3 Experiment Setup

The experiment was performed on Jessey Studturn mechanical lathe. The vibration in the direction of the tool is generated using a stack piezoelectric actuator Model P-225-10, PI. The cutting tool used in the experiment has the parameters shown in Table 1.

Table 1. Cutting tool pieces

Materials	Coating	Rake angle (°)	Clearance angle (°)	Rc
Carbide	MEGACOAT NANO	0	7	0.4

The workpiece material used in the experiment is SUS304 with a diameter of 30 mm and the cutting speed is selected in Table 2. Cutting parameters such as cutting speed, cutting depth and tool amount are fixed. Tool parameters are changed in sequence, frequency F = 0.25; 0.5; 1 kHz and amplitude A = 2.5 μm.

Table 2. Parameters of the cutting process

Vc (m/min)	ap (mm)	s (mm)	F (KHz)	A (μm)
16	0.1	0.03	0.25–1	2.5

4 Result and Discussion

According to the result in Fig. 4, surface roughness of traditional processing is Ra = 0,61. With the cutting speed fixed and the amplitude of the vibration is constant, when increasing frequency, the vc/F ratio decreases or the wavelength decreases, which causes the area where the machining surface is polished. The surface roughness decreases and it helps the machining surface quality to be improved. Surface roughness value at frequency F = 1 kHz (vc/F = 0.26) improved better than F = 0.5 kHz (vc/F = 0.52) and F = 0.25 kHz (vc/F = 1.04) when oscillating at the same amplitude.

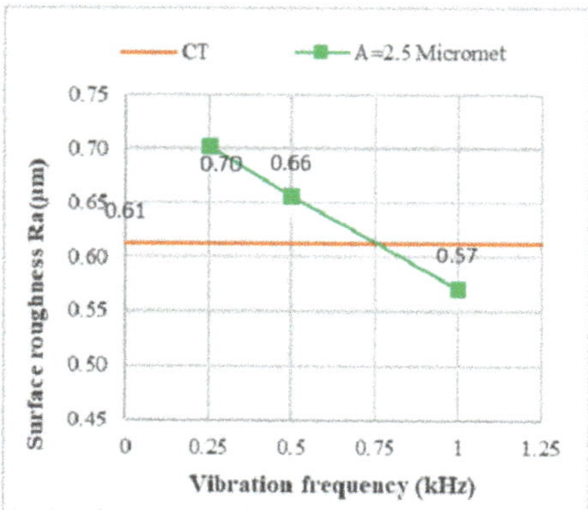

Fig. 4. Effect of vibration frequency on roughness

Besides, the reason why these smooth surfaces are created is that the conventional turning process is transformed into the turning process with high frequency vibration, thereby reducing the natural vibration created by the machining process. Another problem with traditional machining is that the chip may be trapped in the cutting area. It will scratch the machining surface and increase the surface roughness. Machining with assisted vibration does not observe the phenomenon of chip jam, because it is a discontinuous cutting process so the chip is always broken (Fig. 5).

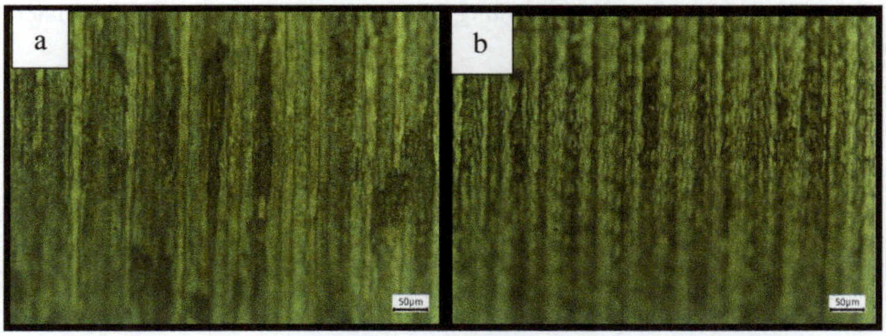

Fig. 5. Microscope picture of machining surface: a. traditional processing; b. Machining with vibration (V = 16 m/min, f = 0.03 mm, F = 1 kHz, A = 2.5 μm)

5 Conclusion

The fundamental analysis of the influence of the frequency of the vibration-assisted turning is presented in this paper. Based on the theory and experiments, it was successfully proved on SUS304 material and the following conclusions are drawn.

- Applying VAM with low frequency F = 0.25 − 1 kHz and small amplitude A = 2.5 μm in longitudinal direction in lathe machining can achieve better surface brightness than traditional 8–18%.
- The roughness of the machined surface is inversely proportional to the frequency of assisted vibrations. A smaller V/F ratio will give better roughness values.
- Higher frequencies and amplitudes would be potential in order to improve the quality of the cut surface.

Acknowledgement. This work belongs to the project in 2021 funded by the Ministry of Education and Training and hosted by Ho Chi Minh City University of Technology and Education, Vietnam.

References

1. Nestler, A., Schubert, A.: Surface properties in ultrasonic vibration assisted turning of particle reinforced aluminium matrix composites. Procedia CIRP **13**, 125–130 (2014)
2. Jiao, F., Liu, X., Zhao, C.Y., Zhang, X.: Experimental study on the surface micro-geometrical characteristics of quenched steel in ultrasonic assisted turning. Adv. Mater. Res. **189**, 4059–4063 (2011)

3. Babitsky, V.I., Kalashnikov, A.N., Meadows, A., Wijesundara, A.A.H.P.: Ultrasonically assisted turning of aviation materials. Mater. Process. Technol. **132**, 157–167 (2003)
4. Nosouhi, R., Behbahani, S., Amini, S., Khosrojerdi, M.R.: An experimental study on the cutting forces, surface roughness and the hardness of Al 6061 in 1D and 2D ultrasonic assisted turning. Appl. Mech. Mater. **680**, 224–227 (2014)
5. Schubert, A., Nestler, A., Pinternagel, S., Zeidler, H.: Influence of ultrasonic vibration assistance on the surface integrity in turning of the aluminium alloy AA2017. Mater. Sci. Eng. Technol. **42**, 658–665 (2011)

Study of Vibration-Assisted Injection Molding: System Design

Hoang Trung Kien[1(✉)], Pham Son Minh[1], Ta Nguyen Minh Duc[1], and Dang Nguyen Nhan[2]

[1] Faculty of Mechanical Engineering, Ho Chi Minh City University of Technology and Education, Ho Chi Minh City 700000, Vietnam
kienhtr@hcmute.edu.vn
[2] Faculty of Mechanical Engineering Technology, Cao Thang Technical College, Ho Chi Minh City 700000, Vietnam

Abstract. Plastic injection molding is one of the most popular and important manufacturing processes in human history. This paper presents a new technology namely the vibration-assisted injection molding using non-ultrasonic vibration which was used to improve the efficiency of the molding process. In this technology, a compliant mechanism is designed to transfer the vibration generated by a piezoelectric actuator to the pressing chamber. The vibration process causes periodic pressure deviations in the pressing chamber which improve the flow of melting plastic. A parallel flexure hinge was used for vibration and pressing chamber coupling. Hydraulic piston and motor are used to generate pressure up to 100 bar. The experiment conducted with PP plastic showed that with vibration, the flow rate and the filled volume have been improved. In this paper, the design of the pressing chamber and the vibration-assisted injection molding system will be described. The experimental results will also be presented and analysed.

Keywords: Injection molding · Vibration-assisted technology · Non-ultrasonic vibration

1 Introduction

Injection molding is one of the most popular methods in producing polymer products. With this method, we can make every plastic product at any shape depend on our intended use. This capability is made possible by modern control systems and precision mold systems. But over time, with the development of technology, the higher the demand of people is, the better the quality of the final product must be. And the conventional injection molding cannot handle this, due to some defects in the final product such as: sinks and voids within material, warpage and weld lines. Because these defects adversely affect the appearance as well as the mechanicals properties of the product, the work of finding a solution to intervene deep inside the structure of the product has become urgent in recent decades.

Several attempts were carried out and many solutions were proposed. One of the most effective methods is using vibration to assist into the injection molding so it is

B. T. Long et al. (Eds.): MMMS 2020, LNME, pp. 45–51, 2021.
https://doi.org/10.1007/978-3-030-69610-8_6

named the VAIM methods (vibration assisted injection molding) [1–4]. All of these researches and others have shown many positive effects of vibration on the quality of the final product at the molecular level. The appearance of vibration in forming plastic products (especially in two stages of filling and packing) will change the nature of the process as well as interfere directly with the arrangement of the plastic molecule. In this research, a molding system was designed to produce small size polypropylene specimens by VAIM method.

2 Vibration-Assisted Injection-Molding

The basic idea behind VAIM is to enable enhanced control of rheological behavior of plastic flow during processing, which could facilitate the altering the mechanical response of molded components to the desired level.

Firstly, it was proved that VAIM processing can significantly strengthen molded polymer samples and at also improve remarkedly the consistency of molded products from cycle to cycle [5]. When the product birefringence was studied, it was found that VAIM processing significantly alters the residual stress distribution throughout final products. However, the material density in the products does not change. VAIM is an effective processing tool for the structure development control of semi-crystallines such as HDPE [6].

Secondly, the introducing of pressure variation caused by vibration in the mold during the filling stage decreased the viscosity of polymer flow so that the filling time could be shortened. During packaging stage, the vibration postponed the gate frozen time so that more materials could be packed into the mold making the mean cavity pressure higher than conventional injection molding [7].

In this research, high frequency vibration is used. The vibration is transferred in to the heating and pressing chamber through the compliant mechanism which helps maintaining the vibration direction and protect the vibration device (Fig. 1).

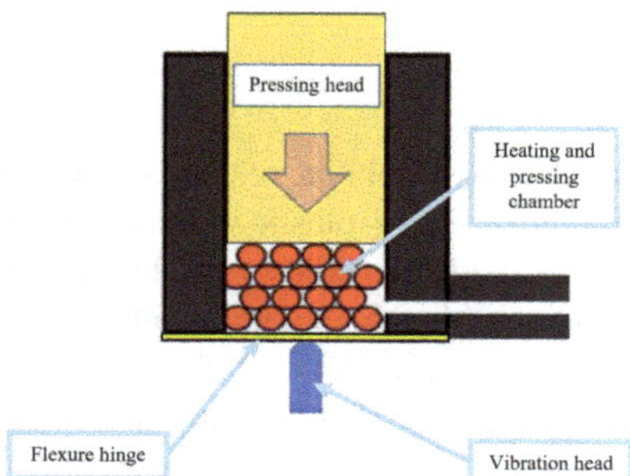

Fig. 1. VAIM configuration

3 Flexure Hinges Theory

3.1 Material

Some of the materials which are both flexible and strong are well suited for compliant mechanisms. Therefore, we look for materials with a high strength and a low Young's modulus. Materials with large σy/E values are generally good candidates for compliant mechanisms. If the σy/E value is high, the material is said to be stronger and flexible [8]. Table 1 shows the list of yield stress to young's modules ratio and resilience of different materials.

Table 1. List of yield stress to Young's modules ratio and resilience of different materials

Material	E (GPa)	S_y (MPa)	S_y.1000/E
Steel (1010 hot rolled)	207	179	0.87
Aluminum 6061	68.9	55.2	0.8
Aluminum 7075	71.7	275.9	4.0
Titanium (Ti-35A annealed)	114	207	1.8
Titanium (Ti-13 heat treated)	114	1170	10
Beryllium copper (CA170)	128	1170	9.2
Brass	97	200	2.06

As given in Table 1, Aluminum 7075–T6 is the material which has the lowest modulus of elasticity E of 71.7 GPa and the yield stress is high enough for the flexure mechanism, only lower than Titanium heat treated and Beryllium copper. Besides, Aluminum 7075–T6 can be purchased easily at a reasonable price. This makes Aluminum 7075–T6 the most appropriate material.

3.2 Flexure Hinge

Figure 2 shows the configuration of a commonly used circular flexure hinge. Different methods have been developed to derive satisfactory compliance/stiffness equations of flexure hinges that can be found in Castigliano's second theorem [9], inverse conformal mapping [10] and empirical equations formed from finite-element-analysis (FEA) [11]. Depends on the t/R ratios of circular flexure hinges, the accuracy of the provided method could be different.

The stiffness-based load-deformation equation is given by [9]:

$$\begin{Bmatrix} F_x \\ F_y \\ M_z \end{Bmatrix} = \begin{bmatrix} K_{x-F_x} & 0 & 0 \\ 0 & K_{y-F_y} & K_{y-M_z} \\ 0 & K_{y-M_z} & K_{\theta_z-M_z} \end{bmatrix} \cdot \begin{Bmatrix} u_x \\ u_y \\ \theta_z \end{Bmatrix} \tag{1}$$

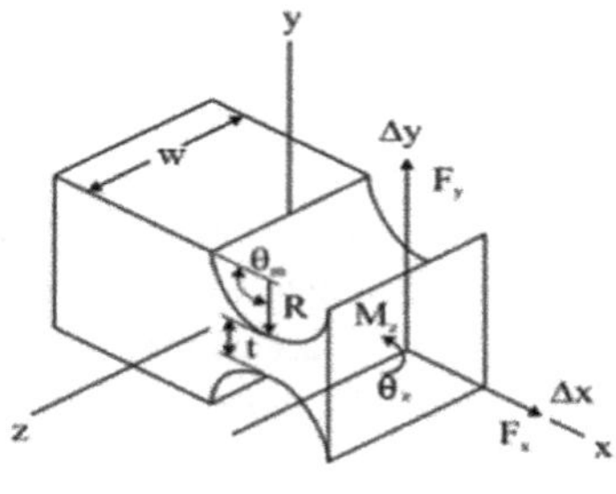

Fig. 2. Geometric parameters of flexure hinge

The stiffness matrix is:

$$[K_1] = \begin{bmatrix} K_{x-F_x} & 0 & 0 \\ 0 & K_{y-F_y} & K_{y-M_z} \\ 0 & K_{y-M_z} & K_{\theta_z-M_z} \end{bmatrix} \tag{2}$$

Simplified equations given by Paros and Weisbord [12] is:

$$\frac{\theta_z}{M_z} = \frac{9\pi R^{\frac{1}{2}}}{2Ewt^{\frac{5}{2}}} = \frac{1}{K_{\theta_z M_z}} \tag{3}$$

Where: t is The neck thickness (mm); R is The radius of the circular hinge (mm); E is the Elastic modulus (N/mm^2); w is the width of the hinge (mm); θ_z, Δ_x, Δ_y are the rotation angle, x-deformation and y-deformation, respectively; M_z, F_y, F_x are the moment and forces acting on the corresponding axes.

3.3 Flexural Prismatic Joint

Figure 3 shows a flexural prismatic joint which can be considered as a parallel structure with an infinitely rigid platform B_1B_2 and two similar limbs A_1B_1 and A_2B_2 [13]. Each limb A_iB_i (i = 1, 2) is a serial chain with two flexure hinges A_i, B_i and one link A_iB_i.

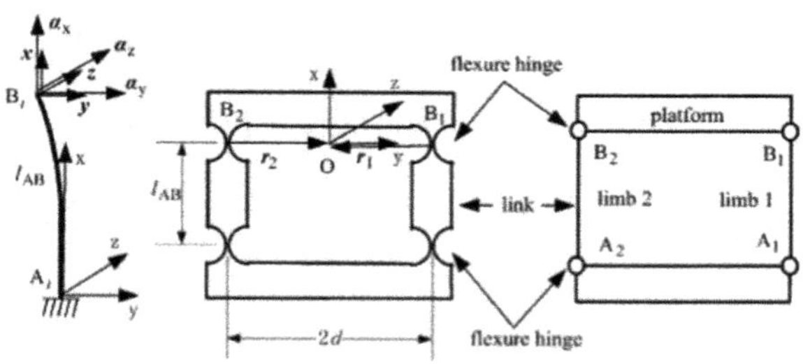

Fig. 3. Flexural prismatic joint [13]

3.4 Circular Prismatic Mechanism

In this study, the circular prismatic mechanism (Fig. 4) is used to transfer vibration into the pressing chamber. Divide the circular prismatic mechanism into a finite number of elements which can be considered as a serial chain with two flexure hinges and one link. The width of each element is equal to dw. The flexural hinge is now a system of a finite number of parallel prismatic joints. The stiffness of an element is as follow [14]:

$$K_S = \frac{4Et^{5/2}}{9\pi l^2 R^{1/2}} dw \tag{4}$$

Assume that the thicknesses of each element are equal at two hinges, the stiffness of the circular prismatic mechanism will be then determined by:

$$K_S = \frac{4Et^{5/2}}{9\pi l^2 R^{1/2}} \int_0^{2\pi} r d\theta \tag{10}$$

Where r is the average radius of the circular grooves. The stiffness of the circular parallel flexure hinge should be designed so that it can satisfy the following condition:

- Maintain the dynamical behavior of the system
- Maintain the vibration direction and vibration parameters.

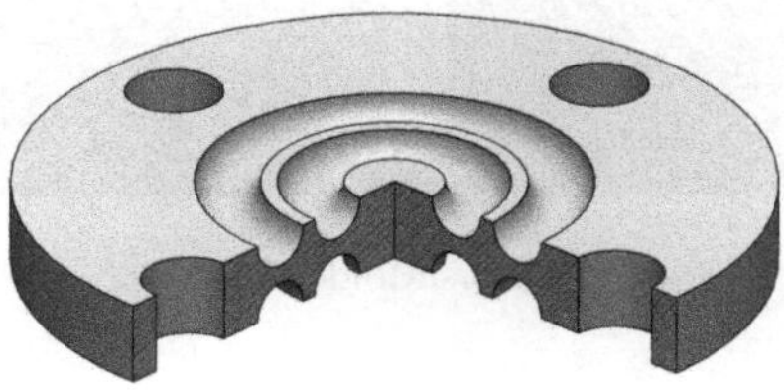

Fig. 4. Flexure hinge 3D model with cross section

The final dimensions of the flexure hinge are given in Table 2.

Table 2. Flexure hinge parameters

R (mm)	l (mm)	t (mm)	K_s (N/mm)
2	5.5	1.14	31250

3.5 Experimental Results

The experimental setup is shown in Fig. 5. The piezoelectric actuator can create frequency ranging from 0 to 4 kHz with amplitude ranging from 0 to 15 μm which depends on the load pressing on the PZT. The hydraulic system can generate pressure up to 100 bar. The heating system can generate up to 1000 °C. The mold product is shown in Fig. 6.

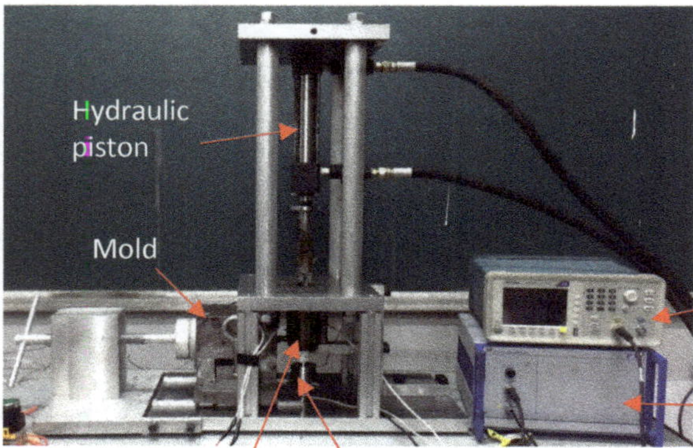

Fig. 5. Experimental setup

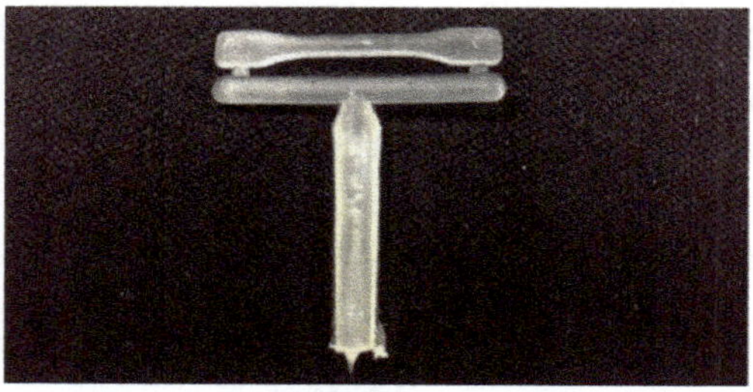

Fig. 6. Molded part

4 Conclusion

This paper presented the development of a vibration-assisted injection molding system. The system includes the vibration system, the pressing system, the heating system and the mold. The vibration system is designed using a compliant mechanism to transfer vibration generated by a piezoelectric actuator. The compliant mechanism which is a circular prismatic joint, has a stiffness of 31250 N/mm and was made from 7075 aluminum alloy. The high frequency vibration is generated by a piezoelectric actuator which helps increasing the plastic flow in the molding system. The system worked successfully with better results compared to conventional molding process.

Acknowledgement. This work belongs to the project in 2021 funded by Ministry of Education and Training, hosted by Ho Chi Minh City University of Technology and Education, Vietnam.

References

1. Alan, M.T., Akihisa, K., John, P.C.: Experimental Determination of Optimized Vibration-assisted Injection Molding Processing Parameters for Atactic Polystyrene, Specialized Molding Techniques, pp. 245–252. William Andrew Publishing, Oxford (2001)
2. Kikuchi, A., Coulter, J.P., Angstadt, D., Pinarbasi, A.: Localized material and processing aspects related to enhanced recycled material utilization through vibration-assisted injection molding. ASME. J. Manuf. Sci. Eng. **127**, 402–410 (2005)
3. Yin, X., He, G., Jinping, Q.: Response of cavity pressure to vibration parameters during VAIM process. Polym. Bull. **57**, 243–251 (2006)
4. Youbing, L., Kaizhi, S.: Self-reinforced high-density polyethylene prepared by low frequency vibration-assisted injection molding. II: microstructure investigation. J. Macrom. Sci. Part B **49,** 242–249 (2010)
5. Kikuchi, A., Coulter, J.P., Gomatam, R.R.: Assessing the effect of processing variables on the mechanical response of polysytrene molded using vibration-assisted injection molding process. J. Appl. Polym. Sci. **99**, 2603–2613 (2006)
6. Youbing, L., Yongheng, L., Xueqin, G., Yi, Y., Wenti, K., Kaizhi, S.: Structure and properties of polyethylene prepared via low- frequency vibration-assisted injection molding, wiley periodicals. Inc. J. Polym. Sci. Part B Polym. Phys. **43**, 13–21 (2005)
7. Xiaochun, Y., Guangjian, H., Qu, J.: Response of cavity pressure to vibration parameters during VAIM process. Polym. Bull. **57**, 243–251 (2006)
8. Roopa, R., Navin, K.P., Kulkarni, S.M.: Effect of flexure beam geometry and material on the displacement of piezo actuated diaphragm for micropump. IOP Conf. Ser. Mater. Sci. Eng. **310** (2018)
9. Lobontiu, N.: Compliant Mechanisms: Design of Flexure Hinges. CRC, Boca Raton (2003)
10. Tseytlin, Y.: Notch flexure hinges: an effective theory. Review of Scientific Instruments (2002)
11. Smith, S.T., Chetwynd, D., Bowen, D.: Design and assessment of monolithic high precision translation mechanisms. J. Phys. E **20**, 977–983 (1987)
12. Smith, S.T.: Flexures: Elements of Elastic Mechanisms, pp. 180–183. CRC Press, Baco Raton (2010)
13. Pham, H.-H., Chen, I.-M.: Stiffness modeling of flexure parallel mechanism. Precis. Eng. **29**, 467–478 (2005)
14. Smith, S.T., et al.: Design and assessment of monolithic high precision translation mechanisms. J. Phys. E Sci. Instrum. **20**, 977 (1987)

Impact of Building Envelope Design on CO2 Footprint Industrial Buildings: A Case Study in Vietnam

Tuan Tran(✉)

University of Architecture Ho Chi Minh City (UAH), 196 Pasteur Dist. 3, Ho Chi Minh City, Vietnam
tuan.tranhuuanh@uah.edu.vn

Abstract. As part of economic growth, the number of industrial buildings in many developing countries like Vietnam has rapidly increased. Energy consumption of industrial buildings accounts for a significant amount of total energy consumption of all economic sectors. Among many building components or systems, building envelope protects the indoor environment and provides indoor comfort against the adverse environment, plays an important role in reducing building energy consumption, thus minimizing CO2 footprint impact to the environment. In this paper, building envelope design alternatives for a conceptual design of an industrial project in Vietnam are proposed and analyzed using building energy simulation. The paper recommends designing future building envelopes for industrial buildings, especially in countries with similar climate type like Viet Nam.

Keywords: Building envelope · Carbon footprint · Industrial projects · Hot and humid climate · Modeling

1 Introduction

Building envelope maintains a comfortable indoor environment from outside environment thanks to balancing deployment of passive design as well as mechanical systems. Building envelopes and their components are designed to meet user's comfort with respect to functional, aesthetic, environmental, and social requirements [1]. Technically, sustainable design's utmost goal focuses on reducing building energy consumption without compromising on user comfort.

Most industrial buildings in developing countries such as Vietnam have excessively high energy consumption. Energy demand for the industrial sector is the highest among other sectors in the economy, accounting for 40% of the total energy demand of Vietnam [2]. There is a significant growth in factory and warehouse projects in Vietnam during recent years. In 2029, the total built industrial projects in Vietnam Northern region and Southern region has increased by 25.2% and 18%, respectively [3].

This paper aims to study the impact of building envelope on building performance, especially highlight the role of building envelope features of industrial buildings in sustainability in terms of energy savings and carbon footprint impact. A case study with

B. T. Long et al. (Eds.): MMMS 2020, LNME, pp. 52–60, 2021.
https://doi.org/10.1007/978-3-030-69610-8_7

a conceptual design of a factory in Vietnam has been carried out by testing building envelope design alternatives using whole energy modeling analysis.

2 Literature Review

2.1 Building Envelope of Industrial Projects

Building envelope is one of four systems comprised of a building [4]. The term "envelope" has a function to address natural forces and human values. Building envelope is defined as building components separating interior spaces from the outdoor environment, including below-grade basement walls, foundation, and floor slabs [5]. Specifically, building envelope systems include below-grade construction, exterior wall (both structural and nonstructural component), fenestration, roofs, and atria.

A building envelope plays a complex role with many functions. For example, a wall might have following functions: structural, water peneration, air infiltration, condensation, differential movement, energy conservation, sound attenuation and transmission, fire safety, security, maintenance, constructability, durability, aesthetics, and economy.

Industrial buildings are mostly built with steel structure and steel-concrete structure, which is different from civic buildings. Industrial building with large span, massive roof areas require high thermal resistance. Industrial building also has many internal loads from industrial processes beside heat gains from human bodies, electronic devices, lighting devices [6].

2.2 Building Envelope and Sustainability

Building envelope or building enclosure is the main component of a building to protect the indoor environment and provide indoor comfort against the adverse environment. Building envelope has an important role in regulating the amount of external heat gain and losss, sizing HVAC systems and load, thus the amount of energy consumption to ensure thermal comfort and other operation purposes. Such energy consumption causes a certain amount of GHG emissions to the natural environment. Energy efficient performance assessment of building envelope with sustainable design alternatives confirmed the significant role of the sustainable design concept to building sustainability and the capability of sustainable building envelope design to achieve building sustainability [7].

Building envelope also can be divided into two types: (1) single skin façade and (2) double or multiple skin façade. Single skin building envelope is a basic type of enclosure which simply consists of walls with openings and roof with skylight. Double skin or multiple skin façade has several properties that enhance the performance of the building such as higher thermal resistance, noise barrier, natural ventilation, daylighting…Some multiple skin façade building types are box window, shaft box, corridor, and multi-story double-skin façade [1].

2.3 Hot and Humid Climates

Climates of the earth are regulated by complex interactions among many factors such as the sun, ocean, atmosphere, clouds, mountains, forests…Climate can be classified according to temperature and precipitation. A well-known climate classification by ASHRAE which has been widely used in validated energy modeling, divide climate into very hot-humid (1A), very hot-dry (1B), hot-humid (2A), hot-dry (2B), warm-humid (3A), warm-dry (3B), warm-marine (3C), mixed-humid (4A), mixed-dry (4B), mixed-marine (4C), cool-humid (5A), cool-dry (5B), cool-marine (5C), cold-humid (6A), cold-dry (6C), very cold (7) and subarctic (8). In this system, Vietnam's climate is considered as hot-humid (1A) [8].

2.4 CO2 Footprint

Every human's daily activities directly or indirectly cause greenhouse gas (GHG) emissions. GHGs exist naturally in the Earth's atmosphere at equilibrium state. However, excessive human activities are increasing the levels of GHG's in the atmosphere, causing greenhouse gas effect and global warming. Among many GHGs, carbon dioxide (CO2) is the most common GHG emitted by human activities. However, it is noted that other GHGs last differently, absorbing different amount of heat [9]. The environmental impact called global warming potential" (GWP) of some GHGs as follows (Table 1):

Table 1. Greenhouse gas's Global Warming Potential (GWP) [9]

No.	Greenhouse gas	Global Warming Potential (GWP)
1	Carbon dioxide (CO_2)	1
2	Methane (CH_4)	25
3	Nitrous oxide (N_2O)	298
4	Hydrofluorocarbons (HFCs)	124–14,800
5	Perfluorocarbons (PFCs)	7,390–12,200
6	Sulfur hexafluoride (SF_6)	22,800
7	Nitrogen trifluoride (NF_3)[3]	−17,200

GHG carbon footprint is a measurement of carbon emissions caused by certain activities. The calculation of total carbon footprint of an activity is quite complex because of inadequate information about complex processes of storing and releasing of carbon dioxide. Wright, Kemp, and William (2011) proposed that carbon footprint is the total amount of carbon dioxide equivalent using the relevant 100-year global warming potential (GWP100) [10].

3 Methods

This paper used energy modeling with Integrated Environmental Solutions Virtual Environment (IESVE) software to virtually predict the whole energy consumption of an industrial project in Vietnam. IESVE is a well-known modeling tool for whole building performance simulation which is qualified and recommended by many green building certification systems such as LEED (US) and BREAM (UK). Hourly weather database to be fed into the simulation tool is the Typical Meteorological Year weather data of Tan Son Nhat Airport (VNM_SVN_Ho.Chi.Minh-Tan.Son.Nhat.Intl. AP.489000_TMYx.2003-2017.epw).

3.1 Building Information

The industrial project is a factory consisting of 03 main buildings: Factory (Production, Warehouse, and Utility), Office (Office and Lab), and Canteen (Canteen and Kitchen based on the design at the conceptual design stage. All space areas are summarized in Table 2.

Table 2. Space summary

Building use	Conditioned area (m^2)	Un-conditioned area (m^2)	Total area (m^2)
Production area	6,994	0	6,994
Warehouse	1,763	7,860	9,624
Utility	1,768	0	1,768
Office & Lab	1,132	0	1,132
Canteen	515	0	515
Totals	**12,173**	**7,860**	**20,033**

Construction types and U-values of the building envelopes such as exterior walls, roofs are defined in Table 3.

Table 3. Baseline building envelope's U-values

Material name	Description	U-Value (W/m^2K)	Remark
Exterior wall type 1	Aluminum cladding panel 150 mm	2.269	Non air-conditioned warehouse areas
Exterior wall type 2	Conventional fired clay brick 210 mm	1.865	Office, Lab, Kitchen and Utility areas
Exterior wall type 3	PU panel 50 mm THK	3.555	Production and air-conditioned warehouse areas
Window	Double clear glazing w/alum. Frame	U = 2.952 SHGC = 0.7	Canteen and office (assumed WWR = 40% for office & lab exterior walls)
Roof type 1	Metal roof w/polynum sheet	4.024	Factory area
Roof type 2	R.C concrete slab	0.731	Office and Canteen

The internal loads such as occupancy, internal lighting power density, plug load, and HVAC… are presented in Table 4 as follows:

Table 4. Baseline internal loads

Building use	Total area (m^2)	No. of people	LPD (w/m^2)	Plug load (w/m^2)
Production area	6,994	82	14	5
Warehouse	9,624	–	7	5
Office	1,132	30	24	25
Lab		5	24	25
Canteen	515	117	16	25
Kitchen		10	16	5
Utilities	1,768	–	8	5
			Area-weighted average: **10.72**	Area-weighted average: **5.03**

3.2 Mechanical Information

The load schedule is based on the detailed head count calculation. There are 3 working shifts (8:00–16:00, 16:00–24:00, 0:00–8:00). The Production area load is at 100% capacity at Shift 1 but reduced to 30% at Shift 2 and Shift 3 (only GMP-2 line works at Shift 2 & Shift 3). The Lab has 3 shifts at full capacity (100%). Office and Canteen works Shift 1 only. The load schedule is modeled in the energy models and graphically presented in Fig. 1.

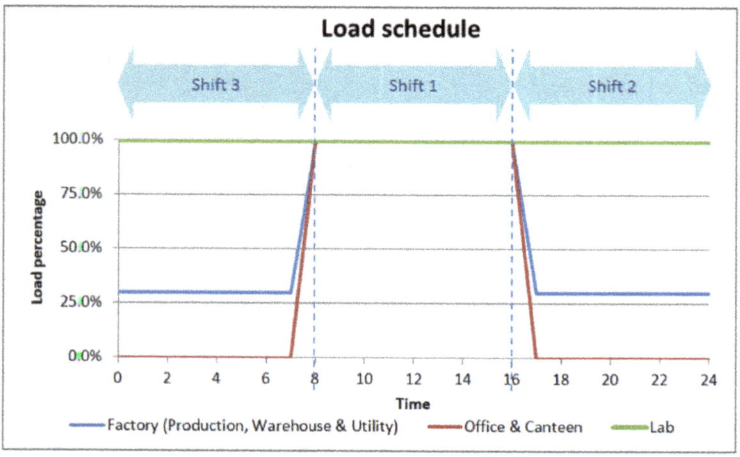

Fig. 1. Load schedules

MVAC system of the factory includes VAV chiller HVAC is used for the Production area and air-conditioned warehouse (Raw warehouse and FG warehouse area). Direct expansion (DX) coil split system is used for Office and Canteen. The systems' efficiencies of the HVAC systems are assumed to comply with ASHRAE 90.1 standard. The infiltration rate of the building is assumed to be 0.1ACH, the ventilation rates (mechanical ventilation) are 5ACH for office and canteen building, 10ACH for factory building. Exterior lighting power is 1.0w/sqm of landscape area.

3.3 Energy Efficiency Design Analysis

The estimated energy consumption in Sect. 3 is based on inputs of the building envelope, internal loads of the factory at the current conceptual design called the Baseline. The Energy efficiency design analysis is to reduce the building energy consumption of the Baseline by potential energy saving design alternatives such as reducing internal heat gains through building envelopes and internal loads. The energy modeling of energy efficient design analysis of the following alternatives will be carried out (Table 5).

Table 5. Improved building envelope thermal resistance alternatives

Material	Baseline	Alternative 1	Alternative 2
Exterior wall	PU panel U-3.555	AAC block 200mm THK U-0.704	PU panel U-3.555
Roof	Metal roof w/polynum sheet U-4.024	Metal roof w/polynum sheet U-4.024	Metal roof w/glass wool U-0.379

4 Results

The total annual energy consumption of the whole factory (including Factory, Office, and Canteen) is **5,577.27 Mwh** per year. Since the GFA of the project is 20,033 m^2, the Site Energy Use Intensity (EUI) is **278.40 kwh/m²** (5,577.27 Mwh/20,033 $m^2 \times 1{,}000 = 278.40$ kwh/m^2).

The results of energy consumptions of Alternatives are being compared against Baseline. Alternative 2 has the lowest reduction percentage (1.2%) while Alternative 4 yields the most savings (4.9%) (Fig. 2).

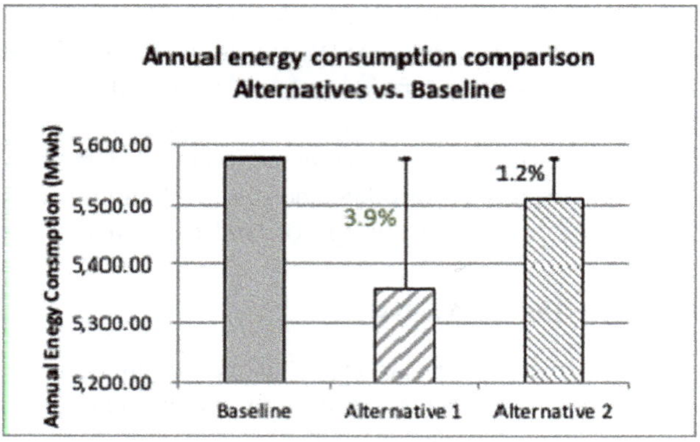

Fig. 2. Annual energy consumption of Alternative vs. Baseline

5 Discussions

The estimation of indirect GHG emissions is based on estimated annual electricity consumption for building services and manufactory process of the factory and the carbon emission rate of grid electricity of Vietnam. The annual electricity consumption of the factory of the Baseline design is 5,577.27 Mwh per year. The carbon emission rate is adopted from Vietnam's average electricity grid emission factor of 0.8154 tCO_2eq/MWh[1]. The annual carbon emission is 5,577.27 Mwh $\times$ 0.8154 tCO_2eq/MWh = **4,547.70 tCO2e** per year.

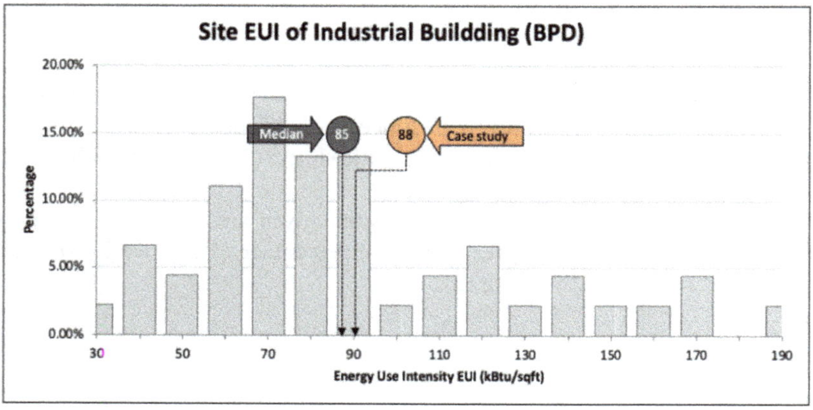

Fig. 3. Energy Use Intensity (EUI) of industrial building rank

[1] Source: World Bank report 2018 - Report No: ICR00004208.

The predicted annual energy consumption of the project is 5,577.27 Mwh per year, resulted in the CO2 emissions footprint of 4,547.70 tCO2e per year. The calculated Site Energy Use Intensity (EUI) of 278.40 kwh/m^2 (88.25 kBtu/sqft), is below the EUI median (85 kBtu/sqft) of industrial building stock obtained from BPD[2] shown in Fig. 3. Two design alternatives related to thermal resistance improvement of the building envelope were proposed and the resultant percentages of energy reduction are compared to against that of the baseline. Implementation of thermal resistance improvement could reduce energy consumption as well as GHG emissions to about **4.0%**. It is clearly that savings from improvement of roof's thermal resistance is higher that of improvement of external walls. That is because, roof surfaces are usually the most dominant factor among other building envelopes of industrial buildings.

6 Conclusions

This study has shown that building envelope of industrial buildings significantly affects building energy consumption and CO2 emissions to natural environment. While many sustainable design strategies on building envelope should be carefully considered, depending on specific site conditions and configurations, improvement of roof's thermal resistance is one of the most effective strategies for its dominant size and thermally interacting with natural forces under a hot and humid climate. Due to resource constraint, this paper only focused on thermal insulation, many other studies should be further investigated such as building envelope shapes, types and sizes of transparent components of building envelopes such as fenestration, clerestories, and multi-façade. Validation of energy simulation with actual performance will be also considered for further study.

References

1. Kumar, G., Raheja, G.: Design determinants of building envelope for sustainable built environment: a review. Int. J. Built Environ. Sustain. **3**(2), 111–118 (2016)
2. Danish Energy Agency: Vietnam Energy Outlook Report 2017
3. CBRE Vietnam: The Growing Prominence of Supporting Industry – Vietnam's Emerging Hubs and the Rise of Second Tier Cities, 2020 Vietnam Industrial Real Estate Market (2020). https://www.cbrevietnam.com/Vietnam-Property/pressrelease/vietnam-industrial-real-estate-market
4. Rush, R.D.: The Building Systems Integration Handbook, p. 464. Elsevier Science & Technology, Amsterdam (1991)
5. Arnold, C.: Building Envelope Design Guide – Introduction. National Institute of Building Science, November 2016. https://www.wbdg.org/guides-specifications/building-envelope-design-guide/building-envelope-design-guide-introduction. Accessed 01 Feb 2020
6. Wang, Y., Cao, Z.: Industrial building environment: old problem and new challenge. Indoor Built Environ. **26**(8), 1035–1039 (2017)

[2] Source: Building Performance Database (https://bpd.lbl.gov/).

7. Iwaro, J., Mwasha, A.: The impact of sustainable building envelope design on building sustainability using Integrated Performance Model. Int. J. Sustain. Built Environ. **2**, 153–171 (2013)
8. ANSI/ASHRAE/IES Standard 90.1-2010, Energy Standard for Building Except Low-Rise Residential
9. Matthew, B.: Green House Gases, CO2, CO2e, and Carbon: What do all these terms mean? Ecometria (2012)
10. Wright, L., Kemp, S., Williams, I.: Carbon footprinting': towards a universally accepted definition. Carbon Manage. **2**(1), 61–72 (2011)

Simulation and Optimization Study on Shear Thickening Polishing of Complex Surfaces

Duc-Nam Nguyen(✉), Van-Minh Pham, Thanh-Danh Lam, and Truong-Giang Nguyen

Faculty of Mechanical Engineering, Industrial University of Ho Chi Minh City, Ho Chi Minh City, Vietnam
nguyenducnam@iuh.edu.vn

Abstract. The complex surfaces have been applied in industrial production fields such as aeronautical engineering, mold engineering, artificial knee joints and bearings. In order to improve the applicability of these complicated surfaces, it must be machined to achieve the high quality with a small surface roughness. For that reason, a non-Newtonian fluid polishing method is studied and carried out to polish the complicated surfaces. The combination of the abrasive slurry and the non-Newtonian fluid will create a shear thickening zone for cutting process. As a results, the quality of the workpiece surface is greatly improved. This machining process is affected by polishing parameters such as inclination angle (*I*), height from the workpiece to the bottom of the tank (*H*), and velocity (*V*). In this paper, a simulation process is performed to investigate the effect of machining parameters on the spherical titanium alloys of Ti-6Al-4V through the ANSYS software. The pressure distribution will be presented and analyzed. In addition, the multi-responses optimization is used to optimize the maximum pressures on the workpiece surfaces in polishing process. According to the simulation and optimization results, the best machining parameters were selected for increasing the maximum pressure area on the workpiece surface in machining process. As a results, the optimal machining parameters such as the height of 11.5 mm, velocity of 2.5 m/s, and inclination angle of 16° were established for machining process.

Keywords: Shear thickening polishing (STP) · Non-Newtonian fluid · Complex surface · Inclination angle · Velocity · Optimization

1 Introduction

The complex surfaces have been fabricated for industrial applications such as automotive, optics, hip and artificial knee joints, robotics and bearings. In order to improve the applicability of these complicated surfaces, it must be machined to achieve the high quality with a small surface roughness. There are many different cutting methods that applied for machining the products for various industrial application [1, 2]. In addition, the lapping and polishing process with abrasive slurry was also studied to improve the surface roughness of complicated surfaces [3–6]. However, these methods are mainly used in the cutting process of the non-complex surfaces.

B. T. Long et al. (Eds.): MMMS 2020, LNME, pp. 61–69, 2021.
https://doi.org/10.1007/978-3-030-69610-8_8

For increasing the cutting performance and surface quality of complicated product, the shear thickening polishing methods is carried out and used to polish these surfaces [7–10]. The polishing fluid is a combination of non-Newtonian fluid and abrasives. It will become a solid when it is moved in the polishing process. The scratches on the workpiece surfaces were removed by the pressure of the non-Newtonian polishing fluid. The polishing conditions such as velocity, abrasive sizes, inclination angle, workpiece position and viscosity greatly affect the cutting efficiency, pressure value and quality of workpiece surfaces. Therefore, the optimization of machining parameters is one of the important problem to find the best polishing conditions. The Taguchi method has been widely applied in the optimization of machining process parameters [11–13]. The surface roughness of workpiece was explored by analysis of variance ANOVA method. Moreover, the mathematical models of surface roughness using response surface method and the genetic algorithm were also applied to find out the optimum cutting parameters [14, 15].

In this paper, a simulation process is performed to investigate the effect of machining parameters such as inclination angle (*I*), height from the workpiece to the bottom of the tank (*H*), and velocity (*V*) on the spherical titanium alloys of Ti-6Al-4V through the ANSYS software. The pressure distribution on the workpiece will be presented and analyzed in the simulation process. In addition, the multi-responses optimization is used to optimize the maximum pressures on the workpiece surfaces. Based on the simulation and optimization results, the best machining parameters were selected for increasing the maximum pressure zone on the workpiece surface during the polishing process.

2 Simulation Models and Results

2.1 Simulation Conditions

For simulation process, three parameters were selected including the inclination angle (*I*), height from the workpiece to the bottom of the tank (*H*), and velocity (*V*), that strongly affects the generated pressure on the workpiece surfaces. The simulation model is built as shown in Fig. 1. The radius of the polishing tank, diameter of workpiece and distance between the workpiece and the inner wall of tank are 150 mm, 40 mm and 10 mm respectively.

The boundary conditions of polishing fluid flow are defined by the non-Newtonian power law. The consistency and viscosity index of non-Newtonian polishing fluid inlet are installed of 0.62 and 1.5, respectively [16]. The simulation model is meshed with a total of 121129 nodes and 616277 elements in simulation process. The element meshing model is shown in the Fig. 2.

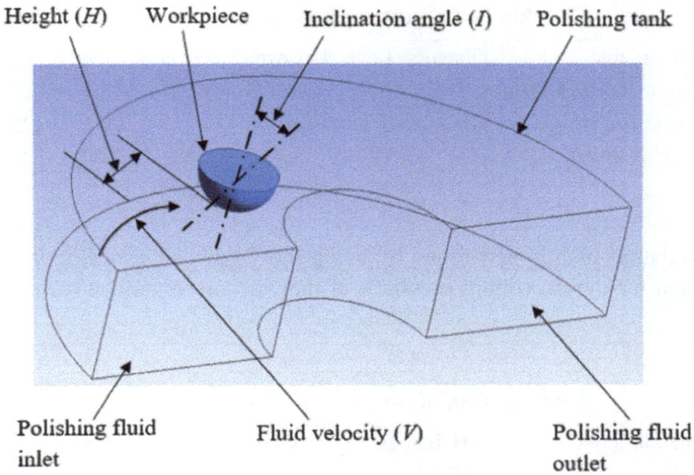

Fig. 1. Simulation model

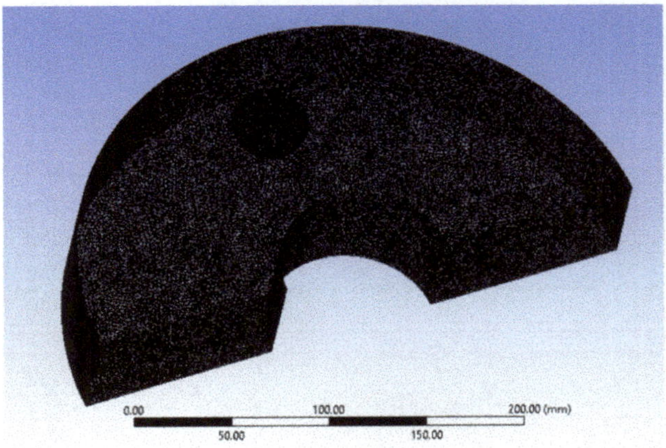

Fig. 2. Picture of element meshing

2.2 Simulation Results

The simulation process is performed with the change of three input parameters such as inclination angle (*I*), height from the workpiece to the bottom of the tank (*H*), and velocity (*V*). The maximum pressure values corresponding to the change of input parameters are collected under the simulation conditions. The orthogonal array for mixed level design for three factors are inclination angle (*I*), height (*H*), and velocity (*V*), as indicated in Table 1.

Table 1. Parameters factors and their level

Design variable	Factors	Level 1	Level 2	Level 3	Level 4	Level 5
Inclination angle: I (deg.)	A	0	4	8	12	16
Height: H (mm)	B	5	10	15	20	25
Velocity: V (m/s)	C	1.0	1.4	1.8	2.2	2.5

The simulation plan is performed by using the Taguchi L_{25} orthogonal array. The simulated data for the maximum pressures at the workpiece surface were collected, as in Table 2.

Table 2. Plan of simulations and their responses

No	Inclination angle I (degree)	Height H (mm)	Velocity V (m/s)	Maximum pressure P (Pa)
1	0	5	1.0	4287
2	0	10	1.4	6213
3	0	15	1.8	7062
4	0	20	2.2	10110
5	0	25	2.5	11100
6	4	5	1.4	4513
7	4	10	1.8	9117
8	4	15	2.2	11460
9	4	20	2.5	6234
10	4	25	1.0	3559
11	8	5	1.8	8056
12	8	10	2.2	11300
13	8	15	2.5	8563
14	8	20	1.0	2482
15	8	25	1.4	5813
16	12	5	2.2	9961
17	12	10	2.5	14650
18	12	15	1.0	11550
19	12	20	1.4	6369
20	12	25	1.8	4782
21	16	5	2.5	12380
22	16	10	1.0	3367
23	16	15	1.4	6862
24	16	20	1.8	8132
25	16	25	2.2	7329

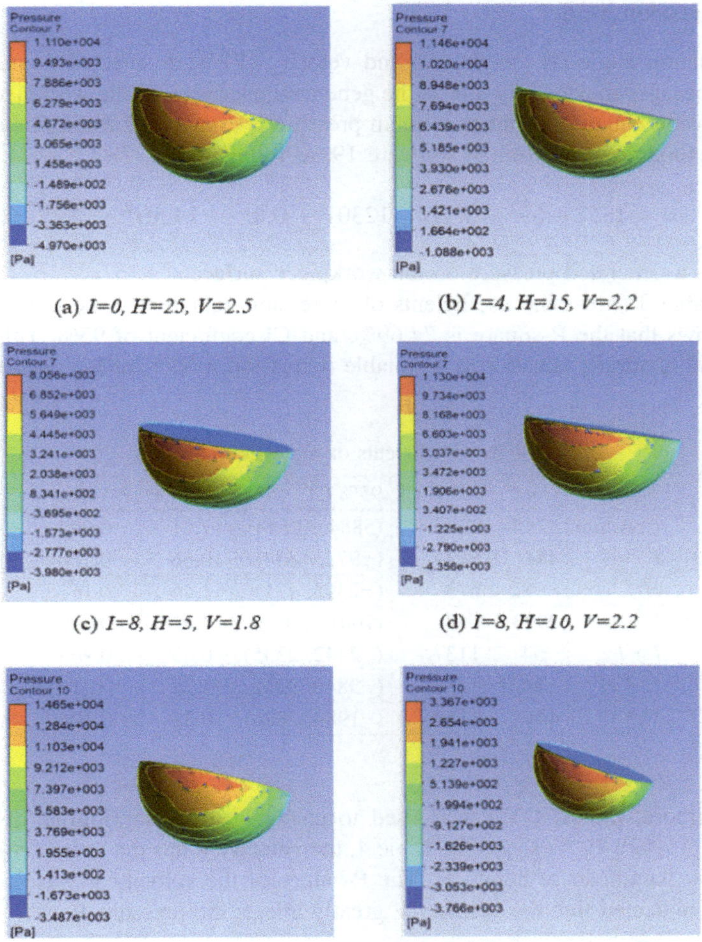

(a) $I=0, H=25, V=2.5$ (b) $I=4, H=15, V=2.2$

(c) $I=8, H=5, V=1.8$ (d) $I=8, H=10, V=2.2$

(e) $I=12, H=10, V=2.5$ (f) $I=16, H=10, V=1.0$

Fig. 3. Simulation results with different machining parameters

Several simulation results with different parameters of inclination angle, height and velocity are shown in Fig. 3.

3 Optimization Results and Discussion

As a simulation results, the low surface roughness of workpiece can be achieved in the zone of maximum pressure. It is nescessary to optimize the polishing parameters to generate the maximum pressure on the workpiece surface during polishing process.

3.1 Regression Model

The inclination angle (*I*), height (*H*) and velocity (*V*) were selected as main design variables because they greatly affects the generated pressures on the workpiece surface. The regression equation of the maximum pressure is established through the response surface methodology in Minitab software 19, as follow:

$$P = 1357 + 69I + 337H + 1230V - 0.8I^2 - 14.5H^2 + 789V^2 \tag{1}$$

where *P* is a generated pressure on the workpiece surface.

The Table 3 shows the coefficients of regression equation of the pressure *P*. The results shows that the R-square is 74.69%, and CI coefficient of 95%. This confirms that the full quadratic equation is a suitable regression model for the pressure *P*.

Table 3. Coefficients of regression equation

Term	Coef	SE Coef	95% CI	T-Value	P-Value
Constant	8214	1109	(5884, 10544)	7.41	0.000
I	441	673	(−972, 1854)	0.66	0.520
H	−982	673	(−2395, 431)	−1.46	0.161
V	2993	663	(1601, 4386)	4.52	<0.0001
I * *I*	−54	1137	(−2442, 2335)	−0.05	0.963
H * *H*	−1452	1137	(−3840, 936)	−1.28	0.218
V * *V*	444	1158	(−1989, 2876)	0.38	0.706

Furthermore, the ANOVA was used to calculate the contributions of polishing factors on pressure *P*. As shown in Table 4, the velocity *V* has the largest contribution, followed by parameter of height *H*. The P-values of the velocity *V* was smaller than 0.05. This indicated that the velocity *V* greatly affects the pressure *P*.

Table 4. The value of the regression model on pressure *P*.

Source	DF	Seq SS	Contribution	Adj SS	Adj MS	F-Value	P-Value
Model	6	140020062	57.91%	140020062	23336677	4.13	0.009
Linear	3	129954752	53.75%	129796707	43265569	7.65	0.002
I	1	2431013	1.01%	2431013	2431013	0.43	0.520
H	1	12052086	4.98%	12052086	12052086	2.13	0.161
V	1	115471653	47.76%	115313608	115313608	20.40	<0.0001
Square	3	10065310	4.16%	10065310	3355103	0.59	0.627
I * I	1	12596	0.01%	12596	12596	0.00	0.963
H * H	1	9222378	3.81%	9222378	9222378	1.63	0.218
V * V	1	830336	0.34%	830336	830336	0.15	0.706
Total	24	241778659	100.00%				

3.2 Optimal Results

This study aims to maximum the value of generated pressure on the workpiece surface in shear thickening process. The multi-response optimization process can be outlined as below:

Find: Variables (inclination angle *I*, height *H*, and velocity *V*).

Maximum the pressure P:

$$\text{Max P } (I, H, V) \tag{2}$$

s.t.

$$\begin{cases} 0^\circ \le I \le 16^\circ \\ 5\text{ mm} \le H \le 25\text{ mm} \\ 1.0\text{ m/s} \le V \le 2.5\text{ m/s} \end{cases} \tag{3}$$

Table 5 shows the optimal parameter results of polishing process include the height *H* of 11.5 mm, velocity *V* of 2.5 m/s, and inclination angle *I* of 16°. The optimal pressures on the workpiece surface was achieved about 12204.2 Pa. The prediction interval for responses were shown in Table 6. The confident interval of 95% is indicated that the optimal solutions are satisfactory. As a results, a good desirability of 0.8 was established to calculate the optimization problem.

Table 5. Optimal machining parameter results

Optimal parameters			Optimal result of *P* (Pa)	Composite desirability
Height, *H* mm	Velocity, *V* m/s	Inclination angle, *I* (degree)		
11.5	2.5	16	12204.2	0.799001

Table 6. Optimal results of response *P*

Response	Optimal solution	95% Prediction Interval (PI)
P	12204	(6407, 18001)

4 Conclusion

The simulation studies and the multi-responses optimization were carried out to investigate the effect of machining parameters such as inclination angle *(I)*, height from the workpiece to the bottom of the polishing tank *(H)*, and velocity *(V)* on the spherical workpieces. The main results are summarised as follows:

The velocity *V* has a great effect on the maximum pressure which distributed on the workpiece surface. However, the height from the workpiece to the bottom of the polishing tank has little effect on the maximum pressure. Meanwhile, the inclination angle only affects the distribution of generated pressure on the workpiece surfaces during machining process.

The multi-responses optimization was used to optimize the maximum generated pressure on the workpiece surfaces in polishing process. The maximum pressure achieves 12204 Pa under conditions of the height of 11.5 mm, velocity of 2.5 m/s and inclination angle of 16°. This optimal results are very useful for establishing experimental conditions for future works.

Acknowledgments. This research is funded by the Industrial University of Ho Chi Minh City, Vietnam through the University Research Foundation (No. 20/1.1CK03).

References

1. Li, B.X., Zhang, S., Wang, R.W., Fang, Y.J.: Toward understanding of metallurgical behaviours in dry machining of hardened steel: phase transformation and surface oxidation. J. Mater. Res. Technol. **8**, 3811–3821 (2019)
2. Sun, S., Brandt, M., Dargusch, M.S.: Thermally enhanced machining of hard-to-machine materials - a review. Int. J. Mach. Tools Manuf. **50**, 663–680 (2010)
3. Nguyen, D.N.: Study on improving the precision of form surface produced in elastic deformation molding process. Int. J. Adv. Manuf. Technol. **93**, 3473–3484 (2017)
4. Nguyen, D,N., Yuan, J.L., Wu, Z.: Experimental study on elastic deformation molding process for generating aspheric surface glass. Int. J. Adv. Manuf. Technol. **82**, 859–866 (2016)
5. Nguyen, D.N., Lv, B.H., Yuan, J.L., Wu, Z., Lu, H.Z.: Experimental study on elastic deformation machining process for aspheric surface glass. Int. J. Adv. Manuf. Technol. **65**, 525–531 (2013)
6. Nguyen, D.N., Chau, N.L., Dao, T.P., Prakash, C., Singh, S.: Experimental study on polishing process of cylindrical roller bearings. Measure. Control **52**, 1272–1281 (2019)
7. Nguyen, D.: Simulation and experimental study on polishing of spherical steel by non-Newtonian fluids. Int. J. Adv. Manuf. Technol. **107**, 763–773 (2020)
8. Li, M., Lyu, B., Yuan, J., Dong, C., Dai, W.: Shear-thickening polishing method. Int. J. Mach. Tools. Manuf. **94**, 88–99 (2015)
9. Lyu, B.H., He, Q.K., Chen, S.H., Shao, Q., Chen, Y., Geng, Z.Y.: Experimental study on shear thickening polishing of cemented carbide insert with complex shape. Int. J. Adv. Manuf. Technol. **103**, 585–595 (2019)
10. Nguyen, D.N., Dao, T.P., Prakash, C., Singh, S., Pramanik, A., Krolczyk, G., Catalin, P.: Machining parameter optimization in shear thickening polishing of gear surfaces. J. Mater. Res. Technol. **9**, 5112–5126 (2020)
11. Sanjeev, K., Lalta, P., Sandeep, K., Vinay, K.: Physico-mechanical and Taguchi-designed sliding wear properties of Himalayan agave fiber reinforced polyester composite. J. Mater. Res. Technol. **8**, 3662–3671 (2019)
12. Nguyen, D.N., Ho, N.L., Dao, T.P., Chau, N.L.: Multi-objective optimization design for a sand crab-inspired compliant microgripper. Microsyst. Technol. **25**, 3991–4009 (2019)
13. Shin, H., Yoo, D., Lee, J., Lee, S., Yoon, Y.: Optimized mix design for 180 MPa ultra-high-strength concrete. J. Mater. Res. Technol. (8), 4182–4197 (2019)

14. Pramanik, A., Islam, M.N., Basak, A.K., Dong, Y., Littlefair, G., Prakash, C.: Optimizing dimensional accuracy of titanium alloy features produced by wire electrical discharge machining. Mater. Manuf. Processes **34**, 1083–1090 (2019)
15. Prakash, C., Kansal, H.K., Pabla, B.S., Sanjeev, P.: Multi-objective optimization of powder mixed electric discharge machining parameters for fabrication of biocompatible layer on β-Ti alloy using NSGA-II coupled with Taguchi based response surface methodology. J. Mech. Sci. Technol. **30**, 4195–4204 (2016)

Optimal Design Structure of Sleeper Bus to Enhance Injury Safety of Human in Frontal Collision

Thanh-Tam Nguyen[1], Cong-Thanh Nguyen[1], Van-Sy Nguyen[1], Thanh-Danh Lam[2], and Duc-Nam Nguyen[2](✉)

[1] Faculty of Automotive Engineering Technology, Industrial University of Ho Chi Minh City, Ho Chi Minh City, Vietnam
[2] Faculty of Mechanical Engineering, Industrial University of Ho Chi Minh City, Ho Chi Minh City, Vietnam
nguyenducnam@iuh.edu.vn

Abstract. The models of driver and passenger injuries in frontal collision were built and investigated in this study. Based on the ECE R94 and R66 European safety standards, the influence of factors on human injuries in frontal collision will be simulated by the HYPERWORKS and MADYMO softwares. The simulation results indicated that the high acceleration causes people to be seriously injured in the head. The reason is that the structure of the car has been designed to absorb the less energy. Based on the analysis of the passenger injury mechanism, the structure of the sleeper bus will be designed and improved for increasing the ability to absorb collision energy by changing the shape and thickness of the chassis structure. As a result, the injury values of passengers will be significantly reduced and meet the requirements of the ECE R94 standard. Finally, the structure and restraint system of sleeper bus were optimized to improve the safety of passengers. According to the optimal results, the injury value of HIC passenger and the weight of the optimized structures was decreased by 18% and 5.4%, respectively.

Keywords: Sleeper bus · Frontal collision · Energy absorption · Optimal structure · ECE R94 standards · ECE R66 standards

1 Introduction

In recent years, a growing number of passengers are choosing sleeper coaches for long trips because of its comfort and convenience. However, there are many serious crashes involving the drowsiness of the driver, those accidents have injured and killed a lot of people. That's why researchers have done a lot of research on how to optimize frame and chassis of the vehicle, and passive safety features. In order to improve safety for passengers, the vehicle frame is designed which can absorb collision energy based change of shapes and sizes of chassis bars at the front end of the vehicle. In which collisions related to buses are a major cause of many serious casualties. In many previous studies, Thanh-Tam Nguyen [1] has applied LS-DYNA software to design and optimization of the frame structure to improve the structure of Bus Rollover. Khac-Huan Nguyen used

B. T. Long et al. (Eds.): MMMS 2020, LNME, pp. 70–77, 2021.
https://doi.org/10.1007/978-3-030-69610-8_9

ANSYS software to conduct the inspection of Transinco K46 passenger car frame according to ECE R66 standard [2]. As a results, this frame does not violate the safe space and has a lighter weight. In other research, the dynamics of the car and the durability of the skeleton when the car encountered a direct collision has been studied and calculated by Liu Wen Qing [3]. In addition, Yu C.L et al., has a research and design on the safety structure of passenger cars as a direct collision occurred [4]. Based on the results, the improvement process was conducted on the following two options: increasing the thickness of the vehicle and attaching an Honeycomb structure collision absorbers ensure a safe space for drivers and passengers. Zhang Weigang has an improvement on the design of the passenger car's skeleton and chassis in the frontal collision by designing a steel pipe that can absorb the energy of acceleration and stress [5]. A model simulating on the collision process of the car's front frame with LS-DYNA and Hypermesh software was investigated by Liu and Wang [6]. Muhammand Aamir Hassan also studied and analyzed the structural deformation and damage of occupants in the case of frontal, wall and column collisions, but the reduction of injuries to passengers has not been mentioned in this study [7]. In the same studies, Tian Fang and Hailiang Wang using the finite element method to analyze the static states of passenger car body structure [8, 9]. Beside that Liu Jiang optimizes the structure of passenger car under structural vibration analysis to improve the applicability of vehicle structure [10]. This studies applied the model design technique using SolidWorks software, built a finite element model with Hypermesh software in MADYMO environment, conducted simulation of frontal collision process of passenger cars. Assessing the extent of passenger injury through body damage assessment indicators. From that results, the front end to absorb collision energy combined with seat belts and airbags will be studied in this paper. The results of this study are not only practical and economical to ensure human safety but also meet the requirements of European standards ECE R94.

2 Simulation Models

Based on the 3D CAD model of the sleeper car from the manufacturer, using Hypermesh software in the LS-DYNA working environment, meshing and assigning the mass value of the parts such as the engine, fuel tank, air conditioning systems were conducted. The material for the skeletal structure and the floor-frame chassis structure are Q235 and Q345, respectively. Table 1 shows the material properties of these structure which used in the simulation process.

Table 1. Material properties

Material	Young's Modulus (GPa)	Poisson's ratio	Density (kg/mm^3)	Ultimate strength (MPa)
Q235	210	0,3	$7{,}85.10^{-6}$	235
Q345	210	0,3	$7{,}85.10^{-6}$	345

After setting the initial basic parameters, proceed to give the value of center coordinates, mass and moment of inertia for each chassis component has been separated as a working database of Hypermesh software with MADYMO environment (as shown in Fig. 1).

$$\begin{bmatrix} M & 0 & 0 & 0 & MZ & -MY \\ 0 & M & 0 & -MZ & 0 & MX \\ 0 & 0 & M & MY & -MX & 0 \\ 0 & -MZ & MY & I_{xx} & -I_{xy} & -I_{xz} \\ MZ & 0 & -MX & -I_{yx} & I_{yy} & -I_{yz} \\ -MY & MX & 0 & -I_{zx} & -I_{zy} & I_{zz} \end{bmatrix}$$

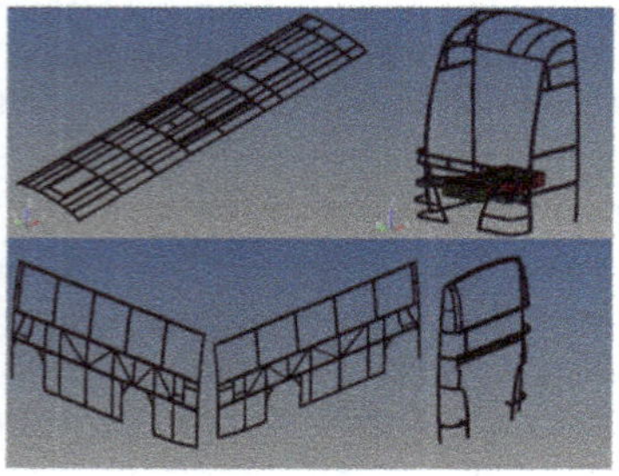

(a) (b)

Fig. 1. Moment of inertia (a) and structural model (b)

Carrying out a test to assess the possibility of violating safety space according to ECE R66 standard before participating in simulation of Dummy Hybrid III 50th and passive safety systems of straps, airbags for both drivers and passengers. Change the thickness of the chassis bars to satisfy ECE R66 safety space (Fig. 2).

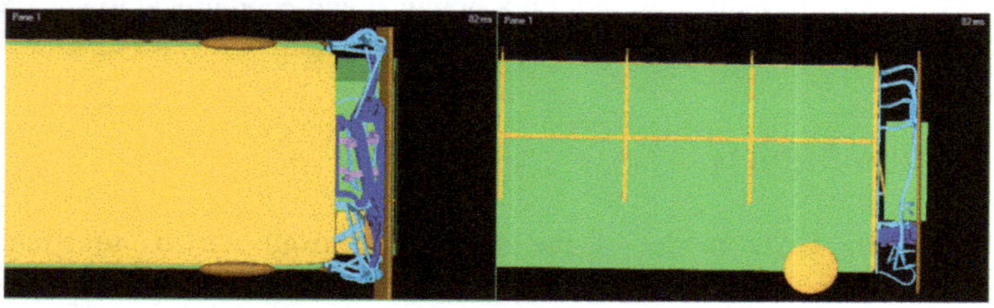

Fig. 2. The tested is displayed at a poor performer in a frontal offset crash

Considering that the safety of space is violated due to the structure of the chassis bars when the collision is heavily deformed because the thickness is not enough strong and the shape of the bars when colliding tends to be badly deformed. So, the thickness of the chassis was changed to 11 mm and kept the same material as the original. Safety of space results were guaranteed, from which the thickness of chassis of 11 mm is used for the simulation process (Fig. 3).

When increasing the thickness of the bars to 11 mm, a safe space is guaranteed, but besides watching the car's chassis deformed, the iron bars do not absorb energy well when collided adverse form. Therefore, we have an idea of changing the structure chassis to be able to absorb collision energy while ensuring a safe space (Fig. 4).

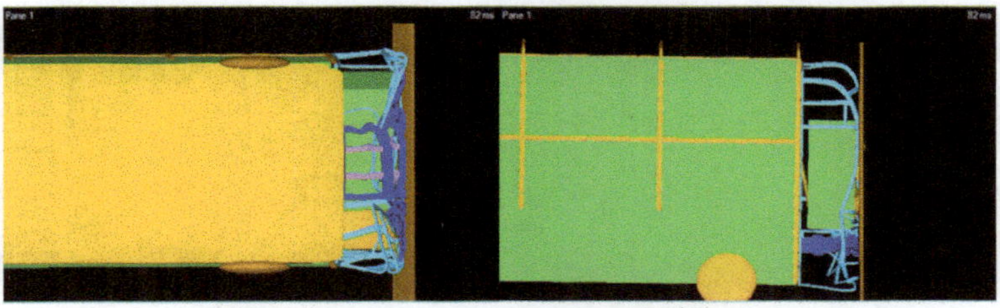

Fig. 3. Chassis conforms to safety standards

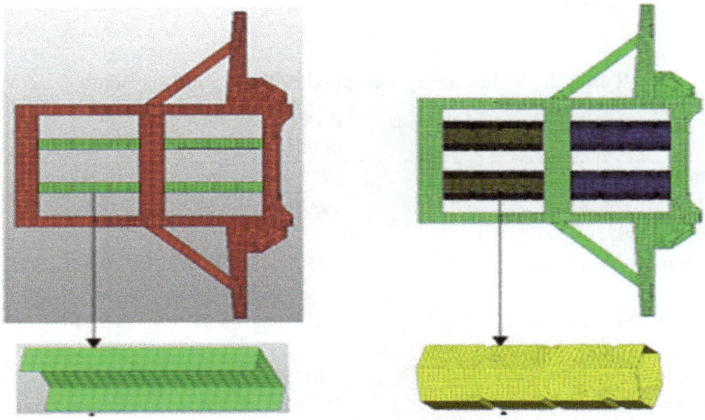

Fig. 4. FE modeling improved texture for crash analysis

3 Results and Discussion

Using MADYMO Protocol Rating software to evaluate the damage of passengers and drivers according to ECE R94. Through HIC head injury indicators (Fig. 5 and Tables 2, 3).

$$HIC = max\left[\frac{1}{t_2 - t_1}\int_{t_1}^{t_2} a(t)\mathrm{d}t\right]^{2,5}(t_2 - t_1) \tag{1}$$

t1, t2: time of the change in maximum acceleration value.

t2 − t1: maximum range 36 ms.

a: acceleration over time at the center of the dummy's head.

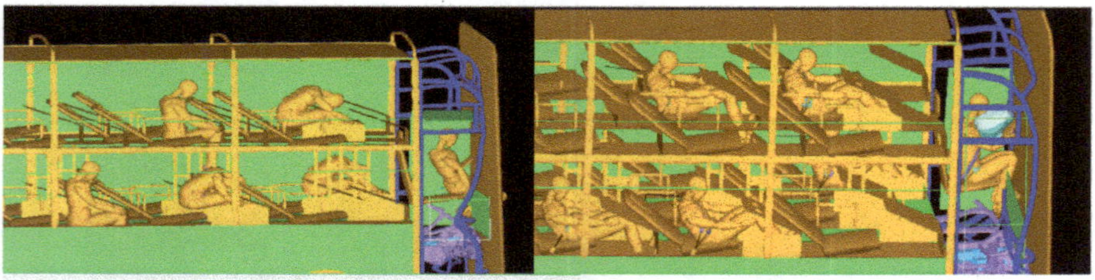

(a) (b)

Fig. 5. Simulation results without safety system (a), Collision simulation results in case of equipping with safety systems (b)

Table 2. Value injury equipped without safety systems

Index		Standards index	PAX 1 details	PAX 2 details	PAX 3 details	Driver details
Head	HIC36ms	1000	3777	3685	3588	4019
	Resultant Acc. 3 ms exceedence (g)	80	148.47	147.17	150.28	225.31
Neck	Pos Shear level exceeded (kN)	3,1	0.55	0.56	0.57	1.64
	Neg Shear level exceeded (kN)	3,1	1.21	1.12	0.94	1.64
	Tension level exceeded (kN)	3,3	6.16	6.10	6.02	2.34
	Extension (Nm)	57	159.17	157.21	158.17	116.62
Chest	Compression (mm)	50	63.8	61.2	54.8	78.9
	Viscous Criterion (m/s)	1	1.24	1.10	0.90	2.14

Table 3. Value injury equipped with safety systems

Index		Standards index	PAX 1 details	PAX 2 details	PAX 3 details	Driver details
Head	HIC36ms	1000	890	838	850	1965
	Resultant Acc. 3 ms exceedence (g)	80	77.45	73.92	77.68	128.65
Neck	Pos Shear level exceeded (kN)	3,1	0.35	0.32	0.29	0.47
	Neg Shear level exceeded (kN)	3,1	1.81	1.41	1.58	1.90
	Tension level exceeded (kN)	3,3	2.91	2.91	2.87	3.55
	Extension (Nm)	57	43.15	41.85	43.96	70.97
Chest	Compression (mm)	50	49.2	53.5	47.2	38.7
	Viscous Criterion (m/s)	1	0.37	0.37	0.28	0.21

In order to increase safety for passengers and drivers in the event of a collision. From studying the theoretical basis of collision on energy absorption, we proceeds to change parameters: the angle of setting the belt buckle, the height of the shoulder belt slider, the thickness of the bars improved chassis and timing of airbag activation in a collision through the variables X1, X2, X3, X4. In this experiment. The values of the variables are changed to reduce the damage, ensuring that the damage values are smaller than the allowable values and meet the European safety standard ECE R94.

Using regression analysis method, conduct analysis of orthogonal test results (Figs. 6 and 7):

$$y = a_0 + \sum\nolimits_{i=1}^{s} a_i x_i + \sum\nolimits_{i=1}^{s} a_{ii} x_i^2 + \sum\nolimits_{i<j}^{s} a_{ij} x_i x_j \qquad (2)$$

Where, $i = 1, 2, .., j$.

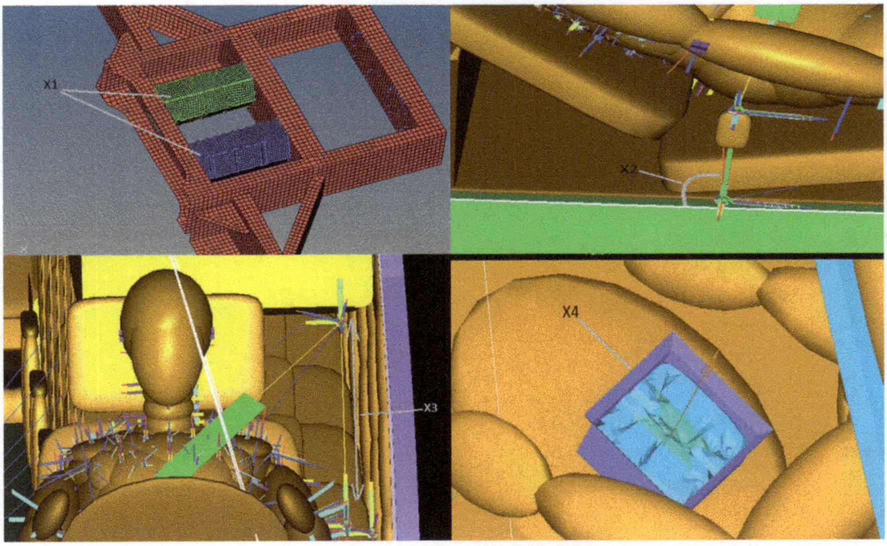

Fig. 6. Variables X1, X2, X3, X4 are selected for optimal experiments

	x1	x2	x3	x4	x1x1	x2x2	x3x3	x4x4	x1x2	x1x3	x1x4	x2x3	x...	x...	HIC36	a3ms	epnguc
1	8.0	90.0	50.0	.030	64.00	8100.0	2500.0	.0009	720.00	400.0	.240	4500.0	2.7	1.5	788.0	78.5	41.6
2	9.0	95.0	70.0	.040	81.00	9025.0	4900.0	.0016	855.00	630.0	.360	6650.0	3.8	2.8	583.0	70.3	27.3
3	9.5	100.0	90.0	.050	90.25	10000.0	8100.0	.0025	950.00	855.0	.475	9000.0	5.0	4.5	1696.0	120.0	34.7
4	7.5	105.0	40.0	.030	56.25	11025.0	1600.0	.0009	787.50	300.0	.225	4200.0	3.2	1.2	783.0	76.2	40.1
5	8.5	110.0	60.0	.050	72.25	12100.0	3600.0	.0025	935.00	510.0	.425	6600.0	5.5	3.0	1782.0	125.0	33.7
6	7.0	115.0	80.0	.040	49.00	13225.0	6400.0	.0016	805.00	560.0	.280	9200.0	4.6	3.2	453.0	59.9	33.9

Fig. 7. Combining linear regression analysis methods to set up regression equations

Regression equations of conditional indicators:

$$HIC = 8671.611 - 135.591x_3 - 0.040x_2^2 - 44157.608x_4^2 - 6.581x_1x_2 + 2664.049x_3x_4 \quad (3)$$

$$a_{3ms} = -127.189 + 5.941x_3 + 0.003x_2 - 4338.099x_4^2 + 0.287x_1x_2 - 114.059x_3x_4 \quad (4)$$

Using the R coefficient to assess the reliability of the regression equation, the reliability of the equations HIC, a_{3ms} are 1, 1 respectively. Therefore, it can be assumed that the regression equations order reaction surface satisfies exact requirement.

Next, we optimize the variables to bring out the most beneficial values of safety by GA moving algorithm method in Matlab, the results X_1 = 8.8988; X_2 = 104.9590; X_3 = 79.2861; X_4 = 0.037 (Table 4).

Table 4. Value injury of the optimal results

Index		Standards index	PAX 1 details	PAX 2 details	PAX 3 details	Driver details
Head	HIC36ms	1000	761	806	735	440
	Resultant Acc. 3 ms exceedence (g)	80	71.71	74.12	66.18	52.15
Neck	Pos Shear level exceeded (kN)	3,1	0.32	0.33	0.41	0.27
	Neg Shear level exceeded (kN)	3,1	1.65	1.75	1.64	0.82
	Tension level exceeded (kN)	3,3	2.62	2.81	2.47	0.85
	Extension (Nm)	57	42.54	42.87	47.22	43.12
Chest	Compression (mm)	50	48.6	47.2	47.2	24.8
	Viscous Criterion (m/s)	1	0.32	0.35	0.31	0.12

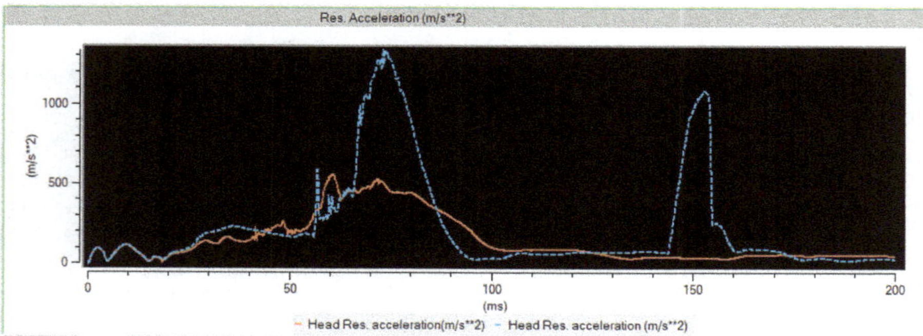

Fig. 8. Chart of driver's head acceleration

Carrying out improvements in shapes of chassis bars at the front of the vehicle for the purpose of absorbing most collision energy which makes passengers injured. The simulation results showed that the passengers safety improved and met the requirements of European ECE R94 standards (as shown in Fig. 8).

4 Conclusions

The injuries of passengers and drivers when a car collided head-on has been simulated and analyzed in this study. Proposing solutions to improve, test and evaluate the effectiveness of each plan to minimize injuries to passengers and drivers. In addition, an optimal and most suitable plan to ensure the safety of the driver and passengers on the vehicle according to ECE R94 standard has been proposed. The simulation results show that the driver's injuries and fatalities have been significantly reduced when the parameters of the belt parts, airbags and the front structure of car structures are optimized and improved. The driver's injuries at the head and chest are always within the safety value range specified by ECE R94.

References

1. Cao, L., Nguyen, T.T., Huang, X., Pang, J.: Study on Bus Rollover Crashwordthiness Based on Tube Filling Method (2011)
2. E/ECE/324 Regulation No. 66, Uniform technical prescriptions concerning the approval of large passenger vehicles with regard to the strength of their superstructure (1995)
3. Liu, W.Q.: Design of Experiments, 224, Tsinghua University (2008)
4. Yu, C.L., Hong, C.N.: Structural design optimization of the body section using the finite element method. SAE Technical 2006-01-0954 (2006)
5. Zhang, W.: Simulation of bus safety body structure (2006)
6. Liu, C.K., Song, X.P., Wang, J.: Simulation analysis of car front collision based on LS-DYNA and hyper works. J. Transp. Technol. **4**(2014), 337–342 (2014)
7. Muhammad, A.H.: Comparison of structural damage and occupant injuries corresponding to a vehicle collision onto a pole versus a flat barrier, University of Engineering and technology Karachi, Pakistan (2002)
8. Tian, F., Wang, T., Shi, Q.: Finite element analysis for monocoque bus/coach body structure. Bus Coach Technol. Res. **1**, 17–19 (2012)
9. Wang, H., Jin, X.L., Lin, Z.: FEM Static and Dynamic Analysis of the Body Structure of SK6120 Low Floor City Bus. SAE Technical 2002-01-0813 (2002)
10. Liu, J., Gui, L., Wang, Q., Fan, Z.: Multi-objective Optimization on the Body Structure of Integral Bus. Automot. Eng. **30**, 170–173 (2008)

Influence of Thermal-Mechanical Process on the Shape Memory Effect of CuAl9Fe4Ni2 Alloys

Nguyen Duong Nam[1], Vu Anh Tuan[1], and Pham Mai Khanh[2(✉)]

[1] Viet Nam Maritime University, Hai Phong, Viet Nam
{namnd.khcs, anhtuan.cscb}@vimaru.edu.vn
[2] School of Materials Science and Engineering, HUST, Ha Noi, Viet Nam
khanh.phammai@hust.edu.vn

Abstract. In recent years, the CuAl9Fe4Ni2 alloy has been studied and applied in the manufacturing. In recent years, scientists have initially found and demonstrated about shape memory effect of this alloy. In addition to increasing, the martensite transformation has increased the strength of the alloy, also creates a shape memory effect by reversing the crystal structure in the aluminum-copper alloy which added nickel, zinc, or iron elements. Martensite can easily be formed when the amount of deformation is about 8%. The crystalline structure of martensite can be monoclinic or orthorhombic. In this paper, we present the research results about the phase transformation process of CuAl9Fe4Ni2 alloy when mechanical-heat-treatment to increase the shape memory effect of this alloy. The research results show that: This alloy is heated at 900 °C to create a martensite phase with large grain size; hot rolling at 800 °C to facilitate the process of shape memory effect. Then, this alloy continued quenching for a second time at 900 °C to create martensite with small grain size and a convenient distribution for the shape memory effect. Continue to conduct cold deformation to energize the martensite sheets, forming the detwinner martensite to facilitate the shape memory effect. Then proceed to reheat at 570 °C to form the process of reversing the martensite phase back to the initial state of the alloy created ($\alpha + \gamma_2$). SEM analysis proved the existence of the martensite phase β' which has hcp structure with Cu_3Al formula is the same as the phase β which has bcc structure but exists in alloys at room temperature (martensite after quenching, residual martensite after tempering and after shape memory effect).

Keywords: Detwinner martensite · Shape memory · Mechanical-heat treatment · SEM · Phase transformation

1 Introduction

Shaped memory alloys are applied to groups of metal materials that can return to a defined shape or size before being destroyed by the heat treatment process. In general, these materials can deform at relatively low temperatures and can return to their original shape (before being deformed) when treated at higher temperatures. Such groups of materials are called one-dimensional image retention. Some materials can

B. T. Long et al. (Eds.): MMMS 2020, LNME, pp. 78–84, 2021.
https://doi.org/10.1007/978-3-030-69610-8_10

return to shape at low temperatures; Such groups of materials have two-dimensional memory capacity.

The condition for the shape memory effect is that when cooled, the austenite (β) phase changes to the martensite (β′) phase and vice versa, along with shifting and non-diffusion. Phase transformation can also occur at a temperature higher than M_s. If there is an external force, that is the case of the elastic dummy ($T_3 > M_s$) and plastic dummy ($T_2 > M_s$), but the temperature to occur elastic dummy must be higher because the tendency to change phase $\alpha \rightarrow \beta$ here is greater.

Besides, the martensite transformation increased strength for the alloy, in addition to creating a memory effect by reversing the transformation of the martensite phase in copper aluminum alloys with nickel, zinc, or iron.

The twinned martensite has 24 variants, meaning that these 24 variants differ from the orientation of crystallography. When the martensite is created by stress, 24 twinned martensite variants become one. The crystal is then oriented in the direction of stress [1].

Assuming initially, in the austenite state, the alloy has a "root shape". When cooled, under variable temperature, the shape memory of alloy has a martensite structure. In the martensite phase, it can be bent into many different shapes. When the alloy is heated to a certain temperature, the atoms arrange themselves in compact structures and, as a result, create a cube phase which we call austenite, a shape-shifting memory alloy. The martensite forms into austenite and restores its "original shape".

Because this alloy has many applications in military engineering, medicine, manufacturing safety devices, fire alarms, manufacturing robots, in manufacturing household appliances (thermal sensors, oscillators in tools,…), mechanics, aviation, space,… [2–7].

According to the research results of A. Ibarra et al. 2007 [8] on martensitic germs in shape memory of Cu-Al-Ni alloy. The research results show that the nucleation on the defects has been detected. Through transmission electron microscope proved the formation of martensite. On defects of phase, phase β can be the germ for γ'_3 and martensite β'_3. The martensite sheets will be the seed on the defects and passed on to the L21 phases.

Copper-shaped memory alloys are of special interest because of their low fabrication costs and good memory properties. The Cu-Al-Ni-shaped memory alloys can be used at temperatures near 200 °C. [4, 6, 7, 9].

However, there are not many research about shape memory effect of Cu-Al which added Fe and Ni. In the article, we present about some results of CuAl9Fe4Ni2 shaped memory alloys and martensite phase transformation in this alloy.

2 Materials and Method

2.1 Sample Preparation

This alloys with 9% Al, 4% Fe, 2% Ni were selected for research. Pure Cu, Al, Fe, and Ni are calculated with 300g of alloys, melted in a reactor with protective Ar gas.

2.2 Thermal Mechanical Processes

The rods of size 100 mm are cut from the original ingot. They were treated at 1173 °K in the β phase for 6 h with protected Ar gas and quenched at room temperature water. The sample will be rolled to 1 mm thick at 1073 °K. These rolling samples were treated at 1173 °K for 15 min then cooling in 373 °K hot water and finally added to the water to room temperature.

Using microscopy, and the Xray and SEM method to identify and observe the alloy structure.

After hot-rolling process, the samples are subjected to a bending test at 309 °K with 10% deformation levels to determine post-deformation shape resilience for the alloy by comparing the initial bending angles and final.

2.3 Shape Memory Effect

Hot-rolled specimens with a thickness of 1mm (after being heat-treated and rolled in the above procedure (*) above) are subjected to deformation recovery by a bending test. All samples were initially deformed with 10% levels. These sheets are curled around the U-shaped mandrel (in the Fig. 1).

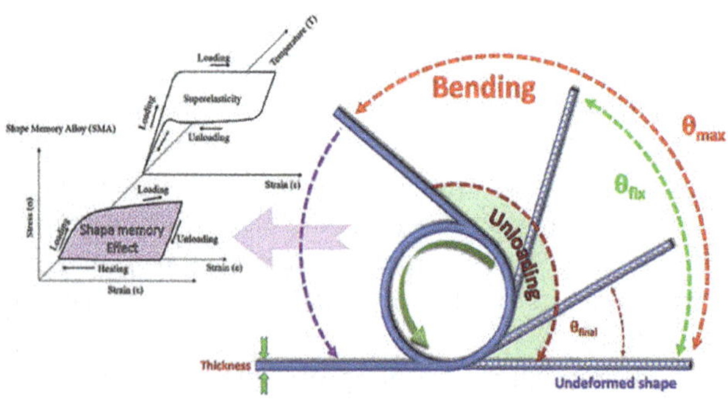

Fig. 1. Testing shape memory effect

Bend these samples at such angles (10% deformation) at room temperature (below M_f). The samples were then heated again to 773 °K (above A_f). We will measure the angle when bending (before heating) and the angle after heating. Samples before and after the deformation determined to remember the shape are measured angled by the profiler with an accuracy of ±0.001.

3 Results and Discussion

In the Fig. 2, The alloy is heat-treated at 900 °C in the α + β phase area and after cooling in water to change from phase β β′, there may exist phase α.

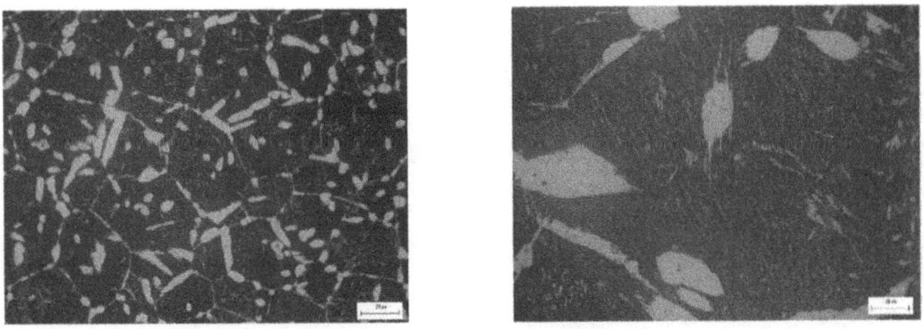

Fig. 2. Microstructure of sample after quenching in hot water at 1073 °K

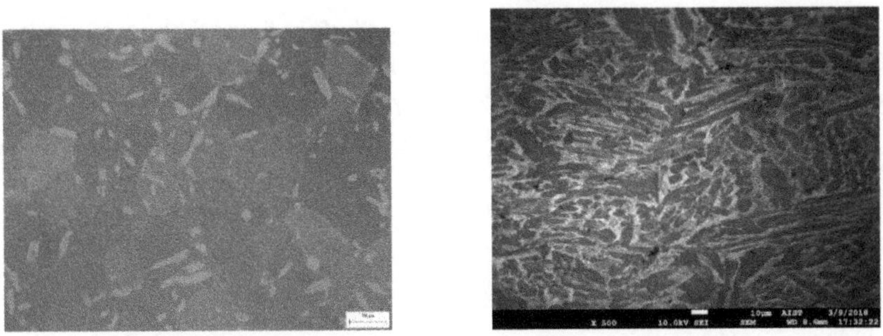

Microscopy optical SEM

Fig. 3. Microstructure after deformation with 10%

After rolling the microstructure of sample show that the disordered phase β′ disappears, the martensite plates are no longer intertwined in the matrix, it seems that only the dark phase β; light phase α and the arrangement of particles of the phase so that the dimensions are more suitable after casting (Fig. 3).

Fig. 4. Microstructure of alloy after deformation (10%) and heat treatment

In the Fig. 4, After rolling, the alloy is heated at 900 °C and quenching in water, then tested to bend, the results in Fig. 4 show phase β′ martensite (Cu_3Al) specific here has become the phase again β β′, which appears at the boundary of the α light phases.

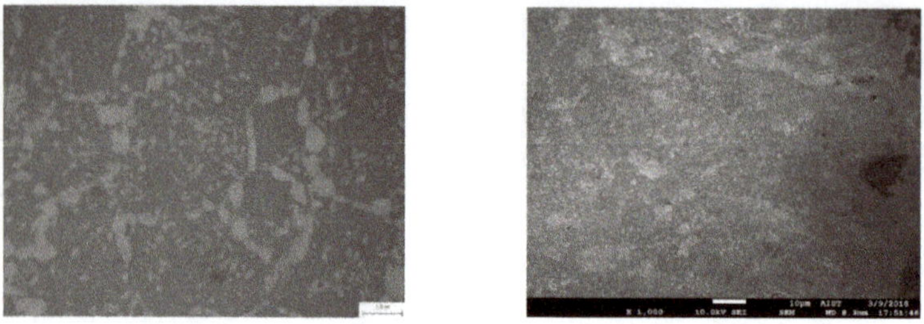

OM SEM

Fig. 5. The microstructure of alloy after SME

After the end of the shape memory test shown in Fig. 5, the β-martensitic plates on the boundary between the α phases have not due breaking during heat-treatment process. At this time, there is variable β′ (α + γ_2), besides, the microstruce has Fe-rich iron phase (δ), similar to the original structure at the casting of the alloy, but here the phase is secreted finely, dispersed evenly. in the structure. At the boundary of the phases, there appears a structure of NiAl.

XRD:

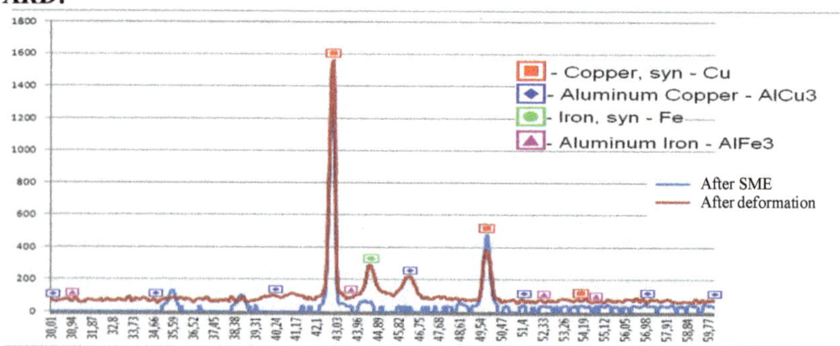

Fig. 6. XRD before and after SME of CuAl9Fe4Ni2

In the Fig. 6, analysis of XRD alloy before and after shape memory effect: For the alloy before shape memory in the structure, there is the formation of martensite phases but when the calcination process does not appear the martensite or the peaks are very low undetectable. This proves that there has been a transition from the martensitic phase with the structure β′ to the α and γ_2 phases.

Analysis of alloy structure before and after image recognition shows: Before shape memory, the ratio of light-phase ratio only about 10%; however, after shape memory when baked at 570 °C, the alpha phase fraction increased to 52% (Fig. 7). The increase in the alpha phase ratio is explained by the decomposition of the martensite phase to increase the memory capacity of the alloy system. Thus, the sample with Ni shows that the ability to remember the image of the alloy is much larger than the non-alloyed sample.

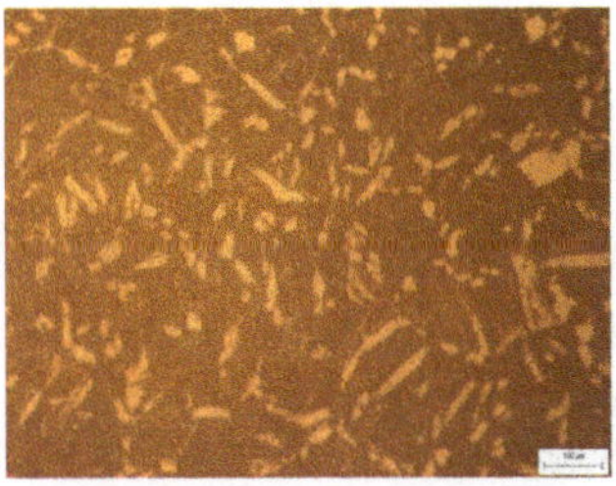

Before shape memory is 10%

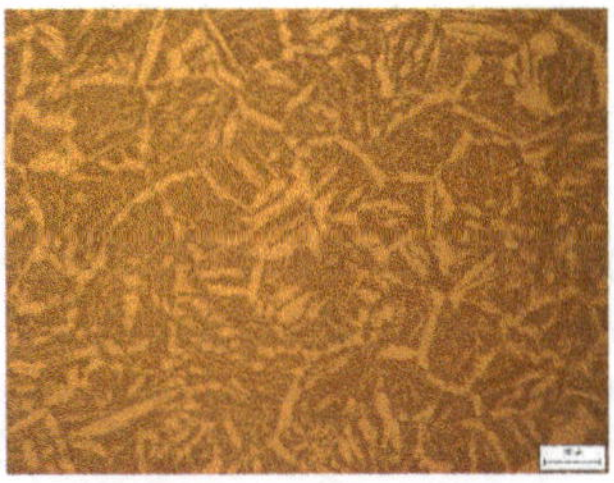

After shape memory is 52%

Fig. 7. The ratio of the α phase

The hot rolling process has the effect of providing additional energy to convert into β, which makes the actual amount of will be higher. This further boosts the number of actual martensite transformations, resulting in a cumulative effect in increasing memory retention (Table 1).

Table 1. The amount of shape memory of CuAl9Fe4Ni2

The angle before the heat	The angle after heat	The amount of deformation (%)	The amount of shape memory
90	84	5	7
30	16	5	47
30	6	10	80

Bending angle to remember the large image of racquet through local plastic deformation at a curved position. When the heating changes from β′ to β, the recovery amount only for structure deviations is smaller than the structure parameters. Distorted distortions that exceed the size of the structure parameters will not be able to recover.

4 Conclusion

The paper has identified the shape memory mechanism of CuAl9Fe4Ni2 alloy. This mechanism is explained by the phase transformation process under heat-mechanical treatment and the reversal martensite phase. The thermal-mechanical process of alloy is: The alloys were treated at 1173 °K in the β phase for 6h with protected Ar gas and quenched at room temperature water. The sample will be rolled to 1mm thick at 1073 °K. These rolling samples were treated at 1173 °K for 15 min then cooling in 373 °K hot water and finally added to the water to room temperature. Moreover, the intermetallic of this alloys is Fe_3Al.

Acknowledgements. The authors would like to thank the Viet Nam Maritime University for their support.

References

1. Stanciu, S., Bujoreanu, L.G.: Formation of β′1 stress-induced martensite in the presence of γ-phase, in a Cu–Al–Ni–Mn–Fe shape memory alloy. Mater. Sci. Eng. A **481–482**, 494–499 (2008)
2. Zhang, X., Qian, M.: Ferromagnetic shape memory alloys: foams and microwires. In: Shape Memory Alloys - Fundamentals and Applications, pp. 03–36. Intech (2017)
3. Kihara, T., et al.: Magnetocaloric effects in metamagnetic shape memory alloys. In: Shape Memory Alloys - Fundamentals and Applications, pp. 59–79. Intech (2017)
4. Nizamani, A.M., Daudpoto, J., Nizamani, M.A.: Development of faster SMA actuators. In: Shape Memory Alloys - Fundamentals and Applications, pp. 106–126. Intech (2017)
5. Nasakina, E.O., et al.: Applications of nanostructural NiTi alloys for medical devices. In: Shape Memory Alloys - Fundamentals and Applications, pp. 81–104. Intech (2017)
6. Sutou, Y., Omori, T., Wang, J.J., Kainuma, R., Ishida, K.: Characteristics of Cu-Al-Mn-based shape memory alloys and their applications. Mater. Sci. Eng. A **378**(1–2), 278–282 (2004)
7. Lojen, G., Gojić, M., Anžel, I.: Continuously cast Cu-Al-Ni shape memory alloy - properties in as-cast condition. J. Alloy. Compd. **580**, 497–505 (2013)
8. Ibarra, I.A., Caillard, D., San Juan, J., Nó, M.L.: Martensite nucleation on dislocations in Cu–Al–Ni shape memory alloys. Appl. Phys. Lett. **90**, 101907-1–3 (2007)
9. Akselsen, O. M.: Joining of shape memory alloys. Intechopen.com (2010)

Case Study: Prototyping a Low-Cost Integrated Automation System in Footwear Industry for Small and Medium Enterprises in Vietnam

Minh D. Tran, Toan H. Tran, Diem T. H. Vu, Thang C. Nguyen, Vi H. Nguyen, and Thanh T. Tran(✉)

Vietnamese-German University, Thu Dau Mot, Binh Duong, Vietnam
thanh.tt@vgu.edu.vn

Abstract. Automation is the key for small and medium enterprises (SMEs) to gain profit and win market. Investment required for automation projects are normally high and would cost extra whenever upgrade is needed. Sometimes, present automation system cannot adapt to the change and must be replaced. These matters cause SMEs to consider whether to invest on automation or not. This paper proposes a concept for an automation system which can be integrated into larger systems with reasonable prices. A case study in footwear industry is analyzed and prototype of automated conveying system is fabricated to demonstrate the opportunity of this concept. The result of implementation proven that SMEs can save up to 50% of the cost if applying the proposed concepts.

Keywords: Smart factory · Automated production system · Low-cost automation · SME · IoT · PLC

1 Introduction

The production SMEs is more complicated than the large corporations since their paradigm is for high mix low volume (HMLV). However, to fulfill customer demand on time and improve product and process quality, automatize process is a must.

The rapid development of technology enable the automation system to adapt with new changes, making them cheaper to implement and easier to control (digitalization for instance). The concept of Cyber Physical Production System (CPPS) help SMEs to gradually apply automation and improve the system when needed. The Internet of Things (IoT) allows machines to communicate with each other and managers to manage the production efficiently. Digital Twin (DT) enhance the planning phase to save cost on real implementation.

With the rapid development of technology, automation is now possible for SMEs. With the philosophy of Agile Manufacturing applied, automation system can be flexible enough to adapt with changes in products cluster of the SMEs.

B. T. Long et al. (Eds.): MMMS 2020, LNME, pp. 85–93, 2021.
https://doi.org/10.1007/978-3-030-69610-8_11

To help the SMEs with application of automation into production, a methodology including concepts of: digitalization, on-hand components usage, outsourcing technology and design for change.

2 Automation and SMEs

2.1 About the Extremely Small, Small and Medium Enterprises

SMEs played an important role in Vietnam economy. Accounted for 98% of total businesses, contributed 40% total GDP and offered 50% jobs on total employment and marked 100,000 business in 2016 ([1] in [2]). Especially in recent years, start-up accelerating policies from Vietnamese government expand the capital invested into SMEs [3, 4]. The majority of these SMEs are registered as manufacturers while less than 1.26% of which functioning as service providers. The manufacturing field of these SMEs are wide range from Food and Beverage to Office Supplies and remanufacturing [5].

Despite the rapid growth of SMEs models, 72% of total SMEs are extremely small (lower than 10 employee) and 22% of which are small (10–50 employee). During the period of 2013–2015, survive ratio of SMEs in Vietnam is 83.3% and as survey, 83% of the survived enterprises having problem with their current state of business [5], yet SMEs must compete with large corporations to win the share of market.

Large corporations have advantages in huge resource which they can invest into R&D for technology breakthrough, or fully automation production lines to reduce manufacturing cost. On the contrary, SMEs also going for technology 4.0 with limited shares to R&D, which may cause failure to take advantages of the new technology and lead to the disruption in business [6].

2.2 Automation

Automation is defined as “the technology by which a process or procedure is performed without human assistance”. Automation systems consist of Power, Program and control system [7]. In real practice, automation are focusing on automatic workpieces supply or position. Automatize production offer many benefits to firms for better competition on the market [8].

There is a wide variety of benefits which automation can offer SMEs. Some notable advantages when apply automation would be: machine will do the heavy, hazardous tasks which human incapable of. The replacement of labor when automatize will save an amount of labor cost. Moreover, machines are more stable with high and precise repeatability. Last but not least, automation can improve process quality (process more stable since machine cannot call in sick) and product quality [9]. Another significant advantage of automation is that automation helps implementing LEAN philosophy. Many improvement from LEAN cannot be achieved without automation such as transporting system for one-piece-flow [10].

Despite the huge benefits from automation, there are also some challenges which needed to be noticed. Some tasks might be too complex for the current technology or the technology readiness level is not met yet. Automation requires maintenance to achieve the highest profit, which means maintenance teams and programs are also required. Last but not least, automation demands high initial cost [9], which is the greatest barrier to prevent SMEs from retrieving the benefits of automation.

2.3 Set up Constrains for Automation to Be Applied to Production of SMEs

SMEs can use their strength to overcome the challenges listed above, some notable strength for SMEs toward innovation are: Quick decision making; less management barrier; face-to-face communication between engineers and managers; dynamic, flexibility to change when required [11].

Based on the strength to cope with the challenges, some constrains are set to proceed on building up concept for automation which can be feasibly applied to the SMEs (also referenced from [6]).

- Flexibility is required, especially for assembly stations.
- CPPS is advised, where human collaborate with robot to complete the task.
- Initial cost must be low enough to be affordable to the extremely small enterprises.
- Conveying and positioning mechanism is preferred, especially production feeder.
- Using simulation for the planning and evaluation of the automation system.

3 Building Concept for a Low-Cost Automation System

3.1 Digitalization

The key for going IoT is digitalization. Most devices nowadays can communicate with each other using digital signal. Using digital signal output components in automation system will allow the system to adapt with changes or integrate to information system easily with minimum cost.

Digital signal is electrical pulse with the amplitude of approximately 0.3 A at the output port of the device. Variety of sensors using digital output which can be connected easily to the controllers. If sensors used as caliper, the output is analog, then an Analog-to-Digital converter (ADC) is required. The principle of an ADC is to set the point on analog band where to send digital signal (send a pulse to port 1 when temperature reach 25 °C for instance).

Digitalization also refer to the simulation of a real system. In implementation phase of the project of new product (automation system) development, a low fidelity prototype normally built for project assessment. In term of digitalization, a DT version of the product is created on the computer for evaluation. This DT not only simulate the real physical prototype but also save cost for implementation.

3.2 On-Hand Components Usage

This concept is basic and commonly aware of by majority of engineer. However, it is normally forgotten when new automation project initiated. Available components inside factory can save a lot for SMEs.

Nowadays, devices can communicate with each other easily using digital signal. Therefore, components from a project can be used for another one.

Using internal available components can also be considered as remanufacturing, which is a key feature in sustainable manufacturing. Moreover, systems which meet the end of lifecycle can also be upgraded using minimum investment.

3.3 Outsourcing Technology

For SMEs, normally the required investment is too much for procuring technology. On the other hand, self-develop technology usually make the project exceed constrains of time and cost. Therefore, outsourcing the technology of other firms in terms of using their products.

Globalization and Internet of Everything transform the tradition way of outsourcing to the online searching for needed technology. The development of digital market and logistic contribute significantly to the drop of price for outsourcing technology.

One remarkable note for the SMEs is the notice on technologies which are not licensed. Normally these technology developed by free-lancers, but with appropriate use, they can be useful for the automation systems with minimum cost.

3.4 Design for Changes

The concepts of Design for changes is advisory for the SMEs due to the small scale of investment and high flexibility. The automation system should be easily adapt to changes and integrate to new system 255± or upgradable. Using hardware with a long lifecycle, which allow the hardware to be upgradable or communicate with other devices for longer time. For controller, SMEs should choose PLC for the multi-purposes of it, or else select the relay modules which allowed to receive signal from sensors directly with the AND algorithm (these hardware can be outsourced easily). For actuators, choosing the standard components is important for interchangeability.

4 A Case Study and Implementation of the Concepts

4.1 Case Study

A medium company in Vietnam working in footwear industry and planning to invest into Lean and Automation. They make the sole part of the shoes using Injection Molding machines. Some soles require molding for multiple times, post-processing and packaging. The semi-finished parts are pile up at workstation making it hard for a one-piece flow. The operators after making some soles will have to carry the semi-finished parts to the next workstation. The labor efficiency decreases require the company to have extra carriers or a conveying system.

Different tasks were made by different workstation, but most workstation can follow the same sequence: workstation 1 finished part are cooled down at room temperature, then transferred to the next work station which is 1.8 m away. The next workstation can be a robot arm for extra tasks, then the product is transferred through a defect-checking camera then drop at packaging.

The semi-product which needed to be transferred is very light-weight but bulky. Normally the conveyors have to transport sole and sometime shoes boxes with maximum dimension of 130 × 140 × 340 mm.

The products needed to stay on the conveyors for at least 3 s to cooldown. At the end of the conveyor have to be a stop switch in order stop the products dropped on the floor and to positioning for the robot arm or operator from the next workstation to pick up the semi-product. Engine speed needed to be adjustable for different products and quality checking camera.

The company cannot afford to invest at once for information system, so the transporting system must be able to be integrated easily with Enterprise Resource Planning (ERP) system or upgradable. New technologies application is preferred with reasonable cost, IoT for instance.

4.2 Design

4.2.1 Conveyors

Based on the dimension of sole and shoes boxes, the width of conveyors is chosen. The conveyor belt (since the workpieces very light-weight so PVC belt is used) size must be wide enough to hold the shoes boxes, but the width of belt would affect the cost of material and shop floor capacity. In this case, standard belt size of 300 mm is chosen.

The height and length of conveyors is chosen follow ergonomical rules. The height should be 0.8 to 1.5 m high (less than 1 m if the operators are sitting), the gap between 2 workstations is 1.8 m so the length of each conveyor should be 1.5 m within the reach of each operator.

Belt should run at maximum speed 0.5 m/s. The roller diameter can be chosen by using available steel or stainless steel pipe in the factory. In this case a pipe of ϕ110 mm is available, which make the motor run at maximum speed of 43 rpm and can be adjusted. Motors commonly run at 1500 rpm and connected to a portable gearbox or a power inverter which are costly. In this case a motor with internal gearbox ratio of 1/36 is used. The speed of motor shaft is now 42 rpm. To adjust the speed of conveyor, a power dimer is used.

The bases of conveyors are made of square steel pipes. Industrial formed aluminum and UCP bearing are advisory to save machining cost.

4.2.2 Controller

To program the system, instead of using PLC which is costly, a set of relays are used to replace the PLC in logical control of the automation system. A set of sensors will recognize the object, then send digital signal to relays to run or stop the conveyor. By sending digital signal, sensors can also trigger the program of the robot arm. A wifi device is used to control the system from another place using voice command.

In this case study, the electronic devices are mostly outsourced in form of End-of-Line product procurement. The main controller is set of relay modules using 5VDC power supply (component 1 of Fig. 1). This module using AND logic: relay change status when there is power supply and "ON" signal from the sensor. The sensors are Infrared (IR) sensors using the same 5VDC power supply with relay module. Instead of using expensive AC sensors, the DC sensors are cheaper with same function. The sensors that are using have the code: E18-D80NK, with the price of 2.5 USD/unit, this sensor can stand vibration and heat which is suitable for industrial use. The power supply for sensors define the selection of relay since the "Ground" input of sensor and relay must be connected. The digital signal of this sensors are 0.3 A electric pulse. The timer relay (component 2 of Fig. 1) also installed to set auto-off time for conveyor 2 since conveyor 2 don't have to run the whole shift. Finally, the IoT solution is outsourced. Instead of self -developing the wifi module from ESP8266 which cost more investment and time, the wifi control module (component 3 of Fig. 1) is bought from another SME which provide the system to be remotely controlled and managed. The module also supports voice control by connecting with Google Assistant and an application on Android and IOS platform with the total cost of 12 USD.

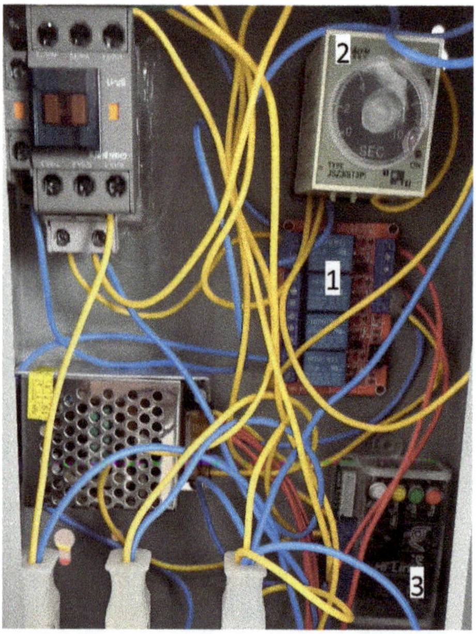

Fig. 1. Inside the controlling box of the system

4.3 Result

By applying the new concepts, initial cost of this project is dropped significantly to the price which extremely small enterprises can invest. In terms of technology 4.0, the prototype can integrate with information systems such as ERP or controller system such as PLC. If connected to PLC, further development of the HMI is also available (Fig. 2).

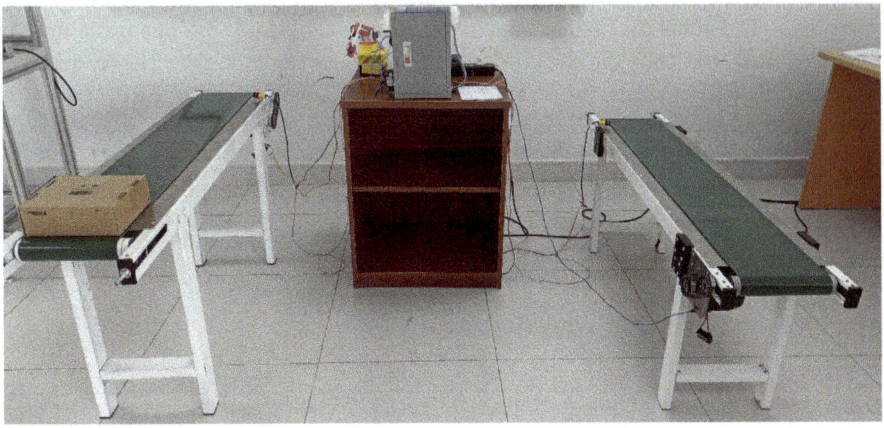

Fig. 2. Prototype of the conveying system

For the mechanical part, standard parts are used in case the process change and require justification for the system. Formed aluminum beam is advised for the base due to the practical application of the component. If there is a change in production line and conveyor need a change in dimension, the parts can be adjusted to meet new requirements or changed quickly. The UCP type bearing, with the housing easy for maintenance, can be assembly on the aluminum beam and suitable for installing a conveyor belt tensioning mechanism (Fig. 3).

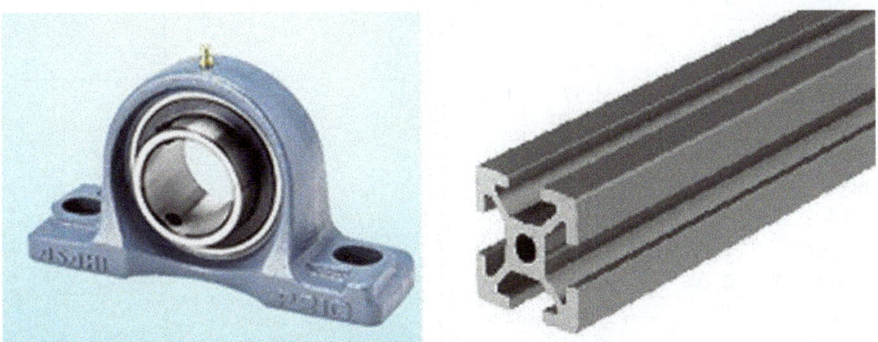

Fig. 3. UCP bearing (left) and formed aluminum can be found at almost all store

For electrical hardware, technology and products are outsourced from local store (can be found at any electronic store) and assembly to control the conveyor. With the right programing with relays for the 2 conveyors and the robot arm, the automation system is basically finished.

The total cost for the automated conveying system which consist of 2 conveyors and controller are less than 500 USD, which can be affordable to extremely small enterprises. The function of this system can help establishing the one-piece-flow or integrated with the fully production system for footwear industry. The system can connect with ERP system by installing counter to send signal to ERP system, or connect with digital ports of PLC. When manufacturing product change, the conveyors can also be calibrated or upgrade to fit with the new product.

The price list in Table 1 given a comparison between the price of making the automated system with (solution 1) and without applying the proposed concepts (solution 2), the data is taken for Vietnamese market.

Table 1. Price list of 2 solutions

Solution 1				Solution 2			
Name	Quantity	Price (USD)	Cost (USD)	Name	Quantity	Price (USD)	Cost (USD)
Belt	2	10	20	Conveyors	2	364.38	728.76
Bearing	8	1.5	12	PLC S7-200	1	128.61	128.61
AC motors	2	25	50	Sensors	2	15.09	30.18
Bases materials	1	150	150	Beacon	2	1	2
Machining cost	1	25	25	Circuit breaker	1	2.57	2.57
Conveyor Aux	1	25	25	Aux	1	25	25
Relay module	1	2.96	2.96				
Timer relay	1	3.43	3.43				
Wifi module	1	11.57	11.57				
Power supply 5VDC	1	4.07	4.07				
Contactor	1	7.29	7.29				
Sensors	2	3.43	6.86				
Circuit Breaker	1	2.57	2.57				
Beacon	2	1	2				
Electronic Aux	1	20	20				
Total			**342.75**	Total			**917.12**

5 Conclusion and Further Research

With the result retrieved from implementation of the conveyors, the can assume that the concepts proposed are feasible. In order to compete on market and win a share, SMEs must invest in automation to reduce cost and increase process efficiency. The investment should follow the concepts suggested for SMEs in order to have the good result.

Further research on information system (especially ERP system) or Virtual Production Integration to be applied in context of automation is recommended. The research on low-cost scara robot arm or fully automation system for SMEs are also beneficial for SMEs in Vietnam.

References

1. Romero, J., Ransden, N., Japhta, R., Huynh, H.M.: Women-owned enterprises in Vietnam: perceptions and potential (2017)
2. Nguyen, T.A., Nguyen, T.T., Dang, T.T.H.: Employment and quality of employment in Vietnam: the roles of small firms, formalization and education (2016)
3. Tran, V.T., Do Quang, H., Minh, H.L.: Entrepreneurial human capital, role models, and fear of failure and start-up perception of feasibility among adults in Vietnam. Int. J. Eng. Bus. Manag. **11**(start-up models), 1–11 (2019)
4. Nguyen, H.X., Tran, D.C., Pham, H.Q.: Solution to develop innovative start-up enterprise in Vietnam. Eur. Sci. Rev. **1–2/2019**(innovative start-up) (2019)
5. Rand, J., et al.: Characteristics of the Vietnamese Business Environment: Evidence from a SME Survey in 2015. Ha Noi (2015)
6. Grube, D., Malik, A.A., Bilberg, A.: Generic challenges and automation solutions in manufacturing SMEs. In: DAAAM International Symposium on Intelligent Manufacturing and Automation, vol. 28, pp. 1161–1169 (2017)
7. Groover, M.P.: Fundamentals of Modern Manufacturing: Materials, Processes, and Systems, 5th edn. Wiley, Hoboken (2012)
8. Taylor, J.: The right way to start robotic automation. Industrial Automation Asia, Bryton House, p. 24, August 2019
9. Lamb, F.: Industrial Automation: Hands On. McGraw-Hill Professional, New York (2013)
10. Protzman, C., McNamara, J., Protzman, D.: One-Piece Flow vs. Batching: A Guide to Understanding How Continuous Flow Maximizes The core idea of one-piece or continuous flow is to maximize customer value while eliminating waste. Simply put, it means creating more value for customers using fewer res, 1st edn. Taylor & Francis Group, Boca Raton (2016)
11. Mrugalska, B., Wyrwicka, M.K.: Towards lean production in industry 4.0. Procedia Eng. **182**, 466–473 (2017)

A Comparative Investigation on Theoretical Models for Forming Limit Diagram Prediction of Sheet Metals

Quoc Tuan Pham[1,2(✉)], Trung Nguyen-Thoi[1,2], Duc Toan Nguyen[3], and Young Suk Kim[4]

[1] Division of Computational Mathematics and Engineering, Institute for Compuational Science, Ton Duc Thang University, Ho Chi Minh City, Vietnam
phamquoctuan@tdtu.edu.vn
[2] Faculty of Civil Engineering, Ton Duc Thang University, Ho Chi Minh City, Vietnam
[3] School of Mechanical Engineering, Hanoi University of Science and Technology, Hanoi, Vietnam
[4] School of Mechanical Engineering, Kyungpook National University, Daegu, Korea

Abstract. This study provides a coherent comparison among four theoretical models including Swift's instability, Hill's localized neck, Storen-Rice bifurcation analysis, and Hora's MMFC criteria for explicitly estimating the forming limit diagram (FLD) of sheet metals. These explicit solutions for FLD are derived based on a power hardening law coupling with J_2 flow theory. Accordingly, the advantages and limitations of each model are discussed from the points of view of metal physics, which appears to be a useful suggestion for selecting a proper theoretical model to evaluate FLD of sheet metals without conducting experimental tests. To verify the effectiveness of those suggestions, the predicted FLD of AA6016 and DP590 sheet materials are compared with experimentally measured data from a series of Nakazima tests.

Keywords: Sheet metals · Forming limit diagram · Swift's instability · Hill's localization · Bifurcation analysis · Modified maximum force criterion

1 Introduction

The formability of sheet metals is usually evaluated through a forming limit curve (FLC) which theoretically separates the strain space into safe and unsafe forming regions. The concept is widely applied in automotive industries to design and optimize parts. Therefore, theoretical and experimental methods have been developed in the decades to determine FLC for sheet metals.

Nakazima test and Marciniak-Kuczynski test are the most used methods to determine FLC experimentally. However, conducting these experimental methods requires high computational time and expensive experimental costs. Thus, different theoretical methods have been presented to determine FLC efficiently. Swift's instability [1] and

B. T. Long et al. (Eds.): MMMS 2020, LNME, pp. 94–100, 2021.
https://doi.org/10.1007/978-3-030-69610-8_12

Hill's localization [2] are well-known as pioneering works for this task. Both methods are developed based on an assumption that deformations of material are observed under proportional loading paths. These theories consequently resulted in explicit expressions of the critical strains, which originate the computational FLC. Alternatively, Storen and Rice [3] assumed the discrepancy between the rates of deformation inside and outside the neck, which yields two expressions of the critical strains based on bifurcation analysis. The first solution is the same with that derived by Hill; whereas, the second solution well-known as the bifurcation analysis solution differs from the other. Recently, Hora et al. [4] introduced a modified maximum force criterion to determine FLC considering the transition of strain path after diffuse neck. Within an assumption of proportionally imposed loading paths, the criterion provides an explicit formulation for the critical strains [5].

This study presents an overview of the theoretical basis of these before-mentioned necking criteria under a general and unified framework which consequently provides a comparison of their governing equations. Advantages and drawbacks of each criterion are discussed based on their application on estimating FLC for two automotive sheet metals: DP590 and AA6016-T4 material [5, 6].

2 Necking Criteria

Swift Diffuse Necking Criterion

Suppose a sheet patch subjected to a biaxial in-plane stretching under the forces F_1 and F_2. The force are given by $F_1 = \sigma_1 A_1$ and $F_2 = \sigma_2 A_2$, where $A_{1,2}$ denote areas of the cross-section normal to $F_{1,2}$, respectively. Swift [1] stated a so-called diffuse necking criterion of which the critical strains are observed as soon as the loading forces reach to their maximum, that means

$$dF_1 = d\sigma_1 A_1 + \sigma_1 dA_1 = 0 \text{ and } dF_2 = d\sigma_2 A_2 + \sigma_2 dA_2 = 0 \tag{1}$$

Using the assumption of incompressibility leads to the following constrains

$$d\sigma_1 = \sigma_1 d\varepsilon_1 \text{ and } d\sigma_2 = \sigma_2 d\varepsilon_2 \tag{2}$$

Under the plane-stress condition, the equivalent stress is calculated from two principal stress components, i.e. $\overline{\sigma} = \overline{\sigma}(\sigma_1, \sigma_2)$. Let's define the stress ratio $\alpha = \sigma_2/\sigma_1$ and strain ratio $\beta = \frac{d\varepsilon_2}{d\varepsilon_1} = \frac{\partial\overline{\sigma}/\partial\sigma_2}{\partial\overline{\sigma}/\partial\sigma_1}$ based on the associated flow rule theory. Hence, the changing rate of the yield criterion can be calculated as follows

$$dF = d\overline{\sigma} - dH(\overline{\varepsilon}) = \frac{\partial\overline{\sigma}}{\partial\sigma_1} d\sigma_1 + \frac{\partial\overline{\sigma}}{\partial\sigma_2} d\sigma_2 - \frac{\partial H}{\partial\overline{\varepsilon}} d\overline{\varepsilon} \tag{3}$$

where $H(\overline{\varepsilon})$ denotes the hardening law of the tested material. Substituting the condition of diffuse neck expressed in Eq. 2 into this equation in conjunction with flow rule leads to the following condition for material's instability

$$\frac{\partial H}{\partial \bar{\varepsilon}} = \left(\frac{\partial \bar{\sigma}}{\partial \sigma_1}\right)^2 \sigma_1 + \left(\frac{\partial \bar{\sigma}}{\partial \sigma_2}\right)^2 \sigma_2 \tag{4}$$

Using Euler's theorem for yield function leads to

$$f(\alpha) = \frac{\bar{\sigma}}{\sigma_1} = \frac{\partial \bar{\sigma}}{\partial \sigma_1}(1 + \alpha\beta) \tag{5}$$

Where $f(\alpha) = \bar{\sigma}/\sigma_1$ is a pre-defined auxiliary function. Substituting this equation into Eq. 4 leads to

$$\frac{\partial H}{\partial \bar{\varepsilon}} = \left(\frac{\partial \bar{\sigma}}{\partial \sigma_1}\right)^2 \bar{\sigma}/f(\alpha)\left[1 + \alpha\beta^2\right] \tag{6}$$

Rearranging this equation with a notice that $\bar{\sigma} = H(\bar{\varepsilon})$ yields to

$$\frac{\partial H/\partial \bar{\varepsilon}}{H(\bar{\varepsilon})} = f(\alpha)\frac{1 + \alpha\beta^2}{(1 + \alpha\beta)^2} \tag{7}$$

Hill's Localized Necking Criterion

From a different point of view the, force per unit width in the major direction of the sheet patch is defined according to the thickness of the sheet t, as $f_1 = \sigma_1 t$. Hill [2] suggested that a localized neck is formed on the sheet patch when the force per unit width in the major direction reaches to its maximum, for example

$$df_1 = d\sigma_1 t + \sigma_1 dt = 0 \tag{8}$$

Again, using incompressible assumption leads to the following constrain

$$d\sigma_1 = \sigma_1(d\varepsilon_1 + d\varepsilon_2) \tag{9}$$

Following the same procedure used in Swift analysis with an additional assumption that the stress ratio remains a constant during plastic deformation, i.e. $d\sigma_2/d\sigma_1 = \sigma_2/\sigma_1 = \alpha$, the changing rate of yield function leads to the following condition for localized neck

$$\frac{\partial H/\partial \bar{\varepsilon}}{H(\bar{\varepsilon})} = f(\alpha)\frac{1 + \beta}{1 + \alpha\beta} \tag{10}$$

Table 1. Expression of the critical major strain based on the necking criteria.

Criterion	Critical major strain
Swift	$\varepsilon_1^* = \frac{2n(2-\alpha)\left(1-\alpha+\alpha^2\right)}{4\alpha^3-3\alpha^2-3\alpha+4}$
Hill	$\varepsilon_1^* = \frac{n}{1+\beta}$ with $\phi = arctan\left(\sqrt{-\beta}\right)$, $-0.5 \le \beta \le 0$
Storen-Rice	(i) $\varepsilon_1^* = \frac{n}{1+\beta}\frac{1}{\frac{1-n}{2}+\sqrt{\frac{(1+n)^2}{4}-\frac{n\beta}{(1+\beta)^2}}}$ with $\phi = arctan\left(\sqrt{-\beta}\right)$, $-0.5 \le \beta \le 0$ (ii) $\varepsilon_1^* = \frac{3\beta^2+n(2+\beta)^2}{2\left(1+\beta+\beta^2\right)(2+\beta)}$ with $\phi = 0$, $-0.5 \le \beta \le 1$
MMFC	$\varepsilon_1^* = n + \frac{3\beta^2}{2(2+\beta)\left(1+\beta+\beta^2\right)}$

Storen-Rice Bifurcation Analysis

For studying the necking behavior of sheet metal, Storen and Rice [3] assumed that the incremental velocity quantities of the flow field inside and outside of a neck are permitted to take on different values. The difference in velocity leads to the following controlling equations used in the traditional bifurcation analysis

$$\eta_1(\Delta\dot{\sigma}_{11} - \sigma_{11}g_1\eta_1) + \eta_2(\Delta\dot{\sigma}_{21} - \sigma_{22}g_2\eta_2) = 0 \tag{11a}$$

$$\eta_1(\Delta\dot{\sigma}_{12} - \sigma_{22}g_1\eta_2) + \eta_2(\Delta\dot{\sigma}_{22} - \sigma_{22}g_2\eta_2) = 0 \tag{11b}$$

Employing the deformation theory of plasticity (J_2D), explicit expressions of the difference between the fields inside and outside of the neck, i.e. $\Delta\dot{\sigma}_{11}, \Delta\dot{\sigma}_{12}, \Delta\dot{\sigma}_{22}$ are achieved [3]. Substituting these expressions into Eq. 11 obtains a system of linear homogeneous equations in term of $g_{1,2}$. Non-zero functions $g_{1,2}$ can be found which lead to the condition of neck formation as follows

$$\left(\frac{H'}{n\overline{\sigma}}\right)^2\left\{4n - 3(1-n)\left(\frac{\partial\overline{\sigma}}{\partial\sigma_{tt}}\right)^2\right\} - \left(\frac{H'}{n\overline{\sigma}}\right)\left[3\frac{\partial\overline{\sigma}}{\partial\sigma_{tt}}\overline{\sigma}^2 + (1-n)\left(\overline{\sigma} - \frac{(\sigma_1+\sigma_2)}{2}\frac{\partial\overline{\sigma}}{\partial\sigma_{tt}}\right)\right] - \left(1 - \frac{(\sigma_1+\sigma_2)}{2\overline{\sigma}}\frac{\partial\overline{\sigma}}{\partial\sigma_{tt}}\right) = 0 \tag{12}$$

where σ_{tt} denotes the stress component normal to the neck direction. Solving this equation with respect to H' obtains the limited strains calculated at the pre-defined angle ϕ. For a given loading path represented by a specific value of β, a critical angle ϕ^* minimized the limited strains should be determined in a numerical manner. This is not easy to achieve an analytical solution for FLC with arbitrary constitutive equations. Considering a special case of deformation theory based on Mises yield function, Storen and Rice derived two possible solutions to satisfy the above governing equation as follows [3]:

(i) $\frac{\partial\overline{\sigma}}{\partial\sigma_{tt}} = 0$ corresponding to a zero extension along the localized neck
(ii) $\frac{\partial\overline{\sigma}}{\partial\sigma_{tt}} = \frac{\partial\overline{\sigma}}{\partial\sigma_{22}}$ corresponding to minimum extension along the localized neck

Therefore, the calculated FLC based on bifurcation analysis is the minimum of these solutions, which are shown in Table 1 as comparisons.

Modified Maximum Force Criterion
Based on experimental observation of strain path transitions after diffuse necking, Hora et al. [4] introduced a modified maximum force criterion (MMFC) as following

$$d\sigma_1 = \frac{\partial\sigma_1}{\partial\varepsilon_1}d\varepsilon_1 + \frac{\partial\sigma_1}{\partial\beta}\frac{\partial\beta}{\partial\varepsilon_1}d\varepsilon_1 \geq \sigma_1 d\varepsilon_1 \tag{13}$$

Under uniform deformations, the strain path is proportional which leads to a constant value of β. Therefore, the second term of the left-hand side of the equation has vanished; and, the criterion reduces to the Swift's instability. After the initiation of diffuse neck, the strain path gradually changes making the evolution of β, which contributes towards the formation of localized neck.

Again, using a power hardening law coupling with the Mises yield function leads to an explicit expression of the critical major strain [5], which is expressed in Table 1.

3 Comparative Results

The previous section presents governing equations and explicit formulation of limited strains determined by different necking criteria. The applications of these formulations for automotive sheet metals are clarified in this section. To this end, experimentally measured forming limit strains of two sheet metals including DP590 and AA6016-T4 materials, which have been reported in the author's previous studies [5, 6], are employed to compare with theoretical predictions and plotted in Fig. 1a and Fig. 1b, respectively.

Table 2. Conditional and sufficient assumptions of these necking criteria.

Assumption	Swift	Hill	Storen-Rice	MMFC
Conditional	$d\sigma 1 = \sigma 1 d\varepsilon 1$	$d\sigma 1 = \sigma 1(d\varepsilon 1 + d\varepsilon 2)$	$\Delta\vartheta_\alpha = f_\alpha(\eta_\beta x_\beta)$	$d\sigma_1 = \left(\frac{\partial\sigma_1}{\partial\varepsilon_1} + \frac{\partial\sigma_1}{\partial\beta}\frac{\partial\beta}{\partial\varepsilon_1}\right)d\varepsilon_1$
Sufficient	$d\sigma_2/d\sigma_1 = \alpha\beta$	$d\sigma_2/d\sigma_1 = \alpha$	$\Delta\dot{\sigma}_{ij}$ based on J_2D	$\sigma_1 = \partial\sigma_1/\partial\varepsilon_1$

Within a proper hardening law, all of these necking criteria provide correct predictions for the lowest point of FLC which corresponds to the critical strain of plane-strain forming mode. Hill's localized necking criterion provides acceptable predictions for limited strains in the negative ranges of strain ratio. However, Swift model seems to be underestimated the experimentally measured data because the diffuse neck is a purely theoretical assumption. In addition, the experimentally critical strains are conventionally determined at the onset of necking localization. Therefore, the transition of strain path beyond the diffuse neck should be considered in FLC calculation, as recommended in MMFC. Consequently, MMFC model gives the best prediction for both two examined materials. The results derived from Storen-Rice model are closed to those of MMFC model in the right-hand side of FLC, although a remarkable discrepancy is observed. However, their prediction for the left-hand side of FLC is lower than those of Hill model. Consequently, it seems to underestimate the experimental data. Table 2 summarizes the conditional and sufficient conditions imposing in each calculation framework. On the one hand, both Swift and Hill models were developed based on an assumption of the proportional loading path. The expression of governing equations is straightforward under the flow theory of plastic deformation. Introducing the evolution of the loading path as shown in MMFC model may improve the accuracy of FLC calculation. However, all of these models were developed using principal stress components that limit the accuracy of the calculated FLC. Future studies on this topic should take the shear stress component into account. On another hand, Store and Rice included the shear stress component in their analysis. However, the increment of stress components was estimated based on J_2D under an assumption of unchanged strain paths. Integrating the strain path change in the model may improve its accuracy of prediction.

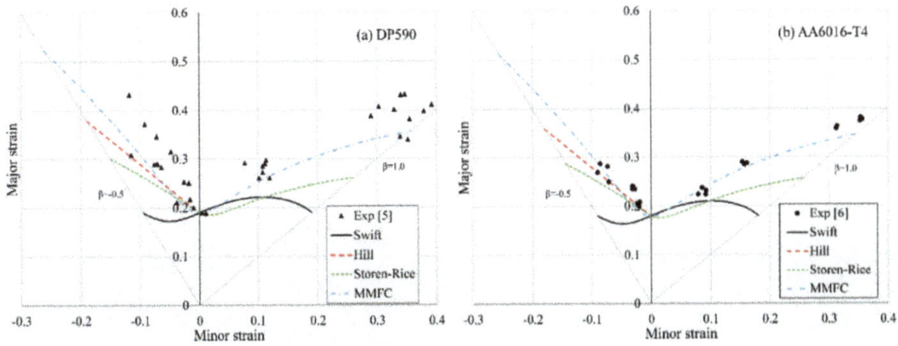

Fig. 1. Comparison between theoretical predictions and experimentally measured forming limit curve for examined materials (a) DP590 (b) AA6016-T4.

4 Conclusion

This study presents details of the theoretical formulation of four explicit necking criteria. Their applications on estimating the FLC of two automotive sheet metals (DP590 and AA6016-T4) are compared with experimentally measured data. The comparison reveals the importance of involving the train path change after diffuse neck and shear stress component in the theoretical prediction of FLC based on these criteria.

References

1. Swift, H.: Plastic instability under plane stress. J. Mech. Phys. Solid. **1**, 1–18 (1952)
2. Hill, R.: On discontinuous plastic states, with special reference to localized necking in thin sheets. J. Mech. Phys. Solid. **1**, 19–30 (1952)

3. Storen, S., Rice, J.: Localized necking in thin sheets. J. Mech. Phys. Solid. **23**, 421–441 (1975)
4. Hora, P., Tong, L., Berisha, B.: Modified maximum force criterion, a model for the theoretical prediction of forming limit curves. Int. J. Mat. Form. **6**, 267–279 (2013)
5. Pham, Q.T., Kim, J.J., Luyen, T.T., Nguyen, D.T., Kim, Y.S.: Application of a graphical method on estimating forming limit curve of automotive sheet metals. Int. J. Automot. Technol. **20**, 3–8 (2019). https://www.scopus.com/source/sourceInfo.uri?sourceId=27571&origin=resultslist
6. Pham, Q.T., Lee, B.H., Park, K.C., Kim, Y.S.: Influence of the post-necking prediction of hardening law on the theoretical forming limit curve of aluminium sheets. Int. J. Mech. Sci. **140**, 521–536 (2018)

Mechanical Responses of Single-Layer Borophene Under Nanoindentation Using Molecular Dynamics

Thi-Bao-Tien Tran[1,2], Te-Hua Fang[1(✉)], Van-Thuc Nguyen[1], and Van-Trung Pham[1]

[1] Department of Mechanical Engineering, National Kaohsiung University of Sciences and Technology, Kaohsiung 80778, Taiwan
fang@nkust.edu.tw
[2] Department of Mechanical Engineering, Nha Trang University, Khanh Hoa, Viet Nam

Abstract. This study surveyed the mechanical characteristics of a single-layer borophene under nanoindentation by using molecular dynamics. These simulations used a spherical and a conical diamond as indenters, the indenters moved at a velocity of 10 m/s and a temperature of 300 K. Some results indicated that the spherical indenter causes greater force, deformation, shear strain, and von Mises than the conical one. With the spherical indenter, the maximum indentation force is 24.05 nN. While the maximum indentation force of the conical tip is 7.62 nN. Both the spherical and conical indenters generated the same crack shape that develops along the armchair direction first and then continues with the zigzag direction of the borophene film. However, the conical indenter created a larger crack than the spherical one. Notably, under the indentation process of a spherical tip, after the initial appearance of the crack, the von Mises stress experienced a dramatic reduction. These results shed more light on the mechanical responses of the borophene film.

Keywords: Molecular dynamics · Borophene · Indentation · Conical indenter · Spherical indenter

1 Introduction

In recent decades, two-dimensional (2D) materials (such as graphene, WS_2, silicene, black phosphorus, and borophene) have attracted the intensive interest of researchers [1–3]. They have exhibited novel properties such as excellent physical, chemical, thermal conductivity, electronic, optical properties, and especially the superconductivity characteristics [4, 5]. These advanced properties could lead to extraordinary applications in electronic devices, strain-dependent optoelectronics, sensors composite materials, energy storage, and utilization [6–8]. Among them, borophene appears as a promising 2D material as it poses advantages properties of thermal conductivity, mechanical properties, atomic adsorption, optical properties [9, 10].

Nanoindentation is a popular method that can survey the fracture behavior of 2D materials. However, the mechanical responses of a single-layer of borophene under

B. T. Long et al. (Eds.): MMMS 2020, LNME, pp. 101–106, 2021.
https://doi.org/10.1007/978-3-030-69610-8_13

indentation have not been studying yet. Nanoindentation can be conducted by experiment or simulation method. By using the simulation method, the deformation can be recreated, observed, and analyzed at a nanoscale.

In this paper, molecular dynamics (MD) simulations are used to survey the indentation process and deformation behaviors of a borophene monolayer. An indenter with a spherical or a conical shape would indent the borophene monolayer. The deformation, forces, atomic strain, and von Mises of this process were carefully analyses.

2 Modelling and Method

The MD simulation model consists of a single-layer borophene film and a spherical or conical diamond indenter, as shown in Figs. 1(a)–(b). The borophene sheet has dimensions of (200 × 200) Å along x-axis (zigzag direction) and y-axis (armchair direction) and it was fixed at four edges. The conical indenter has a cone angle of 30° and a bottom diameter of 40 Å, while the spherical indenter has a diameter of 40 Å. Both of them were treated as rigid bodies and placed 5 Å above the film surface. The indenter velocity and temperature were chosen at 10 m/s and 300 K, respectively. The Lennard-Jones (LJ) potential function was used to illustrate the interaction between the carbon atoms of the diamond tip and the borophene atoms [11]. The bonding among the borophene atoms was imitated by implying the Tersoff potential function [12]. Moreover, all indentation processes were done by applying the LAMMPS code (Large-scale Atomic/Molecular Massively Parallel Simulator code) [13]. Finally, the simulation results were observed and analyzed by the Open Visualization Tool (OVITO) software [14].

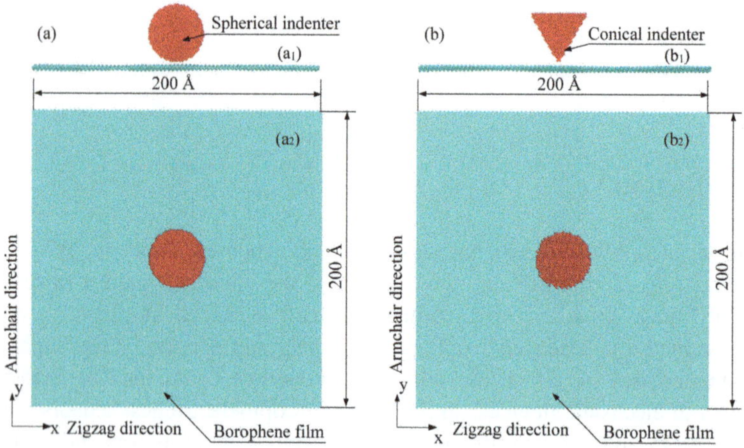

Fig. 1. The simulation models of Borophene (a) using a spherical indenter, (b) using a conical indenter.

3 Results and Discussion

Figure 2(a)–(b) displays the snapshot of the deformation process of borophene film indented by a spherical or conical diamond indenters at 10 m/s and 300 K. Before indenting, a thermal balance was performed to hold the atoms in a steady state. Initially, the indenters moved to the borophene film and touched it, as illustrated in Figs. 2(a_1)–(b_1). Then, Figs. 2(a_2)–(b_2) show the maximum displacement of the perfect borophene film. While Figs. 2(a_3)–(b_3), 2(a_4)–(b_4) indicate the crack-beginning and crack-development of the broken stage. Interestingly, the indenters also created a deformation wave on the borophene film during the indenter process, as illustrated in Fig. 2(a_5)–(b_5). Finally, the indenters passed through the borophene and left a fracture on borophene film, as shown in Figs. 2(a_6)–(b_6). Fig. 2(c) shows the maximum elastic displacement, h_{me}, of borophene film under various indenter shapes. This is the distance between the lowest borophene atom and the fixed borophene atoms. The value of the spherical indenter was about 22 Å, which was larger than that of 12 Å of the conical one.

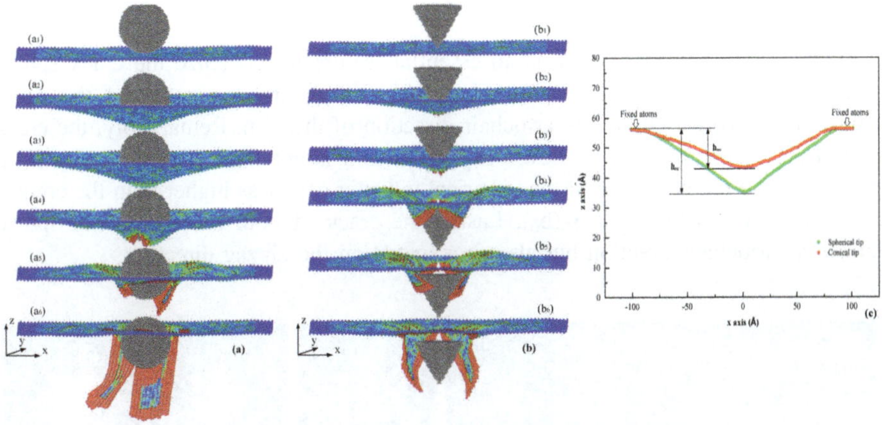

Fig. 2. Deformation development of borophene membrane under indentation process with (a) spherical tip, (b) conical tip, (c) the maximum elastic displacement.

In order to investigate the mechanical properties of the borophene sheet under two indenters, a force-displacement curve was created, as shown in Figs. 3(a)–(b). At the first elastic stage deformation of (I)–(II), the indentation force rapidly increases from zero to the highest value of about 24.05 nN (with spherical tip) and 7.62 nN (with conical tip). Then, with the spherical tip, there was a dramatically drop at the (II)–(III) stage when the crack appeared and developed. Thereafter, the force gradually reduced to zero as the indenter passed through the film. With the conical tip, there are two dropping stages of (II)–(III) and (IV)–(V), interleaved with an increasing stage at (III)–(IV), before dropping to zero at (V)–(VI) stage at the end of the loading process.

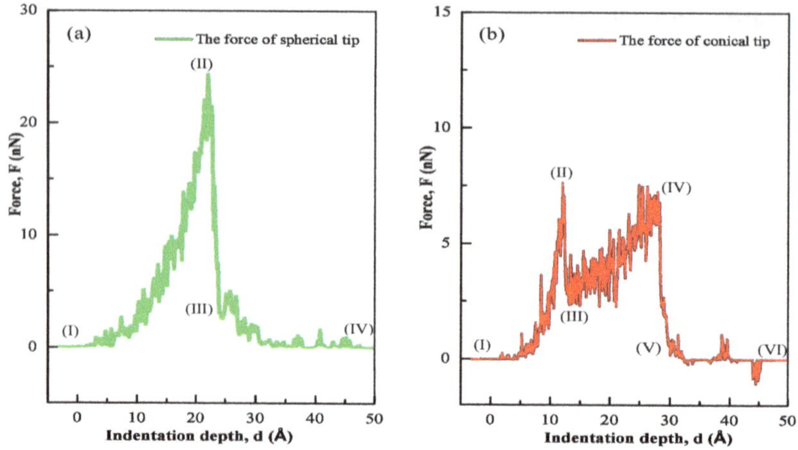

Fig. 3. Indentation force-displacement curve of the single-layer borophene film at T = 300K and v = 10 m/s: (a) spherical tip, (b) conical tip.

The shear strain distribution is an essential feature to recognize the crack spread progress in materials, as shown in Figs. 4(a)–(b). At the beginning of the crack appearance, it spread along the armchair direction of the film. Remarkably, the crack made by the conical indenter developed faster and larger than the spherical one. It means that the recovery rate of the spherical indenter case was higher than the conical case, as shown in Figs. 4(a_6)–(b_6). Lastly, the crack of both cases not only spread across the armchair direction but also propagated on the zigzag direction.

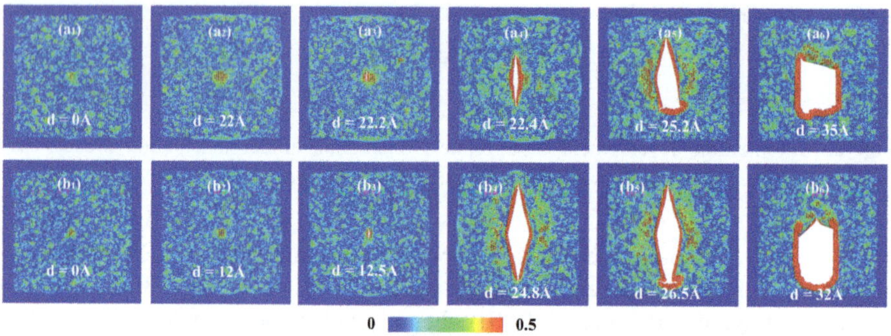

Fig. 4. Shear strain distribution of borophene film under a (a) spherical indenter, (b) conical indenter.

Figure 5(a)–(b) show the development of the atomic von Mises stress of the specimen at 300 K and 10 m/s. The atomic stress distribution of the monolayer borophene under the nanoindentation process was calculated as below Eq. (1):

$$\sigma_{\alpha\gamma} = \frac{1}{\Omega}\sum_{i}^{N}\left(-m_i v_{i\alpha} v_{i\gamma} + \frac{1}{2}\sum_{i \neq j} r_{ij\gamma} f_{ij\alpha}\right) \tag{1}$$

where $\sigma_{\alpha\gamma}$ is the atom average virial stress components i. The symbol α and γ present the contents of the Cartesian coordinate system. Ω is the total volume of a membrane and the whole number of atoms is N. The symbol m_i, $v_{i\alpha}$, and $v_{i\gamma}$ indicate the mass and velocity of the ingredient of atom i. The symbol $f_{ij\alpha}$ is the α-component force between atom i and atom j. The symbol $r_{ij\gamma}$ are the separation distance of the γ-component between atoms i and j [15].

Generally, the high-stress area gradually appeared and strongly developed at an early stage. Then, the stress dropped dramatically as the crack appeared as following the arrows on Figs. 5 (a_4), (a_5)–(b_4), (b_5) when the deformation was released. Eventually, when the indenter passed through the film, there was still some residual stress still existed on the borophene film, as shown in Figs. 5 (a_6)–(b_6). Notably, the spherical indenter created higher stress during the indentation process and even residual stress value than the conical one.

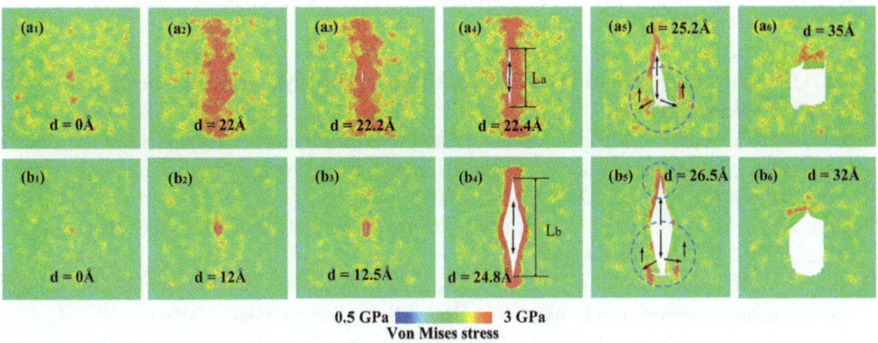

Fig. 5. Von Mises stress distribution of borophene film under a (a) spherical tip, (b) conical tip.

4 Conclusion

The deformation behavior of borophene film was studied by using conical and spherical indenters by conducting MD simulations. The following conclusions were obtained:

1) The conical indenter produces a larger crack than the spherical one. The crack creates by the conical indenter develops on both armchair and zigzag directions, while the spherical one only creates an armchair crack.

2) The spherical indenter products a much higher maximum force than the conical one. Interestingly, the conical indenter creates the larger crack-length.
3) Right before the crack appears, the spherical indenter causes more von Mises stress than the conical one. However, after the indentation process, the borophene film indented by the spherical indenter appears lower stress than the conical one.

Acknowledgments. The authors acknowledge the support by the Ministry of Science and Technology, Taiwan under grant numbers MOST 109-2622-E-992-011-CC3, and MOST 109-2221-992-009-MY3.

References

1. Geim, A.K.: Graphene: status and prospects. Science **324**(5934), 1530–1534 (2009)
2. Radisavljevic, B., Radenovic, A., Brivio, J., Giacometti, V., Kis, A.: Single-layer MoS_2 transistors. Nat. Nanotechnol. **6**(3), 147–150 (2011)
3. Georgiou, T., Jalil, R., Belle, B.D., Britnell, L., Gorbachev, R.V., Morozov, S.V., et al.: Vertical field-effect transistor based on graphene-WS_2 hetero structures for flexible and transparent electronics. Nat. Nanotechnol. **8**(2), 100–103 (2013)
4. Fang, T.H., Chang, W.J., Lin, K.P., Shen, S.T.: Stability and wrinkling of defective graphene sheets under shear deformation. Curr. Appl Phys. **14**(4), 533–537 (2014)
5. Fang, T.H., Chang, W.J., Yang, J.C.: Molecular dynamics study on mechanical properties of graphene sheets with a hole. Dig. J. Nanomater. Biostruct. **9**, 1207–1214 (2014)
6. Huang, Y., Wu, J., Hwang, K.C.: Thickness of graphene and single-wall carbon nanotubes. Phys. Rev. B **74**, 245413 (2006)
7. Lu, Q., Huang, R.: Nonlinear mechanics of single-atomic-layer graphene sheets. Int. J. Appl. Mech. **1**, 443–467 (2009)
8. Fang, T.H., Shen, C.Y., Fan, Y.C., Chang, W.J.: Fracture characteristics of silicene nanosheet with a crack under tension estimated using molecular dynamics simulation. Superlattices Microstruct. **129**, 124–129 (2019)
9. Sun, H., Li, Q., Wan, X.G.: First-principles study of thermal properties of borophene. PCCP **18**(22), 14927–14932 (2016)
10. Moon, W.H., Son, M.S., Lee, J.H., Hwang, H.J.: Molecular dynamics simulation of C60 encapsulated in boron nitride nanotubes. Basic Solid State Physics (2004)
11. Tersoff, J.: Modeling solid-state chemistry: interatomic potentials for multicomponent systems. Phys. Rev. B **39**(8), 5566 (1989)
12. Plimpton, S.: Fast parallel algorithms for short-range molecular dynamics. J. Comput. Phys. **117**(1), 1–19 (1995)
13. Stukowski, A.: Visualization and analysis of atomistic simulation data with OVITO–the open visualization tool. Model. Simul. Mater. Sci. Eng. **18**(1), 015012 (2009)
14. Pham, V.T., Fang, T.H.: Anisotropic mechanical strength, negative Poisson's ratio and fracture mechanism of borophene with defects. Thin Solid Films **709**, 138197 (2020)
15. Diao, J., Gall, K., Dunn, M.L.: Atomistic simulation of the structure and elastic properties of gold nanowires. J. Mech. Phys. Solids **52**(9), 1935–1962 (2004)

Modeling and Simulating Working Processes of the Main Turbocharged Marine Diesel Engine

Do Duc Luu[1(✉)], Nguyen Quang Vinh[2], and Bui Hong Duong[2]

[1] Vietnam Maritime University, 484-Lach Tray, Hai Phong, Viet Nam
luudd@vimaru.edu.vn

[2] Ho Chi Minh City University of Transport, 02-Vo Oanh, Ho Chi Minh, Viet Nam

Abstract. The modern marine diesel (MDE) always is designed to work together with a turbocharger (TC). The paper presents an integral model (IM) and simulation of the working processes of a complex of the MDE and TC. The MDE as a main engine (ME) drives a propeller on the motor vessel (MV). The IM was received on the basis of the integration of the theoretical and experimental researches to improve the model accuracy. The simulation algorithm was programmed in MatLab/Simulink environment and verified to 8MAK43 engine installed on the PhucHung MV. The received simulation was conducted to study the influences of nozzle cross-section to the engine performances and to optimize the nozzle cross-section in accordance with the deteriorated states of the MDE–TC complex.

Keywords: Modeling working process of marine diesel engine · Modeling working process of turbocharger · Model of the MDE–TC complex · Simulation of MDE-TC complex in MatLab/Simulink

1 Introduction

On medium and large-size diesel engines such as on the MDE, the axial TCs were widely used because their efficiency is higher than radial TC at the same working conditions. The actual working and the standard (designed) conditions of the MDE are quite different, besides that the hull resistance and weather condition affected much on engine operation. After a long operation period, the engine performance indicators would have been reduced. One of the significant causes is the un-enough scavenging air in accordance with the engine loads. The scavenging air (pressure parameter p_i, bar) depends on mainly the turbocharger working.

Watson and Janota [1] given the basic equations for the axial flow turbine. Ranke [2] researching experimentally on a 2.3L engine with an axial turbine at transient regimes shown that with the same centrifugal compressor, the axial turbine attained full boost 25 ÷ 40% faster than the TC with the radial turbine. Pesiridis [3] using MatLab to carry out the thermodynamic analysis of the axial turbine showed that the total to static efficiency (of TC) reached 74.94%.

B. T. Long et al. (Eds.): MMMS 2020, LNME, pp. 107–112, 2021.
https://doi.org/10.1007/978-3-030-69610-8_14

The method of this research is to use the integration idea of the theoretical and experimental studies [4] to improve the IM for investigating the working processes of the MDE –TC complex and to optimize the TC works by adjusting the cross-sectional area of the turbine nozzle.

2 Modeling Working Processes of the MDE–TC Complex

2.1 MDE and TC Relationship

The MDE includes combustion chambers (CC), a Scavenging air collector (SAC), and an exhaust gas block (EGB). The TC includes a turbine and a compressor. The working relationship is shown in Fig. 1.

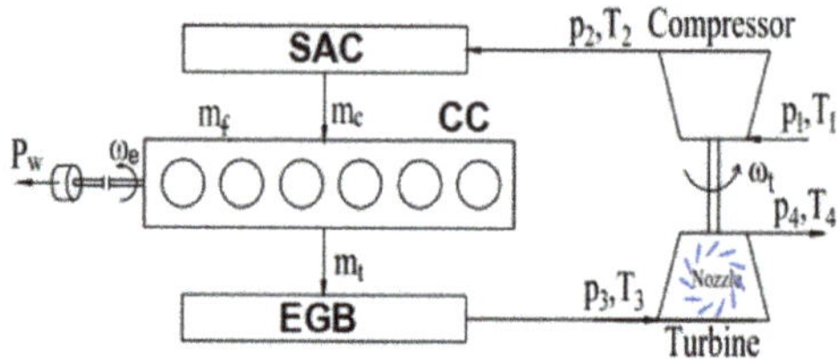

Fig. 1. Relationship between MDE and TC (*m_f, m_c(g/cycle): the amount of fuel provided for a cycle; ω_t, ω_e(rad/s): the turbine and engine speed; p(bar), T(K): pressure and temperature; subscript 1: before the compressor, 2: after the compressor, 3: before the turbine, 4: after the turbine*)

2.2 Models of the TC Working Process

The velocity triangle of the turbine blade is used to make the mathematic model. Figure 2 shows the positions and velocity triangles of the nozzle and working stages of the turbine.

Before and after the working blade: pressure (bar) p3, p4; temperature (°K) T3, T4.

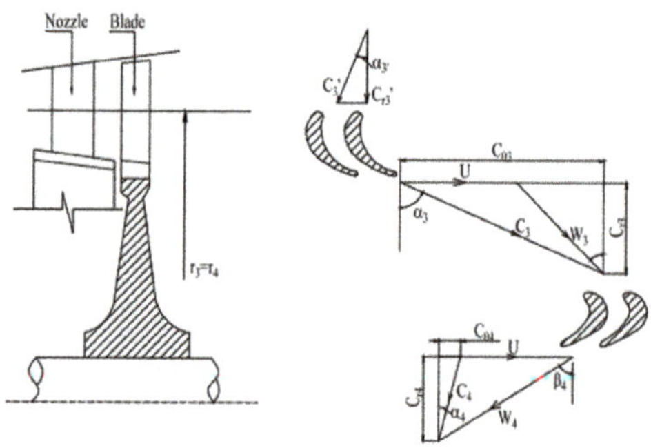

Fig. 2. Velocity triangles of the turbine [1]

The torque produced by the exhaust gas at turbine blades:

$$\tau = \dot{m}_t(C_{\theta 3}r_3 + C_{\theta 4}r_4) \tag{1}$$

The energy transfer:

$$\dot{W} = \omega_t \dot{m}_t (C_{\theta 3} r_3 + C_{\theta 4} r_4); \quad \text{axial turbine}: r_3 = r_4$$

$$\dot{W} = \dot{m}_t U (C_{\theta 3} + C_{\theta 4}) = \dot{m}_t U (C_3 \sin \alpha_3 + C_4 \sin \alpha_4) \quad (2)$$

Where, U (m/s) – speed of rotor; α_3, α_4 - angles of the velocities C_3, C_4; $\dot{m}_t$-mass flow.

C_3 is the outlet velocity of the nozzle, from the first energy conservation law:

$$\dot{m}_t c_{pe} (T_3 + T_{3'}) = \frac{1}{2} \dot{m}_t C_3^2 \quad (3)$$

Assuming the isentropic process in the nozzle we have an isentropic relationship:

$$T_3 / T_{3'} = (p_3 / p_{3'})^{(k_e - 1)/k_e} \quad (4)$$

$$C_3 = \sqrt{2 c_{pe} T_3 \left[1 - (p_{3'}/p_3)^{(k_e-1)/k_e}\right]} = \sqrt{2 c_{pe} T_3 \left[1 - \pi_{t,noz}^{(k_e-1)/k_e}\right]}; \pi_{t,noz} = p_{3'}/p_3 \quad (5)$$

The outlet velocity of the turbine blade can be obtained as below:

$$C_4 = \sqrt{2 c_{pe} T_3 \left[1 - (p_4/p_3)^{(\gamma_e-1)/\gamma_e}\right]} = \sqrt{2 c_{pe} T_3 \left[1 - \pi_t^{(\gamma_e-1)/\gamma_e}\right]}; \pi_t = p_4/p_3 \quad (6)$$

Where, π_t,(−) is the pressure ratio through the turbine; $\pi_{t,noz}$(−) is the pressure ratio through the nozzle; γ (−) is the specific heat capacity ratio; k_e (−) is the isentropic exponent index; c_{pe} (J/kg.K) is the specific heat at constant pressure.

The mass flow through the turbine depends on the gas states before the turbine, nozzle cross-sectional area, and expansion ratio, and modeled as below [1]

$$\dot{m}_t = \frac{p_3}{\sqrt{T_3 R_e}} A_T f(\pi_t) \quad (7)$$

Where R_e (J/kg.K) - exhaust gas constant, and the function $f(\pi_t)$ is modeled below [5]:

$$f(\pi_t) = \sqrt{1 - \pi_t^{k_\pi}} \quad (8)$$

Where k_π(−) is the tuning parameter, according to [5], $k_\pi \approx 2$.

2.3 Models of the Diesel Engine Working Processes

Working processes in the combustion chambers of the diesel cylinders are modeled in accordance with the thermodynamic law, the mass conservation law, Weibe equations for the fuel oil burning and spread heating in the firing process, Woschni equations for the heat exchanging with the cylinder walls. The mathematic models with corrected coefficients according to the relative real processes are conducted and shown in [4].

3 Simulating the MDE-TC Models in MatLab/Simulink

The blocks in Simulink for TC are shown in Fig. 3, and in Fig. 4 – for MDE-TC.

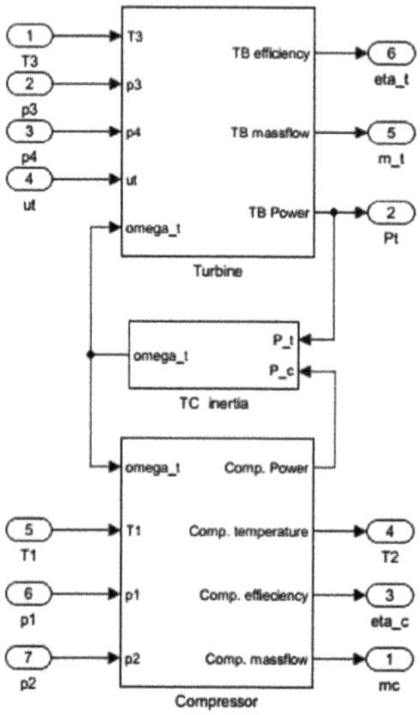

Fig. 3. Block model of the TC in MatLab/Simulink

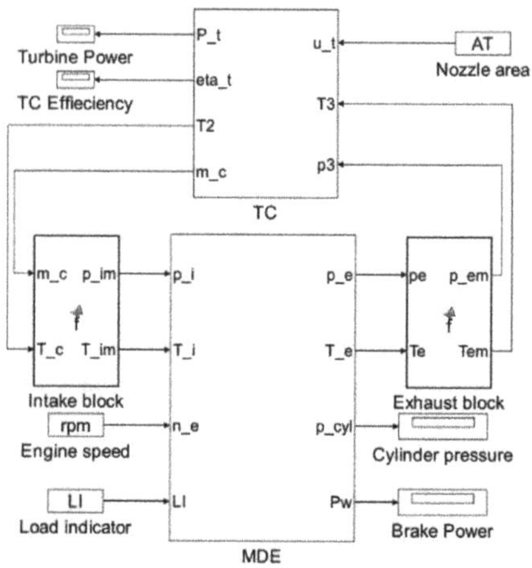

Fig. 4. Block model of the MDE-TC in MatLab/Simulink

The simulation object is a 4-stroke MDE, installed on the MV PhucHung, GLS Co. Ltd (Viet Nam). The features of the object are given in Table 1.

Table 1. Engine specifications [6]

Parameters	E.Unit	Giá trị
Bore x stroke	mm	430 × 610
Nominal speed	rpm	500
Number of cylinders	–	8
SFOC at nominal	g/kW.h	175–178
Nominal power	kW	7200
Turbocharger type	–	TPL76C (ABB)
Turbine blade diameter	mm	520
Number of nozzles	–	24

The Interface Design of the Simulation. In MatLab/Simulink environment, the interface design consists of: input controls, load (%), engine speed (rpm), and nozzle area (%)) and get results (Indicator diagram (p-φ), brake power, specific fuel consumption (g_e), exhaust temperature…). The interface is shown in Fig. 5 after running the simulation.

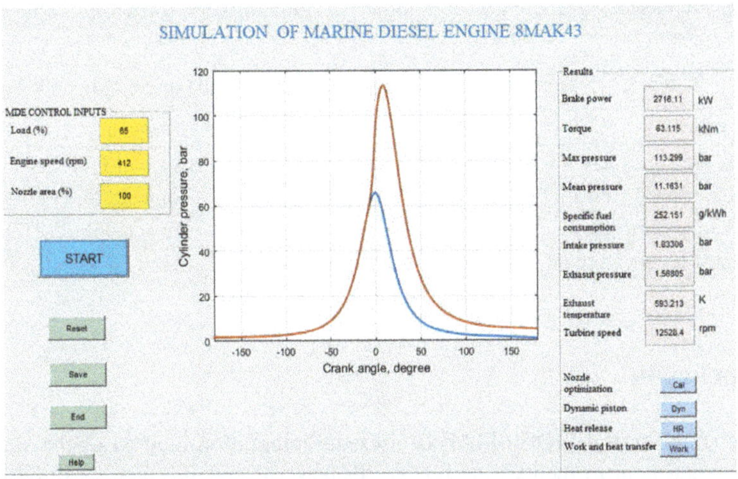

Fig. 5. Simulation program (at speed 410 rpm, load 65%)

This is simulated for the MDE at working condition, the input parameters gathered from engine room (pressure and temperature values), and some key output parameters (exhaust temperature, intake, and maximum pressure) can be compared with the ones in the ship diary for tuning parameters of the modelling.

The optimization of the nozzle area is determined from the simulation as below:

According to Fig. 6 and Fig. 7, the optimization point of the nozzle area is around 91%A_{Tmax}. At this point, the nozzle area was optimized.

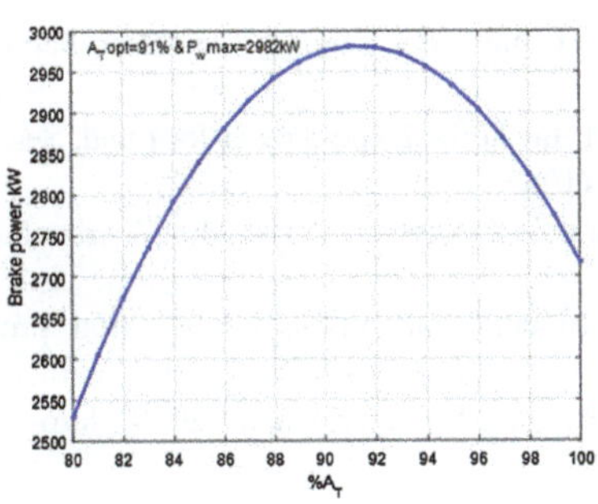

Fig. 6. Varies engine power P_w with nozzle area (%A_{Tmax})

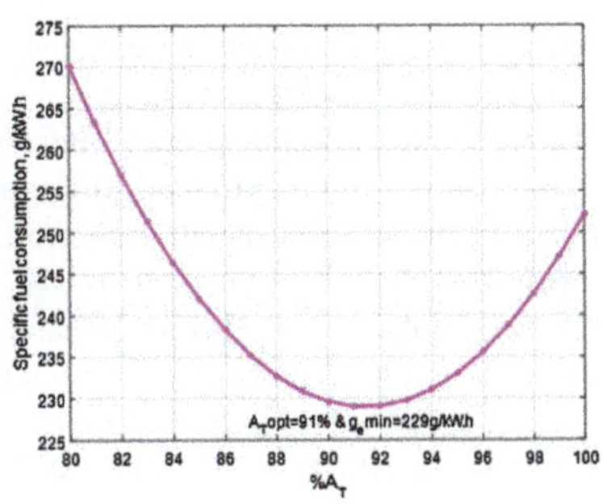

Fig. 7. Varies engine g_e with nozzle area (%A_{Tmax})

4 Experiments on the MDE 8MAK43

The MDE of the MV PhucHung after a long time in using, its technical parameters were gradually worse, especially SFOC and exhaust temperatures. Based on the above simulation results, the turbine nozzle ring was removed and repaired.

After repairing the nozzle ring was re-assembled and tested. The quality working of the MDE increased significantly, and some parameters were shown in Table 2.

Table 2. Evaluation of nozzle adjustment (65% load, speed 411 rpm)

Parameters	Before repairing	After repairing	Changed
Max. pressure (bar)	114	127	Δp_z = 13; (11.3%)
Turbine speed (rpm)	11165	13341	Δn_t = 2176; (19.5%)
Intake pressure (bar)	1.83	2.29	Δp = 0.46; (25.1%)
Exhaust temperature (°C)	379,1	343,6	ΔT = 35.5 °C; (9.4%)

5 Conclusion

The case study on the MDE 8MAK43 uses the integration method of the simulation and real experiment to increase its power, decrease its specific fuel consumption, and exhaust temperature of the old MDE in practice. The simulation research conducts the optimization of the nozzle area to carry out the experimental study for repairing the nozzle ring of the TC in practice. The experiment was carried out and has positive results, with adjusted nozzle area, engine performance improved significantly (See in Table 2, increase max. Pressure with + 11.3%, turbine speed + 19.5%, intake pressure + 25.1%, and decrease exhaust gas temperature −9.4%). This is a relatively simple method and affordable cost which can be applied in the maintenance and improvement of the marine diesel engine field.

References

1. Watson, N., Janota, M.: Turbocharging the Internal Combustion Engine. The Macmillan Press Ltd., London and Basingstoke (1982)
2. Rahnke, C.: Axial Flow Automotive Turbocharger. In: Turbo Expo: Power for Land, Sea, and Air. American Society of Mechanical Engineers (1985)
3. Pesiridis, A., et al.: Conceptual Design of a Variable Geometry, Axial Flow Turbocharger Turbine (2017)
4. Luu, D.D., Vinh, N.Q.: Theoretical and experimental integration for working process simulation on marine diesel engine. In: Fujita, H., et al. (eds.): Proceeding of the International Conference, ICERA 2018. LNNS, vol. 63, pp. 580–588. Springer, Cham (2019). https://doi.org/https://doi.org/10.1007/978-3-030-04792-4_75
5. Eriksson, L., Nielsen, L.: Modeling and Control of Engines and Drivelines. Wiley, Hoboken (2014)
6. Motoren, C.: MaK Marine Propulsion Engines (2010)

Hardening Effect of the Interstitial Free Steel Due to Cold Rolling

Anh-Hoa Bui(✉), Cao-Son Nguyen, and Thi-Thu-Hien Tran

School of Materials Science and Engineering,
Hanoi University of Science and Technology,
No. 1, Dai Co Viet Road, Hai Ba Trung District, Hanoi, Vietnam
{hoa.buianh, hien.tranthithu}@hust.edu.vn

Abstract. In this research, hardening effect of a 1.6 mm thickness commercial interstitial free (IF) steel has been investigated by tensile test and micro-hardness measurement. Microstructural characterization using optical microscopy revealed the single ferrite phase in all the steels. As-received steel had a homogeneous microstructure with grains size of about 40 μm, whereas the cold-rolled steel shown two different microstructures which included finer grains near the deformed surface and the same with as-received steel in the midsection. The tensile tested results show that the strength and micro-hardness have increased with increasing the deformed reduction, but decreased strongly in the annealed steel. It was found that increase of the YS was mainly contributed by dislocation density of the steels after each cold-rolling passes. Dislocation densities of the steels were calculated as 4.67×10^{13} m^{-2} for as-received steel and achieved maximum as 8.95×10^{14} m^{-2} for the rolled steel. The IF steel can be strongly hardened by cold rolling in case the application requires the strength instead of the ductility.

Keywords: Interstitial free steel · Hardening · Microstructure · Dislocation

1 Introduction

Interstitial free (IF) steel usually contains carbon (C) content less than 0.005 %mass and small amount of other elements such as manganese (Mn), silicon (Si), titanium (Ti) or niobium (Nb). Because of having excellent forming characteristics and an increased strength that is obtained during paint baking or galvanizing of the final product, this steel can be widely applied in the automotive and food industries [1–3]. The sheets used for making products required a good formability should have high ductility and low strength. However, several researches have been implemented to increase strength and improve press formability of the steel and subsequently reduce thickness of the products. For example, Galan *et al.* studied on improving strength and dent resistant capacities with bake hardenable ULC steels in order to fulfill the requirements of thinner sheet steel for automotive applications [4]. Jiang *et al.* studied on the IF steel sheet produced by different compact strip production and concluded that a compromised strength of the sheets could be obtained if hot rolling conducted at Ac_1 temperature [5].

B. T. Long et al. (Eds.): MMMS 2020, LNME, pp. 113–117, 2021.
https://doi.org/10.1007/978-3-030-69610-8_15

As far as the IF steels are produced, its mechanical properties have been studied by earlier research works [4–8]. Since the demand for IF steel with higher tensile has strongly increased, most research has been concerning on the annealed or bake hardened IF steel because of improving the drawability and the strength. The compromise resulting from these processes can help the manufacturers to reduce the thickness of the products and minimize the environmental pollution. Besides, the use of cold-rolled IF steel sheets is popular to produce a part with high strength and coating for the purpose of increasing corrosion resistance. Other applications of the cold-rolled IF steel have been known so far in case of that the products require a good strength instead of high ductility [9]. Furthermore, behavior of the cold-rolled IF steel is also important for the sequent annealing step in order to get the proper characteristics of the steel. Therefore, this study investigates mechanical properties such as tensile strength and micro-hardness of a commercial IF steel sheet and discussed in the correlation with the microstructure.

2 Experimental

As-received steel containing 0.0025%C, 0.0808%Mn, 0.0155%Si, 0.0275%Ti, 0.0066%S and 0.0076%P was cut from a commercial IF steel sheet with 1.6 mm thickness, followed by five passes of cold rolling to 0.8 mm thickness. The tensile specimens (gauge length 30 mm and width 6 mm) were prepared in accordance with the Standard ASTM E8, and tested using a MTS 809 machine. Micro-hardness of the steels was averagely determined from 5 measurements by using the Indentamet 1106. Some samples were cut from as-received and deformed sheets in longitudinal rolling direction for microstructural observation by using an optical microscopy (Axiovert 40MAT). Then, the sample was mounted in an epoxy resin, grinded, polished and etched with the nital solution. Grain size was measured by the linear intercept approach, in which a line was superimposed over the optical micrograph.

3 Results and Discussion

Figure 1 shows the optical micrographs of as-received, cold-rolled and annealed steels. Recrystallized grains of the ferrite were observed in all the steels. A homogeneous microstructure including the grains which average size was measured as about 40 μm was seen in the as-received steel (Fig. 1a). However, Fig. 1b shown a bit different microstructure in the cold-rolled steel. Microstructure of the midsection was almost the same with the as-received steel while smaller grains were located near the deformed surface of the cold-rolled steel. This result was in agreement with Guo *et al.* [10] who explained that the sub-surficial area was usually deformed more strongly than the midsection. Thus, it could be concluded that the present cold-rolling process did not make the midsection deform. In addition, a bit coarsen microstructure was seen in the steel after being cold-rolled and annealed at 800 °C for 20 min (Fig. 1c). It can be emphasized that the processing condition affected the microstructure that in turn changed mechanical properties of the steel.

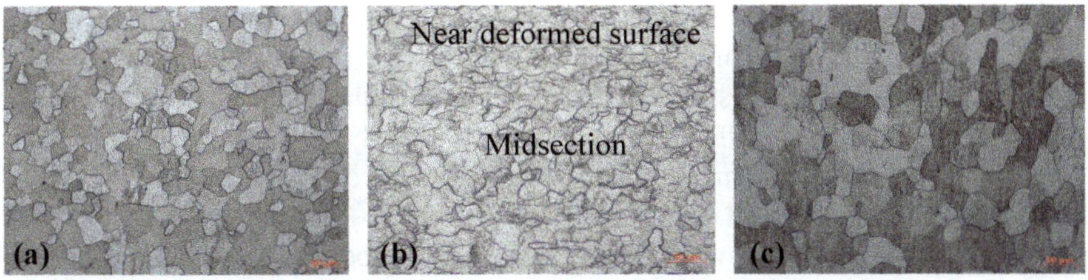

Fig. 1. Microstructure of (a) the as-received, (b) cold-rolled and (c) annealed steels

Effect of the cold-rolling and the annealing on ultimate tensile strength (UTS), yield strength (YS), elongation (EL) and micro-hardness of the steels is illustrated in Fig. 2. UTS and YS of the as-received steel are 278 and 182 MPa, respectively. A very high EL of 42% and low micro-hardness of 168 HV, i.*e.* good ductility, will be a benefit for the consequent deformation of this steel grade. The low strength is attributed to a small amount of strengthening elements such as Si and Mn. In the IF steel, Si plays as a substitution solute in ferrite and has remarkably effect in generating dislocation during quasi-static tensile deformation; resulting in increasing dislocation density and hence, increasing strength of the steel [11]. Mechanical properties of the as-received steel were consistent with the microstructures in Fig. 1a. The fact that the hardened steels were obtained after cold-rolling as seen in Fig. 2 can be simply explained due to attribution to the residual stress in the steel after cold deformation, or in another way that increase in dislocation density with increase in rolling level and as well as in strain rates.

Clearly, the cold deformation caused an increase of the strength and a decrease of the ductility compared with as-received steel. It is seen that the hardening effect increased and the ductility decreases with increasing number of the cold-rolling passes, after each which higher density of dislocation and residual stress has been accumulated in the steel. The YS increment has been reported to be increased with the deformation level, but decreased when the reduction is too large; in the latter case, fine cracks may occur in the steel due to the severe stress [1, 12]. Beside the deformation, precipitation is thought to contribute in strengthening that is often influenced by the size and the ratio of precipitates existing in the steel. Once no precipitate was found in the microstructures of the steels, it has been concluded that the contribution to the YS difference was mainly by the dislocation strengthening.

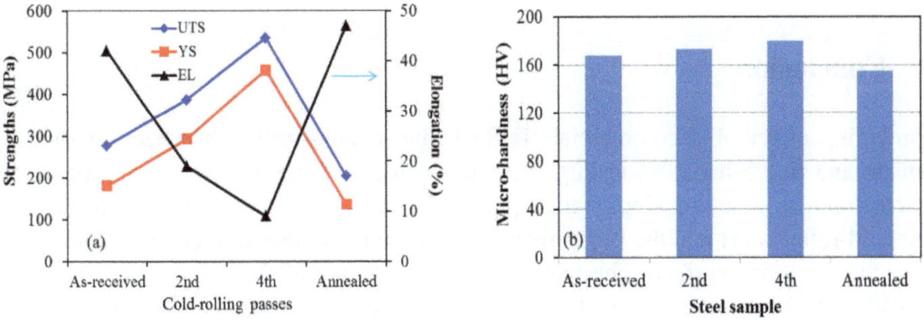

Fig. 2. Variation of the strength (a) and micro-hardness (b) of the steels

Generally, it is accepted that dislocation plays a role in hardening effect of steel [12–15]. Barragan *et al.* confirmed that a stored energy was derived from the dislocation generated during plastic deformation, and calculated the dislocation density based on micro-hardness difference between cold-rolled steel and the prior [15]. Lu *et al.* used X-ray diffraction measurement to calculate the dislocation density in the low carbon martensitic steel, and found the result as 1.64×10^{15} m^{-2} for the cold rolled specimen [14]. In similarity with Takaki *et al.* who proposed that dislocation density has been raised by the refinement of microstructure [13], Antoine *et al.* studied effect of the microstructure on the strain-hardening behavior of a Ti-IF steel grades and concluded that decreasing the grain sizes resulted in higher dislocation densities, which caused the remained stress increment in all the specimens after work hardening [5]. The fact that strength of the annealed steel was lower than as-received steel indicated that the YS depended on the microstructural parameters. In the purpose of quantitative calculation of the dislocation density (ρ), the following correlation can be used for cold rolled ferritic steel [13].

$$YS[GPa] = 0.1 + 1.2 \times 10^{-8} \rho^{1/2} \quad (1)$$

Table 1. Calculated dislocation density of the steels

	As-received	2nd rolling	4th rolling	Annealed
ρ (m^{-2})	4.67×10^{13}	2.61×10^{14}	8.95×10^{14}	9.51×10^{12}

The calculated result is listed in Table 1, in which the dislocation density increased from 4.67×10^{13} m^{-2} (as-received steel) to 8.95×10^{14} m^{-2} (cold-rolled steel) with increasing the reduction, i.e. the ductility decreased as the number of dislocation increased. This is in agreement with the other findings, after which a decrease of the residual stress associated with the removal and rearrangement of dislocation was responsible for a decrease in the mechanical properties of the deformed steel [14, 15]. The dislocation density decreased to 9.51×10^{12} m^{-2} for the annealed steel because the dislocation was removed and the grain size increased during the recrystallization. This result consolidated the explanation of that the residual stress increased the strengths of the cold-rolled steels. TEM observation of the dislocation showed that mobile dislocation changed the arrangement from unstable to stable distribution under the deformation [13].

4 Conclusion

Hardening effect of a commercial IF steel due to cold-rolling has been studied by tensile and micro-hardness testing method. As-received steel shown a homogeneous microstructure including single ferrite phase of 40 μm in average grain size, meanwhile the cold-rolled steel had the same microstructure in the midsection but a smaller grains near the deformed surface. The fact increased YS from 182 MPa for as-received steel to 459 MPa for the cold-rolled steel corresponded to the increased dislocation densities from 4.67×10^{13} to 8.95×10^{14} m^{-2}; indicating that the hardening effect was mainly contributed by the dislocation. A drastically decrease in the strength and micro-hardness in the annealed steel was attributed to the removal of dislocation and recrystallization. Mechanical properties of the cold-rolled steel were concluded to be good for the product preferred the strength to the drawability.

Acknowledgements. This research is funded by Hanoi University of Science and Technology (HUST) under grant number T2018-PC-224.

References

1. Kim, S.I., Lee, Y.: Influence of cooling rate and boron content on the microstructure and mechanical properties of hot-rolled high strength interstitial-free steels. Met. Mater. Int. **18**, 735–744 (2012)
2. Balke, P., De Hosson, J.: Orientation imaging microscopic observations of in-situ deformed ultra low carbon steel. Scripta Mater. **44**, 461–466 (2001)
3. De, A.K., Vandeputte, S., De Cooman, B.C.: Kinetics of strain aging in bake hardening ultra-low carbon steel – a comparison with low carbon steel. J. Mater. Eng. Perform. **10**, 567–575 (2001)
4. Galan, J., Samek, L., Verleysen, P., Verbeken, V., Houbaert, Y.: Advanced high strength steels for automotive industry. Revista de Metallurgia **48**, 118–131 (2012)
5. Antoine, P., Vandeputte, S., Vogt, J.B.: Effect of microstructure on strain-hardening behavior of a Ti-IF steel grades. ISIJ Int. **45**, 399–404 (2005)
6. Ono, Y., Funakawa, Y., Okuda, K., Seto, K., Ebisawa, N., Inoue, K., Nagai, Y.: Roles of solute C and grain boundary in strain aging behavior of fine-grained ultra-low carbon steel sheets. ISIJ Int. **57**, 1273–1281 (2017)
7. Bui, A.H., Le, H.: Strength and microstructure of cold-rolled IF steel. Acta Metallurgica Slovaca **22**, 35–43 (2016)
8. Matoba, R., Nakada, N., Futamura, Y., Tsuchiyama, T., Takasi, S.: Effect of duplex-grained structure on yield stress of IF steels. Tetsu to Hagane J. Iron Steel Inst. Jpn. **93**, 513–517 (2007)
9. Bhagat, A.N., Baek, S.J., Lee, H.C.: A simple method for prediction of shelf life of bake hardening steels. ISIJ Int. **48**, 1781–1787 (2008)
10. Guo, Y.H., Wang, Z.D., Xu, J.S., Wang, G.D., Liu, X.H.: Texture evolution in a warm-rolled Ti-IF Steel during cold rolling and annealing. J. Mater. Eng. Perform. **18**, 378–384 (2009)
11. Yadav, V., Singh, A.K., Sahoo, G.: Effect of tempering temperature on micrsotructure and mechanical properties of low alloy high strength steel. J. Metals Mater. Miner. **28**, 16–21 (2018)
12. Kukla, D., Balkowiec, A., Grzywna, P.: Evaluation of microstructural changes of S235steel after rolling on the basis of microscopic observations and eddy current non-destructive method. Adv. Mater. Sci. **14**, 40–48 (2014)
13. Takaki, S., Ngo-Huynh, K.L., Nakada, N., Tsuchiyama, T.: Strengthening mechanism in ultra low carbon martensitic steel. ISIJ Int. **52**, 710–716 (2012)
14. Lu, Z.Q., Qian, L.H., Liu, S.A., Zhan, L., Qin, S.J.: Preparation and mechanical behavior of ultra-high strength low-carbon steel. Materials **13**, 459–462 (2020)
15. Barragan, J., Diaz, R., Martinez, M., Jimenez, S., Islas, J.: Effect of isothermal treatment on microstructure and mechanical properties of cold-deformed IF steel. Adv. Mater. Sci. Eng. **2019**, 1–13 (2019)

Optimization of Predicting Spring-Back When V-Bending for SS400 Steel Plate

Vuong Gia Hai[1,2], Nguyen Thi Hong Minh[1], Hoang Van Quy[1,2], and Nguyen Duc Toan[1(✉)]

[1] Hanoi University of Science and Technology, Ha Noi, Viet Nam
toan.nguyenduc@hust.edu.vn
[2] Hai Phong University, Hai Phong, Viet Nam

Abstract. In this paper, the Taguchi method and regression analysis used to evaluate the influence of technological parameters: punch radius, temperature and punch stroke to the spring-back angle after V-bending for SS400 steel plate output. The experiments were designing with a mixed orthogonal array and conducted using the L9 (3 × 3) on the bending and heating testing machines modern system. Analysis of variance (ANOVA) was used to determine the effects of input parameters on the output parameters. The analysis results show that the temperature was the biggest factor affecting the springback angle after forming. Nonlinear regression analysis used a quadratic polynomial function to predict experimental results. The predicted values according to construction and experimental mathematical models were vere close to each other, this shows that the built mathematical model has high reliability.

Keywords: SS400 · V-bending · Springback · Elevated · Taguchi method

1 Introduction

Nowadays the steel plate are widely used in the processing of steel structure and shipbuilding [1, 2] because of the reasonable cost while ensuring the requirements of the structure and equipment. Among of the products made from steel plate, the parts with curved profiles machined by bending technology account for a large of proportion. The spring-back of forming processes is greatly affects to the accuracy of the products. And the spring-back of bending processes are often greatly influenced by technology parameters such as the bending radius, the punch radius, the thickness of the plate, the velocity of bending punch, heating when bending, etc. Among of the technological parameters affecting the spring-back after forming deformation, the temperature is a very greatly influential parameter, there were many studies [3–7] that had shown that when the temperature is the higher, then spring-back is the lower. These studies have used simulation and experiment methods to evaluate the effect of temperature on the spring-back after forming deformation.

In this study, the influence of technical parameters on the spring-back angle after V-bending made by SS400 steel plate was investigating. The experimental method according to orthogonal array Taguchi L9 is used in this study. To determine the technological parameters (punch radius, temperature and punch stroke) optimal for

B. T. Long et al. (Eds.): MMMS 2020, LNME, pp. 118–125, 2021.
https://doi.org/10.1007/978-3-030-69610-8_16

V-bending, processing analyzes the influence ratio of technological parameters according to the Taguchi method is used. By the nonlinear regression method based on experimental results, a mathematical model was built to describe the relationship between the spring-back angle after forming deformation and the selected input parameters. Comparing experimental and calculations results through mathematical models has shown that the results are close to each other, this shows the reliability of the built mathematical model.

2 Experimental Design

2.1 Materials

The experimental samples used in this study has long × width × thickness as: 110 mm × 35 mm × 5 mm Fig. 1. The material is SS400 steel plate, according to JISG 3101 [8] with a chemical composition as Table 1.

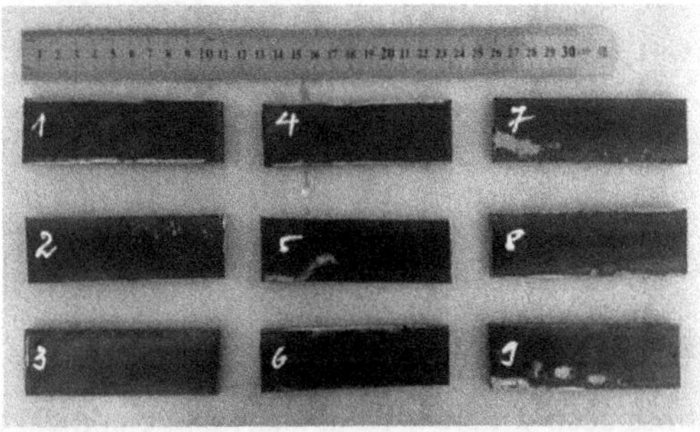

Fig. 1. The experimental samples used for Taguchi method

Table 1. Chemical composition of SS400 steel sheet

C	Si	Mn	P	S	Cr
0,19–0,21	0,05–0,17	0,4–0,6	0,04	0,05	$\leq 0,3$

The mechanical properties of tested material as show in Table 2 [9].

Table 2. Mechanical properties of tested material [9]

Parameters	Temperature (°C)	Poisson's ratio	Density, kg/m^3	Young's Modulus, MPa	Yield stress, (σ_y, MPa)
Levels	32	0,3	7,850	213.000	348
	300	0,3	7,850	184.410	199,3
	600	0,3	7,850	107.640	72,43

2.2 Experimental System Setup

The experimental system used in this study is designed as Fig. 2 and setup as Fig. 3. To maintain stable temperature during V-bending, all punch, die bending and workpiece is placed in the insulation box. The system works correctly at the required temperatures owing to the temperature controller installed on the system, so that the required temperature is maintained stable during bending. The velocity of the punch bending is 2 mm/s for experiments.

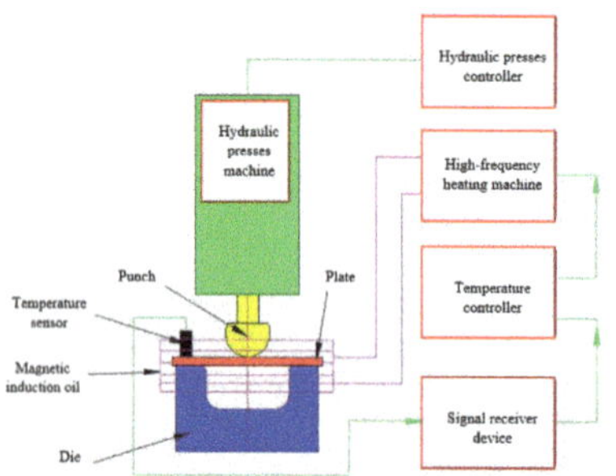

Fig. 2. The diagram of experimental bending system with heating

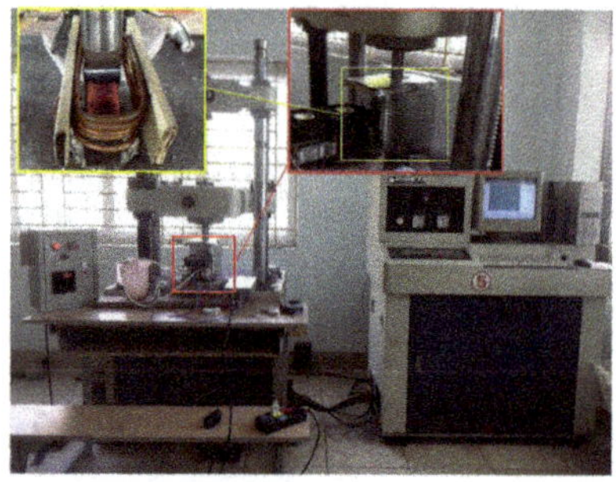

Fig. 3. Setup the experimental system of V-bending with heating

2.3 Taguchi Orthogonal Array

The Taguchi method has been widely used in analyzing empirical data of engineering disciplines [10, 11]. This method has many advantages such as: effective experiments, simple and easy application, it need only a minimum number of experiments can evaluate the best results, thus bringing high economic efficiency.

The purpose of this study is to adjust the input parameters to control the desired output parameters. Therefore, the Taguchi method was chosen to allow the minimum of experiments needed to study the effect of parameters on an output characteristic and quickly adjust the input parameters to optimize the fastest output value.

Table 3. Bending parameters and their levels

Parameters	Unit	Levels		
		1	2	3
Temperature (T)	°C	32	300	600
Punch radius (R_{punch})	mm	10	15	20
Punch stroke (H)	mm	10	16	22

Table 4. Full factorial design with orthogonal array of Taguchi L9 (3^2)

No	T (°C)	R_{punch} (mm)	H (mm)
1	1 (32)	1 (10)	1 (10)
2	1 (32)	2 (15)	2 (16)
3	1 (32)	3 (20)	3 (22)
4	2 (300)	1 (10)	2 (16)
5	2 (300)	2 (15)	3 (22)
6	2 (300)	3 (20)	1 (10)
7	3 (600)	1 (10)	3 (22)
8	3 (600)	2 (15)	1 (10)
9	3 (600)	3 (20)	2 (16)

In this study, the effect of the parameters: temperature (T - 0 °C), punch radius (R_{punch} - mm) and punch stroke (H - mm) on spring-back after forming the bending process has been to investigate. A set of three parameters is chosen with three different levels of values given in Table 3. The S/N noise ratio selected according to smaller is better conditions (Eq. 1).

$$\text{S/N} = -10\text{Log}_{10}(MSD) \tag{1}$$

$$MSD = \frac{(y_1^2 + y_2^2 + \ldots + y_n^2)}{n} \tag{2}$$

Where, MSD is the mean squared deviation from the target value of the quality characteristic, and y_1, y_2, …, y_n are the observed data at the ith experiment, n is the number of observations of the experiment.

The order of experiments using the orthogonal array method L9(3^2) is built according to the Table 4.

3 Analysis and Evaluation of Experimental Results

3.1 Analysis of the Signal-to-Noise (S/N) Ratio and ANOVA

The results of measuring the experimental samples for 9 experiments after conducting experiments using the orthogonal array method Taguchi as designed in Table 4 are shown in Fig. 4 and Table 5. The S/N ratio of each experiment is presented as in Table 5. In which, the ratio of S/N with the quality characteristic of spring back (θ_{SP}) angle the smaller is better to be determined by Eqs. (1) and (2).

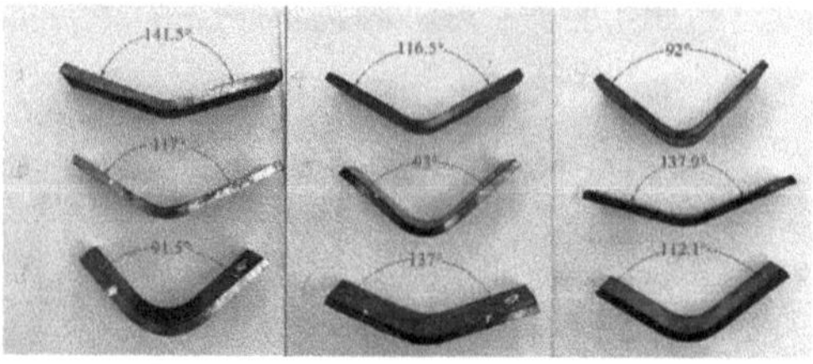

Fig. 4. Result of experiments when using the orthogonal array Taguchi method

Table 5. The results according to Taguchi method

No	T (°C)	R_{punch} (mm)	H (mm)	Bending angle (°)	Experiment angle (°)	Springback angle (θ_{SP}) (°)	S/N
1	32	10	10	138	141,5	3,5	−10,88
2	32	15	16	112	117	5,0	−13,98
3	32	20	22	86	91.5	5,5	−14,81
4	300	10	16	114	116,5	2,5	−7,959
5	300	15	22	88	93	5,0	−13,98
6	300	20	10	134	137	3,0	−9,542
7	600	10	22	90	92	2,0	−6,021
8	600	15	10	136	137,9	1,9	−5,575
9	600	20	16	110	112,1	2,1	−6,444

The S/N ratio of each control parameter with three levels as Table 6 and the analysis of their influence on the spring-back angle during V-bending with heated detail is shown in Fig. 5. The results show that the optimal technological parameter set for the smallest spring-back angle is:

$$\text{T3R1H1 respectively } T = 600\,^{\circ}\text{C},\ R_{punch} = 10\ \text{mm}, H = 10\ \text{mm}$$

To analyze variance (ANOVA) for the spring-back angle, the study uses Minitab 17 software to assist in the data analysis process. The ANOVA analysis results are shown in Table 6.

Table 6. ANOM and ANOVA table of data affecting the springback

Factor	Average by level			Sum of squares	Contribution (%)
	1	2	3		
T	−13,233	−10,494	**−6,013***	10,7222	65,726
R_{punch}	**−8,287***	−11,178	−10,265	2,6289	16,115
H	**−8,666***	−9,461	−11,602	2,9622	18,158
Err.				0,0956	0,001
Sum				16,4089	100

From Table 6 and Fig. 5, we see that the temperature has the largest influence 65,726%. on the spring-back angle (θ_{SP}). Next it is the percentage of influence of the radius of the punch radius (R_{punch}) and the punch stroke (H) are quite similar and respectively: 16,115% and 18,158%. And the percentage distribution of error is quite small, only 0,001%. Thereby, we can see that, with the selected technological parameters when bending with heating, the temperature is the decisive parameter and most strongly affects the spring-back angle (θ_{SP}).

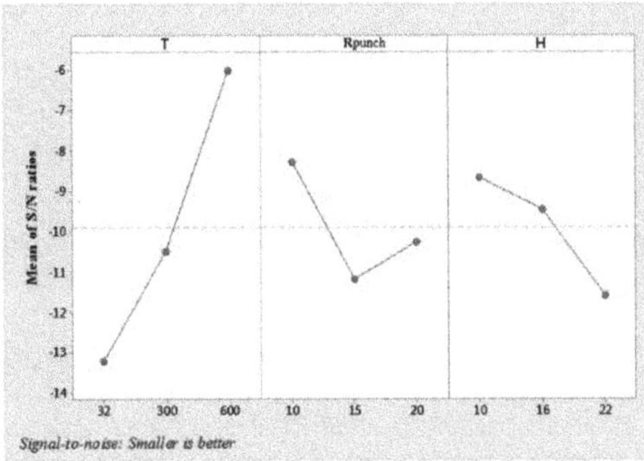

Fig. 5. Main effects plot for means

3.2 Regression Analysis of Springback When V-bending

Regression analysis is used for the modelling and analyzing of one or several variables with dependent relationships with several independent variables [12]. In this paper, the dependent variable is the spring-back angle (θ_{SPC}) and the independent variables used are: Temperature (T), punch radius (R_{punch}) and punch stroke (H) as Eq. (3).

Equation (3) is used to predict the dependency relationship between spring-back angle θ_{SP} and input parameters. The predictive equation used in this study is a quadratic polynomial function. The predicted equation for the quadratic polynomial function of the spring-back angle (θ_{SPC}) Eq. (3) was formulated using the Fitting Regression tool of Minitab 17 software as follows:

$$\theta_{SPC} = g_{(T, R_{punch}, H)} \tag{3}$$

$$\begin{aligned} \theta_{SPC} = & -2,778 - 0,006440.T + 1,113.Rch - 0,2219.H + 0,000139.T.R \\ & + 0,000023.T.H - 0,000001.T^2 - 0,03551.R^2 + 0,01077.H^2 \end{aligned} \tag{4}$$

$$\text{R-sq} = 99,42\% \qquad \text{R-sq}(adj) = 97,67\%$$

The accuracy of the model for calculating the spring-back angle (θ_{SPC}) when V-bending part with heating according to Eq. (4) is evaluated by error when compared with experimental data such as Table 7. In which, the error angle of spring-back $\Delta\theta_{SP}$

Table 7. Compare results experimental and calculated by the formula of springback

No	T (°C)	R_{punch} (mm)	H (mm)	θ_{sp} (°)	θ_{SPC} (°)	$\Delta\theta_{SP}$ (%)
1	32	10	10	3,5	3,504	0,107
2	32	15	16	5	5,005	0,107
3	32	20	22	5,5	5,507	0,126
4	300	10	16	2,5	2,513	0,525
5	300	15	22	5	5,013	0,269
6	300	20	10	3	3,017	0,567
7	600	10	22	2	2,045	2,274
8	600	15	10	1,9	1,950	2,645
9	600	20	16	2,1	2,149	2,358

was determined from Eq. (5) compared with experimental data obtained from 9 experiments using orthogonal array method Taguchi.

$$\Delta\theta_{SP} = \left|\left(\frac{\theta_{SPC} - \theta_{SP}}{\theta_{SP}}\right)\right| \times 100\% \tag{5}$$

The results from Table 7 show that, the error between the mathematical model determining the spring-back angle calculated by Eq. (4) and the results 9 experiments are quite small. The maximum error value is only 2,645% in experiment number 8. This shows that the construction mathematical model determines the spring-back angle when V-bending part with heating has high reliability.

4 Conclusion

In this study, the Taguchi orthogonal array method was used to determine the optimal parameter set for the smallest spring-back angle when V-bending parts fabricate from SS400 steel plate. Experimental results were evaluated using variance analysis (ANOVA). The results of the study showed that: The optimal parameter set is T3R1H1 corresponding to T = 600 °C, R_{punch} = 10 mm, H = 10 mm will give the smallest spring-back angle is 0,41°. And the heating temperature (T) is the parameter that influences the spring-back angle with the distribution is 65,726%. A quadratic polynomial regression model describing the relationship between the spring-back angle and the input parameters has been developed and it also shows reliability when compared to empirical data. The biggest error between experimental and calculation models is only 2,6%. Developed quadratic regression model demonstrated a very good relationship with high correlation coefficient (θ_{SP} = 0,9767) between the measured and predicted values for.

References

1. Bích, V.N.: KẾT CẤU TÀU THỦY. Hồ Chí Minh (2007)
2. Đăng, T.H.: Nghiên cứu ảnh hưởng của các thông số công nghệ cơ bản khi lăn ép đến khả năng tạo hình tấm dày có biên dạng phức tạp ứng dụng trong công nghệ đóng tàu (2016)
3. Stachowicz, F., Trzepieciński, T., Pieja, T.: Warm forming of stainless steel sheet. Arch. Civil Mech. Eng. **10**(4), 85–94 (2010). https://doi.org/10.1016/S1644-9665(12)60034-X
4. Nguyen, D.T., Park, J.G., Kim, Y.S.: A study on yield function for Ti–6Al–4V titanium alloy sheets at elevated temperatures. Trans. Indian Inst. Met. **69**(7), 1343–1350 (2016). https://doi.org/10.1007/s12666-015-0687-5
5. Bruni, C., Forcellese, A., Gabrielli, F., Simoncini, M.: Air bending of AZ31 magnesium alloy in warm and hot forming conditions. J. Mater. Process. Technol. **177**(1–3), 373–376 (2006). https://doi.org/10.1016/j.jmatprotec.2006.03.222
6. Wang, A., et al.: Springback analysis of AA5754 after hot stamping: experiments and FE modelling. Int. J. Adv. Manuf. Technol. **89**(5–8), 1339–1352 (2017). https://doi.org/10.1007/s00170-016-9166-3
7. Duc-Toan, N., Seung-Han, Y., Dong-Won, J., Tien-Long, B., Young-Suk, K.: A study on material modeling to predict spring-back in V-bending of AZ31 magnesium alloy sheet at various temperatures. Int. J. Adv. Manuf. Technol. **62**(5–8), 551–562 (2012). https://doi.org/10.1007/s00170-011-3828-y
8. JIS G3101 SS400 steel plate/sheet for general purpose structural steels, p. 86011881
9. Vuong, G.H., Nguyen, D.T.: Studies on predicting spring-back and verifying the effects of temperature, sheet thickness and punch speed on forming force of V-bending for SS400 steel plate. In: Advance Materials, Springer Proceedings in Materials, vol. 6, (2020). https://doi.org/10.1007/978-3-030-45120-2_9
10. Ng, B., Hoàng, V.: Effect of feed rate, toolpath and step over on geometric accuracy of freeform suface when 3 axis CNC milling, vol. 9, no. May, pp. 576–584 (2018)
11. Asiltürk, I., Akkuş, H.: Determining the effect of cutting parameters on surface roughness in hard turning using the Taguchi method. Measurement: J. Int. Meas. Confed. **44**(9), 1697–1704 (2011). https://doi.org/10.1016/j.measurement.2011.07.003
12. Kivak, T.: Optimization of surface roughness and flank wear using the Taguchi method in milling of Hadfield steel with PVD and CVD coated inserts. Measurement: J. Int. Meas. Confed. **50**(1), 19–28 (2014). https://doi.org/10.1016/j.measurement.2013.12.017

Using the Essential Oils Extracted from Vietnamese Plants as Antibacterial Agents for Nonwoven Polyethylene Fabric

Luu Thi Tho[2], Vu Thi Hong Khanh[1(✉)], and Tran Thi Phuong Thao[1]

[1] School of Textile – Leather and Fashion (STLF), Hanoi University of Science and Technology (HUST), Hanoi, Vietnam
khanh.vuthihong@hust.edu.vn

[2] Garment Technology and Fashion Design (GTF), Hanoi University of Industry (HaUI), Hanoi, Vietnam

Abstract. In recent years, many works showed that Essential oils (EOs) extracted from aromatic medicinal plants have the good antimicrobial effects against bacteria, yeasts, filamentous fungi, and viruses. Many studies have shown that many EOs extracted from Vietnamese plants have antibacterial activity (AA). The purpose of the paper is to evaluate the AA of polyethylene nonwoven fabric treated with four Vietnamese EOs: Cinnamon; basil; lemongrass, and mint. Two types of bacteria were used: Escherichia coli (E. coli) and Staphylococcus aureus (Staph. aureus). Antibacterial evaluation method was according to AATCC method 147. According to the AA of the treated fabrics, these four EOs could be ordered as follows: The best oil is cinnamon, basil oil ranks the second, the next is the mint oil, and the lowest effective is the lemongrass oil. The results show that these Vietnamese natural EOs could be used as antibacterial agents for disposable textile products such as wet paper tow or medical textiles.

Keywords: Cinnamon · Basil · Lemongrass · Mint · Polyethylene nonwoven fabric

1 Introduction

Textile and garment products have a large surface area and high moisture absorption, which is a favorable condition for the growth of microorganisms [1]. The growth of microorganisms on the textiles can degrade the product quality, possibly even destroy the product. Moreover, they also affect the health of textile product users [2]. Consequently, the demand for anti-microbial textiles has increased significantly in recent times [1, 2]. Antimicrobial methods for textiles are classified mainly based on the antimicrobial compounds used. The main antimicrobial compounds that have been used to provide antimicrobial properties to textile products are quaternary ammonium compounds, triclosans, metal salts, polybiguanides or even natural polymers [1–4]. These compounds are not only effective antimicrobial agents, but must also be non-toxic to the consumer and the environment [1, 3, 4]. This need has spurred the studies on using natural-based compounds, which have revealed potential AA without harmful

B. T. Long et al. (Eds.): MMMS 2020, LNME, pp. 126–137, 2021.
https://doi.org/10.1007/978-3-030-69610-8_17

effects to the surrounding medium to make antimicrobial textile products [1]. Among natural antibacterial compounds, EOs extracted from medicinal plants have been used in many studies as antimicrobial agents. The results showed that they are resistant to microorganisms such as bacteria, yeasts, filamentous fungi, and viruses [5]. EOs are defined as "The highly concentrated volatile substances extracted from various parts of certain plant species, each with specific therapeutic and energetic effects" [6, 7]. The use of EOs as antimicrobial agents has been described qualitative for many years in many works for example: In work [6] (1996), 51 EOs were examined the AA against a range of Gram-positive and Gram-negative bacteria. The tests with the drop diffusion method showed that all 51 oils were able to inhibit the growth of at least one of the seven microorganisms used, 13 oils have been shown to be active against at least 5 types of microorganisms. In work [8] (2000), the 6 volatile oils were assessed for AA against 25 different genera of bacteria. The results show that these oils have a significant inhibitory effect on all the organisms tested. However, their main constituents have shown the different growth inhibition abilities. In work [9] (2001), the AA of 14 EOs and their major constituents in the gaseous state was evaluated against Haemophilus influenza, Streptococcus pneumoniae, Streptococcus pyogenes and Staph. aureus. Work [5] (2009) shows that 24 EOs tested exhibited a high anti-Helicobacter activity. Moreover, results also show the antiviral activity of the 30 EOs tested could be clearly demonstrated for enveloped DNA and RNA viruses. Thanks to their properties including antibacterial, odorous, and friendly with environment and human, natural EOs have been used in many areas: perfumes, cosmetics, soaps and other products, foods and drinks (to add flavor) as well as incense and household cleaning products (to add scents) [7].

In the textile industry, EOs have also attracted the attention of many researchers [1, 10]. Although there are many benefits [7, 11], the practical application of natural EOs for textiles is limited because of being volatile, having poor water and chemical solubility and being unstable in the presence of air, light and high temperature [11, 12]. However, the application of EOs for antimicrobial effect on textiles have increased in recent times because they are highly efficient when applied using the right technology [10]. With the aforementioned limitations, the use of natural EOs for disposable textile products is appropriate. In the work [13], 100% Viscose Hydroentangled nonwoven was treated with 5 EOs by pad-dry method, the drying was at 60 °C for 35 min. The results showed that the nonwoven fabric treated with clove oil even at a concentration of 1% had very good antibacterial properties. Thanks to its high absorbency and antibacterial properties, this product can have a wide range of medical applications such as wipes, wound care bandages, bed sheets, pillow cases, surgical absorbent composites and others. To overcome the volatile nature of natural EOs for textiles, researchers have applied various methods. In the work [11], Tea tree oil (TTO) was encapsulated into β-CD, PVA and GA as wall materials. Then, the prepared capsules were applied to the 100% viscose nonwoven fabric samples through padding method. The results show that the fabric samples, which were treated with prepared TTO capsules gained AA at the different levels depending on the capsule walls and bacteria species. All of the treated samples had high AA against Staph. aureus bacteria, the TTO/PVA capsules applied samples had the highest AA. This method can be applied to produce disposable antibacterial textile surfaces that can be used as medical or cosmetic textiles. In the work [14], the woven and knitted fabric were treated with chitosan–lavender EO and monochlorotriazinyl-β-cyclodextrin–lavender EO systems by Pad-dry-cur method. All treated samples present antibacterial properties against both gram positive (Staph. aureus) and gram negative (E. coli) bacteria. However, the best results are obtained in the case of the monochlorotriazinyl-β-cyclodextrin/lavender EO system. In work [15], the woven cotton fabric was treated with 10% aqueous solution of microencapsulated geranium powder by pad-dry-cur technique. The result show that the cotton microcapsulateds fabric have antibacterial effects against both Staph. aureus and E. coli. Furthermore, this method also helps to retain up to 50–60% of the fragrance after 10 washings. In work [16], the treated textile materials were finished with herbal extracts and EOs by using pad-dry-cure method and micro-spraying method. The result show that the products are equipped with good AA. These products also showed good result in the evaluation of the health care performance. In the work [17], the untreated cotton, carboxymethyl cotton and cationized cotton fabrics were treated with a mixture of emulsions with different concentrations. The composition of the emulsion includes Neem and Tulsi EOs and Silicon Micro Emulsion (SiME) or Nonionic softeners. The treated fabrics were encouraged for use in the medical textile sector for their excellent antimicrobial properties and softness. The investigation [12] is to study the combined effect of fragrance and antimicrobial finishing on cotton fabric by lavender EO with the use of β-Cyclodextrin, Chitosan citrate and β-Cyclodextrin/Grafted Chitosan through pad-dry method. The results revealed that cotton treated with β-Cyclodextrin/Grafted Chitosan is better alternative to conventional finishing. This novel technique can be use for design and development of customized aroma fabric and give new idea on textile fragrance.

Although many works have shown the good AA of EO treated textile materials, the exact mechanism/mode of action of their synergies or purified components on microorganisms when applied in textiles have not been explicitly described. Presently the generally accepted mechanism of antimicrobial action is the sequential inhibition of the common biochemical pathway, protective enzymes and the destruction of the cell wall of the microbe enhancing quick uptake of the antimicrobial agent as well as cell lysis [10].

In Vietnam, there are many kinds of natural EOs that have attracted many researchers but mainly in the fields of health and food. In 2013, Thanh, N.T.T., et al. [18] studied and compared the AA of Vietnamese cinnamon EO and its chemotype (trans-cinnamaldehyde) with the purpose of using them in food preservation. In 2015, Lan-Phi, N et al. [19] studied about the chemical composition, antioxidant and antibacterial activities of peels' EOs of different pomelo varieties in the South of Vietnam. More recently, the chemical composition and the inhibitory activity of Vietnamese Citrus peels EOs also were studied in the work [20]. These studies show that many of the EOs extracted from Vietnamese plants are antibacterial but have not been used as antimicrobial agent for textile fabric. Therefore, in this study, four Vietnamese natural oils (cinnamon, basil, lemongrass, and mint) that have antimicrobial property (their antibacterial abilities have been confirmed in many studies: lemongrass [6, 21], cinnamon [6, 22], basil [6, 23] and mint [6, 24])), were used as antibacterial agents for polyethylene nonwoven fabric. The novelty of this study is the use of Vietnamese natural EOs in the antibacterial treatment for textiles. The purpose is to clarify the possibility of using these EOs as antibacterial agents for textile fabrics. To achieve this goal, the study includes three sections: Investigate the antibacterial ability of polyethylene nonwoven fabric treated with pure EOs (cinnamon, basil, lemongrass, and mint); The effect of these four EOs on the color of EO treated non-woven fabric; The effect of the concentration of these EOs on the AA of the EO treated fabric.

2 Research Method

2.1 Materials

Substrate: Nonwoven fabric 100% polyethylene; Surface mass of fabric: 26 g/m^2; Thickness: 0.3 mm, which was made in Malaysia and imported into Vietnam by Joint Stock Company Diana Unicharm in Bac-Ninh province, Vietnam.

Essential oils: Pure cinnamon, basil, lemongrass, and mint EOs supplied by the farmers in the Northern mountain of Vietnam.

The chemical composition of the oils:

The main ingredients of cinnamon bark EO include cinnamaldehyde (55%–76%), eugenol (5%–18%) and saffron (up to 2%) [25, 26].

In the EO of basil, there are 25 different ingredients, of which the main ingredients are methyl eugenol (39.3%) and methyl chavicol (38.3%) [27].

In the EO of lemongrass, the percentages of geranial and neral were highest, at 45.95% and 31.13%, respectively [28].

In peppermint essential oil a total of 63 different ingredients have been identified. However, the proportions of carvone and limonene accounted for (40.8% ± 1.23%) and (20.8% ± 1.12%), respectively [29].

The molecular formula of the main components of cinnamon, basil, lemongrass, and mint oils are presented in Table 1.

The bacterial strains

Gram positive: Staphylococcus Aureus according to AATCC – 6538.

Gram negative: Escherichia coli according to AATCC – 25922.

2.2 Methods

Treatment of the Nonwoven Fabric with the EOs

Nonwoven fabrics were treated with finishing mixtures by the padding method.

The finishing mixtures were prepared with one of the four EOs at different concentrations. The padding (finishing) mixtures were prepared as follow: The distilled water and the selected oil were added according to the prescribed composition, then, the mixture was stirred with a magnetic stirrer until a uniform oil-in-water emulsion was reached. The composition of all padding mixtures are presented in Table 2.

Table 1. The molecular formula of the main components in four EOs.

Type of oil	Molecular formula of the main components in EOs		Ref.
Cinnamon	Cinnamal-dehyde 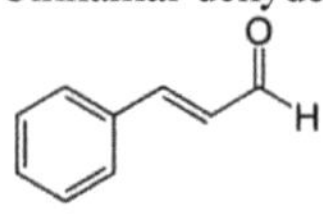	Methyl cinnamate: A methyl ester of cinnamic acid 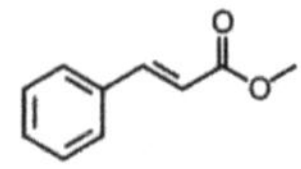	[26]
Basil	Methyl eugenol 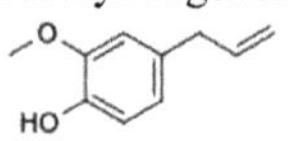	Methyl chavicol 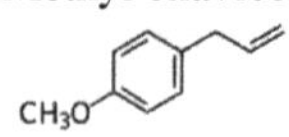	[27]
Lemongrass	Geranial 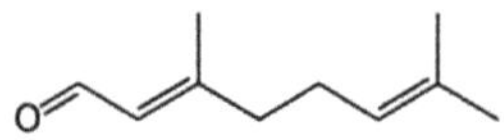	Neral 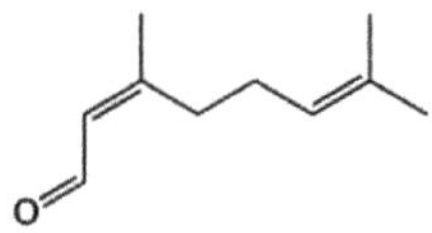	[28]
Mint	Carvone: a member of a family of chemicals called terpenoids 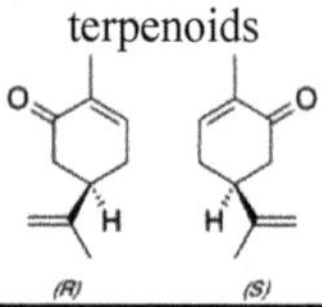	Limonene: a colorless liquid hydrocarbon classified as a cyclic terpene 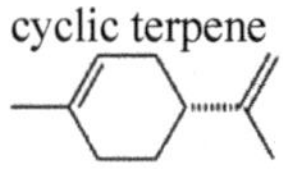	[29]

Table 2. Composition of the padding mixtures

Type of oil	Composition of the padding mixtures (%): Oil/distilled water				
	100/0	1/99	3/97	5/95	10/90
Cinnamon	X	X	X	X	X
Basil	X	X	X	X	X
Lemongrass	X	X	X	X	X
Mint	X	X	X	X	X

The non-woven fabric samples 35 cm × 35 cm were impregnated in the finishing mixture, then padded with the wet pick-up of approximately 375% by padder SDL D394A. The treated samples were then stored in a polyethylene bag to prevent the evaporation of the EO prior to any test.

The experiments were carried out in the Chemical-Textile lab (CTL) (STLF – HUST).

Antibacterial Testing Method for Fabric

Agar diffusion test according to AATCC test method 147 – 1998 is used in this research.

The agar solution was prepared with nutrient then poured into the sterilized petri dish. The microbes, that were prepared according to AATCC test method 147 – 1998 are inoculated in 5 parallel lines on the agar surface without refilling the loop. After that, the nonwoven samples with the size 20 × 50 mm were placed on the agar surface, across 5 parallel inoculated lines, ensuring intimate contact with the agar surface. The petri dishes were incubated for 24 h at 37 ± 2 °C. The incubated plates are examined for interruption of growth along the streaks of inoculums beneath the specimen and for the clear zone of inhibition beyond its edge. AA of EO treated samples was evaluated by comparing their area of inhibition zone with that of original nonwoven fabric samples (non-treated sample or control sample).

All these antibacterial tests were carried out at Proteomics lab of Center for Research and Development in School of Biotechnology and Food Technology (SBFT – HUST).

Color Testing Method for Fabric

Color of fabric has been measured according to ISO 105 - A02 – 1993 using light box and D65 light. The test was carried out in the CTL (STLF – HUST).

The changing of fabric color was evaluated by comparing the color of fabric before and after treatment with oils.

3 Results and Discussion

3.1 Antibacterial Activity of the Nonwoven Fabric Treated with Pure EO

The photos of the incubated plates with control samples were presented in Fig. 1. The results show that both Staph. aureus and E. coli have been developed on the agar after 24 h incubation, all inoculated lines were become bigger than before incubation, and there were not inhibited zones around the border of the fabric. That means the control samples have not demonstrated any AA.

(a) Staph. aureus (b) E. coli

Fig. 1. The incubated plates with control samples and 2 trains of bacteria.

Figure 2 shows the antibacterial properties of nonwoven fabric samples treated with the pure EOs. According to this result, the nonwoven fabric sample treated with the pure cinnamon oil has the highest AA for both types of bacteria. The agar surfaces were completely transparent and there are no colonies on the plates (photos in the first column from the left). This result shows that Vietnamese cinnamon EO, when applied on the fabrics, retains the antimicrobial properties as in other applications [18, 22].

For nonwoven fabric samples treated by the pure basil EO (two photos in the last column), the zone of inhibition appeared, this zone was fairly large, the agar surfaces transparent, but there were still some of colonies developed at the beginning of the inoculated lines. That means the fabric treated with pure basil EO also has strong AA against both strains Staph. aureus and E. coli, however its AA is slightly weaker in comparison with the fabric treated by cinnamon oil.

Nonwoven fabric treated with lemongrass and mint EOs had the effect only on Staph. aureus, the agar surfaces were completely transparent, does not appear colonies. On the other side, on the agar dish with E. coli, the colonies were still growing but only a few and size of each is smaller than that with the control sample (Fig. 1). These results are also consistent with the published papers about the AA of mint and lemongrass EOs [5, 6]. This means that although these four EOs were extracted from plants grown in Vietnam, they have the similar AA as EOs from these plants grown elsewhere. When EOs are applied on fabrics, their AA does not change.

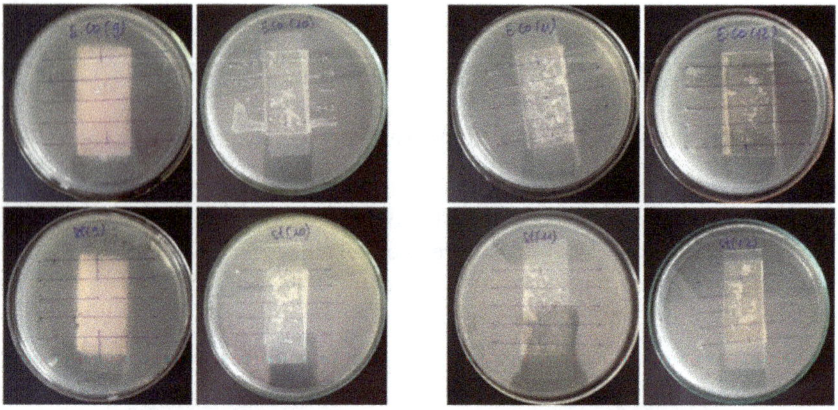

Fig. 2. The incubated plates with the samples treated with 4 pure EOs, from left to right: cinnamon, lemongrass, mint, and basil. In the top row, the plates inoculated with E. coli, in the bottom row, the plates inoculated with Staph. Aureus.

3.2 The Effect of These 4 EOs on the Color of White Non-woven Fabric

Since lemongrass and mint oils are pale green, their impact on the color of treated polyethylene nonwoven fabric was hardly visible, i.e. the color of fabric changed very slightly. However, because the cinnamon oil has a dark brown color, after being applied on the fabric, the white nonwoven fabric has been transformed into a pinkish brown color. Similarly, basil oil is brown in color, when it was applied on the white fabric, the fabric was yellowed.

3.3 The Effect of the EO Concentration on the AA of the EO Treated Non-woven Fabric

The above research showed that nonwoven fabric samples treated with pure cinnamon and basil EOs have the effect on both Staph. aureus and E. coli while the nonwoven fabric samples treated with pure mint and lemongrass EOs have clear effect on only Staph. aureus. From these results, in this study, the samples treated with a mixture of 1%, 3%, 5%, and 10% cinnamon oil or basil oil were tested for AA against both bacteria Staph. aureus and E. coli. Meanwhile, the fabric samples treated with the padding mixtures containing 1, 3, 5, 10% of mint or lemongrass oil were tested for AA only with Staph. aureus. The results of these tests are presented in the Figs. 3, 4, 5 and Table 3.

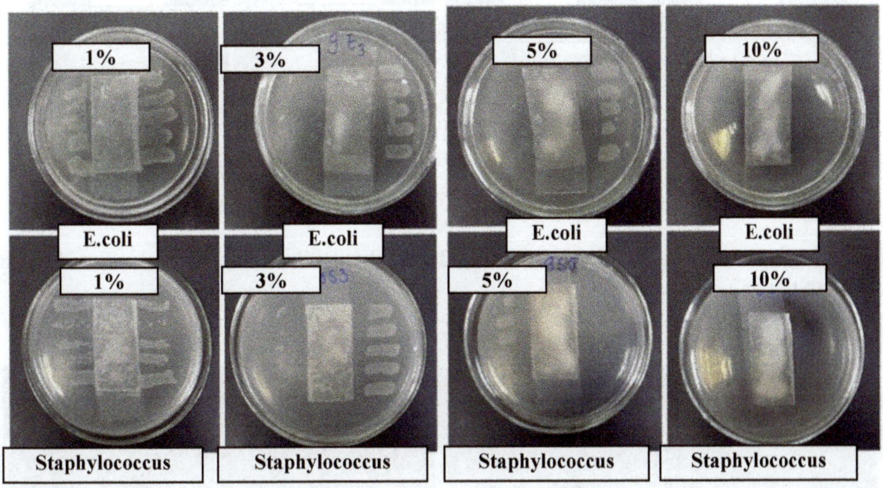

Fig. 3. Incubated plates with samples treated with the mixtures containing cinnamon oils at different concentrations. (From left to right: concentrations of oils are 1, 3, 5, and 10%, respectively. In the top row are the plates inoculated with E. coli. In the bottom row are the plates inoculated with Staph. aureus).

Figure 3 shows the AA of the samples treated with the diluted cinnamon oils. The results show that the AA of these samples depends on the oil concentration of the padding mixture. The sample treated with the mixture containing 1% cinnamon oil demonstrated virtually no AA against both E. coli and Staph. aureus. However, the sample treated with the oil-in-water emulsion containing 3% of the oil showed some AA. On the agar surface, the width of the inhibition zone is about 10 mm for both strains. It increased to approximately 12–13 mm when the oil concentration increased to 5%. At the highest oil concentration (10%), there were no colonies on the agar surface. More importantly, the cinnamon oil-in-water emulsions at low concentrations only presented a little effect on the color of the treated fabric. At the concentration of 10%, the treated nonwoven samples were light yellow brown by the naked eye. The padding mixtures with the lower concentrations of oil (i.e. 1%, 3% and 5%) almost did not change the color of the treated samples in comparison with the untreated fabric. Indeed, the color change could not be determined by naked eye. Thus, it is possible to use the cinnamon oil-in-water emulsions with the low concentrations (from 3% or higher) as an agent against both E. coli and Staph. aureus.

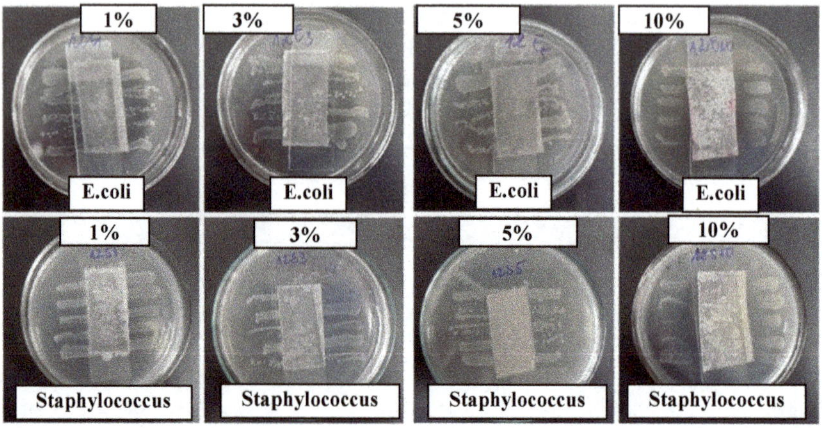

Fig. 4. Incubated plates with samples treated with basil oil-in-water emulsions at different concentrations.

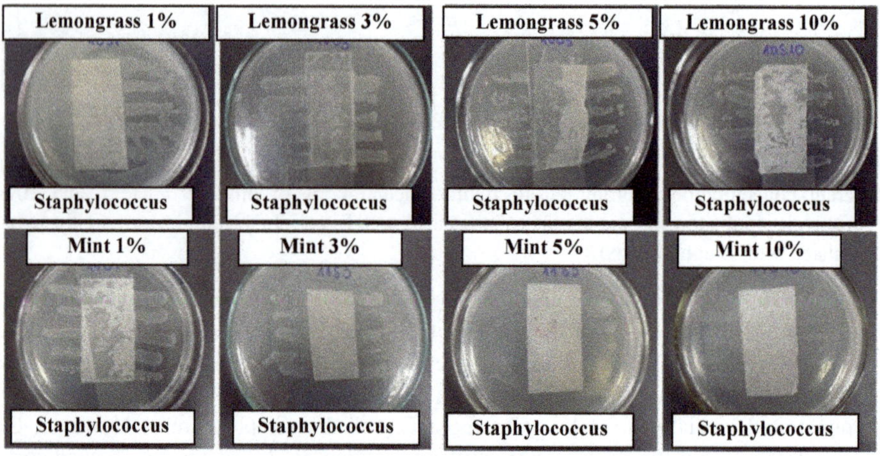

Fig. 5. The incubated plates with samples treated with lemongrass or mint oils at different concentrations. (From left to right: concentrations of oils are 1, 3, 5, and 10%, respectively. In the top row, the plates with fabrics treated with diluted lemongrass oils. In the bottom row, the plates with fabrics treated with diluted mint oils).

Figure 4 shows the AA of the samples treated with basil oil-in-water emulsions at different concentrations. Figure 2 and Table 3 show that the fabric treated with pure basil EO has quite good AA against both strains (E. coli and Staph. aureus). However, the AA of the fabrics treated with the diluted basil EOs has been significantly reduced (Fig. 4). Even at the concentration of 5%, the treated fabric did not show any impact on the colonies. The zone of the inhibition only appeared when the concentration reached 10%. On the agar surface, the width of the inhibited zone was about 7–8 mm for E. coli and got slightly larger (about 10mm) for the Staph. aureus. This indicates that the concentration of basil EO should be at least 10% or higher so that the treated fabric can be antibacterial.

(From left to right: concentrations of oils are 1, 3, 5, and 10%, respectively. In the top row, the plates inoculated with E. coli. In the bottom row, the plates inoculated with Staph. aureus).

Antibacterial properties of fabrics treated by four EOs at different concentrations are summarized in Table 3.

Table 3. The antibacterial activities of fabric treated by 4 essential oils at different concentrations.

Type of oils	Concentration of oils (%)	The width of inhibited zone (mm)	
		Against E. coli	Against Staph. aureus
Cinnamon	1	No zone of inhibit	No zone of inhibit
	3	10 mm	10 mm
	5	12–13 mm	12–13 mm
	10	No colonies, the width of inhibited zone > 20 mm	No colonies, the width of inhibited zone > 20 mm
	100	No colonies, the width of inhibited zone > 20 mm	No colonies, the width of inhibited zone > 20 mm
Basil	1	No zone of inhibit	No zone of inhibit
	3	No zone of inhibit	No zone of inhibit
	5	No zone of inhibit	No zone of inhibit
	10	7–8 mm	10 mm
	100	18 mm	No colonies, the width of inhibited zone > 20 mm
Lemongrass	1	Not studied	No zone of inhibit
	3	Not studied	No zone of inhibit
	5	Not studied	No zone of inhibit
	10	Not studied	No zone of inhibit
	100	No zone of inhibit	No colonies, the width of inhibited zone > 20 mm
Mint	1	Not studied	No zone of inhibit
	3	Not studied	7–8 mm
	5	Not studied	15 mm
	10	Not studied	No colonies, the width of inhibited zone > 20 mm
	100	No zone of inhibit	No colonies, the width of inhibited zone > 20 mm

4 Conclusion

From the results, some conclusions could be drawn:

- All four EOs that have been studied in this paper can be used as antibacterial agents for fabric.
- The antibacterial properties of treated fabrics with these four types of oil are different in both the type of bacteria that they can influence and the level of their AA. According to the AA of the treated fabrics, these four EOs could be ordered as follows: The best oil is cinnamon, basil oil ranks the second, the next is the mint oils and the lowest effective is the lemongrass.
- Cinnamon and basil oil are dark in color, but when they are used in concentrations of 10% or lower, their color does not significantly affect the whiteness of the treated fabric.
- These results suggest that these oils can be used for disposable textile products that require fragrance and antibacterial properties such as wet paper tow or medical textiles, etc. However, how to use the natural EOs so that they can become effective antibacterial agents for antibacterial textiles that require the antibacterial washing durability? this is always a challenge for researchers. This content will be interested in our further research.

Acknowledgement. This study has been carried out in the framework T2015-232. The authors would like to thank the HUST for funding this project.

The authors also wish to extend the thanks to the SBFT of HUST for allowing the authors to carry out this research in the Proteomics lab of Center for Research and Development in Biotechnology.

References

1. Morais, D.S., Guedes, R.M., Lopes, M.A.J.M.: Antimicrobial approaches for textiles: from research to market. Materials **9**(6), 498 (2016)
2. Shahidi, S., Wiener, J., Bobbarala, V. (ed.): Antibacterial Agents in Textile Industry, pp. 387–406. InTech (2012)
3. Duong, P.D.: Study on antimicrobial treatment for cotton fabrics used in apparel. Ph.D. thesis, Hanoi University of Science and Technology, Hanoi, Vietnam (2012)
4. Tho, L.T.: Study on utilization of vietnamese chitosan as antibacterial agent for cotton fabric. Ph.D. thesis, Hanoi University of Science and Technology. Hanoi Vietnam (2016)
5. Reichling, J., et al.: Essential oils of aromatic plants with antibacterial, antifungal, antiviral, and cytotoxic properties–an overview karger. Complement. Med. Res. **16**(2), 79–90 (2009)
6. Hili, P., Evans, C., Veness, R.: antimicrobial action of essential oils: The effect of dymethylsulphoxide on the activity of cinamon oil. Lett. Appl. Microbiol. **24**(4), 269–275 (1997)
7. Butnariu, M., Sarac, I.J.J.O.B.: Essential oils from plants. J. Biotechnol. Biomed. Sci. **1**(4), 35 (2018)
8. Dorman, H., Deans, S.G.J.J.o.a.m.: Antimicrobial agents from plants: antibacterial activity of plant volatile oils. J. Appl. Microbiol. **88**(2), 308–316 (2000)

9. Inouye, S., Takizawa, T., Yamaguchi, H.J.J.o.a.c.: Antibacterial activity of essential oils and their major constituents against respiratory tract pathogens by gaseous contact J. Antimicrob. Chemother. **47**(5), 565–573 (2001)
10. Tawiah, B., et al.: Advances in the development of antimicrobial agents for textiles: the quest for natural products. Review. Fibres Text. Eastern Europe (2016)
11. Beşen, B.S.J.F.: Production of disposable antibacterial textiles via application of tea tree oil encapsulated into different wall materials. Fibers Polym. **20**(12), 2587–2593 (2019)
12. Singh, N., et al.: Sustainable fragrance cum antimicrobial finishing on cotton: indigenous essential oil. Sustain. Chem. Pharm. **5**, 22–29 (2017)
13. Chinta, S., et al.: Medical textiles – application of essential oil as antimicrobial agent on nonwoven. Glob. J. Bio-Sci. Biotechnol. **1**(1), 75–80 (2012)
14. Angela, C., et al.: Textile materials functionalised with natural biologically active compounds. Rom. Biotechnol. Lett. **15**(5), 5538 (2010)
15. Thilagavathi, G., Kannaian, T.: Combined antimicrobial and aroma finishing treatment for cotton using micro encapsulated geranium (pelagonium graveolens L' Herit. ex Ait). Indian J. Nat. Products Resour. **1**(3) (2010)
16. Chandrasekaran, K., Anbumani, N., Priyal, V.V.: Analysis of eco-friendly medicinal herb extracts and esential oil application on textile products for healthcare application. Indian J. Tradit. Knowl. **14**(3) (2015)
17. Zaghloul, S., et al.: Herbal textile finishes – natural antibacterial finishes for cotton fabric. Egypt. J. Chem. **60**(2), 161–180 (2017)
18. Thanh, N.T.T., et al.: Comparison of the antibacterial activity of Vietnamese Cinnamon essential oil and its chemotype (trans-cinnamaldehyde) in tryptone soya broth (TSB) and in an oil in water emulsion containing TSB: consequences for its use in food preservation Scientific works volume LX Food science. Fngineering Technol. 18–19 (2013)
19. Lan-Phi, N., Vy, T.J.I.F.R.J.: Chemical composition, antioxidant and antibacterial activities of peels'essential oils of different pomelo varieties in the south of Vietnam. Int. Food Res. J. **22**(6), 2426 (2015)
20. Dang, N.H., et al.: Chemical composition and α-glucosidase inhibitory activity of vietnamese citrus peels essential oils. J. Chem. (2016)
21. Pattnaik, S., Subramanyam, V., Kole, C.J.M.: Antibacterial and antifungal of ten essential oils in vitro. Microbios **86**(349), 237 (1996)
22. Le, T.B., et al.: In vitro anti-leishmanial activity of essential oils extracted from vietnamese plants. Molecules **22**(7), 1071 (2017)
23. Gaio, I., et al.: Antibacterial activity of basil essential oil (Ocimum basilicum L.) in Italian-type sausage. J. Consum. Prot. Food Saf. **10**(4), 323–329 (2015)
24. Saeed, S., Naim, A., Tariq, P.J.P.J.o.B.: In vitro antibacterial activity of peppermint sabahat saeed, asma naim and perween tariq. Pakistan J. Botany **38**(3), 869 (2006)
25. Haddi, K., Faroni, L., Oliveira, E.E.: Cinnamon oil, in Green Pesticides Handbook: Essential Oils for Pest Control, pp. 117–150. CRC Press (2017)
26. Rao, P., Gan, S.: Cinnamon: A Multifaceted Medicinal Plant. Hindawi Publishing Corporation (2014)
27. Joshi, R.K.J.A.S.o.L.: Chemical composition and antimicrobial activity of the essential oil of Ocimum basilicum L. (sweet basil) from Western Ghats of North West Karnataka, India. Ancient Sci. Life **33**(3), 151 (2014)
28. Tajidin, N., et al.: Chemical composition and citral content in lemongrass (cymbopogon citratus) essential oil at three maturity stages. Afr. J. Biotech. **11**(11), 2685–2693 (2012)
29. Snoussi, M., et al.: Mentha spicata essential oil: chemical composition, antioxidant and antibacterial activities against planktonic and biofilm cultures of vibrio spp. Strains. Molecules **20**(8), 14402–14424 (2015)

Study of the Mechanisms of Chip Formation in Turning of 304 Austenitic Stainless Steel

Viet-Hoi Tran[1(✉)], Van-Bong Pham[1], and Van-Dich Tran[2]

[1] Hanoi University of Industry, Hanoi, Viet Nam
hoitv@haui.edu.vn
[2] Hanoi University of Science and Technology, Hanoi, Viet Nam

Abstract. During the machining process, the morphological mechanism of the chip also shows the factors affecting the machining process such as cutting parameters, materials, physical phenomena, cutting force, vibration, ext. This study deals with the mechanisms of chip formation in turning of 304 austenitic stainless steel, analyzing chip forms formed in different cutting parameters. The study also assessed the influence of chip formation on surface roughness after machining. The investigated chip formation can cause a significant improvement in surface integrity.

Keywords: Austenitic stainless steel · Chip formation · Surface roughness

1 Introduction

Chip geometry is an important parameter for evaluating a material's machinability [1]. The chip deformation in the cutting process depends on many factors such as characteristics of the machining material (chemical composition, the physical properties of the material), machining conditions (machining parameters, machining conditions, ext.), changes in cutting area characteristics, changes in the tool-workpiece interact area. The morphology of the chip formed during the machining has influenced on the properties of removing material from the cutting zone. The chip displacement speed also affects the shearing force, vibration, heat, and mechanical properties of the tool-workpiece interact area, the tool life and the quality of the machining surface [1, 2]. Research of author S. Sun [3] studies chip formation and shear force when machining titanium alloy. The study also shows that the chip formation size depends on cutting speed and cutting depth. The increasing shearing force also increases chip segmentation. Author Xiaobin Cui [4] studies surface roughness and chip formation when milling AISI H13 steel. The research shows that the surface roughness obtained at different cutting speeds can be estimated by observing the variation in chip shape and different chip colors. Some other studies on surface roughness and chip formation, such as [5, 6] showed that the surface quality after work is affected by chip formation. This study experimentally assesses surface roughness and chip formation when turning AISI 304 and the influence of chip generation on surface roughness. The results of the study help manufacturers adjust the cutting parameters to achieve the best productivity and quality.

B. T. Long et al. (Eds.): MMMS 2020, LNME, pp. 138–146, 2021.
https://doi.org/10.1007/978-3-030-69610-8_18

2 Experimental Procedure

2.1 Machining Experimental

The experiments were conducted on MORI SEIKI-SL253 CNC lathe, the capacity of 4.5KW, maximum spindle rotation number is 4500 rpm, and surface roughness is measured Mitutoyo SV-2100 machine as Fig. 1, chips are analysis on FE-SEM (S4800, Hitachi).

2.2 Material and Tools

Cutting tool Sanvik DCMT 11 T3 04 - MF 2220 coated with CVD Ti (C, N) + Al2O3 + TiN, workpiece AISI 304 size 50 × 200 mm (Fig. 2).

Materials: Researching to use AISI 304 is a type of steel with better corrosion resistance and heat resistance than ordinary carbon steel. AISI 304 stainless steel is the most used steel in austenitic stainless steel, about 72%. AISI 304 is difficult to machining due to its low thermal conductivity and high mechanical and physical properties, Table 1 shows the material's chemical composition properties.

Table 1. Chemical composition of materials

AISI 304	C	Cr	Ni	Si	Mn	P	S
	0.07	18.49	8.15	0.57	0.76	0.03	0.009

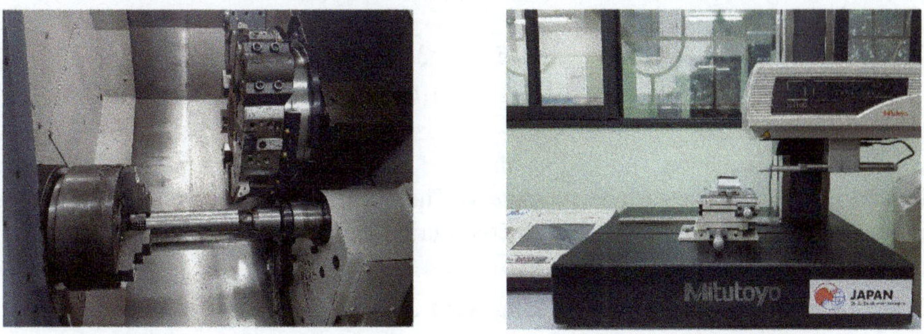

a) CNC Lathe b) Test surface roughness.

Fig. 1. Machining experimental

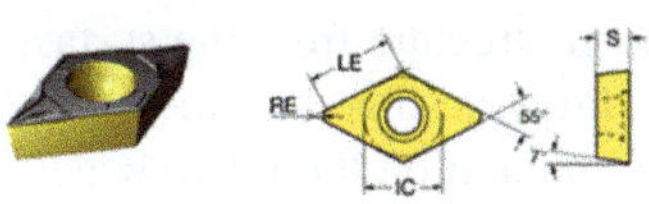

Fig. 2. Cutting tool

Table 2. Cutting ranges

Cutting parameters	Level		
	1	2	3
V (m/min)	230	260	290
f (mm/rev)	0.08	0.14	0.20
t (mm)	0.10	0.25	0.50

3 Results and Discussion

3.1 Influence of Cutting Parameters on Surface Roughness

Experiments were performed according to the parameters given in Table 2, selected research with 3 factors, 3 levels and conducted according to the Box-Behnken method including 15 experiments, the least number of experiments to the second-order polynomial regression model. Experimental results of measuring AISI 304 stainless steel's surface roughness when turning on a CNC machine are shown in Table 3. Figure 3 shows the roughness of the machining detail according to the experiments and observations. Experiments 1, 4, 10 have a higher feed rate and depth, have higher surface roughness and experiments 6, 9, 12 have a lower feed rate and the depth cut is smaller. The study used Minitab 18 software to analyze the cutting parameters to Ra by the Box-Behnken method to evaluate the effect of cutting parameters on surface roughness. Regression product of Ra to (V, f, t), this analysis shows the R Square value is 99.49%. Analysis of the influence coefficients of surface roughness of workpieces with technological parameters (V, f, t) is given in Eq. (1):

$$\begin{aligned} \mathrm{Ra} = 12,11\ - 0,0818\mathrm{V}\ - 11,57\mathrm{f}\ - 3,69\mathrm{t}\ + 0,000149\mathrm{V2}\ + 64,68\mathrm{f2} \\ + 1,460\mathrm{t2}\ + 0,0079\mathrm{V} * \mathrm{f}\ + 0,01002\mathrm{V} * \mathrm{t}\ + 2,27\mathrm{f} * \mathrm{t} \end{aligned} \tag{1}$$

The above equation shows that the feed rate and cutting depth are the most influencing factors on surface roughness. The shear factor also influences surface roughness, but with the cutting speed considering less influence.

3.2 Chip Formation When Turning AISI 304

The chip formed by turning AISI 304 is shown in Fig. 4. Chips with string chips are divided into surfaces: the outer surface of the chip is called the free face; the inner side that comes into interaction with the tool is called the inside. The surface interact with the tool due to friction; temperature makes the surface smooth and shiny. The free surface has an entirely different structure from the surface in interact with the tool. In addition to the two basic surfaces on the chip, there are two outer edges. The inside edge of the chip and the deformation on these two edges have a significant difference on the chip's free surface (Fig. 5).

Table 3. Experimental design and cutting conditions

Run	V (m/min)	f (mm/rev)	t (mm)	Ra (μm)
1	290	0.2	0.25	1.58
2	260	0.14	0.25	0.73
3	260	0.14	0.25	0.73
4	230	0.2	0.5	1.72
5	230	0.14	0.1	0.93
6	260	0.08	0.5	0.45
7	260	0.2	0.1	1.55
8	260	0.14	0.25	0.73
9	260	0.08	0.1	0.44
10	230	0.2	0.25	1.66
11	290	0.14	0.1	0.87
12	290	0.08	0.25	0.48
13	230	0.14	0.5	0.85
14	230	0.08	0.25	0.64
15	290	0.14	0.5	1.02

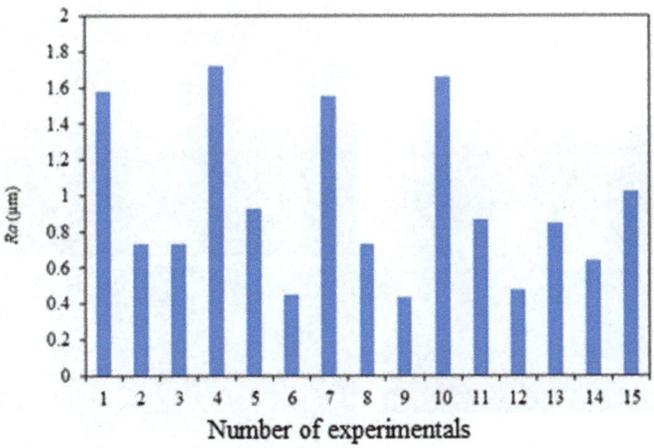

Fig. 3. Surface roughness measured

Figure 5 is a snapshot of the AISI 304 chip-free surface SEM when turning. There are two zones on the free chip surface: the inner chip area and the outer chip area. It is observed that the inner chip area on the free surface is more compressed, so the chip surface appears small and thick chip layers together. When formed, the outer area is less creaking, so the larger chip layer is formed and divided into layers.

Thus, in the turning AISI 304, the chip formed is subject to different deformations, so the two above areas have other surface structures. Specifically, when observing the SEM image of the chip's free surface, it was found that the outer chip area on the free surface is not subject to the interact tool-workpiece that should be heavily deformed to form

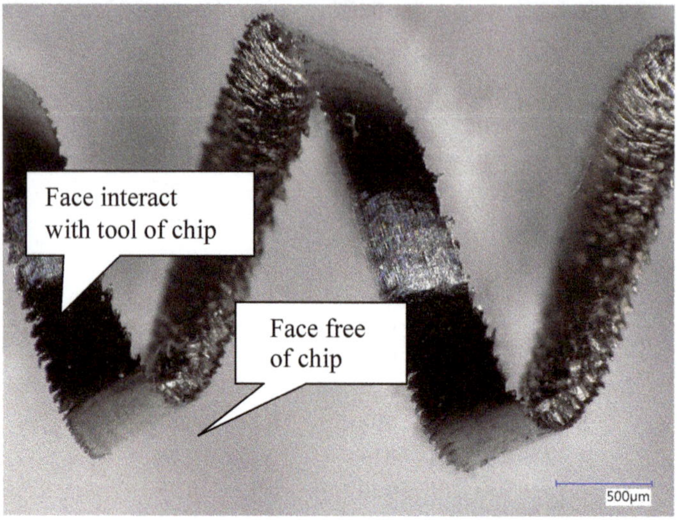

Fig. 4. Chip surfaces form when turning AISI 304

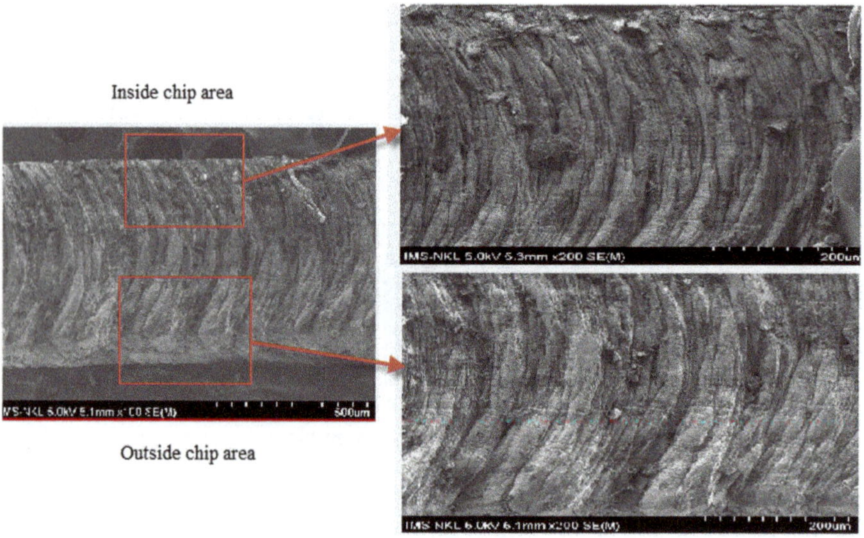

Fig. 5. Free chip surface layer structure

trapezoidal layers (tooth saw). The inner part of the chip is influenced by the contact between the tool and the rest of the workpiece, the metal particles in this area are almost compressed and it is found that the structure of this chip differs from the outer chip area. The layers in the inner chip area tend to tilt along the tool's inner edge, small in size and evenly distributed (Fig. 6). Simultaneously, in the outer chip area, the forming layers are observed on the chip's surface. The segmentation on the outer edge of the chip when turning AISI 304 depends on the properties of the material and the cutting parameters.

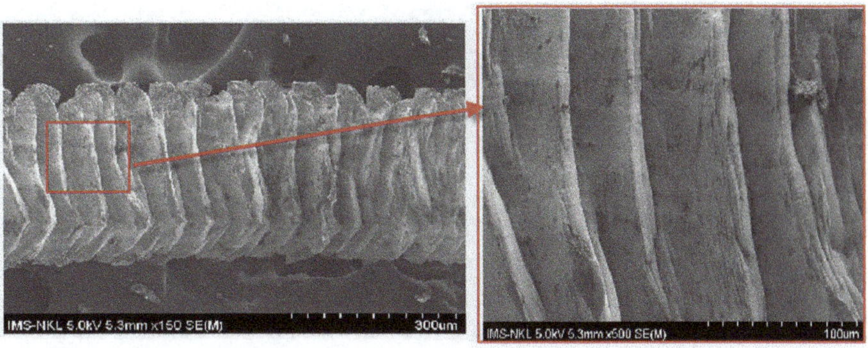

Fig. 6. Chip layer formed on the free surface in experiment 9

3.3 Effect of Chip Form and Surface Roughness of Workpiece

Chip formation and construction also affect the surface roughness of the workpiece. Experimental study on measuring detailed surface roughness after turning AISI 304, the results are shown in Table 2. In experiment 4 with V = 230 m/min, f = 0.2 mm/rev and t = 0, 5mm surface roughness is measured to be 1.72 m. This is the test with the largest results among 15 experiments. A photo of SEM surface texture after detailed machining is shown in Fig. 7a. The corresponding chip form of the experiment is shown in Fig. 7b. Largely formed chip widths are 659 um wide and free surface pattern spacing is on average 34 μm (Fig. 8).

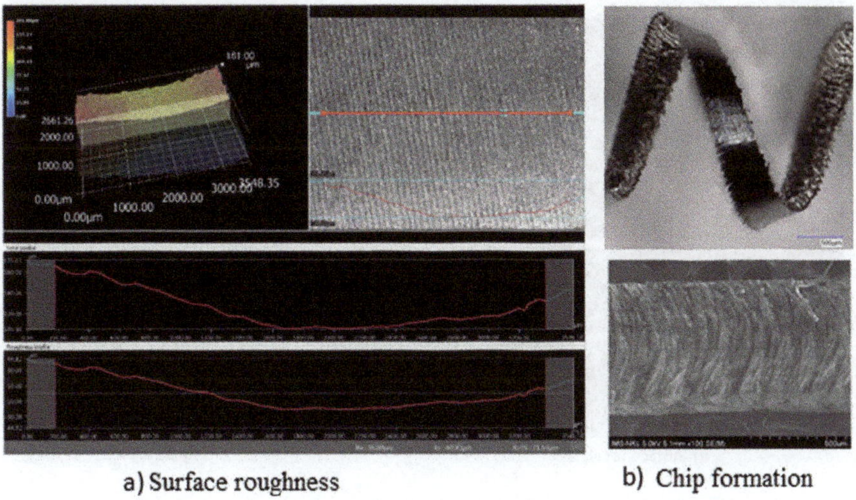

Fig. 7. Surface roughness and chip formation in experiment 4

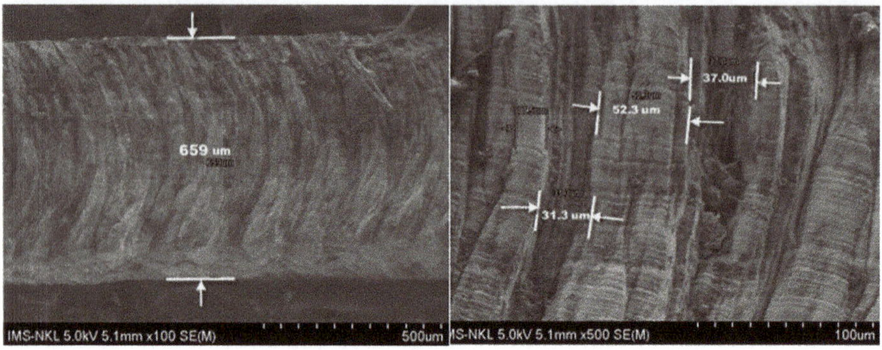

Fig. 8. Chip widths in experiment 4

In experiment 9 (V = 260 m/min, f = 0.08 mm/rev and t = 0.1 mm), the measured surface roughness was 0.44 µm, which is the smallest roughness in the experiments. The surface roughness of post-machined workpieces is shown in Fig. 9. In this experiment, the chip formed is a wire chip with small chip width and chip twist, as shown in Fig. 9b. SEM imaging can also measure a chip width of 362 µm and the pattern formed on the free surface of the pronounced mark. Thus, under the condition of high speed, small thickness and low feed rate, chip forms in the form of wire chips, making the surface of the details smooth. The SEM image also shows the chip layers formed on the free surface of this experiment quite clearly and quite similar in both the inner and outer parts of the chip (Fig. 10).

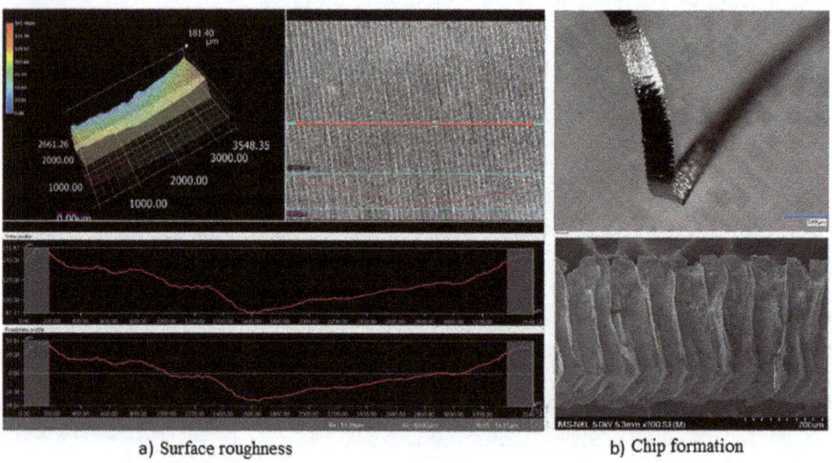

Fig. 9. Surface roughness and chip formation in experiment 9

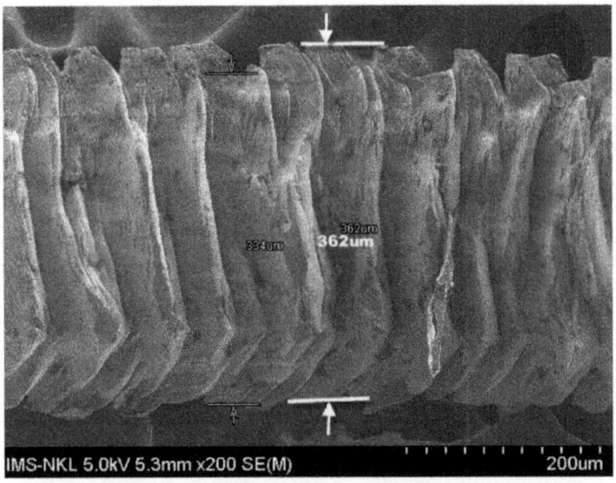

Fig. 10. Chip widths in experiment 9

Thus, increasing the cutting speed and depth, the low chip feed rate forms the wiring chip and the chip deformation is clearly shown by the chip layer on the free surface layer. Under this condition, the workpiece's surface roughness also reaches the smallest value in the experimental area.

4 Conclusion

This study experimentally assesses the effect of surface roughness and chip formation when turning AISI 304. The study draws the following conclusions:

- The amount of toolpath and cutting depth is the most influencing factor on the surface roughness, the cutting element also affects the surface roughness, but with the speed zone being considered, the effect is less.
- Formed chip affects the surface roughness of the workpiece. On the other hand, chip shape, width and free surface layers are also dependent on different cutting speeds.

References

1. Rathi, A., Mahor, A., Ranjan, R., Gajbhiye, A., Rehman, A., Krishna, C.M.: Characterization of chip morphology for aluminum metal matrix composites in end milling machining. Univers. J. Mech. Eng. **2**(7), 240–247 (2014)
2. Hua, J., Shivpuri, R.: Prediction of chip morphology and segmentation during the machining of titanium alloys. J. Mater. Process. Technol. **150**(1–2), 124–133 (2004)
3. Sun, S., Brandt, M., Dargusch, M.S.: Characteristics of cutting forces and chip formation in machining of titanium alloys. Int. J. Mach. Tools Manuf. **49**(7–8), 561–568 (2009)
4. Cui, X., Zhao, J., Jia, C., Zhou, Y.: Surface roughness and chip formation in high-speed face milling AISI H13 steel. Int. J. Adv. Manuf. Technol. **61**(1–4), 1–3 (2012)

5. Wang, M., Xu, B., Zhang, J., Dong, S., Wei, S.: Experimental observations on surface roughness, chip morphology, and tool wear behavior in machining Fe-based amorphous alloy overlay for remanufacture. Int. J. Adv. Manuf. Technol. **67**(5–8), 1537–1548 (2013)
6. Lalwani, D.I., Mehta, N.K., Jain, P.K.: Experimental investigations of cutting parameters influence on cutting forces and surface roughness in finish hard turning of MDN250 steel, vol. 6, pp. 167–179 (2007)

Damage Detection Using Local Dominant Frequency of Pulse-Echo Laser Ultrasonic Waves

Truong Thanh Chung(✉) and Nguyen-Anh-Vu Le

Nha Trang University, Nha Trang, Vietnam
{chungtt, vulna}@ntu.edu.vn

Abstract. This paper presents the Local Resonance Frequency Map (LRFM) algorithm that provides an instant interpretation of structural features and defects using an estimation of local dominant frequency of the pulse-echo laser ultrasonic waves at a fine grid of spatial sampling points. A Q-switched laser and a laser Doppler vibrometer are used for generation and sensing of laser ultrasonic waves, and a two-axis translation stage is used for raster scanning of the combined generation and sensing laser beams. Experiments show that LRFM works well for damage detection in various structures from metallic to composites specimens.

Keywords: Local resonance frequency map · Pulse-echo laser ultrasonic waves

1 Introduction

Laser ultrasonic techniques for Non-Destructive Testing (NDT) of aeronautic industry have been developed for several decades. Initial studies in the 70s and 80s [1–3] showed strong potential for industrial applications. The pulse-echo mode of ultrasonic waves is showed to be the most effective method for NDT since it requires only access to one surface of the structure, in contrast to the through-transmission mode where two sides of access is needed; although the latter is more effective for detection of adhesively disbond [4]. In the 90s, several groups investigated the use of laser NDT for composite structures in cooperation with organizations such as Lockheed Martin and US Air Force [4–7]. Today, there are several companies provide commercial laser NDT: PaR Systems [8], Tecnar Automation [9], iPhoton Solutions [10], and Tecnatom [11]. In academic, this topic is also attracted much interested with several laboratory scale systems [12–14].

The key idea of the mentioned laser NDT systems [8–14] is to combine an extreme short pulse laser excitation with a sensing laser beam; and use a scanning mechanism to scan the combined beam: such as 5-DOF-axis robot [12], mirror scanner [8–11], and mechanical scanner XY stage [13, 14]. Each scanning mechanism has its own advantages as well as limitations. The use of a 5-DOF-axis robot for scanning allows inspection of very complex shaped objects such as small radius of curvature structures; however, the scan speed is slow with the max scan speed in [13] is 30 Hz. Mirror

B. T. Long et al. (Eds.): MMMS 2020, LNME, pp. 147–151, 2021.
https://doi.org/10.1007/978-3-030-69610-8_19

scanner provides extremely fast scan speed (up to kHz range), also it is very compact. On the other hand, one issue of mirror scanner is that when the scan area is relatively large, the angle of incidence of the sensing beam may significant large near the edge of the scan area, which causes the signal quality is not uniform across the scan structure. In the case the structure is flat or mildly contoured, the mechanical scanner XY stage avoids the issue with high angle scan though it is much bulkier compared to the mirror scanner [14].

In this paper, a novel algorithm is presented for imaging damages of structure so-called Local Resonance Frequency Map (LRFM). It maps the dominant frequency of the pulse-echo laser ultrasonic waves at each point in the scan area. The result is a 2D image instead of 3D movie format in a C-scan movie. This algorithm provides an instant interpretation of structural features and damages since no input is necessary for LRFM, meaning that the result is shown immediately after finish scanning without any further input from user. Also LRFM has physical meaning instead of an arbitrary color-scale in a C-scan movie.

This paper is organized as follows. The next section presents the Local Resonance Frequency Map algorithm. Next, we apply this algorithm to detect damages of metallic and composite specimens. The final section is the conclusions.

2 Local Resonance Frequency Map Algorithm

When the laser focusses on the material, the laser irradiation generates bulk waves penetrated through the thickness h of the structure. The bulk waves reflected back and forth between two surfaces has the dominant frequency $\hat{f} = h/2v$ where v is the longitudinal velocity of the ultrasonic waves [14]. Local Resonance Frequency Map (LRFM) algorithm maps the dominant frequency of the pulse-echo laser ultrasonic waves at each point in the scan area. The flowchart of the algorithm is shown in Fig. 1. First, the time-space domain data $u(x,y,t)$ is transformed to frequency-space domain data $U(x,y,f)$ using 1D- Fourier transform. Next, for each frequency-domain signal from each scanning point, the maximum frequency is calculated. Finally, the 2D image containing all maximum frequency of the scan area is plotted.

$$u(x,y,t) \xrightarrow{\text{1D Fourier Transform}} U(x,y,f) \xrightarrow[f]{\arg\max(U)} \hat{f}(x,y)$$

Fig. 1. Flowchart of Local Resonance Frequency Map (LRFM) algorithm

3 Experiments and Results

3.1 Aluminum Specimen with Thinning Area

The Pulse-Echo Ultrasonic Propagation Imager (PE-UPI) system [14] was used for inspection of a 4-mm aluminum plate with thinning area (100 mm × 200 mm) at 50%

thickness reduction. Picture of the specimen is shown in Fig. 2(a). The scan process were performed on the opposite side of the wall-thinning. The parameters of the experiment were set as follows. The laser was set to 5.85 mJ per pulse and 500 Hz pulse repetition rate. The standoff distance from the scanning laser heads to the specimen was 840 mm. The Q-switched laser beam impingement on the specimen had a diameter of 2.23 mm. The scanning interval of the translation stage was set to 0.52 mm on both the horizontal and vertical axes. The sensitivity of the LDV was set to 50 mm/s/V.

Figure 2(b)(c) show C-scan results at different time points. The damage area can be clearly visualized; however, there is no conclusion about the thickness of the damage area compared to the intact area. Figure 2(d) shows the LRFM result. The LRFM shows that in the thinning area, the dominant frequency is approximately 1.5 MHz and in the intact area, the dominant frequency is approximately 0.75 MHz. This result is showed to be perfectly agreed with the theory as presented in Sect. 2 where the thinning area has 50% thickness reduction.

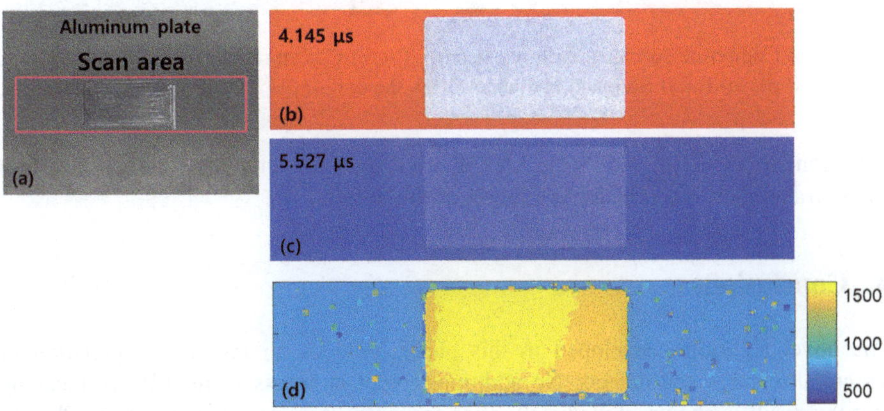

Fig. 2. (a) Aluminum plate of 4-mm thickness with 50% thickness reduction at thinning area, (b) (c) C-scan results captured at 4.145 μs and 5.527 μs, (d) Local Resonant Frequency Map, the unit of color bar at the right side is kHz

3.2 Composite Specimen

The PE-UPI system was used for inspection of a composite specimen with ply defects made during its manufacturing. Picture of the specimen is shown in Fig. 3(a). The parameters used for testing this specimen was identical with the previous test with the aluminum specimen, except the laser energy is set at lower level 2.5 mJ per pulse to avoid surface damage of the composite specimen.

Figure 3(b)(c) show the C-scan results at different time points. The C-scan image at 10.700 μs only shows one ply defect at the top-right corner. The C-scan image at 78.400 μs shows all three ply defects; however the image of the defects is not clear since the ultrasonic waves is highly diminished after multiple back-and-fourth

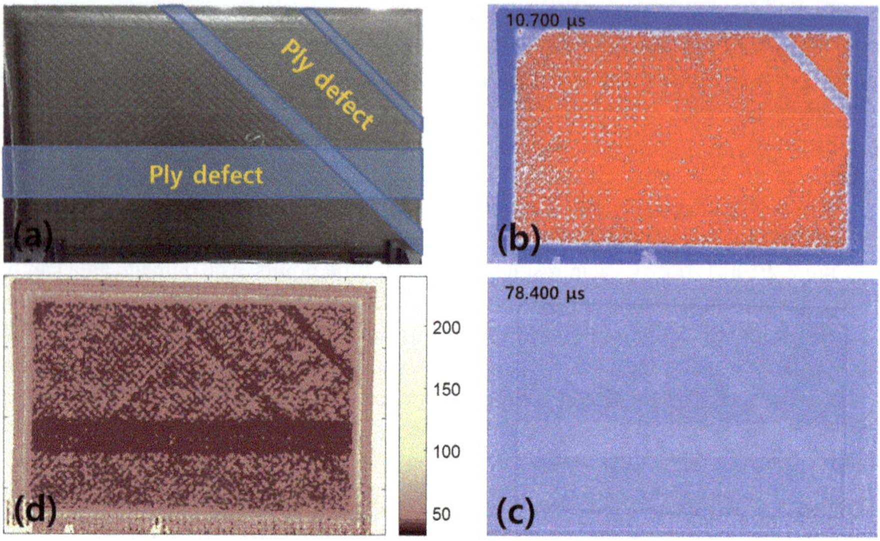

Fig. 3. (a) Composite specimen with ply defects, (b) (c) C-scan results captured at 10.700 μs and 78.400 μs, (d) Local Resonant Frequency Map, the unit of color bar at the right side is kHz

reflections. Figure 3(d) shows the LRFM result, which captured all ply damage in a single frame with much better resolution of the defects.

4 Conclusions

The novel algorithm developed in this paper provides an instant interpretation of structural features and defects. No additional input is necessary for LRFM, meaning that the result is shown immediately after scanning. It gives the users of the PE-UPI system the preliminary result before proceed to use other algorithms such as VTWAM [15], or UWI [16] for deeper analysis. Another advantage of the LRFM algorithm is that it has physical meaning instead of arbitrary color-scale in a C-scan movie, which allows possible evaluation of structural thickness. Experiments show that LRFM works well in various structures from metallic to composite specimens.

Acknowledgement. We express a deep gratitude to Prof. Jung Ryul Lee (OESL, KAIST) for providing the experimental data for this study.

References

1. Calder, C.A., Wilcox, W.W.: Noncontact material testing using laser energy deposition and interferometry. In: Proceedings of 6th International Conference on Experimental Stress Analysis, European Permanent Committee for Stress Analysis, Munich, Germany (1978)

2. Bourkoff, E., Palmer, C.H.: Noncontact material testing using low-energy optical generation and detection of acoustic pulses. In: Review of Progress in Quantitative Nondestructive Evaluation, vol. 5A, pp. 659–667. Plenum Press, New York (1986)
3. Monchalin, J.P., Aussel, J.D., Bouchard, P., et al.: Laser-ultrasonics for industrial applications. In: Thompson, D.O., Chimenti, D.E. (eds.) Review of Progress in Quantitative NDE, vol. 7B, p. 1607. Plenum Press, New York (1988)
4. Chang, F.H., Drake, T.E., Osterkamp, M.A., et al.: Laser-ultrasonic inspection of honeycomb aircraft structures. In: Review of Progress in QNDE, vol. 12, pp. 611–616. Plenum Publishing Corporation (1993)
5. Tittmann, B.R., Linebarger, R.S., Addison, R.C.: Laser-based ultrasonic on Gr/Epoxy composite. J. Nondestr. Eval. **9**(4), 229–238 (1990)
6. Padioleau, C., Bouchard, P., Heon, R., et al.: Laser ultrasonic inspection of graphite epoxy laminates. In: Review of Progress in QNDE, vol. 12, pp. 1345–1352. Plenum Publishing Corporation (1993)
7. Thomas, E., Drake, Jr., Kenneth, R.Y, et al.: Affordable NDE of aerospace composites with laser ultrasonics. In: Review of Progress in QNDE, vol. 17, pp. 587–593. Plenum Publishing Corporation (1998)
8. Osterkamp, M.A., Kaiser, D.L.: Application of laser ultrasonics for the non-destructive inspection of complex composite aerospace structures. In: 1st International Symposium on Laser Ultrasonics: Science, Technology and Applications (2008)
9. Bentouhami, F., Campagne, B., Cuevas, E., et al.: LUCIE - a flexible and powerful laser ultrasonic system for inspection of large CFRP components. In: 2nd International Symposium on Laser Ultrasonics (2010)
10. Monchalin, J.P., Padioleau, C., Neron, C., et al.: A laser-ultrasonic inspection system for large structures fabricated by automated fiber placement. In: International Workshop on Smart Materials & Structures NDT in Aerospace (2011)
11. Esmeralda, C., Miquel, L., Monica, G.: Ultrasonic techniques and industrial robots: natural evolution of inspection systems. In: 4th International Symposium on NDT in Aerospace (2012)
12. Vandenrijt, F.J., Walter, J., Brouillette, T., et al.: Performances comparison of a laser ultrasonic system using 10.6 μm infrared or 532 nm visible generation beam for the investigation of CFRP. In: 5th International Symposium on Laser-Ultrasonics and Advanced Sensing, Linz, Austria (2016)
13. Pelivanov, I., Shtokolov, A., Wei, C.W., et al.: A 1 kHz A-scan rate pump-probe laser-ultrasound system for robust inspection of composites. IEEE Trans. Ultrason. Ferroelectr. Freq. Control **62**(9), 1696–1703 (2015)
14. Hong, S.C., Abetew, A.D., Lee, J.R., et al.: Three dimensional evaluation of aluminum plates with wall-thinning by full-field pulse-echo laser ultrasonic. Opt. Lasers Eng. **99**, 58–65 (2016)
15. Lee, J.R., et al.: Laser ultrasonic anomalous wave propagation imaging method with adjacent wave subtraction: algorithm. Opt. Laser Technol. **44**(5), 1507–1515 (2012)
16. Truong, T.C., Kang, D.H., Lee, J.R., Farrar, C.R.: Comparative study of laser Doppler vibrometer and capacitive air-coupled transducer for ultrasonic propagation imager and the new development of an efficient ultrasonic wavenumber imaging algorithm. Strain **51**(4), 332–342 (2015)

Analyzing the Effect of Some Welding Parameters TIG to Properties and Microstructure of Titanium Materials

Sai Manh Thang[1], Nguyen Duong Nam[2(✉)], Pham Van Truong[3], and Pham Mai Khanh[4(✉)]

[1] Military Institute of Science and Technology Ha Noi, Ha Noi, Viet Nam
smtbk77@yahoo.com
[2] School of Mechanical Engineering, Viet Nam Maritime University, Hai Phong, Viet Nam
namnd.khcs@vimaru.edu.vn
[3] Factory X56, Hai Phong, Viet Nam
[4] School of Materials Science and Engineering, HUST, Ha Noi, Viet Nam
khanh.phammai@hust.edu.vn

Abstract. The article presents about TIG welding technology for titanium material with a thickness of 3 mm. TIG welding experiment of 17 samples on titanium material with 3 mm thickness billet in the rolling state, annealed with variable input parameters (current (I), protective gas flow (Q), and electrode distance e). After welding the samples were analyzed, measured mechanical properties about strength limit, hardness, rigorous planning development showed that the optimal welding technology parameters were: I = 126.3 A; Q = 16.6 L/min; e = 3.1 mm. On the selected optimization mode, the results of mechanical properties achieved σ_b = 488,2 Mpa; hardness = 287,96 HV, elongation ε = 18,1%. Compared to the sheet sample which was not welded: σ_0 = 492,7 Mpa; hardness = 257,7 HV; elongation ε_0 = 18,5%. The microstructure of welding was homogeneous and not defects of welding as porosity. The microstructure of welding has large particle sizes, was larger than the boundary of welding and the base metal. The uneven composition in the areas of light and dark patches was indicated an uneven distribution of the composition when cooled. The grain size is 50–200 μm. The particles of matrix metal were small and uniform.

Keywords: TIG welding · Titanium material · Titan welding technology · Microstructure · Mechanical properties

1 Introduction

Welding is an important technology to create joints for structures. Titanium and most industrial titanium alloys are capable of welding. Can be used for melting welding, resistance welding, laser, electron beam, diffusion welding, and pressure welding … And is applied to create joints for titanium and titanium alloys. Gas electrode arc welding (GTAW) - also known as TIG (Tungsten Inert gas) welding and Gas metal arc

B. T. Long et al. (Eds.): MMMS 2020, LNME, pp. 152–161, 2021.
https://doi.org/10.1007/978-3-030-69610-8_20

welding (GMAW) - commonly called MIG (Metal Inert gas) welding is commonly used for titanium welding [1]. In which TIG welding is widely used to weld titanium alloys with other materials which has high-quality, suitable for welding materials from thin to thick, manual welding, semi-automatic, or fully automatic. Titanium is the most active metal especially in the liquid state and at high temperatures, titanium easily reacts with elements H, O, N, C, with a small amount of these elements also creating compounds. Brittle and rapidly reduce the material properties, so when welding needs to be protected with inert gases such as Ar or He. TIG titanium welding using tungsten electrodes should be suitable so that the molten welding pool is not contaminated with tungsten.

Grade 2 non-alloy titanium (according to US standard UNS R50400 with "C 0,1; Fe $\leq$ 0,3; H $\leq$ 0,015; O $\leq$ 0,25; N $\leq$ 0,03 remaining is Ti") [3] is non-alloy engineering titanium, combining properties such as durability, good corrosion and weldability. Therefore, it is commonly used for manufacturing body frames and aircraft engine parts. Corrosion-resistant marine applications, tubes and pressure vessels, medical applications such as bone and orthopedic implants, condenser tubes, heat exchangers, etc.

In the world, titanium welding technology has been studied and achieved great achievements for titanium alloy welding methods, developing various welding technologies, applied in many fields [4]. In Vietnam, some titanium welding technologies have been transferred in some fields such as oil and gas, shipbuilding, and medicine, but the number of researches on titanium welding in the country has been published. This paper studies TIG welding technology on grade 2 titanium plate material with 3 mm thickness. Assess the influence of some welding technology parameters such as current, protection gas flow, distance from the tip of the welding needle to the weld, the mechanics, and the microstructure of the weld. These are some of the main parameters in welding operations and greatly affect the quality of the weld.

2 Materials and Methods

Intending to examine the influence of welding technology parameters are current intensity (I), protection gas flow (Q), electrode distance (e) on weld mechanics. Using the Box-Behnken method of experimental planning for the three parameters above, the results need to perform 17 welding samples with different modes of amperage, gas flow, distance from the electrode to the weld to determine the optimal value, 3 welding parameters are in Table 2.

Welding experiment on grade 2 titanium plates; 3 mm thickness with composition in Table 1:

Table 1. Chemical grade 2 titanium plate

Al	Zr	Mo	V	Si	Mn	Cr	Ni	Fe	Cu	W	O	N	Ti
0.004	< 0.003	0.0005	< 0.001	0.012	< 0.001	0.014	0.002	0.12	0.003	< 0.002	0.072	< 0.002	99.8

Select TIG welding technology as follows: Welding machine ZX7 - 200, titanium plate thickness: 3 mm. The number of welding layers: 2. Welding current varies from 70–130 A. Welding voltage: 15V. WT20 tungsten electrode (0.8–4.2% ThO2) diameter: 3.2 mm. The distance from the electrode to the welding metal surface varies from 1,5 ÷ 4,5 mm. Agon protection gas 99.99% purity, the amount of protection gas weld direction: 12 ÷ 20 L/min, the back: 5 L/min. Spacing maintaining jig, welded plate model, shown in Figs. 1 and 2.

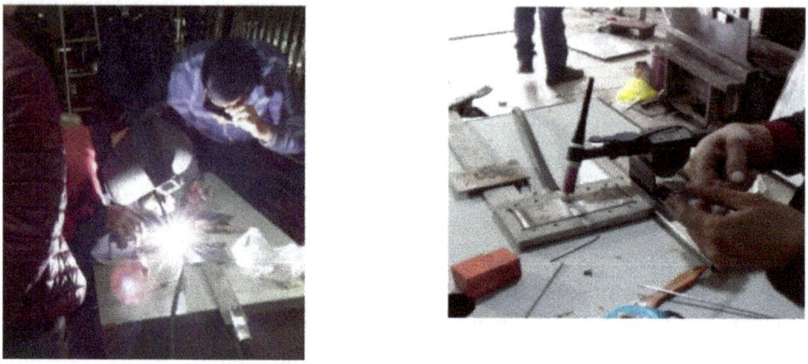

Fig. 1. TIG welding process

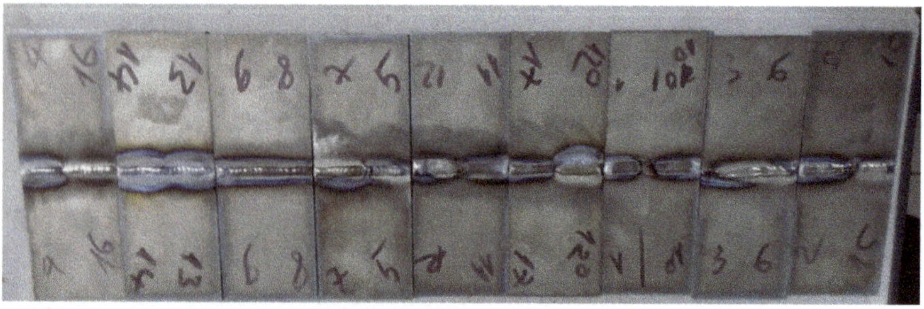

Fig. 2. After welding

After welding, the tensile strength test piece, weld stiffness is measured, determining the influence of welding parameters to select the best welding mode under test conditions.

3 Determining the Optimal Account for Welding Mechanics

The sample after welding is cut wire, milling by the position of the weld, The size of the tensile sample (according to TCVN 5403-1991 standard) as shown in Fig. 3. Mechanical measurement results include hardness, durability, elongation given in Table 2.

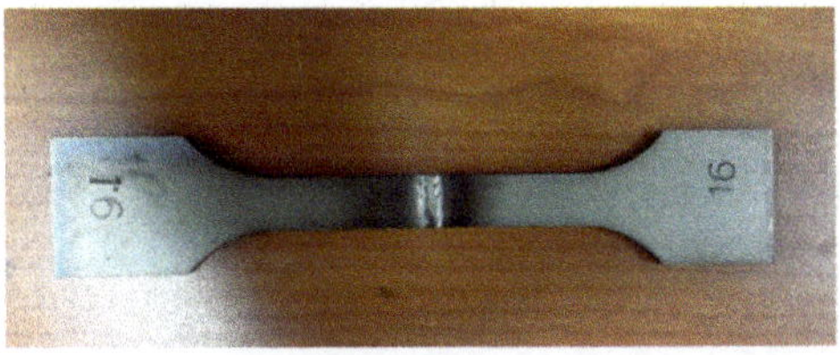

a) Sample after welding

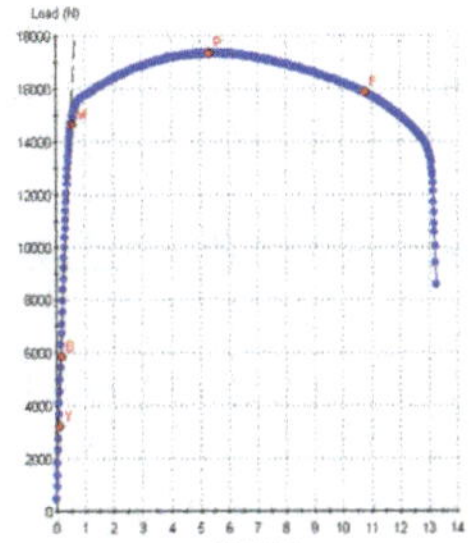

b) Tensile diagram

Fig. 3. Sample and tensile value

Table 2. Experimental welding parameters and results of mechanical

No	Sample	Electric current. A	Gas flow l/min	Electrode distance mm	Hardness of welding HV	Tensile MPa	Elongation, %
1	Sample 1	130	16	1,5	257,7	463,7	16,50
2	Sample 2	100	12	1,5	295,2	450,3	12,63
3	Sample 3	100	16	3	290,5	450,3	17,25
4	Sample 4	70	16	4,5	294,6	448,6	15,88
5	Sample 5	100	12	4,5	271,6	469,1	13,50
6	Sample 6	100	16	3	252,1	437,8	16,00
7	Sample 7	100	20	4,5	265,5	410,8	18,13
8	Sample 8	100	16	3	240,1	416,4	9,88
9	Sample 9	100	16	3	292,6	451,3	17,50
10	Sample 10	130	20	3	246,9	402,4	17,50
11	Sample 11	130	12	3	231,8	370,7	16,00
12	Sample 12	130	16	4,5	256,6	488,2	13,75
13	Sample 13	70	12	3	220,3	405,4	6,75
14	Sample 14	70	20	3	251,8	428,4	12,75
15	Sample 15	100	16	3	254,5	457,1	7,38
16	Sample 16	70	16	1,5	243,3	379,4	13,00
17	Sample 17	100	20	1,5	295,,1	452,6	6,75
18	Sample 0	0	0	0	257,7	492,7	18,5

In which sample 0 is a sample of grade 2 welded metal sheet 3 mm thickness.

4 Simulation and Selection of the Optimal Range

Using the Box-Behnken method of experimental planning is not comprehensive with 3 levels. This method is effective when estimating coefficients of quadratic polynomials. Experiments in the Box-Behnken method are taken from matrix blocks orthogonal to the midpoint of the design domain [2].

The mathematical model of the second-order response surface is described as follows:

$$y = \beta_0 + \sum\nolimits_{i=1}^{k} \beta_i x_i + \sum\nolimits_{i=1}^{k} \beta_{ii} x_i^2 + \sum\nolimits_{i=1}^{k-1} \sum\nolimits_{j=i+1}^{k} \beta_{ij} x_i x_j + \varepsilon \quad (1.1)$$

Where k is a factor, ε is the error and parameters; β_i; β_{ii} and β_{ij} are called regression coefficients.

Result of calculating the objective function:

Tensile:

$$y_1 = -244,51033 - 1,04088.I + 79,30.Q + 83,38372.e - 0,096646.I.Q + 8,11111.10^{-3}.I.e - 1,55042.Q.e + 0,017548.I^2 - 1,909.Q^2 - 8,87622.e^2 \quad (1.2)$$

The precision jaw tensile function: $R_{Squared} = 0,9934$.

Hardness:

$$y_2 = -475,82889 - 5,62528.I + 37,52250.Q + 106,68556.e + 0,016875.I.Q - 0,24722.I.e + 0,40417.Q.e - 0,023964.I^2 - 1,21672.Q^2 - 13,60788.e^2 \quad (1.3)$$

The precision jaw tensile function: $R_{Squared} = 0,9921$.

Optimal value: I = 126,31 A; Q = 16,6 l/min; e = 3,09 mm.

Value of the target function at the optimal point: $y_1 = 299,32$; $y_2 = 588,2$

The construction chart from empirically constructed: Fig. 4, 5, 6, 7, 8 and 9.

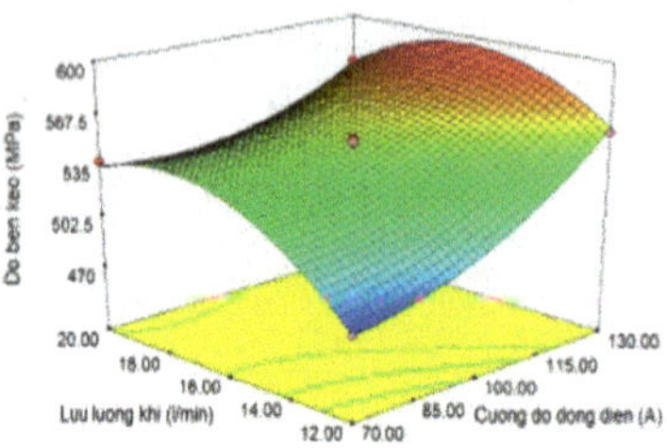

Fig. 4. Effect of electric current (I) and protection gas flow (Q) on tensile strength

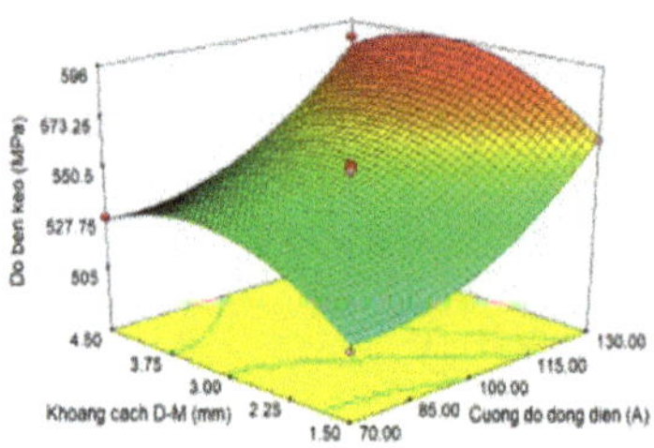

Fig. 5. Effect of electric current (I) and electrode distance (e) on tensile strength

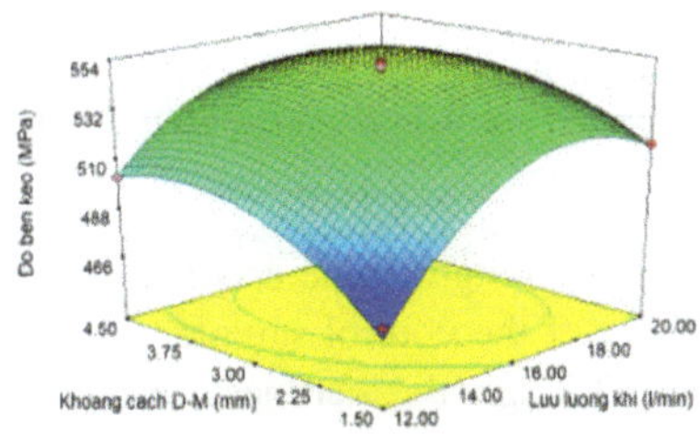

Fig. 6. Effect of protective gas flow (Q) and electrode distance (e) on tensile strength

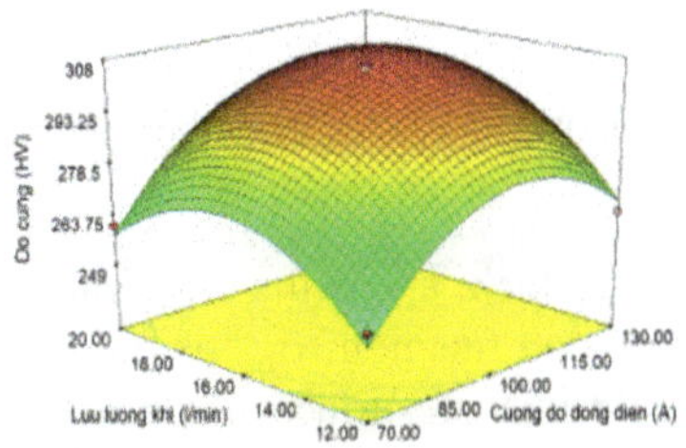

Fig. 7. Effect of electric current (I) and protection gas flow (Q) on hardness

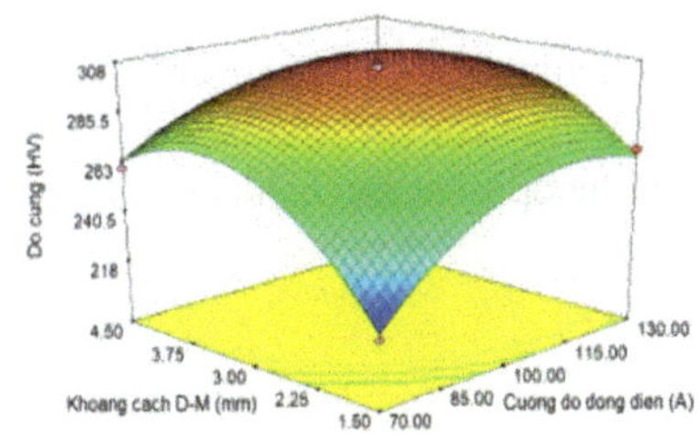

Fig. 8. Effect of electric current (I) and electrode distance (e) on hardness

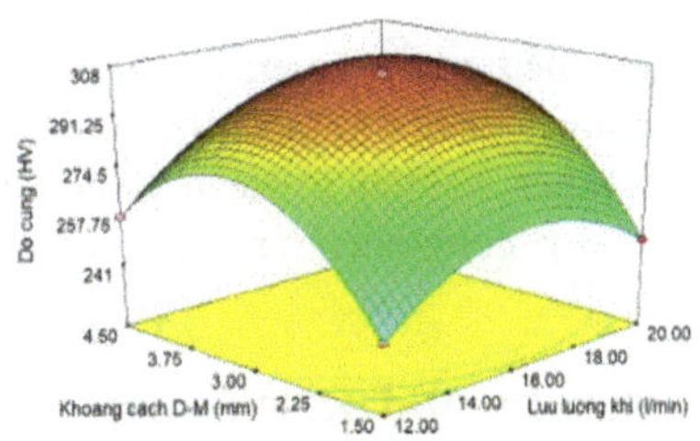

Fig. 9. Effect of protective gas flow (Q) and electrode distance (e) on hardness

The optimal zone for welding results is the maximum tensile strength and stiffness. In this experiment, based on the built-in graphs: the electric current (I) is near the highest level, indicating that the greater the current, the greater the heat of the weld. Increase the ability to fill the welding slot. The optimal protective gas flow (Q) is near the average of the survey interval, indicating that the protective gas flow is too much or too little. The optimal electrode distance is near the average of the survey range.

The optimal values of the input variables identified are:

- Optimum amperage: 126.29 A.
- Optimal protection air flow: 16.65 L/min.
- Optimal electrode distance: 3.12 mm.
- The value of the tensile objective function follows the optimal input values: σ_{bmax} = 488,2 Mpa
- Target function surface hardness value according to the optimal input values: Hmax = 287.96 HV

Perform welding to create a prototype according to the optimal regime of mechanical properties surveyed. Evaluate and test the mechanical properties of titanium material welds at the calculated optimal value, compare the error with the optimal value:

Result of tensile strength: σ_b^{kc} = 476,2 MPa.

$$\text{Error}: \quad \varepsilon_{\sigma_b} = \frac{\left|\sigma_b^{hq} - \sigma_b^{kc}\right|}{\sigma_b^{hq}} = \frac{|488,2 - 476,2|}{488,2} \approx 0,0246 = 2,46\% \quad (1.4)$$

In which: $\varepsilon_{\sigma b}$ is the error value; σ_b^{hq} is the tensile strength value according to the regression function; σ_b^{kc} is the tensile strength value of the control sample.

Results of surface hardness measurement, k) Through the hardness test of the control sample: Hkc = 292.3 HV.

$$\text{Error}: \quad \varepsilon_H = \frac{\left|H_{hq} - H_{kc}\right|}{H_{hq}} = \frac{|287,96 - 292,3|}{287,96} \approx 0,0151 = 1,51\% \quad (1.5)$$

In which: ε_H is the error value; H_{hq} is the surface hardness value according to the regression function; H_{kc} is the surface hardness value of the control sample.

Conduct welding to create prototypes according to the optimal mechanical mode of the survey, component analysis and microstructure.

5 Results and Discussion

5.1 Component Analysis

Table 3. The composition of the sample analysis at the welding position of the sample test the optimal mode

Al	Zr	Mo	V	Si	Mn	Cr	Ni	Fe	Cu	W	O	N	Ti
0.005	0.004	< 0.0003	< 0.001	0.012	< 0.001	0.013	0.002	0.12	0.001	< 0.002	0.12	< 0.002	99.7

Comparing the chemical composition of the weld in optimal mode (Table 3) with grade 2 sheet metal (Table 1), it can be seen:

- Some elements have changed and reduced components such as Mo, Cr, Cu. This is the phenomenon of alloy loss during welding.
- Some elements have increased content such as Al, Zr, O. The reason is due to the impurities phenomenon during welding. For Titanium, impurities have a great influence on the weld properties, especially O, N, H but still within the allowed range of grade2 materials.

5.2 Microstructure

Analysis of verified welds in optimum mode, the base sample results in Fig. 10.

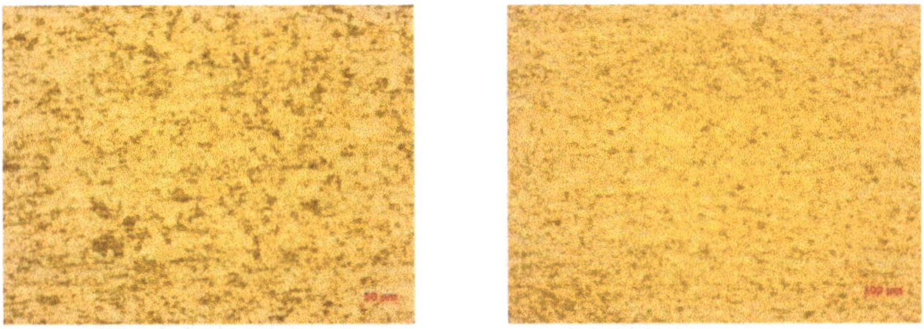

Fig. 10. Microstructure of welded grade 2 titanium plates

The microstructure of matrix metal shows uniform, small particles, no defects. Particle size 10–20 μm. This is a microstructure of the rolled sheet which is processed to make small particles and then annealed.

Fig. 11. Microstructure of welding

In the Fig. 11, The microstructure of the welding area with large particle sizes, irregular edges, larger than the weld border areas, and the base metal. The microstructure of cold welding similar to cold casting. The uneven composition in the areas of light and dark patches, indicating an uneven distribution of the composition when cooled. The particle size of 50–200 μm. No pore defects appear defects.

Fig. 12. Microstructure welding area boundary

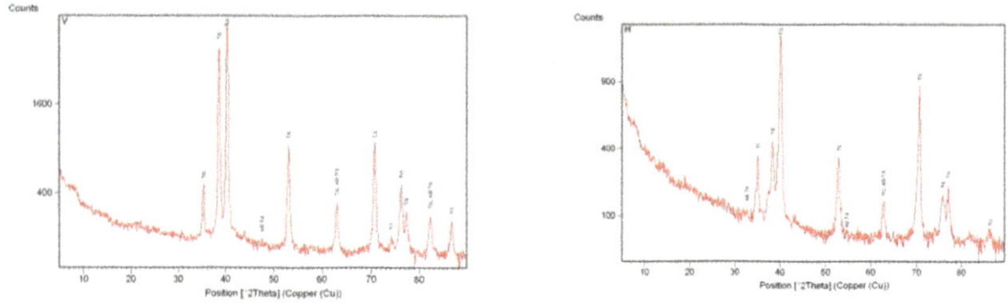

a- Welded plate b- Welding area

Fig. 13. XRD at welding boundary

The microstructure of metal at the junction between molten weld metal and non-melting metal substrates indicates a clear demarcation of welding. Part metal particles small uniform matrix (Fig. 12). The hot and cold-welded metal parts with large coarse particles, irregular components appearing under etching corrosion are areas of light and dark array showing the uneven distribution of the composition, more impurities. Large, irregular, multi-faceted particles may vary in size from a few dozen to hundreds of μm. No pore or corrosion holes appear to show good fill and weld protection.

5.3 The Results of Xray

Xray analysis of weld parts compared to the matrix of microstructure (Fig. 13).

Comparing the position, the number, the intensity of the peaks, shows that the chemical composition of the welded area and the welded metal sheet are the same. There are no phases other than Ti, especially the inter-metallic phases that reduce the mechanical properties. This proves that the weld is well protected. The weld metal component is the same as the base metal component.

6 Conclusion

Titanium has good welding properties, TIG welding technology elements are similar to TIG welding of other materials. TIG welding of Titanium materials requires special attention to cleaning the welding position before welding and the weld must be protected from both the front and back of the weld, both the titanium weld area is cooled below 2000C. Welding of Titanium structures requires suitable fixtures to ensure good weld protection.

TIG welding experiment of 17 samples on Titanium Grade 2 material with 3mm thickness billet in the rolling state, annealed with variable input parameters (current I, gas protection Q, and electrode distance e). After welding, the samples were analyzed, measured mechanical properties of durability, rigidity, developed a rigorous plan showing the optimal welding technology parameters are: I = 126.3 A; Q = 16.6 L/min; e = 3.1 mm.

Performing the test on the selected optimization mode, welding organization analysis, and analysis in the optimal mode, the results achieved $\sigma_b = 488{,}2$ Mpa; H = 287,96 HV, elongation $\varepsilon = 18.1\%$. Compared to the unfilled sheet sample: $\sigma_0 = 492{,}7$ Mpa; $H_0 = 257{,}7$ HV; elongation $\varepsilon_0 = 18.5\%$. Holding non-porous welds, components are guaranteed.

Acknowledgements. The authors would like to thank the Viet Nam Maritime University for their support.

References

1. Boyer, R., Welsch, G., Collings, E.W.: Materials Properties Handbook: Titanium Alloys. ASM International. Fourth Printing, p. 100, August 2007
2. Otani, T.: Titanium Welding Technology. Nippon Steel Technical Report, no. 95, January 2007
3. Vagi, J.J., Monroe, R.E., Evans, R.M., Martin, D.C.: Welding proceduces for titanium and titanium alloys. Nasa Technical, October 1965
4. Kramár, T., Michalec, I., Kovačócy, P.: The laser beam welding of titanium grade 2 alloy. GRANT J. (2012). ISSN 1805-062X, 1805-0638 (online), ETTN 072-11-00002-09-4

Asymmetry in Material Flow Patterns and Mechanical Properties Along Friction Stir Welding Interface of Dissimilar Metal T-lap Joint

Duong Dinh Hao[1(✉)], Tran Hung Tra[1], and Masakazu Okazaki[2]

[1] Department of Engineering Mechanics, Nha Trang University,
02 Nguyen Dinh Chieu str., Nha Trang, Vietnam
{dinhhao,tra}@ntu.edu.vn

[2] Department of Mechanical Engineering, Nagaoka University of Technology,
1603-1 Kamitomioka, Nagaoka, Niigata 940-2188, Japan
okazaki@mech.nagaokaut.ac.jp

Abstract. In this work, macro and microstructures along friction stir welding interface in the dissimilar T-lap joints between AA5083 and AA7075 were investigated by several parallel and perpendicular sections to the welding line. The experimental results showed that the asymmetry in material flow patterns and mechanical properties along the welding interface was found. The bonding strength of interface at the hook and kissing bond defects was much lower than that at the mixed zone. Mixed zone with the periodicity of material flow patterns played the important role in this strength.

Keywords: Friction stir welding · Dissimilar T-lap joints · Material flow patterns · Bonding strength · Welding interface

1 Introduction

In the recent years, the manufacturer has been facing with the big challenges from the fuel efficiency. In order to solve this difficulty, the use of lightweight alloys is considered as a suitable solution to replace the more heavy materials such as steel. However, the weldablity of them by applying the traditional welding techniques is obtained the low mechanical properties [1]. In addition, the greatest challenger of these methods lies in AA2xxx and AA7xxx series [2]. To deal with these problems, Friction Stir Welding (FSW) technique has been emerging, that is known as a solid-state joining [3]. As evidence, the success of FSW butt-joint has been attained to the high strength by the previous research [4].

The FSW process produced the asymmetry in both microstructure and strength of the joints. This result can be correlated to the asymmetric flow, plastic deformation, and welding temperature during FSW process. A few former studies have focused on the asymmetric mechanical properties of butt-joint. D. Rao et al. [5] presented that the steep change in hardness value and tensile strength was found in the retreating side (RS) compared to the advancing side (AS).

B. T. Long et al. (Eds.): MMMS 2020, LNME, pp. 162–167, 2021.
https://doi.org/10.1007/978-3-030-69610-8_21

Dissimilar to butt-joint, the strength of the T-joints is evaluated by both the skin- and stringer-directional loading tests. It is true that the poor strength of the FSWed T-lap joints was induced by the stringer-directional loading. As shown in literature [6, 7], the joints were cracked along interface between the skin and stringer plate. The weak bonding of interface was reason for these results. This means that the interface characteristic has become the important key for making a success of the T-lap joints and need to be clarified to enhance the joint efficiency.

This work focused on the investigation of the interface characteristic of the T-lap joints. A dissimilar joint was fabricated to observe the material flow patterns and measure tensile strength distribution along interface. In addition, the fracture behavior of the joints was also investigated.

2 Experimental Procedures

Two aluminum alloys of 5083 (AA5083) and 7075 were chosen to fabricate the T-lap joints (Fig. 1(a)). The dimensions of skin and stringer plates are 300 × 150 × 3.0 mm and 300 × 50 × 8.2 mm, respectively. The shoulder diameter of the welding tool was 28.0 mm. The length of pin was 4.4 mm and tapered from 11.0 mm to 6.8 mm. Two corner fillets were manufactured with the radius of 3.0 mm.

For investigating microstructure and metal flow patterns, the specimens were extracted perpendicular to welding line. All of them were polished and etched with Kroll's reagent for 10 s. The macro- and micro-structures were observed by mean of optical microscope. To evaluate effect mixing material on mechanical properties, the small tensile specimens were extracted from three various locations (i.e. specimens at AS, Center, and RS) as shown in Fig. 1(b, c).

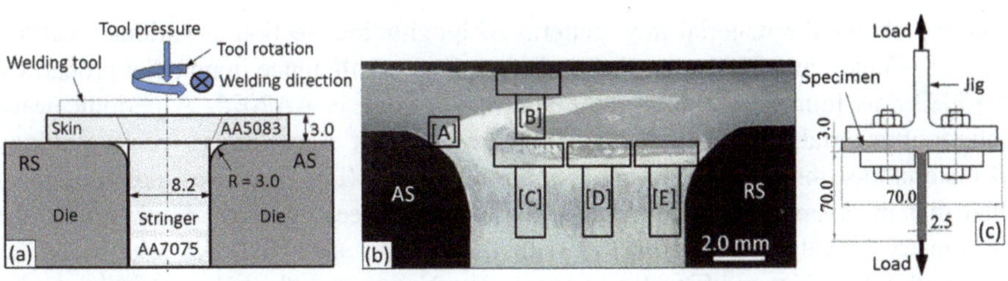

Fig. 1. (a) Geometry of T-lap joint, (b) Evaluation of local interface bonding strength, and (c) test system to evaluate local mechanical properties (dimensions in mm).

3 Results and Discussion

3.1 Morphological Cross-Section of Interface

The macro and microstructures of cross-section along welding interfaces are showed in Fig. 2. It can be seen that the asymetry in metal flow during FSW process was found by the various formation of mixing zone between two alloys. Here, the material flow trended from the RS to the AS and produced a hook defect (Fig. 2(a)), as presented in previous work [7]. The kissing bond defects (KBs) was also found at two corner defects. Note that the mixing zone was not formed at these defects. The diffusion of two aluminum alloys reduced from the RS to AS. There was a difference in the formation of the joint along interface such as the severe diffusion of materials in the RS (Fig. 2(b)), the onion ring formation in the center (Fig. 2(c)), and the laminar flow in the AS (Fig. 2(d)).

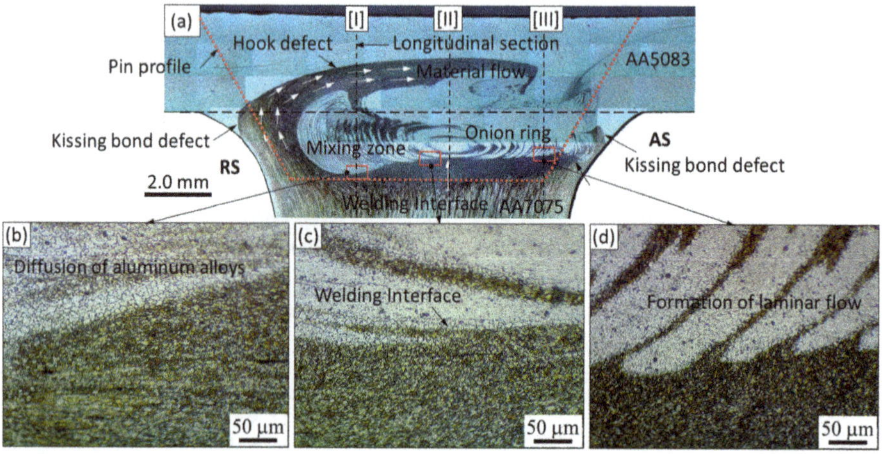

Fig. 2. Macro and microstructures along welding interface.

3.2 Morphology of Longitudinal Section of Interface

Figure 3 shows the material flow patterns of longitudinal section at various locations (i.e. AS, Center, and RS), as marked in Fig. 2. Due to difference in material properties, AA5083 was found as brighter zone and darker zone is AA7075. A periodic wave pattern of material flow was detected in all three sections. According to the morphology of material distribution, the mixing zone in the [I] area (Fig. 3(a)) was more dominant than that in others, resulting in the formation of perfect interface. Furthermore, the asymmetry in velocity direction between the [I] area and [III] area was found, as observed in Fig. 3(a) and Fig. 3(c), respectively. Noting that, the heights of mixed zone reduced from the RS to the AS. This difference was also found in cross-section. It can be seen that the combination between downward and upward flow, and the movement of welding tool induced the periodic metal flow along interface [8].

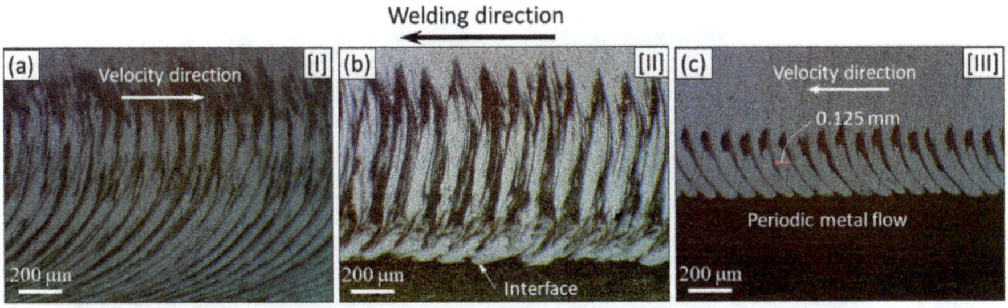

Fig. 3. Periodicity of material flow patterns in longitudinal section along welding interface.

3.3 Mechanical Properties Distribution

The bonding strength along the interface is shown in Fig. 4. The maximum bonding strength was heterogeneous. The strength at the KBs region [A] is nearly zero [7]. The bonding at the KBs region [A] is lower than that at the hook defects [B]. The strength of [C]-[E] zones was much higher than other regions. The bonding strength is significantly affected by the KBs and hook defects.

The fracture surfaces of the specimen at the [D] position that cracked at mixed zone (Fig. 5(a, b)) were analyzed under the SEM, and are showed in (Fig. 5(c, d)). As showed in low magnification (Fig. 5(c)), the rough surface was found. The heterogeneity was caused by the mixture between two aluminum alloys that contain the different properties. Figure 5(d) indicates the presence of them due to difference from characteristic fracture behavior. Based on analyzed EDS with the presence of Zinc element, it was found that the appearance of dimples on fracture surface of AA5083 indicated a ductile behavior while the brittle behavior was obtained at AA7075. Furthermore, a large number of the dimples were detected in Fig. 5(e) due to disappearance of AA7075.

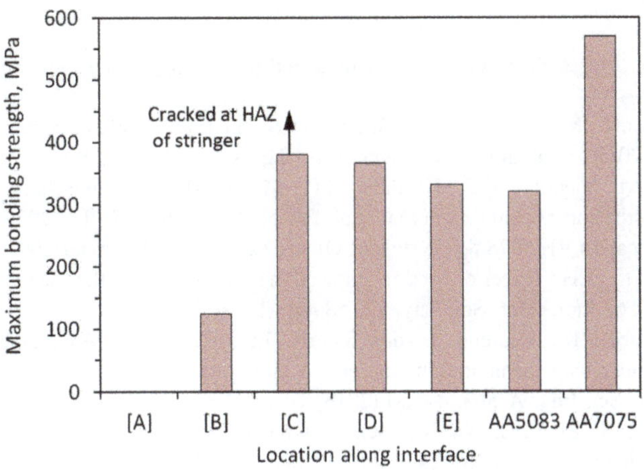

Fig. 4. Bonding strength distribution along welding interface.

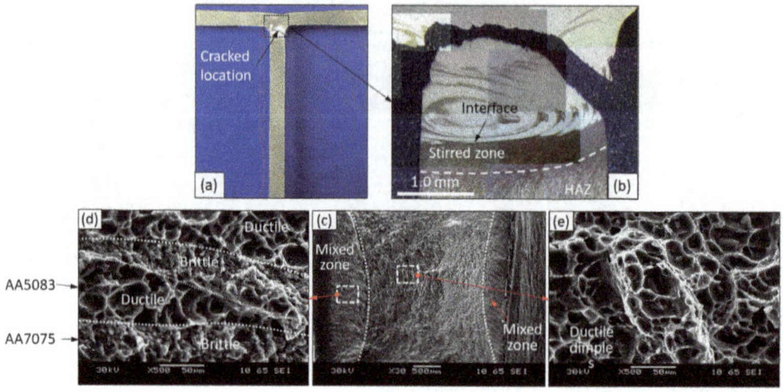

Fig. 5. Fracture location (a, b) and fracture surface (c–e) of specimen at [D] position.

4 Summary

The FSWed T-lap joints of dissimilar aluminum alloys were fabricated and analyzed. The attention was focused on the welding interface in terms of material flow patterns and bonding strength distribution. The following findings were derived:

1. The asymmetry in material flow patterns was found along interface in both cross-section and longitudinal section.
2. Mixed zone with the periodicity of material flow patterns played the important role in the bonding strength of the interface.
3. The bonding strength at the hook defects was the lowest while this strength was dominant at mixed zone.

References

1. Mishra, R.S., Ma, Z.Y.: Friction stir welding and processing. Mater. Sci. Eng. R. **50**, 1–78 (2005)
2. Lu, Z.J., Evans, W.J., Parker, J.D., Birley, S.: Simulation of microstructure and liquation cracking in 7017 aluminum alloy. Mater. Sci. Eng. A **220**, 1–7 (1996)
3. Thomas, W.M., Nicholas, E.D., Needham, J.C., Murch, M.G., Temple-Smith, P., Dawes, C. J.: Friction-stir butt welding. International Patent Application PCT/GB92/02203 and GB Patent Application 9125978.8, UK Patent Office, London, 6 December 1991
4. Hao, D.D., Tra, T.H.: Effect of friction stir welding parameters on the mechanical properties of AA7075-T6. Arch. Mat. Sci. Eng. **77**, 58–64 (2016)
5. Rao, D., Huber, K., Heerens, J., dos Santos, J.F., Huber, N.: Asymmetric mechanical properties and tensile behaviour prediction of aluminium alloy 5083 friction stir welding joints. Mater. Sci. Eng. A **565**, 44–50 (2013)
6. Cui, L., Yang, X., Zhou, G., Xu, X., Shen, Z.: Characteristics of defects and tensile behaviors on friction stir welded AA6061-T4 T-joints. Mater. Sci. Eng. A **543**, 58–68 (2012)

7. Hao, D.D., Okazaki, M., Tra, T.H.: Effect of welding parameters on mechanical properties of friction stir welded T-lap dissimilar metal joints between 7075 and 5083 aluminum alloys. JSME-Mech. Eng. J. **6** (2019). https://doi.org/10.1299/mej.19-00091
8. Jain, R., Pal, S.K., Singh, S.B.: Finite element simulation of pin shape influence on material flow, forces in friction stir welding. Int. J. Adv. Manuf. Technol. **94**, 1781–1797 (2018)

Dynamic Response and Vibration of Shear Deformable Sandwich Plate with GPL-RC Face Sheets and a Core Layer with Negative Poisson's Ratios Under Blast Loading

Vu Dinh Quang, Nguyen Van Quyen, Vu Thi Thuy Anh, and Nguyen Dinh Duc(✉)

Advanced Materials and Structures Laboratory, VNU Hanoi, University of Engineering and Technology (UET), Hanoi 100000, Vietnam
ducnd@vnu.edu.vn

Abstract. The impact threats and blast have led to a demand for new structure in recent years. One of the ways is to add the smart materials layers for creating the new, which is called a sandwich structure. This work presents the dynamic response and vibration of sandwich composite plate supported by elastic foundations in thermal environment. This present takes advantage of the sandwich-structured configuration with three layers: two graphene platelet-reinforced composite (GPL-RC) face sheets and an auxetic core with negative Poisson's ratios to analyze the dynamic and vibration problem.

Keywords: Dynamic · Blast loads · Sandwich · GPL-RC · Auxetic · Thermal environment

1 Introduction

In recent years, the number of studies about sandwich structure has increased significantly, due to the applications of sandwich structures in terms of bearing to mechanical and thermal loads, as well as the benefits of cost optimization or structure weight. Notably, as well as most outstanding are sandwich structures with cores layer made from materials with negative Poisson's ratios- auxetic materials.

Auxetic, as is known as a new material with a negative Poisson's ratios. Thanks to this material parameter, auxetic has special properties such as high energy absorption and useful in applications such as body armor, robust shock absorbing material and especially for structures subjected to blast loading. However, controlling the behavior of auxetic structures is still a big challenge. This requires much more research on auxetic as well as the behavior of structures made of this type of material, especially the structure of sandwich auxetic. Within the scope of the report's research about the dynamic and vibration response, from open source documents, a few studies on this area can be listed as [1–5].

Recently, issues related to the behavior of auxetic - graphene structures have been paid much attention by scientists [6, 7]. Can see that the research in this regard is still a completely new field and has not received much attention. There are still a lot of issues

B. T. Long et al. (Eds.): MMMS 2020, LNME, pp. 168–176, 2021.
https://doi.org/10.1007/978-3-030-69610-8_22

that need further research. Base on the FSDT, this report presents the dynamic response and vibration of sandwich composite plate with three layers: two GPL face sheets and an auxetic core layer supported by elastic foundations in thermal environment. Numerical results are provided to explore the influences of geometrical parameters, blast and mechanical loads on the nonlinear dynamic behavior and vibration of sandwich plate resting on elastic foundation and in thermal environment.

2 Mathematical Model

2.1 The Configuration of Sandwich Plate with Auxetic Core Layer

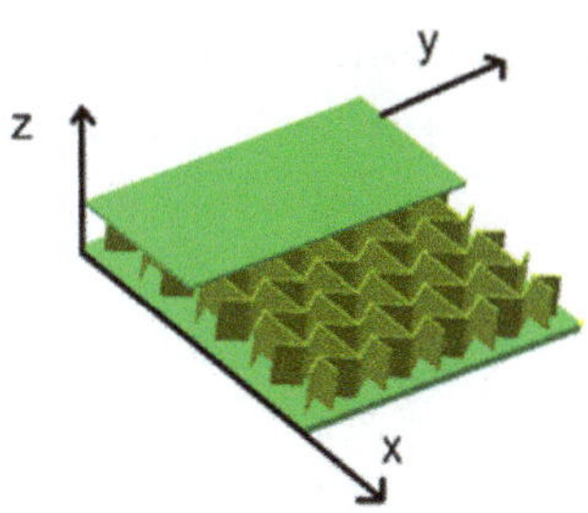

Fig. 1. Model of sandwich plate

Consider a sandwich plate on elastic medium with two graphene platelet – reinforced composite (GPL) face sheets and an auxetic core layer (Fig. 1). With such a combination of layers, the total plate thickness is $h = h_f + h_{AU} + h_b = 2h_{GRA} + h_{AU}$ (with h, h_f, h_{AU}, h_b- the thickness of plate, top layer, auxetic layer and bottom layer, respectively, and $h_f = h_b = h_{GRA}$, h_{GRA} - the thickness of GPL layer).

The reaction–deflection relation represents for the interaction of the elastic medium is given by [4]:

$$q_e = K_W w - K_P \nabla^2 w \tag{1}$$

with K_W, K_P - the foundation stiffness of Winkler foundation and Pasternak model; w, ∇ - the deflection of the plate, and the Laplace operator.

2.2 Material Properties of Face Sheets

The functions of non – dimensional thickness coordinate:

$$Q_{11}^{GRA} = Q_{22}^{GRA} = \frac{E^{GRA}}{1 - \left(v_{12}^{GRA}\right)^2}, \; Q_{12}^{GRA} = v_{12}^{GRA} Q_{22}^{GRA}, \; Q_{66}^{GRA} = G_{12}^{GRA}, \; Q_{44}^{GRA} = G_{23}^{GRA}, Q_{55}^{GRA} = G_{13}^{GRA}. \tag{2}$$

in which, the Halpin - Tsai model are selected to calculate the elastic modulus of the GPL layer, E_{GPL}, E_m - the Young's modulus, V_{GPL}, V_m - the volume fractions of GPLs and metal, $a_{GPL}, b_{GPL}, h_{GPL}$ - the average length, with and thickness of GPLs, and

$$E^{GRA} = \frac{E_m}{8(1 - \eta_L V_{GPL})}\left[3\eta_L\left(\frac{1}{\eta_L} + \frac{2a_{GPL}}{h_{GPL}} V_{GPL}\right) + 5\eta_T\left(\frac{1}{\eta_L} + \frac{2b_{GPL}}{h_{GPL}} V_{GPL}\right)\right], \tag{3}$$

$$\eta_L = \frac{h_{GPL}(E_{GPL} - E_m)}{E_{GPL} h_{GPL} + E_m 2a_{GPL}}, \quad \eta_T = \frac{h_{GPL}(E_{GPL} - E_m)}{E_{GPL} h_{GPL} + 2b_{GPL} E_m}.$$

The mass density and Poisson's ratio of the GPL layer are calculated by applying the rule of mixture follows as [13]:

$$\nu_{12}^{GRA} = \nu_{21}^{GRA} = V_m(\nu_m - \nu_{GPL}) + \nu_{GPL},\ V_{GPL} = 1 - V_m,$$
$$\alpha_1^{GRA} = \alpha_2^{GRA} = \frac{V_{GPL}E_{GPL}\alpha_{GPL} + E_m\alpha_m - V_{GPL}E_m\alpha_m}{V_{GPL}E_{GPL} + E_m - V_{GPL}E_m}.$$

with α_{GPL}, α_m - the thermal expansion coefficient of GPLs and metal, ρ_{GPL}, ρ_m - the mass density of GPLs and metal, ν_{GPL}, ν_m - the Poisson coefficient of GPLs and metal.

2.3 Material Properties of Honeycomb Core

The functions of non – dimensional thickness coordinate and elastic modulus of the auxetic layer are calculated by [12] (Fig. 2).

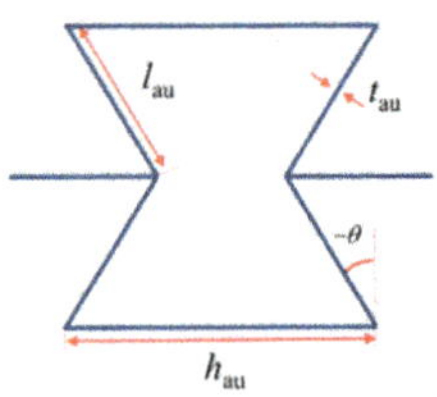

Fig. 2. Geometric of the cell of honeycomb core (t_{au}, l_{au}, h_{au} - the thickness of cell wall, length of the inclined cell rib and length of the vertical cell)

$$Q_{11}^{AU} = \frac{E_m\eta_3^3(\eta_1 - \sin\theta)}{\cos^3\theta\left[1 + (\tan^2\theta + \eta_1\sec^2\theta)\eta_3^2\right]\left(1 - \nu_{12}^{AU}\nu_{21}^{AU}\right)};\ \nu_{12}^{AU} = -\frac{\sin\theta(1 - \eta_3^2)(\eta_1 - \sin\theta)}{\cos^2\theta\left[1 + (\tan^2\theta + \sec^2\theta\eta_1)\eta_3^2\right]};$$
$$Q_{22}^{AU} = \frac{E_m\eta_3^3}{\cos\theta(\eta_1 - \sin\theta)(\tan^2\theta + \eta_3^2)\left(1 - \nu_{12}^{AU}\nu_{21}^{AU}\right)};Q_{12}^{AU} = \nu_{12}^{AU}Q_{22}^{AU};\nu_{21}^{AU} = \frac{-\sin\theta(1 - \eta_3^2)}{(\tan^2\theta + \eta_3^2)(\eta_1 - \sin\theta)};$$
$$Q_{44}^{AU} = \frac{G_m\eta_3\cos\theta}{\eta_1 - \sin\theta},\ Q_{55}^{AU} = \frac{G_m\eta_3}{2\cos\theta}\left[\frac{\eta_1 - \sin\theta}{1 + 2\eta_1} + \frac{\eta_1 + 2\sin^2\theta}{2(\eta_1 - \sin\theta)}\right],\ Q_{66}^{AU} = \frac{E_m\eta_3^3}{\eta_1(1 + 2\eta_1)\cos\theta}. \quad (4)$$

$$\rho^{AU} = \frac{t_{au}\rho_m(h_{au} + 2l_{au})}{2l_{au}\cos\theta(h_{au} - l_{au}\sin\theta)};\ \alpha_1^{AU} = \alpha_m\frac{\eta_3\cos\theta}{\sin\theta + \eta_1};\ \alpha_2^{AU}$$
$$= \alpha_m\frac{t_{au}(h_{au} + l_{au}\sin\theta)}{l_{au}(2h_{au} + l_{au})\cos\theta}.$$

2.4 Fundamental Formulation

In this report, the FSDT is used to establish the basic equations and investigate the nonlinear vibration of the plate with negative Poisson's ratio in auxetic honeycombs.

The strain-displacement relations across the plate thickness at a distance from the mid-plane taking into account the von Karman nonlinear terms and are defined as [2].

$$\begin{bmatrix} \varepsilon_x \\ \varepsilon_y \\ \gamma_{xy} \end{bmatrix} = \begin{bmatrix} \varepsilon_x^0 \\ \varepsilon_y^0 \\ \gamma_{xy}^0 \end{bmatrix} + z \begin{bmatrix} \chi_x \\ \chi_y \\ \chi_{xy} \end{bmatrix} = \begin{bmatrix} u_{,x} + 0.5(w_{,x})^2 \\ v_{,y} + 0.5(w_{,y})^2 \\ u_{,y} + v_{,x} + w_{,x}w_{,y} \end{bmatrix} + z \begin{bmatrix} \phi_{x,x} \\ \phi_{y,y} \\ \phi_{x,y} + \phi_{y,x} \end{bmatrix}; \quad \begin{bmatrix} \gamma_{xz} \\ \gamma_{yz} \end{bmatrix} = \begin{bmatrix} w_{,x} + \phi_x \\ w_{,y} + \phi_y \end{bmatrix}, \tag{5}$$

For auxetic core layer, the relationship between stress and strain components with the effect of temperature are expressed as.

$$\sigma_x^{(k)} = Q_{11}^{(k)}\varepsilon_x + Q_{12}^{(k)}\varepsilon_y - \left(Q_{11}^{(k)}\alpha_1^{(k)} + Q_{12}^{(k)}\alpha_2^{(k)}\right)\Delta T; \quad \sigma_{xy}^{(k)} = Q_{66}^{(k)}\gamma_{xy}; \quad \sigma_{xz}^{(k)} = Q_{55}^{(k)}\gamma_{xz};$$

$$\sigma_y^{(k)} = Q_{12}^{(k)}\varepsilon_x + Q_{22}^{(k)}\varepsilon_y - \left(Q_{12}^{(k)}\alpha_1^{(k)} + Q_{22}^{(k)}\alpha_2^{(k)}\right)\Delta T; \sigma_{yz}^{(k)} = Q_{44}^{(k)}\gamma_{yz}. \tag{6}$$

in here k = "f", "AU", "b".

The force and moment resultants of the GPL – auxetic sandwich plate are expressed as

$$\begin{aligned} (N_i, M_i) &= 2\left(N_i^{GRA}, M_i^{GRA}\right) + \left(N_i^{AU}, M_i^{AU}\right), \ i = x, y, xy, \\ (Q_i) &= 2\left(Q_i^{GRA}\right) + \left(Q_i^{AU}\right), \ i = x, y, \end{aligned} \tag{7}$$

The Airy's stress function $f(x, y, t)$ is introduced to simplify the problem as.

$$N_x = \frac{\partial^2 f}{\partial y^2}, N_y = \frac{\partial^2 f}{\partial x^2}, N_x = \frac{\partial^2 f}{\partial x \partial y}, \tag{8}$$

Based on the analytical approach, the selected variables are stresses and two basic equations are the equilibrium equations using the Hamilton principle and the geometrical compatibility equation. These equations are defined as in [11].

3 Solution Procedure

Two cases of boundary conditions will be considered [11]: Case 1: The imperfect plate edges are simply supported and freely movable (FM), Case 2: The imperfect plate edges are simply supported and immovable (IM). The analytical solutions are assumed to have the form as.

$$\begin{aligned} &w(x, y, t) = W(t)\sin(\alpha x)\sin(\beta y); \ \phi_x(x, y, t) = \Phi_x(t)\cos(\alpha x)\sin(\beta y); \\ &\phi_y(x, y, t) = \Phi_y(t)\sin(\alpha x)\cos(\beta y), \end{aligned} \tag{9}$$

where $\alpha = m\pi/a,\ \beta = n\pi/b,\ W(t)$ is the time dependent total amplitude and $\Phi_x,\ \Phi_y$ are the amplitudes which are functions dependent on time.

For initial imperfection, assume that the function w^* has the same form

$$w^*(x, y, t) = \mu h \sin(\alpha x) \sin(\beta y), \tag{10}$$

and μ $(0 \le \mu \le 1)$ is a quantity that is characteristic of the structural imperfection.

The form of stress function is obtained as

$$f(x, y, t) = \mathrm{T}_1(t)\cos(2\alpha x) + \mathrm{T}_2(t)\cos(2\beta y) + \mathrm{T}_3(t)\sin(\alpha x)\sin(\beta y) + \frac{N_{x0}y^2}{2} + \frac{N_{y0}x^2}{2}, \tag{11}$$

$$\mathrm{T}_1 = -\left(W^2 + 2W\mu h\right)\left(\Delta\beta^2/32A_{11}\alpha^2\right),\ \mathrm{T}_2 = -\left(W^2 + 2W\mu h\right)\left(\Delta\alpha^2/32A_{22}\beta^2\right),\ \mathrm{T}_3 = D_1\Phi_x + D_2\Phi_y,$$

Through transformations, and by applying the Bubnov - Galerkin method, the new equations written for the imperfect plate are rewritten as follows.

$$\begin{gathered}
h_{11}\mathrm{W} + h_{12}\Phi_x + h_{13}\Phi_y + h_{14}(\mathrm{W} + \mu h)\Phi_x + h_{15}(\mathrm{W} + \mu h)\Phi_y + h_{16}(\mathrm{W} + \mu h), \\
+ h_{17}W(\mathrm{W} + \mu h)(W + 2\mu h) + F(\mathrm{W} + \mu h) + h_{18}q = I_0 W_{,tt}, \\
h_{21}\Phi_x + h_{22}\Phi_y + h_{23}(\mathrm{W} + \mu h) + h_{24}W(W + 2\mu h) = \left[I_2 - \left(I_1^2/I_0\right)\right]\Phi_{x,tt}, \\
h_{31}\Phi_x + h_{32}\Phi_y + h_{33}(\mathrm{W} + \mu h) + h_{34}W(W + 2\mu h) = \left(I_2 - \left(I_1^2/I_0\right)\right)\Phi_{y,tt}.
\end{gathered} \tag{12}$$

3.1 Nonlinear Dynamic Analysis with Effect of Pre-loaded Axial Compression

Assume that the plate is loaded under uniform compressive loads P_x and P_y (Pascal) on the edges $x = (0, a)$ and $y = (0, b)$(FM case) with: $N_{x0} = -P_x h,\ N_{y0} = -P_y h$.

The natural frequencies of the plate can be determined directly by solving determinant.

$$\begin{vmatrix}
h_{11} + h_{16} + \left(P_x\alpha^2 + P_y\beta^2\right)h + I_0\omega^2 & h_{12} & h_{13} \\
h_{23} & h_{21} + \rho_1\omega^2 & h_{22} \\
h_{33} & h_{31} & h_{32} + \rho_1\omega^2
\end{vmatrix} = 0. \tag{13}$$

3.2 Nonlinear Dynamic Analysis in Thermal Environment

The effect of temperature on the structure will be considered in case 2 (IM), the natural frequencies of the plate under thermal loads can be determined directly by solving determinant.

$$\begin{vmatrix} h'_{11} & h_{12} & h_{13} \\ h_{23} & h_{21}+\rho_1\omega^2 & h_{22} \\ h_{33} & h_{31} & h_{32}+\rho_1\omega^2 \end{vmatrix} = 0, \quad \begin{pmatrix} h'_{11} = h_{11} - \left(c_{13}\alpha^2 + c_{23}\beta^2\right)\left(\Phi_1 + \Phi_1^a\right) + I_0\omega^2 + \\ -\left(\alpha^2\Phi_{1x}^s + \beta^2\Phi_{1y}^s\right) - \left(\alpha^2 A_{11}^a + \beta^2 A_{12}^a\right) \end{pmatrix} \tag{14}$$

3.3 Nonlinear Dynamic Analysis Using Simplified Assumption

By choosing the right parameters, the fundamental frequencies of the plate can be determined approximately by explicit expression:

$$\omega_{mn} = \sqrt{-(a_1 + a_2)/I_0} \tag{15}$$

The frequency-amplitude relation of nonlinear forced vibration is obtained.

$$\alpha^2 - \left(1 - \frac{8}{3\pi}MA + \frac{3}{4}NA^2\right) + \frac{R}{A\omega_{mn}^2} = 0, \ \alpha = \frac{\Omega}{\omega_{mn}}. \tag{16}$$

And the frequency-amplitude relation of free nonlinear vibration is obtained.

$$\omega_{NL}^2 = \omega_{mn}^2\left(1 - \frac{8}{3\pi}MA + \frac{3}{4}NA^2\right) \tag{17}$$

4 Results and Discussion

4.1 Comparison Studies

The differences between the first and second natural frequency parameters of isotropic plate which obtained in this report and by Hashemi et al. in [8, 9] are summarized in Table 1. It can be seen that a good agreement is obtained in this comparison.

Table 1. Comparison study of frequency $\beta = \left(\omega a^2/h\right)\sqrt{\rho_c/E_c}$ for isotropic plates.

Material		Mode 1			Mode 2		
		h/a			h/a		
		0.05	0.1	0.2	0.05	0.1	0.2
Al_2O_3	Present	**5.9317**	**5.8117**	**5.3957**	**14.677**	**13.98**	**11.941**
	[8]	0.177%	0.639%	1.859%	0.417%	1.348%	2.807%
	[9]	0.174%	0.634%	1.856%	0.417%	1.341%	2.798%
Al	Present	**3.0192**	**2.9581**	**2.7463**	**7.4704**	**7.1158**	**6.0778**
	[8]	−0.45%	0.02%	1.28%	−0.21%	0.76%	2.29%
	[9]	0.18%	0.64%	1.85%	0.41%	1.35%	2.80%

4.2 Nonlinear Dynamic and Vibration Analysis

The effect of pre-loaded and elastic foundations on natural frequencies $\omega(s^{-1})$ of sandwich plate is showed in Table 2. The effects of imperfection coefficient μ on the amplitude – time curves of sandwich plate (mode $(m, n) = (1, 1)$) are shown in Fig. 3.

Table 2. Effect of pre-loaded and coefficient K_W $(GPa/m), K_P$ $(GPa.m)$ on natural frequencies $\omega(s^{-1})$ of sandwich stiffened plate, $\beta_{om} = \omega \times a^2 \times \sqrt{\rho_m/E_m}/h$

(K_W, K_P)	Mode	$P_x(MPa)$				
		0	100	200	300	500
(0, 0)	1	12.6517	12.1698	11.6681	11.1437	10.0130
	2	31.0434	30.2630	29.4619	28.6384	26.9159
(0.2, 0.03)	1	18.6458	18.3223	17.9929	17.6574	16.9665
	2	36.8286	36.1732	35.5057	34.8254	33.4233

The influence of a/b, on the dynamic behavior of the structure are surveyed and indicated through Fig. 4 below. And the influence of the elastic foundation on the amplitude – time curves of the sandwich plates is examined and shown in Fig. 5.

Impact of blast and pre-loaded axial compressive force and the influence of parameter Ts on the amplitude – time curves of sandwich plate arre shown in Fig. 6, 7, 8. The influence of uniform temperature rise and the effect of GPL volume fraction on the amplitude – time curves of the sandwich plates are shown in Fig. 9, 10 with UD types of GPL distribution.

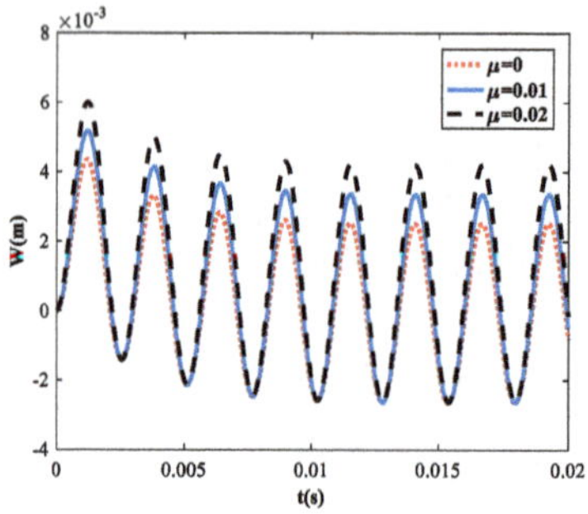

Fig. 3. Influence of coefficient μ on the amplitude – time curves of plate.

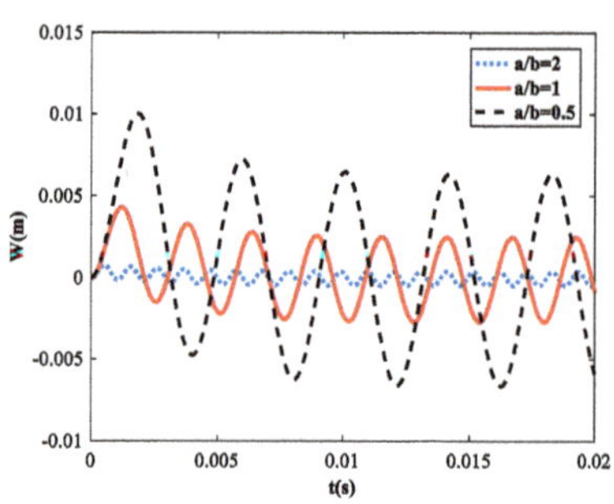

Fig. 4. Influence of ratio a/b on the amplitude – time curves of the plate.

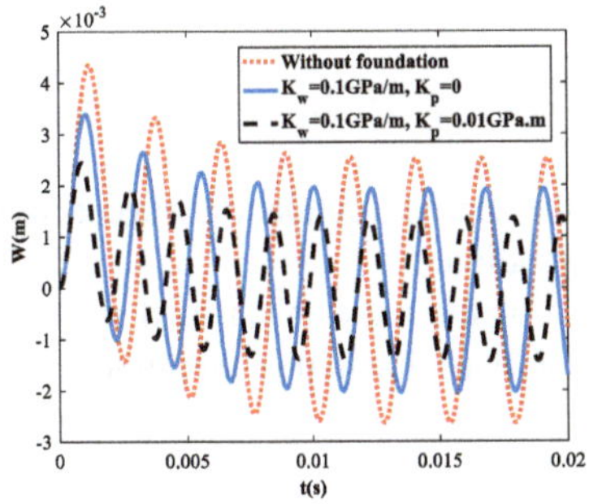

Fig. 5. Influence of K_W, K_p on the amplitude – time curves of the plate

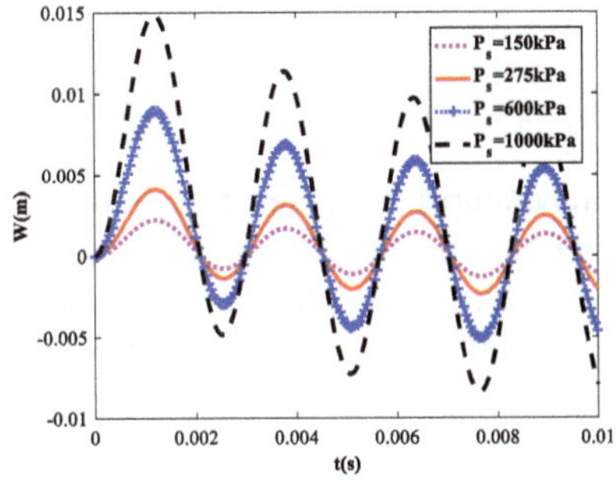

Fig. 6. Influence of blast loading on the amplitude – time curves

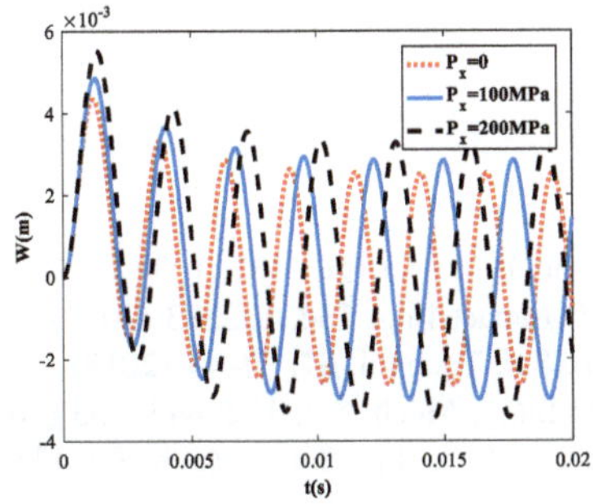

Fig. 7. Influence of compression load on the amplitude – time curves of the plates

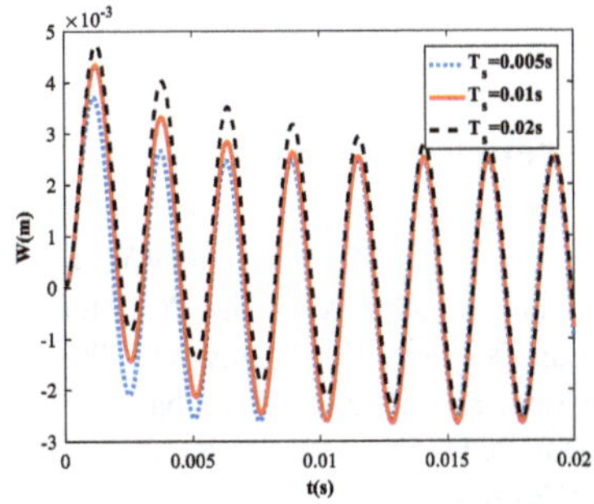

Fig. 8. Influence of Ts on the amplitude – time curves of the sandwich plate.

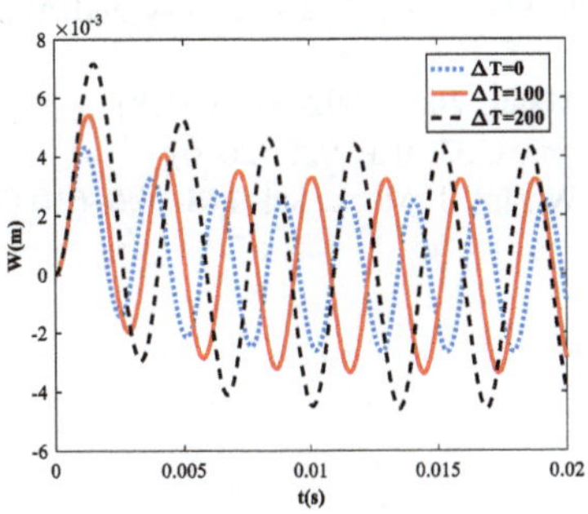

Fig. 9. Influence of ΔT on nonlinear displacement-time response.

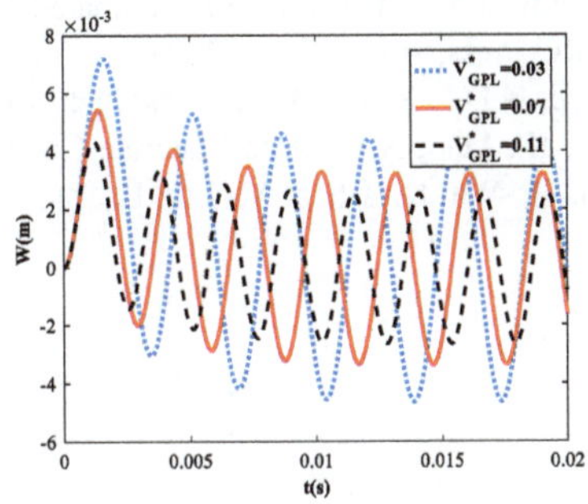

Fig. 10. Influent of V^*_{GPL} on the amplitude – time curves of the sandwich plates.

5 Conclusion

A few comments are drawn:

- Temperatures have negative impact while the elastic foundation have a positive impact on the amplitude - time curve and vibration.
- The change in amplitude under the influence of the mechanical load is not much difference, then when under the influence of blast load, this amplitude changes significantly.
- The material parameters, parameters of structure and boundary conditions greatly influence on the amplitude of the dynamic response of the sandwich structure.

Acknowledgement. This research is funded by Vietnam National Foundation for Science and Technology Development (NAFOSTED) under grant number 107.02-2018.314. The authors are grateful for this support.

References

1. Li, Y., Chen, Z., Xiao, D., Wu, W., Fang, D.: Int. J. Impact Eng. **137**, 103442 (2020)
2. Zhang, J., Zhu, X., Yang, X., Zhang, W.: Int. J. Impact Eng. **134**, 103383 (2019)
3. Lan, X., Feng, S., Huang, Q., Zhou, T.: Aerosp. Sci. Technol. **87**, 37–47 (2019)
4. Cong, P.H., Long, P.T., Nhat, N.V., Duc, N.D.: Int. J. Mech. Sci. **152**, 443–453 (2019)
5. Duc, N.D., Seung-Eock, K., Cong, P.H., Anh, N.T., Khoa, N.D.: Int. J. Mech. Sci. **133**, 504–512 (2017)
6. Winczewski, S., Rybicki, J.: Carbon **146**, 572–587 (2019)
7. Oh, J.H., Kim, J.S., Hiep, N.V., Oh, I.K.: Composites Part B: Eng. **1861**, 107817 (2020)
8. Hashemi, S.H., Fadaee, M., Atashipour, S.R.: Int. J. Mech. Sci. **53**(1), 11–22 (2011)
9. Hashemi, S.H., Arsanjani, M.: Int. J. Solids Struct. **42**(3–4), 819–853 (2005)
10. Reddy, J.N.: Mechanics of Laminated Composite Plates and Shells: Theory and Analysis. CRC Press, Boca Raton (2004)
11. Duc, N.D., Cong, P.H.: J. Sandwich Structures Mater. **20**(6), 692–717 (2018)
12. Di, K., Mao, X.B.: Acta Materiae Compositae Sinica. **33**, 910–920 (2016)
13. Li, Q., Wu, D., Chen, X., Liu, L., Yu, Y., Gao, W.: Int. J. Mech. Sci. **148**, 596–610 (2018)

Modeling of the Effect of Cutting Parameters on Surface Residual Stress When Turning of 304 Austenitic Stainless Steel

Viet-Hoi Tran[1(✉)], Van-Bong Pham[1], and Van-Dich Tran[2]

[1] Hanoi University of Industry, Hanoi, Vietnam
hoitv@haui.edu.vn
[2] Hanoi University of Science and Technology, Hanoi, Vietnam

Abstract. Residual stress is induced in a workpiece's surface layer by machining, which badly affects the components' static strength, fatigue strength, and corrosion resistance. In this study, experiments are conducted based on a response surface methodology (RSM) using Box - Behnken experiment design. Surface residual stress measured by the X-ray diffraction method and the test results were analyzed using variance analysis. The paper proposes a mathematical model of the influence of cutting parameters such as cutting speed, feed rate, depth of cut on surface residual stress.

Keywords: Austenitic stainless steel · Residual stress · Response surface methodology

1 Introduction

Stainless steel is widely used in life as home appliances, medical, food, aerospace industry. It has good mechanical and physical properties, such as beautiful shape, high hardness, resistance to corrosion and good heat resistance. However, stainless steel also has many disadvantages such as high shear force, high cutting heat, high hardness and easy to lead to BUE and BUL, which is the cause of low productivity and increased wear, poor surface quality. Stainless steel is typical of difficult material to machining.

The term 'surface quality' was introduced in the 60s of the last century and is increasingly focused on production worldwide [1]. Surface quality has a significant impact on part performance, including the state and properties of a machined surface. In general, surface quality is determined by the mechanical, physical, chemical and topological characteristics of the surface properties such as changes in roughness, hardness, structure and residual stress, ext [2]. Surface roughness and residual stress are often considered one of the most critical surface quality [3].

In general, residual stresses in conventional machining processes are caused by three sources, including thermal and mechanical loads and metallurgical transformations [4]. Mechanical loads are created by contact and compression between tool and workpiece and lead to mechanical plastic deformation on workpiece machining. These deformations produce residual compression stress. Besides, the plastic deformation created by the thermal load is responsible for generating the residual tensile stress.

B. T. Long et al. (Eds.): MMMS 2020, LNME, pp. 177–183, 2021.
https://doi.org/10.1007/978-3-030-69610-8_23

Residual stress prediction has been a research topic since the 1950s. Most of the initial research efforts to determine the effect of machining on residual stress are experimental, learn about the fundamental mechanism that generates residual stress when machining, and model the relationship between machining conditions, material properties with machining residual stress [5]. Investigations were presented in [2, 6, 7], the authors report the residual stress on the surface after machining for highly soft steel is primarily the tensile stress. Capello [8] analyzed the effect of feed pitch, tooltip radius, cutting depth and inlet angle on residual stress formation along the axial direction on different steels, concluding that feed and radius tooltips are the main parameters for controlling residual stress at turning. In this case, the cutting speed is not taken into account in the experiments. The effect of shear parameters in the principal stress direction is also not evaluated. The same stainless steels were analyzed by Outeiro et al. [9] but again, their assumption of the direction of stress in terms of the sheer speed and the vehicle's pitch as to the primary stress direction. It is not clearly demonstrated in the case of surface residual stress. On the other hand, they found that the external stress in the direction of the cutting rate as the cutting rate increases, indicating a trend opposite to the movement of other authors such as Jang et al. [6] stainless steel AISI 304 and Navas et al. in [10] in AISI 4340 steel, among others.

As mentioned above, not only the effects of machining parameters on residual stress have not been studied extensively, but the studies also give different results in previous works. It can say that the machining residual stress state of AISI 304 varies significantly under various machining conditions. Also, there are no comprehensive studies to evaluate the effect of this alloy's machining parameters on residual stress state.

Therefore, the purpose of this study is to evaluate the residual surface stress upon turning AISI 304.

2 Experimental Procedure

2.1 Machining Experimental

The experiment was conducted on MORI SEIKI-SL253 CNC lathe, the capacity of 4.5 KW, maximum spindle rotation number is 4500 rpm (Fig. 1).

Residual stresses were determined through the XRD method. X-ray powder diffraction (XRD) patterns were obtained on a Rigaku D/Max 2500/PC diffractometer with a Cu Kα radiation. Samples after machining on CNC lathes. After being processed, these samples will be analyzed for X-ray analysis using Anode radiation of Cu - Kal with a wavelength of l = 1.54051 Å. The obtained XRD diagrams will be calculated and considered to determine the residual stress. From XRD analysis schematic. Data were processed by MDI Jade 6.5 software to get angle 2, intensity Peak, distance d, and Width Scherrer (FWHM).

2.2 Material and Tools

Cutting tool Sanvik DCMT 11 T3 04 - MF 2220 coated with CVD Ti (C, N) + Al2O3 + TiN, workpiece AISI 304 size 50 × 200 mm (Fig. 2).

The material used in this experiment is a type of steel with better corrosion resistance and heat resistance than ordinary carbon steel, austenitic stainless steel AISI 304. This is a difficult material to process due to its low thermal conductivity, high mechanical and physical properties. The chemical compositions are listed in Table 1.

Table 1. Chemical composition of materials

AISI 304	C	Cr	Ni	Si	Mn	P	S
	0.07	18.49	8.15	0.57	0.76	0.03	0.009

a) CNC Lathe

b) Goniometer

Fig. 1. Machining experimental

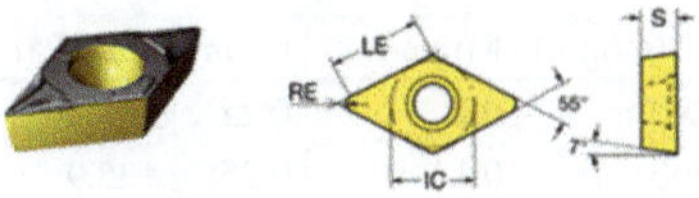

Fig. 2. Cutting tool

Table 2. Cutting ranges

Cutting parameters	Level		
	1	2	3
V (m/min)	230	260	290
f (mm/rev)	0.08	0.14	0.20
t (mm)	0.10	0.25	0.50

3 Results and Discussion

Experiments were performed according to the parameters given in Table 2, selected research with 3 factors, 3 levels and conducted according to the Box-Behnken method including 15 experiments, the least number of experiments to the second-order polynomial regression model.

X-ray analysis results of an AISI304 stainless steel sample processed by XRD MDI Jade 6.5 data processing software.

From XRD analysis schematic. Data were processed by MDI Jade 6.5 software to obtain parameters such as angle 2, intensity Peak, distance d, Width Scherrer (FWHM) (Fig. 3).

Calculation results of ressidual stress of AISI 304 stainless steel when turning on CNC Lathe are shown in Table 3.

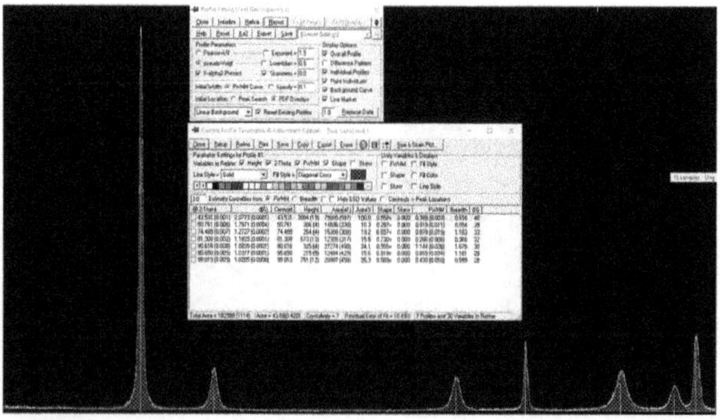

Fig. 3. X-ray analysis results of an AISI304 stainless steel sample processed by XRD MDI Jade 6.5 data processing software.

Table 3. Experimental design and cutting conditions

Run	V (m/min)	f (mm/rev)	t(mm)	σ (MPa)
1	290	0.2	0.25	201,6
2	260	0.14	0.25	125,9
3	260	0.14	0.25	125,9
4	230	0.2	0.5	240,8
5	230	0.14	0.1	136,3
6	260	0.08	0.5	143,1
7	260	0.2	0.1	233,3
8	260	0.14	0.25	125,9
9	260	0.08	0.1	131,7
10	230	0.2	0.25	204,5
11	290	0.14	0.1	172,5
12	290	0.08	0.25	166,7
13	230	0.14	0.5	226,5
14	230	0.08	0.25	143,2
15	290	0.14	0.5	148,3

The study used Minitab 18 software to analyze the cutting parameters to surface residual stress by the Box-Behnken method to evaluate the effect of cutting parameters on surface residual stress. The R-sq = 91,61% for the quadratic and power model is high enough to obtain reliable estimates. Analysis of the influence coefficients of the surface residual stress of workpieces with cutting parameters (V, f, t) is given in Eq. (1):

$$\sigma = 1557 - 11,97V - 664f + 1066t + 0,0260V^2 + 7515f^2 + 484t^2 \\ - 2,40V * f - 4,52V * t - 1176f * t \quad (1)$$

Results of ANOVA in Fig. 4 indicate that feed rate is the statistically significant turning process parameters that affect the response variables chosen for this experiments or overall performance characteristics.

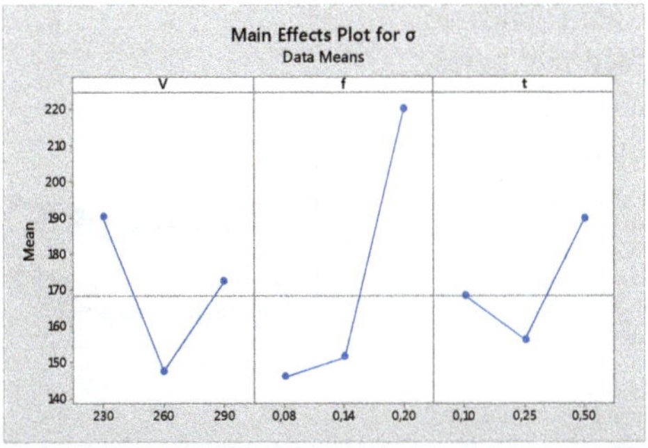

Fig. 4. Mean Plot effect for residual stress

Modeling of the influence of cutting parameters on surface residual stress is shown in Fig. 5, 6 and 7.

Increasing the feed rate and cutting depth in three cases brings about an increase in residual stresses.

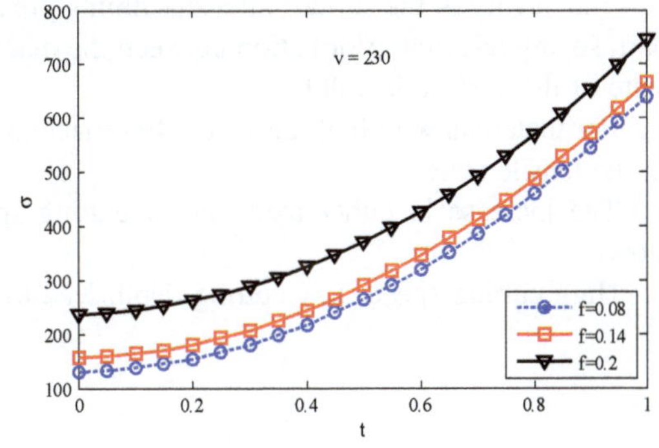

Fig. 5. Residual stress at v = 230 mm

Increasing the cutting speed has the same effect which is nevertheless inferior to that of feed rate.

Figures 5 and 6 show that if the depth of cut increases $t \geq 0.8$ mm it will give residual stress value σ 500 Mpa. According to the specification of AISI304 in [11], the material's tensile strength below 515 Mpa will result in surface damage.

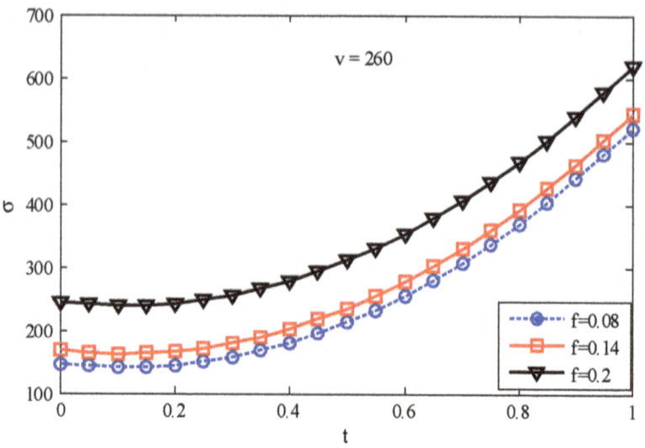

Fig. 6. Residual stress at = 260 mm

When the depth of cutting t is low $(0.1 \text{ mm} \leq t \leq 0.25 \text{ mm})$, the residual surface stress tends to decrease in all three feed rate cases, while when t is high from $t \geq 0.25$ mm, the residual increases with increasing feed rate.

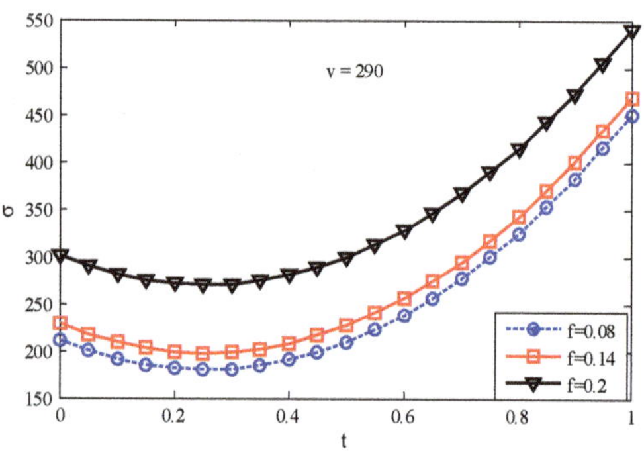

Fig. 7. Residual stress at v = 290 mm

4 Conclusion

This study experimentally assesses the effect of cutting parameters on surface residual stress when turning AISI 304. The study draws the following conclusions:

The quality of the surface after machining greatly influences the performance of the part, so any relevant information between the machining parameter and the mechanical state of the surface is valid.

For materials with high ductility after machining, the measured stresses are usually in the tensile state.

The increase in either feed rate or cutting speed affects the increase in residual stress.

High cutting speed, low cutting depth lead to decreased surface residual stress.

References

1. Wu, Q.R., Lu, J.P., Chen, X.P., Jiao, S.C.: The influence of cutting parameters on residual stress distribution during turning of 20Cr2Ni4 steel. In: IEEE International Conference on Industrial Engineering and Engineering Management, pp. 1574–1578, December 2016. https://doi.org/10.1109/IEEM.2016.7798142
2. Jang, D.Y., Liou, J., Cho, U.: Study of residual stress distribution in the machined stainless steel components. Tribol. Trans. **37**, 37–41 (1994). https://doi.org/10.1080/10402009408983334
3. Umbrello, D., Filice, L.: Improving surface integrity in orthogonal machining of hardened AISI 52100 steel by modeling white and dark layers formation. CIRP Ann. Manuf. Technol. **58**, 73–76 (2009). https://doi.org/10.1016/j.cirp.2009.03.106
4. Davim, J.P.: Machining Fundamentals and Recent Advances (2008)
5. Su, J.: Residual stress modeling in machining processes. Mechanical Engineering, Ph.D., p. 186 (2006)
6. Jang, D.Y., Watkins, T.R., Kozaczek, K.J., et al.: Surface residual stresses in machined austenitic stainless steel. Wear **194**, 168–173 (1996). https://doi.org/10.1016/0043-1648(95)06838-4
7. Matsumoto, Y., Barash, M.M., Liu, C.R.: Effect of hardness on the surface integrity of AISI 4340 steel. J. Manuf. Sci. Eng. Trans. ASME **108**, 169–175 (1986). https://doi.org/10.1115/1.3187060
8. Capello, E.: Residual stresses in turning: Part I: influence of process parameters. J. Mater. Process. Technol. **160**, 221–228 (2005). https://doi.org/10.1016/j.jmatprotec.2004.06.012
9. Outeiro, J.C., Dias, A.M., Lebrun, J.L., Astakhov, V.P.: Machining residual stresses in AISI 316L steel and their correlation with the cutting parameters. Mach. Sci. Technol. **6**, 251–270 (2002). https://doi.org/10.1081/MST-120005959
10. Navas, V.G., Gonzalo, O., Bengoetxea, I.: Effect of cutting parameters in the surface residual stresses generated by turning in AISI 4340 steel. Int. J. Mach. Tools Manuf. **61**, 48–57 (2012). https://doi.org/10.1016/j.ijmachtools.2012.05.008
11. Steels, A.: Stainless Steel Grade Datasheets (2013)

Evaluation of the Influence of Electrical Parameters and Powder Concentration to the Content of Tungsten on the Surface of SKD61 Steel in EDM Process

Van Tao Le[1(✉)] and Tien Long Banh[2]

[1] Advanced Technology Center,
Le Quy Don Technical University, Hanoi, Vietnam
taoitd@yahoo.com
[2] Hanoi University of Science and Technology, Hanoi, Vietnam

Abstract. In this study, the authors study the influence and evaluation of technological parameters, as follows: The peak current, the pulse time on and the powder concentration affects the content of tungsten on the surface of SKD61 steel after the Electrical Discharge Machining (EDM) process, which mixes the tungsten carbide powder (PMEDM) into the dielectric fluid. Using the least square method to build the relationship between the content of tungsten element and technological parameters, thereby assessing the impact of each technological parameter to see which technological parameter factors are the main influence on the content of tungsten element of surface. Through analyzing and assessing the influence of each technological parameter. At the small peak current, the short pulse on time and high powder concentration, the content of tungsten penetrating the surface is best.

Keywords: EDM · Surface modification · Tungsten

1 Introduction

Electrical discharge machining (EDM) is a non-traditional process and highly powerful for machining arduous materials such as SKD61 steel. The required characteristics of the material surface and material removal can be manipulated by appropriate parameters of EDM process. The added powder in the dielectric medium of EDM, called powder mixed electrical discharge machining (PMEDM), This method can improve the machined ability such as material removal rate (MRR), tool wear rate (TWR), and modification surface. In this research, the domination of tungsten carbide powder in the EDM process is investigated to explore the change in the surface of SKD61 steel. Recently, researchers have been interested in improved surface quality, some of the published works are mentioned as follows:

The SiC powders were studied by Santosh Kumar Sahu et al. [1] in the EDM process on the Inconel 718 Super Alloys. The authors pointed out that adding the powder significantly improved roughness of workpiece surface, the MRR, surface structure was improved better than conventional EDM. Micro-MoS2 powder combined

B. T. Long et al. (Eds.): MMMS 2020, LNME, pp. 184–188, 2021.
https://doi.org/10.1007/978-3-030-69610-8_24

with ultrasonic vibration has been experimented by Gunawan Setia Prihandana et al. [2]. Experimental results are reported to have significantly improved the MRR and improved the quality of surface after machining. Talla et al. [3] studied the effects of three powders including: aluminum powder (Al), graphite powder and Silicon powder (Si) in the EDM process on Inconel 625. The authors point to the effect of each type of powder to improve the quality of surface and the MRR. Graphite powder has the highest effect in MRR, silicon powder has the highest effect in micro hardness of surface. Marashi et al. [4] studied the effect of TiO2 powder in EDM on the surface properties of AISI D2 steel. As a result, the craters are shallower, the cracks are gradually reduced on the surface of workpieces. Amorim et al. [5] investigated surface modification after EDM machining with the addition of molybdenum powder on AISI H13. The research results of the report showed a great influence on the recast layer, the microhardness of surface, molybdenum powder size has a great influence on the recast layer, the micro hardness of surface. Within the framework of the paper, the authors investigated the penetration of tungsten content into the surface of SKD61 steel. The least square method was used to model the relationship between tungsten content of SKD61 steel with technological parameters, the model were used to evaluate the role of each technological parameters, that influences the penetration of tungsten element into the surface of SKD61 steel.

2 Materials and Methods

Steel SKD61 was used in the experiment (supplied by Daido corporation of Japan). The nominal chemical composition of steel SKD61 are 0.38C-1Si-5Cr-1.25Mo-1V-0.4Mn (wt.%). The dielectric medium is Shell Oil EDM Fluid 2. The tungsten carbide powder is used in this experiment by Praxair Surface Technologies with code WC-727-6. The chemical composition are 5.56C-11.9Co-0.02Fe-82.5W- 0.02 other components (wt.%) and the particle size of less than 31 μm.

The machine of electrical discharge is supplied by Aristech Corporation, with the commercial code is CNC-460 EDM, was used to remove the SKD61 workpiece. The copper material was the tool electrode, was negative. In this experiment, the tungsten carbide powder was mixed into the dielectric fluid.

The chemical composition of surface was measured by the EDX method with JSM6610LA-JEOL (JAPAN). The content of each element on surface is averaged among the three surface region on the surface of samples (Table 1).

Table 1. Experimental conditions

Condition	The current (I_p)	The pulse on (T_{on})	The pulse off (T_{off})	The current voltage (V)	Powder concentration (C_p-g/l)
Detail	1; 2; 3 A	16; 32; 50 μs	50 μs	80–120V	20; 40; 60

3 Results and Discussion

3.1 Establishing the Relationship Between Technological Parameters and Tungsten Content

Experimental effect of EDM technology parameters and powder concentration (Cp) on the content of tungsten element of surface SKD61 steel ($f_{w\%}$) were selected with 3 elements and 3 levels of (T_{on}, I_p, C_p). The levels and factors constitute 11 experiments.

Base on the experimental data was obtained and the experimental data points are represented by the graphical method, the authors proposed the regression function in the form of an exponential function:

$$f_{W\%} = K.I_p^a \cdot T_{on}^b N_{bt}^c \ (\%) \tag{1}$$

Using the least square method to construct regression functions in the experimental planning calculations, the system of equation is solved by Maple software, the empirical regression function of the model is:

$$f_{W\%} = 755.37 \cdot I_p^{-0.8818} \cdot T_{on}^{-1.9325} \cdot N_{bt}^{0.6632} \ (\%) \tag{2}$$

With the reliability of empirical regression function: r = 0.922.

3.2 Analysis of the Influence of the Technological Parameters Affecting the Content of Tungsten Element ($f_{w\%}$)

According to function (2), it find that the role of each technological parameter affecting the content of tungsten element is very different, as follows:

* T_{on}: According to function (2) the pulse on time have a negative exponent and the maximum absolute value is 1.9325, so T_{on} have the strongest impact of the three input parameters to affect the content of tungsten element on the surface of workpiecess. The pulse on time increases resulting in the reduction of the content of tungsten element on the surface of workpieces. According to the graphs of Fig. 2, 3, when the pulse on time increases, the content of tungsten element in the surface decreases. So, in theory [6, 7] this is entirely appropriate.

*I_p: According to function (2) the peak current have a negative exponent and the second largest absolute value are 0.8818, so the peak current have the second strongest impact on the three input parameters to affect the content of tungsten element. The peak current increases resulting in the reduction of the content of tungsten element on the surface of workpieces, as shown in Fig. 1, 3.

* C_p: According to function (2) the powder concentration have a positive exponent and the absolute value are 0.6632, so C_p has the weakest impact on the three input parameters to the surface roughness. The C_p increases resulting in the increasing the content of tungsten element on surface of SKD61 steel, as shown in Fig. 2, 3.

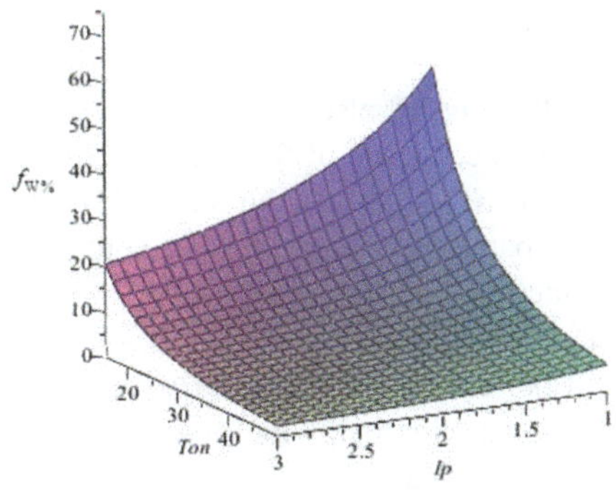

Fig. 1. The effect of pulse on time (T_{on}) and peak current (I_p) to the content of tungsten element

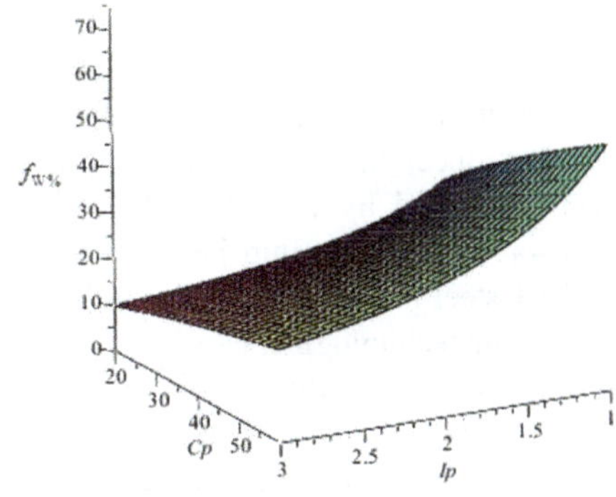

Fig. 2. The effect of peak current (I_p) and powder concentration (C_p) to the content of tungsten element

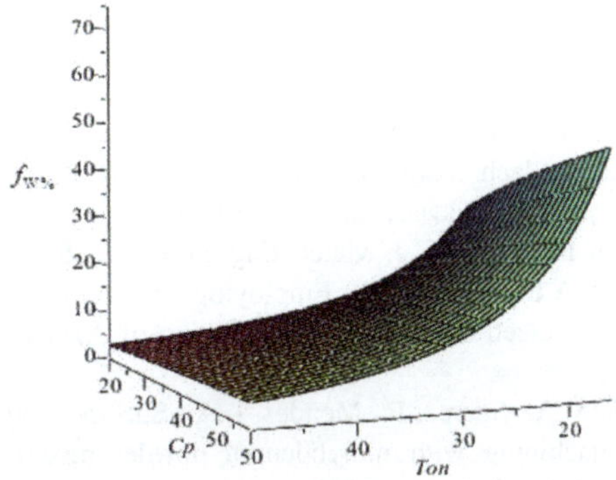

Fig. 3. The effect of pulse on time (T_{on}) and powder concentration (C_p) to the content of tungsten element

Through analyzing and assessing the influence of each technological parameter by function (2) and graphs in Fig. 1, 2, 3, they show the role of each technological parameter to the tungsten content entering the surface through the PMEDM process. At the parameters as: the small peak current, the short pulse on time and all concentrations, the high content of tungsten has been penetrated into the surface of SKD61 steel. Otherwise, at the parameters as: the large peak current, the long pulse on time and all concentrations, the low content of tungsten has been penetrated into the surface of SKD61 steel. Especially at the small peak current, the short pulse on time and high powder concentration, the content of tungsten penetrating the surface is best. From there, it serves as a basis for controlling technological parameters for improving surface quality, which requires a lot of tungsten content on the surface.

4 Conclusions

In this current study, through the evaluation effecting of electrical parameters and powder concentration to the content of tungsten on the surface of SKD61, which have been machined by the PMEDM method. The main results can be summarized as follows: The relationship between technological parameters and tungsten content of SKD61 steel has been built, according to the function (2). The strongest and weakest among the technological parameters have been identified. Specifically, T_{on}, I_p, C_p have an impact in descending to the tungsten content present on the surface. By evaluating and analyzing the role of each technology parameter, this helps the process of controlling tungsten content penetration the surface more clearly.

References

1. Kumar Sahu, S., Jadam, T., Datta, S., Dhupal, D., Nandi, G.: Application of SiC power added in kerosene dielectric media for electro-discharge machining of Inconel 718 super alloys: effect of powder concentration. Mater. Today Proc. **5**, 20297–20305 (2018)
2. Prihandana, G.S., Mahardika, M., Hamdi, M., Wong, Y.S., Mitsui, K.: Effect of micro-powder suspension and ultrasonic vibration of dielectric fluid in micro-EDM processes — Taguchi approach. Int. J. Mach. Tools Manuf. **49**, 1035–1041 (2009)
3. Talla, G., Gangopadhyay, S., Biswas, C.K.: Effect of powder-suspended dielectric on the EDM characteristics of Inconel 625. J. Mater. Eng. Perform. **25**, 704–717 (2016)
4. Marashi, H., Sarhan, A.A.D., Hamdi, M.: Employing Ti nano-powder dielectric to enhance surface characteristics in electrical discharge machining of AISI D2 steel. Appl. Surf. Sci. **357**, 892–907 (2015)
5. Amorim, F.L., Dalcin, V.A., Soares, P., Mendes, L.A.: Surface modification of tool steel by electrical discharge machining with molybdenum powder mixed in dielectric fluid. Int. J. Adv. Manuf. Technol. **91**, 341–350 (2017)
6. Schumacher, B.M.: About the role of debris in the gap during electrical discharge machining. CIRP Ann. **39**, 197–199 (1990)
7. Erden, A., Bilgin, S.: Role of impurities in electric discharge machining. In: Proceedings of 21st International Machine Tool Design and Research Conference, pp. 345–350. Palgrave, London (1980)

Improving Mechanical Properties of Concrete Utilizing Crushed-Sand Combined with River Sand

Duc-Trong Nguyen[1] and Duy-Liem Nguyen[2(✉)]

[1] Campus in Ho Chi Minh City, University of Transport and Communications, Ho Chi Minh City, Vietnam
trongnd_ph@utc.edu.vn
[2] Faculty of Civil Engineering, Ho Chi Minh City University of Technology and Education, Ho Chi Minh City, Vietnam
liemnd@hcmute.edu.vn

Abstract. The extreme exploitation of river sand in Mekong Delta region has seriously caused landslides sweeping away a large area, even affecting traffic movement on national highway. Thus, utilizing crushed-sand, a co-product in aggregate making, to partially replace river sand is really useful and urgent. In this investigation, the effect of crushed-sand fraction, in hybrid sand systems of river sand blended with crushed-sand, on compressive and flexural strength of concrete was experimentally studied. The percentage ratio of river sand/crushed-sand were used at 100/0 (M100), 60/40 (M060), 50/50 (M050), 40/60 (M040), 30/70 (M030) and 0/100 (M000) by weight. The compressive specimen was cubic-shaped with dimension of 100 × 100 × 100 mm while the flexural specimen was 100 × 100 × 400 mm subjecting to four-point bending loading. The testing results revealed that the maximum flexural strength and compressive strength were produced from the M040 and M030, respectively. The effectiveness of the hybrid sands were also evaluated.

Keywords: Crushed-sand · Artificial sand · River sand · Environmental impact · Landslide

1 Introduction

Concrete is well-known to be the most popular construction material for civil infrastructure. Recently, the quick economic development of Vietnam requires an increasing demand of sand to make concrete for a large number of constructions throughout the country. However, the extreme exploitation of river sand in Mekong Delta region has caused landslide phenomenon at river bank more and more seriously. Some important highways near the rivers also were destroyed by the landslide [1]. To minimize the landslide situation in the Mekong Delta, finding an artificial sand to replace the river sand has gotten essential. Crushed-sand is a non-natural sand made from waste production in manufacturing process of coarse aggregate. Some references of crushed-sand for making concrete in Southern Vietnam are available [2, 3]. In these references, the crushed-sand, produced in An Giang, a province of Mekong Delta region, was entirely

B. T. Long et al. (Eds.): MMMS 2020, LNME, pp. 189–195, 2021.
https://doi.org/10.1007/978-3-030-69610-8_25

replaced the river sand in making concrete, the such crushed-sand was evaluated to be practicable in both cost and technical conditions. In this study, the authors tried to blend the river sand with the crushed-sand, it is highly expected that the hybrid sand will produce a favorable effect in improving mechanical properties of concrete. The main target of this study is to explore the favorable non-natural sand in making concrete while the study objective is to explore the best fraction amount of crushed-sand in hybrid sand system to produce the highest compressive strength and flexural strength.

2 Experimental Test

2.1 Materials and Preparation of Specimens

There were six types of the hybrid sands including river sand and crushed-sand blended. The percentage ratio of river sand/crushed-sand were used at 100/0 (M100), 60/40 (M060), 50/50 (M050), 40/60 (M040), 30/70 (M030) and 0/100 (M000) by weight. The river sand was from Tan Chau District, An Giang Province while the crushed-sand was produced from Hoa An Company, Dong Nai Province. Table 1 provides the particle size fractions of the river sand and crushed sand while Table 2 provides the properties of hybrid sands. As shown in Table 2, the weight total of river sand and crushed-sand was kept constant. Figure 1 shows the photos of river sand and crushed-sand used. The compositions of the studied concretes were supplied in Table 3. The compressive specimen was cubic-shaped with size of 100 × 100 × 100 mm while the flexural specimen was a prism with size of 100 × 100 × 300 mm (width × depth × span-length), its full length of 400 mm, subjecting to four-point bending loading. The water for curing specimen was controlled at 28–32 °C temperature. After that the specimens were removed from the curing water and dried at room temperature before testing.

Table 1. Particle size fractions of the river sand and crushed sand

Sieve size (mm)	Cumulative percent retained on sieve (%)	
	River sand (*)	Crushed-sand (+)
7.50	0.00	0.00
5.00	0.00	8.96
2.50	0.35	39.14
1.25	1.06	64.64
0.63	3.01	80.95
0.315	53.32	91.62
0.14	89.90	95.51
<0.14	100.00	100.00

(*) River sand from Tan Chau District, An Gaing Province
(+) Crused-sand from Hoa An, Dong Nai Province

Table 2. Properties of hybrid sands

Name of concrete	Density	Fineness modulus	Cumulative percent retained on sieve of 0.14 mm
	(g/cm^3)	Mdl	(%)
M100	1.665	3.72	4.49
M060	1.599	2.36	7.86
M050	1.614	2.61	7.35
M040	1.655	2.81	6.74
M030	1.660	3.04	6.15
M000	1.390	1.48	10.1

(a) (b)

Fig. 1. The hybrid sand system used in the study: (a) River sand, (b) Crushed-sand

Table 3. Compositions of the studied concretes

Name of concrete	Cement (*) (kG)	Aggregate size 1x2 (kG)	River sand (+) (kG)	Crushed-sand (kG)	Water (kG)	Total weight (kG)
M100	468	1058	734	0	203	2463
M060	468	1058	440	294	203	2463
M050	468	1058	367	367	203	2463
M040	468	1058	294	440	203	2463
M030	468	1058	220	514	203	2463
M000	468	1058	0	734	203	2463

(*) Cement PCB40 of INSEE

2.2 Experiment Setup

A testing machine with displacement control was used to test all specimens. The applied displacement speed during the test was constantly 1 mm/min. Each tested series was examined using three specimens then averaged their peak loads. The compressive strength ($f_c' = P_{\max}/A$) and the flexural strength ($f_r = (P_{\max}S)/(bh^2)$) were computed using the peak load applied ($P_{\max}$). In these equations, A is the sectional area of the compressive specimen, S is the span length of the flexural specimen, b is the width and h is the depth of rectangular section of the flexural specimen.

3 Experiment Results and Discussions

3.1 Comparative Mechanical Properties of the Studied Concretes

There were three parameters in the comparison of the studied concretes including slump, compressive strength, flexural strength, and, Table 4 supplies their average values. The compressive strength was explored at 3-day, 7-day and 28-day age while the flexural strength was only measured at 28-day age. The M040 produced the highest flexural strength of 5.26 MPa, while the M030 produced the highest compressive strength of 51.65 MPa at 28-day age. In addition, the slump measured was observed to be agreed with the flexural strength in the trend, as shown in Table 4, the highest slump of 4.3 cm was also produced from M040 (crushed-sand/total of sand of 0.6). Figure 2 provides the SEM (Scanning Electron Microscope) images of the surfaces of the river sand and crushed-sand particle used. As shown in Fig. 2, the surface of crushed-sand partical was rougher than that of river sand, this helped increase the friction between the embedded particles in the concrete mixture, and thus resulted the better performances in strength [4, 5].

Table 4. Mechanical properties of the studied concretes

Concrete type	Slump, S_n (cm)	Compressive strength, f_c' (MPa)			Flexural strength at 28 days, f_r (MPa)
		3 days	7 days	28 days	
M100	2.7	30.22	35.05	46.04	4.50
M060	3.5	32.09	37.47	48.24	4.90
M050	4.0	32.64	38.13	49.45	5.06
M040	4.3	33.63	39.56	51.43	5.26
M030	4.0	33.52	39.34	51.65	5.24
M000	3.3	33.19	38.90	51.32	5.14

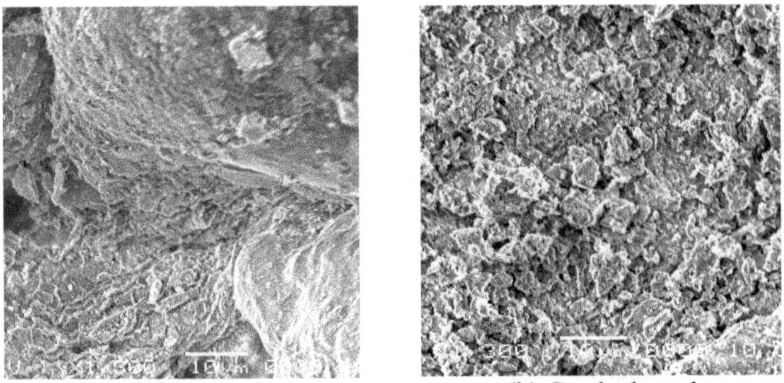

(a) River sand (b) Crushed-sand

Fig. 2. SEM images of the river sand and crushed-sand particle

3.2 Effective Amount of Crushed-Sand Used in the Hybrid Sand

Although the Table 4 reveals the best amount of crushed-sand was 60–70%, the changes of both compressive strength and flexural strength were not much as the ratio crushed-sand/total of sand was beyond 60%. In the detail, as the crushed-sand increased from 60% to 100% (increased 1.667 times), the flexural strength decreased from 5.26 MPa to 5.14 MPa (decreased 1.023 times), the compressive strength decreased from 51.43 MPa to 51.32 MPa (decreased 1.002 times). This means the favorable content of crushed-sand was from 60–100% in the hybrid sand system, and the crushed-sand could be used to entirely replace the river sand. As the crushed-sand entirely replacing the river sand (100% crushed-sand), the improvements were observed as follows: 1.114 times for compressive strength and 1.142 times for flexural strength, in comparison between M000 and M100 matrix. Furthermore, using one sand type was simpler, the more utilization of crushed-sand led to the less harmful environmental impact.

3.3 Testing Age-Dependent Compressive Strength of the Concretes

The mechanical strengths of concrete have been reported to increase with increasing of testing age since the cement hydration gradually occurs for a long time [6, 7]. For estimating the $f_c^{'}$ according to time (t), a relationship between them would be mathematically analysed. A hyperbolic profile with equation $f_c^{'}(t) = t/(at + b)$ was proposed in order to fit the experimental curves. The hyperbolic equation was wholly satiated with the boundary requirement: the compressive strength was equal to zero at the testing age of zero, and it approached a constant value when the testing age was infinity. The constants a and b would be achieved with the least square method and linear regression. Figure 3 displays the regression analysis of relationship between compressive strength and testing age.

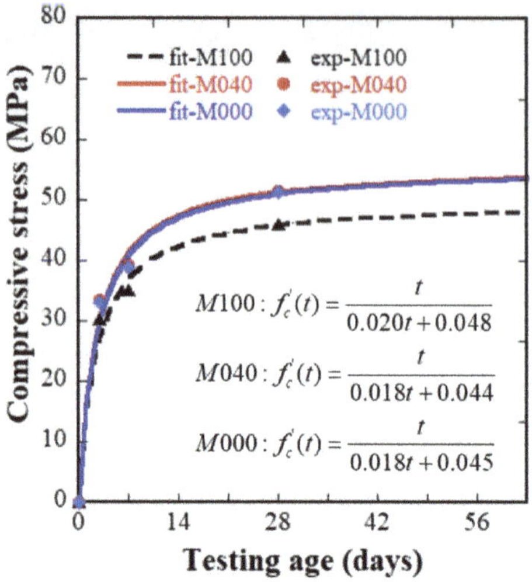

Fig. 3. Regression analysis for compressive strength versus testing age

3.4 Correlation Between Flexural and Compressive Strength of the Concretes

Table 5. Derivation of the scale coefficient in relationship between flexural and compressive strength of the studied concretes

Concrete type	Compressive strength using cube specimen, f_c' (MPa)		Flexural strength of using prism specimen, f_r (MPa)		Scale coefficient, ψ
	100 mm	150 mm	100 mm	150 mm	
M100	46.04	41.90	4.50	4.29	0.66
M060	48.24	43.90	4.90	4.67	0.70
M050	49.45	45.00	5.06	4.82	0.72
M040	51.43	46.80	5.26	5.01	0.73
M030	51.65	47.00	5.24	4.99	0.73
M000	51.32	46.70	5.14	4.90	0.72

The flexural strength (f_r) of a normal concrete is often correlated with the compressive strength (f_c') under the square root scale regarding ACI 318 [6], given by equation $f_r = \psi\sqrt{f_c'}$. The scale coefficients (ψ) for normal concrete is 0.62. The standard specimens for scale coefficient were used as follows: using cylindrical specimen with a diameter × height of 150 × 300 mm for compressive strength, using prism specimen with a width × depth × span-length of 150 × 150 × 450 mm (full length of 600 mm) for flexural strength. For the studied concretes in this research, the f_r and f_c' were firstly multiplied with 0.91 and 0.95 [7], respectively, for conversing to standard specimens due to size effect on strength of concrete [8], as presented in Table 5. The ψ value of the studied concretes would be experimentally determined through experimental results. Table 5 also supplies the derivation of the scale coefficient in relationship between flexural and compressive strength of the studied concretes. All ψ values were ranged from 0.66 to 0.73 and their average value of six concrete types was 0.71, which was higher than that of normal concrete 1.146 times (0.71/0.62).

4 Conclusion

The summary of the testing research can be listed as follows:

1) Crushed-sand is possibly combined with river sand in making concrete for buildings and other civil infrastructures. The much reduction of river sand and the increase of crushed-sand utilized could bring greatly meaningful effect on nature.
2) Crushed-sand amount for the best performance in compressive strength and flexural stregnth were 70% and 60% of the total of sand.
3) Crushed-sand could be used to entirely replace the river sand with the improvements of 1.114 times for compressive strength and 1.142 times for flexural strength.

References

1. Information on https://vietnamnet.vn/en/sci-tech-environment/experts-worry-about-the-fate-of-mekong-delta-566111.html
2. Nguyen, D.-L., Duong, M.-T.: Compressive resistance of environmental concrete using fly ash and fine aggregate for replacing traditional sand. Appl. Mech. Mater. **889**, 283–288 (2019). © 2019 Trans Tech Publications, Switzerland. https://doi.org/10.4028/www.scientific.net/AMM.889.283
3. Nguyen, D.-L., Nguyen, V.-T., Tran, M.-P., Duong, M.-T.: Comparative performances of reinforced beams using concrete made from crushed sand and fly ash. In: Reddy, J.N., et al. (eds.) Lecture Notes in Civil Engineering, vol. 80. ICSCEA 2019, pp. 437–446. https://doi.org/10.1007/978-981-15-5144-4_39. ISBN 978-981-15-5143-7
4. Hong, L., Gu, X., Lin, F.: Influence of aggregate surface roughness on mechanical properties of interface and concrete. Constr. Build. Mater. **65**, 338–349 (2014). https://doi.org/10.1016/j.conbuildmat.2014.04.131
5. Rao, G.A., Prasad, B.K.R.: Influence of type of aggregate and surface roughness on the interface fracture properties. Mater. Struct. **37**(5), 328–334 (2004)
6. ACI 318-14, Building code requirements for structural concrete (2014)
7. TCVN 8218:2009, Hydraulic concrete - Technical requirements. Vietnamese code (2009)
8. Bažant, Z.P., Planas, J.: Fracture and Size Effect on Concrete and Other Quasibrittle Materials. CRC Press LLC, United States (1997)

An Investigation on the Impact of the Interpass-Cooling Time on the Metallurgy of Wire-Arc-Additive-Manufacturing SS308L Components

Van Thao Le(✉) and Van Tao Le

Advanced Technology Center, Le Quy Don Technical University, Hanoi, Vietnam

Abstract. In this paper, we focused on observing the influence of the interpass-cooling time on the microstructure of stainless steel (SS308L) walls fabricated by the wire and arc additive manufacturing (WAAM) process. For this purpose, two SS308L walls were produced by WAAM according to the unidirectional depositing strategy with an interpass-cooling time of 30 (s) and 60 (s) between two consecutively deposited layers. The findings indicate that in terms of microstructures there is no significant difference between two as-built thin walls. The microstructure of WAAM SS308L walls contains dominant austenite phases and a small amount of ferrite, which exists in the boundaries of austenite grains. However, when the interlayer idle time increases the secondary dendrite arm spacing (SDAS) becomes narrow, resulting in an augmentation in the microhardness and tensile properties of as-built components.

Keywords: Wire-arc-additive-manufacturing (WAAM) · Interpass-cooling time · 308L steel · Metallurgy

1 Introduction

WAAM - wire arc additive manufacturing is known as a metal-based additive manufacturing (AM) technology, in which the arc is used as the energy source to melt metal wire and fabricates metal parts layer-by-layer. In comparison to other metal AM technologies, this technique features high rate of production and low costs of device investment [1]. Therefore, WAAM is considered as an excellent-technological option to build large-scale components. However, the main drawback of such a technique is that the heat accumulation in WAAM parts is very high. This is due to the high heat input generated by the arc source and reheating cycles during the WAAM process [2]. Therefore, high thermal distortion and stress is one of the most crucial issues related to WAAM as-built components. To reduce the heat accumulation, the interpass-cooling time between successive deposits is generally used [3–5]. The interpass-cooling time enables cooling down the workpiece before the deposit of the next layer.

In the current paper, the impact of the interpass-cooling time on metallurgical characteristics of stainless steel (SS308L) walls built by WAAM was explored. From

B. T. Long et al. (Eds.): MMMS 2020, LNME, pp. 196–201, 2021.
https://doi.org/10.1007/978-3-030-69610-8_26

the findings obtained in this study, we can select an effective interpass-cooling time for the production of WAAM SS308L components.

2 Experiments

In the current work, two stainless steel (SS308L) walls had been manufactured from the commercial SS308L wire by using a robotic welding system (Fig. 1a) according to the unidirectional depositing direction strategy (Fig. 1b). The wire has a diameter of 1 mm, and its chemical composition contains (19.5–21)% Cr and (9.0–11.0)% Ni. The percentage of C, P, and S is inferior to 0.03%. The processing parameters employed to produce the walls were presented in Table 1 [6].

Table 1. Processing conditions employed in experiments.

Process parameter	Value and unit
Welding current, I	122 (A)
Voltage, U	20 (V)
Traveling speed, v	0.368 (m/min)
Flow rate of the shielding gas	15 (L/min)
Interpass-cooling time	30 (s) for the first wall, 60 (s) for the second wall

The length of each deposit is 120 mm and each wall consists of 18 deposited layers. During the depositing process, a gas (99.9% of argon) was employed to protect the molten pool against the ambiance. The flow rate of the shielding gas is 15 l/min. In addition, an interpass-cooling time (T_{id}) of 30 s and 60 s between two successive deposits was applied during the build of the first and the second wall, respectively, for the free cooling purpose.

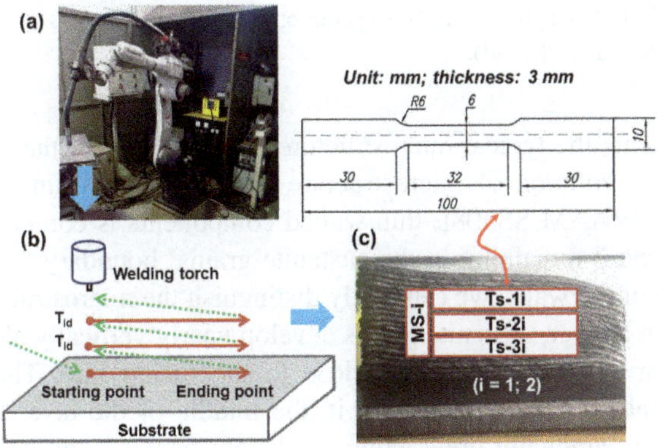

Fig. 1. The welding robot (a), the unidirectional depositing strategy (b), and (c) the built wall with the locations to extract metallurgical and tensile samples.

To compare the impact of different interpass-cooling times on the metallurgical properties, the material of as-built walls in the middle region was taken into consideration. In each case, a metallurgical sample (MS-i) and 03 tensile samples (Ts-1i, Ts-2i, and Ts-3i) in the horizontal direction had been prepared (Fig. 1c), where i = {1; 2} for the first wall and the second wall, respectively.

To observe the microstructure of the walls, an AXIO imager A2M was used. The micro-hardness was tested by a Vickers micro hardness tester (Duramin-2 of Struers). Moreover, a tensile tester (Instron 3T, Model 5967B12620) was employed to implement all tensile tests. The tests of tensile strengths had been conducted at ambient temperature. The traveling speed of the crosshead was set by 10 mm/min. In order to evaluate the defects of components, all metallurgical samples (MS-i) and tensile samples were tested through the X-ray CT tests.

3 Findings and Discussion

The X-ray CT photographs of MS and tensile samples are shown in Fig. 2, in which defects generally appear in black color, whereas the parts appear in bright color. It is found that there are no black zones observed in the microstructure testing samples and the tensile samples. Therefore, we can consider that the walls were successfully produced by WAAM without major defects such as macro porosity and cracks.

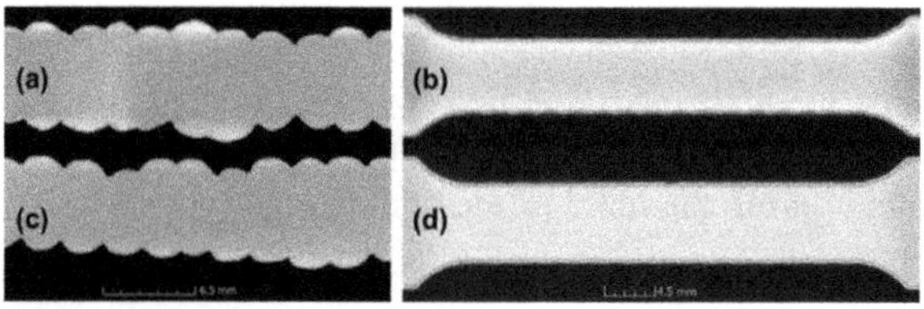

Fig. 2. X-Ray CT photographs of the MS specimen and tensile samples of the first wall (a–b) and those of the second wall (c–d).

Figure 3 presents the typical microstructure in the middle of the walls. It is firstly observed that the evolution of microstructure in two walls is almost identical. The microstructure of WAAM SS308L thin-walled components is composed of austenitic phases γ and ferrite δ remaining in the austenite grains' boundary.

In the middle of the walls, we can easily distinguish the microstructure evolution of each layer. Within a layer, austenite grains develop nearly vertically along the building direction. Cellular structures appeared close to the fusion line. They subsequently developed into finer columnar structures in the middle of the layer (Fig. 3a–c). The columnar structures gradually grew along the height of the layer. They finally transformed into coarser granular grains in the layer's upper part (Fig. 3b). The secondary-dendrite arms are also clearly observed (Fig. 3a).

To analyze the impact of interpass-cooling times on the microstructure of the as-built walls, we measured the secondary dendrite arms' spacing (SDAS) at the same

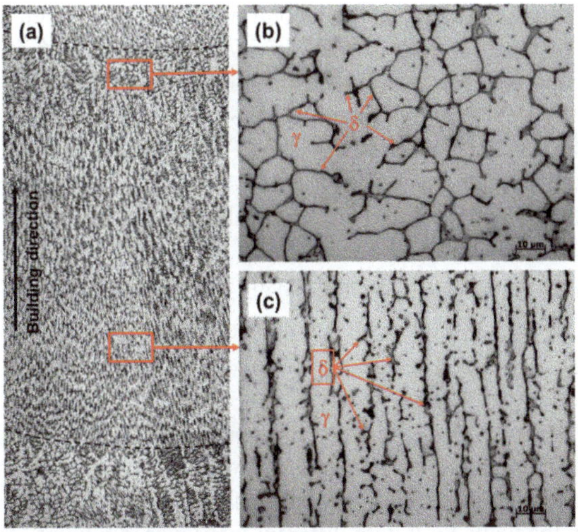

Fig. 3. Overall microstructures in the middle region (a) at low magnification (x100) of the as-built walls, and microstructure in the upper portion (b) and in other portions of a layer (c) in the middle region of the walls at high magnification (x500).

layer by using ImageJ software. In both two cases, it is found that the average SDAS augments in the height direction of the layer. In the case of the first wall (T_{id} = 30s), the average SDAS is 8.20 ± 0.49, 10.52 ± 0.33, and 12.64 ± 0.86 μm in the bottom, the middle, and the top of the layer, respectively. Moreover, the increase in the interpass-cooling time from T_{id} = 30 s to T_{id} = 60 s results in a decrease in SDAS. For instance, in the middle of the layer, the average SDAS is 10.52 ± 0.33 μm and 9.62 ± 0.54 μm for the first and the second wall, respectively. Indeed, due to the interlayer idle time increases, the interlayer temperature becomes lower, and the cooling rate increases [5, 7, 8]. As a result, the SDAS increases [7, 9].

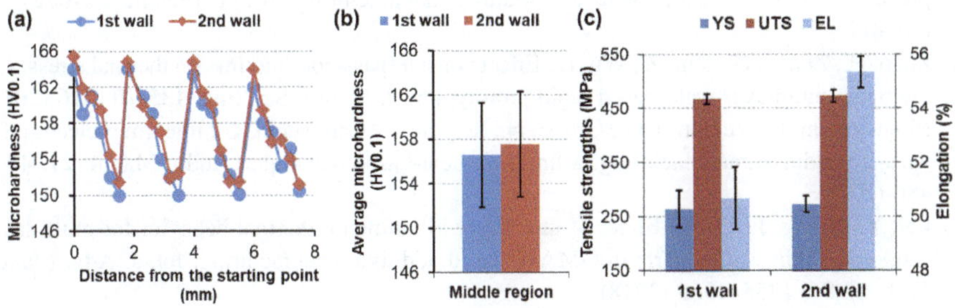

Fig. 4. Microhardness in the middle portion (a, b), and the tensile property (c) of the WAAM as-built material.

Figure 4a shows the microhardness distribution in four adjacent layers of the walls in the middle, and a comparison on the average hardness values between the walls was described in Fig. 4b. It is found that the second wall features an average hardness that is higher than that of the first wall (157.62 ± 4.78 HV0.1 vs. 156.58 ± 4.71 HV0.1). Similarly, the second wall features the tensile strength that is superior to that of the first wall (Fig. 4c). The YS (yield strength), UTS (ultimate tensile strength), and EL (elongation) of the second wall are 275.67 ± 15.53 MPa, 472.33 ± 11.06 MPa, and 54 ± 1%, respectively, whereas YS = 263.67 ± 33.65 MPa, UTS = 467.67 ± 9.29 MPa, and EL = 50.67 ± 1.15% for the first wall. The reason is that the SDAS of the first wall was greater than that of the second one. Therefore, in agreement with the Hall-Petch relationship [10], the hardness and tensile strength of the second wall are higher.

4 Conclusion

In this article, the impact of the interpass-cooling time on material properties of stainless steel 308L manufacture by WAAM was addressed. The obtained findings indicated that no remarkable difference in terms of the microstructure evolution was observed between the walls built with different interlayer idle times. The microstructure of WAAM thin-walled SS308L components is mainly composed of austenite phases and residual ferrite in the austenite boundaries. The microhardness and tensile strength of as-built thin-walled components can be improved by increasing the interpass-cooling time between successive deposited layers. However, it is noted that the increase in the interpass-cooling time causes a loss in productivity.

Acknowledgements. This research is funded by Vietnam National Foundation for Science and Technology Development (NAFOSTED) under grant number 107.99-2019.18.

References

1. Wu, B., Pan, Z., Ding, D., et al.: A review of the wire arc additive manufacturing of metals: properties, defects and quality improvement. J. Manuf. Process. **35**, 127–139 (2018)
2. Cunningham, C.R., Flynn, J.M., Shokrani, A., et al.: Invited review article: strategies and processes for high quality wire arc additive manufacturing. Addit. Manuf. **22**, 672–686 (2018)
3. Zhao, H., Zhang, G., Yin, Z., Wu, L.: Effects of Interpass-cooling time on thermal stresses in multipass multilayer weld-based rapid prototyping. J. Manuf. Sci. Eng. **135**, 011016 (2013)
4. Montevecchi, F., Venturini, G., Grossi, N., Scippa, A., Campatelli, G.: Idle time selection for wire-arc additive manufacturing: a finite element-based technique. Addit. Manuf. **21**, 479–486 (2018)
5. Lei, Y., Xiong, J., Li, R.: Effect of inter layer idle time on thermal behavior for multi-layer single-pass thin-walled parts in GMAW-based additive manufacturing. Int. J. Adv. Manuf. Technol. **96**, 1355–1365 (2018)
6. Le, V.T., Mai, D.S., Doan, T.K., Hoang, Q.H.: Prediction of welding bead geometry for wire arc additive manufacturing of SS308L walls using response surface methodology. Transp. Commun. Sci. J. **71**, 431–443 (2020)

7. Wang, L., Xue, J., Wang, Q.: Correlation between arc mode, microstructure, and mechanical properties during wire arc additive manufacturing of 316L stainless steel. Mater. Sci. Eng. A **751**, 183–190 (2019)
8. Yang, D., Wang, G., Zhang, G.: Thermal analysis for single-pass multi-layer GMAW based additive manufacturing using infrared thermography. J. Mater. Process. Technol. **244**, 215–224 (2017)
9. Yin, H., Felicelli, S.D.: Dendrite growth simulation during solidification in the LENS process. Acta Mater. **58**, 1455–1465 (2010)
10. Singh, K.K., Sangal, S., Murty, G.S.: Hall-Petch behaviour of 316L austenitic stainless steel at room temperature. Mater. Sci. Technol. **18**, 165–172 (2002)

A Brief Review of Renewable Energy: Sustainable Energy Production, Integration, and Application

V. L. Trinh[1(✉)], C. K. Chung[2(✉)], X. C. Nguyen[3], and T. S. Nguyen[1]

[1] Manufacturing Technology Division, Faculty of Mechanical Engineering, Hanoi University of Industry, Hanoi, Vietnam
longtv@haui.edu.vn
[2] Department of Mechanical Engineering and Center for Micro/Nano Science and Technology, National Cheng Kung University, Tainan 701, Taiwan
ckchung@mail.ncku.edu.tw
[3] Vietnam-Japan Centre, Hanoi University of Industry, Hanoi, Vietnam

Abstract. Energy is really important in the human social development. The energy consumption need significantly rises in over the world. Conventional energy comes from fossil fuels or related-fossil material. These methods created a serious environmental pollution and ambient destruction by its gas emissions and harmful pollutants. Renewable energy is produced from renewable energy sources. The sustainable energy shows good characteristics of non-pollutant, non-gas emissions. This paper reviews current methods of energy conversion from renewable energy sources to useful energy. Current methods and technologies have been used to production, integration, and application of sustainable energy.

Keywords: Energy · Renewable energy · Sustainable energy · Fuel · Integration

1 Introduction

Energy demand is extremely huge in over the world. Conventionally, energy is created from fossil fuel resources or related-fossil materials by burning methods or power plants. These methods remained a serious environmental pollution and ambient destruction by its gas emissions, and toxic pollutants [1]. Fuel burning processes have been wasted lots of pollutants and greenhouse gases like: methane gas, nitrous oxide, sulphur dioxide, carbon dioxide, nitrogen oxides, black carbon, and poly-aromatic hydrocarbons. The carbon dioxide emissions rose 1.9% in 2018 as the highest rising rate from 2013 [2]. Conventional energy resources are become dwindled by rising of energy need in over the world. The BP statistics give evidence of the worldwide primary energy consumption increased up to 2.9% in 2018 [3]. Traditional energy production methods are reason of the climate change and global warming problems. Renewable energy is produced from renewable energy resources such as: wind, sunlight, rain, geothermal heat, bioenergy, tides, or waves [4]. Renewable energy is not created from fossil resources, or inorganic waste things. Figure 1 shows renewable

B. T. Long et al. (Eds.): MMMS 2020, LNME, pp. 202–207, 2021.
https://doi.org/10.1007/978-3-030-69610-8_27

energy resources [4]. These energy resources were changed to useful energies such as: electricity, transportation, and heat [5]. Technologies uses to change renewable energy resources to useful energy like: solar photovoltaic, hydro, wind, biofuels. Renewable energy contributed 23% of global electrical energy with 17% of variable renewables and 70% by hydropower, in 2014 [2].

Renewable energy has been discovered and integrated in power grids with good characteristics such as: observability, controllability, self-diagnosis, and healing [6]. Nations have valued 20% electrical energy from renewable energy resources in 2019 and estimated about 100% renewable energy in the future for some areas [7]. The paper reviews renewable energy conversion methods, integration, and applications of sustainable energy.

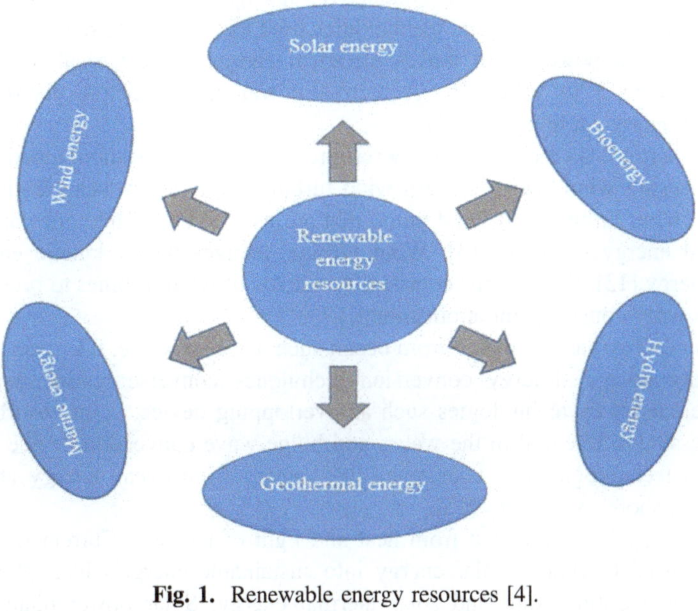

Fig. 1. Renewable energy resources [4].

2 Sustainable Energy Production

Sustainable energy production is the top priority in the modern social because of its advantages of non-climate change, non-gas emissions, and non-pollutants. Sustainable energy development will be contributed to replace traditional energy with lots of disadvantages of reasons of climate change, gas emissions, and industrial pollutants. Many technologies have been applied to produce sustainable energy or renewable energy from renewable resources such as solar, hydro, wind, geothermal, maritime, or biomass energy.

Biomass is a renewable energy resource come from animal material or plant to make heat, electrical, and biofuel energy [4]. Bio-refinery and biotransformation technology have been developed to convert biomass resources into renewable energy via electrochemical, thermochemical, chemical, and biochemical methods [8]. Figure 2 is the technology of biomass conversion into renewable energy by processing and synthetic operations of renewable energy production [9].

Hydro energy creates useful renewable energy from water flow. Turbines generally use to change hydropower into electrical energy. Dams were generally built to control water flow through turbines for generating electricity as shown in Fig. 3 with a general hydropower plant and its basic parts [10]. In that, running water rotates rotor mounted on the turbine of the electromagnetic generator. With electromagnetic principle, the electromagnetic rotor rotates inly the stator to create electric energy. The electric power is integrated to the electrical grid for daily applications and sustainable development.

Geothermal energy is produced and stored in the Earth. Geothermal energy resources consist of deep aquifer, conductive, and hydrothermal sources. Electricity, heat, or combination methods are technologies used to change into renewable energies [11]. Geothermal power plant makes electrical energy via electrical turbine under moving of steam or hot water. The geothermal power stations include dry steam, flash steam, and binary cycle plants [12].

Wind energy is a renewable energy resource. Wind power station generate electric energy by using wind energy to turn wind turbine. Conversion technologies include wind mill, wind turbine, and wind pump that are used to create the renewable energy from wind energy resource [13]. Wind turbine changes windy kinetic energy into electric energy [13]. Wind farms consist of hundreds of wind turbines to produce more power for daily consumption application.

Marine energy has originated from ocean such as ocean wave, tide, ocean thermal, and salinity energy. Energy conversion techniques converse marine energy into renewable energy by technologies such as overtopping device, point absorber, oscillating, pressure differential in the water, and bulge wave conversion device [14]. The conversion technologies can take renewable energy from ocean energy at different areas of nearshore to offshore [15].

Solar energy has originated from heat and light of the Sun. Current technologies have been used to change solar energy into sustainable energies like: photovoltaic, heating, concentrating solar and solar thermal energy. Solar power plant converse sunlight into electrical power directly via photovoltaic device or indirectly via solar concentration technique [16]. Photovoltaic technology directly changes sunlight into electric power.

3 Integration and Applications

Integration of renewable energy and power grid permits renewable energy serving easily in practical applications. Renewable energy penetration gives a role key to replace traditional energy by renewable energy to get off carbon dioxide and industrial pollutants. Group of renewable substations, wind farms, solar farms gathered single stations to enhance power capacity and quality of electrical power grid [17]. Renewable

Fig. 2. The biomass energy conversation technology [9].

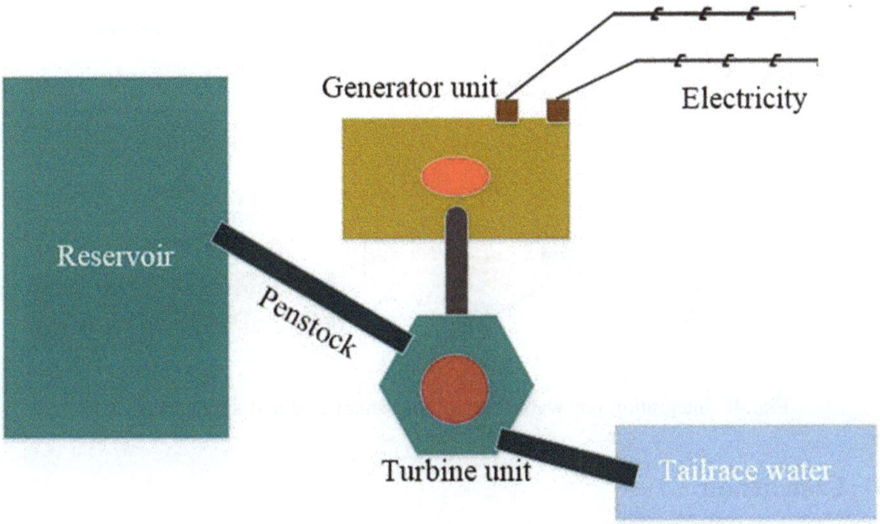

Fig. 3. A common hydro plant and its basic parts [10]

energy is penetrated to the power grid by an alternating current or direct current. Smart grids are modern systems to enhance, reliability, efficiency, and safety as shown in Fig. 4 [18]. The grid has full of morden communication technologies to connect devises, smart sensors to monitor operating and condition of equipments, and computing ability to distribute and control the system. Smart grid is a perfect technique to integrate renewable energy resources into electric power grid. Recently, renewable energy is applied successfully in home consumption, factories, transport vehicles sustainable production, sustainable manufacturing, flexibility of manufacturing systems.

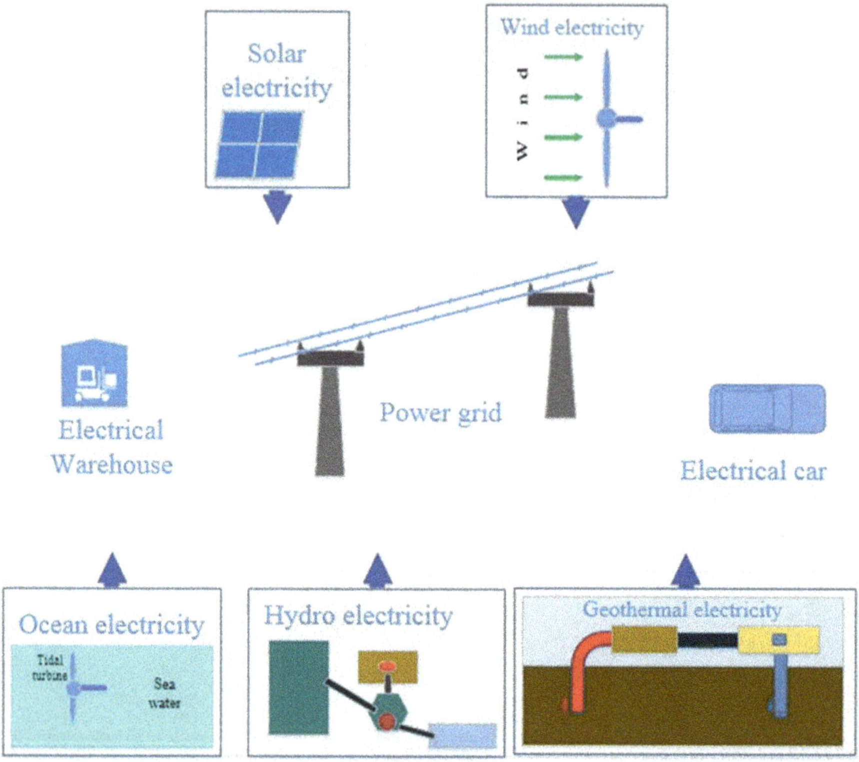

Fig. 4. Integration renewable energy into smart grid and application [18]

4 Conclusion

Traditional energy sources are limited. Environmental pollution is really harmful to our health and surrounding environment. Renewable energy is the best answer for sustainable development. The paper shortly reviews current technologies in renewable energy conversion. Renewable energy is integrated in the power grid and applied in many practical application as transportation, factories, home consumption, manufacturing systems, and sustainable systems. Renewable energy gradually replaces conventional energy resources with lots of gas emissions and industrial pollutants that will bring sustainable development in the future.

Acknowledgement. This work was partially sponsored by Department of Mechanical Engineering, National Cheng Kung University, Taiwan. It was also supported in part by Hanoi University of Industry, Vietnam. The work was also supported in part by Faculty of Mechanical Engineering, Hanoi University of Industry, Vietnam.

References

1. Ali, M.B., et al.: A review on emission analysis in cement industries. Renew. Sustain. Energy Rev. **15**, 2252–2261 (2011)
2. World Energy Outlook, IEA: IEA, World Energy Outlook (2019). www.iea.org
3. BP Statistical Review of World Energy, June 2019. bp.com/statisticalreview
4. Ellabban, O., et al.: Renewable energy resources: current status, future prospects and their enabling technology. Renew. Sustain. Energy Rev. **39**, 748–764 (2014)
5. REN21, Renewables 2019 Global Status Report, September 2010. www.ren21.net
6. Cecati, C., et al.: Combined operations of renewable energy systems and responsive demand in a smart grid. IEEE Trans. Sustain. Energy **2**, 468–476 (2011)
7. IEA: Technology Roadmap: Smart Grids (2019). www.iea.org
8. Kundu, K., et al.: Thermochemical conversion of biomass to bioenergy: a review, pp. 235-268. Springer, Singapore (2018)
9. Klass, D.L.: Biomass for renewable energy and fuels. Encyclopedia of Energy, vol. 1, pp. 193–212 (2004)
10. Kaunda, C.S., et al.: Hydropower in the context of sustainable energy supply: a review of technologies and challenges. ISRN Renewable Energy (2012). Article ID 730631
11. Hammons, T.J.: Geothermal power generation worldwide. In: 2003 IEEE Bologna Power Tech Conference Proceedings, vol. 1, p. 8 (2003)
12. Zhou, Y., et al.: Optimization and thermodynamic performance analysis of a power generation system based on geothermal flash and dual-pressure evaporation organic rankine cycles using zeotropic mixtures. Energy **194**, 116785 (2020)
13. Herbert, G.M.J., et al.: A review of wind energy technologies. Renew. Sustain. Energy Rev. **11**, 1117–1145 (2007)
14. Farrok, O., et al.: Electrical power generation from the oceanic wave for sustainable advancement in renewable energy technologies. Sustainability **12**, 2178 (2020)
15. Zhou, Z., et al.: Developments in large marine current turbine technologies – a review. Renew. Sustain. Energy Rev. **71**, 852–858 (2017)
16. Singh, G.K.: Solar power generation by PV (photovoltaic) technology: a review. Energy **53**, 1–3 (2013)
17. Liang, X.: Emerging power quality challenges due to integration of renewable energy sources. IEEE Trans. Ind. Appl. **53**, 855–866 (2017)
18. Gungor, V.C., et al.: Smart grid technologies: communication technologies and standards. IEEE Trans. Industr. Inf. **7**, 529–539 (2011)

Determination of Thermal Spectra in High Pressure Die Casting by Finite Difference Method

Luu Thuy Chung[1,3(✉)], Dinh Thanh Binh[2], Nguyen Thi Phuong Mai[3], and Vu Thanh Tung[3]

[1] Vinh University of Technology Education, Vinh, Nghe An, Vietnam
[2] Defense Industrial College, Phu Tho, Phu Tho, Vietnam
[3] Hanoi University of Science and Technology, Hanoi, Vietnam

Abstract. Thermal fatigue is one of the most importance failure forms of high pressure casting dies. The numerical calculation to predict the damage with the aim to select the materials and coating layers on the surface die is one of great interest to researchers. In this study, the authors used a finite difference method to determine the thermal spectra in the die walls with different materials to evaluate the effect of thermal conductivity on the thermal spectra. Calculations were shown that the coefficient of thermal conductivity affects the rate of temperature stability and the thermal gradient in high pressure die.

Keywords: Heat transfer · Finite difference method · Temperature spectrum · Thermal gradient spectrum

1 Introduction

Thermal fatigue was one of the major failures of in high pressure die casting (HPDC) mold, which was a widespread and unavoidable phenomenon [1–6]. This failure form depends on variety factors [3–8]: Material, surface quality, surface engineering. Thermal stress in HPDC dies includes:

- The stress by the difference of coefficient of thermal expansion between the coating and substrate: When temperature increases, due to the difference in the coefficient of thermal expansion, there is a difference between the expansion of the coating and the substrate, which leads to stress formation in the coating.
- The stress by the temperature gradient in the dies: Thermal gradients lead to differences in thermal expansion between material layers, which causes the material layers to tensile or compress and form stresses.
- Mold temperature in casting process: As the temperature rises, the mold has thermal expansion, but due to space constraints the mold cannot expand freely, which causes compression stresses in the mold.

Research [9] shows that thermal gradients cause tensile/compressive stresses in the material and affect the durability of HPDC dies. Research [9, 10] shows that both stress amplitude and stress mean affect the fatigue strength of materials.

Some studies using simulation tools have calculated the thermal stress in a HPDC die as well as predicted the life of the die [11, 12]. Research on thermal and thermal

B. T. Long et al. (Eds.): MMMS 2020, LNME, pp. 208–213, 2021.
https://doi.org/10.1007/978-3-030-69610-8_28

gradient spectra is essential to assess the impact of casting temperature on the durability of mold materials. This study focuses on the influence of coefficient of thermal diffusivity (CTD) of coating material on the thermal gradients of the HPDC mold and the fluctuation of thermal gradient on the mold's surface.

2 Research Method

2.1 Finite Differential Equation

Fourier equation for heat transfer in the case of 1D heat transfer is presented as Eq. 1 [13]. Where u(x, t) is the temperature; t is time; x is the coordinate axis in the Cartesian coordinate system along the heat transfer direction; *a* is CTD.

$$\frac{\partial u}{\partial t} = a\frac{\partial^2 u}{\partial x^2} \quad (1)$$

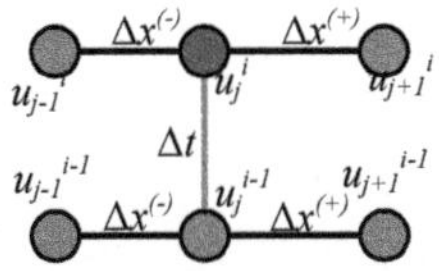

Fig. 1. Difference model in 1D heat transfer equation

The finite difference equation of differential Eq. 1 with the heat transfer model as shown in Fig. 1 is shown as Eq. 2. Equation 2 can be shown as Crank-Nicolson equation [13] (Eq. 4) when $\Delta x^{(-)} = \Delta x^{(+)} = \Delta x$. The value of *r* and *r'* are shown as Eqs. 3, 5. The finite difference equations (Eqs. 2, 4) are hidden differential equations, which makes it difficult to calculate the values of the differential grid, but these equations ensure the unconditional convergence of the differential equation.

$$\begin{aligned}&\left(1 + r\Delta x^{(-)} + r\Delta x^{(+)}\right)u_j^i - r\Delta x^{(-)}u_{j+1}^i - r\Delta x^{(+)}u_{j-1}^i \\ &= r\Delta x^{(-)}u_{j+1}^{i-1} + r\Delta x^{(+)}u_{j-1}^{i-1} + \left(1 - r\Delta x^{(-)} - r\Delta x^{(+)}\right)u_j^{i-1}\end{aligned} \quad (2)$$

$$r = \frac{2.a.\Delta t}{\Delta x^{(-)}.\Delta x^{(+)}(\Delta x^{(-)} + \Delta x^{(+)})} \quad (3)$$

Crank-Nicolson equation:

$$(1 + 2r')u_j^i - u_{j+1}^i - u_{j-1}^i = u_{j+1}^{i-1} + u_{j-1}^{i-1} + (1 - 2r')u_j^{i-1} \quad (4)$$

$$r' = \frac{a.\Delta t}{2.(\Delta x)^2} \quad (5)$$

2.2 Modeling

To simplify, consider the mold walls to be flat and wide enough, the outer is cooled continuously. As an approximation, the heat transfer process in a mold can be considered to be an unstable 1D heat transfer process. The temperature in the mold is a solution of the differential equations Eq. (2) or Eq. (4). All models are 50 mm thick. Mold materials in all cases are SKD61 with CTD a = 6.78 (mm^2/s). One model has no coating, the other models all have 10 μm thickness coating: The details of coatings are shown in Table 1.

Initial conditions: The casting process is carried out from cold mold, initial temperature of mold is ambient temperature, initial condition $u° = 22$ °C.

Boundary conditions: This study use Dirichlet boundary condition by using the results from other experiments. The line chart in Fig. 2a is provided by the attachment of the HPDC machine. The line graphs in Fig. 2b, c are graphs of the regression function from the data of Fig. 2a, which is similar to other research results [8, 11, 12].

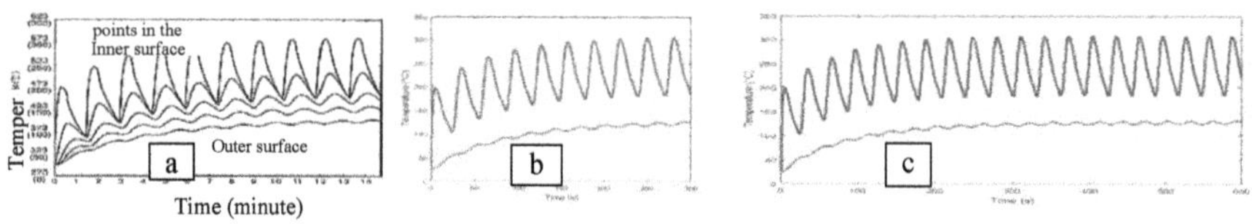

Fig. 2. Temperature variation at points in the mold over time: *(a) real charts and (b, c) regression charts in 300 s and 600 s*

Table 1. Thickness and CTD of coating material in investigation model

Model no	1	2	3	4	5	6	7	8
Coating material	–	CrN	TiN	TiC	WC	Cr	W	CrN/Cr
Thickness *(μm)*	–	10	10	10	10	10	10	CrN:8/Cr:2
CTD a *(mm^2/s)*	–	2.68	4.67	5.64	13.27	20.85	64.03	2.68/20.85

Time step and differential grid: In this study, the differential equation has solved by the time step Δt = 0.0025 s and differential grid in x axis Δx = 1 μm.

3 Results and Discussions

3.1 Results of None Coating Model

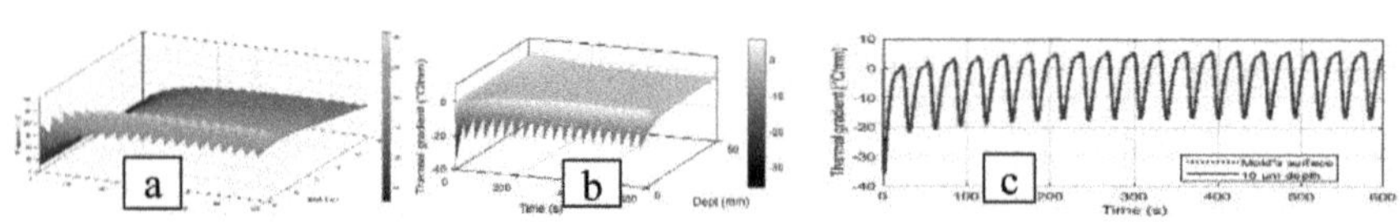

Fig. 3. Result of solving differential equations of SKD61 in the first 600s *(a)- Temperature spectrum; (b) Thermal gradient spectrum; (c)-Thermal gradient on surface and in 10 μm depth point*

Figure 3 shows the calculation results for the thermal and spectral gradient spectra of the HPDC die without coating. The thermal spectrum of pressure die is almost stable from the 10th cycle. The amplitude of fluctuation of the thermal gradient in the first cycle is 18.2925°/mm, in the regular cycle is 11.303 °C/mm. The mean value of the thermal gradient in the first cycle is −9.704 °C/mm, in the regular cycle is −2.465 °C/mm. Thermal gradients have almost negative values.

3.2 Temperature Spectra

The thermal spectra in HPDC coating molds are shown in Fig. 4. The temperature spectra in Fig. 3a and Fig. 4 show that the coating has almost no effect on the thermal spectrum of the HPDC die.

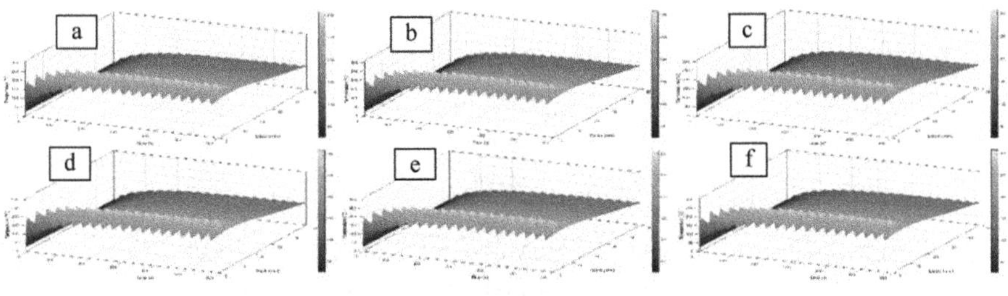

Fig. 4. Temperature spectra of mold with 10 μm coating *a- CrN coating; b-TiN coating; c-TiC coating; d-Cr coating; e-WC coating and f- W coating*

3.3 Thermal Gradient Spectra

The thermal gradient spectrum of HPDC coated dies are shown in Fig. 5. The coating has almost no effect on the thermal gradient of the mold material (SKD61), the thermal gradient of the surface layer (coating) varies markedly by the value of the CTD of coating material. Figure 6 is the value of the thermal gradient on the mold surface and thin film/substrate interface. This figure also shows that the mold surface's thermal gradient is highly dependent on the coating's CTD, the mold material thermal gradient is almost unaffected by the coating. Table 2 is the amplitude and mean values of thermal gradient on the mold surface of the coating cases. All thermal gradient spectra

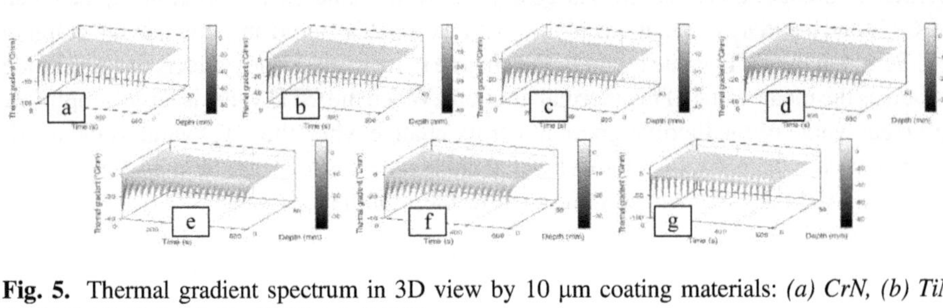

Fig. 5. Thermal gradient spectrum in 3D view by 10 μm coating materials: *(a) CrN, (b) TiN, (c) TiC, (d) WC, (e) Cr, (f) W, (g) CrN/Cr*

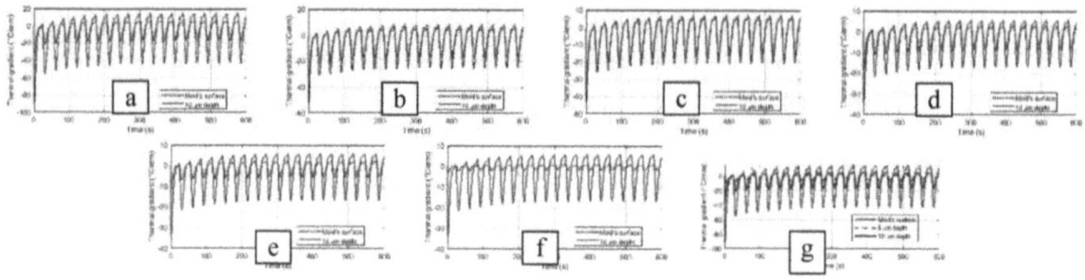

Fig. 6. Thermal gradients on surface and 10 μm depth point of 10 μm coating mold by material: *(a) CrN, (b) TiN, (c) TiC, (d) WC, (e) Cr, (f) W, (g) CrN/Cr*

Table 2. Amplitude and Mean of thermal gradient in coating material

Coating material	Thermal gradient in the 1st cycle		Thermal gradient in normal cycle	
	Amplitude	Mean	Amplitude	Mean
CrN	46.13	−24.53	28.54	−6.23
TiN	26.53	−14.09	16.40	−3.58
TiC	21.98	−11.67	13.58	−2.96
WC	9.35	−4.96	5.78	−1.26
Cr	5.954	−3.15	3.679	−0.80
W	1.938	−1.03	1.18	−0.26
CrN/Cr	CrN: 46.17 Cr: 5.93	CrN: −24.63 Cr: −3.17	CrN: 28.55 Cr: 3.666	CrN: −6.24 Cr: −0.80

show that, most of the time in the casting cycle, the thermal gradients of the material have negative values, the Mean value of the thermal gradients are less than zero. In the first cycle, the absolute values of the thermal gradients (both amplitude and mean value) are the greatest, from the second cycle, the oscillation cycle of the thermal gradient quickly reaches a stable value.

Tungsten is the highest CTD coating material (Table 1), the amplitude of thermal gradient of this material is smallest and the absolute of the mean values is smallest too (as shown in Table 2). Chromium nitride is the exact opposite. Chromium nitride is the smallest CTD coating material, the amplitude and the absolute of the mean values of thermal gradient of this material is largest (first row in Table 2). The relationship between the CTD and the thermal gradients of the coating on HPDC mold are shown in the graph in Fig. 7. When the CTD value is small, the change in its value greatly affects the amplitude as well as the mean value of the thermal gradient. When the CTD value is large, amplitude and mean value of the thermal gradient hardly change when the CTD value is changed.

In the case of a CrN/Cr multi-layer coating, the thermal gradient of the coating material is virtually unchanged compared to the monolayer case.

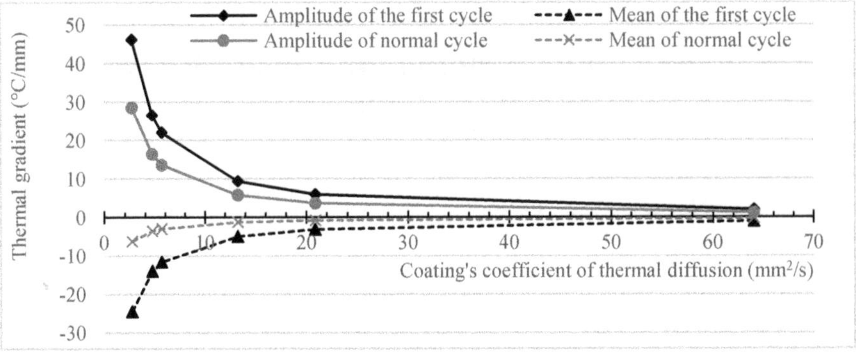

Fig. 7. Thermal gradients in coating of HPDC mold via CTD of coating material

4 Conclusion

Calculation results on mathematical models show that, in all cases, the thermal gradients are almost negative. The coating hardly affects the thermal spectrum and thermal gradient spectrum of the mold material. However, the coefficient of thermal diffusion of the coating material greatly affects the thermal gradient of the coating. The higher the CTD is (W: 64.03 mm^2/s), the smaller the amplitude (1.938 °C/mm in the first cycle and 1.18 °C/mm in regular cycle) and absolute value of Mean of thermal gradient are (1.13 °C/mm and 0.26 °C/mm) and vice versa (CrN: 2.68 mm^2/s – Amplitude: 46.13 °C/mm, 28.54 °C/mm; Mean: −24.53 °C/mm and −6.23 °C/mm).

References

1. Jhavar, S., Paul, C.P., Jain, N.K.: Eng. Failure Anal. **34**, 519–535 (2013)
2. Chen, C., Wang, Y., Ou, H., Lin, Y.-J.: Int. J. Fatigue **92**, 166–178 (2016)
3. Srivastava, A., Joshi, V., Shivpuri, R., Bhattacharya, R.: Satish Dixit – Surf. Coat. Technol. **163–164**, 631–636 (2003)
4. Nunes, V., Silva, F.J.G., Andrade, M.F., Alexandre, R., Baptista, A.P.M.: Surf. Coat. Technol. **332**, 319–331 (2017)
5. Persson, A., Hogmark, S., Bergström, J.: Surf. Coat. Technol. **191**, 216–227 (2005)
6. Muhič, M., Tušek, J., Kosel, F., Klobčar, D., Pleterski, M.: Metalurgija **49**, 9–12 (2010)
7. Persson, A., Bergström, J., Burman, C., Hogmark, S.: Surf. Coat. Technol. **146–147**, 42–47 (2001)
8. Fazarinc, M., Muhič, T., Kugler, G., Terčelj, M.: Eng. Fail. Anal. **25**, 238–249 (2012)
9. Hetnarski, R.B.: Encyclopedia of Thermal Stresses. Springer Reference (2014)
10. Stephens, R.I., Fatemi, A., Stephens, R.R., Fuchs, H.O.: Metal Fatigue in Engineering, 2nd edn. Wiley-Interscience Publication, New York (2001)
11. Srivastava, A., Joshi, V., Shivpuri, R.: Wear **256**, 38–43 (2004)
12. Long, A., Thornhill, D., Armstrong, C., Watson, D.: Appl. Therm. Eng. **44**, 100–107 (2012)
13. Fadugba, S.E., Edogbanya, O.H., Zelibe, S.C.: Int. J. Appl. Math. Modeling IJA2M **1**(3), 8–23 (2013)

Advantages Analysis of Composite Materials Used as Bus Automobile Covers Body According to Vibrations

Nguyen Thanh Quang[1(✉)] and Nguyen Quang Thai[2]

[1] Department of Automobile Technology, Hanoi University of Industry, Hanoi, Vietnam
nguyenthanhquang@haui.edu.vn
[2] Kettering University, Michigan, USA

Abstract. This paper presents the following contents: Replacing thin sheet metal with composite materials on Bus automobile. Applying finite element method for vibration analysis in the vehicle body with both original sheet metal and composite material. The vibration parameters of the vehicle include a vibration frequency range, vibration shape modes, and the largest displacement of the body vehicle. Using this information, we can design the body to avoid the frequency band causing the resonance oscillation of the body. One of the design methods is to choose the material type or thickness of the composite sheet. The results of the study have shown the effectiveness of composite materials for automobile body and suggested a replacement for some plate parts on the vehicle.

Keywords: Bus vehicle · Composite materials · Vibration

1 Introduction

Composites are used in automobile industries due to the following reasons: Reduce weight, due to which there is an improvement in fuel efficiency; Attractive styling; High resistance to corrosion, bio-degradation and extreme environmental conditions; High thermal insulation and sound deadening properties; Easy repairing and maintenance free structural elements. Also, the structural elements are mostly less stressed and hence glass fibers composites with polyester or vinyl ester resins are used for these applications. Manufacturing is an issue for composites in the automotive sector when one considers the high production volumes required. For small quantity, mostly glass reinforced composite products are used due to low cost and easy availability. The processing techniques are mostly handmade technique and compression moldings techniques for making these products. Automatic aspects that have to be considered are tooling costs, scrap production and cycle time for high production volumes.

B. T. Long et al. (Eds.): MMMS 2020, LNME, pp. 214–219, 2021.
https://doi.org/10.1007/978-3-030-69610-8_29

2 Base on Change Materials in Body of Bus

2.1 Bus Body Structural Analysis

Body structural of Bus automobile analysis is an approximation for stress in an integral can be obtained by using the bending loads are carried by the side plates. The side door apertures are critical areas as they represent large cut-outs in the structural surfaces. Shear force at the center of each side plate is assumed to be shared equally between the cant rail and sole bar according to their respective stiffnesses. In the structure, ED coating from 50–75 microns was used to primer for all of the plates. Main assemblies were together, these comprising floor, roof, side front and rear plates assemblies of spotwelded steel pressings. The loading acts in the plane of the section the elements are chosen can be plane rectangles and the relationships must ensure continuity of strain across the nodal boundaries. Choosing the optimum mesh density (size and distribution) of elements is a skill which is gradually learned which experience. The calculate several values at various key points where stresses vary rapidly in a region. The body structure defined as a thin sheet metal is limited by two parallel planes with a plate thickness, the ratio between the plate thickness and the minimum plate dimension is within the limit: $1/20 \leq t/b \leq 1/5$ [1]. The front and rear plate are complex form will be manufacturing by heavy stamping machines with high investment cost. In order to proactively implement technology and reduce the cost of machinery and investment, the can be instead composite materials for sheet metal on the body of the bus.

2.2 Selection Materials

Materials selection criteria for vehicle body are mechanical strength, low weight, shock/Energy absorption capability, cost, and ease of manufacturing. Composite materials like CFRP-Epoxy, aramid Kevlar 49 and Boron fiber having tensile strength more than steel and aluminum and had best flexural properties. Whereas in plastic composites, SMC have fair tensile properties but shows better results in flexural cases. Plastic composites are having less weight as compared tosteel as its density is less. The research selected composite material like comparing CFRP-Epoxy, [2] and GFRP-Epoxy, [3] for Bus body.

$$C_b = \sqrt{\omega p k C_1} \tag{1}$$

In which ω is nature frequency of a plate; ρ is material density (in Table 1); C1 is the velocity of the elastic compressive wave, which is about 5100 m/s for steel, and κ is the ratio of the rotational moment of inertia to the cross-section area, which is only dependent on plate thickness [4].

$$k = \frac{h}{\sqrt{12}} \tag{2}$$

Applied equation resemble to Ohm's law, with the stress and particle velocity corresponding to voltage and current in a circuit.

The relationship between the displacement to the natural vibration depends on the physical data section specification material of plate.

Total displacement of the plate, at the time T after the start of the forces.

$$u = (p.k.C_1)^2 \int_{o}^{T} \omega^2 dt \tag{3}$$

3 Results FEA Analysis

3.1 The Front Plate of Body

The plates on the vehicle body usually need local stiff́ness strengthening by adding extra plates or beams or being stamped with corrugated shapes or other complex treatment to attain the effect of local stiffness strengthening. The modal frequencies and modal shapes of complex plates are usually obtained by using the finLte element method or real testing. The modal frequencies corresponding to the peaks on the spectrum of acceptance are dependent on mode stiffness and mode mass, whereas the peaks depend on mode damping. If a specific mode frequency must be increased, the usual approach is to increase stiffness. The front of vehicle is subjected to the accidents which are most dangerous among all kinds of accidents since it is a directly attack. It is also a component that is directly Influenced by air resistance causing vibration noise. The front plate FE model is used with shell elements to model of three types for compare the sheet metal and CFRP-Epoxy, GFRPEpoxy composite components. The two types of shell elements used are quadrilateral and triangular.

The natural frequencies of the vibration modes of a front plate are all of the analysis can be obtained by using the finite element method shown in Fig. 1. This graph shows the difference between mode verses frequency to locate at each mode shape how many frequencies generated on steel and CFRP-Epoxy, GFRP-Epoxy front plate. In modal analysis, the results calculated by ANSYS in the form of total deformation up to ten mode shapes show the values in the form of frequency at each mode for front plate carrying uniformly undistributed load. The equation shown on graph using trend line equation is regression analysis shown in Fig. 2.

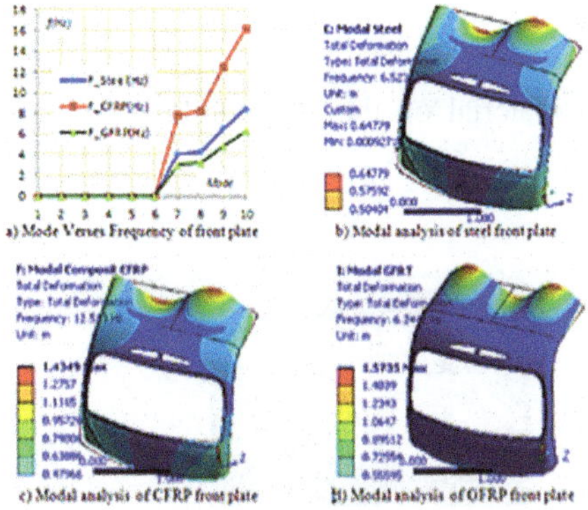

Fig. 1. Mode Frequency of front plate

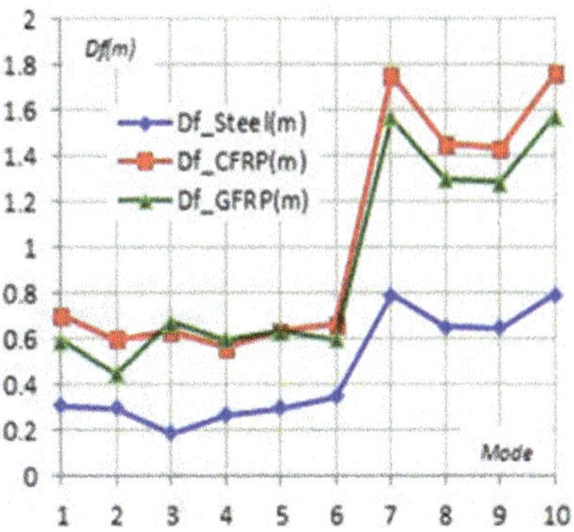

Fig. 2. Total deformation of front plate

3.2 The Assembling Body

The body assembled from six plates in Fig. 1 above formed a compartment with an interior volume of 3.6 cubic meters, the mode numbers within the bandwidth of 150 Hz centered at 1200 Hz are as much higher as single each plate. When mode density reaches a certain high level, the vibration problem is no longer a deterministic problem with certainty instead it is a non-deterministic problem having statistical characteristics. Therefore, the experimental statistical approach must be used to quantify and characterize for ensure accuracy. The natural frequencies of the vibration modes for body are all of the analysis can be obtained by using the finite element method shown in Fig. 3. This graph shows the difference between mode verses frequency to locate at each mode shape how many frequencies generated on steel and CFRP-Epoxy, GFRP-Epoxy all of the plates.

In modal analysis the results calculated by ANSYS in the form of total deformation up to ten mode shape which shows the values in the form of frequency at each mode for all body assembled carrying uniformly undistributed load. The equation shown on graph using trend line equation is regression analysis shown in Fig. 4.

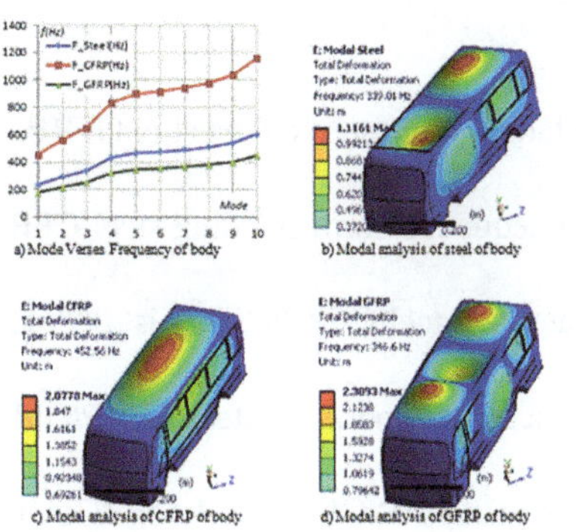

Fig. 3. Mode Frequency of body

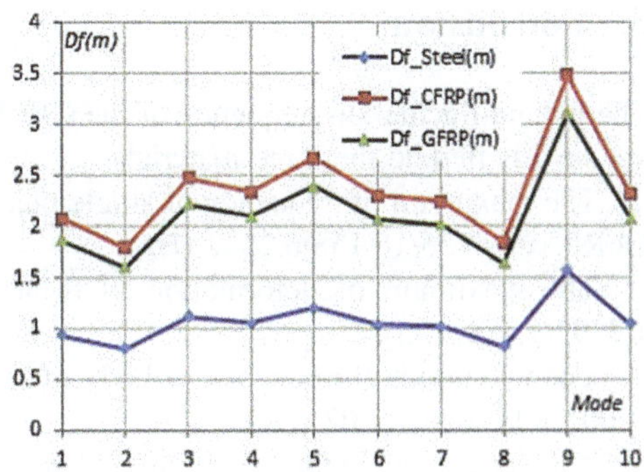

Fig. 4. Total deformation of body

4 Experiment

Replacement of sheet metal for composite material on the automobile body is conducted by two experiments.

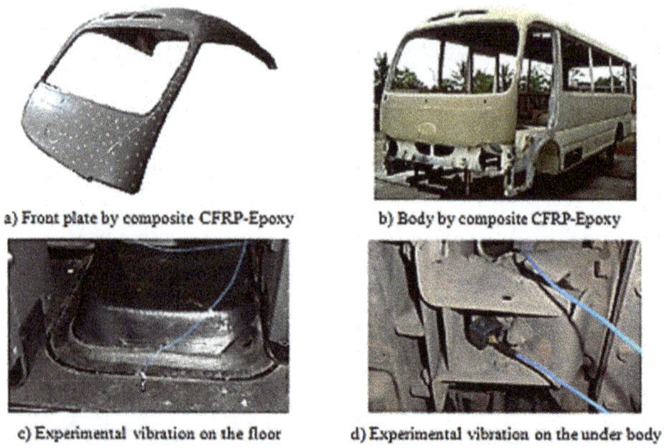

Fig. 5. Experimental replacement of plate for composite material on the bus body

The first experimental is to replace the front plate only while the rest of the body retains 1 mm thick sheet steel original. The front plate has the 3 mm thickness, the material is CFRP composite a continuous and aligned glass fiber-reinforced composite consists of 40 vol% of glass fibers having a modulus of elasticity of 69 GPa and 60 vol % of a polyester resin that, when hardened, displays a modulus of 3.4 GPa. The front plate is shown in Fig. 5(a). The second experiment is conducted by replacement of sheet metal for CFRP composite material on the automobile body. The structure of bus body is shown in Fig. 5(b). Ten experiments were performed with different modes. Vibration measurement equipment installed on the floor is shown in Fig. 5(c) and on the vehicle body is shown in Fig. 5(d) to record and analyze the obtained results.

5 Conclusion

The natural fluctuation frequency of the CFRP plate is highest, the lowest is GFRP, of the steel is middle the three materials.

The maximum of frequency of single plate of sequence corresponding steel/CFRP/GFRP is 8.4235/16.158/6.2442 (Hz).

The maximum of deformation of front plate of steel/CFRP/GFRP is 0.79421/1.7592/1.5735 (m).

The maximum of frequency of body of sequence corresponding steel/CFRP/GFRP is 603.1/1156.9/447.07 (Hz).

The maximum of deformation of body of steel/CFRP/GFRP is 1.5567/3.4702/3.1038 (m).

The steel plates are better than composite plate's materials, because they give minimum total deformation and frequency in modal analysis as compared.

However, composite panels are smaller in weight than steel plates, reducing overall vehicle weight. Besides, it is easier to make the body of the composite body, thus reducing the manufacturing cost. The results show that CFRP-Epoxy composite should be used.

References

1. Fenton, J.: Handbook of vehicle design analysis. Society of Automotive Engineers Inc., Warrendale (1996)
2. Laranjeira, E., Carvalho, L.H., Silva, S.M.L., Almeida, J.R.M.D.: Influence of fiber orientation on the mechanical properties of polyester/ jute composites. J. Reinf. Plast. Compos. **25**, 1269–1278 (2006)
3. Yunfu, O., Zhu, D., Zhang, H., Huang, L., Yao, Y., et al.: Mechanical characterization of the tensile properties of glass fiber and its reinforced polymer (GFRP) composite under varying strain rates and temperatures. Polymers **8**, 196 (2016)
4. Gang, S.: Vehicle noise, vibration and sound quality. SAE International (2012). ISBN 978–0–7680–3484–4.SAE Order No. R-400. https://doi.org/10.4271/R-400
5. Zalka, K.A.: A simplified method for calculation of the natural frequencies of wall-frame buildings. Eng. Struct. **23**, 1544–1555 (2001)
6. Quang, N.T.: Finite element analysis in automobile chassis design. Appl. Mech, Mat. **889**, 461–468 (2019). ISSN:1662–7482. https://doi.org/10.4028/www.scientific.net/AMM.889.461
7. Tucker, N., Lindsey, K.: An Introduction to Automotive Composites Shawbury, Shrewsbury, SY4 4NR, United Kingdom (2002). ISBN: 1–85957–279–0

Effect of Tool Offset on the Mechanical Properties of the Dissimilar Friction Stir Welding Cu/Al

Tran Hung Tra(✉), Quach Hoai Nam, and Duong Dinh Hao

Department of Engineering Mechanics, Nha Trang University, 02 Nguyen Dinh Chieu Street, Nha Trang City, Vietnam
{tra, namqh}@ntu.edu.vn, dinhhao@ntu.vn

Abstract. Pure copper plate C1100 and aluminum 1050 plate were joined by friction stir welding. The influence of the tool offset on the macrostructure and the strength of the joints were explored. At various tool offsets, the tunnel defects seemed to be in-avoidable. However, at the high offsets, the tunnel size is small. The tunnel size is increased with the reduction of the tool offset. The tensile and bending strength reached to a highest value at a welding regime with the tool offset around 3.0 mm. The strength bonding between copper and aluminum could be obtained about 80% of the base aluminum 1050.

Keywords: Dissimilar joint · Friction stir welding · Defects · Tensile strength

1 Introduction

Friction stir welding (FSW) has been emerging as a green technique in metal joining. It is suitable solution for welding the dissimilar metals recognized to be un-weldable by conventional fusion welding methods. The fundamental theory of the FSW process is simple. The rotation of shoulder and pin of a welding tool produces a heating around metal. Consequently, a solid-state joint is established.

Recently, with the development of electric vehicles and equipment, several researchers have addressed to join the dissimilar electrical materials, *i.e.*, copper to aluminum. Since the significant different properties between copper and aluminum alloy, the bonding between these two metals becomes a big challenge. Several researched works [1–5] have been conducted to join these metals by FSW technique. However, the macrostructure and the mechanical properties of the bonding were quite sensitive to the welding parameters. The non-uniform materials in the welded zone during the stirring process must be one of the main culprits. In this work, we try to fabricate the FSW joints between pure copper C1100 and aluminum 1050 in the Friction welding Lab of Nha Trang University. The influence of the tool offset on the strength of the joints was performed.

B. T. Long et al. (Eds.): MMMS 2020, LNME, pp. 220–224, 2021.
https://doi.org/10.1007/978-3-030-69610-8_30

2 Experimental Procedures

The pure copper C1100 plate and aluminum 1050 plate with 5.0 mm thickness were joined by applying FSW method with supporting the CNC milling machine. The diameter and length of cylindrical pin was 8.0 mm and 4.8 mm, respectively. The tilt angle of tool was 2.0 deg. To reveal the influence of the tool offset on the strength of the joints, all the joints were fabricated at constant welding regimes except for the tool offset. Tool offset is understood as the distance from the interface to the tool axis line (the tool is offseted to aluminum side).

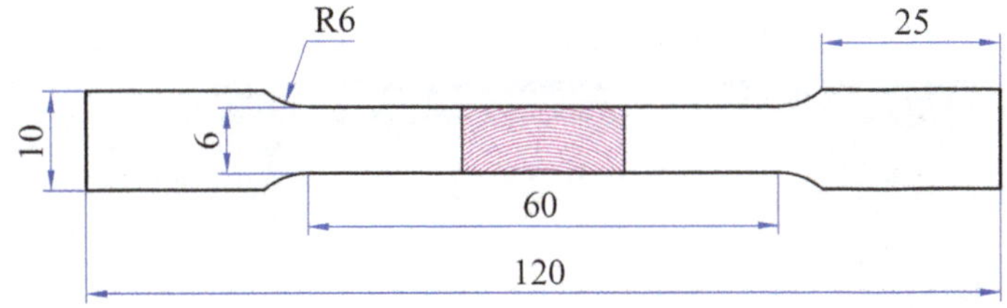

Fig. 1. Geometry of tensile specimen.

The macrostructure was observed by an Olympus microscope. The tensile specimens were machined from the welded plate with a thickness of 5.0 mm and the geometry based on ASTM E08 (see Fig. 1). The tensile specimen loading direction was cut perpendicular to the welding line. The tensile tests were performed by means of the Instron 3366, at a constant strain rate of 5 mm/min.

3 Results and Discussion

The macrostructure of the joints at various welding regimes was summarized in Table 1. In these macrostructures in Table 1, the right side is the aluminum, and the left side is copper.

The bonding between copper 1100 and aluminum 1050 could be obtained advantageously by the FSW process. The tensile strength could be gained up to 98.2 MPa (≈80% tensile strength of base aluminum 1050). The tunnel defects seemed to be dominant in the joints, see Table 1. The tunnel size was quite large at the low tool offset. In all welding regimes, the copper material was stirred into the aluminum and obviously seen (Table 1) and Fig. 2. The macrostructure tunnel in the joint was magnitude in the 3rd column of Table 1.

The tensile strength of the joint at various tool offsets was shown in Fig. 3. The strength of the joint seemed to gradually rise with the increasing of the tool offset. The highest strength of the joint was obtained with the tool offset of 3.0 mm. Then, it was reduced with increasing the tool offset. In the bending view, the bending behavior was quite similar to that of tensile. The bending strength was improved with the increasing of the tool offset. The bending strength reached to the peak value at the tool offset 3.3 mm. Then, the bending was drop with the increasing of the tool offset.

In all cases of the welding regimes, the tensile fracture and bending fracture took place at the tunnel defects (see Figs. 4 and 5) except for the case of the joints was fabricated at 3.0 mm of the tool offset (see Figs. 4(b) and 5(b)). In this case, the joint was fractured at the interface. As an overview, the joint fabricated at the tool offset around from 3.0 mm to 3.3 mm showed a best strength.

Table 1. Macrostructure and defects in the joint at various tool offset regimes

Offset, mm	Maccrostructure	Tunnel defects
3.6	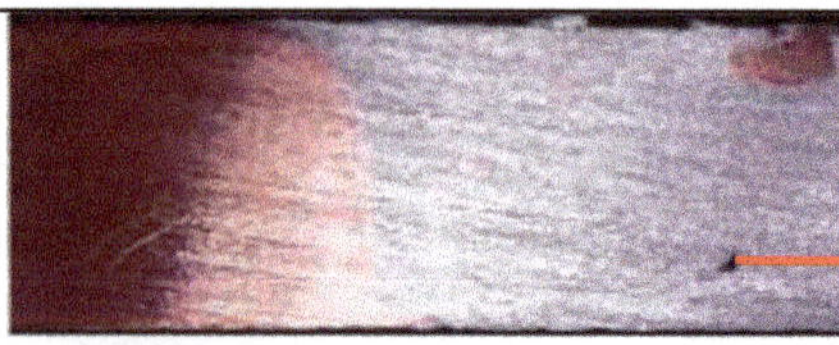	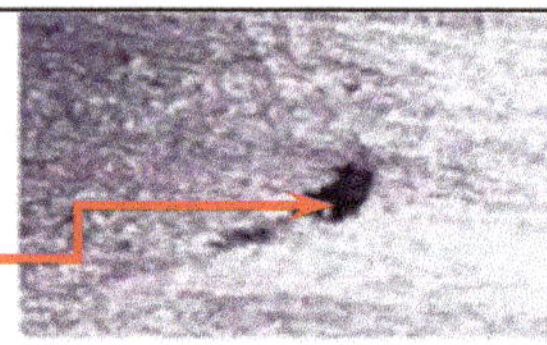
3.3	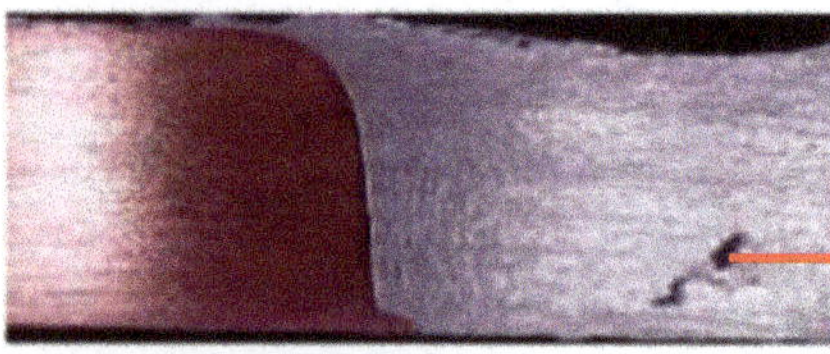	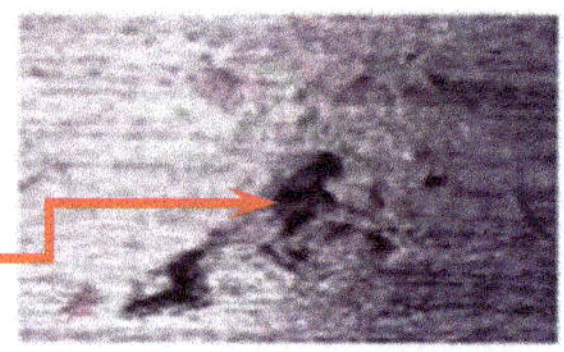
3.0	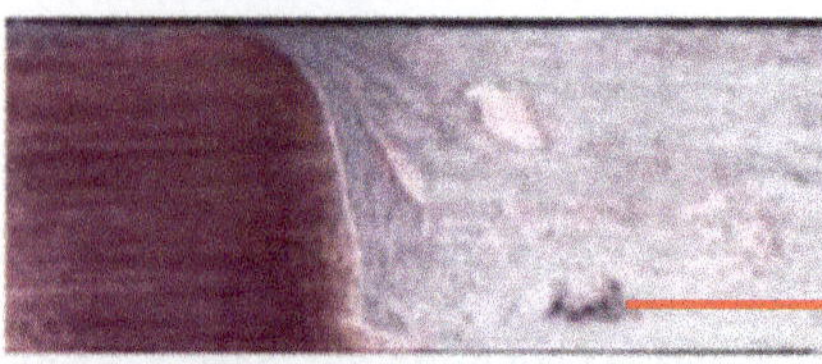	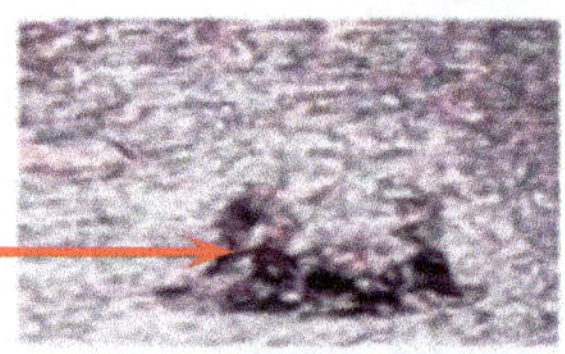
2.7	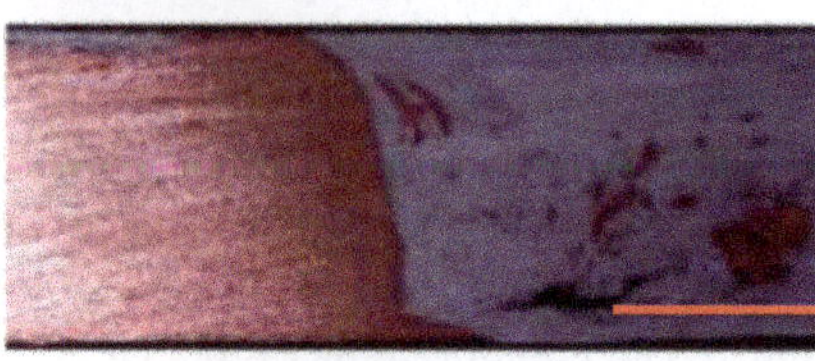	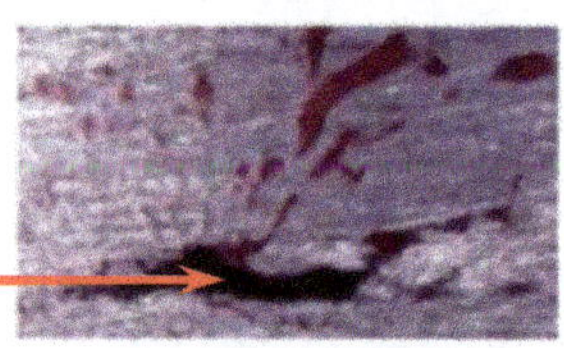
2.4		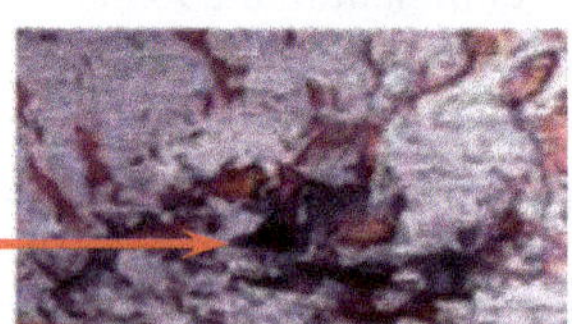

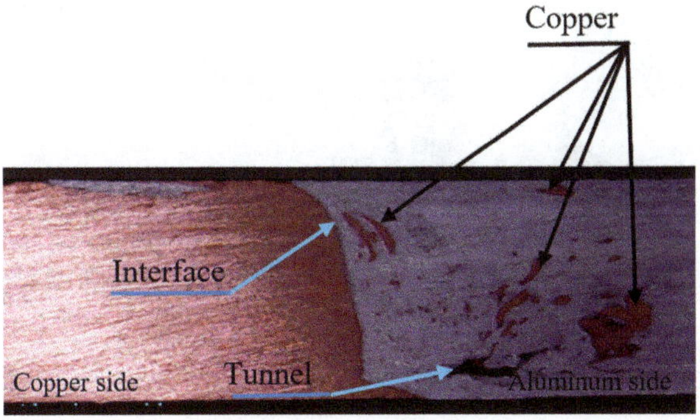

Fig. 2. The mixed copper in to aluminum (Offset 2.7 mm).

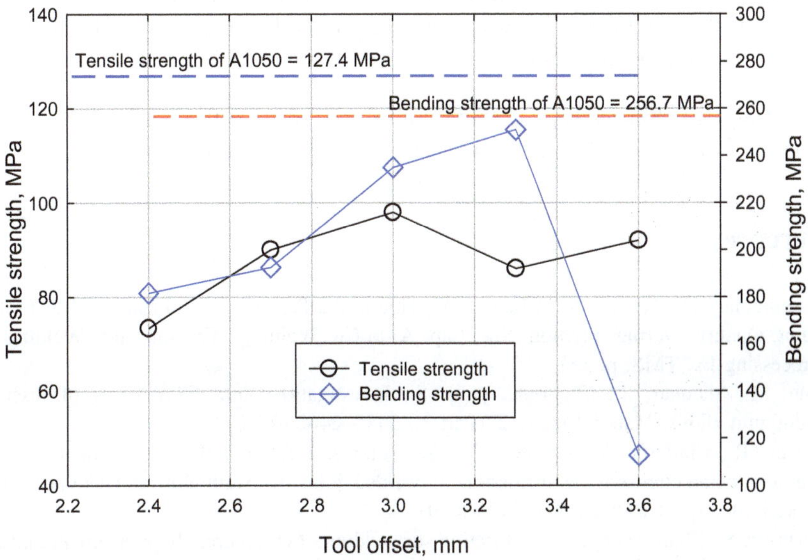

Fig. 3. Tensile and bending strength of the joints at various tool offset regimes.

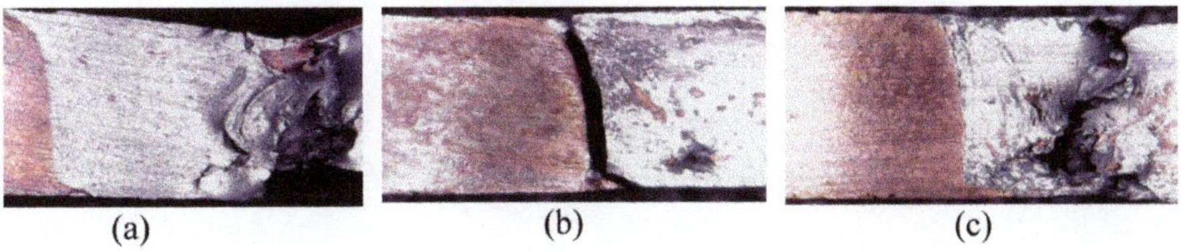

Fig. 4. Tensile fracture of the specimens with offset (a) 3.6 mm, (b) 3.0 mm, (c) 2.4 mm.

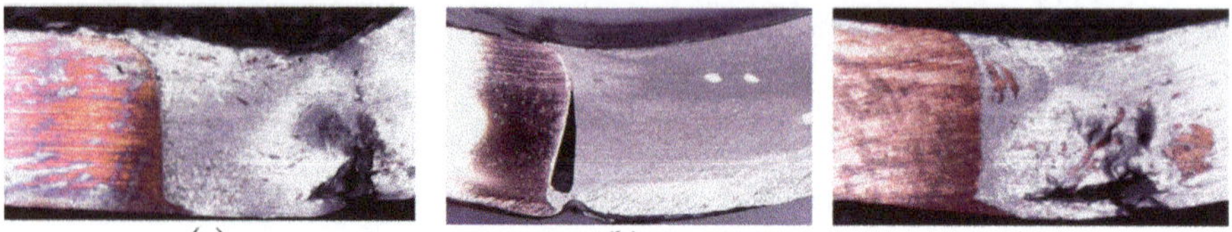

(a) (b) (c)

Fig. 5. Bending fracture of the specimens with offset (a) 3.3 mm, (b) 3.0 mm, (c) 2.7 mm.

4 Conclusion

Copper C1100 and aluminum 1050 was joined by FSW. The strength of the joint could be reached to 80% that of the base aluminum 1050. The interface bonding between copper and aluminum was free defects, but the tunnel defects in the stir zone seemed to be unavoidable. The tunnel size was quite large at the low tool offset. The strength of the joints reached to peak values at the tool offset around 3.0 mm to 3.3 mm.

Acknowledgment. This research is funded by Vietnam National Foundation for Science and Technology Development (NAFOSTED) under grant number: **13/2020/TN**.

References

1. Parningotan, D., Tarrant, M., Chen, Z.W., Hilton, A., Pasang, T.: Influence of Stir Flow on Joint Quality During Friction Stir Lap Al-to-Cu Welding, Friction Stir Welding and Processing IX, TMS, p. 157
2. Sen, M., Shankar, S., Chattopadhyaya, S.: Investigations into FSW joints of dissimilar aluminum alloys. Materialstoday **27**(Part 3), 2455–2462 (2020)
3. Rzaev, R., Chularis, A., Smirnov, V., Semyenova, L.: The influence of the friction stir welding parameters on the formation of welded joint of Aluminum and Copper Alloys. Materialstoday **11**(Part 1), 534–542 (2019)
4. Shankar, S., Chattopadhyaya, S.: Friction stir welding of commercially pure copper and 1050 aluminum alloys. Materialstoday **25**(Part 4), 664–667 (2020)
5. PatrickMubiayi, M., Akinlabi, E.T., Makhatha, M.E.: Current state of friction stir spot welding between aluminium and copper. Materialstoday **5**(9, Part 3), 18633–18640 (2018)

Adaptive Variable Stiffness Vibration Absorber Using Magnetorheological Material

Xuan Bao Nguyen(✉) and Phung Minh Tung

The University of Danang, University of Technology and Education, Danang, Vietnam

Abstract. Magnetorheological elastomer (MRE) is a class of elastomer that consist of rubber with embedded ferromagnetic particles. As a result, the mechanism of these materials can be controlled under the different magnetic field levels. The systems that use MRE-based devices can efficiently tune natural frequencies to avoid resonance. Recently, the use of this material in intelligent vibration systems has attracted many scientists. In this study, MREs were used in an intelligent suspension system. The stiffness controllable elastomer is first fabricated, and then the static and dynamic properties are tested under different magnetic fields. Second, an on-off semi-active controller was used to reduce the vibration of a system. The control algorithm scheme is based on the Lyapunov stability theory to maximize vibration energy absorption. The effectiveness of the developed controller has been evaluated by experiment.

Keywords: Magnetorheological elastomer · Semi-active control · Vibration system

1 Introduction

The development of vibration control technology is an important issue to prevent the destruction and damage of mechanical structures and ensure human safety and comfort. Vibration control methods are classified into three classes: passive type, active type, and quasi-active type. The passive type is generally widely used where the damper is the device that dissipates the system's vibrational energy. In contrast to the passive type, an active type equipped with an actuator suppresses the vibration by the generated force. Recently, research and application of a semi-active vibration system are the subjects of many researchers [1–3]. The semi-active vibration system has been widely used in the automotive, construction, and mechanical systems industries. Many controllers for semi-active suspension systems have been developed, such as Skyhook controllers, adaptive controllers, the fuzzy controller [4–7].

Magnetorheological fluid (MR fluid), which can change viscosity in the response under an external magnetic field, is well known as functional fluids. Therefore, a wide range of applications in the engineering field, such as dampers with adjustable damping force, are being investigated. Problems such as aggregation, aggregation, and deterioration over time are listed as issues to be solved. A magnetorheological elastomer (MRE) has attracted scientists as smart material whose stiffness can be varied under a magnetic field. Recently, semi-active control systems using MRE have appeared as

B. T. Long et al. (Eds.): MMMS 2020, LNME, pp. 225–232, 2021.
https://doi.org/10.1007/978-3-030-69610-8_31

smart devices that have been widely applied in vibration control (especially suspension systems, earthquake systems) [8–11]. The behavior and model of MRE has also been fully presented [12].

In this work, static and dynamic tests were conducted for MR elastomer, static restoring force, and the dynamic viscoelastic characteristics according to the magnetic field were determined. Besides, a semi-active vibration control using an on-off type variable stiffness control law was performed. A model structure with one degree of freedom is constructed using this MR elastomer, and the stiffness is switched between two modes of minimum and maximum values. The system using MR elastomer achieved effectiveness for vibration control.

2 Magnetorheological Elastomer (MRE)

2.1 MRE and Fabrication of the Material

Magnetorheological elastomers (MREs) are solid, consisting of a rubber matrix with nano-sized ferromagnetic particles such as carbonyl iron. As a result of this synthetic microstructure, these materials' mechanical properties can be controlled using magnetic fields. The change in the material structure is influenced by magnetism; as shown in Fig. 1, the iron particles are polarized by magnetic force. A link is created between the particles to form chains of iron particles. Therefore the mechanical properties of the material are changed. This change can be controlled by adjusting the strength of the magnetic field passing through the material.

The two components were used for fabrication consist of Silicone of brand Shin-Etsu and the iron particles (particle size about 20 μm) of the brand BASF SG-BH. The silicone and particle volume ratio was 10:3 and was mixed to become a homogeneous mixture by a centrifugal mixer. The material mixture was cured at room temperature for 24 h under the presence of a magnetic field. The elastomer's cross-sectional shape was standardized as a 25 × 25 mm with a thickness of 10 mm, as shown in Fig. 2(b).

2.2 Mechanical Characteristics of MREs

To investigate the change in shear force concerning MR elastomer's magnetic field, the measurement device is shown in Fig. 2. During the measurement, the room temperature was always kept at 26 °C. It is expected that when a current is applied to the electromagnet, clusters of magnetic particles are formed along the lines of magnetic force in the MR elastomer. As a result, the magnetic particles' binding force increases, and the amount of deformation in the shear direction decreases. In this experiment, we examined the changes in static restoring force characteristics when the applied current value was 0 to 4 A in 0.5 A steps. As shown in Fig. 3, the applied current value and the stiffness can be regarded as nearly linear.

To determine the material's stiffness, the shaker performs harmonic vibrations with different frequencies $x = x_0 \sin(\omega t)$. Then the force response is also a harmonic function $F = F_0 \sin(\omega t)$. The stiffness and the damping of the material are determined as,

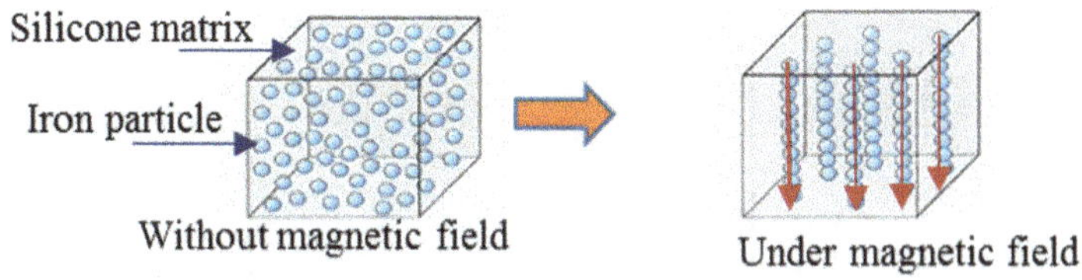

Fig. 1. The change in structure of the MRE under magnetic field

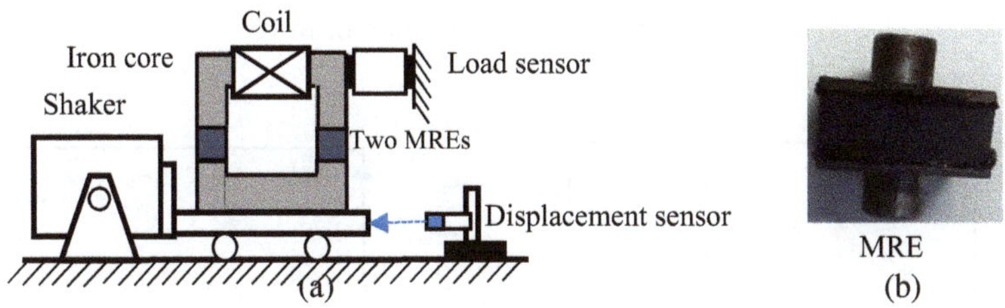

Fig. 2. Experiment setup to measure MRE mechanical properties: (a) schematic, and (b) photograph of MRE sample.

$$k = \frac{F}{x} = \frac{F_0}{x_0}, c = \frac{\Delta E}{2F_0 x_0} \tag{1}$$

where x_0 is the amplitude of the curve vibration measured by the displacement sensor, F_0 is the amplitude of force measured by the force sensor, and ΔE is the loss of energy per cycle calculated by the area of the hysteresis loop. The value of the amplitude is determined by the maximum value measured from the sensors.

The equivalent stiffness and damping coefficient increase as frequency increases, as shown in Fig. 4. Equivalent stiffness increases with a large value when the material works at low frequencies (1–15 Hz) and increases with a small value in the high-frequency range (>15 Hz). The damping coefficient increases continuously as the frequency increases. The mechanical properties of materials in the magnetic field domain are analyzed in detail. The material's stiffness increases significantly when the applied current turns up from 0 A to 4 A, and then the stiffness increases slowly and reaches the saturation value at 6 A. The damping increases slightly when applying a magnetic field. It is a similar trend in different frequencies. It can be summarized that mechanical properties are nonlinear and depend on excitation frequency and applied magnetic field.

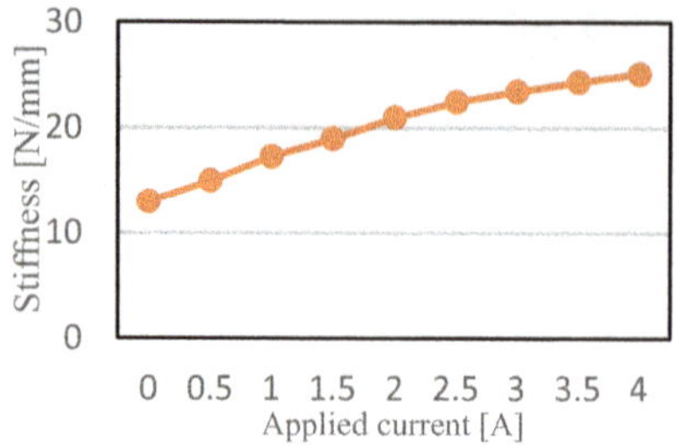

Fig. 3. Static stiffness variation of MRE

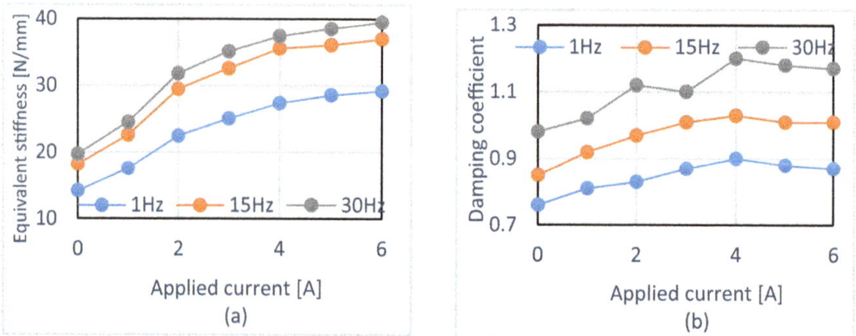

Fig. 4. Mechanical characteristic of MRE in magnetic field domain for different frequencies: (a) the equivalent stiffness and (b) the damping coefficient

3 Semi-active Control of 1-Dof System

3.1 1-DOF Vibration System

The model of the 1-DOF vibration system is shown in Fig. 5. The dynamic model of the system is described as:

$$m\ddot{x} = -k_0(x - u) - c_0(\dot{x} - \dot{u}) - F_{MRE}, \tag{2}$$

$$m\ddot{x} + k_0 x + c_0 \dot{x} + F_{MRE} = k_0 u + c_0 \dot{u} \tag{3}$$

where u, x is the excitation displacement and mass displacement, respectively, m is the mass, k_0 and c_0 are the nominal stiffness and nominal damping coefficient of MRE when no magnetic field is applied.

The force of MRE is given by:

$$F_{MRE} = k_m(x - u) + c_m(\dot{x} - \dot{u}), \tag{4}$$

k_m and c_m are the increment stiffness and increment damping coefficient when the magnetic field is applied. To simplify the equation, let $k = k_0 + k_m$ and $c = c_0 + c_m$.

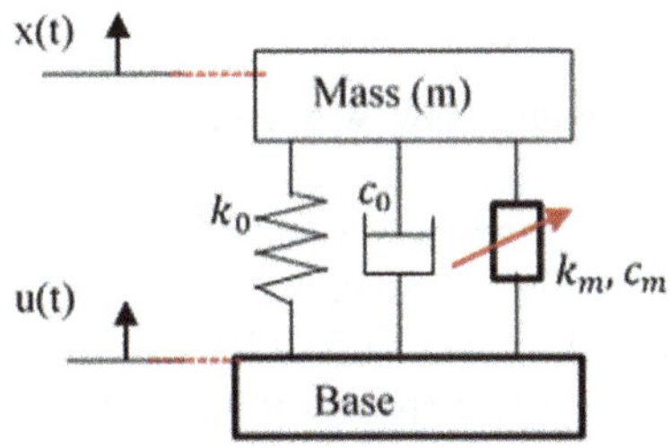

Fig. 5. 1-DOF vibration system

A control algorithm based on the Lyapunov stability theory has been applied. The function for the system (Eq. 2) can be rewritten to as,

$$m\ddot{x} + c_0(\dot{x} - \dot{u}) + k_0(x - u) + k_m(x - u) = 0, \tag{5}$$

Choose a Lyapunov function

$$V(x, \dot{x}) = \frac{1}{2}\left(k_0 x^2 + m\dot{x}^2\right) \tag{6}$$

The derivative of Lyapunov function combined Eq. (2) is deployed as:

$$\begin{aligned}\dot{V}(x, \dot{x}) &= k_0(x - u)\dot{x} + m\dot{x}\ddot{x} \\ &= c_0(\dot{u} - \dot{x})\dot{x} - k_m(x - u)\dot{x} \\ &= -c_0\dot{x}^2 + (c_0\dot{u} + k_0 u)\dot{x} - k_m(x - u)\dot{x}.\end{aligned} \tag{7}$$

The semi-active control algorithm below is proposed for the function $\dot{V}(x, \dot{x})$ to achieve the minimum value. As analyzed in Sect. 2, the MRE's variable stiffness can adjustable from $k_m = 0$ to maximum value k_m^{max} corresponds to the applied dynamic range 0–4 A. The following control scheme, which defines the stiffness of MRE, can minimize the (7)

$$k_m = k_m^{max} H((x - u)\dot{x}) \tag{8}$$

where k_m^{max} is the maximum increase in stiffness of MRE corresponding to the maximum value of applied current (I = 4 A), $x_r = x - u$ is the relative displacement between mass and base excitation H is Heaviside step function. From Eq. (8), the applied current is defined as,

$$I = I_{max} H((x - u)\dot{x}) \tag{9}$$

3.2 Experimental Results

Since the variable stiffness control law was considered for the effectiveness of the vibration system. An experimental with semi-active variable stiffness control was performed. The experimental equipment was used for the viscoelasticity test. As shown in Fig. 6, the upper core, including the coil, was regarded as the mass, and two MREs placed between the upper and the lower parts of the electromagnet. Two parts above are assumed to emulate a mass and two springs for the system. The value of the spring stiffness reaches the maximum or minimum value by applying a corresponding current 0 A or 4 A, respectively. A coil with a diameter of 1 mm, 800 turns to create magnetic field strength in the iron core and through the MRE. The lower iron was rigidly attached to the vibrating table to create base excitation. The shaker generates random excitation with a cutoff of 50 Hz. Both the displacement of the mass and the base vibration system were measured with a laser displacement. The vibration isolation was evaluated by obtaining the transfer function between both displacements, which corresponds to displacement's transmissibility, with an FFT analyzer. The displacement signals are sent to the digital signal controller. The applied current is provided by using a DC power supply so that the stiffness of MRE achieve max-min values, respectively.

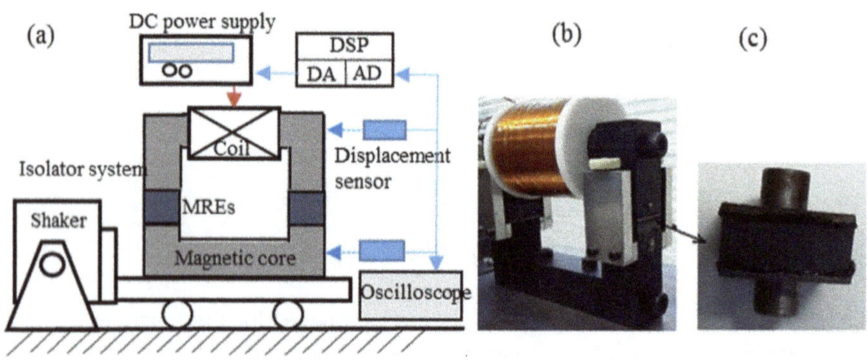

Fig. 6. Experiment setup for 1-dof vibration system using MRE: (a) schematic, (b) isolator system photo and (c) MRE samples

The experiment results are shown in Figs. 7 and 8. The results are compared between the two passive systems (applied current values of 0A, 4A) and the semi-active controller system. Passive systems with zero applied current have the lowest natural frequency (16 Hz). Passive systems with 4 A current applied to have the highest specific frequency (27 Hz), as shown in Fig. 7. These results are following the MRE material characteristics mentioned in Sect. 2. The transmissibility displacement response has been significantly reduced when using a on-off semi-active controller, as shown in Fig. 8. We performed each case over a long period of time and calculated the root mean square (RMS) for comparison. The RMS for the cases is: the case of passive (0A) RMS = 1.08; case of passive on (4A) RMS = 0.96; case of the on-off semi-active control RMS = 0.75. So, the system using MRE material with a semi-active controller is effective in reducing vibration.

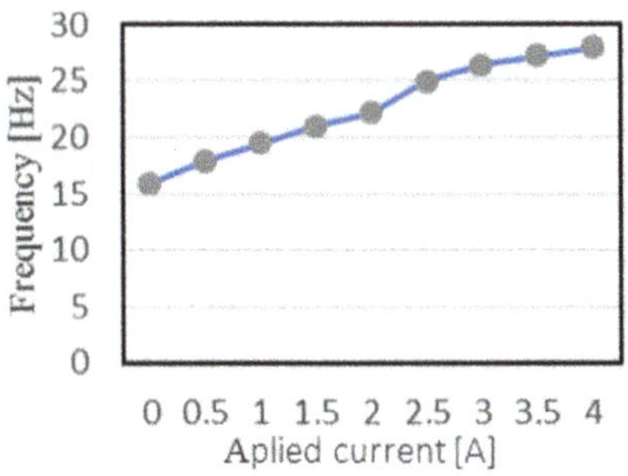

Fig. 7. Natural frequency variation

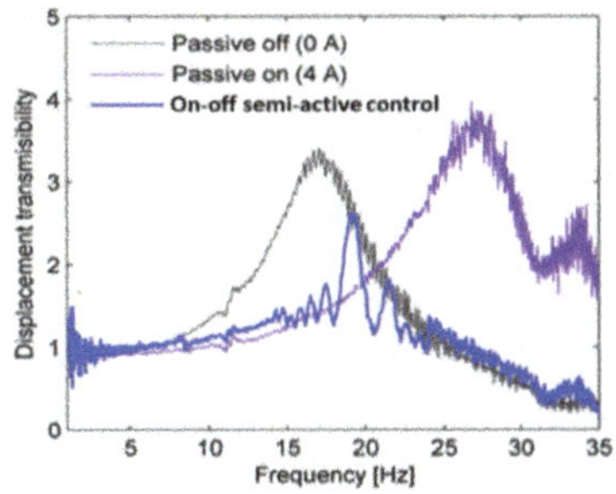

Fig. 8. The transmissibility by frequency

4 Conclusion

This research used MR elastomer as new functional material and evaluated its static and dynamic properties experimentally. This elastomer was applied to semi-active vibration control of a one-degree-of-freedom vibration system subjected to displacement excitation. The effectiveness of the vibration system was evaluated by combining variable stiffness of MRE and on-off control.

Acknowledgements. This research is funded by Funds for Science and Technology Development of the University of Danang under project number B2019-DN06–16.

References

1. El Majdoub, K., Ghani, D., Giri, F., Chaoui, F.Z.: Adaptive semi-active suspension of quarter-vehicle with magnetorheological damper. J. Dyn. Syst. Measurement Control **137**(2) (2015)
2. Nie, S., Zhuang, Y., Liu, W., Chen, F.: A semi-active suspension control algorithm for vehicle comprehensive vertical dynamics performance. Vehicle Syst. Dyn. **55**, 1099–1122 (2017)
3. Nguyen, X.B., Komatsuzaki, T., Iwata, Y., Asanuma, H.: Robust adaptive controller for semi-active control of uncertain structures using a magnetorheological elastomer-based isolator. J. Sound Vib. **343**, 192–212 (2018)
4. Nguyen, X.B., Komatsuzaki, T., Iwata Y., Asanuma H.: Fuzzy semi-active control of multi-degree-of-freedom structure using magnetorheological elastomers ASME Proceedings. No: ASME 2017 Pressure Vessels and Piping Conference Volume 8: Seismic Engineering, 2017, pp. 1–10 (2017)
5. Opie, S., Yim, W.: Design and control of a real-time variable modulus vibration isolator. J. Intell. Mater. Syst. Struct. **22**(2), 113–125 (2011)
6. Liao, G.J., Gong, X.L., Xuan, S.H., Kang, C.J., Zong, L.H.: Development of a real-time tunable stiffness and damping vibration isolator based on magnetorheological elastomer. J. Intell. Mater. Syst. Struct. **23**(1), 25–33 (2011)

7. Nguyen X. B., Komatsuzaki T., Iwata Y., Asanuma H.: Fuzzy Semiactive Vibration Control of Structures Using Magnetorheological Elastomer, Shock and Vibration, Hindawi, Volume 2017. Pages: 15p.
8. Jansen, L.M., Dyke, S.J.: Semi-active control strategies for MR dampers: a comparative study. J. Eng. Mech. ASCE **126**(8), 795–803 (2000)
9. Yi, F., Dyke, S., Caicedo, J., Carlson, J.: Experimental verification of multi input seismic control strategies for smart dampers. J. Eng. Mech. **127**, 1152–1164 (2001)
10. Nguyen, X.B., Komatsuzaki, T., Iwata, Y., Asanuma, H.: Modeling and semi-active fuzzy control of magnetorheological elastomer-based isolator for seismic response reduction. Mech. Syst. Signal Proces. **101**, 449–466 (2018)
11. Nguyen, X.B., Komatsuzaki, T., Zhang, N.: A nonlinear magnetorheological elastomer model based on fractional viscoelasticity, magnetic dipole interactions, and adaptive smooth Coulomb friction. Mech. Syst. Signal Process. **141**, 106438 (2020)

Reconstruction of Defect Configuration in Plane Construction Using Numerical Modeling and Neural Networks at Positional Scanning

A. N. Soloviev[1,2], I. A. Parinov[1], and A. V. Cherpakov[1,2(✉)]

[1] Southern Federal University, Rostov-on-Don, Russia
[2] Don State Technical University, Rostov-on-Don, Russia

Abstract. The results of modeling a plane construction with defect are presented. The aim of the work is to identify the parameters of the defect based on the use of finite element modeling (FEM) and artificial neural network (ANN). The properties and configuration of the defect differ from the layer in which it is located. The calculation of construction oscillations is carried out during its impulse loading at a certain point of the construction surface. The shape of the defect is considered as identifiable parameter of the defect. Neural networks are used as a tool for analysis. Surface waves excited by impulse loading are considered as input parameters. The results of the restoration of defect parameters obtained on the base of a combination of FEM and ANN are presented.

Keywords: Layer construction · FEM · Impact load · Surface waves · Defect · ANN

1 Introduction

In connection with the development of approaches to researching construction infrastructure facilities, as well as organizing support for project activities, assessing sites for the construction of environmentally hazardous structures, it is impossible without information on the structure of their geological foundation. Into framework of the applied problem, it is supposed to construct an optimal design scheme of the geological foundation. A map for constructing an optimal structure in practice is obtained using expensive methods of physical examination of soils, namely, drilling tests, laboratory studies of the composition and properties of soils. Deformable elastic structures in the elastic statement of problem can be considered as models of geological structures. Soil foundations in the most cases are modeled as layered homogeneous structures. As a simple model, a set of layers in the horizontal area with elastic couples can be used.

By considering particular cases of layered constructions, problems with the presence of inhomogeneities and defects of various structures can be considered. For design purposes, simulation can most often be applied in the form of certain simplified models of the defects in the foundation in the form of cracks, delamination, inclusions and void.

B. T. Long et al. (Eds.): MMMS 2020, LNME, pp. 233–237, 2021.
https://doi.org/10.1007/978-3-030-69610-8_32

Spectral analysis and multichannel analysis of surface waves are among the methods for diagnosing the parameters of layered structures. The Spectral Analysis of Surface Waves (SASWs) method is discussed in [1, 2] and the Multichannel Analysis of Surface Waves (MASWs) method is described in [3–6]. These methods are based on the study of the dependence of the dispersion characteristics of the measured waves of the medium. The analysis discusses the processing vibrations measured at specific points in an extended medium.

Methods for solving the problems of restoring the parameters of a layered structures are presented in [7–10] and based on the use of analytical and numerical solutions obtained into framework of inverse problems. To solve the problem of restoring properties, information on the wave parameters of a certain area of the structure is required. The time range of the vibrational response of the structure should include the areas requiring restoration of the properties.

The application of artificial neural networks (ANNs) is considered as an analysis tool [11]. In works [11–16] some solutions of the reconstruction problems of the damage state of structural elements are considered. In [18], the authors pointed out the advantages of identification methods which did not require preliminary statement of a mathematical model of the research object.

Application of FE modeling is sufficiently effective [19–21]. The parameters of natural vibration modes as well as angular values of the vibration curves showed fairly good correlation with the size of the defect in the simulated structure.

2 Statement of Problem

The problem is solved in a plane statement using the finite element method (FEM). In this work, we use the linear equations of the theory of elasticity taking into account an energy dissipation, adopted in the ANSYS software. The goal of the research is to restore information about the configuration of a defect based on the analysis of vibration parameters.

3 Description of the Model

A plane structure is considered and modeling is carried out in the FE ANSYS software. A construction considered has inner layer 1 and outer layer 2. The elastic characteristics of the layers are the same: Young's modulus $E = 0.1$ GPa, density $\rho = 2000$ kg/m^3, and Poisson's ratio $\nu = 0.33$. The properties of the defect: $E = 1$ GPa and the other properties are identical to the base layer. Layer 1 has a lower damping coefficient $\xi_1 = 0.001$. At the same time, layer 2 has a higher damping coefficient $\xi_2 = 0.1$. This distinction is required to fixing the passage of reflected waves from the side and bottom sides of the structure. All layers are rigidly coupled to each other (Fig. 1).

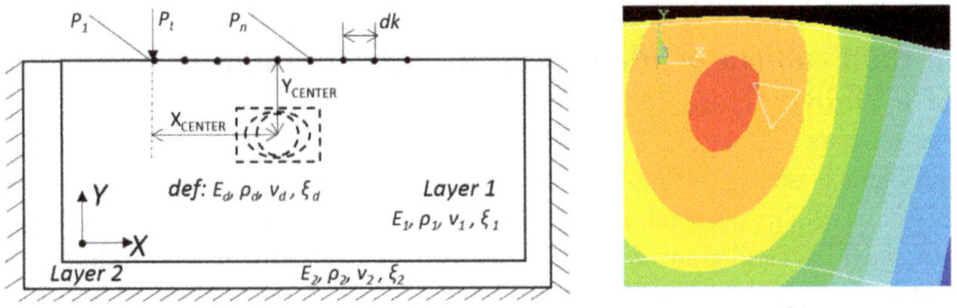

(a) (b)

Fig. 1. (a) Model of plane structure with defect; (b) example of displacement field Y at time t = 0.06 s in the area of triangular defect based on the results and modeling in the FE complex Ansys

Table 1. Configuration of defective area

Cluster no.			
1	2	3	4
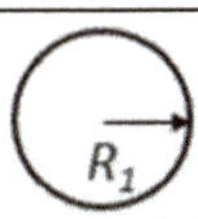	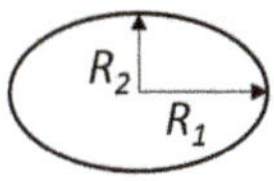	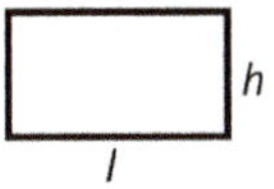	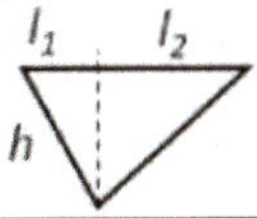

The defect is located at a depth of Y_{center} = 1.5 m from the surface and at a distance of X_{center} = 2 m from the point P_0 of impact loading. The displacement field of the plane structure is calculated as a result of a short-term impulse load. The shape of the defect area is considered as the variable parameter under study. The shape of the defect area can be as follows: (i) circle of radius R, (ii) ellipse-like structure R_1–R_2, (iii) rectangular structure h–l, (iv) triangular structure h-l_1-l_2. Thus, 4 clusters of defective areas are modeled. The variation of the corresponding size parameters (R_i, l_i, h) lies in the range 0–0.5 m. It is assumed that under dynamic loading of the structure, the cluster number of the structure and its size parameters (R_i, l_i, h) depending on the value of the vibration responses will be found. The positional scanning is applied and the vibration responses are measured at the corresponding points P_n of the structure (Fig. 1).

Modeling the layered structure (with a width of 10 m and a depth of 7 m) with defect was carried out in the FE ANSYS software: triangular and quadrangular finite elements of PLANE82 type with plane deformation were used. During the simulation, the structure was broken by more than 3000 finite elements. The estimated time was chosen so that the impulse from external boundaries of the structure did not return back. The impact load was applied at the central point P_1 of the structure surface.

In the result of impulse loading, transverse and longitudinal wave displacements of points occur both as on the construction surface as and throughout its entire volume. Control points (43 points) are located on the surface at a distance of d_k = 0.1 m from each other (see Fig. 1). The first point is located at a distance of 0.1 m from the impact site.

4 Application of Neural Network Technologies in the Problem of Identifying the Cluster Number of a Hidden Defect

The problem is stated on determining the cluster number for classifying the defect shape. Let us establish a relationship between the vibration amplitudes obtained on the base of the FE calculation of oscillations in a layered structure. The defect variant number is taken as an output parameter of the ANN. Based on the obtained data, a training set is formed at the stage of FE modeling. The constructed model of the neural network is trained with its help. To identify the layer thickness, we used a fully coupled multilayer ANN model in the Matlab software. The neural network model, containing 10 hidden layers, consisted of 43 input neurons and 1 output neuron. Rectified Linear Unit (ReLU) was selected as the input activation function, and Sigmoid was stated as the output activation function. For the training and test sampling, consisting of 150 vectors, it was found that 5 training epochs are enough to achieve the required level of ANN performance, in the case of using the amplitude values in the training set. Figure 2 shows the results of the mean square error (MSE) by training the ANN and the estimated data by testing a sample of random vectors with oscillation data. As the analysis performed showed, the accuracy of MSE was 99%.

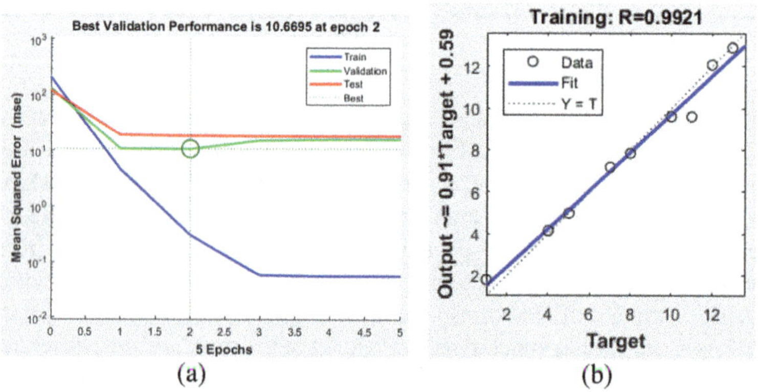

Fig. 2. (a) Change in mean square error (MSE) of neural network during its training based on the values of amplitudes; (b) output parameters of test sampling

5 Conclusion

The results of modeling vibrations of plane structure with single defect are presented. Positional scanning of vibration parameters is performed. The analysis of restoration of the cluster configuration of the defect was considered. An ANN with 10 hidden layers was used as an analysis tool. The calculation results showed that the use of ANN allowed one to reconstruct the configuration of the defect region with a sufficiently high degree of correlation $R = 0.99$.

Acknowledgment. Research was financially supported by Southern Federal University, grant No. VnGr-07/2020–04-IM (Ministry of Science and Higher Education of Russia).

References

1. Kim, D.-S., Seo, W.-S., Lee, K.: NDT and E Int. **39**(2), 143–154 (2006)
2. Badsar, S., Schevenels, M., Haegeman, W., Degrande, G.: Geophys. J. Int. **182**(3), 1493–1508 (2010)
3. Park, C.B., Miller, R.D., Xia, J.: Geophysics **64**(3), 800–808 (1999)
4. Xia, J., Miller, R.D., Park, C.B.: Geophysics V **64**, 691–700 (1999)
5. Jaganathan, A.P.: J. Appl. Geophys. **180**, 104132 (2020)
6. Park, C.B., Miller, R.D.: J. Environ. Eng. Geophys. **13**(1), 1–11 (2008)
7. Morabie, K.M.: J. Appl. Math. Mech. **76 I 6**, 732–737 (2012)
8. Lowe, M.J.: IEEE Trans. Ultrasonics Ferroelectrics Frequency Control **42**(4), 525–542 (1995)
9. Vatul'jan, A.O.: Inverse Problems in Solid Mechanics, p. 224. Moscow, Fizmatlit (2007)
10. Brigante, M., Sumbatyan, M.A.: Russ. J. Nondestruct. Test. **49**(4), 185–195 (2013)
11. Haykin, S.: Neural Networks - A Comprehensive Foundation. Prentice Hall. (1998). 842 p.
12. Krasnoshchekov, A.A., Sobol, B.V., Solov'ev, A.N., Cherpakov, A.V.: Russ. J. Nondestruct. Test. **47**(6) (2011)
13. Waszczyszyn, Z., Ziemianski, L.: Comput. Struct. **22**(79) (2001)
14. Solov'ev, A.H., Kurbatova, P.S., Saprunov, N.I., Shevcov, S.N. Materialy, X.: Mezhdunar-odnoj konferencii "Sovremennye problemy mehaniki sploshnoj sredy. In: Proceedings of the 10th International Conference Modern Problems of Continuum Mechanics. Rostov-on-Don (2006)
15. Liu, S.W., Huang, J.H., Sung, J.C., Lee, C.C.: Comput. Meth. Appl. Mech. Eng. 191 (2002)
16. Solov'ev, A.N., Cherpakov, A.V., Vasil'ev, P.V., Parinov, I.A., Kirillova, E.V.: Adv. Eng. Res. **20**(3), 205–215 (2020). (In Russ.) https://doi.org/10.23947/2687-1653-2020-20-3-205-215
17. Cherpakov, A.V., Vasiliev, P.V., Soloviev, A.N., et al.: Advanced Materials. Proceedings of the International Conference on "Physics and Mechanics of New Materials and Their Applications", PHENMA 2019, pp. 439–447 (2020)
18. Fang, X., Luo, H., Tang, J.: Comput. Struct. 85 (2005)
19. Cherpakov, A.V., Shilyaeva, O.V., Grigoryan, M.N., Polyakova, T.V.: IOP Conf. Ser. Mat. Sci. Eng. **698**(6), 066021 (2019)
20. Akopyan, V.A., Solov'Ev, A.N., Cherpakov, A.V., Shevtsov, S.N.: Russ. J. Nondestruct. Test. **49**(10), 579–583 (2013)
21. Cherpakov, A.V., Shlyakhova, E.A., Egorochkina, I.O., Kokareva, Y.A.: Materials Sci. Forum **931** MSF 373–378 (2018)

Influence of Machining Parameters on the TPIF Formability for Aluminum Sheet at Room Temperature

Ma Van Viet[1], Nguyen Truong Thinh[1], and Le Van Sy[2](✉)

[1] Department of Mechanical Engineering, HCMC University of Technology and Education, Ho Chi Minh City, Vietnam
{vietmv.ncs, thinhnt}@hcmute.edu.vn
[2] PetroVietnam University, Ba Ria city, Vietnam
sylv@pvu.edu.vn

Abstract. Two Point Incremental Forming (TPIF) technology is very flexible forming technology, cost-effective, and potential advantages (high flexibility, simplicity, less-time consumption). This technology uses a hemispherical-end tool controlled by CNC milling machine to deform a metal sheet which is fixed on the jig/fixture. The jig/fixture is firmly positioned on the CNC milling machine table which includes clamped plate, movable plate base plate and semi-support. The metal sheet material is clamped between movable plate and clamp plate by 8 bolts and can move down along 4 guide bars. The sheet material is deformed layer by layer by the hemispherical-end tool following CNC tool path. This paper aims influence of machining parameters on the formability of aluminum sheet at room temperature which is deformed by TPIF technology. The investigated parameters are depth step, feed rate, tool diameter, and spindle speed through Design of Experiment (DOE). The formability has been known the ability of metal sheet to deform until happen failure through the achieved forming angle.

Keywords: Incremental sheet forming · TPIF · Formability · Maximum forming angle · Metal sheet material · Design of Experiment

1 Introduction

Incremental sheet forming technology is invented by Leszak in 1967 [1]. After that, it is researched by many studying groups. ISF technology uses a head forming tool deform a sheet material layer by layer to complete product. The forming tool is controlled by a toolpath which is exported from Creo 6.0 software. A sheet material is fixed on a simple frame which is fixed on the table of 3-axis CNC milling machine. The ISF has got two sorts of Single Point Incremetal Forming (SPIF) and TPIF. The TPIF technology is moldless machining technology with many advantages (high flexibility, simplicity, less-time consumption, and etc.). It uses more and more in the other countries. A review of TPIF studies such as.

B. T. Long et al. (Eds.): MMMS 2020, LNME, pp. 238–245, 2021.
https://doi.org/10.1007/978-3-030-69610-8_33

Attanasio [4] do experiments on a car door handle cavity for evaluating geometrical and dimensional errors, and surface finishing, between TPIF and SPIF with the same working parameters. TPIF assures the achievement of a better dimensional accuracy and surface finishing.

Hani Mostafanezhad [2] studied experimental study based on response surface methodology (RSM) was carried out to analyze effect of wall angle, tool nose diameter, initial sheet thickness and step down on thinning ratio and forming force during TPIF of AA1050 truncated cone. A series of experiments was carried out based on Box-Bhenken experimental design and mathematical models of responses are developed by means of RSM and analysis of variances. Response surface methodology, optimal parameter setting regarding minimum thinning ratio and forming force. Asghari [3] were optimate by grey relational analysis with response factors (min thickness, springback, surface roughness) optimize parameters were 15 mm tool nose diameter, 63° wall angle, 800 r/min spindle speed and 0.2 mm deep step with cone shape, analyze formability of aluminum 1050 in TPIF.

Although there are a lot of teams study geometrical accuracy, surface quality, compare forces in SPIF and TPIF, and etc. on TPIF, the studys of sheet material formability have not been investigated clearly. This research uses a step frustum cone shape (every step with 1°) (Fig. 3) to investigate influence of machining parameters on formability of aluminum sheet at room temperature by TPIF technology. Survey limited angle is from 65°–85°. This investigated model is a new model in study influence of machining parameters on the formability by TPIF technology which it has never been used the last researches. The machining parameters are considered their effects on the formability (depth step, feedrate, tool diameter, and spindle speed) through DOE strategy. The achieved results show that it can put into practice in production.

2 Experimental Equipment

In this equipment, a sheet material with dimension of 400 × 400 mm^2, 1.5 mm in thickness is fixed between movable plate and clamp plate by 8 bolts and can move down along 4 guide bars on the jig/fixture. The jig/fixture (Fig. 1) is firmly positioned on the CNC milling machine table. The sheet material is deformed layer by layer by the hemispherical-end tool with 6 mm, 12 mm, 18 mm diameter following CNC tool path. The end of a round steel bar and a steel ball (good hardness and wear resistance) are welded with each other to create the forming tool (Fig. 3). The forming tool is checked inversion. Solid graphite powder and lithium grease with scale 1:1 and lubrication oil (multi 20W-50) are mixed each other to create a mixed lubrication (Fig. 2).

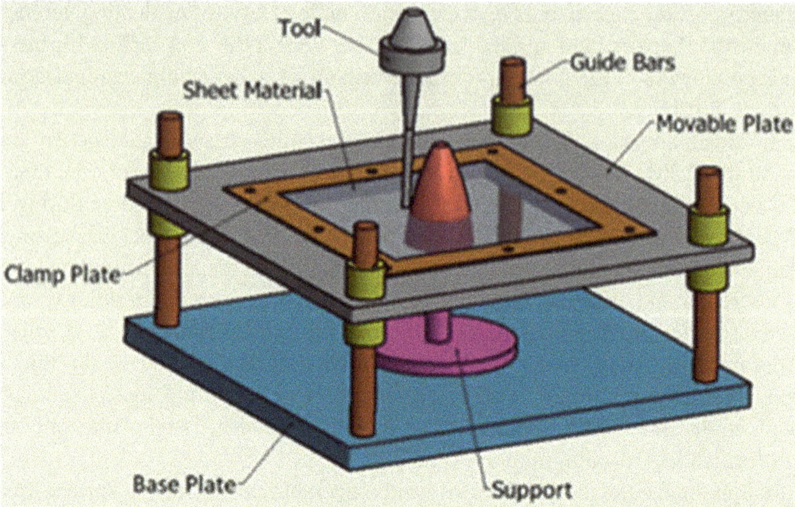

Fig. 1. Jig and fixture system

Fig. 2. Experiment system for TPISF process

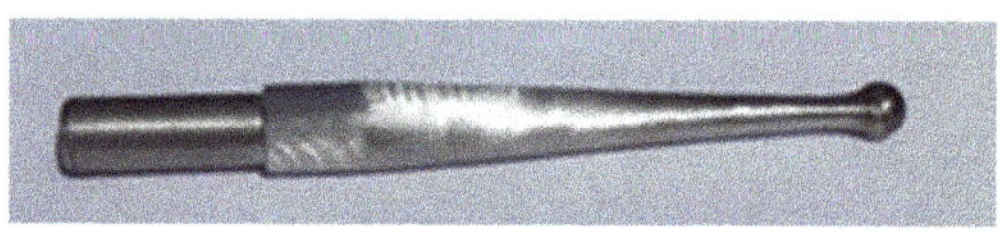

Fig. 3. Forming too

3 Experimental Design

A step frustum cone shaped (Fig. 4) is built to investigate forming angle from 65° to 85°. Formed product is investigated about maximum wall angle through DOE to determine the influence of the machining parameters on the TPIF formability for aluminum sheet at room temperature. The Box-Behnken experiment mode is used 4 factors, 5 center points, 29 base runs (Table 2), with factor level (Table 1). Response parameter of DOE is the maximum wall angle in the experiments. The formed products are measured height from head of product to any mechanical failure (Fig. 6). Every height value is converted into a wall angle and inserted into the design matrix (Fig. 5).

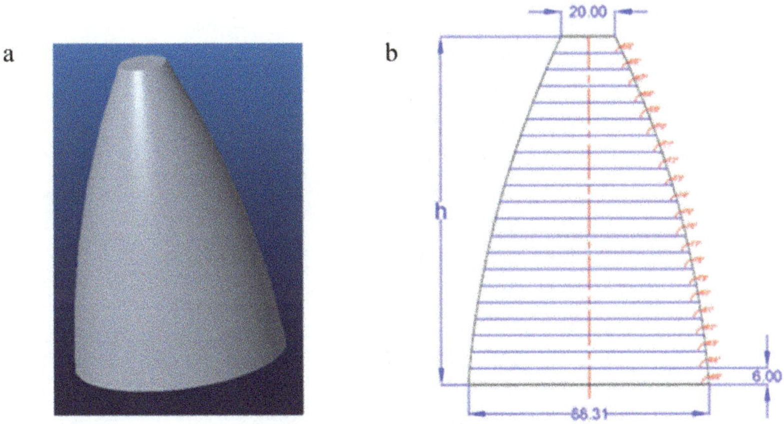

Fig. 4. A (a) CAD model of forming product, (b) Geometric profile of forming product.

Table 1. Machining parameters for experimental design

No	Machining parameters	Unit	Range of values	
			Low level	High level
1	Depth step (z)	mm	0,1	1.5
2	Feed rate (V_{xy})	mm/minute	300	1500
3	Tool diameter (D)	mm	6	18
4	Spindle speed (n)	rpm	300	1800

Fig. 5. The formed products

Fig. 6. Measuring the forming height

Table 2. Design of experiments and results for forming angle

Run	z	V_{xy}	n	D	Wall angle (degree) α
1	0.8	900	1050	12	82
2	0.8	1500	300	12	83
3	0.8	900	1800	18	81
4	1.5	900	300	12	82
5	0.1	900	300	12	80
6	0.8	300	1050	18	80
7	0.8	900	1050	12	82
8	0.8	900	1050	12	82
9	0.8	900	1050	12	82
10	1.5	300	1050	12	81
11	0.8	900	1050	12	82
12	0.8	300	300	12	81
13	0.8	1500	1800	12	83
14	0.1	900	1050	18	80
15	0.8	300	1050	6	83
16	0.8	900	300	6	83
17	0.8	1500	1050	6	84
18	0.8	900	300	18	81
19	1.5	900	1050	18	80
20	1.5	1500	1050	12	83
21	0.8	900	1800	6	83
22	0.8	300	1800	12	82
23	0.1	1500	1050	12	81
24	0.1	900	1800	12	80
25	1.5	900	1050	6	83
26	0.8	1500	1050	18	82
27	0.1	900	1050	6	80
28	0.1	300	1050	12	79
29	1.5	900	1800	12	82

4 Results and Discussion

The model summary (Table 3), the experiments model is 96.25% obtain statistically significant data. According to Table 3, we can analyze the effect of machining parameters on formability of aluminum sheet at room temperature. A relationship among machining parameters and response.

$\alpha = 79.17 + 7.374z + 0.000556\ V_{xy} + 0.000111n - 0.0863D - 2.451\ z^2 - 0.1786zD + 0.000069\ V_{xy}\ D.$

A relationship among processing parameters and response, it shows that the formability is proportional to feed rate (V_{xy}), spindle speed (n), and inverse proportion to tool diameter (D).

These machining parameters are effect on formability and product quality. So we can optimizing to have good results. In this experiment, optimized machining parameters are depth step 1.29 (mm), feed rate 1500 (mm/minute), tool diameter 6 (mm), spindle speed 1800 (rpm).

According to the analysis of variance (ANOVA) result (Table 3), we can see the percent influence of parameters and the interaction parameters such as depth step (z) 22.49%, feed rate (V_{xy}) 18.59%, tool diameter (D) 26.77%, spindle speed (n) 0.19%, interaction parameter z * z 22.63%, interaction parameter zD 5.02%, interaction parameter V_{xy} D 0.56% (Fig. 7).

Table 3. Analysis of variance (ANOVA)

Source	DF	Adj SS	Adj MS	F-Value	P-Value
Model	8	43.1462	5.3933	64.15	0.000
Z	1	10.0833	10.0833	119.94	0.000
V_{xy}	1	8.3333	8.3333	99.13	0.000
n	1	0.0833	0.0833	0.99	0.331
D	1	12.0000	12.0000	142.74	0.000
z*z	1	10.1462	10.1462	120.69	0.000
z*D	1	2.2500	2.2500	26.76	0.000
Error	20	1.6814	0.0841		
Total	28	44.8276			
S	R-sq	R-sq(adj)	R-sq(pred)		
0.289946	96.25%	94.75%	88.96%		

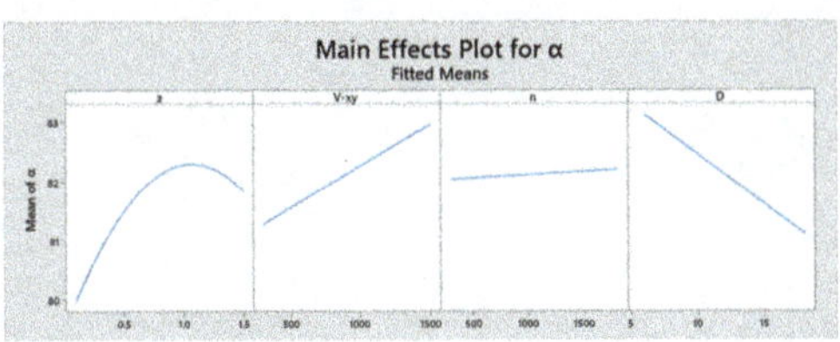

Fig. 7. Main effects

Through residual analysis (Fig. 8) carried out a model adequacy examination. Experimental data are the same with the predicted results of the constructed model. Figure 9 shows all residual plots for formability. Figure 8 (normal probability plot) shows the residuals concentrate in a straight line approximately. So, the error distribution is normal, the observed results are the same predicted results. The plot of the residual versus fitted values show Fig. 8 (versus fits), a confidence interval with 95% for all values. In Fig. 8 (histogram), the histogram of residuals is distributed normally. In Fig. 8 (versus order), the residuals are independent. According to the analysis of residuals, the constructed model is suitable for predicting formability, and all residuals are falling within control limits.

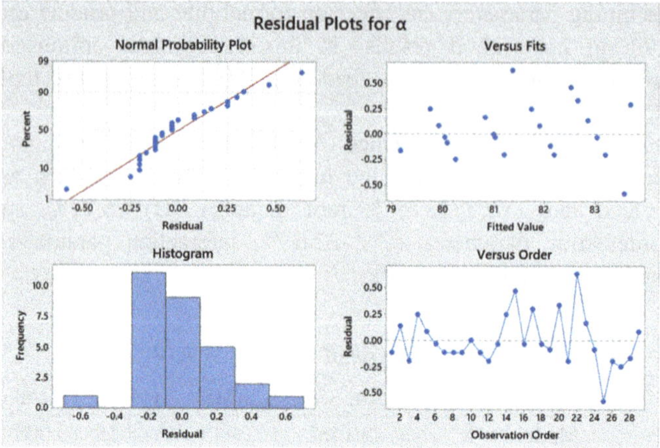

Fig. 8. Residual Plots for α

The interactive influences of independent parameters on formability show the 3D surface plot of formability (Fig. 9). It base on a relationship among machining parameters and responses. These graphs show change of two factors. The roughness of the surface exposed to the forming tool is smaller than the other surface, due to many different reasons such as contact conditions, lubrication and machine parameters. Therefore, we can control the best surface quality.

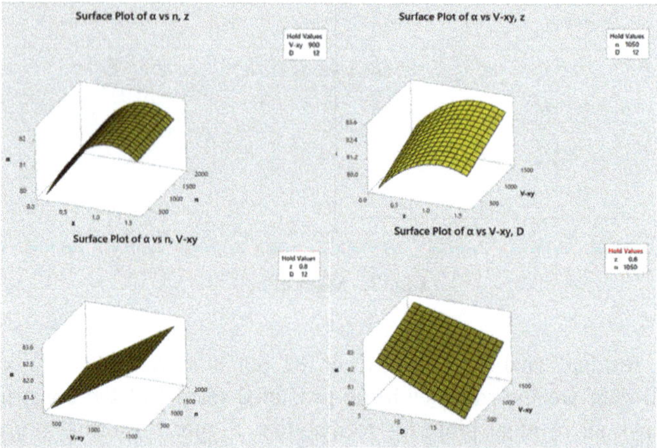

Fig. 9. Surface Plot of α vs n, z; α vs Vxy,z; α vs n, Vxy; α vs Vxy, D

5 Conclusion

In order to have good results, the jig and fixture system for TPIF process is important to reach good results. It is designed very stable to create some results as follows.

According to the analysis results, the experiments model is 96.25% with investigated machining conditions.

The percent influence of parameters and the interaction parameters such as depth step (z) 22.49%, feed rate (V_{xy}) 18.59%, tool diameter (D) 26.77%, spindle speed (n) 0.19%, interaction parameter zz 22.63%, interaction parameter zD 5.02%, interaction parameter $V_{xy}D$ 0.56%.

The formability is proportional to feedrate (V_{xy}), spindle speed (n), and inverse proportion to tool diameter (D).

Optimized machining parameters are depth step 1.29 (mm), feed rate 1500 (mm/minute), tool diameter 6 (mm), spindle speed 1800 (rpm) to obtain maximum formability $\alpha = 84.36°$ by TPIF technology for aluminum sheet A1050 H14, 1.5 mm thickness at room temperature.

Lubrication is an important factor to obtain high quality of product surface. The surface roughness exposed to the forming tool is smaller than the other surface.

Acknowledgments. This research is supported by Key Laboratory of Digital Control and System Engineering (DCSELAB), HCMUT, VNU-HCM under grant number TX2020-20b-01.

References

1. Sy, L.V.: modeling of single point incremental forming process for metal and polymeric sheet. University of Padua, Department of innovation in Mechanics and Management (2009)
2. Mostafanezhad, H., Menghari, H.G., Esmaeili, S., Shirkharkolaee, E.: Optimization of Two-point incremental forming process of AA1050 through response surface methodology. Measurement **127**, 21–28 (2018)
3. Asghari, S.A.A., Sarband, A.S., Habibnia, M.: Optimization of multiple quality characteristics in two-point incremental forming of aluminum 1050 by grey relational analysis. Institution of Mechanical Engineering Science, vol. Proc IMechE Part C (2017)
4. Attanasio, A., Ceretti, E., Mazzoni, L., Giardini, C.: Use of TPIF or SPIF for prototype productions. In: Conference on Material Forming, American Institute of Physics, pp. 978–0–7354–0414–4 (2007)

Deep Level Defects in GaN Grown at Low Temperature by Metalorganic Vapor Phase Epitaxy

Nguyen Hoang Thoan, Nguyen Trung Do, Tran Thien Duc, and Nguyen Ngoc Trung(✉)

School of Engineering Physics,
Hanoi University of Science and Technology, Hanoi, Vietnam
{thoan.nguyenhoang, trung.nguyenngoc}@hust.edu.vn

Abstract. Most of GaN-based light emitting diodes (LED) uses the structure of InGaN/GaN multiple quantum well (MQW) as the active layer in which GaN barrier layer has been grown at lower temperature compared with that of n-GaN and p-GaN. However, it also raises more defects in GaN layer mainly related to the large lattice mismatch and the difference of growth temperature of InN and GaN. These defects could introduce deep levels in the bandgap. To characterize the deep level defects in 400 nm-thick GaN barrier layer grown on top of n-GaN. Deep level transient spectroscopy (DLTS) revealed five deep levels in barrier layer which locate at $E_1 = E_C - 0.21$ eV; $E_2 = E_C - 0.25$ eV; $E_3 = E_C - 0.51$ eV, $E_4 = E_C - 0.69$ eV, and $E_5 = E_C - 1.06$ eV. The total trap density of the low temperature GaN is about 4.0×10^{15} cm^{-3}, which is one order of magnitude higher than GaN template.

Keywords: GaN · Deep traps · Deep level transient spectroscopy

1 Introduction

Nowadays, III-V nitride semiconductor materials have been used widely in fabricating LED and laser diode (LD) in the range of short wavelength. In 1995, Nakamura et. al. made the single quantum well (SQW) of InGaN to fabricate super-bright green LED of 520 nm wavelength [1]. Then, these have been many studies to improve the output power and efficiency. The most potential way is to improve the quality of InGaN/GaN MQW structure[2, 3]. The growth of GaN light emission diodes (LED) need a compromising of growth temperature to obtain high quality InGaN MQW and GaN barrier layer is important because InGaN is grown at relatively low temperature while higher growth temperature of subsequent GaN barrier could degrade the quality of InGaN QW.

To justify the quality of the GaN barrier layer in InGaN/GaN MQW, characterization of the GaN barrier layer in the LED is necessary to improve quality of this layer. In order to study deep level defects in the GaN barrier layer, deep level transient spectroscopy (DLTS) method is used. DLTS, invented by D.V Lang, is a common method for studying deep level traps in semiconductors [4]. The DLTS technique

B. T. Long et al. (Eds.): MMMS 2020, LNME, pp. 246–251, 2021.
https://doi.org/10.1007/978-3-030-69610-8_34

measure capacitances at various condition, therefore, it normally require a Schottky diode or p-n junction. In this experiment, we growth a 400 nm thick layer of low temperature GaN on GaN template. Then the Schottky diode is formed in such layer. We report on the deep level trap characterization of GaN barrier layer in the GaN/InGaN LED. There are five deep level traps observed in the measurement. The most dominant level is $E_c - 0.51$ eV traps and density of about 4.0×10^{15} cm^{-3}.

2 Experiment

Figure 1 showed the structure of samples fabricated in this report in which an Ohmic contact of Ti/Al/Ni/Au was deposited by e-beam evaporator on n^+ type GaN. Ohmic contact was annealed at 700 °C for 1 min in air ambient by using rapid thermal anneanling (RTA) method. Ni/Au layers were used for Schottky contact. The n^+-GaN was doped by Si to archieve electron conentration of 5×10^{18} cm^{-3}. An additional low temperature GaN layer with electron concentration in the order of 10^{17} cm^{-3} was grown on n^+-GaN layer. The net donor concentration of the low temperature GaN layer is $N_D \sim 2.0 \times 10^{17}$ cm^{-3} for **sample 1** growth at 800 °C and $N_D \sim 4.5 \times 10^{17}$ cm^{-3} for **sample 2** growth at 760 °C.

DLTS measurements were performed with the temperature scanning from 80 K to 450 K. The sample was applied under a reverse bias of 0.5 V. The filling pulse amplitude is of 0.5 V and the filling pulse width of 10 ms was used and the rate window was 10 ms/25 ms. After each filling pulse, the emission of carriers trapped in deep level in depletion region creates the capacitance transient which was recorded by DLTS system. Samples were measured in dark condition to reduce an influence of light on emission process. The DLTS data were investigated by using double boxcar at different rate windows for a temperature increasing rate of 0.1 °C. Activation energy and capture cross section of deep level traps were extracted from DLTS spectra.

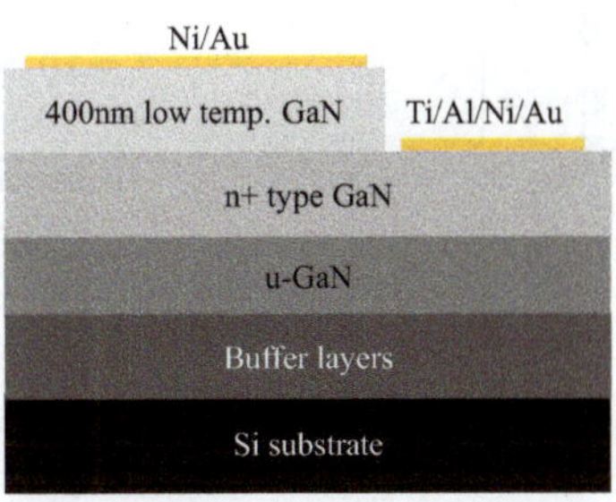

Fig. 1. Structure of the GaN Schottky layers for characterization.

3 Results and Discussion

The n-GaN Schottky diode was characterized by DLTS to obtain the deep trap parameters. The DLTS spectra of the two samples using a rate window of $t_1/t_2 = 10$ ms/25 ms are shown in the Fig. 2. It was found that there are five deep levels (E1 − 5) in two samples. Observing the DLTS peak amplitudes of two samples, the trap E1 and E2 in sample 2 has higher concentration than in sample 1.

Assuming capture cross-section of defects in the n-type GaN σ_n independent on temperature T, which is normally true for GaN material, the following equation is used to identify energy level of the deep level traps [5]:

$$ln\left(\frac{T^2}{e_n}\right) = (E_C - E_T)\frac{1}{k_B T} - ln(\sigma_n \gamma_n) \quad (1)$$

where e_n is emission coefficient of the carrier from deep level traps, T is the temperature at the peak of DLTS spectrum, E_C is conduction band energy, E_T is the trap energy, σ_n is the capture cross-section and $\gamma_n = gN_C v_{th}$ is constant. In the calculation, emission coefficient e_n is calculated using following equation [5]:

$$e_n = \frac{ln\frac{t_2}{t_1}}{t_2 - t_1} \quad (2)$$

where t_1/t_2 is selected rate window in the measurement.

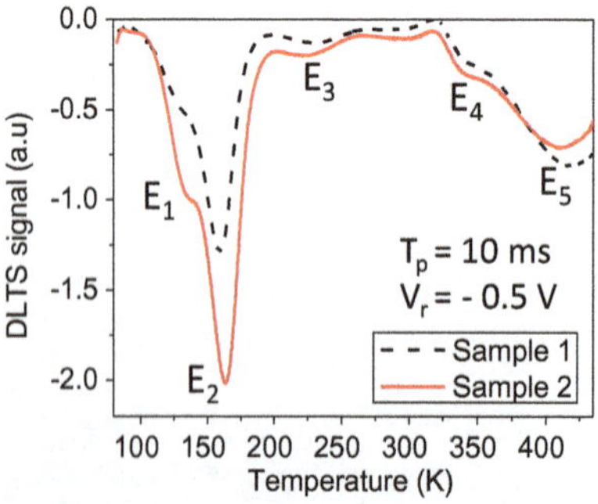

Fig. 2. DLTS spectra of the GaN Schottky diode for GaN samples 1 growth at 800 °C and sample 2 growth at 760 °C.

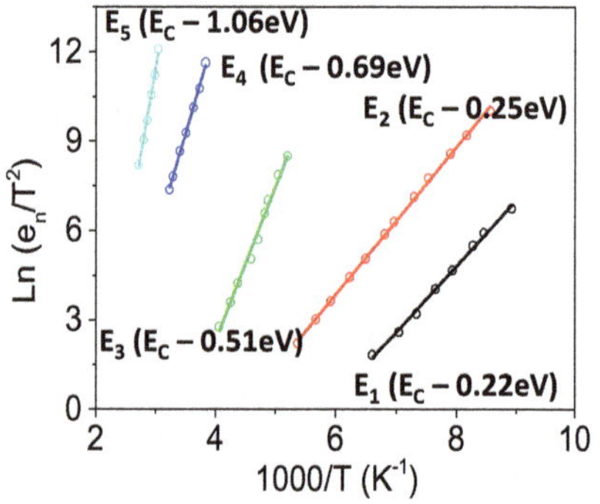

Fig. 3. Arrhenius plot calculated from DLTS measurement to show energy level.

By drawing a plot of relationship of $ln(e_n/T^2)$ versus 1000/T (K^{-1}), Arrhenius plot was obtained (Fig. 3). The slope of the Arrhenius plot give an information of the energy level and the capture cross section. They are located at E_C − 0.21eV; E_C − 0.25 eV; E_C − 0.51 eV; E_C − 0.69 eV; and E_C − 1.09 eV.

The density of the deep level traps is calculated using following equation [5]:

$$N_t \approx 2N_D \frac{\Delta C}{C_0} \quad (3)$$

where N_t is the trap density, N_D is the doping of the GaN layer, shown in the Fig. 2 (b), ΔC is DLTS signal and C_0 is the capacitance in DLTS pulsing condition. Table 1 summarizes the peak position and density of the deep level traps.

As shown in the Table 1, the total density of all the traps is about 4.0×10^{15} cm^{-3}, which is about one order higher than standard GaN [6]. The most dominant traps are E2 located at $E_C - 0.25$ eV. This trap is similar to the most common level $E_C - 0.21$ to 0.26 eV observed in GaN materials [6–9]. The first trap E1 has capture-cross section falls within $8.4 \times 10^{-17} - 1.4 \times 10^{-15}$ cm^{-2}. Fang et al. suggested this trap to be attributed to Ga and N vacancy defects based on a correlation with thickness reduction or dislocation increase [6, 8]. In freestanding GaN, this trap is also a dominant trap [7]. The second trap of E3, which located at $E_C - 0.51$ eV, was also normally appeared in GaN epi-layer. These trap is similar with the level in range of $E_C - 0.51$ to 0.59 eV, it has capture-cross section falls within 1.8×10^{-16} to 2×10^{-12} cm^2 [10, 11]. Tokuda et al. attributed this trap to dislocation-related defects as the trap is found to have shown logarithmic dependence on filling pulse width [9]. In an earlier report, Fang et al. also proposed the same conclusion [8]. However, Haase et al. [10] concluded that this trap is related to be an intrinsic point defect as its concentration changes little with nitrogen implantation and annealing. He postulated that N antisite could be the associated defect. The E3 level is the dominant trap in GaN grown on Sapphire and SiC [9]. The third trap $E5 = E_C - 1.16$ eV is similar with the trap located at $E_C - 0.93$ to 1.02 eV in GaN materials [9, 11]. This has capture-cross section falls within 9.4×10^{-15} to 5.3×10^{-15} cm^{-2}. Fang et al. attributed this trap to threading dislocation based on a logarithmic dependence of the DLTS signal with the filling pulse width [12].

Table 1. Summary of deep level traps and their density in GaN layer growth at low temperature.

Deep levels	Energy level, eV	Capture cross section, cm^2	Defect density, cm^{-3}	
			Sample 1	Sample 2
E1	$E_C - 0.22$ eV	6.24×10^{-16}	3.0×10^{14}	4.0×10^{14}
E2	$E_C - 0.25$ eV	9.05×10^{-17}	1.9×10^{15}	3.0×10^{15}
E3	$E_C - 0.51$ eV	1.39×10^{-13}	1.0×10^{14}	2.0×10^{14}
E4	$E_C - 0.69$ eV	4.32×10^{-15}	4.0×10^{14}	5.0×10^{14}
E5	$E_C - 1.16$ eV	2.36×10^{-11}	1.1×10^{15}	1.3×10^{15}

Beside the common traps observed in GaN, new levels can be seen in low temperature grown GaN which not seen before in GaN materials are $E_{T1} = E_C - 0.22$ eV and $E_{T4} = E_C - 0.69$ eV. The origin of those levels is not clear yet. From DLTS measurement, we can see that the growth temperature affects density of deep level traps in GaN. The GaN layer grow at lower temperature (sample 2) has higher total density but the same five deep levels as compared with that of sample 1. In general, GaN barrier layer in GaN LED grown at low temperature has one order of magnitude higher than trap density than GaN template but the trap density still two order of magnitude lower than doping in that layer ($\sim 10^{17}$ cm^{-3}). These results to support that the quality of GaN barrier layer still decent for high performance of GaN LED.

4 Conclusion

In this paper, deep level traps of the GaN barrier layer was characterized by DLTS characterization. The measurement revealed five deep levels in the GaN barrier layers with the total deep level defect density of about 4×10^{15} cm^{-3}. Among five deep levels found in those GaN samples, there are three deep levels are commonly seen in GaN on sapphire, Si or SiC substrate. They are $E_2 = E_C - 0.25$ eV; $E_3 = E_C - 0.51$ eV; and $E_5 = E_C - 1.16$ eV. They nature of these levels was mentioned in other papers. Two new levels are found in these samples, they are $E_1 = E_C - 0.22$ eV and $E_4 = E_C - 0.69$ eV. The total trap density is one order of magnitude higher than that of standard GaN layer but still two order of magnitude lower than doping in that layer. Therefore, that low temperature growth GaN barrier still results in reasonable material quality for high performance of GaN LED.

Acknowledgements. This research was supported by Hanoi University of Science and Technology under grand number T2020-PC-058.

References

1. Nakamura, S., Senoh, M., Iwasa, N., Nagahama, S., Yamada, T., Mukai, T.: Superbright green InGaN single-quantum-well-structure light-emitting diodes. Jpn. J. Appl. Phys. **34**, L1332–L1335 (1995)
2. Meel, K., Mahala, P., Singh, S.: Design and Fabrication of Multi Quantum well based GaN/InGaN Blue LED. IOP Conf. Ser. Mater. Sci. Eng. **331**(1), 012008 (2018). https://doi.org/10.1088/1757-899X/331/1/012008
3. Kim, D.J., Moon, Y.T., Song, K.M., Park, S.J.: Effect of barrier thickness on the interface and optical properties of InGaN/GaN multiple quantum wells. Jpn. J. Appl. Phys. **40**, 3085–3088 (2001). https://doi.org/10.1143/JJAP.40.3085
4. Lang, D.V.: Deep level transient spectroscopy: A new method to characterize traps in semiconductors. J. Appl. Phys. **45**, 3023 (1974)
5. Schroder, D.K.: Semiconductor Material and Device Characterization. Wiley-IEEE Press (2006)
6. Fang, Z.-Q., Look, D.C., Jasinski, J., Benamara, M., Liliental-Weber, Z., Molnar, R.J.: Evolution of deep centers in GaN grown by hydride vapor phase epitaxy. Appl. Phys. Lett. **78**(3), 332 (2001). https://doi.org/10.1063/1.1338970
7. Fang, Z.-Q., et al.: Deep centers in a free-standing GaN layer. Appl. Phys. Lett. **78**(15), 2178 (2001). https://doi.org/10.1063/1.1361273
8. Fang, Z.-Q., Look, D.C., Wang, X.-L., Han, J., Khan, F.A., Adesida, I.: Plasma-etching-enhanced deep centers in n-GaN grown by metalorganic chemical-vapor deposition. Appl. Phys. Lett. 82(10), 1562 (2003). https://doi.org/10.1063/1.1560562
9. Tokuda, Y.: Traps in MOCVD n-GaN studied by deep level transient spectroscopy and minority carrier transient spectroscopy. In: CS MANTECH Conference, no. C, pp. 19–24 (2014)

10. Haase, D., et al.: Deep-level defects and n-type-carrier concentration in nitrogen implanted GaN. Appl. Phys. Lett. **69**(17), 2525 (1996). https://doi.org/10.1063/1.117727
11. Ito, T., Nomura, Y., Selvaraj, S.L., Egawa, T.: Comparison of electrical properties in GaN grown on Si (111) and c-sapphire substrate by MOVPE. J. Cryst. Growth **310**(29), 4896–4899 (2008)
12. Fang, Z.-Q., Look, D.C., Lu, C., Morkoç, H.: Electron and hole traps in GaN p-i-n photodetectors grown by reactive molecular beam epitaxy. J. Electron. Mater. **29**, L19–L23 (2000)

Material Applications for Sustainable Development

Experimental Vibration Diagnostics of the Floor Plate Set of Building Construction

A. N. Soloviev[1,2], I. A. Parinov[1], A. V. Cherpakov[1,2(✉)], and Yu. V. Esipov[2]

[1] Southern Federal University, Rostov-on-Don, Russia
[2] Don State Technical University, Rostov-on-Don, Russia

Abstract. The problem of the damage diagnostics of reinforced concrete floor plates in full-scale construction is considered. Vibration analysis of forced vibrations of building elements was carried out using a mobile measuring set-up. A calibrated impact loading was applied to a certain point of the construction. The experimental approach allowed one to fix the frequencies and forms of natural vibrations of floor plates. Analysis of the data obtained identified damaged objects. A method for assessing damage in beam-type structures is presented.

Keywords: Reinforced concrete floor plates · Vibration analysis · Forced vibrations · Impact load · Mobile measuring set-up · Acceleration · Natural frequencies

1 Introduction

The study of the damage of reinforced concrete goods in building constructions is of great sound, especially for buildings with an expired service life. This should be taken into account in the implementation of the parameters of the reliability of constructions during their operation. It is known that any operating constructions gradually lose their exploitation characteristics over time. This is due to the inevitable destruction and worsening properties of materials under the influence of different external factors.

The restoration of the integrity and reliability of the bearing elements of buildings is carried out in the course of repair and restoration work, which is necessarily preceded by scheduled technical inspections. It should be noted that there are insufficient methods and criteria in normative and technical literature for assessing the quality of repair work. It is assumed that there is mainly visual control of the repaired areas, the reliability and objectivity of which is low. Along with this, specialized methods of analysis can be used, such as vibration diagnostics [1–4].

To control the current state and identify defects in the structure of concrete constructions, along with standardized methods of non-destructive testing, the method of vibration diagnostics is widely used [5–8].

The works [9, 10] reflect the issues of mathematical and finite element modeling, statement of tests and experimental results in laboratory conditions. Dependences of the frequency of natural vibrations of a concrete construction on the size and type of structural defects have been established [11, 12].

B. T. Long et al. (Eds.): MMMS 2020, LNME, pp. 255–260, 2021.
https://doi.org/10.1007/978-3-030-69610-8_35

The authors of [13] proposed integral diagnostic (ID) sign for rod structures. It was shown that on its base with the help of modal analysis and using the finite-element (FE) method, it is possible to solve the ID problem for elements of rod structures with an increased degree of reliability.

2 Algorithm for Assessing the State of Floor Plates

2.1 The Aim of the Study

The aim of the study is to develop a technique for assessing the damage of floor plates using vibration diagnostics using a mobile vibration diagnostic set-up developed in the laboratory of Mechanics and Physics of Advanced Materials and Devices of the I. I. Vorovich Mathematics, Mechanics and Computer Science Institute of the Southern Federal University.

2.2 Description of the Identification Algorithm

This algorithm is implemented as part of a set-up for vibration diagnostics of multi-parametric identification of damage in rod structures.

The basic block of the algorithm is the creation of a bank (array) of data on the dynamic state of the construction. The parameters of the state of the construction can be obtained by combining the computational modeling and experimental approach [14].

To assess the parameters of vibrations of floor plates and implement the method for assessing damage, a vibration-diagnostic set-up, shown in Fig. 1, was used.

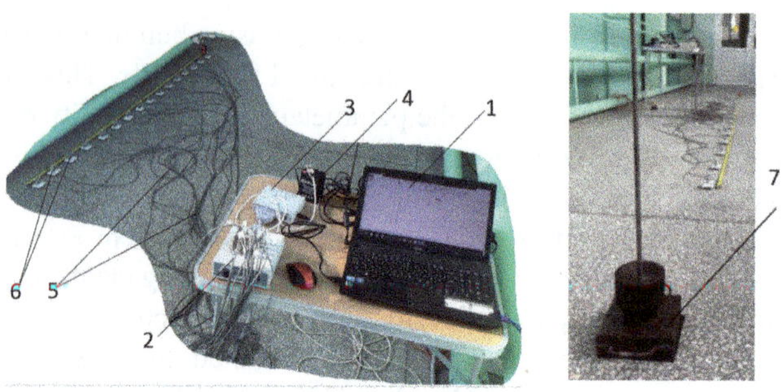

Fig. 1. Set-up of vibration diagnostic equipment: 1 – PC; 2 – matching device; 3 – ADC; 4 – power supply unit of the matching device; 5 – transmitting path; 6 – sensors-accelerometers; 7 – impact loading device

The damage identification algorithm procedure is described as a sequence of the following steps:

(i) collecting the hardware part of the diagnostic system: design - sensors - transmitting path - matching device - ADC - device for collecting and processing information (PC);
(ii) excitation of structure vibrations and collection of information on natural frequencies and vibration modes;
(iii) processing of data on oscillations, obtainement of measurement results and modal calculations.

2.3 Research Object

Three floor plate sections of the transitional gallery of one of the buildings of the Don State Technical University, Rostov-on-Don (see Fig. 2(a)) were studied. Accelerometers were installed along the length of the floor plates; impact loading of the structure was performed and transverse oscillations defined. Comprehensive comparative analysis allows one to draw a conclusion about the identification of "weak", damaged floor slabs that require further observation.

(a) (b)

Fig. 2. (a) Transition gallery; (b) Layout of test zone in the inner part of the transitional gallery

Figure 2(b) shows the layout of sensors. Vibration excitation is produced by dropping the load on a fixed platform at points V1–V3 of the tested construction. The parameters of the displacements at the points D1–D16 in the vertical direction are taken along the lines of the location of the sensors L1, L2, L3 along the floor plates (D1–D16 are the sites of fixing the displacement sensors). Analysis of the parameters of construction is carried out on the base of mathematical processing of vibration parameters.

With the help of vibration-diagnostic equipment at impact excitation of test floor plates without damage, the amplitudes of vibration accelerations were measured and the frequencies of natural vibrations were calculated. The first oscillation frequency was taken as the basic calculated one.

An example of the measured amplitudes of accelerations in the vertical direction of points on the floor plate, taken from the accelerometers, is shown in Fig. 3(a).

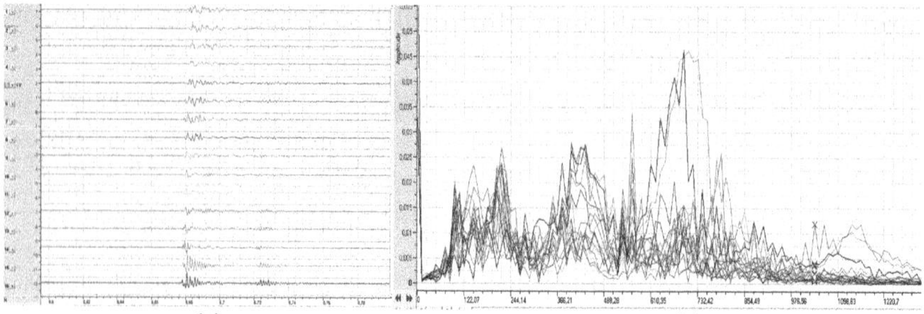

Fig. 3. (a) Signal recorded from sensors (D1–D16) with vibration excitation; (b) results in the form of spectrum of natural frequencies obtained by testing floor plate No. 3

The analysis consists of the following stages: (i) construction of Fourier-images of natural and forced oscillations of construction; (ii) definition of oscillation modes; (iii) obtainment of phase characteristics of the oscillations of construction and its elements. As the time ranges, are selected certain periods of the exploitation of construction, taking into account damage formation and accumulation of internal stressed sections. Based on the data obtained, the deviation of natural frequencies is revealed and the vibration modes are refined with the identification of the damage zone. Figure 3 (b) presents an example of frequency spectrum of floor plate No. 3.

At the next stage, the first harmonics of vibrations of floor plates were calculated. Table 1 shows the first three harmonics measured for three floor plates.

Table 1. Natural frequencies of floor plates

Floor slab	Natural frequency, Hz		
No	ω_1	ω_2	ω_3
1	122	134	207
2	122	146	195
3	97	146	219

Analysis of natural vibration frequencies of three floor plates shows that for the 1st and 2nd plates the first vibration mode has the first frequency $\omega_1 = 122$ Hz. Floor plate No. 3 has the first frequency $\omega_1 = 97$ Hz. What can we say about the damage to the plate? Further analysis should be based on fixing the modes of natural vibrations of the plates.

3 Conclusion

An experimental mobile vibration-diagnostic set-up was tested. A vibration measurement of natural frequencies at various points was made for three floor plates. The analysis showed the deviation of the first natural frequency for plate No. 3 from the other two. This indicator may speak on the presence of a damage state in a reinforced concrete plate, and require its further detailed examination. A more detailed analysis can be made on the basis of a comprehensive analysis of studies of modal vibration modes and phase characteristics of non-stationary vibrations.

Acknowledgments. Research was financially supported by Southern Federal University, 2020 grant number VnGr-07/2020–04-IM (Ministry of Science and Higher Education of the Russian Federation).

References

1. Shevtsov, S.N., Soloviev, A.N., Parinov, I.A., Cherpakov, A.V., Chebanenko, V.A.: Piezoelectric Actuators and Generators for Energy Harvesting - Research and Development, vol. 182. Springer, Cham (2018)
2. Kotlyarevsky, V.A.: Features of vibration diagnostics in construction. J. Monit. Sci. Secur. **3** (11), 44–54 (2013)
3. Sohn, H., Farrar, C.R., Hemez, F.M., Shunk, D.D., Stine-Mates, D.W., Nadler, B.R., Czamecki, J.J.: A review of structural health monitoring literature: 1996–2001, Los-Alamos National Laboratory, Report LA-13976-MS, Los Alamos, NM, USA (2004)
4. Rainieri, C., Fabbrocino, G.: Operational Modal Analysis of Civil Engineering Structures, vol.322. Springer, New York
5. Cherpakov, A.V., Soloviev, A.N., Gricenko, V.V., Goncharov, O.U.: Damages identification in the cantilever-based on the parameters of the natural oscillations. Def. Sci. J. **1**, 44–50 (2016)
6. Cherpakov, A., Egorochkina, I., Shlyakhova, E., Kharitonov, A., Zarovny, A., Dobrohodskaya, S.: Using technique vibration diagnostics for assessing the quality of power transmission line supports repairs, MATEC Web Conf. **106**, 04009 (2017)
7. Ilgamov, M.A., Khakimov, A.G.: Diagnosis of damage of a cantilever beam with a notch. Russ. J. Nondestr. Test. **45**(6), 430–435 (2009)
8. Sobyanin, K.V., Shardakov, I.N., Shestakov, A.P., Glot, I.O.: Dynamic deformation interaction of elements of the "drummer–gasket–reinforced concrete beam" system. PNRPU Mech. Bull. **4**, 119–126 (2018)
9. Soloviev, A.N., Cherpakov, A.V., Egorochkina, I.O., Shlyakhova, E.A., Parinov, I.A.: Analytical modeling influence of defects in the base of transmission tower on the parameters of its own transverse vibrations. In: Proceedings of the 2016 International Conference on Physics and Mechanics of New Materials and Their Applications Nova Science Publishers NY, pp. 439–446 (2017)
10. Akopyan, V.A., Rozhkov, E.V., Soloviev, A.N., Shevtsov, S.N., Cherpakov, A.V.: Identification of damages in elastic structures: approaches, methods, analysis. Rostov-on-Don Southern Federal University Press (2015)

11. Egorochkina, I.O., Shlyakhova, E.A., Cherpakov, A.V., Soloviev, A.N.: Analysis of the influence of defects in the base of the transmission line support on the parameters of natural transverse vibrations based on an analytical model. Eng. J. Don 4 (2015)
12. Vankov, Yu.V., Kazakov, R.B., Yakovlev, E.R.: Natural frequencies of the product as an informative sign of defects EJ Technical Acoustics 5 (2003)
13. Akopyan, V.A., Solov'ev, A.N., Cherpakov, A.V., Shevtsov, S.N.: Russian Journal of Nondestructive Testing **49**(10) (2013)
14. Cherpakov, A.V., Akopyan, V.A., Rozhkov, E.V., Soloviev, A.N.: Identification of damage parameters in an elastic rod using finite element and experimental analysis of bending vibration modes. Vestnik Don State Technical University **3**, 312–318 (2011)
15. Volkovas, V., Petkevicius, K.: Modeling and identification of building's construction defects. In: Proceedings of the World Congress on Engineering, WCE-2011, vol. 3, pp. 2518–2522 (2011)

Research on the Effect of Geometric Parameters of Billet and Punch on Upsetting Technology of Hollow Transmission Details

Nguyen Dac Trung[1(✉)] and Phan Thi Ha Linh[1,2]

[1] Hanoi University of Science and Technology, Hanoi, Vietnam
trung.nguyendac@hust.edu.vn
[2] Hung Yen University of Technology and Education, Hung Yen, Vietnam

Abstract. In gearboxes, drives or mechanical devices, there are many hollow transmission parts, which are usually made by forging technology from solid billet and then finished by cutting process. Such technology causes waste of forming energy and materials due to the removal of large quantities of excess metals. The volume of material, that needs to remove, can reach up to 35% of billet volume. In order to save time and effectively making of parts, the hollow billets such as tubes are proposed to use and the technology undergoes the main tasks: first is heating of billet to achieve high temperature, then local upsetting of tube to create an appropriate intermediate billet, and after that forging in closed die. However, the upsetting of tube in hot state is often difficult because the tube is unstable. This leads to low upsetting ratio and appearance of folding defects in the billet. To avoid the instability of tube billet during deformation, it is necessary to calculate the suitable initial dimensions of tube billet and determine the appropriate shape of punch (upper die). This paper presents the research results to determine the time, which the billet is destabilized depending on the ratios height/diameter and thickness/diameter of billet based on numerical simulation. With the target of strain bigger than 40% in the thickness of tube, this research also deals with development of a model for tube upsetting with conical punch, establishment of the relationship between the strain depending on billet geometry and tilt angle of punch based on numerical simulation and experiments.

Keywords: Upsetting · Hot forming · Defect in upsetting · Tube forming

1 Introduction

There are many axis symmetrical and hollow transmission parts such as gears, couplings, shafts, which are small size, made of carbon steel such as C45, and widely used in gearboxes, drives or mechanical devices. They are often fabricated by bulk forming technology from initial solid billet due to many advantages in comparison to cutting technology for example high productivity, saving production costs, improving mechanical properties and product quality. But after such bulk forming, drilling process is necessary to create a hollow part [1]. To save material by using bulk forming from tube billet, the technology needs to undergo three main tasks: billet heating, upsetting, and die forging (Fig. 1).

B. T. Long et al. (Eds.): MMMS 2020, LNME, pp. 261–266, 2021.
https://doi.org/10.1007/978-3-030-69610-8_36

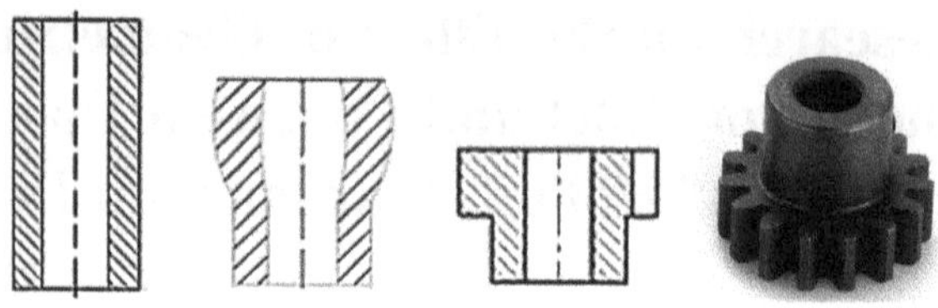

Fig. 1. Forming steps from initial tube billet, intermediate billet (workpiece) after upsetting and gear product after forging in closed die

To ensure a successful forming process from tube billet, the upsetting plays important role to make intermediate workpiece with the appropriate shape and size. However, the tube upsetting is often difficult (when the ratio of height and diameter of initial tube $H_0/D_0 > 0.6$ and the ratio of thickness and diameter of initial tube $S_0/D_0 < 0.3$), because tube is prone to instability and the folding defect can be occurred in workpiece [2] (Fig. 2). Moreover, due to the small deformation by upsetting, it is not able to create the desired thickness and height of the workpiece after upsetting. Therefore, the initial dimension of tube billet must be calculated exactly, but it is not easy to implement by theoretical calculation or experiment [3]. This research will be proposed a method based on numerical simulation combined with experiment to solve this problem.

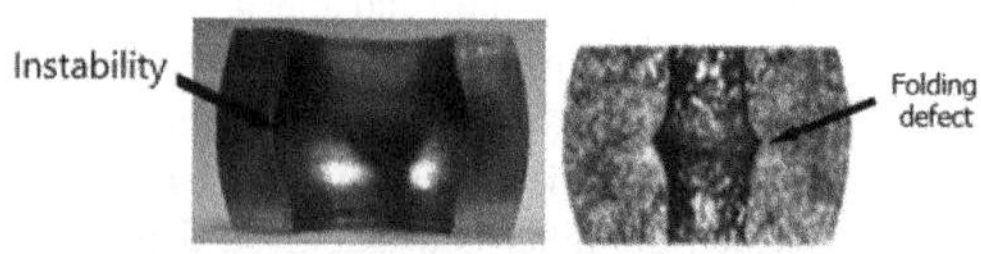

Fig. 2. Instability and folding defect during upsetting

2 Establish a Model of Local Upsetting

2.1 Determination of Instability During Upsetting by Numerical Simulation

Research target is creating an appropriate intermediate workpiece by upsetting, so that a model is built (shown in Fig. 3) [2] and applied for numerical simulation. H_0, H_1, D_0, D_1, d_0, d_1, S_0, S_1 are the height, outer diameter, inner diameter and thickness of initial tube and intermediate workpiece after upsetting. During upsetting, the punch/upper die goes down to compress the tube billet, so that height of tube decreases, the cross section or thickness increases, respectively. It is necessary to determine the strain or upsetting ration ε, as well as starting point for appearance of instability and folding defect in workpiece.

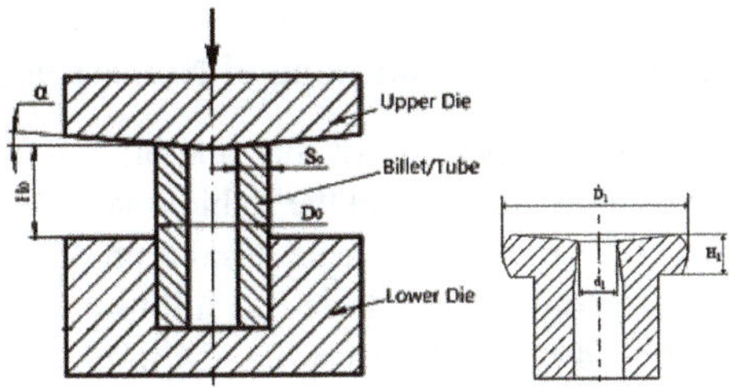

Fig. 3. Model of local upsetting and workpiece after upsetting

Numerical simulation will be implemented with the following forming conditions: material: C45; initial temperature of tube billet for hot forming: T = 1150 °C; Crankshaft press speed: 40 strokes/minute; Friction coefficient: 0,4 [4].

The geometry model (tube billet with H_0 = 32 mm, D_0 = 40 mm, S_0 = 8 mm and dies with α = 0°), all forming conditions are imported into Deform software to simulate and analysis of upsetting process. The simulation is implemented at each step of punch travel and unstable condition of workpiece is checked. If the instability occurs, the simulation process will stop to avoid the defects in workpiece. The time of instability and appearance of folding defect can be determined by images or graphs of simulation results such as deformation mesh/flow net, displacement according to the coordinates and strain rate of the defined points in workpiece.

Figure 4 shows the flow net and displacement of 5 points on the inside surface of workpiece.

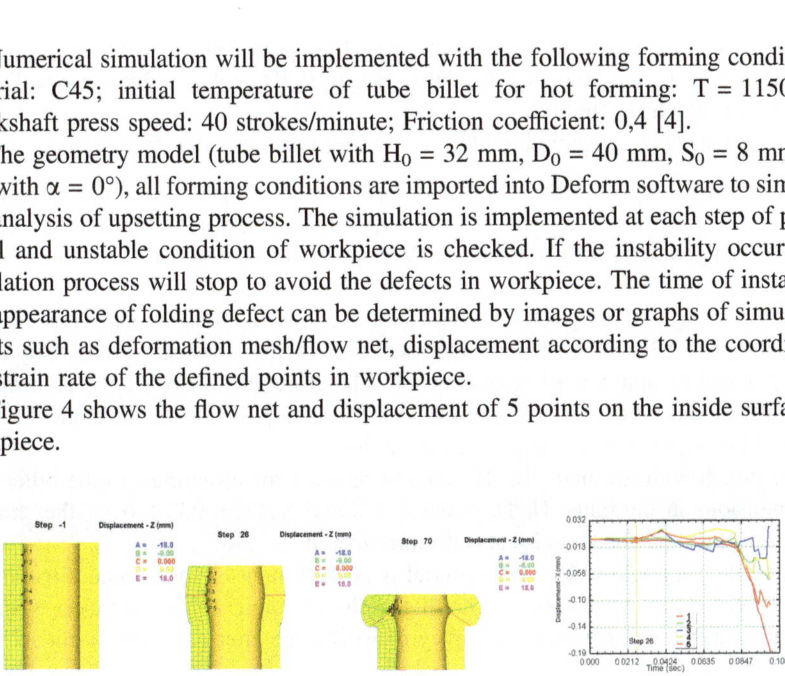

Fig. 4. Flow nets and displacements at beginning and steps 26, 70

Based on the increasing of the mesh's deviation according to the compress stroke of the punch, it is easy to realize, that the net at step 70 is broken at step 70 and the folding defect is appeared on the inside surface of workpiece (Fig. 5). Based on displacements in X-direction of 5 consecutive points (P1, P2, P3, P4, P5), it can determine that step 26 is the starting time of the tube instability. Corresponding to this step, the strain or upsetting ratio reaches 27,3%.

2.2 Determination of Strain Depending on the Initial Geometry of Tube Billet and Shape of Punch by Analyzing of Numerical Simulation

The determination of the initial dimension of the tube billet must be ensured that the upsetting can be reach the maximum strain in the stable state of workpiece and meet the criteria suitable with closed die forging to create desired product: $S_1/S_0 \geq 1{,}4$ and $d_1/d_0 \geq 0{,}8$. The initial billet geometry is selected as follows:

$D_0 = 40$ mm, $H_0 = 16 \div 40$ mm and $S_0 = 8 \div 14$ mm.

The dimension ratios of the billet: $H_0/D_0 = 0{,}4 \div 1{,}0$ and $S_0/D_0 = 0{,}2 \div 0{,}35$.

Punch has the conical bottom with values of tilt angle for investigation: $\alpha = 0° \div 5°$.

Based on simulation results for upsetting with difference shape of punch, the graphs are buit to discribe the influence of relative height H_0/D_0, relative thickness S_0/D_0 and tilt angle of punch on strain.

The graphs (Fig. 5) show the effects by using conical punch as follows:

- In the range $H_0/D_0 = 0{,}4 \div 0{,}6$ and $S_0/D_0 = 0{,}3 \div 0{,}35$, the effect to improve strain by using of conical punch is not bigger than using of flat punch $\alpha = 0°$. With the initial geometry of tube billet in this range, it is recommended to select an angle of $\alpha = 5$–$10°$.
- In the ranges $H_0/D_0 = 0{,}4 \div 0{,}6$ and $S_0/D_0 < 0{,}3$ or $H_0/D_0 > 0{,}6$ và $S_0/D_0 = 0{,}3 \div 0{,}35$, the workpiece is unstable. However, the risk of instability could be reduced by applying of conical punch. Therefore, the strain and the deformation ability of the tube is raised more than 40%. The punch with tilt angle $5 \div 10°$ should be used in those ranges of tube billet.
- The punch with tilt angle 10–15° should be used for upsetting of tube billet with dimensions in the range $H_0/D_0 = 0{,}8 \div 1{,}0$ and $S_0/D_0 = 0{,}2 \div 0{,}25$, the strain is significantly enhanced and reached more than 40%.
- When the tilt angle is 20°, the material is pushed intensely horizontal direction, so the effect of strain increasing is not high. The tilt angle 20° is only used when the requirement for dimensional accuracy of workpiece after upsetting is not so high.

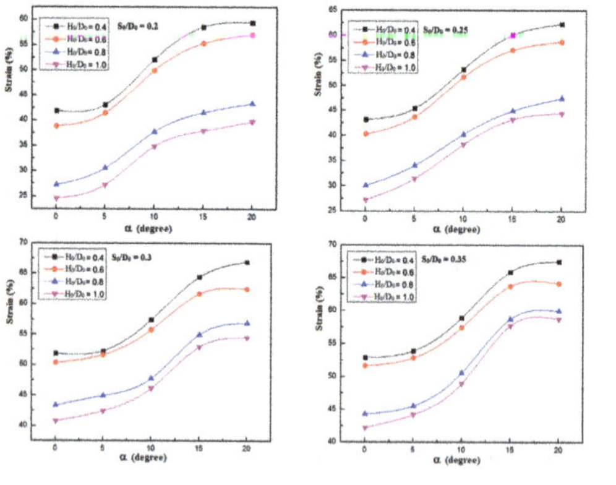

Fig. 5. Influence of the tube parameters and tilt angle on the strain

3 Experiment and Comparison with Numerical Simulation Results

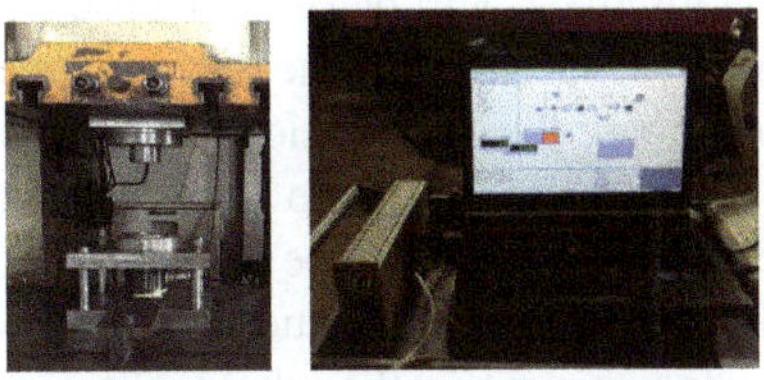

Fig. 6. Experimental system

The investigation results by numerical simulation are important for applying of the appropriate technology design. However, it is necessary to do the experiments to compare the results with simulation cases to evaluate and adjust input parameters of the research model. The experimental system was built on Crankshaft press FCP-160 with nominal pressure 160 tons (Fig. 6). All sensors are connected from die, machine to the signal collector and the computer. For each experiment, the results will be stored in the computer, analyzed, and extracted data supported by dedicated software DasyLab 11.

To compare the results between numerical simulation and empirically with high reliability, each range (above mentioned) of the initial dimension of tube billet will be selected from 2 to 3 case studies. Each specific case is conducted with 10 samples and calculated the average results to ensure the objectivity and randomness of the experiments. After finishing of experiments, the measurement data are stored in the computer, the products are cleaned and cut for measurement of shapes, dimensions and comparison with numerical simulation.

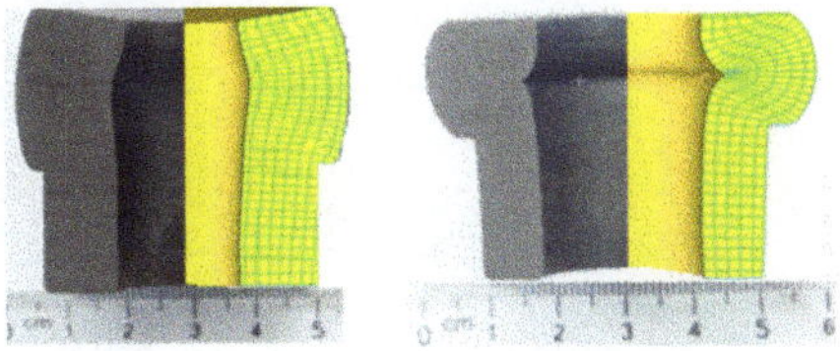

Fig. 7. Comparison of workpiece shape between experiment and simulation

Figure 7 shows the comparison of workpiece shape between experiment and simulation on the steps of starting instability and the step of folding defect appearance.

The steps began to appear unstable and the folding defect at the same deformation level, shapes of workpieces are quite similar and the thickness of tube after upsetting is approximately equal. Based on comparison with experiment, it can be affirmed that numerical simulation of upsetting process with established boundary conditions is consistent with reality, numerical simulation results are reliable, simulated results can be used for research, technology design and practical manufacturing.

4 Conclusion

The manufacture of hollow, symmetrical transmission parts can be achieved by bulk forming technology from tube billet. It is important to select the initial tube billet to ensure the shape of the intermidiate workpiece after upsetting without folding defect. Based on numerical simulation, the initial dimension of tube billet is determined and met the desired strain. To increase the deformation ability of tube bigger than 40%, it is possible to use the conical punch. The results of numerical simulation were compared with experiment. The research results will be applied to the calculation and optimization of the initial billet geometry to produce the hollow transmission parts with enhance forming process efficiency as well as the quality of final products.

References

1. Politis, N.J., Politis, D.J., Davies, C.M., Lin, J., Dean, T.A.: An experimental and numerical investigation into forming force reduction in precision gear forging. Key Eng. Mater. **623**, 165–173 (2017)
2. Poursina, M., Parvizian, J.: Simulation of folding defect in forging, Nimiform pp. 486–491 (2004)
3. Su, Y.L., Yang, W.Z., Wang, C.P.: Upsetting process analysis and numerical simulation of metal pipe's end. Appl. Mech. Mater. **364**, 488–492 (2013)
4. Li, Y., Onodera, E., Chiba, A.: Friction coefficient in hot compression of cylindrical sample. Mater. Trans. **51**(7), 1210–1215 (2010)

Multi-variable Regressive Models for Diagnostics of the Unbalances on Rapid Rotor in Shop Dynamic Balance

Duc Do Luu[1] and Luu Minh Hai[2(✉)]

[1] Vietnam Maritime University, 484-Lach Tray, Hai Phong City, Viet Nam
luudd@vimaru.edu.vn
[2] Naval Academy, Nha Trang City, Khanh Hoa, Viet Nam

Abstract. The paper presents creating a multi-variable regression model (MRM) for estimating the residual unbalanced mass based on the simulations of dynamic balancing rotor that is placed on a dynamic balancing machine (DBM). The model shows the relationship between the unbalanced Index (UI) that monitors the DBM and the model's factors of the actual unbalanced masses (x_1, x_2) and the balancing position (angle x_3, x_4 and radius x_5, x_6). The MRM was built on the base of the made Design of the Experiments (DoE), regression analysis. The DoE is $\mathbf{2^{k-p}}$ fractional factorial design, with k = 6, p = 4, main independent factors ($x_1 \ldots x_4$). $N_p = 2^4 = 16$ experiments, The simulation studying for creating the MRM was verified with the low-pressure turbine rotor on military ships gas turbine engine. The obtained regressive models were tested with 99% confidence. The received MRM would be used for improving the software of the dynamic balancing machines.

Keywords: Rotor unbalances · Multi-variable regressive model · Rotor dynamic balancing · Unbalance diagnostics

1 Introduction

The rotating parts as well as the rotor machine often work at high speed, even very high, so it is required to dynamic balance before assembling the rotor and putting the machine into operation [2]. Usually, the rotors as well as the details of the ship's rotors are short and are considered as rigid rotors, which are dynamically balanced on the supports of the dynamic balancing machine (DBM) at the workshop. Dynamic balancing for the rotor at the workshop is carried out on one or two balancing planes depending on the configuration of the rotor. When dynamic balancing the rotor on two planes, it is necessary to measure the vibrations at the soft support of the dynamic balancing machine in 3 times and simultaneously at the two bearings. And for dynamic balancing on one plane, only measures vibration on one bearing and two experiments [3].

Dynamic balancing has a strict procedure, requiring knowledge and skilled implementers. In education and training courses for operating marine engines at universities as well as electro-mechanical engineers for the Navy, the simulation software is important to improve the quality and efficiency of the specialized training courses [3, 4].

B. T. Long et al. (Eds.): MMMS 2020, LNME, pp. 267–272, 2021.
https://doi.org/10.1007/978-3-030-69610-8_37

Dynamic balance simulation software (SSDB) for rotor-mounted robots has been developed at Vietnam Maritime University, which is part of the National Project [3]. The study evaluating the homogeneity of the results obtained in the unbalanced data indicator when using SSDB gives 99% confidence level according to the Schi statistical standard [4]. The algorithm and SSDB give an imbalance indicator for a rotor that is balanced at each plane according to two parameters: the unbalanced mass (grams) and the unbalanced position (angle that deviates from placement of the test mass, rotation angle), and the radius is considered to be fixed from before. In fact, the indicator that SSDB gives is a predictable result, there is always an error compared to the actual (expected) value of the rotor to be balanced. If the forecast results are not enough accurate, it will lead to the implementation of the equilibrium which must be repeated many times, leading to an increase in costs, sometimes not achieving the expected results. Research [4] developed a regression model for the diagnosis residual unbalanced of rotor in military ship GTE, providing two tasks to determine the actual unbalanced mass (regardless of the angle where the unbalanced masses are placed) that SSDB indicates. The two regression models were written for 2 unknowns, achieving 99% confidence, based on DoE Full $2^2 = 4$, plus 8 experiments with the assumption of 3 levels of random error of 5% from the equilibrium radius, total TNE = 12.

For better application of dynamic balancing, unbalanced diagnostic regression models need to simultaneously indicate mass quantities and angular positions. This paper will address this case study.

2 Materials and Methods

2.1 Designing Experiments for Making Regressive Models

The total number of factors for the DB in two planes is 6: actual unbalanced masses (x_1, x_2) and the balancing position (angle x_3, x_4 and radius x_5, x_6). In the cases study, we consider that the main Independent factors are (x_1, x_2), (x_3, x_4), and the dependent (generated) factors (x_5, x_6). The simulation study is conducted in accordance with the $\mathbf{2}_{\mathrm{IV}}^{6-2}$ Fractional Factorial Design [1], that is shown in Table 1

Table 1. A $\mathbf{2}_{\mathrm{IV}}^{6-2}$ Fractional Factorial Design [1]

Run	F_0	Main Independent Factors				Generated factors	
		$A = x_1$	$B = x_2$	$C = x_3$	$D = x_4$	$x_5 = ABC$	$x_6 = BCD$
1	**1**	1	1	1	1	1	1
2	**1**	−1	1	1	1	−1	1
…							
15	**1**	1	−1	−1	−1	1	−1
16	**1**	−1	−1	−1	−1	−1	−1

Table 1 shows the DoE of the 2^{6-2} fractional factorial design, with the 4 main indepent factors A, B, C, D and 2 interacted E = x_5 and F = x_6 as the generators [1]: E = ABC; F = BCD. Therefore, we create a experiment design **F** with N = 2^4 = 16 total number of Experiments (TNE).

2.2 Defining Coefficients and Testing the Confidence of the Regressive Models

The regression models finding in the matrix form are the linear Eqs. (1). In (1) factors $x_1 \div x_6$, which is typical for the input of the experiment, has a value of ± 1, corresponding to the unbalanced value of unbalanced mass input (m_1, m_2), angular position and radius (tt_1, tt_2), (R_1, R_2) at maximum and minimum values on balanced planes I and II: $x_{1\ldots6} = [-1, 1]$; $U_k = [U_{kO} - \Delta U_k, U_{kO} + \Delta U_k]$; k = 1, 2; $U_{1(2)} = m_{1(2)}$; $U_{3(4)} = tt_{1(2)}$; $U_{5(6)} = R_{1(2)}$ also observed values of Y (indicator from SSDB) are values of imbalancing $V_{1(2)} = (m_1, m_2)$, angular position $V_{3(4)} = (tt_1, tt_2)$ and radius set to $V_{5(6)} = (R_1, R_2)$.

$$\begin{bmatrix} \hat{y}_1 \\ \hat{y}_2 \\ \ldots \\ \hat{y}_6 \end{bmatrix} = \begin{bmatrix} A_{01} & A_{11} & \ldots & A_{51} & A_{61} \\ A_{02} & A_{12} & \ldots & A_{52} & A_{62} \\ . & . & \ldots & . & . \\ A_{06} & A_{16} & \ldots & A_{56} & A_{66} \end{bmatrix} * \begin{bmatrix} 1 \\ x_1 \\ \ldots \\ x_6 \end{bmatrix} \Leftrightarrow \mathbf{Y} = \mathbf{A.X}$$
$$\hat{y}_k = A_{0k} + \sum_{j=1}^{6} A_{kj}.x_j = \mathbf{A}_k.\mathbf{X}, k = 1 \ldots 6 \tag{1}$$

Let us to consider the every indicator V_k of the following parameters (m_1, m_2, tt_1, tt_2, R_1, R_2), in the common form are written in the below way. In the every experiment m, m = 1…16, the diagnostic sign $\mathbf{V}_k$ is written in the form:

$$V_k(m) = \mathbf{f}(m).\mathbf{A_k};\ m = 1 \ldots 16; \mathbf{A_k} = [a_k(0) \ldots a_k(6)]^T; k = 1 \ldots 6$$

$$\begin{bmatrix} V_k(1) \\ \ldots \\ V_k(16) \end{bmatrix} = \begin{bmatrix} 1 & 1 & 1 & 1 & 1 & 1 & 1 \\ . & . & . & . & . & . & . \\ 1 & -1 & -1 & -1 & -1 & -1 & -1 \end{bmatrix} .\mathbf{A_k} = \mathbf{F}.\mathbf{A_k} \tag{2}$$

Where, **F** – expanded information matrix is defined from the DoE $\mathbf{2_{IV}^{6-2}}$.
We define the coefficient vector of the diagnostic signs:

$$\begin{aligned} \mathbf{FA}_k = \mathbf{V}_k \Rightarrow \mathbf{F^TF.A}_k = \mathbf{F^TV}_k \Rightarrow \mathbf{A}_k = (\mathbf{F^TF})^{-1}.\mathbf{F^TV}_k \\ \text{Put: } \mathbf{G} = \mathbf{F^TF} \Rightarrow \mathbf{A}_k = \mathbf{G}^{-1}.\mathbf{F^TV}_k \end{aligned} \tag{3}$$

The confidence testing of the RM for every diagnostic sign is executed statistically in accordance with the F-criterion [5].

The inverse task: using the received regressive Eqs. (3) and the observed indicator parameters that are shown by SSDB (vector $\mathbf{V} = [\mathbf{V}_1 \ldots \mathbf{V}_6]^T$) in the cases study), we receive the diagnostic regressive models for unbalances on the balancing rotor via the indicators shown by SSDB:

$$\begin{aligned} \mathbf{X} = (\mathbf{B}^T\mathbf{B})^{-1}\mathbf{B}^T(\mathbf{V} - \mathbf{A_0}) = \mathbf{H}_1(\mathbf{V} - \mathbf{A_0}) = \mathbf{X}_0 + \mathbf{H}_1\mathbf{V} = \mathbf{H}[1\ \mathbf{V}]^T; \\ \mathbf{A_0} = [\mathbf{A}_{01} \ldots \mathbf{A}_{06}]^T; \mathbf{B} = \{A_{p.q}; p, q = 1 \ldots 6\}; \mathbf{H}_1 = (\mathbf{B}^T\mathbf{B})^{-1}\mathbf{B}^T; \mathbf{X}_0 = \mathbf{H}_1\mathbf{A_0}; \mathbf{H} = [\mathbf{X}_0\ \mathbf{H}_1] \end{aligned} \quad (4)$$

2.3 Simulating and Data Acquisition for the Regressive Models

Let us assumption, the initial unbalances of the balancing rotor in the two-planes are: $\mathbf{m_0} = [m_{10}, m_{20}] = [8, 6]$ (gams); $\mathbf{tt_0} = [tt_{10}, tt_{20}] = [0, 45]$ (deg); $\mathbf{R_0} = [R_{10}, R_{20}] = [95, 160]$ (mm), and present mean values with the simulation errors: (0 ÷ 5)%. We would design variable interval of every $U_0(k) = [U_0(k) - \Delta U_0(k), U_0(k) + U_0(k)]$, where $k = 1 \ldots 6$.

The study simulates dynamic balancing rotor of the Low Pressure Turbine (LPT) of a Marine Gas Turbine Engine [4]. The main parameters for balancing are shown in Fig. 1 and Table 2. The designed factors of this cases study are shown in Table 3 and selected database for regressive models are shown in Table 4.

Table 2. Features of the LPT rotor

No	Input parameters	Unit	LPC
1	Operate speed	Rpm	10000
2	Mass of rotor	Kg	47,5
3	Balanced Standard (ISO 1940–1)		G2.5
4	Model		CF1
5	Balanced Speed	rpm	900
6	A	mm	75
7	B	mm	565
8	C	mm	740
9	R_1 (Balanced radius)	mm	95
10	R_2 (Balanced radius)	mm	160

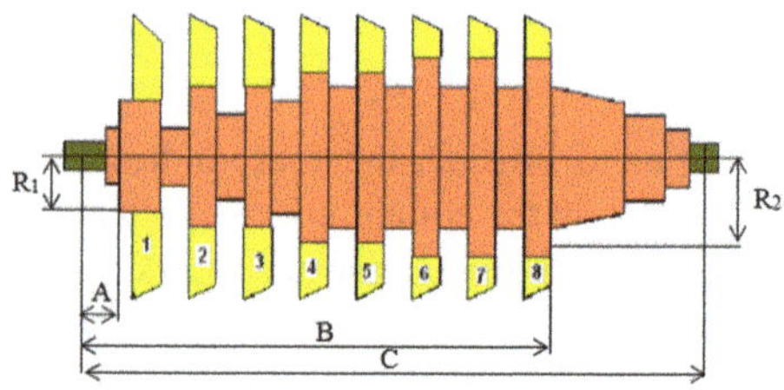

Fig.1. Configuration of LPT rotor on DBM

Table 3. Designed factors for unbalance study

Value	U_{10} = 8 g		U_{20} = 6 g		U_{30} = 30°		U_{40} = 60°		U_{50} = 95 mm		U_{60} = 160 mm	
	Min	max	min	max	min	Max	min	max	Min	max	min	max
U	7.6	8.4	5.7	6.3	28.5	31.5	57	63	90.3	99.8	152	168
x	−1	1	−1	1	−1	1	−1	1	−1	1	−1	1

Table 4. Selected database for regressive models ($\mathbf{x_5 = x_1 \cdot x_2 \cdot x_3}$; $\mathbf{x_6 = x_2 \cdot x_3 \cdot x_4}$)

N°E	x_1	x_2	x_3	x_4	x_5	x_6	V_1 gam	V_2 gam	V_3 deg	V_4 deg	V_5 mm	V6 mm
1	1	1	1	1	1	1	8.18	6.27	31.5	63	99.8	168
2	−1	1	1	1	−1	1	7.34	6.28	31.5	63	90.3	168
3	1	−1	1	1	−1	−1	8.18	5.66	31.5	63	90.3	150
4	1	1	−1	1	−1	−1	8.10	6.25	28.5	63	90.3	150
5	1	1	1	−1	1	−1	8.27	6.28	31.5	57	99.8	150
6	−1	−1	1	1	1	−1	7.40	5.66	31.5	63	99.8	150
7	−1	1	−1	1	1	−1	7.33	6.25	28.5	63	99.8	150
8	−1	1	1	−1	−1	−1	7.46	6.29	31.5	57	90.3	150
9	1	−1	−1	1	1	1	8.13	5.65	28.5	63	99.8	168
10	1	−1	1	−1	−1	1	8.26	5.69	31.5	57	90.3	168
11	1	1	−1	−1	−1	1	8.19	6.28	28.5	57	90.3	168
12	−1	−1	−1	1	−1	1	7.30	5.67	28.5	63	90.3	168
13	−1	1	−1	−1	1	1	7.41	6.28	28.5	57	99.8	168
14	−1	−1	1	−1	1	1	7.47	5.69	31.5	57	99.8	168
15	1	−1	−1	−1	1	−1	8.25	5.67	28.5	57	99.8	150
16	−1	−1	−1	−1	−1	−1	7.43	5.68	28.5	57	99.8	150

In Table 3, the input parameters for building the regression model are following: U_{10}, U_{30} and U_{50} – three parameters of the unbalanced mass on the Ist balancing plane (mass; angle position and radius, respectively). Similarly, on the IInd balancing plane, there are parameters: U_{20}, U_{40}, and U60.

3 Results

We code in LabView to find the MRM in the matrix form A and to test the confidence for every regressive equations. Figure 2 shows the received results matrix **A** of the multi-variable regressive models and confidence vector relatively with color lamps. The matrix **H** (see Eq. 4) also is defined and is shown in Fig. 2.

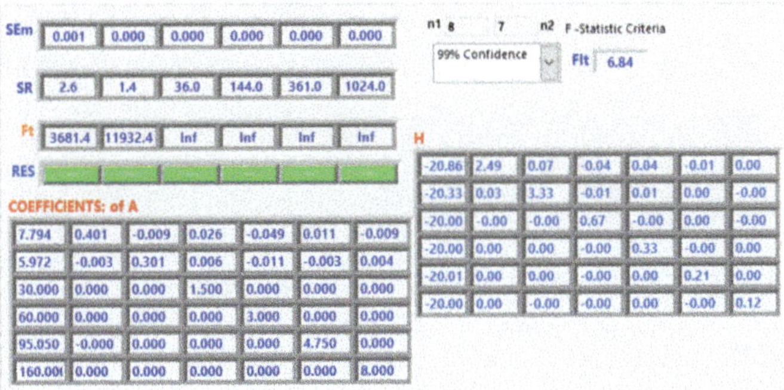

Fig. 2. Received coefficients and results of confidence testing of the MRM

The **H** matrix contains the constant vector $\mathbf{X}_0$ and influenced coefficients of the unbalanced indicators $V_1 \ldots V_6$. The original mass factors are estimated via SSDB indicators in the following regressive models:

$$\begin{aligned} x_1 &= -20.86 + 2.49V_1 + 0.07V_2 - 0.04V_3 + 0.04V_4 - 0.01V_5; \\ x_2 &= -20.33 + 0.03V_1 + 3.33V_2 - 0.01V_3 + 0.01V_4 \end{aligned} \quad (5)$$

4 Conclusion

The cases study on the LPT of the gas turbine engine gives the linear multi-variable regressive models that are written by the matrixes of coefficients **A** and **H**. The matrix **H** shows that the radius (R_1, R_2) and angle (tt_1, tt_2) indicate exactly values as original values of the unbalance parameters on the two planes. The unbalance masses (original values) are quite dependent of the 5 indicators $V_1 \ldots V_5$, that are shown by SSDB (see Eq. 5). In the model of the x_1, the coefficient before the indicator V_6 is zero, and in the model of the x_2, the coefficients before the indicator V_5 and V_6 are zeros.

References

1. Douglas, C.M.: Chapter 8 – Two levels fractional factorial designs. In: Design and Analysis of Experiments, 9th edn., pp.. 343–419. Wiley (2017)
2. ISO 1940–1. ISO 1940/1, Balance Quanlity requirements of Rigid Rotors
3. Nho, L.C., Luu, D.D.: Researching, simulating main propulsion plant and main switch board for marine cargo ship, Content 20: Researching, and simulating vibrations and rotor balancing. National Science-Technology Project, ID ĐT.ĐLCN-14/15 managed by Vietnam Maritime University
4. Hai, L.M., Luu, D.D.: Regressive models for diagnosing the unbalances of the rapid rotor using the simulation software for dynamic balancing rotor placed on the balancing machine, J. Transp. Commun., 71–80 (2020). ISSN 2354–0818
5. Corn, G., Corn, T.: Mathematical handbook for scientific researchers and engineers Publisher «Land» (2003). 831 p. (In Russian).

Effect of Crosslinking Agent and Chitosan Molecular Weight on the Mechanical Properties of Chitosan-Treated Cotton Fabric

Luu Thi Tho[1] and Vu Thi Hong Khanh[2](✉)

[1] Garment Technology and Fashion Design, Hanoi University of Industry, Hanoi, Vietnam

[2] School of Textile – Leather and Fashion, Hanoi University of Science and Technology, Hanoi, Vietnam

khanh.vuthihong@hust.edu.vn

Abstract. The aim of this study is to clarify the influence of the chitosan molecular weight and the crosslinking agents on the mechanical properties of the chitosan treated cotton fabric. In this study, the cotton fabric was treated with 2 types of chitosan with molecular weight of 187 kDa and 2.6 kDa. Two types of crosslinking agents (citric acid (CA) and Arkofix NET (DMDHEU)) have been used in combination with the above two types of chitosan. The following mechanical properties of treated and untreated fabrics were tested: Fabric drapery was tested according to NF G07–109 standard, the fabric's wrinkle recovery angle has been tested in accordance with ISO 2313 standard, the tensile strength and breaking elongation of the fabrics were tested in accordance with TCVN 1754 standard, the surface roughness of the fabric was assessed by Kawabata method and by the surface observation method using the SEM images with the magnification of 1500 times. The results showed that the low molecular weight chitosan solution has low viscosity, which makes the antimicrobial treatment more favorable. Furthermore, the treated fabric is also softer. Fabric treated with CA has higher tensile strength but lower wrinkle recovery angle than fabric treated with cross-linking agent Arkofix NET.

Keywords: Chitosan · Crosslinking agent · Molecular weight · Mechanical properties

1 Introduction

Chitosan is the deacetylated derivative of chitin, when the degree of deacetylation (DD) higher than 60%. Differ with chitin, chitosan can be easily dissolved in acid solution and showed good antimicrobial activity depending on its DD, molecular weight and pH [1–3]. The amine groups in chitosan backbone can attach to negatively charged residues in the cellulose surface of bacteria and thereby changed the structure of cellulose membrane, leading to leakage of intracellular substances, then inhibit the growth even kill the bacteria. Thus chitosan acts like a natural biocide [4]. Chitosan has been intensively studied and used in many areas, such as food processing, medicine, industry, cosmetics and pharmaceutics [5]. In textile industry, chitosan has been

B. T. Long et al. (Eds.): MMMS 2020, LNME, pp. 273–281, 2021.
https://doi.org/10.1007/978-3-030-69610-8_38

applied as the friendly textile auxiliary, that can be replaced for traditional toxic textile chemicals in finishing treatment [6].

Cotton fabric has many valuable properties such as: high moisture absorption, sweat escape, ensuring hygiene, and creating comfort for the wearer [9, 10]. However, cotton is an invaluable environment for bacteria to grow, so antibacterial treatment of cotton fabrics is essential. Chitosan, a biological antibacterial agent, has been used in many studies for antibacterial treatment of cotton fabrics [2–4, 7, 9, 11–13, 19].

To improve antimicrobial laundering durability, chitosan has been cross linked to cotton by using chemicals such as dimethylol dihydroxy ethylene urea (DMDHEU) [4, 7], citric acid [7, 13], or 1, 2, 3, 4-Butane tetra carboxylic acid (BTCA) [4, 11, 15]. These chemicals crosslink chitosan to cotton through hydroxyl groups, this bonding requires a curing step at high temperatures (170–180 °C), which might decrease the tensile or tear strength of cellulose fabrics [4]. However, this bonding could also increase the wrinkle recovery angle of the chitosan-treated cotton fabrics [4]. In addition, due to the high viscosity of the chitosan solution, the surface properties of the treated fabrics such as drapery and hand-touch may also change because of the formation of a coating on the fiber surface [3, 10, 12, 16].

In this study, the cotton fabric was treated with 2 types of chitosan with the molecular weight of 187 kDa and 2.6 kDa. Two types of crosslinking agents CA and Arkofix NET (DMDHEU) have been used in combination with the above two types of chitosan. The tensile strength, drapery, wrinkle recovery angle, and surface characteristics of the untreated and treated cotton fabrics were compared to clarify the effect of the molecular weight of chitosan and the cross-linking agent on the mechanical properties of the cotton fabric.

2 Research Method

2.1 Materials

2.1.1 Fabric Characterization

The cotton fabric that was supplied by Nam Dinh Textile Company, was desized, scoured and bleached, its technical characteristics was demonstrated in [17, 18] are showed in the Table 1.

Table 1. The characteristics of fabric

Fiber content	Construction	Count of yarn (Ne)		Density of the yarns (yarn/10 cm)		Surface mass of fabric (g/m^2)
		Warp	Weft	Warp	Weft	
100% cotton	Twill weaves	34	16	410	175	230

2.1.2 Chitosan

Chitosan with the molecular weight of 187 kDa, the degrees of deacetylation (DD) about 75% was purchased from the Chitosan Vietnam One member Limited Company. This chitosan was irradiated to obtain chitosan with a low molecular weight of 2.6 kDa. The irradiation was at Department of Radiation Technology, Hanoi Irradiation Center The chitosan (CTS) samples denoted are CTS1 and CTS2 with molecular weights of 2.6 kDa and 187 kDa respectively [17, 18].

2.1.3. Cross-linking agent given by Citric acid (CA) and Arkofix NET were purchased from HUNTSMAN Ltd Co.

2.2 Methods

2.2.1 Fixation of Chitosan on the Cotton Fabric

Fabric samples (35 cm × 35 cm) were washed in the non-ionic water with pH 7, dried at room temperature, then impregnated with the finishing solution, and padded two dips with the wet pick-up of 80% on weight of fabric (o.w.f). The padded samples were dried at 100 °C for 3 min and cured at 160 °C for 2 min. Next, they were washed under running water for 5 min to remove all residual chemicals. Finally, they were dried at room temperature. All treatment processes were carried out at chemical-textile LAB of School of Textile - Leather and Fashion design of HUST and at the lab of Textile Research Institute.

These finishing solutions were prepared as followed: Chitosan (2.6 or 187 kDa): 0.3% (o.w.f); Citric acid (CA): 7% (o.w.b); Natri Hypophostphite: SHP-CA mole ratio 1:1; Hostapal MRN Liquid was supplied by Clariant: 0.1% (o.w.b).

And: Chitosan (2.6 or 187 kDa): 0.3% (o.w.f); Arkofix NET: 100 g/l; Catalyst NKC: 30 g/l; CH_3COOH: 2 g/l; Hostapal MRN Liquid was supplied by Clariant: 0.1% (o.w.b).

2.2.2 Testing the Mechanical Properties of the Untreated and Treated Cotton Fabrics

- The drapery of the cotton fabric was measured according to the NF G07–109 standard using the drapery coefficient Hr (the higher the Hr, the stiffer the fabric).
- The crease resistance of fabrics was evaluated using the wrinkle recovery angle of fabrics, which measured on according to ISO 2313–1972.
- The tensile strength and breaking elongation of the fabrics were tested in accordance with TCVN 1754 standard.
- The surface roughness of the fabric was assessed by Kawabata method

3 Results and Discussions

3.1 Effect of the Crosslinking Agent and Molecular Weight of Chitosan on the Drapery of the Cotton Fabric

The drape coefficient of fabric is determined according to NF G07–109 standard as shown in Table.

Table 2. Drape coefficient of the untreated and cotton fabrics treated with chitosan and cross-linking agents (CA, Arkofix NET)

Side of fabric	Parameters	Untreated cotton fabric	Cotton fabrics treated with chitosan and cross-linking agents			
			CA		Arkofix NET	
			2.6 kDa	187kDa	2.6 kDa	187 kDa
		Hr (%)				
Wrong side of fabric	Hr	69.88	71.15	74.61	75.46	79.15
	CV of Hr (%)	0.53	1.68	0.70	2.93	0.37
Right side of fabric	Hr	66.00	69.68	73.16	73.46	75.33
	CV of Hr (%)	0.40	0.60	1.00	2.67	2.19
Average		67.94	70.42	73.88	74.46	77.24

Where: Hr: drape coefficient (%).

As indicated in Table 2, the Hr coefficient of four cotton samples treated with chitosan were higher than those of the untreated fabric. These results suggested that chitosan-treated fabrics are stiffer. It is possible that the cross-linking bonds have been formed between the cellulose molecules that are responsible for this phenomenon. However, Table 2 also show that two samples treated with Arkofix NET had a higher Hr coefficient than the samples treated with CA. With the same cross linking agent, the higher the molecular weight of chitosan, the higher the hardness of the treated fabric is. This phenomenon has also been observed in the other published studies [4, 9]. This effect may be attributed to the high viscosity of chitosan. Furthermore, a higher MW chitosan results in a higher viscosity solution, which reduces the penetration of chitosan solution inside the cotton fiber and then increases the amount of chitosan on the fiber surface making the fabric stiffer.

3.2 Effect of Cross-Linking Agent and Molecular Weight of Chitosan on the Creasing Resistance of the Cotton Fabric

The wrinkle recovery angle of the samples determined with ISO 2313–1972 standard are presented in Table 3.

Table 3. The wrinkle recovery angles of the untreated sample and samples treated with chitosan and cross-linking agents (CA, Arkofix NET)

Fabric samples	Warp				Weft				$\bar{\alpha}$ (degree)	
	5 mins		30 mins		5 mins		30 mins			
	$\bar{\alpha}$ of 5 times (degree)	CV (%)	$\bar{\alpha}$ of 5 times (degree)	CV (%)	$\bar{\alpha}$ of 5 times (degree)	CV (%)	$\bar{\alpha}$ of 5 times (degree)	CV (%)	5 mins	30 mins
untreated	51.2	0.30	62.1	0.70	46.5	0	52.5	2.20	48.85	57.30
2.6 - CA	52.7	1.88	64.3	0.90	46.5	1.37	54.0	0.52	49.60	59.15
187 - CA	55.6	0.97	68.4	0.89	47.1	0.45	56.2	2.77	51.35	62.30
2.6 - NET	91.6	0.25	102.2	0.53	84.9	0.78	97.9	0.09	88.25	100.05
187 - NET	96.4	0.56	108.8	0.10	91.3	0.22	100.9	0.09	93.85	104.85

$\bar{\alpha}$: are average of the wrinkle recovery angles of fabric samples.

Table 3 show that the creasing recovery angles of four cotton fabrics treated with chitosan were higher than those of the untreated fabric. This can be seen that the crease recovery angles of the samples treated with CA were insignificantly increased (3.3 and 8.7% corresponding with the chitosan having MW of 2.6 and 187 kDa). However, the crease recovery angles of the cotton fabrics treated with Arkofix NET at the same conditions were much increased (74.7 and 83.1% corresponding to the fabric samples treated with chitosan of 2.6 and 187 kDa). The crease recovery angles of the treated cotton fabrics were insignificantly increased with the increasing of MW of chitosan. This result is similar to the report on the cotton fabrics treated with chitosan and CA [19]. Though the crease recovery angles of the fabrics treated with Arkofix NET is much higher than that treated with CA, however, the antibacterial activity and antibacterial durability of the cotton fabrics treated with CA were higher than those of the cotton fabrics treated with Arkofix NET as reported in our previous study [20]. Possibly, the cellulose - Arkofix NET- cellulose reaction is easier to produce than the cellulose - CA - cellulose reaction. This results in a higher wrinkle recovery angle of the Arkofix NET treated fabric. In addition, the chitosan - CA - cellulose reaction may be more easily formed than the chitosan - Arkofix NET - cellulose reaction, resulting in higher antibacterial activity and antibacterial durability of CA-treated fabrics. Therefore, there is a difference in the role of the two cross-linking agents (CA and Arkofix NET) in the antimicrobial finishing of chitosan cotton fabrics: CA plays an important role in the cross-linking between cellulose and chitosan, and it can also improve the cross-linking between cellulose molecules. Meanwhile, compared with CA, Arkofix NET is easier to create cross-linking between cellulose molecules, and harder to cross-link between cellulose and chitosan. These results are consistent with the previous studies mentioned above.

3.3 Effect of Cross-Linking Agent and Molecular Weight of Chitosan on the Tensile Strength and Breaking Elongation of the Cotton Fabric

The tensile strength and breaking elongation of the cotton fabric determined with TCVN 1754 standard are presented in Table 4.

Table 4 show that the tensile strength and breaking elongation of all the treated samples are reduced in both warp and weft direction. These results are in agreement with the findings that have also been reported in the literatures [4, 9, 19]. In addition, the decreases in the tensile strength and breaking elongation of the treated samples can be attributed to the curing in temperature of 160 °C or the hydrolysis of cellulose in the acidic environment.

Table 4. Tensile strength and breaking elongation of the untreated sample and samples treated with chitosan and cross-linking agents (CA and Arkofix NET)

Fabric samples	Tensile strength – P_d (N)		Loss of the tensile strength in warp (%)	Loss of the tensile strength in weft (%)	Breaking elongation – E_d (mm)		Loss of the breaking elongation in warp (%)	Loss of the breaking elongation in weft (%)	Intensity (yarn/10cm)	
	Warp	Weft			Warp	Weft			Warp	Weft
Cotton untreated	797.66	550.19	-	-	28.35	38.73	-	-	410	175
2.6kDa - CA	571.66	421.66	28.33	23.36	22.79	27.61	19.61	28.71	410	175
2.6kDa - NET	404.45	262.60	49.30	52.27	20.68	21.66	27.05	44.07	410	175
187kDa - CA	581.92	403.63	27.05	26.64	21.89	28.13	22.79	27.37	410	175
187kDa - NET	481.23	285.82	39.67	48.05	22.82	22.53	19.51	41.83	410	175

Where:

Pd: Tensile strength of sample (N).
Ed: Breaking elongation of sample (mm).

Similar to the crease recovery angles, the tensile strength and breaking elongation of the cotton treated with chitosan changed obviously with different cross-linking agents: the tensile strength and breaking elongation of the cotton fabric samples treated with chitosan and CA were higher than those treated with chitosan and Arkofix NET.

For each cross-linking agent, the effects of MW of chitosan on the tensile strength and breaking elongation were different. With CA, there was insignificant difference in the decrease of tensile strength and breaking elongation when using the chitosan with molecular weight of 2.6 kDa and 187 kDa. However, in case of Arkofix NET, the tensile strength and breaking elongation significantly increased when integrating with chitosan 2.6 kDa in comparison with the case integrating with chitosan 187 kDa. This phenomena may be explained that the solution of high molecular weight chitosan having high viscosity, so it could be more on the surface of the fabric, whereas the low molecular weight chitosan solution with low viscosity could be penetrated deeply into the fiber structure of the cotton fabric, which increased the cross-linking between the cellulose molecules of the cotton fabric. This result was also according to our previous results on the drapery and the crease recovery angles of the cotton fabric treated with chitosan.

3.4 Effect of Cross-Linking Agent and Molecular Weight of Chitosan on the Surface Characteristics of the Cotton Fabric

The surface properties of the fabric were evaluated using KAWABATA method:

The surface characteristics of the cotton samples were assessed through the measurements of the surface friction coefficient, the average deviation of surface friction coefficient. The average deviation of the roughness was measued by using the KAWABATA equipment, and the results are presented in Table 5.

Table 5. The surface characteristics of the untreated sample and samples treated with chitosan and cross-linking agents (CA and Arkofix NET)

Characteristic	Times	Untreated cotton fabric			Cross-linking agents											
					CA						NET					
					2.6 kDa			187 kDa			2.6 kDa			187 kDa		
		Warp	Weft	Average	Warp	Weft	Average	Warp	Weft	Average	Warp	Weft	Average	Warp	Weft	Average
MIU	Average	0.223	0.219	0.221	0.216	0.223	0.220	0.238	0.223	0.231	0.218	0.198	0.208	0.164	0.160	0.162
	CV(%)	2.017	0.693	0.686	0.655	0.969	0.811	1.159	1.062	1.023	0.432	0.990	0.261	0.831	0.011	0.009
MMD	Average	0.006	0.011	0.009	0.006	0.009	0.008	0.007	0.012	0.010	0.006	0.011	0.009	0.007	0.013	0.010
	CV(%)	6.244	0.057	0.057	5.813	4.457	0.891	2.815	7.454	4.243	3.898	4.598	3.303	4.694	2.946	1.692
SMD (μm)	Average	1.637	2.525	2.081	1.668	2.675	2.172	1.747	2.725	2.236	1.788	2.887	2.338	1.613	2.503	2.058
	CV(%)	4.694	3.921	1.947	6.077	3.853	1.743	3.197	2.078	1.542	5.868	1.715	3.284	3.891	5.294	4.730

Table 5 shows that the fabric treated with chitosan 2.6 kDa has the average surface friction coefficient of 0.22 corresponding to the standard deviation of 0.008; and for the fabric treated with chitosan 187 kDa, the average surface friction coefficient is 0.223 corresponding to the average standard deviation of 0.01. Thus, chitosan with a molecular weight of 2.6 kDa did not affect the roughness of treated cotton fabric, while the chitosan with a molecular weight of 187 kDa made the fabric rougher. This may be explained that the high viscosity of the high molecular weight chitosan solution can make it more difficult to penetrate the fiber, and the high MW chitosan solution is may be only on the fabric surface causing roughness and stiffness to the treated fabric.

For the cotton fabric treated with chitosan and Arkofix NET, the average surface friction coefficient was lower than that of the untreated one. This means that the surface of the treated cotton fabric was smoother than the untreated one.

4 Conclusion

- The effect of cross-linking agents on the mechanical properties of chitosan-treated fabrics is quite obvious. The chitosan-treated sample with CA has higher tensile strength than the sample treated with Arkofix NET. However, the chitosan-Arkofix NET-treated fabric has a higher crease recovery angle than the chitosan-CA-treated fabric. This trend was repeated with both chitosan.

– Comparing the influence of the molecular weight of chitosan on the mechanical properties of the treated fabric, it can be seen that the 187 kDa-chitosan-treated cotton fabric has a higher wrinkle recovery angle and a higher tensile strength.

References

1. Lim, S.H.: Synthesis of A fiber-reactive Chitosan Dericative and Its Application to Cotton Fabric as an Antimicrobial Finish and a Dyeing-imprvoing Agent, Doctor of Philosophy in Fiber and Polymer Science of North Carolina State University (2002)
2. Fouda, M.M.G.: Use of Natural Polysaccharides in Medical Textile Applications, in Chemistry Department, Univerdity of Duisburg-Essen (2005)
3. Sandra Flinčec Grgac, A.T.: The Chitosan Implementation into Cotton and Polyester/Cotton Blend Fabrics, Tihana Dekanić, Witold Sujka and and Z. Draczyński. Materials **13**(1616), 1–19 (2020)
4. Khaled, F., El-tahlawy, M.A.E.-B., Elhendawy, A.G., Hudson, S.M.: The antimicrobial activity of cotton fabrics treated with different crosslinking agents and chitosan. Carbohydrate Polymers **60**, 421–430(2005)
5. Vellingiri, K., Senthilkumar, T.R.M.: Eco-friendly application of nano chitosan in antimicrobial coatings in the textile industry. Nanosci. Nanotechnol. **3**(4), 75–89 (2013)
6. Enescu, D.: Use of chitosan in surface modification of textile materials. Roumanian Soc. Biol. Sci. **13**(6), 4037–4048 (2008)
7. Fu, X., Shen, Y., Jiang, X., Huang, D., Yan, Y.: Chitosan derivatives with dual-antibacterial functional groups for antimicrobial finishing of cotton fabrics. Carbohyd. Polym. **85**, 221–227 (2011)
8. Anahita Rouhani Shirvan, M.S., Bashari, A.: Recent advances in application of chitosan and its derivatives in functional finishing of textiles, the Impact and Prospects of Green Chemistry for Textile Technology, Copyright © 2019 Elsevier Ltd. All rights reserved, pp. 107–133 (2019
9. Shin, Y., Yoo, D.I., Jang, J.: Molecular weight effect on antimicrobial activity of chitosan treated cotton fabrics. J. Appl. Pol. Sci. **80**, 2495–2501 (2001)
10. Rahman Bhuiyan, M.A., Hossain, M.A., Zakaria, M., Islam, M.N., Zulhash Uddin, M.: Chitosan coated cotton fiber: physical and antimicrobial properties for apparel use. J. Pol. Environ. **25**, 334–342 (2017)
11. A Edwin Sunder, G.N.: Principal, croslinking of chitosan with cotton using polycarboxylic acids. Int. J. Eng. Res. Technol. **3**(4), 1769–1774 (2014)
12. Kim, H.W., Kim, B.R., Rhee, Y.H.: Imparting durable antimicrobial properties to cotton fabrics using alginate–quaternary ammonium complex nanoparticles. Carbohyd. Polym. **79**, 1057–1062 (2010)
13. Periolatto, M., Ferrero, F.: Antimicrobial chitosan finish of cotton and silk fabrics by UV-curing with 2-hydroxy-2-methylphenylpropane-1-one. Carbohyd. Polym. **88**, 201–205 (2012)
14. Lim, S.-H., Hudson, S.M.: Synthesis and antimicrobial activity of a water-soluble chitosan derivative with a fiber-reactive group. Carbohyd. Res. **339**, 313–319 (2004)
15. Hebeish, A., Abdel-Mohdy, F.A., Fouda, M.M.G., Elsaid, Z., Essam, S., Tammam, G.H., Drees, E.A.: Green synthesis of easy care and antimicrobial cotton fabrics. Carbohyd. Polym. **86**, 1684–1691 (2011)

16. Wu, Y., Bian, Y., Yang, F., Ding, Y., Chen, K.: Preparation and properties of Chitosan/Graphene modified bamboo fiber fabrics. Polymers **11**(1540), 1–13 (2019)
17. Tho, L.T., Hong Khanh, V.T., Thong, N.V.: Effect of molecular weight of chitosan on antibacterial activity of cotton fabrics treated by chitosan. J. Sci. Technol. Tech. Univ. **93**, 27–33 (2013)
18. Tho, L.T., Hong Khanh, V.T., Thong, N.V.: Effect of molecular weight of chitosan on the antibacterial durability of cotton fabrics after treatment by chitosan. In: The 15th International Symposium on Eco-materials Processing and Design, ISEPD 2014, Ha noi University of Science and Technology, Hanoi, Vietnam, pp. 23–28 (2014)
19. Chung, Y.-S.: Durable press and antimicrobial finishing of cotton fabrics with a citric acid and chitosan treatment. Text. Res. J. **68**(10), 772–775 (1998)
20. Hong Khanh, V.T., Tho, L.T., Thong, N.V.: Effect of cross-linker on the antibacterial durability of cotton fabrics after treatment by chitosan. In: Proceedings of The 89th Textile Institute World Conference Wuhan, P. R. China, 2–6 November, pp. 488–492(2014)

Manufacturing and Heat Treatment of High-Precision Ring Gauge

Trinh Van Trung and Nguyen Van Thanh(✉)

Hanoi University of Science and Technology, No. 1, Dai Co Viet, Hai Ba Trung, Hanoi, Viet Nam

Abstract. Two high-precision gauges (Ø162-Go and Ø162-Not Go) for assessing shaft components with a nominal diameter of 162 μm were machined from SKD11 steel and heat-treated using some technologies, such as quenching, cryogenic treatment, and aging, to improve hardness, wear resistance, and dimensional stability. The hardness, abrasion intensity, microstructure, phase analysis, and change in size of the gauge were observed using a Metrolog hardness tester, the pin on disc abrasion equipment, Axiovert microscopy, X-Ray Diffraction X pert PRO, and 3D measurement FUSION 9106 machine, respectively. Results showed that the hardness of gauges is approximately 62 HRC after quenching at 1030 °C for 90 min and reduces to around (59 ÷ 59,5) HRC after applying heat treatment technology in M0 mode (quenching and aging at 150 °C for 48 h). The hardness of gauges is nearly (62.5 ÷ 63) HRC, which is 2 HRC higher than the value in M1 mode (quenching + cryogenic treatment at − 80 °C/6 h + aging at 150 °C/48 h), after deep cryogenic treatment at − 160 °C in M2 mode (quenching + cryogenic treatment at − 160 °C/6 h + aging at 150 °C/48 h) due to carbides formation. The gauge after heat treatment in M2 mode demonstrated the smallest change in size (ΔL < 3 μm) in the temperature range of 20 °C–40 °C compared with M0 and M1 modes.

Keywords: High-precision gauge · Cryogenic treatment · Aging · SKD11 steel

1 Introduction

Gauge is a tool used for examining components in a limited size range. The gauge fails to determine the exact size of components and its purpose is to classify between standard and nonstandard components in mass production rapidly.

The evaluation results showed whether the size of test components is within the tolerable range. At least two gauges are required to examine a certain nominal size, wherein both upper and lower tolerances are within the tolerance of the component. For example, two cylindrical shape gauges were used in accordance with the size and tolerance of the hole to be tested when we assess the size of a circular hole of a component. The first gauge will pass through the hole (the gauge with the lower tolerance is called “GO”), the second gauge will not pass (the gauge with the upper tolerance is called “NOT GO”), and the hole size of the component is qualified. If the manufactured hole is smaller than required, then the “GO” gauge will not pass. If the

B. T. Long et al. (Eds.): MMMS 2020, LNME, pp. 282–288, 2021.
https://doi.org/10.1007/978-3-030-69610-8_39

hole is excessively large, then the "NOT GO" gauge will pass and the component is unqualified.

High-precision or high-quality gauges demonstrate high hardness (60 ÷ 65 HRC) and high dimensional stability (with a small thermal expansion coefficient of around $10^{-5} \div 10^{-6}$ /°C) [1].

An important issue in gauge manufacturing is choosing suitable heat treatment processes to minimize retained austenite and increase hardness and dimension stability of gauges. Gauges need to be quenched to achieve high hardness. Our previous study [2] showed that quenched SKD11 steel at 1030 °C demonstrates the maximum hardness results because of the low amount of carbides dissolved in austenite when austempering at low temperatures (980 °C) leading to the low hardness of obtained martensite after quenching; however, the increase in both grain size and retained austenitic content leads to reduced hardness when the quenching temperature is excessively high (1080 °C). Short soaking time (60 min) in the quenching process led to nonuniform hardness at the surface and in the core of the component, whereas the long soaking time (150 min) that coarsens the grain size led to reduced hardness. The maximum hardness of 62 HRC is achieved after quenching at 1030 °C for 90 min.

Manufacturing high-quality gauges is a serious problem because it requires precise machining operation and maintenance of dimensional stabilization of gauges under test conditions. High-alloy steels, such as SKD11, commonly contain a large amount of retained austenite after quenching; hence, the hardness and abrasion resistance of SKD11 steel fail to reach the maximum level. Furthermore, dimension changes easily occur in retained austenite due to the 4% volume expansion when austenite transforms into martensite [3]. After quenching, subzero or cryogenic treatment can help reduce this retained austenitic content and improve hardness and dimensional stability [4–6]. Apart from selecting steel with a small thermal expansion coefficient, such as SKD11, microstructure stability is important in improving the dimensional stability of the gauge. The microstructure stability can be improved by applying an aging process (steel or component is heated at a low temperature range of less than 150 °C for long periods of time, such as 1 to 2 days) after quenching. This work focuses on the manufacturing and application of heat treatment technologies, such as cryogenesis and aging, for two high-precision ring gauges (Ø162-Go and Ø162-Not Go), which are used to determine dimensions of cylindrical components with a nominal diameter of 162 mm.

2 Experiment

The diagram of manufacturing and heat treatment processes for gauges is shown in Fig. 1.

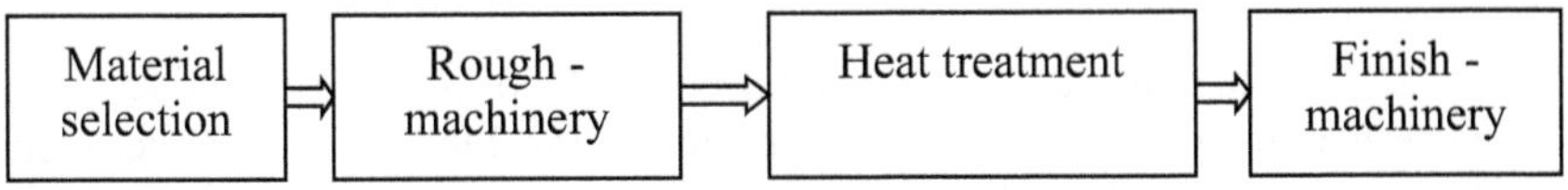

Fig. 1. Basic stages of manufacturing and heat treatment processes for high-precision gauges

SKD11 steel (chemical composition: Fe - 1.5% C - 12.1% Cr - 0.9% Mo and low thermal expansion coefficient: 10.3×10^{-6} /°C) is an appropriate choice for the requirement of high hardness and dimensional stability. The uniform microstructure and hardness of 200 HB as-received SKD11 steel makes it suitable for rough-machinery process without preheating.

SKD11 steel billet is machined via CNC machine (MvP10) to make ring gauges. Machined ring gauges with a nominal diameter of 162 mm are used to determine the diameter of the shaft or cylinder and manufactured according to the design drawing shown in Fig. 2 (Ø162-Go ring gauge) and 3 (Ø162-Not Go ring gauge) with two working sizes of D1 ($\varnothing 162_{0}^{+0.005}$) and D2 ($\varnothing 162_{-0.025}^{-0.02}$), respectively. The temperature range of 20 °C–40 °C is suitable for the testing temperature in mechinary factories.

Machined ring gauges require heat treatment via quenching at 1030 °C for 90 min in a single-chamber vacuum quenching furnace (Turbo^{2}Treater, Ipsen, Germany) with compressed air pressure of 6 bars and then heat-treated in three modes, namely, M0 (aging at 150 °C for 48 h), M1 (cryogenic treatment at − 80 °C/6 h + aging in 150 °C/48 h), and M2 (cryogenic treatment at − 160 °C/6 h + aging at 150 °C/48 h), to achieve high hardness of 60 ÷ 65 HRC and dimensional stability.

The finish machinery (profile cutting) step was carried out using a wire cutter (30P) to achieve the working size. Specifically, Ø162-Go gauge reached diameter to inspect product within the range (162,000 ÷ 162,005) mm; Ø162-Not Go gauge reached diameter to inspect product within the range (161,975 ÷ 161,980) mm after finish machinery. In addition, surface roughness of gauges (Ra = 0.63) is required.

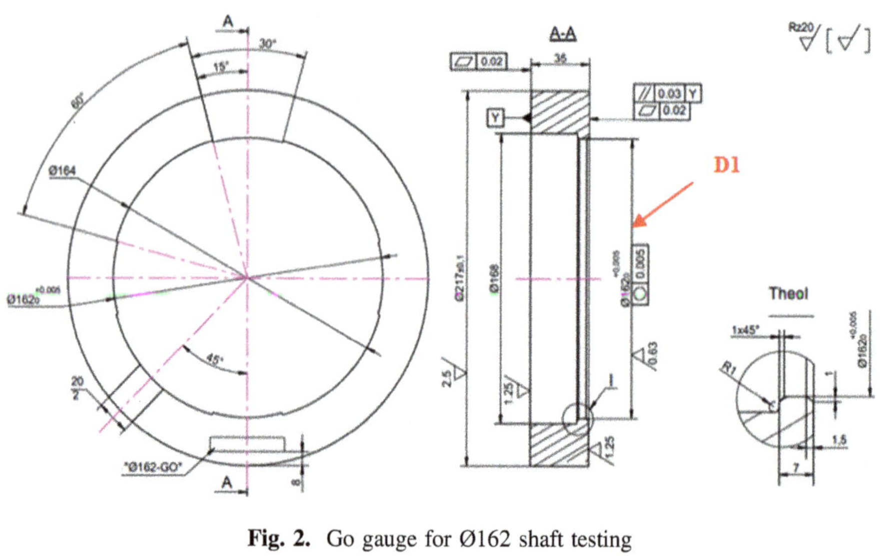

Fig. 2. Go gauge for Ø162 shaft testing

The microstructure image was surveyed using an Axiovert optical microscope. The quantitative measurement was analyzed using Image-Pro Plus software. Color-etching method with etching solution (HCl + picric acid) was used to determine the percentage of retained austenite and the total percentage of carbides and austenite, and an etching solution of NaOH and K_3FeN_6 was utilized to calculate the carbide percentage. Measurements were carried out at five locations on the sample to minimize errors. Phase of heat-treated SKD11 steel was investigated using X-ray diffraction (XRD) equipment. The hardness, wear resistance, and dimensions of the gauge were calculated using a Metrolog hardness tester, Pin-on-disc equipment, and FUSION 9106 3D-dimensioners.

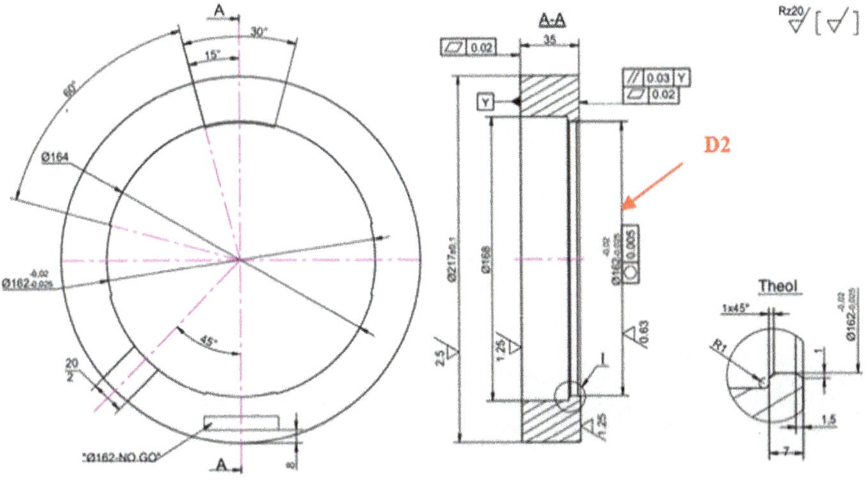

Fig. 3. Not Go gauge for Ø162 shaft testing

3 Results and Discussion

Figure 4 shows the fabricated Ø162-Go and Ø162-Not Go ring gauges (with working sizes: D1 and D2) fabricated after the machining operation shown in Fig. 1.

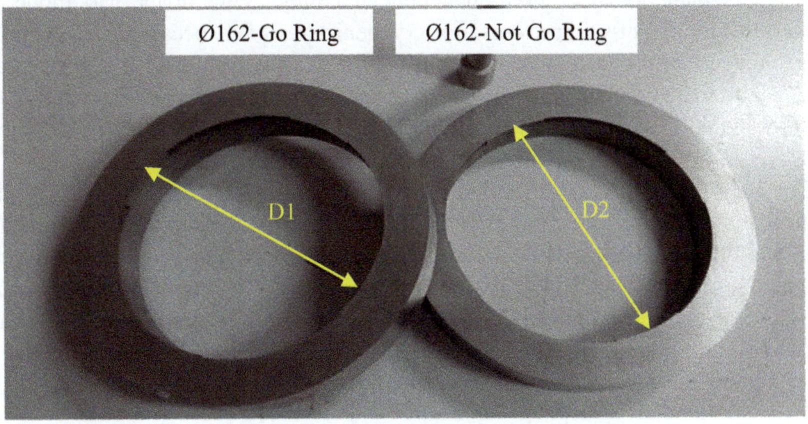

Fig. 4. As-machined ring gauges (Ø162-Go and Ø162-Not Go)

As-machined gauges must be quenched to improve their hardness and continuously subjected to cryogenic treatment technology and/or aging for dimensional stability. Both gauges (Ø162-Go and Ø162-Not Go) achieved a hardness value within the range of design requirements after heat treatment in different modes (Table 1).

Table 1. Hardness test of two gauges (Ø162-Go and Ø162-Not Go) at different heat treatments

Hardness (HRC)	M0	M1	M2	Hardness according to design requirements
Ø162-Go	59.0	60.5	62.5	58 ÷ 63 HRC
Ø162-Not Go	59.5	61.0	63	58 ÷ 63 HRC

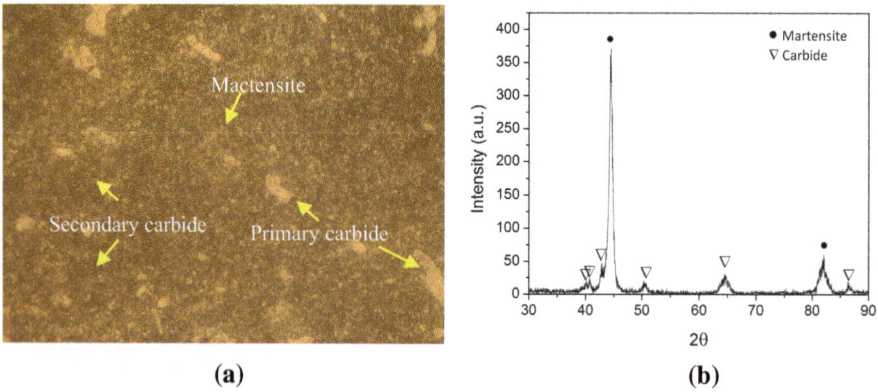

Fig. 5. (a) Microstructure and (b) XRD pattern of SKD11 steel after quenching + cryogenic treatment at − 160 °C/6 h + aging at 150 °C/48 h (M2 mode)

The microstructure image and XRD analysis (Fig. 5) show that martensite and carbide (primary and secondary carbides) phases were detected. The difference in hardness between the two gauges is approximately 0.5 HRC. The hardness decreased to a value less than 60 HRC after quenching and only aging in the M0 mode. The hardness is significantly improved after cryogenic treatment because this technology helps to reduce the retained austenitic content significantly from 9.65% after quenching to 1.85% after treating in the M2 mode. Cryogenic treatment at deep negative temperatures in the M2 mode obtains a hardness of 2 HRC higher than the value in the M1 mode.

The improvement of hardness is extremely important in the field of tool manufacturing in general and the production of high-precision gauges in particular because it will help improve wear resistance and therefore increase the longevity of the product. High hardness enhances the wear resistance. The results of wear intensity measurement of three heat treatment modes (M0, M1, and M2) showed that the minimum wear intensity (highest hardness after deep cryogenic treatment at − 160 °C in the M2 mode) obtains the maximum wear resistance (Fig. 6).

Working sizes D1 of Ø162-Go and D2 of Ø162-Not Go gauges after heat treatment in M0, M1, and M2 modes are listed in Table 2. The results showed that all the measured working sizes of both gauges after heat treatment in the M0, M1, and M2 modes are within the range of nominal dimension.

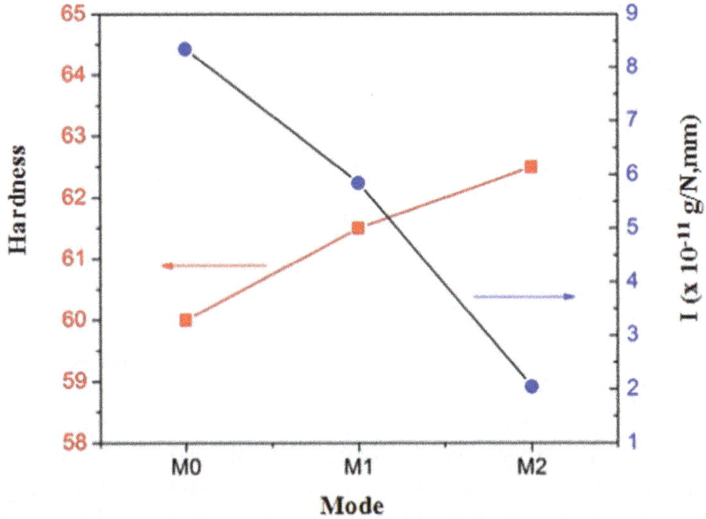

Fig. 6. Effect of M0, M1, and M2 heat treatment modes on the hardness and wear resistance of the manufactured SKD11 steel gauge

Table 2. Results of measuring the working size of two gauges at 25 °C after heat treatment in M0, M1, and M2 modes

Gauge	Nominal dimension, D	Working size	M0	M1	M2
Ø162-Go	$\varnothing 162_{0}^{+0.005}$	D1	162.0019	162.0021	162.0018
Ø162-Not Go	$\varnothing 162_{-0.025}^{-0.02}$	D2	161.9775	161.9786	161.9793

The dimensional stability of gauges (Table 3) were investigated by surveying the change in the working size D2 of the Ø162-Not Go gauge at four temperature ranges of 20 °C–25 °C, 25 °C–30 °C; 30 °C–35 °C, and 35 °C–40 °C. The results showed that the working size of the gauge treated in the M2 mode demonstrates the minimum size change (<3 µm) in all the temperature ranges from 20 °C to 40 °C compared with treatments in M0 and M1 modes. Hence, the cryogenic treatment at a temperature of –160 °C obtaining the minimum content of austenitic phase (1.85%) in the microstructure of SKD11 steel significantly improves the dimensional stability of the gauge in the M2 mode.

Table 3. Dimension change of working size of the heat treated gauge in M0, M1, and M2 modes in relation to the temperature of the testing environment

Temperature range of the testing environment (°C)	Dimension change (ΔL) of working size, μm		
	M0	M1	M2
20–25	0.1	0.1	0.4
25–30	4.5	1.3	2.1
30–35	7.3	2.4	2.5
35–40	10.8	5.1	2.7

4 Conclusion

Two ring gauges (Ø162-Go and Ø162-Not Go) were manufactured to examine shaft components with a nominal diameter of 162 μm. Both gauges were applied in three various modes of heat treatment technologies, namely, M0 (quenching + aging at 150 °C/48 h), M1 (quenching + cryogenic treatment at − 80 °C/6 h + aging at 150 °C/48 h), and M2 (quenching + cryogenic treatment at − 160 °C/6 h + aging at 150 °C/48 h). The results showed that (1) two high-precision gauges are manufactured within the required design dimensions and (2) the application of cryogenic treatment and aging processes improves the hardness, wear resistance, and dimensional stability of gauges.

Acknowledgment. This research is funded by the Hanoi University of Science and Technology (HUST) under project number T2020-SAHEP-040.

References

1. Kalpakjian, S., Schmid, S.R.: Manufacturing Engineering & Technology, 4th edn., Prentice Hall (2001)
2. Van Trung, T., Van Thanh, N.: Study on heat treatment for SKD11 steel for high presise gauges using for mechanical assembly. J. Sci. Technol. Metal **87**, 33–39 (2019)
3. Thelning, K.E.: Steel and Its Heat Treatment, 2nd edn. Butterworths (1984). p.199
4. Barron, R.F.: Cryogenic treatment of metals to improve wear resistance. Cryogenics **22**, 409–413 (1982)
5. Baldissera, P., Delprete, C.: Deep cryogenic treatment: a bibliographic review. Open Mech. Eng. J. **2**, 1–1 (2008)
6. Yang-Yu, S., Chiu, L.-H., Chen, F.-S., Lin, S.-C., Pan, Y.-T.: Residual stresses and dimensional changes related to the lattice parameter changes of heat-treated JIS SKD 11 tool steels. Mater. Trans. **55**, 831–837 (2014)

Computer Simulation of Alumina and Aluminosilicate at Amorphous and Liquid State

Mai Thi Lan(✉)

School of Engineering Physics, Hanoi University of Science and Technology, No. 1 Dai Co Viet Road, Hai Ba Trung District, Hanoi City, Vietnam
lan.maithi@hust.edu.vn

Abstract. The structural characteristics of $Al_2O_3.2SiO_2$ and Al_2O_3 at amorphous (600K) and liquid (3500K) states have been investigated by Molecular Dynamics (MD) simulation. MD models contain 5500 atoms in cubic box with periodic boundary and application of the Born-Mayer potentials. Most of the Al and Si atoms are 4-coordinated by O atoms in both states to form SiO_4 and AlO_4 tetrahedra units. The links between two adjacent SiO_x units and between AlO_x and SiO_x units (x = 3 ÷ 5) in AS2 at 600K and 3500K is mainly corner-sharing and edge-sharing bonds. We found that there is the first speak splitting of Al-Al pair radial distribution function in AS2 at amorphous state. This does NOT appear in AS2 at liquid state as well as in Al_2O_3 at both states. The origin of the first peak splitting relating to the existence of number of edge-sharing bonds will be explained and discussed in detail.

Keywords: MD · Structure · The first peak splitting · Alumina · Aluminosilicate

1 Introduction

Al_2O_3-SiO_2 system is the main components in the Earth's mantle as well as in glass and ceramic materials [1–3]. Al_2O_3-SiO_2 is also known as the main component of Geopolymer material which the family of inorganic polymers [4]. Studying the structure and dynamics of them in both liquid and amorphous states is essential for understanding the physical and chemical changes in the earth as well as controlling and finishing of the processes of material manufacturing technology. Many experimental and simulation studies were performed for Al_2O_3-SiO_2 system [1–3, 5–13]. These studies showed that the chemical ordering in Al_2O_3-SiO_2 is very complicated. To understand its structure, it is necessary to clarify the local structure, network structure around Al, Si, O atoms. The studies also indicated that the NBO and tri-clusters may play an important role in dynamic properties of Al_2O_3-SiO_2 melts [1, 2, 13]. In the work [1] showed that there is the first peak splitting of Al-Al pair radial distribution function (RDF) in $Al_2O_3.2SiO_2$ at amorphous state. However, this problem is still explained clearly. In this work, we used MD simulation for $Al_2O_3.2SiO_2$ systems which the simple pseudo-binary silicate (denoted as AS2) and compared with Al_2O_3 system. The network structure, dynamics properties and the origin of the first peak splitting of Al-Al pair RDF will be clarified and discussion in detail.

B. T. Long et al. (Eds.): MMMS 2020, LNME, pp. 289–295, 2021.
https://doi.org/10.1007/978-3-030-69610-8_40

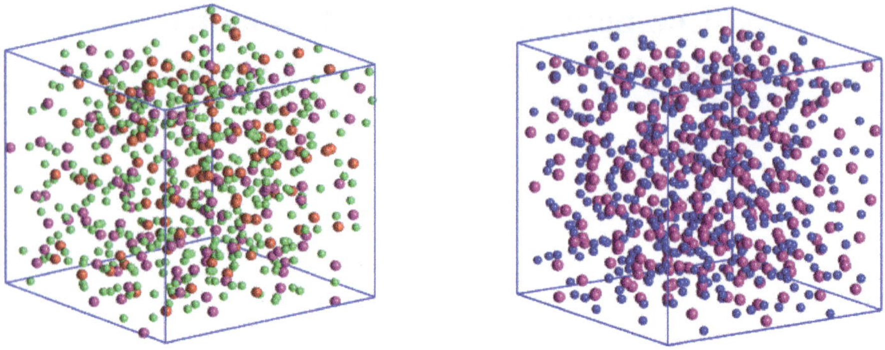

a) AS2 model (Pink spheres are Si, green spheres are O and red spheres are Al atoms)

b) Al_2O_3 model (blue sphere are O and pink sphere are Al atoms)

Fig. 1. Illustration of MD initial configuration for AS2 and Al_2O_3 systems

Table 1. Structural characteristics of AS2 and Al_2O_3 at amorphous (600K) and liquid (3500K) states; r_{ij} (Å) - position of first peak of pair RDF $g_{ij}(r)$; Al_x, Si_y (%) - the fraction of structural unit AlO_x, SiO_y ($x = 3 \div 5$; $y = 4 \div 5$)

System	AS2		Al_2O_3	
T (K)	600 K	3500 K	600 K	3500 K
$r_{Al\text{-}Al}$	3.16(pre-peak: 2.74)	3.18	3.14	3.18
$r_{Al\text{-}O}$	1.64	1.64	1.72	1.72
$r_{O\text{-}O}$	2.64	2.64	2.84	2.84
$r_{Al\text{-}Si}$	3.16	3.16	-	-
$r_{Si\text{-}Si}$	3.18	3.18	-	-
$r_{Si\text{-}O}$	1.60	1.60	-	-
Al_3	13.11	35.72	0.40	6.10
Al_4	75.24	57.80	72.35	64.60
Al_5	11.23	26.23	26.23	27.20
Si_4	89.49	98.93	-	-
Si_5	9.97	1.07	-	-
< O-Al-O >	105° (pre-peak: 85°)	105°		
< O-Si-O >	105°	105°	-	-

2 Calculation Method

We used MD simulation method and employed the Born-Mayer potential to construct MD models for Al_2O_3 and AS2 systems. The MD initial configuration is created by randomly placing 5500 atoms in cubic box with periodic boundary condition. MD model for AS2 system (5500 atoms with 1000 Si, 3500 O, 1000 Al) and for Al_2O_3 system (5500 atoms with 2200 Al, 3300 O) is presented in Fig. 1. The MD model constructing can be refer in our previous work [14]. To study structural characteristics, the obtained samples are relaxed in constant temperature and pressure (NPT model). The structural properties of the obtained models determined by averaging over 1000 configurations. The time consumed to obtain these is about 10^6 MD steps.

Table 2. Distribution of O atoms forming types of linkages in Al_2O_3 and AS2 at 600 K and 3500 K

T (K)	AS2		Al_2O_3	
	600 K	3500 K	600 K	3500 K
O-Al2 (%)	5.46	6.71	18.58	36.21
O-Al3 (%)	10.51	9.86	78.61	61.06
O-Al4 (%)	1.51	1.40	2.82	2.61
O-Al5 (%)	0.11	0.09	0.00	0.00
O-Si2 (%)	25.11	29.89	-	-
O-Si3 (%)	0.06	0.00	-	-
Si-O-Al (%)	25.03	33.66	-	-
Si-O-Al2 (%)	20.00	14.63	-	-
Si-O-Al3 (%)	2.66	1.11	-	-
Si-O-Al4 (%)	0.14	0.00	-	-
Si2-O-Al (%)	8.69	2.63	-	-
Si2-O-Al2 (%)	0.69	0.03	-	-
Si3-O-Al (%)	0.03	0.00	-	-

3 Result and Discussion

The structural properties of AS2 and Al_2O_3 at amorphous (600K) and liquid (3500K) states are shown in Table 1 and Table 2. The simulation results show a good agreement with the published experimental data and simulation in the first peak position, coordination number and bond angle [1, 2, 5]. The first peak position of the radial distribution function (RDF) for all pairs (Al-O, O-O, Al-Si, Si-Si, Si-O) is almost not dependent on temperature. At both states, the position of the first peak of RDF for Si-O pair is 1.60 Å. This know the strong covalent nature of the Si-O bond. The Al-O bond length in AS2 (1.64 Å) is smaller than one in Al_2O_3 (1.72 Å) at both states. There is significantly difference between the first peak position of RDF for Al-Al pair in AS2 and Al_2O_3 at 600K. The first peak of RDF for Al-Al pair in AS2 splits up in two peaks at r_1 = 2.74 Å (pre-peak) and r_2 = 3.16 Å (main peak) at 600 K. Moreover, it also

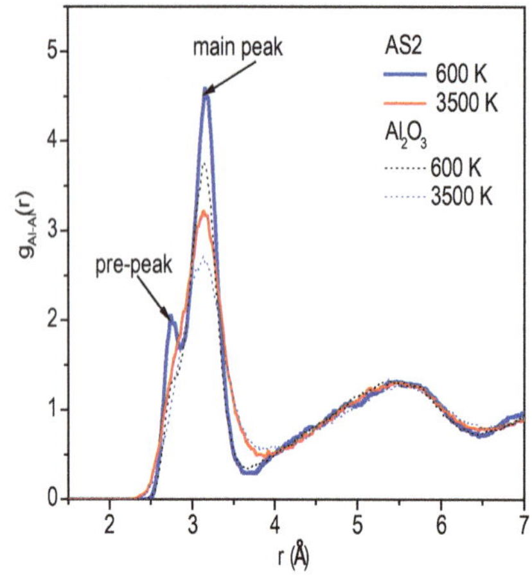

Fig. 2. Radial distribution function for Al-Al pair.

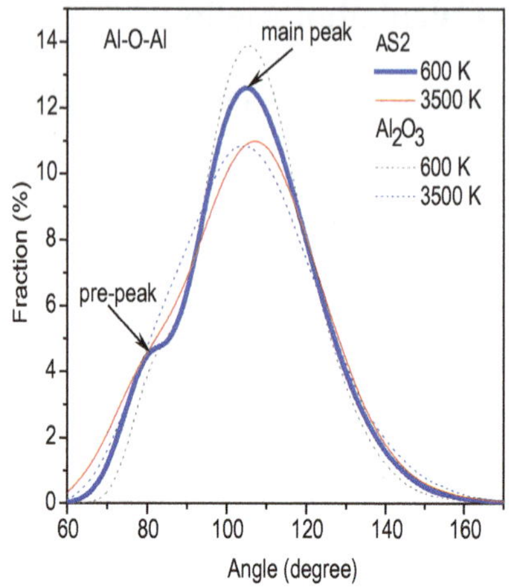

Fig. 3. Distribution of bond angle O-Al-O in AlO_4 unit.

exhibits the pre-peak in the O-Al-O bond angle distribution function (pre-peak at $\alpha_1 = 85^0$, main peak at $\alpha_2 = 105^0$). It is showed in Fig. 2 and 3. Whereas the RDF for Al-Al pair in Al_2O_3 have only single peak at 3.14 Å. The first peak splitting is NOT presence in Al_2O_3 and AS2 at liquid state. We found that the pre-peak relates to the existence of the edge-sharing bonds and will be clarified in the following. From Table 1, the results show that the structure of AS2 is comprised by TO_x basic structural units (TO_x means that T atom is surrounded by x O atoms at the nearest-neighbor distance; T = Si, Al; $x = 3 \div 5$). Fraction of SiO_4 units is 89.49% and 98.93% at 600K and 3500K, respectively. Fraction of AlO_4 units is mainly in AS2 (75.24 and 57.80% at 600K and 3500K, respectively) and Al_2O_3 (72.35% and 64.60% at 600K and 3500K, respectively). From Table 2, it shows that most O atoms is surrounded by y T atoms at the nearest-neighbor distance to form OT_y units (T = Al, Si; $y = 2 \div 5$). For Al_2O_3 system, fraction of OAl_3 units is dominant in both states (78.61% at 600 K and 61.06% at 3500 K). For AS2 system, most of OT_y is OAl_3 (9–11%), OSi_2 (25–30%), OSiAl (25–34%) and $OSiAl_2$ (14–20%) in which OSi_2 and OSiAl is dominant. It means that the structure of Al_2O_3 system at both amorphous and liquid state is mainly build by AlO_4 tetrahedral network structure which linked to each other via O-Al_3 linkages. The structure of AS2 comprises of both SiO_4 and AlO_4 tetrahedral units which linked each to other via O-T_2 or O-T_3 linkages to form network structure in three-dimensional space. The results show that the adjacent TO_x units can be connected by one O atom to forming corner-sharing bond, or two O atoms to forming edge-sharing bond, or three O atoms to forming face-sharing bond. The statistics of the types of bonds between two adjacent TO_x units are showed in Table 3.

Table 3. Distribution of the types of bonds between two adjacent SiOx units, two adjacent AlO_x units and between AlO_x and SiO_x in AS2 at 600 K and 3500 K

T (K)	SiO_x-SiO_x		AlO_x-AlO_x		AlO_x-SiO_x	
	600 K	3500 K	600 K	3500 K	600 K	3500 K
Corner-sharing	1216	1138	1319	1081	2243	2014
Edge-sharing	0	0	340	274	200	59
Face-sharing	0	0	0	8	3	0

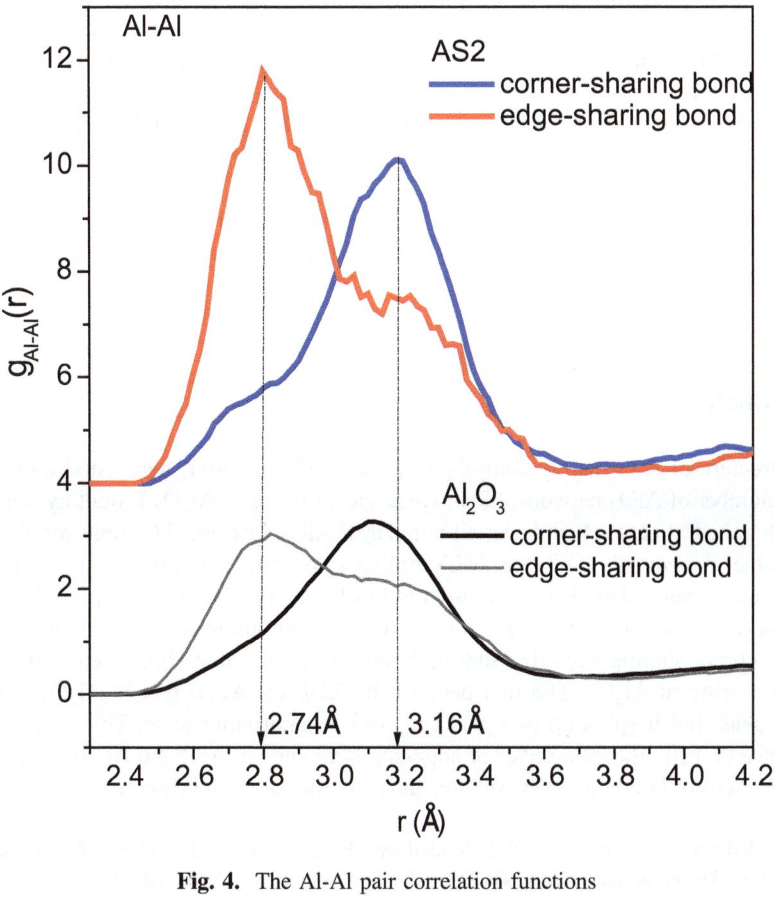

Fig. 4. The Al-Al pair correlation functions

For AS2, links between two adjacent SiO_x units are corner-sharing bonds. Links between two adjacent AlO_x units or between AlO_x and SiO_x units are mainly corner- and edge-sharing bonds. For Al_2O_3, links between two adjacent AlO_x units are mainly corner- and edge-sharing bonds meanwhile the number of edge-sharing links is not significant compared to the corner-sharing. The number of face-sharing bonds is very small, almost zero. As mentioned above, the structure of AS2 is mainly build by AlO_4 and SiO_4 tetrahedra units and the first peak in the RDF for Al-Al pair in AS2 splits up in two peaks at r_1 = 2.74 and r_2 = 3.16 Å at 600 K. The distribution of bond angle Al-O-Al in AlO_4 tetrahedra units for AS2 at 600 K have two peaks at $\alpha_1 = 85^0$ and $\alpha_2 = 105^0$. We found that these two angles relate to the distance r_1 and r_2 between two nearest neighboring Al atoms in the RDF for Al-Al pair. When the distance between the two nearest neighboring Al atoms is r_1 where AlO_4 tetrahedra link to each other by edge-sharing bond, the Al-O-Al angle in the AlO_4 structure unit corresponds to α_1. And when the distance between the two nearest neighboring Al atoms is r_2 where AlO_4 tetrahedra link to each other by corner-sharing bond, the angle of Al-O-Al in the AlO_4 structure corresponds to the angle α_2. This result showed that the main peak (r_2) in the RDF for Al-Al pair may be related to the corner-sharing bonds and the sub-peak (r_1) may be related to the edge-sharing bonds. To clarify this, we have calculated the RDF for the set Al-Al pairs with corner- and edge-sharing bonds (see Fig. 4). It can be seen that the RDF for set Al-Al pairs with corner-sharing bond has the first peak at a location of approximately 3.16 Å. Meanwhile, the RDF for the set of Al-Al pairs with edge-sharing bonds has peak at 2.74 Å. The Al-Al bond length in corner-sharing is much longer than the one in edge-sharing bonds. This demonstrates that the pre-peak of the RDF for Al-Al pair in Fig. 2 is due to the edge-sharing bonds. Beside the RDF for the set Al-Al pairs with edge-sharing bonds in Al_2O_3 has a smaller amplitude and the full width at half maximum (FWHM) of the first peak is much broader than in AS2, so there is NOT splitting peak in Al_2O_3.

4 Conclusion

The structure of AS2 mainly comprises of both SiO_4 and AlO_4 tetrahedral units and a small number of AlO_3 network. Meanwhile, the structure of Al_2O_3 is built by dominant AlO_4 tetrahedral network and a small number of AlO_5 network. TO_x units are linked to each other via mainly O-T_2 or O-T_3 linkages to form network structure in three-dimensional space. The degree of structural order in AS2 is lower than in Al_2O_3. The links between two adjacent TO_x units in AS2 at both amorphous and liquid states is mainly corner-sharing and edge-sharing bonds. The AlO_x units link to each other by a corner-sharing in Al_2O_3. The first peak of the RDF for Al-Al pair in Al_2O_3 only has single peak, but it splits up two peaks in AS2 at amorphous state. This is due to the existence of a significant number of edge-sharing bonds in AS2 and Al-Al bond length in edge-sharing bonds is much smaller than that in corner-sharing bonds.

Acknowledgements. This research is funded by Vietnam National Foundation for Science and Technology Development (NAFOSTED) under grant number 103.05–2018.37.

References

1. Anke, W., Jurgen, H., Walter, K., Kurt, B.: Structure and diffusion in amorphous aluminum silicate: A molecular dynamics computer simulation. J. Chem. Phys. **120**, 384 (2004)
2. Patrick, P., Jurgen, H., Kurt, B.: Structure and transport properties of amorphous aluminium silicates: Computer simulation studies. Chem. Geol. **229**, 186–197 (2006)

3. Hoang, V.V.: Dynamical heterogeneity and diffusion in high-density Al2O3.2SiO2 melts. Physica B **400**, 278–286 (2007)
4. Mustafa, A.M., Muhammad, F.M.T., Sandu, A.V., Alida, A., Salleh, M.A.A., Ruzaidi, C.M.: Microstructure studies on different types of geopolymer materials. Appl. Mech. Mat. **421**, 384–389 (2013)
5. Wilding, M.C., Benmore, C.J., Weber, J.K.R.: High-energy X-ray diffraction from aluminosilicate liquids. J. Phys. Chem. B **114**, 5742–5746 (2010)
6. Poe, B.T., Romano, C., Zotov, N., et al.: Compression mechanisms in aluminosilicate melts: Raman and XANES spectroscopy of glasses quenched from pressures up to 10 GPa. Chem. Geol. **174**, 21–31 (2001)
7. Sen, S., Youngman, R.E.: High-resolution multinuclear NMR structural study of binary aluminosilicate and other related glasses. J. Phys. Chem. B **108**, 7557 (2004)
8. Okuno, M., Zotov, N., Schmucker, M., Schneider, N.: Structure of SiO2-Al2O3 glasses: combined X-ray diffraction, IR and Raman studies. J. Non-Crystalline Sol. **351**, 1032–1038 (2005)
9. Bindera, K., Jurgen, H., Anke, W., Walter, K.: Modeling glass materials. Ceram. Int. **31**, 713–717 (2005)
10. Takei, T., Kameshima, Y., Yasumori, A., Okada, K.: Calculation of metastable immiscibility region in the SiO2–Al2O3 system using molecular dynamics simulation. J. Mater. Res. **15**, 186–193 (2000)
11. Schmucker, M., Schneider, H.: New evidence for tetrahedral triclusters in aluminosilicate glasses. J. Non-Cryst. Solids **311**, 211–215 (2002)
12. Allwardt, J.R., Stebbins, J.F., Schmidt, B.C., et al.: Aluminum coordination and the densification of high-pressure aluminosilicate glasses. Am Mineralogist **90**, 1218–1222 (2005)
13. Hoang, V.V.: Composition dependence of static and dynamic heterogeneities in simulated liquid aluminum silicates. Phys. Rev. B **75**, 174202 (2007)
14. Lan, T.M., Yen, V.N., Hong, V.N., Hung, K.P.: Visualization-based analysis of structure and dynamics of liquid aluminosilicate under compression. Phys. Chem. Liq. **55**, 62–84 (2017)

The Effect of Approach in the Contact Problem for Two Elastic Parallel Circular Cylinders

German Zhuravlev[1(✉)], Yuri Drobotov[2], and Tatiana Andreeva[2,3]

[1] LLC "SCB Mechanics SFEDU", Rostov-on-Don, Russia
[2] Southern Federal University, Rostov-on-Don, Russia
[3] Southern Mathematical Institute of VCS RAS, Vladikavkaz, Russia

Abstract. In contact mechanics, there are two opposite viewpoints on the way in which the radii of two circular cylinders effect on the deformation of these bodies. While some of the approaches consider its universality to be a theoretically proven fact, the others state this influence to be absent. And these groups of methods are both applied in mechanical engineering without discussing the reasons for contradictions to take place. The present paper considers the selective effect of contact approach and provides its theoretical and numerical verification.

Keywords: Contact mechanics · Elastic cylinders · Approach · Effect of approach

1 Introduction

For prolonged periods of time, a misleading belief about two fundamentally different principles of estimating the effect caused by the radii of two parallel elastic circular cylinders under power contact on the approach of their axes has been maintained in theory and practice of contact calculations. Several researchers (see, for instance, [1–3]) used to attribute such an effect to the strictly validated theoretical implication of a nonlinear relationship between the contact approach and loading. However, the method [4], based on experimental results, demonstrates the total lack of this impact, which implies that the cylinders' contact deformation is depended linearly on the loading. While considering the grouping [1–4], one can see that the methods [2, 3] are based on the solution of N. I. Muskhelishvili's problem [5], namely the one on a cylinder's plane deformation under the influence of two point forces which are equal in absolute value and oppositely directed. The application of the solution [5] in [2, 3] distinguishes these studies by the lack of serious simplifications that are immanent in other methods, for instance, the methods of the type [1].

To analyze the inconsistency of these solutions, the effect of approach, or EA for short, is estimated. EA states the contact approach to be selectively independent on the smaller radius of the cylinders in contact. The EA's selectivity is that it occurs only in case of great residual of the cylinders' radii.

B. T. Long et al. (Eds.): MMMS 2020, LNME, pp. 296–301, 2021.
https://doi.org/10.1007/978-3-030-69610-8_41

2 Fundamental Solution of the Approach Problem

EA could be revealed through analyzing of any sufficiently accurate method, but the one which was used is the solution [3] of the approach problem. Consider the deformation of an isotropic elastic circular cylinder of radius R due to the self-balanced unit loads $q = R\int_{-\varphi}^{\varphi} p(\theta)d\theta$, which are distributed over the surface of the cylinder as shown in Fig. 1 on the left.

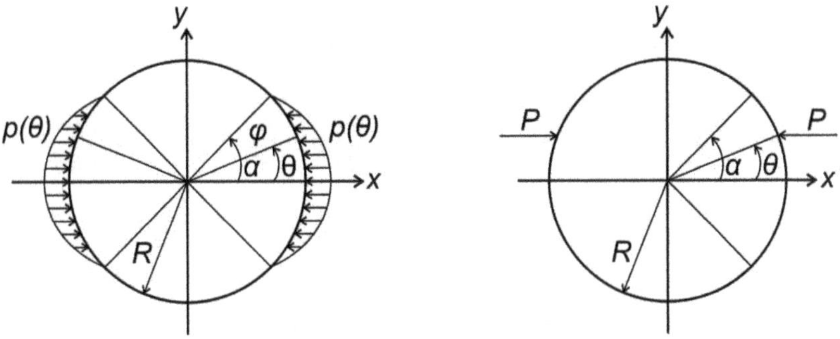

Fig. 1. Deforming of a cylinder with a system of self-balanced loads (on the left) and with two equal opposite concentrated forces (on the right).

The plane problem on deforming a cylinder with two equal opposite concentrated forces P, applied as shown in Fig. 1 on the right, was solved in [9]. The deflections of surface points in the x-axis direction are

$$u(R\cos\alpha, R\sin\alpha) = \tfrac{2}{\pi}\eta P\left[\ln tg\tfrac{|\alpha-\theta|}{2} + \cos\theta\cos\alpha\right], \quad -\tfrac{\pi}{2}\leq\alpha\leq\tfrac{\pi}{2}, \tag{1}$$

where $\eta = (1-\mu^2)/E$; μ and E are, respectively, Poisson's ratio and modulus of elasticity of the cylinder's material. The loads $p(\theta)$ are assumed to be distributed over the cylinder's surface, and the deflection of the surface points is considered as a sum of displacements caused be the elementary forces $p(\theta)R\cdot d\theta$, so (1) becomes

$$u(R\cos\alpha, R\sin\alpha) = \frac{2}{\pi}\eta R\int_{-\varphi}^{\varphi} p(\theta)\left[\ln tg\frac{|\alpha-\theta|}{2} + \cos\theta\cos\alpha\right]d\theta.$$

The following relationship defines the elastic displacement w of the points of a straight line $x = R$, $y = 0$:

$$w = |u(R,0)| = -\frac{2R}{\pi}\eta \int_{-\varphi}^{\varphi} p(\theta)\left[\ln tg\frac{|\theta|}{2} + \cos\theta\right]d\theta.$$

Assume that the unit load q is distributed over the cylinder's surface according to the law of an elliptical cylinder. Then

$$p(\theta) = \frac{2q}{\pi b}\sqrt{1-\left(\frac{\theta}{\varphi}\right)^2},\ p(\pi-\theta) = -p(\theta),\ -\varphi \le \theta \le \varphi,$$

where $b = R\varphi$. Given that $p(\theta) = p(-\theta)$ and by defining $t = \theta/\varphi$, one can obtain

$$w = -\frac{8\eta q}{\pi^2}\int_0^1 \sqrt{1-t^2}\left[\ln tg\left(\frac{bt}{2R}\right) + \cos\frac{bt}{R}\right]dt = \frac{8\eta q}{\pi^2}\left[I_0\left(\frac{b}{R}\right) + I\left(\frac{b}{R}\right)\right]. \quad (2)$$

The integral $I_0(b/R)$ is easy to calculate [9]:

$$I_0\left(\frac{b}{R}\right) = \int_0^1 \sqrt{1-t^2}\cos\frac{bt}{R}dt = \frac{\pi R}{2b}J_1\left(\frac{b}{R}\right),$$

whereas for calculating $I_1(b/R)$ one can use the series expansion of ln tan $\frac{bt}{2R}$:

$$\begin{aligned} I\left(\frac{b}{R}\right) &= \int_0^1 \sqrt{1-t^2}\ln tg\left(\frac{bt}{2R}\right)dt \\ &= -\frac{\pi}{4}\left[\ln\frac{2R}{b} + \frac{1}{2} + \sum_{n=1}^{\infty}(-1)^{n+1}\frac{(2^{2n-1}-1)(2n-1)!!B_{2n}}{2^n n(n+1)!(2n)!}\left(\frac{b}{R}\right)^{2n}\right]. \end{aligned}$$

Thus, after several obvious transformations one can express the approach of elastic parallel circular cylinders' axes from (2) by the following formula:

$$\begin{aligned} \delta &= \tfrac{2q}{\pi}\sum_{i=1}^{2}\eta_i\left\{\ln\tfrac{2R_i}{b} - \tfrac{3}{2} + \ln 2 + \tfrac{2R_i}{b}J_1\left(\tfrac{b}{R_i}\right) - \Sigma_i\right\}, \\ \Sigma_i &:= \sum_{n=1}^{\infty}\tfrac{(-1)^{n+1}(2^{2n-1}-1)}{n!^2 n(n+1)}\left(\tfrac{b}{2R_i}\right)^{2n}B_{2n},\quad i = 1,\ 2. \end{aligned} \quad (3)$$

Note that the series Σ_i is convergent. First, the following estimation is to be held to prove this fact:

$$\Sigma_i = \sum_{n=1}^{\infty} \frac{(-1)^{n+1}(2^{2n-1}-1)}{n!^2 n(n+1)} B_{2n} \left(\frac{b}{2R_i}\right)^{2n} \leq \sum_{n=1}^{\infty} \frac{2^{2n}(-1)^{n-1}(2^{2n}-1)}{2^{2n} n!^2 n^2} B_{2n} \left(\frac{b}{2R_i}\right)^{2n}.$$

As $(2n)!/(2^{2n}n!^2n^2) = (2n-1)!!/(2^n n!\, n^2) = \mathrm{B}\left(\frac{1}{2}, n\right)/(\pi n^2)$, from applying the extended first mean-value theorem to the beta-function, if follows that $\mathrm{B}\left(\frac{1}{2}, n\right) \leq \pi n^2$, which implies that $(2n)! \leq 2^{2n} n!^2 n^2$. The latter has the following estimations of each series Σ_i, $i = 1, 2$, as a consequence:

$$\Sigma_i \leq \frac{b}{2R_i} \sum_{n=1}^{\infty} (-1)^{n-1} \frac{2^{2n}(2^{2n}-1)}{(2n)!} B_{2n} \left(\frac{b}{2R_i}\right)^{2n-1} = \frac{b}{2R_i} \tan \frac{b}{2R_i}.$$

The restriction $b < \pi \min\{R_1, R_2\}$, which is due to the estimations used, holds within the Hertz theory, as it assumes the half-width of the contact area to be small in comparison to the radii of the cylinders in contact.

3 Conclusion

EA can be considered as one of the validation criteria for formulating and solving the problem on approach of two elastic parallel circular cylinders under power contact, and related problems of specific types. Namely, if the problem is set and solved correctly, EA holds and illustrates the behavior of the approach function within varying the ratio between the radii of the cylinders in contact, irrespectively of calculation errors.

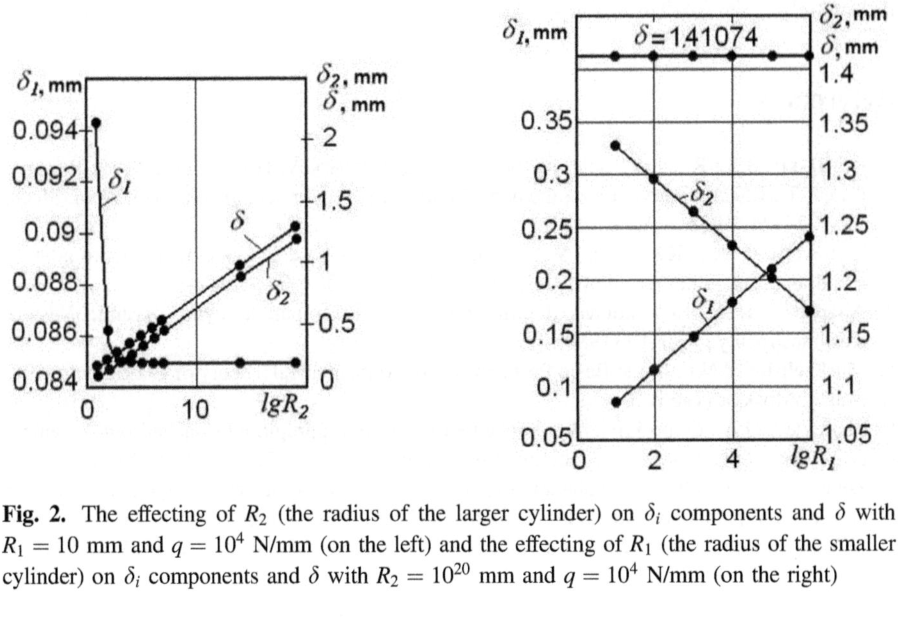

Fig. 2. The effecting of R_2 (the radius of the larger cylinder) on δ_i components and δ with $R_1 = 10$ mm and $q = 10^4$ N/mm (on the left) and the effecting of R_1 (the radius of the smaller cylinder) on δ_i components and δ with $R_2 = 10^{20}$ mm and $q = 10^4$ N/mm (on the right)

If the cylinders' material properties are constant, the width parameter b of the contact area is a function of both the two radii R_i and the load q, determining the complicated impact of these variables on each component of the approach δ_i.

The plots on Fig. 2 (with a logarithmic scale of abscissa) represent some of the calculation results from [6, 7], obtained through applying the formula (3), with $q =$ const for two steel ($\mu_i = 0.3$ and $E_i = 2.15 \times 10^5$ MPa, $i = 1, 2$) cylinders with $R_1 \neq R_2$. These results reveal the mechanism of convergence for elastic bodies.

The increasing of the second cylinder's deformation δ_2 (Fig. 2, on the left) or the first cylinder's deformation δ_1 (Fig. 2, on the right) is taking place with the increasing of the corresponding cylinder's radius – R_2 or R_1, respectively. When the greater cylinder's radius R_2 (Fig. 2, on the left) increases, the approach $\delta = \delta_1 + \delta_2$ also increases. As the radius R_1 of the smaller cylinder increases (Fig. 2, on the right), the impact of the larger cylinder's radius on the approach value decreases while the load-approach relation tends to linearity.

Note that even in this case there is a profound effect of cylinders' radii on the changes in the values of δ_1 and δ_2; however, the function $\delta_1 = \delta_1(R_1)$ is increasing (provided that $q =$ const and $\eta_i =$ const) while the function $\delta_2 = \delta_2(R_1)$ is decreasing (Fig. 2, on the left). While $R_2 \to \infty$ and the cylinder is virtually in contact with a half-space (for instance, when $R_2 = 10^{20}$), the radius R_1 has almost no contribution to the value $\delta = \delta_1 + \delta_2$ (Fig. 2, on the right); this fact is consistent with the results of the researches in [8]. Constancy of the approach (which has been revealed earlier only by experiment) for two bodies leaded to the illusory and untrue conclusion about the lack of cylinders radii's effect on their deformation.

Acknowledgements. Research was financially supported by Southern Federal University, grant No. VnGr-07/2020–04-IM (Ministry of Science and Higher Education of the Russian Federation).

References

1. Koval'skij, B.S.: Raschet detalej na mestnoe szhatie. Kharkov, HVKIU (1967). (in Russian)
2. Loo, T.T.: Effect of curvature on the hertz theory for two cylinders in contact. J. Appl. Mech. **25**, 122–124 (1958)
3. Zhuravlev, G.A., Karpenko, V.A.: Sblizhenie uprugih tel, modeliruemyh krugovymi cilindrami.,Tekhnika mashinostroeniya **6**(28), 51–55 (2008). (in Russian)
4. Ajrapetov, E.L.: Kontaktnaya deformaciya cilindrov s parallel'nymi osyami. Vestnik mashinostroeniya **6**, 6–10 (1988). (in Russian)
5. Muskhelishvili, N.I.: Some Basic Problems of the Mathematical Theory of Elasticity, p. 708. Nauka, Moskva (1966)
6. Zhuravlev, G.A.: O roli krivizny v kontaktnom sblizhenii uprugih tel s nachal'nym kasaniem po pryamoj linii. Visnik East Ukrainian Natl. Univ. **5**(39), 19–22 (2001). (in Russian)
7. Zhuravlev, G.A.: Physical explanation of the valuation inconsistency of contact approach of elastic round parallel cylinders. In: Morozov, N.F. (ed.) Proceedings International Conference "XVIII Session of Int. School on Models of Continuous Media, pp. 127–129 (2007). (in Russian)

8. Hoeprich, M.R., Zantopulos, H.: Line contact deformation: a cylinder between two flat plates. ASME J. Lubr. Technol. **103**, 21–25 (1981)
9. Zwillinger, D., Moll, V., Gradshteyn, I.S., Ryzhik, I.M.: Table of Integrals, Series, and Products, Table of Integrals, Series, and Products (Eighth Edition). Academic Press (2014). https://doi.org/10.1016/B978-0-12-384933-5.00007-2

Experimental Research and Evaluate the Result of Models of Engine Mount by Natural Rubber (NR) and Butadiene Rubber (SBR) When Replacing Diesel Engines for Gasoline Engines on ZIL-130, ZIL-131 Vehicles

Nguyễn Huy Trưởng[1(✉)], Lê Ngọc Tú[1], and Đào Đình Nam[1,2]

[1] Institute of Military Mechanical Engineering, Caugiay, Hanoi, Vietnam
nguyenhuytruonghd@gmail.com
[2] School of Mechanical Engineering and Automation, Northeastern University, Shenyang 110819, China

Abstract. In recent years, ZIL-130, ZIL-131 vehicles used for transport, it has been researched by the Institute of Military Mechanical Engineering to change from using gasoline engines to using diesel engines. To reduce the impact of vibration when the engine works on the body, the author has researched and manufactured engine mounts from natural rubber (NR) and butadiene rubber (SBR) to replace the engine mounts that have been used before.

This article has presented the results of research, testing and evaluation. At the same time, it also provides optimal parameters of engine mounts when replacing using diesel engines for using gasoline engines on ZIL-130 and Zil-131 vehicles.

Keywords: Engine mounts · Natural rubber (NR) · Butadiene rubber (SBR) · Anti-vibration rubber · Metal-rubber mount

1 The Experimental Model

The engine is integrated to the frame of the car by rubber-metal mount. We can build dynamic model of engine vibrations on the pillow elastic suspension as Fig. 1. This model reflects that the engine has six degrees of freedom in which three degrees of freedom allow the engine relatively shift under the shaft ox(X), oy(Y), oz(Z), 3 degrees of freedom allow the motor spin deflection around the shaft ox, oy, oz.

When working the engine suffers the major external effects of inertia forces of the moving structure and torque generated at the crankshaft. The combined inertial force vectors acting on the motor body are positioned to deviate from the dynamic cluster heart by a distance, therefore the main oscillation of the engine cluster is the vertical axis transposition oscillation (Oz) and the pendulum oscillation on the horizontal axis of the motor (Ox). Thus in, the calculation of the article, when building computational models, the authors have bypassed vertical axis pendulum oscillations of the motor

B. T. Long et al. (Eds.): MMMS 2020, LNME, pp. 302–310, 2021.
https://doi.org/10.1007/978-3-030-69610-8_42

(Oy), vertical axis pendulum (Oz) and transposition of the Ox and Oy axis. Experimental model (Fig. 2) is referred as a flat model, meaning that the engine has one front engine mount and two rear engine mounts that have hardness value equivalent the

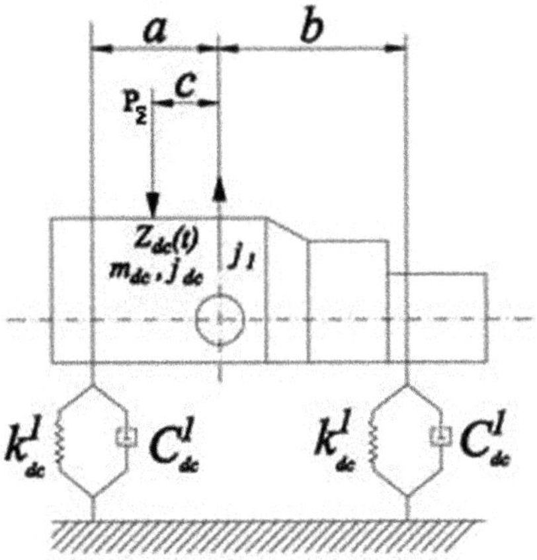

Fig. 1. Model for calculating of engine vibrations [3]

value of one engine mount.

In which [6]:

m_{dc} – The volume mass of engine

z_{dc} – Acceleration of the engine

C^1_{dc}, C^2_{dc}, k^1_{dc}, k^2_{dc} – Coefficient value and hardness coefficient value of front and rear engine mount

J_{dc} – Inertia moment at centre of gravity of vehicle body

P_Σ – General force acting on the engine body

a, b – The distance from the bearing to the center of gravity of the engine

c – The distance between the location of general force and coordinates the center of gravity of the engine.

D1, D2, D3: Sensor mounting positions (Fig. 3).

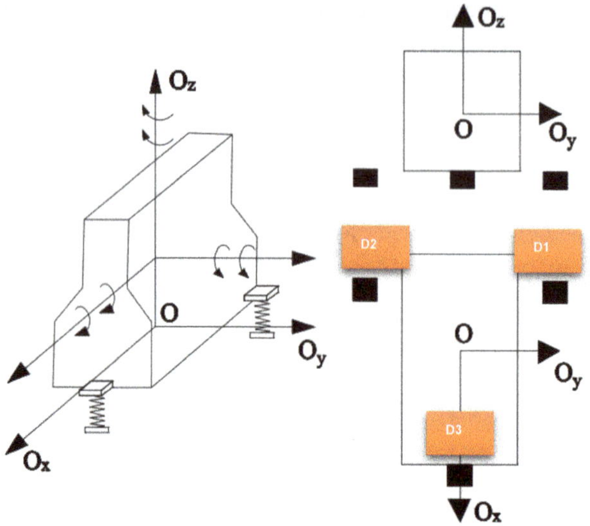

Fig. 2. Experimental model of engine mount

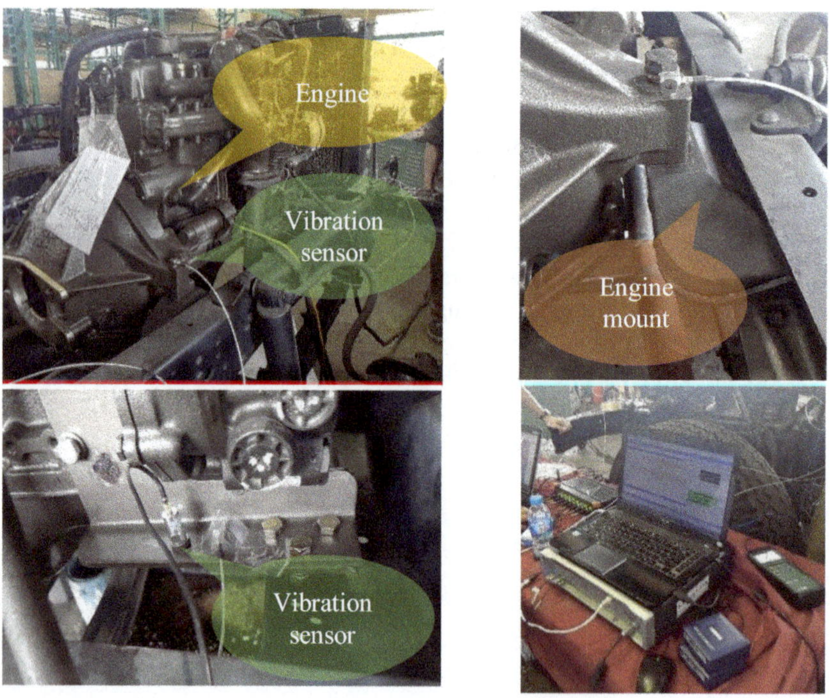

Fig. 3. Real experiment layout diagram

2 Results of Calculation and Experiment

2.1 Vibrate Calculation

Spectral density value vibrational transposition is calculated according to the formula [1, 2]:

$$S_z(\omega) = |W_z(j\omega)|^2 \cdot S_q(\omega), \tag{1}$$

In which:

- $P_\Sigma(t)$: The total working force of the engine
- $S_q(\omega)$: The spectral density linear deformation.

Spectral density value vibrational acceleration is calculated according to the formula:

$$S_{\ddot{z}}(\omega) = |W_{\ddot{z}}(\omega)|^2 \cdot S_q(\omega) = \omega^4 |W_z(\omega)|^2 \cdot S_q(\omega), \tag{2}$$

The average squared values of vibrational acceleration [1, 5] (Fig. 4):

$$\ddot{z}_{CK} = \sqrt{\int_{-\infty}^{\infty} S_{\ddot{z}}(\omega) d\omega}, \tag{3}$$

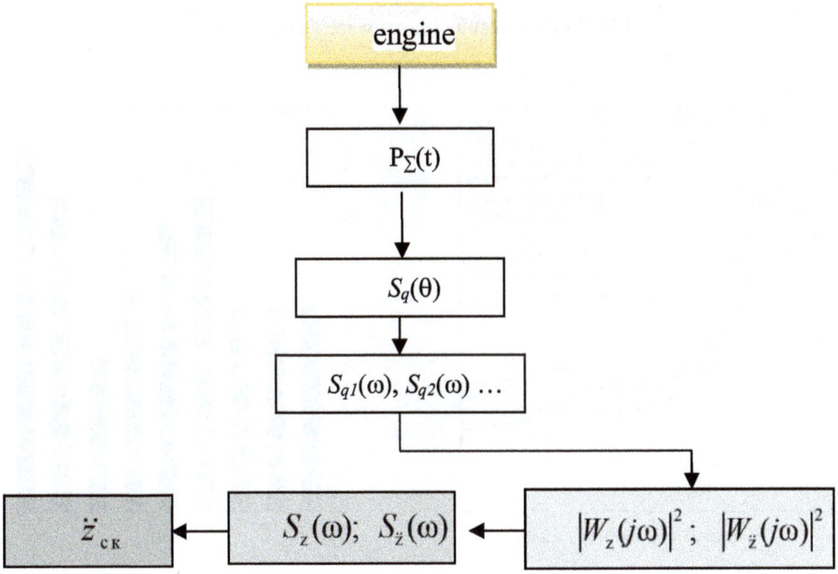

Fig. 4. Calculation diagram of system of vibrating differential equation [1]

2.2 Compare Experiment Results

Tests were conducted on 3 engine mounts with different ingredients: Butadiene rubber 5% Silica (Model 1); Butadiene rubber 10% Silica (Model 2), Butadiene rubber 15% Silica (Model 3) and placed in 3 positions: right rear (Sensor 1), left rear (Sensor 2) and front (Sensor 3).

The calculation and test results are shown on the following graphs (Figs. 5, 6, 7, 8, 9 and 10):

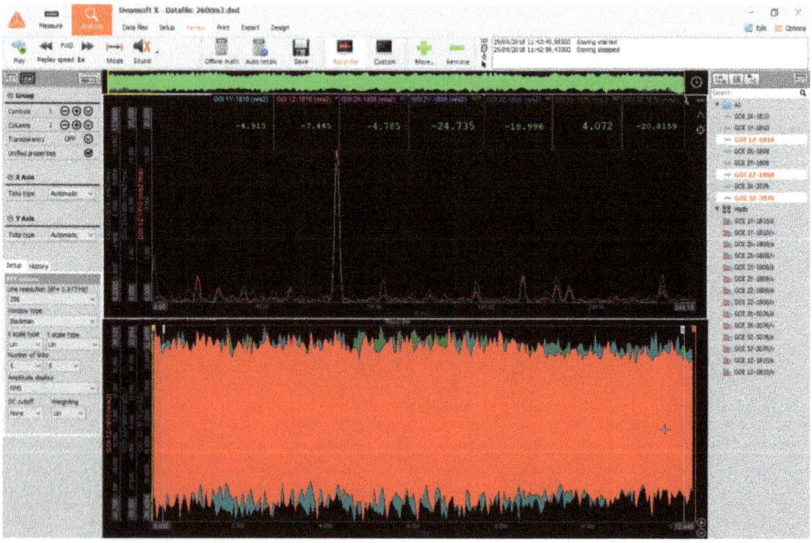

Fig. 5. Test results on software Dewesoft X

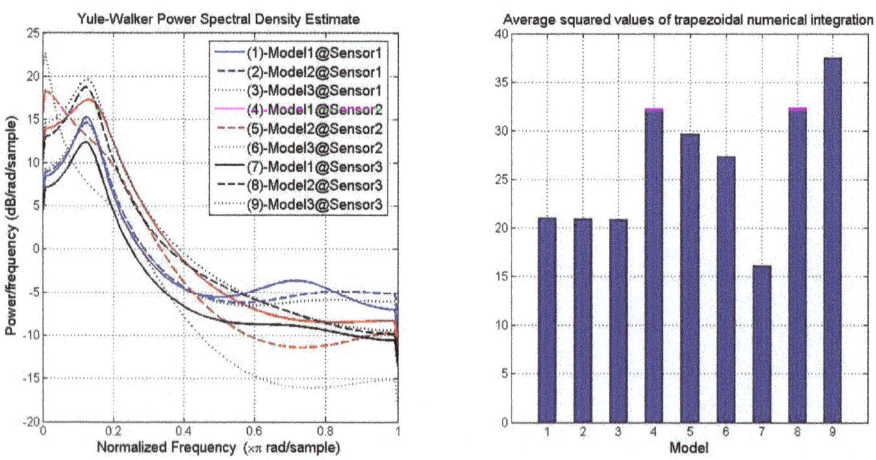

Fig. 6. Compare power spectral estimate values vibrational acceleration and average squared values of trapezoidal numerical integration at 700 r/min of engine speeds

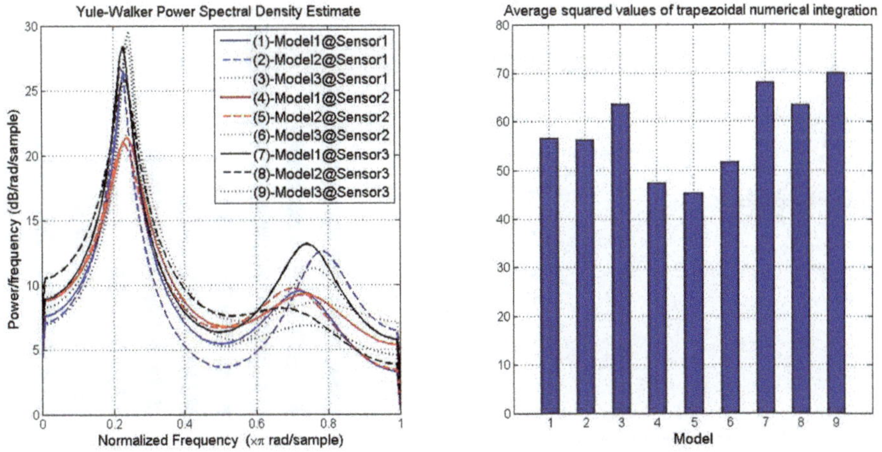

Fig. 7. Compare power spectral estimate values vibrational acceleration and average squared values of trapezoidal numerical integration at 1600 r/min of engine speeds

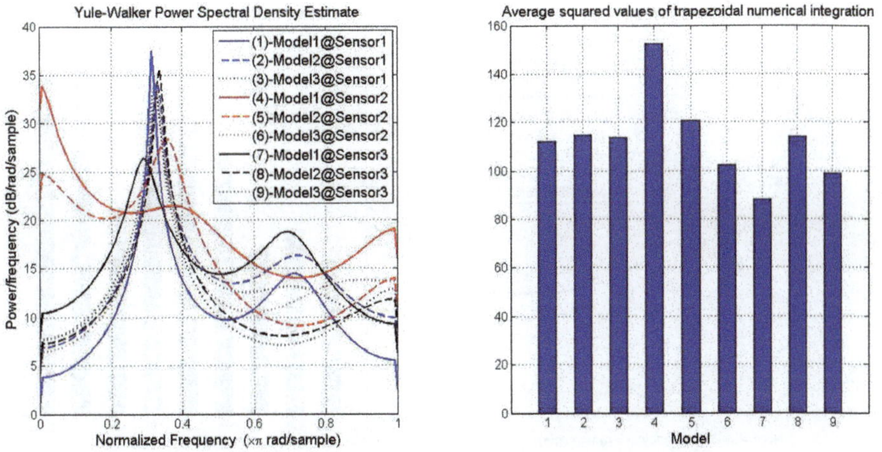

Fig. 8. Compare power spectral estimate values vibrational acceleration and average squared values of trapezoidal numerical integration at 2600 r/min of engine speeds

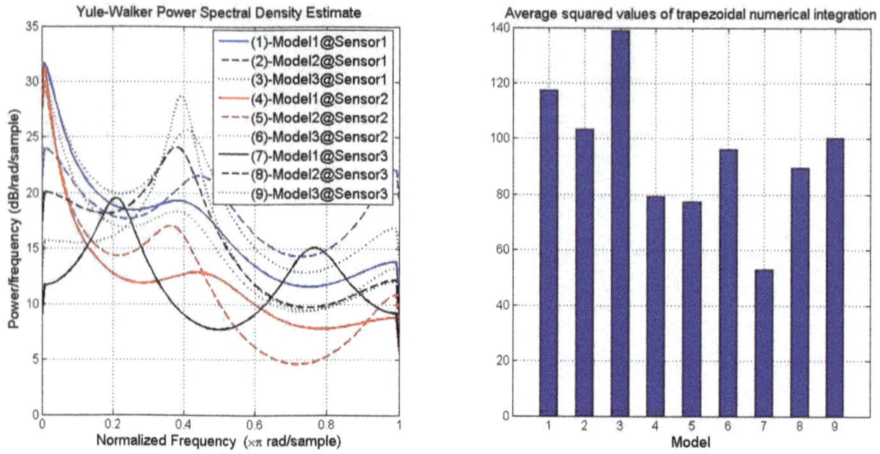

Fig. 9. Compare power spectral estimate values vibrational acceleration and average squared values of trapezoidal numerical integration at 700 r/min to 2600 r/min

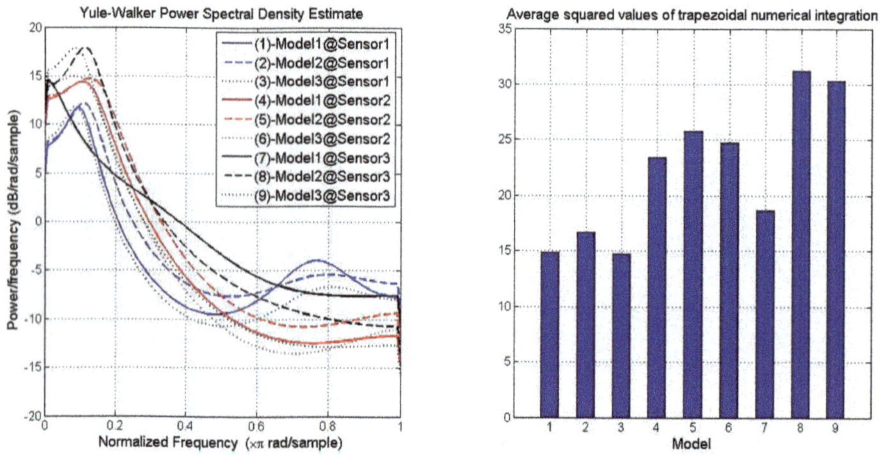

Fig. 10. Compare power spectral estimate values vibrational acceleration and average squared values of trapezoidal numerical integration at stop engine

Based on the value of the bar graph, the table compares the results in Table 1:

Table 1. Compare the results average squared values of trapezoidal numerical integration

Number	Test case	Model 1	Model 2	Model 3
For right rear engine mount (Sensor 1)				
1	700 r/min	3	2	1
2	1600 r/min	2	1	3
3	2600 r/min	1	3	2
4	700-2600 r/min	2	1	3
5	Stop engine	3	2	1
Summary		**11**	**9 (minimum)**	**10**
For left rear engine mount (Sensor 2)				
1	700 r/min	3	2	1
2	1600 r/min	2	1	3
3	2600 r/min	3	2	1
4	700-2600 r/min	2	1	3
5	Stop engine	1	3	2
Summary		**11**	**9 (minimum)**	**10**
For front engine mount (Sensor 3)				
1	700 r/min	1	2	3
2	1600 r/min	2	1	3
3	2600 r/min	1	3	2
4	700-2600 r/min	1	2	3
5	Stop engine	1	3	2
Summary		**6 (minimum)**	**11**	**13**

Note: In the instance shown the values of Average squared values of trapezoidal numerical integration: 1-minimum; 2-medium; 3-largest. Sum up the columns in 5 test cases: at engine speed 700 r / min, 1600 r/min, 2600 r/min, speed engine increases from 700 r/min to 2600 r/min and stop engine. Whichever column result is the smallest, that model is the most suitable.

Based on the results, we found that Model 2 (Butadiene rubber 10% Silica) will be suitable as the rear mount of the engine and Model 1 (Butadiene rubber 5% Silica) will be suitable as the front engine mount.

3 Conclusion

In the paper, the authors have given a model to test the vibration of the engine mounts.

In this experiment, engine has one front engine mount and two rear engine mounts, and there are 3 vibration sensors located at 3 bearing positions. The engine mounts are replaced respectively 3 types: Model 1 (Butadiene rubber 5% Silica); Model 2 (Butadiene rubber 10% Silica); Model 3 (Butadiene rubber 15% Silica). The results obtained were modified with the help of MATLAB software transform with formulas (1), (2), (3), we have table comparing the results average squared values of trapezoidal numerical integration (Table 1).

With the smallest total results in Table 1, we found that Model 2 (Butadiene rubber 10% Silica) will be suitable as the rear mount of the engine and Model 1 (Butadiene rubber 5% Silica) will be suitable as the front engine mount. This is also consistent with the theoretical calculations presented in the references [3, 4].

References

1. Nguyen Huy Truong. Разработка метода снижения виброакустической нагруженности полноприводного легкового автомобиля путём оптимизации упругих параметров силового агрегата и трансмиссии. Дисс. канд.техн.наук. – М (2008)
2. BOSCH: Automotive Handbook. 5th edn. Robert Bosch GmbH, Gerlingen (2004)
3. Le, N.T., Nguyen, H.T.: Method of calculating optimization of metal-rubber bearing parameters of engine by using Gomory graph. In: The 5th World Conference on Applied Sciences, Engineering and Technology (5th WCSET 2016) (2016)
4. Le, N.T., Chu, C.H., Nguyen, H.T., Dang, N.S.: On the impact of silica and black carbide in improving the anti-vibration of the rubber blends based on natural rubber (NR) and styrene butadiene rubber (SBR). J. Sci. Technol. **56**(24), 17–23 (2018)
5. Harris, C.M.: Shock an Vibration Handbook, 6th edn. McGraw Hill, New York (2010)
6. Кузнецов, А.С.: Руководство по техническому обслуживанию и ремонту автомобилей ЗИЛ 130, 431410, 131. М.: Издательский Дом Третий Рим (2004)

Investigation on the Effect of Nanofluid-MQL Parameters on Surface Roughness During Milling of Hardened 9CrSi Steel

Quoc-Manh Nguyen[1], Huu-That Nguyen[2], Nguyen-Anh-Vu Le[2], Thi-Nguyen Nguyen[3], Thi-Huong-Thao Dang[1], and The-Vinh Do[3](✉)

[1] Hung Yen University of Technology and Education, Hải Dương, Vietnam
[2] Nha Trang University, Nha Trang, Vietnam
[3] Thai Nguyen University of Technology, Thái Nguyên, Vietnam
thevinh8880@tnut.edu.vn

Abstract. Nanofluid MQL method has been used in the hard milling process to replace conventional milling for overcoming environmental and economical problems. In the cutting process, the selection of nanofluid MQL parameters is of great significance in improving surface roughness. In this study, the milling of hardened 9CrSi steel under different Al_2O_3 nanofluid MQL parameters including nanoparticle concentration, pressure, and fluid flow was accomplished. A series of experiments were designed by use of the Taguchi method. The signal to noise (S/N) ratio and ANOVA were applied to find the optimal values of nanofluid MQL parameters for minimizing surface roughness. The result shows that nanofluid MQL condition with the nanoparticle concentration of 4 wt%, the pressure of 3 kg/cm^2, and the fluid flow of 90 ml/h presents the better surface roughness. Furthermore, the nanoparticle concentration is the most influential parameter on the surface roughness followed by the pressure.

Keywords: Nanofluid · MQL · Hard milling · Taguchi method · Al2O3 nanoparticle · 9CrSi steel

1 Introduction

Cutting fluid plays an important role in metal cutting with three main functions including heat transfer, providing lubrication, and removing metal debris from the cutting area [1]. In conventional cutting, cutting fluid is supplied to the cutting area in large quantities that flooding the entire cutting area. This cooling lubrication process is called flood cutting or wet cutting. The biggest drawback of wet cutting is its impact on the environment and operator health [2–4]. Currently, there are several outstanding methods that have gradually replaced wet cutting such as dry cutting and MQL. The development of materials science, including cutting tool materials, has resulted in cutting tools that are capable of cutting in a state of high temperature, severe scratching. This is the premise for a complete cutting fluid removal method called dry cutting. Dry cutting, obviously, is a very environmentally friendly machining method. Also, this method has the advantage of reducing costs by saving on the costs of coolant and cleaning [3, 5]. However, the high temperature during the cutting process that causes

B. T. Long et al. (Eds.): MMMS 2020, LNME, pp. 311–316, 2021.
https://doi.org/10.1007/978-3-030-69610-8_43

low tool life and high roughness is a limitation of dry cutting [6]. A "win-win" solution that helps both minimizing the amount of cutting fluid required and enhancing the efficiency of the cutting process is the MQL method. MQL is a solution in which a very small amount of cutting fluid (less than 50 ml/h) is used [7–9]. The effectiveness of MQL in reducing roughness, improving durability, cutting heat, and reducing machining costs has been reported in many studies [10–12]. Currently, a new trend that is called nanofluids is the addition of nanometer-sized solid particles to cutting fluids applied in MQL machining. Nanofluid offers high efficiency in machining such as improving roughness, reducing force and cutting heat, enhancing tool life, and reducing tool wear [13–15].

In the cutting, the roughness index is an important indicator in assessing the quality of processed products. The cooling condition as well as the processing mode greatly affects the surface roughness. Therefore, choosing the appropriate cutting and cooling mode is the key to reducing roughness. The major concern in cutting is to find out the effect of cutting parameters and cooling conditions thereby improving the roughness.

In this research, the Taguchi methodology was applied to find the effects of parameters of the cooling condition such as nanoparticle concentration, pressure, and fluid flow on surface roughness. The optimal values of nanofluid MQL parameters for minimizing roughness were determined by signal to noise (S/N) ratio. Also, the parameter having the most impact on roughness was presented by using ANOVA.

2 The Experimental Procedure

In this research, an experimental organization is carried out according to Taguchi's design of experiment. Three parameters of nanofluid MQL condition are nanoparticle concentration, pressure, and fluid flow. Each parameter has three levels. Therefore, the orthogonal array L9 is suitable for conducting experiments. The levels of the parameters were selected based on the results of previous research by the author [4, 9, 14]. The cooling parameters and levels are indicated in Table 1.

Table 1. Parameters selected for input

Parameters	Levels		
	Low	Middle	High
Nanoparticle concentration *c* (wt%)	0	2	4
Fluid flow *f* (ml/h)	30	60	90
Pressure *p* (kg/cm^2)	3	4	5

All experiments were conducted by using a CNC machining center. 9CrSi steel workpiece has dimensions of 150 × 100 × 50 mm, hardness of 50 HRC. The Al_2O_3 nanoparticles added into the cutting oil have a size of 20 nm. A magnetic stirring device was used for mixing the nanoparticle and cutting oil. The cutting tools are ϕ10 TiAlN coated end-mill. The surface roughness data was collected by using the SJ - 201 instrument of Mitutoyo. Each test was repeated 3 times to maximize the accuracy of the results.

3 Results and Discussions

Table 2 indicates the results of the milling process of hardened 9CrSi steel and S/N ratio. The input parameters of the nanofluid MQL condition consist of nanoparticle concentration (c), fluid flow (f), and pressure (p). The goal of this work is to optimize the nanofluid MQL condition parameters to get minimum surface roughness (Ra). Therefore, "the smaller is the better" S/N type was selected and determined in the Eq. (1):

$$S/N = -10log\frac{1}{n}\left(\sum_{i=1}^{n} y_i^2\right) \quad (1)$$

Where: y_i is the observed data, n is the total of experimental runs.

As shown in Table 2, the roughness index is in the range of 0.162 μm to 0.407 μm. The roughness has a maximum value of 0.407 μm in experiment 1 (Run 1) and has a minimum value of 0.162 μm in experiment 9 (Run 9). The S/N ratio is calculated by using the Minitab 17 software.

Table 2. The experimental result

Run	*c* (wt%)	*f* (ml/h)	*p* (kg/cm²)	Ra (μm)	S/N
1	0	30	3	0.407	7.808112
2	0	60	4	0.34	9.370422
3	0	90	5	0.392	8.134279
4	2	30	4	0.338	9.421666
5	2	60	5	0.311	10.14479
6	2	90	3	0.191	14.37933
7	4	30	5	0.364	8.777972
8	4	60	3	0.209	13.59707
9	4	90	4	0.162	15.8097

Table 3. Response for S/N ratios

Level	*c*	*f*	*p*
1	8.438	8.669	11.928
2	11.315	11.037	11.534
3	12.728	12.774	9.019
Delta	4.291	4.105	2.909
Rank	1	2	3

The response table for S/N ratios is shown in Table 3. The ranked highest of each level of parameters is the third level for nanoparticle concentration, the third level for fluid flow, and the first level for pressure. Therefore, the optimal parameters of the experiment are 3-3-1. The rank of the nanoparticle concentration as "1" means that the concentration is the most influential parameters in the surface roughness. The second factor affecting the roughness is the fluid flow. The least influencing factor for roughness is the pressure.

The S/N response analysis is expressed in Fig. 1. Figure 1 presents that the optimal parameters of nanofluid MQL condition are nanoparticle concentration of 4 wt%, fluid flow 90 ml/h, and pressure of 3 kg/cm^2. To get the best roughness, the nanofluid MQL condition applied is the higher concentration, the higher fluid flow, and the lower pressure. It is explained that a higher amount of nanoparticles added into the cutting oil result in an improvement in the heat transfer and lubrication characteristic of the cutting fluid. In other words, the excellent effect of adding nanoparticles to cutting fluids was demonstrated when compared to conventional MQL. The higher fluid flow provides a greater amount of cutting fluid into the cutting area resulting in increased lubrication and cooling efficiency. The pressure of 3 kg/cm^2 is suitable for the cooling process. Under greater pressure, large amounts of cutting fluid will be ejected from the interface between the cutting tool and the machined surface. That reduces the efficiency of lubrication and cooling.

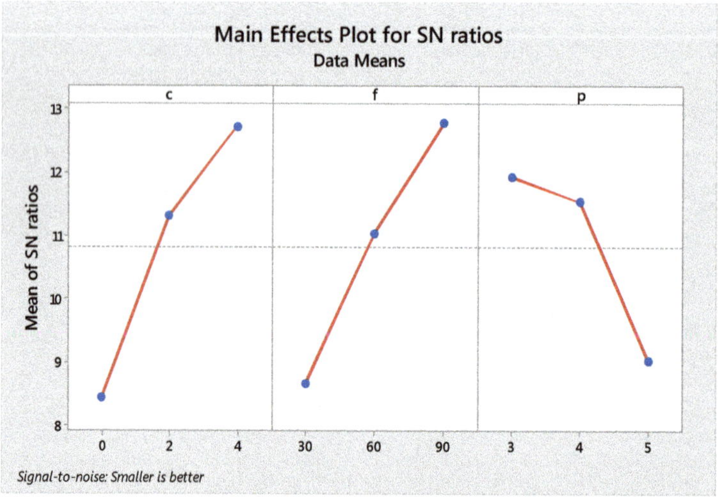

Fig. 1. The S/N response analysis

Table 4. ANOVA for SN ratios

Source	DF	Adj-SS	Adj-MS	F-value	P-value	%C
Regression	3	0.060552	0.020184	17.38	0.004	91.3
c	1	0.027203	0.027203	23.43	0.005[a]	41.0
f	1	0.022083	0.022083	19.02	0.007[a]	33.3
p	1	0.011267	0.011267	9.70	0.026[a]	17.0
Error	5	0.005806	0.001161			8.7
Total	8	0.066358				100.0
R-sq = 91.3%						

[a]significant

Table 4 indicates the analysis of variance for the signal to noise. Based on ANOVA, concentration is the factor that has the greatest impact on roughness. The impact of this factor accounts for 41% of the total impact on roughness. The second biggest factor is fluid flow. It contributes 33.3% of the total influence. The smallest impact factor is pressure. It accounts for only 17% of the total impact on roughness. The P index of all inputs is less than 0.05. That means that the impact of these inputs is statistically significant. With the coefficient of determination R-sq = 91.3%, it shows that 91.3% of the variation of the response is explained by the inputs selected in the study.

A verification test with optimized nanofluid MQL condition (i.e., concentration of 4 wt%, fluid flow 90 ml/h, and pressure of 3 kg/cm^2) was carried out to examine the result of this research. The roughness obtained from the verification test is 0.153 μm. This value is smaller than value of surface roughness of the experiment 9 (Run 9) having the smallest value of 0.162 μm shown in Table 2. That proves the reliability of the research results.

4 Conclusion

In this work, the Taguchi method was used to optimize nanofluid MQL condition for minimizing the roughness index in milling of hardened 9CrSi steel. It can be concluded that as the following:

- The nanofluid MQL condition including nanoparticle concentration of 4 wt%, fluid flow 90 ml/h, and pressure of 3 kg/cm^2 is the optimal condition to obtain the minimum surface roughness value.
- The nanoparticle concentration is the most influential factor in roughness followed by the fluid flow. They contribute 41% and 33.3% of the total affecting roughness, respectively.
- The excellent effect of adding nanoparticles to cutting fluids has been shown when compared to the usual MQL process.

Acknowledgement. The authors would like to thank Thai Nguyen University of Technology. The study was supported by Thai Nguyen University of Technology.

References

1. Senevirathne, S., Punchihewa, H.: Comparison of tool life and surface roughness with MQL, flood cooling, and dry cutting conditions with P20 and D2 steel. In: IOP Conference Series: Materials Science and Engineering (2017)
2. Sreejith, P., Ngoi, B.: Dry machining: machining of the future. J. Mater. Process. Technol. **101**(1–3), 287–291 (2000)
3. Do, T.V., Nguyen, Q.M., Pham, M.T.: Optimization of cutting parameters for improving surface roughness during hard milling of AISI H13 steel. In: Key Engineering Materials. Trans Tech Publications (2020)
4. Do, T.-V., Hsu, Q.-C.: Optimization of minimum quantity lubricant conditions and cutting parameters in hard milling of AISI H13 steel. Appl. Sci. **6**(3), 83 (2016)
5. Klocke, F., Eisenblätter, G.: Dry cutting. Cirp Ann. **46**(2), 519–526 (1997)
6. Gupta, M.K., Sood, P., Sharma, V.S.: Optimization of machining parameters and cutting fluids during nano-fluid based minimum quantity lubrication turning of titanium alloy by using evolutionary techniques. J. Clean. Prod. **135**, 1276–1288 (2016)
7. Walker, T.: The HANDBOOK A Guide to Machining with Minimum Quantity Lubrication, V 1.0. 3. Unist. Inc. (2013)
8. Do, T.-V., Le, N.-A.-V.: Optimization of surface roughness and cutting force in MQL hard-milling of AISI H13 steel. In: Fujita, H., Nguyen, D., Vu, N., Banh, T., Puta, H. (eds.) Advances in Engineering Research and Application. ICERA 2018. Lecture Notes in Networks and Systems, vol. 63. Springer, Cham (2019). https://doi.org/10.1007/978-3-030-04792-4_58
9. Do, T.-V., Vu, N.-C., Nguyen, Q.-M.: Optimization of cooling conditions and cutting parameters during hard milling of AISI H13 steel by using Taguchi method. In: 2018 IEEE International Conference on Advanced Manufacturing (ICAM). IEEE (2018)
10. Boswell, B., et al.: A review identifying the effectiveness of minimum quantity lubrication (MQL) during conventional machining. Int. J. Adv. Manuf. Technol. **92**(1–4), 321–340 (2017)
11. Weinert, K., et al.: Dry machining and minimum quantity lubrication. CIRP Ann. **53**(2), 511–537 (2004)
12. Dhar, N.R., Islam, S., Kamruzzaman, M.: Effect of minimum quantity lubrication (MQL) on tool wear, surface roughness and dimensional deviation in turning AISI-4340 Steel. Gazi Univ. J. Sci. **20**(2), 23–32 (2007)
13. Sayuti, M., Sarhan, A.A., Salem, F.: Novel uses of SiO2 nano-lubrication system in hard turning process of hardened steel AISI4140 for less tool wear, surface roughness and oil consumption. J. Clean. Prod. **67**, 265–276 (2014)
14. Do, T.-V.: Empirical model for surface roughness in hard milling of AISI H13 steel under nanofluid-MQL condition based on analysis of cutting parameters. J. Mech. Eng. Res. Dev. **43**(2), 89–94 (2020)
15. Sayuti, M., Sarhan, A.A., Hamdi, M.: An investigation of optimum SiO_2 nanolubrication parameters in end milling of aerospace Al6061-T6 alloy. Int. J. Adv. Manuf. Technol. **67**(1–4), 833–849 (2013)

Effect of Dielectric Liquid on the Diameter of Nickel Microspheres Produced by Aided Electric Discharge

Nguyen Tien Dung(✉) and Le Van Diem

School of Mechanical Engineering, Vietnam Maritime University, Haiphong, Vietnam
{dungnt, diemlv.mtb}@vimaru.edu.vn

Abstract. Nickel microspheres were produced in many different methods. In this paper, the method of aided electric discharge (EDM) was applied to produce hollow and solid nickel microspheres. We used three types of dielectric liquids, such as deionized water, kerosene, and EDM fluid. The effects of different dielectric liquids on the diameter of nickel microspheres were studied, using scanning electronic microscopy (SEM). We have utilized smile view software and origin software to measure the diameter of the microspheres and to analyses the diameter distributions of the nickel microspheres, respectively. Our findings show that while the dielectric liquid is deionized water, the diameter range of nickel microspheres ranges mainly from 0 to 5 μm, the percentages of the total number of microspheres of this range is 73.4%, even appearing nickel microspheres with nanometer size and hollow nickel microspheres. The dielectric liquid is kerosene and EDM fluid, the diameter range of nickel microspheres is mainly from 15 μm to 20 μm, the percentages of the total number of microspheres of this range are 30% and nickel nanospheres are not found.

Keywords: EDM · Dielectric fluid · Nickel microspheres · Nickel nanospheres · Hollow microspheres

1 Introduction

Currently, studying new materials is one of the most pressing issues. Many scientists are working to find new materials to increase the properties of the material. X.D. Pham et al. [1] based on the theory of spinodal decomposition, after heat treatment, to change microstructure and hardness of Cu-15Ni-8Sn alloy. A.T. Hoang et al. [2] has been studying heat treatment to increase the quality and longevity of weld. M.K. Pham et al. [3] increased the content of vanadium into high manganese steel to increase the particle size and hardness of the alloy. The results showed that, when increasing 1% vanadium, the impact toughness of the alloy is the highest.

Nanomaterials are also one of the new material technologies. Nanomaterials are used extensively in the chemical field [4], medical field [5–7], physics field [8–12]. To make these nanoparticles, there are many methods such as: using beads as templates method [13, 14], hydrothermal method [15], vapor condensation method [16], aided electric discharge method [17, 18].

B. T. Long et al. (Eds.): MMMS 2020, LNME, pp. 317–323, 2021.
https://doi.org/10.1007/978-3-030-69610-8_44

There are many studies to use the method of aided electric discharge (EDM) to produce metal microspheres. Nersessian [19] used EDM to produce hollow and solid magnetostrictive particulate composites under both dielectric liquid nitrogen and water conditions. Then these hollow and solid nickel microspheres epoxy composites and their magnetostrictive properties are measured. These spherical Terfenol-D particles would give strains that exceed the values currently achievable with nonaligned composites. A. E. Berkowitz [17, 18] used liquid nitrogen as the dielectric liquid to produce hollow and solid Ni microspheres. Vasudevamurthy [20] also used an electric sparking method to fabricate SUS 304 stainless steel particles, analyzing the effect of the technological parameters on the particle diameter. Liu Yifan [21–24] has combined EDM with ultrasound to produce nickel particles. Study the effect of dielectric liquid on the diameter of microspheres, then studied the effects of several electrical parameters on the sizes of the microspheres at an ultrasound frequency of 40 kHz. C.R. Sanghani [25] studied the effect of various dielectric fluids on the performance of EDM. However, the study of the effect of technological parameters on particle size when machining with EDM has not been mentioned.

In this paper we use the EDM to produce solid nickel microspheres, then study the effect of dielectric liquid on the diameter of nickel microspheres, ever after find the dielectric liquid for the diameter of microspheres is the smallest.

2 Materials and Methods

Digital controller machine - E46PM containing machining box made of stainless steel with 260 mm × 190 mm × 170 mm in size is used to perform the experiments. The ultrasound systems of 40 kHz are mounted on the machining box, the machining tool with 10 mm × 10 mm in square and electrode are both of 99.99% nickel. Due to the magnetic properties of nickel, magnetic magnets are used to absorb the nickel particles after processing (Fig. 1).

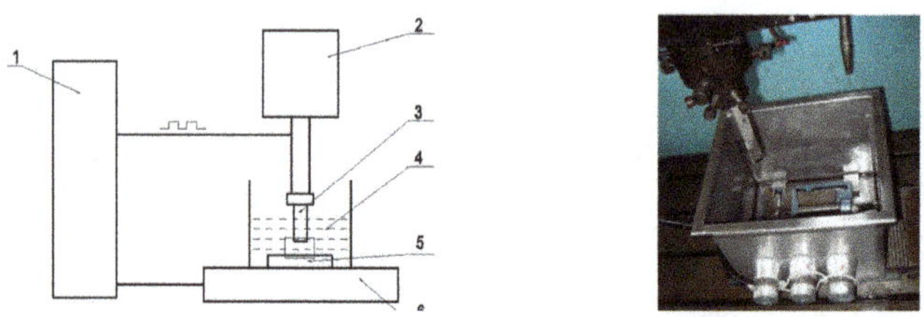

Fig. 1. The principle diagram of the machining process of aided electric discharge 1-Power source; 2-Main axis control system; 3-Electrode; 4-Machining liquid; 5-Machine parts; 6-Machining table

Figure 1 shows the operation principle of the machining process of aided electric discharge, where the electrode (3) and machine parts (6) is connected to the power source (1) through the main axis control system (2). The electrode (3) move downward and approach to machine parts (6) till the gap between the electrode and the machine parts are small enough, the discharge phenomena occur through the machining liquid (4).

The tests were performed in an E46PM digital control EDM machine, with current varying from 1.5 to 60 A, pulse width varied from 2 to 1200 μs, and the gap voltage can vary between 30 and 120 V. In this study, we conducted many groups of experiments, which differ in pulse width, current and voltage, in each group only changes the dielectric liquid. It is worth noting that we found that the results from all experimental groups had very similar tendencies and properties, so we only had one group to explain the tendency and changing properties of the differentiation particle size.

In this experiment, we used three types of dielectric liquids, which are changed corresponding EDM fluid, kerosene, and deionized water. However, the unchanged electric parameters such as current (I) is 15 A, electrical discharge voltage (U) is 45 V, the pulse widths (t_i) is 300 μs, and machining time (t) is 60 min. Experimental parameters of the machining process of aided electric discharge while changed dielectric liquids are given in Table 1.

Table 1. Experimental parameters of the machining process of aided electric discharge

Number	Dielectric liquid	Pulse width (μs)	Current (A)	Voltage (V)
A	EDM fluid	300	15	45
B	Kerosene	300	15	45
C	Deionized water	300	15	45

3 Results and Discussion

Nickel particles are cleaned after machining by alcohol and checked under the SEM microscope. We can see two types of nickel particles: solid nickel particles and hollow nickel particles. Hollow nickel particles are formed with the following mechanism:

- During the freezing process, the outermost metal is frozen and shrunk, this shrinkage of volume will be offset by its inner elements closely, then these elements are also frozen and shrunk, this shrinkage of volume will continue to be offset by its inner elements closely so the center of the seed is finally frozen but not compensated, so create a hollow particle.
- At high temperatures, the liquid metal dissolves a certain amount of gas, but when the metal freezes, this gas is released. Because the outer surface of the particle will freeze first, this amount of gas will be trapped inside the particle to form a hollow particle.

This article used smile view software to measure the diameter of the microspheres and to analyses the diameter distributions of the nickel microspheres. The results of three experiments (Exp) are shown in Table 2 and Fig. 2.

Exp A *Exp B* *Exp C*

Fig. 2. The size of the nickel particles under a SEM microscope

Table 2. Percentage of dispersion of diameter nickel particles with the change of dielectric liquid

Diameter (μm)	Exp A (%)	Exp B (%)	Exp C (%)
0–5	0	0,54	73,4
5–10	2,15	7,22	16,1
10–15	17,13	22.,48	7,12
15–20	27,52	30,26	2,62
20–25	17,43	18,8	0,75
25–30	14,06	10,63	0
30–35	6,11	7,36	0
35–40	3,36	1,5	0
40–45	4,59	0,95	0
45–50	3,36	0,26	0
50–55	2,45	0	0
55–60	0,61	0	0
60–65	0,92	0	0
65–70	0,31	0	0

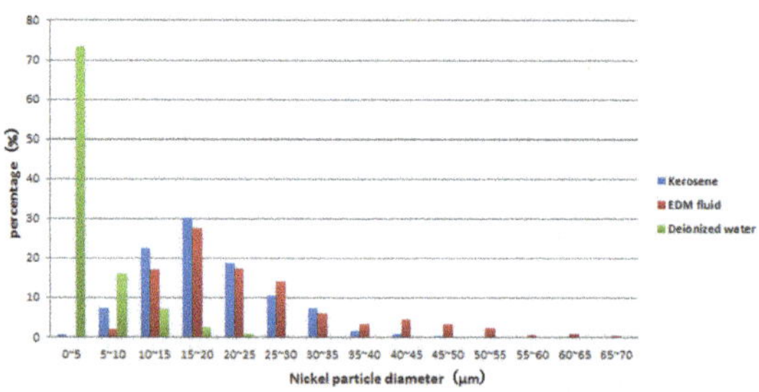

Fig. 3. Dispersion percentage of diameter nickel particles with the change of dielectric liquid

Figure 2 and Fig. 3 show that the dielectric liquid is kerosene and EDM fluid, the diameter range of nickel particles is mainly from 15 μm to 20 μm, the percentages of the total number of microspheres of this range are 30%, nickel nanospheres are not found and maximum and minimum diameters were 69.3 μm and 90 nm. Conversely, when the dielectric liquid is deionized water, the diameter range of nickel microspheres ranges mainly from 0 to 5 μm, the percentages of the total number of microspheres of this range is 73.4%, maximum and minimum diameters were 24.3 μm and 60 μm, appearing nickel particles with nanometer size and hollow nickel particles (Fig. 4). These results are consistent with the results of other studies [20, 22, 23].

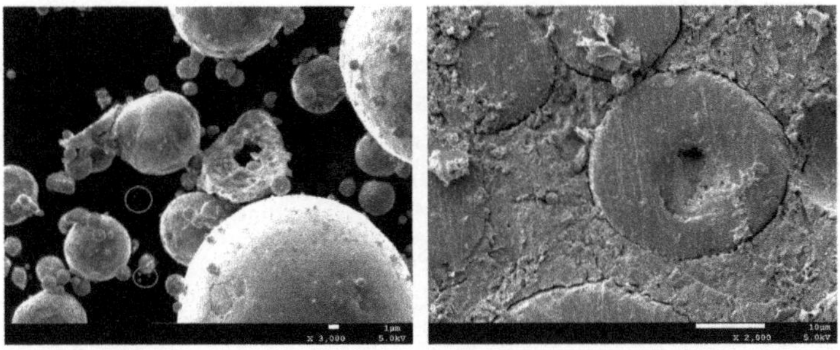

Fig. 4. The size of the nickel particles under SEM Microscope when zoomed 3000 times and 2000 times

In the machining process, the processing area has a very high temperature causing the metal to melts and evaporates, they penetrate the dielectric liquid then freeze. During freezing, metallic elements may evaporate together creating larger metal particles. Due to the kinematic viscosity and the insulation performance of the deionized water is much smaller than that of the EDM fluid and kerosene. When the dielectric liquid is deionized water, the ability to disperse the metal is easier, the evaporating metal particles can hardly accumulate together to form larger particles so that the diameter of particles is smaller.

Otherwise, under the same processing conditions, the discharge gap is larger when water is used as the working fluid, the compression effect on the discharge channel is relatively poor, and the amount of erosion is relatively small. When kerosene and EDM fluid are used as the working fluid, the kerosene and EDM fluid can be cracked to produce free carbon under the high temperature and high-pressure environment of the discharge channel (During the experiment, it can be observed the discharge starts for several tens of seconds, the EDM fluid and kerosene became black. Then the liquid is settled for a few weeks, the upper kerosene and EDM fluid was again turned into a transparent yellowish color). The addition of carbon will increase the distance of sparking, will increase the discharge channel, and the discharge of the electrode is scattered, makes the discharge power is dispersed. This is also the reason that metal particles cannot go far, nickel particles easily accumulate together to form larger particles, so the diameter of the particles is larger.

When deionized water is used as the working fluid, the thermal conductivity of deionized water is high, the time required to cool the molten metal, and the metal vapor is short so that the diameter of particles is smaller. On the contrary, when kerosene and EDM fluid are used as the working fluid, the thermal conductivity of kerosene and EDM fluid is low, which causes the molten metal and metal vapor to cool for a long time. Therefore, the molten metal and the metal vapor are more easily combined when solidified, so the diameter of the particles is larger.

4 Conclusion

Our research has given the relationship of dielectric liquid with dispersion percentage of diameter nickel particles. We can see that when dielectric liquid is deionized water, the diameter range of nickel microspheres ranges mainly from 0 to 5 μm, the percentages of the total number of microspheres of this range is 73.4%, even appearing nickel microspheres with nanometer size and hollow nickel microspheres. Compared with the experiments in kerosene and EDM fluid, the diameter of nickel microspheres when dielectric liquid is deionized water is the smallest.

Acknowledgements. The authors would like to thank the Vietnam Maritime University for their support.

References

1. Pham, X.D., Hoang, A.T., Nguyen, D.N.: Int. J. Mech. Mechatron. Eng. **18**(4), 27–34 (2018). Corpus ID: 165160084
2. Anh Tuan, H., Nam Nguyen, D., Viet Pham, V.: Int. J. Mech. Eng. Technol. **9**(6), 496–506 (2018). Article ID: IJMET_09_06_057
3. Pham, M., Nguyen, D., Hoang, A.T.: Int. J. Mech. Mechatron. Eng. **18**(2), 141–147 (2018). Corpus ID: 201742605
4. Li, X., Xiong, Y., Li, Z., Xie, Y.: Inorg. Chem. **45**, 3493–3495 (2006). https://doi.org/10.1021/ic0602502
5. Caruso, F.: Chem. - A Eur. J. **6**(3), 413–419 (2000). https://doi.org/10.1002/(sici)1521-3765(20000204)6:3%3c413::aid-chem413%3e3.0.co;2-9
6. Gao, X., Zhang, J., Zhang, L.: Adv. Mater. **14**(4), 290–293 (2002). https://doi.org/10.1002/1521-4095(20020219)14:4%3c290::AID-ADMA290%3e3.0.CO;2-U
7. Koktysh, D.S., et al.: Adv. Funct. Mater. **12**(4), 255 (2002). https://doi.org/10.1002/1616-3028(20020418)12:4%3c255::AID-ADFM255%3e3.0.CO;2-1
8. Tartaj, P., Morales, M.D.P., Veintemillas-Verdaguer, S., Gonzalez-Carreno, T., Serna, C.J.: J. Phys. D Appl. Phys. **36**(13), R182–R197 (2003). https://doi.org/10.1088/0022-3727/36/13/202
9. Tartaj, P., González-Carreño, T., Serna, C.J.: Adv. Mater. **13**(21), 1620–1624 (2001). https://doi.org/10.1002/1521-4095(200111)13:21%3c1620::AID-ADMA1620%3e3.0.CO;2-Z
10. Wang, Y., Zhu, Q., Zhang, H.: J. Mater. Chem. **16**(13), 1212–1214 (2006). https://doi.org/10.1039/B517176H
11. Bang, J.H., Suslick, K.S.: J. Am. Chem. Soc. **129**(8), 2242–2243 (2007). https://doi.org/10.1021/ja0676657

12. Bao, J., Liang, Y., Xu, Z., Si, L.: Adv. Mater. **15**(21), 1832–1835 (2003). https://doi.org/10.1002/adma.200305315
13. Chen, Z., et al.: Adv. Mater. **16**(5), 417–422 (2004). https://doi.org/10.1002/adma.200306445
14. Li, L., Ding, J., Xue, J.: Chem. Mater. **21**(15), 3629–3637 (2009). https://doi.org/10.1021/cm900874u
15. Feng, X., Yang, L., Liu, Y.: Mater. Lett. **64**(24), 2688–2691 (2010). https://doi.org/10.1016/j.matlet.2010.09.006
16. Gurav, A.S., Kodas, T.T., Wang, L.-M., Kauppinen, E.I., Joutsensaari, J.: Chem. Phys. Lett. **218**(4), 304–308 (1994). https://doi.org/10.1016/0009-2614(93)E1491-X
17. Berkowitz, A.E., et al.: Appl. Phys. Lett. **85**(6), 940–942 (2004). https://doi.org/10.1063/1.1779962
18. Berkowitz, A.E., Hansen, M.F., Parker, F.T.: J. Magn. Magn. Mater. **254–255**, 1–6 (2003). https://doi.org/10.1016/S0304-8853(02)00932-0
19. Nersessian, N., Or, S.W., Carman, G.P., Choe, W., Radousky, H.B.: J. Appl. Phys. **96**(6), 3362–3365 (2004). https://doi.org/10.1063/1.1775303
20. Vasudevamurthy, G., Knight, T.W.: Mater. Lett. **61**(27), 4872–4874 (2007). https://doi.org/10.1016/j.matlet.2007.03.070
21. Liu, Y., Li, X., Li, Y., Chen, J., Zhao, W., Bai, F.: Adv. Powder Technol. **27**(6), 2399–2408 (2016). https://doi.org/10.1016/j.apt.2016.08.016
22. Liu, Y., Li, X., Bai, F., Chen, J., Wang, Y., Liu, N.: Particuology **17**, 36–41 (2014). https://doi.org/10.1016/j.partic.2013.11.002
23. Liu, Y., Li, X., Li, Y., Zhao, Z., Bai, F.: Appl. Phys. A **122**(3), 174 (2016). https://doi.org/10.1007/s00339-016-9698-2
24. Bai, F., Li, X., Liu, Y., Wang, Y., Liu, N., Ma, Y.: Adv. Mater. Res. (2015). https://doi.org/10.4028/www.scientific.net/AMR.887-888.126
25. Sanghani, C.R., Acharya, G.D.: Trends Mech. Eng. Technol. **6**(1), 55–71 (2016). Corpus ID: 148562649

On Increasing Informativity and Reliability of the Results of Condition Diagnostics of Materials and Goods

I. P. Miroshnichenko[1(✉)] and I. A. Parinov[2]

[1] Don State Technical University, Rostov-on-Don, Russia
ipmir@rambler.ru
[2] Southern Federal University, Rostov-on-Don, Russia

Abstract. The article proposes the results of the development of new high-precision measuring means for solving applied measuring problems when diagnosing the state of structural materials and products in operation. New contactless optical measuring means based on modern laser technologies and new methods of optical interferometry are described. The original software for preparation, measurement and processing of the obtained measurement results is proposed. The developed promising optical measuring complex is described, the use of which for solving measuring problems during diagnostics of the state of structural materials and products makes it possible to increase the information content and reliability of measurement results up to 30%, as well as improve the quality of measurement results up to 40%.

Keywords: Construction materials · Condition diagnostics · Optical measuring means · Displacement measurement · Software · Simulation of diagnostic processes

1 Introduction

Ensuring safe operation and preventing emergencies during the operation of products for various purposes is possible through periodic diagnostics of the state of structural materials and the marked products themselves. This makes it possible to systematically monitor the state of structural materials and products in operation, timely assess the residual resource of their operation and promptly make decisions either about the possibility of further operation or about its termination.

In connection with the above, it becomes relevant to increase the information content and reliability of the results of diagnostics of the state of structural materials and products in operation.

The development of new contactless optical measuring means based on modern laser technologies and new methods of optical interferometry is one of the most relevant and promising areas for achieving an increase in the information content and reliability of the results of diagnostics of the state.

B. T. Long et al. (Eds.): MMMS 2020, LNME, pp. 324–329, 2021.
https://doi.org/10.1007/978-3-030-69610-8_45

The use of such measuring means can significantly increase the information content, reliability and quality of measurement results, and, respectively, the results of diagnostics of the state of structural materials and goods in operation.

2 Optical Measuring Complex and Its Description

To solve this problem, a new optical measuring set-up has been proposed. Its structure is shown schematically in Fig. 1.

The components of the set-up are a control system (1), a new measuring mean (2), a device for protecting a measuring mean from the influence of internal destabilizing influences and correction of measurement results (3), a device for protecting a measuring mean from the influence of external destabilizing influences and correction of measurement results (4), a system for *a priori* analysis and preparation for measurements (5), registration of measurement results (6) and a system for processing and *a posteriori* analysis of measurement results (7).

The control system (1) provides control and preparation for measurements, registration and correction of measurement results during their performance and processing of the obtained measurement results after their completion.

Measuring mean (2) (optical interference device for contactless measurement of small spatial displacements of control surfaces) provides contactless measurements of the small spatial displacements of control object surfaces.

The proposed measuring mean (2) is based on the design of a laser two-way interferometer with aligned branches.

The measuring mean (2) can be implemented in design options for measuring small linear displacement, small linear and angular displacements, small linear and all angular (spatial) displacements of the control object surface.

Each of the design options can be used both as part of stationary and as part of mobile diagnostic station using active acoustic methods for monitoring and diagnosing the state of materials in instrument making, mechanical engineering, shipbuilding, aircraft construction, the fuel and energy complex, etc.

All possible design options for measuring mean (2), as well as devices for protection (3) and (4) are scientifically grounded, tested, passed trial operation, and protected by patents of the Russian Federation for inventions [1–6].

The system of *a priori* analysis and preparation for measurements (5) makes it possible to carry out preliminary predictive modeling of the dependences of changes in small displacements of the surfaces of test objects for the used active acoustic method of non-destructive testing and distributions of the intensity of optical fields of interference patterns during measurements.

The system for recording measurement results (6) makes it possible to register the intensity of the optical field of the interference pattern in specified areas during measurements.

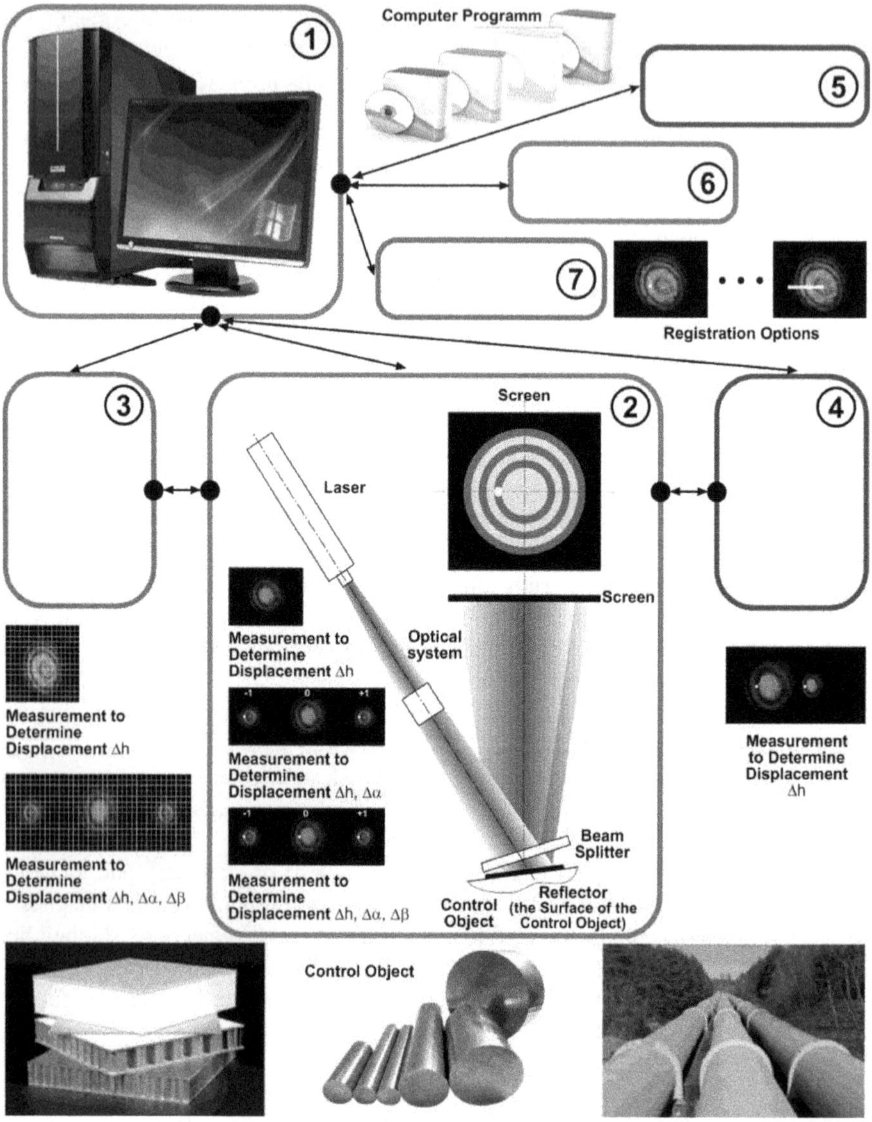

Fig. 1. Scheme of optical measuring complex

The system for processing and *a posteriori* analysis of measurement results (7) allows processing the measurement results after their completion and recognition of the obtained measurement results using the results of preliminary predictive modeling the dependences of changes in small displacements of the control object surfaces for the used active acoustic method of non-destructive testing, obtained during *a priori* analysis.

To ensure the functioning of the proposed set-up, original software has been developed, which is included in systems (5)–(7).

The pointed software contains:

(i) computer programs for modeling the intensity distributions of optical fields of interference patterns for preliminary predictive modeling of the measurement process, which take into account the geometric characteristics of the measuring circuit, the type of polarization and the type of beam splitter;
(ii) computer programs for modeling the dependencies of changes in small displacements of the control object surfaces for the used active acoustic method of non-destructive testing for preliminary predictive modeling of the state of the control object, which allow taking into account the spatial temporal distribution of sources of probing influence, the material type of the control object, its physical and mechanical characteristics, types of elastic waves propagating in a given material and the features of the arising wave processes;
(iii) computer programs for recording and processing the intensity of optical fields from specified areas of the interference pattern;
(iv) computer program for determining the total intensity of the optical field of the interference pattern and correcting the measurement results of the intensity during measurements;
(v) computer program for correcting measurement results from the influence of external destabilizing influences.

All above-mentioned computer programs are promised contactless means designed to measurements of optical displacements [7] and protected by certificates of the Russian Federation on state registration of computer programs [8–18].

3 Conclusion

Summarizing the results described above, we can draw the following conclusions:

1. New high-precision optical measuring means have been developed for solving applied measuring problems when diagnosing the state of structural materials and goods in operation, which are functionally combined into an optical measuring set-up.
2. The proposed new technical solutions and original software included in the set-up are scientifically grounded, tested, and also protected by patents of the Russian Federation for inventions and certificates of the Russian Federation on state registration of computer programs, respectively.
3. The use of the proposed complex for solving measuring problems during diagnostics of the state of materials and products allows increasing the information content and reliability of measurement results up to 30%, as well as improving the quality of measurement results up to 40%.

Acknowledgement. The study was performed under partial support of the Russian Foundation for Basic Research (No. 19-08-00365).

References

1. Alekhin, V.E., Miroshnichenko, I.P., Serkin, A.G., Sizov, V.P.: Method of recording displacements by optical sensors Patent RU No. 2343403/10.01.2009
2. Miroshnichenko, I.P., Parinov, I.A., Rozhkov, E.V., Serkin, A.G.: Optical device for measuring displacements Patent RU No. 2373492/20.11.2009
3. Miroshnichenko, I.P., Serkin, A.G., Sizov, V.P.: Measuring method of linear and angular displacements Patent RU No. 2388994/10.05.2010
4. Miroshnichenko, I.P., Parinov, I.A., Rozhkov, E.V., Serkin, A.G.: Optical device for measuring displacements Patent RU No. 2407988/27.12.2010
5. Miroshnichenko, I.P., Nesterov, V.A., Serkin, A.G., Sizov, V.P., Shevtsov, V.A.: Interference meter of small displacements Patent RU No. 2410642/27.01.2011
6. Miroshnichenko, I.P., Parinov, I.A., Sizov, V.P.: Measuring method of linear and angular components of the small displacements of control object surfaces Patent RU No. 2606245/10.01.2017
7. Miroshnichenko, I.P., Parinov, I.A.: Development of a new optical measuring set-up. In: IOP Conference Series: Materials Science and Engineering, vol. 374, p. 012052 (2018)
8. Miroshnichenko, I.P., Shevtsov, V.A.: Determination of the intensity of the optical field of interference pattern created by laser two-way interferometer with aligned branches Certificate of the Russian Federation on state registration of computer program RU No. 2014662261/26.11.2014
9. Miroshnichenko, I.P., Parinov, I.A.: Determination of the intensity of the optical field of interference pattern created by laser interferometer for beam splitter in the form of phase sinusoidal grating with uniform period Certificate of the Russian Federation on state registration of computer program RU No. 2015610921/21.01.2015
10. Miroshnichenko, I.P., Parinov, I.A.: Determination of the intensity of the optical field of interference pattern created by laser interferometer for beam splitter in the form of phase zone plate Certificate of the Russian Federation on state registration of computer program RU No. 2015611078/23.01.2015
11. Miroshnichenko, I.P.: Computer program for visualization and processing the distribution of the optical field intensity of interference pattern created by laser interferometer Certificate of the Russian Federation on state registration of computer program RU No. 2017614710/ 26.04.2017
12. Miroshnichenko, I.P.: Computer program for visualization and statistical processing the optical field intensity distribution in the selected areas of interference pattern Certificate of the Russian Federation on state registration of computer program RU No. 2017614715/ 02.05.2017
13. Miroshnichenko, I.P.: Computer program for visualization and statistical processing the optical field intensity distribution in the regions of interference pattern limited by the geometric characteristics of photodetectors Certificate of the Russian Federation on state registration of computer program RU No. 2017614764/26.04.2017
14. Miroshnichenko, I.P.: Computer program for visualization and processing the dependences of the intensity of optical field in the selected area of interference pattern Certificate of the Russian Federation on state registration of computer program RU No. 2017614861/ 02.05.2017
15. Miroshnichenko, I.P.: Computer program for correcting the results of measurements of the intensity of the optical field of interference pattern Certificate of the Russian Federation on state registration of computer program RU No. 2017614900/27.04.2017

16. Miroshnichenko, I.P., Sizov, V.P.: Modeling displacements and stresses in construction layer under impulse loading along its surface Certificate of the Russian Federation on state registration of computer program RU No. 2018664987/27.11.2018
17. Miroshnichenko, I.P.: Modeling displacements and stresses in conjugate layers of construction under impulse loading on surface Certificate of the Russian Federation on state registration of computer program RU No. 2019663431/16.10.2019
18. Miroshnichenko, I.P.: Modeling displacements and stresses in set of conjugate layers of construction under impulse loading on surface Certificate of the Russian Federation on state registration of computer program RU No. 2019663432/16.10.2019

An Experimental Study of 3 Axes CNC Milling for a Blade of the Axial Flow Pump Impeller

Ngoc Tuyen Bui[1(✉)] and Duc Luong Ngo[1,2]

[1] Hanoi University of Science and Technology, Hanoi, Vietnam
tuyen.buingoc@hust.edu.vn
[2] Institute for Pump and Water Resource Machine, Hanoi, Vietnam

Abstract. Impeller is an important part of the axial flow pump. Basing on main characteristics of the pump as pump head, flow rate and efficiency, impeller blades are designed with free form surface to meet requirements of hydrodynamical functions. The paper presents a complete process from designing an impeller blade by Vozexexki - Pekin method, modeling the blade by Solidwork to manufacturing it by 3 axes CNC milling. Through an empirical research for assessing influence levels of three factors as tool path type, side step, diameter of tool on the geometrical accuracy of the free - surface forming of the impeller blade, the authors have selected appropriate technological parameters to machine the blade surface. By the technological parameters, the impeller blade was completely machined by 3 axes CNC milling.

Keywords: 3 axes CNC milling · Impeller blade · Tool path · Side step · Diameter of tool

1 Introduction

Nowadays, pumps in general and axial pumps in particular have had wide applications in the national economy, industry, agriculture as well as in life.

The impeller is an important part of the axial pump. The impeller blade is designed to have a free form surface to meet the hydrodynamic functions required. Currently, impellers are usually made by casting method, then grinding by hand, so it takes a lot of time, large costs and often have low geometrical accuracy. This paper presents a complete process from design, experimental research to choose the optimal technological parameters to machine and manufacture impeller blades on 3 axis CNC milling machines. By the experiment, influence levels of the three factors as tool path type, side step, tool diameter to the geometric accuracy of the free form surface of the impeller blade were defined. A set of the most appropriate parameters to machine the impeller blade on the 3 axis CNC milling machine was found out, too.

The research method here is a combination of theoretical and empirical researches. The design method of Vozexexki - Pekin, that commonly used in Russia and SNG countries [1–3] was used to design the impeller blade profile. On the basis of the designed profile grid, the 3D impeller blade model was built by using SOLIDWORK software. Three - axis CNC milling with ball - end cutter was used to machine the impeller blade surface with the free form surface shape. In the machining process, there

B. T. Long et al. (Eds.): MMMS 2020, LNME, pp. 330–336, 2021.
https://doi.org/10.1007/978-3-030-69610-8_46

were many factors that affected to the geometric accuracy of the blade surface such as cutting regime parameters, cutting tool parameters, materials, etc. Three parameters, which had direct effects on the geometric accuracy of the forming process, were chosen for this study including of tool path type, side step and tool diameter. The Taguchi method combined with variance analysis (ANOVA) [4] with the support of the Minitab software was used for experimental design and processing of experimental data.

2 Experiments, Results and Discussion

2.1 Design and Modeling of Impeller Blade

Input data for impeller design was supplied by the Institute of Pump and water resource machine from the demand for axial pump design of the South Ma River pumping station, specifically as follows:

Calculated flow, m^3/h: 490
Calculated water column, m: 2.60
Number of revolutions of the engine, rev/min: 900
Number of characteristic revolutions of the pump n_s, rev/min: 598
Pump performance η%: ≥ 72.

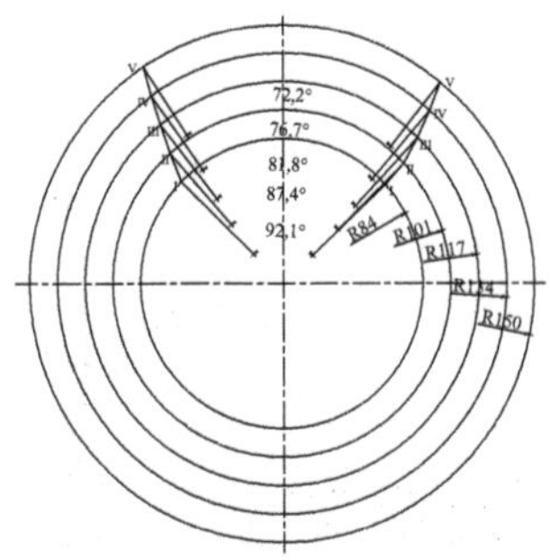

Fig. 1. Plan view of the impeller blade and the location of sections

The impeller blade of the axial pump was designed by the design method of Vozexexki - Pekin. The basic content of this method is to replace the effect of profile on the fluid flow by the vortices distributed on the nucleus of the profile according to a defined rule. The profile grid will be identified, then. The detailed solution of this problem is introduced in the specialized documents. The results of the design calculation were obtained in a file of coordinates of points, plan view drawings, profiles at cross sections and profiles' string drawings (Figs. 1, 2 and 3).

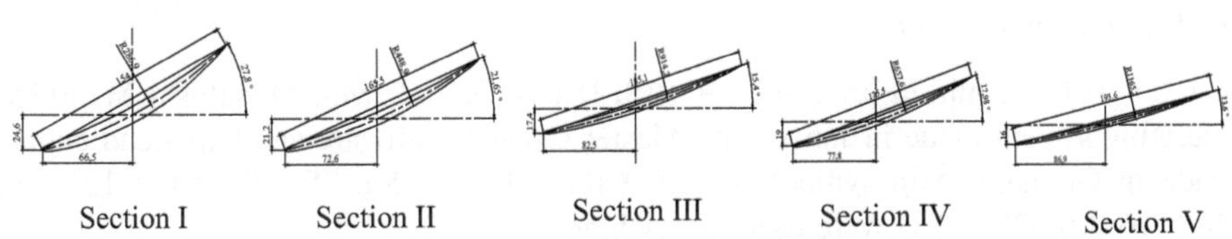

Fig. 2. Profiles of the blade at the sections

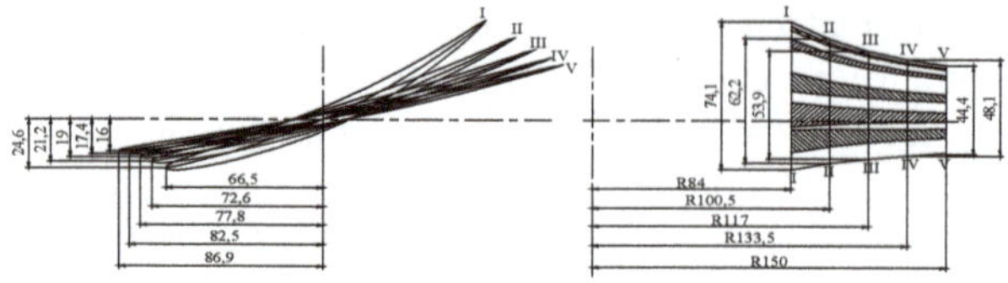

Fig. 3. Views of the profiles string of the blade

The 3D model of the impeller was built from the profiles at cross sections by using Solidworks software (Fig. 4).

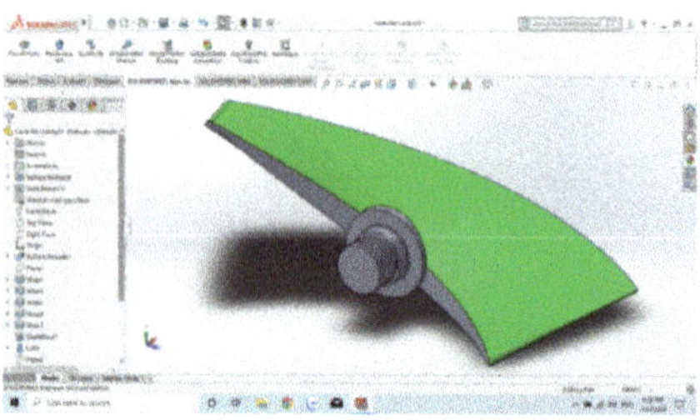

Fig. 4. 3D modeling of the impeller blade

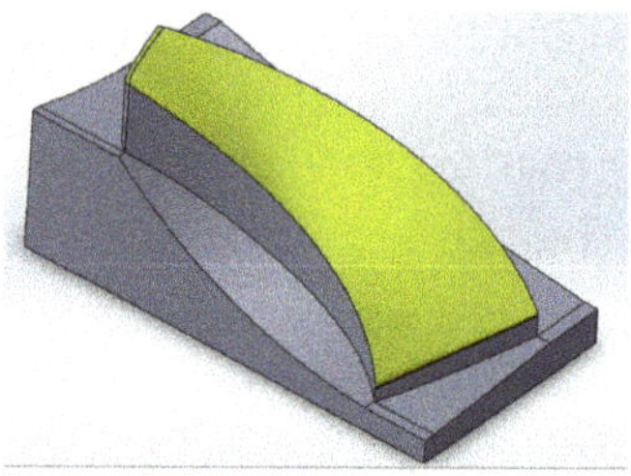

Fig. 5. Experimental model

2.2 Experiment of Machining Free-Form Surface with Impeller Blade Shape

The purpose of the experiment is the assessing the influence level of three factors as tool path type, side step, tool diameter to the geometric accuracy of the machined free form surface with the impeller blade shape when machining it on a 3 axis CNC milling machine by ball - end cutters [5]. From there, finding out an optimal parameter set to machine impeller blades.

- *Experimental conditions*

Three axis CNC milling machine MORISEIKI NVX5060 with Mitsubishi M730BM operating system made in Japan with Master CAM X software and ball - end cutters made in Germany with symbols as: 2F * R2 * 4 * 8 * 50; 2F * R3 * 6 * 12 * 50; R4 * 16 * 60-2F; ALU were used to machine.

HSCAN331 Handheld 3D Scanner with the accuracy up to 0.03 mm, speed reaches 265,000 measures/s and Geomagic Studio were used to assess accuracy of the machined surface.

Experimental model with machined surface taken from the upper surface of the impeller was designed by Solidwork (Fig. 5).

Workpieces from A6061 material having dimensions of 190 × 100 × 90 (mm).

- *Experimental plan, results and assessment*

Rough and semi - finish milling were performed and left an allowance of 0.1 mm. Finish milling was performed according to L9 Taguchi design. Three input variables as tool path type (T), Side step (S), tool diameter (D) were selected. Each variable has 3 selected levels as in Table 1. Other technology parameters were kept constant for all experiments during the machining process: cutting speed v = 138 m/min, cutting depth t = 0.1 mm, feed rate f = 549 mm/min, Emunxi cool fluid, etc.

Table 1. Factors and their levels

Factors	Levels		
	1	2	3
Tool-path style T	Zigzag	One way	Spiral
Side step S (mm)	0.10	0.45	0.80
Tool diameter D (mm)	4	6	8

The output target was the average deviation of the sample surface. After processing, according to the experimental plan above, 9 samples were obtained (Fig. 6). Using the HSCAN331 Handheld 3D Scanner, we had 9 surface point cloud files. These point clouds were registered with the designed model surface to evaluate surface geometric errors, through Geomagic Quality 2003 software. The test results of two samples 1 and 4 were presented in Fig. 7. The criterion of evaluating forming accuracy was the average deviation defined by a formula: average dev. = (upper dev. − lower dev.)/2. Average deviations of the samples were presented in Table 2.

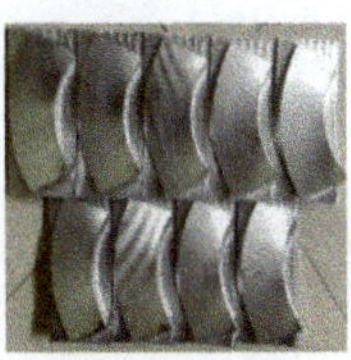

Fig. 6. Nine machined samples

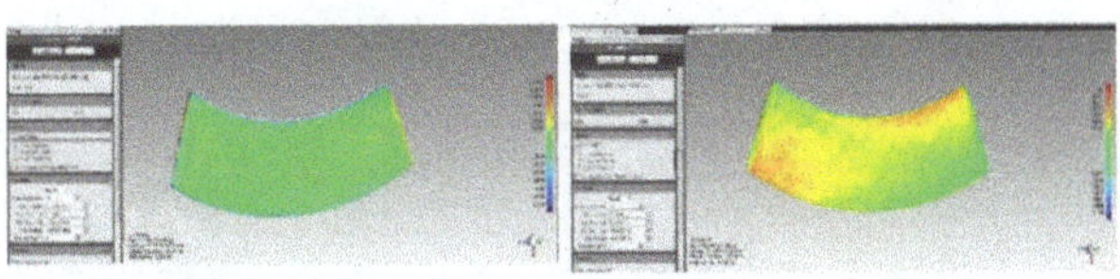

Fig. 7. Results of checking samples 1 and 4

Table 2. Experimental layout using the orthogonal array Taguchi L9 (3 ^ 3)

No.	End milling machining parameters			Average dev. (μm)	MSD	S/N ratios
	Tool path	Side step (mm)	Tool diameter (mm)			
1	Zigzag	0.10	4	30.450	927.20	−29.67
2	Zigzag	0.45	6	38.280	1465.36	−31.66
3	Zigzag	0.80	8	43.650	1905.32	−32.80
4	One way	0.10	6	30.500	930.25	−29.69
5	One way	0.45	8	33.050	1092.30	−30.38
6	One way	0.80	4	31.980	1022.72	−30.10
7	Spiral	0.10	8	36.450	1328.60	−31.23
8	Spiral	0.45	4	41.260	1702.39	−32.31
9	Spiral	0.80	6	50.820	2582.67	−34.12

The average S/N ratio for each of the cut parameters at levels 1, 2 and 3 can be calculated by averaging the S/N ratios for the respective experiments. Main effects of each factor on average deviation of machined surface are defined by Minitab as shown in Fig. 8. The S/N ratio is higher for smaller surface variances around the desired value. The Analysis of Variance (Table 3) shows that the tool path and side step are the most important cutting parameters. Tool path contributes nearly 49.65%, which has the most significant influence on the geometric accuracy of the machined surface. Next are the side step and then the cutting tool diameter with the contribution of 36.78% and 10.60%, respectively. With the S/N value as large as possible as shown in Table 4, we can choose an optimal set of parameters as $T_2S_1D_1$. The optimum cutting parameters for the best geometric accuracy of the machined surface are the One way tool path type, side step with S = 0.1 mm and tool diameter D = 4 mm.

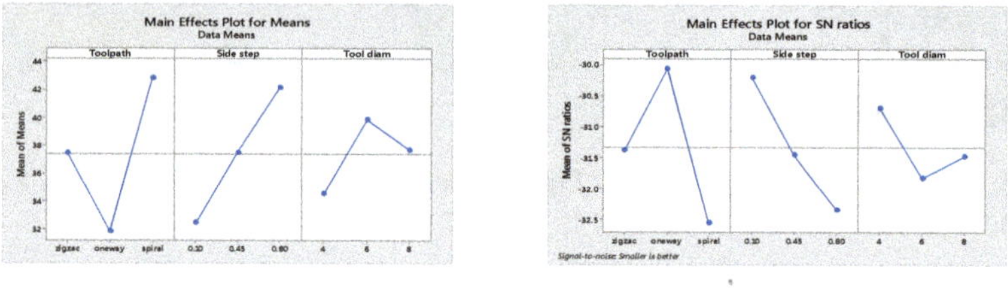

Fig. 8. Main effects of each factor on the average deviation of the machined surface

Table 3. ANOVA results for the deviation of the surface

Source	DOF	Sum of squares	Mean squares	F-ratio	Contribution (%)
A:T	2	9.380	4..690	16.72	49.65%
B:S	2	6.950	3.475	12.39	36.78%
C:D	2	2.003	1.002	3.57	10.60%
Error	2	0.561	0.280		2.97%
Total	8	18.895			100%

Table 4. Summary of S/N ratios of factors with their levels

Factor level	Tool path (T)	Side step (S)	Tool diameter (D)
	−31.38	−30.20*	−30.69*
2	−30.06*	−31.45	−31.82
3	−32.56	−32.34	−31.47

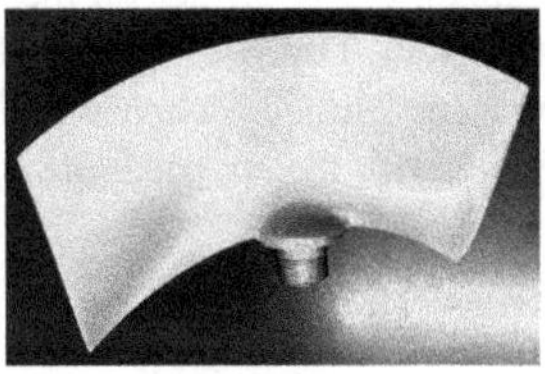

Fig. 9. The machined impeller blade

And an impeller blade was completely machined by the set of the parameters (Fig. 9).

3 Conclusion

The impeller blade of an axial pump has been calculated and designed based on the demand and the input parameters by the design method of Vozexexki – Pekin. The blade has been completely modeled by Solidworks software.

The experimental study of the effect of 3 factors as tool path, horizontal step and tool diameter to the geometric accuracy of the impeller blade surface when machining on a 3 axis CNC milling machine by ball - end cutters give us following results:

The geometric accuracy of the pump impeller blade surface milled on a 3 axis CNC machine depends on three machining parameters (tool path type, side step and tool diameter) with different degrees. The parameter that has the greatest influence on the geometric accuracy of the machined surface is the tool path type (49.65%), followed by the side step (36.78%) and finally the tool diameter (10.60%).

A set of optimal parameters for the best geometric accuracy of the machined surface under test conditions is: side step (S) of 0.1 mm, tool path type (T) of one way and tool diameter (D) of 4 mm. Finally, the impeller blade has been machined.

The results of this research can be extended to apply to the design and manufacture of other types of axial impellers in the future.

References

1. Lo-Ma-Kin, A.A.: Centrifugal pump and axial direction pump. Translations from Russian language. Translators: Phu, L., Tung, L.D., Thi, D.X., Publishing Scientific and Technical, Hanoi (1971)
2. Lien, L.D.: Pumps, fan blades. Bach Khoa Publishing House – Hanoi, Hanoi (2014). Code: 128-2014/CXB/61-01/BKHN
3. Scientific report of independent state-level scientific research project on large flow irrigation pump: Research, design and test of axial pump models of 20,000 m^3/h and 36,000 m^3/h. Project manager: Pham Van Thu, Hanoi (1996)
4. Taguchi, G.: A primer on the Taguchi method - Joyce Cary, TS156.R69 (1990)
5. Quy, H.V., Tuyen, B.N.: Effect of feed rate, tool path and step over on geometric accuracy of freeform surfaces when 3 axis CNC milling. In: Applied Mechanics and Materials, vol. 889 (2019). ISSN: 1662-7482

Applied Theory of Bending Vibration of Magnetoelectroelastic Bimorph

Do Thanh Binh[1], A. N. Soloviev[1,5(✉)], V. A. Chebanenko[2], E. Kirillova[3], and Truong Huu Dong Ha[4]

[1] Don State Technical University, Rostov-on-Don, Russian Federation
solovievarc@gmail.com
[2] Southern Scientific Center of RAS, Rostov-on-Don, Russian Federation
[3] RheinMain University of Applied Sciences, Wiesbaden, Germany
[4] Air Force - Air Defence Technical Institute, Hanoi, Vietnam
[5] Southern Federal University, Rostov-on-Don, Russian Federation

Abstract. Transverse steady-state vibrations of a two-layer plate, consisting of a piezoelectric layer and a layer with piezomagnetic properties, are considered. On the basis of hypotheses about the distribution of mechanical quantities, electric and magnetic potential over the thickness of the plate, the equations and boundary conditions of the applied theory of bending vibrations of the plate are obtained using the variational principle. The first natural vibration frequencies were found, the values of which showed conformity with their calculations in the ACELAN finite element package.

Keywords: Applied theory · Piezoelectrics · Piezomagnetics · ACELAN

1 Introduction

An urgent task in the development of a number of electronic devices (sensors, measuring systems, cell phones, etc.) is their energy independence. In this case, it is sufficient to use low-power energy sources that can be charged from the external influence of the environment. At the moment, such energy sources can be various types of piezoelectric generators (PEG). In the case when a PEG oscillates or experiences mechanical loads, a large number of devices are presented and their effectiveness has been studied in the literature. If the PEG is in an alternating magnetic field caused by some reasons (for example, moving permanent magnets…), then the presence of a piezomagnetic layer leads to oscillation of the device and the generation of an electric current due to deformation of the piezoelectric layer. In this work, we consider a device that is a two-layer plate (with piezoelectric and piezomagnetic layers), which performs transverse vibrations under the action of an alternating magnetic field. An applied theory of oscillations has been developed, according to which the dynamic characteristics of the device (natural frequencies) are calculated. The results are compared with the calculations using the FEM in the ACELAN package.

B. T. Long et al. (Eds.): MMMS 2020, LNME, pp. 337–342, 2021.
https://doi.org/10.1007/978-3-030-69610-8_47

2 Problem Formulation

The motion of a piezomagnetoelectric environment is described by a system of differential equations [1]:

$$\nabla \cdot \boldsymbol{\sigma} + \rho \mathbf{f} = \rho \ddot{\mathbf{u}}, \quad \nabla \cdot \mathbf{D} = \sigma_\Omega, \quad \nabla \cdot \mathbf{B} = 0. \tag{1}$$

$$\begin{aligned} \boldsymbol{\sigma} &= \mathbf{c} : \boldsymbol{\varepsilon} - \mathbf{e}^T \cdot \mathbf{E} - \mathbf{h}^T \cdot \mathbf{H} \\ \mathbf{D} &= \mathbf{e} : \boldsymbol{\varepsilon} + \boldsymbol{\kappa} \cdot \mathbf{E} + \boldsymbol{\alpha} \cdot \mathbf{H} \\ \mathbf{B} &= \mathbf{h} : \boldsymbol{\varepsilon} + \boldsymbol{\alpha}^T \cdot \mathbf{E} + \boldsymbol{\mu} \cdot \mathbf{H}. \end{aligned} \tag{2}$$

$$\boldsymbol{\varepsilon} = \frac{1}{2}\left(\nabla \mathbf{u} + (\nabla \mathbf{u})^T\right), \quad \mathbf{E} = -\nabla \varphi, \quad \mathbf{B} = -\nabla \phi. \tag{3}$$

Where, $\boldsymbol{\sigma}$ - stress tensor, $\boldsymbol{\varepsilon}$ - strain tensor, $\mathbf{D}$ - electric displacement vector, $\mathbf{E}$ - electric field vector, $\mathbf{B}$ - magnetic induction vector, $\mathbf{H}$ - magnetic field vector, ρ - material density, $\mathbf{c}$ - elastic coefficients tensor, $\mathbf{e}$ - piezoelectric coefficients tensor, $\mathbf{h}$ - piezomagnetic coefficients tensor, $\boldsymbol{\kappa}$ - dielectric permittivity tensor, $\boldsymbol{\alpha}$ - magnetoelectric permeability tensor, $\boldsymbol{\mu}$ - magnetic permeability tensor, $\mathbf{f}$ - mass forces density vector, σ_Ω - volume density of electric charges, $\mathbf{u}$ - displacement vector, φ and ϕ - electric and magnetic potentials.

Equations (1)–(3) should be supplemented with mechanical boundary conditions and boundary conditions for the magnetic and electric fields. Note that when considering only a piezoelectric body or only a piezomagnetic body in (1)–(3), the corresponding field connectivity coefficients should be zeroed.

3 Finite Element Analysis

To construct an applied theory of bending vibrations of the PEG under consideration, it is necessary to know the distribution of mechanical quantities, magnetic and electric potential over the thickness of the device, on the basis of which the corresponding hypotheses can be accepted.

To determine these distributions, a finite element modeling in ACELAN of a two-layer plate with piezoelectric and piezomagnetic layers with the corresponding thicknesses H_1 and H_2 was carried out. The layers are polarized by thickness. Since the equations for electric and magnetic potential are qualitatively the same, such modeling can be carried out in ACELAN using a quadratic triangular piezoelectric element with the corresponding constants for piezoelectric material (PZT-4) and piezomagnetic material $CoFe_2O_4$ [2, 3]. Material constants of these materials are given in Tables 1, 2.

Table 1. Properties of piezoceramics PZT-4

C_{11}^E	C_{12}^E	C_{13}^E	C_{33}^E	C_{44}^E	e_{31}	e_{33}	e_{15}	k_{11}/ε_0	k_{33}/ε_0
139	77.8	74.3	115	25.6	−5.2	15.1	12.7	730	635

Where, C_{ij}^{E} (GPa) - elastic coefficients, e_{ij} (C/m^2) – piezoelectric coefficients, k_{ij} and ε_0 - dielectric permittivity, $\varepsilon_0 = 8,85 \times 10^{-12}$ F/m, density of piezoceramics PZT-4: $\rho = 7500$ kg/m^3.

Table 2. Properties of piezomagnetic element $CoFe_2O_4$

C_{11}^{M}	C_{12}^{M}	C_{13}^{M}	C_{33}^{M}	C_{44}^{M}	Q_{31}	Q_{33}	Q_{15}	λ_{11}	λ_{33}
286	173	170	269.5	45.3	580.3	699.7	550	590	157

Where, C_{ij}^{M} (GPa) - elastic coefficients, Q_{ij} (N/Am) - piezomagnetic coefficients, $\lambda_{ij}\left(10^{-6} N \cdot s^2 / C^2\right)$ – magnetic permeability, density of piezomagnetic element $CoFe_2O_4$: $\rho = 5290$ kg/m^3.

The plate, the length of which is 10 mm, is pinned at both ends. The front surfaces of the piezolayer are covered with electrodes. The magnetic potential is zero on the inner surface of the piezomagnetic layer (condition 1) and is equal to zero at the fixing points (condition 2).

a)
b)
c)
d)
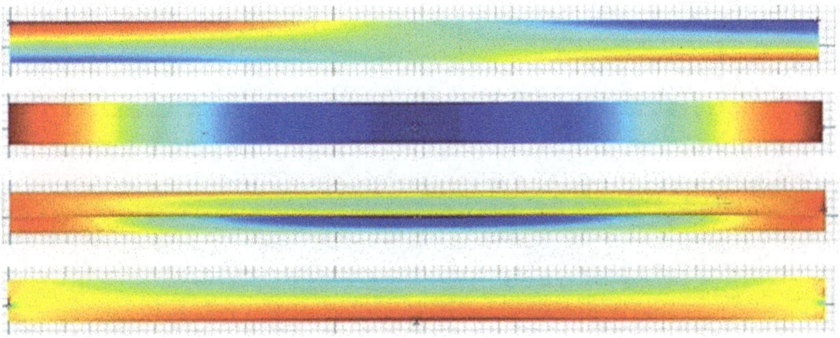

Fig. 1. Distribution of: (a) horizontal displacements, (b) vertical displacements, (c) electric and magnetic potentials, (d) longitudinal strain.

For condition 2 at the frequency of the first bending mode, Fig. 1a shows the distribution of horizontal displacements, and Fig. 1b shows the vertical displacements in the area, Fig. 1c shows the distribution of electric and magnetic potentials, Fig. 1d shows longitudinal deformation. The thickness of the electroelastic layer is $H_1 = 0.3$ mm, of the magnetoelastic is $H_2 = 0.2$ mm.

Figures 2a, 2b, 2c indicate the distribution over the bimorph thickness of horizontal displacement, electric and magnetic potentials in the first case at $H_1 = H_2 = 0.2$ mm and in the second case at $H_1 = 0.3$ mm, $H_2 = 0.2$ mm at the frequency of the first resonance, respectively. When accepting hypotheses about the distribution of mechanical magnetic and electrical quantities, which will be guided by their behavior shown in Figs. 1 and 2. The longitudinal displacement is assumed to be linear, the transverse displacement over the thickness is constant.

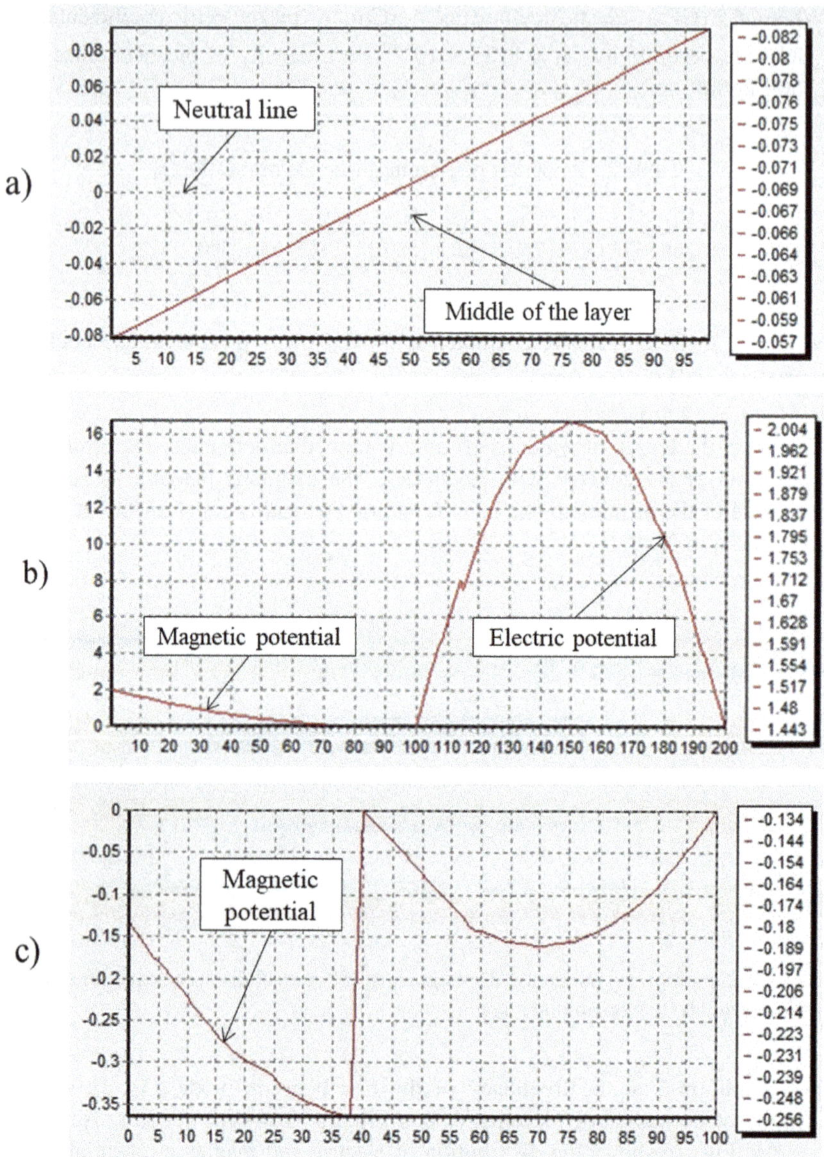

Fig. 2. Distribution of: (a) horizontal displacements, (b) electric and magnetic potentials in the first case at $H_1 = H_2 = 0.2$ mm, (c) electric and magnetic potentials in the second case at $H_1 = 0.3$ mm, $H_2 = 0.2$ mm.

The electric potential of the electroelastic layer is described by quadratic law in range from 100 to 200 in Fig. 2b. While range from 0 to 100 in Fig. 2b and in range from 0 to 40 in Fig. 2c shows that the magnetic potential over the thickness is linear.

Because there is a significant difference in elastic coefficients for piezoelectric and piezomagnetic materials with their equal thicknesses (Fig. 2a), the neutral line is shifted from the middle of the plate which is marked by number 50 on the horizontal axis.

4 Applied Theory

Using the variational principle and the hypotheses described above, an applied theory of bending vibrations of a two-layer bimorph is constructed for the first case of boundary conditions for the magnetic potential:

$$\begin{gathered}
\frac{2e_{31}H_1}{3}\left(\frac{d^2}{dx^2}\Phi(x)\right)+\frac{\left(H_1^3c_{p11}+H_2^3c_{m11}\right)}{3}\left(\frac{d^4}{dx^4}w(x)\right)-\omega^2\left(H_1\rho_p+H_2\rho_m\right)w(x)\\
+\frac{\omega^2\left(H_1^3\rho_p+H_2^3\rho_m\right)}{3}\left(\frac{d^2}{dx^2}w(x)\right)-p_1(x)-p_2(x)-\frac{h_{31}H_2}{2}\left(\frac{d^2}{dx^2}\Xi(x)\right)=0.\\
\frac{8\varepsilon_{11}H_1}{15}\left(\frac{d^2}{dx^2}\Phi(x)\right)+\frac{8\varepsilon_{33}V_1}{3H_1}-\frac{16\varepsilon_{33}\Phi(x)}{3H_1}+\frac{8\varepsilon_{33}V_3}{3H_1}+\frac{2H_1e_{31}}{3}\left(\frac{d^2}{dx^2}w(x)\right)=0.\\
\frac{\mu_{11}H_2}{3}\left(\frac{d^2}{dx^2}\Xi(x)\right)-\frac{h_{31}H_2}{2}\left(\frac{d^2}{dx^2}w(x)\right)+\frac{\mu_{33}(M_1-\Xi(x))}{H_2}=0.
\end{gathered} \tag{4}$$

Thus, the problem is reduced to a system of three ordinary differential equations for the deflection of the neutral line, the electric potential in the midline of the piezoelectric layer, and the distribution of the magnetic potential on the outer surface of the piezomagnetic field. The equation includes V_1 and V_2 electric potentials on the electrodes, if there is a potential difference, then the distribution of the electric potential is close to linear, if one of them is unknown, then an additional condition for the current in the piezoelectric layer is formulated to find it.

$$\begin{aligned}
Q_x &= -\tfrac{2e_{31}H_1}{3}\left(\tfrac{d}{dx}\Phi(x)\right)-\tfrac{\left(H_1^3c_{p11}+H_2^3c_{m11}\right)}{3}\left(\tfrac{d^3}{dx^3}w(x)\right)\\
&\quad-\tfrac{\omega^2\left(H_1^3\rho_p+H_2^3\rho_m\right)}{3}\left(\tfrac{d}{dx}w(x)\right)+\tfrac{h_{31}H_2}{2}\left(\tfrac{d}{dx}\Xi(x)\right).\\
M_x &= \tfrac{e_{31}H_1V_1}{6}-\tfrac{5e_{31}H_1V_3}{6}+\tfrac{2e_{31}H_1\Phi(x)}{3}\\
&\quad+\tfrac{H_1^3c_{p11}+H_2^3c_{m11}}{3}\left(\tfrac{d^2}{dx^2}w(x)\right)+\tfrac{h_{31}H_2(M_1-\Xi(x))}{2}.\\
D_1 &= -\tfrac{8\varepsilon_{11}H_1}{15}\left(\tfrac{d}{dx}\Phi(x)\right).\\
B_1 &= -\tfrac{\mu_{11}H_2}{3}\left(\tfrac{d}{dx}\Xi(x)\right).
\end{aligned} \tag{5}$$

To satisfy the boundary conditions, it is necessary to use the obtained relations for the force factor (force and moment) and the average values over the thickness of the longitudinal components of electric induction and magnetic induction vectors.

To confirm the adequacy of the applied theory, a comparison of the results of solving the inhomogeneous problem (frequency 100 Hz, load referred to the neutral line constant 1000 Pa) was carried out. So the maximum deviation value of error in calculating the deflection, electric potential, magnetic potential and resonance frequency are presented in Table 3 for different layer thicknesses.

Analysis of the results in Table 3 shows that if the neutral line is at the interface between the layers, the applied theory gives an adequate description of the stress-strain electric and magnetic states. The error in determining the first resonant frequency of 10 kHz for the case of a thickness ratio of 2:3 (column 4 of Table 3) does not exceed 1%.

Table 3. Error comparison

Thickness of the piezomagnetic and piezoelectric layers, respectively (mm)	0.2–0.2	0.2–0.4	0.2–0.3
Deflection (%)	2.6	3.7	0.7
Electric potential (%)	1.8	3.4	0.07
Magnetic potential (%)	10	8	2.4
The first resonant frequency (%)	1.3	1.5	0.3

5 Conclusion

On the basis of hypotheses confirmed by preliminary finite element analysis, an applied theory of bending cylindrical vibrations of a two-layer plate with two active piezoelectric and piezomagnetic layers is constructed. The proposed theory has shown its adequacy in calculating the characteristics of mechanical, electric and magnetic fields in the vicinity of the frequency of the first flexural vibration mode. So the error in determining the first natural frequency does not exceed 1%.

Acknowledgement. The second author is grateful for the support of the grant from the Government of the Russian Federation No. 075-15-2019-1928.

References

1. Kurbatova, N.V., Nadolin, D.K., Nasedkin, A.V., Oganesyan, P.A., Soloviev, A.N.: Finite element approach for composite magneto-piezoelectric materials modeling in ACELAN-COMPOS package. In: Analysis and Modelling of Advanced Structures and Smart Systems. Series "Advanced Structured Materials", vol. 81, pp. 69–88 (2018)
2. Kim, J.-Y.: Micromechanical analysis of effective properties of magneto-electro-thermoelastic multilayer composites. Int. J. Eng. Sci. **49**, 1001–1018 (2011)
3. Challagulla, K.S., Georgiades, A.V.: Micromechanical analysis of magneto-electro-thermoelastic composite materials with applications to multilayered structures. Int. J. Eng. Sci. **49**, 85–104 (2011)

A Study on Vibration During Heat Assisted End Milling of SKD11 Tool Steel

Thi-Bich Mac[1(✉)], The-Thanh Luyen[1], Tien-Long Banh[1,2], and Duc-Toan Nguyen[1,2]

[1] Department of Mechanical Engineering, Hung Yen University of Technology and Education, Hungyen, Vietnam
[2] School of Mechanical Engineering, Hanoi University of Science and Technology, Hanoi, Vietnam

Abstract. The paper presents the study carried on a vertical machining center (VMC) to determine the vibration amplitude of SKD11 steel in the heat assisted end milling by electromagnetic induction. The experiments of analyzing the frequency response function and the vibration amplitude in the idle state are first performed to control the resonance phenomenon causing instability during machining. After that, experiments are conducted to evaluate the influence of technology parameters on vibration amplitude, showing that the cutting speed is inversely proportional to the vibration amplitude. Meanwhile the feed speed and cutting depth are proportional to vibration amplitude. The effect of the heating process on vibration is studied, showing that the vibration amplitude much decreases during machining at elevated temperatures.

Keywords: SKD11 tool steel · Heat assisted end milling · Vibration amplitude

1 Introduction

SKD11 steel is widely used in the automotive industry for its excellent properties. This material is favored for its unique metallurgical properties. SKD11 steels are machined by advanced machining methods like electrostatic discharge, diamond grinding wheels, etc. However, they are limited because of low material removal rate, expensive cutting tools and fast wear [1–4]. These issues are the basis for studying methods of improving the machining properties of SKD11 steel, while maintaining their properties.

In order to reduce vibration during machining, the heat processing technology has been studied by various studies. Amin et al. [5] studied modeling and optimizing vibration and surface roughness of heated milling process of SKD11 tool steel using ball nose tool. Research showed that the heat-assisted process is conducive to achieving low vibration amplitude and surface roughness. The results of the optimal analysis showed that the appropriate temperature is from 400–500 °C. Additionally, cutting temperature and speed have the greatest influence on surface roughness and produce maximum vibration amplitude.

The effect of heating process on the processing of high hardness materials Ti-6Al-4V was studied by Ginta et al. [6]. The experiments were performed on a vertical machining

B. T. Long et al. (Eds.): MMMS 2020, LNME, pp. 343–348, 2021.
https://doi.org/10.1007/978-3-030-69610-8_48

center. An online vibration monitoring system was used to capture the vibration during cutting process. The results showed that vibration decreased from 53% to 86% when working at high temperatures of 650 °C. This reduction in vibration amplitude significantly reduced the bouncing effects on the cutting tool, thereby reducing the tool wear rate.

The main purpose of this study is to study the effect of cutting parameters and elevated temperatures on the vibration. The frequency response function and the vibration amplitude at idle state are established that the basis for analyzing and controlling the resonance phenomenon that causes instability in the machining process.

2 Experimental Procedure

2.1 Workpiece Material

The study uses material of SKD11 tool steel. The dimension workpiece of 30 × 70 80 mm. The chemical composition of SKD11 steel material is presented as Table 1.

Table 1. Chemical composition of SKD11 (wt%) [7]

C	Cr	Mo	Si	Mn	Ni	V
1.4–1.6	11–13	0.7–1.2	$\leq$0.6	$\leq$0.6	–	0.15–0.3

2.2 Experimental Setup

Milling experiments were conducted on vertical machining center (MC500). Face milling cutter's diameter is 40 mm which has hard alloy milling insert coated with titanium of PRAMET firm (Model: 1604DR-GM). All experiments were carried out in the absence of coolant. The device for mesuring vibration of Bruel & Kjaer Danish company includes: LAN - XI data acquisition module which have 4 inputs and 2 outputs with frequency of 51.2 kHz; PULSE FFT 7770 analysis module, 1–3 channels; Triaxial DeltaTron 3-direction accelerometer with TEDS 4525-B-001 model, shell material made from Titanium. The cutting conditions selected for the experiment are shown in Table 2.

Table 2. Cutting condition for experiments

Cutting parameters	Values
Cutting speed (m/min)	190, 235, 280
Feed rate (mm/min)	230, 305, 380
Cutting depth (mm)	0.5, 1.0, 1.5
Temperature (°C)	25, 200, 300, 400

3 Results and Discussion

3.1 Experimental Analysis of Frequency Response Function

Frequency response function analysis experiment is the process of determining the kinetic properties of a system through natural frequencies, damping coefficients and shape modes and then using those properties to construct mathematical model which are called modal models of the system, and information of properties is called modal data.

In order to ensure that responses are correctly obtained and to avoid wasting time analyzing a set of false response data, the empirical modal analysis method before use needs to be tested. The test is simply done on obtained response graphs or graphs constructed from obtained graphs. In this study, the author uses modal analysis on the frequency domain and uses frequency response peak selection method. The specific frequency and the amplitude of the oscillation of the system are determined to evaluate the dynamics during the machining process. Use modal analysis to determine the specific oscillation frequency of the cutting tools used in the experimental model. These frequencies will be mixed with the vibration spectrum measured during the machining. This allows the elimination of extraneous frequencies (if any) in the vibration spectrum of the cutting process.

3.2 Analyzing Machine Vibration at Idle State

In order to analyze vibration during machining correctly, it is necessary to control the frequency resonance phenomenon that leads to erroneous analysis. In addition to the analysis of individual vibrations, the study experimentally analyzed machine vibrations in the no-load state, also known as background vibration.

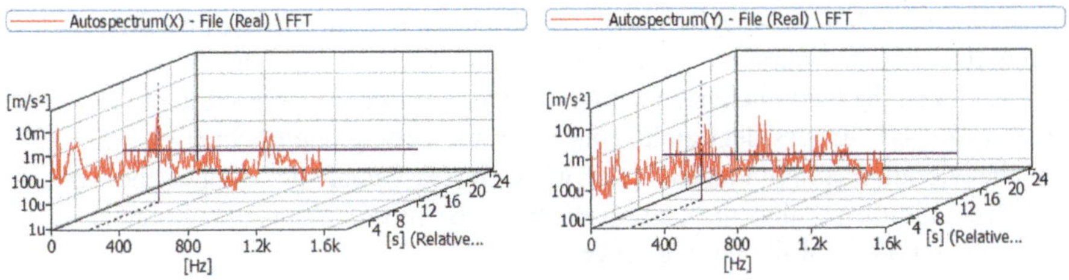

Fig. 1. Results of analysis of machine vibration at idle state

Figure 1 shows the results of background vibration frequency spectra with vibrating signals converted from real-time domain to frequency domain (Fourier FFT - Fast Fourier Transformation). Thus, vibration is described through two values: amplitude and frequency. Amplitude describes the level of vibration. Frequency describes the speed at which vibration is vibrated. This is also the basis for analyzing and controlling the resonance phenomenon that causes instability in the machining process.

3.3 Effects of Technology Parameters on the Vibration Amplitude

The vibration amplitude of machining process is the sum of vibration amplitude in the two directions X and Y determined by formula 1.

$$A_{XY} = \sqrt{A_X^2 + A_Y^2} \qquad (1)$$

Where A_{XY} is the sum of the amplitude of vibration of the cutting process; A_X, A_Y are in order of vibration amplitude in the X and Y directions.

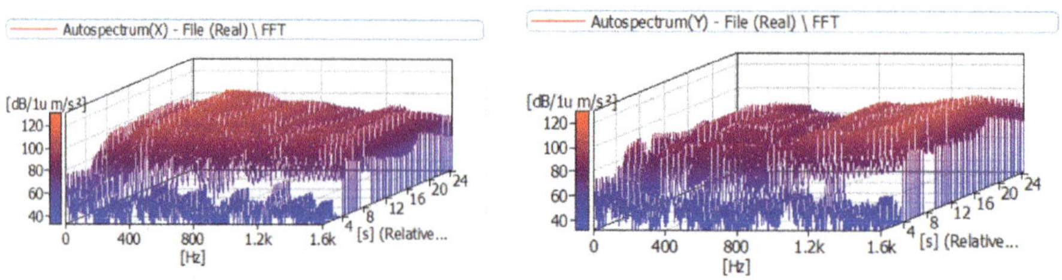

Fig. 2. Vibration during machining with cutting condition V = 235 (m/min), f = 305 (mm/min), t = 1.5 (mm) and T = 200 °C.

Figure 2 shows the results of vibration during machining with cutting condition V = 235 (m/min), f = 305 (mm/min), t = 1.5 (mm) and elevated temperature of 200 °C. To analysis the efficiency of the cutting speed on the vibration, the experiments are designed with increasing cutting speed and other cutting conditions unchanged. The measurement results of average amplitude of vibration are presented as Fig. 3a. The results show when the speed of cutting increases, the cutting force decreases. However, the amount of reduction is negligible of 2.3 dB. Similarly, the effects of the feed rate and cutting depth on the vibration amplitude are also studied that are shown in Fig. 3b and Fig. 3c. When the feed rate changes from 230 to 380 mm/min, the vibration amplitude increases by 1.5 dB was negligible. But, when the cutting depth increases from 0.5 to 1.5, the vibration amplitude increases with greater value, the amount increases by 10.3 dB. This means that the cutting depth is the most influence parameters on the vibration amplitude of the cutting process.

3.4 Effect of Elevated Temperature on Vibration Amplitude

Heat assisted end milling is a solution for machining high hardness materials. To evaluate the effect of elevated temperature on vibration amplitude during machining SKD11 steel, the experiments are designed respectively at room temperature and elevated temperature with unchanged cutting mode. Figure 4 shows the results of vibration amplitude when machining different temperature conditions. The results show that the vibration amplitude decreases significantly by 26.5 dB when machining at 200 °C. The reason is that under the effect of elevated temperatures, the bonds between metal particles decrease, the cutting process is made more easily. When the temperature

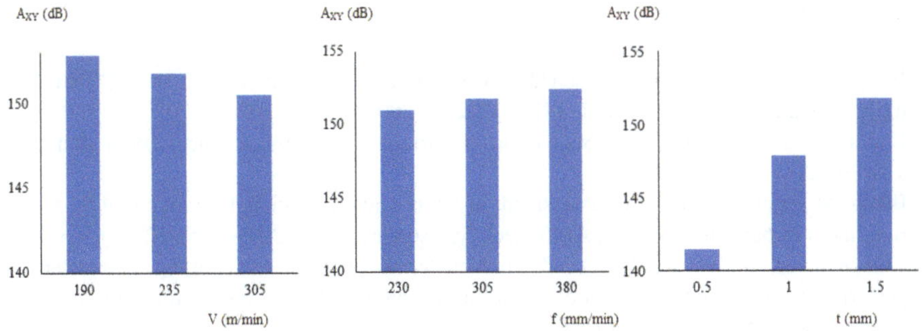

(a) (b) (c)

Fig. 3. Average vibration amplitude at different machining conditions

continue to increase to 300 and 400 °C, the vibration amplitude continue to decrease but the amount decreased slightly. This means that rising temperatures have little effect on the vibration amplitude.

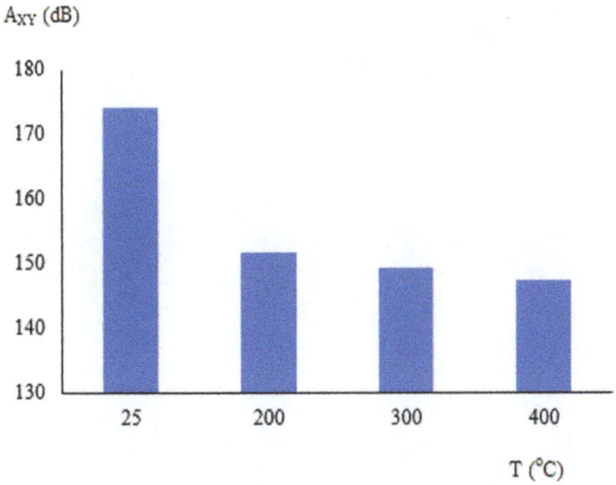

Fig. 4. Vibration amplitudes when machining at different temperatures

4 Conclusion

A study of the vibration when heat assisted end milling SKD11 steel was performed in this study. Analyzing the frequency response function and the vibration amplitude at idle state were the basis for analyzing and controlling the resonance phenomenon that causes instability in the machining process. The effects of the cutting parameters on the vibration analyzed showed that the vibration increases with increasing feed rate and cutting depth. Vibration decreases with increasing cutting speed. The efficiency of the heating process has also been verified, the vibration is significantly reduced when machining at 200 °C in comparing with it in machining at room temperature.

References

1. Benga, G., et al.: Turning of hardened 100 cr6 bearing steel with ceramic and PCBN cutting tools. J. Mater. Process. Technol. **143**, 237–241 (2003)
2. Bartarya, G., Choudhury, S.K.: State of art in hard machining. Int. J. Mach. Tools Manuf. **53**, 1–4 (2012)
3. Elbah, M., et al.: Comparative assessment of wiper and conventional ceramic tool on surface roughness in hard turning AISI 4140 steel. Measurements **46**, 3041–3056 (2013)
4. Chinchanikar, S., et al.: Effect of work material hardness and cutting parameters on performance of coated carbide tool when turning hardened steel: an optimization approach. Measurements **46**, 1572–1584 (2013)
5. Nurul Amin, A.K.M., et al.: Modeling & optimization of surface roughness & vibration amplitude in heat assisted end milling of SKD 11 tool steel using ball nose tool. In: Advanced Materials Research, vol. 538–541, pp. 799–803 (2012)
6. Ginta, T.L., et al.: Thermally-assisted end milling of titanium alloy Ti-6Al-4V using induction heating. Int. J. Mach. Mach. Mater. **14**, 194–212 (2013)
7. Wang, C., et al.: Modeling and simulation of the high-speed milling of hardened steel SKD11 (62 HRC) based on SHPB technology. Int. J. Mach. Tools Manuf. **108**, 13–16 (2016)

Analysing Two Crashed Vehicles to Determine the Right Material for Body Panels

Nguyen Thanh Quang(✉)

Faculty of Automobile Technology, Hanoi University of Industry, Hanoi, Vietnam
nguyenthanhquang@haui.edu.vn

Abstract. In designing buses, composite materials are replacing traditional body's steel sheets, with criteria of reducing vehicle weight and fuel consumption, increasing vehicle speed. However, in case of crash, it is subject to passenger safety and body structure durability. This paper presents results of analyzing the crash process of two buses. Low carbon steel, high carbon steel and composite panels are used in vehicle body. Linear momentum equations and Ansys workbench software are used to simulate the crash process. The results of analysing displacement, equivalent deformation, stresses in body structure of each material show that vehicle body structure using composite materials has largest deformation. Thus, it is necessary to enhance the rigidity of the composite sheets in vehicle's body.

Keywords: Bus crash · Deformation · Equivalent Elastic · Equivalent Stress

1 Introduction

While running, if a vehicle crashes with another object, it will cause safety problems for passengers and the vehicle. For passengers in the vehicle, there should be a safe enough space to avoid immediate crash as well as long pressure causing danger. For vehicles, it's body should absorb maximum impact force of the crash to reduce the subsequent impacts, but need to be hard enough to have minimum deformation. The crash theory has presented types, characteristics and methods of solving crash problems. Regulations for assessing automobile crash conditions have been developed for application at the manufacturing and operating stages of vehicles, including popular Euro NCAP standards, which assess vehicle's safety. Vehicle safety regulations and socio-economic costs are related [1]. Vehicle crashs are very complicated and diversified, which are described by models: Front, side and rear crashs. These models are relative and do not represent all crashs cases [2]. Crash testing is very costly, so using simulation models will replace some real test samples which can save a lot of cost [3]. Analysing vertical and horizontal movements of the vehicle's center of gravity determines the vehicle's trajectory after the crash, involving the crash location and vehicle's speed [4, 5]. Studying the durability of the body in a crashs by numerical simulation, load analysis, stress analysis, deformation, displacement of the body structure based on dynamics equations [5].

B. T. Long et al. (Eds.): MMMS 2020, LNME, pp. 349–354, 2021.
https://doi.org/10.1007/978-3-030-69610-8_49

2 Scientific Basis

In a crash, two vehicles are deformed and transmit energy to each other. Using full linear momentum Eq. (1), one can describe the energy transmit ion in two vehicles, [6].

$$V_1^2 m_1 + V_2^2 m_2 = V_3^2 m_1 + I_1 \omega_1^2 + I_2 \omega_2^2 + V_{ekv1}^2 m_1 + V_4^2 m_2 + V_{ekv2}^2 m_2 \tag{1}$$

where: m_1 and m_2 are the mass of the first and second vehicle; V_1, V_2 are the vehicles velocities at the start time of the crash; V_3, V_4 are the vehicles velocities after the crash; I_1, I_2 are the moments of inertia of a vehicles after the start of a crash; ω_1, ω_1 are the angular accelerations of vehicles after the start of the crash; $V_{kev(1,2)}$) are the velocities at the time of crash which are determined by formula (2)

$$V_{ekv_{(1,2)}} = \sqrt{\frac{2E_{(1,2)}}{m_{(1,2)}}} \tag{2}$$

$E_{(1,2)}$ is the crash energy corresponding to the deformation in the trunk body, determined by formula (3).

$$E_{(1,2)} = \sum_{i=1}^{n=1} k_i \left(1 + tg^2\theta\right) \left(\frac{B}{6} \cdot \frac{S_{i+1}^3 - S_i^3}{S_{i+1} - S_i} + \frac{A}{2}(S_{i+1} + S_i) + (n-1)G\right) \tag{3}$$

where: k_i is a coefficient depending on the width of the impact area and greater than 1; S_i is the depth of impact area (mm); A_i is the horizontal hardness coefficient of the structure relative to the contact force at the impact surface; B_i is the vertical hardness coefficient of the structure relative to the force at the point of impact depth; The total hardness coefficient of the structure is determined by formula (4), [5, 7].

$$G_i = \frac{A_i^2}{2B_i} \tag{4}$$

The crash process occurs in two stages. The first stage occurs when two vehicles come into contact and crash points appear in each car S1, S2. During this period, a portion of the motion energy of each vehicle is converted into potential and heat. The second stage is the process of elastic deformation of the body. At this time, the two contact surfaces stick together. The potential of elastic deformation becomes their kinetic energy. The deformation crash area is limited by two points A and B. The area of intersection on two vehicles is called the average displacement area, which is also the total displacement of the structure body. The line connecting two crash points S1, S2 is the direction of impact and distance S1S2 is the crash depth shown in Fig. 1.

Depth of impact, velocity of impact and force of impact at the point of impact are determined by (5), (6) and (7).

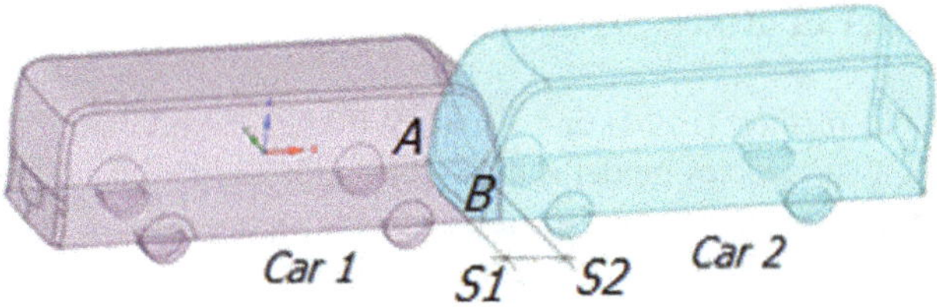

Fig. 1. Analysing model of vehicles front crash

$$x_s(t) = S_1S_2(t) \tag{5}$$

$$\dot{x}_s(t) = dx_s(t) = \frac{dx_s}{dt} \tag{6}$$

$$F_s(t) = k_s x_s(t) + h_s \dot{x}_s(t) \tag{7}$$

Where: x_s(t) is depth, dx_s(t) is velocity, F_s(t) is force of crash over time; k_s, h_s is the coefficient of hardness and elasticity of material at displacement element.

In the simulation, the following assumptions have been used: (1) Static load including vehicle weight at full load is 6475 kg [10]; (2) The entire vehicle body, consisting of body cover plates, a frame, and sub-plates, assuming full force; (3) Links of the assemblies in the vehicle are absolutely hard, regardless of abnormal deformation there. The body cover and frame materials in the two vehicles are of the same type. The properties of materials listed in Table 1.

Table 1. Properties of materials of body structure

	Type of materials	Density (kg/mm^3)	Young's modulus (MPa)	Poisson's ratio	Tensile yield strength (MPa)	Tensile ultimate strength (MPa)
1	Low alloy steel, 300M (low carbon)	7.833e−006	2.106e+005	0.3265	1595	1958
2	Low alloy steel, 300M (high carbon)	7.833e−006	2.106e+005	0.3265	1667	2030
3	Composite, Epoxy/glass fiber, UD prepreg, QI	1.904e−006	19970	0.3065	479.9	479.9

3 FEM Model of Body Structure

The finite element method (FEM) and Ansys Workbench software to analyses crash model used. The two vehicles model is drawed in grids for the linear problem with the result of three-dimensional elements having average diagonal size of 30 mm. Using the default transition ratio of 0.272 for 5 layers according to increasing speed of 1.2, we get 148256 nodes and 148357 elements.

The boundary conditions of model are displacement units the crash direction S1S2 allowing free movements for all edges in the two vehicles bodies, the first vehicle has 228 edges and the second vehicle has 456 edges.

Set analytical mode according to analytical ability of computer. We set the number of analytical cycles 1e + 07 cycles with very small crash time 7e−04 (s), linearity 0.2. The number of points for analyzing is 100 points in the results area, Fig. 2.

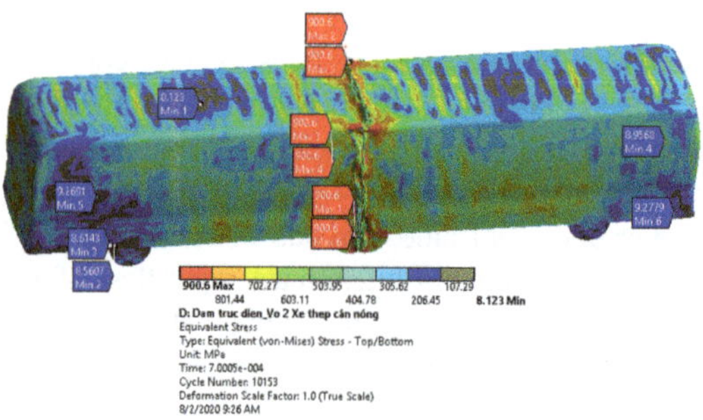

Fig. 2. Cycle modes, time and number of analytical selected points

4 Analysis of Results

The consider the body displacement in the case of different materials. The simulation results include: Directional Deformation; total deformation in three directions x, y, z; Equivalent Elastic Strain; Equivalent Stress distribution in the impact area. In the simulation, we can take the values of graphs in the numerical table over time to evaluate in detail the time of the survey cases.

The displacement in the direction of crash df (mm) gives the result of evaluating the depth of the displacement in the direction of front crash. The depth of displacement area of the composite body is the largest, 0.5 m, while it is 0.33 m of high carbon body cover structure and 0.17 m of low carbon. The total displacement in three directions has the same value as the front direction. The depth of crash area will increase steadily without mutation. Thus, in front crash of x direction, the displacement of the remaining z and y directions does not much affect. Body displacement results are presented in Fig. 3.

The equivalent deformation of EEs (mm/mm) depends on the coefficient of hardness and elasticity of the material in formula (4) above, Fig. 4. The body cover structure of composite material is very much deformed and will be completely destroyed (broken) at the first ¼ time (about 21 s) of the 2 vehicles contact.

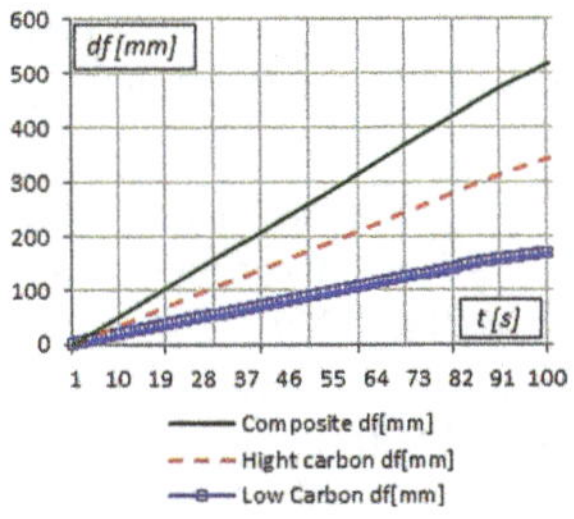

Fig. 3. Body cover structure displacement of different materials

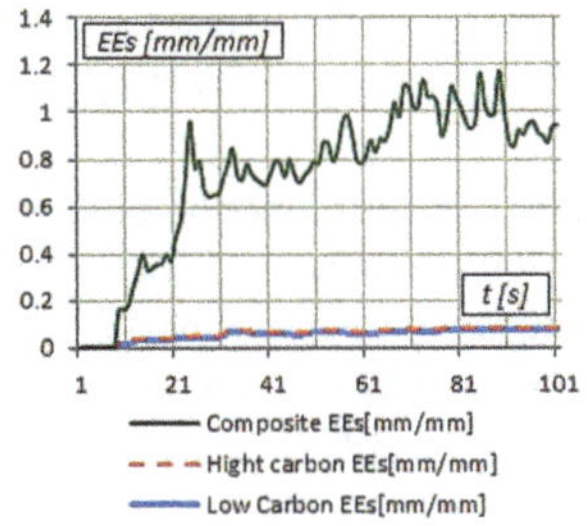

Fig. 4. Equivalent deformation in vehicle body of different materials

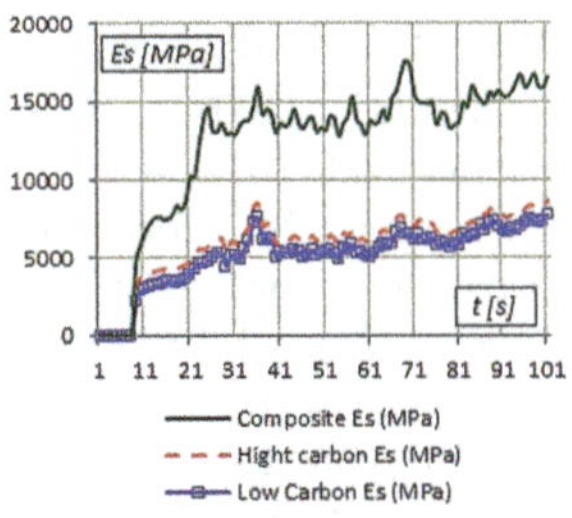

Fig. 5. Stress distribution over impact area

The stress distributed over the crash area Es (MPa) reaches maximum value also at about first ¼ time of the crash process. In body cover structure of composite material, the stress is three times higher than that of metal materials, Fig. 5.

5 Conclusions

The paper simulated frontal crash of two Buses with body cover structure materials of low carbon steel, high carbon steel and composite. The results of stress, displacement and deformation analysis in the impact area showed that the body cover structure of composite material has largest deformed, quickly and completely damaged.

From these results, choosing composite for body structure should include a solution to enhance body hardness to increase the body durability and ensure a safe space for passengers in case of crash. We advise designers and manufacturers to consider the weight carefully when adding reinforce bars to composite panels in body structure.

References

1. European Commission: TTIP – car safety analysis in the EU and US in relation to US and EU regulatory standards on crash testing, Industrial Transformation and Advanced Value Chains Automotive and Mobility Industries (2016)
2. European Commission: Vehicle safety, European Road Safety Observatory (2018)
3. Standards by ISO/TC 22/SC 36, Safety and impact testing

4. Tautkus, A., Bazaras, Z.: The investigation of car movement trajectories after contact in the case of collisions with stationary safety barriers. WIT Trans. Built Environ. (2007). Urban Transport XIII: Urban Transport and the Environment in the 21st Century 557. https://doi.org/10.2495/UT070521
5. Tautkus, A., Bazaras, Z.: Modelling and investigation of car collisions. Transport XXI **I**(4), 279–283 (2007). https://doi.org/10.1080/16484142.2007.9638142. ISSN 1648-4142 print/ISSN 1648-3480
6. Gidlewski, M., Prochowski, L.: Analysis of motion of the body of a motor car hit on its side by another passenger car. In: IOP Conference Series: Materials Science and Engineering, vol. 148, p. 012039 (2016). https://doi.org/10.1088/1757-899X/148/1/012039
7. Neptune, J.A., Flynn, J.E.: A method for determining accident specific crush stiffness coefficients. SAE paper No. 940913 (1994). https://doi.org/10.4271/940913. ISSN: 0148-7191. e-ISSN: 2688-3627
8. Neptune, J.: A comparison of crush stiffness characteristics from partial overlap and full–overlap frontal crash tests. SAE Paper 1999 (1999). https://doi.org/10.4271/1999-01-0105. ISSN: 0148-7191. e-ISSN: 2688-3627. Paper 1999-01-0105
9. Prentkovskis, O., Sokolovskij, E., Bartulis, V.: Investigating traffic accidents: a collision of two motor vehicles. Transport **25**(2), 105–115 (2010). https://doi.org/10.3846/transport.2010.14. ISSN 1648-4142 print/ISSN 1648-3480
10. https://www.hyundai.com/et/en/find-a-car/county/specification

Morphology and Microstructure of Cr - Al - Fe Coating by Electric Arc Spraying on JIS SKD61 Steel

Hong-Nga Thi Pham(✉) and Van-Thuc Nguyen

Faculty of Mechanical Engineering, Ho Chi Minh City University of Technology and Education, Ho Chi Minh City 71307, Vietnam
hongnga@hcmute.edu.vn

Abstract. Metal spraying is one of the modification surface methods that is widely used to increase the hardness, durability, resistance to abrasion, and corrosion of the surface. This method can be used to rehabilitate worn and corroded parts. This paper applies the electric arc spraying method to generate a 25%Cr - 5%Al - Fe coating on an SKD61 steel substrate. The results show that the coating surface of 25%Cr - 5%Al - Fe has a large undulating, bright, and relatively uniform coating. The coating structure of the samples includes the coating area, bonding region, heat affected area, and bottom steel. The hardness after quenching and tempering increases significantly from 252.5 $HV_{0.1}$ to about 595 $HV_{0.1}$. The average hardness is 240.6 $HV_{0.1}$, 549.5 $HV_{0.1}$, and 772.7 $HV_{0.1}$ correspondings to the steel substrate, the heat-affected zone, and the coating, respectively. The hardness of the coating is about 3 times higher than that of the steel substrate.

Keywords: Electric arc spraying · Cr - Al - Fe coating · SKD61 steel · Microstructure

1 Introduction

In recent years, the demand for improving product quality and prolong the service life of machine parts is strongly concerned. For example, hot stamping dies, which requires the material surface to work in heat-etching conditions, high hardness, toughness, resistance to abrasion. Every time the parts are worn and need to change, the replacement cost is high. Recently, surface treatment technologies have received enormous attention as it can effectively improve the quality of the surface. The surface can be modified by thermal spraying to increase its hardness, wear resistance, corrosion resistance, and heat resistance [1–3]. Among some popular surface modification processes, such as vacuum plasma spraying (VPS), wire arc spraying (WAS), atmospheric plasma spraying (APS), high velocity oxygen fuel (HVOF) spraying, and detonation gun spraying (DGS), metal spraying has a lot of superiority characteristics. This method not only creates a new surface that can work under critical conditions but also be able to recover and repair the failure surface [4–6]. Spray forming is a flexible manufacturing method, it can be applied for both thin protective coating and bulk

B. T. Long et al. (Eds.): MMMS 2020, LNME, pp. 355–361, 2021.
https://doi.org/10.1007/978-3-030-69610-8_50

material. Gas atomized spray forming with a high melting point has the disadvantage of being very difficult to control and design the injector. The electric arc spraying has more advantages, there is a voltage difference is applied between two alloy conductors that are driven together to form an electric arc. The arc melts the conductors and stream metal droplets are deposited on the substrate [7]. Electric arc spraying is the process of spraying metal or alloy wire into particles that can stick to the surface of the object to be sprayed. The particles are sprayed under high-pressure gas flow, and arc formed by two metal wires with opposite charges that need to be sprayed. Arc spraying is very beneficial as it can avoid the problems associated with melting technology. This technology can provide the spray layer with excellent adhesion ability and minor porosity. This method can be used to coat relatively thin metal coatings in the 100–1500 nm range to resist corrosion and abrasion [8]. Popular alloys such as Fe-based alloys have received great attention for their good mechanical properties and resistance to corrosion. There are many authors reported about Fe-based alloys spraying. However, there is not much research survey on the effect of microscopic structures, such as particle size and grain boundary characteristics, on the properties of Fe-Cr alloys. Both mechanical properties and corrosion resistance are closely related to the grain structure [9, 10].

2 Experimental

2.1 Materials and Experimental Method

The sample size of steel SKD61 used for the experiment is 10 × 12 × 10 mm. The atomic emission spectroscopy analysis result showed that the elemental composition of the steel sample used in the experiment was in accordance with the SKD61 steel standard. The spraying wire materials is Cr-Al-Fe with 25% Cr, 5% Al, and 70% Fe.

Heat Treatment Process Before Spraying: Heat treatment is implied at Minh Hanh Heat treatment Company, Bien Hoa City, Dong Nai. The selected quenching temperature is 1030 °C, the total heating time is 9.6 min [11]. The samples are then tempered with an average temperature of 500 °C to achieve a hardness suitable for hot stamping [12]. The tempering time is 1.5 h, after tempering, the samples are cooled in static air.

Surface treatment Before Spraying: Before metal spraying, the samples are roughened by spraying brown coridon abrasive. Sandblasting parameters: Abrasive particle: brown coridon of Hai Duong abrasive fabrication joint-stock company, spray distance: 200 mm, spray pressure: 8–10 atm.

Electrical Arc Spraying: The samples are coated at the Institute of Tropical Technology, Vietnam Academy of Science and Technology, Hanoi. Parameters of electric arc spray for coating 25%Cr - 5%Al - Fe: gas pressure: 4.5 atm, arc voltage: 29–30 V, spray distance: 150 mm, feeding material shape: wire.

2.2 Methods for Determining the Mechanical Properties

Microstructure Analysing Method: The microstructure analyse is made at Nhu Ngoc precision processing company. Initially, the sample cutting is done on wire cutters, 10 × 10 × 10 mm in size, with the cross-section perpendicular to the direction of the spray gun. After cutting, the samples were molded to create epoxy samples with Blueler special machines. Then, the samples are etched by applying a solution having 3% HNO_3 in alcohol. Finally, they are observed on a microscope.

Hardness Testing Method: The samples are tested at D.V.G company, Ho Chi Minh city by Vickers testing method.

3 Result

3.1 Microstructure of Initial SKD61 Steel

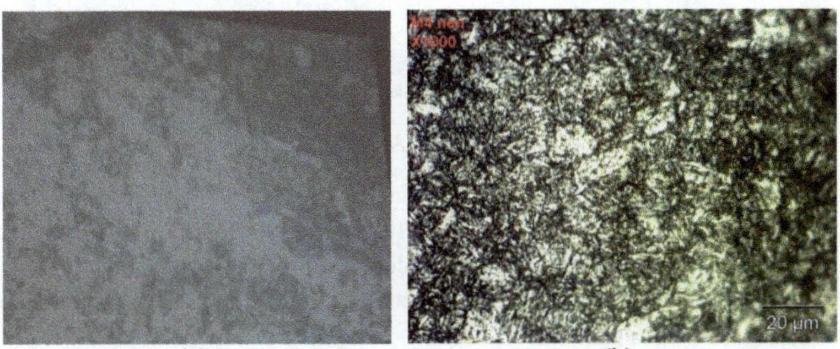

(a) (b)

Fig. 1. Microstructure of SKD61 steel a) Initial SKD61 steel (x400) b) SKD61 samples after quenching and tempering

Figure 1 shows microstructure of SKD61 steel of initial SKD61 steel (x400) and SKD61 samples after quenching and tempering. The structure has spherical carbide particles and a ferrite matrix. Moreover, the microstructure content of fine particles, which are regularly distributed. In the microstructure of the SKD61 steel substrate after quenching and tempering, the particles in the microstructure are smaller and more evenly distributed than the original one. When the steel temperature reaches a temperature of 400 °C, ferrite and cementite phase appears. At the temperature reach 500 °C, a mixture of ferrite and cementite in a granular shape or tempered troostite is formed [13]. At that temperature, the structure with fine and uniform grains is generated, having good synthetic mechanical properties.

3.2 Sample Morphology Analyses

Figure 2 shows the shape of the 25%Cr - 5%Al - Fe coating. The coating surface has an undulating, uniform, and thin coating. Between the coating and the steel substrate, there is a black boundary. The existence of this boundary due to the technical parameters, treatment technology conditions, the coating materials, the state of the substrate. Additionally, the metal substrates and the coating materials have a large difference in thermal expansion factors. Therefore, during the cooling process, there will be residual stress on the boundary.

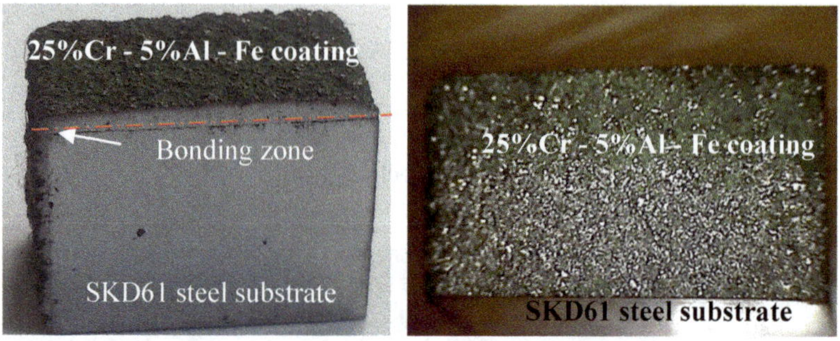

Fig. 2. The 25%Cr - 5%Al - Fe coating sample

Figure 3 illustrates the microstructure of the cross-section of the sample. The coating has an undulating surface, dark color, and regularly distributed. No cracks could be found in the coating. Thus, the electric arc coating method can create a high-quality coating on steel substrate SKD61.

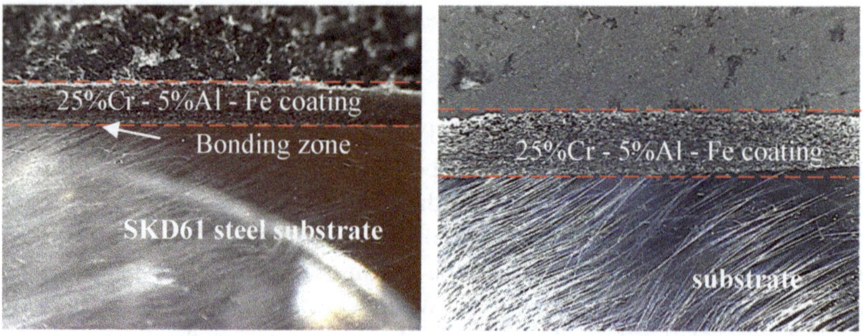

Fig. 3. Morphology of the 25% Cr - 5% Al - Fe coating

3.3 The Morphology of the Coating

Figure 4 represents the microstructure of the cross-section of the sample. The coating structure consists of four distinct areas: the coating area, the bonding area, the heat affected area, and the steel substrate. The microstructure the coating has the typical shape of a coating made by electro-arc spraying with wire material: between the lamellas, there are some holes and specific impurities of aluminium oxide. Student et al. [4] shows that the basically phase of the coatings is made of solid solutions of Cr and Al in Fe, and a small amount of Fe_2O_3, Al_2O_3 and Cr_2O_3. In the crystallized lamellas, the dispersion phase is separated from the largest droplets. In addition, in the melt spraying process, large droplets with different chemical composition were also observed. In addition, A.P. Newbery et al. [10] also indicated that the particles were splashed during spraying. Hedges et al. [8] argued that the unfused layer contained the splashed droplets that were deflected by the thickening 'step' of the material under the spray footprint. The relatively fine, rapidly condensing droplets from the droplets splashed onto the deposited equiaxed/fused splat layer previously deposited occurred at a random angle. Therefore the coating has a relatively high porosity.

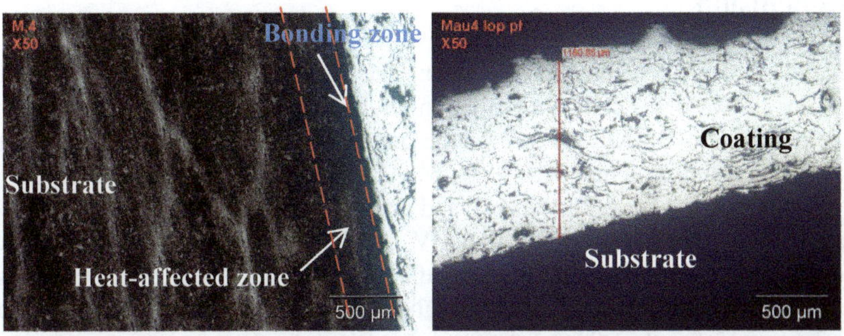

Fig. 4. Microstructure of the 25%Cr - 5%Al – Fe sample cross-section

3.4 Micro-Hardness of the Coating

The average hardness of the initial SKD61 sample and the heat-treated samples are 252.5 $HV_{0.1}$ and 595 $HV_{0.1}$, respectively. After quenching and tempering, the hardness of the SKD61 samples is more than two times higher than the initial samples. Figure 5 shows the hardness distribution of the sample. The average hardness at different positions of the sample is 602.6 $HV_{0.1}$, 549.5 $HV_{0.1}$, and 772.7 $HV_{0.1}$ correspondings to the substrate steel after quenching and tempering, the heat-affected zone, and the coating respectively. In general, from the substrate steel to the top of the coating, the hardness mostly increases. The coating hardness is greatly higher than that of the substrate steel after tempering and tempering. Additionally, the hardness of the coating is about 3 times higher than that of the initial SKD61 steel.

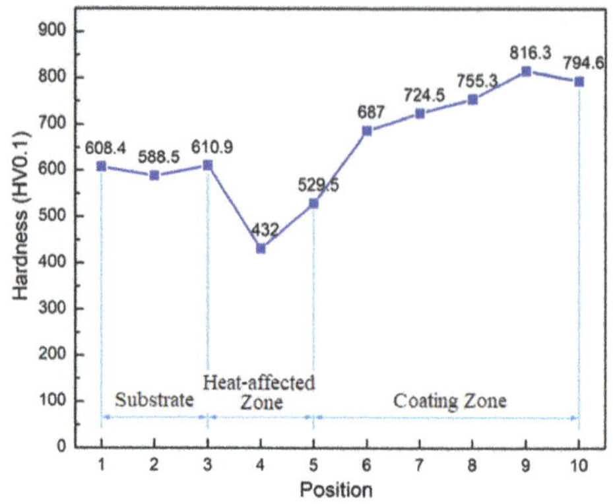

Fig. 5. Hardness distribution of the coating sample

4 Conclusion

The hardness after quenching and quenching increases significantly from 252.5 $HV_{0.1}$ to 595 $HV_{0.1}$, two times compared to the initial hardness of SKD61 steel. The coating surface of 25%Cr - 5%Al - Fe has a large undulating, bright, and relatively uniform coating. The coating structure of 25%Cr - 5%Al - Fe consists of 4 clear regions: coating area, bonding region, heat affected area, and steel substrate. The coating layer has some porous. The average hardness is 240.6 $HV_{0.1}$, 549.5 $HV_{0.1}$, and 772.7 $HV_{0.1}$ correspondings to the steel substrate, the heat-affected zone, and the coating, respectively. The hardness of the coating is greatly higher than that of the steel substrate, and is about 3 times higher than that of the steel substrate.

References

1. Sarikaya, O., Anik, S., Aslanlar, S., Okumus, S.C., Celik, E.: Al-Si/B_4C composite coatings on Al-Si substrate by plasma spray technique. Mater. Des. **28**(9), 2443–2449 (2007)
2. Sun, H., Ma, S.: Microstructural characteristics and protection performances of plasma-sprayed Al_2O_3-Al composite coatings on AZ91D magnesium alloy. Sci. Eng. Compos. Mater. **23**(5), 467–474 (2016)
3. Skopp, A., Kelling, N., Woydt, M., Berger, L.M.: Thermally sprayed titanium suboxide coatings for piston ring/cylinder liners under mixed lubrication and dry-running conditions. Wear **262**(9–10), 1061–1070 (2007)
4. Student, M., Dzioba, Yu., Hvozdets'kyi, V., et al.: High-temperature corrosion of electric-arc coatings sprayed from powder core wires based on the Fe-Cr-B-Al system. Mater. Sci. **44**(5), 93–97 (2008)
5. Wang, Z.J., Zhu, G.M., Yin, F.S.: Study on the microstructure of FeCrAl/aluminum bronze composite coatings prepared by supersonic electric arc-spraying. Appl. Mech. Mater. **217–219**, 1346–1349 (2012)
6. Konstantinov, V.M., Gubanov, A.S.: Effect of alloying elements of steel wire on the structure and properties of coatings in electric arc spraying. Weld. Int. **22**(2), 113–117 (2008)

7. Newbery, A.P., Jordan, R.M., Singer, A.R.E., Cantor, B.: Electric arc spray forming of an Ni_3Al based alloy. Scripta Mater. **35**(1), 47–51 (1996)
8. Hedges, M.K., Newbery, A.P., Grant, P.S.: Characterisation of electric arc spray formed Ni superalloy IN718. Mater. Sci. Eng.: A **326**(1), 79–91 (2002)
9. Zhou, Z.C., Wei, J.N., Han, F.S.: Influences of heat treatment and grain size on the damping capacity of an Fe-Cr-Al alloy. Physica Status Solidi (A) **191**(1), 89–96 (2002)
10. Newbery, A.P., Grant, P.S.: Oxidation during electric arc spray forming of steel. J. Mater. Process. Technol. **178**(1–3), 259–269 (2006)
11. Jia, Z.X., Li, J.Q., Liu, L.J., Liu, Y.W., Wang, Y.Q., Li, H.L.: Influence and application of laser parameters on unit of H13 steel by laser remelting process. Int. J. Adv. Manuf. Technol. **79**(1–4), 551–568 (2015)
12. Yeh, S.H., Chiu, L.H., Pan, Y.T., Lin, S.C.: Relative dimensional change evaluation of vacuum heat-treated JIS SKD61 hot-work tool steels. J. Mater. Eng. Perform. **23**(6), 2075–2082 (2014)
13. Cui, X.H., Wang, S.Q., Wei, M.X., Yang, Z.R.: Wear characteristics and mechanisms of H13 steel with various tempered structures. J. Mater. Eng. Perform. **20**(6), 1055–1062 (2011)

Bamboo – The Healthy Material Reducing Carbon and Preventing the Coronavirus in Interior Design

Nguyen Thi Bich Van(✉)

University of Architecture Ho Chi Minh City, Ho Chi Minh City, Vietnam
van.nguyenthibich@uah.edu.vn

Abstract. In folk architecture, bamboo is obviously a traditional material. Bamboo attaches itself to the minds of the Vietnamese people with villages, thatched roofs and bamboo roofs that are present in everything that belong to the Vietnamese environment and living environment. Bamboo is associated with community activities and creates many values in interior architecture design. The bamboo constructions are increasingly favored because it is not only a work but also a precious yield of art, a work of humanity and sustainability. However, this article only emphasizes the value and how to put bamboo materials into the components of the interior space: walls, ceilings and floors. The inherent properties of bamboo and advances in manufacturing techniques have resulted in high-value consumer products such as bamboo flooring and bamboo furniture to replace laminate flooring and wood furniture, offering more benefits to society and the environment. The impact of the Covid-19 pandemic is huge and comprehensive globally, especially on development issues and development trends. The impact of the Covid-19 epidemic had a lot to do with the way of life and working of many organizations, individuals, many industries, including interior design. Research and use of healthy, non-viral materials in interior design and construction are very interested in the current period. In addition to industrial materials manufactured with antibacterial technology, bamboo materials are considered to be the optimal solution to exploit and put into use in the interior, especially they are used to make walls and floors as a Practical solution for limiting the influence of viruses. Therefore, the use of environmentally friendly materials, healthy materials such as bamboo is essential.

Keywords: Healthy material · Low carbon material · Bamboo wall · Bamboo floor · Bamboo ceiling · Laminated bamboo flooring – bamboo · Block pressed bamboo flooring · Plant bamboo

1 Introduction

Bamboo can grow up to 40 m high and lives in the tropics and subtropics, it can serve human interests in many ways. However, the role of Vietnamese bamboo in the market has not had a place compared to other trees and is still considered the tree of the poor today, while the world has been enriched from bamboo even though they are not intact materials are as rich as in Vietnam. Today, society is more developed; people live in

B. T. Long et al. (Eds.): MMMS 2020, LNME, pp. 362–368, 2021.
https://doi.org/10.1007/978-3-030-69610-8_51

their own space such as houses, schools, hospitals, offices, hotels ... almost 90% of people experience their lives inside space. In particular, the emergence of the Coronavirus is seen as a great human event - the Covid19 epidemic, causing many severe losses to humanity. It is a wake-up call for people to live simpler and closer to nature. People are almost 100% encouraged to stay at home. Therefore, the use of materials with health benefits and environmental protection, such as bamboo should be promoted. In interior design, bamboo has been and is a valuable material in terms of aesthetics, sustainability, environmental friendliness, and many positive influences on the emotions and spirits of users.

Bamboo is a plant that has a positive impact on the soil surface, preventing erosion and saltwater intrusion. Bamboo planting is the most reasonable and easy solution to improve the soil and bring nutrients to the soil. In all sectors: food industry, textile, construction ... bamboo is involved and is meant as a sustainable material. Bamboo has the ability to absorb large amounts of carbon dioxide, which means that growing bamboo will contribute to preventing the greenhouse effect, creating a good environment for the residential area. There have been many studies on bamboo, but this article wants to strongly emphasize the inclusion of bamboo in the solutions of designing and constructing human space from walls, ceilings, and products is very necessary. When environmental problems seriously affect human health and life, it is time for people to return to traditional bamboo materials and renovate them, produce them in the direction of sustainable materials, protect health. People and create a good environment and landscape. Designers should consider the sustainable value of this material to be given special attention in interior design and as a sustainable Vietnamese material.

2 Bamboo is a Sustainable and Eco-Friendly Material with Many Significant Benefits

Bamboo has the potential to improve the ever-increasing depletion of natural resources around the world, and in particular, it is capable of effectively replacing natural wood. Some of the advantages and benefits of bamboo are as follows:

- Fast growing: Bamboo grows rapidly and can be harvested within 3 to 5 years of planting, compared with hardwoods that take up to 40 years to mature and give good quality to harvest.
- Abundant natural bamboo resources: There are 37 million hectares of bamboo forests in the world. Therefore, there is an abundant supply of raw materials that do not need much cost to grow and do not take care of [1].
- Benefits for the poor: Some bamboo forests are allocated to poor people for management and exploitation, so advances in the bamboo industry will bring more benefits
- Sustainability: Bamboo can be harvested annually and regenerates itself; in fact, planned harvesting contributes to the sustainability of the forest as well as increases productivity in the future.
- Soil protection: Planting bamboo forests helps to improve soil fertility and prevent erosion. The remaining bamboo roots in the soil after harvest help retain nutrients

and moisture for the next crop. Bamboo also provides limited protection from natural disasters such as landslides.

- High rate of material utilization: Not only the trunk, but all other parts of the bamboo can also be used for a variety of uses such as bamboo shoots for food, leaves for fodder, and the branches used to make brooms and firewood.
- Absorbs greenhouse gases: Bamboo forests absorb greenhouse gases. Bamboo absorbs carbon dioxide and emits 35% more oxygen into the atmosphere, more so than hardwood.
- No need for necessary fertilizers, pesticides, or herbicides: Unlike most industrial crops, bamboo does not need fertilizer to thrive. Also unlike other plants, bamboo does not need careful care and care does not create excess chemicals for the environment [2].

3 Solution to Use Bamboo in the Design of Decorative Wall Cladding Art

Mobile interior space is a new model for advanced countries, minimizing energy consumption. This space can be arranged from the composite panels to create new features for the interior (Fig. 1).

Fig. 1. Bamboo wall [3]

- Bamboo fiber panels can be used to design beautiful, simple mobile interiors that are both environmentally friendly, protect the health, and enhance the use-value of bamboo through new products
- Bamboo material woven together with bamboo fiber sheets from many types of bamboo has the advantages of meeting the aesthetic factors, at the same time creating jobs for people. Bamboo fiber is an ideal substitute for traditional materials in sustainable construction. Can be used as reinforcement for composites.
- Bamboo fiber material sheet has thermal conductivity from 0.976–2.5 W/mK. The coefficient of thermal conductivity represents a material's ability to conduct heat. The greater the coefficient of thermal conductivity, the better the ability to conduct

heat. The coefficient of thermal conductivity is not dependent on thickness but on the properties of the material [4].

- Bamboo fiber material sheets have sound absorption coefficients from 0.02–0.06 with frequencies from 500–2000 Hz, indicating that only 2–28% of the sound is transmitted through the material. Thus, it is possible to produce soundproof plywood products from bamboo fibers for commercialization.

4 Using Bamboo in Ceiling and Floor Cladding

Bamboo flooring - good material for everyone's health, families with young children should use bamboo flooring instead of carpet, ceramic tile. Its antibacterial and prevent bacteria, dirt on the bamboo floor is very good. This material is very suitable for the ceiling, and the floor in the interior of houses.

4.1) Bamboo flooring is very suitable for the Vietnamese climate: The first thing that many customers and users worry about is the suitability with the hot and humid tropical climate of Vietnam. So it can be confirmed immediately that the bamboo floor is very suitable for the weather and climate conditions of Vietnam, even when it is hot in the northern climate. First, the bamboo fiber itself is less water permeable due to its special structure. Indoor bamboo flooring is not designed for use in submerged conditions but may make better use of laminate flooring in high humidity conditions.

4.2) Bamboo flooring has bright colors for many years: Unlike flooring using color paint, bamboo flooring is treated with color right from the raw material stage and coated with transparent paint. Bamboo that is boiled and carbonized to remove substances such as lignin will be very stable over time and not destroyed by the environment. The color of the bamboo floor that we perceive is the true color of the bamboo. Any scratch on the floor surface, just varnish, and the floor will be as bright as new.

4.3) Bamboo flooring is very durable: Bamboo is also considered as "soft iron". Not only is it durable, but also has high stability and fewer scratches.

4.4) Bamboo floor makes it easy to decorate your house: For a long time there have been thousands of handicraft products and interior decoration made from bamboo, now bamboo flooring helps you to have more new choices. Bamboo is very easy to combine with other materials used for decoration. Not only is the flooring material, but the bamboo floor also makes the house more beautiful and luxurious. Bamboo flooring is also very easy to combine with other items and decorations in your home to create a friendly and warm space [5].

4.5) Bamboo flooring is suitable for every home space: Whether your house is in a quiet, peaceful place or in a noisy and noisy place, bamboo flooring is the best choice to create a space. Goodness and a sense of balance, lightness. Bamboo flooring can be used for any space in the house from the living room to the bedroom, from the dining room to the kitchen.

4.6) Bamboo flooring is an ecological product: Not only using bamboo which is eco-material with fast regeneration speed, other materials, and additives used to make bamboo flooring are also selected and meet the standards of safe to use and friendly with the environment. Formaldehyde-free sealants and low lead paints are tested to European and American standards.

4.7) The beauty of bamboo flooring is not outdated: Like bamboo handicrafts and decorations, bamboo flooring retains its sophistication, uniqueness, and elegance that is not easy to replace. With the deep philosophical values from bamboo, with the historical significance of the bamboo tree and the closeness to the culture of the Vietnamese people, the longer the homeowner uses the bamboo floor, the longer he loves it, thus the more attentive it is. a little bit for it, the more attractive its beauty.

4.8) Inexpensive Bamboo Flooring: With the availability of fast-regenerating natural materials, bamboo flooring brings high economic efficiency to producers and consumers. Bamboo flooring is not as expensive as a natural wood floor, but it has many outstanding technical properties similar to natural wood [5].

5 Classification of Bamboo Floors

5.1 Laminated bamboo flooring - bamboos, after heat-denatured treatment, are square planed and carefully selected according to a certain specification, then the bamboos are assembled with specialized glue - creating a product with high elasticity that shows the natural beauty of the bamboo tree and its distinctive pattern due to the difference in color between the bamboos and the body of the bamboo tube. With a hardness of 1200–1600 on the Janka hardness scale, the laminate bamboo floor is suitable for installation in areas with moderate travel intensity such as households, hotel rooms, gyms, aerobic. In some cases, it can be installed for offices, showrooms to display products (Fig. 2).

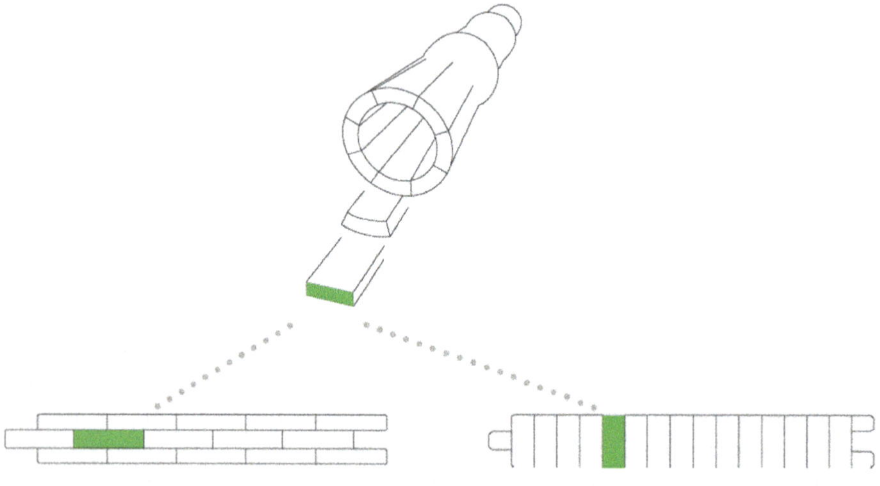

Fig. 2. Laminated bamboo flooring – bamboos [5]

5.2 Block pressed bamboo flooring is produced by crushing bamboo bars into fibers and pressed under the high pressure of a 2,500-ton press. The product uses European E1 safety glue, bamboo sticks stick together to form blocks like natural wood. The specific weight of block laminate flooring reaches 1,100–1200 kg/m^3 and is 2 times harder than Oak. Diamond'Ali bamboo flooring is suitable for use in places with high traffic density such as offices, restaurants, hotels, showrooms. Studying to design interior architecture with bamboo resources of Vietnam to create a simple, economical, ecological and cultural living environment. Based on the biodiversity of bamboo with different shapes, colors, sizes, flexibility, and physical properties to give initial and specific solutions (Fig. 3).

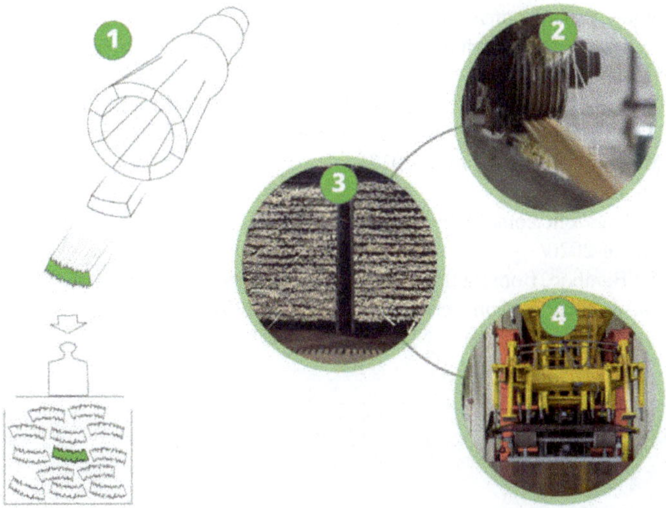

Fig. 3. Block pressed bamboo [5]

1. Harvest and select the right quality bamboo materials.
2. Bamboo chips are coarsely shredded to remove the outer green shell and the soft inner lining as well as made of bamboo clips.
3. Crushing bamboo sticks into fibers.
4. Process denaturation and natural color by heat. Pressing technology in the production of pressed bamboo.

6 Conclusion

Bamboo has many advantages and high applicability. It can be seen that bamboo has been and is a sustainable material suitable for all environments and climates. Using bamboo materials in the housing space also contributes to protecting human health, increasing feelings of closeness to nature. The emergence of Covid 19 changes people's living habits, they tend to work from home and organize their home office spaces and feel the values of living from within their homes [6]. They will consider more materials selected for their home. Therefore, the message of the article wants everyone to have a better understanding of living a more balanced life, preserving the environment, growing bamboo, taking bamboo as the saying Vien Dong says, "Better to eat without meat, a family without bamboo" [7].

References

1. Troya, F.A., Xu, C.: Plantation management and bamboo resource economics in China. Ciencia y Tecnología **7**(1), 1–12 (2014)
2. https://sangonguyenkim.com/san-go-tre/
3. https://www.michaelfreemanphoto.com/media/ec83cce6-3aa1-11e0-b729-eb19eeb9cb75-bamboo-wall-house
4. Nguyen, D.M., Grillet, A.-C., Diep, T.M.H., Woloszyn, M.: Influence of the inorganic absorbents on hygrothermal properties of bamboo insulation boards. Construction and Building Materials (under review) (2019)
5. https://bambooali.com/
6. https://colombogazette.com/2020/07/29/how-covid-19-is-changing-interior-and-architectural-design-trends-of-2020/
7. Chele, E.-S.: Bamboo, from traditional crafts to contemporary design and architecture (2012). https://www.sciencedirect.com/science/article/pii/S1877042812033770

Multi-objective Optimization of Cutting Force, Surface Roughness and Vibration Amplitude During End Milling SKD11 Steel

Thi-Bich Mac[1(✉)], Vinh-Hung Tran[2,3], and The-Thanh Luyen[1]

[1] Department of Mechanical Engineering, Hung Yen University of Technology and Education, Hungyen, Vietnam
bich.utehy@gmail.com
[2] Faculty of Mechanical Engineering and Mechatronics, PHENIKAA University, Hanoi 12116, Vietnam
[3] PHENIKAA Research and Technology Institute (PRATI), A&A Green Phoenix Group JSC, No. 167 Hoang Ngan, Trung Hoa, Cau Giay, Hanoi 11313, Vietnam

Abstract. The paper studied about optimization machining parameters when milling SKD11 tool steels with multi - objective based on grey relational analysis in Taguchi method. Firstly, the experimental design method of Taguchi was used to establish the experimental set. Parameters that directly affect the machining process include cutting depth, feed rate and cutting speed optimized based on multi - objective like cutting force, surface roughness and vibration. Then, a GRG (grey relational grade) is determined from the grey analysis. The influence level of the input parameters on the output one shows that the cutting depth was the most influence parameter on the total performance, followed by the cutting speed, the feed rate having the least effect. Finally, the optimal technology parameter to get the minimum cutting force, surface roughness and vibration is A3B1C1. In which, the cutting speed, feed rate, cutting depth are corresponding 280 m/min, 230 m/min, 0.5 mm.

Keywords: SKD11 material · Grey relational analysis · Multi-objective optimization · End milling

1 Introduction

In mechanical processing, input parameters including cutting speed, feed rate and cutting depth are the main factors affecting the output parameters such as cutting force, surface roughness, chip formation, cutting temperature, and vibration, etc. Many scientists studies on the effect of input parameters on output parameters [1–3]. These are important parameters for evaluating the quality of a machining process.

Improving product quality is the top task of the manufacturer. One of the solutions to that task is to find the optimal set of technological parameters for machining, especially multi-parameter optimization. There are several methods of multi-parameter optimization such as: weighted principal component [4], combined Taguchi method and neuro-fuzzy based model [5], GRA in Taguchi method [6].

B. T. Long et al. (Eds.): MMMS 2020, LNME, pp. 369–374, 2021.
https://doi.org/10.1007/978-3-030-69610-8_52

In this study, optimization is performed simultaneously for 3 output parameters including cutting force, surface roughness and vibration based on orthogonal array and grey relational analysis. The experimental set was established using Taguchi's experimental design method. Order and influence level of input parameters were analyzed based on ANOVA variance analysis method. Optimal parameter levels are determined based on grey relational grade with the goal of achieving minimum shear force, surface roughness and shear vibration simultaneously.

2 Experimental Procedure

2.1 Workpiece Material

In this study, the SKD11 steel is used that has the chemical composition as shown in Table 1. It is the steel used in the mold industry. Billet size 30×70×80 mm. Do not use coolant during machining.

Table 1. Chemical composition of SKD11 (wt%) [7]

C	Cr	Mo	Si	Mn	Ni	V
1.4–1.6	11–13	0.7–1.2	≤ 0.6	≤ 0.6	–	0.15–0.3

2.2 Experimental Setup

The machining schematic diagram is shown as Fig. 1. The CNC milling machine with code of MC500 is used to perform machining. To measure cutting force, the study uses 3-component cutting force measuring equipment from Kisler - Switzerland. This device uses dynamometer 9257B - Kisler with force measuring range F_X, F_Y, F_Z of 1500, 1500, 5000 (N), respectively. Surface roughness after machining is measured by Mitutoyo SV - C3200 roughness tester combined with Formtracepak software. The device to measure the vibration made in Bruel & Kjaer Denmark that includes a LAN-XI data acquisition module, PULSE FFT and PULSE FFT 7770 analysis module. The triaxial accelerometer of Triaxial DeltaTron with type TEDS 4525-B-001 and its shell material is made from Titan.

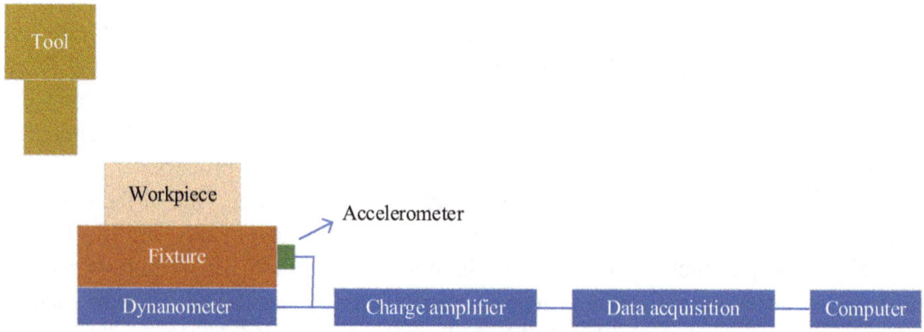

Fig. 1. Schematic diagram of end milling SKD11 steel

2.3 Control Paramters and Levels

Taguchi orthogonal arrays were used to design experiments with three parameters and three levels. The three inputs are feed rate, cutting speed and cutting depth. The benefit of the Taguchi method is optimizing design the number of experiments to study the entire parameter space [8]. Control parameters and levels are presented as Table 2. The results of experimental for the cutting force, vibration amplitude and surface roughness for orthogonal arrays L9 are presented in Table 3.

Table 2. Parameters and the levels

No.	Factors	Unit	Symbols	Level 1	Level 2	Level 3
1	Cutting speed	m/min	A	190	235	280
2	Feed rate	mm/min	B	230	305	380
3	Cutting depth	mm	C	0.5	1.0	1.5

Table 3. L9 orthogonal array and experimental results

Trial No.	Actual value			F	Ra	$A_{X\text{-}Y}$
	A	B	C	(N)	(μm)	(dB)
1	190	230	0.5	135.980	0.177	158.386
2	190	305	1	298.685	0.208	174.741
3	190	380	1.5	434.063	0.221	176.233
4	235	230	1	213.651	0.161	158.354
5	235	305	1.5	360.168	0.184	174.080
6	235	380	0.5	160.462	0.157	163.347
7	280	230	1.5	262.667	0.148	157.329
8	280	305	0.5	118.768	0.128	150.093
9	280	380	1	239.342	0.165	166.246

3 Results and Discussion

3.1 Experimental Analysis of Frequency Response Function

To specify the optimal technological parameters for both cutting force, surface roughness and vibrations amplitude, the study used the grey relation analysis (GRA) method. The output target of the above parameters is "lower is better". Therefore, the exact form of the original sequence is shown in the following formula:

$$S_{kj} = \frac{maxX_{kj} - X_{kj}}{maxX_{kj} - minX_{kj}} \tag{1}$$

Trong đó: S_{kj} is value after GRA; j = 1 ÷ 9; k = 1 ÷ n, n is the number of process responses; $maxX_{kj}$ and $minX_{kj}$ are the largest and smallest value of X_{kj} for the kth response, respectively.

$$GZ_{kj} = \frac{S_{kmin} - \xi S_{kmax}}{S_{kj} + \xi S_{kmax} kmax} \tag{2}$$

Trong đó: GZ_{kj} is the coefficient of the GRA of the kth response in the jth sample; ξ is the distingushing coefficient, ξ = 0.5 in this case.

G_j is a coeficient describes the weighted sum of the gray relation which represents overall evaluation of experimental data for multi-objective optimization.

$$G_j = \frac{1}{n}\sum_{k=1}^{n} GZ_{kj} \tag{3}$$

Standard values of F, Ra and A_{XY}, process responses and grades of gray relation are shown in Table 4. The combination of quality characteristics is closer to the desired level, the grey relational grade is higher. So, the better process parameters are corresponding with a better process responses. Figure 2 presents the grey relational grade at different experiments. Table 4 and Fig. 2 show that experiment 8 has the highest relational grade grey value. Therefore, in the 9 experiments of the orthogonal array L9, the eighth experiment is the first prefered order. The first experiment and the sixth one are corresponding the second and the third prefered order.

Table 4. Grey relational analysis results

Exp. No.	S for F	S for Ra	S for A_{XY}	GZ for F	GZ for Ra	GZ for A_{XY}	G	Order
1	0.104	0.593	0.335	0.827	0.314	0.599	0.580	2
2	0.712	0.889	0.947	0.413	0.265	0.346	0.341	7
3	1.000	1.000	1.000	0.333	0.250	0.333	0.306	9
4	0.453	0.420	0.334	0.525	0.352	0.600	0.492	4
5	0.856	0.665	0.923	0.369	0.300	0.351	0.340	8
6	0.232	0.374	0.527	0.683	0.364	0.487	0.511	3
7	0.612	0.266	0.293	0.449	0.395	0.630	0.492	5
8	0.000	0.000	0.000	1.000	0.500	1.000	0.833	1
9	0.541	0.465	0.637	0.480	0.341	0.440	0.421	6

The order of influence of the cutting parameters on the GRA analysis results is presented as Table 5. The influence percentage is described in Table 6 and graphed as Fig. 3. The results indicated that the cutting depth hold the greatest effect on total efficiency with 63.21%, cutting speed has second infulence level with percentage effect of 25.95%. The feed rate has the least effect on total efficiency with a percentage effect of 10.84%.

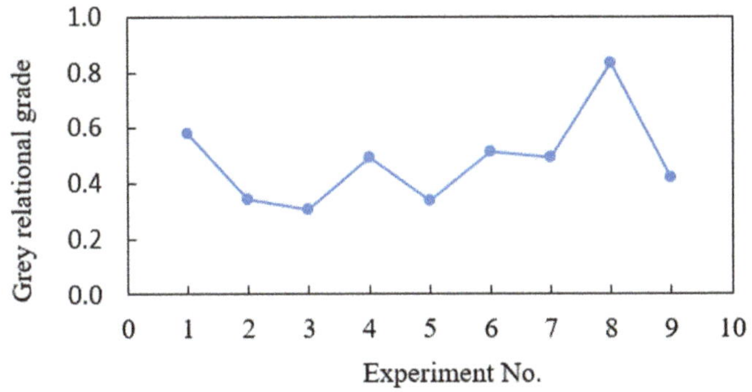

Fig. 2. The grey relational grade at different experiments.

Table 5. Results of Grey relational analysis

Symbol	Factor	Grey relational grade			Max-Min	Rank
		1	2	3		
A	V	0.409	0.448	0.582	0.173	2
B	f	0.521	0.505	0.412	0.109	3
C	t	0.641	0.418	0.379	0.262	1

Table 6. ANOVA table

Symbol	Factor	Df	Sum of squares	Mean square	Contributiton
A	V	2	0.0165	0.0082	25.95
B	f	2	0.0069	0.0034	10.84
C	t	2	0.0401	0.0201	63.21

Table 4 presents the influence of the levels of each control parameter on the grey relational grade. The results show that the optimal control parameters for the 3 target output of the cutting force, vibration and surface roughness are A3B1C1. In which, V, f, t are corresponding to 280 m/min, 230 mm/min and 0.5 mm (Fig. 4).

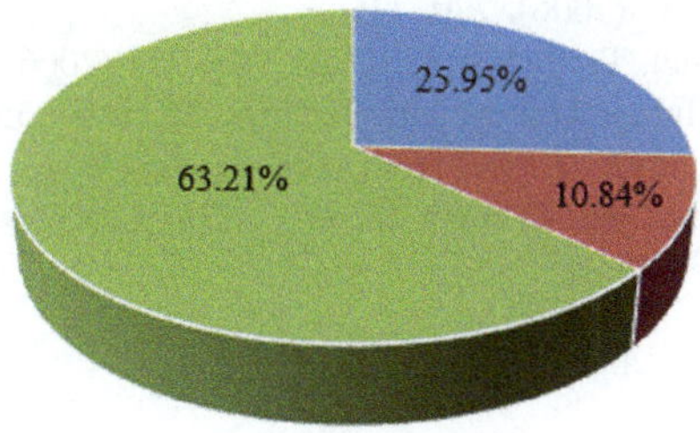

Fig. 3. Percentage contributions of parameters on the grey relational grade

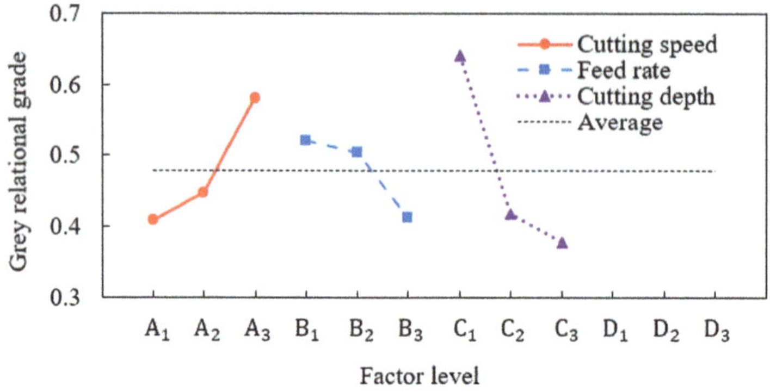

Fig. 4. Effects of process parameters on the grey relational grade

4 Conclusion

A multi-parameter optimization study has been performed in this paper. The Taguchi experimental design method, the ANOVA variance analysis method and the GRA method were used respectively and the research results were obtained as follows:

- The cutting depth is the parameter has most influencing on determining the multiple performance characteristics. The cutting speed has second effect and the feed rate has a trivial effect on the total performance.
- The percentage effects of the cutting speed, feed rate and cutting depth on the total efficiency are 63.21%, 25.95% and 10.84% respectively.
- The set of optimal control parameters for 3 output targets including the cutting force, surface roughness and vibration is A3B1C1. In wich, the cutting speed, feed rate and cutting depth are corresponding to 280 m/min, 230 mm/min and 0.5 mm.

References

1. Makwana, R., et al.: Int. Conf. Multidiscip. Res. Pract. **1**, 488–489 (2014)
2. Amin, A.K.M.N., et al.: Adv. Mater. Res. **541**, 799–803 (2012)
3. Chang, C., et al.: Int. J. Mach. Tools Manuf. **47**, 141–147 (2007)
4. Hung-Chang Liao, Vol. 27 (2006), 720–725.
5. Antony, J., et al.: J. Manu. Tech. Manag. **17**(6), 908–925 (2006)
6. Noorul Haq, A., et al.: Int. J. Adv. Manuf. Technol. **37**, 250–255 (2008)
7. Wang, C., et al.: Int. J. Mach. Tools Manuf. **108**, 13–16 (2016)
8. Palanikumar, K.: J. Reinf. Plas. Compo. **25**, 1739–1751 (2006)

Effect of Calcination Temperature on the Structure and Characteristics of Cuprous Oxide (Cu_2O-ONPs)

Nguyen Thi Tuyet Mai[1(✉)], Nguyen Kim Nga[1], Dang Thi Minh Hue[1], Truong Thi May[2], Nguyen Cong Tu[2], Trinh Xuan Anh[1], Ta Ngoc Dung[1], Luu Thi Hong[3], Huynh Dang Chinh[1], and Luu Thi Lan Anh[2]

[1] School of Chemistry Engineering, Hanoi University of Science and Technology, No. 1 Dai Co Viet, Hanoi, Vietnam
mai.nguyenthituyet@hust.edu.vn

[2] School of Engineering Physics, Hanoi University of Science and Technology, No. 1 Dai Co Viet, Hanoi, Vietnam

[3] Viet Nam Institute for Building Materials, No. 235 Nguyen Trai, Thanh Xuan, Hanoi, Vietnam

Abstract. This study synthesized cuprous oxide nanoparticles (Cu_2O-ONPs) from the green precursors $CuSO_4.3H_2O$ and Na_2SO_3. The materials samples were studied at prepared temperatures of 70 °C, 80 °C, 90 °C and 100 °C. The structure and characteristics of the materials were determined by methods such as Raman shift, UV-vis absorption spectrum, SEM, XRD. The results showed that, Cu_2O-ONPs were crystallized the best in octahedral form at temperature of 90 °C. The absorption spectrum of the materials was in the visible light region $\lambda \approx$ 500–600 nm with the band gap energy defined as Eg $\approx$ 1.78–2.1eV. At calcination temperatures of 200 °C the materials retained the octahedral shape, but the absorption spectrum was shifted strongly in visible light region $\lambda \approx$ 600–700 nm with Eg $\approx$ 1.94eV. At the firing temperature of 300–400 °C the octahedral shapes of the materials were slightly distorted. At 500 °C the materials particles were inflated, broken and no longer retaining the original octahedral shape. Within this high temperature range of 300–500 °C the absorption spectra of the materials samples were shifted more strongly in visible light region $\lambda \approx$ 800–900 nm, the optical band gap energy was reduced to a value Eg $\approx$ 1.37–1.46eV which was due to the existence of two Cu_2O-CuO phases. These will be desirable of material fabrication for applications in photochemical reactions in visible light region and other specific commercial applications that need further research and exploitation.

Keywords: Cu_2O · CuO · Octahedral crystals · Nanoparticles · Green precursors

B. T. Long et al. (Eds.): MMMS 2020, LNME, pp. 375–383, 2021.
https://doi.org/10.1007/978-3-030-69610-8_53

1 Introduction

Cuprous oxide Cu_2O is a p-type semiconductor that has attracted research interest in many fields of application in recent years because of its non-toxic properties, abundant raw materials, low-cost manufacturing processes and reasonable, good optical-electrical properties [1–7]. Copper oxide is known to exist in two semiconductor phases as Cu_2O and CuO (cuprous oxide Cu_2O and cupric oxide CuO). In which, the Cu_2O direct semiconductor has a theoretical optical gap band of 2.0–2.2 eV [8, 11, 12]. Potential applications of Cu_2O have been used in the chemical industry, manufacturing highly efficient solar cells (theoretical efficiency $\sim$18%), biosensors, Li-ion batteries, photocatalysts, luminescent, optoelectronics, gas sensors. [5–24]. Cu_2O nano materials had been fabricated by many different methods such as: simple reduction method of CuO suspension by $NH_2OH.HCl$ [25], micro-emulsion method [16, 26], hydrolysis of a Cu(I)-triethylamine (Cu(I)-TEA) complex at room temperature [27], liquid phase synthesis [28], magnetron sputtering [29], solvothermal scheme [30], method of super saturation [31]. These different fabrication methods have been given Cu_2O products with great differences in structure, morphology including crystal particle shape, crystal particle size and phase transition (e.g., phase transition of Cu_2O and CuO), which all contribute to a significant change in the properties of the materials. X. Deng *et al.* prepared Cu_2O octahedra and Cu_2O/CuO heterogeneous and showed that the methyl orange (MO) dye degradation efficiency of the Cu_2O octahedra was higher than that of Cu_2O/CuO heterogeneous [11]. W. Sun *et al.* fabricated Cu_2O with the reduction of particle size to nanometer size and showed that the photocatalytic efficiency for MO dye decomposition of Cu_2O particles in small size was higher than that of Cu_2O particles in large size [15]. The studies above indicate the need for further in-depth studies of single phase Cu_2O or Cu_2O/CuO heterogeneous or etc. materials to achieve high efficiency for each specific commercial application.

In this study, we performed the synthesis of octahedral Cu_2O nanoparticle (Cu_2O-ONPs) at the synthetic temperature of 70, 80, 90, 100 °C. Next, the best crystallinity Cu_2O sample was further calcinated at temperatures of 200, 300, 400, 500 °C. Study on the effect of temperature on the formation of the Cu_2O crystalline phase as well as the structure and characterizations of Cu_2O-ONPs materials samples.

2 Experimental Procedure

Pure chemicals (Merck) used as raw materials in the study include: $CuSO_4.5H_2O$, Na_2SO_3, NaOH, doubly distilled water.

A mixture of 25 ml of $CuSO_4$ solution (0.1M), 100 mL of Na_2SO_3 solution (0.25M) and 50 mL of NaOH solution (0.1M) was stirred evenly on a magnetic stirrer with a constant gentle stirring speed (250 rpm) for 20 min. The mixed solution was blue color and gradually decreases with stirred time. Then, the mixed solution was set at different temperatures 70, 80, 90, 100 °C, respectively. Kept the constant heating temperature until the mixed solution had red-brown and a red-brown precipitate gradually appeared. Continue to stir the mixed solution for another 30 min. Then samples were cooled and deposited a red-brown precipitate which was Cu_2O crystals.

The precipitate was filtered, centrifuged several times with doubly distilled water, and a red brown powder product was obtained. At different heating temperatures (called sample synthesis temperature) of 70, 80, 90, and 100 °C, red brown powder products of different degrees of color are obtained and samples were denoted as Cu_2O-70, Cu_2O-80, Cu_2O-90 and Cu_2O-100, respectively. Next, select the sample at the synthetic temperature that has the best crystallinity of Cu_2O for further heating at the high heating temperatures of 200, 300, 400 and 500 °C. The resulting products were the black powders and were denoted as Cu_2O-200, Cu_2O-300, Cu_2O-400 and Cu_2O-500, respectively.

The microRaman spectra were determined by a Renishaw Invia Raman Microscope (laser power: 6.25mW). The solid Uv-vis spectra were determined by a JASCO V-750 (scan rate: 200nm/min). The morphologies of Cu_2O-ONPs were measured by a scanning electron microscope (HITACHI TM4000Plus). The X-ray diffraction spectra were determined by a X'pert Pro (Cu-Kα, 1.54056 Å, $2\theta \approx$ 25–75°).

3 Results and Discussion

The raman spectra of Cu_2O-70, Cu_2O-80, Cu_2O-90 and Cu_2O-100 samples (Cu_2O samples were prepared at the synthetic temperature of 70, 80, 90, 100 °C, respectively) were shown in Fig. 1a. The raman spectra in Fig. 1a showed that, the samples synthesized at 70°C did not appear the frequencies of the normal modes of vibrations of Cu_2O. The samples synthesized at 80 °C and 90 °C appeared peaks at the vibration modes of 220, 520 and 642 cm^{-1} which confirmed the frequencies of the vibration modes of Cu_2O crystalline phase [10, 13, 22–24]. In which, the raman vibrations peaks of the Cu_2O samples synthesized at 90 °C were stronger than that of the Cu_2O sample synthesized at 80 °C. The samples synthesized at 100 °C appeared peaks at the vibration modes of 220, 642 cm^{-1} which were the vibration modes of Cu_2O phase. In addition, there appeared more raman vibrations at 282 cm^{-1} and 342 cm^{-1} which reported for the CuO crystalline phase [10, 22–24]. From the analytical results on the raman spectra could be indicated that, Samples synthesized at 80 °C and 90 °C had the crystallization of the Cu_2O single crystalline phase. The sample synthesized at 100 °C had been converted a portion of Cu_2O crystalline phase to the CuO crystalline phase. The surface morphology of Cu_2O-70, Cu_2O-80, Cu_2O-90 and Cu_2O-100 samples were shown in Fig. 1b. From Fig. 1b could seen that, the Cu_2O-70 sample showed that the particles were spherical, their distribution on the sample surface was relatively uniform with particles sizes ranged in about ~ 400–500 nm. The Cu_2O-80 sample had the particles in the octahedral shape (Cu_2O-ONPs).

The distribution of particles on the sample surface was uniform, particles sizes ranged in about 200–300 nm. The Cu_2O-90 sample had also the particles in the octahedral shape (Cu_2O-ONPs), the distribution of particles on the sample surface was uniform with a denser density than the Cu_2O-80 sample, particles sizes were about 200–250 nm. The Cu_2O-100 sample showed that the particles no longer retain their octahedral shape, the crystals had been slightly distorted. Among them, there were some particles had broken away and had grown larger to be about 500–700 nm in size.

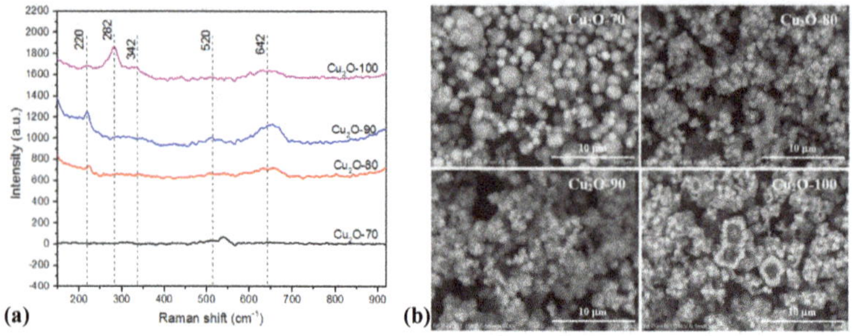

Fig. 1. (a)- The raman shift spectra of Cu_2O-70, Cu_2O-80, Cu_2O-90, Cu_2O-100 samples; (b)- The SEM of Cu_2O-70, Cu_2O-80, Cu_2O-90, Cu_2O-100 samples

The solid Uv-vis absorbance spectra results of Cu_2O-70, Cu_2O-80, Cu_2O-90 and Cu_2O-100 samples were shown in Fig. 2. The spectra in Fig. 2 showed that, the Cu_2O-70 sample had a spectral peak with a low intensity at 492 nm. The Cu_2O-80, Cu_2O-90 samples had a spectral peak at 520 nm, in which the absorption peak of Cu_2O-90 the was higher than that of the Cu_2O-80. The Cu_2O-100 sample had an absorbance peak shifted to a wavelength position of 705 nm. The absorption peak positions at 492 nm and 520 nm were confirmed to be absorption peak of Cu_2O crystalline phase, and the absorption peak position at 705 nm were corresponding to CuO crystalline phase [1, 25, 27, 31]. These results to be agreed with the analysis raman spectra results above.

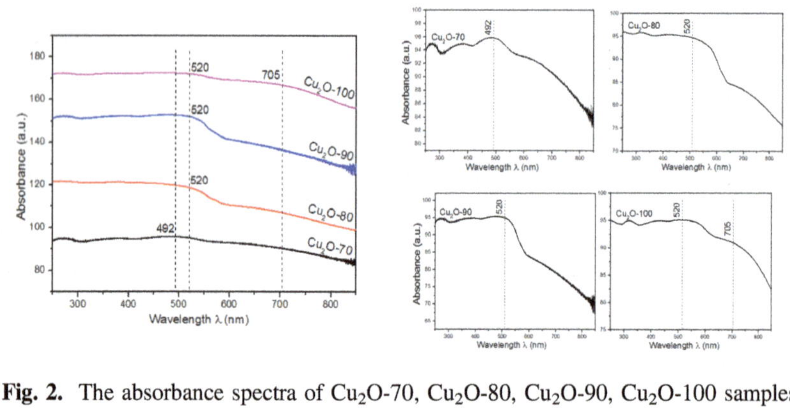

Fig. 2. The absorbance spectra of Cu_2O-70, Cu_2O-80, Cu_2O-90, Cu_2O-100 samples

Select the Cu_2O-90 sample which had the best crystallinity Cu_2O single phase to continue heating at high temperatures of 200, 300, 400, 500 °C (denoted as Cu_2O-200, Cu_2O-300, Cu_2O-400 and Cu_2O-500 samples, respectively). Followed by investigations on the influence of heating temperature on the structure, characterizations and crystallinity and the phase transition of Cu_2O crystals.

The Fig. 3a was the raman spectra of Cu_2O-200, Cu_2O-300, Cu_2O-400, Cu_2O-500 samples. The Fig. 3a showed that, the Cu_2O-200 sample had peaks of the vibration modes at 149, 220, 520 and 625 cm^{-1}, which attributed to the vibration modes of Cu_2O crystalline phase [10, 13, 22–24]. These samples had the vibration peaks shifted to a vibration position of the frequencies at 280, 292, and 340 cm^{-1} which were corresponding to CuO crystalline phase [10, 22–24]. This could be seen, the Cu_2O-200 sample retained the stability of the Cu_2O crystalline phase. But when the sample was calcined at the higher temperature to 300–500 °C, the Cu_2O crystalline phase in sample was partially converted to the CuO crystalline phase. The Fig. 3b presented the scanning electron microscope of Cu_2O-200, Cu_2O-300, Cu_2O-400 and Cu_2O-500 samples. The Fig. 3b showed that, Cu_2O-200 sample had the particles in the octahedral shape (Cu_2O-ONPs), the distribution of particles on the sample surface was uniform, particles sizes ranged in about ~1000–1600 nm. The Cu_2O-300, Cu_2O-400 samples showed that, the octahedral Cu_2O crystals had been slightly distorted, tended to stick together, cluster together and the large particles were about ~1200–2000 nm. The Cu_2O-500 sample showed that, the octahedral crystalline particles had broken apart, tended to agglomerated into a cluster and did not had a certain shape.

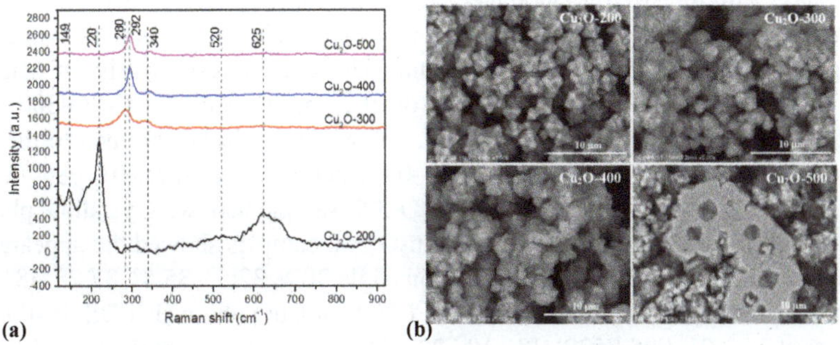

Fig. 3. (a)- The raman spectra of Cu_2O-200, Cu_2O-300, Cu_2O-400, Cu_2O-500 samples; (b)- The SEM of Cu_2O-200, Cu_2O-300, Cu_2O-400, Cu_2O-500 samples

The Fig. 4 was the solid Uv-vis absorbance spectra results of Cu_2O-200, Cu_2O-300, Cu_2O-400 and Cu_2O-500 samples. The absorbance spectra of Fig. 4 showed that, Cu_2O-200 sample had the absorption peak at ~506 nm, which were confirmed to be absorption peak of Cu_2O crystalline phase [1, 25, 27, 31]. The Cu_2O-300, Cu_2O-400 and Cu_2O-500 samples had the absorbance peak shifted to a wavelength position of ~706 nm, which were confirmed to be absorption peak of CuO crystalline phase [1, 25, 27, 31]. These were could seen that, the Cu_2O-200 sample retained the stability of the Cu_2O crystalline phase. The higher temperature to 300–500 °C, the samples had a partial phase transition from Cu_2O phase to CuO phase. The X-ray diffraction spectra results of the Cu_2O-90, Cu_2O-200, Cu_2O-300, Cu_2O-400 and Cu_2O-500 samples were presented in Fig. 5. The X-ray diffraction spectra (Fig. 5) showed that, the Cu_2O-90 and Cu_2O-200 samples had the diffraction peaks at $2\theta \approx 29.6°$, 36.5°, 42.45° and 62° which could be ascribed to the (110), (111), (200) and (220) planes.

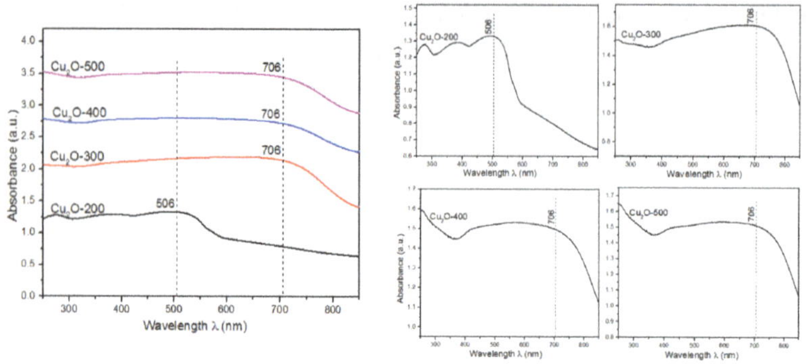

Fig. 4. The Uv-vis spectra of Cu_2O-200, Cu_2O-300, Cu_2O-400, Cu_2O-500 samples

These diffraction peaks were the octahedral Cu_2O phase (JCPDS 05–0667) [5, 6, 8, 31]. In addition, there had not obvious peaks of impurity as well as of CuO phase which were detected in the XRD spectra of the Cu_2O-90 and Cu_2O-200 samples. This indicated that, the Cu_2O-90 and Cu_2O-200 samples both obtained octahedral Cu_2O single phase with the size of nanometer crystalline particles (Cu_2O-ONPs). The Cu_2O-300 sample had the diffraction peaks at $2\theta \approx 29.6°$, 36.5°, 42.45° and 62° which were corresponding to the (110), (111), (200) and (220) planes, respectively, of octahedral Cu_2O phase (JCPDS 05–0667). Besides, on the spectrum of the Cu_2O-300 sample there had other peaks at $2\theta \approx 35.5°$, 37.2°, 48° which could be assigned to the (002), (111) and (-202) of CuO phase (JCPDS 45–0937 and ICDD card No.01–085-7520) [8, 10, 22, 23]. This demonstrated that, the Cu_2O-300 sample had two crystalline phases of Cu_2O and CuO. Next, the Cu_2O-400 and Cu_2O-500 samples showed the appearance of diffraction peaks at the position of diffraction angle $2\theta \approx 32.5°$, 35.5°, 37.2°, 48° which were corresponding to the (002), (111) and (-202) of CuO phase (JCPDS 45–0937 and ICDD card no. 01–085-7520) [8, 10, 22, 23]. These two samples still have low intensity diffraction peaks at the position of $2\theta \approx 62°$ corresponding to the (220) lattice plane of Cu_2O crystals. This proved that, two samples the Cu_2O-400 and Cu_2O-500 samples both had existed in the form of CuO crystalline phase and mixed with a small part as Cu_2O crystalline phase. The average crystalline particle size of the samples were calculated by the Debye-Scherrer equation: $D = k\lambda/\beta\cos\theta$ (1) [30, 31]. In which, D was the average crystalline particles size, k was the Scherrer's constant related to the shape of the crystals, λ was the X-ray diffraction wavelength, β was the full width at halfmaximum (FWHM), and θ was the diffraction angle at the position of the most intense diffraction peak of the crystalline phase for each fabricated sample (determined for the crystalline lattice planes Cu_2O (111) and CuO (111)). The calculated results the average crystalline particles size of the samples were: 16.9 nm; 18.4 nm; 20.05 nm; 18.8 nm; 19.3 nm. The optical band gap (Eg) of the samples was determined by Tauc's equation: $(\alpha h\nu)^2 = B(h\nu\text{-}Eg)$ (2) [8, 25, 27]. Where B is a constant, hν is the photon energy, Eg is the optical band gap.

The results of calculating the optical band gap (Eg) of the samples were shown in Table 1. From the results in Table 1 showed that, the optical gap band energy of the

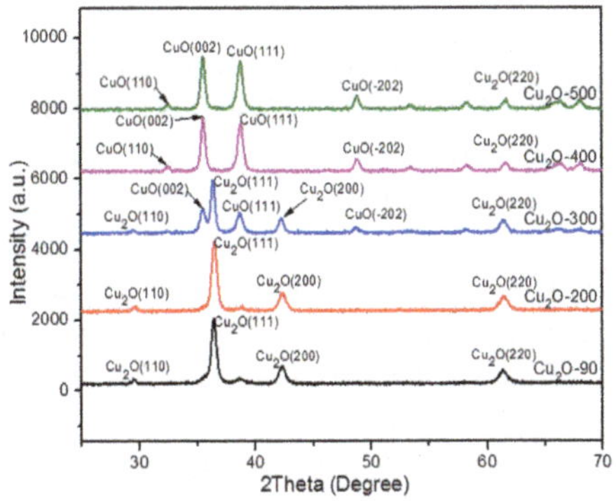

Fig. 5. The XRD of Cu_2O-90, Cu_2O-200, Cu_2O-300, Cu_2O-400, Cu_2O-500 samples

samples at the synthetic temperatures of 70–90 °C was in the range of 1.78–2.1eV. The optical gap band energy of the Cu_2O-200 sample reached 1.94eV. These optical gap band energy was smaller than that of bulk Cu_2O sample (2.2eV) [30]. Meanwhile, the band gap energy of the samples at the heating temperature of 300–500 °C was reduced to reach 1.37–1.46eV. This was also suitable for the phase conversion from the Cu_2O phase to the CuO phase at high heating temperatures of samples (the cause could be attributed to the CuO optical band gap being low, reaching around 1.2–1.4eV) [8, 34].

Table 1. The optical band gap (Eg) of the Cu_2O samples at the different synthetic temperatures of 70–100 °C and the different heating temperatures of 200–500 °C

Samples	Cu_2O-70	Cu_2O-80	Cu_2O-90	Cu_2O-100	Cu_2O-200	Cu_2O-300	Cu_2O-400	Cu_2O-500
λ (nm)	–	–	–	–	640	852	900	905
Eg (eV)	1.78	2.1	1.8	1.75	1.94	1.46	1.38	1.37

4 Conclusion

The cuprous oxide nanoparticles (Cu_2O-ONPs) have been successfully synthesized with the green precursors $CuSO_4.3H_2O$ and Na_2SO_3. The effects of the synthetic temperatures as well as the heating temperatures on the phase formation, the crystalline size development, the structure and the characterizations of the materials samples were investigated. The sample synthesized at 70°C only started the crystallization of spherical Cu_2O crystalline particles with the low intensity absorption peak at 492 nm with the particle sizes were about 200–300 nm. The Cu_2O-80, Cu_2O-90 samples had the well crystallization of the octahedral Cu_2O single crystalline phases (Cu_2O-ONPs) with the absorption peak at 520 nm and particle sizes were about 200–300 nm. In which, the Cu_2O-90 sample had the crystallization of octahedral crystalline particles the denser, the sizes were smaller than that of the Cu_2O-80 sample. The Cu_2O-100 sample had the octahedral Cu_2O crystals which were slightly distorted with two absorbance peaks at 520 and 705 nm with the large particles sizes ~500–700 nm and had been converted a portion of Cu_2O crystalline phase to the CuO crystalline phase.

The Cu_2O-200 sample retained the stability of the Cu_2O crystalline phase with the absorption peak at ~506 nm and had the nanoparticles in the octahedral shape (Cu_2O-ONPs) with sizes of ~1000–1600 nm. The Cu_2O-300, Cu_2O-400, Cu_2O-500 samples had the octahedral Cu_2O nanoparticles slightly distorted, tended to cluster together into large particles in size of ~1200–2000 nm with the absorbance peak shifted to a wavelength position of ~706 nm. The Cu_2O crystalline phase in these samples was partially converted to the CuO crystalline phase. The optical band gap energy of the samples at the synthetic temperatures of 70–90 °C was in the range of 1.78–2.1eV. The optical band gap energy of the Cu_2O-200 sample reached 1.94eV. These optical band gap energy values were smaller than that of bulk Cu_2O sample (2,2eV). The band gap energy of the samples at the high heating temperatures of 300–500 °C was reduced to reach 1.37–1.46eV which were suitable for the phase conversion from the Cu_2O phase to the CuO phase.

Acknowledgments. This work was funded by Hanoi University of Science and Technology (HUST) under the project number T2018-PC-227.

References

1. Sawant, S.S., Bhagwat, A.D., Mahajan, C.M.: J. Nano Elec. Phys. **8**(01035), 5 (2016)
2. Xiong, L., Yu, H., Ba, X., Zhang, W., Yu, Y.: App. and Mechanism, Nanomater for Environ. Protection. Wiley, Hoboken (2014)
3. Gao, Y., Wu, Q., Liang, X., Wang, Z., Zheng, Z., Wang, P., Liu, Y., Dai, Y., Whangbo, M. H., Huang, B.: Adv. Sci. **1902820**, 1–7 (2020)
4. Kim, M.C., Kim, S., Han, S.B., Kwak, D.H., Hwang, E.T., Kim, D.M., Lee, G.H., Choi, H. S., Park, K.W.: J. of Mater. Chem. A **1–3**, 1–9 (2013)
5. Li, S., Lin, Q., Liu, X., Yang, L., Ding, J., Dong, F., Li, Y., Irfana, M., Zhang, P.: Roy. Soc. Chem Adv. **8**, 20277–20286 (2018)
6. Zou, X., Fan, H., Tianb, Y., Yan, S.: Royal Soc. Chem. **16**, 1149–1156 (2014)
7. Helaïli, N., Bessekhouada, Y., Bouguelia, A., Trari, M.: J. Hazar. Mater. **168**, 484–492 (2009)
8. Mikami, K., Kido, Y., Akaishi, Y., Quitain, A., Kida, T.: Sensors **19**(211), 1–14 (2019)
9. Lan, T., Fallatah, A., Suiter, E., Padalkar, S.: Sensors **17**(1944), 1–10 (2017)
10. Cretu, V., Postica, V., Mishra, A.K., Hoppe, M., Tiginyanu, I., Mishra, Y.K., Chow, L., de Leeuw, N.H., Adelungc, R., Lupan, O.: J. Mater. Chem. A **4** (2016) 6527–6539
11. Deng, X., Wang, C., Shao, M., Xu, X., Huang, J.: Royal Soc. Chem. Adv. **7**, 4329–4338 (2017)
12. Liu, C., Zhou, X., Chen, S., Zhao, X., Dai, S., Xu, B.: Adv. Sci. **6**(1801169), 9 (2019)
13. Kardarian, K., Nunes, D., Sberna, P.M., Ginsburg, A., Keller, D.A., Pinto, J.V., Deuermeier, J., Anderson, A.Y., Zaban, A., Martins, R., Fortunato, E.: Solar Energy Mater. Solar Cells **147**, 27–36 (2016)
14. Kumara, B.R., Hymavathi, B., Rao, T.S.: Mater. Today Proc. **4**, 3903–3910 (2017)
15. Sun, W., Sun, W., Zhuo, Y., Chu, Y.: J. Solid State Chem. **184**, 1638–1643 (2011)
16. Suzuki, K., Tanaka, N., Ando, A., Takagi, H.: Am. Ceram. Soc. **94**(8), 2379–2385 (2011)
17. Yu, X., Huang, L., Wei, Y., Zhang, J., Zhao, Z., Dai, W., Yao, B.: Mater. Res. Bull. **64**, 410–417 (2015)

18. Xiong, L., Yu, H., Nie, C., Xiao, Y., Zeng, Q., Wang, G., Wang, B., Lv, H., Lia, Q., Chen, S.: Roy. Soc. Chem. **7**, 51822–51830 (2017)
19. Xu, L., Xu, H., Wu, S., Zhang, X.: App. Surface Sci. **258**, 4934–4938 (2012)
20. Janczarek, M., Kowalska, E.: J. Catal. 7(317), 1–26 (2017)
21. Liua, Y., Xueb, J., Chic, C., Liang, W., Li, G.: Energy Environ. Mater. **743–744**, 623–628 (2013)
22. Dharmalingam, K., Bordoloi, D., Kunnumakkara, A.B., Anandalakshmi, R.: J. App. Polymer Sci. **e49216**, 1–14 (2020)
23. Deng, Y., Handoko, A.D., Du, Y., Xi, S., Yeo, B.S.: Am. Chem. Soc. Catal. **6**, 2473–2481 (2016)
24. Balkanski, M., Nusimovici, M.A., Reydellet, J.: Solid State Commun. **7**, 815–818 (1969)
25. Yang, M., Zhu, J.J.: J. Cryst. Growth **256**, 134–138 (2003)
26. He, P., Shen, X., Gao, H.: J. Colloid Interface Sci. **284**, 510–515 (2005)
27. Singha, M., Jampaiaha, D., Kandjani, A.E., Sabri, Y.M., Gaspera, E.D., Reineck, P., Judd, M., Langley, J., Cox, N., Emdben, J.V., Mayes, E., Gibson, B.C., Bhargava, S.K., Ramanathan, R., Bansal, V.: Roy. Soc. Chem. **1–3**, 1–3 (2017)
28. Wang, X., Liu, D., Li, J., Zhen, J., Zhang, H.: NPG Asia Mater. 7(e158), 7 (2015)
29. Fortunato, E., Figueiredo, V., Barquinha, P., Elamurugu, E., Barros, R., Gonçalves, G., Park, S.H.K., Hwang, C.S., Martins, R.: App. Phys. Lett. **96**(192102), 3 (2010)
30. Khan, M.A., Ullah, M., Iqbal, T., Mahmood, H., Khan, A.A., Shafique, M., Majid, A., Ahmed, A., Khan, N.A.: Nanosci. Nanotech. Res. **3**(1), 16–22 (2015)
31. Rastabi, S.A., Moghaddam, J., Eskandarian, M.R.: J. Ind. Eng. Chem. **22**, 34–40 (2015)
32. Shahmoradi, Y., Souri, D.: Ceram. Int. **45**, 7857–7864 (2019)
33. Wang, S., Lian, J.S., Zheng, W.T., Jiang, Q.: App. Surface Sci. **263**, 260–265 (2012)
34. Zaman, S., Zainelabdin, A., Amin, G., Nur, O., Willander, M.: J. Phys. Chem. Solids **73**, 1320–1325 (2012)

An Atomistic Study of Indentation and Scratching Characteristics of Cu_xZr_{100-x} Metallic Glasses

Anh-Son Tran(✉)

Faculty of Mechanical Engineering, Hung Yen University of Technology and Education, Khoai Chau District, Hung Yen, Vietnam

Abstract. In present study, the combination of the indentation and scratching process was developed to analyze the mechanical behaviors of the CuZr amorphous alloys (MGs) using molecular dynamics (MD) simulation. The mechanical behaviors of the Cu-Zr MGs are appraised through the surface morphology, pile-up height, hardness, machining forces, resistance coefficient, and radial distribution function (RDF) diagram. The influences of component concentration in the Cu-Zr MGs are investigated. The results exhibit the significant differences in mechanical properties of the Cu-Zr MGs as changing the percentages of the Cu and Zr element.

Keywords: Cu-Zr MGs · Indentation · Scratching · Resistance coefficient

1 Introduction

Recently, the material sciences and technologies are developing very strongly, many new materials types with many novel characteristics and special properties have been discovered such as composite, metallic glasses (MGs), graphene, amorphous polyethylene, and so on. Among them, MGs reveal outstanding mechanical characteristics such as superior corrosion resistance, very good strength, and great hardness [1–3]. So, MGs attracted great attention from researchers.

Deformation behaviors and mechanical properties are very typical and exceedingly important factors used to evaluate the characteristics of the materials. These factors directly influence the workability of the materials. Therefore, the investigation and evaluation on the mechanistic characteristics of materials is very necessary. With the strong development of computer technology, molecular dynamics (MD) simulation method is an appropriate choice in simulating and evaluating the properties of nanomaterials. Nowadays, the Cu-Zr MGs system has appeared as the promising immiscible alloy type for applications in electrical engineering, magnetic- sensing, chemical, and structural materials [4]. The indentation and scratching processes are usually performed to explore the material's deformation phenomena and properties, however, the combination of these two processes is scarce, especially with Cu-Zr MGs.

By the content mentioned above, the mechanical dynamic responses properties of the Cu-Zr MGs systems are analyzed and assessed via the combination of indentation and scratching processes using MD simulation in this work. The machining processes are simulated under different testing conditions in the portion of the Cu-Zr component.

B. T. Long et al. (Eds.): MMMS 2020, LNME, pp. 384–388, 2021.
https://doi.org/10.1007/978-3-030-69610-8_54

2 Methodologies

Figure 1 shows the machining system model for the machining process. The machining system includes a Cu_xZr_{100-x} MGs specimen and a sphere diamond indenter.

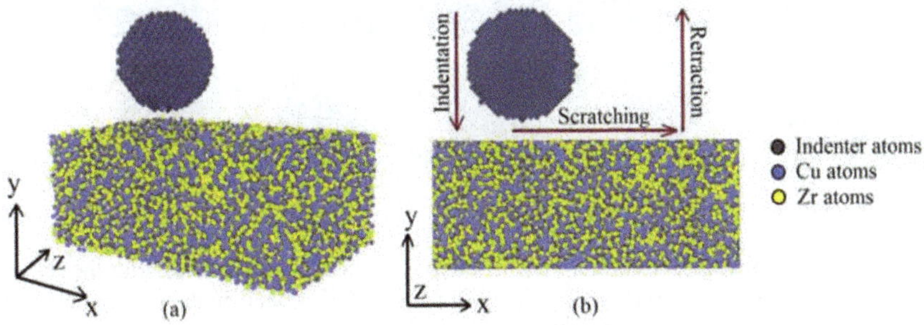

Fig. 1. The Cu_xZr_{100-x} MGs model system for the the machining process.

The machining process is divided into three stages including indentation, scratching, and retraction. The indenter is considered as an ideal rigid body with the radius of 2.5 nm. The dimensions of the Cu_xZr_{100-x} MGs specimen respectively are 15 nm (length) × 6 nm (height) × 10 nm (width) corresponding to x-, y-, and z-axis. The free boundary is specified along the y-axis. The x-, and z-axis are assigned the periodic boundary conditions. In all MD simulations, the NVT (canonical ensemble) is used. The initial distance between the surface of the specimen and the indenter is 1 nm. The machining process begins by the indentation stage with a machining depth of 2 nm and the indentation velocity of 50 m/s along the y-axis. Then, the scratching stage is performed with a scratch distance of 5 nm and a scratch velocity of 50 m/s along the x-axis. Finally, the indenter retracts to the original position at a retraction velocity of 100 m/s.

In present work, the Cu-Zr atomic interactions are depicted using the EAM potential [5]. The Lennard-Jones (LJ) potential is enrolled to delineate the atomic interaction of the indenter and Cu-Zr MGs [6]. The C-C atomic interaction of the indenter is ignored due to the indenter is considered as a rigid body.

The series of the testing simulations are executed through the LAMMPS simulator. Besides, the OVITO software is employed to represent the data obtained through the MD simulations.

3 Results and Discussion

Five Cu_xZr_{100-x} samples are prepared for the machining process at a temperature of 300 K, a depth of 2 nm, and a machining speed of 50 m/s for the purpose of investigating the effect of the component concentrations on the mechanical behaviors.

Figure 2 shows the lateral cross-sectional-view of the pile-up and groove formed after the retraction of the indenter (a) and the surface morphology (b) of the Cu_xZr_{100-x} MGs in the machining process.

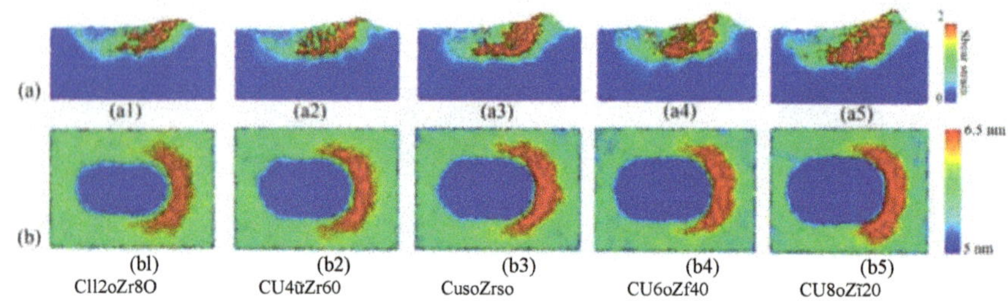

Fig. 2. (a) The lateral cross-sectional-view of the pile-up and groove formed after the retraction of the indenter and (b) the surface morphology of the Cu_xZr_{100-x} MGs in the machining process.

The significant changes in deformation behavior are obviously observed with the changing composition of Cu and Zr components. The height of the chip increases and the deformed area of the groove is shaped more broadly as increasing the Cu component, which can be seen in Fig. 2(a1)–(a5). The significantly larger surface morphology is observed in Fig. 2(b1)–(b5) to confirm the strong influence of Cu components in the Cu_xZr_{100-x} MGs system. So, the substrate is more susceptible to deformation and the percentage of high shear strain atoms increases when the Cu content increases. The maximum pile-up height and hardness of the Cu_xZr_{100-x} MGs systems in the machining process are shown in Fig. 3.

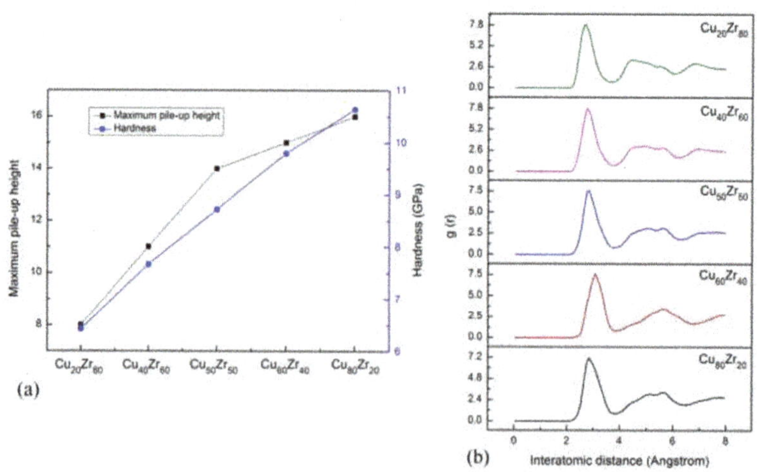

Fig. 3. (a) The maximum pile-up height and hardness and (b) the RDF diagram of the Cu_xZr_{100-x} MGs in the machining process.

The maximum pile-up heights are 8, 11, 14, 15,and 16 Â; the hardness values are 6.43, 7.67, 8.72, 9.80, and 10.64 GPa corresponding to component of Cu values are 20,40, 50, 60 and 80%, respectively. As increasing Cu component in the Cu_xZr_{100-x} MGs systems, the brittleness of the material prevails, leading to the material becomes more rigid and the resilience of the material becomes more difficult during the machining process. In addition, the radius of the Cu atom is smaller than that of Zr atom, resulting in the structure of the Cu_xZr_{100-x} MGs systems become more sustainable as increasing the Cu component. So, the deformed area of groove, the maximum pile-up height and the hardness value of the Cu_xZr_{100-x} MGs system rise up as the Cu component increases.

The RDF diagram supplies valuable knowledge to prove the effectiveness of the different element contributions about the material's constructional stability. The peak of RDF value is the highest for the $Cu_{20}Zr_{80}$ specimen and is the lowest for the $Cu_{80}Zr_{20}$ specimen. Meanwhile, the $Cu_{50}Zr_{50}$ MGs represents the intermediate value.As we can see, the $Cu_{20}Zr_{80}$ MGs specimen possesses the highest RDF peak and the lowest hardness value. In contrast, the lowest RDF peak and the highest hardness value belongs to the $Cu_{80}Zr_{20}$ MGs specimen. These mean that as the Cu component concentration increases, the specimen becomes harder and more crunchy, while the specimen seems to be relatively more flexible and softer with the decrease of the Cu component concentration. The $Cu_{50}Zr_{50}$ MGs specimen exhibits the mechanical behavior that lies in between both specimens. Imran et al. [7] also found the similar result with the Cu-Zr bulk MGs specimens under the tensioning process in the front study.

Figure 4 shows the normal (a) and tangential force diagram (b) of the Cu_xZr_{100-x} MGs during the indentation, scratching, and retraction process.

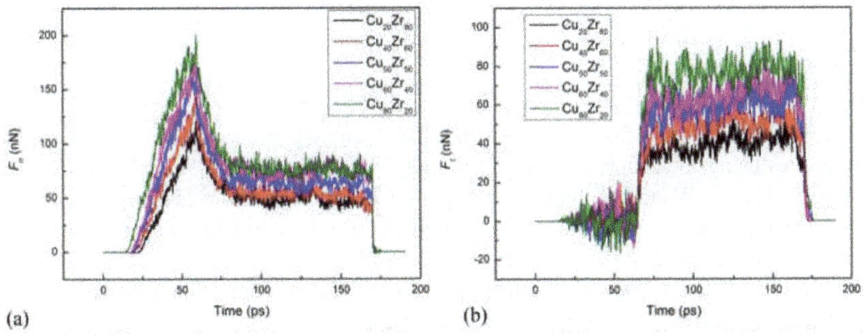

Fig. 4. (a) Normal and (b) tangential force diagram of the Cu_xZr_{100-x} MGs during the indentation, scratching, and retraction process.

Based on the reasons explained above, it is easy to understand that the changing trend of both *Fn* and *Ft* of the Cu_xZr_{100-x} MGs in machining processes with changing the components of Cu and Zr. The machining force increases linearly as increasing the Cu component in the Cu_xZr_{100-x} MGs. The proportional increase between *Fn* and *Ft* causes the values of resistance coefficient of the Cu_xZr_{100-x} MGs under the scratching process seem to be slightly different with the changes in the Cu component distribution, which is observed in Fig. 5.

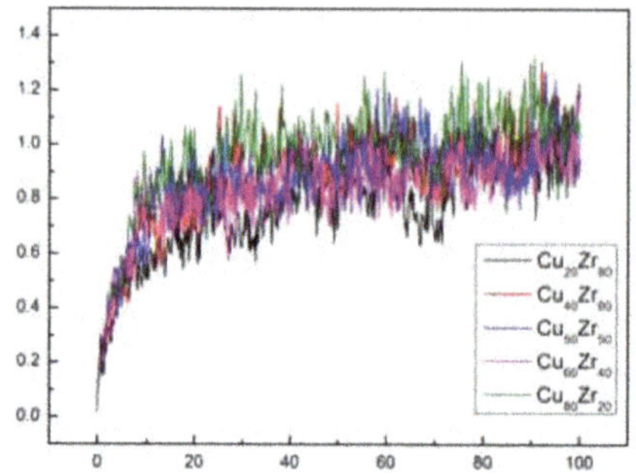

Fig. 5. Resistance coefficient of the Cu_xZr_{100-x} MGs under the scratching process. The values of *Fn* and *Ft* change markedly, however, the resistance coefficient slightly increases with the increase of the Cu component. That means the changes of Cu and Zr components in Cu_xZr_{100-x} MGs strongly affect the *Fn* and *Ft* but slightly affect the resistance coefficient.

4 Conclusions

The deformed area of the groove, the maximum pile-up height, the hardness value, machining force of the Cu_xZr_{100-x} MGs system rise up as the Cu component increases, however, the resistance coefficient slightly increases. The $Cu_{80}Zr_{20}$ substrate is relatively harder and more crunchy, the $Cu_{20}Zr_{80}$ substrate is more flexible and softer.

References

1. Tran, A.S., Fang, T.H.: Size effect and interfacial strength in nanolaminated Cu/Cu_xTa_{100-x} composites using molecular dynamics. Comput. Mater. Sci. **184**, 109890 (2020)
2. Liang, X., He, A., Wang, A., Pang, J., Wang, C., Chang, C., Liu, C.T.: Fe content dependence of magnetic properties and bending ductility of FeSiBPC amorphous alloy ribbons. J. Alloy. Compd. **694**, 1260–1264 (2017)
3. Wu, F.F., Chan, K.C., Jiang, S.S., Chen, S.H., Wang, G.: Bulk metallic glass composite with good tensile ductility, high strength and large elastic strain limit. Sci. Rep. **4**(1), 1–6 (2014)
4. Inoue, A., Nishiyama, N.: New bulk metallic glasses for applications as magnetic- sensing, chemical, and structural materials. MRS Bull. **32**(8), 651–658 (2007)
5. Mendelev, M.I., Kramer, M.J., Ott, R.T., Sordelet, D.J.: Molecular dynamics simulation of diffusion in supercooled Cu-Zr alloys. Phil. Mag. **89**(2), 109126 (2009)
6. Lin, T., Bian, X.F., Jiang, J.: Relation between calculated Lennard-Jones potential and thermal stability of Cu-based bulk metallic glasses. Phys. Lett. A **353**(6), 497–499 (2006)
7. Imran, M., Hussain, F., Rashid, M., Cai, Y., Ahmad, S.A.: Mechanical behavior of Cu-Zr bulk metallic glasses (BMGs): a molecular dynamics approach. Chin. Phys. B **22**(9), 096101 (2013)

Determining the Stability Chart of the High-Speed CNC Milling Machine Based on the Vibration Criterion

Phan Van Hieu(✉)

School of Mechanical Engineering, Hanoi University of Science and Technology, Hanoi, Vietnam
hieu.phanvan@hust.edu.vn

Abstract. In the field of mechanics, improve product quality and productivity Cutting is a requirement to be placed on top. To improve productivity tools but still ensure product quality, we have two approaches: Improving the quality of cutting tools such as the use of tools coated.., or research to build a sustainable chart of processes, thus helping to select the best cut mode. This paper presents the research for constructing the stability chart of the High-Speed CNC Milling Machine based on vibration criteria. The object was shaped plane Details, C45 steel, the experiments are performed on a High-Speed CNC Milling Machine "Super MC 500", the cutting tool is hard alloy milling cutters (NACHI GS MILL 4 GS 10 -LIST9384), 4 bits of Japan, data is measured experimentally on a three-dimensional vibration measuring device (BR 4513B-002). The author builds a stability chart experimentally to show the correlation between the spindle rotation speed, axial depth of cut with vibration amplitude.

Keywords: Stability lobe diagram · Vibration · High-Speed milling · Axial depth of cut · Spindle rotation speed

1 Introduction

Scientists have shown that during milling, vibration can be generated from a combination of cutting speed and depth of cut (Fig. 1). The stability chart is a function of the spindle rotation speed with shaft cutting depth used to make decisions about the cutting mode to avoid vibrations when the feed rate increases [1, 2].

With the help of the stabilization chart, one can select the cutting modes to perform cutting operations with free vibration. They are interested in studying the methods of stabilization available especially for many machine combinations, jigs, cutting tools, and detailed materials.

Vibration during machining includes forced vibration, individual vibration, and auto vibration. Thereby finding that vibration is an intermediate factor generated while processing affects the surface quality of the part for good control and optimization of cutting parameters affecting the quality of machining surfaces. The study of the effect of cutting mode on vibrations during high-speed milling is necessary and is the basis for constructing the stability chart of high-speed three-axis CNC milling with vibration criterion [3–5].

B. T. Long et al. (Eds.): MMMS 2020, LNME, pp. 389–395, 2021.
https://doi.org/10.1007/978-3-030-69610-8_55

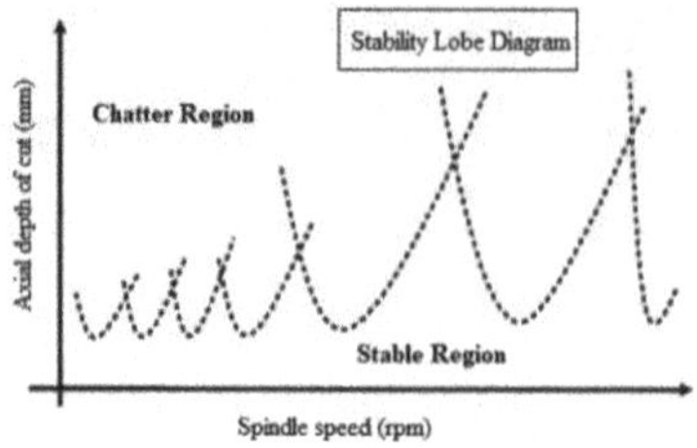

Fig. 1. Stability chart

This article will address an empirical approach to constructing a stabilization chart, thereby helping us see the difference between steady and unstable cutting modes and choosing a dark combination. The area under investigation is divided into some points along with a network, and we obtain a vibration chart by taking measurements at the points identified in this evaluation area. The remaining score is calculated using interpolation. For milling, the stability chart is relative to the areas on the vibration chart. If the stability chart is considered an area, then it can be divided into small regions with an element grid, and we define the stability chart by applying vibration charting. This approach will be verified by comparing the stability chart generated by the vibration chart with the stability chart generated by the intermediate values of the tap test method [6, 7].

2 Determine the Stability Chart of the "Super MC 500" High-Speed CNC Milling Machine Based on Vibration Criteria

2.1 Vibration Mapping to Determine the Stability Chart During Milling

In a milling operation, excess metal is removed from the workpiece by rotating the tool and translating it into the toolpath at a certain speed (Fig. 5). Blade collision stimulates the machine tool/tool holder/cutter/part system, thereby generating vibrations. A short initial period will determine whether the process is stable or not (Fig. 2).

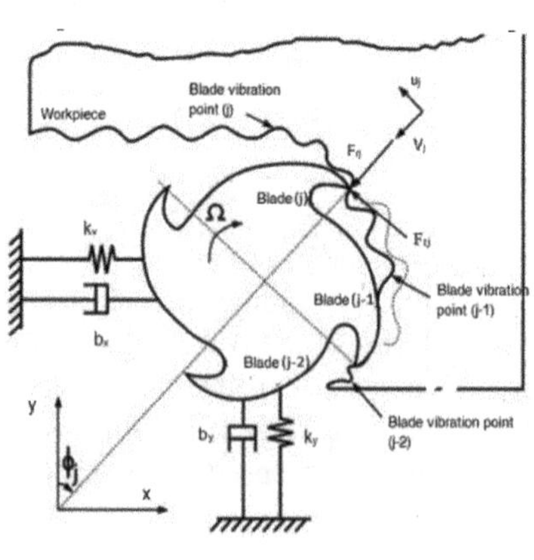

Fig. 2. Stable survey model

Stable or unstable machining is closely related to vibration amplitude. If the machining mode is not stable, then the vibration chart will not converge (amplitude of vibration becomes larger with time). If the machining mode is stable, the vibration chart will converge (vibration amplitude will decrease with time) (Figs. 3 and 4).

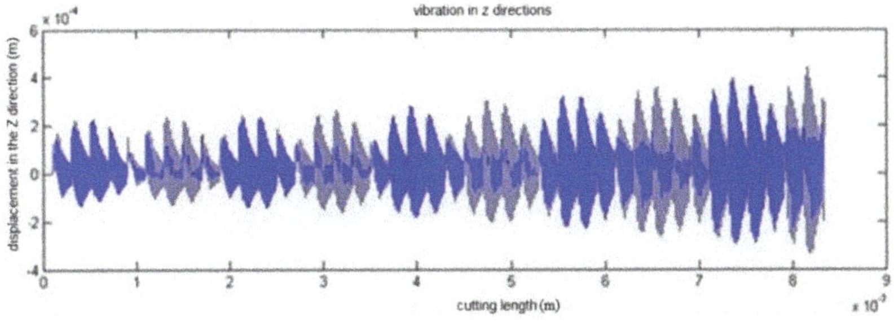

Fig. 3. Vibration chart with unstable machining mode

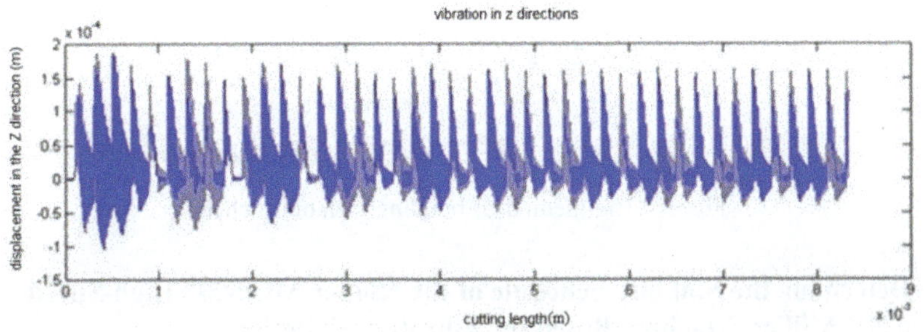

Fig. 4. Vibration chart with steady machining mode

So we can rely on the amplitude of vibration to evaluate whether the machining mode is stable or unstable.

Vibration graphs are constructed using two simple methods: by sampling, through analysis of the resulting vibration amplitude, and then digitizing these analyzes or by simulating. Vibrate again in a certain area. They are all based on performing calculations thanks to the advancement of electronic computers, and both have the advantage of minimizing costs for data acquisition and saving time in building vibration charts.

In the confines of the milling task and to focus on the problem of vibration, if you consider the stability chart as an area and the area is divided into multiple meshed regions, a graph can be constructed using the vibration charting method. The graph shows the boundary of stability where the horizontal axis represents the rotation speed of the spindle, and the vertical axis is the radius of the depth of cut. In this study, the stability chart was determined from 190 points on a grid consisting of 19 rotational speeds and 10 different depths of cut. Vibration amplitudes were analyzed, and the frequencies of vibration were determined. The above method is shown in Fig. 5. Upstairs. On the chart, it is not difficult to observe a stable border and then to choose the right set of technological parameters for the milling process [7].

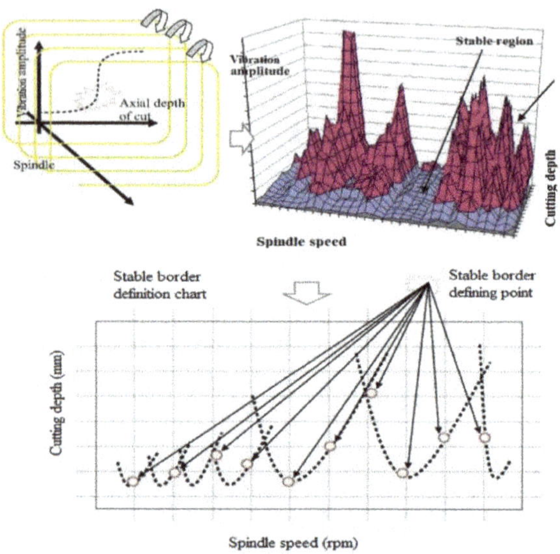

Fig. 5. The method of building a stability chart

2.2 Determine the Stability Schedule of the "Super MC 500" High-Speed CNC Milling Machine Based on Vibration Criterion

The objective of the experiment is to determine the stability chart of the high-speed CNC milling machine by measuring the vibration in the directions when machining with each cutting mode. After obtaining the above-measured parameters, based on the theory of milling process stability presented above, the author will proceed to build an experimental stability chart (Fig. 6).

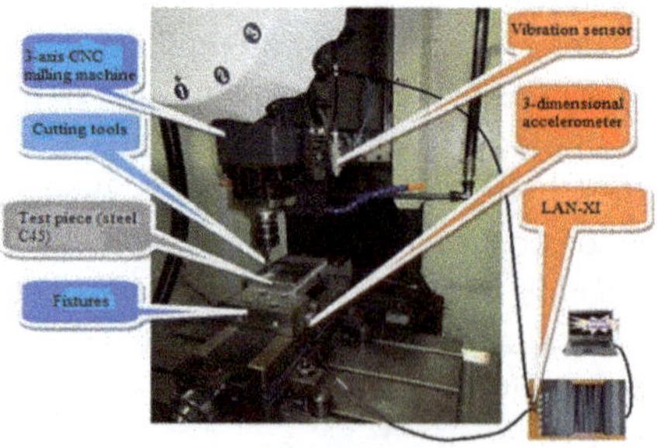

Fig. 6. Experimental model

2.2.1 Experimental Model

The experiment was carried out under the following conditions:

Processing machines: CNC high-speed milling machine Super MC 500 - Spindle speed from 100 ÷ 20000 rpm; BT 30; FANUC control system.

Cutting tools: Hard alloy milling cutter (NACHI GS MILL 4 GS 10-LIST9384), 10 mm, 4 teeth, Cutting tool holder length: 40 mm; Total length: 70 mm; section length: 22 mm; can cut steel with hardness 50–55 HRC.

Test samples: 10 samples of C45 steel with the size 40 × 100 × 250, mounted on a vise.

Microphone: Vibration sensor (microphone 4189) mounted in the engine room.

Sensor for vibration measurement: 3-dimensional accelerometer.

LAN-XI: Data acquisition and signal amplification equipment with 4 inputs, 2 outputs up to 51.2 Kz of the Danish company.

2.2.2 Stability Chart of the Surveying Machine System

The author conducted testing different cut modes. The number of experimental cuts is 190.

Traverse along the width of the workpiece with the same depth and change the speed per turn, increases the speed to 1000 (rpm).

For each cutting thickness, we change the rotation speed from 2000 to (rpm), with the step change of 1000 (rpm) (19 speeds).

With each knife, we change the cutting thickness from 0.15 mm with the step change of 0.15 mm each, to the cutting thickness of 1.5 mm. (10 cutting depths).

With one cutting thickness, cutting until severe tool wear or breakage occurs, stop and move on to machining at the next cutting thickness. Develop a stable chart of the cutting process on "SUPER MC 500" by experiment (Figs. 7 and 8).

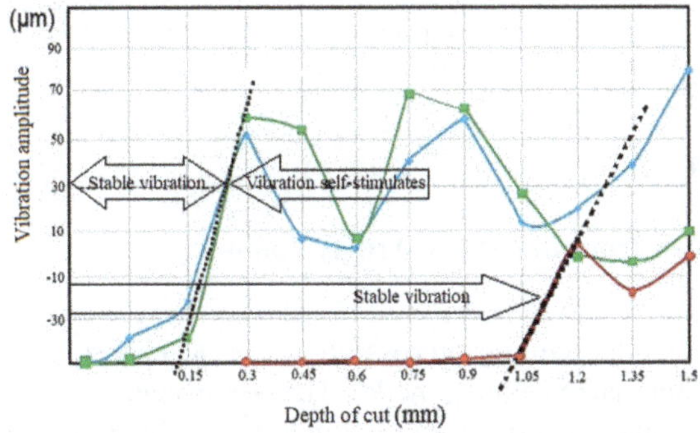

Fig. 7. An experiment of an unstable cutting operation

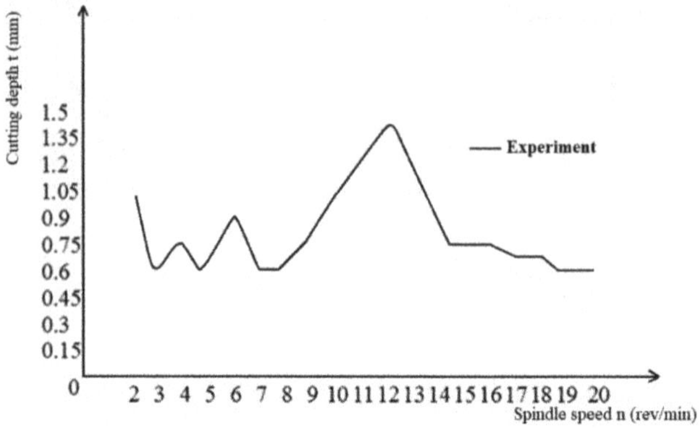

Fig. 8. Construction stability chart by experiment

3 Discussion

From the stability chart of the cutting process on the "SUPER MC 500" high-speed milling machine, we have the following comments:

When processing C45 steel material with a finger mill ngón10; 4 cutting teeth, the tooling material is a hard alloy (NACHI GS MILL 4 GS 10-LIST9384), with a cantilever length of 40 mm on the Super MC 500 Maximum depth of cutting the machine can cut. The machining mode for the highest productivity and surface quality is guaranteed at n = 12000(rpm) and cutting depth t = 1.5 mm.
With the "SUPER MC 500" machine, the actual machining mode cuts the maximum milling depth of 0.5 mm. So studying the stability chart of the machining process, we have chosen the cutting mode (reasonable cutting speed) that the depth of cutting reaches 1.5 mm, helping us to increase the cutting productivity of the machine by three times. But the surface quality of the workpieces is still guaranteed.

4 Conclusion

Conclusions Drawn from the Study Are as Follows

Conclusion 1

The stability chart of the cutting process allows us to choose the most appropriate machining mode (optimal machining mode). The best machining mode must satisfy the following requirements: great depth of cut (high productivity), quality of workpieces, and physical phenomena such as vibration, cutting heat, and shearing force.

Conclusion 2

In this paper, the information about the vibration signal of the milling process is analyzed to build a stable chart. The timed vibration signal is analyzed off-line to determine the frequency of vibrations, and a vibration map is built for the milling process inspection. The purpose is to obtain a stability chart. The analysis of the frequency and amplitude of the milling vibrations, based on the fast Fourier transformation of the vibration signal over time, has yielded good results and allows us to achieve further directions. A precision approach to understanding milling problems through vibration.

Conclusion 3

The biggest drawback of this method is that it takes a lot of experiments, and it takes time and money to gather enough data to build a stable chart through table vibration. Once the stability chart has been completed, it is possible to choose an appropriate cutting mode to prevent vibration for a milling task. Extensive testing, as just presented, has corroborated the results of many previous surveys, to this day still topical and needs further attention and research.

References

1. Altintas, Y., Budak, E.: Analytical prediction of stability lobes in milling. CIRP Ann. Manuf. Technol. (1995)
2. Al - Regib, E., Ni, J., Lee, S.: Programming spindle speed variation for machine tool chatter suppression. Int. J. Mach. Tools Manuf. **43**(12), 1229–1240 (2003)
3. Gradis, J.: On stability prediction for milling. Int. J. Mach. Tools Manuf. **45** (2005). Elsevier Ltd
4. Faassen, R.: Chatter Prediction and Control for High-Speed Milling. University Press Facilities, Eindhoven (2007)
5. Long, B.T., Phan, V.-H., Sinh, H.V.: Study on parameters affecting machining stability on high-speed CNC milling "DNM400". Vietnam Mech. J. (2010)
6. Long, B.T., Phan, V.-H., Sinh, H.V.: Development of method and measurement diagram for determination of hardness "K" and damping coefficient "ς" of high-speed CNC milling machines. Vietnam Mech. J. (2011)
7. Long, B.T., Phan, V.-H., Sinh, H.V.: Constructing stability chart of high-speed milling machine "DNM400" based on surface wave strength criteria by Matlab software. Vietnam Mech. J. (2011)

Study of the Effect of Friction on the Errors of the Stopping Position of the Automatic Tool Changer on CNC Machine Under Variable Relative Humidity Conditions

Pham Van Hung, Nguyen Van Khoa, Nguyen Van Quyen, and Nguyen Thuy Duong(✉)

School of Mechanical Engineering, Ha Noi University of Science and Technology, No. 1 Dai Co Viet Stress, Hai Ba Trung District, Hanoi City 100000, Vietnam
duong.nguyenthuy@hust.edu.vn

Abstract. An automatic tool changer (ATC) is an important module on computer numerical control (CNC) machine, it determines the level of automation and flexibility of the machine. Accurate stopping at the position of the ATC during the tool change process would minimize the damage of the taper hole and ensure the machining quality of the product. In a CNC machine, the motion of ATC to the tool change position is performed by a pneumatic cylinder. Friction, in which its behavior is complex and depends on many factors such as pressure, velocity lubrication, etc., always embarrasses controlling of the position of the ATC. This paper presents the effect of friction on the stopping position error of the ATC on the CNC machine under conditions of relative humidity (RH) varying from 51% to 99%. The study results showed that the stopping position error of the pneumatic cylinder depends on the number of tools stored in the ATC and tends to increase when the relative humidity and the travel speed increased.

Keywords: Position control · Pneumatic positioning system · Friction · Humidity relative · Automatic tool change · CNC machine

1 Introduction

CNC machine is a modern mechatronics system which is commonly used in mechanical manufacturing with many advantages such as high precision, machining quality stable, short preparation and processing time, high economic efficiency, etc. The most important part constitutions of the CNC machine would be ATC, it determines the level of automation and flexibility of the machine. The movement carrying the ATC into the tool change area is often performed by a pneumatic cylinder (as shown in Fig. 1). During the automatic tool change processed, the concentric of the tool holder to the spindle center line must be maintained. The stopping position error of the pneumatic piston during the movement of the tool charging would directly produce the concentric error of the tool holder to the spindle center line. This concentric error is

B. T. Long et al. (Eds.): MMMS 2020, LNME, pp. 396–404, 2021.
https://doi.org/10.1007/978-3-030-69610-8_56

one of the main reasons of wearing and collision of the main spindle taper hole in CNC machine (as shown in Fig. 1b, c), therefore reducing the machining accuracy [1, 2].

The sliding friction in a pneumatic cylinder embarrasses the control of the precise position and stability of the tool changer. In order to reduce the wear and collision of the spindle taper hole, and to reduce the radial spin imbalance, the improvement of the movement and the precise control of the position of the tool changer is necessary basing on the friction property in the pneumatic cylinder.

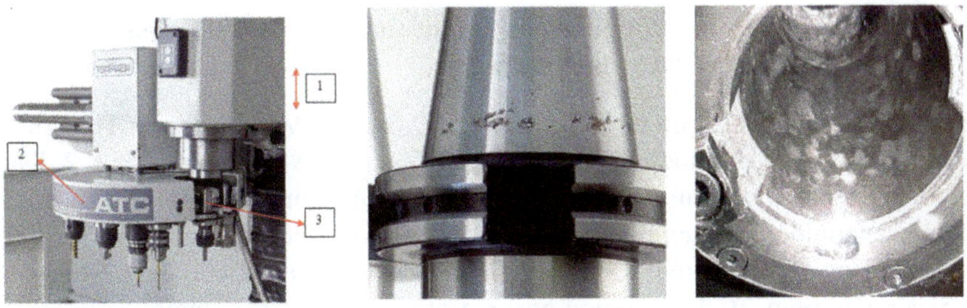

a. b.

Fig. 1 a. ATC of a CNC machine. b. Surface damages of the tool holder and the spindle taper hole.

Many previous studies presented the friction property in pneumatic cylinders, however it mainly focused on a single factor such as pressure (p), velocity (v), lubrication material, material of gasket sealing, and friction force. In fact, the environment is also a significant factor in friction property of the pneumatic cylinder, particularly, the humid tropical climate environment in Vietnam. The previous studies indicated that the friction properties depend on the humidity change, such that, the friction reduces when the relative humidity increases [3–8]. Studies on the influence of the humidity on the friction in pneumatic cylinder indicated that the friction force reduces when the RH increased from 51% to 99%, and the friction force was experimentally identified as a formula of RH and velocity v [9, 10].

Some current studies aimed to enhance the accuracy position of the pneumatic cylinder, as Bashir M.Y. Nouri [11, 12], indicated that the reason of the difficulty in the position controlling of the pneumatic cylinder is the friction force during working. Therefore, the modeling of the positional stepless control system necessarily includes the friction force during the pre-sliding and the complete sliding process. Kuo-Ming Chang et.al. Introduced the fuzzy controller with the friction compensation which applied for the accurate position control of the pneumatic cylinder with the position accuracy less than 30 nm [13]. Tran Xuan Bo et al. indicated that the simulation results of the motion of the pneumatic cylinder from the revised LuGre model (RLG model) were better than the results from the steady-state friction model (SS model) and the LuGre model (LG model) [14]. To improve the position accuracy, the control problem including the friction Leuven, LG, and RLG models which depended mainly on p and v, however, the environmental factor commonly has not been mentioned. Therefore, the influence of the environment with high humidity on the position accuracy of the pneumatic piston-cylinder is not clear.

In this paper, we investigated the influence of the friction in a pneumatic cylinder on the position error of automatic tool changer in a CNC machine under condition of the RH varying.

2 Modeling of the ATC motion by a pneumatic cylinder

2.1 Error Between the Stopping Position of the ATC and the Spindle Center Line in a CNC Machine

This study was conducted on the ATC on a CNC machine V30 Leadwell. The principal scheme of the automatic tool change of the CNC machine is shown in Fig. 2. During the tool changing, the pneumatic cylinder carries the ATC into the tool change position which is determined by the position sensor LS_2. This procedure needs to be sure of the concentric between the spindle and the tool holder. However, due to the CNC working on the air environment, the stopping position of the pneumatic cylinder when carrying the tool changer to the tool change position would be influenced by the impact of the varying friction. Therefore, the stopping position of the pneumatic cylinder also changes and oscillates the bilateral of the spindle center. The scheme presents the error of the stopping position of the ATC in comparison with the LS_2 position as below:

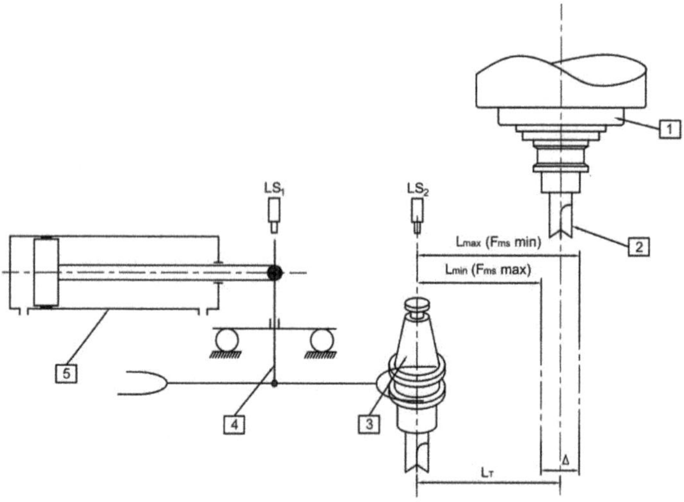

Fig. 2. The scheme identifies the errors of the stopping position and adjusting the ATC 1 - Spindle, 2 – Tool, 3 - Tool holder; 4 - The ATC, 5- Pneumatic cylinder

The tolerance of the stopping position Δ of the tool changer is presented as a formula below:

$$\Delta = L_{max} - L_{min} \quad (1)$$

In which:
L_{max} – The maximum error of the ATC when stopping.
L_{min} – The minimum error of the ATC when stopping.
L_T – The correction interval of the ATC position.
LS_1, LS_2 – The stopping position sensors.
Assume that the distribution of the stopping position error is normality distribution, the spindle center tolerance would provide the minimum stopping position error.
The specification of the ATC in the CNC machine is shown in the Table 1 [10].

No	Specification of the ATC	Value
1	ATC type	Drum
2	Quantity	20
3	Weight (kg)	7
4	Time of tool changing (sec)	8
5	Velocity of the piston mm/s	30–100
6	Stroke of piston (mm)	250
7	Humidity	51%–99%

2.2 Simulation of the Stopping Position Error of the ATC

The simulation scheme of the movement of the ATC with the piston carrying the tool changer to the tool change position is in Fig. 3.

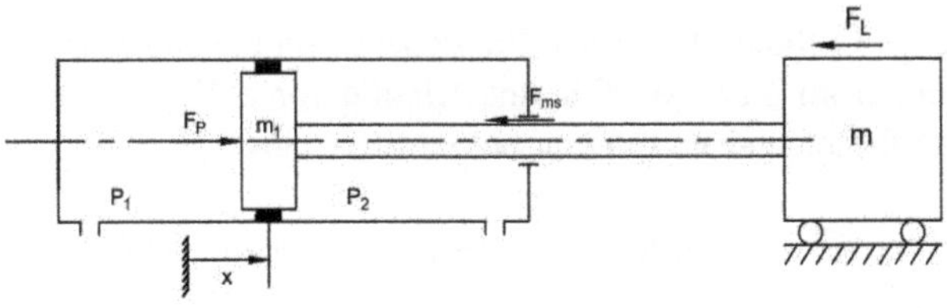

Fig. 3. Scheme of the movement of the piston carrying the ATC to the tool change position.

Consider the stopping process of the ATC, the motion equation can be written below:

$$M\ddot{x} = -F_L - F_{ms}(0 \leq x \leq L) \quad (2)$$

$$Mv\frac{dv}{dx} = -F_L - F_{ms} \Rightarrow dx = \frac{Mv}{-(F_L + F_{ms})}dv$$
$$\Rightarrow \int_0^L dx = \int_V^0 \frac{Mv}{-(F_L + F_{ms})}dv$$

In which:
v – Velocity of the ATC during stopping process;
L – The position error of the ATC when stopping;
m – The weight of the ATC;
M – The total weight of the ATC system; $M = m + m_1$.
F_{ms} - The friction force of the pneumatic cylinder.
F_L – The friction force of the rolling guideline, $F_L = f.m.g$;
f - The friction coefficient of the guideline.

The position error L of the tool changer of the milling CNC machine when stopping is calculated by the formula:

$$L = \int_0^V \frac{Mv}{(f \cdot m \cdot g + F_{ms})}dv \quad (3)$$

Thus, the position error when stopping L depends on the friction in the pneumatic cylinder, the weight of the ATC, and the velocity at the beginning of the stopping process.

3 Result and Discussion

3.1 The Position of the Pneumatic Cylinder when Stopping

In the motion Eq. (3), the friction force F_{ms} depends on the air environment. When the humidity varying from 51% to 99%, the friction force F_{ms} can be experimentally identified basing the recursive equation below [9]:

$$F_{ms}(RH, v) = p_{00} + p_{10}.RH + p_{01}.v + p_{20}.RH^2 + p_{11} \cdot RH \cdot v + a.e^{(-b.v)} \quad (4)$$

In which, coefficients (with 95% confidence bounds):
$a = 49.05$; $p_{00} = 14.41$; $p_{01} = 0.05426$; $p_{10} = 0.006261$; $p_{11} = 0.0007671$; $p_{20} = -0.0007448$; $b = 0.123$.

Substituting F_{ms} from the Eq. (4) into the Eq. (3), we can find the over-distance interval L during the stopping process of the ATC power by the pneumatic piston-cylinder when considering the RH factor by the equation follow:

$$L = \int_{0}^{V} \frac{Mv}{(f.m.g + p_{00} + p_{10}.RH + p_{01}.v + p_{20}.RH^2 + p_{11} \cdot RH \cdot v + a.e^{(-b.v)})} dv \quad (5)$$

From the working condition of the ATC of the CNC machine, we may have the following cases: the tool changer with the maximum and minimum weights; the highest and lowest humidity; the maximum and minimum velocities. The dependence of the position error when stopping *L* on *M*, *RH*, and v is shown in Fig. 4.

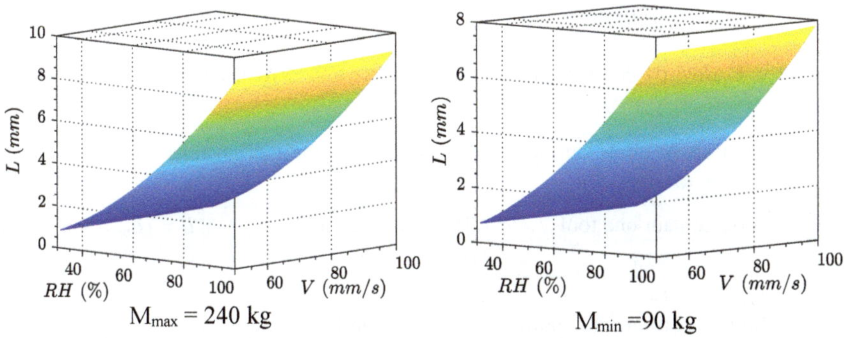

Fig. 4. The dependence of *L* on *RH*, and *v*

From Fig. 4, the error of the stopping position of the pneumatic cylinder in RH varying from 51 to 99% condition significantly affects the stopping position of the tool changer of the CNC machine, especially when m changing. The error of stopping position tends to increase when the velocity, the RH and the weight of the ATC increase.

Effect of the weight of ATC on the error of stopping position and the tolerance of the error with each velocity is shown in Fig. 5. The error of the stopping position tends to decrease when the weight of the tool changer is large. The tolerances (Δ) of the error of the stopping position at velocities 30mm/s, 50mm/s, and 100mm/s, and weights are shown in Table 1.

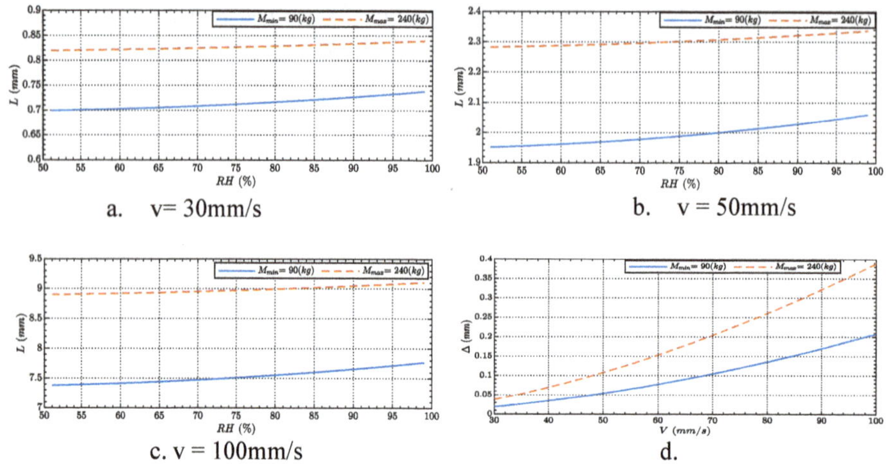

Fig 5. a,b,c. *L when the RH changes at different v.* Δ *on velocity when* the weight of the ATC

Table 1. Synthesis of errors on the stopping position of the tool changer when the humidity varying from 51% to 99%.

v (mm/s)	ATC contain one tool			ATC contain 20 tools			$L = (L_{min1} + L_{max20})/2$
	51%	99%		51%	99%		
	L_{min1} (mm)	L_{max1} (mm)	Δ_1 (mm)	L_{min20} (mm)	L_{max20} (mm)	Δ_{20} (mm)	
30	0.7	0.74	0.04	0.82	0.84	0.02	0.77
50	1.96	2.06	0.1	2.28	2.34	0.06	2.15
100	7.4	7.8	0.4	8.9	9.15	0.25	8.28

In which:
L_{min1} – The minimum error of the tool changer with one tool when stopping.
L_{max1} – the maximum error of the tool changer with one tool when stopping.
Δ_1 – the tolerance of the error of the tool changer with one tool when stopping.
L_{min20} – the minimum error of the tool changer with all tools when stopping.
L_{max20} – the maximum error of the tool changer with all tools when stopping.
Δ_{20} – the tolerance of the error of the tool changer with all tools when stopping.
L – the adjustment interval of the position of the tool changer.

The impact of the humidity of the specific climate in Vietnam with RH varying from 51 to 99% could lead to the error of the stopping position of the tool changer using the pneumatic cylinder. The tolerance interval Δ and L are showed in Table 1. These error values are quite high and are exactly the non-concentric of the spindle and the tool holder on the tool changer. These errors are the main reason for the wearing and collision of the spindle taper hole on the milling CNC machine. Therefore, it is necessary to adjust the center position of the tolerance interval Δ coinciding with the centerline of the spindle taper hole (refer to the L dimension in Table 1) according to the *RH* of the environment. This adjustment needs to implement during the run-up test and correction of the machine in the assembly process.

On the other hand, the varying weight of the tool changer also affects the stopping position error, as well as the tolerance interval. When carrying a maximum or a minimum number of tools, the error of the stopping position is up to 11–17%. Therefore, when operating the CNC machine, besides the selection of the number of tools storing on the tool changer in accord with the processing instruction, the total weight of tools is necessarily considered to reduce the error of the tolerance of the stopping position. In the industrial fact, the use of air-conditioning for the whole factory is not feasible and low efficiency. Therefore, it is appropriate to equip a desiccant system to maintain stable humidity. This is also a condition to adjust L_T interval when assembling and testing CNC machines.

4 Conclusion

Effect of the friction on the error of the stopping position of the tool changer of the CNC-V30 machine using the pneumatic cylinder when the humidity varying was investigated. From the rule of frictional behavior of pneumatic cylinder in changing humidity conditions with Vietnam's climatic characteristics, we quantified the error of the stopping position of the tool changer with the following suggestions:

1. The error of the tool changer of the CNC machine depends on the RH, velocity, and number of tools storing on the tool changer.
2. The error of the stopping position of the tool changer is maximum at the area of RH = 99% and minimum at the area of RH = 51%. In the stroke of the pneumatic piston-cylinder, the investigation indicated that the higher velocity produces a larger position error.
3. The error of the stopping position of the tool changer reaches to the maximum value at the RH 99%, full 20 tools storing, and the high velocity of the pneumatic piston-cylinder. In contrast, this error is smallest at RH 51%, one tool storing, and the small velocity.
4. The tool feed position correction interval defined in Table1 can be used for calibration and run-up testing when installing CNC milling machines in Vietnamese climatic conditions with velocities from 30 to 100 mm/s. This will limit the deterioration of CNC machine accuracy, in general, and the assembly accuracy of the spindle cluster and tool holder, in particular, when operating in Vietnam.

References

1. https://www.cattspindlegrinding.com/evaluating-your-spindle-taper/
2. de Lacalle, L.N.L., Lamikiz; A.: machine tools for high performance machining (2009)
3. Chowdhury, M.A., Helali, M.: The effect of frequency of vibration and humidity on the coefficient of friction. Tribol. Int. **39**, 958–962 (2006). https://doi.org/10.1016/j.triboint.2005.10.002
4. Imada, Y., Nakajima, K.: Effect of humidity on the friction and wear properties of SN. J. Tribol. **117**, 737–744 (1995). https://doi.org/10.1115/1.2831545
5. Imada, Y.: Effect of humidity and oxide products on the friction and wear properties of mild steel. J. Jpn. Soc. Tribol. **114**, 131–140 (1996)
6. Lancaster, J.K.: A review of the influence of environmental humidity and water on friction, lubrication and wear. Tribol. Int., 371–389 (1990). https://doi.org/10.1016/0301-679X(90)90053-R
7. Chowdhury, M.A., Helali, M.: The effect of relative humidity and roughness on the friction coefficient under horizontal vibration. Open Mech. Eng. J. 2, 128–135 (2008). https://doi.org/10.2174/1874155X00802010128
8. Chowdhury, M.A.: Effect of sliding velocity and relative humidity on friction coefficient of brass sliding against different steel counterfaces. Int. J. Eng. Res. Appl. (IJERA) **2**(2), 1425–1431 (2012)
9. Nguyen, T.D., Pham, V.H.: Study of the effects of relative humidity and velocity on the friction characteristics of pneumatic cylinders. Int. J Modern Phys. B 34 2(020)
10. Pham, V.-H., Nguyen, T.-D., Bui, T.-A.: Behavior of friction in pneumatic cylinders with different relative humidity; tribology in industry, vol. 42, no. 2 (2020)
11. Nouri, B.M.Y., Al-Bender, F., Swevers, J., Vanherck, P., Van Brussel, H.: Modeling a pneumatic servo positioning system with friction. In: Proceedings of the ACC 2000, pp. 1067-1071 (2000)
12. Nouri, B.: Modelling and control of pneumatic servo positioning systems (2001)
13. Chang, K.M., Liu, Y.T., Hsieh, S.J.: Fuzzy control with friction compensation for a pneumatic positioning system. In: Proceedings of the 13th International Conference on Informatics in Control, Automation and Robotics (ICINCO 2016), vol. 1, pp. 471–476 (2016)
14. Tran, X.B., Nguyen, V.L., Tran, K.D.: Effects of friction models on simulation of pneumatic cylinder. Mech. Sci. **10**, 517–528 (2019)

Buckling and Vibration Analysis Based on a Novel Unified Model for Laminated Composite Beams

Quoc-Cuong Le[1] and Ba-Duy Nguyen[2(✉)]

[1] Faculty of Engineering – Technology, Thu Dau Mot University, 6 Tran Van On Street, Thu Dau Mot City, Binh Duong, Vietnam
[2] Faculty of Architecture, Thu Dau Mot University, 6 Tran Van On Street, Thu Dau Mot City, Binh Duong, Vietnam
duynb@tdmu.edu.vn

Abstract. Based on fundamental equations of the elasticity theory, a novel unified beam model is developed for laminated composite beams. In this model, the displacement field is selected in a unified form which can be recovered to that of existing shear deformation beam theories available in the literature. Based on Lagrange's equations, the governing equations of the present theory are derived. They are then solved for natural frequencies and critical buckling loads of composite beams under different boundary conditions and lay-ups by using the Ritz approach with novel hybrid trigonometric functions. Various examples are also presented to verify the accuracy and generalization of the present theory, as well as investigate the influences of fiber angle on the behavior of composite beams under different boundary conditions and lay-ups.

Keywords: Vibration and buckling · Laminated composite beams · The Ritz method

1 Introduction

Laminated composite materials are commonly used in spacecraft, aircraft, mechanical engineering, construction and other different engineering fields due to their excellent mechanical properties including high strength, high stiffness and lightweight. The widespread applications of these structures led to the development of different computational models to predict their behaviours.

A general review of theories to analyze the laminated composite beams can be found in the previous works [1–5]. Generally, their behaviours can be captured based on either 2D beam theories or 3D theory of elasticity. The first approach is more popular due to its simplicity, whilst the second one is complicated to implement although it can analyze exactly response. Based on the first method, a huge number of shear deformation models have been developed for composite beams. The simplest one is the first-order shear deformation beam theory (FSBT) which requires a shear correction factor to compensate for the inadequate distribution of shear stress. The higher-order shear deformation beam theory (HSBT) with higher-order variations of axial

B. T. Long et al. (Eds.): MMMS 2020, LNME, pp. 405–419, 2021.
https://doi.org/10.1007/978-3-030-69610-8_57

displacement gives slightly improved predictions compared with FSBT. However, it involves more number of unknowns and thus is more complicated than FSBT. Another shear deformation beam theory developed based on the first approach is quasi-3D theory where both axial and transverse displacements are approximated as high-order variations through the beam thickness. Therefore, quasi-3D theory can predict the behavior of composite beams more accurately than HSBT. However, quasi-3D theory is more complicated than HSBT because it involves more numbers of unknowns. Both HSBT and quasi-3D theory do not require the shear correction factor. However, their accuracy strictly depends on a choice of shear functions. The development of shear functions for these theories is therefore an interesting topic that has attracted many researches with different approaches. Various types of shear functions have been developed for composite plates such as polynomial [6–10], trigonometric [11–17], exponential [18], hyperbolic [19, 20], and hybrid [21, 22]. For laminated composite beams, only some representative references are herein cited. For example, Khdeir and Reddy [23, 24] derived closed-form solutions of Reddy's theory for critical buckling loads and natural frequencies of cross-ply composite beams. Chandrashekhara and Bangera [25], Shi and Lam [26, 27], Murthy et al. [28] and Manur and Kant [29] investigated vibration behaviours of composite beams via finite element method (FEM). Karama et al. [30, 31] examined static, buckling and free vibration responses of composite beams based on trigonometric theory. Aydogdu [32–34] used polynomial, hyperbolic and exponential theories and the Ritz method to explore the behavior of composite beams. Shao et al. [35] presented various HSBTs for the free vibration of composite beams. Vo and Thai [36–39] presented FEM solutions of both HSBT and quasi-3D theory for the structural analysis of composite beams. Mantari and Canales [40, 41] developed both Ritz and FEM solutions for composite beams by using the quasi-3D theories with a third-order, polynomial and hybrid polynomial-trigonometric theories. Matsunaga [42] considered free vibration responses of composite beams by means of quasi-3D theory and Navier procedure. By using Carrera Unified Formulation (CUF) [43], Carrera et al. [44–47] and Arruda et al. [48] analyzed laminated composite beams. Vidal et al. [49] recently developed higher-order beam elements to model composite beams with generic cross-section. This paper aims to propose a novel unified HSBT which can be recovered to existing HSBT by changing the shear functions. The Ritz solution method with novel hybrid shape functions is employed to develop approximate solutions for critical buckling loads and natural frequencies of laminated composite beams under various boundary conditions and lay-ups. In order to validate the accuracy of the proposed theory, several numerical examples in vibration and buckling are considered. In addition, the effects of fibre orientation on the behavior of laminated composite beams are also examined.

2 Theoretical Formulation

2.1 A General Framework of Higher-Order Displacement Field

For the simplicity purpose, it is supposed that the effects of the transverse normal strain is neglected and all formulations in this section are performed under the assumption of isotropic materials. The stress-strain relationships of the beam are therefore given by:

$$\varepsilon_x = \frac{1}{E}\sigma_x \tag{1a}$$

$$\gamma_{xz} = \frac{1}{G}\sigma_{xz} \tag{1b}$$

in which E and G are Young's modulus and shear modulus, respectively. Moreover, the linear elastic relationships of strains and displacements are expressed by:

$$\varepsilon_x = u_{,x} \tag{2a}$$

$$\gamma_{xz} = u_{,z} + w_{,x} \tag{2b}$$

in which u and w are respectively axial and transverse deflections. Substituting Eq. (2) into Eq. (1) leads to:

$$u_{,x} = \frac{1}{E}\sigma_x \tag{3a}$$

$$u_{,z} + w_{,x} = \frac{1}{G}\sigma_{xz} \tag{3b}$$

The transverse shear stress $\sigma_{xz}(x, z)$ can be expressed in terms of the transverse shear force $Q_x(x)$ as follows:

$$\sigma_{xz}(x, z) = g(z)Q_x(x) \tag{4}$$

where $g(z)$ is a higher-order shear function that satisfies the traction free conditions at the bottom and top surfaces of the beam, i.e. $g(z = \pm h/2) = 0$ with h being the beam thickness. Substituting Eq. (4) into Eq. (3b) and taking into account the hypothesis of $w(x, z) = w_0(x)$, the axial displacement of the beam can be expressed by:

$$u(x, z) = u_0(x) - zw_{0,x} + f(z)\frac{Q_x(x)}{G}\theta \tag{5}$$

where $f'(z) = g(z)$.

Therefore, a general formulation of the displacement field of the proposed theory is obtained as follows:

$$u(x,z) = u_0(x) - zw_{0,x} + f(z)\frac{Q_x(x)}{G}\theta \tag{6a}$$

$$w(x,z) = w_0(x) \tag{6b}$$

It can be noticed that when $f(z) = \frac{3}{2h}\left(z - \frac{4z^3}{3h^2}\right)$, Eq. (6) yields to the zeroth-order shear deformation theory developed by Shimpi [50] for plates. Moreover, if the shear force is expressed in the form:

$$Q_x(x) = G\theta_0(x) \tag{7}$$

Where θ_0 is the beam rotation, the displacement field in Eq. (6) becomes a common form of the HSBT, which has been widely used by many authors due to its simplicity:

$$u(x,z) = u_0(x) - zw_{0,x} + f(z)\theta_0(x) \tag{8a}$$

$$w(x,z) = w_0(x) \tag{8b}$$

Furthermore, if the transverse shear force is written in terms of the rotation and the derivative of transverse deflection as follows [7]:

$$Q_x(x) = \frac{5Gh}{6}(\theta_0 + w_{0,x}) \tag{9}$$

Equation (6) becomes a general formulation of the HSBT as follows:

$$\begin{aligned} u(x,z) &= u_0(x) + \left(\frac{5hf}{6} - z\right)w_{0,x} + \frac{5hf}{6}\theta_0(x) \\ &= u_0(x) + (\Phi - z)w_{0,x} + \Phi(z)\theta_0(x) \end{aligned} \tag{10a}$$

$$w(x,z) = w_0(x) \tag{10b}$$

where $\Phi(z) = \frac{5h}{6}f(z)$.

The displacement field in Eq. (10) can be considered as an unified HSBT that many theories can be obtained. For example, the HSBT proposed by Reissner [7], Shi [26] and Reddy [9] can be recovered with $f(z) = \frac{3}{2h}\left(z - \frac{4z^3}{3h^2}\right)$ and $f(z) = \frac{6}{5h}\left(z - \frac{4z^3}{3h^2}\right)$, respectively.

2.2 Strain and Stress Fields

The non-zero strains associated to the displacements in Eq. (10) are therefore given by:

$$\varepsilon_x(x,z) = u_{0,x}(x) + (\Phi - z)w_{0,xx} + \Phi(z)\theta_{0,x} \tag{11a}$$

$$\gamma_{xz}(x,z) = \Phi'\left(\theta_0 + w_{0,x}\right) \tag{11b}$$

The non-zero stresses at the kth-layer of the beam (Fig. 1) associated to the strains in Eq. (11) are given by [5]:

$$\sigma_x(x,z) = \overline{Q}_{11}\varepsilon_x = \overline{Q}_{11}\left[u_{0,x}(x) + (\Phi - z)w_{0,xx} + \Phi(z)\theta_{0,x}\right] \tag{12a}$$

$$\sigma_{xz}(x,z) = \overline{Q}_{55}\gamma_{xz} = \overline{Q}_{55}\left[\Phi'\left(\theta_0 + w_{0,x}\right)\right] \tag{12b}$$

where

$$\overline{Q}_{11} = Q_{11} - \frac{Q_{13}^2}{Q_{33}},\ \overline{Q}_{55} = Q_{55} \tag{13a}$$

$$Q_{11} = C'_{11} - \frac{C'_{16}C'_{2} - 2C'_{12}C'_{16}C'_{26} + C'^2_{12}C'_{66}}{C'^2_{26} - C'_{22}C'_{66}} \tag{13b}$$

$$Q_{13} = C'_{13} - \frac{C'_{16}C'_{22}C'_{36} + C'_{12}C'_{23}C'_{66} - C'_{16}C'_{23}C'_{26} - C'_{12}C'_{26}C'_{36}}{C'^2_{26} - C'_{22}C'_{66}} \tag{13c}$$

$$Q_{33} = C'_{33} - \frac{C'^2_{36}C'_{22} - 2C'_{23}C'_{26}C'_{36} + C'^2_{23}C'_{66}}{C'^2_{26} - C'_{22}C'_{66}} \tag{13d}$$

$$Q_{55} = C'_{55} - \frac{C'^2_{45}}{C'_{44}} \tag{13e}$$

where C'_{ij} are the elastic stiffness coefficients at the kth-layer of the beam (see [51] for more details).

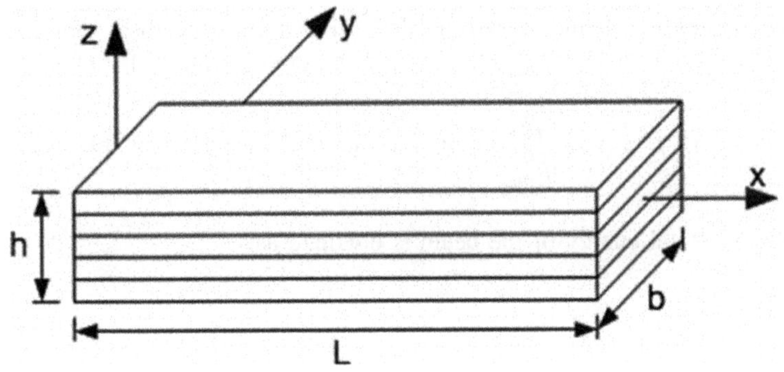

Fig. 1. Geometric dimensions of a typical laminated composite beam.

2.3 Governing Equations of Motion

The strain energy of the proposed beam model can be written as:

$$\begin{aligned} U &= \frac{1}{2}\int_V (\sigma_x \varepsilon_x + \sigma_{xz}\gamma_{xz})dV \\ &= \frac{1}{2}\int_0^L \Big[Au_{0,x}^2 + 2Bu_{0,x}w_{0,xx} + Dw_{0,xx}^2 + 2B^s u_{0,x}\theta_{0,x} \\ &+ 2D^s w_{0,xx}\theta_{0,x} + H^s\theta_{0,x}^2 + A^s\left(\theta_0^2 + w_{0,x}^2 + 2\theta_0 w_{0,x}\right) \end{aligned} \tag{14}$$

where (A, A^s, B, B^s, D, D^s and H^s) are the stiffness coefficients defined as:

$$(A,\ B,\ D,\ B^s,\ D^s,\ H^s) = \int_{-h/2}^{h/2} \left(1, \Phi - z, (\Phi - z)^2, \Phi, (\Phi - z)\Phi, \Phi^2\right)\overline{Q}_{11}(z)bdz \tag{15}$$

$$A^s = \int_{-h/2}^{h/2} \Phi'^2 \overline{Q}_{55}(z)bdz \tag{16}$$

The external work done by the axial compression N_0 is obtained as follows:

$$V = -\frac{1}{2}\int_0^L N_0 w_{,x}^2 dx \tag{17}$$

The kinetic energy of the proposed beam model is calculated by:

$$\begin{aligned} K &= \frac{1}{2}\int_V \rho(z)\left(\dot{u}^2 + \dot{w}^2\right)dV \\ &= \frac{1}{2}\int_0^L \left[I_0\dot{u}_0^2 + 2I_1\dot{u}_0\dot{w}_{0,x} + I_2\dot{w}_{0,x}^2 + 2J_1\dot{\theta}_0\dot{u}_0 + 2J_2\dot{\theta}_0\dot{w}_{0,x} + K_2\dot{\theta}_0^2 + I_0\dot{w}_0^2\right]dx \end{aligned} \tag{18}$$

in which dot-superscript indicates time derivative t; ρ is the mass density; and the moment of inertia terms I_0, I_1, I_2, J_1, J_2 and K_2 are the inertia coefficients, defined by:

$$(I_0, I_1, I_2, J_1, J_2, K_2) = \int_{-h/2}^{h/2} (1, \Phi - z, (\Phi - z)^2, \Phi, (\Phi - z)\Phi, \Phi^2)\rho bdz \tag{19}$$

Finally, the total energy of the beam is obtained as:

$$
\begin{aligned}
\Pi = & \frac{1}{2}\int_0^L \Big[Au_{0,x}^2 + 2Bu_{0,x}w_{0,xx} + Dw_{0,xx}^2 + 2B^s u_{0,x}\theta_{0,x} \\
& + 2D^s w_{0,xx}\theta_{0,x} + H^s\theta_{0,x}^2 + A^s\left(\theta_0^2 + w_{0,x}^2 + 2\theta_0 w_{0,x}\right)\Big]dx - \frac{1}{2}\int_0^L N_0 w_{,x}^2 dx \\
& - \frac{1}{2}\int_0^L \Big[I_0\dot{u}_0^2 + 2I_1\dot{u}_0\dot{w}_{0,x} + I_2\dot{w}_{0,x}^2 + 2J_1\dot{\theta}_0\dot{u}_0 + 2J_2\dot{\theta}_0\dot{w}_{0,x} + K_2\dot{\theta}_0^2 + I_0\dot{w}_0^2\Big]dx
\end{aligned}
\tag{20}
$$

Based on the Ritz method [51], the displacement variables in Eq. (20) can be approximated as:

$$u_0(x,t) = \sum_{j=1}^{m}\psi_{j,x}(x)u_j e^{i\omega t} \tag{21a}$$

$$w_0(x,t) = \sum_{j=1}^{m}\varphi_j(x)w_j e^{i\omega t} \tag{21b}$$

$$\theta_0(x,t) = \sum_{j=1}^{m}\psi_{j,x}(x)\theta_j e^{i\omega t} \tag{21c}$$

where ω is the frequency, $i^2 = -1$ the imaginary unit; (u_j, w_j and θ_j) are unknown and need to be determined; $\psi_j(x)$ and $\varphi_j(x)$ are shape functions which should be chosen to satisfy the boundary conditions [51]. If they do not meet these conditions, a penalty function method can be applied to recover the boundary conditions [52]. Practically however, this approach leads to an increase in the size of the stiffness and mass matrices and thus causing computational costs. Moreover, it is known that the equations of motion of the beams are expressed under the fourth-order differential ones, hence the primary solutions are given in terms of exponential functions. It is numerically observed that the approximation of field variables based on the elastic homogeneous solution will lead to an accuracy and fast convergence of the solution. Therefore, the present research proposes novel hybrid shape functions given in Table 1. These functions are composed of the exponential functions and admissible trigonometric ones that satisfy automatically various boundary conditions of the beam given in Table 2 in which S-S, C-F and C-C are represented for simply supported, clamped-free and clamped-clamped, respectively.

Table 1. Approximate shape functions.

BCs	$\frac{\psi_j(x)}{e^{-j\frac{x}{L}}}$	$\frac{\varphi_j(x)}{e^{-j\frac{x}{L}}}$
S-S	$\cos\frac{\pi x}{L}$	$\sin\frac{\pi x}{L}$
C-F	$\sin\frac{\pi x}{2L}$	$1-\cos\frac{\pi x}{2L}$
C-C	$\sin\frac{\pi x}{L}$	$\sin^2\frac{\pi x}{L}$

Table 2. Boundary conditions of the proposed beam model

Beam	$x = 0$	$x = L$
S-S	$w_0 = 0$	$w_0 = 0$
C-F	$u_0 = 0,\ w_0 = 0,\ \theta_0 = 0,\ w_{0,x} = 0$	
C-C	$u_0 = 0,\ w_0 = 0,\ \theta_0 = 0,\ w_{0,x} = 0$	$u_0 = 0,\ w_0 = 0,\ \theta_0 = 0,\ w_{0,x} = 0$

The governing equations of motion can be obtained in the form of Lagrange's equations by substituting Eq. (21) into Eq. (20).

$$\frac{\partial \Pi}{\partial d_j} - \frac{d}{dt}\frac{\partial \Pi}{\partial \dot{d}_j} = 0 \tag{22}$$

with d_j representing the values of $\left(u_j,\ w_j,\ \theta_j\right)$.

Equation (22) can be written in the matrix forms as:

$$\left(\begin{bmatrix} \mathbf{K}^{11} & \mathbf{K}^{12} & \mathbf{K}^{13} \\ {}^T\mathbf{K}^{12} & \mathbf{K}^{22} & \mathbf{K}^{23} \\ {}^T\mathbf{K}^{13} & \mathbf{K}^{23} & \mathbf{K}^{33} \end{bmatrix} - \omega^2 \begin{bmatrix} \mathrm{M}^{11} & \mathrm{M}^{12} & \mathrm{M}^{13} \\ {}^T\mathrm{M}^{12} & \mathrm{M}^{22} & \mathrm{M}^{23} \\ {}^T\mathrm{M}^{13} & {}^T\mathrm{M}^{23} & \mathrm{M}^{33} \end{bmatrix}\right) \begin{Bmatrix} \mathbf{u} \\ \mathbf{w} \\ \boldsymbol{\theta} \end{Bmatrix} = \begin{Bmatrix} 0 \\ 0 \\ 0 \end{Bmatrix} \tag{23}$$

or under the compact form:

$$\left([\mathbf{K}] - \omega^2[\mathrm{M}]\right)\mathrm{d} = \{0\} \tag{24}$$

where the components of the stiffness matrix **K**, mass matrix **M** and force vector **P** are given by:

$$\begin{aligned}
&K_{\mathrm{ij}}^{11} = A\int_0^L \psi_{i,x}\psi_{j,x}dx,\ K_{\mathrm{ij}}^{12} = B\int_0^L \psi_{i,x}\varphi_{j,xx}dx,\ K_{\mathrm{ij}}^{13} = B^s\int_0^L \psi_{i,x}\psi_{j,x}dx, \\
&K_{\mathrm{ij}}^{22} = \int_0^L \left(D\varphi_{i,xx}\varphi_{j,xx} + A^s\varphi_{i,x}\varphi_{j,x} - N^0\varphi_{i,x}\varphi_{j,x}\right)dx, \\
&K_{\mathrm{ij}}^{23} = \int_0^L \left(D^s\varphi_{i,xx}\psi_{j,x} + A^s\varphi_{i,x}\psi_j\right)dx,\ K_{\mathrm{ij}}^{33} = \int_0^L \left(H^s\psi_{i,x}\psi_{j,x} + A^s\psi_i\psi_j\right)dx \\
&M_{\mathrm{ij}}^{11} = I_0\int_0^L \psi_i\psi_j dx,\ M_{\mathrm{ij}}^{12} = I_1\int_0^L \psi_i\varphi_{j,x}dx,\ M_{\mathrm{ij}}^{13} = J_1\int_0^L \psi_i\psi_j dx \\
&M_{\mathrm{ij}}^{22} = \int_0^L \left(I_0\varphi_i\varphi_j + I_2\varphi_{i,x}\varphi_{j,x}\right)dx,\ M_{\mathrm{ij}}^{23} = J_2\int_0^L \varphi_{i,x}\psi_j dx,\ M_{\mathrm{ij}}^{33} = K_2\int_0^L \psi_i\psi_j dx
\end{aligned} \tag{25}$$

It is noted from Eq. (24) that the buckling responses by solving the equation $|\mathbf{K}| = 0$, and vibration responses based on the equation $\left|\mathbf{K} - \omega^2\mathrm{M}\right| = 0$.

3 Numerical Results

To demonstrate the accuracy of the proposed beam model, two shear functions f (z) involved in Eq. (10) are considered: $f(z) = \frac{3}{2h}\left(z - \frac{4z^3}{3h^2}\right)$ [7] (HSBT1) and $f(z) = \frac{6}{5h}\left(z - \frac{4z^3}{3h^2}\right)$ [13] (HSBT2). Several cases in vibration and buckling of laminated composite beams are carried out to validate the proposed theory. The following materials are used [32]: $E_2 = E_3,\ G_{12} = G_{13} = 0.6E_2, G_{23} = 0.5E_2,\ \nu_{12} = \nu_{13} = \nu_{23} = 0.25$.

For convenience, the following normalized terms are used:

$$\varpi = \frac{\omega L^2}{h}\sqrt{\frac{\rho}{E_2}} \tag{26}$$

$$\overline{N}_{cr} = N_{cr}\frac{L^2}{E_2 bh^3} \tag{27}$$

The Ritz method has been widely used for bending, vibration and buckling analysis of the structures. The static responses are related to the equilibrium problem while the buckling and vibration behaviours correspond to the eigenvalue ones. It is from many previous works that this method yields upper bound to the exact values for the natural frequencies and buckling loads, and lower bound for the displacements in average sense (see Refs. [53–55] for more details). In order to verify the convergence of the present solutions, a (0°/90°/0°) composite beam for different boundary conditions with E_1/E_2 = 40 and L/h = 5 has been considered and its results are shown in Table 3. It can be seen that the present solutions converge when the number of series m = 12. Therefore, this value will be used herein for the following numerical examples. It is also observed from Table 3 that the solutions of C-F and C-C beams converge slower than that of S-S beam due to essential boundary conditions.

Table 3. Convergence studies for (0°/90°/0°) composite beams (L/h = 5, HSBT1, E_1/E_2 = 40).

Beam	Number of terms in series (m)						
	2	4	6	8	10	12	14
Fundamental frequency							
S-S	9.3215	9.2080	9.2062	9.2062	9.2062	9.2062	9.2062
C-F	4.3362	4.2373	4.2309	4.2304	4.2298	4.2295	4.2297
C-C	12.6368	11.7766	11.6348	11.6094	11.6052	11.6047	11.6043
Critical buckling load							
S-S	8.7232	8.6116	8.6093	8.6092	8.6092	8.6092	8.6092
C-F	4.9610	4.7075	4.7035	4.7034	4.7029	4.7028	4.7025
C-C	11.8736	11.6514	11.6484	11.6482	11.6478	11.6475	11.6471

Table 4. Normalized fundamental frequencies of (0°/90°/0°) and (0°/90°) composite beams ($E_1/E_2 = 40$)

BCs	Theory	0°/90°/0°			0°/90°		
		5	10	50	5	10	50
S-S	HSBT1	9.206	13.607	17.449	6.125	6.940	7.297
	HSBT2	9.223	13.608	17.449	6.149	6.948	7.297
	HSBT [4]	9.208	13.614	17.462	6.128	6.945	7.302
	HSBT [28]	9.207	13.611	–	6.045	6.908	–
	HSBT [56]	9.208	13.614	–	6.128	6.945	–
	Quasi-3D [42]	9.200	13.608	–	5.662	6.756	–
	Quasi-3D [40]	9.208	13.610	–	6.109	6.913	–
	Quasi-3D [5]	9.208	13.610	17.449	6.140	6.948	7.297
C-F	HSBT1	4.229	5.490	6.260	2.380	2.540	2.602
	HSBT2	4.242	5.492	6.259	2.386	2.542	2.602
	HSBT [4]	4.234	5.498	6.267	2.383	2.543	2.605
	HSBT [28]	4.230	5.491	–	2.378	2.541	–
	HSBT [56]	4.234	5.495	–	2.386	2.544	–
	Quasi-3D [40]	4.221	5.490	–	2.375	2.532	–
	Quasi-3D [5]	4.223	5.491	6.262	2.382	2.543	2.604
C-C	HSBT1	11.605	19.727	37.656	10.024	13.665	16.418
	HSBT2	11.607	19.823	37.666	10.169	13.741	16.424
	HSBT [4]	11.607	19.728	37.679	10.027	13.670	16.429
	HSBT [28]	11.602	19.719	–	10.011	13.657	–
	HSBT [56]	11.603	19.712	–	10.026	13.660	–
	Quasi-3D [40]	11.486	19.652	–	9.974	13.628	–
	Quasi-3D [5]	11.499	19.672	37.633	9.944	13.664	16.432

The comparisons of the normalized fundamental frequencies are presented in Tables 4, whilst the comparisons of the normalized critical buckling loads are presented in Tables 5. The comparison is made for different beam configurations, i.e., (0°/90°/0°), (0°/90°), (0°/θ°/0°) and (0°/θ°) under various span-to-thickness ratios, fibre orientations and boundary conditions.

It is noted that although the bifurcation buckling phenomena does not occur for laminated composite beams with unsymmetric layers, in order to verify the accuracy of the present beam theory in comparison with the earlier researches, the critical buckling loads of laminated composite beams with layers (0°/90°) and (0°/θ°) are still considered. The validation of these results are compared with those derived from the HSBTs ([4, 24, 28]) and quasi-3D theories ([5, 40, 42]). It can be seen that the consistencies of the solutions are again found.

Table 5. Normalized buckling loads of (0°/90°/0°) and (0°/90°) composite beams (E_1/E_2 = 40)

BCs	Theory	0°/90°/0°			0°/90°		
		5	10	50	5	10	50
S-S	HSBT1	8.609	18.814	30.859	3.902	4.935	5.398
	HSBT2	8.640	18.817	30.858	3.934	4.947	5.398
	HSBT [4]	8.613	18.832	30.906	3.907	4.942	5.406
	HSBT [56]	8.613	18.832	–	–	–	–
	Quasi-3D [40]	8.585	18.796	–	3.856	4.887	–
	Quasi-3D [5]	8.613	18.822	30.860	3.921	4.946	5.398
C-F	HSBT1	4.703	6.762	7.871	1.233	1.322	1.353
	HSBT2	4.703	6.760	7.880			
	HSBT [4]	4.708	6.772	7.886	1.236	1.324	1.356
	HSBT [56]	4.708	6.772	–	–	–	–
	Quasi-3D [40]	4.673	6.757	–	1.221	1.311	–
	Quasi-3D [5]	4.699	6.762	7.874	1.233	1.324	1.354
C-C	HSBT1	11.648	34.434	114.205	8.668	15.605	21.337
	HSBT2	11.710	34.556	113.530			
	HSBT [4]	11.652	34.453	114.398	8.674	15.626	21.372
	HSBT [56]	11.652	34.453	–	–	–	–
	Quasi-3D [40]	11.502	34.365	–	8.509	15.468	–
	Quasi-3D [5]	11.652	34.452	114.260	8.615	15.693	21.371

The influence of fibre orientation on the normalized fundamental natural frequencies and critical buckling loads of (0°/θ°/0°) and (0°/θ°) laminated composite beams are displayed in Figs. 2 and 3, respectively, for S-S beams. It can be seen that the increase of fibre angle leads an decrease in the buckling loads and fundamental frequencies. The significant deviations on the results of asymmetric beams are observed from θ = 5° to θ = 70°, which is the same response with the transverse displacement.

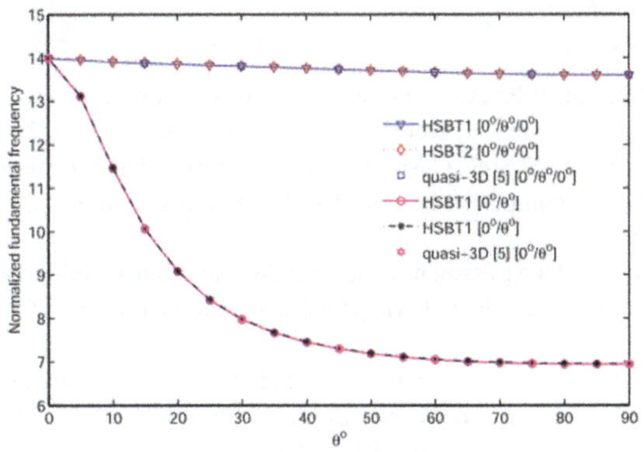

Fig. 2. Influences of fibre orientation on normalized fundamental frequencies of [0°/θ°/0°] and [0°/θ°] beams (L/h = 10, E_1/E_2 = 40).

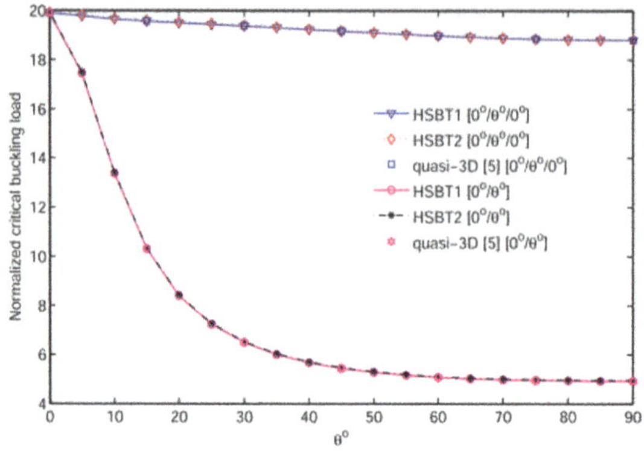

Fig. 3. Influences of fibre orientation on normalized critical buckling loads of [0°/θ°/0°] and [0°/θ°] beams (L/h = 10, E_1/E_2 = 40).

4 Conclusion

A novel unified beam theory has been proposed for the buckling and free vibration analysis of composite beams. Based on the fundamental equations of elasticity theory, the displacement field of the proposed model was derived in a unified form which can be recovered to that of existing HSBTs available in the literature. The Ritz method with a new shape function under hybrid form has been employed to solve the governing equations of the proposed beam model for buckling loads and natural frequencies. A comprehensive verification study has been conducted, and the numerical results demonstrated that the proposed beam model can well predict the free vibration and buckling responses of composite beams under different boundary conditions and arbitrary lay-ups. In addition, the comprehensive result presented in this paper can also be served as a reference solution for the development of composite beam models in the future.

References

1. Ghugal, Y.M., Shimpi, R.P.: A review of refined shear deformation theories for isotropic and anisotropic laminated beams. J. Reinf. Plast. Compos. **20**(3), 255–272 (2001)
2. Aguiar, R., Moleiro, F., Soares, C.M.: Assessment of mixed and displacement-based models for static analysis of composite beams of different cross-sections. Compos. Struct. **94**(2), 601–616 (2012)
3. Zhen, W., Wanji, C.: An assessment of several displacement-based theories for the vibration and stability analysis of laminated composite and sandwich beams. Compos. Struct. **84**(4), 337–349 (2008)
4. Nguyen, T.-K., Nguyen, N.-D., Vo, T.P., Thai, H.-T.: Trigonometric-series solution for analysis of laminated composite beams. Compos. Struct. **160**, 142–151 (2017)
5. Nguyen, N.-D., Nguyen, T.-K., Vo, T.P., Thai, H.-T.: Ritz-based analytical solutions for bending, buckling and vibration behavior of laminated composite beams. Int. J. Struct. Stab. Dyn. **18**(11), 1850130 (2018)

6. Ambartsumian, S.: On the theory of bending of anisotropic plates and shallow shells. J. Appl. Math. Mech. **24**(2), 500–514 (1960)
7. Reissner, E.: On transverse bending of plates, including the effect of transverse shear deformation. Int. J. Solids Struct. **11**(5), 569–573 (1975)
8. Levinson, M.: An accurate, simple theory of the statics and dynamics of elastic plates. Mech. Res. Commun. **7**(6), 343–350 (1980)
9. Reddy, J.N.: A simple higher-order theory for laminated composite plates. J. Appl. Mech. **51** (4), 745–752 (1984)
10. Nguyen, T.N., Thai, C.H., Nguyen-Xuan, H.: On the general framework of high order shear deformation theories for laminated composite plate structures: a novel unified approach. Int. J. Mech. Sci. **110**, 242–255 (2016)
11. Murthy, M.: An Improved Transverse Shear Deformation Theory for Laminated Anisotropic Plates. National Aeronautics and Space Administration, Washington DC (1981)
12. Stein, M.: Nonlinear theory for plates and shells including the effects of transverse shearing. AIAA J. **24**(9), 1537–1544 (1986)
13. Touratier, M.: An efficient standard plate theory. Int. J. Eng. Sci. **29**(8), 901–916 (1991)
14. Arya, H., Shimpi, R., Naik, N.: A zigzag model for laminated composite beams. Compos. Struct. **56**(1), 21–24 (2002)
15. Thai, C.H., Ferreira, A., Bordas, S., Rabczuk, T., Nguyen-Xuan, H.: Isogeometric analysis of laminated composite and sandwich plates using a new inverse trigonometric shear deformation theory. Eur. J. Mech. A/Solids **43**, 89–108 (2014)
16. Mantari, J., Oktem, A., Soares, C.G.: A new trigonometric shear deformation theory for isotropic, laminated composite and sandwich plates. Int. J. Solids Struct. **49**(1), 43–53 (2012)
17. Nguyen, V.-H., Nguyen, T.-K., Thai, H.-T., Vo, T.P.: A new inverse trigonometric shear deformation theory for isotropic and functionally graded sandwich plates. Compos Part B: Eng **66**, 233–246 (2014)
18. Aydogdu, M.: A new shear deformation theory for laminated composite plates. Compos. Struct. **89**(1), 94–101 (2009)
19. Soldatos, K.P.: A transverse shear deformation theory for homogeneous monoclinic plates. Acta Mech. **94**(3), 195–220 (1992)
20. Akavci, S.S., Tanrikulu, A.H.: Buckling and free vibration analyses of laminated composite plates by using two new hyperbolic shear-deformation theories. Mech. Compos. Mater. **44** (2), 145–154 (2008)
21. Mantari, J., Oktem, A., Soares, C.G.: A new higher order shear deformation theory for sandwich and composite laminated plates. Compos. Part. B: Eng. **43**(3), 1489–1499 (2012)
22. Thai, C.H., Kulasegaram, S., Tran, L.V., Nguyen-Xuan, H.: Generalized shear deformation theory for functionally graded isotropic and sandwich plates based on isogeometric approach. Comput. Struct. **141**, 94–112 (2014)
23. Khdeir, A., Reddy, J.: Free vibration of cross-ply laminated beams with arbitrary boundary conditions. Int. J. Eng. Sci. **32**(12), 1971–1980 (1994)
24. Khdeir, A., Redd, J.: Buckling of cross-ply laminated beams with arbitrary boundary conditions. Compos. Struct. **37**(1), 1–3 (1997)
25. Chandrashekhara, K., Bangera, K.: Free vibration of composite beams using a refined shear flexible beam element. Comput. Struct. **43**(4), 719–727 (1992)
26. Shi, G., Lam, K., Tay, T.: On efficient finite element modeling of composite beams and plates using higher-order theories and an accurate composite beam element. Compos. Struct. **41**(2), 159–165 (1998)
27. Shi, G., Lam, K.: Finite element vibration analysis of composite beams based on higher-order beam theory. J. Sound Vib. **219**(4), 707–721 (1999)

28. Murthy, M., Mahapatra, D.R., Badarinarayana, K., Gopalakrishnan, S.: A refined higher order finite element for asymmetric composite beams. Compos. Struct. **67**(1), 27–35 (2005)
29. Marur, S., Kant, T.: Free vibration analysis of fiber reinforced composite beams using higher order theories and finite element modelling. J. Sound Vib. **194**(3), 337–351 (1996)
30. Karama, M., Harb, B.A., Mistou, S., Caperaa, S.: Bending, buckling and free vibration of laminated composite with a transverse shear stress continuity model. Compos. Part B: Eng. **29**(3), 223–234 (1998)
31. Karama, M., Afaq, K., Mistou, S.: Mechanical behaviour of laminated composite beam by the new multi-layered laminated composite structures model with transverse shear stress continuity. Int. J. Solids Struct **40**(6), 1525–1546 (2003)
32. Aydogdu, M.: Vibration analysis of cross-ply laminated beams with general boundary conditions by ritz method. Int. J. Mech. Sci. **47**(11), 1740–1755 (2005)
33. Aydogdu, M.: Buckling analysis of cross-ply laminated beams with general boundary conditions by ritz method. Compos. Sci. Technol. **66**(10), 1248–1255 (2006)
34. Aydogdu, M.: Free vibration analysis of angle-ply laminated beams with general boundary conditions. J. Reinf. Plast. Compos. **25**(15), 1571–1583 (2006)
35. Shao, D., Hu, S., Wang, Q., Pang, F.: Free vibration of refined higher-order shear deformation composite laminated beams with general boundary conditions. Compos. Part B Eng. **108**, 75–90 (2017)
36. Vo, T.P., Thai, H.-T.: Static behavior of composite beams using various refined shear deformation theories. Compos. Struct. **94**(8), 2513–2522 (2012)
37. Vo, T.P., Thai, H.-T.: Vibration and buckling of composite beams using refined shear deformation theory. Int. J. Mech. Sci. **62**(1), 67–76 (2012)
38. Vo, T.P., Thai, H.-T., Nguyen, T.-K., Lanc, D., Karamanli, A.: Flexural analysis of laminated composite and sandwich beams using a four-unknown shear and normal deformation theory. Compos. Struct. **176**, 388–397 (2017)
39. Vo, T.P., Thai, H.-T., Aydogdu, M.: Free vibration of axially loaded composite beams using a four-unknown shear and normal deformation theory. Compos. Struct. **178**, 406–414 (2017)
40. Mantari, J., Canales, F.: Free vibration and buckling of laminated beams via hybrid ritz solution for various penalized boundary conditions. Compos. Struct. **152**, 306–315 (2016)
41. Mantari, J., Canales, F.: Finite element formulation of laminated beams with capability to model the thickness expansion. Compos. Part B Eng. **101**, 107–115 (2016)
42. Matsunaga, H.: Vibration and buckling of multilayered composite beams according to higher order deformation theories. J. Sound Vib. **246**(1), 47–62 (2001)
43. Erasmo Carrera, M.P., Gaetano, G.: Beam structures: classical and advanced theories. Wiley (2011)
44. Carrera, E., Filippi, M., Zappino, E.: Laminated beam analysis by polynomial, trigonometric, exponential and zig-zag theories. Eur. J. Mech. A/Solids **41**, 58–69 (2013)
45. Giunta, G., Biscani, F., Belouettar, S., Ferreira, A., Carrera, E.: Free vibration analysis of composite beams via refined theories. Compos. Part B Eng. **44**(1), 540–552 (2013)
46. Filippi, M., Carrera, E.: Bending and vibrations analyses of laminated beams by using a zig-zag-layer-wise theory. Compos. Part B Eng. **98**, 269–280 (2016)
47. Pagani, A., de Miguel, A., Petrolo, M., Carrera, E.: Analysis of laminated beams via unified formulation and legendre polynomial expansions. Compos. Struct. **156**, 78–92. 70th Anniversary of Professor J.N. Reddy (2016)
48. Arruda, M., Castro, L., Ferreira, A., Garrido, M., Gonilha, J., Correia, J.: Analysis of composite layered beams using carrera unified formulation with legendre approximation. Compos. Part B Eng. **137**, 39–50 (2018)

49. Vidal, P., Giunta, G., Gallimard, L., Polit, O.: Modeling of composite and sandwich beams with a generic cross-section using a variable separation method. Compos. Part B Eng. **165**, 648–661 (2019)
50. Shimpi, R.P.: Zeroth-order shear deformation theory for plates. AIAA J. **37**(4), 524–526 (1999)
51. Reddy, J.N.: Mechanics of Laminated Composites Plates: Theory and Analysis. CRC Press, Boca Raton (1997)
52. Nguyen, T.-K., Nguyen, T.T.-P., Vo, T.P., Thai, H.-T.: Vibration and buckling analysis of functionally graded sandwich beams by a new higher-order shear deformation theory. Compos. Part B Eng. **76**, 273–285 (2015)
53. Courant, R.: Variational methods for the solution of problems of equilibrium and vibrations. Bull. Am. Math. Soc. **49**, 1–23 (1943)
54. Courant, R., Hilbert, D.: Methods of Mathematical Physics. Wiley, Hoboken (1989)
55. Leissa, A.W., Shihada, S.M.: Convergence of the ritz method. Appl. Mech. Rev. **48**(115), 590–595 (1995)
56. Khdeir, A., Reddy, J.: An exact solution for the bending of thin and thick cross-ply laminated beams. Compos. Struct. **37**(2), 195–203 (1997)

Influence of the Preload on the Starting Torques of a Pair Angular Contact Bearings of CNC Lathes Spindle

Duong Nguyen Thuy, Duc Nguyen Tri, and Hung Pham Van(✉)

School of Mechanical Engineering,
Ha Noi University of Science and Technology,
No 1 Dai Co Viet Stress, Hai Ba Trung District, Hanoi City 100000, Vietnam
{Duong.nguyenthuy,Hung.phamvan}@hust.edu.vn

Abstract. In machine tool spindles, train and automobile wheel hubs, gearboxes in heavy industries and aircraft widely used the single row angular contact ball bearings, because It can withstand a combination of high axial and radial loads. All above applications require precision rotary motion with a higher degree of stiffness and positional accuracy. So that, the preload must be applied during assembly to remove excess play of bearing. The preload of bearing increases the stiffness of the assembly, decreases noise and improves rotational accuracy, but also increases starting torque. The starting torque of bearing relates directly to power loss and stability of a rotating system. This paper presents the influence of the preload on the starting torque for a pair single row angular contact ball bearings on the spindle unit of lathe CNC in laboratory conditions. The results show that the loss due to starting friction on the spindle bearing is significant and varies up to 75% during the operating cycle of the machine. The starting torque of the spindle of the mini CNC lathe will be reduced by 40% when the shifting hierarchy creates preload.

Keywords: Spindle unit · Preload · Starting torque · Bearing preload · CNC lathe spindle

1 Introduction

There may be some applications as: machine-tool head spindles, hypoid-gear pinion shaft in automotive differential, shaft in electric motors … and other similar where the bearing arrangement need to be often preloaded. This preload increase bearing stiffness and reduce as far as possible undesirable bearing displacement when loads applied. Machine tool spindles widely used a pair single row angular contact ball bearings, because It can withstand a combination of high axial and radial loads and have got the precision rotary motion with a higher stiffened and positional accuracy. So that, the preload must be applied during assembly to remove excess play of bearing in head spindles of machine-tool. But also, the preload of bearing increases starting torque, run torque, temperature, power loss and reduce the stability of a rotating spindles systems of machine-tool.

B. T. Long et al. (Eds.): MMMS 2020, LNME, pp. 420–427, 2021.
https://doi.org/10.1007/978-3-030-69610-8_58

The amount of preload depends on the bearing series, the contact angle, the internal geometry, surface quality of rolling elements, cooling, lubrication, misalignment, size of the bearing and applies to bearing sets in back-to-back or face-to-face or tandem arrangements. Bearings series are manufactured to four different preload classes: Class A, extra light preload; Class B, light preload; Class C, moderate preload; Class D, heavy preload. In applications where high speeds take precedence over the degree of rigidity, the following additional preload classes are available: Class L, reduced light preload for asymmetrical bearing sets; Class M, reduced moderate preload for asymmetrical bearing sets; Class F, reduced heavy preload for asymmetrical bearing sets [1].

The studying about the starting torque, run torque, and stiffness of CNC machine-tool head spindles depended on preload was done. Momir Šarenac [2] has chosen the suitable structure and preload for the machine tool spindle unit to achieve the stiffness to maintain and improve the machining accuracy. Tri Prakosa et al. [3] has studied in-depth on how to set the preload, the drive distance, and the number of drives to increase the stiffness of main spindles. It increases the machining accuracy of machine tools. The study to determine the relationship between the stiffness and the vibration of the main spindles was done by Holroyd et al. [4] based on choosing the integrated bearings of them. Rastegari. A et al.'s research [5] showed that the wear of the main spindles reduced preload displacement, which was the cause of reduced stiffness of the machine tool spindle unit. Van-Canh Tong et al. [6] investigates the running torque of TRBs with angular misalignment under a constant displacement preload. The angular misalignment increases the running torque. Chi Zhang et al. [7] was presented an equation for calculating the frictional torque of a dry-lubricated tapered roller bearing for two representative preload methods: axial force preload and axial displacement preload. H. Matsuyama and S. Kamamoto [8] have experimented with the frictional torque generated in the raceway contacts of tapered roller bearings. The result has been compared with the prediction from the EHL theory. As a result, a new formula for calculating frictional torque has been proposed. The above studies only mention that increasing preloads will increase stiffness, maintain, and improve machining accuracy, calculating frictional torque without considering that the increase of frictional torque of a bearing relates directly to the power loss of a machine-tool spindles system.

The head spindles of lathes CNC mini have got a pair single row angular contact ball bearings 7210C, which are mounted Back-to-Back arrangement (DB) shown in Fig. 1a. Axial preload in an adjusted bearing arrangement is produced by displacing one bearing ring axially, relative to the other, by an amount corresponding to the desired preload force. This paper presents research on starting torque behavior of head spindles of mini CNC lathes when preload changes with the DB arranged bearing. The relationship between the starting torque and preload by axial displacement is also discussed below.

2 Method and Experimental Apparatus

2.1 Preload Method for Head Spindle of Machine Tool

There are two mains types of preloading methods: Preload with a constant force and Preload by axial displacement.

Preload with a constant force [9, 10]: The constant preload force (Spring Preload) is very important within precision, high-speed applications (Fig. 1b). The calibrated linear springs are typically used between the bearing outer ring and housing shoulder. In this case, the bearing arrangement has got a lower stiffened than an arrangement using axial displacement to set the preload. The spindles on internal grinders machine used often spring preload method. The using spring preload forces don't increases axial rigidity and operating temperatures at high rotational speed.

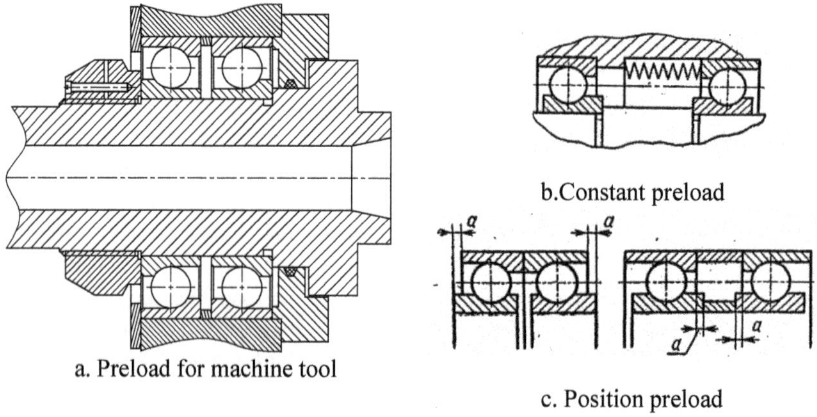

Fig. 1. Preload method

Preload by axial displacement [9, 10]: This preload method (Position Preload) is usually obtained by adjusting: The bearing rings relative to each other in the axial direction or the thickness of spacers rings between two bearings arranged back-to-back or face-to-face, when used in sets of two or more bearings (Fig. 1c). The main advantage of this type is that the degree of rigidity is much higher. Fix preload method is widely used for head spindle of machining centres, milling machines, lathes and drills. It is usually applied to improve the stiffness and precise axial guidance especially when alternating axial loads occur. During operation, an additional increase in preload can also be perhaps grown.

2.2 Calculating Method the Frictional Moment with Preloaded Bearing

The model [9] for calculating the frictional moment uses equation:

$$M = M_s.Z sin\alpha \; (mN.m) \tag{1}$$

Where: M_s : Spin friction for contact angle a centered on the shaft.

$$M_s = \frac{3}{8}.\mu_s.Q.a.E(k) \; (mN.m)$$

μ_s: Contact-surface slip friction coefficient;
Q: Load on rolling elements (N);
a: (1/2) of contact-ellipse major axis (mm);
$E(k)$: With $k = \sqrt{1 - \left(\frac{b}{a}\right)^2}$

As the population parameter, second class complete ellipsoidal integration.

b: (1/2) of contact-ellipse minor axis (mm);
Z: Number of ball;
α: Contact angle (°)

As a general rule, a medium preload should be adopted for spindle of lathers requiring rigidity. The starting torque of 7210 C bearing of head spindles for mini CNC lathes is determined [1]: $M_{ST} \sim$ 40 mN.m while a preload should be 590 N.

2.3 Measurement Method for Starting Torque

Measurement method for starting torque is common. It is possible to stop adjustment when a frictional moment corresponding to the desired preload has been reached. This method obtained by measuring the tangential force of spindle by either using a spring balang or rotary torque devise as showed in Fig. 2a [10]. It can be influenced by the oil film formation in the contact area. It is best suite to machining centres, milling machines, lathes and drills. Which have got high preload.

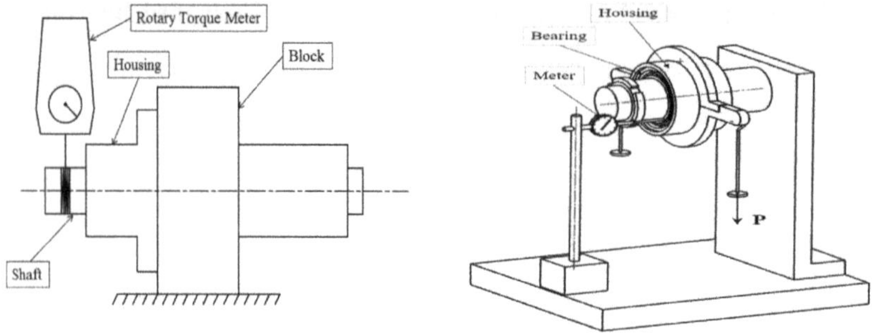

a. Measuring the tangential force of spindle by either using a spring balang or rotary torque devise[10]

b. Starting torque measurement method for Duplex angular contact bearing is done by DB 7210C with counterbalance

Fig. 2. Measurement method for starting torque

2.4 Experimental Apparatus

The basing on Fig. 2a can built the starting torque measurement schema for 7210 C bearing of mini CNC lathes, it's shown in Fig. 2b. The experimental apparatus use a pair single row angular contact bearings 7210 C, which are mounted DB arrangement same as head spindle of mini CNC lathes (show in Fig. 1a). Axial displacement to set the preload performed by tightening the nut on the inner ring and displacement value determined by the dial indicator, with an accuracy of 0.001 μm. The starting tangential force of bearing is determined by the weight of the weight.

Test sequence: Tightening nut A to make the inner ring move a certain distance, applying the tangential load to rotate the bearings. The values of the weights are the starting tangential force of them. Each experiment repeated three times to ensure the measurement accuracy.

3 Results and Discussion

The measurement result of starting torque for the main spindle of lathe depends on preload by axial displacement, are listed in Table 1.

Table 1. The starting torque of bearing in spindle of the mini CNC lathe and step of axial displacement.

Position preload δ(μm)	Experiment 1		Experiment 2		Experiment 3		Average
	P(g)	MST (mN.m)	P(g)	MST (mN.m)	P(g)	MST (mN.m)	MST (mN.m)
0	15	2.1	10	1.4	20	2.8	2.1
8	40	5.6	50	7	35	4.9	5.8
16	90	12.6	100	14	110	15.4	14
24	200	28	190	26.6	220	30.8	28.5
32	340	47.6	350	49	370	51.8	49.5
40	540	75.6	560	74.2	580	81.2	78.4

Based on the data in Table 1, a graph of the variation of the fictional starting torque and the axial displacement of bearings in lathes spindles is shown in Fig. 6.

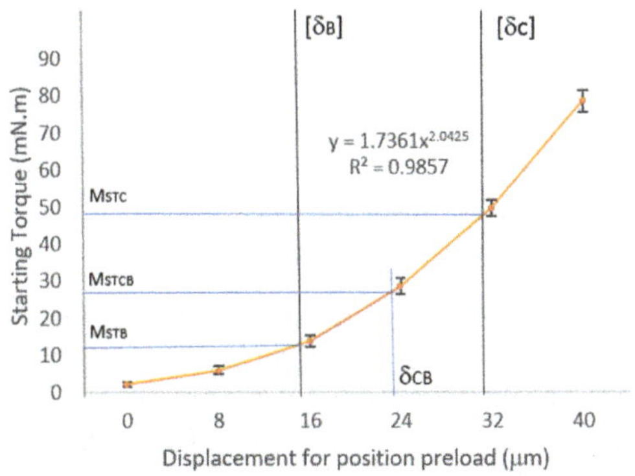

Fig. 6. The relationship of the starting rotary torque of bearing in lather spindles and step of axial displacement

Creating preload by moving the relative rings in the bearing is consistent with the structure of the CNC machine spindle unit. However, it is difficult to adjust precisely the displacement amount to a few μm, and an adjustment jump that exceeds the requirements may occur. If the preload is larger than necessary, abnormal heat generation increased friction torque, reduced fatigue life, and so on may occur. The amount of the preload should be determined carefully considering the operating conditions and purpose of the preload.

Figure 6 shows that the frictional starting torque of bearings always nonlinearly increases with the position preload. Comparing the experimental results with the calculation indicates that the calculated starting friction torque value is only ~80% of the test friction torque.

The mini CNC lathes when using 7210 C, position preload max is $[\delta_C] = 31 \mu m$. During working, the main bearings are wear lead to the reduction of the amount of axial displacement. When the displacement amount is reduced to about 15 μm = $[\delta_B]$ (~ Class B), it is necessary to adjust the position preload again to restore Class C. Thus, the starting torque changes to 75%. In the common case, adjust the bearings once after a running cycle when axial wear is up to 16 μm. Hence the friction loss on main bearings is significant, and it changes significantly during the machine's operating cycle at the adjustable preload level.

Figure 6 illustrates the effect of the axial displacement preload on the frictional starting torque of main spindles. It is possible to reduce the starting friction on the main spindles but still ensure the stiffness within the class C range. Divide the class C position preload into two ranges of values:

$$\delta_{CB} = ([\delta_C] - [\delta_B])/2 = 8\,\mu m \tag{2}$$

Where

$[\delta_C]$ - preload by axial displacement for Class C;
$[\delta_B]$ - preload by axial displacement for Class B.

During the cycle of the spindles, bearings adjust the preload twice. The first time to fix the displacement is δ_{CB} = 8 μm + δ_B, the second time is 8 μm. Thus, based on the graph in Fig. 6, the starting friction torque of the spindle of the mini CNC lathes is reduced to 40%, and the fluctuation is about ~30%. Hence, the time meet controls will decline, and the spindles will have a more stable speed.

In case of classifying the group of rough-machines and finishing machines: The finishing machines, adjusted to create a preload at the upper range in accordance with small cutting deep, cutting power, high accuracy and rigidity requirements. Rough-Machines which adjusted to lower range of preloaded that is suitable for large cutting deep, cutting power, and does not require high accuracy, rigidity. Then, the machine efficiency will be best due to the compensation of friction and cutting power.

4 Conclusions

The research results on starting rotary torque of bearing on spindles of mini CNC lathes when preload by axial displacement changed have given the following conclusions:

1. The starting rotary torque of bearing on the spindles determination system has been developed based on preload by axial displacement. The preload by axial displacement generation is easier to implement than the preload with a constant force.
2. Fictional starting torque of bearing on spindles depends on the nonlinearity of the preload by axial displacement. With class C for bearings on spindles on mini CNC lathes, starting rotary torque is significant, preload needs to be adjusted under long cycle operating conditions.
3. Proposing a solution to reduce the friction loss of the mini CNC lathe spindle unit by segmenting preload by axial displacement into two parts on a class B basis. Then, a low-level adjustment will be performed twice in the machine's duty cycle with the corresponding amount of travel.
4. In case it is necessary to maintain the stiffness of the spindle unit stable in a higher region suitable for precision machining, it is possible to segment the preload by axial displacement in the opposite direction. Then the main bearing friction losses are in a large area.
5. Preload by axial displacement can be adjusted according to the rough and finishing groups to satisfy the machine accuracy and productivity while increasing machine efficiency.

References

1. Rolling bearings for industrial machinery; NSK (2016)
2. Šarenac, M.: Stiffness of machine tool spindle as a main factor for treatment accuracy. Sci. J. Facta Universitatis **1**(6), 656 (1999)
3. Prakosa, T., Wibowo, R.I.A.: Optimizing static and dynamic stiffness of machine tools spindle shaft, for improving machining product quality. J. KONES Powertrain Transp. **20**(4), 363–370 (2013)
4. Holroyd, P., Crinela and Ford, Geoffrey, D.G.: Determination of stiffness and damping sensitivity for computer numerically controlled machine tool drives. Proc. Inst. Mech. Eng. Part C J. Mech. Eng. Sci. 1165–1177 (2003). ISSN 0954-4062
5. Rastegari, A., Mobin, M., Archenti, A.: Condition based maintenance of machine tools: vibration monitoring of spindle units. In: IEEE 63nd Annual Reliability and Maintainability Symposium (2017)
6. Canh Tong, V., et al.: The effect of regular misalignment on the running torques of tapered roller bearings. Tribol. Int. **95**, 76–85 (2016)
7. Zhang, C., Gu, L., Mao, Y., Wang, L.: Modeling the frictional toque of a dry -lubricated tapered roller bearing considering the roller skewing. Friction **7**(6), 551–563 (2019)
8. Matsuyama, H., Kamamoto, S.: Analysis of frictional torque in raceway contacts of tapered roller bearings. Koyo Eng. J. English Edition No 159E (2001)
9. López de Lacalle, L.N., Lamikiz, A.: Machine tool for high pèrformance machining. Springer (2009)
10. Machine tool spindle bearing selection & mounting guide, NSK

Enhancement of UV Resistance and Thermal Stability of the Unsaturated Polyester Material by Introducing MHHPA Fragment into the Molecular Skeleton

Nang Xuan Ho[1,2(✉)], Tuan Anh Pham[1,2], Huong Thu Ha[3], Thuc Quang Dong[3], Nam Hoai Nguyen[3], and Dieu Vinh Tran[1,2]

[1] Phenikaa University, Yen Nghia Street, Ha Dong District, Hanoi 10000, Vietnam
nanghx@phenikaa.com

[2] Phenikaa Research and Technology Institute (PRATI), A&A Green Phoenix Group JSC, 167 Hoang Ngan Street, Cau Giay District, Hanoi, Vietnam

[3] Polymer Center, A&A Green Phoenix Group JSC, 167 Hoang Ngan, Cau Giay District, Hanoi 10000, Vietnam

Abstract. To impove the ultraviolet (UV) resistance and thermal stability of unsaturated polyester resin (UPR), methylhexahydrophthalic anhydride (MHHPA) – a saturated cyclic fragment was added into molecular skeleton of UPR. To get the best objectives mentioned above, replaced phthalic anhydride with MHHPA in different rates of 25%, 50%, 75% and 100% and in sequence, UPRs products were named as UPR-M25, UPR-M50, UPR-M75 and UPR-M100 respectively. In the process of UV radiation with the presence of oxygen, carbonyl groups had been formed gradually in the structure of the unsaturated polyester polymer. These carbonyl groups conjugated with neighboring doubles lead to form chromophore bearing color groups. The total changes of color ΔE* were determined according to the CIE L*a*b* method. After 1000 h exposure under UV radiation, the results showed that the increment ΔE* of the original sample Ortho – UPR is higher than UPR-M100, the increase of ΔE* value was from 7.36 of UPR-M100 to 21.82 of UPR-M0, respectively. The thermal stability of cured UPR-M100 and UPR-M0 was investigated by TGA. The results indicated that the mass loss of UPR-M100 happened from the higher temperature at 391 °C, whilst UPR-M0 from the lower temperature at 378 °C.

Keywords: Unsaturated polyester resin (UPR) · Methylhexahydrophthalic anhydride (MHHPA) · UV resistance · Weathering resistance · Thermal stability

1 Introduction

Unsaturated polyester resins are typically synthesized from the condensation process of unsaturated and saturated acids/anhydrides and diols or oxides, then mixed with styrene monomer as the both of the dilute and the crosslinking agent to form a rigid three-dimension network [1]. UPR materials are the most versatile types of thermoset

B. T. Long et al. (Eds.): MMMS 2020, LNME, pp. 428–436, 2021.
https://doi.org/10.1007/978-3-030-69610-8_59

material, play an important role in the outdoor and indoor applications such as the automobile, building construction, coating, marine, aerospace industries. UPR materials have been widely used individually and in the matrix of glass fiber reinforcement composites because of its competitive price, easy processing, high chemical resistance and high strength [2–4]. Besides, UPR materials also be affected by the weathering conditions including UV radiation, heat, moisture, etc. resulting in degrading its surface and mechanical properties and reducing the lifetime of these products [5]. Therefore, it seriously limits the further commercial utilization in the outdoor field.

To improve the weathering resistance of UPR materials, Min Jae Shin et al. used the photo-stabilized compounds for the unsaturated polyester to protect against UV exposure consisting of 2-hydroxy4-acryloyloxybenzophenone (HABP), 2-[3-(2H-benzotriazol-2-yl)-4-hydroxyphenyl]ethyl methacrylate (BTEM), and 4-acryloyloxy-1,2,2,6,6 pentamethylpiperidine (APMP). The results were estimated using a color difference meter and showed that HABP, BTEM and APMP have a synergistic effect with the best ratio was 29.4:68.6:2, respectively to increment the UV stability of UPR materials [6]. Recently, nanoparticles have been utilized as an advanced filler to improve the anti-ultraviolet aging performance of UPR-based composites as well as the stiffness, harness, and so on. Nano TiO_2 appears as a potential candidate for nanoparticle, which was the surficial modification with titanate in the UPR by the experiments of Chaohua Jiang et al. [7]. The results presented that TiO_2 nanoparticles were comparable with UPR matrix. The best coupling ratio of TiO_2 and titanate are 4% and 3% enhancing the anti-ultraviolet aging properties of the composites.

In this work, MHHPA fragment was used for the enhancement of the UV resistance and thermal stability properties.

2 Materials and Methods

2.1 Materials

Anhydride maleic (AM) 99.8%, NanYa Plastic, (Taiwan); Ortho-Anhydride phtalic (AP) 99.85%, Continental Petrochemicals, Thailand; Methylhexahydrophthalic Anhydride (MHHPA) 99.9%, Linda Chemicals, England; 1,2 propyleneglycol (PG) 99.8%, Shell Chemicals, Malaysia; Hydroquinone (HQ) 99,5%, Anhui Jinao Chemical, China; Styrene monomer 99.9%, Wee Tee Tong Chemicals, Singapore; Dibutyltin oxide (DBTO) 99,9%: Ackoss, Belgium; Tert butyl peroxybenzoate (TBPB) and Cobalt (II) octoate 10%, AkzoNobel, Belgium.

2.2 Synthesis of Modified UPRs

Ortho-UPR was synthesized in the one-stage process by the condensation of polycarboxylic acids AM and AP with propylene glycols (PG). In a 1000 ml three-neck flask equipped with a mechanical stirrer, thermo meter, the mixture was reacted in the molar ratio of reactants including AP: AM: PG was 0.60: 0.40: 1.05, respectively under the nitrogen gas atmosphere. The reaction took place by hot melting solution method at the range of temperature 150 ÷ 190 °C for 12 h in the presence of 0.2% wt. of DBTO to accelerate the esterification rate and inert gas nitrogen to prevent discoloration. The progress of the polymerization reaction was conventionally monitored by means of acid number. The reaction was finished and nitrogen gas was stopped until acid number below 25 mg KOH/g. The condensation product was diluted by reactive styrene monomer and was added 0.02% wt. of HQ as an inhibitor to prevent gelation. Technical parameters of Ortho-UPR: acid number 20 ÷ 25 mg KOH/g; styrene content 33 ÷ 35% wt.; viscosity of 630 ÷ 650 cps at 23 °C.

MHHPA-modified UPR samples were synthesized by similar mentioned method in which AP was replaced by MHHPA in the formula and showed in Table 1. Technical parameters of modified UPR: acid number 20 ÷ 25 mg KOH/g; styrene content 33 ÷ 35% wt.; viscosity of 630 ÷ 650 cps at 23 °C.

Table 1. The molar ratios between AP and MHHPA were used in the modified UPR formulation.

Samples	Molar ratios			
	AP	MHHPA	AM	PG
UPR-M0	0.60	0.00	0.40	1.05
UPR-M25	0.45	0.15	0.40	1.05
UPR-M50	0.30	0.30	0.40	1.05
UPR-M75	0.15	0.45	0.40	1.05
UPR-M100	0.00	0.60	0.40	1.05

2.3 Preparation of the Specimens

The UPR composition was prepared with 1% wt. of TBBP as an initiator and 0.01% wt. of Co^{2+} as an accelerator. Each sample composition then was stirred and poured into mold with dimensions 120 mm × 12 mm × 2 mm for the following tests. For all samples, the curing process occurred in the isothermal heating at 80 °C for 45 min, followed by post-cured at 120 °C in 60 min. The maximum degree of the reaction was reached at the curing conditions.

2.4 Accelerated Weathering Test/UV – Ageing Exposure Methods

The accelerated weathering test was carried out in the Atlas UV-Test™ Fluorescent UV/Condensation Weathering Instrument according to ASTM G154 – 06 standard. The non-modified and modified UPR samples were set up for UV exposure at 500 h and 1000 h.

2.5 Color Measurement

The color changes of the original UPR and modified UPRs were measured at the initial time, 500 h and 1000 h exposing UV radiation using a *Ci7800.*

Benchtop Spectrometer equipped with a CIE Lab system in the reflection mode. Three important parameters are L*, a* and b* determined by CIE Lab; where L* assigned to lightness, that has a range from 100 (white) to 0 (black), and a* and b* are the chromaticity indices, with +a* assigned to the red direction, –a* as green, + b* as yellow, and –b* as blue.

The total change in color (ΔE*) was calculated based on ASTM D2244 using the Euclidean formula as presenting in the following the Eq. (1):

$$\Delta E* = \sqrt{(\Delta L^*)^2 + (\Delta a^*)^2 + (\Delta b^*)^2} \quad (1)$$

Where ΔL*, Δa*, Δb* are the differences in the initial and specific time.

2.6 Thermogravimetric Analysis

Thermogravimetric analysis (TGA) was employed to evaluate the thermal properties of UPR samples. The analysis took place using *Setaram Labsys TG* in the range of 20 to 900 °C and the heating rate at 10 °C/min. The weight of each sample was used to analyze about 10 ± 5 mg in a platinum sample pan under an inert atmosphere.

3 Result and Discussion

3.1 The Color Changes

The value of delta E for each UPR sample was calculated following the Eq. (1) after 500 h and 1000 h exposure UV radiation. Table 2 shows the L*, a*, b* and calculated value ΔE* presenting the color changes of UPR samples.

Table 2 shows the changes of color index constituents L*, a*, b* inducing the shift of ΔE* values. It indicates that accelerated weathering exposure promoted the photodegradation process, which may be the result of various mechanisms [8]:

Photolysis: the chromophore of the UPRs or impurities can absorb the UV radiation leading to form the photoproducts absorbing light. These photoproducts may or may not relate to the intermediate radicals. Oxygen was assumed not to involve in the photochemical process to affect the color changes of the UPR samples.

Direct photo-oxidation: the chromophore of the polyester structure can absorb the UV radiation leading to the radicals react with oxygen causing the oxidation processes. After the process, the products that have absorbing light ability can be generated.

Indirect photo-oxidation: the chromophore of defects or impurities that does not belong to the polyester structure can absorb the UV radiation to start the oxidation reactions and form the products that have absorbing light ability.

From the color measurement results exhibited in Fig. 1, we can see the ΔE* value decreases in accordance with increasing of the concentration of MHHPA fragment in UPR samples at 500 and 1000 h after the exposure of UV radiation. Besides, the lightless index slightly decreases and the color index a* also lightly reduces for all samples after 500 and 1000 h exposing UV radiation, which expresses that the color fading of the samples. However, the change of color index b* increases nearly the same with the trend of total change in color ΔE* value, which means that the surface transforms blue to yellow [1]. The tendency of L* values inclines the black, the intensive increase of b* values and a slight decrease of a* values confirmed that the yellow-color products are generated during the 1000 h exposing UV time [9].

Table 2. The three parameters L*, a*, b* and total change in color ΔE* of UPRs

Samples	Exposing time	L*	a*	b*	ΔE*
UPR-M0	0 h	90.30	−1.24	5.65	0.00
UPR-M25		95.14	−0.99	3.54	0.00
UPR-M50		93.97	−1.18	4.30	0.00
UPR-M75		93.66	−1.17	4.15	0.00
UPR-M100		94.35	−0.98	3.50	0.00
Samples	Exposing time	L*	a*	b*	ΔE*
UPR-M0	500 h	89.30	−3.43	19.55	14.11
UPR-M25		89.49	−3.15	16.07	13.91
UPR-M50		91.39	−2.92	13.21	9.44
UPR-M75		89.30	−2.84	11.99	9.12
UPR-M100		90.40	−2.20	8.18	6.24
Samples	Exposing time	L*	a*	b*	ΔE*
UPR-M0	1000 h	88.69	−3.67	27.27	21.82
UPR-M25		87.51	−3.64	23.02	21.09
UPR-M50		89.24	−3.45	18.48	15.12
UPR-M75		87.46	−2.81	13.27	11.15
UPR-M100		90.20	−2.29	9.44	7.36

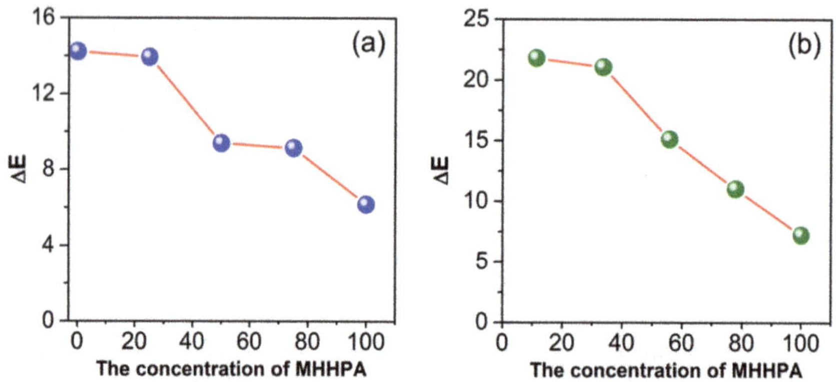

Fig. 1. The relationship between the total color change ΔE* and the concentration of MHHPA fragment in the UPRs at (a) 500 h and (b) 1000 h after exposing UV radiation.

Among the UPR samples at 1000 h exposing time, it is important to note that the smallest change belongs UPR-M100 sample with ΔE* value 7.36 and the biggest change in UPR-M0 with ΔE* value 21.82, the decrease of 66.27%. This observation suggests that the effect of the alicyclic structure from MHHPA fragment reduced the absorption of UV radiation comparing with phthalate groups in the original UPR possessing responsible for primary absorption [10].

From the analysis, it may be concluded that comparing with the non-modified UPR, the MHHPA fragment in the modified UPR samples significantly attributed to improve the weathering resistance after 1000 h of time exposure.

3.2 Thermal Stability of UPR Samples

Figure 2 and Fig. 3 illustrate the TGA and DTG curves of UPR samples that provided the information on thermal stability and the occurred reactions. The curve of UPR-M0 showed the weight loss below 200 °C due to the moisture evaporation and the unreacted reactants of the samples. It can be observed that the chemical transformation of UPR samples occurred in separated two stages, which was exhibited on the TGA curves. As Yolanda Bautista et al. [3] presented in their work, the decomposition of the first group began at 200 °C and was relatively rapid from 300 to 450 °C; the total mass loss was determined more than 90% of the original mass. The peak can be seen with the asymmetric shape, which means the results of several reaction. The second group took place above 450 °C and accounted for the remaining mass loss. It indicates that the thermal degradation process with the presence of oxygen producing the combustion products. Anderson et al. [11] investigated the pyrolysis of UPRs in air and found the mainly degradable products were generated including benzaldehyde, unsaturated hydroxyesters, phthalic anhydride and the combustion products like carbon dioxide, hydrogen and a little methane gas. Learmonth et al. [4] proposed that the first stage related the clearage of styrene crosslinking, possibility by peroxidation to benzaldehyde. Then, the decomposition of major polyester chains occurred to form pyrolysis products.

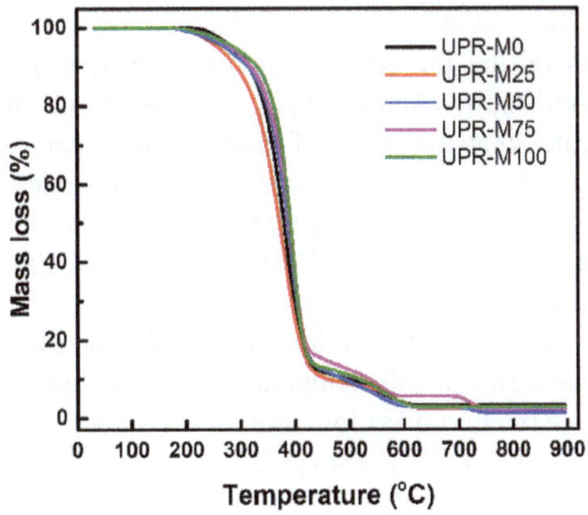

Fig. 2. The TG curves of UPR samples in the air

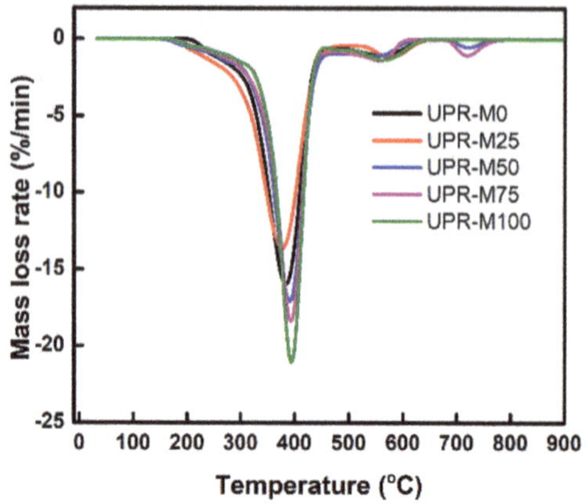

Fig. 3. The DTG curves of UPR samples in the air

Table 3 shows the data obtained from TGA and DTG analysis, that were characterized in the weight loss temperature. It proved that the temperature at 10% weight loss assigned to initial decomposition temperature increased with increasing the content of MHHPA in UPR samples, from 263.80 °C of UPR-M0 to 331.92 °C of UPR-M100, the increase of 25.82%. It means that the original UPR degraded earlier than the modified UPRs. Moreover, T_{1max} of UPR samples which possess much more than 25% of content MHHPA, also was higher than the original UPR sample. Besides, it is observed that the char residues at 800 °C of UPR-M100 remarkably grew compared with that of UPR-M0, from 0.83 to 2.76%, respectively. It demonstrates that the UPR samples modified with the MHHPA component were more stable than the neat UPR sample.

The results obtained from the TGA and DTG measurements revealed that the MHHPA component introduced to the samples affected the thermal decomposition behavior of UPR. The initial temperature decomposition of UPR samples was noticeably raised with raising the concentration of MHHPA and reach 331.92 °C at 100% of MHHPA modification. And the T_{1max} was found to be higher in the modified samples, which indicates that the chain of modified polymers was more stable than non-modified polymer chains. Furthermore, the char residues acquired at 800 °C exhibits high thermal stability at the high temperature of the modified UPR samples. According to the obtained results, the MHHPA modified UPR samples showed the enhancement of thermal stability.

Here, T_{10} – the temperature of 10% loss decomposition (set T_{10} as the beginning decomposition temperature), T_{50} – the temperature of 50% loss decomposition, T_{1max} – the temperature at the maximum mass loss rate for the stage 1, T_{2max} – the temperature at the maximum mass loss rate for the stage 2, R_{max} – the decomposition speed at the decomposition peak.

Table 3. The TG and DTG data of the UPR samples under the air condition.

Specimen	Temperature (°C)			R_{max}	Char residues at 800 °C (wt%)
	T10	T_{50}	T_{1max}/T_{2max}		
UPR-M0	263.80	378.14	386.89/560.90	16.9474/1.3295	0.83
UPR-M25	293.58	371.57	376.24/575.77	13.6188/1.1197	2.51
UPR-M50	315.55	386.69	391.19/556.93	17.0623/1.0747	1.25
UPR-M75	321.30	390.28	393.25/558.82	18.3619/1.3954	2.17
UPR-M100	331.92	391.77	394.18/566.22	21.0319/1.3371	2.76

4 Conclusion

An unsaturated polyester resin was successfully modified by the saturated cyclic fragment of MHHPA. The discoloration of UPR samples was studied during the degradation and photo-oxidation by 1000 h of accelerated wearing test with the CIE Lab system. The results indicated that the UV resistance of UPR samples improved with increasing concentration of MHHPA in the resin formulations and reached the decrease ΔE^* value of 66.27% in the total change in color using MHHPA as a modification substance replacing phthalic anhydride (PA) at the rate of 100%. All the obtained test results from the TGA data revealed that the best thermal resistance was acquired from the UPR modified by MHHPA replacing 100% phthalic anhydride. Overall, the UPR samples modified by MHHPA can enhance the weather resistance and the thermal stability as well as reduce char residues at the high – temperature region.

Acknowledgements. This work is supported by Phenikaa Group and Phenikaa University.

Declaration of Competing Interest. We declare that we have no financial and personal relationships with the other people or organizations that could inappropriately influence (bias) their work or state.

References

1. Jia, Z., Li, X., Zhao, Q.: Effect of artificial weathering on surface properties of unsaturated polyester (UP) resin. Mater. Chem. Phys. **121**, 193–197 (2010)
2. Guo, Q., Zheng, H., Zheng, S., Mi, Y., Zhu, W.: Phase behaviour, mechanical properties and thermal stability of thermosetting polymer blends of unsaturated polyester resin and poly (ethylene oxide). J. Mater. Sci. **34**, 123–128 (1999)
3. Bautista, Y., Gozalbo, A., Mestre, S., Sanz, V.: Thermal degradation mechanism of a thermostable polyester stabilized with an open-cage oligomeric silsesquioxane. Materials **11**, 22 (2018)
4. Learmonth, G.S., Nesbit, A.: Flammability of polymers. v. thermal volatilisation analysis of polyester resin compositions. Br. Polymer J. **4**, 317–325 (1972)

5. Stamatakis, G., Knuutinen, U., Laitinen, K., Spyros, A.: Analysis and aging of unsaturated polyester resins in contemporary art installations by NMR spectroscopy. Anal. Bioanal. Chem. **398**, 3203–3214 (2010)
6. Shin, M.J., Shin, Y.J., Moon, Y.U., Shin, J.S.: Polymerizable ultraviolet stabilizers for unsaturated polyester-based bulk molding compounds. J. Elastomers Plast. **46**, 569–576 (2014)
7. Jiang, C., Jin, C., Wei, M., Yan, S., Chen, D.: Mechanical and thermal properties improvement of unsaturated polyester resin by incorporation of TiO_2 nanoparticle surface modified with titanate. Mater. Res. Express **5**, 115008 (2018)
8. Salim, M.S., Ariawan, D., Ahmad Rasyid, M.F., Mat Taib, R., Ahmad Thirmizir, M.Z., Mohd Ishak, Z.A.: Accelerated weathering and water absorption behavior of kenaf fiber reinforced acrylic based polyester composites. Front. Mater. **7**, 1–5 (2020)
9. Müller, U., Rätzsch, M., Schwanninger, M., Steiner, M., Zöbl, H.: Yellowing and IR-changes of spruce wood as result of UV-irradiation. J. Photochem. Photobiol. B **69**, 97–105 (2003)
10. Michaille, S., Khalil, Z., Lemaire, J., Arlaud, P.: Photolysis and photo-oxidation of unsaturated polyesters. Makromol. Chem. Macromol. Symp. **25**, 263–269 (1989)
11. Anderson, D.A., Freeman, E.S.: The kinetics of the thermal degradation of the synthetic styrenated polyester, laminac 4116. J. Appl. Polym. Sci. **1**, 192–199 (1959)

Research and Design of Rubber Extrusion Molds

Vinh Hung Tran[1,2](✉) and Khac Khanh Bui[3]

[1] Faculty of Mechanical Engineering and Mechatronics,
PHENIKAA University, Hanoi 12116, Vietnam
hung.tranvinh@phenikaa-uni.edu.vn
[2] A&A Green Phoenix Group JSC,
PHENIKAA Research and Technology Institute (PRATI),
No. 167 Hoang Ngan, Trung Hoa, Cau Giay, Hanoi 11313, Vietnam
[3] Department of Mechanical Engineering,
Hung Yen University of Technology and Education, Hung Yen, Vietnam

Abstract. In rubber manufacturing, the shape and size of rubber details after extrusion are often different from the mold because of its characteristics. In order to achieve required shape and size, it is necessary to determine the adjustment coefficient and extrusion mold profile. Experimental planning is one of the effective methods to determine this coefficient. The authors combine theoretical research with use of design software and experience to determine correction coefficients and extrusion profiles. Obtained results of adjustment coefficients and extrusion profiles for rubber seals on B80HN wagon ensure technical requirements.

Keywords: Experimental · Extrusion moulds · Rubber

1 Introduction

In industrialization and modernization of our country, production industry of rubber products is increasingly developing to meet increasing demand for quantity, various kinds as well as high quality. These kinds of products are applied in the automobile manufacturing industry, construction, health care, home appliances, etc. with increasing demand. In fact, due to the characteristics of this kind of product, the shape and dimension of detail after demolding differs from the mold. Therefore, the design and manufacture of molds to create products with expected shape and dimension are necessary. Due to effects of extrusion temperature and speed, profile model, rubber type, etc. design and manufacture of rubber extrusion mould in right profile is very difficult. In reality, manufacturer has to test extrusion and adjust many times. With aim of finding the extrusion mould profile with the least calibration and best result [1], the research team found the profile correction coefficients of rubber extrusion moulds by means of experimental planning method [1, 2].

B. T. Long et al. (Eds.): MMMS 2020, LNME, pp. 437–446, 2021.
https://doi.org/10.1007/978-3-030-69610-8_60

2 Content

Based on actual experimental planning in the extrusion process, we set relationship between temperature, speed and profile with the variation of product profile by mathematical relationship [2, 3].

$$y = f(x_1, x_2, \ldots, x_j; \alpha_1, \alpha_2 \ldots \alpha_n) = a_0 + a_1x_1 + a_2x_2 + \ldots + a_nx_n + a_{12}x_1x_2 + \ldots + a_{(n-1)n}x_{(n-1)}x_n \quad (1)$$

In which:

y: is the correction profile
xi: is influencing factors
a: is correction coefficient

Thereby, we can determine the optimal mode of the product extrusion process and the profile correction coefficient of rubber extrusion mould in comparison with design profile. Implemention process follows the diagram: Fig. 1.

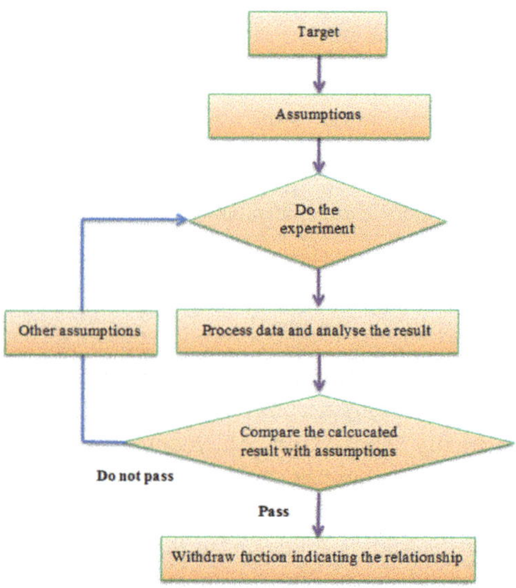

Fig. 1. Definition diagram of mould coefficient

2.1 Design Drawing

The target is extrusion of product with the most exactly or nearly exactly design profile as Fig. 2.

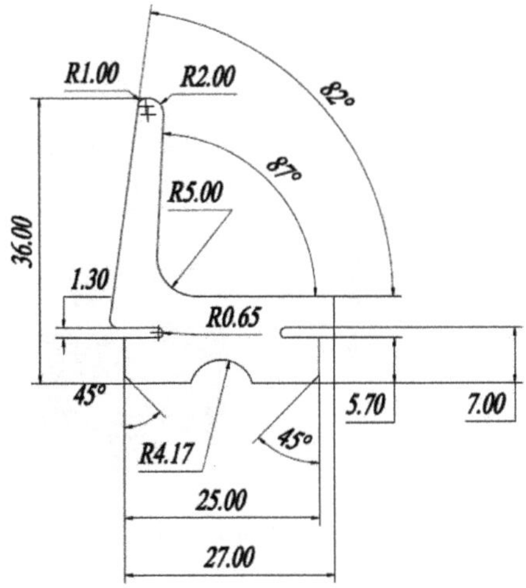

Fig. 2. Profile of rubber felt seal needed to design

2.2 Experiment

Do N = 2^k experiment [4–6]. In which k is an experiment factor (extrusion speed, temperature, profile). Do experiment according to "experiment table with 3 factors" as Table 1.

Table 1. Experiment table with 3 factors

Factor	Experimental points							
	(1)	(2)	(3)	(4)	(5)	(6)	(7)	(8)
X_1	−1	+1	−1	+1	−1	+ 1	−1	+1
X_2	−1	−1	+1	+1	−1	−1	+1	+1
X_3	−1	−1	−1	−1	+1	+1	+1	+1
y	y_1	y_2	y_3	y_4	y_5	y_6	y_7	y_8

In which:

X_1: Extrusion speed;
X_2: Extrusion temperature;
X_3: Profile model;
X_j = 1 is corresponding with high level of factor j;
X_j = −1 is corresponding with low level of factor j.

Profile of experiment product is measured on specialized machine Micro-Hite® DCC.

Fig. 3. Coordinates measurement machine Micro-Hite® DCC

Technical characteristics of the machine:

There are high-speed direct drives; Measured dimensions of samples are 22 × 30 × 19 in (559 × 750 × 483 mm); The working table is made by granite stone with high accuracy and stability; High measurement speed: 40 measuring points/minute; The accuracy of machine 0.1 μm; Air cushions reduce friction for axes when moving; The machine structure allows improving accuracy of measurement when the machine is operating in environments with large temperature variations; Measured results are checked for homogeneity and put into experimental problems;

After drawing, according to requirements of user, the program can put into the drawing frame and the descriptions stored in a **DWG file** outside.

2.3 Apply Experimental Results in Software to Calculate the Profile Adjustment Coefficients of Extrusion Molds

2.3.1 Profile Adjustment Coefficient Calculation Software of the Extrusion Molds

The software is written in **Autolisp** programming language, embedded in Autocad environment [7, 8].

The software application is written in Autolisp language, so whole program is a collection of functions divided into functional groups as Table 2 with the distribution diagram as shown in Fig. 4:

Table 2. Function groups of programs

Graphic user interface				
Excel loader	Definition file loader	State machine	Calculation	Drawing
Definition files Autolisp file (dialogs, menus)				

Drawing process is implemented as follows:

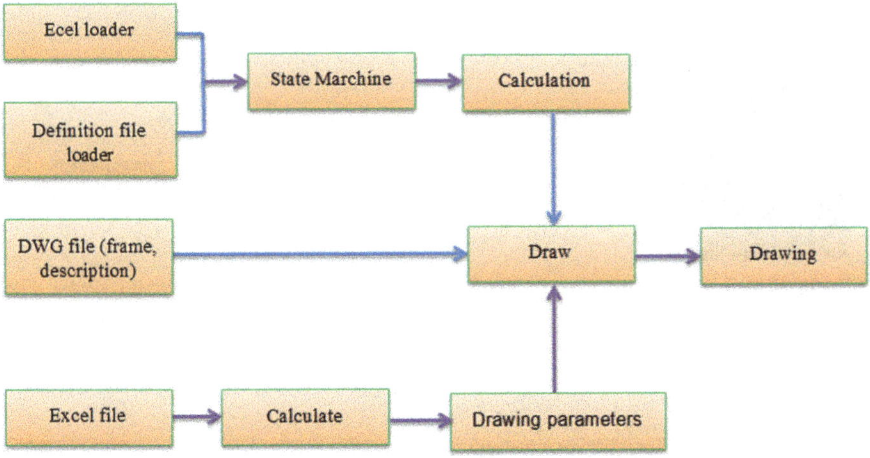

Fig. 4. Distribution diagram of function group

In which:

User interface:

Includes **dialog boxes** that allow users to input data and control commands. The **excel loader** group has the function of reading data from excel file and writing data to **excel file**. The **definition file loader group** has the function of reading and processing information in definition files. The **state machine** group is capable of handling strings read from a definition file or input from the keyboard to extract drawing commands and parameters of drawing commands. The calculated values are then recorded in excel file. The **drawing** group performs the drawing of profile into an Autocad drawing. Drawing parameters are imported from excel file and processed before calculating. Draw commands can be imported from the definition file or directly from the keyboard. Each drawing command is processed via **state machine** to extract the commands and parameters. After drawing parameters and commands are entered and processed, the program starts to draw each command.

After installing on the Menu of Autocad software [7, 9, 10], additional Menu **"B2007–04-53TD"** (circled) appears as in Fig. 5.

Fig. 5. Menu B2007-04-53TD appears on Autocad

And this time, the software interface is shown as Fig. 6.

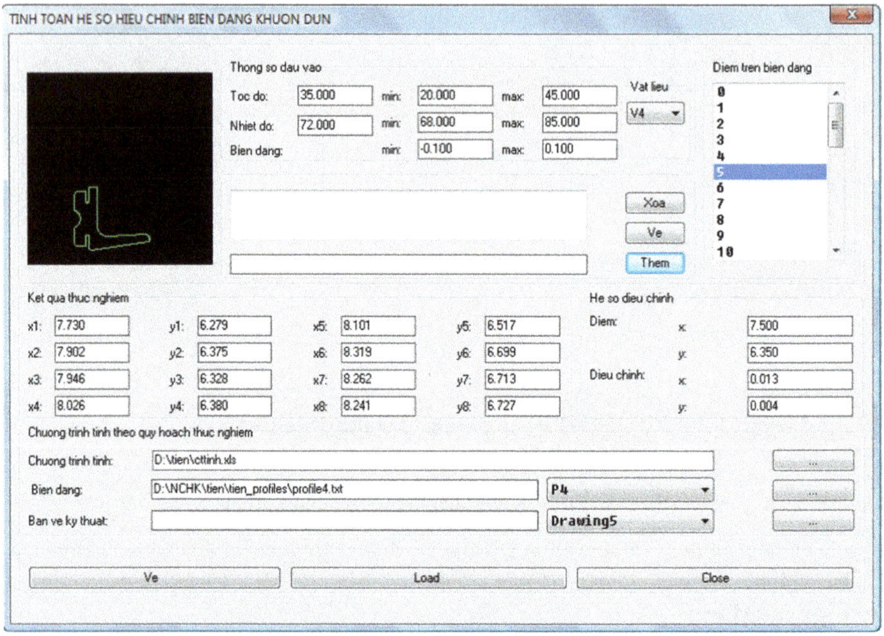

Fig. 6. Software interface after adding Menu [7, 9]

2.3.2 Explanation on Software Interface

The interface allows the mould designer to enter the extrusion mode (input parameters) for the type of rubber to be extruded as shown in Fig. 7.

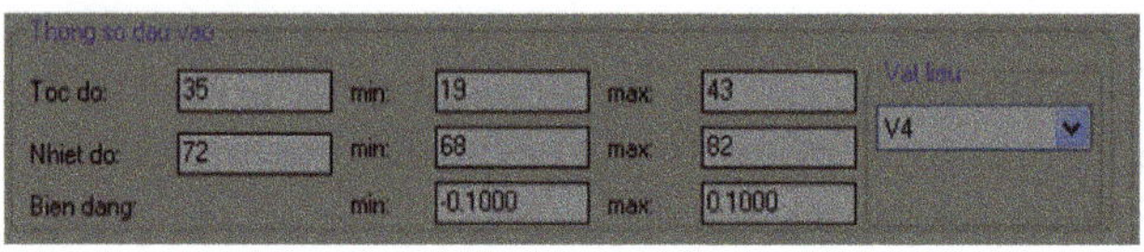

Fig. 7. Software interface after adding Menu

Speed: Enter the extrusion speed which is not exceeded the **min** and **max** limit (The **min** and values are speed limit that extrusion of machine can be guaranteed, depending on each machine and rubber type). We can choose extrusion rubber in the material section, which will automatically give **min** and **max** temperature.

Temperature: Enter extrusion temperature which is not exceeded the **min** and **max** limit (limit value of extrusion temperature).

After entering the input parameters and importing the experiment **file**, Click as Fig. 8.

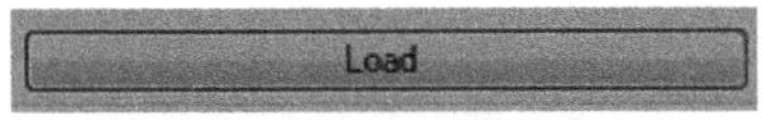

Fig. 8. Load icon

The program will automatically calculate and retrieve the coordinates of the points on selected mould profile.

Profile: Allows users to select mould profile model to edit, implementation method:

Click on the P1 icon as Fig. 9 and select profile model to edit, and then, the profile will appear in the upper left of interface.

Fig. 9. P1 icon

Points on Profile:
They contain points on the surveyed mould profile. We can view coordinates and correction result of each point by **click** on each point them as Fig. 10.

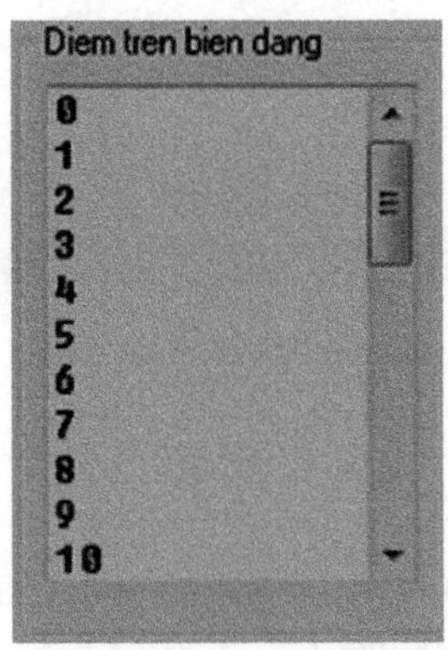

Fig. 10. Points on mould profile

Experiment Result: Allow designer to check or enter fluctuations of each surveyed point on profile in experiments such as Fig. 11.

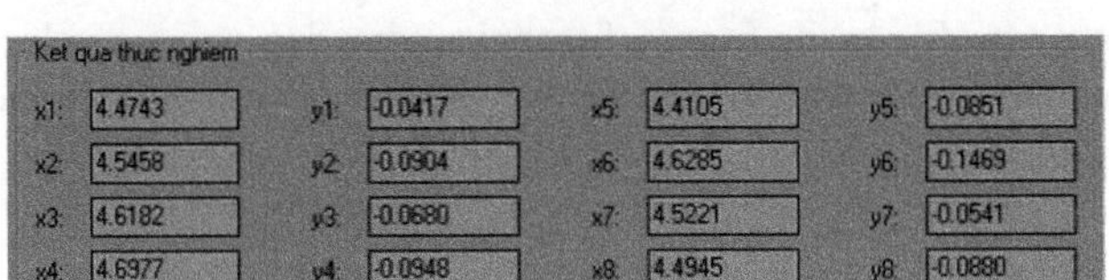

Fig. 11. Experimental result

Correction Factor: shown design coordinates of points on mould profile and size to be adjusted for those points as Fig. 12.

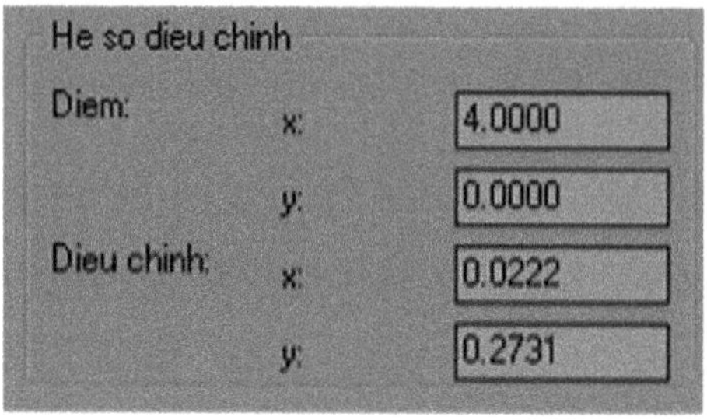

Fig. 12. Calibration result

If there is no **profile** file: We can draw the mould profile directly using **Autocad** drawing commands in the dialog box as Fig. 13:

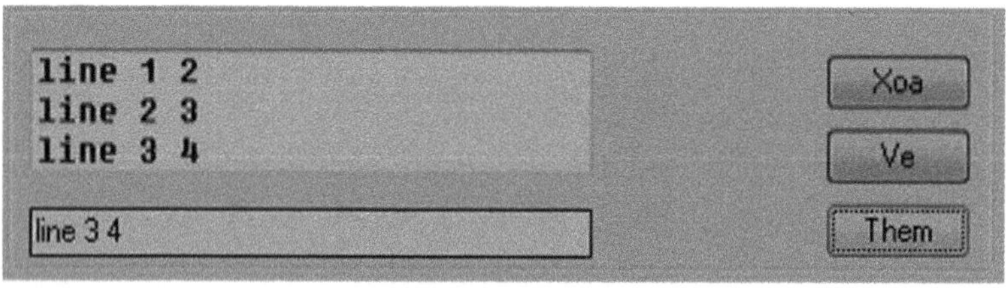

Fig. 13. Dialog box to draw mould profiles

Then press **"Them".**

Experiment:

Need to draw a line connecting point 1 to point 2: **Line 1 2** → **Them**. Need to draw the arc going through the three points 1, 2, 3: **arc 1 2 3** → **Them**. Allow user to import files containing experimental results as Excel. After calculating, this **File** also contains the correction results of each point on profile. Implementation method [10, 11] and [12]:

Click icon 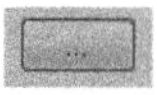 and come to archive folder **File. -> Open.**

Draw: After entering all the information and profile model, click on "Ve", the program will automatically draw a correction mould profile and design mould profile. In order to draw a profile in the frame of technical drawing, we can select **"ban ve ky thuat"** section which contains a number of predefined drawing frames and name frames as Fig. 14 [11], [12].

Fig. 14. Dialog box of technical drawing on interface

After drawing: The result is Fig. 15: Red is correction mould profile, black is design mould profile.

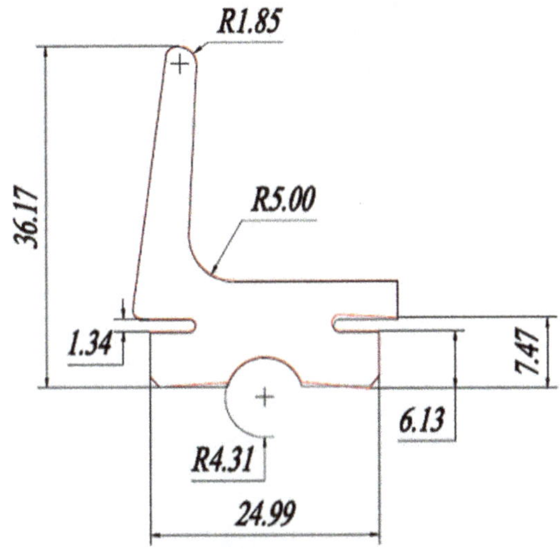

Fig. 15. Result of mould after drawing

3 Conclusion

The change in profile of rubber feat seal under extrusion technology depends on many factors of which mould temperature, extrusion speed, and profile model are chosen for research on Acrylonitril Butadien Rubber (NBR). Research results have set up correction coefficients for different types of felt seal. When fabrication deviations of seal compared to design profile are within (0.1 ÷ 0.2) mm, we can save time and moulding cost. The accuracy of seals will save installation time, improving the tightness and water permeability of cars.

References

1. NIST. Engineering Statistics Handbook (2006)
2. Myers, R.H., Montgomery, D.C., Anderson, M.C.: Cook, Response Surface Methodology: Process and Product Optimization Using Designed Experiments, 3rd edn. Wiley (2009)
3. Mathew, P.G.: Design of Experiments with Minitab. ASQ Quality Press (2005)

4. Antony, J.: Design of Experiments for Engineers and Scientists. Elsevier Science & Technology Books (2003)
5. Montgomery, D.C.: Design And Analysis of Experiments, 5th edn. (2000)
6. Box, G.E.P., Hunter, J.S.: Multifactor experimental designs for exploring response surfces. Annals Math. Stat. **28**, 195–241 (1957)
7. Ridder, D.: Autocad 2000 im Maschinenbau (2000)
8. Kwiatkowski, D., Modławski, M., Jaruga, T.: Design of flash pocket inserts in an extrusion blow mould based on the results of numerical simulations. In: XXI International Polish-Slovak Conference "Machine Modeling and Simulations 2016", pp. 127–134 (2017)
9. Razeghiyadaki, A., Zhang, D., Wei, D., Perveen, A.: Optimization of Polymer Extrusion Die Based on Response Surface Method. Licensee MDPI, Basel (2020)
10. Sikora, J.: Design of Extrusion Heads. Department of Polymer Processing Lublin University of Technology
11. Ouwerkerk, G.V.: CAD implementation of design rules for aluminium extrusion dies. University of Twente, ISBNs9789036528146 (2009)

Construct the Stable Domain of the High-Speed CNC Milling Machine Based on the Surface Wave Criterion

Phan Van Hieu(✉)

School of Mechanical Engineering,
Hanoi University of Science and Technology, Hanoi, Vietnam
hieu.phanvan@hust.edu.vn

Abstract. To improve productivity tools but still ensure product quality, we have two approaches: Improving the quality of cutting tools such as the use of tools coated.., or research to build a sustainable chart of processes, thus helping to select the best cut mode. In this article, the author studies and determines the stable domain of the Super MC 500 High-speed CNC milling machine, to improve the machining productivity on this machine. To assess stability during the cutting process, we can rely on many criteria such as: Based on the system's vibration (vibration measurement), based on sound (measuring sound intensity), based on surface waves, or based on the sheer force chart. The author chooses the surface wave criterion to determine the stable domain. The method of determining the stable domain of the cutting process based on surface wave criteria of Super MC 500 high-speed CNC milling machine is using Matlab software, ensuring high reliability, saving time and money. Testing fee (test cut). During the research process, the author has determined the stable domain by machining and experimental measurement to test and re-evaluate the reliability of the above method.

Keywords: Stable domain · Surface wave · Axial depth of cut · Spindle rotation speed

1 Introduction

Vibration during the cutting process reduces the quality of the machining surface, causing a lot of noise, which is also known as cutting instability. To identify the unstable machining mode, the following methods are available: acoustical (noise) detection [1], force spectral detection, and vibration detection (vibration amplitude diagram). In this paper, the author focus on the method of identifying the unstable machining mode by assessing the surface quality, namely, the surface wave evaluation (Fig. 1).

B. T. Long et al. (Eds.): MMMS 2020, LNME, pp. 447–455, 2021.
https://doi.org/10.1007/978-3-030-69610-8_61

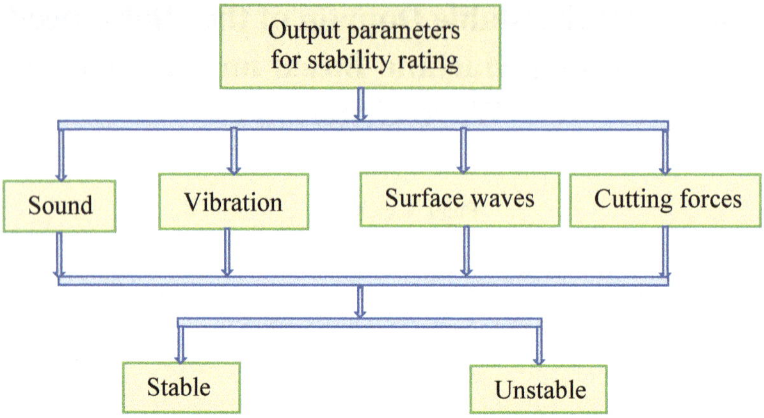

Fig. 1. Indicators are used to assessing stability during cutting

The criteria to evaluate the machining surface error include surface roughness, surface wave, and geometry error. The scales with a ratio between the roughness step (p) and the length (h) of less than or equal to 50 (p/h $\leq$ 50) belong to the surface texture. The ridges (50 < p/h $\leq$ 1000) belong to the surface waves. The vibrations (1000 < p/h) belong to the misalignment (Fig. 2).

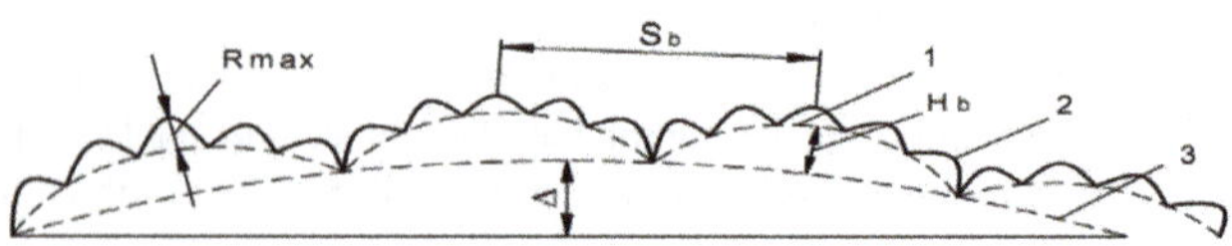

Fig. 2. Geometrical elements of the surface layer 1 - Wave surface, 2 - undulating surface, 3 - shape error

The surface wave is the period of the unevenness of the surface of the workpiece being observed to the extent greater than the surface roughness. Surface waves usually appear when machining with technology system vibrations, cutting is not continuous, cutting tools are reversed… Usually, waveforms appear when machining large and medium-sized parts. Through means, milling, and grinding.

Stable or unstable machining is closely related to the surface wave. If the machining mode is not stable, the surface waveform will not converge (surface waves get bigger with time). If the machining mode is stable, the waveform will converge (surface waves will decrease with time) [2, 3] (Figs. 3 and 4).

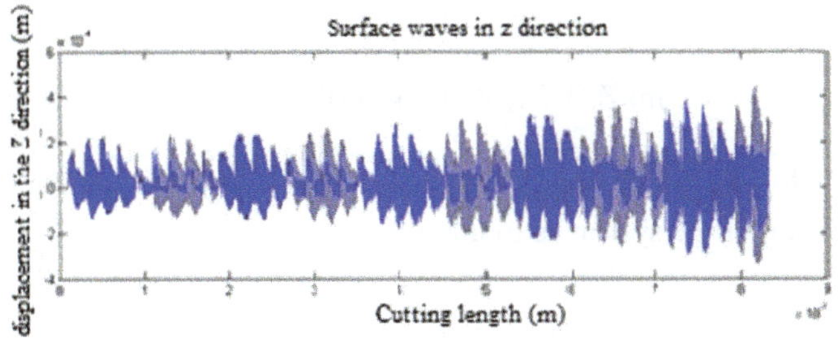

Fig. 3. Surface wave graph when machining with unstable mode

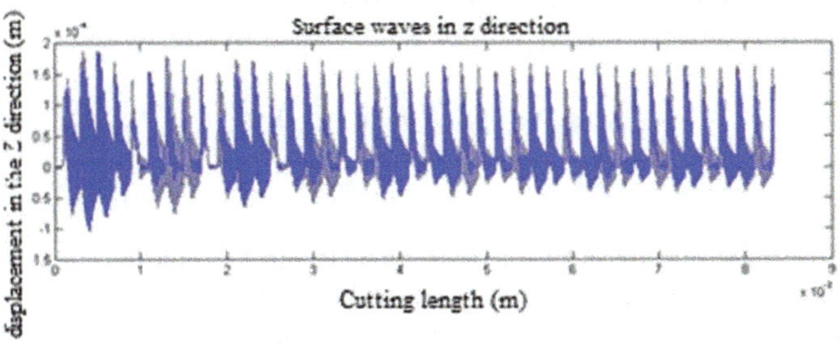

Fig. 4. Surface wave graph when machining with a stable model

So we can rely on the surface wave to evaluate whether the machining mode is stable or unstable.

2 Constructing Surface Wave Equations

Stability survey model based on surface wave index (Fig. 5).

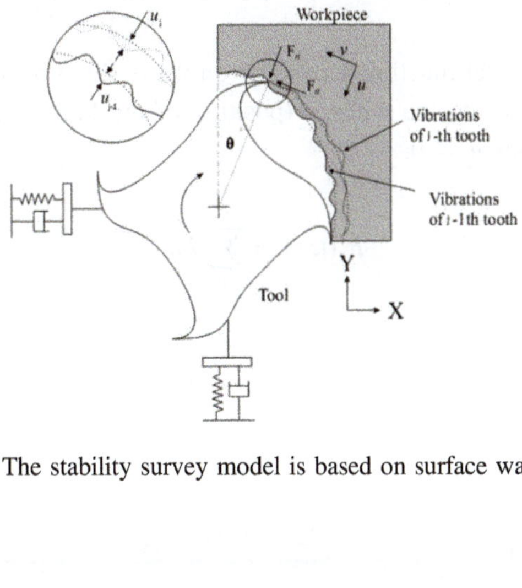

Fig. 5. The stability survey model is based on surface wave index

Equation of forced oscillation of the system

$$m_z.\ddot{Z}(t) + c_z.\ddot{Z}(t) + k_z.Z(t) = F_z(t, \tau) \tag{2.1}$$

Divide both sides by m_z and put $\omega_{0z}^2 = \frac{K_z}{m_z}, \xi_z = \frac{C_z}{2.\sqrt{K_z.m_z}}$ We get the equation for forced oscillation of the system as follows:

$$\ddot{Z}(t) + 2.\omega_{oz}\xi_z.\dot{Z}(t) + \omega_{oz}^2.Z(t) = \frac{F_z(t,\tau)}{m_z} \tag{2.2}$$

With F_z: Axial force in the axial direction of the cutter.
According to the empirical formula: $F_z = 0.5\ F_t$.
F_t: a tangential shear force for j^{th} tool is calculated by the formula:

$$F_{tj} = g\left[\phi_j(t)\right].K_t.h\left[\phi_j(t)\right].a \tag{2.3}$$

Inside:

+ m_z: The converted mass of machine math in the z-direction
+ C_z: Damping coefficient of the system in the z-direction
+ K_z: The hardness of the system in the z-direction
+ t: Time, τ is the period of external force.
coefficient of tangential shear force (depends on workpiece material).
+ a: depth of cut (m)
+ $\phi_j(t)$ is the instantaneous position of the jth tooth
+ $h\left[\phi_j(t)\right]$ The cutting thickness includes the static thickness (due to motion) and the thickness (due to vibration)

$$h\left[\phi_j(t)\right] = h_{static}\left[\phi_j(t)\right] + h_{dynamic}\left[\phi_j(t)\right]$$

$$h\left[\phi_j(t)\right] = f_z.sin\left[\phi_j(t)\right] + \left[sin\left[\phi_j(t)\right].\cos\left[\phi_j(t)\right].\begin{bmatrix} x_t - x_{t-T} \\ y_t - y_{t-T} \end{bmatrix}\right] \tag{2.4}$$

+ $g\left[\phi_j(t)\right]$ is a logical function depending on the position of the jth tooth at time t When the knife feeds into the embryo (position of the knife $\phi_{st}(t) < \phi_j(t) < \phi_{ex}(t)$) $g_j = 1$, and other cases $g_j = 0$.

$$F_z(t,\tau) = \sum_{j}^{N} F_{tj}$$

So the wave equation for milling has the following form:

$$\ddot{Z}(t) + 2.\omega_{oz}\xi_z.\dot{Z}(t) + \omega_{oz}^2.Z(t) = \frac{1}{m_z}\left[0,5.\sum\nolimits_j^N g[\phi(t).K_t.h(\phi).a]\right] \tag{2.5}$$

After setting up the wave level equation, we realize that the wave degree equation is the differential equation with second-order delay. Here we use Matlab software to solve and give the graph form of the wavelength in the milling process with specified input parameters [4, 5].

3 Determining the Stable Domain by Using Matlab Software

The calculates of the stability of the cutting process is shown in Fig. 6.

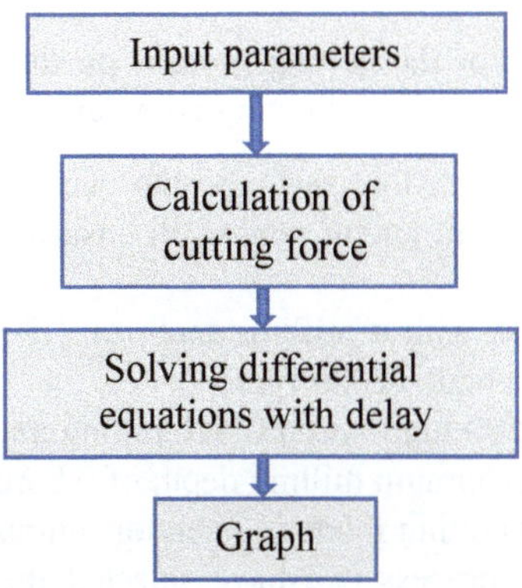

Fig. 6. Block diagram calculates the stability of the cutting process

Input parameters of the stability problem of the cutting process are input parameters including Specifications of the machine: These parameters are determined by the experimental measurement method on the Super MC 500 high-speed CNC milling machine, with a 4-flute alloy milling cutter, the length of the tool holder 20 mm is. The converted mass in the z-direction is: $m_z = \frac{k_z}{\omega^2} = \frac{0,45.10^8}{(2.\pi.745)^2} = 2(\text{kg})$, the rigidity of the machine system in the z-direction is k_z = 0,45.108 and the damping coefficient of the machine system in the z-direction is ξ_z = 0,03.

Coefficients of shear force: k_t = 1495 (N/mm2) and k_n = 538,7 (N/mm2) (the coefficients are selected based on the machined material C45 steel - document [1]).

Variables run for different machining modes include: Change range of cutting speed 1000 ÷ 20000 (rpm) with step speed change of 1000 (rpm).

Range of cutting depth 0 ÷ 5 (mm), with a variable interval of 0,2 (mm).

The result of a stable region cutting on the Super MC 500 high-speed CNC milling machine by Matlab software. The stable region is the lower area of the graph in Fig. 7.

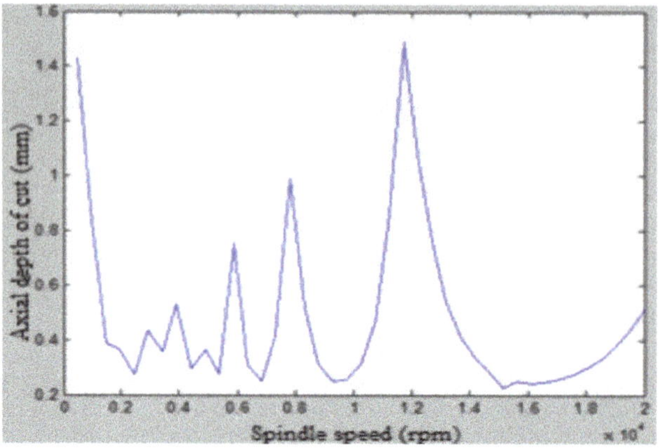

Fig. 7. Stability chart built with Matlab software

From the stability chart of the cutting process on the Super MC 500 high-speed CNC milling machine we have the following comments:

+ The largest cutting depth that the machine can cut with a 4-flute hard alloy milling machine and 60 mm length while still ensuring a good machining surface quality is 1.5 mm.
+ To be able to process with a milling depth of 1.5 mm, we have to adjust the speed of the spindle to about 12000 rpm.
+ With the Super MC 500 high-speed CNC milling machine, the actual machining mode can only cut the maximum milling depth of 0.5 mm (the largest milling depth achieved by the current cutting modes - operator figures). So studying the stability chart of the machining process, we have selected the cutting mode (reasonable cutting speed) that the depth of cutting reaches 1.5 mm, helping us to increase the cutting productivity of the machine by 3 times. But the surface quality of the workpieces is still guaranteed [5–7].

4 Determine the Stable Domain by the Experimental Cut Method

To re-evaluate the reliability of the software stability chart construction method. The author conducts a test cut on the Super MC 500 high-speed CNC milling machine and builds the experimental stability chart.

The surface wave level is still used as an assessment indicator of stability. With stable machining, the surface wave is shaped as in Fig. 8.

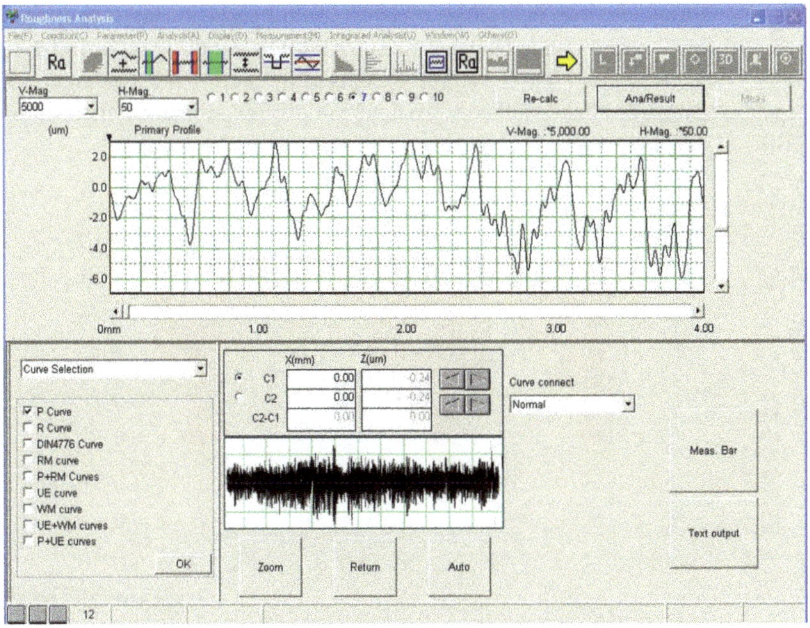

Fig. 8. Surface wave experimental chart with stable machining

The surface wave unstable machining is shown in Fig. 9.

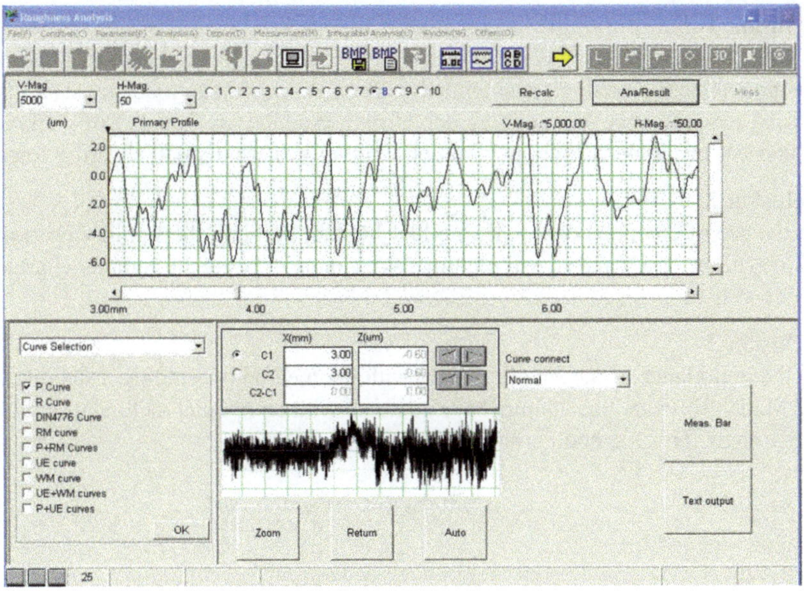

Fig. 9. Chart experimental surface wave with unstable machining

The test cutting process is conducted with a cutting speed range from 2000 to 20000 (rpm), each cutting mode changes 1000 (rpm). The depth of cut is variable 0.15 (mm) one in the range from 0 to 2.1 (mm). Results of measuring the machining surface profile and based on the stability theory we have built the experimental chart as follows (Fig. 10).

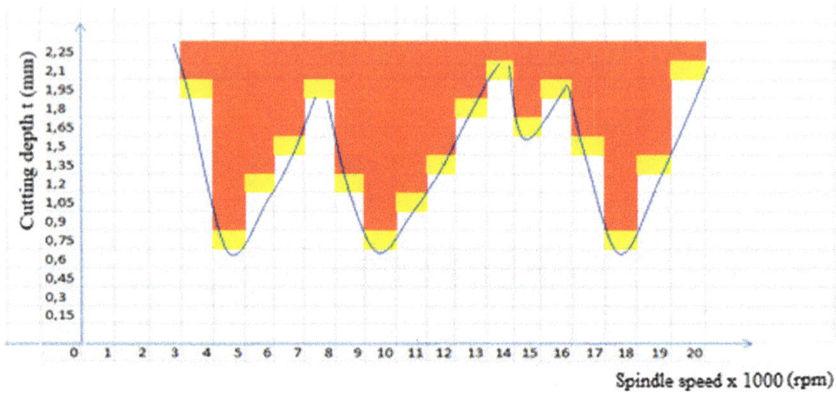

Fig. 10. The chart of experimental stability

5 Conclusion

Conclusions drawn from the study are as follows

Conclusion 1
The stability chart of the cutting process allows us to choose the most appropriate machining mode (optimal machining mode). The best machining mode must satisfy the following requirements: great cutting depth (high productivity), quality of workpieces, and physical phenomena such as vibration, cutting heat, and small shearing force.

Conclusion 2
With the use of computational software like Matlab, building the stability chart of the machine will be easier and save more time. We can quickly get the stability chart of the machine system.

Conclusion 3
With the importance of the stability chart, with the method of building a stability chart using Matlab software, the authors have come up with an optimal method of building a stability chart. This method applies to all machine systems.

References

1. Altintas, Y., Budak, E.: Analytical prediction of stability lobes in milling. CIRP Annals-Manufacturing Technology (1995)
2. Al-Regib, E., Ni, J., Lee, S.: Programming spindle speed variation for machine tool chatter suppression. Int. J. Mach. Tools Manuf. **43**(12), 1229–1240 (2003)
3. Gradis, J.: On stability prediction for milling. Int. J. Mach. Tools Manuf. **45**, 1229–1240 (2005)
4. Faassen, R.: Chatter Prediction and Control for High-Speed Milling. University Press Facilities, Eindhoven (2007)
5. Long, B.T., Van-Hieu, P., Sinh, H.V.: Study on parameters affecting machining stability on high-speed CNC milling "DNM400". Vietnam Mech. J. (2010)
6. Long, B.T., Van-Hieu, P., Sinh, H.V.: Development of method and measurement diagram for determination of hardness "K" and damping coefficient "ς" of high-speed CNC milling machines, Vietnam Mech. J. (2011)
7. Long, B.T., Van-Hieu, P., Sinh, H.V.: Constructing stability chart of high-speed milling machine "DNM400" based on surface wave strength criteria by Matlab software. Vietnam Mech. J. (2011)

Research on the Dynamic Load of Tractor Semi-trailer When Braking on a Round Road

Nguyen Thanh Tung[1(✉)] and Vo Van Huong[2]

[1] Vinh Long University of Technology Education, 73, Nguyen Hue Street, Vinh Long City, Vietnam
tungnt@vlute.edu.vn

[2] Ha Noi University of Science and Technology, 01, Dai Co Viet Road, Ha Noi City, Vietnam

Abstract. The paper presents the research results on the dynamic load of tractor semi-trailer on a round road with grip coefficient $\varphi_{xmax} = 0.8$ at a speed of 50 km/h, being braked with different braking intensities. The authors used the method of structural separation and the Newton-Euler equation to set up the system of dynamic equations of tractor semi-trailer by 3D dynamic model. Matlab-simulink software was used to examine the effect of braking intensity on the dynamic load of tractor semi-trailer. The results showed that, when the tractor semi-trailer moved on the left roundabout with the turning radius of R = 100 m, grip coefficient $\varphi_{xmax} = 0.8$ at a speed of 50 km/h, being braked with 6 different levels, the dynamic load coefficient of the left wheels decreased to about 0.2 to 0.65, and the dynamic load coefficient of the right wheels increased to about 1.35 to 1.8; When braking with a torque of $M_B = 70\%M_{Bdm}$, the braking distance, braking acceleration meet the requirements of ECE-R13 standard.

Keywords: Brake · Dynamic load · Dynamic load coefficient · Tractor semi-trailer

1 Introduction

The braking level greatly affects the dynamic load, braking efficiency and stability of tractor semi- trailer, especially when moving on a round road. Therefore, the driver must choose an appropriate braking level to ensure movement safety and increase braking efficiency. In this paper, the authors present how to construct a dynamic model of tractor semi- trailer and investigate the effects of braking levels on dynamic load, braking efficiency and stability of of tractor semi- trailer when operating on round road at a speed of 50 km/h.

B. T. Long et al. (Eds.): MMMS 2020, LNME, pp. 456–461, 2021.
https://doi.org/10.1007/978-3-030-69610-8_62

2 Dynamics Model

To study dynamic load of tractor semi-trailer, the authors chose the spatial model as shown in Fig. 1. The tractor semi-trailer is a system of many objects with complex structures, so the authors used the method of structural separation to set up the model as follows: Analyzing the structure of the tractor semi-trailer; definition of coordinate systems; splitting the structure of tractor semi-trailer and determining internal force; writing Newton-Euler equations for objects and connecting objects according to modules; solving differential equations using numerical methods.

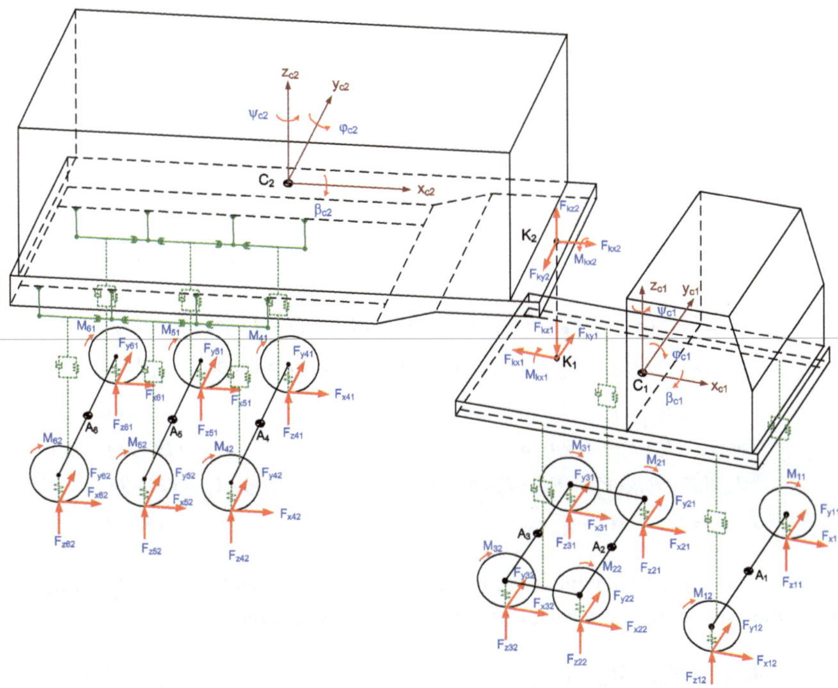

Fig. 1. Dynamic model of the tractor semi-trailer

The Newton-Euler equation was used to set up the system of dynamic equations of the tractor semi-trailer, comprising 42 equations. The system of dynamic equations of the tractor semi-trailer in the road plane is as follows:

$$m_{c1}\ddot{x}_{c1} = F_{xij} - F_{wx1} - F_{kx1} \quad (i = 1-3) \quad (1)$$

$$m_{c1}\ddot{y}_{c1} = F_{yij} - F_{ky1} \quad (i = 1-3) \quad (2)$$

$$J_{zc1}\ddot{\psi}_{c1} = F_{x1j}l_1 + F_{x1j}b_1 + F_{ky1}l_{k1} + (F_{xi2} - F_{xi1})b_i - F_{yij}l_i \quad (3)$$

$$m_{c2}\ddot{x}_{c2} = F_{xij} + F_{kx2} \quad (i = 4-6) \quad (4)$$

$$m_{c2}\ddot{y}_{c2} = F_{ky2} + F_{yij} \quad (i = 4-6) \quad (5)$$

$$J_{zc2}\ddot{\psi}_{c2} = (F_{xi2} - F_{xi1})b_i + F_{ky2}l_{k2} - F_{yij}l_6 \quad (i = 4-6) \quad (6)$$

The motion equations of the tractor semi-trailer in a vertical plane are as follows:

$$m_{c1}\ddot{z}_{c1} = F_{Cij} + F_{Kij} - F_{kz1} \quad (i = 1 \div 3) \quad (7)$$

$$J_{yc1}\ddot{\varphi}_{c1} = (F_{C1j} + F_{K1j})l_i + F_{kz1}l_{k1} - F_{kx1}(h_{c1} - h_{k1}) + M_{ij} \quad (i = 1 \div 3) \quad (8)$$

$$m_{c2}\ddot{z}_{c2} = F_{Cij} + F_{Kij} + F_{kz2} \quad (i = 4 \div 6) \quad (9)$$

$$J_{yc2}\ddot{\varphi}_{c2} = -(F_{Cij} + F_{Kij})l_i + F_{kx2}(h_{c2} - h_{k2}) + F_{kz2}l_{k2} + M_{ij} \quad (i = 4 \div 6) \quad (10)$$

Horizontal dynamic equations of the tractor semi-trailer are as follows:

$$J_{xc1}\ddot{\beta}_{c1} = (F_{Ci2} + F_{Ki2} - F_{Ci1} - F_{Ki1})w_i + M_{kx1} \quad (i = 1 \div 3) \quad (11)$$

$$J_{xc2}\ddot{\beta}_{c2} = (F_{Ci2} + F_{Ki2} - F_{Ci1} - F_{Ki1})w_i - M_{kx2} \quad (i = 4 \div 6) \quad (12)$$

$$m_{Ai}\ddot{z}_{Ai} = F_{CLij} + F_{KLij} - F_{Cij} - F_{Kij} \quad (i = 1 \div 6) \quad (13-18)$$

$$m_{Ai}\ddot{y}_{Ai} = F_i + F_{yij} \quad (i = 1 \div 6) \quad (19-24)$$

$$J_{Axi}\ddot{\beta}_{Ai} = (F_{Ci1} + F_{Ki1} - F_{Ci2} - F_{Ki2})w_i + (F_{CLi2} - F_{CLi1})b_i - F_{yij}(r_{ij} + \xi_{Aij}) \quad (25-30)$$

$$J_{Ayij}\ddot{\varphi}_{ij} = M_{Aij} - M_{Bij} - F_{xij}r_{dij} \quad (31-42)$$

The tractor semi-trailer has 6 axles, and each has 2 wheels, so 12 dynamic equations of the wheels in the vertical plane are shown as follows:

$$J_{Ayij}\ddot{\varphi}_{ij} = M_{Aij} - M_{Bij} - F_{xij}r_{dij} \quad (31-42)$$

3 Survey Results and Discussions

Based on ISO 19747: 2011, let a tractor semi-trailer move on a left roundabout with a radius of R = 100 m, at a speed of V_o = 50 km/h, the road has a grip coefficient of 0.8. The survey was conducted with 6 braking levels of M_B = [0, 50, 60, 70, 80, 90]%$M_{Bđm}$.

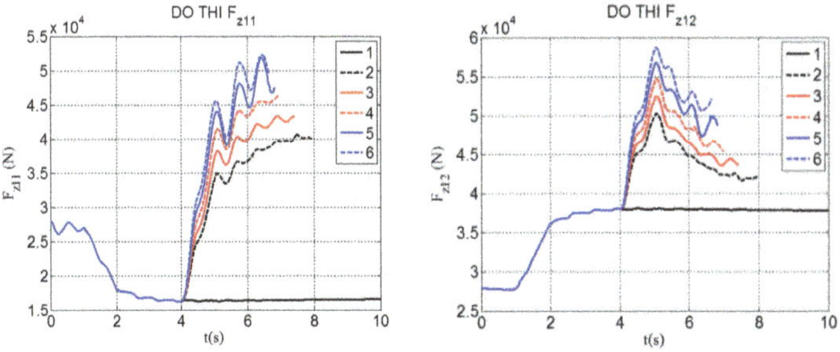

Fig. 2. The vertical force of wheel load on the front axle

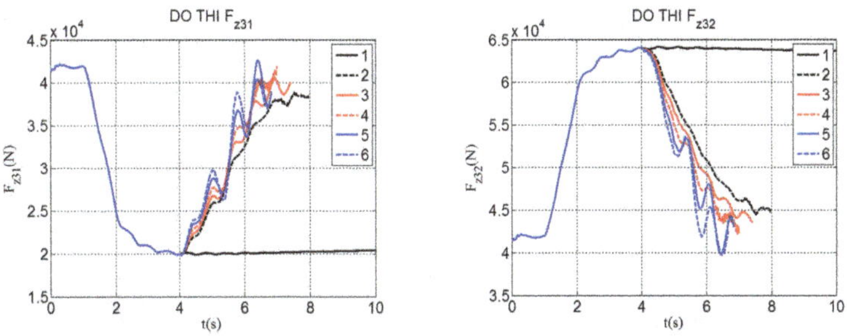

Fig. 3. The vertical force of wheel load on the middle axle

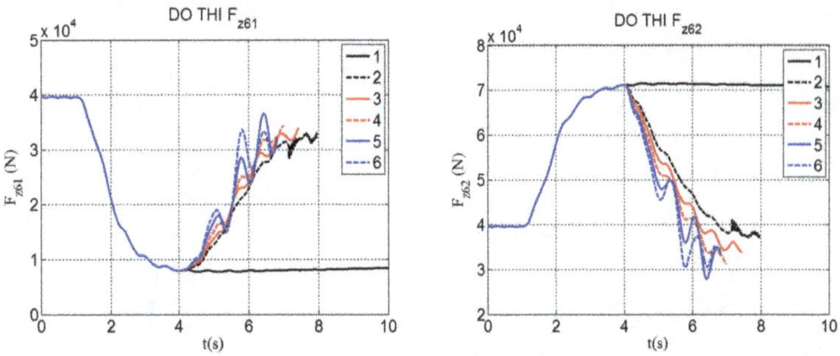

Fig. 4. The vertical force of wheel load on the rear axle

Figure 2, 3, 4 is a graph of vertical force of wheel load on the left of F_{zi1} and on the right of F_{zi2} of the front axle (i = 1), the middle axle (i = 3), the rear axle (i = 6). From the graphs, it can be seen when the rotation is on the left (t = 1 ÷ 4 s), the vertical force of wheel load on the left of F_{zi1} (i = 1 ÷ 6) has a decreasing value, whereas the vertical force of wheel load on the right of F_{zi2} (i = 1 ÷ 6) has an increasing value. When braking (t = 4 ÷ 7 s) due to vertical inertial forces in the same direction of motion, the weight of the tractor semi-trailer is cornered forward, so the load on the wheels of the front axle increases, but the rear axle decreases.

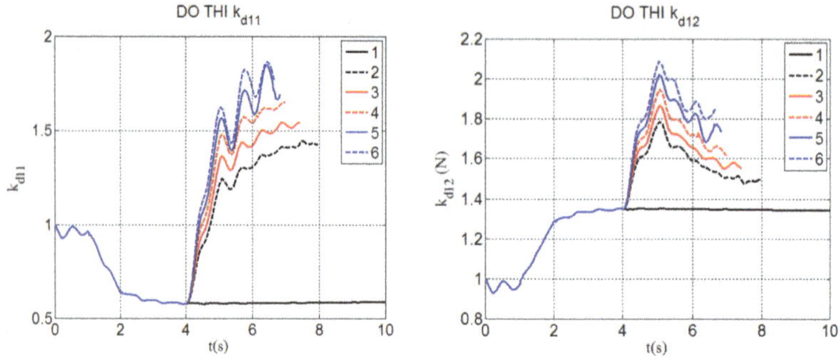

Fig. 5. Dynamic load corresponding to 6 braking levels of the front axle

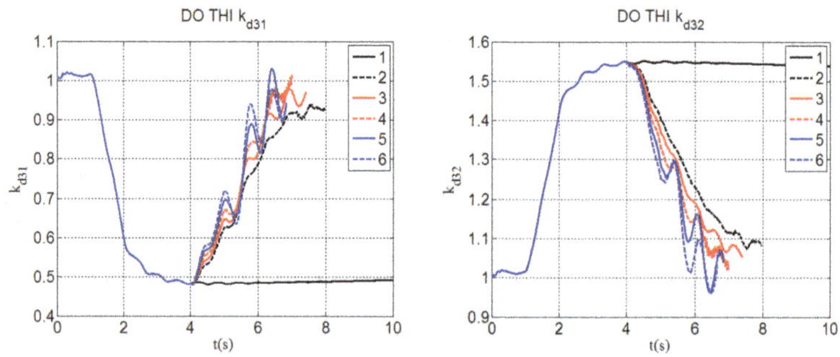

Fig. 6. Dynamic load corresponding to 6 braking levels of the middle axle

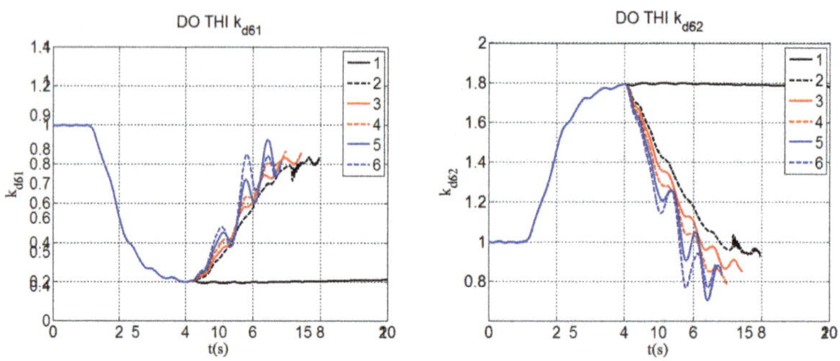

Fig. 7. Dynamic load corresponding to 6 braking levels of the rear axle

Figure 5, 6, 7 is a graph of dynamic load coefficient $k_{dij} = \frac{F_{zij}}{F_{Gij}}$ of the tractor semi-trailer while the rotation is on the left at a speed of V_o = 50 km/h and then braking with 6 different levels. When the left rotation is stable (t = 2 ÷ 4 s), the dynamic load coefficient of the left wheels decreases $k_{di1} = 0.2 \div 0.51$, but the right increases $k_{di2} = 1.35 \div 1.8$. When braking (t = 4 ÷ 7 s), the dynamic load coefficient of the front wheels increases $k_{d12} = 2.0$, but the rear decreases $k_{d62} = 0.8$.

4 Conclusion

The tractor semi-trailer when moving on the left roundabout with a radius of R = 100 m, at a speed of V_o = 50 km/h on the road φ_{xmax} = 0,8, and then braking with 6 levels of M_B = [0; 50; 60; 70; 80; 90]%$M_{Bđm}$, the dynamic load coefficient of the left wheels decreased to about 0.2 to 0.51, and the dynamic load coefficient of the right wheels increased to about 1.35 to 1.8; When braking with a torque of M_B = 70% $M_{Bđm}$, the braking distance, braking acceleration meet the requirements of ECE-R13 standard.

References

1. Emery, L.: Design and construction of a variable dynamic vehicle. In: Proceedings of the 16th International Technical Conference on the Enhanced Safety of Vehicles, Canada (1998). Paper number 98-S2-P-27
2. The international braking regulation of the Economic commission for Europe in Geneva - ECE-Regulation No. 13, ECE-R13, Vehicle Regulations (2004)
3. International standard ISO 14794:2011 (2011)
4. Jazar, R.N.: Vehicle Dynamics Theory and Application. Springer, Cham (2017)
5. Jazar, R.N.: Advanced Vehicle Dynamics. Springer, Cham (2019)

An Experimental Research on Converting Gasoline Engines Using a Distributor Ignition System into a Direct Ignition System

Nguyen Thanh Tung(✉)

Vinh Long University of Technology Education,
73, Nguyen Hue Street, Vinh Long City, Vietnam
tungnt@vlute.edu.vn

Abstract. The paper presents the research results on converting gasoline engines using a distributor ignition system into a direct ignition system. The researcher calculates the parameters of ignition system, thereby calculating to choose a suitable direct ignition system to replace a distributor ignition system; checking and evaluating exhaust concentration, fuel consumption, power, torque of Toyota 5A-F engine using a distributor ignition system and a direct ignition system. The results show that when converting Toyota 5A-F engine using a distributor ignition system into a direct ignition system, it will help the engine increase its power by about 3.1%, increase the torque by 6.6%, save about 17.0% fuel, CO concentration decreases by 18% and HC concentration decreases by 10.4%.

Keywords: Gasoline engines · Distributor ignition system · Direct ignition system

1 Introduction

According to Vietnam Register, as of December 2019, Vietnam has 3.673.065 cars in circulation, many of them still use distributor ignition system. The disadvantage of the gasoline engine using the distributor ignition system is to consume a lot of fuel and cause environmental pollution. Therefore, the researcher calculates and suggests the method to convert gasoline engines using a distributor ignition system into a direct ignition system (DIS); evaluates environmental pollution, speed, power, fuel consumption of Toyota 5A-F engine using a distributor ignition system and DIS.

2 Conversion Method

In order to convert the engine using distributor ignition system into DIS, the researcher performs the following procedures: (i) Calculate the specifications of the ignition system being used to choose the appropriate DIS; (ii) Disassemble the distributor ignition system; (iii) Install sensors and DIS. The primary current of the ignition system at the time transistor off (I_{ng}) is determined by the formula (1):

B. T. Long et al. (Eds.): MMMS 2020, LNME, pp. 462–466, 2021.
https://doi.org/10.1007/978-3-030-69610-8_63

$$I_{ng} = \frac{U}{R_1}\left(1 - e^{\frac{-R_1}{L_1}\Delta t}\right) \quad (A) \tag{1}$$

The inductance L_1 of the primary coil is determined by the formula (2):

$$L_1 = \frac{R_1 \Delta t}{\ln\left(1 - \frac{R_1 I}{U}\right)} \quad (mH) \tag{2}$$

The stored energy W_{dt} in the primary coil is determined by the formula (3):

$$W_{dt} = \frac{L_1 I_1^2}{2} \quad (mJ) \tag{3}$$

The ignition frequency of a 4-stroke engine is determined by the formula (4):

$$f = \frac{nZ}{120} \quad (Hz) \tag{4}$$

The ignition period T is the time between two spark occurrences determined by the formula (5):

$$T = \frac{1}{f} = t_d - t_m \quad (S) \tag{5}$$

Note: t_d - *transistor time of saturation led power;* t_m - *transistor time of breaker capacity.*

The researcher experiments on Toyota 5A-F engine as shown in Fig. 1. The engine has the following basic parameters: 4 cylinders; 16-valve DOHC; Working volume of engine 1498 cc; Fuel system using carburettor; Maximum power 85HP (63 kW)/ 6000 rpm; Maximum torque of 122 Nm/3600 rpm; Compression ratio 9.5: 1; cylinder diameter 78.7 mm; Piston stroke 77 mm; integrated ignition assembly (IIA) has $R_1 = 1.4\ \Omega$, $I_1 = 5$ A, $\Delta t = 6$ mS, $U = 12$ V, $L_1 = 9.655$ mH, $W_{dt} = 121.25$ mJ, $I_{ng} = 4.5$ A.

Fig. 1. Toyota 5A-F engine using integrated ignition assembly – IIA

Through the research, the researcher finds that Toyota 5E-FE engine (using direct ignition system 2 - DIS2), Toyota 1SZ-FE engine (using direct ignition system 1 - DIS1) has a primary coil resistance of 0.4 Ω ÷ 0.5 Ω and a secondary coil resistance of 10 KΩ ÷14 KΩ. Calculation results of basic specifications of DIS of Toyota 5E-FE and 1SZ-FE engine are as follows: R_1 = 0.45 Ω, I_1 = 7 A, Δt = 4 mS, U = 12 V, L_1 = 6.0 mH, W_{dt} = 147 mJ, I_{ng} = 7A.

Through the calculation above, it can be seen that the stored energy in the primary coil winding for IIA of Toyota 5A-F engine is W_{dt} = 121.25 J and DIS of Toyota 5E-FE, 1SZ-FE engine is W_{dt} = 147 J, is about 21% higher than that of IIA, ensuring enough reserve energy to burn well the mixture of air gasoline. Therefore, the researcher chooses DIS2 of Toyota 5E-FE engine and DIS1 of Toyota 1SZ-FE engine to replace IIA of Toyota 5A-F engine.

There are 2 options for converting Toyota 5A-F engine using IIA into DIS as follows: (i) Option 1: Using DIS2, as shown in Fig. 2; (ii) Option 2: Using DIS1, as shown in Fig. 3.

Fig. 2. Direct ignition system 2

Fig. 3. Direct ignition system 1

3 Evaluation of Specifications

To check the exhaust concentration of Toyota 5A-F engine using IIA, DIS2 and DIS1, the researcher uses vehicle diagnostic tool SCA 3500 of SUN firm, as shown in Fig. 4.

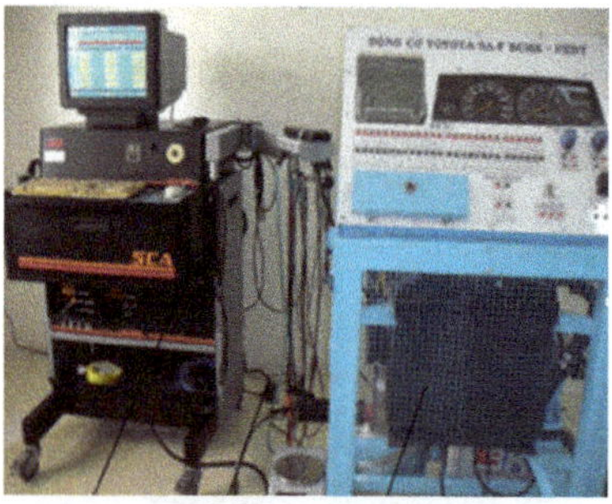

Fig. 4. Diagnostic tool SCA 3500

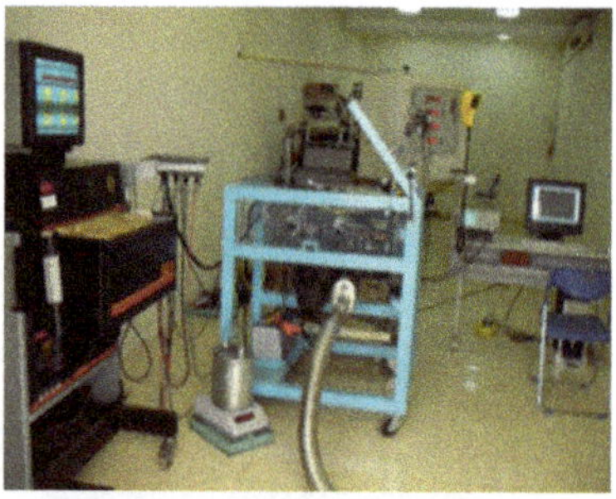

Fig. 5. Engine test equipment MP-100S

Emission measurement results of Toyota 5A-F engine using IIA, DIS2 and DIS1 at the speed n = 850 rpm is summarized in Table 1. Empirical data show that when the engine uses IIA does not meet emission standards at idle speed. The engine using DIS meets Vietnamese standards (VNS).

Table 1. Emission concentration of Toyota 5A-F

	IIA	DIS 2	DIS 1	VNS
CO (%)	5.38	4.46	4.41	4.5
HC (ppm)	1293	1174	1158	1200
CO_2 (%)	8.54	7.73	6.82	–

Measuring the fuel consumption, power, torque of Toyota 5A-F engine with Weinlich's engine power test equipment MP-100S, as shown in Fig. 5. Graph of fuel consumption in 1 min of Toyota 5A-F engine using IIA, DIS2 and DIS, as shown in Fig. 6. From this figure, it can be seen that when converting Toyota 5A-F engine using IIA into DIS2 and DIS1, the average fuel consumption decreases by 13% and 17% respectively.

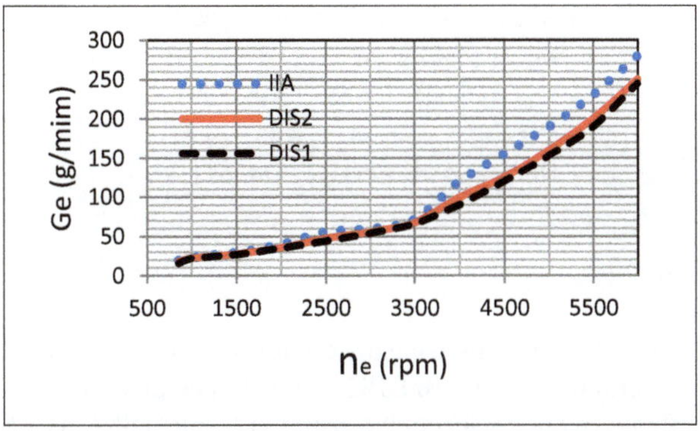

Fig. 6. Fuel consumption of Toyota 5A-F engine

Graph of power and torque of Toyota 5A-F engine using IIA, DIS2 and DIS1 are presented in Fig. 7. It can be seen that the power and torque of the engine when using DIS is higher than when the engine uses IIA.

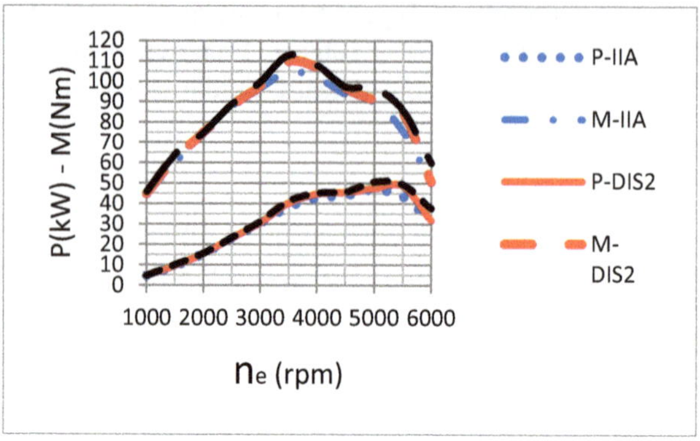

Fig. 7. M, P graph of 5A-F engine

A summary of the specifications of the Toyota 5A-F engine using IIA, DIS2 and DIS, as shown in the Table 2.

Table 2. Specifications of Toyota 5A-F engine

	IIA	DIS 2	DIS 1
n_{max} (rpm)	6200	6200	6200
N_{emax} (kW)	48	49	49.5
M_{emax} (Nm)	106	111	113
G_{nltb} (g/min)	106.58	93.08	89
CO (%)	5.38	4.46	4.41
HC (ppm)	1293	1174	1158

4 Conclusion

The conversion of gasoline engines using distributor ignition system into DIS, depending on the specific conditions, we can use DIS2 or DIS1. The plan to use DIS1 is the most optimal. Converting Toyota 5A-F engine using IIA into DIS1 will help the engine increase the power by about 3.1%, increase the torque by 6.6%, save about 17.0% fuel, CO concentration decreases by 18% and HC concentration decreases by 10.4%.

References

1. Tien, N.T.: Principles of Internal Combustion Engines. Educational Publishing House (2000)
2. MP 100S engine test bed with MP computer operating instructions, Weinlich Steuerungen (2005)
3. The modular gas analyser MGA-1200 operator's manual, Sun electric Europe (2005)
4. Garrett, T.K.: The Motor Viehicle. Ceng, FIMechE, MRAeS (2006)
5. Stone, R.: Introduction to the Internal Combustion Engine. Dept of Engineering Science, University Oxford (2009)

Effect of Different Tool on Surface Roughness for Aluminium Cookwares Polishing

Truong Dinh Luan[1(✉)], Nguyen Trong Doanh[1], and Bui Qui Luc[2]

[1] Department of Mechatronics, School of Mechanics, HUST, Hanoi, Vietnam
[2] School of Mechanics, HUST, Hanoi, Vietnam

Abstract. In mechanical polishing, there are many factors affecting into product surface roughness such as: technical factors and other factors. The technical factors are as velocity, force, time, contact angle, tool move, polishing materials; polishing tools. Other factors as: shape of workpiece, hardness of workpiece, additives, etc. Polishing of aluminium cookwares has two goals: achieving fin surface roughness and having good esthetic looks. This paper produces the experimental result about the effect of felt and fiber tool on the surface roughness in the same conditions: velocity, force, time for aluminum cookwares polishing. The goal of study is finding a optimal functions for aluminium polishing process.

Keywords: Polishing · Aluminum · Felt · Fiber · Tool · Functions

1 Introduction

– There are many input parameters that affect surface roughness. Through studying documents and a priori information, the most influence parameters are the velocity of polishing tool n (rpm), contact force F (N) and time t (s)
– Through screening experiment, determine the value domain as follows:
    Velocity n(rpm): $2000 \leq n \leq 2800$
    Force F(N): $100 \leq F \leq 300$
    Time t(s): $30 \leq t \leq 90$.

2 Building Experimental

2.1 Experimental Equipment

– Sample for experiment: aluminum pot Ø160 (Fig. 1).
– Roughness measuring machine SJ-301 (Fig. 2).

B. T. Long et al. (Eds.): MMMS 2020, LNME, pp. 467–473, 2021.
https://doi.org/10.1007/978-3-030-69610-8_64

Fig. 1. Sample for experiment

Fig. 2. Roughness measuring machine SJ-301

- Polishing tools felt and fiber Ø150 (Fig. 3 and Fig. 4).

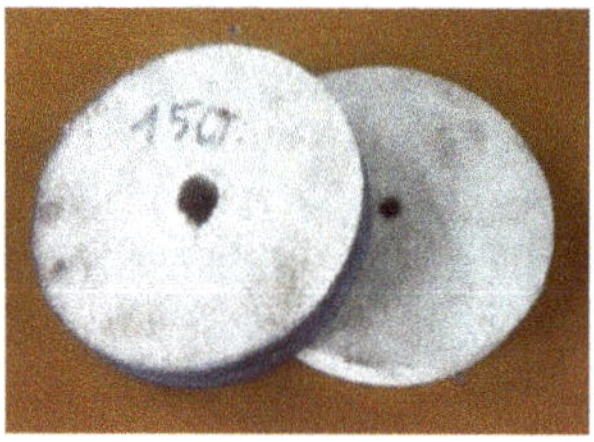

Fig. 3. Felt

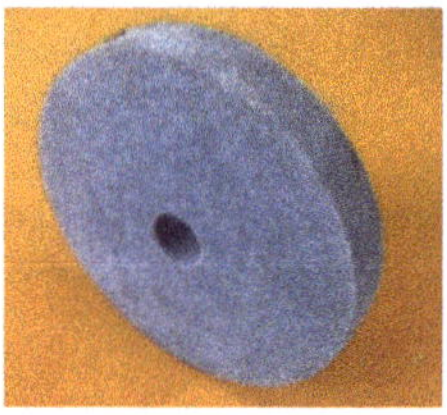

Fig. 4. Fiber

- Turning T18A (Use to carry aluminium pot, velocity 110 rpm).
- 3-phase electric motor, velocity 2850 rpm (carry polishing tool).
- Indication HPW501.
- Inverter 1-phase/3-phase to adjust the velocity of a 3-phase electric motor (Fig. 5).
- Loadcell measure force, load 50 kg, accuracy Class – C3 (Fig. 6).

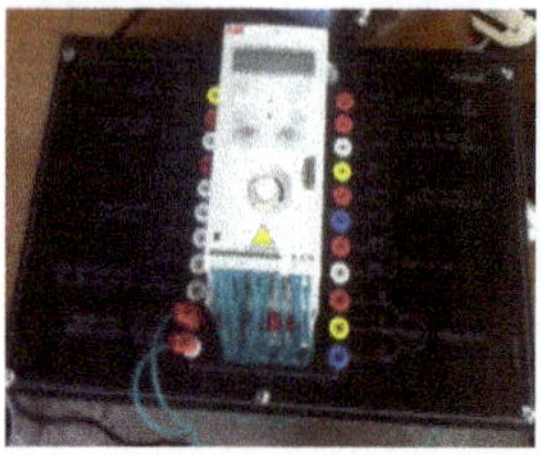

Fig. 5. Inverter 1phase/3phase

Fig. 6. Loadcell Keli PST

2.2 Identify the Domain of Technology Parameters

Through experimental screening, shown in the following Table 1.

Table 1. Value of experimental technology parameters

Technology parameters	The levels		
	High	Medium	Low
Velocity n(rpm)	2800	2400	2000
Force F(N)	300	200	100
Time t(s)	90	60	30

2.3 Experimental Planning

According to [3], the total number of experiments was: $N + n_0 = 8 + 4 = 12$

N: Number of experiments; n_0: number of experiments at heart

Choose

$$n_{min} = 2000\,\text{rpm}; F_{min} = 100\,\text{N}; t_{min} = 30\,\text{s}$$

$$n_{max} = 2800\,\text{rpm}; F_{max} = 300\,\text{N}; t_{max} = 90\,\text{s}$$

Inferred: $n_j^0 = 2400\,\text{rpm}; F_j^0 = 200\,\text{N}; t_j^0 = 60\,\text{s}$

Variable range:

$$\gamma_I^n = 400\,\text{rpm}; \gamma_I^F = 100\,\text{N}; \gamma_I^t = 30\,\text{s}$$

- The surface roughness selected linear model has the following form:

$$\tilde{y} = b_0 + b_1x_1 + b_2x_2 + b_3x_3 + b_{12}x_1x_2 + b_{23}x_2x_3 + b_{13}x_1x_3 + b_{123}x_1x_2x_3 \quad (1)$$

Include: b_0: regression coefficient; b_1, b_2, b_3: Linear coefficients

b_{12}, b_{23}, b_{13}, b_{123}: interaction coefficients (Table 2)

Table 2. The experimental variables

Elements affecting	Velocity n (rpm)	Force F (N)	Time t (s)
Variables	Z_1	Z_2	Z_3
Z_{jmin}	2000	100	30
Z_j^0	2400	200	60
Z_{jmax}	2800	300	90
ΔZ_j	400	100	30

Z_{jmin}, Z_{jmax}: Minimum and maximum values of input variables

Z_j^0: Value at the center of the variable; ΔZ_j: Variable range; j: variable index (j = 1, 2, 3)

– Encode variables and create experimental matrix (Table 3):

Encode variables according to the formula:

$$x_j = \frac{Z_j - Z_j^0}{\Delta Z_j} \tag{2}$$

Table 3. Experimental matrix

No	x_0	x_1	xx_2	x_3	x_1x_2	x_2x_3	x_1x_3	$x_1x_2x_3$	Z_1	Z_2	Z_3
1	1	−1	−1	−1	1	1	1	−1	2000	100	30
2	1	1	−1	−1	−1	1	−1	1	2800	100	30
3	1	−1	1	−1	−1	−1	1	1	2000	300	30
4	1	1	1	−1	1	−1	−1	−1	2800	300	30
5	1	−1	−1	1	1	−1	−1	1	2000	100	90
6	1	1	−1	1	−1	−1	1	−1	2800	100	90
7	1	−1	1	1	−1	1	−1	−1	2000	300	90
8	1	1	1	1	1	1	1	1	2800	300	90
9	1	0	0	0	0	0	0	0	2400	200	60
10	1	0	0	0	0	0	0	0	2400	200	60
11	1	0	0	0	0	0	0	0	2400	200	60
12	1	0	0	0	0	0	0	0	2400	200	60

x_0: Virtual variable that adds to the matrix

2.4 Experimental Results

a. When polishing by felt tool (Table 4).

Table 4. Value of surface roughness when polished by felt tool (Rz) (Each experiment was measured 12 times)

No	1 (μm)	2 (μm)	3 (μm)	4 (μm)	5 (μm)	6 (μm)	7 (μm)	8 (μm)	9 (μm)	10 (μm)	11 (μm)	12 (μm)	Rz (μm) average
1	4.95	5.32	6.24	5.28	5.74	6.02	6.13	5.64	4.33	4.51	5.93	6.00	5.508
2	6.19	3.47	2.98	3.68	3.74	2.87	5.19	6.93	2.96	4.64	7.45	4.01	4.509
3	5.28	2.96	4.84	5.50	4.47	4.69	4.94	5.23	4.96	5.30	4.58	5.00	4.813
4	2.64	3.38	3.96	5.52	2.85	2.12	4.80	3.10	4.48	4.58	3.33	3.61	3.698
5	2.60	2.82	2.72	1.76	2.58	2.25	2.89	3.65	2.56	2.61	3.45	2.70	2.716
6	2.78	2.82	2.73	2.05	2.22	1.91	1.99	3.30	2.64	2.25	1.79	1.79	2.356
7	1.40	1.58	2.14	1.22	1.86	1.63	2.11	2.31	1.67	1.79	1.62	2.40	1.811
8	1.98	2.36	2.14	2.21	1.60	1.00	0.62	2.01	2.10	1.29	0.88	0.63	1.485
9	3.12	3.38	2.11	2.77	3.11	1.87	2.82	2.53	2.91	2.33	2.04	2.59	2.632
10	1.58	2.85	6.78	2.72	4.59	1.94	2.26	6.24	5.39	2.26	1.89	2.21	3.423
11	2.45	2.43	5.46	3.84	1.83	2.83	1.72	3.20	2.79	2.30	2.42	2.36	2.802
12	4.59	4.84	4.61	3.14	2.49	2.64	3.13	1.71	5.70	3.47	2.79	3.95	3.588

According to [3] after calculating the b coefficient, replace Eq. (1) [6]

$$\begin{aligned}\tilde{y} = 3.362 - 0.35x_1 - 0.41x_2 - 1.27x_3 - 0.01x_1x_2 \\ - 0.038x_2x_3 + 0.179x_1x_3 + 0.019x_1x_2x_3\end{aligned} \tag{3}$$

- Check the meaning of the coefficient b in the equation:

Choose $\alpha = 0.2$, degrees of freedom $n_0 - 1 = 4 - 1 = 3$. Look up tables *Student* (page 171 [2]), $t_\alpha = 0.978$. So b_1, b_2, b_3 signify (due to $t_1, t_2, t_3 > t_\alpha$), coefficients $b_{12}, b_{23}, b_{13}, b_{123}$ is removed. The linear regression Eq. (3) is rewritten as follows:

$$\tilde{y} = 3.362 - 0.35x_1 - 0.41x_2 - 1.27x_3 \tag{4}$$

- Check the compatibility of regression equation with experiment.

$$\hat{F} = \frac{s^2_{du'}}{s^2_{ts}} = \frac{0.2677}{0.6498} = 0.412$$

Choose $\alpha = 0.05$, numerator is 8 − 4 = 4, denominator is $n_0 - 1 = 4 - 1 = 3$. Look up tables *Fisher* (page 165 [2]), have $F_\alpha = 9.12$.

Because $\hat{F} < F_\alpha$ so the 1-linear equations suitable to the experiment.

Converts (4) to variable Z_j.

$$\begin{aligned}\tilde{y} &= 3.362 - 0.35\frac{Z_1 - 2400}{400} - 0.41\frac{Z_2 - 200}{100} - 1.27\frac{Z_3 - 60}{30} \\ &= 8.822 - 0.000875Z_1 - 0.0041Z_2 - 0.0423Z_3\end{aligned} \tag{5}$$

Consequence:

$$\boldsymbol{R_z} = 8.822 - 0.000875\boldsymbol{n} - 0.0041\boldsymbol{F} - 0.0423\boldsymbol{t} \tag{6}$$

b. When polishing by fiber tool (Table 5).

Similarly 2.4.1, we also calculate the value of response function as follows:

$$\begin{aligned}\tilde{y} = 4.901 + 0.386x_1 - 0.555x_2 - 0.494x_3 + 0.026x_1x_2 - 0.086x_2x_3 \\ - 0.051x_1x_3 + 0.021x_1x_2x_3\end{aligned} \tag{7}$$

Table 5. Value of surface roughness when polished by fiber tool (Rz) (Each experiment was measured 12 times)

No	1 (μm)	2 (μm)	3 (μm)	4 (μm)	5 (μm)	6 (μm)	7th (μm)	8 (μm)	9 (μm)	10 (μm)	11 (μm)	12 (μm)	Rz (μm) average
1	4.42	4.75	4.82	4.69	5.21	5.82	5.24	5.85	4.18	4.47	4.72	5.82	4.999
2	6.38	5.48	5.93	6.50	5.33	6.19	5.27	7.78	6.07	5.26	8.59	6.77	6.296
3	4.96	4.03	5.23	4.09	4.22	4.62	5.70	4.76	4.99	5.83	6.59	6.23	5.104
4	6.51	5.92	6.75	5.04	5.72	5.37	6.84	7.94	6.02	7.30	7.86	6.05	6.433
5	3.03	3.92	3.50	4.10	3.72	4.28	3.96	3.88	5.19	5.09	3.60	3.62	3.991
6	4.98	3.95	3.59	4.04	4.17	5.36	5.76	5.41	4.30	5.44	5.54	5.75	4.858
7	3.12	3.71	3.89	3.85	3.71	3.03	3.63	2.84	3.30	3.73	3.06	3.57	3.453
8	4.07	4.69	3.75	3.60	3.62	4.65	4.77	5.74	5.05	5.31	4.71	5.23	4.599
9	4.84	3.98	5.27	5.35	4.72	3.53	6.30	4.73	3.99	3.92	4.69	5.74	4.755
10	5.15	3.86	4.96	4.31	5.45	5.08	5.07	5.68	4.34	4.27	3.64	4.85	4.722
11	5.11	4.78	3.88	4.35	4.11	4.57	3.42	4.30	3.45	4.40	4.05	3.51	4.244
12	4.61	6.97	5.74	5.59	5.29	5.55	5.89	5.12	5.63	4.28	4.99	4.63	5.358

- Check the meaning of the coefficient b in the equation:
  Choose $\alpha = 0.05$, degrees of freedom $n_0 - 1 = 4 - 1 = 3$. Look up table *Student* (page 171 [2]), $t_\alpha = 2.35$. So b_1, b_2, b_3 signify (due to $t_1, t_2, t_3 > t_\alpha$), coefficient $b_{12}, b_{23}, b_{13}, b_{123}$ is removed. The linear regression (7) is rewritten as follows:

$$\tilde{y} = 4.901 + 0.386x_1 - 0.555x_2 - 0.494x_3 \tag{8}$$

- Check the compatibility of regression equation with experiment.

Choose $\alpha = 0.05$, numerator is 8 – 4 = 4, denominator is $n_0 - 1 = 4 - 1 = 3$. Look up table *Fisher* (page 165 [2]), have $F_\alpha = 9.12$.

Because $\hat{F} < F_\alpha$ so the 1-linear equations suitable to the experiment.

Converts (8) to variable Z_j.

$$\begin{aligned}\tilde{y} &= 4.901 + 0.386\frac{Z_1 - 2400}{400} - 0.555\frac{Z_2 - 200}{100} - 0.494\frac{Z_3 - 60}{30} \\ &= 4.683 + 0.000965Z_1 - 0.0056Z_2 - 0.0165Z_3\end{aligned} \tag{9}$$

Consequence:

$$\boldsymbol{R_z} = 4.683 + 0.000965\boldsymbol{n} - 0.0056\boldsymbol{F} - 0.0165\boldsymbol{t} \tag{10}$$

3 Conclusions

- When polishing with felt, if n is higher then Rz is lower (shown in Eq. (6)), the high glossy
- When polishing with fiber, if n is higher then Rz is higher (shown in Eq. (10)), the low glossy
- In both cases (i.e. polished with felt or fiber) when F, t is bigger, Rz is lower or the glossy is proportional to F, t (shown in Eqs. (6) and (10)).

References

1. Que, N.T., Huong, N.T.L., Yen, P.T.N.: Basis of Measurement Techniques (2009)
2. Tri, B.M.: Statistical Probability & Experimental Planning (2011)
3. Praiapati, N.K., Patel, S.M.: Adv. Eng. Res. Stud., 2, 1 (2012)
4. Dev Singh, D., Yadav Raju, N.: Eng. Appl. Manage., 4, 3 (2018)
5. Selvam, M.D., Saik Dawood, A.K., Karupussami, G.: Eng. Sci. Technol., 2, 4 (2012)
6. Luan, T.D., Doanh, N.T., Luc, B.Q.: VietNam Mechanical Eng. J., 70–75 (2018). ISSN0886-7056

Material Machining, Industrial Engineering and Renewable Energy for Efficient Systems

Influence of Surface Layers' Defects of Gear Teeth's Working Surfaces on the Tension of Steel Thermochemically Reinforced Novikov Gear

V. I. Korotkin[1], N. P. Onishkov[2], and E. M. Kolosova[1](✉)

[1] Southern Federal University, Rostov-on-Don, Russia
[2] Don State Technical University, Rostov-on-Don, Russia

Abstract. The article is dedicated to the study of several factors that must be taken into account when assessing the hardness of the reinforced Novikov gears surface. Previously, it was established that, as the plastic properties of steel decrease due to thermal or thermochemical hardening, the contact stress at the periphery of the instantaneous contact spot (ICS) can exceed the tension at its center and initiate alternative types of failures (for example, observed tooth fracture). The effect of tensile stresses at the ICS boundary and the inevitable defects of thermochemical treatment (TCT) on the level of transmission load capacity are evaluated. The optimum hardness of the working surfaces of the contact is determined.

Keywords: Novikov gears · Thermochemical treatment · Hardened layer · Surface defects · Strength criteria · Hardness · Ductility · Friction conditions

1 Introduction

There are practically no classical types of contact fractures of the working surfaces of the Novikov gears teeth (pitting). All failures usually have the character of end chips and fractures, and the localization of primary surface defects in the core near the boundaries of the contact spot and, in some cases, close to a flat fracture surface indicated the influence of shear stresses.

Stresses on the surface of a quasielliptical contact area, characteristic of Novikov gear teeth, compressing everywhere, except for the region of its boundary. There, stresses of equal magnitude arise - radial (tensile) and circumferential (compressive). The shear stress is maximum for the circular contact area and decreases as the ellipticity coefficient decreases $\overline{\beta} = b_H/a_H$, where b_H, a_H- the minor and major semiaxes of the contact ellipse, respectively, this stress decreases. The criterion (equivalent strength according to the accepted hypothesis) stresses also decrease at the boundaries of the MPC and increase in the center of the spot. For normalized and heat-treated steels, the equivalent stresses σ_e and Poisson's ratio 0.3 for the circular region of the ICS of radius a (Fig. 1) at the border they exceed those operating in the center by approximately 12.5% [1, 2].

B. T. Long et al. (Eds.): MMMS 2020, LNME, pp. 477–482, 2021.
https://doi.org/10.1007/978-3-030-69610-8_65

The increase in hardness, especially when H > (450… 500) HB, leads to an increase in the difference in the limits of tensile strength ($\sigma_{в+}$) and compression ($\sigma_{в-}$). The role of shear stresses in the process of fracture decreases with increasing brittleness of the material, and the fracture becomes brittle or mixed (brittle-plastic) depending on the stiffness of the load. Under these conditions, the use of the Tresca – Mises plasticity criteria in assessing the strength leads to incorrect results. It was shown in [2, 3] that the most objective for such cases was the use of the Lebedev – Pisarenko criterion for the limiting state for a structurally inhomogeneous material.

The degree of participation of shear deformations in material microfracture is reflected by the plasticity parameter $\chi = \sigma_+/\sigma_-$, where σ_+, σ_-- stresses of failure under uniaxial stretching and compression. In particular, for hardened alloy structural steels with surface hardness $H_o \approx (60\ldots62)HRC_э$ and plasticity parameter $\chi \approx 0{,}7$ the relative tension of the border zone according to the Lebedev-Pisarenko criterion increased by approximately another 12–13% (Fig. 1). It is important to bear in mind that the level of permissible stresses in the center of the ICS is conditioned by the compressive strength of the material, while on the edge of the ICS - by tensile strength, so it decreases proportionally to χ [1].

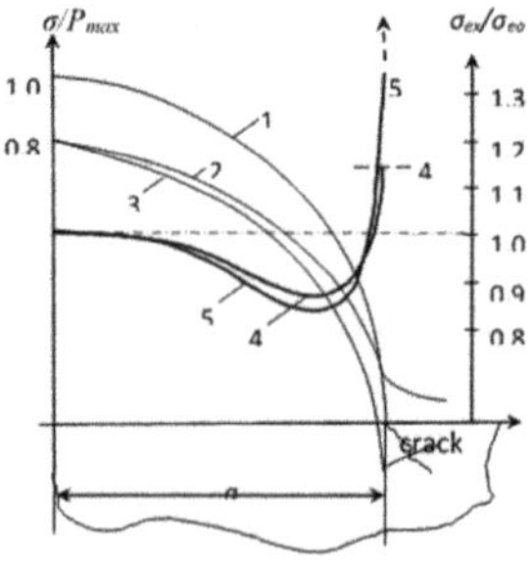

Fig. 1. Plots of stresses on the surface and along the axis of symmetry with a hertz distribution of pressures in a circular region: P_{max} – max pressure MPa; 1 – normal stress σ_n; 2 – circumferential stress σ_θ; 3 – radial stress σ_r; 4,5 – relative tension σ_{ex}/σ_{e0}, where σ_{ex} – equivalent stresses at a distance x from the central axis, σ_{e0} – equivalent stresses in the center of ICS (4 – von Mises criterion, 5 – Lebedev – Pisarenko criterion).

In [2], it was shown that in the case of an elliptical contact area at values $\overline{\beta} \geq 0{,}2$ the stress at the ends of the major semiaxis of the ICS exceed those in the center of the spot, which increases the risk of end chips; at $\overline{\beta} \geq 0.35$, the criterion stresses at both boundaries of the ICS semiaxes become decisive for the danger of a cut. Note that for normalized and heat-treated steels with hardness up to (400… 450) HB χ = 0.95… 1.0; for tool and bearing steels hardened to 60… 6 The analysis of gears with initial contours of DLZ 1.0–0.15, DLZ 0.7–0.15, according to GOST 30224–96 showed: in unprocessed gears, the ellipticity coefficient was in the interval (0.1–0.4), decreasing as increasing the gear ratio and angle of inclination of the tooth line and reducing the number of gear teeth. 4 HRC χ = 0.6… 0.7. The run-in in Novikov's gears [4, 5] makes adjustments to the size and configuration of the ICS, but does not change the nature of Stress-strain state at its borders. As a result, various friction conditions are implemented within the ICS. In the central zone of the spot (where the maximum pressures), a stable elastohydrodynamic regime prevails, and the run-in is predominantly plastic. However, as we approach the ICS boundary and with decrease of specific pressures, the effect of plastic run-in of microroughness decreases, and a boundary friction regime with increased wear is likely at the periphery of the ICS.

2 Results

Let us analyze the effect of typical defects of the surface layer of TCT-hardened samples on their stress-strain state (Fig. 2). When cementing in a thin (not more than several tens of microns) surface area, defects are inevitable as a result of decarburization of the solid solution, and so-called "internal oxidation" in the form of a troostite mesh along grain boundaries (Fig. 2a). This is the result of direct interaction of saturated with oxygen media with alloying elements of the material having a higher oxidation potential than iron. With nitrocarburizing, the so-called the dark component is observed (Fig. 2b), which is detected on the surface in the form of a dark dotted grid. The nature of the dark component is a matter of discussion, but its influence is extremely negative.

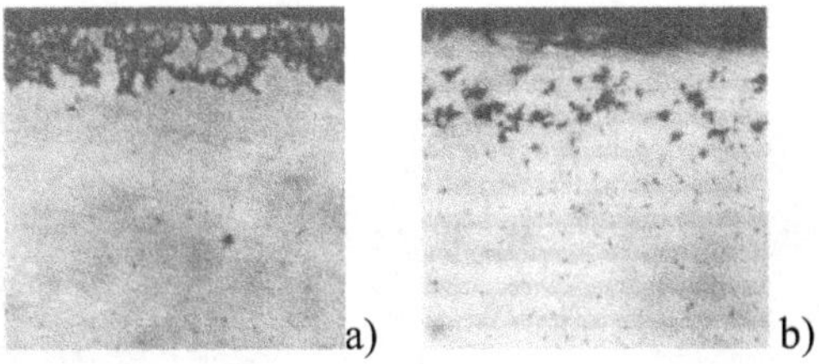

Fig. 2. Defects of the diffusion layer on the samples [6]: a) cementation, steel 20XH2M, a troitite mesh is visible; б) nitrocarburizing, 25KhGT steel, the "dark component" is visible.

Defects in the surface zone sharply reduce the strength properties of the part. In [8], cemented to effective depth $h_{te} \approx 1.3\ldots\ 1.4$ mm (to hardness $H_e = 550$ HV) involute wheels (option 1) are described, which showed a fatigue limit of more than 30% lower than cemented to a depth of $h_{te} \approx 0.3\ldots\ 0.4$ mm (option 2) [7]. But after shot-blasting the fractured area, the samples of both variants showed comparable durability. Metallographic studies revealed the presence of defects in the form of a troostite mesh in a thin surface zone. As you know, defective inclusions are formed at the saturation stage, and subsequent types of heat treatment do not change the depth of the defective zone. The exposure time of the samples of option 1 was 12 h, and the depth of surface oxidation reached 26 μm. For option 2, with a holding time of 3 h, the oxidation extended to a depth of 14 μm.

The influence of defects of the near-surface zone on the flexural strength of Novikov gears and involute gears doesn't have principal differences, but the quantitative difference is significant. Novikov's gears are oriented to work with increased specific loads, which require, according to the criterion of deep contact strength, a greater thickness of the hardened layer. This increases the duration of the saturation process and the depth of the defective zone, which reduces the resistance of the part to cuts and fractures.

The experiments [4, 5], which were discussed, along with the confirmation of the high load capacity of gears with the initial circuit according to GOST30224-96, posed a number of questions. The microstructure of the hardened layer and core by martensite, residual austenite, dark component, and carbonitrides in general corresponded to OST23.4.52–83. However, despite the fact that the gears were made of steel of the same supply and TCT passed simultaneously, the scatter of hardness values reached $\Delta H \approx 150$ HV within the same tooth and $\Delta H \approx 250$ HV for the same points of different samples. And in all cases, in the near-surface zone at depths of 0.18… 0.20 mm, the presence of defective troostomartensitic structures was noted due to depletion of the zone solid solution by alloying elements; a local decrease in hardness compared to the maximum reached 100… 120 HV.

As a result of running-in, a decrease in roughness was noted: by 17… 26 μm, and the hardness of the near-surface zone increased, on average, by 40… 60 HV.

As already indicated, all the fractures of the teeth were in the nature of end chips and fractures with the localization of primary defects in the core near the boundaries of the contact patch.

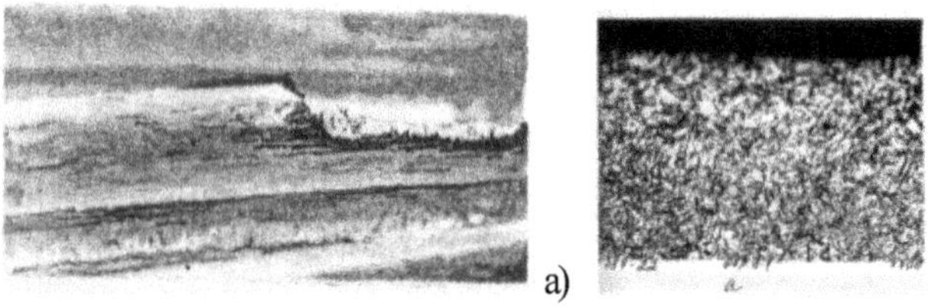
a) b)

Fig. 3. Experimental Novikov gear wheel, operating time - $33.84 \cdot 10^6$ cycles; a) type of destruction; b) the microstructure of the surface zone.

Figure 3 is a typical example of refusal is given. The fracture occurred along the tooth head, where the influence of the bending moment is negligible. The fracture surface is relatively flat. Areas characteristic of fatigue fracture are not clearly expressed, the area of slow crack propagation is poorly smoothed. The zone of accelerated development is velvety, with a small, ruddy structure characteristic of fragile destruction. Troostomartensite structure - as in most samples, but deeper: up to 0.32… 0.34 mm - medium-acicular martensite with needles up to 10 μm long (points 5.6). Note that there were always signs of brittle fracture in the presence of medium-acicular martensite in the structure of the samples. There are many other examples of edge chips along the tooth tip of Novikov's wheels.

In [2], a comparative analysis of changes in the maximum permissible equivalent, according to the Lebedev-Pisarenko criterion, stresses in the center and at the boundary of the IPC, depending on the hardness of the working surfaces, is given. Design model – "sphere-plane", material - case hardened alloy steel. With a surface hardness up to H0 ≈ 750 HV, an increase in P_o up to 2236 MPa was observed. With a further increase in H0, a decrease in P_o took place, which generally corresponds to the previously obtained results for a linear contact [8]. But at the slick boundary, a progressive decrease in bearing capacity with an increase in H0 increased already from the level H0 = 650 HV by 8… 9% at H0 = 700 HV and up to 35% at H0 = 840 HV. In this sample, the surface hardness reached 64-65HRC (i.e. 850–880 HV). The microstructure of the nitrocarburized layer at depths of up to 0.32–0.36 mm consisted of coarse-needle (with needles up to 16 microns - 8 points) martensite and residual austenite, which indicated significant overheating.

For involute transmissions under the conditions of deep contact fracture (GKR), the recommended optimum hardness is not less than 750 HV at a depth of 0.1 module [9]. But since when testing Novikov's gears at contact stresses σkmax up to 2300 MPa, GCR were not limiting, then one should focus on alternative types of failure (cut, break), for which the level of hardness H0 = 650… 700 HV seems rational.

3 Conclusions

The conducted research allows us to draw the following conclusions. 1. Areas with tensile stresses at the boundaries of the local contact area are preserved after running-in. 2. As a result of TCT, a layer is formed on the surface with a troostite mesh of up to 30 μm and a troostomartensitic structure and crypto-needle martensite with significantly reduced mechanical properties. 3. As you approach the border of the ICS, the effect of plastic break-in of microroughness decreases and regime close to boundary friction is realized, which leads to an extremely negative combination of "hard" stress-strain state and TCT defects. 4. An increase in surface hardness above 650 HV practically does not lead to an increase in hardness in the central zone of the ICS, but progressively reduces it at the periphery, therefore it is advisable to limit the maximum hardness of the working surfaces to 650–700 HV. 5. It is necessary to evaluate the stress-strain state characteristics of the contact area of the teeth of the Novikov gears formed as a result of running in. 6. When assessing the influence of the microstructure of diffusion layers on the level of their bearing capacity, it is advisable to focus on the recommendations of [9], which are not yet fully reflected in the technical standards.

Acknowledgement. Research was financially supported by Southern Federal University, grant No. VnGr/2020-04-IM (Ministry of Science and Higher Education of the Russian Federation).

References

1. Beskopylny, A., Meskhi, B., Onishkov, N., Korotkin, V.: Limit-state criteria and their use in conditions of the significant structural heterogeneity of the gear steel. In: Popovic, Z., Manakov, A., Breskich, V., (eds.) VIII International Scientific Siberian Transport Forum. TransSiberia 2019; Advances in Intelligent Systems and Computing, vol. 1115, pp. 904–912. Springer, Cham (2020). https://doi.org/10.1007/978-3-030-37916-2_89.
2. Korotkin, V.I., Onishkov, N.P.: Assessment of load capacity of chemically-thermally hardened gears with local contact of teeth. Russian Engineering Research №8 (2020)
3. Pisarenko, G.S., Lebedev, A.A.: Deformation and Strength of Materials Under Complex Stress State. Naukova dumka, Kiev (1976). 415s.
4. Korotkin, V.I., Onishkov, N.P., Kharitonov, Y.D.: Novikov Gearing: Achievements and Development, p. 249p. Nova Science Publishers, Inc., Hauppauge (2011)
5. Beskopylny, A., Onishkov, N., Korotkin, V.: Bending strength assessment of chemically-heat strengthened Novikov gearing. In: MATEC Web of Conferences Electronic Edition C. 02033
6. Zinchenko, V.M.: Engineering of the Surface of Gears by Means of Chemical-Thermal Treatment. Monograph. Bauman MSTU Publishing House, p. 303 (2002)
7. Anzinger, M.: Werkstoff-und Fertugungseinflisse auf die Zahnfuβ tragfahigkeit, insbesondere im hohen Zeitfestigktitsgebiet./ Diss. Munchen (1991). 129p.
8. Onishkov, N.P., Korotkin, V.I.: Evaluation of contact fatigue life of chemically heat-strengthened gear wheels. N. P. Advanced Engineering Research, T. 17 №3, pp. 5–13 (2017)
9. Rudenko, S.P., Val'ko, A.L.: Contact, fatigue of gear wheels of transmissions of power-packed machines. Belarusian Science, Minsk (2014). 126s. ISBN 978-985-08-1694-8

Modelling and Simulation of the Hydraulic System on EODM Machine

Dat Nguyen Duy[1], Dat Chu Van[1], and Sy Le Van[2(✉)]

[1] Le Quy Don Technical University, Hanoi, Viet Nam
[2] PetroVietnam University, Ba Ria, Ba Ria - Vung Tau, Viet Nam
sylv@pvu.edu.vn

Abstract. The remote-controlled explosive ordnance disposal machine (EODM) is an important device using in the military, which had been designed and developed by many countries in the world. One of the most important components in this machine is hydraulic system that drives the operating units. This paper focuses on modelling and dynamic simulating of the hydraulic system during the bomb laying process. The main target of this study is to meet the requirements for controlling an EODM with high specific power, low hysteresis, high precision control, which ensures precision operation and safety. The hydraulic system was modeled in mathematical formula and then was simulated by using LMS Amesim software to evaluate the dynamic response of the system.

Keywords: Hydraulic system · LS-PC control · EODM · Bomb clearance

1 Introduction

The activities of bomb/mine clearance and explosive ordnance disposal are an important task of each country after the war finished. Many technologies have been developed in many decades for both semi-automation and full automation systems. In the recent year, explosive ordnance disposal technicians are using remote-controlled machines called EODM (remote-controlled explosive ordnance disposal machine) to help detect and defuse improvised explosive devices. It is equipped with integrated operational devices such as radar detectors, digging devices, bomb pickers, high-pressure water jets and hydraulic hammers. Moreover, EODM has been developed with more intelligent and multi-functional operations, multi-purpose with full controlled system and high-resolution cameras [1]. The smart EODM can replace single functional machines which are using for cleaning the bomb and explosive ordnance disposal such as: single-bucket excavators, high-pressure water jetting devices and bomb picking devices. The application of this machine in the practical activities allows to save the greatly operating time and technician effort with easy-to-use, high accuracy, safety for operators. Many commercial EODMs have designed and developed by well-known companies UBIM (Russia), Wisent (Germany), Kodiak (UK), GCS (Switzerland), … etc. [12, 13]. Although there are many similarities in the designed function and using purpose of EODMs among the manufacturers, but significant differences of control system found in their design. Specially, the design of hydraulic systems which

B. T. Long et al. (Eds.): MMMS 2020, LNME, pp. 483–488, 2021.
https://doi.org/10.1007/978-3-030-69610-8_66

drives operating units (1. Base machine, 2. Booms, 3. Arm, 4. Gripper bomb mechanism, shown on Fig. 1a) to carry out soil excavation, bomb picking, bomb transportation onto specialized vehicles and other tasks is a typical design of each manufacturers. The hydraulic drive system, in comparison with the electric and pneumatic drive system, has a high specific power, low hysteresis, high precision control, which ensures the speed and movement of the working equipment. In addition, the hydraulic system meets an excellent agreement for the explosive ordnance disposal activities with high precise control which is a very important factor contributing to ensuring safety for operator and properties [4, 5].

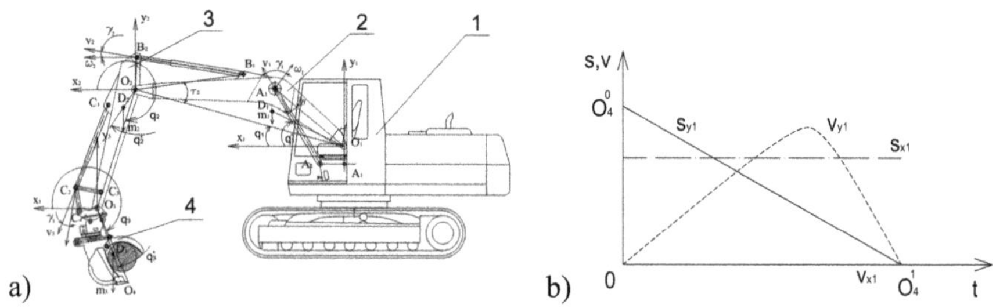

Fig. 1. a) Main components; b) Required displacement, speed.

This paper presents the dynamics model of working equipment and the hydraulic drive system during the bomb laying process, in order to meet the requirements set out when the bomb is placed on a specialized vehicle. The hydraulic system used on EODM are analyzed and modelled in this study. In this machine, the hydraulic system was modeled fully from the LS-PC control pump using mechanical feedback to the control system which consists of proportional pressure reducing valve and main valve with integrated pressure compensation valve. The model of working equipment during the laying bomb process was fully built. All hydraulic components, lifting equipment, as well as input control signals are simulated by using LMS Amesim software to make the dynamic assessments for the system.

2 Mathematical Modelling

2.1 Main Components of EODM

The EODM is developed and manufactured by many manufacturers in the world. It was included the main units such as base machine, boom, arm, excavation/gripper bomb mechanism as shown on Fig. 1a. These units are controlled by the hydraulic system which needs to ensure the exact movement and good energy characteristics. Many previous researches have shown that the throttle-controlled hydraulic drive system is eliminated due to low efficiency. Hydraulic system with volumetric control has high efficiency, but it has some disadvantages such as complicated structure, large volume and bulky [4, 5, 8, 9]. In recent year, the hydraulic drive systems combined with pump - valve control has been equipped for EODM which can overcome these disadvantages such as LS (Load sensing), FS (Flow sharing) and FM (Flow matching), FS hydraulic systems. In this study, the flow sharing (FS) hydraulic drive system will be considered which has high efficiency, no saturation of the pump flow, and low hysteresis [2, 6, 11]. The pump provides a fluid flow for the operating units with the pressure to be equal to the load pressure plus the pressure drop in the control devices (from 0.8 to 1.5 MPa). To operate the working units, the EODM uses electric-proportional pressure reducing valves to control the main valves.

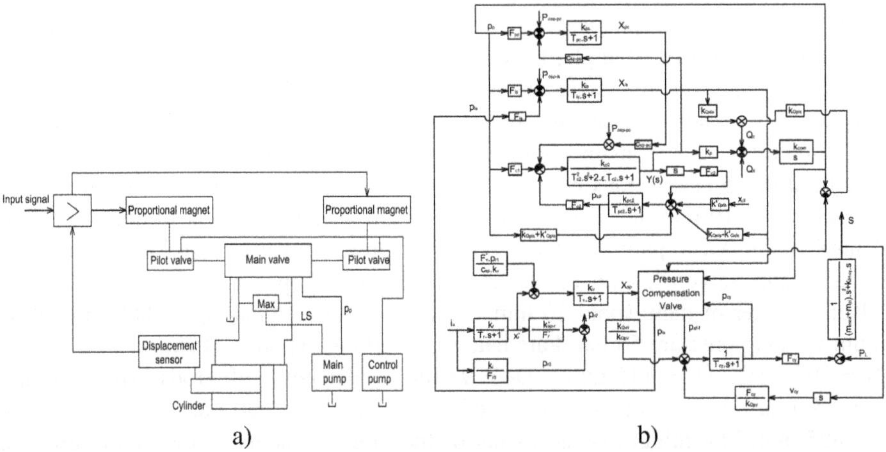

Fig. 2. a) Block diagram and b) Mathematic model of the hydraulic drive

2.2 Mathematical Model

The dynamic equations of the system used Newton – Euler's equation approach was carried out for each component such as the booms, arm, gripper bomb mechanism, and excavation gripper mechanism. These equations have used popularly in the previous researches relating to EODM. The static and dynamic components of acting forces during the bomb laying process were converted into the force P_L which applied to the rod of piston by using moment equation and momentum conservation.

The study of hydrodynamics for hydraulic system driving the operating components in EODM used Newton-Euler equation for each unit in the system. The mathematic relations among hydraulic component of the system included flow rate, pressure based on the Pascal's law, flow rate conservation and working fluid compressibility. In addition, the hydraulic system was modeled fully from the LS-PC control pump using mechanical feedback to the control system which consists of proportional pressure reducing valve and main valve with integrated pressure compensation valve.

Based on the Block diagram of hydraulic system, mathematic models for each component in the hydraulic system are established in this section. The dynamic analyzing and calculating in detail for EODM were presented in [1] and block diagram of the hydraulic drive system shown in Fig. 2a. This paper summaries the main formula of transfer function and present full structure diagram of hydraulic system controlling working equipment in Fig. 2b. This diagram is used for inputting the simulating software so called LMS Amesim.

3 Simulation Setting and Results

Because the process of calculating with a very large volume, to simplify, the process of studying the dynamics properties when laying bomb, is carried out by the laying method of the boom, not moving the arm. This method will ensure the accuracy of the bomb placement when the distance from the starting position to the laying surface of the bomb is small (for example, lowering the lifting cylinder 0.1 m).

To study the dynamics characteristics during bomb laying using LMS Amesim software. Components as pump, control valves are designed from elements in the Amesim library. Some main parameters of the system: HPV 95 hydraulic pump: Specific flow 95 cc/rev, maximum working pressure $p_{max} = 350$ bar; diameter of the PC, LS valves 10 mm, normally open mode; Pump use mechanical feedback control. Control pump: Gear pump, specific flow 36 cc/rev; Maximum working pressure is 30 bar. Main valve: 25 mm diameter of the spool. Proportional pressure reducing valve: diameter of main spool 10 mm, electrical control signal $i_{max} = 800$ mA, maximum working pressure 50 bar. Parameters of working cylinder: Diameter of cylinders $D_1 = 105$ mm, $D_2 = 115$ mm, $D_3 = 95$ mm, diameter of rods $d_1 = 70$ mm, $d_2 = 75$ mm, $d_3 = 65$ mm, stroke of cylinders $l_1 = 990$ mm, $l_2 = 1175$ mm, $l_3 = 885$ mm. The initial displacements of the three cylinders – the boom, arm and excavation – gripper bomb mechanism were 0.818 m, 0.84 m, 0.71 m, respectively. Masses of working equipment: $m_1 = 936$ kg, $m_2 = 410$ kg, $m_3 = 955$ kg. Mass of bomb MK-84: 907 kg.

In Fig. 4, the displacement of the piston has not reached the required value (from 0.818 to 0.718 m), this can be explained in Fig. 3. Specifically, when the value of the pressure to control the main valve reaches the value of 1.5 bar at 3 s, with this pressure value, the opening of the main valve close to zero and the piston of the boom does not move any further. In order to achieve the displacement value, the laying velocity of the lifting cylinder need a suitable proportional pressure relief valve control signal, so that the valve control pressure is maintained to decrease to 4 bar, to maintain the small opening of the main valve, and when the necessary displacement is reached, immediately give this value down to 1.5 bar (or control signal to 0 mA).

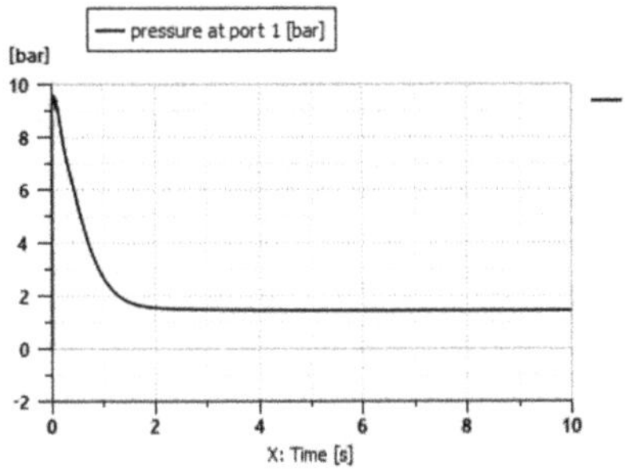

Fig. 3. Control pressure at the main valve

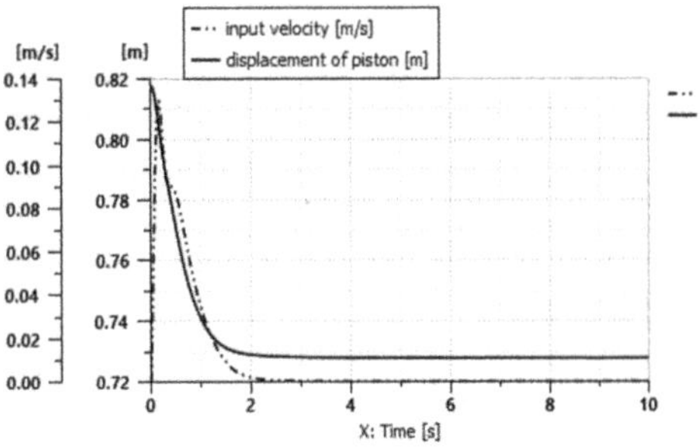

Fig. 4. Speed of the boom piston

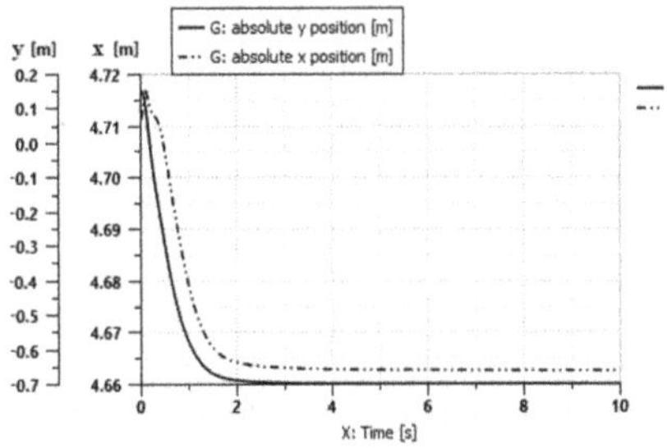

Fig. 5. Movement of center gravity of excavation – gripper bomb mechanism

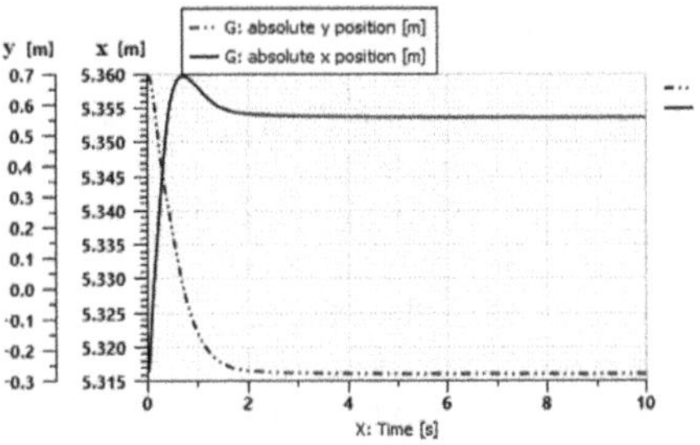

Fig. 6. Movement of the center gravity of the gripper bomb mechanism (when extending arm)

According to Fig. 5, with the displacement of 0.09 m in the direction of the laying of the boom piston, the displacement of the centroid of the excavation – gripper bomb mechanism (together with the bomb) in the y direction is 0.85 m (from 0.15 m to −0.7 m), while the movement in the x direction is only 0.053 m (from 4.716 m to 4.663 m).

In Fig. 6, when extending arm of 0.16 m, the displacement in the y direction of the center gravity decreases 0.98 m (from 0.7 m to −0.28 m), while the movement in the x direction of the center gravity increased by 0.039 m (from 5.315 m to 5.354 m). Thus, the position deviation in the x direction has decreased to less than 0.05 m.

4 Conclusion

The paper has completed the dynamics modeling of working equipment with the hydraulic system on EODM for the laying bomb process. With high requirements on accuracy, safety, the process of laying bombs must be conducted with important notes. Firstly, when laying bomb near the base vehicle, just combine the lowering boom and extending arm, when laying bomb far away from the base vehicle (arm is extended), laying bomb by the method of laying boom (due to cameras for observation support). Therefore, in practice, the application of method of laying bombs by laying boom method will be simple, while ensuring accuracy. Secondly, to ensure the proper laying bomb according to the distance and velocity requirements, need control signal of the pressure reducing valve to meet the requirements mentioned above, to overcome the delay of the main valve, as well as the actual delay of the pressure reducing valve.

References

1. Internal report: Design and Manufacturing EODM, Institute of military engineering, Hanoi (2017)
2. Popov, D.N.: Dynamic and regulation of hydraulic and pneumatic systems, Moscow (1987)
3. Gimadiev, A.G.: LMS Imagine.lab Amesim As an effective tool for modeling dynamic processes in mechatronic systems, Samara (2014)
4. Burks, B.L., Killough, S.M., Thompson, D.H., Dinkins., M.A.: Explosive Ordnance Disposal Technology Demonstration Using the Telerobotic Small Emplacement Excavator, Oak Ridge National Laboratory (1994)
5. Terwelp, C.R.: Remote Control of Hydraulic Equipment for Unexploded Ordnance Remediation, Virginia Polytechnic Institute (2003)
6. Shop manual PC120, 130-6
7. Vaseliu, N., Puhalschi, R.: Simulation of fluid power systems with Simcenter Amesim, Siemens, Germany (2018)
8. Yang, Y., Chen, Q.: Modeling and Analysis of Full Hydraulic Pressure Drive System of the Wheel Loader Based on AMEsim, Guizhou University, China (2013)
9. Wei, C., Lian, J., Li, J.: The modeling analysis and dynamics simulation of load-sensing hydraulic systems, Taiyuan University of Science and Technology, China (2011)
10. Casoli, P., Riccò, L.: Modeling and verification of an excavator system – Axial Piston Pump, Kinematics and Load Sensing Flow Sharing Valve Block, Linköping, Sweden (2013)
11. Rexroth, M.: Hydraulic, Proportional and servo valve technology, Germany
12. https://www.army-technology.com/projects/wisent-2-armoured-support-vehicle/
13. https://btvt.info/1inservice/ubim158.htm

Corrective Maintenance of Manufacturing System Using Artificial Intelligence Techniques Support for Engineering Process

Ngoc-Hoang Tran(✉)

University of Technology and Education - The University of Danang, Danang, Vietnam
tnhoang@ute.udn.vn

Abstract. Corrective Maintenance (CM) take an important role on automated manufacturing process in many industrial areas. However, CM process depends mainly on operator's knowledge and their decision of locating failure components or its causes in order to make corrections on their equipment. The common knowledge - based system as Failure Mode and Effect Analysis (FMEA) has especially used since many years, but it has required an integrated approach which exploit production data to generate a more effective maintenance procedure in recent periods. In this context, this paper proposes an integrated protocol of diagnosis process by combining expert's system with Pearson correlation (R) model and using AI technique: Bayes Learning. This protocol toward exploiting historical production data problem to identify and diagnosis drift production or bad quality product for dealing good correction decision. Therefore, our key contribution to support maintenance tool machine for increasing life-cycle engineering.

Keywords: Corrective Maintenance · Life cycle engineering · Machine process control · Bayesian network · Decision support

1 Introduction

Today, manufacturing system become more complex because of the increasing of production lines, mix – product and high manufacturing volume. While operator requires always accuracy of manufacturing equipment operations, that make a planned maintenance become so important for keeping maximum utilization performance of equipment. However, a huge production data is not easily to extract on Industry 4.0 context. In fact, its complex level is increasing because of development of new 4.0 technology as IoT, data analyzing, blockchain and intelligence solution.

Furthermore, it's more difficult because the signal from sensors cannot show all production state since its position on the manufacturing process for technical reasons. Therefore, it's necessary a data driven approach which exploit maximum potential of historical data for supervising equipment state and provide correction actions on real-time. In that context, this work proposes a way of dealing with finding root cause problem for supporting corrective maintenance process by using Artificial Intelligence Technique: Bayesian learning on historical production data, which performed on

B. T. Long et al. (Eds.): MMMS 2020, LNME, pp. 489–494, 2021.
https://doi.org/10.1007/978-3-030-69610-8_67

MATLAB environment. Our approach will be applied to monitor and diagnosis failure causes of a critical CNC machine component: Spindle Tool in our workshop. The principal idea of this paper is to demonstrate a model for automatic identification of variable, establish collected data table for learning and final produce a protocol consists of the classes researches (observed/latent).

2 Literature Review

Industrial production system is represented by large volume of production, multi production lines, high complexity and unscheduled risks that represent in required characteristics for the common machine learning methods:

(+) A significant number of input parameters, sensors signal from manufacturing process (cycle time, metrology, motor's speed, etc.).
(+) Various natural parameters type: discrete, continuous, qualitative or quantitative.
(+) Strong causality between the events on equipment and product: failure, effects, drift function, bad product quality.
(+) Uncertain or incomplete data in some cases.

2.1 K-Nearest Neighbors (KNN)

This method is traditional classification learning methods but they are not producing any model from learning examples data. By observing the nearest class to find the similar characterizes of them, they decide the class in which this new observation belong. In order to predict the class of a new variable, the KNN algorithm searches for the K nearest neighbors of this variable and predicts by the most frequent response (by correlation coefficients, such as Pearson and Spearman) of them. This method works on continuous data. It is also possible to take into account binary data (discrete variable: Yes/No; 0 or 1), but not multinomial (discrete variable with more than 2 modalities) [1]. In addition, KNN's classes are mainly represented by the K nearest neighbors, so that it is difficult to find the case of large classes. Also, there are no priori between classes and the calculation time is too high because this method needs a lot of learning examples.

2.2 Artificial Neural Network (ANN)

The ANN is structured based on a classic neural network compose of two principal classes neurons: input and output layer. In order to combine these layers, the ANN take role as a black box with these hidden layers. They analyze and process information using connectionist approach to computation [2]. The objective of an ANN is to find out the relationship between input and output patterns. Normally, they define the input/output of ANN model according to system's parameters or variables but it's often performed depend on their experiences without a certain mathematical formula. Therefore, it's difficult for determining the reasonable number for this model.

Furthermore, learning with missing data is important problem if without necessary certain conditionally in ANN method [3]. The main advantage of ANN is allowed modeling of nonlinear multivariate processes and defining classes's priori of its structure by synaptic weights. However, choosing and taking account the synaptic weight into ANN model is another difficult problem that requires a logical approach to calculation.

2.3 Bayes Network (BN)

A BN is a system which represent knowledge and allow the calculation of conditional probabilities providing solutions to different kinds of problems: classification, identification, modelling relationship [4]. In general, S = {S_1, S_2..., S_n} is a set of nodes of BN, the joint distribution function P(S) is written as follows:

$$P(S) = \prod_{i=1}^{n} P(S_i/Pa(S_i))$$

With Pa(Si) is set of parents of Si nodes.

One of advantage of BN is dealing with both the case of discrete variables and continuous variables. Bayes network is represented by visual graph that is vertices and edges have meaning gives us valuable information about conditional dependence between the variables (cause – effect). Moreover, the priori of causality is represented by conditional probabilistic. In addition, BN can represent many variables (by nodes) and hierarchical classes (with naive hierarchical BN). The probability calculation is performed by the optimal estimate algorithm (Likelihood Maximation and Expectation-Maximization (EM)) can represent correlations between nodes. Furthermore, BN allow taking into account of the temporal causality to Dynamic Bayesian networks [5]. However, its complex level of computational process depends on the amounts of learning data.

In summary, the BN is generally simpler than Neural network, with many decisions about hidden layers, and topology and variants parameters.

3 Toward to Model Diagnosis Protocol by Bayes Learning Approach in Case Study

3.1 Modelling Bayesian Network

Modelling Bayesian network is generally composed by two principal steps:

(i) Define the structure of Bayesian model under a graph form. Our case study in this paper is located on a CNC machine. In this situation, all variables of machine's data are automatically transformed correspond to Bayesian nodes. These variables are defined by "DOOSAN" default manufacturer (describe on top part of Fig. 1a). Causality of each variable is defined at first by the expert charge the operating of this machine. Then, we used the Pearson Correlation Coefficient [6] for verifying and approving these causal relationship (below part of Fig. 1a). This coefficient value can range between −1.00 and 1.00 corresponding to mean the relationship between the variables is negatively or positively through increases or decreases. As mentioned, each variable corresponds to a node. We develop a cause and effect type model; its generic structure is presented in Fig. 1b. It is structured by three levels: The essential component of CNC functions ("RC" class), the "F" class represent the regular fault of system and the parameter/sensors signal ("X" class).

(ii) Estimation of the probability tables is performed by the EM learning algorithm, in below equation. The values of the node are observed from historical production data. We used this dataset which recorded in 6 months from 9/2019.

$$\hat{P}\left(X_i = x_k/parents(X_i) = X_j\right) = \hat{\theta}^{MV}_{i,j,k} = \frac{N_{i,j,k}}{\sum_k N_{i,j,k}}$$

$N_{i,\,j,\,k}$ is the number of occurrences where the variable X_i takes the state x_k and the parents of X_i take the state x_j.

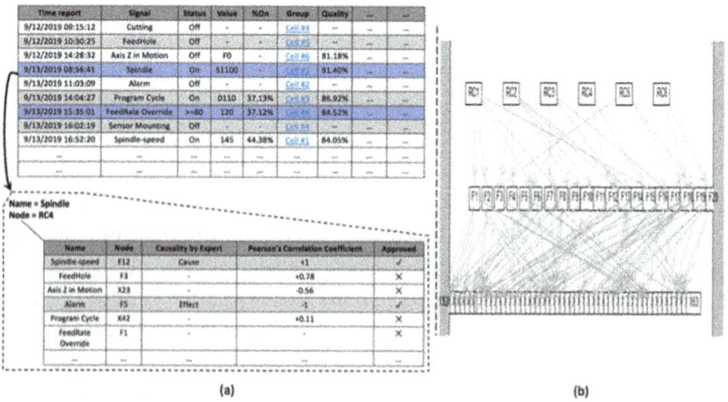

Fig. 1. (a) Historical data for learning & Causal table of variable "Spindle". (b) Our Bayes Network is modeled in MATLAB environment.

3.2 Diagnosis Results for Corrective Maintenance (CM) Suggestion

Figure 1b show us the result of probabilistic learning phase based on historical production data in MATLAB environment with BN network of 79 nodes.

In order to test validation phase, we used CNC data (show in Fig. 2a) in which the increasing of Spindle Speed that effect to low performance of production quality of Spindle Tool (represented by RC_4 node). Based on established BN, the remaining step is only a probabilistic calculation matter. The probability table for some considered variable is performed and show in Fig. 2b. It represents the conditional probability of RC_4 with its considered components including of Spindle Speed (F_{12} node), Alarm (F_5 node), spindle motor rate (X_{32})…

The result is detected by examining their evolution of conditional probability P (variables | RC_4) in two situations of presence/absence of “Low Performance of Spindle” event in system. In which, the most significant evolution focus on P(F_{12} | RC_4) represents the primary cause “increasing of Spindle Speed” who is also shown on our data (Fig. 2a). Whereas the evolution of probability of another variables is not evidently found. This result implies that this BN model produces an acceptable result when the operator requires a diagnosis for CM actions by learning historical production data.

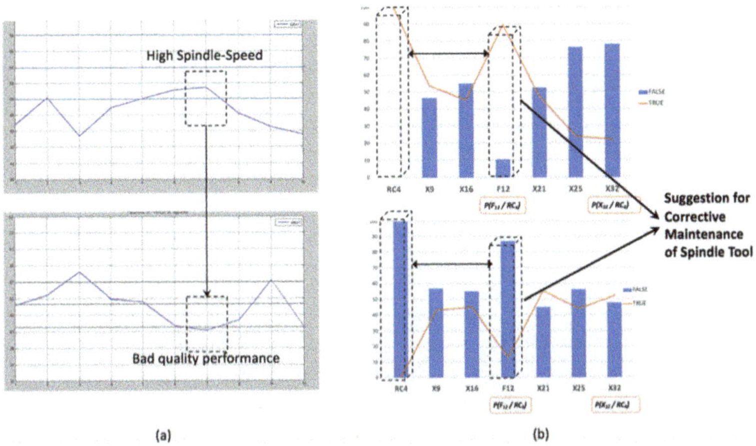

Fig. 2. (a) Curve of Spindle - Speed and Production Quality. (b) Diagnosis results for supporting Corrective Maintenance actions.

4 Conclusions

In this paper, our work provides a supporting process for building Corrective Maintenance activities by Bayes learning algorithm based on historical production data. The author integrates also expert’s knowledge system and Pearson coefficient to build Bayes structure and testing it on an example of CNC machine. The diagnosis results demonstrate that our model work successfully for supporting maintenance decision in order to improve life production cycle.

References

1. Cover, T.M., Hart, P.E.: Nearest neighbor pattern classification. IEEE Trans. Inf. Theory **13**, 21–27 (1967)
2. Yilmaz, I.: Landslide susceptibility mapping using frequency ratio, logistic regression, artificial neural networks and their comparison: a case study from Kat landslides (Tokat—Turkey). Comput. Geosci. **35**(6), 1125–1138 (2009)

3. Dreyfus, G., Martinez, J.-M., Samuelides, M., Gordon, M., Badran, F., Thiria, S., Hérault, L.: Réseaux de neurones: Méthodologie et applications. Eyrolles, 2ème édition (2004)
4. Tran, N.H., Sébastien, H., Zamaï, E.: Generic and configurable diagnosis function based on production data stored in manufacturing execution system (2016)
5. Roos, J., Gavin, G., Bonnevay, S.: A dynamic Bayesian network approach to forecast short-term urban rail passenger flows with incomplete data. Transp. Res. Procedia **26**, 53–61 (2017)
6. Pearson, K.: Correlation coefficient. In: Royal Society Proceedings, vol. 58, p. 214 (1895)

Generate the Ultimate Longitudinal Strength of Damaged Container Ship Considering Random Geometric Properties Under Combined Bending Moment

Huynh Van-Vu[1(✉)] and Cho Sang-Rai[2]

[1] Nha Trang University, Nha Trang, Vietnam
vuhv@ntu.edu.vn
[2] University of Ulsan, Ulsan, Korea
srcho@ulsan.ac.kr

Abstract. In this paper the ultimate longitudinal strength of a container ship in grounding and collision cases is generated by Monte Carlo simulation with 10000 samples, the effect of randomness in an extent damaged dimension of collision and grounding scenarios, yield strength and plate thickness of midship section will be considered. For each sample, the ultimate longitudinal strength is obtained by using the simplified incremental-iteractive method under combined vertical and horizontal bending moments. The comparison results of non-dimensional combined ultimate bending moments M_u/M_p are performed at the angle 0°, 90°, 180° and 270° of heeling angles in midship section. The ultimate strength results of both intact and damaged conditions have differences between deterministic and randomness simulation. The COV correspond the neutral axis angles at 0° and 360° is always smaller than at 180°, the values of ultimate strength where angles of neutral axis between 90° to 180° and 270° to 360° are further away from the intact one for collision, and opposite phenomenon for grounding.

Keywords: Ultimate longitudinal strength · Container ship · Bending moments · Randomness variables

1 Introduction

The ultimate longitudinal strength is the most important criteria for the structural design of ship hulls. Smith [1] was the first to propose a method to account for the behavior of an each individual element in calculation of ship ultimate longitudinal strength. Based on procedure of Smith method, a computer program ULSAN [7] was developed to determine the ultimate longitudinal strength of ship's hull girders. Many papers published in the open literature for vertical bending moment. However, ships are affected by combined vertical and horizontal bending moments in a rough sea. Gordo [4] calculated the ultimate collapse load of ship hull girders under combined load effect in intact condition. Luís [9] studied the residual longitudinal strength of two double hull Suezmax tankers after grounding or collision accidents under combined bending

B. T. Long et al. (Eds.): MMMS 2020, LNME, pp. 495–501, 2021.
https://doi.org/10.1007/978-3-030-69610-8_68

moment. Beside the combined ultimate bending moment results, the coefficients were proposed for the interaction equation for the collapse damages. Khan [10] calculated the residual longitudinal strength under combined vertical and horizontal bending moments with grounding or collision damages. The interaction equations and reliability analysis using limit-state function by this equation were obtained. Cho [8] modified the program ULSAN to estimated the residual longitudinal strength under combined vertical and horizontal bending moments for various types of damaged ships. Most studies aforementioned assumed materials and geometric properties of ships are deterministic, but all of those may be random in nature. Suhas [11] studied the effects of randomness in yield strength and initial imperfection of bulk carrier and tanker to calculated the ultimate hull girders strength under vertical bending moment. The statistic of ultimate hull girder strength capacity was generated through Monte Carlo simulation with 100 samples for each ship. In the present study, the container in ISSC 2000 [5] (the midship section is shown in Fig. 1) were generated the ultimate longitudinal strength under combined vertical and horizontal bending moments on both intact and damaged conditions by using ULSAN program. Besides, the effects of randomness in plate thickness, yield stress and extent of damaged area due to collision or grounding accidents were considered and there were generated by the Monte Carlo simulation with 10000 samples.

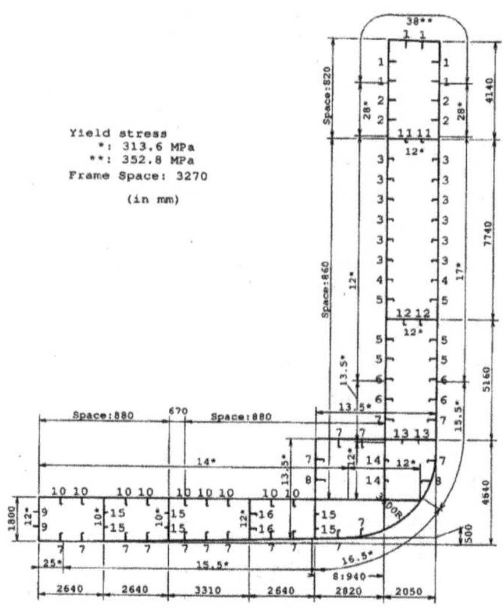

Fig. 1. Midship section of container ship in ISSC 2000 [5].

2 Extent of Damages

The damages due to collision or grounding is the most common results of destruction of ship structure. In this study, the collision or grounding damages calculation are performed at mid-ship location (1/2 length of ship) and extent of damages according to the ABS 1995 [3].

3 Combined Ultimate Bending Moment

The combined vertical and horizontal bending moment is calculated changing a neutral axis at a given angle in relation to the horizontal axis. The angle of neutral axis adopted from 0° to 360° in steps of 10°.

4 Effects of Randomness in Material and Geometric Properties

In nature, the material and geometric properties of ship structures may be random. In this study, the plate of thickness, the yield stress of each structural member (plates, stiffeners or hard corner) and the extent of damages are considered the uncertainties, another parameters remain are considered the deterministic. In this paper, the plate thickness variable is lognormal with mean value equal original thickness and COV 4% according to Lee et al. [2]. The yield stress is also lognormal with COV 8% and mean value of 1.2 times the nominal value according to Cheon et al. [12]. The extent of damages area is assumed lognormal with COV 30% and mean value is two values which one is equal to one time of ABS 1995 [3] and another one is equal to two times of ABS 1995 (Table 1). All those random variables are independence.

Table 1. Mean value of extent damages (m)

C-ABS	C-2ABS	G-ABS	G-2ABS
5.5	10.6	5.3	11.3

5 The Monte Carlo Simulation

Let X is a lognormal random variable [6] with mean μ_X and standard deviation σ_X. To generate a sample value x_i, it begin by generating a sample value u_i of uniformly distributed random number such that $0 \leq u_i \leq 1$. The ultimate longitudinal strength in probabilistic case is generated by the Monte Carlo simulation (MCS). The process of MCS is following as below:

- Determine the number of simulated values. Choosing $S_i = 10000$.
- Generate S_i uniform random variables for simulating of plate thickness utp for each plate, S_i uniform random variables of yield stress u_y for each member element, S_i uniform random variables of extent of collision damaged u_c and also for grounding damaged u_g.
- Generate S_i values of plate thickness t_p, S_i values of yield stress σ_Y for member element, and also S_i value of collision or S_i value of grounding extents.
- Applied the ULSAN program to calculate S_i value of vertical ultimate bending moment M_{uv}, horizontal ultimate bending moment M_{uh} and value of ultimate combined bending moment M_u.

From this time forth, the value of ultimate longitudinal strength in deterministic case have suffix notation is -Det and randomness case is -Ran, respectively. Therefore, for intact condition will be denoted Intact-Det, Intact-Ran, for collision condition with mean value equal to one time ABS is C.-ABS-Det, C.-ABS-Ran, mean value equal to two times ABS is C.-2ABS-Det, C.-2ABS-Ran, and also for grounding condition is G.-ABS-Det, G.-ABS-Ran, G.-2ABS-Det, G.-2ABS-Ran, respectively.

6 Results and Discussions

The comparison results of non-dimensional combined ultimate bending moments M_u/M_p, where M_p is a fully plastic bending moment in intact case, is performed in Fig. 2 and Fig. 3.

In the figures, the angle 0°, 90°, 180° and 270° of neutral axis are performed at middle right, middle bottom, middle left and middle top, respectively. The position of angle 360° is the same with angle 0°. A half right pattern from 90° to 270° of neutral axis angles is performed the hogging condition and remain half left pattern is performed the sagging condition. The outcome of the analysis is the non-dimensional ultimate vertical bending moment M_v/M_p on horizontal axis, the non-dimensional ultimate horizontal bending moment M_h/M_p on vertical axis and each current point on the pictures is a non-dimensional ultimate combined bending moments M_u/M_p. From the figures, it can be show that the results in intact condition are largest than the other

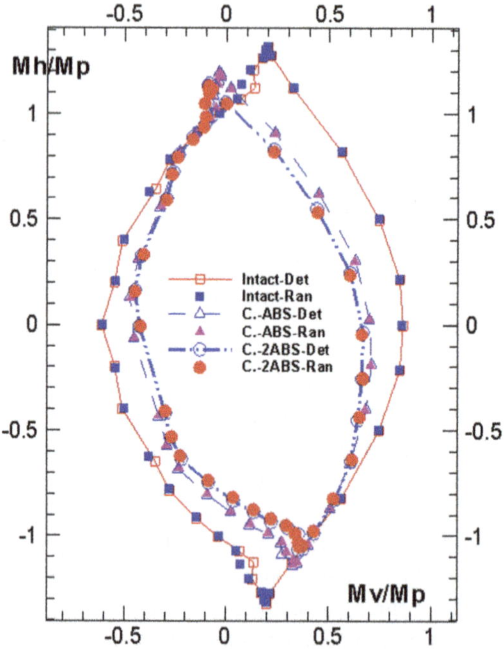

Fig. 2. The non-dimensional combined ultimate bending moment due to collision damaged

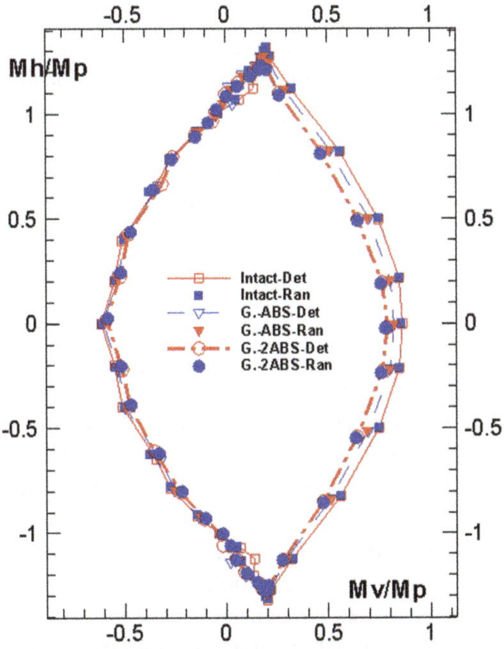

Fig. 3. The non-dimensional combined ultimate bending moment due to grounding damaged

one and the values are symmetry through a horizontal axis. For unsymmetrical collision, the values of ultimate longitudinal strength on the lower left and upper right of the figures, which correspond respectively to the angles of neutral axis between 90° to 180° and 270° to 360°, it can see that they are further away from those of the intact than the values obtained for the other angles. Khan [10] observed that the horizontal load combined with the vertical one is very dangerous from a structural safety point. From these present figures, however, it can be observe in interval from −0.5 to 0.5 of non-dimensional horizontal bending moment, the vertical bending moment slightly decreases. Therefore, the value of ultimate longitudinal strength briefly change when the neutral axis angle small.

Table 2. The COV (%) of residual longitudinal strength in probabilistic simulation

Angle of NA	Intact	C.-ABS	C-2ABS	G.-ABS	G.-2ABS
0°	1.76	2.47	2.55	1.98	2.96
90°	1.46	2.41	4.78	1.57	1.56
180°	2.89	3.11	2.83	2.80	2.82
270°	1.46	1.90	3.27	1.91	3.48

In Table 2 performed the COV of ultimate longitudinal strength in 10000 samples simulation are showed at 0°, 90°, 180° and 270° of neutral axis angles. It can be observe the COV in intact case is smallest and in collision case which mean equal to two times of ABS case (C.-2ABS) is largest. Moreover, the COV at 0° is smaller than at 180° accept in G.-2ABS case, and most of COV at two positions are smaller than 3%. At 90° and 270° of neutral axis angles, in case mean equal to one time ABS, the COV of collision (C.-ABS) is smaller than grounding (G.-ABS). However, in case mean equal to two times ABS, the COV of collision (C.-2ABS) is larger than grounding (G.-2ABS). That meaning is if the extent of damages is longer, the convergence of probabilistic result will be larger.

7 Conclusions

In this work, the combined ultimate bending moment of container ship has been obtained using Smith's method. The effects of randomness of material (yield stress of each member element) and randomness of geometry (plate thickness and extent of damages) have been considered under combined horizontal and vertical bending moments in both intact and damaged conditions through Monte Carlo simulation. It observed that values of ultimate strength where angles of neutral axis between 90° to 180° and 270° to 360° (or 0°) they are further away from intact one for collision damaged. For grounding damaged, it occurred between 0° to 90° and 270° to 360° of neutral axis angles. Besides, residual longitudinal strength briefly change when neutral axis angle is small. On the other hand, ultimate values of both intact and damaged conditions have small differences between deterministic and probabilistic simulation, and convergence of probabilistic result (measured by COV) depend on extent of damages. The COV at 0° and 180° of neutral axis angles are smaller than 3%.

References

1. Smith, C.S.: Influence of local compressive failure on ultimate longitudinal strength of ship's hull. In: Proceedings of International Symposium on Practical Design in Shipbuilding, Tokyo, pp. 73–79 (1977)
2. Lee, J.S., Yang, P.D.C.: Reliability assessment against ultimate bending moment of ships' hull girder. Trans. Soc. Naval Archit. Korea **29**(1), 103–112 (1992)
3. American Bureau of Shipping ABS (1995)
4. Gordo, J.M., Soares, C.G.: Collapse of Ship hulls under combined vertical and horizontal bending moments. In: Proceeding of 6th International Symposium on Practical Design of Ships and Mobile Units PRADS 1995, 17–22 September 1995 (1995)
5. ISSC: Report of special task committee VI.2 (Ultimate hull girder strength). In: Proceedings of 14th ISSC, Nagasaki, Japan, vol. 2, pp. 91–321 (2000)
6. Nowak, A.S., Collins, K.R.: Reliability of Structures. McGraw-Hill Publisher, New York (2000)
7. Cho, S.R.: Manual of ULSAN program. University of Ulsan, School of Naval Architecture and Ocean Engineering (2001)
8. Cho, S.R., Lee, S.H.: Residual longitudinal strength analysis of damaged ships. In: Proceedings of the Annual Autumn Meeting, SNAK, Yongin, 3–4 November, pp. 405–412 (2005)

9. Luís, R.M., Hussein, A.W., Soares, C.G.: On the effect of damage to the ultimate longitudinal strength of double hull tankers. In: Proceeding of 10th International Symposium on Practical Design of Ships and Other Floating Structures PRADS, Houston, Texas, USA (2007)
10. Khan, I.A., Das, P.K.: Reliability analysis of intact and damaged ships considering combined vertical and horizontal bending moments. Ships Offshore Struct. **3**(4), 371–384 (2008)
11. Suhas, V., Baidurya, B.: Ultimate Strength analysis of Ship hull girder under random material and geometric properties. In: Proceeding of the ASME 28th International Conference on Ocean, Offshore and Arctic Engineering OMAE 2009, Honolulu, Hawaii, USA, 31 May–5 June (2009)
12. Cheon, S.G.: Ultimate strength assessment for foundation of bow chain stopper. Master thesis, School of Naval Architecture and Ocean Engineering, University of Ulsan (2010)

Using the Brake Torque to Redistribute the Engine Power Transmitting to the Left and Right Drive Wheels

Phan Tan Tai[1,3](✉), Tran Van Nhu[2], and Truong Quoc Dung[3]

[1] School of Engineering and Technology, Tra Vinh University, Tra Vinh, Vietnam
phantantaickd@tvu.edu.vn
[2] Faculty of Mechanical Engineering, University of Transport and Communications, Hanoi, Vietnam
[3] University of Transport and Communications, Hanoi, Vietnam

Abstract. The automotive differential is a mechanism that splits the engine power to the two drive wheels into the left and right sides of the drive axles and allows the driven wheels of a vehicle to have different speeds while they are turning. When a vehicle moves straight on the same road friction at the two drive wheels, the differential applies the same amount of power to each drive wheel. However, if one of the drive wheels rolls on the skid road, the differential splits more the engine power to this wheel. As this wheel slip on the roads, the engine power is dissipated due to the friction between the road and tire. The solutions to this problem are using the limited-slip differential, the active differential. They allow more power to transmit to the non-slipping wheel. In this paper, the authors use the braking torque applied to the slipping wheel to redistribute the engine power on two drive wheels and design a controller based on the PID controller for traction control. The simulation result shows the efficacy of the designed controller.

Keywords: Automotive differential · Drive wheels · Braking torque · Traction control

1 Introduction

The automotive differential is a mechanism that splits the engine power to the two drive wheels into the left and right sides of the drive axles and allows the driven wheels of a vehicle to have different speeds while they are turning. When a vehicle moves straight on different road friction at the two drive wheels, one of the drive wheels rolls on the slippery road and the differential splits more the engine power to this wheel. Besides, as the wheel slip on roads, the engine power is dissipated due to the friction between the road and tire. To improve the traction performance, some passive approaches such as the self-locking differentials, limited-slip differentials… and some active approaches such as the active/semi-active differentials, the traction control systems using differential braking forces… have been proposed. For instance, in [3–5, 8, 10], the authors use active or semi-active differentials to improve the yaw stability by unsymmetrical

B. T. Long et al. (Eds.): MMMS 2020, LNME, pp. 502–507, 2021.
https://doi.org/10.1007/978-3-030-69610-8_69

traction force distribution in the left and right sides of the drive axle. To achieve the same goal, the authors use the differential braking forces [3, 6]. To improve the traction performance, in [2, 7, 9] the authors use breaking force or/and engine torque to control wheel slip ratios during acceleration on slippery roads or to directly drive motor torque of electric vehicles [1, 11]. However, the wheel slip ratio is difficult to measure or estimate. In this paper, the authors present a new method for improving traction performance in the case of a drive wheel moving on a slippery road.

2 Automotive Power Transmission System Modeling

Considering the one-dimension model of an automotive transmission shown in Fig. 1. Where: J_0 is the inertial moments of the drive gear of final drive; J_1, J_2, J_3, J_4, J_5, and J_6 are the inertial moment of the driven gear of final drive, planetary gears, sun gears at the left and right sides of the differential, and left and right drive wheels, respectively; M_0 is the torque on the drive shaft of the final drive; M_{p1} and M_{p2} are the braking torques applied to the left and right drive wheels, respectively. The angular speed $\omega_0, \omega_1, \omega_2, \omega_3, \omega_4, \omega_5$ and ω_6 are speeds of the drive gear and driven gear of final drive, planetary gears, sun gears at the left and right sides of the differential, and drive wheels at the left and right sides, respectively; k_3 and k_4 are stiffness coefficients of the half shafts; c_3 and c_4 are the viscous drag coefficients of the left and right haft shafts; M_{c1} and M_{c2} are road friction torques at the left and right drive wheels, respectively. The road friction torques in this research are modeled as follows: $M_{c1} = \mu_1\omega_5$; $M_{c2} = \mu_2\omega_6$. Here μ_1, μ_2 are constants depending on the state of the road surface on the left and right wheels.

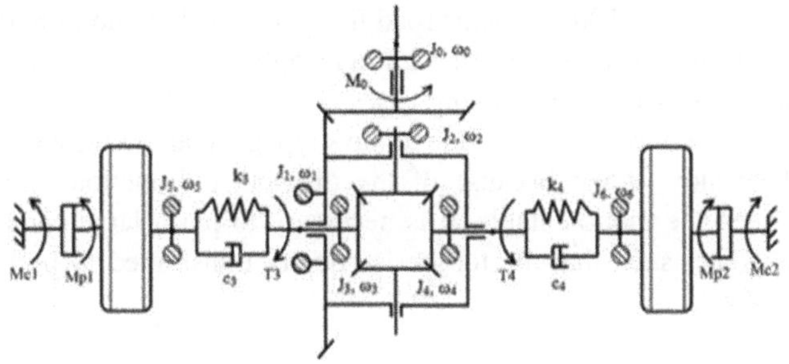

Fig. 1. The dynamic model of the power transmission system

The power transmission mathematical model has six variables. There are three dependent variables: $\omega_0 = \omega_1 . a_1$, $\omega_1 = (\omega_3 + \omega_4)/2$, $\omega_2 = (\omega_4 - \omega_3) . a_2/2$. Therefore, the mathematical model with four differential equations is shown in the Eq. (1)

$$\begin{aligned}
\ddot{\varphi}_3 &= (A(0.5M_0a_1 - T_3) - B(0.5M_0a_1 - T_4))\tfrac{1}{A^2-B^2} \\
\ddot{\varphi}_4 &= (B(0.5M_0a_1 - T_3) - A(0.5M_0a_1 - T_4))\tfrac{1}{B^2-A^2} \\
\ddot{\varphi}_5 &= \tfrac{1}{J_5}\left(T_3 - M_{c1} - M_{p1}\right); \\
\ddot{\varphi}_6 &= \tfrac{1}{J_6}\left(T_4 - M_{c2} - M_{p2}\right)
\end{aligned} \quad (1)$$

Assuming that the road friction at the left drive wheel is less than that at the right one, we investigate the angular speeds of the drive wheels for the cases of keeping the value μ_2 at 100 and varying the value μ_1 from 0 to 100. The simulation results are shown in Fig. 2.

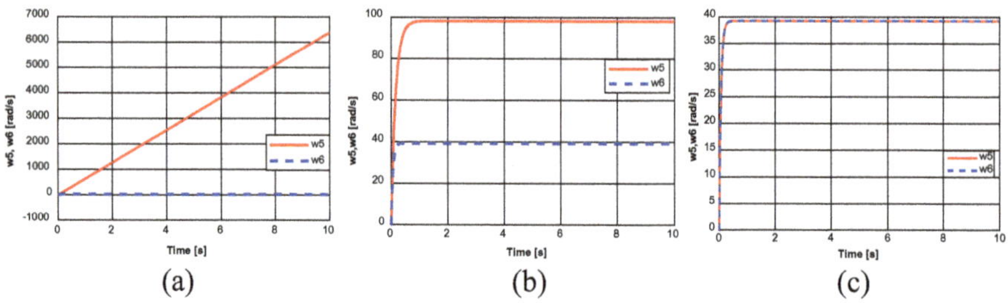

Fig. 2. The angular speeds of the left and right wheels: (a) $\mu_1 = 0$, $\mu_2 = 100$, (b) $\mu_1 = 40$, $\mu_2 = 100$; (c) $\mu_1 = 100$, $\mu_2 = 100$.

Figure 2 shows that when the left wheel completely slips the speed of the right wheel is almost zero (Fig. 2(a)). When increasing the road friction coefficient at the left drive wheel (in the case of $\mu_1 = 40$), the speed of the right wheel increases (Fig. 2(b)). In the case of $\mu_1 = \mu_2 = 100$ (the same road friction at the left and right drive wheels), there is no speed difference between the left and right wheels (Fig. 2(c)).

For similar surveys in the remaining cases, it is found that when the wheel has a lower grip coefficient, it rotates faster due to the slippage, making the angle speed of the remaining wheel slow down because of the distributive function of a differential. Therefore, to keep the motion stable, it is necessary to put a large enough resistance torque on this wheel such that the torque is equally transmitted to both sides of the drive wheels.

3 Controller Design

Considering the case of a vehicle getting bogged down when a drive wheel moves on the slippery road, the drive wheel on the good road can't turn around and the other one on the slippery road is skidded. As a result, the engine power is dissipated due to the friction between the road and the tire. To improve the traction performance, in this paper, a controller is designed to apply a braking torque to the slipping wheel to obtain the speeds of two drive wheels equally. Thereby, the engine power is redistributed on

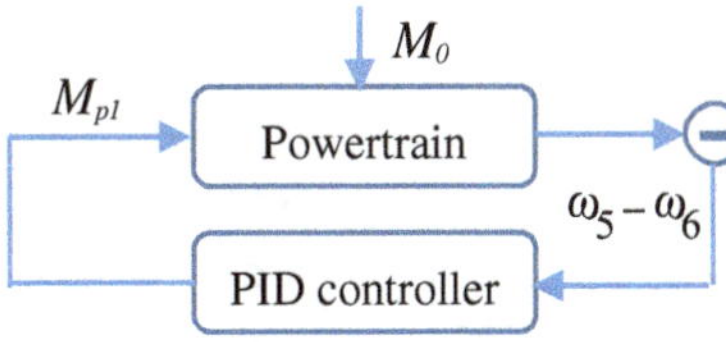

Fig. 3. Diagram of braking torque control by PID controller

two drive wheels. The controller is designed based on the PID (Proportional – Integral – Derivative) controller as shown in Fig. 3. Assuming the left drive wheel moving on the slippery road, the control algorithm is expressed by the Eq. (2).

$$M_{p1} = P(\omega_5 - \omega_6) + I\int(\omega_5 - \omega_6)dt + D(\dot{\omega}_5 - \dot{\omega}_6) \quad (2)$$

4 Simulation Results

Assuming that when the left wheel moves on the road with a low friction coefficient, it slips, and thereby, its angular speed increases over the right wheel. At this time, the controller calculates an appropriate braking torque (M_{p1}) to apply to this wheel. Thus, the angular speed of the left wheel decreases while the angular speed of the right wheel increases, resulting in that the angular speeds of both drive wheels become equal at a certain time (Fig. 4(a)). The angular speed difference between the two drive wheels with various simulation conditions is shown in Fig. 4(b). The applied braking torques are also shown in Fig. 4(c).

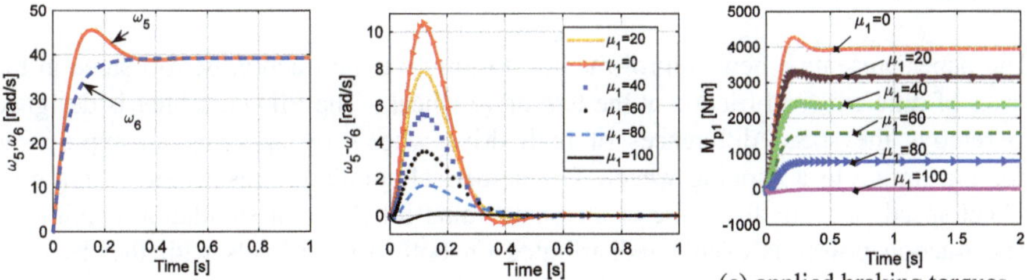

(a) wheel speeds ($\mu_1 = 40$, $\mu_2 = 100$) (b) different speed of drive wheels (c) applied braking torques

Fig. 4. Left and right drive wheel speeds with PID controller

Figure 4(a) shows that, at the beginning of the simulation, the speed of the left drive wheel is higher than that of the right drive wheel. Then, both wheel speeds are rapidly stable at the same value of roughly 39 rad/s. With the different simulation conditions, the angular speeds of the drive wheels always become stable at the simulation time of

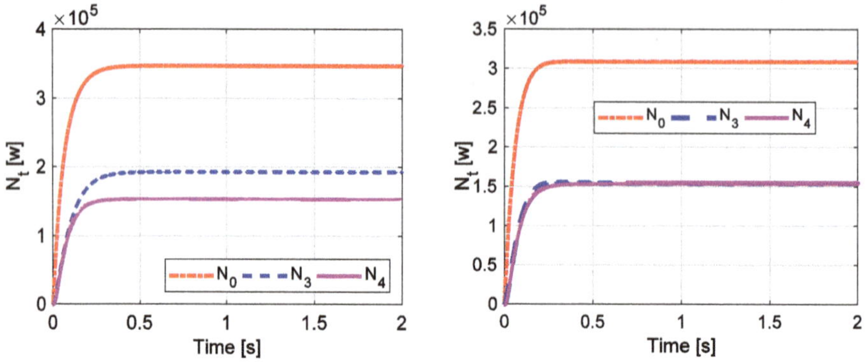

(a) (b)

Fig. 5. Traction power distribution on two drive wheels (case μ_1 = 80): (a) without the PID controller, (b) with the PID controller

roughly 0.5 s, confirming the effectiveness of the designed controller. Figure 4(c) shows the applied braking torques to the left drive wheel for different cases. It is observed that the maximum braking torque is roughly 4000 Nm in the case of $\mu_1 = 0$ when the road friction at the left drive wheel is zero.

Figure 5 shows the input power (N_0) distributing on the left (N_3) and right (N_4) drive wheels with control (Fig. 5(b)) and without control (Fig. 5(a)). Figure 5(a) presenting the simulation results without control in the cases of $\mu_1 = 80$ and $\mu_2 = 100$, the traction power is distributed more on the left drive wheel. However, in the same simulation conditions, the simulation results with the designed PID controller indicates that the traction power is equally distributed on both drive wheels as seen in Fig. 5(b).

5 Conclusion

The paper presents a new control method to improve the traction performance in the cases of different road friction at the two drive wheels. The PID controller is designed to remove the speed differences of both drive wheels by applying an appropriate braking torque to a slipping wheel. The numerical simulation results show that both wheel speeds are rapidly stable at the same value for different simulation conditions. The traction power is evenly redistributed on both drive wheels with the designed controller.

Acknowledgment. This research was fully funded by Tra Vinh University under grant contract number 226/HĐ.HĐKH&ĐT-ĐHTV.

References

1. Saha, S., Ikkurti, H.P., Saha, S.: A robust slip based traction control of an electric vehicle under different road conditions (2015). Paper ID: 66. https://doi.org/10.1049/cp.2015.1618
2. Liu, G., Jin, L.: A study of coordinated vehicle traction control system based on optimal slip ratio algorithm. Math. Probl. Eng. **2016**, 1–10 (2016)
3. Kim, H., Lee, S., Hedrick, J.K.: Active yaw control for handling performance improvement by using traction force. Int. J. Autom. Technol. **16**(3), 457–464 (2015). https://doi.org/10.1007/s12239-015-0047-9
4. Annicchiarico, C., Ranchi, M., Pellari, S., Capitani, R.: Design of a Semi-Active Differential to Improve Vehicle Dynamics. in Volume 1: Applied Mechanics; Automotive Systems; Biomedical Biotechnology Engineering; Computational Mechanics; Design; Digital Manufacturing; Education; Marine and Aerospace Applications, Copenhagen, Denmark, p. V001T02A006 July 2014. https://doi.org/10.1115/ESDA2014-20157
5. Cheli, F., Pedrinelli, M., Resta, F., Travaglio, G., Zanchetta, M., Zorzutti, A.: Development of a new control strategy for a semi-active differential for a high-performance vehicle. Veh. Syst. Dyn. **44**(Suppl. 1), 202–215 (2006)
6. Chou, H., D'andréa-Novel, B.: Global vehicle control using differential braking torques and active suspension forces. Veh. Syst. Dyn. **43**, 261–284 (2005)
7. Li, H.-Z., Li, L., He, L., Kang, M.-X., Song, J., Yu, L.-Y., Wu, C.: PID plus fuzzy logic method for torque control in a traction control system. Int. J. Autom. Technol. **13**(3), 441–450 (2012)
8. Canale, M., Fagiano, L., Milanese, M., Borodani, P.: Robust vehicle yaw control using an active differential and IMC techniques. Control Eng. Pract. **15**(8), 923–941 (2007)
9. Park, Y., Kwak, B.: Slip controller design for a traction control system. Int. J. Autom. Technol. **1**(1), 48–55 (2000)
10. Rubin, D., Arogeti, S.: Vehicle yaw stability control using rear active differential via Sliding mode control methods. In: 21st, Mediterranean Conference on Control and Automation, Platanias, Chania - Crete, Greece, pp. 317–322, June 2013
11. Nam, K., Hori, Y., Lee, C.: Wheel slip control for improving traction-ability and energy efficiency of a personal electric vehicle. Energies **8**(7), 6820–6840 (2015). https://doi.org/10.3390/en8076820

Mathematical Modeling of Thin Layer Drying of Moringa Oleifera by a Combined Microwave and Heat Pump Drying

Le Anh Duc[1(✉)], Tran Dinh Khai[2], and Nguyen Hay[1]

[1] Nong Lam Univerisity Ho Chi Minh City, Ho Chi Minh City, Vietnam
{leanhduc,ng.hay}@hcmuaf.edu.vn
[2] HCMC University of Technology (HUTECH), Ho Chi Minh City, Vietnam

Abstract. The study was performed to determine the mathematical modeling of thin layer drying model, the effective moisture diffusivity and the activation energy of Moringa during a combined microwave and heat pump drying process. The thin layer drying tests were conducted at three different of drying temperature levels of 35, 40, and 45 °C and microwave power levels of 270, 385 and 525 W. Eight different thin layer drying models were fitted to the experimental data.

The Page model was the most adequate model for describing the thin layer drying of Moringa. Effective moisture diffusivity during drying of Moringa varied from 0.71×10^{-8} to 1.63×10^{-8} m^2/s. The activation energy values varied from 34.99 to 47.01 kJ/mol over the range of temperature and microwave power.

Keywords: Moringa oleifera · Microwave · Heat pump · Thin layer drying

1 Introduction

Drying is a simultaneous process of both heat transfer and moisture transfer in a material and between its surfaces and its surroundings [1]. Many models have been used to describe the drying process, among them the most widely used is the experiment and quasi-experiment.

Many studies on thin layer drying have been conducted on products such as rice, corn, beans, etc. However, there are a few of studies on thin layer drying on Moringa leaves. The drying process causes moisture to move from inside the drying material to the surface and then it evaporates. A drying process consists of three drying stages: the heating phase of drying material, the constant rate drying phase and the gradually decreasing rate drying phrase. For many types of drying materials, the gradually decreasing rate drying phrase is primary. In this case, the moisture flow in the material takes place by diffusing the liquid and/or steam during the drying process [2].

When analyzing the diffusion of water during the gradually decreasing rate drying phase, the moisture diffusion model based on Fick's second diffusion law was considered to evaluate the moisture movement. Many products such as rice, sunflower [3], peanuts [4], green beans [5] have been successfully predicted by the second law of Fick. Bhattacharya et al., [6] studied the thin layer model of abalone mushrooms on the principle of convection and convection drying combined with microwaves with drying

B. T. Long et al. (Eds.): MMMS 2020, LNME, pp. 508–519, 2021.
https://doi.org/10.1007/978-3-030-69610-8_70

temperature of 50–70 °C. The results showed that, when only using convection drying, the moisture diffusion coefficient increased from 2.87×10^{-8} to 3.17×10^{-8} m^2/s, the diffuse energy value was 21.998 kJ/mol. When using convection drying combined with microwave, the diffusion coefficient increased from 11.09×10^{-8} to 12.39×10^{-8} m^2/s and the diffuse energy value was 17.126 kJ/mol. Alibas [7] studied energy analysis and drying kinetics for celery leaves when using convection drying combined with microwave, the results showed that when increasing the output power of the microwave from 200 to 900 W, moisture diffusion coefficient increased from 4.46×10^{-8} to 39.63×10^{-8} m^2/s.

The prediction on drying rate and moisture content of drying material through thin layer drying model is an important and essential factor for effective heat transfer and moisture transfer analysis. The suitability of the drying model depends directly on the accuracy of the thin layer drying equation [8]. These predictions will make the drying process more efficient as well as contribute to energy reduction.

The objective of this research is to determine the appropriate thin layer drying model, moisture diffusion coefficient and activation energy for evaporation process when drying moringa leaves by microwave heat pump drying method.

2 Materials and Methods

Moringa leaves are harvested and stripped from the stalks, washed, drained and stored in a refrigerator at 4 ± 1 °C until experiments are performed [9]. The initial moisture content of fresh moringa leaves determined by oven method is 77 ± 0.5% (wb.).

The thin layer drying process is conducted under controlled temperature and microwave power. The experiments were conducted at 3 drying temperature of 35, 40 and 45 °C and 3 levels of microwave power of 270, 385 and 525 W. Most thin-layer drying models do not include the drying agent velocity and consider it an independent variable [10]. In all experiments, the drying agent velocity was fixed at 0.5 m/s. Different combinations of the 3 temperature levels and the 3 levels of microwave power will produce a total of 9 experiments. The order for performing the experiments is completely random.

Moringa leaves are evenly distributed on the drying tray with a layer thickness of 1 cm. The drying sample is weighed regularly to record weight changes with a weighing interval of 10 min. The change in humidity, relative to the drying time, can be calculated based on the change in leaves' weight. When the samples reach a moisture content of 8%, the experiment is stopped.

3 Results and Discussions

3.1 Processing Experimental Data

Statistical analysis was performed by non-linear regression for each individual drying result. The proposed models were tested with experimental drying data and analyzed by regression method step by step, using Statistical Analysis System (SAS), Ver.9.1. The moisture diffusion coefficients are calculated by the moisture diffusion equation for flat plate shapes based on Fick's second law.

Experimental values and values from the model are compared and analyzed statistically to determine the most suitable equation. Coefficient of Determination (R^2), Standard Error of Estimate (SEE) and Residual Sum of Squares (RSS) are used to determine the quality of conformity [1, 11]. The SEE and RSS coefficients are determined by Eqs. (1) and (2) respectively.

$$\mathrm{SEE} = \sqrt{\frac{\sum_{i=1}^{n}(\mathrm{M} - \mathrm{M_S})^2}{\mathrm{df}}} \tag{1}$$

$$\mathrm{RSS} = \sum_{i-1}^{n}(\mathrm{M} - \mathrm{M_S})^2 \tag{2}$$

Where: M: measured value; M_S: value calculated by the formula; n: number of observed samples; df: freedom.

3.2 Thin Layer Drying Mathematical Model

Numerous mathematical models have been proposed to describe the rate of moisture reduction during thin layer drying of various materials. A few thin layer drying models, which are commonly studied for agricultural products, have been selected for this study and are shown in Table 1.

Table 1. Mathematical models for thin layer drying

Model name	Recipe	Source
Lewis	$MR = \exp(-k.\tau)$	Bruce (1985) [12]
Page	$MR = \exp(-k.\tau^n)$	Page (1949) [13]
Modified Page	$MR = \exp(-k.\tau)n$	White et al. (1981) [14]
Henderson and Pabis	$MR = a.\exp(-k.\tau)$	Henderson, Pabis (1961) [15]
Wang and Singh	$MR = 1 + a.t + b.\tau 2$	Wang, Singh (1978) [16]
Parabolic	$MR = c + a.t + b.\tau^2$	Verma và cs (1985) [17]

MR: moisture ratio:

$$MR = \frac{M(t) - M_e}{M_o - M_e} \tag{3}$$

Where: M(t): moisture content of the drying material at the drying time t (%, dry basis); M_e: equilibrium moisture content of drying material (%, db); M_o: initial moisture content of drying material (%, db); τ: drying time (s or h); k, n, a, b, g, c:

coefficients of the models. The models proposed in Table 1 will be tested with experimental drying data to determine the most suitable thin layer drying model.

3.3 Determining the Thin Layer Drying Model

In order to express the drying curves, the moisture content of the drying material at the dry basis moisture content over the drying time is converted to a dimensionless parameter, the moisture ratio by drying time.

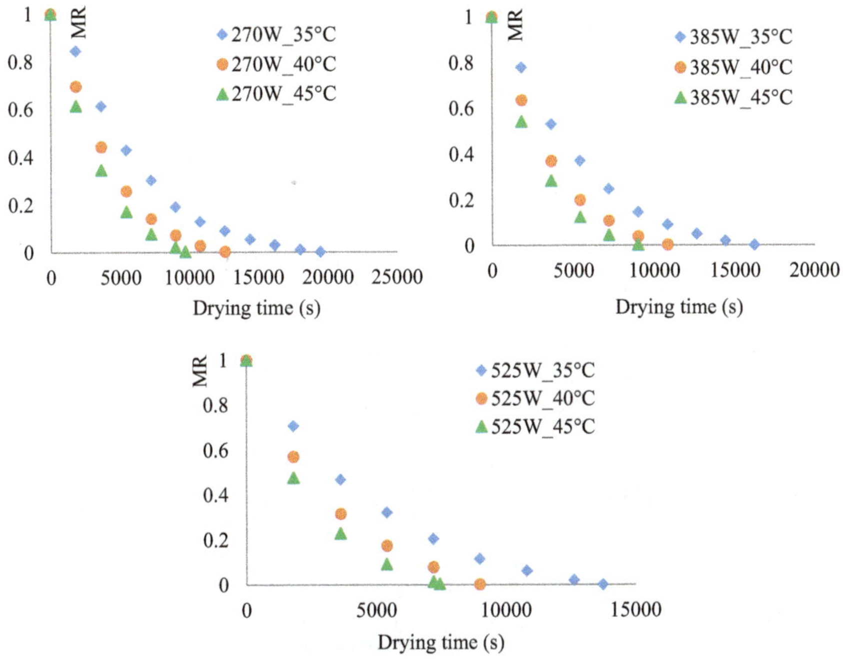

Fig. 1. Moisture ratio versus drying time at three microwave power levels of 270 W, 385 W and 525 W

Figure 1 shows the typical drying curves of the moringa leaves drying process using the heat pump drying method combined with microwave through thin layer drying activity at 9 different drying conditions when combining with 3 levels of temperature and 3 levels of microwave power. The results show that the ratio of moisture reduction depends on both drying temperature and microwave power. The ratio of moisture reduction increases with increased drying temperature as well as with increased microwave power.

To determine the characteristics of the drying process, the drying rate (dM/dτ) is calculated under different drying conditions and plotted in correlation with the moisture content of the drying material. The results are presented in Fig. 2.

The drying rate of moringa leaves shown in Fig. 2 are the same as the results for drying of agricultural products. In particular, the rate of moisture removal is higher in the first stage and gradually decreases in the next stage when the moisture content of the material decreases.

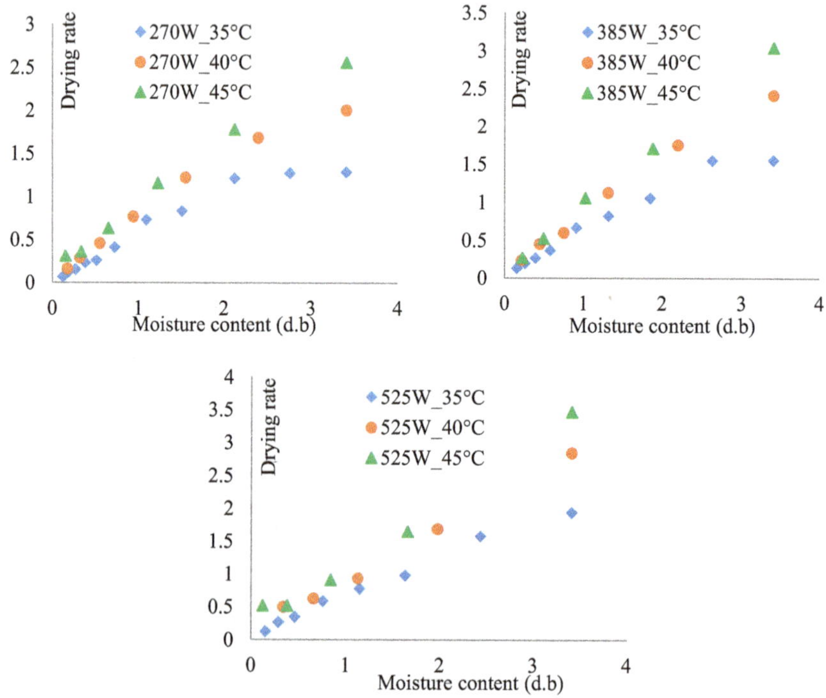

Fig. 2. Drying rate at three microwave power levels of 270 W, 385 W and 525 W

Use statistical analysis to analyze and assess the suitability of experimental values with thin layer drying models shown in Table 1. The change of moisture ratio over time for each experiment is used to calculate the constants in thin layer drying models by non-linear regression method. The coefficients of the models and the suitable assessment index of the models are R^2, SEE and RSS of the six thin layer drying models shown in Table 2.

Table 2. Value of statistical coefficients of thin layer drying models

Temperature (°C)	Microwave power								
	270 W			385 W			525 W		
	R2	RSS	SEE	R^2	RSS	SEE	R2	RSS	SEE
Lewis									
35	0.981	0.002	0.025	0.987	0.002	0.039	0.992	0.001	0.030
40	0.989	0.001	0.038	0.993	0.001	0.031	0.995	0.001	0.026
45	0.991	0.001	0.034	0.995	0.001	0.026	0.995	0.001	0.027
Page									
35	0.999	0.001	0.009	0.999	0.001	0.010	0.998	0.002	0.017
40	0.999	0.001	0.009	0.999	0.001	0.013	0.997	0.002	0.022
45	0.998	0.001	0.016	0.999	0.001	0.015	0.998	0.002	0.022
Modified page									
35	0.981	0.025	0.049	0.987	0.014	0.042	0.992	0.007	0.032
40	0.989	0.010	0.041	0.993	0.006	0.034	0.995	0.003	0.029
45	0.991	0.007	0.037	0.995	0.003	0.029	0.995	0.004	0.030
Henderson and Pabis									
35	0.986	0.018	0.043	0.990	0.011	0.037	0.993	0.007	0.030
40	0.990	0.009	0.038	0.993	0.005	0.033	0.995	0.003	0.029
45	0.992	0.007	0.036	0.996	0.003	0.029	0.995	0.004	0.030
Wang and Singh									
35	0.994	0.008	0.027	0.997	0.004	0.022	0.995	0.005	0.027
40	0.997	0.002	0.020	0.995	0.004	0.028	0.991	0.006	0.039
45	0.997	0.003	0.023	0.993	0.005	0.036	0.992	0.006	0.039
Parabolic									
35	0.994	0.007	0.029	0.997	0.004	0.023	0.996	0.004	0.026
40	0.998	0.001	0.021	0.996	0.003	0.029	0.992	0.005	0.042
45	0.997	0.002	0.024	0.994	0.004	0.038	0.993	0.005	0.042

The results presented in Table 2 show that the SEE and RSS parameters achieved the lowest value and the coefficient R^2 achieved the highest value in the Page model at all temperature and microwave power levels. Thus, the Page model is determined to be the most suitable model to describe the moringa leaves drying process according to the principle of heat pump drying combined with microwave. The coefficients of the Page model are presented in Table 3.

Table 3. Coefficients of Page model in 9 drying modes

Drying temperature, (°C)	Microwave power levels, (kW)	The coefficients of the page	
		k	n
35	270	8.315×10^{-6}	1.338
	385	1.977×10^{-5}	1.261
	525	5.102×10^{-5}	1.171
40	270	2.722×10^{-5}	1.261
	385	5.385×10^{-5}	1.201
	525	0.0001	1.117
45	270	5.020×10^{-5}	1.122
	385	9.858×10^{-5}	1.162
	525	0.0001	1.146

From the regression analysis results for the experimental constants of the Page model, the values of k and n coefficients in the model were determined to be a function of drying temperature and microwave power as the (4) and (5) equations.

$$k = -0.001 - 0.001 \times P + 2.886 \times 10^{-5} \times t_a + 0.0003 \times P^2 - 3.497 \times 10^{-7} \times t_a^2 + 1.484 \times 10^{-5} \times P \times t_a;\ (R^2 = 0.96;\ RMSE = 7.447 \times 10^{-10}) \tag{4}$$

$$n = 3.855 - 2.337 \times P - 0.097 \times t_a + 0.442 \times P^2 + 0.001 \times t_a^2 + 0.037 \times P \times t_a \ (R^2 = 0.986;\ RMSE = 0.0005) \tag{5}$$

Where: t_a: drying temperature, (°C); P: microwave power, (kW).

3.4 Moisture Diffusion Coefficient and Moisture Diffusion Energy

The diffusion of moisture in a material is considered to be a primary mechanism of the process of transferring moisture from within the material to the surface of the material [18]. In thin layer drying of agricultural products, the diffusion coefficient is determined using a simple diffusion model based on Fick's second law [19, 20].

The moisture diffusion coefficient, D_m (m^2/s), is a greatly influential coefficient to the moisture diffusion process in the drying material, this coefficient is usually determined from the experimental results. In case the D_m is constant and coefficient unidirectional moisture diffusion, the moisture diffusion equation as Eq. (6).

$$\frac{\partial M}{\partial t} = D_m \frac{\partial^2 M}{\partial^2 x} \tag{6}$$

The solution to Eq. (6) was solved by Crank [21] provided that the condition for the exchanging of substances took place on both sides of the plane.

$$MR = \frac{8}{\pi^2}\sum_{i=0}^{\infty}\frac{1}{(2i+1)^2}\exp\left[\frac{-(2i+1)D_m\pi^2 t}{4\delta^2}\right] \quad (7)$$

When the drying process have occurred for long enough, the rate of moisture reduction can be calculated as follows:

$$MR = \frac{M - M_e}{M_0 - M_e} = \frac{8}{\pi^2}\exp\left[-\frac{D_m\pi^2 t}{4\delta^2}\right] \quad (8)$$

Take the logarithm of Eq. (8). Fick's model applies to flat sheet [7]:

$$Ln(MR) = Ln\frac{8}{\pi^2} - \left(\frac{\pi^2 D_m}{4.^2}\right).\tau \quad (9)$$

Where: D_m: moisture diffusion coefficient, (m^2/s); δ: thickness of a half flat sheet, (m); τ: drying time (s).

From the experimental data, when building the relationship of ln(MR) in correlation with drying time τ, the graph of the Ln (MR) function as above will form a line from which the moisture diffusion coefficient (D_m) can be calculated from its slope (x_1) as formula (10).

$$x_1 = -D_m\left(\frac{\pi}{2}\right)^2 \quad (10)$$

Moisture diffusion energy (also known as energy activity) is an indicator of the energy required for water molecules to overcome the binding force during the moisture transfer process in materials. The Arrhenius formula is used to calculate the moisture diffusion energy. The formula shows the effect of temperature on the diffusion constant, and the energy for moisture diffusion is determined by the Arrhenius equation [22]:

$$D_m = D_0 e^{\left(-\frac{E_a}{RT_a}\right)} \quad (11)$$

Where: D_o: diffusion constant or Arrhenius coefficient, (m^2/s); T_a: absolute temperature, (K); E_a: activation energy for the moisture diffusion process, (kJ/mol); R: gas constant, (R = 8.3143 kJ/mol.K).

Take the natural logarithm Eq. (7):

$$LnD_m = \left(-\frac{E_a}{R}\right)\frac{1}{T_a} + LnD_0 \quad (12)$$

Equation (12), which expresses the relationship between $Ln(D_m)$ and $1/T_a$ in the form of a graph, is a straight line with the inclination (angle of inclination) expressed as the value of −Ea/R.

To determine the energy for the moisture diffusion process, plot the relationship between $Ln(D_m)$ and $1/T_a$, the graph will be a straight line with an inclination of x_2. The x_2 value is used to calculate the moisture diffusive energy.

$$x_2 = -\frac{E_a}{RT_a} \tag{13}$$

3.5 Determination of Moisture Diffusion Coefficient and Evaporation Energy

From the experimental results and application of Eq. (11), the value of the moisture diffusion coefficient at temperatures of 35 °C, 40 °C and 45 °C combined with 3 levels of microwave power (270, 385 and 525 W) when drying moringa leaves using the heat pump drying method combined with microwave is determined in Table 4.

From the result of calculating D_m, the relational graph between $Ln(D_m)$ and 1/Ta is established as shown in Fig. 3. From the graph, the values of E_a and D_o are determined. The value of the energy needed for moisture diffusion are presented in Table 4.

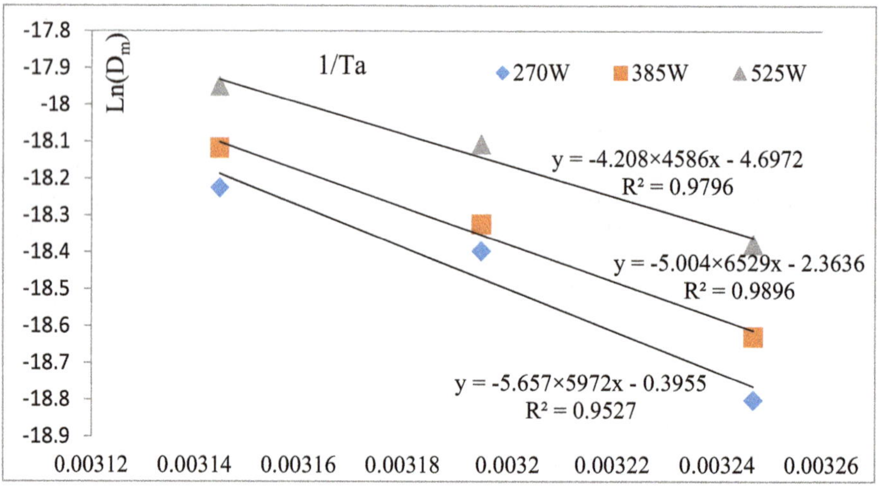

Fig. 3. Relation between $Ln(D_m)$ and 1/Ta

Table 4. Moisture diffusion coefficient and energy required for diffusion process

Parameter	Temperature (°C)	Microwave power (W)		
		270	385	525
Moisture diffusion coefficient ($\times 10^{-8}$ m^2/s)	35	0.71	0.82	1.06
	40	0.95	1.07	1.32
	45	1.26	1.37	1.63
Energy (kJ/mol)		47.04	41.61	34.99

The values of the moisture diffusion coefficient of moringa leaves in Table 4 show that increasing the drying temperature and increasing the microwave power will increase the value of the moisture diffusion coefficient of moringa leaves with values from 0.71×10^{-8} to 1.63×10^{-8} m^2/s. The diffusion energy varies from 34.99–47.04 kJ/mol. The experimental values of the moisture diffusion coefficient and moisture diffusion energy of moringa leaves are similar to the results of previous studies on a number of different agricultural products such as dried abalone mushrooms [6], namely the moisture diffusion coefficient in convection drying is 2.87×10^{-8} to 3.17×10^{-8} m^2/s, the energy is 21.998 kJ/mol. When using convection drying combined with microwave, the moisture diffusion coefficient is 11.09×10^{-8} to 12.39×10^{-8} m^2/s and the energy is 17.126 kJ/mol. When drying chrysanthemum leaves using convection drying combined with microwave [7]: the moisture diffusion coefficient increased from 4.46×10^{-8} to 39.63×10^{-8} m^2/s. When drying sliced potatoes by microwave with the power varies from 90–450 W, the moisture diffusion coefficient is 1.25×10^{-8} to 3.76×10^{-8} m^2/s.

Table 5. Moisture diffusion coefficient by drying mode

Microwave power	Moisture diffusion coefficient, D_m (m^2/s)
270 W	$D_m = 0.666 \times Exp\left(-\frac{5658}{(t_a + 273.2)}\right)$
385 W	$D_m = 0.093 \times Exp\left(-\frac{5005}{(t_a + 273.2)}\right)$
525 W	$D_m = 0.009 \times Exp\left(-\frac{4208}{(t_a + 273.2)}\right)$

Statistical analysis showed that the moisture diffusion of moringa leaves was significantly influenced by drying conditions (temperature, microwave power) over the entire data range (P value < 0.01). The degree of moisture diffusion increases when increasing temperature as well as when increasing microwave power. The higher the moisture diffusion coefficient, the faster the moisture in the drying material diffuses to the outside.

$$D_m = -2.746 \times 10^{-9} + 5.22 \times 10^{-12} \times t_a^2 + 3.589 \times 10^{-10} \times P \times t_a \quad R^2 = 0.990;\ RMSE = 6.914 \times 10^{-19} \tag{14}$$

The overall moisture diffusion coefficient for each respective drying condition is presented in Table 5. The moisture diffusion coefficient is expressed for all drying conditions with any drying temperature and microwave power level is represented by Eq. (14). The results show a linear relationship between the effective moisture diffusivity and the temperature, and that it could be satisfactorily described by an Arrhenius-type equation.

4 Conclusions

Experimental values and values from the model are compared and statistical analysis was performed by non-linear regression for each individual drying result to determine the most suitable equation. The results show that the rate of moisture reduction depends on both drying temperature and microwave power. The ratio of moisture reduction increases with increasing drying temperature as well as with increasing microwave power.

The Page model is determined to be the most suitable model to describe the moringa leaves drying process according to the principle of heat pump drying combined with microwave. The values of the k and n coefficients in the Page model were determined to be a function of drying temperature and microwave power as the (4) and (5) equations.

The moisture diffusion coefficient, D_m (m^2/s) is usually determined from the experimental results calculated as Eq. (10) at temperatures of 35, 40 and 45 °C combined with 3 levels of microwave power (270, 385 and 525 W) when moringa leaves drying using the heat pump drying method combined with microwave is determined. The increasing the drying temperature and increasing the microwave power will increase the value of the moisture diffusion coefficient of moringa leaves with values from 0.71×10^{-8} to 1.63×10^{-8} m^2/s. The diffusion energy varies from 34.99–47.04 kJ/mol over the range of temperature and microwave power.

References

1. Akpinar, E.K., Sarsilmaz, C., Yildiz, C.: Mathematical modeling of a thin layer drying of apricots in a solar energized rotary dryer. Int. J. Energy Res. **28**, 739–752 (2004)
2. Brooker, D.B., Bakker-Arkema, F.W., Hall, C.W.: Drying Cereal Grains. The AVI Publishing Company Inc., Westport, Connecticut (1974)
3. Danae, D., Tzia, K., Gekas, V.: A knowledge base for the apparent mass diffusion coefficient (D_{eff}) of foods. Int. J. Food Prop. **3**(1), 1–14 (2000)
4. Palacios, T.R., Potes, L.B., Montenegro, R.A., Giner, S.A.: Peanut drying kinetics: determination of the effective diffusivity for in-shell and shelled peanuts by applying a short-time analytical model to measured data. In: Proceedings of the 14th International Drying Symposium B, pp. 1448–1455 (2004)

5. Doymaz, I.: Drying behaviour of green beans. J. Food Eng. **69**, 161–165 (2005)
6. Bhattacharya, M., Srivastav, P.P., Mishra, H.N.: Thin-layer modeling of convective and microwave-convective drying of oyster mushroom (Pleurotus ostreatus). J. Food Sci. Technol. **52**(4), 2013–2022 (2015)
7. Alibas, I.: Mathematical modeling of microwave dried celery leaves and determination of the effective moisture diffusivities and activation energy. J. Food Sci. Technol. **34**(2), 394–401 (2014)
8. Basunia, M.A., Abe, T.: Moisture desorption isotherms of medium-grain rough rice. J. Stored Prod. Res. **37**, 205–219 (2001)
9. Dev, S.R.S., Geetha, P., Orsat, V., Gariépy, Y., Raghavan, G.S.V.: Effects of Microwave - assisted hot air drying and conventional hot air drying on the drying kinetics, color, rehydration, and volatiles of Moringa oleifera. Drying Technol. **29**, 1452–1458 (2011)
10. Sinicio, R., Muir, W.E., Jayas, D.S., Cenkowski, S.: Thin-layer drying and wetting of wheat. Postharvest Biol. Technol. **5**, 261–275 (1995)
11. Sacilik, K., Keskin, R., Elicin, A.K.: Mathematical modelling of solar tunnel drying of thin layer organic tomato. J. Food Eng. **73**, 231–238 (2006)
12. Bruce, D.M.: Exposed-layer barley drying: three models fitted to new data up to 150 °C. J. Agric. Eng. Res. **32**, 337–347 (1985)
13. Page, G.E.: Factors influencing the maximum rates of air drying shelled corn in thin layers. M.S. thesis, Department of Mechanical Engineering, Purdue University, Purdue (1949)
14. White, G.M., Ross, J.J., Ponelert, R.: Fully exposed drying of popcorn. Trans. ASAE **24**, 466–468 (1981)
15. Henderson, S.M., Pabis, S.: Grain drying theory. II. Temperature effects on drying coefficients. J. Agric. Eng. Res. **6**, 169–174 (1961)
16. Wang, C.Y., Singh, R.P.: Use of variable equilibrium moisture content in modeling rice drying. Trans. ASAE **11**, 668–672 (1978)
17. Verma, L.R., Bucklin, R.A., Endan, J.B., Wratten, F.T.: Effects of drying air parameters on rice drying models. Trans. ASAE **28**, 296–301 (1985)
18. Beigi, M.: Hot air drying of apple slices: dehydration characteristics and quality assessment. J. Heat and Mass Transfer **52**(8), 1435–1442 (2016)
19. Demiray, E., Tulek, Y.: Thin-layer drying of tomato (Lycopersicum esculentum Mill. CV. Rio Grande) slices in a convective hot air dryer. J. Heat and Mass Transfer **48**(5), 841–847 (2012)
20. Duc, L.A., Han, J.W., Keum, D.H.: Thin layer drying characteristics of rapeseed (Brassica napus L.). J. Stored Prod. Res. **47**, 32–38 (2011)
21. Crank, J.: The Mathematic of Diffusion, 2nd edn. Oxford University Press, Oxford (1975)
22. Duc, L.A., Hong, S.-J., Han, J.-W., Keum, D.-H.: Estimation of effective moisture diffusivity of rapeseed (Brassica napus L.). J. Biosyst. Eng. **33**(5), 296–302 (2008)

Experimental Model and Optimization of Cutting Parameters in Hard Milling SKD61 Steel Under Nanofluid MQL Environment

Minh-Quan Nguyen[1,2], Liangshan Xiong[1], Huu-That Nguyen[2(✉)], and Ngoc-Chien Vu[2]

[1] School of Mechanical Science and Engineering, Huazhong University of Science and Technology, Wuhan, China
[2] Mechanical Engineering Faculty, Nha Trang University, Nha Trang, Vietnam
thatnh@ntu.edu.vn

Abstract. The present paper intends to modeling and optimizing the cutting force and material removal rate (MRR) in hard milling SKD61 steel under the nanofluid MQL environment using a combination of the Kriging model and NSGA-II. Firstly, a set of experiments produced using the Box-Behnken method for four input parameters that combine cutting speed, depth of cut, feed per tooth, and hardness of workpiece. Secondly, the Kriging model is adopted for performing the relation between process parameters and machining responses. Finally, NSGA-II is a tool for optimizing multi-objective for machining responses. The results indicate that feed per tooth and depth of cut are the essential parameters in the cutting force. The Kriging model is highly reliable, with an R^2 coefficient of 96.6%. Employing the Kriging model combines NSGA-II for multi-objective optimization can be shown as the feasible optimal solution with the minimum of cutting force and the maximum of material removal rates.

Keywords: Hard milling · Nanofluid MQL · Kriging · Optimization · Cutting force · MRR

1 Introduction

Over the past few years, hard machining is widely adopted in the manufacturing industry due to its irreplaceable characteristics. The hard machining in dry cutting conditions has the advantages of being environmentally friendly because it not be employing cutting fluids. However, the hard machining in the dry condition results leads to high cutting temperature in the cutting zones, which reduces surface quality and affects the accuracy of surface finished. Additionally, increasing wear on blades of cutting tools over time may lead to tool breakage. Therefore, the employment of lubricants is very important for the hard machining. There is a lot of investigation on the application of lubricants in the machining of hard material [1–4]. They were shown that the use of minimal cutting fluids had outperformed conventional cooling techniques. This technique can be reduced production costs, environmental protection, and human health. The cutting fluids technique mentioned is the MQL.

B. T. Long et al. (Eds.): MMMS 2020, LNME, pp. 520–524, 2021.
https://doi.org/10.1007/978-3-030-69610-8_71

The thermal conductivity of coolant and lubricant in MQL is still weak, so a new technology is emerging as a technique to overcome the disadvantages of MQL. It is called nano-fluids that formed by the nano-particles with the size of 10–100 nm are mixed in cutting fluids to become a homogeneous solution [2].

The effectiveness of using nanofluids MQL has been shown by many studies. Nanofluids can be reducing friction coefficient, reducing energy consumption, significantly extending tool life by diminishing the cutting temperature and enhancing the surface quality [5, 6].

In the hard milling of SKD61 steel, many researchers have optimized the cutting parameters for SKD61 material [3, 7], but there have been no clear studies on the multi-objective with cutting force and material removal rate (MMR). In this study, we conducted experiments, modeling, and optimizing simultaneously with the target to minimum cutting force and maximum MRR by an employed evolutional algorithm.

2 Material and Method

2.1 Material and Method

The material utilized in this study is SKD61 steel that has popularly accepted in the mold industry. The material was chosen because of good characteristics as high hardness, wear resistance after heat treatment, potent hardenability, excellent high-temperature resistance, high-grade dimensional stability.

The workpiece size is 150 mm (L) × 100 mm (W) × 55 mm (H), in which the dimension of 100 mm for performing machining experiments. A 3-axis vertical milling machine (Bridgeport VMC 2216), Ø10 M520 ultra carbide milling cutter with four flutes, and MQL system was utilized for 25 physical experiments. In which the parameters of the MQL system are 90 mL/h for flow rate, 60° for the angle of nozzle, 3 kg/cm^2 for air pressure, and 20 mm for the distance of nozzle.

Cutting fluid was used in the MQL system is cutting oil CT232 that manufactured by LONG LUB-TEK company in Taiwan. Graphite nano-particle with a size of less than 30 nm elected to homogenize with the coolant and lubricant. Nanoparticles are mixed into cutting fluid with 0.5% Vol (%) and was stirred by magnetic stirrer within 48 h to ensure homogeneity of solution.

For design of experiments (DOE), the paper selects four process parameters including cutting velocity, depth of cut, feed per tooth, and hardness. Each process parameter has three levels and is shown in Table 1. The milling performance criteria include cutting force and material removal rate (MRR). In which cutting force was measured through piezoelectric dynamometer.

The approximate Kriging model was used for modeling the relationship between process parameters and response parameters. Then, Nondominated Sorting Genetic Algorithm II (NSGA-II) is an advanced multi-objective optimization technique used for optimization. The Pareto front generated by NGSA-II based on the Kriging model was an exhibition to determine simply the optimal results.

Table 1. Values and levels of factors

Input parameters	Units	Levels		
		1	2	3
Cutting velocity (v_c)	m/min	30	45	60
Feed per tooth (f_z)	mm/tooth	0.02	0.03	0.04
Depth of cut (a_p)	mm	0.2	0.5	0.7
Hardness (H)	HRC	40	45	50

3 Results and Discussion

Table 2. Partial experimental matrix and experimental results

No	v_c (m/min)	a_p (mm)	f_z (mm/tooth)	H (HRC)	F (N)	MRR (mm^3/min)
1	45	0.2	0.02	45	99.6	229.3
2	45	0.2	0.04	45	135.1	458.6
3	45	0.8	0.02	45	236.8	917.2
–	–	–	–	–	–	–
12	60	0.8	0.03	45	291.1	1834.4
13	45	0.5	0.02	40	148.1	573.2
14	45	0.5	0.02	50	187.3	573.2
15	45	0.5	0.04	40	214.6	1146.5
–	–	–	–	–	–	–
22	60	0.5	0.02	45	206.2	764.3
23	30	0.5	0.04	45	274.9	764.3
24	60	0.5	0.04	45	285.5	1528.7
25	45	0.5	0.03	45	234.4	859.9

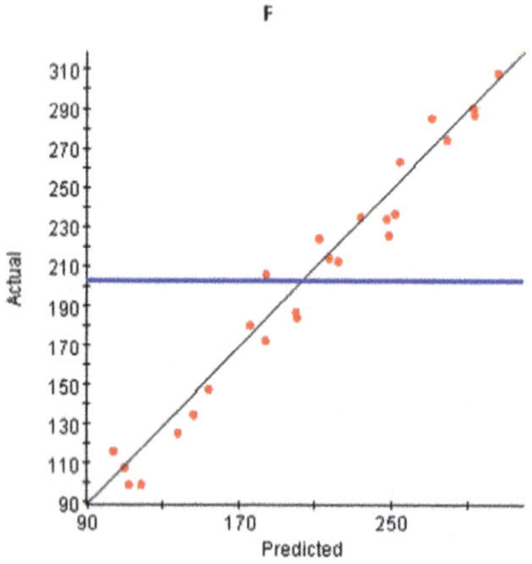

Fig. 1. The evaluation of Kriging model by R^2

The Box-Behnken method was employed for DOE. The results with 25 experiments were listed in Table 2. Subsequently, the Kriging model was rendered the relationship between process parameters and response parameters. Figure 1 demonstrates the evaluation of the Kriging model by the coefficient of determinant ($R^2 = 0.966$). Based on that, it can be inferred that the selection of the Kriging model is proper.

Figure 2 illustrates the three-dimensional (3D) graph generated by the approximate Kriging model that shows the impaction of the input factors

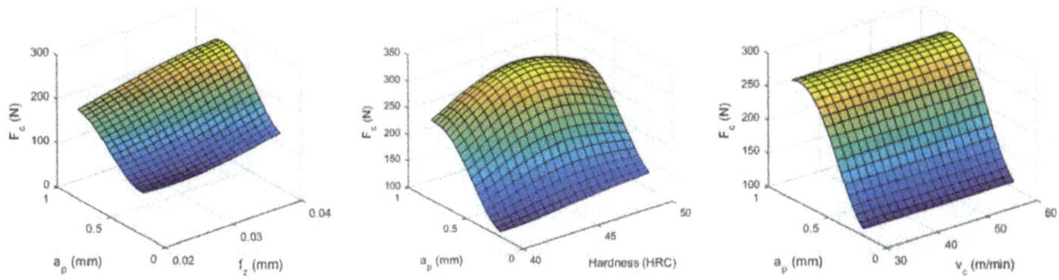

Fig. 2. Cutting force versus process parameters

(v_c, f_z, a_p, and H) on the cutting force. It is worth noting that the increased depth of cut, feed per tooth, as well as the hardness of the workpiece, lead to an enlarge in the cutting force. Meanwhile, the influence of cutting velocity on cutting force is quite small. These results are in satisfactory compromise with previous research [7, 8].

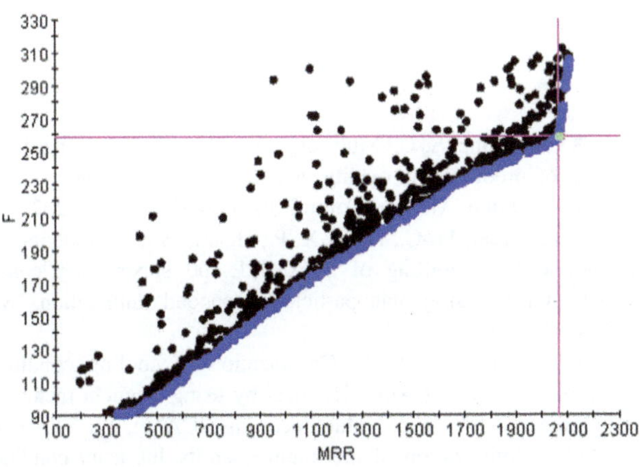

Fig. 3. Pareto front generated by NSGA-II based on Kriging approximate model

After rendering the relationship of input and output parameters by Kriging model, NSGA-II was employed for multi-objective optimization. In Fig. 3, the scatter Pareto plot is produced by NSGA-II based on Kriging approximate model.

Based on Fig. 3, it is easily recognized that rising MRR leads to an increase in cutting force and vice versa. Two goals are conflicting, so these need to be compromised. We also can observe that the slope of the line that collecting the Pareto points gradually increases. When MRR reaches the value of 2067.33 mm^3/min, this slope changes increases exponentially. Therefore, when selecting machining parameters, machine operators should avoid these values to achieve the best optimum value. This is the new important finding that this study finds. It assists to optimize of multi-objective between cutting force and MRR more clearly.

4 Conclusion

This paper investigates hard milling with SKD 61 material under nanofluids condition in terms of cutting force and material removal rate. The combination of the Kriging model and NSGA-II can support the multi-objective optimization problem of the optimization process. The simultaneous raise in MMR and the diminution of cutting force are the main findings of this paper. Some main conclusion can be represented as follows:

1. The Kriging model using to render for the cutting force has R^2 value of 0.966. As a result, this is the suitable model that can be employed to render the relationship between input parameters and technological responses with high accuracy.
2. Based on the Pareto plot, the proper determination of process parameters can support the engineer to enhance productivity and reduce cutting force.
3. The employment of NSGA-II combined with the Kriging model is a powerful procedure for illustrating the association between input and output parameters and for determining machining process optimization.

References

1. Vu, N.-C., Dang, X.-P., Huang, S.-C.: Multi-objective optimization of hard milling process of AISI H13 in terms of productivity, quality, and cutting energy under nanofluid minimum quantity lubrication condition. Measur. Control 0020294020919457 (2020)
2. Hsiao, T.-C., Vu, N.-C., Tsai, M.-C., Dang, X.-P., Huang, S.-C.: Modeling and optimization of machining parameters in milling of INCONEL-800 super alloy considering energy, productivity, and quality using nanoparticle suspended lubrication. Measur. Control 0020294020925842 (2020)
3. Do, T.-V., Vu, N.-C., Nguyen, Q.-M.: Optimization of cooling conditions and cutting parameters during hard milling of AISI H13 steel by using Taguchi method. In: 2018 IEEE International Conference on Advanced Manufacturing (ICAM), pp. 396–398. IEEE (2018)
4. Do, T.V., Hsu, Q.C.: Optimization of minimum quantity lubricant conditions and cutting parameters in hard milling of AISI H13 steel. Appl. Sci. **6**(3), 83 (2016)
5. Khan, A.M., et al.: Multi-objective optimization of energy consumption and surface quality in nanofluid SQCL assisted face milling. Energies **12**(4), 710 (2019)
6. Gupta, M.K., et al.: Performance evaluation of vegetable oil-based nano-cutting fluids in environmentally friendly machining of inconel-800 alloy. Materials **12**(17), 2792 (2019)
7. Vu, N.C., Huang, S.C., Nguyen, H.T.: Multi-objective optimization of surface roughness and cutting forces in hard milling using taguchi and response surface methodology. Key Eng. Mater. **773**, 220–224 (2018)
8. Nguyen, H.T., Hsu, Q.C.: Study on cutting forces and material removal rate in hard milling of SKD 61 alloy steel. J. Chin. Soc. Mech. Eng. **38**(1), 41–51 (2017)

The Effect of Microstructure on the Cutting Forces and Microhardness in the Surface Grinding of Titanium Alloys

Hung Trong Phi[1,2(✉)], Got Van Hoang[3], Tam Ngoc Bui[2,4], Trung Kien Nguyen[2], and Son Hoanh Truong[2]

[1] Mechanical and Power Engineering Faculty, Electric Power University, Hanoi, Vietnam
hungpt@epu.edu.vn

[2] School of Mechanical Engineering, Hanoi University of Science and Technology, Hanoi, Vietnam
tambn@shibaura-it.ac.jp, {trung.nguyenkien, son.truonghoanh}@hust.edu.vn

[3] National Research Institute of Mechanical Engineering, Hanoi, Vietnam
gothv@narime.gov.vn

[4] College of Systems Engineering and Science, Shibaura Institute of Technology, Tokyo, Japan

Abstract. Ti-6Al-4V (Ti64) is the most popular alpha-beta titanium alloy employed in the aerospace industry. Ti64 is developed in many modifications of the basic formulation to gain favorable properties according to different applications. The heating temperature and cooling conditions are the basis of the titanium alloy's heat treatment resulting in various microstructures which influences the machinability of the Ti alloys and other alloys. This research investigates the effect of Ti64 microstructure on the grinding forces and surface roughness in the surface grinding process using a resin-bonded cBN wheel. The grinding process of Ti64 with a fully elongated alpha microstructure (Ti64-Elo) produces smaller grinding forces than those with Ti64 samples with a fully lamellar microstructure (Ti64-La). For the same cutting conditions, the Ti64-La surface shows a gain in hardness compared to the Ti64-Elo one.

Keywords: Ti-6Al-4V titanium alloy · CBN grinding wheel · Grinding force · Microhardness · Microstructure

1 Introduction

Ti-6Al-4V presents unique properties, i.e., high strength-to-weight ratio, superior corrosion resistance, good strength at elevated temperature, and high melting temperature. These properties make Ti-6Al-4V a promising material for industries worldwide, i.e., automotive, aerospace, chemical, marine, and gas turbine technology. Furthermore, Ti-6Al-4V has a languid corrosion speed in the human body, making it suitable for biomedical applications, including the artificial parts of prosthetic hearts and replacements of weight-bearing bone for the spine, shoulder, wrist, knee, and hip. However,

B. T. Long et al. (Eds.): MMMS 2020, LNME, pp. 525–533, 2021.
https://doi.org/10.1007/978-3-030-69610-8_72

the machining of Ti64 is notorious for its high hardness, low thermal properties, work hardening during machining, high chemical reactivity with many cutting tool materials that reduce the machinability of Ti-6Al-4V and make it to be a difficult-to-cut material. In the grinding process, cubic Boron Nitride (cBN) grinding wheels are a promising and a favorable option for cutting difficult-to-process metallic alloys since the grinding with this grinding wheels demonstrates competitive tooling cost, high capacity, high confidence, and low thermal failure. Many researchers have done to study and improve the machining properties of Ti64 in general cutting processes and grinding processes in specially. The cutting force and tool life control were implemented by Armendia et al. [1] to estimate the effect of heat treatment on the machinability of Ti-6Al-4V, Ti6246, and Ti-5Al-4V-0.6Mo-0.4Fe (TIMETAL 54M) titanium alloys. The Ti6246 presented an enormous tool wear rate and cutting force in all the titanium alloys, which could concern its higher strength. TIMETAL 54M alloy, a recently evolved alloy with similar characteristics to the more popularly employed Ti-6Al-4V, indicated the lowest wear rates. The TIMETAL 54M and beta annealed Ti-6Al-4V alloys with highly rough lamellar structure indicated significantly shorter tool life and larger cutting force. Jaffery et al. [2] used turning experiments with H13A class carbide inserts to introduce a new wear map for Ti-64 alloy. This wear map could be utilized to choose compatible cutting parameters that achieved the lowest tool wear rate. Youssef et al. [3] investigated the solution treatment's influence at 1050 °C in 30 min with water cooling, then aging treatment at 500 °C in 24 h, followed by air cooling on the Ti-6Al-4V samples. Using hot swaging and then quick quenching in water from 1050 °C could obtain a more compatible association of tensile properties, hardness and wear resistance.

Ti-6Al-4V is the most primarily employed alloy in all the alpha-beta titanium alloys. Ti-64 includes 4% of vanadium, which stabilizes the β phase, and 6% of aluminum stabilizes the α phase. The aluminum and vanadium additives enhance the strength in elevated temperatures, improves the formability and the oxidation resistance. Alpha (α) titanium is stable up to 880 °C, and it will undergo an allotropic transformation to beta (β) titanium at a higher temperature. The mill annealing of titanium alloys produces an optimal combination of ductility and fatigue strength, while the beta annealing maximizes the fracture toughness and stress corrosion cracking (SCC) resistance. The microstructure of the Ti64 alloy is two-phase consisting of α and β phrase. The lamellar of the α phase are relatively regular and are connected in basket weaving between these phases are thin zones of β phases.

The present work was conducted to investigate the grinding performance of Ti64 samples with different microstructures. The grinding forces and surface roughness were measured and discussed. The effect of workpiece infeed speed on grinding performance was evaluated. The obtained results could be used to select the grinding conditions for the surface grinding of Ti64 alloy.

2 Experimental Setup

The surface grinding experiments of Ti64 samples were performed on a high-speed vertical machining center (HS Super MC500, Fuhong Machinery Co., Taiwan. In the tests, Ti64 alloy specimens with the dimension of 16 mm (L) × 10 mm (W) × 9 mm

(H) were used. The Ti64-Elo samples were obtained at mill anneal condition at 704–788 °C in 2 h and air cool. Ti64-La was a modified version of Ti64 with maximum fracture toughness and SCC Resistance. Ti64-La was beta-annealed at 1066 °C (above transus) for 2 h, and the furnace cooled. Fig. 1 demonstrates the microstructure of Ti64-Elo and Ti64-La samples. The mechanical and thermal properties of the Ti64-Elo and Ti64-La are presented in Table 1. The Ti64-Elo showed a microstructure consisting of equiaxed β-grains and different morphologies of α-phase, as shown in Fig. 1a. The lamellar structure showed various shapes of α, α-grain boundaries, and groups of α structure consisting of lamellar α groups inside the prior β-grains as shown in Fig. 1b.

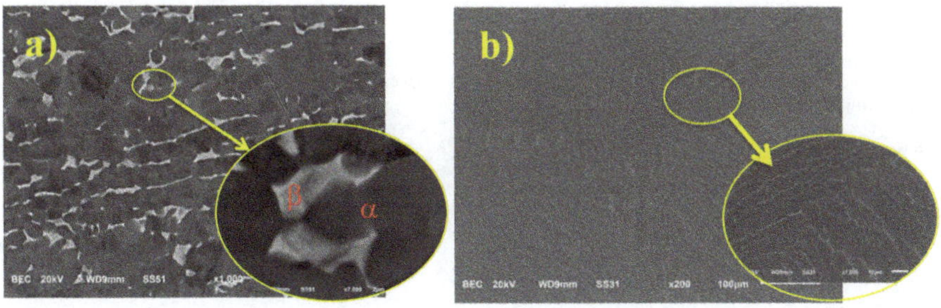

Fig. 1. Microstructure of (a) Ti64-Elo and (b) Ti64-La (α: darker color, β: light color).

Figure 2 schematically describes the grinding experimental setup. The cBN grinding wheel has an outer diameter of 100 mm and the grain size at 120 mesh. The up-grinding mode was applied for all the grinding tests. The grinding parameters are listed in Table 2. The abrasive wheel was dressed with an impregnated diamond dresser. The grinding tests were performed under dry and flood cooling environment. The grinding force signals in surface grinding were captured by the Kistler 9139AA piezoelectric dynamometer. The vibration of the main spindle was measured using a Triaxial DeltaTron Accelerometer to eliminate the effect of vibration on the results. The machined surface's microhardness was measured by IndentaMet 1106 Vicker indentation hardness tester.

Table 1. Mechanical and thermal properties of the Ti64-Elo and Ti64-La

Properties	Ti64-Elo	Ti64-La
Tensile yield strength [MPa]	830	965
Tensile strength [MPa]	900	1035
Compressive yield strength [MPa]	860	1070
Elongation [%]	10	8
Fatigue strength [MPa]	510	700
Modulus of elasticity [GPa]	114	
Thermal conductivity [W/m.K]	6.7	
Melting point [°C]	1604 ÷ 1660	
Specific heat capacity [J/g°C]	0.5263	

Table 2. Parameters of the grinding process

Workpiece material	Ti64-Elo and Ti64-La (16 mm × 10 mm × 9 mm)
Grinding wheel	cBN120 D100 T8 H20 X5 U6
Wheel speed (v_c)	30 m/s
Workpiece speed (v_w)	1000, 3000, 6000, and 10000 mm/min
Depth of cut (a_p)	0.005 ÷ 0.015 mm
Grinding width (b)	4.5 mm
Environment	Dry grinding and 10% emulsified oil in water

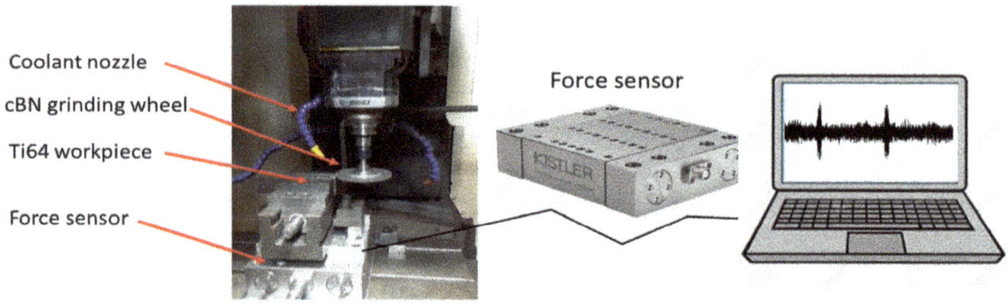

Fig. 2. Experimental set-up for plunge surface grinding (up-grinding).

3 Results and Discussion

3.1 Effect of Grinding Conditions on Grinding Forces

Many factors contribute to cutting forces in the grinding process, i.e., grinding parameters, cutting fluid, wheel condition, and wheel wear. Among these factors, workpiece infeed speed is a significant impact on cutting forces and cutting performance. This section focuses on the relationship between grinding forces and workpiece infeed speed with two types of microstructures of Ti64 workpieces. Fig. 3 plotted normal and tangential grinding forces for Ti64-Elo and Ti64-La alloys at various workpiece infeed speed. The experimental results exhibited that high workpiece speed generally increased grinding forces. In case of dry grinding at a low workpiece speed of 1000 mm/min grinding, Ti64-Elo experienced normal force at 16.45 N and the tangential force at 4.7 N. As the workpiece speed increased to 10000 mm/min, the normal force ascended to 94.92 N while the tangential force raised to 31.02 N. On the other hand, Ti64-La showed slightly higher normal grinding force ranged from 22.16 to 81.86 N as the workpiece infeed speed varied between 1000 and 10000 mm/min, while the tangential grinding force changed from 3.8 to 38.93 N. This was attributed to the fact that the increasing of workpiece infeed speed resulted in the increasing of the undeformed chip thickness, the penetration level of grits into the machined material and the cross-section area of swarfs [4]. As the workpiece approaching the rotating wheel surface rapidly, the grains did not have enough time to plow up the surface layer effectively. This grinding process was the same as slipping motion between sample and wheel surface and created a higher grinding force value. Meanwhile, more grains would take part in the grinding operation creating a more effective plowing action when the table speed was adjusted at a slower degree [5]. The wet cutting showed much reduction in normal grinding force, while the tangential grinding force with the cooling condition introduced moderate degradation. The normal forces in the wet grinding of Ti64-Elo and Ti64-La reached the maximum values of 36.6 and 32.87 N, respectively. Those values for dry grinding were 94.02 and 81.86 N, respectively. The grinding tests with emulsified oil-based fluid generally exhibited lower forces due to emulsion particles' lubricating effect in reducing forces [6]. In wet grinding, the lubricant would penetrate through the interface between the wheel and the Ti64 surface. The lubricant's existence in the contact zone lowered the friction between abrasive grits and cutting surface and delivered better slipping of the grains on the surface [7]. However, the Ti64 alloy is classified as difficult-to-cut material with high cutting temperature; therefore, the advantage of emulsified oil-based fluid in the cutting process was not significant to the machining process of other alloys.

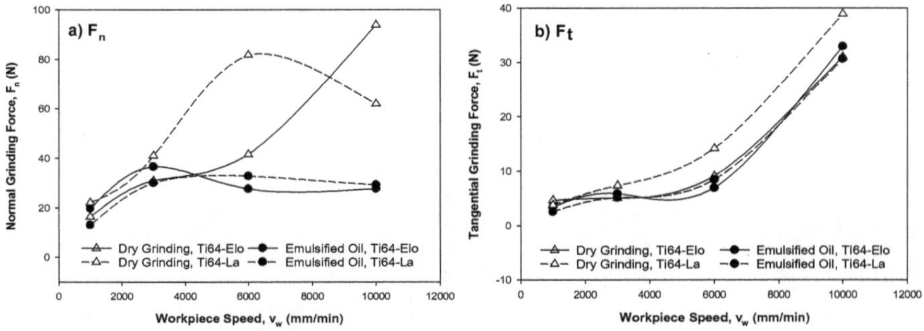

Fig. 3. Effect of workpiece speed and cooling condition on (a) normal force and (b) tangential force in the grinding tests with Ti64-Elo and Ti64-La.

The average grinding forces in Ti64-Elo grinding were slightly lower than their counterparts in Ti64-La grinding as Ti64-Elo showed a reduction in hardness value compared to that of Ti64-La. The higher hardness of material was, the stronger the constraints on its molecules were, the more extensive the grinding forces were. Therefore, a little difference in the hardness of the two Ti64 samples would show grinding force deviation. The Ti64-Elo obtained the lower hardness due to coarse β-grains and the relative inhomogeneity in the α + β structure. The higher hardness value of Ti64-La could be ascribed to the disintegration of the martensite structure, i.e., $\alpha' \rightarrow \alpha + \beta$. The reason is, if the martensite structure was not formed after the beta annealing, the hardness during the aging would increase probably as a result of deposition of the fine α phase from β phase, i.e., metastable $\beta \rightarrow$ fine $\alpha + \beta$ [8]. However, the Ti64 samples' microstructure resulted in less sightly variation of the cutting forces in wet grinding compared to dry grinding. Therefore, lubricant's existence in the contact zone could reduce the effect of the variation in the microstructure of the workpieces.

The tangential forces F_t were less susceptible than normal forces F_n with various workpiece speeds and cooling environments. Fig. 3 (a) and (b) clearly showed that normal forces always had larger values than tangential ones, which was predicted due to the more intensive rubbing and plowing activity by the abrasive grains with great negative rake angle. A slight variance of tangential forces F_t and a distinct change of normal forces F_n could be observed in Fig. 3 as the feed rate varied. At a feed rate of 10000 mm/min, the vibration started appearing on the main spindle, and then both the normal and tangential forces considerably increased regardless of Ti64 microstructures, as shown in Fig. 3. Along with the cooling and lubricating effect of emulsified fluid in the cutting zone, the normal force F_n significantly reduced while the tangential force F_t only showed a little reduction. The wheel's vibration at high workpiece infeed speed also created waviness on the ground surface with both dry and wet cutting conditions. The most massive tangential grinding forces for Ti64-Elo and Ti64-La with emulsified fluid were 32.97 and 30.66 N, while their counterparts in dry grinding were consecutively 31.02 and 38.93 N. The insufficiency of the coolant in dry condition led to immature failure, grain blunting, and the side flow of the material to the grains, then the growth of the forces. The improved temperature control of wet grinding was the reason for the reduction of the cutting forces [9].

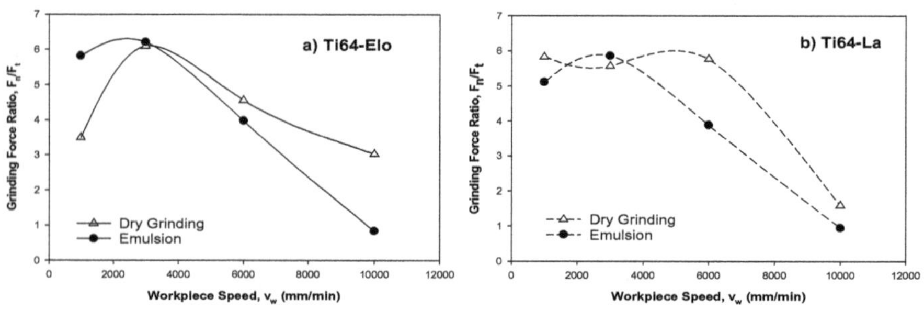

Fig. 4. Effect of workpiece infeed speed on the grinding force ratio F_n/F_t in grinding of (a) Ti64-Elo and (b) Ti64-La.

Figure 4 indicated that the grinding force ratio F_n/F_t in wet grinding is lower than that in dry condition. This demonstrated the better lubricating influence of the cutting oil. It could penetrate deeply into the contact zone and create a stable tribo-membrane to enhance the slipping, plowing, and cutting action of the abrasive grits. It is well recognized that a lower force ratio always points to a sharper grit and a longer tool life. Meanwhile, a higher force ratio usually relates to severe wheel wear [10, 11].

3.2 Effect of Grinding Conditions on Specific Grinding Energy

Specific energy u (J/mm^3) is a crucial index of machinability in the grinding process that describes the necessary power to reject one volume unit of specimen material. It can be computed with using the tangential force F_t and other cutting parameters by Eq. (1) [12]. The undeformed chip thickness (a_{gmax}) of a grinding wheel is described in Eq. (2) as a function of workpiece infeed speed, wheel's diameter, cutting velocity, and the cutting depth [13].

$$u = \frac{F_t.v_s}{v_w.a_p.b} \tag{1}$$

$$a_{gmax} = 2\pi d_s \frac{v_w}{v_c}\sqrt{\frac{a_p}{d_s}} \tag{2}$$

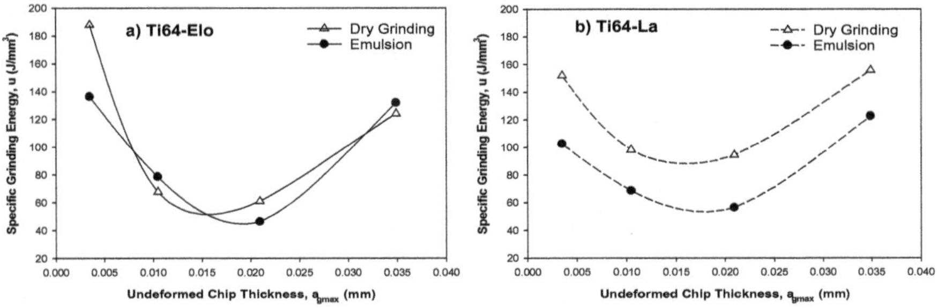

Fig. 5. The undeformed chip thickness and the specific grinding energy in the grinding of (a) Ti64-Elo and (b) Ti64-La.

As observed from the relationship between the specific energy and the undeformed chip thickness shown in Fig. 5 (a), the specific grinding energy of Ti64-Elo and Ti64-La was up to 40 ÷ 200 J/mm^3, which proved that the machinability of these alloys was considerably worse than that of the steels (20 ÷ 80 J/mm^3 [14]). The specific grinding energy generally lowered as the undeformed chip thickness increased. This could be explicated by the "size effect" of grinding wheels. In the instance of a low a_{gmax}, slipping and plowing were the dominant actions in grinding, which energy contribution led to a considerable specific energy. At the cutting with high a_{gmax}, the fraction of slipping and plowing reduced result in lower specific energy [4]. The energy fast declined, then slowly reduced regardless of the cooling conditions and sample materials. In terms of cooling conditions, specific grinding energy under emulsified fluid grinding generally showed lower values compared to that of dry cutting because of a lower tangential force of emulsified fluid condition. However, specific energy when feed rate v_w was 10000 mm/min with various Ti64 versions and cooling conditions was relatively high compared to its counterparts under lower feed rates. The main spindle's vibration at the highest workpiece speed could explain it, and hence the tangential force F_t and the specific energy increased. The average specific energy of Ti64-Elo was generally lower than that of Ti64-La, which was attributed to the higher tangential force F_t of Ti64-La compared to that of Ti64-Elo.

3.3 Effect of Grinding Conditions on Microhardness

The hardening effect is attributed to a high plastic flow rate uniting with the heat creation at the prime cutting region. The microhardness was measured along the cross-section of the Ti64-Elo and Ti64-La specimens under the same cutting condition to assess the impacted zone of residual stress through the machining process. The Vickers $HV_{0.1}$ hardness data was collected between space of two points at 20 μm as shown in Figs. 6a and 6b, while the measuring data was shown in Fig. 6c. The microhardness had the highest value on the surface and reduced in the sub-surface layer indicating that the softening effect did not happen because of the lower grinding zone temperature compared to the β-transus temperature. This was partly attributed to the immense heat dissipation of the cBN wheel that decreased grinding temperature at the cutting interface and discarded thermal failure. The microhardness (~390 HV) of the Ti64-Elo sample gained by 11.8% compared to the bulk material (~349 HV), while that of the Ti64-La surface (~440 HV) was 17% higher than the sample base (~376 HV). The results also confirmed that the enhancing hardness of Ti64-La bulk material in comparison with that of Ti64-Elo due to the difference in their heat treatment. In both microstructures, the hardening depth was less than 130 μm [15, 16], while a slight fluctuation was seen at a depth of 200 μm below the grinding surface. This result indicated that the machining process only affected the Ti64 sample's microhardness at a narrow area from the machined surface.

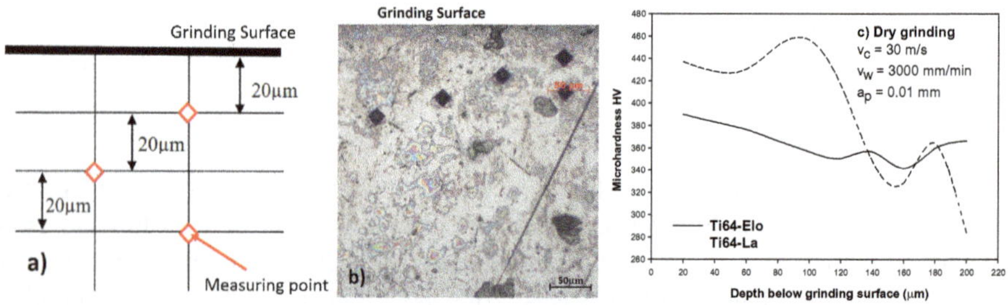

Fig. 6. Microhardness of Ti64-Elo and Ti64-La ground surfaces: (a) Schematic of measuring points, (b) Image of indentations, and (c) Microhardness HV results.

4 Conclusion

The surface grinding experiments were performed on samples of Ti-6Al-4V alloy with fully equiaxed microstructure (Ti64-Elo) and fully lamellar microstructure (Ti64-La) under different cutting conditions. The major conclusions are as follows:

- The high workpiece infeed speed showed the high normal and tangential cutting forces. The grinding forces of Ti64-Elo were lower than those of Ti64-La. The grinding process using water emulsion exhibited lower forces than the value obtained in the dry cutting.
- The wet grinding showed a slight reduction in the grinding force ratio F_n/F_t and specific energy compared to the dry cutting. The specific energy of Ti64-Elo was lower than that of Ti64-La.

– At the same cutting conditions, the Ti64-La surface obtained larger microhardness compared to the Ti64-Elo.

References

1. Armendia, M., Osborne, P., Garay, A., Belloso, J., Turner, S., Arrazola, P.J.: Influence of heat treatment on the machinability of titanium alloys. Mater. Manuf. Process. **27**, 457–461 (2012)
2. Jaffery, S.I., Mativenga, P.T.: Assessment of the machinability of Ti-6Al-4V alloy using the wear map approach. Int. J. Adv. Manuf. Technol. **40**, 687–696 (2009)
3. Mukhopadhyay, M., Kundu, P.K.: Performance evaluation of conventional abrasive wheels for grinding Ti-6Al-4V. Mater. Sci. Eng. **377**, 012–043 (2018)
4. Zhenzhen, C., Jiuhua, X., Wenfeng, D., Changyu, M.: Grinding performance evaluation of porous composite-bonded CBN wheels for Inconel 718. Chin. J. Aeronaut. **27**(4), 1022–1029 (2014)
5. Lin, S.Y., Liu, Y.C., Huang, C.W.: An investigation of surface grinding characteristics for titanium alloy with CBN wheel. Key Eng. Mater. **364–366**, 237–242 (2008)
6. Kundu, A., Mukhopadhyay, M., Mahata, S., Banerjee, A., Mandal, B., Das, S.: Grinding titanium grade 1 alloy with an alumina wheel using soap water. Procedia Manuf. **20**, 338–343 (2018)
7. Guo, G., Liu, Z., Zheng, X., Chen, M.: Investigation on surface grinding of Ti-6Al-4V using minimum quantity lubrication. Adv. Mater. Res. **500**, 308–313 (2012)
8. Youssef, S.S., Ibrahim, K.M., Karim, M.A.: Effect of heat treatment process on tribological behavior of Ti-6Al-4V alloy. Int. J. Mech. Eng. Robot. Res. **2**(4), 385–394 (2013)
9. Manimaran, G., Kumar, M.P.: Effect of cryogenic cooling and sol–gel alumina wheel on grinding performance of AISI 316 stainless steel. Arch. Civ. Mech. Eng. **13**, 304–312 (2013)
10. Li, Z., Ding, W., Liu, C., Su, H.: Grinding performance and surface integrity of particulate-reinforced titanium matrix composites in creep-feed grinding. Int. J. Adv. Manuf. Technol. **94**, 3917–3928 (2018)
11. Zhao, B., Ding, W.F., Dai, J.B., Xi, X.X., Xu, J.H.: A comparison between conventional speed grinding and super-high speed grinding of (TiCp+TiBw)/Ti–6Al–4V composites using vitrified CBN wheel. Int. J. Adv. Manuf. Technol. **72**, 69–75 (2014)
12. Kacalak, W., Lipiński, D., Bałasz, B., Rypina, Ł, Tandecka, K., Szafraniec, F.: Performance evaluation of the grinding wheel with aggregates of grains in grinding of Ti-6Al-4V titanium alloy. Int. J. Adv. Manuf. Technol. **94**, 301–314 (2018)
13. Zhou, H., Ding, W., Liu, C.: Material removal mechanism of PTMCs in high-speed grinding when considering consecutive action of two abrasive grains. Int. J. Adv. Manuf. Technol. **100**, 153–165 (2019)
14. Pal, B., Chattopadhyay, A.K., Chattopadhyay, A.B.: Development and performance evaluation of monolayer brazed cBN grinding wheel on bearing steel. Int. J. Adv. Manuf. Technol. **48**(9–12), 935–944 (2010)
15. Sadeghi, M.H., Haddad, M.J., Tawakoli, T., Emami, M.: Minimal quantity lubrication-MQL in grinding of Ti–6Al–4V titanium alloy. Int. J. Adv. Manuf. Technol. **44**, 487–500 (2009)
16. Changyong, Y., Jiuhua, X., Wenfeng, D., Zhenzhen, C., Yucan, F.: Dimension accuracy and surface integrity of creep feed ground titanium alloy with monolayer brazed CBN shaped wheels. Chin. J. Aeronaut. **23**, 585–590 (2010)

Effect of High-Speed Milling Parameters on Chip Shrinkage and Tool–Chip Junction Length of A6061 Alloy

Pham Thi-Hoa[1(✉)], Doan Thi-Huong[1], Banh Tien-Long[1], Nguyen Duc-Toan[1], and Pham Duc-Thanh[2]

[1] Hung Yen University of Technology and Education, Khoai Chau, Hung Yen, Vietnam
[2] University of Transport and Communications, Hanoi, Vietnam

Abstract. The chip shrinkage coefficient determines the plastic deformation that occurs during high-speed milling. The magnitude of chip shrinkage closely relates to the degree of deformation of the chip. Hence, it depends on the workpiece's properties, the tool's geometry, and the other cutting parameters. During high-speed machining, the separation of metal from the cutting area takes place rapidly over the junction between the tool and workpiece, which increases the friction rapidly. Therefore, the cutting tool's abrasive wear also increases, which leads to a reduction in productivity and surface finish of the cutting process. The chip shrinkage behavior of the A6061 alloy high-speed milling (HSM) is experimentally studied and discussed. Research results provide a basis for selecting suitable cutting conditions for achieving optimal quality of surface finish.

Keywords: Chip shrinkage coefficient · Tool-chip contact length · A6061 aluminum alloy · High-speed milling

1 Introduction

Many experimental and theoretical studies have been conducted to explore optimal cutting setup during HSM of aluminum alloys. Among them, the optimal cutting parameters for improving the cutting tool's life are significant concerns because the cutting tool's durability relates to the cost and productivity of the cutting process [1, 2]. Besides, the optimization studies related to increasing the quality of surface finish, cutting temperature, cutting force, and workpiece vibration were extensively explored [3]. However, there have been very few studies dealing with the chip shrinkage coefficient (CSC) and tool-chip contact length (TCL). Anayet et al. [4] studied the CSC during end milling of Ti6Al4V alloy and showed that the CSC varied within 5% with increasing cutting speed (CS) 20 to 100 m/min. Afonasov and Lasukov [5] investigated the relationship between cutting parameters and longitudinal shrinkage, shear angle, and chip continuity. The TCL is then one of the parameters that affect the friction at the contact zone and the tool wear [6]. In this work, the relationships between cutting

B. T. Long et al. (Eds.): MMMS 2020, LNME, pp. 534–540, 2021.
https://doi.org/10.1007/978-3-030-69610-8_73

parameters (CS, cutting depth, and feed rate) on the CSC and the TCL in high-speed, dry milling of aluminum alloy A6061 were studied. First, an experiment of high-speed, dry milling was conducted using a milling cutter with four carbide inserts. Then, regression formulae for the CSC and the TCL were determined. The effects of cutting parameters on the CSC and the TCL were extensively investigated based on the regression formulae. The obtained results were finally discussed.

2 Experimental Setup

2.1 Milling Machine and Cutter

The experimental milling is carried out on an HS Super MC500 high-speed CNC vertical milling machine. The main parameters of the milling machine are spindle rotation speed from 100 to 30000 (rpm), 15 kW spindle and the table displacement velocity is from 1 to 30,000 (mm/min).

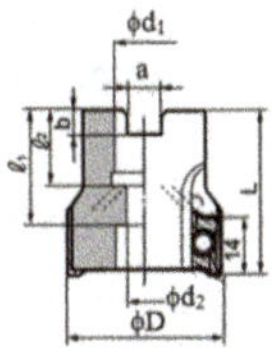

Dimension (mm)							
ϕD	ϕd_1	ϕd_2	a	b	L	l_1	l_2
40	16	9	8.4	5.6	40	28	18

Fig. 1. Geometric parameters of the milling cutter

A 40 mm face milling cutter is used, as depicted in Fig. 1. The milling cutter is attached with four ceramic carbide inserts (code APMT1604PDTR TC300). The rake and the clearance angles of the carbide insert are 5° and 10°, respectively. Dry cuttings are considered for all milling experiments.

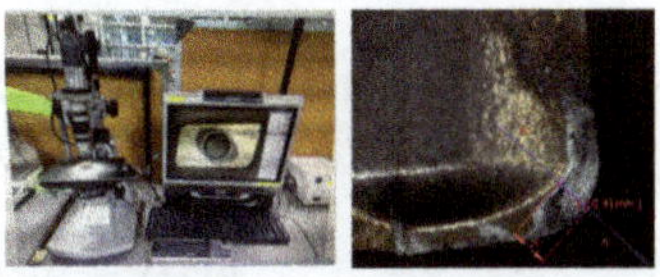

Fig. 2. Measurement of the TCL using VHX-2000E digital microscope

2.2 Tool-Chip Contact Length Measurement

The tool-chip contact lengths (H_s) are measured by a digital microscope (Keyence VHX-2000E), as shown in Fig. 2. The measuring device captures the cutter at a high resolution, from which the TCLs are resolved. Figure 2 shows a sample measured TCL.

2.3 Chip Shrinkage Coefficient Determination

The chip shrinkage coefficient (K) is estimated by dividing the uncut chip length and the actual post-chip length. In HSM A 6061 alloy, the uncut chip lengths are generally challenging to determine. Hence, another approximate method to estimate K is presented by the following formula [6]:

$$K = \frac{1000.Q}{\rho.l.f_v.t} \quad (1)$$

where Q is the mass of the chip (g); ρ is the density of workpiece material (g/cm^3); l is the actual chip length (mm); f_v is feed rate (mm/min); t is the cutting depth (mm). According to Eq. (1), the mass and length of chips need to be measured.

3 Design of Experiment

3.1 Experimental Results

Three cutting parameter levels are selected: CS of 1000, 1128, and 1256 m/min, the feed rate of 800, 1350, and 1800 mm/min, and cutting depth of 0.5, 1.0, and 1.5 mm. A total of 11 experiments follow the design of the experiment method with three replicated experiments at the center point. The experimental results of the TCL and CSC are shown in Table 1.

3.2 Regression Formulae

The following nonlinear equation describes the mathematical formulae of the TCL (Hs) when milling A6061 aluminum alloy:

$$H_S = d_1.V^{d_2}.f^{d_3}.t^{d_4} \tag{2}$$

where d_1, d_2, d_3, and d_4 are the coefficients to be determined. The equation is linearized by performing logarithm transformation to determine these coefficients, as follows:

$$\ln Hs = \ln d_1 + d_2 \ln V + d_3 \ln f + d_4 \ln t \tag{3}$$

The coefficients are found, and the H_s is obtained as follows:

$$H_S = -10.135V^{1.1316}f^{0.1734}t^{0.2925} \tag{4}$$

Table 2 shows the analysis of variance results of $H_{s,}$ indicating that the F value is 0.0002, which is less than 0.5. So, the regression equation is reliable. The R-square value is 0.933, and Fisher, the test of independence result is 32.67, which is greater than the theoretical value of 4.35. Thus, the significance level of the statistics is high. Similarly, the regression formula for K can be found. The analysis of variance results for K is listed in Table 3. The formula for K is as follow

$$K = 0.8899V^{0.009071}f^{0.03395}t^{-0.14026} \tag{5}$$

4 Discussions

The relationships between Hs and cutting parameters (V, f, t) are plotted, as shown in Fig. 3. It can be observed that increasing either V *and* t, or f leads to the increase of H_s. However, the increasing rate of H_s when f increases are relatively small. It is also found that the CS has the highest impact on the TCL owing to increasing the CS leading to the metal removal process faster. Besides, the pressure and temperature increase with increasing V leading to the formation of continuous contact of chips on the front of the cutting tool and adhesion of chip on the cutting tool.

The relationships between K and cutting parameters (V, f, t) are shown in Fig. 4, which is obtained using Eq. (5). The results show that CS only shows very little influence on K. In contrast, the cutting depth (t) is the most influential factor in K. This is because when the depth of cut increases, and the thicker chip. As a result, the curl radii of the chip get reduced, and the chip's deformation becomes more extensive, which is contributed mainly by the higher temperature difference between the sliding and free surfaces of the chip. The effect of feed rate (f) on K is relatively small at low CS, but it increases slightly as the CS increases.

Having obtained formulae of H_s and K, the relationship between the H_s and K can be determined, as shown in Fig. 5. From this figure, the following behaviors are observed: when V, f is fixed and t changes: H_s decreases as K increases; when V and t are fixed and f changes: H_s decreases as K decreases; when t and f are fixed, and V changes: Hs increases rapidly while K does not change.

Table 1. Experimental results of tool-chip contact and CSC

Exp. number	Cutting parameters			Responses	
	V (m/min)	f (mm/min)	t (mm)	H_s (mm)	K
1	1000	800	0.5	0.32	1.294
2	1256	800	0.5	0.27	1.304
3	1000	1800	0.5	0.41	1.322
4	1256	1800	0.5	0.29	1.370
5	1000	800	1.5	0.46	1.144
6	1256	800	1.5	0.37	1.102
7	1000	1800	1.5	0.56	1.137
8	1256	1800	1.5	0.41	1.133
9	1128	1350	1.0	0.37	1.236
10	1128	1350	1.0	0.37	1.236
11	1128	1350	1.0	0.37	1.236

Table 2. Analysis of variance (ANOVA) results for H_s

R	R_2	Adjusted R Square	Standard error	Sum	
0.9661	0.9333	0.9047	0.0631	11	
ANOVA results					
	df	*SS*	*MS*	*F* value	*F*value (theory)
Regression	3	0.3904	0.1301	32.649	0.0002
Error	7	0.0279	0.0040		
Sum	10	0.4183			
Coefficient	Value	Standard error	*t* Stat	*P*-value	
d_1	-10.14	1.4265	-7.105	0.0002	
d_2	1.132	0.1957	5.7814	0.0007	
d_3	0.17	0.0544	3.187	0.0154	
d_4	0.293	0.0403	7.264	0.0002	

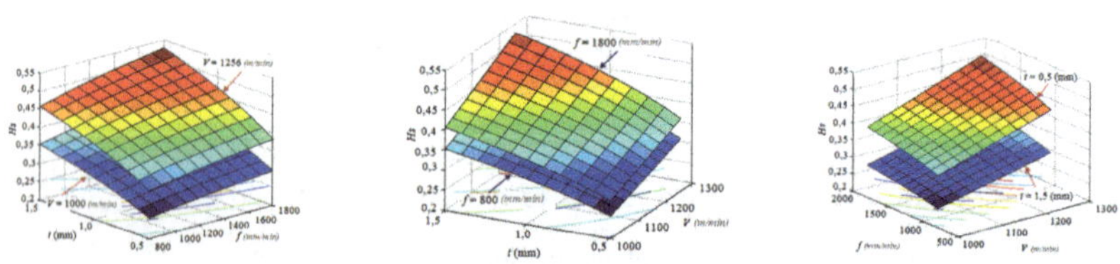

a) fixed values of V b) fixed values of f c) fixed values of t

Fig. 3. Cutting parameters versus the tool – chip contact length

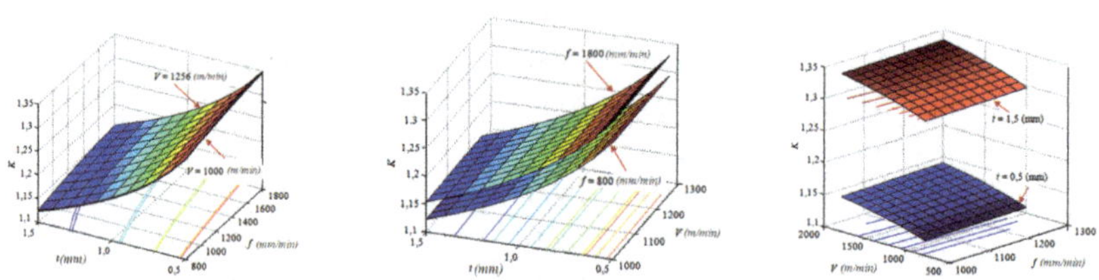

a) fixed values V b) fixed values f c) fixed values t

Fig. 4. Cutting parameters versus the CSC

5 Conclusions

In this study, the influences of the cutting parameters on the CSC and the TCL in HSM A6061 alloy were investigated. Several technical information could be concluded:

- Cutting speed shows the highest effect on TCL. Hence, increasing CS can adversely affect tool life.

Table 3. Analysis of variance (ANOVA) results for *K*

R	R_2	Adjusted *R* Square	Standard error	Sum
0.9673	0.9357	0.9081	0.0221	11

ANOVA results					
	df	*SS*	*MS*	*F* value	*F* value (theory)
Regression	3	0.0496	0.0165	33.953	0.00015
Error	7	0.0034	0.0005		
Sum	10	0.0529			

Coefficient	Value	Standard error	*t* Stat	*P*-value
d_1	0.8899	0.4985	-0.2341	0.8216
d_2	0.00907	0.0684	0.13262	0.8982
d_3	0.03395	0.01902	1.78436	0.1175
d_4	-0.1403	0.01407	-9.9672	2.2E-05

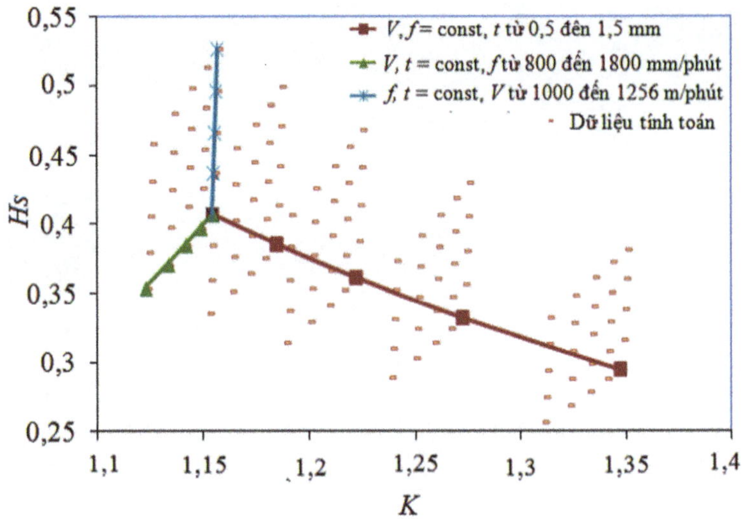

Fig. 5. The relationship between *K* and H_s.

- The CSC is considerably dependent on the depth of cut, subsequently to feed rate. Vice versa, cutting speed does not affect the CSC.

- The CSC and the TCL show a complicated behavior depending on cutting parameters.
- The obtained results provide a fundamental basis for selecting optimal machining parameters in HSM of A6061 alloy.

References

1. Hou, J., Zhou, W., Duan, H., et al.: Influence of cutting speed on cutting forces, flank temperature, and tool wear in end milling of Ti-6Al-4V alloy. Int. J. Adv. Manuf. Technol. **70**, 1835–1845 (2014)
2. Liu, Z.Q., Ai, X., Zhang, H., Wang, Z.T., Wan, Y.: Wear patterns and mechanisms of cutting tools in high-speed face milling. J. Mater. Process. Technol. **129**, 222–226 (2002)
3. Da Silva, R.B., MacHado, A.R., Ezugwu, E.O., Bonney, J., Sales, W.F.: Tool life and wear mechanisms in high speed machining of Ti-6Al-4V alloy with PCD tools under various coolant pressures. J. Mater. Process. Technol. **213**, 1459–1464 (2013)
4. Anayet, M., Patwari, U., Nurul Amin, A.K.M.: Influence of chip serration frequency on chatter formation during end milling of Ti6Al4V. J. Manuf. Sci. Eng. **133**, 011–013 (2011)
5. Afonasov, A.I., Lasukov, A.A.: Elementary chip formation in metal cutting. Russ. Eng. Res. **34**, 152–155 (2013)
6. Pham, T.-H., Mac, T.-B., Tong, V.-C., Banh, T.-L., Nguyen, D.-T.: Simulation and experimental studies to verify the effect of cutting parameters on chip shrinkage coefficient and cutting forces in machining of A6061 aluminum alloy. Adv. Mech. Eng. (2016)

Desulphurization of Hot Metal in Pre-treatment Process with a Flux Containing Basic Oxygen Furnace Slag and CaO

Cao Son Nguyen, Trung Duc Tran, and Anh Hoa Bui(✉)

School of Materials Science and Engineering,
Hanoi University of Science and Technology,
No 1, Dai Co Viet Road, Hai Ba Trung District, Hanoi, Vietnam
hoa.buianh@hust.edu.vn

Abstract. Recycling of the slag is a crucial challenge for the steel-producing companies because of a huge amount of the slag produced in the steel companies. This paper investigated an ability of a basic oxygen furnace slag using in desulfurization of hot metal in pre-treatment process, as a solution in recycling of the slag. Iron and the slag were obtained from a steel company. The iron was completely melt in a high frequency induction furnace to get hot metal, then a flux containing the slag and CaO was embedded into the metal. After 10 min of reaction time, the metal and slag samples were taken out for chemical compositions analysis. The results shown that the S content of the metal decreased from 0.054 to 0.027 mass%. The oxidation of Silicon (Si) strongly influenced the desulfurization of the metal; increase in oxidation of Si caused removing of Sulfur (S) content of the metal to decrease significantly.

Keywords: Basic oxygen furnace slag · Desulphurization · Pre-treatment process

1 Introduction

Basic oxygen furnace (BOF) process is main responsible for steel production. Today, approximately 70% of steel is produced all over the world using the integrated blast furnace-basic oxygen furnace route [1]. The oxygen is injected through a multi-holed, water-cooled lance, into the furnace to react with the hot metal. Consequently, the carbon content significantly reduced to produce steel. Additionally, the slag, a by-product of steelmaking, is produced under an oxidizing atmosphere. The slag mainly contains oxides of FeO, CaO, SiO_2, Al_2O_3 and MgO [2–4]. Because of the high CaO content, the slag is possibility to utilize a large amount of the slag in metallurgical process to replace the flux. Furthermore, the slag also has high economical benefits, which also has been successfully recycled as a construction material, ceramic, fertilizer, etc. [5–7].

However, recycling of the slag is a crucial challenge for the steel-producing companies because of a huge amount of the slag produced in the steel companies; the companies need more landfills for the slag. It hence has significant impact on

B. T. Long et al. (Eds.): MMMS 2020, LNME, pp. 541–546, 2021.
https://doi.org/10.1007/978-3-030-69610-8_74

environment if the total slag is not properly recycled. Environmental requirements will enhance the need for further product development works concerning the ability of the slag recycling. The BOF slag can be a reagent for desulphurization in the pre-treatment process. To study ability of the slag, the role of the slag in the process of pre-treatment has been extensively explored by many researchers [8, 9]. The slag was confirmed to be able to remove Sulphur (S) of the hot metal [8].

In this research, the slag combination with CaF_2 effectively removed S of the hot metal in the pre-treatment process. Hernandez *et al.* reported the rate of desulphurization increased by increasing of MgO content and basicity in slag, decreasing Al_2O_3 and TiO_2 content, and increasing the temperature [9]. In the pre-treatment process, diffusion of S from the iron to the flux containing the steelmaking slag would occur. However, the diffusion of S from the molten iron to the flux could not sufficiently discussed in the previous studies. Additionally, factors affecting desulfurization of the molten iron has not been concerned. Therefore, the objective of this study is to investigate behavior and ability of the slag combination with CaO affecting desulphurization of hot metal in the pre-treatment process.

2 Experimental Method

BOF Slag was obtained from Hoa Phat steel company, then the slag was crushed and grinded to make the homogeneous powder. Composition of slag is given in Table 1. The slag powder and limestone powder were completely mixed at a fixed ratio to make a flux. Meanwhile, iron sample was also obtained from the steel company. Chemical composition of the iron sample was measured using the optical emission spectrometer (METAL LAB 75/80 J MVU). The iron sample contains 3.26%C, 2.29%Si, 0.054%S, 0,086%P and 94.31%Fe in mass.

Table 1. Mineralogical composition of the slag used

	C	MgO	Al_2O_3	SiO_2	P_2O_5	CaO	MnO	Fe_2O_3
Mass%	6.57	6.32	1.18	9.56	1.26	55.24	6.90	15.97

In the experiment procedure, 10 kg the iron sample was melted in a frequency induction furnace. Then the furnace was holding 5 min for homogeneous chemical composition of the molten iron. After holding time, the flux including BOF slag and lime with ratio 1:1 were put into the melt with the weight of 0.3 kg. There were 2 experiments (named M1 and M2) carrying out in the study. In M1 experiment, the furnace with molten iron was closed using a cover and protected by blowing argon gas on the top of the furnace after putting the flux into the furnace. The holding time for reaction between the flux and the molten iron was 10 min. After this period, the power was turned off then the flux and molten iron were taken out for chemical analysis. In the M2 experiment, molten iron reacted with the flux with open furnace. The experimental procedure of M2 was as similar as that of M1 experiment. Finally, S and C concentrations of the iron samples were measured using the optical emission spectrometer.

3 Results and Discussion

3.1 Effect of Melting Condition on Carbon Content of Iron

Figure 1 shows variation of carbon concentrations of iron samples after reactions between the molten iron and the flux. There is a significant drop of carbon content of iron sample from the initial carbon content. The carbon content significantly decreased from the initial value of 3.24 mass% to 1.80 and 1.82 mass% in both cases of M1 and M2 experiments, respectively. Accordingly, the carbon content sample is 1.80 mass% with open furnace.

Meanwhile, in case for protective cover and argon gas blowing, the value of carbon content is 1.82 mass%. It indicated that there was a small different values of carbon content after experiments in both cases. Therefore, the top furnace atmosphere is concluded to have low impact on the decarburization of the hot metal. In the pre-treatment process, the carbon content would decrease in the contact between the flux and the hot metal.

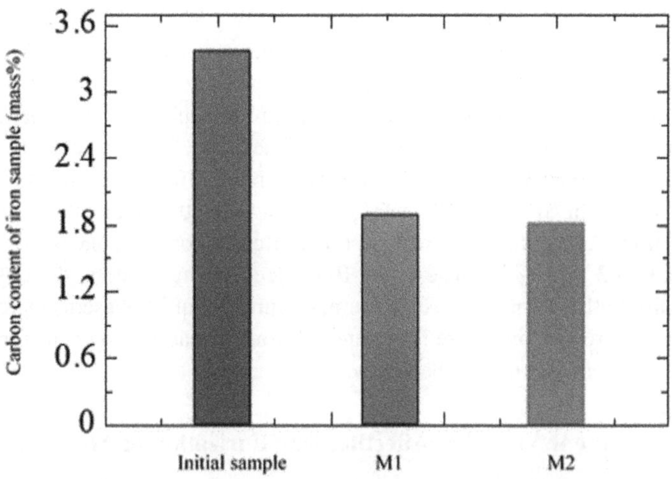

Fig. 1. Variation of carbon content of iron sample

3.2 Silicon Content Variation Under Different Experimental Condition

Figure 2 shows variation of Silicon (Si) content with time after reaction between the flux and the molten iron. The results show that Si contents of the samples dramatically decrease due to different experimental condition, in which Si is oxidized following reaction 1.

$$[Si] + 2[O] = SiO_2 \quad (1)$$

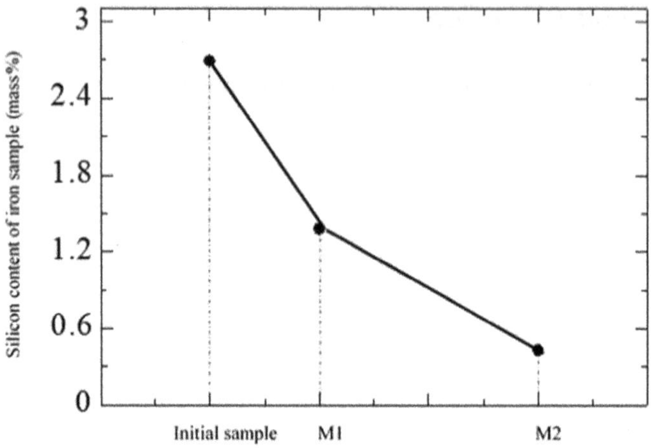

Fig. 2. Variation of Silicon content of iron sample

In M1 experiment, Si of iron sample decreased by 1.31 mass% from initial value of 2.69 mass% with condition of protecting atmosphere as shown in Fig. 3. Meanwhile, in M2 experiment, Si content dramatically decrease by 2.26 mass% (from 2.69 to 0.43 mass%) with open furnace. In the oxidizing atmosphere, the oxidation of Si is indicated to be more sufficient for reactions. Therefore, it is concluded that the Si content is largely affected by the top furnace atmosphere. In this case of the experiment, the lowest content of Si is 0.43 mass% is advantage for the steelmaking process after 10 min reaction. Additionally, in real production condition, Si content of iron should be higher than 0.2 mass% because in the BOF steelmaking process, the oxygen reacts exothermically with silicon to cause the temperature of liquid phase to increase. When Si content of the iron is low, there is not enough for the reaction to occur for increasing the liquid phase temperature in the furnace.

3.3 Silicon Content Variation Affecting Desulfurization of Molten Iron

S content of sample is related to the environment condition will be discussed through the relationship between Si and S content. Those results are useful to evaluate the oxidizing atmosphere affecting the desulfurization of the molten iron. Figure 3 shows the relationship between Si and S contents of iron. The results show that in the M1 experiment, S derogation is 50% (decrease from 0.054 to 0.027 mass%) when variation of Si is 1.31%. It means low oxidizing atmosphere cause ability of S removal to increase. Meanwhile, with high oxidizing atmosphere (M2 experiment), oxidation of Si amount is high, which is able to weaken desulfurization with 20% S removal (decrease from 0.054 to 0.044 mass%). It was revealed that high oxidizing atmosphere causes S removal to decrease.

Sufficient condition of desulfurization includes high basicity, reduced atmosphere and high temperature [7]. High basicity is positive condition for desulfurization. Results show the desulfurization decreases when the amount of SiO_2 increase due to oxidation of Si of the iron. SiO_2 was formed due to oxidation of Si, which is able to cause the basicity of flux to decrease. M2 experiment obtained high amount of SiO_2 is due to high oxidation of Si which causes basicity to decrease. Therefore, decreasing of basicity is one of reasons to decrease the desulfurization of the iron. Besides, oxidation of Si negatively affected S removal which is indicated in result of Stephen *et al.* [8]. The research addressed high content of SiO_2 in slag increase the oxygen potential, which decreased the S removal.

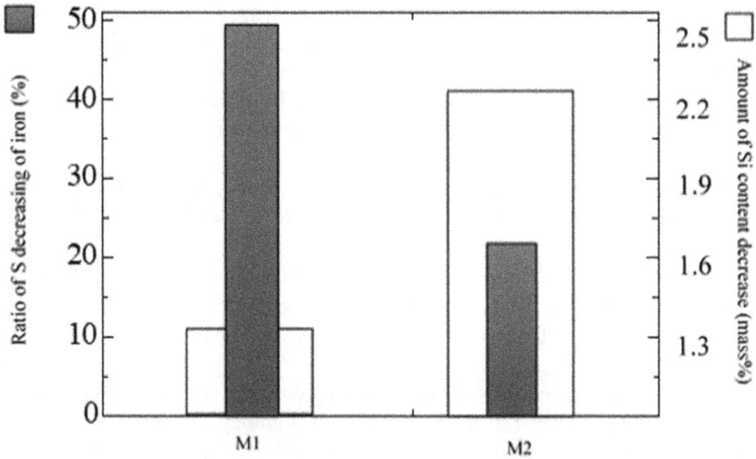

Fig. 3. Relationship between Si and S contents of iron samples

It was thought that the desulfurization was not advantage in case of oxidation of Si. There was a solution of resistance of oxidation by protecting atmosphere, which allowed to improve the desulfurization of iron. In which, method of preventing oxidizing atmosphere in the furnace was concluded to be a suitable solution.

4 Conclusions

The experimental condition was addressed to have low impart on the decarburization of the metal. Variation of Si was large due to the oxidizing atmosphere in the top of the furnace, which strongly influenced the desulfurization of the metal. Accordingly, S content of hot metal decrease with Si variation of iron. The atmosphere in the furnace significantly affected the desulfurization of the metal. Open furnace had a negative effect on desulfurization of the flux, meanwhile, the best condition for desulfurization is protective with cover and argon gas. The slag was concluded to reuse as a flux for desulfurization of hot metal in pre-treatment process with close furnace. In this process, the slag was useful to remove S content of the hot metal.

Acknowledgment. The authors are grateful to the financial support by the Asahi glass foundation under grant number AGF.2020-06.

References

1. Visser, H.-J., Boom, R.: Advanced process modelling of hot metal desulphurisation by injection of Mg and CaO. ISIJ Int. **12**, 1771–1777 (2006)
2. Taguchi, K., Ono-Nakazato, H., Nakai, D., Usui, T., Marukawa, K.: Deoxidation and desulfurization equilibria of liquid iron by calcium. ISIJ Int. **11**, 1705–1709 (2003)
3. Schrama, F., Beunder, E., Berg, B., Yang, Y., Boom, R.: Sulphur removal in ironmaking and oxygen steelmaking. Ironmak. Steelmak. Prod. Appl. **5**, 333–343 (2017)
4. Yang, J., Okumura, K., Kuwabara, M., Sano, M.: Desulfurization of molten iron with magnesium vapor produced in-situ by aluminothermic reduction of magnesium oxide. ISIJ Int. **9**, 965–973 (2001)
5. Hernandez, A., Romero, A., Chavez, F., Angeles, M., Moral, R.D.: Dephosphorization and desulfurization pretreatment of molten iron with CaO-SiO2-CaF2-FeO-Na2O slags. ISIJ Int. **2**, 126–131 (1998)
6. Hur, N.-S., Jung, S.-M., Ban, B.-C.: Dephosphorization and desulphurization pretreatment of liquid iron using CaF2-free fluxes containing BOF slags. Steel Res. Int. **75**, 778–782 (2004)
7. Biswas, A.K.: Principles of Blast Furnace Ironmaking. Cootha Publishing House, Brisbane (1981)
8. Stephen, F.M.: Improvement of the desulfurization process by slag composition control in the ladle furnace. Master Thesis, Luleå University of Technology (2009)
9. Dwarapudi, S., Ghosh, T.K., Shankar, A., Tathavadkar, V., Bhattacharjee, D., Venugopal, R.: Effect of pellet basicity and MgO content on the quality and microstructure of hematite pellets. Int. J. Miner. Process. **99**, 43–53 (2011)

Real-Time Simulation of Electronic Power Steering System on TOYOTA VIOS

Manh Nguyen Huu[1(✉)] and Truong Nguyen Quang[2]

[1] Ly Tu Trong College, Ho Chi Minh City, Vietnam
manhnhdhgt@gmail.com
[2] Central Transport College No.3, Ho Chi Minh City, Vietnam

Abstract. EPS electric power steering system in cars not only brings a comfortable driving feeling but also helps to reduce fuel consumption, easy to repair when damaged. The study of simulation theory, semi-real model testing, and finally the actual model saves the cost of the actual experiment which is very expensive. Thanks to the signals from the sensors, the researchers can input simulation models built on computers. In the scope of the article, the author researches the EPS electric steering system of the TOYOTA VIOS car.

Keywords: TOYOTA VIOS · Electric power steering · Electric power steering · EPS · Steering system

1 Introduction

Nowadays, the field of science and technology develops rapidly and dominates almost all aspects of social life. Many technology products came out with small designs, integrating many functions such as microprocessors, microcontrollers, which have been used in almost all industrial control systems and civil equipment with many advantages to gradually develop to support people.

Due to higher and higher speed requirements, better quality, and reduced vehicle energy consumption requirements. In order to meet these requirements, research, and development in the trend of improving the electronic control system is aimed at further enhancing its functions and properties. A special feature that includes two propositions is to introduce mathematical logic and an intensive steering system to suit the driving environment by changing the power assist to match traffic or road surface conditions to create a sense of sensitivity when driving. The most important issue is the immediate response of the power steering, giving the driver a feeling of attention to the changes caused by the jet. As such, the system provides the driver with information to be aware of in the vehicle's operating conditions, for example, speed and acceleration variation, steering jet, not only improving the relationship between driver and vehicle that can match the driver's feel with the steering system, but the function automatically compensates when the vehicle has uneven changes caused by the disturbance can also be resolved.

B. T. Long et al. (Eds.): MMMS 2020, LNME, pp. 547–557, 2021.
https://doi.org/10.1007/978-3-030-69610-8_75

Electrically Power Steering (EPS) is a complete electrical system that significantly reduces steering system resistance by delivering electric current directly from the electric motor to the steering system. This device includes a vehicle speed sensor, a steering sensor (torque, angular velocity), the ECU electronic control, and a motor. The output signal from each sensor is sent to the ECU which calculates the drive control mode to control the operation of the power motor.

The hydraulic power steering uses the engine power to generate hydraulic pressure and torque, thus increasing the engine load, leading to a waste of fuel, and the electric power system uses an electric motor. Needs engine power and makes for a better fuel economy (Fig. 1).

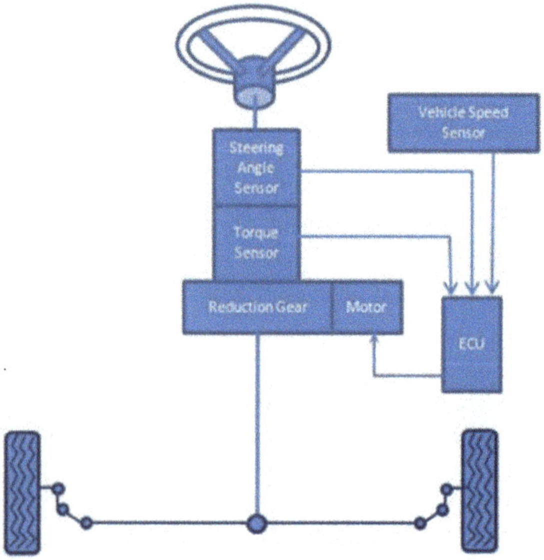

Fig. 1. General structure diagram of the electric power steering system

The purpose of building a model to research and create an electronic controller to control the EPS steering system. The team used STM32 and Arduino communication cards to communicate between the Matlab software and the sensor system set on a semi-experimental model. All are built on the HILS (Hardware-in-the-loop simulation) platform [13]. As a result, the team has built a relationship between the load, the speed, and the rotation of the steering wheel. Semi-empirical model based on electric power steering (EPS) on TOYOTA models (Figs. 2 and 3).

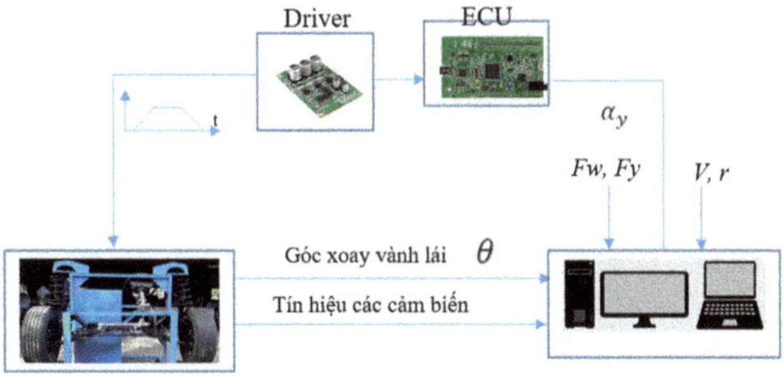

Fig. 2. HILS model

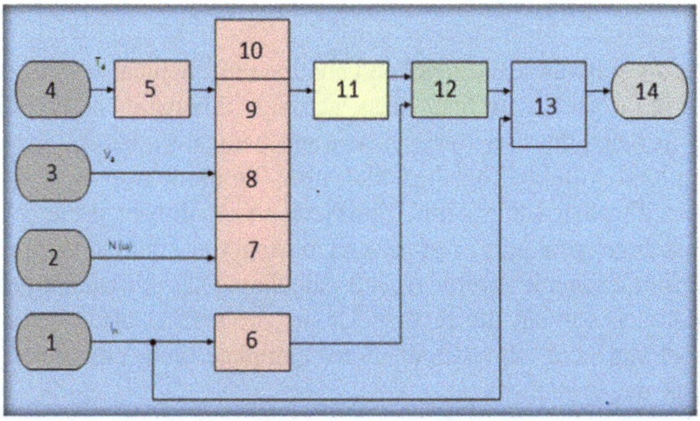

Fig. 3. Block diagram of electric power steering 1- Motor grade current; 2- Motor speed; 3- Motor speed; 4- Driving torque; 6- Motor maximum current control; 7- Control of vibration compensation; 8- Recovery control; 9- Compensation control; 10- Main monitor; 11- Destination line; 12- Limiting the maximum voltage supply current to the motor; 13- Motor supply current control; 14- The current supplied to the motor

2 Building 3D Models and Connected Hardware in Matlab

2.1 Simulation of Steering System on Matlab with SimMechanics

SimMechanics software is a block diagram modeling environment for engineering and simulation design of solid systems and their operations, using Newtonian kinematic standards for force and torque (Fig. 4).

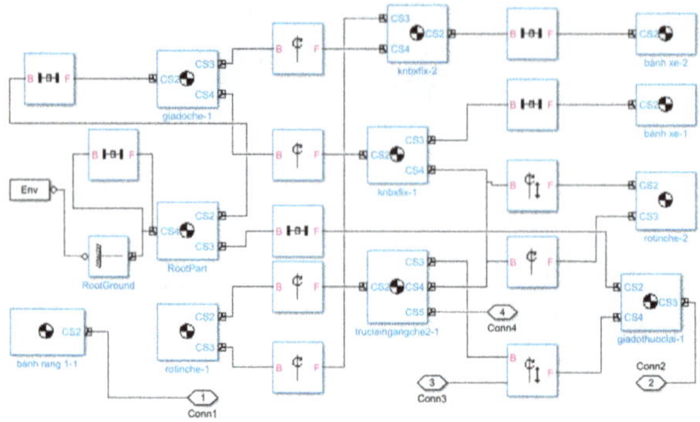

Fig. 4. Dynamic simulation block in Matlab Simulink

SimMechanics software can model and simulate mechanical systems with a system of tools to identify objects with their mass, activity, kinematic and coordinate systems characteristics. It helps the user to describe a mechanical system by means of a block diagram system that is linked together, like other Simulink models, and also designed in conjunction with other subsystems. SimMechanics software presents and performs vividly 3-D machine geometry, before and during the simulation process. Besides simulating by SimMechanics software and calculating the kinematic problem for the model, the approach solving the reverse kinetic problem is also a problem to do to make the modeling and simulation more complete and complete, meeting the requirements set out in reality.

2.2 Controller Design Reads Sensor Data

In-vehicle dynamics, a classic model shown in Fig. 5 is widely used for analysis, control, and controller stability testing [5–12]. This model is the same action from the method is based on the author of after:

- The force exerted on the tire operates in a linear region;
- Cars move on flat/flat surfaces;
- The left and right wheels on the front and rear axles are combined into one wheel at the centerline of the vehicle;
- Constant wheel speed, vertical acceleration: ($ax = 0$);
- Steering angle and sideslip angle are considered very small (≈ 0);
- All wheels are not braked;
- Vehicle center of gravity does not change when vehicle volume changes;
- The two front wheels have the same steering angle;
- The desired lateral slip is assumed to be 0 as the initial state.

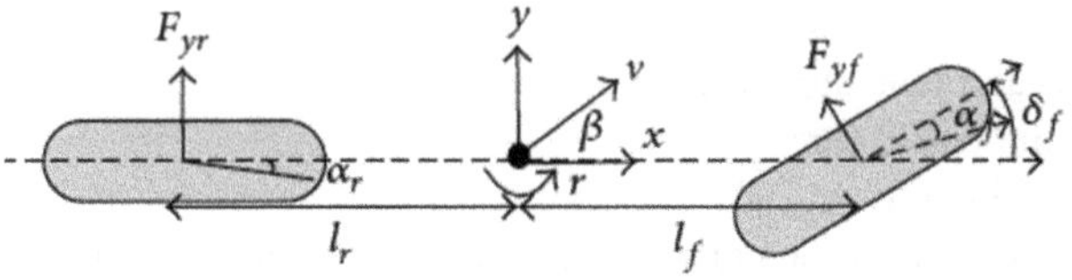

Fig. 5. Single-track 2 DOF model [1]

This is the simplest form of flat motion, this model has 2 DOF that move horizontally and rotate the body. The model is described in the equations below:

Horizontal movement:

$$mv\left(\dot{\beta}+r\right)=\left(F_{yf}+F_{yr}\right)-r$$

Body rotation:

$$I_z\dot{r}=l_fF_{yf}-l_rF_{yr}$$

In this model, the horizontal force of the front and rear tires F_{yf} and F_{yr} represent linear properties respectively and are seen as a linear function of the front and rear stiffness C_f, C_r, as follows:

$$F_{yf}=C_f\alpha_f;\ F_{yr}=C_r\alpha_r.$$

α_f, α_r: The sliding angle of the front and rear tires for linear tire force is shown in the following formulas:

$$a_f=\delta_f-\beta-\frac{l_fr}{v};a_r=-\beta+\frac{l_rr}{v}$$

After arranging and simplifying [2–5], the differential equations of lateral slip and hull angular velocity can be simplified as a linear state-space model as follows:

$$\dot{x}=Ax+Bx$$

$$\begin{bmatrix}\dot{\beta}\\ \dot{r}\end{bmatrix}=\begin{bmatrix}a_{11} & a_{12}\\ a_{21} & a_{22}\end{bmatrix}\begin{bmatrix}\beta\\ r\end{bmatrix}+\begin{bmatrix}b_1\\ b_2\end{bmatrix}u$$

$$\begin{bmatrix}\dot{\beta}\\ \dot{r}\end{bmatrix}=\begin{bmatrix}\frac{-C_f-C_r}{mv} & -1+\frac{C_rl_r-C_fl_f}{mv^2}\\ \frac{C_rl_r-C_fl_f}{I_z} & \frac{-C_fl_f^2-C_rl_r^2}{I_zv}\end{bmatrix}\begin{bmatrix}\beta\\ r\end{bmatrix}+\begin{bmatrix}\frac{C_f}{mv}\\ \frac{C_fl_f}{I_z}\end{bmatrix}\delta_f$$

- $\beta_d = 0$, r are output status variables;
- C_f, C_r: front and rear tire stiffness;
- m: mass of the entire car;
- I_z: moment of inertia;
- l_f, l_r: the distance from the car's center of gravity to the front and rear axles;
- v: car speed;
- δ_f: front wheel steering angle.

The driving angle of the front wheel is the input signal of the model. At this time, the vehicle speed set on the model is constant which means neither accelerating nor depressing the brake. Therefore, only pay attention to horizontal movement and body rotation angle.

In addition, the 2 DOF linear model is also used frequently as the ideal reference model to bring other control models closer to equivalent.

The body rotation angle r and the lateral slip angle β are important parameters in the vehicle's stability control system. As indicated above the diagram on the body rotation angle stability control targets can be classified into three categories:

1. Body rotation speed control;
2. Horizontal slip control;
3. Integration of body rotation speed control and lateral slip control (Figs, 6, 7, 8, 9 and 10).

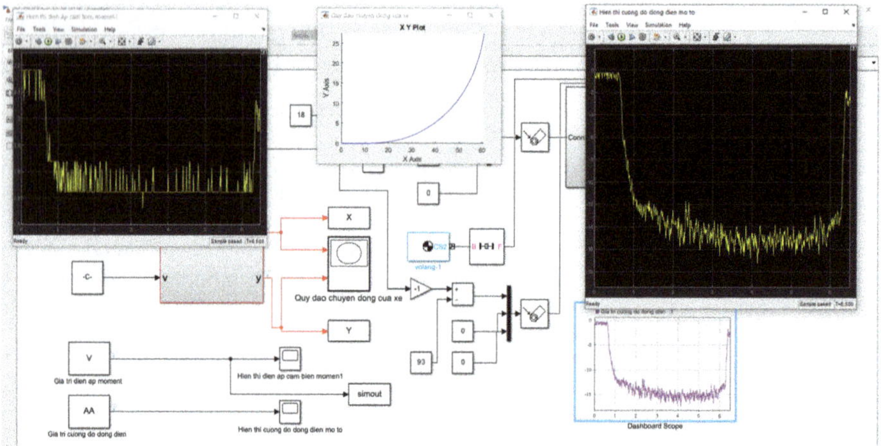

Fig. 6. Interface on the computer

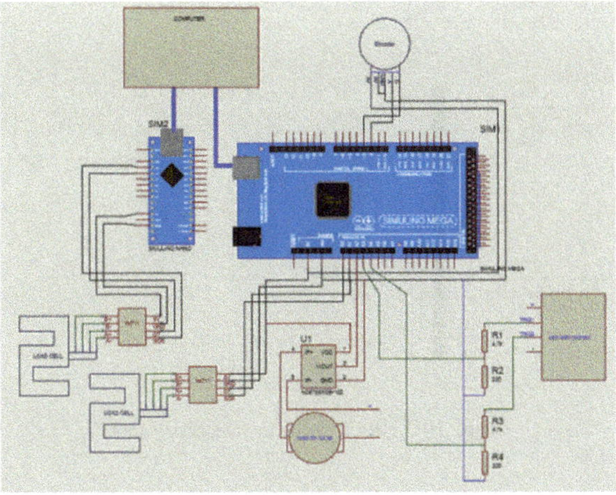

Fig. 7. Sensor reader circuit diagram

Fig. 8. Install the Encoder on the steering shaft

Fig. 9. Install the loadcell under the wheel

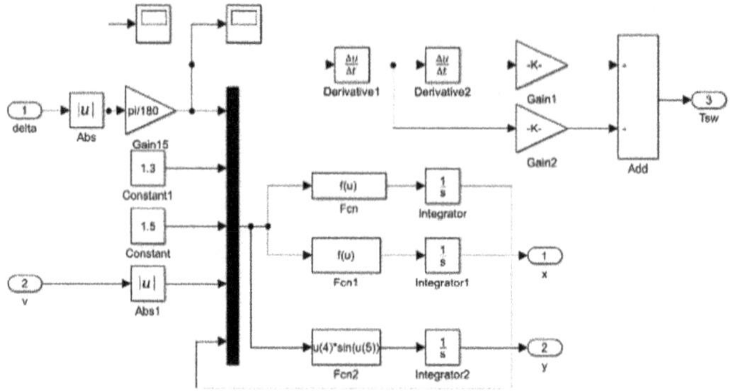

Fig. 10. Block motion trajectory car

3 Data Collection of EPS Electric Power Steering with the "J-Turn" Test

"J-turn" is an experiment used to determine the state of rotation of the car by changing the angle of rotation of the steering wheel [14]. The vehicle's speed before stepping is regulated v = 40 [km/h]. After releasing the accelerator pedal, start steering at a speed of 0.4 [rad/s] (23 [°/s]) until the value $\delta = 90$ [deg]. Other input parameters such as road surface ripple, horizontal air force, and zero braking force, cases set at 2800 [N] (Figs. 11, 12, 13 and 14).

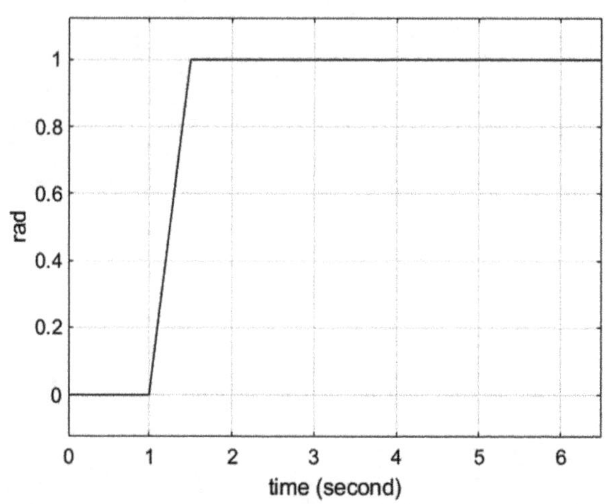

Fig. 11. The theoretical steering wheel rotation angle graph

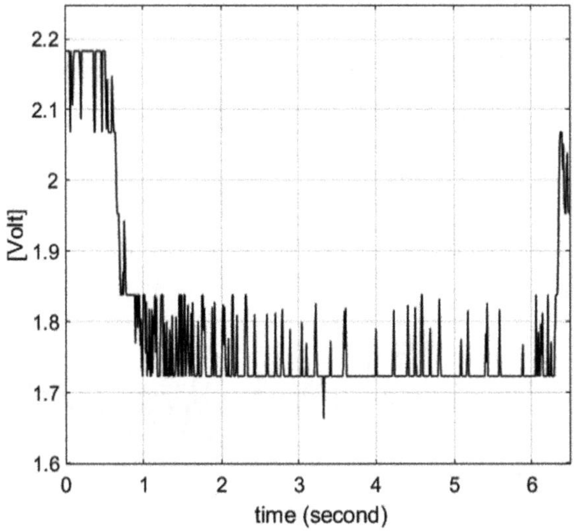

Fig. 12. Torque sensor voltage graph TRQ 1 reality.

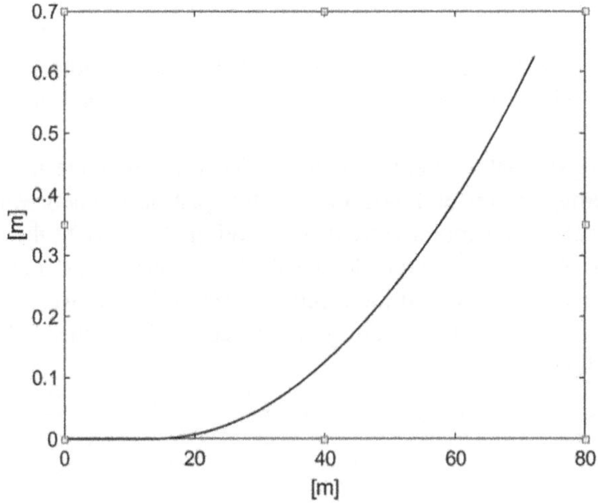

Fig. 13. Graph of vehicle trajectory

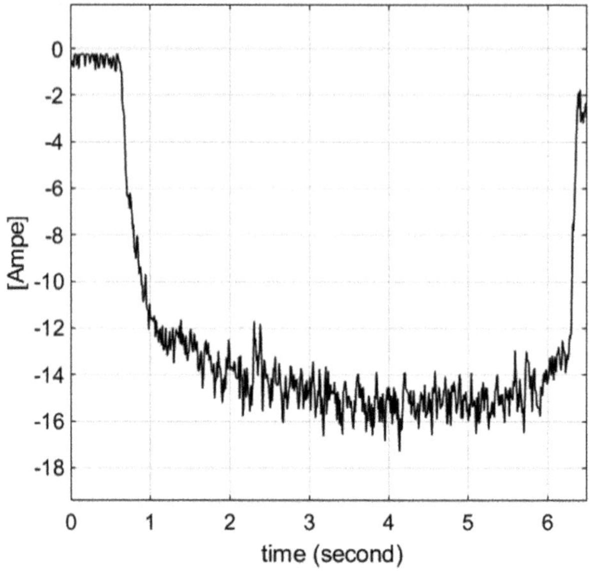

Fig. 14. Graph comparing the electric current of the case of driving left to 90°

The graph shows the electric current of the EPS electric motor (TOYOTA VIOS) with cases set at 2800 [N]. From the above graph at 4th second, the amperage value reaches 16 [A].

The real-time simulation model of the TOYOTA VIOS electric power steering system on the semi-experimental model has built a relationship between the load acting on the guide wheel and the motor current. The motion trajectory of the experiment uses a 1-trace model. With the different load values, the author has built a map of the relationship between the load and the electric current of the power motor.

This is also one of the bases for designing the controller of the EPS steering system on TOYOTA VIOS.

References

1. Trai, N.K.: The Control and Trajectory of Cars. Publisher of Transportation, Vietnam (1997)
2. Du, H., Zhang, N., Naghdy, F.: Velocity-dependent robust control for improving vehicle lateral dynamics. Transp. Res. Part C: Emerg. Technol. **19**(3), 454–468 (2011)
3. Boada, B.L., Boada, M.J.L., Díaz, V.: Fuzzy-logic applied to yaw moment control for vehicle stability. Veh. Syst. Dyn. **43**(10), 753–770 (2005)
4. Yang, X., Wang, Z., Peng, W.: Coordinated control of AFS and DYC for vehicle handling and stability based on optimal guaranteed cost theory. Veh. Syst. Dyn. **47**(1), 57–79 (2009)
5. Ding, N., Taheri, S.: An adaptive integrated algorithm for active front steering and direct yaw moment control based on direct Lyapunov method. Veh. Syst. Dyn. **48**(10), 1193–1213 (2010)

6. Mammar, S., Koenig, D.: Vehicle handling improvement by active steering. Veh. Syst. Dyn. **38**(3), 211–242 (2002)
7. Esmailzadeh, E., Goodarzi, A., Vossoughi, G.R.: Optimal yaw moment control law for improved vehicle handling. Mechatronics **13**(7), 659–675 (2003)
8. Raksincharoensak, P., Mizushima, T., Nagai, M.: Direct yaw moment control system based on driver behavior recognition. Veh. Syst. Dyn. **46**(sup1), 911–921 (2008)
9. Canale, M., Fagiano, L., Milanese, M., Borodani, P.: Robust vehicle yaw control using an active differential and IMC techniques. Control Eng. Pract. **15**(8), 923–941 (2007)
10. Canale, M., Fagiano, L., Ferrara, A., Vecchio, C.: Vehicle yaw control via second-order sliding-mode technique. IEEE Trans. Ind. Electron. **55**(11), 3908–3916 (2008)
11. Falcone, P., Borrelli, F., Asgari, J., Tseng, H.E., Hrovat, D.: Predictive active steering control for autonomous vehicle systems. IEEE Trans. Control Syst. Technol. **15**(3), 566–580 (2007)
12. Falcone, P., Borrelli, F., Tseng, H.E., Asgari, J., Hrovat, D.: Linear time-varying model predictive control and its application to active steering systems: stability analysis and experimental validation. Int. J. Robust Nonlinear Control **18**(8), 862–875 (2008)
13. Galko, C., Rossi, R., Savatier, X.: Vehicle-hardware-in-the-loop system for ADAS prototyping and validation. In: IEEE International Conference on Embedded Computer Systems: Architectures, Modeling, and Simulation (SAMOS XIV), SAMOS Island, Greece (2014)
14. Forkenbrock, G.J., Garrott, W.R., Heitz, M., O'Harra, B.C.: An experimental examination of J-Turn and fishhook maneuvers that may induce on-road, untripped, Light Vehicle Rollover, SAE (2003)

Study and Evaluation the Influence of Rotary Tool on Surface Roughness of SKD61 Steel in EDM Process

Van Tao Le(✉) and Van Thao Le

Advanced Technology Center, Le Quy Don Technical University, Hanoi, Vietnam
taoitd@yahoo.com

Abstract. The research on the processing of Electrical Discharge Machining has much attention. The scientists have found solutions to improve machining efficiency, one of which is the rotary tool method. In this study has investigated the influence of tool rotation on roughness of surfaces in semi-finishing and final-finishing, the role of rotary tool is explained to be clear. In addition, the surface roughness by the rotary tool method is evaluated and compared with the surface roughness by conventional EDM method. The results show that the roughness of surfaces by the rotary tool method is improved as compared to the surface roughness of conventional EDM method, at S_{rt} = 500 Rpm the surface roughness is improved the best, with an improvement of 33.25% at T_{on} = 50 μs; I_p = 2A.

Keywords: EDM · Surface modification · Surface roughness · Rotary tool · Mill EDM

1 Introduction

Electrical discharge technology in mechanical manufacture has been finite due to low production capacity and poor surface quality. The EDM process was revealed about 1770 by J. Priestly when he studied the domination of the electric discharges. Up until the year 1930, endeavors for machined metals was processed by this process. This process was called electrical discharge machining (EDM) [1]. Lazarenko brothers did valuable work on EDM at the university of Moscow in the year 1943. In recent years, many studies have focused on enhancing surface quality. In the methods, which enhance the surface quality, that is the rotary tool method.

Soni et al. [2] used the rotary tool to study the formation of surface properties of recast layer. The authors have reported the properties of the recast layer by the rotary tool method have changed as compared to that of the conventional EDM. Dwivedi et al. [3–5] used the tool rotation to process AISI D3 steel. The authors investigated the influence of the rotary tool on the surface after machining. The results showed that the surface roughness, micro- cracks were improved, the thickness of the recast layer was thinner than the conventional EDM. Thus, as mentioned above, the rotary tool improves the electrical discharge process, which changes the surface properties of the

B. T. Long et al. (Eds.): MMMS 2020, LNME, pp. 558–563, 2021.
https://doi.org/10.1007/978-3-030-69610-8_76

workpiece after machining. To better understand the role of rotary tool. Firstly, let's introduce about the mechanism of the electrical discharge process and the role of debris in the electrical discharge process.

The Mechanism of the Electrical Discharge: To appreciate characteristics of the EDM, the mechanism of the material removal in typical EDM is represented. Typical EDM occurs incessant charging and discharging between the tool electrode and workpiece electrode while sunk in the dielectric medium. Electrical discharge process embraces phases, as follows: (1) preparation, (2) discharge, and (3) interval phases. They are described in [6].

The Role of the Debris in Electrical Discharge Process: The EDM process transforms the electrical energy into thermal energy through a discharge channel. High current density initiates a high temperature in a plasma channel. Melting and evaporation of material occur in a small area of the discharge plasma. Moreover, the pulsating process outcomes in a local mechanical collision. The collision separates melted materials and all other corrode products from the crater to dielectric liquid. Removal of the material would not be possible. If this mechanical repulsion has not existed. Electrical discharges would retrograde into arc, i.e. spark concentration if the two mechanisms have not existed. The pulse generator will produce the non-stop and excruciating arc, which will destroy both the tool and the workpiece if the control system does not act to stop it immediately. Two electrodes frequently short-circuited, it is the other end result of arcing. The arc discharge makes the process unstable. This has heated the pulsating electrode in a relatively large area and replaced the material removal process. The arc discharge does not stop until the control system provides an external force to push the concentrated discharge away from the arc discharge point. Having too much debris in the spark gap is often thought to cause arcing. The debris consisting of small carbon and metal particles should be continuously sucked up through a process of flushing eielectric medium. The debris in the spark discharge gap is always dangerous to the machining process. The debris of the spark discharge is described in the literature [7].

In this study, the authors investigated and evaluated the tool rotation speed affecting the surface roughness. In addition, the surface roughness comparison between the rotary tool method in EDM and conventional EDM, showed the role of the rotary tool in EDM to improve surface roughness.

2 Materials and Methods

Steel SKD61 was used in the experiment (supplied by Daido corporation of Japan). The SKD61 samples are carried out by an electrical discharge machine (CNC-460 EDM) from the Aristech to obtain dimensions asin Fig. 1 and Table 1. The polarity of the copper tool was negative. The roughness of surfaces was measured with a TR200, supplied by TINE Group INC with the resolution of 0.001 μm.

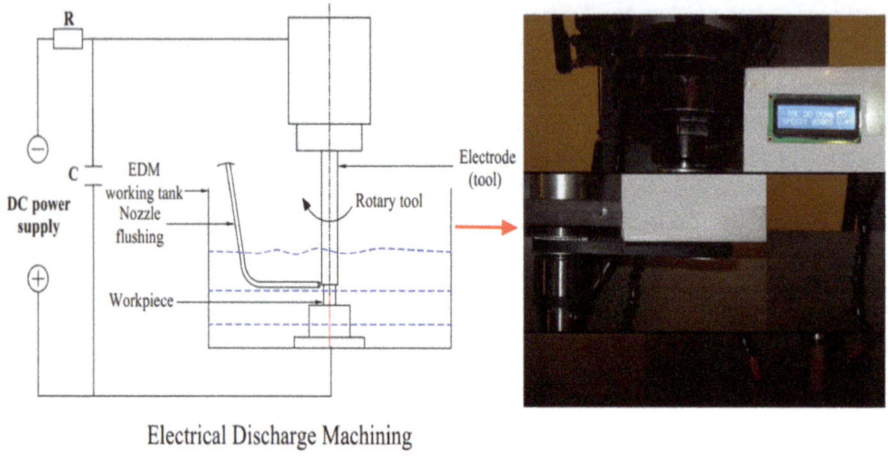

Fig. 1. Setup experimental diagram

Table 1. Experimental Conditions

Deposition Condition	Detail
The current (I_p-A)	1A; 2A
The pulse on (T_{on} - μs)	16 μs; 32 μs; 50 μs
The pulse off (T_{off} - μs)	50 μs
The rotation speed of the tool (S_{rt}- Rpm- Revolutions per minute)	100 Rpm; 300 Rpm; 500 Rpm
The dielectric fluid	Shell EDM Fluid 2
Polarity of Cu-electrode	Negative (−)
Current voltage (V)	80–120V
Dimension of workpiece	D = 19 mm; L = 50 mm
Dimension of workpiece after machining	D = 19 mm; L = 49.7 mm

3 Results and Discussion

The surface roughness is observed in Fig. 2 and 3. It was found that the rotary tool has improved the surface roughness for all peak current modes and pulse on time modes.

According to Fig. 2: With S_{rt} = 100 Rpm at all pulse on time, the improvement of roughness surface is not significant, this is explained by the discharge process combined with pulse on time (T_{on}), the stationary tool in EDM method has done a good job of pushing derbis out of the discharge area the next time, which makes the roughness surface of the rotary tool in EDM changes little as compared to that of the stationary tool in EDM. With S_{rt} = 300 Rpm at all pulse on time, the improvement of roughness surface is significant. With S_{rt} = 500 Rpm at all pulse on time, the improvement of roughness surface is very significant. Thus, through the speed range of the tool is applied to the experiment, it is found that when the tool rotary at a high speed, the surface roughness is significantly improved. Observing Fig. 2, it is found that at the short pulse on time (T_{on} = 16 μs; 32 μs), the improvement in roughness surface between rotary tool and stationary tool is negligible. But at the long pulse on time (T_{on} = 50 μs), the surface roughness has significantly improved. The best roughness surface is improved by 27.22% at T_{on} = 50 μs; I_p = 1A; S_{rt} = 500 Rpm.

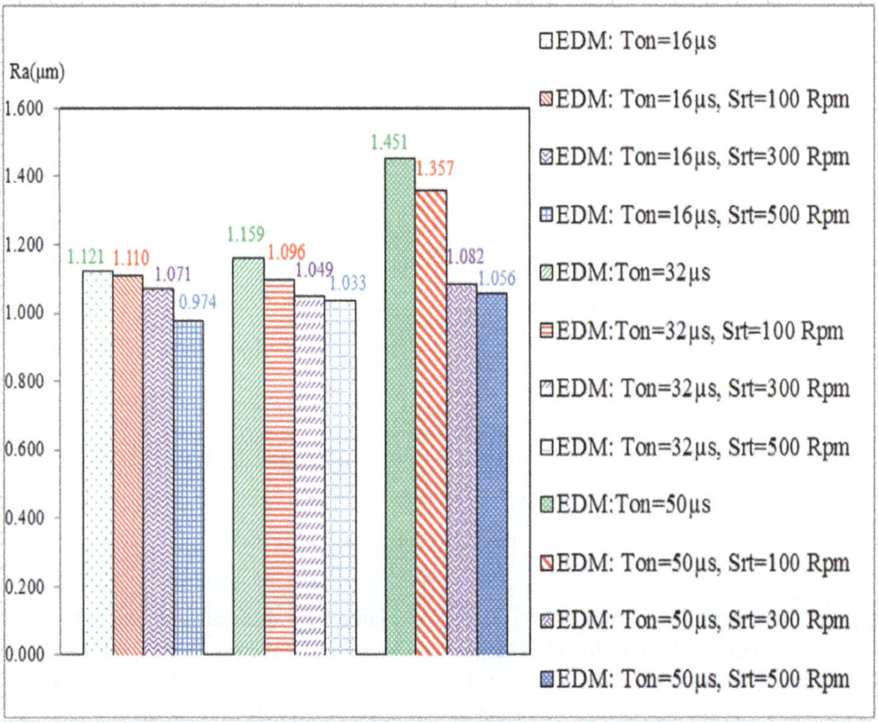

Fig. 2. The surface roughness (R_a) at I_p=1A

According to Fig. 3: With S_{rt} = 100 Rpm at all pulse on time, the surface roughness the rotary tool in EDM is reduced as compared to that of the stationary tool in EDM. The best roughness surface is improved by 14.9% at T_{on} = 32 μs; I_p = 2A. Similarly with S_{rt} = 300 Rpm and S_{rt} = 500 Rpm at all pulse on time, the surface roughness of the rotary tool in EDM is reduced as compared to that of the stationary tool in EDM. With S_{rt} = 300 Rpm, the best roughness surface is improved by 24.58% at T_{on} = 32 μs; I_p = 2A. With S_{rt} = 500 Rpm, the best roughness surface is improved by 33.25% at T_{on} = 50 μs; I_p = 2A. Observing Fig. 3 to see that, there is an improvement in surface roughness at all pulse on time and at the speed range of the tool as compared to that of the stationary tool in EDM.

Through analysis and evaluation, it was found that at the long pulse on time and the large peak current, the effect of the rotary tool on the improvement of surface roughness is most clearly shown. The rotation speed of the tool also strongly affects the

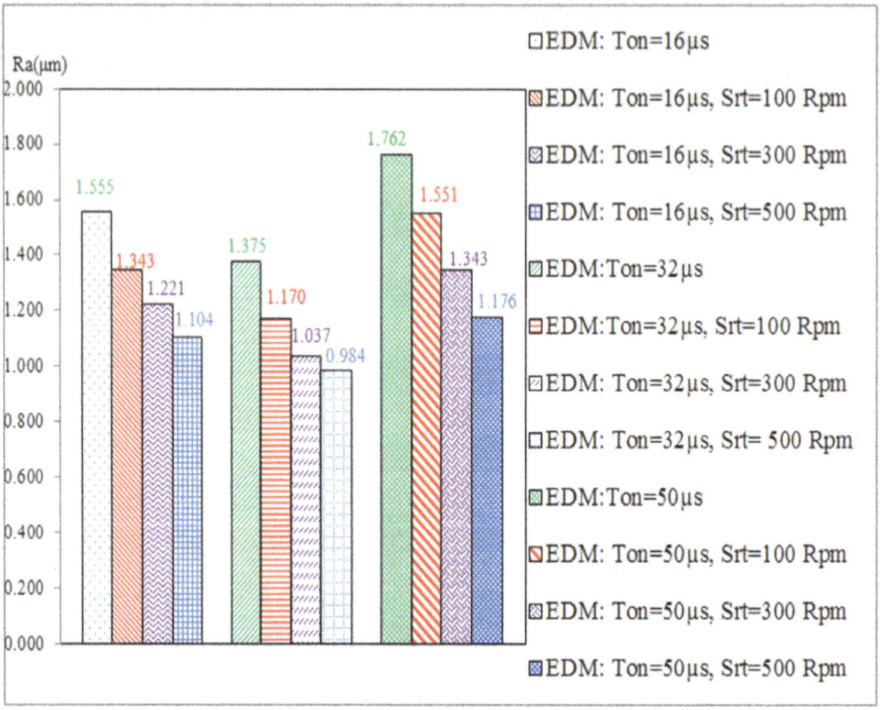

Fig. 3. The surface roughness (R_a) at I_p = 2A

improvement of surface roughness. This is explained by the fact that the rotary tool makes the process of removing the debris due to material removal by the EDM process, so that the discharge process in the next time is more stable. With the rotary tool, the process spark discharges alter the electrical discharges on the following aspects: The involvement of the rotary tool causes debris to be transported away from the next sparking area. The absence or little debris involved in the next launch process leads to a stable discharge [8–10] and does not cause arcing. Similarly, when the tool's rotation speed is large, the surface roughness is improved better than the surface roughness by conventional EDM method with the same electrical technology mode. This is because the large rotation speed creates a large vortex force, which makes the debris escape better than the mode with a small electrical rotation speed. Therefore, the process of the electrical discharge at the next time becomes stable, this makes the surface roughness to be improved better.

4 Conclusions

In this study, the authors proposed the rotatry tool method to improve the surface roughness, the investigation results are summarized as follows: In the investigation rotation speeds, all modes have the effect of improving surface roughness. At S_{rt} = 500

Rpm, the surface roughness is improved the best, with an improvement of 33.25% at T_{on} = 50 μs; I_p = 2A. Through experiments, we can see the impact of the rotation speed of tool in changing the spark discharge process. For small tool rotation speed, the surface roughness is less affected, but the high speed of tool affects the surface roughness more. The role of rotatry tool has been clarified in improving the quality of the electrical discharge.

References

1. Ho, K., Newman, S.: State of the art electrical discharge machining (EDM). Int. J. Mach. Tools Manuf. **43**, 1287–1300 (2003)
2. Soni, J.S., Chakraverti, G.: Experimental investigation on migration of material during EDM of die steel (T215 Cr12). J. Mater. Process. Technol. **56**, 439–451 (1996)
3. Dwivedi, A.P., Choudhury, S.: K, Improvement in the surface integrity of AISI D3 tool steel using rotary tool electric discharge machining process. Proc. Technol. **23**, 280–287 (2016)
4. Dwivedi, A.P., Choudhury, S.K.: Effect of tool rotation on MRR, TWR, and surface integrity of AISI-D3 steel using the rotary EDM Process. Mater. Manuf. Process **31**, 1844–1852 (2016)
5. Dwivedi, A.P., Choudhury, S.K.: Estimation of recast layer thickness in rotary tool EDM process for machining AISI D3 tool steel. Mater. Today Proc. **4**, 10816–10822 (2017)
6. Schumacher, B.M.: After 60 years of EDM the discharge process remains still disputed. J. Mater. Process. Technol. **149**, 376–381 (2004)
7. Luo, Y.F.: The dependence of interspace discharge transitivity upon the gap debris in precision electrodischarge machining. J. Mater. Process. Technol. **68**, 121–131 (1997)
8. Schumacher, B.M.: About the role of debris in the gap during electrical discharge machining. CIRP Ann. **39**, 197–199 (1990)
9. Masuzawa, T., Cui, X., Taniguchi, N.: Improved jet flushing for EDM. CIRP Ann. **41**, 239–242 (1992)
10. Yanatori, K., Kunieda, M.: Study on debris movement in EDM gap. J. Japan Soc. Electr. Mach. Eng **29**, 19–27 (1995)

Natural Radioactivity and Environmental Impact Assessment at Dong Pao Rare Earth Mine, Lai Chau, Vietnam

Nguyen Van Dung(✉) and Vu Thi Lan Anh

Faculty of Environmental, Hanoi University of Mining and Geology, Hanoi 11910, Vietnam
nguyenvandung@humg.edu.vn

Abstract. Dong Pao rare earth mine containing radioactive uranium and thorium in Ban Giang and Ban Hon communes, Tam Duong district, Lai Chau province, with an area of about 12.4 km^2 and total reserves of 11.8 million tons concentrate of rare earth ores, is a mine with large reserves of rare earth metal resources of Vietnam being explored to be put into exploitation in the near future. The activity of natural radionuclides and radiation dose are important parameters in assessing the effect of radioactivity on the environment when the mines go into operation. Investigation soil, water, air, plants and annual radiation dose levels in the region in order to contribute to the management and supervision of the radioactive environment in rare earth mining. The results showed that the radioactive activity of soil and plants increased higher than the permitted standard; The annual radiation dose value results are 15.6 $mSvyear^{-1}$, 6.5 times higher than the world average (2.4 $mSvyear^{-1}$).

The results of the health survey of people living in and outside the mines showed that there was a difference between the abnormal rate of blood counts between inside and outside the radioactive area. The rate of erythrocyte abnormalities (related to anemia) of people outside radioactive anomalies is much lower than that of people living in radioactive anomalies (HST: low rate of 14.3% compared with 36.4 ÷ 38.4%; MCH: 68.5% lower than 24% compared to 93.4 ÷ 96.3%; MCHC: 5.4% lower than 88.5% compared to 93.6 ÷ 97.3%; RED rate is 7.4% lower than 56.4% compared to 68 ÷ 72%). These are evidences on the impact of radiation on the health of workers living and working in mines producing high levels of radioactive material.

Keywords: Radioactive nuclides · Health · Environment · Rare earth mine · Dong pao · Annual effective dose

1 Introduction

The concentration of radioactive substances in the environment varies naturally from region to region. However, people can still live normally in places with high radiation backgrounds (100 times higher than the global average) such as is regions in the with high levels of radioactive uranium, thorium [1–4].

B. T. Long et al. (Eds.): MMMS 2020, LNME, pp. 564–575, 2021.
https://doi.org/10.1007/978-3-030-69610-8_77

According to the assessment of UNSCEAR, in total global radiation is 2.4 $mSvyear^{-1}$, in which the component of external radiation dose from gamma radiation accounts for 20%; the internal projection component due to breathing radon in the air accounts for 59%; composition by cosmic radiation accounts for 15%; the remaining ingredients make up 6% [5, 27].

Vietnam is the country with the third largest reserve of rare earths in the world with about 20 million tons, concentrating in the Northwest of Vietnam, including mines according to analysis results, mineral composition in rare earth. Contains radioactive elements. In mines in Northwest Vietnam, gamma dose exceeding the safe limit is 0.6 μSvh^{-1} [8]. Research results on the concentration of U, Th and radon gas concentration (^{222}Rn) at the radiation level of Dong Pao rare earth mine varied from 0.2 to 3.0 μSvh^{-1}; The concentration of Rn and Tn is 10 times higher than that of the standard of the IAEA [9, 10].

2 Materials and Methods

2.1 The Study Areas

Dong Pao rare earth mine is limited by coordinates 103°33′00″E, 22°18′03″N, in Ban Giang and Ban Hon communes, mine belongs to the fluorcarbonite group with mineral composition mainly bastnaesite, parisite, baryt and fluorite. According to the proportion of minerals divided into ores: rare earth $TR_2O_3 > 5\%$; rare earth - baryt - fluorite with a content of TR_2O_3 0.5 ÷ 30% and rare earth containing fluorite with content of TR_2O_3 1 ÷ 5%. Reliable resource (222) TR_2O_3 = 645,000 tons, estimated resource (333) = 7 million tons; CaF_2 = 9.4 million tons; $BaSO_4$ = 288 million tons (Fig. 1). The content of radioactive minerals ThO_2 ranges from 0.002 to 0.054%, the average is 0.013%; U_3O_8 from 0.002÷0.014%, average 0.007% [11–16].

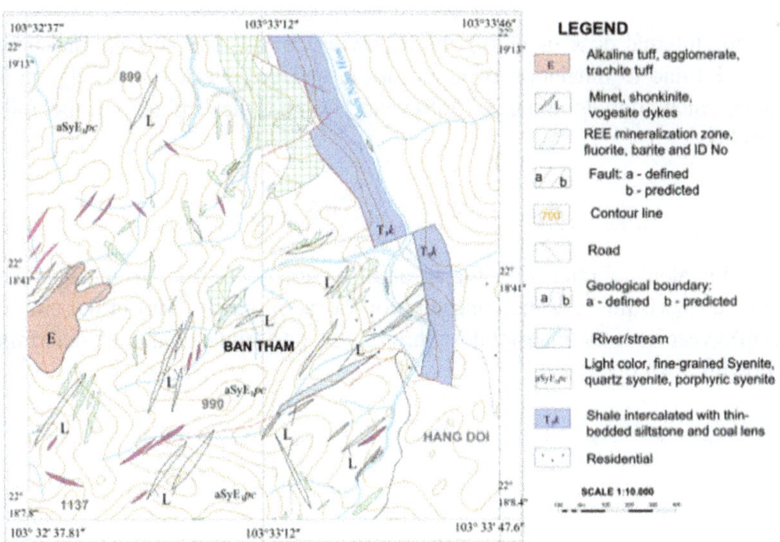

Fig. 1. Geological and mineral characteristics of Dong Pao rare earth mine

Activities of surveying and investigating the natural radioactive environment in the study area include the following contents: measuring the dose rate of external irradiation, the concentration of radioactive gases, determining the activity of natural radioactive substances in the sample. soil, water, plants [8, 14].

Assessing the impact of radiation on people's health in the survey area, the author analyzed hair samples and blood samples from residents living in the mine (*with high irradiation dose*) and residents living outside the mines to assess radiation-related illness symptoms [17].

2.2 Research Methods

Radiation exposure was determined by the effective annual dose H_{eff} ($mSvyear^{-1}$). Radioactive deposits is necessary radiation on the concentrations substances in soil, water and plants. The survey, investigation and assessment of radioactive environmental impacts in mineral deposits are identified as follows [18–21, 29]:

2.2.1 Effective Annual Dose

Which the is determined by formula:

$$H = H_n + H_t, \quad (1)$$

In which:

- Radiation caused by gamma radiation, measured at a position 1 m from the ground, determined by the following formula:

$$H_n = 8,760 \times H_{SL}, \ \left(mSvyear^{-1}\right), \quad (2)$$

where H_{SL} is the dose rate of the external radiation in the surrounding environment (μSvh^{-1});

- H_t is the internal dose including the following components: the internal inhalation dose (H_p) due to inhalation of radon, toron and the internal dose (H_d) (food and water), calculation is equal to ($mSvyear^{-1}$) and is determined by the following formulas:

$$H_p = 0.047 \times C_{Rn}, \ \left(mSvyear^{-1}\right) \quad (3)$$

The equivalent of inhalation for an adult breathes 7,280 cubic meters of air per year; where C_{Rn}(Bqm^{-3}) is the concentration of radon gas (^{222}Rn) in the air.

H_d($mSvyear^{-1}$) is the internal dose through the gastrointestinal tract, determined by the following formula:

$$H_d = \left(6.2 \times 10^{-6} A_K + 2.8 \times 10^{-4} A_{Ra} + 2.3 \times 10^{-4} A_{Th} + 4.4 \times 10^{-5} A_U\right) \times m_d, \quad (4)$$

In which: A_K, A_{Ra}, A_{Th}, A_U respectively the radioactive activity of potassium, radio, thorium, uranium in water (Bq L^{-1}) or food (Bq kg^{-1}); md is the average amount of water or food consumed per person per year (800 L of water and 650 kg of food).

2.2.2 Radiation Activity in Soil Samples

The activity of radioactive determined following formula:

$$A = A_{Ra} + 1.3A_{Th} + 0.085A_K \leq 370, \left(\mathrm{Bqkg}^{-1}\right) \quad (5)$$

where A_{Ra}, A_{Th}, A_K are the activity of the radioactive nuclei Ra, Th, K.

2.2.3 Error Assessment Method

To evaluate the a combination the following methods [18, 19]: 1) Measuring gamma radiation caused by region; 2) Measure the radioactive gas to determine the concentration of radon and toron inhalation; 3) Gamma spectroscopy to determine the activity surface rock;

The error of following:

$$\sigma = \sqrt{\frac{\sum_{i=1}^{n}(x_i - y_i)^2}{2n}} \quad (6)$$

$$\text{Relative error: } \delta = \frac{\sigma}{R} \times 100\% \text{ vớ } R = \frac{1}{2n}\sum_{i=1}^{n}(x_i + y_i) \quad (7)$$

test measurement points; x_i, y_i are main and repeat measurements

2.2.4 Survey the Current Health Status of People Living on Rare Earth Mines

Studying the effects of radiation on people living in the Dong Pao rare earth mine area and surrounding areas, the author has collected and analyzed hair samples, blood samples of people are living in the Dong Pao rare earth mine area and the surrounding area [17].

- Investigating and investigating the health situation, disease characteristics, disease symptoms related to the harmful effects of radiation such as cancer, digestion, respiratory, neurological, birth defects, teratogenicity miscarriage and some genetic diseases of people living in rare earth mines and people giving birth outside the mines;
- *Hair collection and sample:* Each hair sample took 300 mg of hair (5 cm from the root) from different positions of the tip near the scalp and put it in plastic bags with the model number clearly marked. Hair samples were analyzed to determine the content of naturally radioactive substances ^{238}U, ^{232}Th on the ICP-MS device of the Geological Division on Radioactive and Rare Earth Elements.

- *Taking and analyzing blood samples to study the effects of radioactivity on blood cells*: blood is taken from people living in rare earth mines (radioactive anomalous area) and nearby (outside anomalous areas). radioactivity), blood count 100 samples (the samples are selected for subjects according to age and indigenous peoples). The blood sample collection and analysis process is carried out by the Central Institute of Hematology and Blood Transfusion.

3 Results and Discussion

3.1 Evaluation Errors

After measuring and analyzing and processing data, the results of measurement error assessment are given in Table 1.

Table 1. Error assessment of measured results in site

Measured method	Number of moints	Relative error (%)	Absolute error		Limited (%)
			Value	Unit	
Gamma dose measurement					
At the point of 0 m	150	2.24	0.01	μSv h^{-1}	≤ 10
At the point of 1 m	150	2.43	0.02	μSv h^{-1}	≤ 10
Gas measurement					
Radon (^{222}Rn)	30	11.32	6.76	Bq m^{-3}	≤ 30
Toron (^{220}Rn)	30	7.56	18.12	Bq m^{-3}	≤ 30

Results error evaluation show that the errors are within the permissible limits. Therefore, the collected data meets the reliability.

3.2 Characteristics Radiation Rate

Survey results on radiation rate at the Dong Pao rare earth mine showed that: the gamma dose rate value at 0 m position changed from 0.13 to 2.98 μSvh^{-1}, the average of 0.77 μSvh^{-1} (Standard deviation-SD 0.48); at the 1 m position, the values varied from 0.09 to 2.72 μSvh^{-1}, the average of 0.71 μSvh^{-1}(Standard deviation-SD 0.44). Survey results show that the gamma radiation dose rate value less than 0.3 μSvh^{-1} lies in the area outside the rare earth ore body. The value of gamma radiation ranges 0.6 μSvh^{-1} at boundary area between the ore body and the vicinity. In the rare earth ore body, the gamma and the maximum value reached 2.98 μSvh^{-1} at the location of the rare earth ore body with high uranium and thorium content.

According to the standards of Vietnam and Russia [20, 21, 24], the value of gamma radiation dose rate at the Dong Pao rare earth mine exceeds the permitted standard.

3.3 Radon and Thoron Gas Concentrations

Concentration of (^{222}Rn) in the air ranged from 16.8 to 848 Bqm^{-3}, averaging 122.4 Bqm^{-3} (Standard deviation-SD 77.5); The toron gas concentrations (^{222}Rn) ranged from 11.8 to 978 Bqm^{-3}, averaging 133.3 Bqm^{-3} (Standard deviation-SD 161.6). Locations concentrations depend distribution of rare earth on with areas with a Ban Tham, Ban Hon and Ban Giang, radon concentration values are higher than the Vietnamese standard recommended indoor radon concentrations [23]. Vietnamese No. [23], building new houses is not allowed radon radioactive gas concentration exceeds 100 Bqm^{-3}.

3.4 Water

The analysis results of alpha and beta activity in water in the study area showed that: alpha activity values ranged from 0.018 to 0.089 BqL^{-1}, the average was 0.042 BqL^{-1} (Standard deviation-SD 0.016), beta activity ranged from 0.198 to 0.887 BqL^{-1}, averaging 0,501 BqL^{-1} (Standard deviation-SD 0.162). According to Vietnamese standards, the total activity of alpha and beta in water samples is within the allowable limits (alpha < 0.1 BqL^{-1}; beta < 1.0 BqL^{-1}) [25].

3.5 Food Crops

The results analyzing the concentration food samples showed the nuclei food samples grown in Dong Pao mine area (Table 2).

Table 2. Characteristics of radioactivity in food crops

Parameters	Radioactivity ($Bqkg^{-1}$)			
	^{226}Ra	^{238}U	^{232}Th	^{40}K
Min	0.03	0.18	0.24	25.23
Max	2.57	2.78	6.63	159.45
Average	0.93	0.87	1.22	74.34
Standard deviation	0.94	1.02	1.32	9.43
Coefficient of variation (%)	8.31	9.34	8.21	11.89
Total number of samples	20	20	20	20

The results of calculating the average dose in the gastrointestinal tract (H_d) was 0.67 $mSvyear^{-1}$, exceeding the allowable limit compared with the Russian standard ($H_d \leq 0.2$ $mSvyear^{-1}$) [24].

The results analyzing the soil showed in Dong Pao mines. Analysis results of soil samples showed that the radioactivity of limit > 1,000 $Bqkg^{-1}$ [20, 21]. These Dong Pao rare earth mine, where indigenous people use the land for food crops. Therefore, it can be seen that most of the soil on the surface of the Dong Pao mine.

3.6 Effective Annual Dose H_{eff} (MSv Year^{-1}) at Dong Pao Mine

Using Eqs. (1), (2), (3) and (4), the annual effective dose chart for the Dong Pao rare earth mine area was established (Fig. 2).

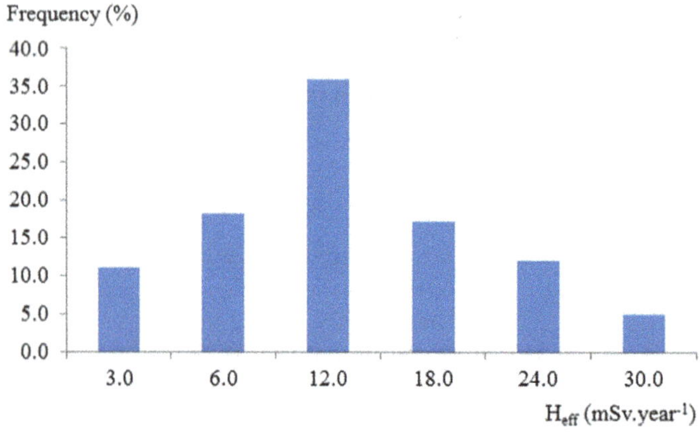

Fig. 2. Diagram of effective annual H_{eff} dose at Dong Pao rare earth mine

From Fig. 2, the annual effective dose characteristic value at Dong Pao rare earth mine is H_{eff} = 15.6 (mSvyear^{-1}), 6.5 times higher than the world average (2.4 mSvyear^{-1}) [5,32].

The on environmental protection, polluted the Dong Pao mine identified, radioactive nuclei specified are still according to the reference documents [18–22].

According to the formulas (1), (2), (3), (4) and (5) the author has calculated the total annual radiation dose for the study area. Comparing with Vietnamese [18, 19, 23, 24] and international standards [20, 21], a zoning map of radioactive pollution in the Dong Pao rare earth mine area has been developed (Fig. 3).

The irradiation dose level shown on the map of radioactive pollution status and zoning in the Dong Pao rare earth mine area on the map divides the areas with different levels of radioactive pollution.

- The area of H_{eff} > 3mSvyear^{-1} is about 20 km^2, covering the entire mine including Tham, Na Khum, Phu Nhieu and Hoa Di Ho villages.
- Three areas of H_{eff} > 10mSvyear^{-1} located completely coinciding with the area with rare earth ores in Tham village, Na Khum village, Phu Nhieu village and northeast Hoa Di Ho village, each area is about 1 ÷ 1.5 km^2. In these areas, people of Tham, Na Khum and Phu Nhieu villages live.
- The area of the control area with potential radiation dose greater than $H_{eff} \geq 6$ mSvyear^{-1} scattered red colored areas in the center of the map, coinciding with areas with rare earth ore bodies in Tham, Na Khum, Phu Nhieu and northeastern Hoa Di Ho village. According to Circular 19/2012-TT-BKHCN [25, 26] this is a control area that requires protection measures and special safety regulations to

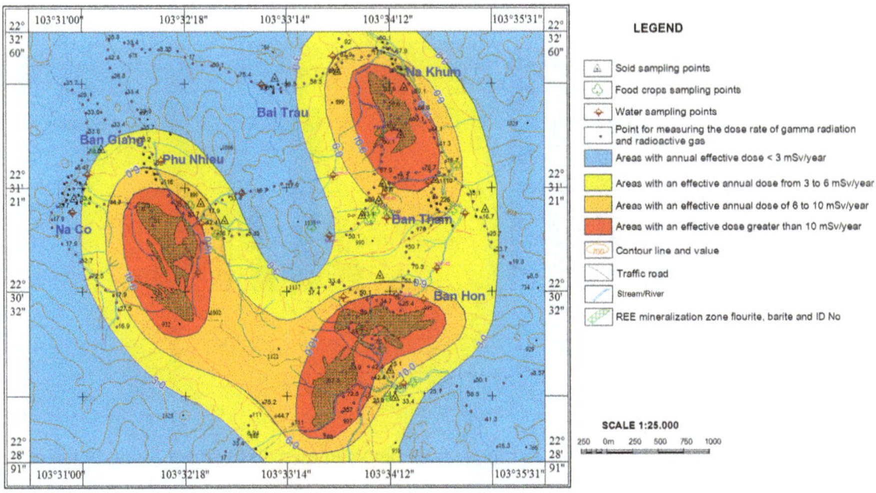

Fig. 3. Diagram of natural radioactive pollution in Dong Pao rare earth mine

control irradiation or prevent radiation contamination from spreading under normal operating conditions, to prevent or limit potential levels of irradiation.

3.7 Results of Epidemiological Survey of People Living in Dong Pao Rare Earth Mine Area

3.7.1 Content of Radioactive Material in Human Hair Samples

The results of analyzing the activity of radioactive elements in human hair samples at Dong Pao rare earth mine are shown in Table 3.

From Table 3, it is clear that in the hair samples of people living mine, average content ^{238}U reached 0.096 ppm, 1.92 times higher than the world's average ^{238}U content (0.05 ppm) [5, 17] The concentration of ^{238}U, ^{232}Th in the hair samples of people born in the rare earth mines is 5 to 10 times higher than their concentrations in the hair samples of people living outside the mines [17].

Table 3. Content of radioactive substances in human hair samples

Parameters	Content of radioactive elements and metals (ppm)				
	Cr	Co	Pb	^{232}Th	^{238}U
Min	0.183	0.025	1.057	0.075	0.062
Max	0.96	0.206	18.238	0.315	0.166
Average	0.570	0.096	7.899	0.212	0.096
Standard deviation	0.231	0.043	2.762	0.021	0.015
Coefficient of variation (%)	4.52	3.47	9.78	2.34	2.76
Number of samples	15	15	15	15	15

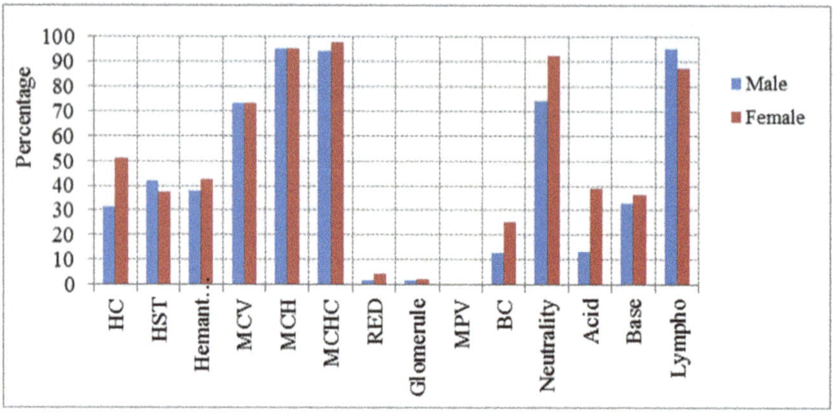

Fig. 4. Abnormal ratio of indicators in blood count between men and women living in the mines

3.7.2 Processing and Synthesizing the Results of Blood Sample Analysis in the Study Area

At the Dong Pao rare earth mine, 100 blood samples were taken and analyzed from people living inside and outside the mine. The criteria for analysis of blood cells include the number of red blood cells, hemoglobin (HST), hermatocrit, average volume of red blood cells (MPV), number of leukocytes, the neutral segment, the acidophilic segment, the basophilic segment, the monocyte, lymphocyte.

To process the results of blood sample analysis, have built graphs of the frequency of abnormal rate in the blood count of men and women living in the mines (Fig. 4); Abnormal ratio in the blood count of people (both men and women) living inside and outside the mines (Fig. 5).

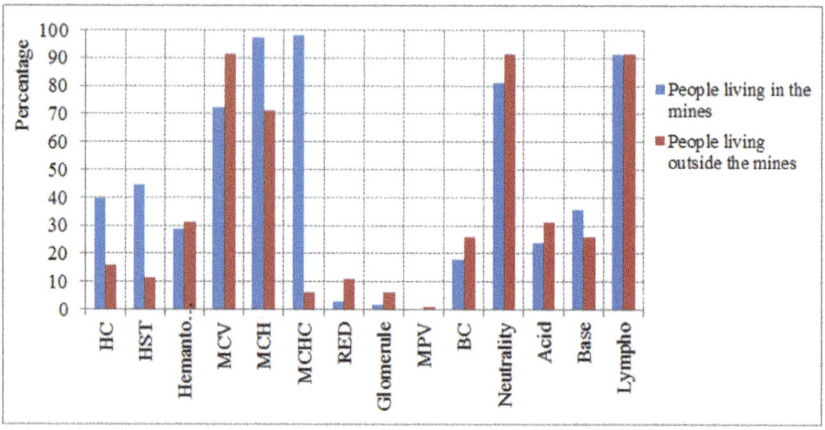

Fig. 5. Abnormal ratio of blood count of people living in the area of radioactive anomalies (*inside the mines*) and outside the area of radioactive anomalies (*outside the mines*)

From the graph in Fig. 4, there is a similarity in the rate of abnormal blood count between men and women in and around the mines, the rate of abnormalities is high in the indicators MCV, MCH, MCHC, RED, BC is neutral, BC Lympho. However, the number of abnormal red blood cells between men and women is markedly different. The rate of abnormal HC in women is 51.6%, while in men it is only 36.5%.

For the total number of people, both men and women, we can see that the rate of abnormal red blood cells and white blood cells in the blood count is high, the platelet abnormality is very low.

From the graph in Fig. 5, we can see that there is a clear difference between the abnormal rate of indicators in blood formula between inside and outside the radioactive anomalous region. The rate of abnormal erythrocytes of the people outside the radiation anomalous area is much lower than the people living in the area of radiation anomalies (HST: low rate of 14.3% compared with 36.4÷38.4%; MCH: 68.5% lower than 24% compared to 93.4÷96.3%; MCHC: 5.4% lower than 88.5% compared to 93.6÷97.3%; RED rate is 7.4% lower than 56.4% compared to 68÷72%).

More than half (55.6% rate) of people with blood disease living in the mine field have identified blood diseases with genetic factors from the previous generation. Blood disease with genetic factors from the previous generation is one of the evidences of the effects of radiation on the health of people living in the mines containing radioactive substances.

The data and evidence of disease manifestations mentioned above have been used as a basis for assessing the effects of radiation on the environment and on the health of people living in radioactive mines.

The harmful effects of radiation may be one of the causes of the high prevalence of anemia symptoms among people living in the Dong Pao rare earth mine area.

Thus, it is initially seen that the level of radioactive pollution (expressed through the annual dose of radiation and the content of radioactive substances in plants, soil and water) affects the health of the people living in the area. rare earth mine in Dong Pao.

4 Conclusion

The study results have determined that the annual radiation dose (H_{eff}) at Dong Pao mine is 15.6 mSvyear^{-1}, higher than the world average (2.4 mSvyear^{-1}) of 6.5 times. The activity of radioactive substances from some food crops such as rice and cassava grown mine shows that the radioactive content of ^{238}U, ^{226}Ra, ^{232}Th, and ^{40}K has high value. Samples with high radioactive substances were grown in areas with rare earth ores in Ban Hon, Ban Giang, Dong Pao. The concentration of radioactive substances ^{238}U, ^{232}Th in hair samples in the study area 2 times higher than the world average.

Study results of blood disease symptoms showed that the majority of people living in the mines suffered from anemia. The rate of abnormal red blood cells such as HST, MCH, MCHC, RED of people living outside the mine (*outside the radioactive area*) is lower than the people living in the mine (*intside the radioactive area*). These are evidences of the health effects of radiation on workers living and working in mines that produce high levels of radioactive material.

The results given the radioactive environment and offering solutions substances ecological environment when the mine goes into exploitation and processing coming time.

References

1. Abdollahi, H.: World high background natural radiationareas and public life: letter to the editor. Radiat. Meas. **59**:288–289 (2013)
2. Ahmed, J.U.: High levels of natural radiation: report of an international conference in Ramsar. IAEA Bull. **2**, 36–38 (1991)
3. IAEA: Extent of environmental contamination by naturally occurring radioactive material (NORM) and technological options for mitigation. IAEA, Vienna (2003)
4. Derin, M.T., Vijayagopal, P., Venkatraman, B., Chaubey, R.C., Gopinathan, A.: Radionuclides and radiation indices of high background radiation area in Chavara-Neendakara placer deposits (Kerala, India). PLoS One **7**(11), e50468 (2012). https://doi.org/10.1371/journal.pone.0050468
5. Naturally-Occurring Radioactive Materials (NORM): World Nuclear Association. http://www.world.nuclear.org/info/Safetyand-Security/Radiation-and-Health/Naturally-Occurring-Radioactive-Materials-NORM/. Accessed 02 Aug 2014
6. Uranium from rare earths deposits, World Nuclear Association. http://www.world-nuclear.org/info/Nuclear-Fuel-Cycle/UraniumResources/Uranium-From-Rare-EarthsDeposits/. Accessed 16 Oct 2014
7. Ives, M,: Boom in mining rare earths poses mounting toxic risks, Yale Environment 360 (2013). http://e360.yale.edu/feature/boom_in_mining_rare_earths_poses_mounting_toxic_risks/2614/. Accessed 16 Oct 2014
8. Nguyen, V.N.: To study the scientific basis to determine the level of environmental pollution of natural radioactive sources in order to develop technological process for detailed assessment of radioactive contamination areas; archives in science and technology information center; ministry of science and technology of Vietnam: Hanoi, Vietnam (2009)
9. Nam, N.V., Hai, D.V., Son, N.T., Long, L.T.: Rare earth deposits in the Northwest of Vietnam considered from the perspective of radioactive environment. J. Geol. A **335**, 47–55 (2013). (in Vietnamese with abstract in English). http://idm.gov.vn/Data/TapChi/2013/A335/a55.html. Accessed 16 Oct 2014
10. Nam, N.V., Son, N.T., Vinh, N.Q.: Studying on the characteristics of natural radiation field for the assessment of radioactive pollution over rare earth mines in the North Vietnam. J. Geol. A **320**, 9–10 (2010). (in Vietnamese with abstract in English). http://idm.gov.vn/Data/TapChi/2010/A320/A270.htm. Accessed 16 Oct 2014
11. Bui, T.H., Trinh, D.H., Nguyen, P.: Overview of rare earth in Vietnam. Geol. J. Ser. A 447–456 (2010)
12. Trinh, D.H.; Nguyen, P.: Develop a methodology for investigating and assessing the environment related to toxic minerals on the basis of existing equipment in Vietnam. J. Geol. Ser. A, 30–38 (2013)
13. Tien, D.M.: Report "Exploration and supplementation of the Dong Pao rare earth - fluorite - barite mine of Ban Hon commune, Ban Giang commune, Phong Tho district, Tam Duong, Lai Chau". General Department of Geology and Minerals of Vietnam (2011)

14. Trong, T.B., Huan, T.D., Phuong, N.: Investigating the current state of radioactive environment on the deposits of Dong Pao, Then Sin-Tam Duong (Lai Chau), Muong Hum (Lao Cai), Yen Phu (Yen Bai), Thanh Son (Phu Tho), An Diem, Ngoc Kinh-Rib Between (Quang Nam), Geological Review, Series A (298), pp. 41–47, Hanoi (2007)
15. Phuong, N., Dung, N.V., Phon, L.K., Dong, N.P., Anh, V.T.L, Huan, T.D., Chau, T.L., Hai, D.V.: A study on the increase of natural radiation field due to the exploration of rare earth ores in the Dong Pao and Nam Xe areas - Lai Chau Province. J. Geol. A **335**, 56–62 (2013). (in Vietnamese with abstract in English). http://idm.gov.vn/Data/TapChi/2013/A335/a62.htm. Assessed 16 Oct 2014
16. Yolanda, F.S.: The mineral industry of Vietnam. U.S. geological survey minerals yearbook 2011 (2013). http://minerals.usgs.gov/minerals/pubs/country/2011/myb3-2011-vn.pdf. Assessed 02 Aug 2014
17. Le Phon, K.: Summary report on scientific topic epidemiological investigation of residents living in rare earth mines area of Phong Tho district, Tam Duong province, Lai Chau province. Code 09.10-DTLC-KY. Archives of Mining University – Geological (2014)
18. TCVN 9414:2012: Investigation, Assessment of Geological Environment Gamma Method. Accessed 30 Aug 2020
19. TCVN 9416: 2012: Investigation, Assessment of Geological Environment Radioactive Air Method. Accessed 30 Aug 2020
20. Ryabukhin, Y.S.: Radiation safety standards (NRB-96). Meditsinskaya Radiologiya i Radiatsionnaya Bezopasnost **42**, 70–73 (1997)
21. National Commission for Sanitary Inspection of Russian Epidemiology. Radiation Safety Standards (NRB-99); Minzdrav Rossii: Moscow, Russia (1999)
22. Wrixon, A.D., Green, B.M.R., Lomas, O.R., Miles, J.C.H., Cliff, K.D., Francis, E.A., et al.: Natural radiation in UK dwellings. National Radiological Protection Board Report R190. NRPB, Chilton (1988)
23. TCVN 7889:2008: Natural Radon activity in Buildings-Levels and General Requirements of Measuring Methods. Accessed 30 Aug 2020
24. International Atomic Energy Agency (IAEA). Technical Reports Series No.295. Measurement of Radionuclides in Food and the Environment. Accessed 28 Sept 2019
25. Ministry of Science and Technology: Circular "Provisions on control and ensure the safety of irradiated and irradiated career public", No.19/2012/TT-BKHCN, Hanoi (2012)
26. The Ministry of Science and Technology - Ministry of Natural Resources and Environment - Joint Circular "Regulations on radiation safety in the exploration, exploitation and processing of radioactive ores", Hanoi (2012)
27. ICRP Publication 82. Protection of the Public in Situations of Prolonged Radiation Exposure; Elsevier Science Ltd.: Amsterdam, The Netherlands (2000)
28. ICRP Publication 103. The 2007 Recommendations of the International Commission on Radiological Protection; Elsevier Science Ltd.: Amsterdam, The Netherlands (2007)
29. UNSCEAR Report Volume I: Sources and effects of ionizing radiation (Annex A: Dose assesment methodologies; Annex B: Exposures from radiation sources). United Nations Scientific Committee on the Effects of Atomic Radiation, United Nations, New York (2000)

Application of IE Techniques to Support Businesses to Make the Adoption of Sustainable Manufacturing Solutions: A Case Study of Adoption of AGV Technology in a Garment Factory

Bui Thi Kim Dung(✉), Thi My Le Anh, and Tran Thien Chuong

Faculty of Automotive Mechanical and Electrical and Electronics Engineering (FAME), Department of Industrial Systems Engineering, Nguyen Tat Thanh University, 331 1A Route, An Phu Dong Ward, 12 District, HCM City, Vietnam
btkdung@nttu.edu.vn, buitkdung@gmail.com

Abstract. In export garment industry, implementation of sustainable manufacturing processes, technology and practices are currently pooled and facilitated by international buyers as Nike; Adidas, Puma. Many contractual advantages and benefits are offered, technical assistants are provided to develop more capable and sustainable suppliers. However, the progress seems to be very slow, ineffective even unsustainable. One important reason is businesses are unable to make such a complex decision as the adoption of sustainable manufacturing solution taking into account even the economic benefits of new technology, let alone considering at the same time other social and environmental aspects. Industrial Engineering used to be considered as a bridge between technology and management is relatively new to Vietnam Industry. Concerned with the design, operation, control and improvement of efficient manufacturing processes integrating human and technology, IE techniques and methods are essential tools to support operation managers to make all decisions regarding manufacturing processes and system. This paper aims at demonstration of how Industrial Engineering techniques can support businesses making difficult and complex technology adoption decisions through a case study of the adoption of AGV technology in a garment factory. Technique as work design and measurement and engineering economics are employed to analyze the cost and benefit and to evaluate the cost effectiveness of AGV technology in a garment factory taking into account the improvement of worker health and the reduction of resources for the same level of output. Also in the case study, it is shown another advantage of the IE techniques. That is a new work method for smooth integration of AGV technology into manufacturing processes is developed for the implementation stage.

Keywords: Technology adoption · Cost-effectiveness · IE techniques: work design measurement · Engineering economics · Sustainable manufacturing processes · practices and systems · AGV technology

B. T. Long et al. (Eds.): MMMS 2020, LNME, pp. 576–583, 2021.
https://doi.org/10.1007/978-3-030-69610-8_78

1 Introduction

Since 2016, Government of Vietnam selected garment industry for a programme implementing sustainable manufacturing and consumption [1]. Earlier, sustainable practices such as those through Better Work Vietnam program; lean manufacturing; green factory; and LEED (Leadership in Energy and Environmental Design) certified factories are implemented in several Vietnam and FDI garment companies [2–4]. In addition, many emerging technologies for more competitive and sustainable garment apparel manufacturing are diffused into Vietnam factories. For examples, flexible automation and IoT systems are used for reduction of human hard work in material handling and in ironing; automated hanger systems; modular sewing cell; automated spreading and cutting machine are used for improvement of process reliability. The implementation of sustainable manufacturing processes, technology and practices are mainly pulled and facilitated by international buyers such as Nike; Adidas, Puma. Many contractual advantages and benefits are offered, technical assistants are provided to develop more capable and sustainable suppliers [5–7].

However, the progress seems to be very slow, ineffective even unsustainable. An important reason is the lack of ability of business manager to make complex decision of adoption of sustainable manufacturing solution that takes into account both the economic benefits of new technology and other involved social and environmental aspects [8–10]. The adoption decision seams even more difficulty for Vietnam Factory Managers who have not been trained and involved in investment decision making but specialized in the management of on-going operations.

Industrial Engineering used to be considered as a bridge between technology and management is relatively new to Vietnam Industry. Concerned with the design, operation, control and improvement of efficient manufacturing processes and systems integrating human and technology, IE techniques and methods are essential tools to support business managers to make all decisions regarding manufacturing processes and system [11].

This paper aims at demonstration of how Industrial Engineering techniques can support business managers making difficult and complex technology adoption decisions through a case study of the adoption of AGV technology in a garment factory.

Technique as work design and measurement and engineering economics are employed to analyze the cost and benefit and to evaluate the cost effectiveness of AGV technology in a garment factory taking into account the improvement of worker health and the reduction of resources for the same level of output.

Also in the case study, it is shown another advantage of the IE techniques. That is a new work method for smooth integration of AGV technology into manufacturing processes is developed for the implementation stage.

2 The Situation

The present case started at a newly developed garment facility of a big company addressed in this paper as XYZ located in Dongnai Province. The company factory was in process of getting LEED certification. XYZ company is a partner of international buyers such as Adidas; Nike, Puma. Prior to the present case, the XYZ company management visited Chinese partners of the buyers in order to learn how flexible automation and IoT are implemented in Garment Industry. China garment manufacturing productivity with automation astonished the visitors. Machines do not replace workers entirely. However, automated ironing machines totally eliminated sweaty jobs. Modular sewing cells with automated hanger system; automated spreading and cutting machines were widely used and especially, AGVs operated for all material, part transportation activities ensuring high productivity and reliable delivery, i.e., ensuring delivery at the right quality, at the right quantity; at the right place and at the right time.

A quick decision was made to implement the exact configuration of China manufacturing system. With the support of the company's clients, China equipment and machines suppliers also offered trial periods for all the automated machines needed with full maintenance and training support at XYZ company facility. The results of the system implementation were however not satisfied, the capacity of the production line was just about 55% of the China's counterpart.

Changes to the system were introduced in order to improve the situation by trial-error: adding more workers, reducing workload, changing the layout, raising the benefits. Operation problems did not seam to disappear but sometime even deepened. After several months of intensive effort, the production line was finally operated at an accepted capacity of 60 shirt/h. It is necessary to add that 60 shirt/h is the capacity accepted for production plan. The capacity of the same line slightly increased in comparison to the first trial but still was just approximate 65% of that at China's factory with the addition of workers. Productivity did not seam to be improved.

The company's management and technicians were suggested many kind of reasons and solutions for the situation, some of the solutions also implemented, but there was not any progress. 65% capacity of China's system seams to be the best to expect.

The situation was revealed to IE experts through an annual internship and a site visit for final year students and lecturers. The company agreed to support a team of lecturers and students to conduct investigation and to develop solution in return to employment for the students and other financial rewards for the team.

3 Methodology and Implementation

Two methodologies were chosen for problem identification and solution development.

3.1 Scientific Method for Problem Exploration [12]

The need for the chosen methodology is to explore and to identify the real problem behind the current "messy situation". Simple steps of the methodology was followed to answer the question why the company management had to accept such low capacity and so low productivity?

After discussion with factory managers and supervisors and observation of ongoing operation in one cut-delivery-sewing section, some issues were noticed.

According to the production plan, three sections: Cutting, Transport and Sewing all operated at the capacity of 60 shirt/h. However records showed that the cutting section could well maintain the capacity of 86 shirt/h and the sewing section could relatively easy to secure the capacity of 72 shirt/h. Besides, there was no record on the capacity of the Transport section. It was only known to management that 2 person allocated to delivery two pallet carts could maintain normally the capacity of 60 shirt/h. It means that a single delivery of parts and materials from the cutting section to the sewing workstations took 2 h.

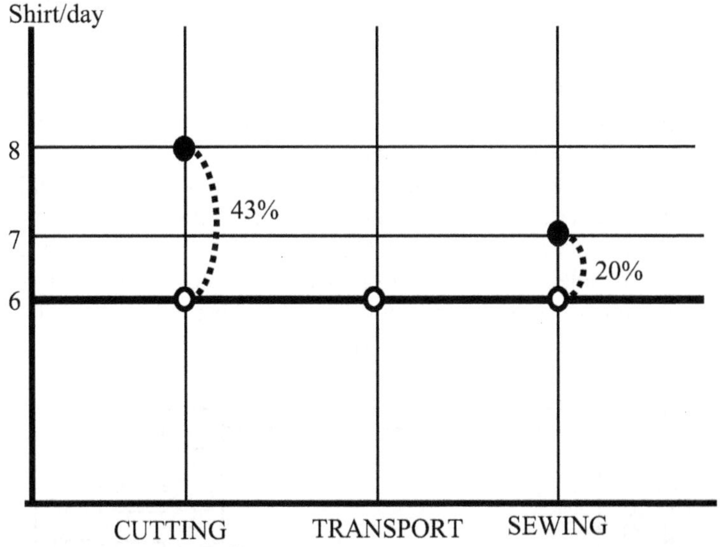

Fig. 1. Capacity and capacity allowance for Cutting, Transport and Sewing sections

Also, observation at the shop floor and discussion with the supervisors of the three sections showed while standardization has been started at the cutting and sewing operations, it had never been applied to the transport. The reason was mentioned that the transport section employed low-skill aged workers for the "simple job". It is clear that non-standardization of operation was definitely the cause for large time and consequently capacity variations. The variation in turn forced the operation managers to use large allowance to secure the production target. It is estimated that, allowance of about 43% of the running capacity (60 shirt/h); 20% of the running capacity and about 100% of the running capacity was allocated to the Cutting; Sewing and Transport sections respectively (Fig. 1).

3.2 Method Engineering Steps: A Systematic Procedure to Develop a Work Center [11]

After the problem of large variation and the consequent large allowance of capacity as the main cause of low productivity in the entire production process was identified,

Method Engineering Steps were applied to develop 2 standardized work centers: one with manual transport and the other with the use of an AGV.

The two work centers were standardized and experimentally implemented on an improved shop layout and with material delivery routes shown in the Fig. 2. A pallet cart was load with bundles of fabric and parts by a transport worker at the cutting section (work center 1 and 2). The pallet cart then moved slowly to the line of individual printing and pressing cells (work centers 3–13) at the distance of 25 m. In front of 4 work centers 3, 5; 6; 9, the worker unloaded the fabrics and part bundles. Then, the pallet cart moved to the line of 4 sewing work centers 14, 15, 16; 17 at which the worker again unloaded the rest of the fabric and part bundles. After the first round, the pallet cart went back to the loading/unloading points in front of the work center 12 and 13 to load the bundles processed by the cells and then it returned and delivered the bundles to the work centers 14, 15, 16; 17.

The process flow of transporting bundles of fabric and parts from the cutting section to sewing work centers was standardized in terms of the followings: Layout of delivery routes, position of loading/unloading points along the delivery route, the mass of each material and part bundle and the number of bundles, number of turns and time to turn, the number and time of slow-stop-speedup around loading/unloading points and time of transport with material or without material.

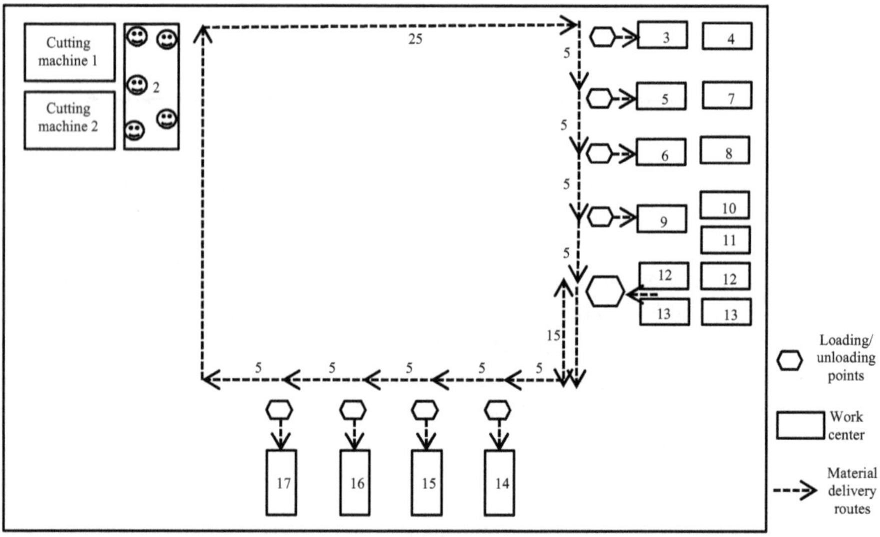

Fig. 2. Improved layout and material delivery routes

The standardized process flow for 2 experiments: one with no AGV and one with AGV was implemented and work measurement was conducted. The results is shown belows (Table 1).

Table 1. The results of experiments on two standardized process flows

No	Elements of Activity	Time (min)	
		AGV	NO AGV
1	Load 1, 72 bundles	9,60	14,40
2	Load 2, 36 bundles	4,80	7,20
3	Transport 1: 25m + 2 turn	2,29	3,67
4	Unload 1; 72 bundles	9,60	14,40
5	Transport 2: 5m	0,42	0,63
6	Transport 3: 5m	0,42	0,63
7	Transport 4: 5m	0,42	0,63
8	Transport 5: 20m + 1 turn	1,47	3,33
9	Unload 5, 36 bundles	4,80	7,20
10	Transport 6: 5m	0,42	0,63
11	Transport 7: 5m	0,42	0,63
12	Transport 8: 5m	0,42	0,63
13	Transport 9: 35m + 1 turn	2,12	4,33
14	Load 3, 36 bundles	4,80	7,20
15	Transport 10: 20m + 1 turn	1,47	3,33
16	Unload 9:	4,80	7,20
17	Transport 11: 5m	0,42	0,63
18	Transport 12: 5m	0,42	0,63
19	Transport 13: 5m	0,42	0,63
20	Transport 14: 40m + 1 turn	2,34	4,67
	Total	51,84	82,56

The figures showed that, the duration for one material delivery is reduced considerably by 57% for delivery with AGV and by 31% for delivery without AGV comparing to 2 h/delivery at the present. The delivery reliability was also improved due to standardization and the method of work measurement.

With AGV, the worker did not perform the hazardous pushing-pulling tasks as a result the work did not affect the worker's health in a negative way. Also, using AGV for transportation reduced delivery time, reduced variation and therefore improved productivity through lower capacity allowances for the same level of output.

4 The Economic Analysis

With the data collected, the economic analysis of the three methods of material transport was performed considering the following data (excluding common elements of the three methods):

- 1 worker salary: 8.000.000 VND/month
- AGV Investment/10 year of use: 250.000.000 VND

- Analysis period: 5 years
- Maintenance cost: 5% of annual depreciation of AGV
- Electricity consumption: 0,2kW/h
- Electricity unit cost: 1388 VND/kWh
- Annual loan interest: 7,5%
- Annual saving (opportunity cost): 6%
- 26 working days/month
- 312 working days/year
- Number of material delivery/day: 4 (maintain the current capacity of 60/h)
- Cost effectiveness = Present Cost (No standardized method)/No AGV method or AGV method. Cost effectiveness shows the relative reduction of cost while maintaining the same delivery or transport capacity (Table 2)

Table 2. The economic analysis of the three methods of material transport

	Cost Effectiveness	Present cost	Year 1	Year 2	Year 3	Year 4	Year 5
No Standar-dised		413200538	96000000	88800000	82140000	75979500	70281038
No AGV	1,45	285108371	66240000	61272000	56676600	52425855	48493916
AGV	2,96	139414807	32390620	29961323	27714224	25635657	23712983

The economic analysis showed high cost effectiveness of the adoption of AGV for material transport in the garment factory. The reduction of cost almost 3 times promised huge economic potential. Especially, the use of AGV offered an opportunity to improve capacity of material transport by raising AGV speed and as a result, to improve the overall capacity and the productivity of the whole production line. Although the analysis did still not take into account factors like inflation, idle time; downtime and inventory and pre-mature part replacement. Additional opportunity for cost reduction and productivity improvement can be expected when standardization can be applied appropriately for the other two sections: the Cutting and Sewing. At the same time, optimized scheduling and capacity planning employing IoT could be the best leverage the whole production line improvement.

5 Conclusion

Vietnam Garment manufacturing companies are in needs of becoming more and more competitive and sustainable suppliers for international buyers. Advanced Technology is widely available at a reasonable cost and with improved services. Sustainable manufacturing processes and practices are successfully tested in Vietnam and in the world. The technology decision is always difficult but Industrial Engineering techniques are convincingly capable to support making such decision. IE techniques are not only able to consider economic concern of businesses but also to take into account the improvement of work environment, and so IE techniques take important role in the development of sustainable manufacturing in Vietnam.

References

1. Decision No 76/QD-TTg, Hanoi, 2016 January 11: "National Programme on sustainable manufacturing and consumption to 2020 and forwards to 2030" (2016)
2. FY18 Sustainable Business Report NIKE, Inc. (2018)
3. Adidas Green Company Report 2019 (2019)
4. Fraunhofer Research: Emerging green technologies for the manufacturing sector UNIDO (2014)
5. Deloit University Press. The smart factory**.** Responsive, adaptive, connected manufacturing, A Deloitte series on Industry 4.0, Digital Manufacturing Enterprises, and Digital Supply Networks (2017)
6. Soosay, C., Nunes, B., Bennett, D.J., Sohal, A., Jabar, J., et al.: Strategies for sustaining manufacturing competitiveness: comparative case studies in Australia and Sweden. J. Manuf. Technol. **27**(1), 6–37 (2016)
7. Bennett, D., Sohal, A., Jabar, J., Winroth, M.M.:.Sustainable and smart manufacturing: an integrated approach. Sustainability **12**(6), 2280 (2020)
8. Abdullahi, T., Abdullah, S.S.: Sustainability considerations in manufacturing and operation management. Int. J. Sci. Eng. Appl. Sci. (IJSEAS), **1**(4), 490–497 (2015)
9. Baier, D., Rausch, T.M., Wagner, T.F.: The drivers of sustainable apparel and sportswearconsumption: a segmented Kano perspective. Sustainability **12**, 2788 (2020). www.mdpi.com/journal/sustainability
10. McKinsey Global Institute Manufacturing the future: The next era of global growth and innovation. November 1. Report (2012)
11. Andris, F., Benjamin, W.N.: Niebel's Methods, Standards,and Work Design 13 edn., The McGraw-Hill Companies, New York (2014)
12. Glazunov, N.M.: Foundations of scientific research. National Aviation University, Kiev, Ukraine, pp. 144–158 (2012)

Evaluation of the Importance of the Blank Holder Pressure in Hydrostatic Forming for Sheet Metal

N. T. Thu(✉), N. D. Trung, and N. T. H. Minh

Hanoi University of Science and Technology, Hanoi, Vietnam
thu.nguyenthi@hust.edu.vn

Abstract. In recent years, hydrostatic forming for sheet metal has emerged as an impressive technology applied in the manufacture of shell parts due to its outstanding advantages. The technology can create products with good surface quality, reduce the number of forming tasks, and improve the rigidity of the product. This technology is suitable for forming complex detailed shapes that are difficult to achieve with conventional technologies. However, this advanced technology has been faced a challenge in accurately shaping the product, especially at concave radius locations of product. Therefore, technological parameters in forming process must be studied to evaluate the product quality. In this paper, one essential technological parameter mentioned is the blank holder pressure and the evaluation is made through experiments. The selected object is a cylindrical part fabricated from DC04 steel with different initial thicknesses (s_0 = 0.8, 1.0, 1.2 mm). The results indicate that this parameter has great influence on the shape of the concave radius at the bottom of the cylindrical product. In addition, the relationship between the blank holder pressure and the forming fluid pressure is also clarified in the technology.

Keywords: Hydrostatic forming · Blank holder pressure · Forming fluid forming

1 Introduction

Hydrostatic forming is an advanced technology for shaping thin shell products. Under the effect of high pressure liquid, the product is formed according to the shape of the die's geometry. This technology has been highly appreciated by researchers as well as manufacture thanks to its suitability when shaping sheet products [1–3]. Along with the development of the automobile industry, this technology has also made significant development in its theory and application through a great number of studies on many aspects of this technology.

One of the most frequently mentioned parameters in hydrostatic forming is the forming fluid pressure, which determines the final shape of the product. If the pressure is too small, the product cannot reach the desired shape while it can be cracked or torn with too high pressure. There have been many studies on the parameter in the forming process. When researching the shaping process of square cup details with 5182 AA alloy, Modi used Taguchi method to optimize the forming liquid pressure [4].

B. T. Long et al. (Eds.): MMMS 2020, LNME, pp. 584–590, 2021.
https://doi.org/10.1007/978-3-030-69610-8_79

Karabegovic et al. have built a mathematical model of fluid forming pressure in hydrostatic forming for welded plates to stabilize the forming process [5]. Liu Wei [6] have illustrated the load curve in the forming a pair of hemispherical parts. This curve illustrated the pressure value zones corresponding to three surface quality of product: wrinkle, fine and crack. Assempour and Emami [7] also studied the control of forming pressure in hydraulic forming.

Another parameter is considered to have big influence on product quality is the blank holder pressure. Modi et al. [4] have showed that it was easy to achieve better minimum radius and thinning with variable blank holding force (BHF) instead of constant BHF tduring forming process. In addition, Fitsum has presented the influence of these technological parameters on the formability of highly strength cryorolled Al alloy sheets hydroforming [8].

Although there has been a lot of research on typical technological parameters including blank holder pressure and forming fluid pressure, the influence of blank holder pressure on the forming process need to be assessed thoroughly.

In this paper, by empirical research, the relationship between the blank holder pressure and the forming fluid pressure was given in the graph, showing the necessary correlation between them. In addition, the effect of the blank holder pressure on the forming ability in terms of the cylindrical bottom radius was also mentioned. From this experimental result, it was possible to evaluate the importance of the blank holder pressure in hydrostatic forming.

2 The Hydrostatic Forming Process

The hydrostatic forming process can be divided into 4 phases, including: to place blank onto die, to accelerate fluid pressure and preform, to calibrate the final fluid pressure and remove product from the die. The schematic diagram of the hydrostatic forming process is shown in Fig. 1.

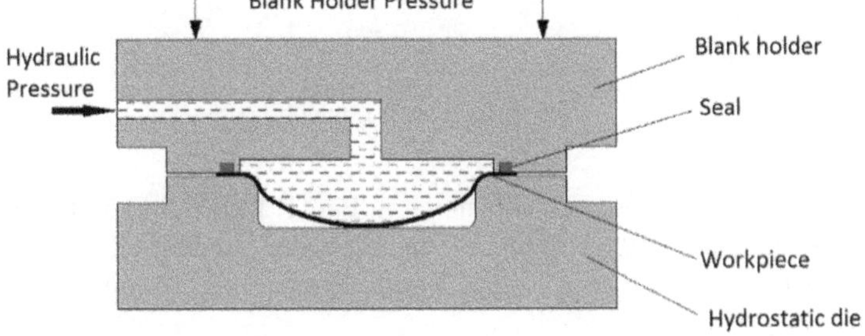

Fig. 1. Scheme of sheet hydrostatic forming process

From the diagram of principle in Fig. 1, the influence of the fluid pressure and the blank holder pressure on product formation can be clearly shown.

On one hand, the fluid pressure acts directly on the workpiece to obtain an expected shape. If the pressure is too low, the product cannot be shaped according to the die's profile, especially at the bottom radius. Meanwhile, if the pressure is too high, the product can be torn apart.

On the other hand, the blank holder pressure in hydrostatic forming has two main purposes, including preventing the instability on the flange of the workpiece and ensuring the tightness (with seals) to maintain a stable fluid pressure during forming. Like the fluid pressure, the blank holder pressure should be selected at a suitable value. If the pressure is too high, the flange will be kept fixed on the die and tearing can occur in forming process. On the contrary, its value should not be too small, because it will destabilize the workpiece's flange and fail to maintain the forming pressure. It can be inferred that the relationship between the blank holder pressure and the forming fluid pressure in the process must be evaluated for high product quality.

3 Research Object and Objectives

In this work, research objectives include determining the relationship of the two mentioned parameters (the blank holder pressure and the forming fluid pressure) and evaluating the influence of the blank holder pressure on the formation of the concave radius on the studied product.

After preliminary investigation, the blank holder pressure range selected in this study is of (80 ÷ 115) bar.

The cylindrical part has been selected as the subject of the study. The material selected was carbon steel DC04 with thickness s_0 of 0.8, 1.0 and 1.2 (mm) as shown in Fig. 2. The properties of this material are given in Table 1.

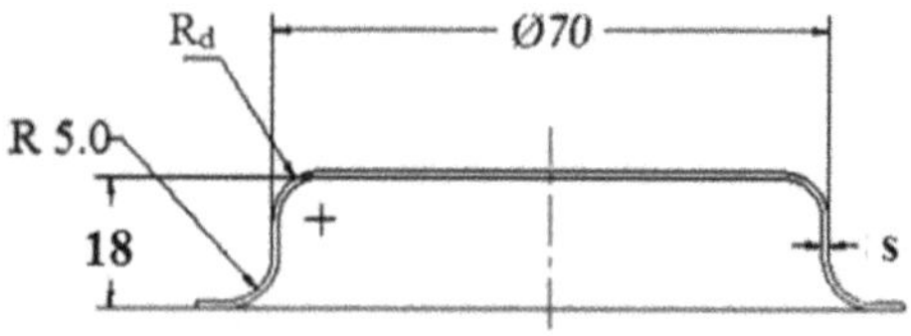

Fig. 2. Research product

The dimensions of the product are shown in Fig. 1. Depending on the intensity of the liquid pressure, the bottom radius of the generated product (R_d) has a different value. Initial workpiece is shown in Fig. 3 with diameter of 110 mm.

Table 1. Properties of carbon steel DC04

Label	Properties		
DC04	Ultimate strength σ_m (MPa)	Yield stress σ_f (MPa)	Specific weight ρ (kg/cm^3)
	270–350	140–210	7.8 * 10^{-3}

Fig. 3. Experimental workpieces

4 Experiments

Experimental system is demonstrated in Ref. [9]. In order to investigate the relationship between the forming fluid pressure and the blank holder pressure, the experiments were carried out with workpieces of different thicknesses. The maximum forming fluid pressure recorded is named the peak pressure P_{peak} and shown in Table 2.

Table 2. Measurement results of the peak pressure P_{peak} (bar)

Initial thickness s_0 (mm)	Blank holder pressure Q (bar)							
	80	85	90	95	100	105	110	115
0.8	355	385	405	448	468	483	488	495
1.0	373	404	425	470	491	506	512	520
1.2	387	420	441	486	507	526	529	537

In order to investigate the impact of the blank holder pressure on the formation of the bottom radius, the bottom radius of the products formed in Table 2 were measured as shown in Fig. 4. Experimental results are collected in Table 3.

Table 3. Results of the bottom radii R (mm)

Initial thickness S_0 (mm)	Blank holder pressure Q (bar)							
	80	85	90	95	100	105	110	115
0.8	6.46	6.25	6.18	6.1	6.13	6.25	6.31	6.5
1.0	7.92	7.66	7.54	7.45	7.51	7.69	7.67	7.92
1.2	11.01	10.68	10.52	10.44	10.38	10.65	11.12	11.06

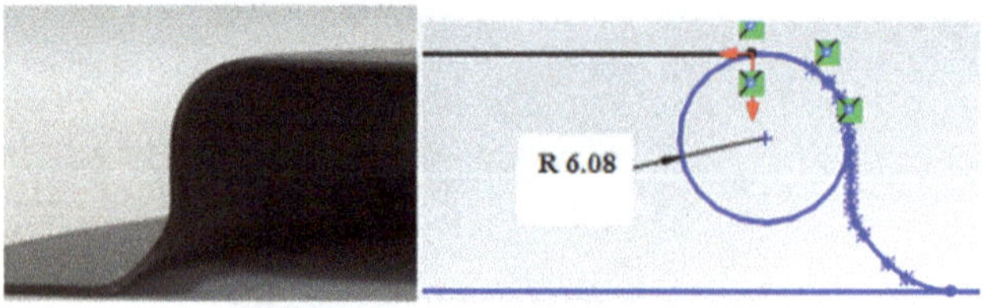

Fig. 4. Measured curve radius R

5 Discussions

5.1 Relationship between the blank holder pressure and the forming fluid pressure.

Based on the results in Table 2, the graph showing the relationship between the blank holder pressure and the peak pressure was shown in Fig. 5.

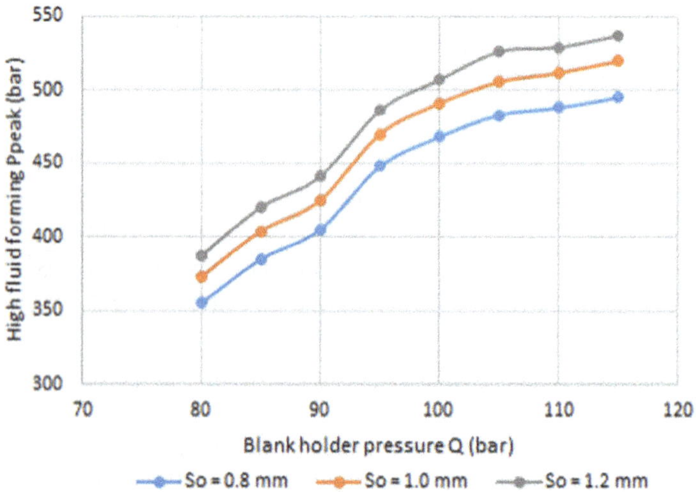

Fig. 5. Relationship between the blank holder pressure and the peak pressure

From experimental process, it can be noted that in each value of the blank holder pressure, only one value of the peak forming pressure can be maintained. If the forming pressure rose above this value, the blank holder would be pulled out and the forming process could not be completed. This is the difference point in the relationship between the blank holder pressure and forming fluid pressure compared with conventional forming.

Figure 5 also shows the relationship between the blank holder pressure and the maximum forming pressure in hydrostatic forming for cylindrical cup with different thicknesses. Within the considered range of the blank holder pressure, it has been found that the larger the blank holder pressure, the larger the forming pressure that could be maintained. It could be explained that, when the blank holder pressure was high, a large binder force was created. This binder force, in addition to preventing the destabilization of the workpiece, was also effective against high pressure liquid leak-age. Therefore, the peak pressure could reach a high value.

The graph also shows that the larger the thickness, the greater the peak pressure. In fact, the forming force was proportional to the thickness of the workpiece. Therefore, in cases the blank holder pressure was sufficiently large, a greater peak pressure could still be maintained to shape workpieces with larger thicknesses.

5.2 Dependence of the bottom radius on the blank holder pressure

Based on the results in Table 3, the graph of the dependence of the radius at the bot-tom of the product on the blank holder pressure was shown in Fig. 6.

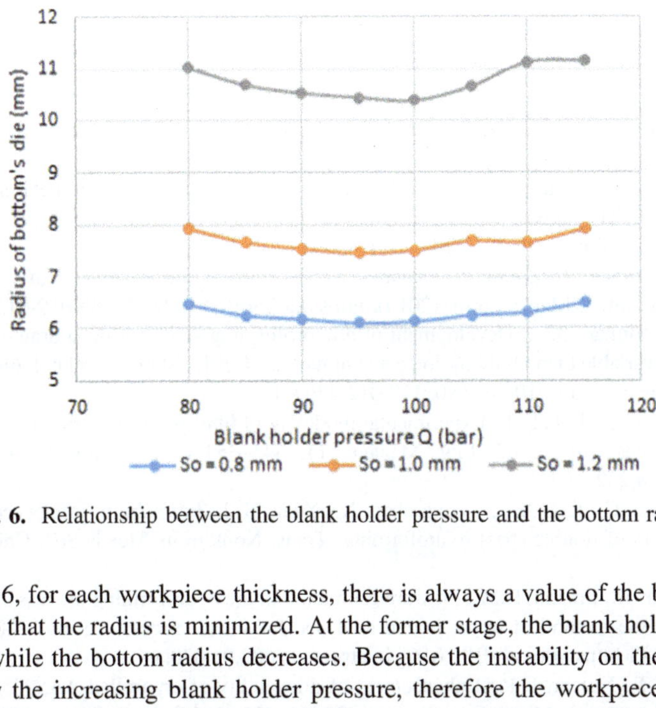

Fig. 6. Relationship between the blank holder pressure and the bottom radius

In Fig. 6, for each workpiece thickness, there is always a value of the blank holder pressure so that the radius is minimized. At the former stage, the blank holder pressure increases while the bottom radius decreases. Because the instability on the flange was reduced by the increasing blank holder pressure, therefore the workpiece was easily pulled into the cavity die and formed well. However, at a later stage, the bottom radius increases with higher blank holder pressure. Because when the blank holder pressure was raised, it was difficult for the workpiece to slide on the die's surface and the blank holder, so it became more difficult to fill the die cavity.

Figure 6 also shows that, with the same blank holder pressure, the radius will be larger if the workpiece thickness is greater. Because then, only one peak pressure could be hold on, therefore the forming was more difficult, especially in the bottom radius.

6 Conclusions

The research results shows that the blank holder pressure plays an important and decisive role in the product shape in the forming process.

On one hand, the forming fluid pressure depended on the blank holder pressure in the forming process. When the blank holder pressure is increased higher, the fluid pressure intensity is also maintained at a higher level. This is a different point from conventional forming.

On the other hand, the blank holder pressure has a great influence on the bottom radius of the product. There existed an appropriate value of the blank holder pressure to reach the minimum radius. Moreover, different thicknesses would also affect the bottom radius formation.

Acknowledgment. This research is funded by Hanoi University of Science and Technology (HUST) under grant number T2020-TT-202.

References

1. Maki, T., Cheng, J.: IOP Conference Series Materials Science Engineering, vol. 418, 012117 (2018)
2. COSTIN, G.A., Afteni, C., Iacob, I., Păunoiu, V.: An overview on sheet metal hydroforming technologies, The Annals of " Dunarea de Jos" University of Galati Fascicle V, Technologies in Machine Building (2018). ISSN 1221-4566
3. Bell, C., Corney, J., Zuelli, N., Savings, D.: A state of the art review of hydro-forming technology. Int. J. Mater. Form. (2019). https://doi.org/10.1007/s12289-019-01507-1
4. Modi, B., Kumar, D.V.: Development of a hydroforming setup for deep draw-ing of square cups with variable blank holding force technique. Int. J. Adv. Manuf. Technol. **66**, 1159–1169 (2013). https://doi.org/10.1007/s00170-012-4397-4
5. Karabegović, E., Poljak, J.: Experimental modeling of fluid pressure during hydroforming of welded plates. Adv. Prod. Eng. Manage. **11**, 345–354 (2016). https://doi.org/10.14743/apem2016.4.232
6. Wei, L., Liu, G., Cui, X.-L., Yong-chao, X., Yuan, S.-J.: Formability influenced by process loading path of double sheet hydroforming. Trans. Nonferrous Metals Soc. China **21**, 465–469 (2011)
7. Assempour, A., Emami, R.M.: Pressure estimation in the hydroforming process of sheet metal pairs with the method of upper bound analysis. J. Mater. Process. Technol. **209**(5), 2270–2276 (2008). https://doi.org/10.1016/j.jmatprotec.2008.05.020.a
8. Feyissa, F.T., Kumar, D.R.: Enhancement of drawability of cryorolled AA5083 alloy sheets by hydroforming. J. Mater. Res. Technol. **8**, 411–423 (2019). https://doi.org/10.1016/j.jmrt.2018.02.012.2
9. Thu Nguyen Thi: Trung Nguyen Dac, Effect of Technological and Geometrical Parameters on Formation of Radius Region at Cylindrical Product Bottom in Hydrostatic Forming. J. Korean Soc. Precis. Eng. **36**, 891–900 (2019). https://doi.org/10.7736/KSPE.2019.36.9.891

Reasonably Use of Accumulated Energy When Moving Downhill During the Operation Cycle of Timber Tractor Complex

Mai Hoang Long[1(✉)], Le Hoang Anh[1], and Bui Hai Trieu[2]

[1] Vinh Long University of Technology Education, Vinh Long, Vietnam
longmh@vlute.edu.vn
[2] University of Transport Technology, Hanoi, Vietnam

Abstract. The paper introduces a hydraulic circuit diagram that controls the anchoring of timber tractor complex when it works on forest land with a steep slope. The paper also presents some research results that simulate the main stages of timber transport technology. The hydraulic anchoring system is supplemented by hydraulic components that help the tractor complex (TC) lift and unload wood, support the departure of moving uphill, gather and load wood at the stops on the slope by accumulated energy. The research results show that if we calculate, determine the size and appropriately arrange the stages, the system only needs to use the accumulated energy in the accumulator after moving downhill, contributing to reducing energy costs for timber transport technology.

Keywords: Tractor complex · Accumulated energy · Hydraulic system

1 Introduction

The cyclical activity of the log transport technology on steep slopes could accumulate the cumulative energy as the TC moves downhill. And this energy is used to supplement departure energy, move uphill or as an energy source to unload wood at the bottom of a slope and collect wood at a stop on the slopes. In several published scientific papers [1–3], the authors have developed a hydraulic system controlling anchor the timber TC. The system can anchor, drag, and drop the TC with the winch force that is almost unchanged when transporting wood on forest land with a slope that is higher than 20°. The hydraulic circuit controlling the anchor has an accumulator that is big enough to recover the downhill kinetic energy of the full-load TC. To properly use the accumulated energy, it is necessary to calculate the size and organize the operation accordingly. The paper introduces the simulation results of the process operations of some assistive working units in timber transport technology using hydraulic energy in the accumulator.

B. T. Long et al. (Eds.): MMMS 2020, LNME, pp. 591–598, 2021.
https://doi.org/10.1007/978-3-030-69610-8_80

2 The Hydraulic Circuit Controls the Anchoring, Loading, and Unloading of Wood

The feature of timber transport technology on sloping land is a circular movement. The TC needs unloading and partial loading, stopping to collect wood and moving downward at full load. Therefore, in addition to hydraulic anchors, other parts need to be equipped to use the accumulated energy in the accumulator. The diagram is a hydraulic cylinder to lift the unloading of wood at the bottom of the slope and a hydraulic motor to collect wood. Single-acting hydraulic cylinders receive hydraulic energy from the accumulator S_2, lift and lower due to VW_7 distribution valve and control speed by V_{Dr3} flow control valve.

The wood-gathering hydraulic motor M_2 also receives hydraulic power from the accumulator S_2, which controls the pull-down of the wood winch with the distribution valve VW_6. In case the accumulator S_2 is not able to provide, the hydraulic motor M_2 can receive hydraulic energy from the pump P through a distribution valve VW_5. Other hydraulic elements have been introduced in [4–9] (Fig. 1).

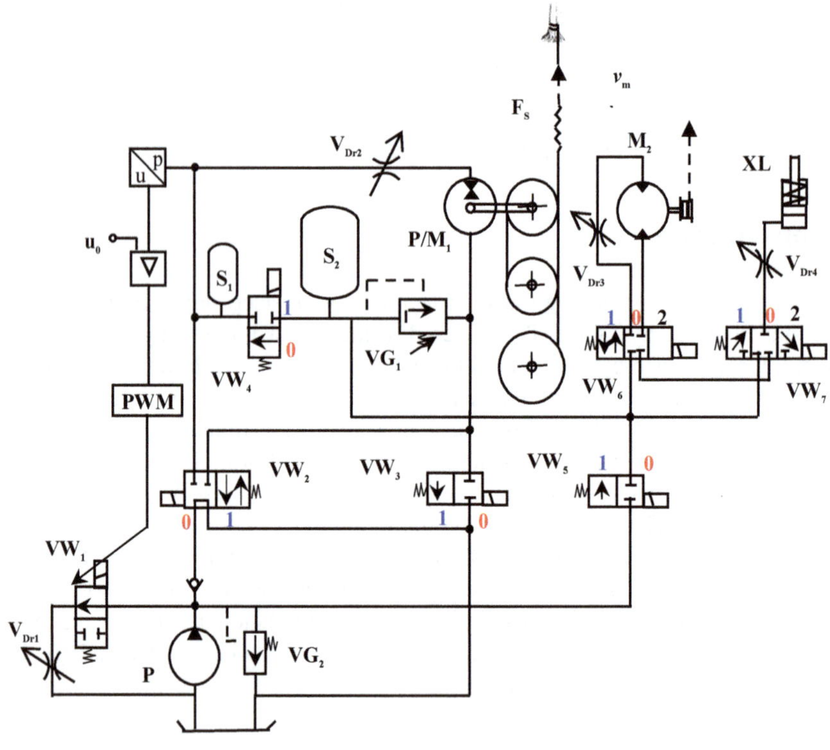

Fig. 1. Diagram of the hydraulic circuit controlling anchors of TC ①. Pressure sensors; ②. Voltage controller; ③. Pulse-width modulation controller; **P.** Hydraulic pump; **P/M$_1$**. Hydraulic Pump/Motor; **M$_2$**. Hydraulic motor pulls wood; **XL.** Cylinder lifting wood; **F$_S$**. Winch tension force, **VW$_1$...VW$_7$**. Distribution valves; **V$_{Dr1}$...V$_{Dr4}$**. Flow control valves; **VG$_1$**. Pressure limiter valve; **S$_1$, S$_2$**. Accummultor.

3 Simulation Model of Hydraulic Elements Using the Accumulated Energy

The hydraulic elements are calibrated so that the use of accumulated energy in the accumulator is reasonable and in effect. Besides, the TC can work smoothly according to the following diagram (Fig. 2).

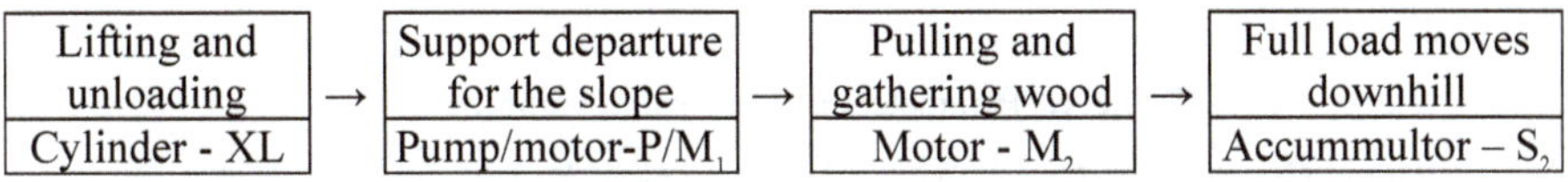

Fig. 2. Working order diagram of the TC on forest land with high slope

3.1 Model of Lifting Timber at the Bottom of the Slope

When the wood is lifted at the bottom of the slope, only the cylinder XL that corresponds to the VW_7 valve is set at position 1 and all the other distribution valves are at position 0. Valve V_{Dr4} is controlled so that the lifting speed hook floor is about 0.02 m/s. From the hydraulic circuit diagram, the floor hook can build a model that simulates the process of unloading the wood due to the accumulated energy (Fig. 3).

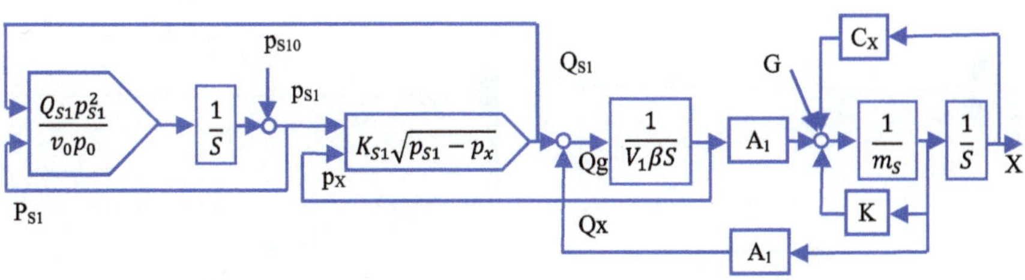

Fig. 3. The model that simulates the floor lifting process

The corresponding differential equation

$$m\ddot{x} = A_1 p_x - G - K\dot{x} - C_x.x \tag{1}$$

$$p_x = \frac{Q_{S_1} - Q_x}{V_1 \beta} \tag{2}$$

With: $Q_{S_1} = K_{S_1}\sqrt{p_{S_1} - p_S}; Q_x = A_1.\dot{x}; p_{S_1} = p_{S_{10}} - \Delta p_S = p_{S_{10}} - \int \frac{p_{S_1}^2 .Q_S}{p_0 V_0} dt$

And: A_1. Cross section of cylinder;
C_X. Rigidity of winch wire;
G. Lifting weight;
K. Hydraulic resistance coefficient of cylinder;

p_X, p_{S1}. The hydraulic oil pressure in the cylinder at the respective positions;
Q_X, Q_{S1}. Oil flow rate at the respective positions of the plunger in the cylinder;
x. The raised height of the cylinder;
β. Compression coefficient of hydraulic oil.

The process of lifting the hook floor is investigated according to the model parameters in Table 1. The initial time was chosen is the time of switching valve VW_7 from position 0 to position 1.

Table 1. Parameters of timber unloading model

Para.	G	K	Cx	A_1	V_O	p_O	p_{S10}	K_{S1}	β	V_1
Units	*N*	*Ns/mm*	*N/mm*	mm^2	dm^3	*bar*	*bar*	$mm^4/s\sqrt{N}$	mm^2/N	mm^3
Value	10^4	10	20	5671	25	60	200	1327,8	10^{-3}	10^6

The simulation result is shown in Fig. 4.

From the pressure result at the end of the unloading stroke, the remaining pressure in the accummultor (p_{S2}) is determined as the input parameter for the next process.

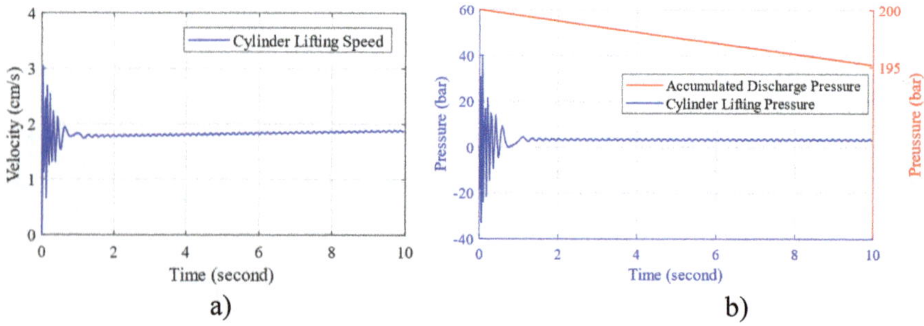

Fig. 4. a) Process of lifting speed of cylinder to unload wood; b) Process of cylinder lift pressure and the pressure of the accumulator;

3.2 Departure Support Model

When starting up the slope with hydraulic energy in the accumulator, only the tension force of the winch makes the TC moving. In order to rotate the winch, the motor P/M_1 is connected to the hydraulic accumulator S_2 through the valve VW_4 in position 1, all other distribution valves need to be in position 0. Valve V_{Dr2} is controlled so that the motor P/M_1 can move the TC at the speed of $v_{max} \approx 1$ m/s. The process is carried out in a period of 0.5 to 1.0 s, then turn the valve VW_4 to position 0, valve VW_2, VW_3 switch to position 1 and set the forward gear and increase the engine speed of the TC.

From the corresponding hydraulic circuit diagram, it is possible to build a simulation model of the departure process of the TC from accumulated energy (Fig. 5).

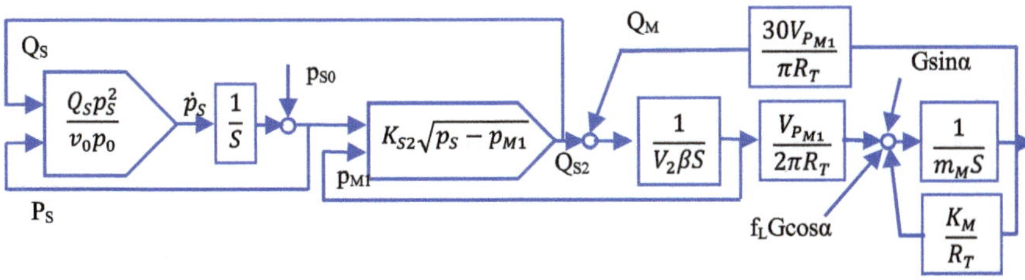

Fig. 5. Block diagram simulating the upward departure process of the TC

The corresponding differential equation

$$\begin{aligned} m_m \ddot{x}_m &= F_S - Gsin\alpha - f_L Gcos\alpha - \frac{K_M}{R_T} \\ with: F_S &= \frac{p_{PM_1} . V_{PM_1}}{2\pi R_T}; \dot{p}_M = \frac{Q_{S2} - Q_M}{V_2 \beta} \end{aligned} \tag{3}$$

$$Q_{S2} = K_{S2}\sqrt{p_{S2} - p_{M1}}; \; Q_M = \frac{30 V_{M1}}{\pi R_T}\dot{x}_m; \; p_{S_3} = p_{S_{30}} - \Delta p_{S_3}$$
$$= p_{S_{30}} - \int \frac{p_{S_3}^2 . Q_{S_3}}{p_O V_O} dt$$

And: α. The slope angle of the road surface;
F_S. Tension of winch wire;
f_L. Rolling resistance coefficient;
K_M. Hydraulic resistance coefficient in the motor;
R_T. The radius of the winch;
x_m. Moving distance of TC.

The departure process is surveyed according to the parameters of the model in Table 2. The simulation result is presented in Fig. 6.

Table 2. Parameters of departure model

Para.	m_M	α	p_{S20}	V_O	p_O	V_{PM1}	R_T	V_2	f_L	K_M	K_{S2}
Units	*Kg*	*degree*	*bar*	*dm*^{3}	*bar*	*cm*^{3}*/rev*	*cm*	*cm*^{3}	–	*Ns/mm*	*mm*^{4}*/ s*$\sqrt{N}$
Value	1400	20	p_{S1}^*	25	60	450,6	10	250	0,1	10	4.10^4

The initial value of p_{S2} is taken from the end value of p_{S1}^*

From the result of the accumulated pressure p_{S2} at the time of switching, VW_4 determines the remaining pressure p_{S1}^* in the accumulator as input for the next process.

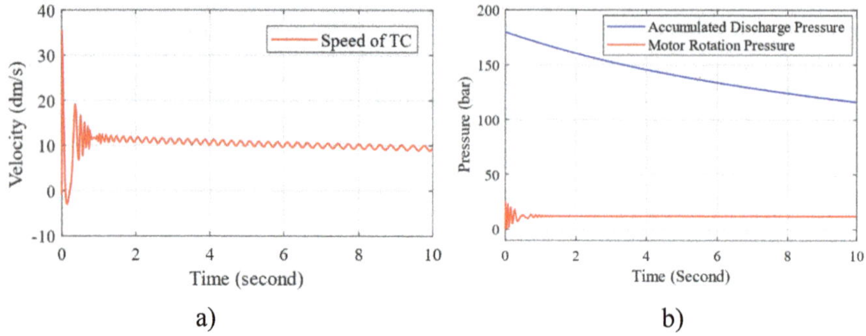

Fig. 6. a) The process of velocity; b) Accumulated discharge pressure and rotating the motor when TC starts up the slope due to the accumulated energy

3.3 Pulling and Gathering Timber Model

The gathering of wood by the accumulated energy at the stop on the slope is done when the VW_6 valve switches to position 1 while the other distribution valves are all at position 0. From the drive circuit diagram for motor M_2, it is possible to build a simulation model of the process of gathering wood by the accumulated energy (Fig. 7).

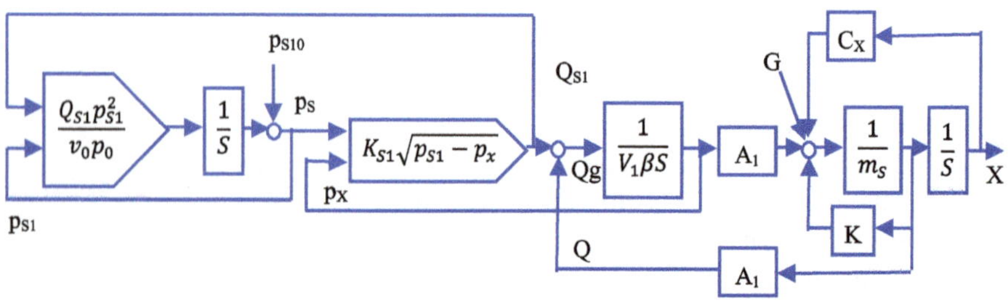

Fig. 7. Block diagram simulating the process of timber gathering.

The corresponding differential equation:

$$m_g \ddot{x}_g = F_{S2} - G_g F_2 - \frac{K_M}{R_2} \dot{x}_g \tag{4}$$

With: $F_{S2} = \frac{p_{M_2}.V_{M_2}}{2\pi R_2}$; $\dot{p}_{M_2} = \frac{Q_{S3} - Q_{M_2}}{V_3 \beta}$; $Q_{S3} = K_{S3}\sqrt{p_{S3} - p_{M_2}}$; $Q_{M_2} = \frac{30 V_{M_2}}{\pi R_2} \dot{x}_g$

And: m_g. Wood weight;
F_{S2}. Traction of wooden winch;
F_2. Coefficient of sliding friction of wood;
R_2. The radius of the winch to the wood;
x_g. Moving distance of wood.

The process of timber gathering was surveyed according to the parameters of the model in Table 3. The simulation results are presented in Fig. 8.

Table 3. Parameters of timber pulling and gathering model

Para.	m_g	p_{S30}	V_O	p_O	V_{M2}	R_2	V_3	f_L	K_{S3}
Units	*kg*	*bar*	dm^3	*bar*	cm^3/rev	*cm*	cm^3	–	$mm^4/\,s\sqrt{N}$
Value	100	p^*_{S2}	5	60	20	10	10	0,2	4.10^4

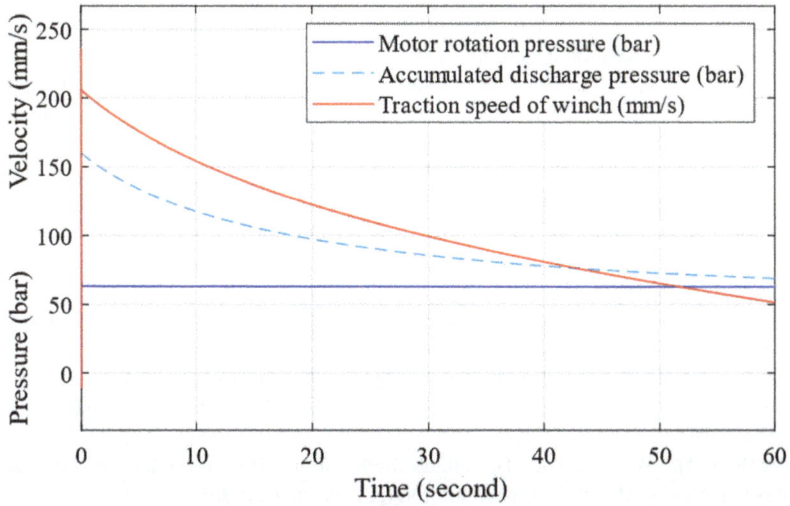

Fig. 8. The process of velocity and discharge accumulated pressure and rotating the motor when gathering timber due to accumulated discharge energy

A 100-kilo log is pulled halfway with a sliding friction coefficient of $f_L = 0.2$ continuously until a minimum pressure of $ps \approx 70$ bar is reached. If all of the accumulated energy is exhausted in case needed wood is not pulled enough, valve VW_5 can be switched to position 1 to receive flow from the pump P.

4 Conclusion

The hydraulic anchoring system not only helps the TC work safely and stably on the slope but also recovers and accumulates energy when going downhill. If stages are logically arranged in timber transport technology, it is possible to use the accumulated energy in the accumulator. In so doing, we can save energy costs in the process of timber transportation. For example, the simulated study of a small-sized timber TC shows that with a 25-l accumulator, a cumulative pressure of 200 bar, when going downhill it could provide enough energy to lift and unload wood at the bottom of the slope, assist the uphill departure, and pull up and gather timber at a stop on the slopes. The simulation study results can be used as a basis for designing hydraulic anchoring system and it also proposes a reasonable operating mode for the timber TC working on forest land with a steep slope.

References

1. Trieu, B.H., Quang, N.T., Long, M.H.: Application of hydrostatic transmission systems for anchoring systems of forestry activities in the uplands. Vietnam Mech. Eng. J. 332–335 (2016)
2. Trieu, B.H., Long, M.H., Thuat, N.C.: Hydraulic transmission primary control for mooring vehicle system operated on upland forest steep slopes. Vietnam J. Agric. Sci. 161–167 (2018)
3. Trieu, B.H., Long, M.H., Thuat, N.C.: Research of the size and structure of the hydraulic circuit for mooring vehicle system of forestry operation on upland. Vietnam Mech. Eng. J. 291–295 (2018)
4. Will, D., Strühl, H., Gebhardt, N.: Hydraulik – Grundlagen, Komponenten. Schaltungen Springer – Verlag, Berlin, Heidenberg (1999)
5. Bombosch, F., Sohns, D., Nollau, R.: Forwarder - Einsatz uber 70% Forst und. Technik **9**(15), 22–23 (2003)
6. Holzleitner, F., Kastner, M., Stampfer, K., Höller, N., Kanzian, C.: Monitoring cable tensile forces of winch-assist harvester and forwarder operations in steep terrain. For. Oper. Eng. Manage. Spec. Issue Publ. For. **9**, 295–307 (2018)
7. Gangwar, G.K., Tiwari, M., Singh, R.B., Dasgupta, K.: Study of different type of hydraulic accumulators, their characteristics and applications. Int. J. Res. Aeronaut. Mech. Eng. **2**(2), 56–63 (2014)
8. Murrenhoff, H., Wallentowitz, H.: Fluidtechnik für mobile Anwendungen, Umdruck zur Vorlesung, Verlag Mainz, Wissenschaftsverlag, Aachen 1 (1998)
9. Long, M.H., Trieu, B.H., Thuat, N.C.: Experimental model of hydraulic anchoring system holds a combination of small timber transport machines. J. Water Resour. Environ. Eng. 149–152 (2019)

Investigation of the Suitability of Energy-Based Skull Fracture Criterion Application for V-dummy Finite Element Pedestrian Model

Hung Anh Ly[1,2](✉) and Bang Cao Vo[1,2]

[1] Department of Aerospace Engineering, Faculty of Transportation Engineering, Ho Chi Minh City University of Technology (HCMUT), 268 Ly Thuong Kiet Street, District 10, Ho Chi Minh City, Vietnam
lyhunganh@hcmut.edu.vn

[2] Vietnam National University, Linh Trung Ward, Thu Duc District, Ho Chi Minh City, Vietnam

Abstract. A kinematically validated modified finite element human model Hybrid III, called V-dummy, was proposed. In this paper, the V-dummy was deeply investigated of its capability of predicting skull fracture using a new energy-based criterion. 27 frontal crash test simulations were reconstructed to discover skull fracture injury risk curve. The application of this new criterion on V-dummy was found correlated based-on a statistical study with another parameter, the head kinematic-based injury criterion (HIC). The study concluded that internal energy can be used for V-dummy to predict skull fracture. The tolerance limit for 50% risk of the injury was correlated to 2000 mJ.

Keywords: Finite element model · Internal energy · Skull fracture · Statistical analysis · V-dummy

1 Introduction

According to World Health Organization's report on road safety, traffic injury took 1.35 million lives in 2018 [1]. In Vietnam, in 2019, there were total 17626 traffic accident cases that caused 7624 deaths and wounded 22152 people. Figure 1 shows the detailed data of cases, death count and injured people in traffic accident from 2015 to 2019 [2]. Arregui et at. investigated hospital data from eight European countries including over 10,000 pedestrians with more than 19,000 injuries [3]. Another study of 5000 patients from pedestrian accidents in Los Angeles County showed that head-neck injuries represent 30% of the total [4].

In the previous literature, various skull fracture criteria were proposed to be the suitable method for diagnosing skull fracture probability via acceleration, stress, strain, internal energy, etc. [5–8]. The head injury criteria (HIC) [9] is widely used over decades which at the value of 1000, it is considered the appearance of skull fracture with severe damage injuries [10]. Vorst firstly introduced the first statistically-based skull fracture criterion (SFC) which is the averaged acceleration over the HIC15 time interval gained from Hybrid III head form impact tests [7]. Chan et al. (2007) used

B. T. Long et al. (Eds.): MMMS 2020, LNME, pp. 599–611, 2021.
https://doi.org/10.1007/978-3-030-69610-8_81

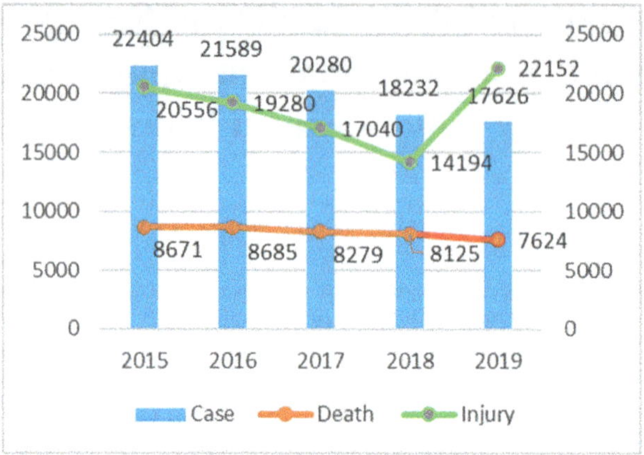

Fig. 1. Number of traffic accident cases, injured people and death 2015–2019.

numerical and statistical method to propose the strain-based SFC15 = 120 g for 15% probability of skull fracture [6]. Later, Sahoo (2015) executed 15 drop tests and reconstructed 70 pedestrians crash tests using finite element method. This study claimed that energy-based has better correlation with skull fracture injury against HIC, SFC and peak force in almost cases [11]. However, because of the variation of material selection for FE head model as well as the geometry configuration, internal energy criterion still leaves many questions because of its sensitivity to other head model with different properties and configurations [12].

The Hybrid III 50th percentiles model mimics the shape, weight, size, of the human body. Since Hybrid III 50th dummy model is built based on 50th percentile weight and height of an American man which are 175 cm and 78 kg [13], respectively, it is needed to have a proper scaled model for Vietnamese and also comply with kinematic laws of motion [14].

The goal of this study is to define whether or not the V-dummy can be used to predict skull fracture with more complex criteria than the acceleration-based one hence the potential of the Hybrid III-based V-dummy FE model can be determined. 27 pedestrian-versus-vehicle crashes was reconstructed using the V-dummy and 3 types of car models provided by National Crash Analysis Center (NCAC). The skull of the V-dummy behaviors using HIC value was extracted and validated by Nguyen Dinh Bao (2019) [15]. The HIC value gathered from these simulations combined with Hertz's method [10] would give the observation of skull fracture outcome. The internal energy of the skull was then collected and statically analyzed to establish the correlation of this parameter to the probability of skull fracture injury.

2 Material and Method

2.1 Finite Element V-dummy Model

An existing validated FE human model base on the Hybrid III [14] was used for the current study to invest its capability of predicting skull fracture using internal energy. In the previous research, the Hybrid III 50th percentile was scaled using 3 scaling techniques [16, 17]:

- Scaling of the dummy geometry
- Scaling of the inertial properties
- Scaling of joint properties

The modified human model is equivalent to a 50th percentile Vietnamese male. The scaling constant result for each direction was $\lambda = 0.9371$ so that the head weigh 3.7 kg for the V-dummy whence it is 4.5 kg for the original Hybrid III head form. Table 1 and Fig. 2 show the comparison between Hybrid III vs V-dummy in terms of weight and height. The detailed for these techniques can be found in [14].

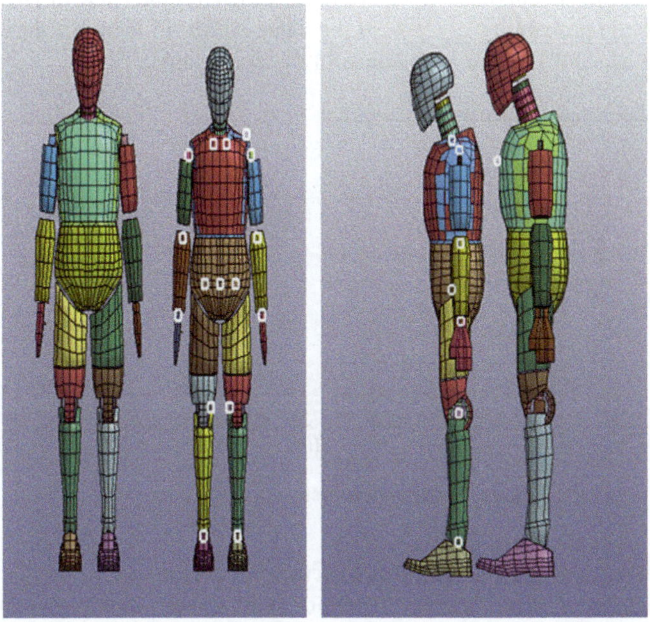

Fig. 2. Result of geometric scaling where Hybrid III 50th is the one with red head. (Color figure online)

Table 1. Weight and height of the Hybrid III and the Vietnamese dummy.

	Hybrid III	V-dummy
Height (cm)	175	164
Mass (kg)	78	58

Due to lack of real-world experiment data, Nguyen Dinh Bao (2019) chose to propose a verification method using simulation result data collected from Autoliv to make a comparison [15].

2.2 Skull Fracture Criteria

2.2.1 The Acceleration-Based Head Injury Criterion (HIC)

The most typical criterion to predict head injuries is the HIC which was adopted by NHTSA for the main head injury criterion for crashworthiness [10]. The HIC value was proposed as the following equation:

$$HIC = \max_{(t_1,t_2)}\left\{(t_2 - t_1)\left[\frac{1}{t_2 - t_1}\int_{t_1}^{t_2} a(t)dt\right]^{2.5}\right\} \tag{1}$$

NHTSA proposed to adopt the duration of 15 ms for all dummy [14].

To determine skull fracture base on HIC15¬ result gained from the simulations, the authors decided to use the skull fracture risk curve provided by Hertz (1993) [10, 21]. The skull fracture risk curve is shown in Fig. 3, which has the value of HIC15 $\geq$ 1000 for skull fracture probability $\geq$ 50%. If the value of HIC15 is higher than 1000, it can be identified as skull fracture appearance in the head model.

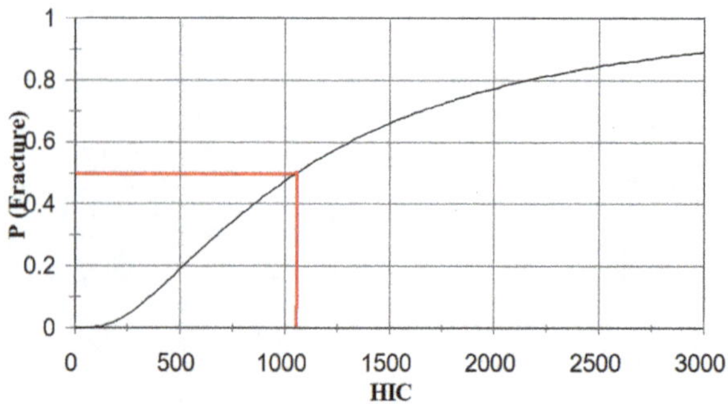

Fig. 3. Skull fracture risk curve by Hertz [37].

2.2.2 Internal Energy-Based Skull Fracture Criterion

A robust method of using energy to predict skull fracture has been developed and proposed in a series of essays by Sahoo (2013–2015) [11, 22–25]. The Strasbourg University Finite Element Head Model (SUFEHM) has been utilized to inspect skull fracture criterion using internal energy, which is the maximum amount of internal strain energy the skull absorbed during the impact before skull starts to fracture [12]. This value can only be calculated via a finite element code LS-DYNA for the whole skull [11]. The simple formula is described in Eq. 2, which can be found in LS-DYNA theory manual [26]:

$$\text{IE(after)} = \text{IE(before)} + \text{sum over all six directions of (stress, incremental strain and volume)} \quad (2)$$

The energy threshold for 50% risk of fracture ranged from 453 mJ to 833 mJ [11, 23, 27]. This criterion was statistically concluded to be the most correlated to skull fracture against other 3 parameters (force, HIC, SFC). A complication to be considered is that the internal strain energy reported in multiple researches are highly diverse. It is supposed that the strain energy values are strongly correlated with the contact area and the skull thickness near the contact area [12].

2.3 Simulation Setup

There were 27 vehicle-to-pedestrian frontal crash tests reconducted using LS-DYNA to compute and LS-PrePost to post-process, both provided by LSTC. All the finite element models in this study, which included vehicle numerical models (Chevy C2500, Dodge Neon, Ford Explorer) are provided by NCAC. The details of these models are comprised in Table 2.

Table 2. Information of the FE models in this study

	V-dummy	Chevrolet C2500	Dodge Neon	Ford Explorer
No of parts	143	251	336	923
No of nodes	7472	66586	283859	724628
No of shells	35	54565	267786	680288
No of beams	6	163	122	196
No of solids	66	3561	2852	33690
No of elements	4299	58313	270768	714205

The simulation set-up configuration was described in [15]. Figure 5 demonstrates the initial positioning and posture of the V-dummy and the Ford Explorer, same method would be applied to other cars. Table 3 shows the simulation matrix (Fig. 4).

In this study, the V-dummy was sustained frontal crashes with vehicles moving at the speed of 40 km/h, 50 km/h and 60 km/h. The CONTACT AUTOMATIC SURFACE TO SURFACE keyword was used between the model's head versus the entire car.

The HIC15 and internal energy were extracted using Ls-PrePost (LSTC) then statistically analyzed using an IBM's software called SPSS.

Fig. 4. Initial posture and positioning of V-dummy and Ford Explorer.

2.4 Statistical Analysis

In order to define the capability of predicting skull fracture of the V-dummy using internal energy as the variable, statistical analyses were conducted for HIC15 and internal energy. The study chose logistic regression to investigate the correlation between "injury" or "no injury" and internal energy [10] which was executed using version 26.0 release the statistical software SPSS.

In logistic regression, the natural logarithm of the odds ratio is equal to a linear function [32]. The probability of skull fracture can be derived as:

$$\log\left(\frac{P}{1-P}\right) = a + bx \tag{3}$$

$$P(\text{skull fracture}) = \frac{e^{a+bx}}{1+e^{a+bx}} \tag{4}$$

The quality of the correlation of internal to skull fracture was assessed via Nagelkerke R^2 statistics. The Hosmer-Lemeshow test and ROC Curve were also performed to evaluate the model performance.

3 Result

3.1 Pedestrian Crash Simulations

27 accident reconstructions have been successfully calculated using LS-DYNA platform. Firstly, the locations of impact were observed in Fig. 5. For the SUV and Pickup scenario, since their high ground clearance and their frontal geometry configuration, the dummy's head tended to hit the upper hood. On the other hand, the head hit the front windshield for the first impact in the Sedan scenario. This first observation is very important for analyzing the final outcome and for the commendation of the influence of

stiffness and shape of the impact surface [25]. The paper summed up the mechanical properties of the Pickup's hood, the Sedan's windshield and the SUV's hood for further discussion (Table 3).

Table 3. Mechanical properties of the Pickup's hood the sedan's windshield and the SUV's hood.

	Pickup's hood	Sedan's windshield	SUV's hood
Density (kg/m^3)	7890	2500	7890
Young's modulus (GPa)	200	76	210
Poisson ratio	0.3	0.3	0.3
Yield stress (MPa)	350	138	200
Material type	Steel	Glass	Steel

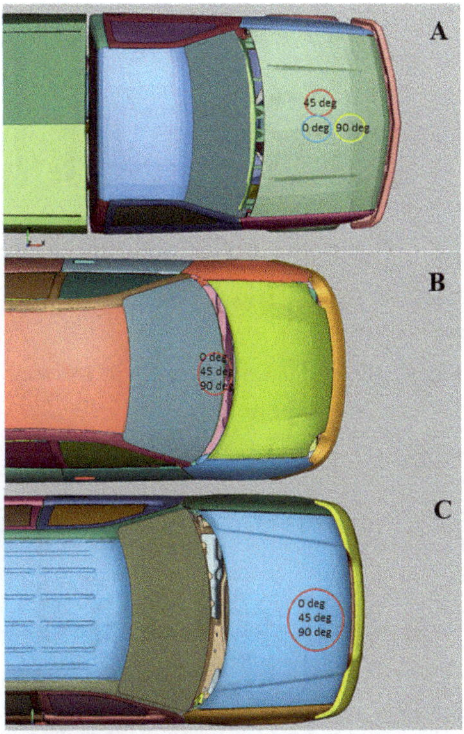

Fig. . Position of the head in the first impact (A: pickup truck; B: sedan; C: SUV).

Secondly, the HIC_{15}, Hertz's method-based binomial skull fracture result and internal energy values of the V-dummy head form are summarized. There are 13 out of 27 cases identified fracture occurred, most of them are in sedan tests (9 out of 9 cases). The result has shown that crashes with the sedan causes higher HIC_{15} (2012-17940) and higher internal energy (1640–9640 mJ) than the other cases.

Thirdly, according to Fig. 6, 7 and 8, the SUV seems to be the friendliest vehicle to pedestrian against the Pickup and the Sedan. For the SUV, there is only one case (60 km/h–0°) that the fracture occurred. The pickup truck also had less injured cases (3 out of 9): 50 km/h–45°, 60 km/h–0° and 60 km/h–45°.

Finally, it should be recognized that with the same vehicle and the V-dummy initial angle (IA) of 90°, the HIC and internal energy values are relatively lower than other two initial angle cases. This search proved that injury risk can distribute differently for distinct regions of the skull.

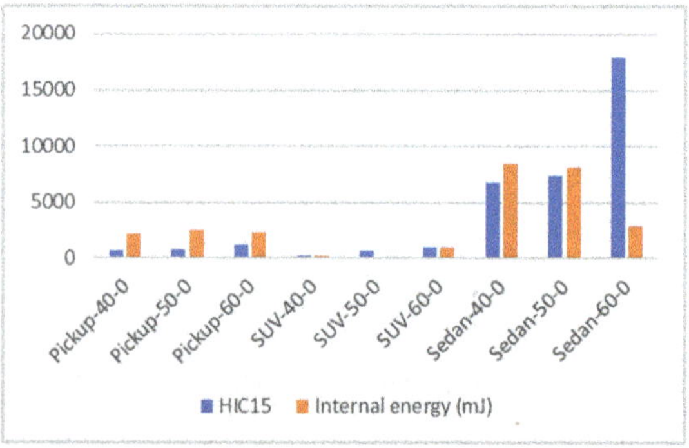

Fig. 6. HIC and internal energy in 0° IA cases.

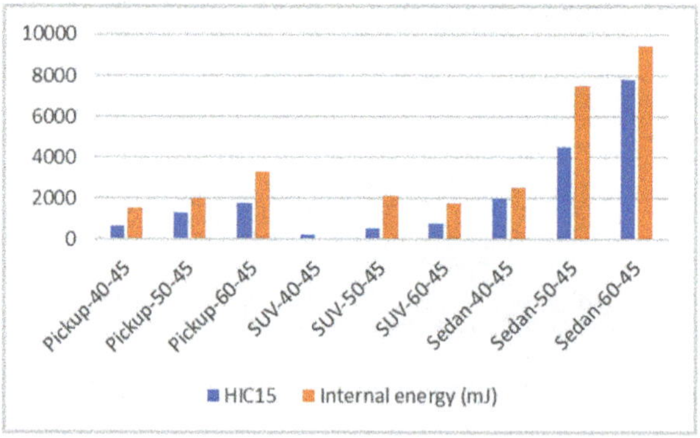

Fig. 7. HIC and internal energy in 45° IA cases.

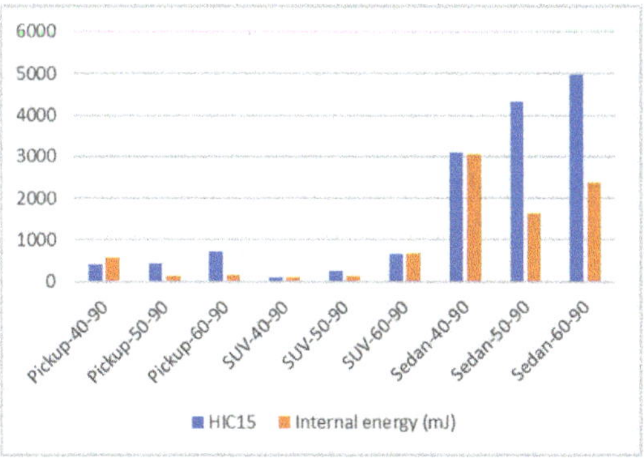

Fig. 8. HIC and internal energy in 90° IA cases.

3.2 Result of Statistical Method

The result of the ROC curve analysis for internal energy gathered from 27 frontal crash simulations are reported in Fig. 9. The area under ROC curve (AUROC) was computed by SPSS which result is shown in Fig. 10. The Nagelkerke R^2, −2log-likelihood, HL test p-value are also obtained in Table 4.

Table 4. Assessment parameters for logistic regression model.

AUROC	Nagelkerke R-square	Hosmer-Lemeshow test p-value
0.923	0.66	0.835

The internal energy presented high performance in 4 tests. The AUROC of 0.923 showed the excellent accuracy of the model. The Nagelkerke R^2 of 0.66 (> 0.5) proved the model we were using is appropriate and correlate. The HL test p-value was also relatively high (p-value = 0.835) demonstrated the model was badly fit to the null hypothesis which made it fit the regression model well, the higher the p-value the better model fit to the data. Finally, injury risk curve for the predictor for skull fracture are plotted in Fig. 10. The dash black line predicts the probability of skull fracture corresponding the value of internal energy and red circles are representative of 27 simulation cases. According to the curve, the internal energy of 2000 mJ is related to 50% risk of skull fracture.

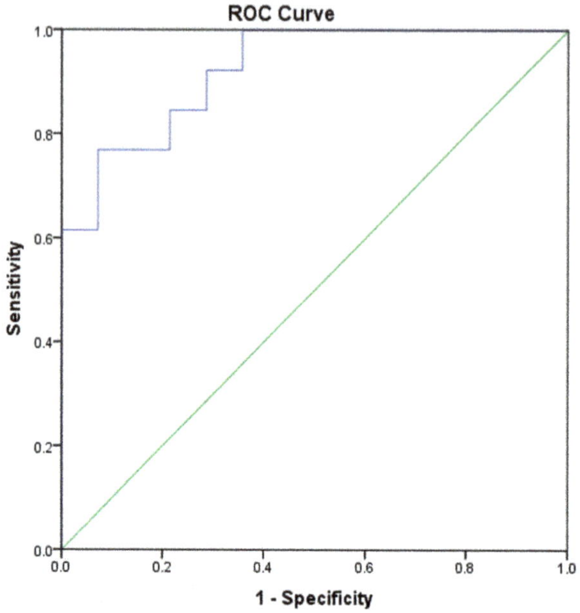

Fig. 9. ROC curve with AUROC = 0.923

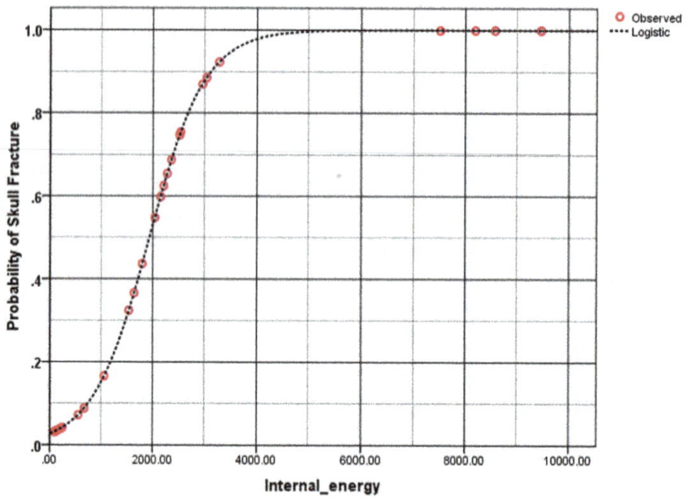

Fig. 10. Injury risk curve for skull fracture based on internal energy computed with the V-dummy. A 50% skull fracture risk is obtained for skull internal energy of 2000 mJ

4 Discussion

27 crashes were reconstructed using a finite element software (LS-DYNA) and a statistical method was conducted using SPSS which had given reliable result due to 3 statistical assessments (Nagelkerke R^2, AUROC and Hosmer-Lemeshow test p-value). All three assessments gave the result of supporting the reliability, goodness of fit to data and accuracy of this regression model.

Despite these optimistic results, there are still several issues that need taken into account. Comparing with other researchers' study (Fig. 11 and 12), the value of internal energy causes fracture is still in wide range and not yet been narrowed [11, 12, 23, 27, 28]. This paper proposed to use a fully developed detailed FE model called THUMS [29] to be a consist, reliable, accurate tools for future tests.

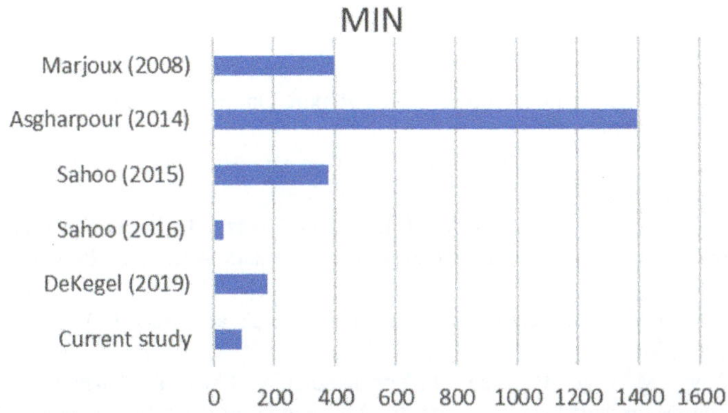

Fig. 11. Comparison of the minimum internal energy causes skull fracture with other researchers.

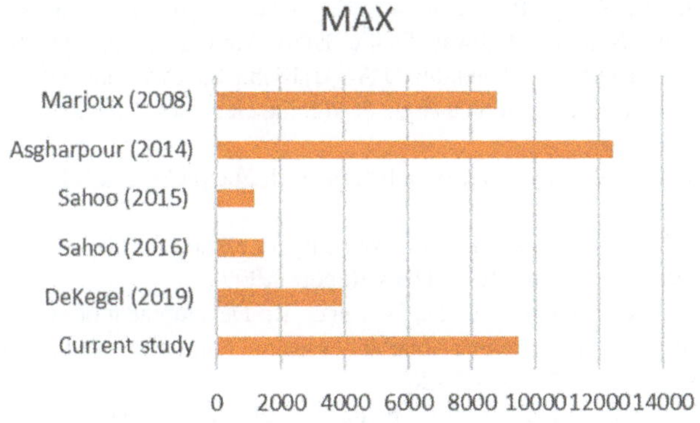

Fig. 12. Comparison of the maximum internal energy causes skull fracture with other researchers.

5 Conclusion

The paper provided an answer of capability of the V-dummy predicting skull fracture when using an up to date criterion. Numerically and statistically, the result supported the potential of this hypothesis. The skull fracture risk curve for V-dummy was also plotted, 50% risk corresponding the internal energy value of 2000 mJ. Logistic regression was performed for the parameter. Based on high statistical assessment result (Nagelkerke R2 = 0.66, AUROC = 0.923, HL p-value = 0835), the skull fracture was then concluded to be well compatible to the V-dummy. There are still some limitations in this study to improve the confidence level for diagnosing skull fracture.

Acknowledgment. This research is funded by Vietnam National University Ho Chi Minh City (VNU-HCM) under grant number C2019-20-04.

References

1. World Health Organization: Global Status Report On Road Safety 2018. World Health Organization, Geneva (2018)
2. Tổng Cục Thống Kê, Niên giám thống kê (Tóm tắt), Hà Nội, Việt Nam: Nhà xuất bản thống kê 2015–2019
3. Arregui-Dalmases, C., Lopez-Valdes, F.J., Segui-Gomez, M.: Pedestrian injuries in eight European countries: an analysis of hospital discharge data. Acc. Anal. Prev. **42**, 1164–1171 (2010)
4. Peng, R., Bongard, F.S.: Pedestrian versus motor vehicle accidents. J. Am. Coll. Surg. **189**, 343–348 (1999)
5. Vander Vorst, M., Chan, P., Zhang, J., Yoganandan, N., Pintar, F.: A new biomechanically-based criterion for lateral skull fracture. In: Association for the Advancement of Automotive Medicine (2004)
6. Vorst, M.V., Stuhmiller, J., Ho, K.: Statistically and biomechanically based criterion for impact-induced skull fracture. In: 47th Annual Proceedings Association for the Advancement of Automotive Medicine, pp. 365–381, 22–24 September 2003
7. Chan, P., Lu, Z., Rigby, P., Takhounts, E.: Development of a generalized linear skull fracture criterion. National Highway Traffic Safety Administration, vols. 07-0227 (2007)
8. Katagiri, M., Katagata, K., Pramudita, J.A., Ujihashi, S.: Development and application of stress-based skull fracture criteria using a head finite element model. J. Biomech. Sci. Eng. **7** (4), 449–462 (2012)
9. Versace, J.: A review of the severity index. In: 15th Stapp Car Crash Conference, Corondo, USA (1971)
10. Eppinger, R., et al.: Development of improved injury criteria for the assessment of advanced automotive restraint systems- II, NHTSA Report (1999)
11. Sahoo, D., Deck, C., Yoganandan, N., Willinger, R.: Development of skull fracture criterion based on real-world head trauma simulations using finite element head model. J Mech. Behav. Biomed. Mater. **57**, 24–41 (2015)
12. Kegel, D.D., et al.: Skull fracture prediction through subject-specific finite element modelling is highly snesitive to model parameters. J. Mech. Behav. Biomed. Mater. **100**, 103384 (2019)

13. Foster, J.K., Kortge, J., Wolanin, M.: Hybid III - a biomehcanically-based crash test dummy. In: 21st Stapp Car Crash Conference (1977)
14. Ly, A.H., Nguyen, B.D., Nguyen, H.A.: Head injury of Vietnamese pedestrian in crash accident with SUV using numerical simulation. In: Bach Khoa Youth Science and Technology Conference, Ho Chi Minh (2019)
15. Nguyen, B.D.: Database construction of vietnamese injury criterion in traffic collision: pedestrian injury criterion responses in collision, Ho Chi Minh (2019)
16. Hyncik, L.: On scaling of human body models. University of West Bohemia (2007)
17. Untaroiu, C.D.: A study of the pedestrian impact kinematics using finite element dummy: the corridors and dimensional analysis scaling of upper body trajectories. Int. J. Crashworthiness **13**(2008), 468–478 (2008)
18. Gadd, C.: National Council Publication **97**, 141–144 (1961)
19. Prasad, P., Mertz, H.J.: The position of the united states delegates to the ISO working group 6 on the use of HIC in the automotive environment. In: SAE, vol. 85, p. 1246 (1985)
20. Hodgson, V.R., Thomas, L.M.: Effect of long-duration impact on head. In: 16th Stapp Car Crash Conference (1972)
21. Mertz, H.J., Prasad, P., Nusholtz, G.: Head injury risk assessment for forehead impacts. SAE Technical Papers (1996)
22. Sahoo, D., Deck, C., Willinger, R.: Development and validation of an advanced anisotropic visco-hyperelastic human brain FE model. J. Mech. Behav. Biomed. Mater. **33**, 24–42 (2013)
23. Sahoo, D., Deck, C., Yoganandan, N., Willinger, R.: Anisotropic composite human skull model and skull fracture validation against temporo-parietal skull fracture. J. Mech. Behav. Biomed. Mater. **28**, 340–353 (2013)
24. Sahoo, D., Yoganandan, N.: Influence of head mass on temporo-parietal skull impact using finite element modelling. Med. Biol. Eng. Comput. **53**(9), 869–878 (2015)
25. Sahoo, D., Deck, C., Yoganandan, N., Willinger, R.: Influence of stiffness and shape of contact surface on skull fractures and biomechanical metrics of the human head of different population under lateral impacts. Accid. Anal. Prev. **80**, 97–195 (2015)
26. Livermore Software Technology Corporation, LS-DYNA Theory Manual. Livermore Software Technology Corporation, Livermore (2007)
27. Marjoux, D., Baumgartner, D., Deck, C., Willinger, R.: Head injury prediction capability of the HIC, HIP, SIMON and ULP criteria. In: IRCOBI Conference - Madrid (Spain) (2006)
28. Asgharpour, Z., Baumgartner, D., Willinger, R., Graw, M., Peldschus, S.: The validation and application of a finite element human head model for forntal skull fracutre analysis. J. Mech. Behav. Biomed. Mater. **33**, 16–23 (2014)
29. Kato, D., Nakahira, Y., Atsumi, N., Iwamoto, M.: Development of human-body model THUMS version 6 containing muscle controllers and application to injury analysis in frontal collision after brake deceleration. In: IRCOBI Conference (2018)
30. Hertz, E.: A note on the head injury criterion (HIC) as a predictor of the risk of skull fracture (1993)

Low-Cost High Voltage Pulse Source Driven by Solar Panel for Green Plasma Generation Used in Environmental Application

Hoa Thi Truong(✉) and Nam Van Nguyen

The University of Danang, University of Technology and Education, Da Nang, Vietnam
tthoa@ute.udn.vn

Abstract. The non-thermal plasma as Dielectric-Barrier Discharges (DBD) plasma is a promising high voltage technology applied in many industrial fields. However, the high cost of the high voltage pulse source used in DBD generation is an issue for the application of DBD-based technologies, especially in the field of environmental treatment. This research presents a low-cost, high voltage pulse source used the characteristic of Silicon Diodes for Alternating Current (SIDAC) accompanied with a high voltage multiplier and the self-termination characteristic of the dielectric barrier discharge (DBD) without external control. The pulse source is designed with a simple modular structure, which requires no control, eliminates transformers requirement, results in reduced dimensions, weight, and high scalability at a low cost. Furthermore, the power supply can be powered directly by the output of the solar panel, which could be a solution for low-cost green DBD plasma generation in an area without power grid connection.

Keywords: Silicon Diodes for Alternating Current (SIDAC) · Dielectric-Barrier Discharge (DBD) · High-voltage pulse source

1 Introduction

Dielectric barrier discharges (DBDs) are plasmas generated in a reactor with an insulating (dielectric) material covering the electrodes. The dielectric is responsible for a self-pulsing discharge. DBD plasma is a typical type of non-thermal plasma at atmospheric pressure [1]. Initially used for the ozone generation, the application of DBD has been opened up to many other fields, e.g., surface treatment, exhaust gas decomposition, water purification, medical treatment, combustion assistance, light source [2, 3]. When the electric field in the discharge gap reaches the breakdown voltage, discharge occurs, then charged particles deposit on the dielectric layers. The charge deposition leads to electric field distortion effects in areas of the surface discharge. Because of electric field distortion effects, the formation of a spark or an arc discharge is prevented, and DBD is kept operating in a non-thermal regime [3–5]. Furthermore, the dielectric layers also work as coupling capacitors. As a result, high-frequency alternating voltage sources or bipolar pulsed power sources are frequently employed. However, these types of power sources are complicated and expensive for widespread application [1–5].

B. T. Long et al. (Eds.): MMMS 2020, LNME, pp. 612–617, 2021.
https://doi.org/10.1007/978-3-030-69610-8_82

In this study, a designation of a high-voltage power source used for a low-cost DBD generation is proposed. This design includes Cockcroft–Walton high voltage multiplier, a SIDAC connection, and load of DBD reactor.

2 Modular Pulse Source

2.1 Voltage Multiplier

Cockcroft-Walton is a voltage multiplier. It is a simple circuit with only capacitors and diodes in a cascade network that multiply the low peak input voltage values of an AC power source or a DC pulse source to an extremely high DC output voltage value. The voltage multiplier shows a lot of advantages over a conventional high voltage power supply with a high voltage transformer used in an application where a high voltage at low current is required. Cockcroft-Walton has been used in electric discharge tubes in traditional televisions, oscillators, photocopiers, X-ray generators where the use of high voltage power supply with low power is necessary [7]. Figure 1 illustrates the operation of a one-level voltage multiplier. In the first half negative cycle of the input voltage wave, the potential on diode D1 is positive, so D1 forward biased, let current passing through and charging capacitor C1 to the peak of the input voltage $V_1 = V_0$. Simultaneously, the potential on diode D2 is negative, then D2 is reverse biased and non-conductive. In the second half positive cycle of the input voltage, the potential on diode D2 is positive, the potential on D1 is negative, so D1 is out of the circuit, D2 is forward biased, let current flow through to charge to capacitor C2. Therefore after several cycles of input voltage, the output voltage of the 1-level multiplier would be stabilized as:

$$V_2 = V_1 + V_0 = 2V_0 \tag{1}$$

If there is a circuit with an N-level multiplier, the output voltage of the circuit is theoretical calculated as:

$$V_{out} = V_0 * N \tag{2}$$

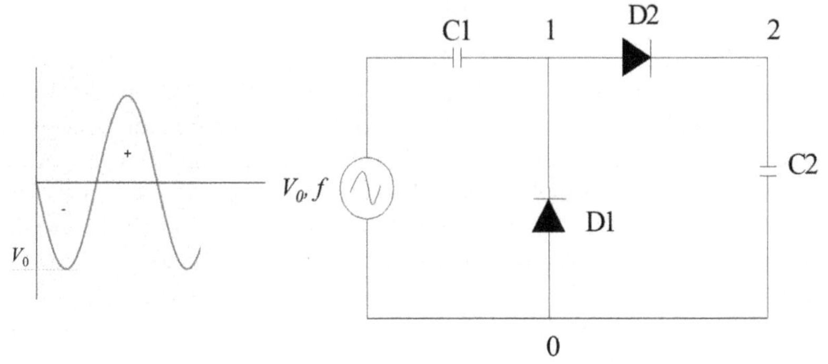

Fig. 1. Circuit diagram of 1-level voltage multiplier

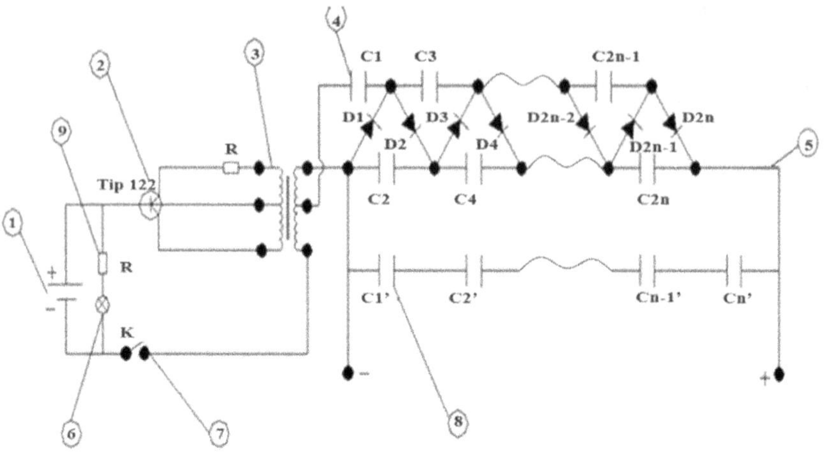

1– DC Input Source (12 V)	4– High Pressure ceramic capacitor 472- 3 KV, C = 4,7nF	7– Switch
2– Transistor Tip122	5– Diode 1N4007 (1 kV, 1A)	8– CBB 223J-2 KV capacitor, C = 0.02 uF
3– Transformers pulse	6– LED Light	9– Resistor

Fig. 2. The Cockcroft–Walton voltage multiplier (VM)

In this research, the Cockcroft–Walton voltage multiplier was designed by using very cheap devices as diodes and capacitors in a simple ladder circuit as Fig. 2. In order to make the circuit suitable for directly connecting to an output of a solar panel, a DC voltage source with a low input voltage of 12 V is used as a primary power supply. This low DC input voltage source can be a conventional DC power supply, batteries, or a solar panel output. The Tip122 transistor and the pulse transformer convert an input DC voltage to output voltage in the range of (300–330 V) with frequency up to approximately 70 kHz. The connection of capacitors (CBB 223J-2 kV) is used to stabilize the output voltage. The voltage multiplier was scaled to convert the pulse transformer's pulses to up to an output voltage of about 16 kV.

2.2 Silicon Diode for Alternating Current

Silicon Diode for Alternating Current (SIDAC) is a bi-direction high-voltage switching device that can be switched from off-state (none conducting state) to the on-state or conductive state with both polarities. The high-speed transition from off-state to on-state is done by simply exceeding the break-over voltage in both directions. It has a high switching voltage and can simply create a pulse up to the kV at a low cost. The V-A characteristic of the SIDAC is shown in Fig. 3; if the magnitude of voltage on SIDAC is less than the value of its break-over voltage, the SIDAC is in off-state (red line). When the voltage equals or higher the break-over voltage (V_{BO}), the SIDAC switches from the off-state to the on-state and jumps to the transition region (green lines) with a nanosecond transition time. The on region (blue line) will be sustained until the current is dropped below the hold current (I_H). If there are N SIDACs in the connection, the break-over voltage of the connection increase N times, but the hold current (I_H) of the connection is kept the same as one's. In this study, the SIDACs (K1V38 (W)) Shindengen Electric Mfg.Co.Ltd.) with switching voltage of about 360–400 V are used. If twenty SIDACs are connected serially, the switching voltage will be increased up to ~4000 V [6, 8].

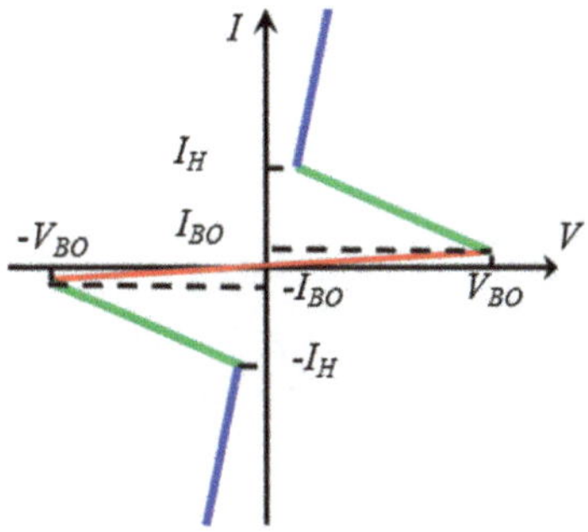

Fig. 3. V-A characteristic of SIDAC

2.3 High Voltage Pulse Generation

The experiment of DBD plasma generation using the proposed designation has been setting up with a circuit diagram shown in Fig. 4. The experiment has been done under the condition shown in Table 1. In this setup, the output of the Cockcroft–Walton voltage multiplier will charge the parallel capacitor C (1nF); this capacitor is used as buffer energy to stabilize the output voltage of the multiplier circuit. A 10 MΩ resistor is connected serially with the buffer capacitor in order to limit current and protect the multiplier circuit. A serial connection of 12 SIDACs plays the main role for pulse generating. The voltage on the SIDAC connection is determined by subtracting voltage on the buffer capacitor and voltage on the reactor. If the result of the subtraction reaches the break-over voltage of the SIDAC connection, the SIDAC is changed their state to become conductive. Simultaneously, a pulse is generated and applied to the DBD reactor. The shunt resistor (10 Ω) with very low inductance is used to observe current flow through the main circuit, a high voltage probe (500 MΩ, 60 Hz) is used to observe voltage on the high voltage electrode of the reactor. Current and voltage signals will be obtained by an oscilloscope (Tektronix TBS 1102B-EDU - 100 MHz).

Table 1. Experimental condition

Output of VM circuit	14 kV
Number of SIDAC	12
Working gas	He
Gas flow	1 L/min
Pressure	Atmospheric pressure

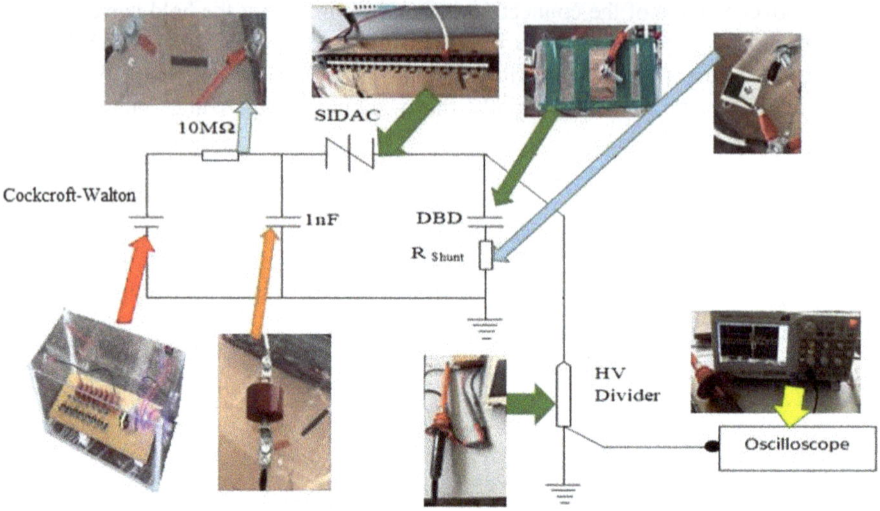

Fig. 4. Schematics of the experimental setup

The DBD reactor was made of two glass plates with a thickness of 0.5 mm. The discharge gap was 1 mm. Helium gas is used as a working gas and blown to the reactor with a flow rate of 1 L/min.

3 Results and Discussion

The proposed high voltage pulse source has been succeeded in generating a high rising rate high voltage pulse train applied on the high voltage electrode of the DBD reactor (shown in Fig. 5). The pulses change from around 10 kV to around 14 kV. As can be seen from this figure that, when SIDAC connection is conductive, the pulse is generated with a frequency of about 80 Hz and changed sharply from around 10 kV to 14 kV.

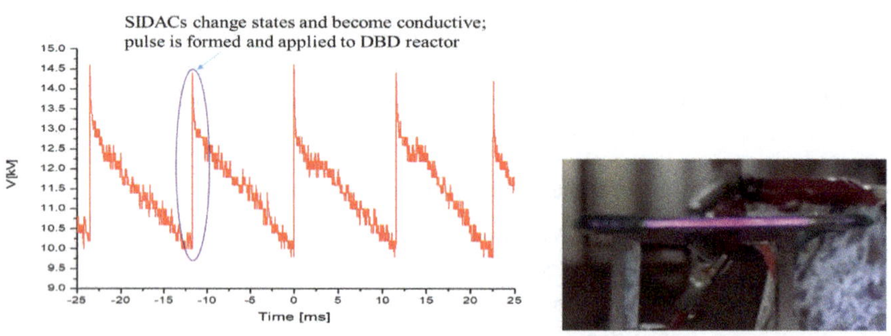

a) Waveform voltage on plasma reactor b) Image of Plasma

Fig. 5. High voltage (HV) pulses' waveform and plasma generated by applying HV pulses

4 Conclusions

In this study, the Cockcroft-Walton Voltage Multiplier with a simple cascade circuit has been employed to convert a low DC input voltage as an output of a solar panel to an extreme high DC voltage value.

The high-speed switching characteristic of SIDAC has been used as pulse manipulation. The connection in the series of the output of voltage multiplier, SIDACs, and a load of plasma reactor has generated a train of high voltage pulses and applied to the plasma generation.

Acknowledgments. This research is funded by Funds for Science and Technology Development of the University of Danang under project number B2019-DN06-17.

References

1. Kogelschatz, U.: Dielectric-barrier discharges: their history, discharge physics, and industrial applications. Plasma Chem. Plasma Process. **23**, 1 (2003)
2. Kogelschatz, U., Eliasson, B.: Ozone generation and applications. In: Chang, J.S., et al. (eds.) Handbook of Electrostatic Processes, vol. 581. Marcel Dekker, New York (1995)
3. Brandenburg, R.: Dielectric barrier discharges: progress on plasma sources and on the understanding of regimes and single filaments. Plasma Sources Sci. Technol. **26**, 053001 (2017)
4. Raizer, Y.P.: Gas Discharge Physics. Springer, Berlin (1991)
5. Roth, J.R.: Industrial Plasma Engineering. Applications to Non-thermal Plasma Processing, vol. 2. CRC Press, Boca Raton (2001)
6. Truong, H.T., Hayashi, M., Uesugi, Y., Tanaka, Y., Ishijima, T.: Novel design of high voltage pulse source for efficient dielectric barrier discharge generation by using silicon diodes for alternating current. Rev. Sci. Instrum. **88**, 065105 (2017)
7. Malik, M.A., Schoenbach, K.H., Abdel-Fattah, T.M., Heller, R., Jiang, C.: Low cost compact nanosecond pulsed plasma system for environmental and biomedical applications. Plasma Chem. Plasma Process. **37**, 59 (2017)
8. Truong, H.T., Uesugi, Y., Tanaka, Y., Ishijima, T.: Effects of dielectric properties on electrical characteristics of dielectric barrier discharge generated by low frequency uni-polar high voltage pulses. Jpn. J. Appl. Phys. **58**, 111001 (2019)

Radon, Thoron Gas Concentration and Level Living in Ban Gie Monazite Mineral Sand Mine Area, Quy Hop District, Nghe An Province, Vietnam

Nguyen Van Dung(✉) and Vu Thi Lan Anh

Faculty of Environmental, Hanoi University of Mining and Geology, Hanoi 11910, Vietnam
nguyenvandung@humg.edu.vn

Abstract. Radon and thoron are inert gases which do not participate in any chemical complexes. The radioactive danger of radon-222 (^{222}Rn) is very high, compared to thoron (^{222}Rn) and radon-219 (^{219}Rn), due to the long half-life of 3.8 days as compared to 55 and 4 s, respectively. As a gas, radon can easily escape from the surfaces of ore minerals, rocks and soils can be dissolved in underground water, and can be transported in long distance from the location of formation. All of these radioactive isotopes emit alpha radiation, but ^{222}Rn is the most important one since it is the major cause of internal dose by respiration with threats to the human health and environment. In this paper, the authors employed the trace detector technique in investigating the indoor in households surrounding the Ban Gie monazite mine, Quy Hop district, Nghe An province.

Level of indoor concentration of in households 25 to 200 Bqm^{-3}, with some outliers higher than 600 Bqm^{-3}, which is much higher radon proposed (EPA).

The results of calculating the annual effective dose caused by radon and thoron show that people in the survey area show that people can be exposed to an average dose of 4.4 mSvyear^{-1}, nearly 3.5 times higher with an average of the world is 1.25 mSvyear^{-1}.

Keywords: Radon · Thoron · Trace detector · Monazite mine · Effective dose

1 Introduction

In nature, radon and its isotopes are decay chains of natural radioactive chains: chain ^{238}U (born ^{222}Rn); chain ^{232}Th (born ^{220}Rn also called thoron); ^{235}U chain (born ^{219}Rn also called actinon, has a half-life of 3.94 s).

Radon levels are highly active and through inhalation of them and their radioactive descendants can cause significant exposure of the people with ionizing radiation [1–3]. Concentrations of air, their of ^{220}Rn they can only move with the therefore are often less concerned than ^{222}Rn. Radon-222 levels can accumulate in surface air, especially in confined spaces such as in houses, caves and mines and can cause significant radiation exposure [4–6].

B. T. Long et al. (Eds.): MMMS 2020, LNME, pp. 618–628, 2021.
https://doi.org/10.1007/978-3-030-69610-8_83

Exposure to ^{222}Rn in now mentioned in medical radioactive gas can cause cancer shows levels of ^{222}Rn and lung cancer [6, 7].

According to the United States Environmental Protection Agency (EPA) [7], radon is the second leading cause of lung cancer after tobacco; Lung cancer due to lethal radon gas ranks second among all cancers in general, about 21,000 Americans die from lung cancer every year from radon gas. In today's civilized age, the demand for quality of life is very concerned, the more society develops, the more attention to radon gas in the home needs more attention.

Since ^{238}U, ^{232}Th and ^{235}U are components of the Earth's crust, radon can be found ^{238}U, ^{232}Th, ^{235}U and therefore radon, especially ^{222}Rn, can accumulate in houses that reach higher concentrations in regular air [4, 6]. Some areas tend to have higher levels of ^{222}Rn indoors in surface air, especially areas of granite, uranium, titanium, monazite, phosphate [5].

Studies on radon are done a lot, because radon is contribution to the internal dose for humans caused by natural radiation is up to 59% (UNSCEAR 2000) [16]. However, in radon measurements, it is often used to measure the total concentration of all isotopes of radon (including ^{222}Rn-radon, ^{220}Rn-thoron, ^{219}Rn-actinon) but not interested in individual components of each isotope. Although the contribution of radon is essential, thoron isotopes also contribute 10 ÷ 15% to radiation doses for humans. With a half-life about 4 s, actinon almost does not contribute to the amount of radon in the air and hardly affects the human body.

Nuclear trace detector (SSNDT) is one of the tools that separates the content of radon and thoron based on the difference in their half life.

The paper presents the technique of detecting nuclear trace (SSNDT LR-115) to determine the concentration of radon, thoron gas and assessing the internal radiation dose caused by radon and thoron in households living in the area of monazite mine Ban Gie, Quy Hop, Nghe An province.

2 Geological Mineral Characteristics of the Monazite Mine of Ban Gie

Ban Gie monazite mine has geographical coordinates 19°46′30″ North latitude, 105°29′ 50″ East longitude in Yen Hop commune, Quy Hop district, in the west of Nghe An province, 130 km from Vinh city. In the field, 05 monazite ore bodies were discovered in valleys with an area of 550,000 ÷ 1,100,000 m^2, with reserves of monazite of 190,356 tons; ilmenite 826,990 tons; The results of mineral composition analysis include: monazite, ilmenite, xenotin, zircon, rutil, silver, uranium and thorium. Chemical composition: monazite = 150 ÷ 4,800 gm^{-3}; ilmenite = 200 ÷ 2,734 gm^{-3}; zircon = 29 ÷ 143 gm^{-3}; Ag = 167 gm^{-3}; U_3O_8 = 0.055 ÷ 0.087%; ThO_2 = 4.62 ÷ 6.61%. At the mine, it causes radioactive anomalies to the area with a radiation intensity of 25 ÷ 170 μRh^{-1}; anomalies of radon gas concentration from 20 ÷ 740 Bqm^{-3} [8–10].

At the monazite mine, the exploration and evaluation of reserves has been conducted and will be put into operation in the future. Activities of mining and processing monazite ore will make the process of dispersing radioactive substances, especially radioactive radon, thoron to the environment more powerful and complicated.

At the monazite mine, people in the area often exploit sand and stone… at the mines to build houses and exploit underground water for daily activities, also makes the ability to spread radon concentration in the house soar.

3 Materials and Methods

3.1 Measuring Technique

Nuclear trace detector (SSNTD): A material known as a nuclear trace detector used to measure charged particles needs to meet two criteria: one is that the resistivity must be greater than 2,100 Ωcm and two is the temperature diffusion must be less than 0.05 cm^2 s^{-1}. SSNDT LR-115 is a type that satisfies both of the above conditions and can be used to record heavy charged particles and alpha particles with energies between 1.7 and 4.2 MeV. SSNDT LR-115 consists of a thin layer of cellulose nitrate and is coated on a 100 μm thick polyester (PET) sheet. When the alpha particle enters the sensitive layer of the trace detector, it will create a hidden mark [11, 12].

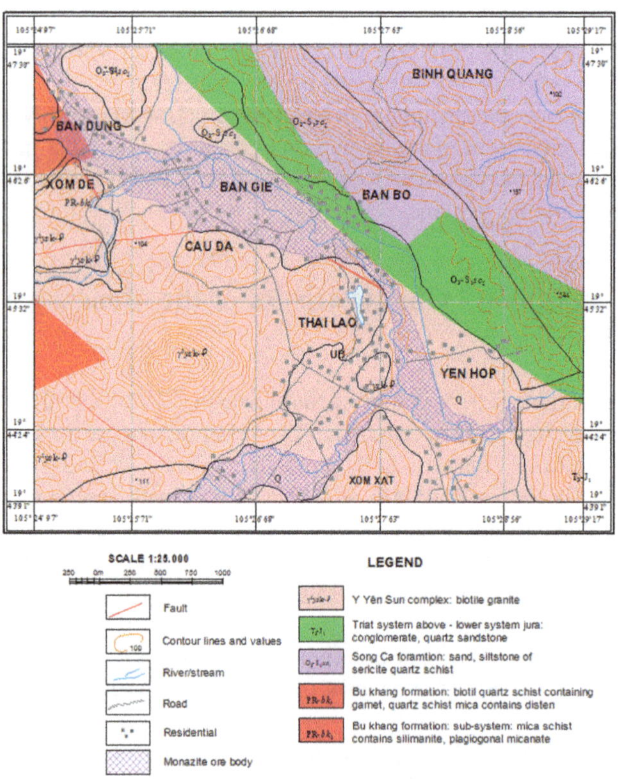

Fig. 1. Geological map of Ban Gie area

SSNTD LR-115 is cut to medium size inside the detector box of size 3 × 3 × 4 cm. The configuration is researched and manufactured by the Nuclear Science and Technology Institute. Each box can attach up to two SSNTD LR-115 sized 3 × 3 × 4 on two box covers, radon gas will enter the inner cavity. Normally, alpha particles produced from radon and descendants in the air cavity to SSNDT LR-115 have relatively large energy so it is impossible to record the interaction of the film in the configuration. Therefore, these SSNTD LR-115 are lined with a very thin aluminum layer (18 μm thick) just in front of the sensitive layer of SSNTD LR-115 [11]. This aluminum foil reduces alpha particle energy to the sensitive energy area (1.7 ÷ 4.2 MeV) before interacting with SSNTD LR-115.

In the area of the monazite Ban Gie mineral deposit, the author set up 70 detector traces (SSNTD LR-115) at the houses, schools and agencies located on the residential areas of Ban Dung village (15 stains), Ban Gie village (20 stains), Yen Hop village (20 stains) and Xom Xat village (15 stains). At each location we placed two SSNTD LR-115 indoors and outdoors to determine the concentration of radioactive (outside the house) and indoors (placed in the bedroom office).

Measurements were made between November and April. Average annual the mean concentration during may estimate because region a tropical climates may not have levels. November to April is a winter time, a time of year when people tend to close their windows at night. The total time for stain detector is 150 days. During the installation, the bar the an altitude to 2 m above. It is then collected and the count is taken and the results are obtained.

3.2 Evaluate the Annual Effective Dose Caused by Radon and Thoron in Households

The internal projection dose is determined by the sum of the effective dose projected in the respiratory tract by radon, thoron gas indoors and outside of the households living in the survey area, determined by the following formulas [7, 13, 14]:

$$E = E_{in(Rn/Tn)} + E_{out(Rn/Tn)} \tag{1}$$

Here E is the effective dose shown in the radon and thoron gases.

In which : $$E_{in(Rn/Tn)} = C_{in(Rn/Tn)} \times F_{in(Rn/Tn)} \times t_{in} \times K_{Rn/Tn} \tag{2}$$

$$E_{out(Rn/Tn)} = C_{out(Rn/Tn)} \times F_{out(Rn/Tn)} \times t_{out} \times K_{Rn/Tn} \tag{3}$$

$C_{in(Rn/Tn)}$ and $C_{out(Rn/Tn)}$ is annual average radon or thoron concentration (Bqm^{-3}) inside ($C_{in(Rn)}$ or $C_{in(Tn)}$) and outside ($C_{out(Rn)}$ or $C_{out(Tn)}$) of the dwellings. $F_{in(Rn/Tn)}$ and $F_{out(Rn/Tn)}$ is the inside and outside equilibrium factors for radon and its progenies or thoron and its progenies. These values were taken from UNSCEAR 2000 [14] with $F_{in\ (Rn)} = 0.4$ and $F_{out\ (Rn)} = 0.6$ for radon, and $F_{in(Tn)} = 0.02$ and $F_{out\ (Tn)} = 0.003$ for thoron; t_{in} and t_{out} is annual spending hours inside (7,000 h) and outside (1,760 h); $K_{Rn/Tn}$ is USCEAR 2000 recommended dose conversion factors, with $K_{Rn} = 9$ nSv per unit integrated radon concentration ($Bqhm^{-3}$) and $K_{Tn} = 40$ nSv per unit integrated thoron concentration (Bqh m^{-3}). The equilibrium coefficient of thoron and its descendants, was taken from UNSCEAR 2000 [14].

4 Results and Discussion

4.1 Radon and Thoron Concentrations

Results of measurement of radon and thoron concentrations inside and outside the area of Ban Gie monazite mineral sand mine, Quy Hop district, Nghe An province are shown in Table 1.

In Ban Gie village, indoor radon concentrations range from 20 to 674 Bqm^{-3}, an average of 168 Bqm^{-3}. Comparing to the worldwide average of indoor radon concentration of 37 Bqm^{-3} as reported by UNSCEAR 2000 [14], the average radon concentration in houses in Ban Gie village is nearly 4.5 times higher compared to the world average. Outdoor radon concentration in Ban Gie village ranges from 12 to 478 Bqm^{-3}, averaging 143 Bqm^{-3} is 14 times higher than the world average of 10 Bqm^{-3} as reported by UNSCEAR 2000 [14].

In Yen Hop village, inside radon concentration ranges from 18 to 161 Bqm^{-3}, the average is 63.3 Bqm^{-3}. Comparing to the worldwide average of inside radon concentration of 37 Bqm^{-3} as reported by UNSCEAR 2000 [14], the average radon concentration in houses in Yen Hop village is nearly 1.7 times higher compared to the world average. Outside radon concentration in Yen Hop village ranges from 8 to 68 Bqm^{-3}, the average of 51.6 Bqm^{-3} is 5 times higher than the world average of 10 Bqm^{-3} as reported by UNSCEAR 2000 [14].

In Xom Xat village, inside radon concentrations ranged from 62 to 413 Bq m^{-3}, an average of 154 Bqm^{-3}. Comparing to the worldwide average of inside radon concentration of 37 Bqm^{-3} as reported by UNSCEAR 2000 [14], the average radon concentration in houses in Xom Xat village is nearly 4.1 times higher compared to the world average. Outside radon concentration in Xom Xon village ranges from 41 to 277 Bqm^{-3}, an average of 125 Bqm^{-3} is 12 times higher than the world average of 10 Bqm^{-3} as reported by UNSCEAR 2000 [14].

In Ban Dung village, inside radon concentrations range from 13 to 614 Bqm^{-3}, an average of 66 Bqm^{-3}. Comparing to the worldwide average of inside radon concentration of 37 Bqm^{-3} as reported by UNSCEAR 2000 [14], the average radon concentration in houses in Ban Dung village is nearly 1.8 times higher compared to the world average. The concentration of outdoor radon in Ban Dung village ranges from 29 to 505 Bqm^{-3}, an average of 66.3 Bqm^{-3} which is 6 times higher than the world average of 10 Bqm^{-3} as reported by UNSCEAR 2000 [14].

Thus, the average concentration of radon inside and outside in the survey areas is higher than the world average as reported by the United Nations Scientific Committee on the Effects of Atomic Radiation (UNSCEAR). This is because all the households live near the monazite deposit containing uranium and thorium radionuclides (Fig. 1).

The level of thoron inside and outside the house in the survey area is not high. The average inside and concentrations are 45 Bqm^{-3} and 23 Bqm^{-3} equivalent to the world average thoron concentration [14]. This may be due to the low thorium content of the

Ban Gie monazite deposit [8–10]. However, the average thoron concentration in Ban Gie and Xom Xat villages was almost 2 times higher than in Yen Hop and Ban Dung villages, with the indoor thoron concentrations (66.45 Bqm^{-3}; 68 Bqm^{-3}, compared with 31 Bqm^{-3}; 35Bqm^{-3}) and outdoors (54 Bqm^{-3}; 59 Bqm^{-3} compared with 26 Bqm^{-3}; 35 Bqm^{-3}). This may be because the households in Ban Gie and Xom Xat villages live closer to the monazite ore body than those in Yen Hop and Ban Dung villages [8–10].

Table 1. Annual average radon and thoron concentrations

No.	House's code	Inside concentration (Bqm^{-3})		Outside concentration (Bqm^{-3})	
		Rn	Tn	Rn	Tn
Ban Gie village					
1	Ban Gie 1	149 ±15	66 ± 3	110 ± 5	17 ± 2
2	Ban Gie 2	20 ± 2	40 ± 2	107 ± 4	21 ± 3
3	Ban Gie 3	67 ± 4	68 ± 3	12 ± 2	26 ± 3
4	Ban Gie 4	76 ± 4	63 ± 4	59 ± 4	67 ± 5
5	Ban Gie 5	208 ± 13	40 ± 5	189 ± 14	73 ± 8
6	Ban Gie 6	199 ± 12	67 ± 7	177 ± 15	30 ± 2
7	Ban Gie 7	190 ± 11	66 ± 6	165 ± 12	79 ± 5
8	Ban Gie 8	220 ± 21	94 ± 8	207 ± 14	65 ± 6
9	Ban Gie 9	188 ± 15	44 ± 5	299 ± 23	103 ± 10
10	Ban Gie 10	299 ± 12	85 ± 9	303 ± 15	30 ± 2
11	Ban Gie 11	674 ± 22	92 ± 8	422 ±2 1	58 ± 4
12	Ban Gie 12	629 ± 19	24 ± 3	478 ± 17	92 ± 7
13	Ban Gie 13	389 ± 24	27 ± 3	351 ± 16	54 ± 4
14	Ban Gie 14	282 ± 22	55 ± 5	238 ± 23	90 ± 8
15	Ban Gie 15	269 ± 18	93 ± 7	247 ± 21	10 ± 2
16	Ban Gie 16	109 ± 9	25 ± 4	49 ± 3	45 ± 3
17	Ban Gie 17	207 ± 21	93 ± 11	100 ± 8	49 ± 4
18	Ban Gie 18	148 ± 12	203 ± 13	97 ± 9	82 ± 7
19	Ban Gie 19	68 ± 5	24 ± 3	79 ± 6	38 ± 3
20	Ban Gie 20	82 ± 7	60 ± 5	108 ± 5	67 ± 7
Min		20 ± 2	24 ± 3	12 ± 2	10 ± 2
Max		674 ± 22	203 ± 13	478 ± 17	103 ± 10
Average		223.65 ± 23	66.45 ± 11	189.85 ± 16	54.8 ± 8
Geometric standard deviation (GSD)		2.24 ± 0.65	1.72 ± 0.34	2.33 ± 0.54	1.85 ± 0.31
Geometric mean (GM)		168 ± 28	57 ± 12	143 ± 23	46 ± 7
Yen Hop village					
1	Yen Hop 1	109 ± 8	13 ± 2	70 ± 6	61 ± 5
2	Yen Hop 2	61 ± 5	28 ± 3	68 ± 7	12 ± 3
3	Yen Hop 3	29 ± 3	55 ± 5	17 ± 3	15 ± 3
4	Yen Hop 4	18 ± 2	67 ± 7	32 ± 4	22 ± 4
5	Yen Hop 5	43 ± 4	16 ± 2	74 ± 9	18 ± 2
6	Yen Hop 6	78 ± 8	10 ± 2	12 ± 3	18 ± 2
7	Yen Hop 7	89 ± 11	36 ± 4	101 ± 12	17 ± 3
8	Yen Hop 8	65 ± 5	15 ± 3	129 ± 13	11 ± 2

(continued)

Table 1. (*continued*)

No.	House's code	Inside concentration (Bqm^{-3})		Outside concentration (Bqm^{-3})	
		Rn	Tn	Rn	Tn
9	Yen Hop 9	102 ± 9	22 ± 4	85 ± 9	13 ± 2
10	Yen Hop 10	66 ± 7	22 ± 3	48 ± 5	17 ± 4
11	Yen Hop 11	90 ± 10	15 ± 4	29 ± 4	18 ± 3
12	Yen Hop 12	71 ± 4	8 ± 2	58 ± 6	18 ± 2
13	Yen Hop 13	68 ± 6	10 ± 2	62 ± 5	34 ± 4
14	Yen Hop 14	49 ± 6	56 ± 6	89 ± 10	34 ± 3
15	Yen Hop 15	79 ± 9	23 ± 3	78 ± 9	65 ± 6
16	Yen Hop 16	82 ± 8	56 ± 7	48 ± 5	23 ± 3
17	Yen Hop 17	18 ± 2	12 ± 2	12 ± 3	17 ± 3
18	Yen Hop 18	106 ± 11	40 ± 3	107 ± 13	19 ± 4
19	Yen Hop 19	56 ± 6	68 ± 11	60 ± 4	26 ± 4
20	Yen Hop 20	161 ± 16	63 ± 8	55 ± 6	67 ± 6
Min		18 ± 2	8 ± 2	12 ± 3	11 ± 2
Max		161 ± 16	68 ± 11	129 ± 13	67 ± 6
Average		72 ± 8	31.75 ± 5	61.7 ± 7	26.25 ± 4
Geometric standard deviation (GSD)		1.73 ± 0.21	2.02 ± 0.32	1.95 ± 0.27	1.69 ± 0.24
Geometric mean (GM)		63.3 ± 6	25.1 ± 5	51.6 ± 6	22.4 ± 3
Xom Xat village					
1	Xom Xat 1	204 ± 15	40 ± 5	41 ± 4	73 ± 7
2	Xom Xat 2	155 ± 11	67 ± 6	138 ± 8	30 ± 4
3	Xom Xat 3	161 ± 14	66 ± 8	48 ± 5	79 ± 8
4	Xom Xat 4	153 ± 12	94 ± 10	72 ± 7	65 ± 6
5	Xom Xat 5	62 ± 7	44 ± 5	111 ± 10	103 ± 11
6	Xom Xat 6	151 ± 13	85 ± 8	138 ± 12	30 ± 4
7	Xom Xat 7	117 ± 9	92 ± 11	55 ± 5	58 ± 5
8	Xom Xat 8	110 ± 10	24 ± 4	221 ± 23	92 ± 9
9	Xom Xat 9	237 ± 21	27 ± 3	245 ± 18	54 ± 6
10	Xom Xat 10	413 ± 18	55 ± 5	277 ± 15	90 ± 11
11	Xom Xat 11	138 ± 12	93 ± 11	192 ± 14	10 ± 2
12	Xom Xat 12	115 ± 9	25 ± 3	234 ± 13	45 ± 6
13	Xom Xat 13	138 ± 12	93 ± 8	122 ± 11	49 ± 4
14	Xom Xat 14	83 ± 8	203 ± 12	76 ± 8	82 ± 8
15	Xom Xat 15	371 ± 23	24 ± 4	266 ± 21	38 ± 5
Min		62 ± 7	24 ± 4	41 ± 4	10 ± 2
Max		413 ± 18	203 ± 12	277 ± 15	103 ± 11
Average		173.8 ± 13	68.8 ± 9	149.1 ± 11	59.8 ± 7
Geometric standard deviation (GSD)		1.62 ± 0.21	1.85 ± 0.24	1.88 ± 0.31	1.79 ± 0.28
Geometric mean (GM)		153.5 ± 23	56.9 ± 6	124.6 ± 12	52.4 ± 7
Ban Dung village					
1	Ban Dung 1	13 ± 3	60 ± 7	29 ± 4	67 ± 6
2	Ban Dung 2	25 ± 4	13 ± 2	39 ± 6	61 ± 8
3	Ban Dung 3	110 ± 8	28 ± 4	43 ± 7	12 ± 2
4	Ban Dung 4	311 ± 21	55 ± 6	427 ± 23	15 ± 3

(*continued*)

Table 1. (*continued*)

No.	House's code	Inside concentration (Bqm^{-3})		Outside concentration (Bqm^{-3})	
		Rn	Tn	Rn	Tn
5	Ban Dung 5	614 ± 19	67 ± 7	505 ± 21	22 ± 5
6	Ban Dung 6	53 ± 5	32 ± 5	50 ± 6	45 ± 5
7	Ban Dung 7	47 ± 5	34 ± 4	88 ± 9	55 ± 6
8	Ban Dung 8	32 ± 4	17 ± 3	52 ± 5	23 ± 5
9	Ban Dung 9	96 ± 9	25 ± 4	39 ± 4	35 ± 4
10	Ban Dung 10	54 ± 6	23 ± 3	63 ± 7	12 ± 2
11	Ban Dung 11	30 ± 5	34 ± 5	65 ± 7	54 ± 4
12	Ban Dung 12	96 ± 11	47 ± 6	34 ± 5	34 ± 3
13	Ban Dung 13	65 ± 8	32 ± 5	54 ± 6	45 ± 7
14	Ban Dung 14	52 ± 6	46 ± 7	54 ± 5	33 ± 6
15	Ban Dung 15	65 ± 8	12 ± 3	55 ± 6	22 ± 3
Min		13 ± 3	12 ± 3	29 ± 4	12 ± 2
Max		614 ± 19	67 ± 7	505 ± 21	67 ± 6
Average		110.9 ± 12	35.0 ± 6	106.5 ± 13	35.7 ± 6
Geometric standard deviation (GSD)		2.53 ± 0.43	1.66 ± 0.22	2.25 ± 0.35	1.75 ± 0.18
Geometric mean (GM)		65.6 ± 8	31.1 ± 5	66.3 ± 9	31.3 ± 6

Table 1 shows that: 34 out of 70 households have a the home >100 Bqm^{-3}, exceeding allowable limit for concentration ≤ 100 Bqm^{-3} in Vietnam standard TCVN 7889:2008.

The results showed were is of the in the area are simply designed with doors and windows allowing air to be exchanged with the outside of the house, similar to the research results on the concentration of radon, thoron indoors and outdoors in households living on mines[13]. Based on the radon and thoron levels measured outdoors and indoors, it can be concluded that the Ban Gie monazite deposit area is considered to be an area with high radon and thoron levels, about 20% of homes have an indoor radon levels 200 Bqm^{-3}.

4.2 Effective Dose

The results of estimation of annual committed effective doses (Eqs. 1–3) originated from the inhalation of radon and thoron in the Ban Gie monazite mine, Quy Hop district, Nghe An province are given in Table 2. It could be seen that all inhabitants in the investigated areas may have been exposed to high annual effective doses.

The annual effective dose value of radon and thoron gas ranges from 0.41 to 21.67$mSvyear^{-1}$, an average of 4.4 $mSvyear^{-1}$. Compared to the worldwide average of 1.25 $mSvyear^{-1}$ (1.15 $mSvyear^{-1}$ for radon and 0.1 $mSvyear^{-1}$ for thoron) given by UNSCEAR 2000 [14], households In the surveyed area, the annual effective dose caused by radon and thoron is 3.5 times higher than the world average.

Table 2. Annual committed effective doses originated from the inhalation of radon and thoron

House's code	Effective dose from radon ($mSvyear^{-1}$)		Effective dose from thoron ($mSvyear^{-1}$)		Total dose from Rn and Tn ($mSvyear^{-1}$)
	Inside	Outside	Inside	Outside	
Ban Gie village					
Min	0.50 ± 0.12	0.11 ± 0.03	0.20 ± 0.05	<0.1	1.87 ± 0.24
Max	16.85 ± 3.21	4.54 ± 1.02	1.70 ± 0.19	<0.1	21.67 ± 5.34
Average	5.59 ± 1.23	1.80 ± 0.23	0.56 ± 0.12	<0.1	7.99 ± 2.54
Geometric standard deviation (GSD)	2.24 ± 0.65	2.33 ± 0.56	1.72 ± 0.21	<0.1	1.94 ± 0.21
Geometric mean (GM)	4.21 ± 1.21	1.36 ± 0.32	0.48 ± 0.11	<0.1	6.47 ± 2.21
Yen Hop village					
Min	0.45 ± 0.12	0.11 ± 0.04	0.06 ± 0.01	<0.1	0.67 ± 0.21
Max	4.02 ± 1.23	1.22 ± 0.13	0.57 ± 0.21	<0.1	5.12 ± 2.12
Average	1.80 ± 0.65	0.58 ± 0.06	0.26 ± 0.09	<0.1	2.67 ± 0.78
Geometric standard deviation (GSD)	1.74 ± 0.21	1.95 ± 0.32	2.02 ± 0.54	<0.1	1.50 ± 0.23
Geometric mean (GM)	1.58 ± 0.22	0.49 ± 0.22	0.21 ± 0.06	<0.1	2.46 ± 0.54
Xom Xat village					
Min	1.55 ± 0.21	0.38 ± 0.07	0.20 ± 0.06	<0.1	3.05 ± 1.32
Max	10.32 ± 5.43	2.63 ± 0.34	1.70 ± 0.32	<0.1	13.48 ± 6.43
Average	4.34 ± 1.23	1.41 ± 0.06	0.57 ± 0.12	<0.1	6.38 ± 2.23
Geometric standard deviation (GSD)	1.62 ± 0.43	1.88 ± 0.23	1.85 ± 0.45	<0.1	1.44 ± 0.23
Geometric mean (GM)	3.83 ± 1.32	1.18 ± 0.34	0.48 ± 0.11	<0.1	5.93 ± 2.32
Ban Dung village					
Min	0.32 ± 0.05	0.27 ± 0.05	0.10 ± 0.04	<0.1	1.14 ± 0.43
Max	15.35 ± 4.45	4.79 ± 1.32	0.56 ± 0.12	<0.1	20.73 ± 8.23
Average	2.77 ± 0.76	1.01 ± 0.07	0.29 ± 0.07	<0.1	4.10 ± 1.22
Geometric standard deviation (GSD)	2.53 ± 0.43	2.25 ± 0.43	1.66 ± 0.43	<0.1	2.15 ± 1.12
Geometric mean (GM)	1.64 ± 0.22	0.63 ± 0.33	0.26 ± 0.08	<0.1	2.73 ± 1.11

From Table shows that, the people in the Ban Gie village may be exposed to dose of 7.99 $mSvyear^{-1}$, nearly 6.4 times higher than the world average. Inhabitants in the Xom Xat village may be exposed to dose of 6.38 $mSvyear^{-1}$, nearly 5.2 times higher than the world average. The people Ban Dung villagers may also be exposed to a dose of 4.1 $mSvyear^{-1}$, nearly 3.3 times higher than the world average. The people in the Yen Hop village may also be exposed to dose of 2.6 $mSvyear^{-1}$, nearly 2.1 times higher than the world average.

5 Conclusion

Survey results of radon and thoron in the Ban Gie monazite sand deposit show that this area has a high concentration of natural radioactive substances. The houses in Ban Gie village, located adjacent to the placer ore body, have the highest mean radon concentrations greater than 200 Bqm^{-3}. Houses in Yen Hop village, Xom Xat, about 250–350 m from the sandstone body, have radon levels of about 100 Bqm^{-3}. Thoron concentration in the survey area is not high and similar to the average concentration of the world.

Estimated results of annual dose derived from inhalation survey area showed that people could an average dose of 4.4 $mSvyear^{-1}$, higher than 3.5 times the world average of 1.25 $mSvyear^{-1}$. It has been concluded that residents living on highly radioactive mineral ores may have been exposed to the effective annual dose caused by high concentrations of radon and thoron in the air.

References

1. Abdollahi, H.: World high background natural radiation areas and public life: letter to the editor. Radiat. Meas. **59**, 288–289 (2013)
2. IAEA: Extent of environmental contamination by naturally occurring radioactive material (NORM) and technological options for mitigation, IAEA, Editor (2003)
3. Naturally Occurring Radioactive Materials (NORM): World Nuclear Association. http://www.world.nuclear.org/info/Safetyand-Security/Radiation-andHealth/Naturally-OccurringRadioactive-Materials-NORM/. Accessed 02 Aug 2014
4. Bossew, P.: Radon: exploring the log-normal mystery. J. Environ. Radioact. **101**, 826–834 (2010)
5. Carvalho, F.P., Reis, M.C.: Radon in Portuguese houses and workplaces. In: International Conference Healthy Buildings, HB 2006 (2006)
6. WHO: Handbook on Indoor Radon: A Public Health Perspective. World Health Organization Press (2009)
7. Environmental Protection Agency: Assessment of Risks from Radon in Homes, EPA 402-R-03-003. Office of Radiation and Indoor Air, United States Environmental Protection Agency, Washington, DC 20460 (2003)
8. Vy, P.X.: Geological report of the search results assessing the northern Bu Khang-Quy Chau-Nghe An gold placer. Archived at the Geological Archives Information Center, Hanoi (2000)
9. Van Dung, N., Anh, V.T.L., Huan, T.D., Chau, T.L.: Radioactive environment status quo in Quy Hop area - Nghe An Province. In: Collection of Summaries of the Scientific Conference. Hanoi Universities of Mining and Geology, pp. 170–171 (2012)

10. Fong-Sam, Y.: The mineral industry of Vietnam. U.S. geological survey minerals yearbook 2011 (2013)
11. Dung, B.D., Van Giap, T., Kovacs, T., Toan, T.N., Cuong, L.D., Minh, T.K., Quyet, N.H., Van Khanh, N.: Estimation of radon and thoron caused dose at extraction and processing sites of mineral sand mining area in Vietnam. J. Radioanal. Nucl. Chem. **299**, 1943–1948 (2014)
12. de Campos, M.P., Martins, E.W.: Calibration of the solid state nuclear track detector CR-39 for radon measurements. In: International Nuclear Atlantic Conference - INAC (2007)
13. Phon, L.K., Dung, B.D., Chau, N.D., Kovacs, T., Van Nam, N., Van Hao, D., Son, N.T., Luan, V.M.: Estimation of effective dose rates caused by radon and thoron for inhabitants living in rare earth field in northwestern Vietnam (Lai Chau province). J. Radioanal. Nucl. Chem. **306**, 309–316 (2015)
14. UNSCEAR: Sources to effects assessment for Rn in homes and workplaces. UNSCEAR Report (2000)

Unmanned Ground Robot for COVID-19 Isolated Area in Vietnam

Xuan Nang Ho[1,2(✉)] and Anh Son Le[1,2(✉)]

[1] Phenikaa Research And Technology Institute, Phenikaa Group, Hanoi, Vietnam
{nang.hoxuan, son.leanh}@phenikaa-uni.edu.vn
[2] Faculty of Vehicle and Energy Engineering, Phenikaa University, Hanoi, Vietnam

Abstract. Covid-19 has a big damage for all of society in the world., it has been taking more than 297,241 lives until now. To protect a society, the isolated area is one of importance solution, which can reduce the spread of the Virus. To keep the distance between the staff and others to prevent the infection, the unmanned robot for online examination is one of a good solution. Realize the importance of the unmanned ground robot, which is limited exit in Vietnam before, we try to develop a supporting doctor robot within a month and apply it to the isolated area. In this manuscript, the design process as well as the algorithm for controlling robot will be shown.Please check and confirm if the authors and their respective affiliations have been correctly identified. Amend if necessary.

Keywords: Unmanned robot · Ground robot · Robot for isolated area

1 Background

From December 2019 up to May 20 2020, there are about 4.735.000 confirmed cases with 316.289 deaths in Coronavirus disease (COVID-19) pandemic [1]. The Covid-19 virus is spreading between peoples via physical contact or close contact with infected person [2]. Respiratory droplet from cough or sneeze of an infected person can be directly injected into healthy peoples, or indirect injected via touching of infected person's hands and their belongings. Due to ease of infection between people and people, the coronavirus disease, started in China, is now spreading to 216 countries, areas or territories all over the world in only 5 months.

To prevent the spreading of virus across nation border, many countries applied a rule of 14 days quarantine in an isolated area for any passage who enters the country. Consequently, there is a new demand of human resource for serving inside isolated areas, and also a need of protecting methods for medical staffs who work there. Essentially, doctors and nurses who directly take care of patients and suspected infected persons will need extra protective suits [3]. However, there are other tasks which can be done by robots and autonomous vehicles, such as food and medicine delivery or medical wastes. They can significantly reduce people-people contact and also reduce working load for medical staffs.

B. T. Long et al. (Eds.): MMMS 2020, LNME, pp. 629–637, 2021.
https://doi.org/10.1007/978-3-030-69610-8_84

Robot were developing for a long time and applying in a lot of area especially in industry area. With many advantages from robot with artificial intelligent, we decided to develop a robot as quick as possible to do some simple jobs includes: carry the food and medicine, assisting the government's social distancing efforts, examination online.

2 Design and Manufacture

2.1 Hardware Design

2.1.1 Driving System

The driving mechanism of the vehicle is common in robotics field, which includes two front driving motor for movement and steering, and two rear driven wheels. Two options of driving DC motor were tested: brushed and brushless motors. The brushless motor (24–36 V input voltage, 350 W max power at 36 V) was selected based on the advantage of timeline, size, as well as the suitable power. The driven wheels are omni wheels, which allow both forward and backward movements, as well as rotation motion.

2.1.2 Vehicle Controller System

Two microcontrollers: 01 F7-ARM controller (low-level controller) for simple outdoor navigation and 01 embedded computer (high-level controller) running Ubuntu were used for vehicle control. The F7-ARM controller was connected to GNSS, IMU sensors while the embedded computer was connected to 2D Lidar sensor for obstacle detection. Two microcontroller boards connect with each other via USB interface to exchange sensors and navigation information.

2.1.3 Communication System

01 main telemetry over 2.4 GHz Wi-Fi or 4G network and 01 secondary telemetry over 915 MHz radio frequency for backup. There is a strong Wi-Fi router (TL-WR941HP) on the rover side and a Wi-Fi receiver (built-in or external) on the side of ground station. Similarly, there are two 915 MHz telemetry modules (Holybro 915 MHz max 500 mW Sik telemetry module) setting up on both side for communication. Frequency range of the Sik telemetry modules is set from 895 MHz to 905 MHz to avoid interference. All two types of telemetry can stream rover data to ground station and control signals from ground station back to rover. Range of the main telemetry system is dependent on the coverage of Wi-Fi or 4G network, while the range of secondary 915 MHz telemetry module is claimed to be 300 m in open air with stock antennas at 100 mw transmitting power.

There is one main video feed over 2.4 GHz Wi-Fi or 4G network and another secondary telemetry over 5.8 GHz analog video feed for backup. Range and quality of the main video feed is dependent on the coverage of Wi-Fi or 4G network, while the analog video feed (TBS Unify Pro V3) was successful tested up to range of 5 km with directional receiver antenna in open-air condition [4].

There is a remote-control system using computer joystick (a PS3 or Xbox joystick) for manual control. The controlling signals is sent via telemetry data and control range is depend on telemetry quality. There is another optional 2.4 GHz radio control system (M12 Pro from Skydroid, 7 km range in open-air condition) which can be used for backup or extended functions.

2.1.4 Ground Station Hardware

The core part of ground station is a personal computer (standard hardware PC) or a tablet for communicate and control the rover. In the case of using a laptop or a desktop as ground station computer, there are many options for connection and control which can be deployed simultaneously, for example using HD video feed via UDP connection and SD video feed via 5.8 GHz analog radio, or using both Wi-Fi and 915 MHz radio telemetries. On the other side, if a tablet is used as ground station, only one of these functions such as analog video feed or 915 MHz telemetry can be deployed due to lack of high-power USB connections (Fig. 1).

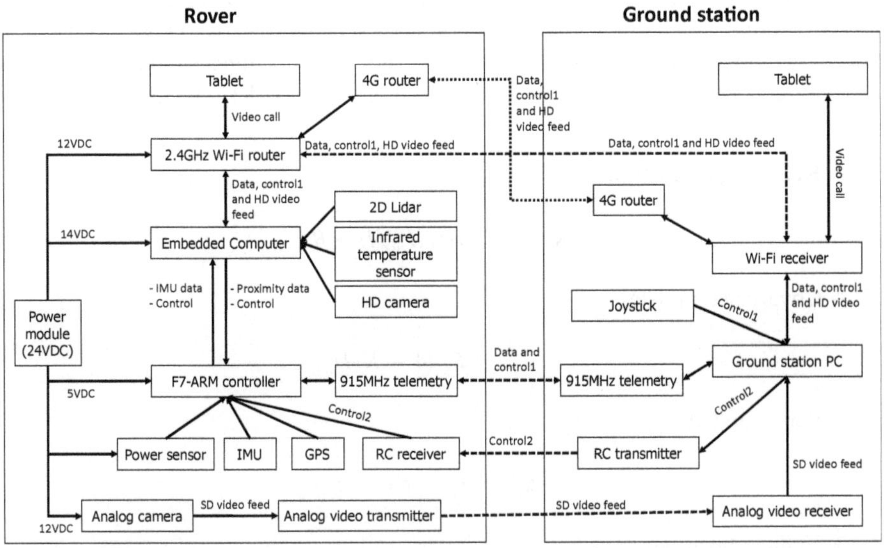

Fig. 1. Hardware design.

2.2 Software Design

Robot Operation System (ROS) and F7-ARM communicate with each other via mavros node, a MAVLink communication node for ROS with a proxy to forward MAVLink data packets via Wi-Fi connection or 4G to Ground Station PC. The other 915 MHz telemetry also can send MAVLink packets directly from F7-ARM controller autopilot to Ground Station PC. Upon mavros node, sensor data from F7-ARM controller such as IMU or Compass can be provided to embedded computer for pose estimation and localization procedures, while the high-level controller in embedded computer can uses these results to guild the low-level controller for indoor and outdoor navigation.

Robot Operation System (ROS) on embedded computer plays a role as the high-level controller, that is used for indoor environment: drive assistance for low-level controller for indoor navigation, indoor remote control with object avoidance, indoor autonomous missions [5]. In fact, ROS is a popular framework for creating autonomous robot software. Using high-performance hardware of embedded computer together with distance sensors such as 360-degree LIDAR, a ROS-based system provides many complex tasks for indoor pose estimation, indoor localization and mapping. Here, ROS was used for detecting obstacle and as well as localization outdoor.

Outdoor navigation with GNSS can be achieved with F7-ARM controller on low-level controller. It uses onboard IMU, compass sensor and GNSS to define its location and movement. Normally, using only standard M8N chip - GNSS using GPS, which provides an average accuracy of 2.5 m [6], is not enough for high-accuracy outdoor navigation.

To improve the accuracy, there are several precious methods by combining with other information such as Takai et al. develop the localization of tractor by using RTK [7], by other way, Sebastian et al. applied the combination IMU and GPS [8]. Here, to reduce the price, the combination of IMU and GPS to enhance the accuracy of localization was applied based on Extended Kalman Filter (EKF) [9–13]. The equation shows below:

+ Kalman gain matrix:

$$K_k = P_k^- H_k (H P_k^- H_k + R)^{-1}$$

+ Covariance update

$$P_k^+ = P_k^- (1 - K_k H_{k)}.$$

+ Status update

$$x_k^+ = x_k^- + K(z - z_p)$$

+ Vehicle position

$$\widehat{x}_k^- = \begin{bmatrix} \left(V_{Long}^-\right)_k \\ \left(a_{Long}^-\right)_k \\ \theta_k^- \end{bmatrix} = \begin{bmatrix} \left(V_{Long}^-\right)_{k-1} + \Delta t.\left(a_{Long}^-\right)_k \\ (ACC_x - g.\sin\left(\theta_{k-1}^-\right)).cos(\theta_{k-1}^-) \\ \theta_{k-1}^- + \Delta t.\left(GYRO_y\right)_k \end{bmatrix} \tag{1}$$

$$z_k = \begin{bmatrix} (V_{GPS})_k \\ (V_{GPS})_k \\ \left(tan^{-1}\frac{V_Z}{V_{XY}}\right)_k \end{bmatrix} \tag{2}$$

$$\widehat{x}_k^- = \left[\widehat{\psi}_k^-\right] = \left[\widehat{\psi}_{k-1}^- + \Delta t.(GYRO_z)_k\right] \tag{3}$$

$$z_k - \left[(\psi_{GPS})_k\right] \tag{4}$$

$$\widehat{x}_k^- = \begin{bmatrix} \widehat{X}_k^- \\ \widehat{Y}_k^- \end{bmatrix} = \begin{bmatrix} \widehat{X}_{k-1}^- + \Delta t.\left(V_{Long}^-\right)_k.\cos(\widehat{\psi}_k^-) \\ \widehat{Y}_{k-1}^- + \Delta t.\left(V_{Long}^-\right)_k.\sin(\widehat{\psi}_k^-) \end{bmatrix} \tag{5}$$

$$z_k = \begin{bmatrix} (X_{GPS})_k \\ (Y_{GPS})_k \end{bmatrix} \tag{6}$$

Where:

$\widehat{x}_0$: *initiallized state vector*
$\widehat{\widehat{P}}_0$: *Covariance matrix*
R: *Measurement covariance*
z: *actual measurement*
z_p: *predicted measurement*
θ: *Pitch angle*
$GYRO_y, GYRO_z$: *angular velocity of y and z-axis*
ACC_x: *x-axis acceleration*
g: *acceleration of gravity*
a_{Long} *and* V_{Long}: *longitudinal acceleration and velocity*
$V_{GPS}, \psi_{GPS}, X_{GPS}, Y_{GPS}$: *velocity, yaw, longitude, and latitude from NMEA of GPS data*
V_Z, V_{XY}: *velocity calculated by variation of altitude and longitude and latitude*

3 Results

3.1 Overview of the Robot

See Fig. 2.

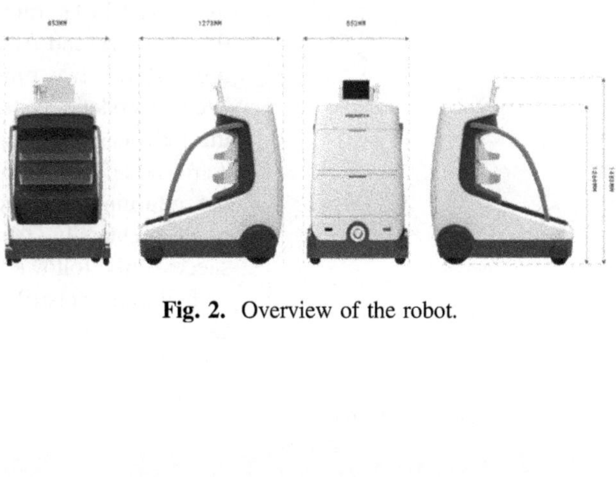

Fig. 2. Overview of the robot.

3.2 Connection Range

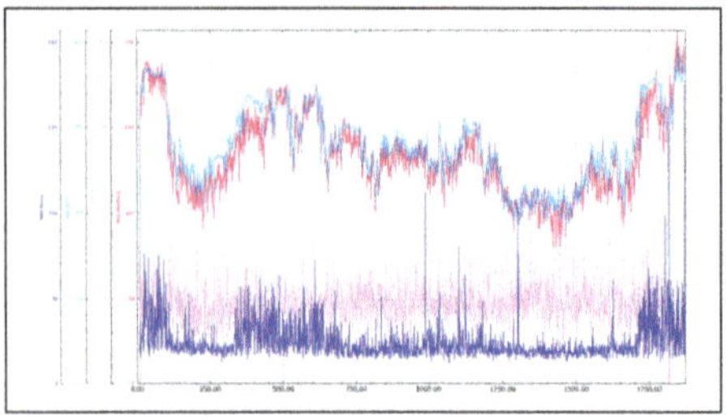

Fig. 3. Signal strength.

The vehicle moved in a close loop around the test field at Phenikaa University and the farthest point from ground station was the opposite corner. The signal strength of 915 MHz 100 mW telemetry transferring data at 64 kbps during the test is shown in Fig. 3. The signal strength was low to $\sim$ -74 dbm at range of $\sim$150 m, which is sufficient for proper remote control14. Noise of vehicle telemetry (pink line) was higher than ground station side (blue line) shown that the 915 MHz telemetry was affected by some interference from other connections on the vehicle (Wi-Fi router or analog video feed).

3.3 Remote Control Performance

Remote control performance was evaluated by manual control ability in indoor and outdoor environment of the prototype. The power of each motor was limited to 30% of maximum designed power. For indoor and outdoor environment, the prototype could do all basic movements (straight forward and backward run, turning by 90° corner, U-turn in 2 m-wide hallway) without any difficulty and collision. Figure 4 shows some movement of the vehicle in hallway.

3.4 Outdoor Autonomous Operation

Outdoor autonomous operation was confirmed by two tests. The first test was conducted using smart Return to Launch mode to check a safety feature of the vehicle, or in other words, to confirm ability of the autonomous following a predetermined way path.

3.4.1 Smart Return to Launch (Smart RTL)

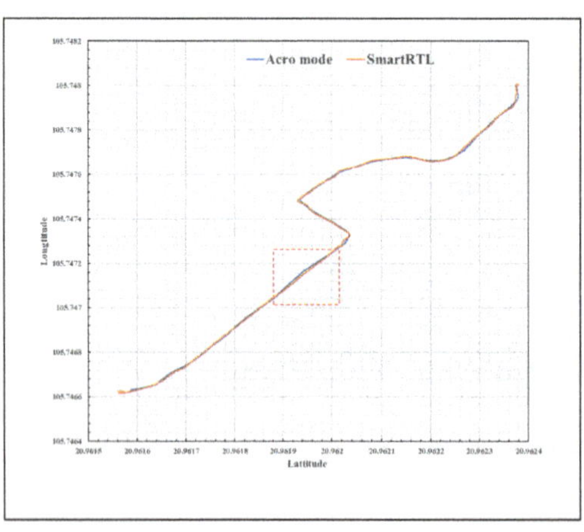

Fig. 4. Acro mode and Smart RTL mode.

In order to do this test, the vehicle was remoted controlled in Acro mode on a 190 m running path with 02 right-turn and 01 left-turn. Figure 4 shows two running paths in World Geodetic System (WGS 84), the red one is for running path in Acro mode, and the blue one is for the returning path in Smart RTL mode. Generally, the vehicle was successfully followed the running path during SmartRTL procedure. One of the largest errors between two paths is shown in Fig. 4. A convert between WGS 84 to SI unit was done by using the calculation distance15. The data shows a maximum error of about 0.7 m with average error of 0.2 m, which is acceptable if using M8N GNSS system for navigation. At the end of the path, Smart RTL mode was activated and the vehicle attempted to return to Home position using the same running path.

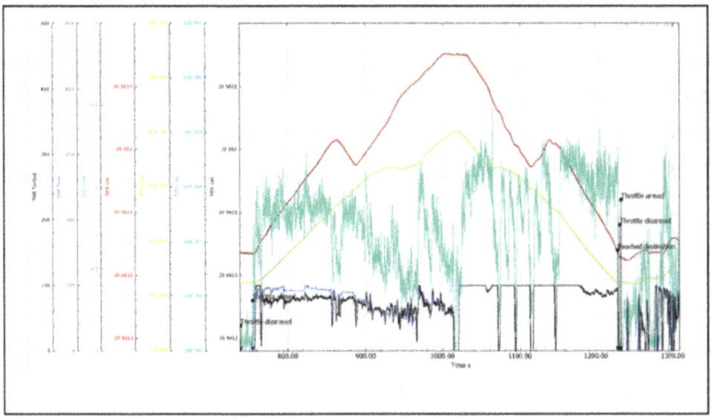

Fig. 5. Vehicle throttle and running speed.

Figure 5 shows vehicle's throttle control and speed data during Acro mode and SmartRTL mode running. In Acro mode, the vehicle mostly followed the throttle command by operation. On the other hand, the vehicle ran mostly with full throttle in SmartRTL mode and the throttle percent only reduced when making a turn. Without load, the cruise speed was about 1–1.1 m/s on asphalted surface, which is close to the prefix cruise speed of 1 m/s of the vehicle.

3.4.2 Autonomous Waypoint Following Mission

The second test to evaluate outdoor autonomous operation was completed in Auto mode. A waypoint following mission was uploaded to the vehicle.

In this test the maximum allowed error to reach each waypoint was set to 0.3 m. The maximum allowed overshoot of each waypoint was set to 0.5 m. Figure 6 shows vehicle autonomous throttle control in Auto mode. The throttle percentage was only reduced to zero in 90° turning angle, otherwise it was kept at 100% to maintain the vehicle speed at 1 m/s on paving stone surface.

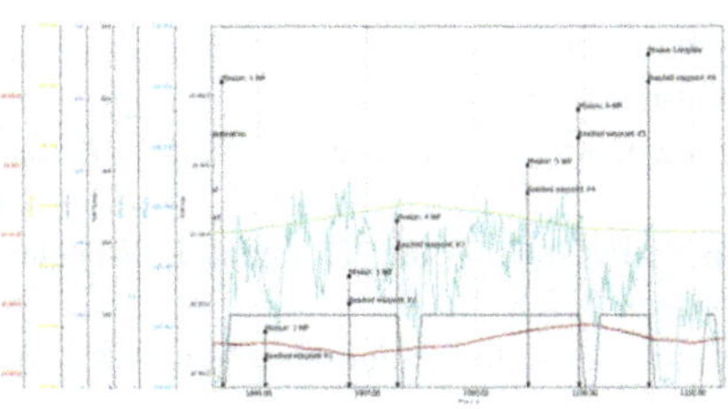

Fig. 6. Auto mode (GWS84)

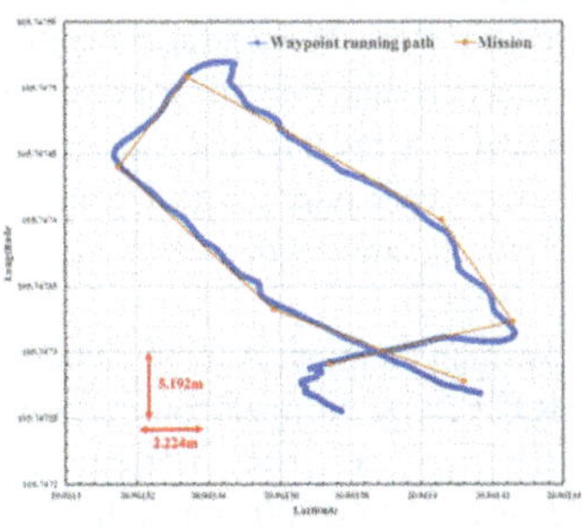

Fig. 7. Path in Auto mode (GW84)

Figure 7 shows running path of the vehicle in Auto mode test. Basically, the vehicle followed the mission. However, due to bad road surface, there were some overshoots up to 1.7 m at sharp turning corner and therefore lead to running path oscillation after. However, because of adding the obstacle avoidance function, there are no worried of the accident.

4 Discussion

In Vietnam, there is no commercial robotic products for hospital market. In fact, there are only a few robotic projects developed for research and marketing purpose in recent years. For example, started from 2018, there were two prototypes of robots in hospitals of Ho Chi Minh city that mainly used for patient greeting and information serve [14]. These robots had limited features and simple movement function. At the beginning of Covid-19 pandemic, there were a few more robot projects used for hospital environment that being introduced in Vietnam [15]. Some of them were simple short-range remote-control robots which had only dozens of meters, while the other were converted from commercial AGVs or former DIY robot projects. Therefore, none of these robotics projects can be mass-produced and commercialized.

The first prototype of Phenikaa robot is designed for goods and food transportation to isolated area within a month. Based on this purpose, this prototype is mainly focused on basic features such as manual remote control with drive assistance and outdoor GNSS autonomous navigation. Thus, it can provide the ability of goods transportation in manual mode between rooms in a hospital's floor and autonomous mode between buildings in hospital. It can do multi autonomous outdoor missions, as well as be easily reprogrammed for new tasks, which is superior when making a comparison with line follower robot and AGV.

References

1. WHO Coronavirus Disease Dashboard (2020). https://covid19.who.int/. Accessed 19 June 2020
2. WHO. COVID-19 advice for the public (2020). https://www.who.int/emergencies/diseases/novel-coronavirus-2019/advice-for-public
3. xWHO. COVID-19 technical guidance (2020) https://www.who.int/emergencies/diseases/novel-coronavirus-2019/technical-guidance/patient-management.
4. TBS. Analog video range calculation. https://maxmyrange.com?freq=5800*pw=800*tx=22*rx=27
5. Morgan Quigley, A.Y.N., Conley, K., Gerkey, B.P., Faust, J., Foote, T., Leibs, J., Wheeler, R.C.: ROS: an open-source Robot Operating System. ICRA (2009)
6. U-blox. NEO-M8 Datasheet. https://www.u-blox.com/sites/default/files/NEO-M8_DataSheet_(UBX-13003366).pdf
7. Takai, R., Yang, L., Noguchi, N.: Development of crawler-type robot tractor based on GNSS and IMU. In: IFAC Proceedings, vol. 1, no. PART 1 (2013). https://doi.org/10.3182/20130327-3-jp-3017.00024

8. Bichler, S., Ogris, G., Kremser, V., Schwab, F., Knott, S., Baca, A.: Towards high-precision IMU/GPS-based stride-parameter determination in an outdoor runners' scenario. Procedia Eng. **34**, 592–597 (2012). https://doi.org/10.1016/j.proeng.2012.04.101
9. Riseborough, P.: InertialNavxhttps://github.com/priseborough/InertialNav
10. Riseborough and InertialNav. Generate Equations 22 states. https://github.com/priseborough/InertialNav/blob/master/derivations/GenerateEquations22states.mx
11. Kim, J., Lee, S.: A vehicular positioning with GPS/IMU using adaptive control of filter noise covariance. ICT Express **2**(1), 41–46 (2016). https://doi.org/10.1016/j.icte.2016.03.001
12. Wei, L., Cappelle, C., Ruichek, Y., Zann, F.: Intelligent vehicle localization in urban environments using EKF-based visual odometry and GPS fusion. IFAC Proceedings, vol. 44, no. 1 PART 1, pp. 13776–13781 (2011). https://doi.org/10.3182/20110828-6-IT-1002.01965
13. Meng, X., Wang, H., Liu, B.: A robust vehicle localization approach based on GNSS/IMU/DMI/LiDAR sensor fusion for autonomous vehicles. Sensors (Switzerland) **17**(9) (2017). https://doi.org/10.3390/s17092140
14. Tính, D.: Cô tấm robot' trong bệnh viện. https://thanhnien.vn/suc-khoe/co-tam-robot-trong-benh-vien-1150531.html
15. Dung. "Cận cảnh robot phục vụ bệnh nhân cách ly của ĐH Bách khoa Đà Nẵng. https://bnews.vn/can-canh-robot-phuc-vu-benh-nhan-cach-ly-cua-dh-bach-khoa-da-nang/151357.html

Study on the Stability of a Wheeled Climbing Robot

Duong Vu[1(✉)], Maged Mikhail[2], and Quy Le Thu[3]

[1] Faculty of Mechanical Engineering, Duy Tan University, 550000 Da Nang, Vietnam
[2] Purdue University Northwest, Hammond, IN 46323-2094, USA
mmikhail@pnw.edu
[3] Key Lab. for Welding and Surface Treatment Technologies/NARIME, Hanoi, Vietnam
quylt@narime.gov.vn

Abstract. Climbing robots are used in many various tasks such as inspection, survey, and even some specific special tasks. The climbing robot uses magnetic adhesion therefore it is necessary to calculate the relative interactions between the magnetic, gravitation (weight), and friction forces in terms of inclination because the robot must be firmly against the surface in every position (adhesion). The greater the adhesive force, the frictional force increases, which limits mobility (locomotion). This paper aims to study this relationship by learning the dynamic design and verification on the physical model to improve the stability of the robot. Specifically, in the hull construction in which most of the robot's movement is on a flat surface.

Keywords: Climbing robot · Wheeled robot · Robot stability · Magnetic wheels · Dynamic stability

1 Introduction

For the wheeled climbing robot to move on the adhesive magnetic surface, it is necessary to evaluate this adhesive force while the robot is moving, being sure to consider obstacle avoidance [1]. Climbing robots require locomotion mechanisms such as wheels and track wheels. Legged and hybrid-type robots use adhesion mechanisms such as magnetic, pneumatic, and electro-adhesives as well as some other methods for stabilizing the robot while moving along a surface. The most common mechanism for the climbing robot is wheeled locomotion [2, 3]. Compared to other locomotion mechanisms, wheeled climbing robots have good mobility, simple drive mechanisms (usually electric), and the ability to achieve high speeds with lower power consumption. However, this mechanism has limitations such as having difficulty with sharp height drops that exceed one third of the wheel's diameter. The wall climbing robots mainly use the tracked locomotion mechanism [2, 4]. Adhesion-locomotion systems determine the operating performance and payload capacity of the wall-climbing robot, which is the focus in this area of research. On one side, the adhesive magnetic force is decreases sharply depending on the gap of air between the magnets and the surface [5].

B. T. Long et al. (Eds.): MMMS 2020, LNME, pp. 638–645, 2021.
https://doi.org/10.1007/978-3-030-69610-8_85

It should be noted, however, that on the other side the magnetic adhesive force affects the anti-sliding and turnover ability of the moving robot [6, 7]. The dynamic design also relates to the friction coefficient and the space position of the robot [8]. One of the solutions to increasing the adhesive ability of the robot, especially in the case of obstacle avoidance on complicate surfaces, is the application of its chain wheel [9]. In terms of configuration, improvement, and upgrading the monitor-driving wheel system, it appears there is a new type of wheel, the Mecanum wheel [10], which allows for in-place rotation and prevents ground friction to a much greater extent, resulting in a lower driving torque. All research efforts are focusing on the relationship between the operating stability of the robot and its dynamic stability of the system in locomotion. Nevertheless, it is difficult to find a unique solution that can exclusively address all the challenges. That is why the aim of this paper is to contribute to the research and methodology of calculating a dynamic stabilization system based on the construction of a variety of robot bodies, magnetic materials, as well as varying friction coefficients all for the purpose of achieving a smooth, stable operation. The theoretical calculations have been approved by the onsite testing at the Vietnam shipyard, Song Thu in the city of Da Nang.

2 Dynamic Stability

In general, a wheel climbing robot uses its 4 wheels at any position on the magnetic adhesive surface at an inclination angle ‘θ’ (Fig. 1). The weight of the robot, ‘W,’ is split between the normal force, W.cos θ, and the tangential force, W.sin θ. The total reaction force is directed from the surface to the robot body, ‘N,’ and starts at the center of gravity, ‘O,’ in the body frame. The magnetic adhesive force is ‘F_m’. The friction force is equal to μ.N, where ‘μ’ is the friction coefficient, which is dependent on the material properties of the wheels and the surface. The wheel has a diameter, ‘D’. The distance between the 2 wheel’s centers is ‘L’ and the distance from the frame body’s center of gravity and the surface is ‘d’.

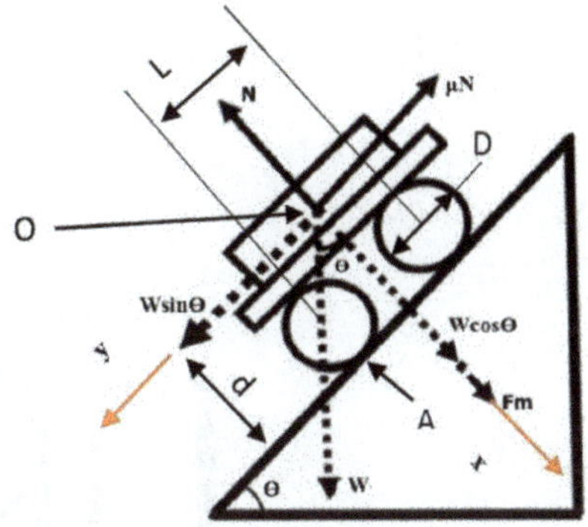

Fig. 1. Dynamic scheme of the robot

2.1 Static Conditions

Sliding Avoidance Conditions for the Robot

From the Static [11], it is known that the slide avoidance conditions to provide stability are based on these equations:

$$\Sigma F_y = W.\cos\theta + F_m - N = 0;$$

Therefore:

$$N = W.\cos\theta + F_m \tag{1}$$

Note that when θ = 0° then N = Fm.

$$\Sigma F_x = W.\sin\theta - \mu.N = 0 \text{ or } = W.\sin\theta/\mu \tag{2}$$

From (1) and (2), it makes:

$$W.\cos\theta + F_m = W.\sin\theta/\mu$$

Therefore:

$$F_m = (W.\sin\theta)/\mu - W.\cos\theta \tag{3}$$

Equation (3) is the condition for slide avoidance of the robot:

$$F_m \geq W.\sin\theta/\mu - W.\cos\theta \tag{4}$$

(where: 'F_m' is the magnetic adhesive force to keep the robot from sliding).

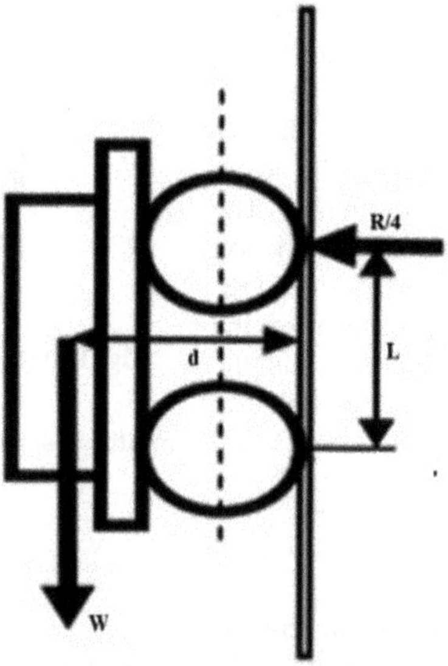

Fig. 2. Vertical position

Figure 2 illustrates one special case when the robot is in a vertical position, where θ = 90°, therefore the sliding avoidance condition formula, Eq. (4), is changed:

$$F_m \geq W/\mu \quad (5)$$

(The total reaction from the surface to the robot is 'R', therefore, R = N, therefore the reaction force onto each wheel is equal to R/4.

Turnover Avoidance Condition

Consider point 'A' (Fig. 1) as the center of rotation and describe all moments based on the above dynamic system. It should be:

$$\Sigma M_A = 0;$$

So: $W.d.\sin\theta + R/2.L - W.L/2.\cos\theta - F_m.L/2 = 0.$

Subsequent, the value of 'F_m' is:

$$F_m = (2.W.d.\sin\theta)/L + N - W.\cos\theta \quad (6)$$

Equation (6) specifies the magnetic adhesive force, 'F_m' to avoid the turnover of the robot.

In specific (critical) cases when the inclination θ = 90°, then:

$$F_m = (2.W.d)/L + N \quad (7)$$

However, due to $N = W.\sin\theta/\mu$, therefore:

$$F_m = (2.W.d.\sin\theta)/L + W.\sin\theta/\mu - W.\cos\theta \quad (8)$$

When Eqs. (4) and (7) are added, it is appearing that the magnetic adhesive force condition, 'F_m', for the robot to satisfy both requirements: slide avoidance and turnover avoidance is as follows:

$$F_m \geq \max[(W.\sin\theta/\mu - W.\cos\theta);\ (2.W.d.\sin\theta/L + W.\sin\theta/\mu - W.\cos\theta)] \quad (9)$$

Equation (9) provides assumptions for the designer with regards to the distance between the two wheel's centers, 'L', and the distance to the surface 'd' for a given adhesion force. The wheel frame is adjustable to increase the length of the robot (body frame) or decrease the distance to the surface (by means of partially decreasing the clearance). In the next section, there will be another situation concerning the torque from drive motor, as well as the obstacle avoidance that partially depends on the diameter of wheel, 'D'.

2.2 Torque Calculation

In Fig. 3: 'T' is the torque of drive motor; 'a' is an acceleration of the robot (in normal condition, a = 0).

For good maneuvering ability, it is necessary to choose the motor with enough torque to overcome all reactive moments. The moment of the robot's weight will deter the rotation of each wheel. This moment is nominated as 'M_w' the radius of the wheel is noted by 'R':

$$M_w = (W.\sin\theta/4).D/2 \tag{10}$$

Moment M_f from the friction force and magnetic adhesive force is calculated as:

$$M_f = ((W.\cos\theta/4) + F_m/4).\,\mu.D/2 \tag{11}$$

From (10) and (11), it can give the formula for calculation motor drive torque:

$$M_d = M_f - M_w = [-W.\sin\theta + \mu.(W.\cos\theta + F_m)].D/8 \tag{12}$$

Substituting the value of the magnetic adhesive force from Eq. (9) with Eq. (12), the final torque value can be calculated. Note that, while inspecting the quality of welding [12], the robot must be moving in all space positions, while the inclination angle, 'θ', changes from 0° to 180°. Using Eqs. (9) and (12) will evaluate the magnetic adhesive force and torque in all critical positions.

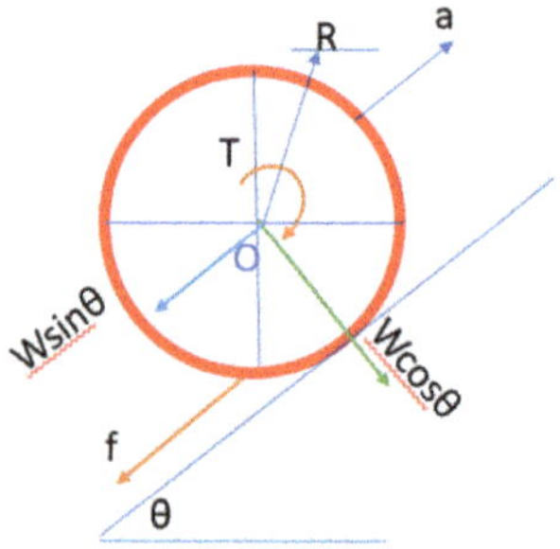

Fig. 3. Torque calculation for drive motor

3 Physical Model

The prototype of climbing robot (Version 1) with 4 wheels has the following measurements (Fig. 4):

- Diameter of wheel = 2.36 in.
- Distance L between front and rear wheel = 3.93 in.
- Width of magnetic wheel = 2 in.
- Center of gravity distance = 2.36 in.
- Cross clearance between two wheels = 2.36 in.
- Full weight of robot = 11 lb.

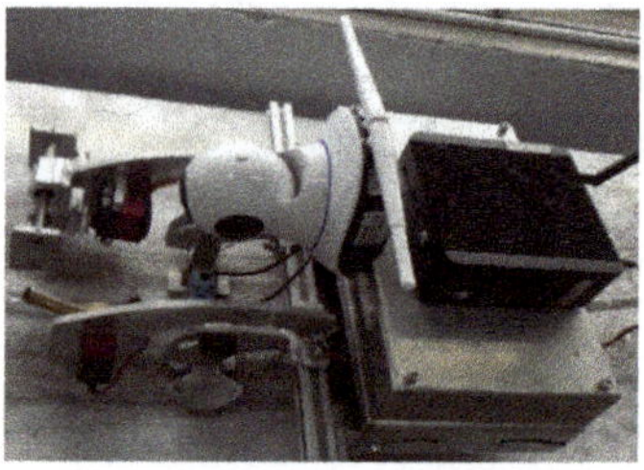

Fig. 4. Prototype of robot

For testing the magnetic adhesion (force), a tension spring balance was been used and it pulled the wheel off from the surface. For evaluation the torque, a torque gauge was used.

Based on Eq. (12), the torque is calculated and equal to 2.36 lb.-in.

The configuration of steel wheel with inserted magnetic pieces covered by rubber layer for friction (Fig. 5) [13]. The dry friction coefficient between rubber and steel is roughly 0.6–0.7 in. And for each wheel, one motor is installed. Therefore, all 4 wheels are actively driven. Motor type DC-DS8911HV has the required torque, which satisfies the value mentioned in the previous section.

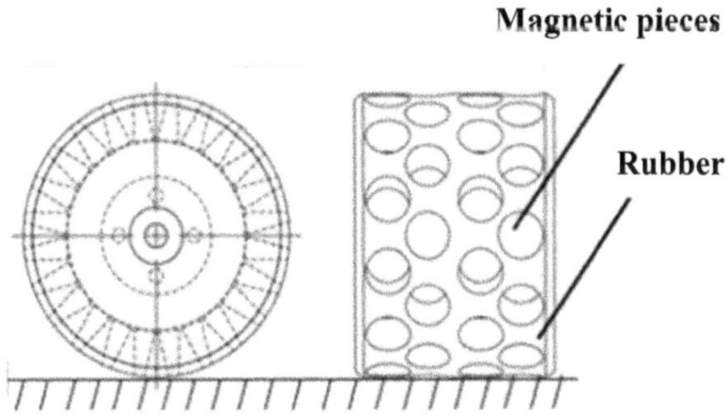

Fig. 5. Magnet wheel

For Version 1 of the robot, the test results are in Table 1, which were collected onsite at the Vietnam shipyard, Song Thu in the city of Da Nang by the engineering research team. The stability of motion (climbing) on the surface of a real hull in all positions (inclinations) and recorded it on video for further analysis were evaluated. The traceability was accessed by the tolerance limit of 0.2–0.3 in. against the width of a real welding seam line. It should be noted that at positions with an inclination of 90° and 120°, the adhesive ability of the robot did not reach the required maximum value. The difference between theoretical and real values set forth the next step of improvement. The actions affecting the slide and turn over avoidance is probably caused by other force acting in the cross direction to the robot frame and causes roll over failure. Another solution is to change the dimensions of the robot body to increase the magnetic adhesive force.

Following this direction, the Version 2 of the robot was created following the dimensions: W = 11 lb.; L = 9.45 in.; d = 2.36 in.; D = 2.36 in.

It was found that with the inclination of 120°, it reaches the maximum value of the magnetic adhesive force. This is contrary to belief as many researchers assumed it would be equal to 90°. In this case, the adhesive magnetic force must be equal to 7 lb. The following value of drive torque is M_d = 0.3 lb.–in. This can help to increase the payload of the robot by an approximate ratio of two.

The theoretical calculations show that in position $\theta = 120°$ (in increments of 30°), the magnetic adhesive force is at its maximum and at least equal to 7 lb. The subsequent value of torque for the motor must be at least equal to 0.3 lb.–in. to provide good stability and locomotion. The test results are in the shipyard (Table 1) specified the improvement of the robot operation was due to its change in distance between the two wheel's centers.

Table 1. Test result on adhesive ability and tracking in shipyard

Inclination, (°)	Robot Version 1		Robot Version 2	
	Adhesive force (%)	Traceability (%)	Adhesive force (%)	Traceability (%)
0	100	100	100	100
30	100	100	100	100
60	100	100	100	100
90	90	100	100	100
120	90	90	98	100
150	100	100	100	100

4 Conclusion

4.1. The theoretical calculations (equations) set fort the preliminary toolset for the primary design of the dynamic part of the magnetic wheel-climbing robot, in terms of firm adhesive attraction on a steel surface. It promises many applications in the industry.

4.2. The main design parameters are magnetic (permanent) adhesive force, body frame length, distance between center of gravity of robot and the surface, diameter of the wheels, the friction coefficient, as well as the full weight of the robot. Case by case, at least one of these parameters should be varied to gain the optimal efficiency, at least with concern to the dynamic stability and locomotion from a viewpoint of the cost of materials.

4.3. The physical test results help the industrial designer with respect to the demand of the client to improve and finalize the product for commercialization, especially for increasing the payload. The data in Table 1 shows the correlation between the theory and physical models of a climbing robot, in terms of stability and locomotion (partially by the traceability). In the critical position (inclination angle of 120°), comparing version 2 with version 1, shows that the increasing of the distance between the front and rear wheel helps to increase the adhesion force from 90% up to 98%, at the same time to improve the dynamic stability (traceability) from 90% up to 100%, as expected in theoretical calculations.

4.4 Future research will be carried on by applying the rare-earth material with more magnetic force, varying all positions of magnetic pieces (not only in the wheels but in the body frame as well), using the passive pair of wheels (non-magnetic) to improve traceability of the robot, and lastly, increase the pay load (for example NDT device).It will be better if there is a simulation model of the climbing robot and a comparison of simulated results with the experiment ones. The authors will compare these results with the simulation as a baseline for research in the future.

References

1. Wakessa Gussu, T., Lin, C.-L.: Geometry based approach to obstacle avoidance of triomnidirectional wheeled mobile robotics platform. J. Sens. (2017). https://doi.org/10.1155/2017/2849537. Article ID 2849537
2. Bisht, R.S., Alecxander, S.J.: Mobile robot for periodic maintenance and inspection of civil infrastructure: a review. In: International and 16th National Conference on Machines and Mechanisms, pp. 1050–1057 (2013)
3. Gradetxky, V.G., Knyazkov, M.M.: Wall climbing robot: mechanics, control and adaptation to environment. In: 23rd International Conference on Robot in Alpe-Adria-Danube Region (RAAD) (2014). https://doi.org/10.1109/RAAD.2014.7002240
4. Nansai, S., Mohan, R.E.: A survey of wall climbing robot: recent advances and challenges. Robotics **5**(3), 1–4 (2016). https://doi.org/10.3390/robotics5030014
5. K&J Magnetics. https://www.kjmagnetics.com/blog.asp?p=magnets-vs-steel
6. Ding, Y., Sun, Z., Chen, Q.: Non-contacted permanent magnetic absorbed wall-climbing robot for ultrasonic weld inspection of spherical tank. In: MATEC Web of Conferences, vol. 269, 02013-IIW2018. (2019). https://doi.org/10.1051/matecconf/201926902013
7. Lauren M.S.: The Tri-Wheel: A novel robot locomotion concept meeting the need for increased speed and climbing capability. Thesis for Degree of Master of Science. Case Western Reserve University (2015)
8. Seriani, S., Scarela, L., Gasparetto, A., Gallina, P.: A new family of magnetic adhesion-based wall climbing robots. Adv. Italian Mech. Sci. 223–230 (2018). https://doi.org/10.1007/978-3-030-03320-0_24
9. Jose, J., Dinakaran, D., Ramya, M.M., Harris Samuel, D.G.: A survey on magnetic wall-climbing robot for inspection. Int. J. Mech. Prod. Res. Dev. (IJMPERD) **8**(6), 59–68 (2018). https://doi.org/10.24247/ijmperddec20186
10. Dhengre, N., Morga, A., Verma, S., Gupta, A.: Design and manufacturing of Mecanum wheel for omnidirectional robot. Int. J. Mech. Prod. Eng. Res. Dev. (IJMPERD), pp. 14–19 (2018). Special Issue
11. Paul, D.R.: Basics of mechanical engineering: Integrating Science, Technology and Common Sense. University of Southern California (2016). (PFD Drive-pdfdrive.com/basics-of-mechanical-engineering-prof-paul-d-ronney-e16452684.html)
12. ISO 17640:2017. Non-destructive testing of weld - Ultrasonic testing - Techniques, testing levels, and assessment
13. Duong, V., Ahmed, A., Ngoc Sy, D.: A robot for weld quality inspection. Int. J. Eng. Res. Innov. **11**(2), 50–55 (2019)

Dynamic of the Shaft in the Hydrodynamic Bearing Examining the Effect of the Centrifugal Force of Lubricating Oil Film for the Short Bearing Type

Nguyen Van Thang(✉)

Faculty of Mechanical Engineering, Ha Noi University of Industry, 298 Cau Dien, Hanoi, Vietnam
nguyenvanthang@haui.edu.vn

Abstract. The paper presents the dynamic of the shaft in the hydrodynamic bearing in case of examining the centrifugal force of the lubricating fluid when the shaft rotates at high speed. In the classical hydrodynamic lubrication theory, this force is often ignored. To simplify, the shaft is considered an absolutely rigid body and the symmetry axes of the shaft and bearing are always parallel to each other, the author in turn sets up the differential equation of motion of the shaft for the short bearing type and investigates its stability of motion through this equation. In this study, the author uses Sommerfeld's hypothesis about the boundary of the lubricating oil film in the bearing. In addition, the author also calculates and compares the force acting of the lubricating oil film on the shaft in two cases: including and excluding the centrifugal force of the lubricating oil film.

Keywords: Floating bush journal bearing · Centrifugal force · Stability of the shaft · Sommerfeld's hypothesis · Lubricating oil film

1 Introduction

In engineering, it is common to see the shaft of a large capacity turbine engine usually supported by the hydrodynamic bearings. In addition, rotating parts require high precision such as the spindle assembly in CNC machines are also supported by this type of bearing. However, one of the disadvantages of this type of bearing is the dynamic instability when the shaft's speed reaches a certain critical value. Instability is manifested in shaft's self-oscillation. To investigate the dynamics as well as the shaft's motion stability in a hydrodynamic bearing, we have to set up the differential equation of motion of the shaft under the acting of the hydrodynamic force. Until now, the analytic expressions for the forces are results of the equations for the oil flow in the gap between shaft and bush (classical Reynolds equation) [1, 7, 8]. When the shaft'speed reaches a certain value, the following phenomena may occur in the bearing: The lubricating oil viscosity decreases due to increased temperature; changes in fluid flow properties; conversion of lubricating fluids from the pre-stressed oil into non Newtonian fluids; the centrifugal force of the lubricating fluid increases significantly. All of

B. T. Long et al. (Eds.): MMMS 2020, LNME, pp. 646–653, 2021.
https://doi.org/10.1007/978-3-030-69610-8_86

these phenomena have been omitted from classical lubrication theory [1, 10]. To explain the phenomenon of shaft's instability when rotating at high speeds, the author has proposed a plan to establish a new form of Reynolds equation examining the effect of the lubricating fluid film's centrifugal force and calculated the hydrodynamic force through this equation [3–5]. In this article, the author continues to research the dynamic of the shaft in the hydrodynamic bearing with the kinematic diagram described in Fig. 1.

The problem of dynamics shaft is closely related to the hydrodynamic force in the bearing acting on it. The investigation of dynamics for the case of common hydrodynamic bearings is conducted in detail in the works [7, 8]. In the document [8] hydrodynamic force and friction force between the lubricating oil film and the surface of the shaft are studied simultaneously. However, in today's industry, in order to increase the stability of the motion, the high speed shaft is often supported by a hydrodynamic bearing with floating bush. The floating bush is a hard metal cylindrical housing placed between the shaft and fixed bush, dividing the lubricating oil field into two separate parts, Fig. 2. Dynamical studies for this type of bearing by the analytical method are complex and not completely resolved [6, 9]. Analytical and experimental methods are applied simultaneously for this type of bearing in the study [2].

2 Research Model

As mentioned above, when considering the dynamic of the shaft in hydrodynamic bearings, two types of bearings are classified: common hydrodynamic bearing (two - solid body model) and hydrodynamic bearing with floating bush (three -solid body model), Fig. 2. To approach these two types of bearings simultaneously when researching the dynamics of the shaft, we consider the general model, Fig. 1.

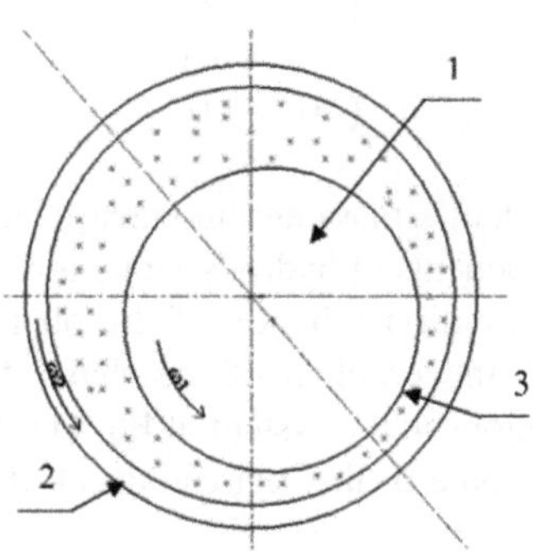

Fig. 1. Model of hydrodynamic bearing: 1 - shaft; 2 - bearing; 3 - lubricating fluid field; ω_1, ω_2 - are the angular speed of elements 1, 2 respectively [4].

The shaft is an absolute rigid cylinder *1*, which rotates free with angular speed ω_1 inside the bearing modeled as a cylindrical shell 2 rotating in the same direction with angular speed ω_2. The rotating axis of the cylindrical shell is fixed. The narrow gap between two solid bodies is filled with a pre-stressed oil with constant dynamic viscosity.

With this general model, research as a lemma to solve the dynamic problem for hydrodynamic bearings with floating bush [2, 6, 9], Fig. 2.

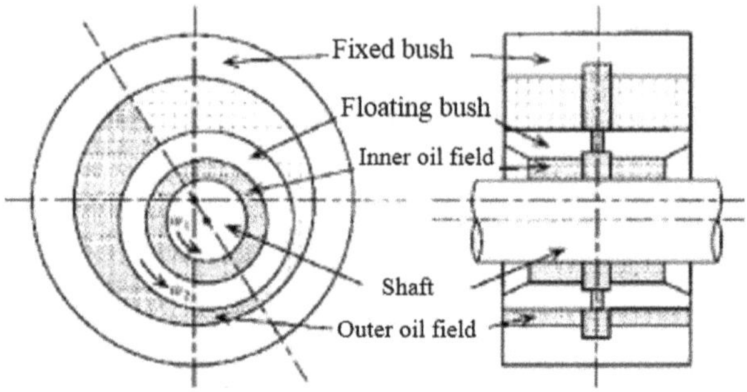

Fig. 2. Model of hydrodynamic bearing with floating bush

In the classical theory of hydrodynamic lubrication for short bearings, the research object is a shaft rotating in a fixed bearing. Ignoring the shaft's deformation and the axial displacement of the shaft, consider the axes symmetry of the shaft and of the bearing always parallel. Because of these assumptions, the problem can be modeled by a 2D model as shown in Fig. 1 and 2.

To investigate the dynamics of the shaft, we must establish the differential equation for its general plane motion in the lubricating fluid field. This equation in the Cartesian coordinate system has the form:

$$\begin{cases} F_x = m\ddot{x} \\ F_y = m\ddot{y} \\ M = I\ddot{\varphi} \end{cases} \tag{1}$$

in which: *m, I* - are the shaft's mass and moment of inertia respectively.

In this equation, the components of hydrodynamic force of the oil film are obtained by integral the pressure distribution function of the oil film over the surface of the rotating shaft [4]. However, the problem of the shaft of motion's stability is only decided by the first two equations in the system of Eq. (1). The third equation describes the resistance of the friction force of the lubricating oil film to the shaft in its motion around its symmetry axis.

3 Differential Equation of the Motion of the shaft's Center

In the Cartesian coordinate system $(O_1\xi, O_1\eta)$, with the direction of the axis $O_1\xi$ corresponding to $\theta = 0$ (coinciding with the center line, Fig. 3), the components of the hydrodynamic force $\mathbf{F}^S$, acting on the shaft has the form [4]:

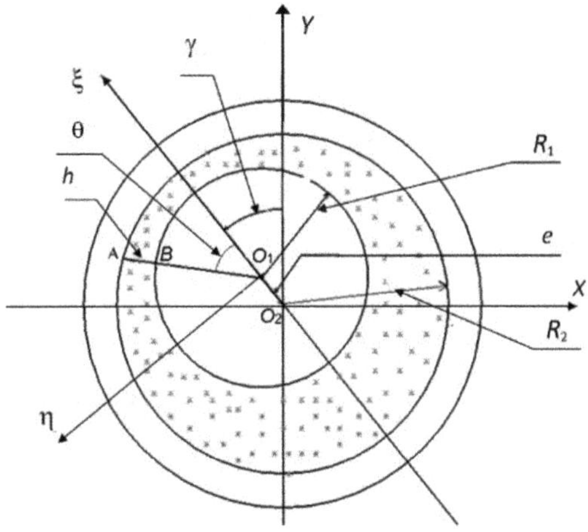

Fig. 3. Model for determining hydrodynamic force: circle (O_1, R_1) - shaft; cylindrical shell (O_2, R_2) - bearing; h = AB - width of the lubricating oil film's gap.

$$\begin{aligned} F_{\xi}^{S} &= -kN_1(\varepsilon)\frac{2\dot{\varepsilon}}{\omega_1+\omega_2}, \\ F_{\eta}^{S} &= -kN_2(\varepsilon)\left(\frac{2\dot{\gamma}}{\omega_1+\omega_2}-1\right); \end{aligned} \tag{2}$$

in which:

$$N_1(\varepsilon) = \frac{1+2\varepsilon^2}{(1-\varepsilon^2)^{5/2}},\ N_2(\varepsilon) = \frac{\varepsilon}{(1-\varepsilon^2)^{3/2}},\ k = \frac{\mu\pi L^3 R_1(\omega_1+\omega_2)}{2h_0^2}. \tag{3}$$

μ, L – are the bearing's dynamic viscosity and length respectively,
γ, e – are the angle associated by the center line O_1O_2 and the vertical direction and the eccentricity: $e = O_1O_2$, $\varepsilon = \frac{e}{h_0}$ – is the relative eccentricity $\varepsilon \in [0; 1].h_0 = R_2 - R_1$,

Assuming a shaft's mass of 2 m is supported by two identical hydrodynamic bearings at both tops. Also, assuming that the shaft is absolutely rigid and homogeneous, its center of gravity coincides with the center of symmetry O_1. Because of the symmetry property, in the plane problem, the differential equation of motion of the shaft's center in the Cartesian coordinate system O_2xy in Fig. 3 has the form:

$$\begin{cases} m\ddot{x} = -F_{\xi}^{S}\sin\gamma - F_{\eta}^{S}\cos\gamma; \\ m\ddot{y} = -mg + F_{\xi}^{S}\cos\gamma - F_{\eta}^{S}\sin\gamma. \end{cases} \tag{4}$$

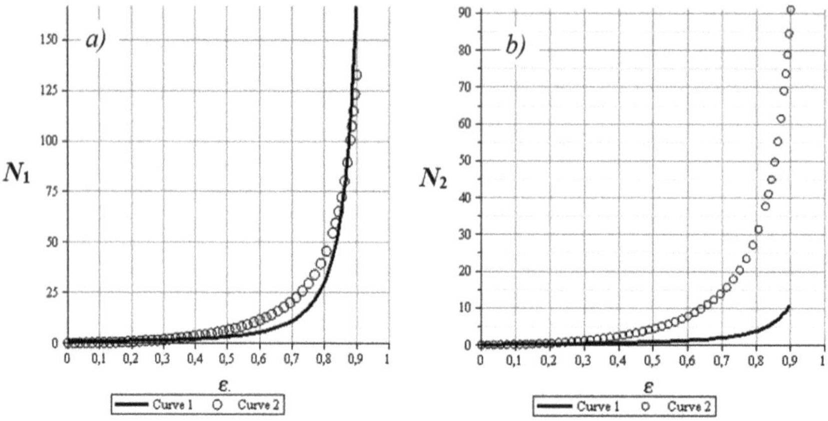

Fig. 4. a) characteristic function $N_1(\varepsilon)$, b) characteristic function $N_2(\varepsilon)$: curve 1 - function graph $N_i(\varepsilon)$ in Eq. (3); curve 2 - function graph $N_i(\varepsilon)$ of Tondl's study for short bearing and does not examine the centrifugal force of the lubricating oil film [11].

In this equation, the components of the hydrodynamic force acting on the shaft $\mathbf{F}^S$ in the case of examining the effect of the centrifugal force of the lubricating oil film, which is represented in formulation (2). We obtain the differential equation of motion of the shaft in polar coordinate system $(\varepsilon, \ \gamma)$:

$$\begin{cases} \ddot{\varepsilon} - \varepsilon\dot{\gamma}^2 = -\frac{g}{h_0}\cos\gamma - \frac{1}{mh_0}kN_1(\varepsilon)\frac{2\dot{\varepsilon}}{\omega_1 + \omega_2}; \\ \ddot{\gamma}\varepsilon - 2\dot{\varepsilon}\dot{\gamma} = \frac{g}{h_0}\sin\gamma - \frac{1}{mh_0}kN_2(\varepsilon)\left(\frac{2\dot{\gamma}}{\omega_1 + \omega_2} - 1\right). \end{cases} \tag{5}$$

Equation (5) is the nonlinear differential equation of motion of the shaft's center in polar coordinate system $(\varepsilon, \ \gamma)$. In which, $N_1(\varepsilon)$, $N_2(\varepsilon)$ are the characteristic functions of the hydrodynamic force acting on the shaft in case of examining the centrifugal force of the lubricating oil film when the shaft rotates at high speed. These characteristic functions depend on the shaft's eccentricity relative to the bearing, as shown in formula (3). Thus, solving the system of Eq. (5) by analytical method and finding out the trajectory of the shaft's center is not easy. However, one of the key problems with shaft's dynamics in a hydrodynamic bearing is investigating the stability of its equilibrium positions. This research can be conducted without solving Eq. (5).

4 Stability of the Shaft's Center at the Equilibrium Positions

To investigate stability at the equilibrium positions of the shaft's center, we must first find these equilibrium positions. Suppose $(\varepsilon^*, \gamma^*)$ is the equilibrium position of the shaft's center. Therefore it must satisfy the equilibrium condition:

$$\dot{\varepsilon}^* - 0,\ \dot{\gamma}^* - 0, \ddot{\varepsilon}^* - 0,\ \ddot{\gamma}^* = 0. \tag{6}$$

Replace this condition into Eq. (5) we obtain equilibrium positions:

$$\begin{cases} \gamma^* = \frac{3\pi}{2}; \\ N_2(\varepsilon^*) = \frac{mg}{k}. \end{cases} \tag{7}$$

See that, if we consider that the lubricating oil film exists and continuously surrounds the shaft - Sommerfeld's hypothesis, the equilibrium positions of the shaft's center are on the horizontal line passing through the bearing's center. Consider $(\varepsilon_1, \gamma_1)$ - is the possible displacement of the shaft's center at its equilibrium position, that is:

$$\varepsilon(t) = \varepsilon^* + \varepsilon_1(t);\ \ \gamma(t) = \gamma^* + \gamma_1(t). \tag{8}$$

Using Poincare's linearization method [13], after replacing the conditions (6) and (7), we obtain the differential equation of possible displacements:

$$\begin{cases} \ddot{\varepsilon}_1 = -\frac{k}{mh_0}\frac{2N_1(\varepsilon^*)}{(\omega_1+\omega_2)}\dot{\varepsilon}_1 - \frac{g}{h_0}\gamma_1; \\ \ddot{\gamma}_1\varepsilon^* = -\frac{k}{mh_0}\frac{2N_2(\varepsilon^*)}{(\omega_1+\omega_2)}\dot{\gamma}_1 + \frac{k}{mh_0}N_2'(\varepsilon^*)\varepsilon_1. \end{cases} \tag{9}$$

Equation (9) is also known as the first-order approximation equation. This is the linear differential equation for possible displacement of the shaft's center around its equilibrium positions. To research the stability of motion of the shaft's center at its equilibrium positions, we can rewrite this equation as the matrix form:

$$\mathbf{M}\ddot{\mathbf{q}} + \mathbf{B}\,\dot{\mathbf{q}} + \mathbf{C}\,\mathbf{q} = \mathbf{0}, \tag{10}$$

in which: generalized coordinate vector - $\mathbf{q} = (e_1, \gamma_1)^T$; **M**, **B**, **C** - are square quadratic matrices whose elements are constants:

$$\mathbf{M} = \begin{pmatrix} m & 0 \\ 0 & mh_0e^* \end{pmatrix};\ \mathbf{C} = \begin{pmatrix} 0 & mg \\ -kN_2'(\varepsilon^*) & 0 \end{pmatrix};\ \mathbf{B} = \begin{pmatrix} \frac{2kN_1(\varepsilon^*)}{(\omega_1+\omega_2)} & 0 \\ 0 & \frac{2kh_0N_2(\varepsilon^*)}{(\omega_1+\omega_2)} \end{pmatrix}. \tag{11}$$

It is easy to see that the damping matrix **B** is a symmetry matrix and the matrix of stiffness **C** is not a symmetry matrix. Therefore, this system is acted by non-conservative force. According to the theory of motion stability, if a system is acted by a non-conservative force, it will not be stable at equilibrium positions [12].

5 Results and Discussion

- The paper has established the differential equation of motion of shaft's center in hydrodynamic bearing in Cartesian coordinates system (4) and polar coordinate system (5).

- The difference of Eq. (4), (5) to previous studies is the expression of hydrodynamic force on the left side. Figure 4 shows the comparison of the characteristic functions $N_i(\varepsilon)$ of the hydrodynamic force in two cases: including and excluding the centrifugal force of lubricating fluid film for the short bearing type. Both characteristic functions $N_i(\varepsilon)$ obtained in this study are smaller than those in the case without excluding the centrifugal force of lubricating fluid film as in Tondl's study [11]. Physically, this can be explained that the force acting on the shaft must actually be the hydrodynamic force of the fluid film acting on the shaft omitting the centrifugal force of the lubricating fluid film. Because when the shaft rotates at high speed, this centrifugal force is quite significant and directed outward to the shaft's surface.
- By Poincare's method, we approximate the nonlinear differential Eq. (5) on the linear quadratic differential equations for possible displacement (9) and prove that this system is unstable at its equilibrium positions. This explains that there is always shaft's self - oscillation in a hydrodynamic bearing.

6 Conclusion

The paper presents establishment of the differential equation of motion of shaft in a hydrodynamic bearing for short bearing type. In which, the hydrodynamic force acting on the shaft of the lubricating oil film is obtained from the new form of Reynolds equation, including the centrifugal force of lubricating oil film. Based on this differential equation, the paper also gives the equilibrium positions of the shaft's center (7) and investigates the stability at its equilibrium positions. In this study, the author uses Sommerfeld's hypothesis about the boundary of the lubricating oil film in the bearing. Application of other hypotheses about the boundary of the lubricating oil film or dynamic research of general floating bush bearings (Fig. 2) including the centrifugal force of lubricating oil film will be presented in further studies.

Acknowledgements. The author would like to thank to support of professor A. K. Belyaev – "Institute of Problems of Mechanical Engineering – Russian Academy of Sciences" for this study.

References

1. Лойцянский Л. Г., Механика жидкости и газа. М: Наука (1987)
2. Dubois, G.B.: Analytical derivation and experimental evaluation of short- bearing approximation for full journal bearing. Cornell Univ. Report, No. 1157, pp. 119–127 (1953)
3. Belyaev, A.K., Krommer, M., Irschik, H., Van Thang, N.: Forces and moments acting on the rapidly rotating floating bearing. Proceedings. of XXXVI Summer School- Conference "Advanced Problems in Mechanics". St. Petersburg, pp. 104–111 (2008)
4. В.Т. Нгуен, Силы и моменты, действующие на ротор в упорном подшипнике скольжения, с учетом гидродинамики смазки и центробежных сил. Научно-технические ведомости СПбГПУ. Физико-математические науки, № 1(116), pp. 116–122 (2011)

5. Văn Thắng, N., Đạc, V.Q.: Về một dạng mới của phương trình Reynolds khi xét đến ảnh hưởng của lực ly tâm trường chất lỏng bôi trơn. Kỷ yếu Hội nghị Cơ học toàn quốc lần thứ IX, Đà Nẵng, Việt Nam, tập 2, 242–246 (2015)
6. Hatakenaka, K.: A theoretical analysis of floating bush journal bearing with axial oil film ruptures being considered. J. Tribol. **124**(3), 494–505 (2002)
7. Пешти Ю. В., Проектирование подшипников скольжения с газовой смазкой. Высшее техническое училище им. Н. Э. Баумана, МВТУ (1973)
8. Максимов С. П., Автоколебания роторов, вызванные масляным слоем подшипников скольжения, Труды ЦКТИ, Котлотурбостроение, № 44, 87–96 (1964)
9. Kirk, A.R.G., Alsaeed, A.A., Gunter, E.J.: Stability analysis of a high-speed automotive turbocharger. J. Tribol. Trans. **50**(3), 427–434 (2007)
10. Gasch, R., Nordmann, R., Pfuetzner, H.: Rotordynamik. Springer (2006)
11. Тондл, А.: Динамика роторов турбогенераторов, Л.: Энергия (1971)
12. Меркин, Д.Р., Введение в теорию устойчивости движения, М.: Наука (1976)
13. Poincaré, H.: Les Méthodes Nouvelles de la Mécanique Célèste II. Dover Publ, New York (1957)

An Application of Electric-Solar Hybrid Energy for Driver Handle Training

Nguyen Lam Tuan[2], Huynh Phuoc Son[2], and Nguyen Phu Thuong Luu[1](✉)

[1] Department of Automotive Engineering, Ho Chi Minh City University of Technology (HUTECH), Ho Chi Minh City, Vietnam
npt.luu@hutech.edu.vn

[2] Department of Automotive Engineering, Ho Chi Minh City University of Technical and Education, Ho Chi Minh City, Vietnam
hpson@hcmute.edu.vn

Abstract. Currently, driving training vehicles in driving schools in Vietnam still use Gasoline and Oil engines, and friction clutches. Researching and applying Electric vehicles and simultaneously changing friction clutches into clutches. This paper aims to save maintenance costs and save the cost of gasoline and oil during the time of learners' driving, and to reduce the environmental pollution caused by the exhaust of Gasoline engines and Oil causes. Radiant energy from the sun is converted by solar panels into electrical energy, and that electrical energy is loaded into batteries (lead-acid batteries) and used to run electric motors so cars move in the direction of front or front. At the same time, the friction clutch will be replaced with the clutch clutch, which saves the repair cost due to the friction clutch because the first time learning to drive is not mature, so the fast friction clutch is worn out or damaged.

Keywords: Electric hybrid cars · Solar car · Photovoltaic solar power · Solar battery

1 Introduction

The solar radiation is the main source of energy for cars, energy from the sun will be absorbed by solar panels and converted to electricity. Therefore, electrical energy is being provided so that the battery is charged and used to run a DC 24 V motor. Chain drive will be applied to transmit motor shaft movement to the rear wheels. It is very important to work properly of the battery. Currently, teaching vehicles in driving schools in Vietnam still use Petrol and Oil engines, and friction clutches. Due to the purpose of saving maintenance and repair costs that will save the gasoline and oil in the training time, also for reducing environmental pollution caused by exhaust gas of gasoline engines and oils.

Today, in Vietnam, many training centers have been researching and experimenting with models for training activities such as [1] designing and manufacturing models of auxiliary electrical systems in cars (AAES). Controlled by a smartphone. The system includes a security system, a safety and comfort system, a lighting system and an information system. This study focuses on lighting systems (front lighting, rear lighting,

B. T. Long et al. (Eds.): MMMS 2020, LNME, pp. 654–660, 2021.
https://doi.org/10.1007/978-3-030-69610-8_87

compartment lighting and signal lights) and safety and comfort systems (windshield wipers and doors). Archived in [2] presents the design and production of a smartphone-controlled model of an air-conditioning system (ACS). This model includes condenser, fan, electric motor, dryer, evaporator, power supply, control panel and compressor. This article is mainly studied on fan, compressor and control panel controls. They are installed into the bodyframe model and connected to an Arduino module. And use Android smartphone to control.

2 Theoretical Basic

Beginning in the 1960s and 70s of the last century, the world faced two major global issues:

- Energy issues: fossil energy sources such as oil and coal are not endless; they are likely to be exhausted and not renewable. Vehicles directly using this source of energy (gasoline, oil) will certainly not exist in the future. While electricity is a very flexible type of energy, it can vary from multiple sources of energy. Endless renewable energy sources like wind, sun, waves,... That the reason why the electric vehicles are the vehicles of the future.
- Environmental issues: It is not difficult to realize that the current environment seriously polluted, but one of the main causes is emissions from vehicles, especially cars. Electric cars are the answer to this problem because they have no emissions at all.

As such, we see that electric cars are the optimal solution to both major problems, which is why it became a special concern in the 20th century, and is becoming a major concern. of the automotive industry and scientists around the world. There have been many studies on this issue such as In Optimizing the Electric Vehicle Solar System on the University Campus in Dhaka, Bangladesh, by Nusrat Chowdhury, Chowdhury Akram Hossain et al. [3]. Due to the rapid decline in gas reserves in the fields coupled with the skyrocketing gas prices worldwide, environmental pollution is caused by the burning of machine fuel. This has made us worry about the exhaustion of energy sources in the future, renewable energy from solar energy, wind energy will provide an optimal and efficient solution for these issues. The goal of this study is to focus on effectively utilizing solar photovoltaic (PV) systems in electric vehicles and minimizing negative environmental impacts in terms of CO2 emissions. This system is also used in the Design of Solar Vehicles by GS U.A. Athavankar, Rajat Shail Singh use solar energy to power electric vehicles with charging facilities and contribute excess energy to the national grid of P. Arivalagan, Manoj mohanan [4]. In the Yogesh Sunil Wamborikar's solar vehicle article, Abhay Sinha [5] had a talk about using solar energy to power cars at the required voltage. A solar vehicle is a type of electric vehicle that provides either total or significant energy using direct solar energy. Typically, photovoltaic (PV) cells in solar panels convert photovoltaic energy directly into electrical energy. The term "Solar vehicle" generally implies that single energy is used to power all or part of a vehicle's propulsion engine. Other Study on Hybrid Solar Vehicle Design Based on Hybrid Energy Systems by A.Spina, R. Jdela Vega et al. [6] studied the issue of the lowest possible energy consumption for vehicle, with the photovoltaic module as the primary source of energy.

In addition, Blaz Luin, Stojan Petelin, and Fouad Al-Mansour [7] also studied the evaluation of electric vehicle energy efficiency in the topic Micro simulation of Electric Vehicle Energy Consumption.

My project basically has two systems, electrical systems and solar systems. The components required for both systems are solar panels, electric charge controllers (Arduino), batteries, motors, motor controllers and inverters, when converting AC to DC current is needed. Regarding the mathematical parameters and specifications of the components, I will use 4pin of 12 V connected in series, the motor controller for the DC motor provides a maximum capacity of 3 kW to move weight vehicle (including rider) 250–350 kg.

2.1 Design Method

When we design a model, there are computational formulas that need to be met and considered in order to produce a sufficient supply of power, performance and safety. The main part is the engine, the higher the capacity the engine has, the more problems that need to be solved for electric vehicles to achieve high efficiency. The first point to determine is how much capacity the engine can provide. Below are the basic approaches and some important and detailed formulas for determining the mechanics required by DC motor power, torque, etc.

2.2 Calculating the Forces Acting on Cars When Selecting an Electric Motor

In order for a car to move, the car needs to overcome the drag acting on it in motion, which is rolling resistance, slope resistance, and wind resistance, inertial resistance.

$$F_{total} = F_f + F_i + F_W + F_j \tag{2.1}$$

$$P_{total} = P_f + P_i + P_W + P_j \tag{2.2}$$

With: total drag of the vehicle.

F_f: rolling resistance, which is the resistance generated by the contact of a tire with the road.

$$F_f = \mathrm{f} * \mathrm{m} * \mathrm{g} \tag{2.3}$$

f: rolling resistance coefficient, depending on the type of road when the speed is less than or equal to 80 km/h

$$P_f = F_f.v \tag{2.4}$$

F_i: slope resistance is the resistance generated by a hill climb or overpass or when traveling on a downhill road. The angle between the ground and the slope of the road expressed in the form of α, we consider in the case of a vehicle running on a slope with a slope of ten degrees = 10 (the slope of 10° is the standard slope angle in driving conditions in Vietnam).

$$F_i = m^*g^* \sin \alpha \quad (2.5)$$

$$\text{Should: } P_i = F_i * V_d \quad (2.6)$$

F_W: Wind resistance is the drag on the air when driving. It is determined primarily by the shape of the vehicle. The formula for calculating aerodynamic drag is given by the following equation:

$$F_W = K * F * V^2 \quad (2.7)$$

$$\mathrm{P}_W = F_W * V \quad (2.8)$$

F_j Inertial drag is positive when the vehicle is accelerating and negative when the vehicle is decelerating.

$$F_j = m * \partial_i * j \quad (2.9)$$

$$P_j = F_j * \mathrm{V} \quad (2.10)$$

When we choose an electric motor, the power of the electric motor must overcome the capacity of rolling resistance, slope resistance, and wind resistance, inertial resistance. Candlestick:

$$P_{total} = P_f + P_i + P_W + P_j \quad (2.11)$$

3 Calculation of Solar Battery Used for the Model

The time it takes the sun to run well at Ho Chi Minh City is from 9 h to 15 h and the solar radiation intensity is about 4.58 $\mathrm{KWh/m^2/day}$. Loss in solar battery system is 1,2. Because only a 2-h test model should be selected, the number of hours is 2.

$$\text{PV panel} = \text{Total electricity consumption} * 1.2 = 5200 * 1.2 = 6240\ \text{Wh/day} \quad (3.1)$$

$$\text{Total Wp of PV panel: } \frac{PVpanel}{4.58} \quad (3.2)$$

With: 4.58 is the level of solar absorption in Vietnam.

$$\text{So the total Wp of PV panel: } \frac{6240}{4.58} = 1362,4\,\text{Wp} \tag{3.3}$$

Choose Wp 345 W type, deduce the number of panels to be used is: 1362.4/345 = 3.94. So we choose 4 panels of power 345 W.

Calculating backup battery capacity for the system:

We choose 4 batteries with 100 Ah capacity and 12 V voltage

$$\text{Battery capacity (Ah)} = \frac{2600}{0.85 * 0.6 * 12} = 424,8\,\text{Ah/day}$$

Battery capacity per 1 h backup: 424.8 / 24 = 17.7 Ah.

Calculating battery time for power supply of electric motors:

T = (48 * 400 * 0.8)/2600 = 5.9 h.

4 Magnetic Clutch Control

- Working principle of magnetic clutch:

When press depressing the clutch pedal: the power supply for pin 3 of Arduino UNO is disconnected, so there is no pulse output PMW (Pulse Width Modulation) at pin 9 to trigger.

- MOSFET HA210N06, resulting in no current Magnetic clutch, resulting in magnetic clutch does not work.
- When releasing the clutch pedal: there is a current to the pin 3 of Arduino UNO, there is a PMW (Pulse Width Modulation) at pin 9 to trigger the MOSFET HA210N06 to work, resulting in a current Magnetic clutch continuous cycle so the magnetic clutch works (Fig. 1).

Fig. 1. Diagram of working principle block of magnetic clutch control circuit.

– Arduino UNO: used to change the pulse width of the output signal cycle by programming using the analogWrite function with low pulse). By programming the Arduino UNO we can change the pulse width to operate continuously.
– PW Programmable pulse generator PWM code for Arduino UNO:

```
        nt rheostatPin = 9;     // rheostat connected to digital pin 9
int analogPin = 3;   // potentiometer connected to analog pin 3
        int val = 0;   // variable to store the read value
void setup()
        {
        PinMode(rheostatPin, OUTPUT);    // sets the pin as output
        }
void loop()
{
        val = analogRead(analogPin);    // read the input pin
 analogWrite(rheostatPin, val / 4);   // analogRead values go from 0 to 1023,
 analogWrite values from 0 to 255
}
```

5 Conclusion

Solar vehicles have solved many environmental problems and are the best way to reduce pollution. We should research and use them to reduce our dependence on fossil fuels. Solar powered vehicles still have some shortcomings that need to be researched and overcome such as low speed, low transport capacity, low operating time because of low battery capacity, manufacturing costs. The car is still tall. In addition, the energy metabolism rate is unsatisfactory (only 17%). However, these shortcomings can be easily overcome by conducting further studies in this area. Solar cars have a huge potential market and we should start using them in everyday life. The study of solar cars in the field of driving practice in Vietnam helps to set the stage for other studies.

References

1. Luu, N.P.T.: Design and manufacture a vehicle auxiliary system model controlled by using smartphone. In: Kahrama, C., Cevik Onar, S., Oztaysi, B., Sari, I, Cebi, S., Tolga, A. (eds.) Intelligent and Fuzzy Techniques: Smart and Innovative Solutions, INFUS 2020. Advances in Intelligent System and Computing, vol. 1197. Springer, Cham (2021). https://doi.org/10.1007/978-3-030-51156-2_187
2. Luu, N.P.T.: A study on vehicle air condition system model controlled by using smartphone. In: Kahrama, C., Cevik Onar, S., Oztaysi, B., Sari, I, Cebi, S., Tolga, A. (eds.) Intelligent and Fuzzy Techniques: Smart and Innovative Solutions, INFUS 2020. Advances in Intelligent System and Computing, vol. 1197. Springer, Cham (2021). https://doi.org/10.1007/978-3-030-51156-2_188

3. Chowdhury, N., Hossain, A., et al.: Optimization of solar energy system for the electric vehicle at university campus in Dhaka Bangladesh. Energies **11**(9), 1–10 (2018)
4. Athavankar, U.A., Singh, R.S.: Design of Solar Powered Vehicle. Project- II Industrial Design Center IIT, Mumbai (2017)
5. Wamborikar, Y.S., Sinha, A.: Solar powered vehicle. In: Proceedings of the Word Congress on Engineering and Computer Science 2010, WCECS 2010, Octorber 20–22, 2010, San Francisco, USA, vol. II (2010)
6. Spina, M.A., dela Vega, R.J., Rossi, S.R., Santillan, G.: Some issues on the design of a solar vehicle based on hybrid energy system. Int. J. Energy Eng. **2**, 15–21 (2012). p-ISSN: 2163-1891 e-ISSN: 2163-1905
7. Luin, B., Petelin, S., Al-Mansour, F.: Microsimulation of electric vehicle energy consumption. Energy **174**, 24–32 (2019). https://doi.org/10.1016/j.energy.2019.02.034

Medical Equipment and Processing Technology for Protecting the Life Environment

Mixture Formation and Pollution Emissions of Ethanol-Gasoline/Ethanol-LPG PI Motorcycle Engines

Van Ga Bui(✉), Le Bich Tram Truong, and Minh Duc Le

Danang University of Science and Technology, The University of Danang, Da Nang, Vietnam
buivanga@ac.udn.vn

Abstract. Mixture formation, soot and HC emissions of a spark-ignition (SI) engine fueled with ethanol-gasoline and with ethanol-liquid petroleum gas (LPG) under effects of intake air temperature and injection methods were investigated numerically. The simulation was performed on an 110cc motorcycle engine with a retrofitted intake system to adopt new fuel supply modes. The results show that ethanol-LPG dual injection (ELDI) mode produces a better homogeneous mixture as compared to that of ethanol-gasoline pre-blended injection (EGBI) mode. The diffusion combustion of liquid fuel droplets in a heterogeneous mixture is the main factor affecting soot emission. Either ELDI or EGBI mode, the higher T_{int} leads to the lower unburned fuel concentration in the exhaust gas. At a low T_{int}, the unburned fuel concentration increases with ethanol content, whereas the rate of this increase is reduced gradually at high T_{int}. At a given engine operating condition, the unburned fuel concentration and the soot volume fraction in the exhaust gas of EGBI mode are nearly 1.3 times and 2.5 times, respectively, higher than those of ELDI mode. The over-fueling is required for EGBI mode, while no start-aids system is needed for ELDI mode under the cold start condition.

Keywords: Ethanol · Alternative fuels · Renewable energy · Soot emission · Air-fuel mixture preparation

Nomenclature

°CA	Degree of crankshaft angle
Cf	Total mole fraction of fuels (%)
Ce	Mole fraction of ethanol (%)
Ex	Gasoline-ethanol blend contains x% ethanol in volume
EGBI	Ethanol-Gasoline pre-Blended Injection
ELDI	Ethanol-LPG Dual Injection
EfR	Partial equivalence ratio of ethanol (%)
fv	Soot volume fraction (m^3 soot/m^3 gas mixture)
LPG	Liquid Petroleum Gas
LPM_v	Liquid Particle Matter (g/m^3)
SI	Spark Ignition
n	Engine speed (rpm)

B. T. Long et al. (Eds.): MMMS 2020, LNME, pp. 663–671, 2021.
https://doi.org/10.1007/978-3-030-69610-8_88

T	Temperature of the mixture in cylinder (K)
T_{int}	Temperature of intake air (K)
φ	Crankshaft angle (°CA)
ϕ	Equivalence ratio
$\phi_{ethanol}$	Equivalence ratio contributed by ethanol content
ϕ_{global}	Global equivalence ratio of the mixture

1 Introduction

Ethanol is an attractive renewable fuel because it has better physical properties, such as higher oxygen content, greater vaporization enthalpy, greater octane number, higher flame speed, and better lean-burn behavior compared to those of other fuels [1–3]. The presence of oxygen in ethanol allows a complete combustion leading to an improvement of engine efficiency [4]. The high latent heat of vaporization of ethanol can cool down the intake charge which allows an increase of the engine volume efficiency [5, 6]. The high flame speed enhances the combustion quality leading to a reduction of harmful emissions. With these advantageous properties, ethanol is widely applied as an additive to the conventional fuels.

The charge cooling due to the high latent heat of ethanol evaporation particularly affects the cold start stage of the engine [7]. To overcome this problem, the fuel injection duration should be increased. For low-volatility fuel, it is important to over-fuel more than 10 times to ensure the reliable start and stable idling at cold conditions [8]. In this case, more fuel remains in the intake port leading to an increase of unburned hydrocarbon emission.

Some techniques boosting the cold start of the ethanol engine, such as intake air preheating or using auxiliary fuels have been introduced. The LPG can be used as an auxiliary fuel to support the starting of the SI ethanol engine in cold condition [9]. The LPG can defuse in the air at a lower inlet temperature compared to that of any other liquid fuels. It leads to faster starting and better combustion with lower unburned emissions of hydrocarbons. In contrast to liquid fuel engines, emission levels of the LPG engine are almost independent with the intake air temperature [9].

During the starting and warming up stages, the evaporating properties of fuel significantly affect the mixture preparation and combustion processes. In the phase of stable combustion, high oxygen in ethanol can shorten the combustion duration as well as promote the completed combustion. Therefore, HC and CO emissions decrease obviously. An addition of 10% ethanol into the gasoline can reduce by about 30% CO emission and 6% HC emission [10]. Rahman [11] has tested the SI engine fueled with LPG enriched E20 and noted that the efficiency of the engine increases in range of 4–6%. As compared to E20, when LPG is inducted to the fuel, the reductions of HC, CO, and CO_2 are in range of 13–14%, 14–15%, and 12–18%, respectively, while NO_x emission increased by 13–16%.

Besides the gaseous pollutants (e.g., HC, CO and NO_x), the emission of soot is also a major problem of exhaust gas control. Due to high latent heat, the evaporation of ethanol reduces the mixture temperature in the combustion chamber. Consequently, the evaporation of the most boiling fuel species can be avoided, leading to the condensation of liquid fuel on the cylinder wall and the mixture becomes more heterogeneous. This creates a high trend in sooting as increasing ethanol content in the fuel mixture [12, 13]. As the SI engine operates under cold condition, the maximum soot gravimetric concentration emission of the E85 fueling mode is about 11 times higher than that of the gasoline fueling mode, while in the warm conditions it is about 8 times higher [12]. While the evaporation properties of ethanol exhibit negative effects on soot emission control, the oxygen composition in the ethanol enhances the combustion thus, reduces soot formation [14]. The addition of ethanol produced a rise in soot emissions particularly when the engine operates under high speeds and/or high load regime [15].

This research focuses mainly on the effects of air temperature, fuel injection methods and ethanol addition on the mixture formation and pollutant emissions of the motorcycle engine fueled with ethanol-gasoline blend and with ethanol-LPG.

2 Material and Method

The research is carried out by numerical simulation on the motorcycle engine (type Honda 110cc) with a modified intake system to adopt new fuel supply modes. This is a four-stroke spark-ignition, single-cylinder engine, which has a bore of 50 mm, a stroke of 55 mm, and a compression ratio of 11:1. The engine produces a rated power output of 6.4 kW at 7500 rpm in gasoline fueling mode. Two fuel couples will be used for the comparison, i.e., ethanol-gasoline and ethanol-LPG. They are inducted into the cylinder by the port injection method through two different modes of fuel supply, i.e., ethanol-gasoline pre-blended injection (EGBI) and ethanol-LPG dual injection (ELDI).

Ethanol or ethanol-gasoline blend is injected through a six-hole injector with the mass flow rate is 6 g/s. LPG is inducted into the intake manifold through a gaseous injector with mass flow rate of 2 g/s. Due to the difference in state of the fuels as they are induced into the intake manifold, for a convenience of comparison between these two fueling modes, the composition of ethanol in the fuel mixture is defined by the partial equivalence ratio of ethanol EfR. It is the ratio between the equivalence ratio contributed by ethanol content divided by the global equivalence ratio of the mixture, $EfR = \phi_{ethanol}/\phi_{global}$.

The CFD Ansys Fluent V18.0 software was used for simulation. The detail of simulation process has been described in our previous works [16–18].

3 Results and Discussion

3.1 Mixture Formation

The Fig. 1a and Fig. 1b illustrate the contours of ϕ and LPM_v concentration on the longitudinal section *xy* and on the cross-section *xz* of the combustion chamber at 330 CA. In the case of ELDI, because the liquid fuel particles evaporate completely earlier, the fuel concentration in the liquid phase at 330°CA has almost vanished. Meanwhile, in the case of EGBI, the droplets continue to evaporate until the end of the compression process. The gap between the minimum and the maximum equivalence ratios in the case of ELDI is very narrow, i.e., from 0.95 to 1.15, while in the case of EGBI, the gap is very large, i.e., from 0.3 to 2.3. In case of EGBI, the distribution of the equivalence ratio on the cross-section *xz* of the combustion chamber is more homogeneous than that on the longitudinal section *xy*. Thus, in this case the location of the spark plug in the section *xy* toward the opposite side of the intake valve is more favorable for ignition. While in case of ELDI, the top center of the combustion chamber is the appropriate position for spark plug location because the mixture is almost homogeneous.

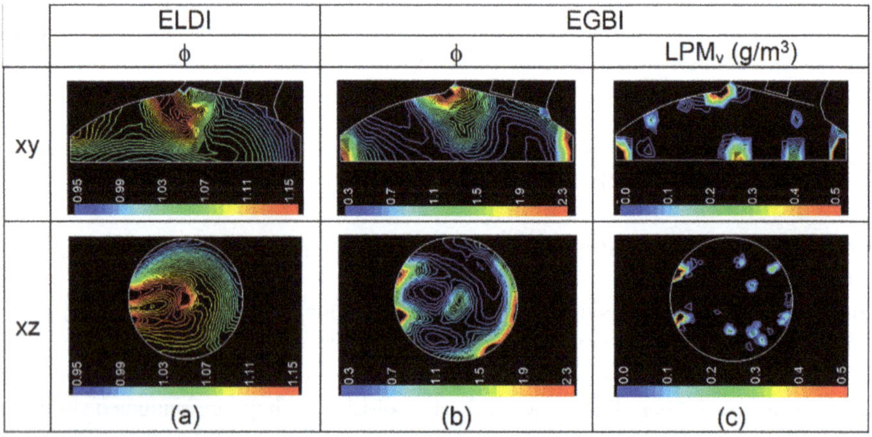

Fig. 1. Comparison of the distribution of the equivalence ratio on the longitudinal section *xy* and cross-section *xz* of the combustion chamber (φ = 330°CA, n = 5000 rpm, T_{int} = 320K): (a) ϕ contours in case ELDI; (b) ϕ contours in case EGBI; (c) LPM_v in case EGBI.

As can be seen from Fig. 1c, the liquid fuel droplets are dispersed in the combustion chamber and the largest droplets concentration areas are closed to the cylinder wall. The presence of liquid fuel droplets in the combustion chamber leads to the diffusive combustion and increases the soot production [14].

The above result indicates that the mixture preparation is greatly influenced by the modes of fuel supply. The quality of mixture preparation in the combustion chamber has a major influence on the combustion process. The homogeneous mixture is a critical aspect of the complete combustion and emission reduction. With the same EfR and intake air temperature, the complete evaporation of liquid fuel droplets in the case of ELDI is achieved much earlier as compared to that in the case of EGBI. The fuel vapor and air have more time to mix together resulting in a homogeneous mixture in the case of ELDI.

3.2 Soot Emission

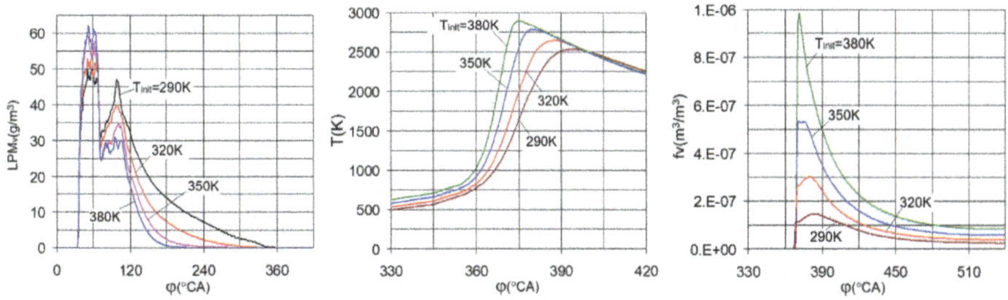

Fig. 2. Effects of intake air temperature on the variation of LPM_v (a), T (b) and fv (c) in function of crankshaft angle (n = 5000 rpm, ϕ_{global} = 1.1, EfR = 40%).

The effects of intake air temperature on the variation of LPM_v concentration as a function of the crankshaft angle in the case of ELDI mode are shown in Fig. 2a. The results show that the higher the intake air temperature, the shorter the existing time of liquid droplets in the cylinder. This is due to the increase of heat transfer coefficient between the air and the fuel droplets with the rise of intake air temperature. At a high intake air temperature, e.g., T_{int} = 380K, ethanol completely evaporates during the intake process. However, at a low intake air temperature, e.g., T_{int} = 290K, the liquid fuel droplets remain longer in the cylinder, until the end of the compression process.

When the complete evaporation of fuel is achieved earlier, the available time for mixture preparation is thus longer leading to a more homogeneous mixture. This improves the combustion efficiency resulting in an increase in maximum combustion temperature as shown in Fig. 2b. When the intake air temperature is about 290K, the evaporation of liquid droplets lasts until the later compression process. This leads to late peaks of pressure and temperature, resulting in a reduction of combustion temperature.

It can be seen from Fig. 2c that with the same injection conditions, the soot concentration in the exhaust gas increases with increasing the intake air temperature. The soot volume fraction yields about $2e^{-8}$ and $9e^{-8}$ corresponding to the intake air temperature of 290K and 380K, respectively. A better preparation of the mixture as rising the T_{int} yielded an increase in the combustion temperature, therefore, accelerates the soot formation consequently.

Figure 3a and Fig. 3b present the effects of EfR to the variation of the soot volume fraction in the exhaust gas according to the intake air temperature in the cases of ELDI and EGBI for the same equivalence ratio of ϕ_{global} = 1.1. It can be seen that the diffusion combustion of liquid fuel droplets in the heterogeneous mixture is the main cause of the soot emission in both cases. The soot concentration increases with increasing the intake air temperature due to the increase of combustion temperature resulted from better evaporation. When the intake air temperature is in the range of 290K to 350K, the slopes of curves $fv(T_{int})$ in both two fueling modes are moderate, on an average of about $4e{-}^{10}K^{-1}$. However, when the intake air temperature is higher than

350K, fv increases significantly. This difference might occur since the high intake air temperature yielded an increase in evaporation rate therefore, the combustion temperature and soot volume fraction are increased consequently. At a given EfR and T_{int}, the soot emission in EGBI mode is about 2.5 times higher than that in ELDI mode as shown in Fig. 3a and Fig. 3b. This is because increasing the content of the liquid fuel leads to an extension of the lifetime of droplets and an increase of the diffusive combustion and soot production [12].

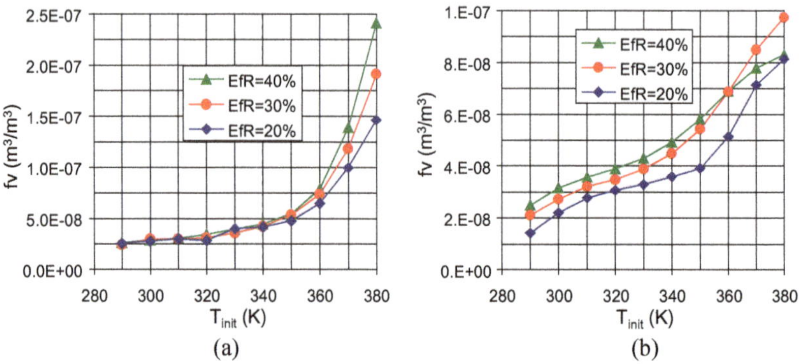

Fig. 3. Effects of EfR on the variation of soot volume fraction as a function of intake air temperature in case of EGBI (a) and in case of ELDI (b) (n = 5000 rpm, ϕ_{global} = 1.1).

It is possible to see in Fig. 3a, in the case of EGBI, when the air temperature is lower than 350K, the EfR does not influence on the soot volume fraction. However, when the temperature is greater than 350K, the soot volume fraction increases strongly with increasing of EfR at a given T_{int}. For instance, at T_{int} = 380K, the soot volume fraction is about $1.4e^{-7}$, $1.8e^{-7}$, and $2.4e^{-7}$ corresponding to the EfR is about 20%, 30%, and 40%, respectively. High values of combustion temperature and high values of ethanol content in the blend are the main reasons for the increase of soot volume fraction with T_{int} and/or EfR. In other words, under the warm conditions, increasing ethanol fraction in blends produces higher soot emissions. The trend of the shoot emissions for the cold condition is less clear in the case of EGBI compared to that in the case of ELDI. The obtained results agree with the experimental data reported by Chen et al. [12]. They claimed that for gasoline and its blends with ethanol, the soot emission under the warm condition is higher than that under the cold conditions.

3.3 HC Emission

Although the soot emission goes up as the intake air temperature increases, the unburned fuel emission decreases because the combustion efficiency is improved. Figure 4a and Fig. 4b exhibit a comparison of the variation of Cf according to the intake air temperature for the cases of ELDI and EGBI under effects of EfR. It is possible to note that at a given T_{int} and EfR, the Cf in the exhaust gas of the EGBI mode is always higher than that of ELDI mode. This is due to the incomplete combustion resulted from unfavorable evaporation conditions of EGBI mode as described previously. Besides, emissions of LPG fuel are practically independent with the intake air temperature [19]. This result is well confirmed by the experimental data proposed by Rahman [11]. Either EGBI or ELDI mode, the higher the intake air temperature, the lower the HC concentration in the exhaust gas. This is occurred due to the improvement of mixture homogeneity as increasing the air temperature which is favorable for the complete combustion.

Figure 4a shows that, in the case of ELDI, the unburned fuel concentration falls from 0.71% to 0.25% when the intake air temperature is raised from 290K to 380K with EfR = 40 and ϕ_{global} = 1.1. In the same conditions, the unburned fuel concentration falls from 0.98% to 0.32% in the case of EGBI. On average, at a given condition of T_{int} and EfR, the unburned fuel concentration in the exhaust gas of EGBI mode is about 1.3 times higher than that of ELDI mode.

Furthermore, at a low intake air temperature, the unburned fuel concentration increases with increasing of EfR. The rate of this increase is reduced gradually as increasing T_{int}. In the case of ELDI, the emission of unburned fuel practically reaches to a remained value since T_{int} is higher than 370K. In the case of EGBI, the emission of unburned fuel decreases as EfR increases when T_{int} is higher than 360K. This is occurred due to the presence of oxygen in the ethanol fuel which allows faster combustion and more efficient combustion as the liquid fuel fully evaporated [12]. This result confirmed the remark of Chansauria et al. [10] which noted that for each 5% of ethanol addition in the blend, there was about 6% of HC emission decreased.

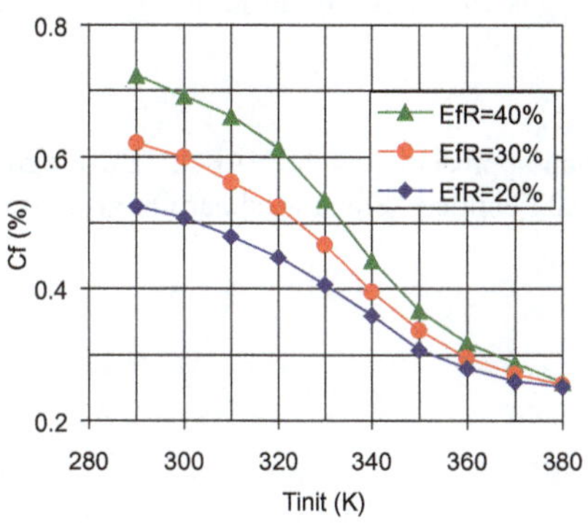

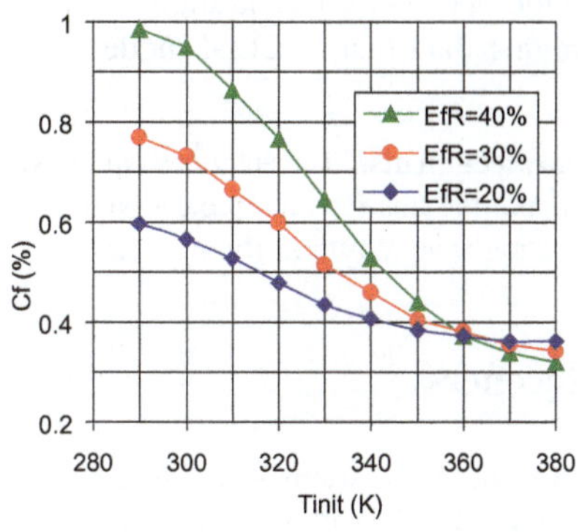

Fig. 4. Effects of EfR on the variation of HC concentration Cf as a function of intake air temperature in case of ELDI (a) and in case of EGBI (b) (n = 5000 rpm, ϕ_{global} = 1.1).

4 Conclusion

The following conclusions are drawn from the results of the present study:

- At fixed values of the EfR and air temperature, ELDI fueling mode produces a more homogeneous mixture as compared to that of EGBI fueling mode. The lifetime of liquid fuel droplets increases with an increase of ethanol content or/and with a decrease in the intake air temperature. The over-fueling is required for EGBI at low intake temperatures, whereas no start-aids system is needed for ELDI mode.
- The diffusion combustion of liquid fuel droplets in the heterogeneous mixture is the main cause of the soot emission. At a given EfR and T_{int}, the soot emission in the EGBI mode is about 2.5 times, on average, higher than that in the ELDI mode. The effect of intake air temperature on the soot concentration is moderate when Tint < 350K, while it is significant when Tint > 350K in both cases of fuel supply mode.
- At a given T_{int}, the soot volume fraction increases with the increase of EfR. The trend of soot volume fraction is more obvious in the ELDI mode as compared to that in the EGBI mode at low T_{int}. At high T_{int}, this sense becomes inverse. The effect of oxygen content in ethanol on the soot formation is obviously clear as using ELDI fuel supply mode at high intake air temperature.
- Either EGBI or ELDI mode, the higher intake air temperature leads to the lower unburned fuel concentration in the exhaust gas. At a low intake air temperature, the unburned fuel concentration increases with increasing EfR, whereas the rate of this increase is reduced gradually at high T_{int}. On average, at a given T_{int} and EfR, the unburned fuel concentration in the exhaust gas of EGBI mode is about 1.3 times higher than that of ELDI mode.

Acknowledgments. The authors wish to express their appreciation to the NAFOSTED Program for supporting this research project NCUD.02-2019.22 entitled "Small Modular On-grid Power Production from Waste in Rural Areas".

References

1. Boretti, A.: Towards 40% efficiency with BMEP exceeding 30 bar in directly injected, turbocharged, spark ignition ethanol engines. Energy Convers. Manag. **57**, 154–166 (2012)
2. Liang, C., Ji, C., Gao, B., Liu, X., Zhu, Y.: Investigation on the performance of a spark-ignited ethanol engine with DME enrichment. Energy Convers. Manag. **58**, 19–25 (2012)
3. Huang, Y., Hong, G., Cheng, X., Huang, R.: Investigation to charge cooling effect of evaporation of ethanol fuel directly injected in a gasoline port injection engine. SAE Technical Paper 2013-01-2610 (2013)
4. Catapano, F., Sementa, P., Vaglieco, B.M.: Air-fuel mixing and combustion behavior of gasoline-ethanol blends in a GDI wall-guided turbocharged multi-cylinder optical engine. Renew. Energy **96**, 319–332 (2016)
5. Celik, M.B., Özdalyan, B., Alkan, F.: The use of pure methanol as fuel at high compression ratio in a single cylinder gasoline engine. Fuel **90**, 1591–1598 (2011)

6. He, B.Q., Wang, J.X., Hao, J.M., Yan, X.G., Xiao, J.H.: A study on emission characteristics of an EFI engine with ethanol blended gasoline fuels. Atmos. Environ. **37**, 949–957 (2003)
7. Lang, K.R., Cheng, W.K.: Effect of fuel properties on first cycle fuel delivery in a SI engine. SAE Trans. **113**, 2095–2105 (2004)
8. Heywood, J.B.: Handbook of Air Pollution from Internal Combustion Engines. Academic Press, New York (1998)
9. Gong, C., Deng, B., Wang, S., Su, Y., Gao, Q., Liu, X.: Investigation on firing behavior of the spark-ignition engine fueled with methanol, liquefied petroleum gas (LPG), and methanol/LPG during cold start. Energy Fuels **22**, 3779–3784 (2008)
10. Chansauria, P., Mandloi, R.: Effects of ethanol blends on performance of spark ignition engine-a review. Mater. Today Proc. **5**, 4066–4077 (2018)
11. Rahman, M.A.: Induction of hydrogen, hydroxy, and LPG with ethanol in a common SI engine: a comparison of performance and emission characteristics. Environ. Sci. Pollut. Res. **26**, 3033–3040 (2019)
12. Chen, L., Stone, R., Richardson, D.: A study of mixture preparation and pm emissions using a direct injection engine fuelled with stoichiometric gasoline/ethanol blends. Fuel **96**, 120–130 (2012)
13. Yang, J., Roth, P., Zhu, H., Durbin, T.D., Karavalakis, G.: Impacts of gasoline aromatic and ethanol levels on the emissions from GDI vehicles: Part 2. Influence on particulate matter, black carbon, and nanoparticle emissions. Fuel **252**, 812–820 (2019)
14. Chen, Y., Ma, J., Han, B., Zhang, P., Hua, H., Chen, H., Su, X.: Emissions of automobiles fueled with alternative fuels based on engine technology: a review. J. Traffic Transp. Eng. **5**, 318–334 (2018)
15. Ratcliff, M.A., Windom, B., Fioroni, G.M., John, P.S., Burke, S., Burton, J., Christensen, E. D., Sindler, P., McCormick, R.L.: Impact of ethanol blending into gasoline on aromatic compound evaporation and particle emissions from a gasoline direct injection engine. Appl. Energy **250**, 1618–1631 (2019)
16. Bui, V.G., Bui, T.M.T.: Soot emission analysis in combustion of biogas diesel dual fuel engine. Int. J. Environ. Sci. Sustain. Dev. **1**, 67–79 (2017)
17. Bui, V.G., Tran, V.N., Nguyen, Q.T., Huynh, T.T.: Evaporation and mixture formation of gasoline-ethanol sprays in spark ignition engines with pre-blended injection and dual injection: a comparative study. IET Renew. Power Gener. **13**, 539–548 (2018)
18. Bui, V.G., Tran, V.N., Bui, T.M.T., Nguyen, Q.T.: Numerical simulation studies on performance, soot and NO_x emissions of dual-fuel engine fuelled with hydrogen enriched biogas mixtures. IET Renew. Power Gener. **12**, 1111–1118 (2018)
19. Gong, C., Liu, Z., Su, H., Chen, Y., Li, J., Liu, F.: Effect of injection strategy on cold start firing, combustion and emissions of a LPG/methanol dual-fuel spark-ignition engine. Energy **178**, 126–133 (2019)

Effects of Flow Rate of Cooling Water on Performance of Electrochemical Water Treatment for Cooling Tower of Water-Cooled Chiller

Le Minh Nhut[1(✉)], Duong Huynh Minh Nhut[2], Nguyen Doan Anh Tai[1], Tran Trung Hao[1], Le Van Nguyen[1], and Tran Trong Nghia[1]

[1] Ho Chi Minh City University of Technology and Education, Ho Chi Minh City, Vietnam
nhutlm@hcmute.edu.vn
[2] College of Technology II, Ho Chi Minh City, Vietnam

Abstract. The main objective of the investigation is to evaluate the effects of the flow rate of cooling water on the performance of the electrochemical water treatment for the cooling tower of the water-cooled chiller. The quality of cooling water of the cooling tower is mainly a parameter of the effect on the coefficient of performance of water-cooled chiller. Therefore, it is necessary to maintain the quality of cooling water during the operation time of the water-cooled chiller. In this study, the electrochemical water treatment for the cooling tower of the water - cooled chiller was designed and built on the campus of Ho Chi Minh City University of Technology and Education for experiments. This system consists of a reactor tank that has a volume of 2 L, three Titanium electrodes connected to DC power, a circulating water pump, a condenser, a cooling tower, and a control box. The experimental results show that the highest performance of the electrochemical water treatment system to maintain the water quality of the cooling tower of water-cooled chiller corresponds to the spacing between of three Titanium electrodes is 2 cm, the current density is 90 A/m^2, and the flow rate of cooling water through the reactor tank is 4 L/min.

Keywords: Electrolysis · Cooling water · Water treatment · Water-cooled chiller · Cooling tower

1 Introduction

In the water-cooled chiller, the cooling tower is used to release the heat from the condenser. The quality of cooling water for the cooling tower is affected by the hard water and it is also the main parameter affecting the coefficient of performance (COP) of the water-cooled chiller [1, 2]. Hard water creates many problems with the water-cooled chiller, such as the precipitated solid of the insoluble salts forms a scale on the heat transfer surface, raises the heat resistance, and thus decreases the efficiency of the heat exchangers, long after the condenser pipe can be blocked [2–4]. There are

B. T. Long et al. (Eds.): MMMS 2020, LNME, pp. 672–679, 2021.
https://doi.org/10.1007/978-3-030-69610-8_89

many methods for softening water and restricting the development of scaling such as the use of chemicals to create chemical precipitation or ion exchange, reverse osmosis, nanofiltration, heating, distillation, etc. However, they have limitations such as increased sludge, high costs, and the formation of sediments on the membrane, need treatment of residual sludge to meet environmental targets [5–7]. The electrochemical technology has many outstanding advantages compared to other technologies in the cooling water treatment of the condenser of water-cooled chiller because this method doesn't use chemicals, softens water quickly, it is easy to use and maintain [8–10]. In electrochemical technology, the precipitation of calcium carbonate is a function of temperature, but it also depends on the ion concentration, the flow rate and the pH index of the cooling water [11]. Henry et al. [12] presented the new method to control hardness, scale, sludge, and reducing water usage for a cooling water system. The polarity reversal in the electrochemical precipitation process for removing scale deposits was suggested by Jin et al. [13]. Mohammad et al. [14] reported the application of the continuous flow electrocoagulation technology for the defluoridation of industrial wastewater. The major objectives of this work are to evaluate the effect of the flow rate of cooling water through the reactor tank on the efficiency of the electrochemical water treatment for the cooling tower of the water-cooled chiller. In addition, the effects of parameters such as electrode spacing, current density, and power consumption on the decrease in total hardness index (such as $CaCO_3$), total dissolved solids, and pH index were also elucidated in this research.

2 Research Methodology

2.1 Theory of Electrochemical Processing

The process of electrolysis of water is the decomposition of water into oxygen and hydrogen gas due to the passage of an electric current, thereby creating an alkaline environment at the cathode that increases the pH in the vicinity of the electrode surface, and promotes precipitation of the calcium hardness in form of $CaCO_3$ according to the reaction as follows [8, 10–16].

Anode reaction (positive electrode), oxidation of water:

$$2H_2O - 4e^- \rightarrow O_2 + 4H^+ \tag{1}$$

Cathodic reaction (negative electrode), decomposition of water:

$$4H_2O + 4e^- \rightarrow 2H_2 + 4OH^- \tag{2}$$

In hard water containing bicarbonate ions a secondary reaction occurs in the anode, secondary anode reactions:

$$H^+ + HCO_3^- + H_2CO_3 \rightarrow CO_2 \uparrow + H_2O \tag{3}$$

In hard water containing bicarbonate ions and calcium ions, two secondary reactions occur in the cathode:

$$OH^- + HCO_3^- \rightarrow CO_3^{2-} + H_2O \quad (4)$$

$$Ca^{2+} + CO_3^{2-} \rightarrow CaC0_3 \downarrow \quad (5)$$

Removal rates of the hardness can be expressed as given by [17]:

$$R = \frac{C_{in} - C_{out}}{C_{in}} \times 100\% \quad (6)$$

Where R(%), C_{in} (ppm) and C_{out} (ppm) are the percentage of hardness removal, the initial concentration and the concentration after treatment, respectively.

2.2 System Description

The schematic diagram of the cooling water treatment system for water-cooled chiller using electrolytic technology is shown in Fig. 1 and Fig. 2. The system is designed and installed on the campus of the Ho Chi Minh City of University of Technology and Education. It consists of a cooling tower with a capacity of 5 RT connected with a condenser, a DC power supply, and voltage controller, measuring devices, an electrochemical reactor tank, and electrodes. The reactor tank has a capacity of 2 L and is cylindrical. The electrodes are made of a titanium mesh, with a reaction area is 145 cm^2 per plate, and are installed in parallel. The anode is coated with lead dioxide (PbO_2), DC power supply can be adjusted to the voltage according to the experiments.

2.3 Experimental Apparatus and Procedure

The quality of the cooling water after the treatment obtained the required water quality for cooling tower operation of water-cooled chiller according to the water quality standards TCXD 232:1999 of Vietnam (the properties of cooling water were: total hardness (such as $CaCO_3$) 280 ppm, pH 7.6; total dissolved solids (TDS) 402–407 ppm). In these tests, the flow rate of the cooling water through the reactor tank varies from 1 L/s to 6 L/s, while the current density varies from 40 A/m^2 to 100 A/m^2, and the spacing of the electrodes ranges from 1cm to 3 cm. The total hardness is measured by a HI3812 (Hanna-USA, uncertainty of 1%), total dissolved solids (ppm) is measured by a TDS-04 m (Total Meter-Taiwan, uncertainty of 2%), while the pH index is measured by a pH-2 m (Mudder, uncertainty of 0.05%), respectively. The energy consumed for experiments (Wh) is measured by a PZEM-002 power meter (uncertainty of 1%), and the flow rate of cooling water through the electrochemical reactor tank measured by a Water Flow Sensor YF-S402B (uncertainty of 1%). The data of the system are recorded during an experimental session and are temporarily stored in an Excel file on the controlling PC and are available for downloading for subsequent calculation.

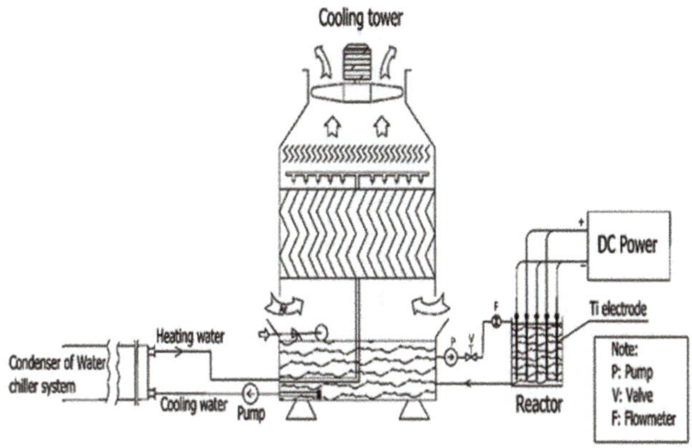

Fig. 1. Schematic diagram of the cooling water treatment system for water-cooled chiller using electrolytic technology

Fig. 2. Components of the experimental system

3 Results and Discussion

Figure 3 indicated the relationship between electrode spacing and total hardness removal. The total hardness steadily dropped in the range of 0 min to 80 min and then was approximately stable in the range of 80 min to 100 min. The efficiency of total hardness removal has obtained the values of 89%, 93%, and 87% corresponding with the electrode spacing at 1 cm, 2 cm, and 3 cm, respectively. The results obtained for the electrode spacing are the same before studies [7, 17].

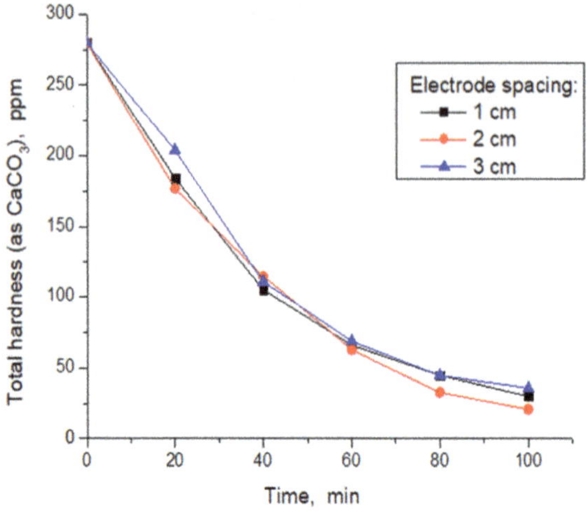

Fig. 3. Relationship between electrode spacing and removal total hardness.

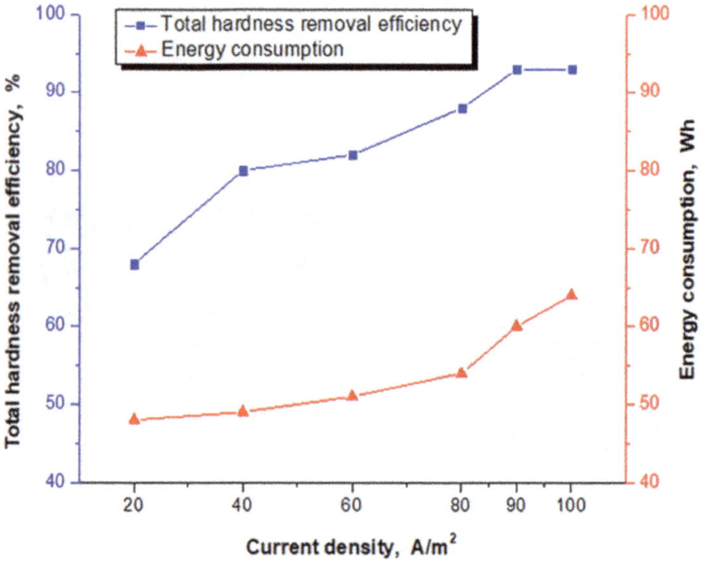

Fig. 4. Effect of current density on the efficiency of total hardness removal and energy consumption.

The effect of current density on the efficiency of total hardness removal and energy consumption is shown in Fig. 4. The efficiency of total hardness removal progressively increased and reached the highest value of 93% at the current density of 90 A/m^2 and then decreased gradually when the current density is greater than 90 A/m^2. The energy consumption of the electrochemical process gradually increased at the backward value of 90 A/m^2 and the forward direction increased rapidly. Therefore, the best value of the current density is chosen at 90 A/m^2.

Figure 5a shows the effect of the flow rate on the efficiency of total hardness removal. When the flow rate of cooling water through the reactor tank ranges from 1 L/min to 6 L/min, the efficiency of total hardness removal strongly increased and obtained the highest value is 23% at the value of flow rate of cooling water is 4 L and then decreased gradually. This result is consistent with studies [11–14]. The effect of the flow rate of cooling water through the reactor tank on the TDS removal efficiency and pH index is shown in Fig. 5b. The highest value of the TDS removal efficiency is 17.5% at the flow rate of cooling water is 4 L/min. The pH index oscillates around between values 7.6 and 8.3 during the experiments. This is demonstrated that the pH index of the cooling water is in the neutral range and helps to limit corrosion and scale formation.

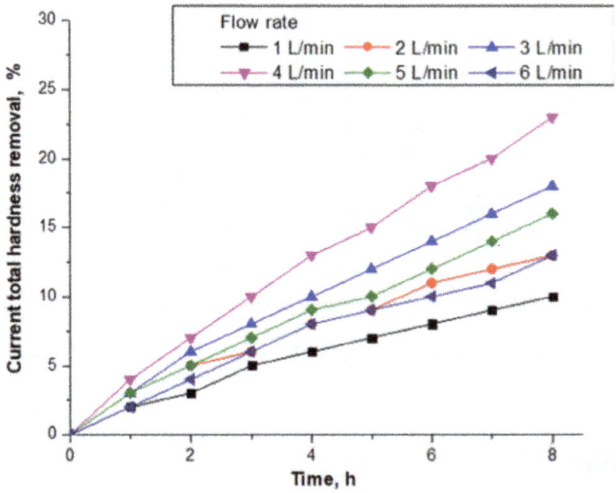

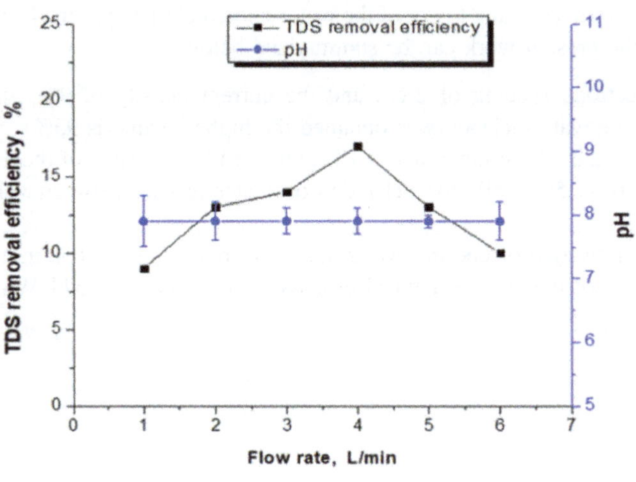

Fig. 5. a. Effect of flow rate on the efficiency of total hardness removal b. Effect of flow rate on the efficiency of TDS removal and pH index.

Figure 6 illustrates the effect of flow rate of cooling water through the reactor tank on the efficiency of TDS removal, total hardness removal efficiency, and ph index. As seen in Fig. 6, when the flow rate of cooling water through the reactor tank ranges from 1 L/min to 6 L/min, both the TDS removal efficiency and total hardness removal efficiency increase gradually until about 4 L/min and then decrease gradually, while the energy consumption of electrochemical process is increased from 311 Wh to 367 Wh.

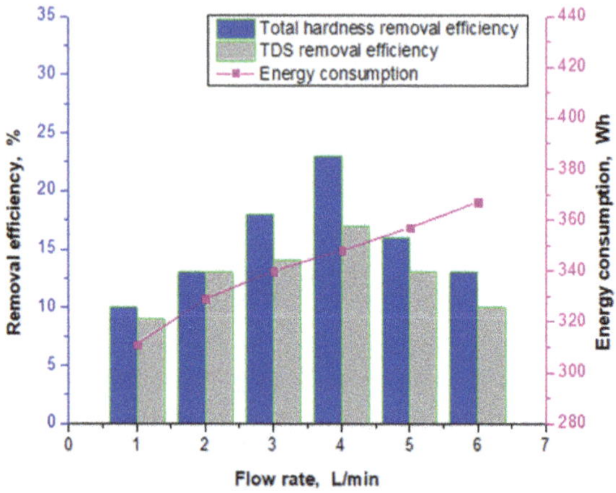

Fig. 6. Effect of flow rate on the efficiency of TDS removal, total hardness removal, and pH index

4 Conclusion

Electrochemical water treatment with three titanium electrodes was used to treat the cooling water from a cooling tower of the water-cooled chiller system. The conclusions drawn from the present work can be summarized below:

- At the electrode spacing of 2 cm and the current density of 90 A/m^2, the total hardness removal efficiency was obtained the highest value is 93%.
- At the flow rate of cooling water is 4 L/min, the highest value of the TDS removal efficiency is 17.5% while the pH index oscillates around between values 7.6 and 8.3.
- When the flow rate of cooling water varies from 1 L/min to 6 L/min, the energy consumption of the electrochemical process is increased from 311 Wh to 367 Wh.

References

1. Frayne, C.: Cooling Water Treatment Principles and Practice, pp. 30–31. Chemical Publishing Co., Inc., New York (1999)
2. Herbert, W., Stanford, I.I.I.: HVAC Water Chillers and Cooling Towers-Fundamentals, Application, and Operation, 2nd edn., pp. 205–206. Taylor & Francis Group, LLC (2012)
3. Rungvavmanee, R., Sayjun, K., Berpant, K., Phalakornkule, C.: Electrolytic treatment with aluminum anodes for reuse of cooling tower water. TIChE International (2011)
4. Gabriellia, C., Maurina, G., Chaussonb, H., Theryb, P., Tran, T.M., Tlili, M.: Electrochemical water softening: principle and application. Desalination **201**, 150–163 (2006)
5. Kabay, N., Demircioglu, M., Ersiiz, E., Kurucaovali, I.: Removal of calcium and magnesium hardness by electrodialysis. Desalination **149**, 343–349 (2002)
6. Liao, Z., Gu, Z., Schulz, M.C., Davis, J.R., Baygents, J.C., Farrell, J.: Treatment cooling tower blowdown water containing silica, calcium and magnesium by electrocoagulation. Water Sci. Technol. **60**, 2345–2352 (2009)
7. Malakootian, M., Mansoorian, H.J., Moosazadeh, M.: Performance evaluation of electrocoagulation process using iron-rod electrodes for removing hardness from drinking water. Desalination **255**, 67–71 (2010)
8. Rungvavmanee, R., Phalakornkule, C.: Reduction of Langelier index of cooling water by electrolytic treatment with stainless steel electrode. Water SA **38**, 825–830 (2012)
9. Cho, Y.I., Lee, S., Kim, W., Suh, S.: Physical water treatment for the mitigation of mineral fouling in cooling-tower water applications. Heat Exchanger Fouling and Cleaning: Fundamentals and Applications Art. 4 (2003)
10. Lumelsky, D.V., Greenberg, G., Pinhas, Y., Semiat, R.: Development of the electrochemical scale removal technique for desalination applications. Desalination **230**, 329–342 (2008)
11. Devos, O., Gabrielli, C., Tlili, M.: Nucleation-growth process of scale electrodeposition: influence of supersaturation. J. Electrochem. Soc. **150**, C494–C501 (2008)
12. Becker, H.A., Cohen, J.J., Zdunek, A.D.: Electrochemical cooling water treatment: a new strategy for control of hardness, scale sludge and reducing water usage. ASHRAE **115**, 399–405 (2009)
13. Jin, H., Yu, Y., Zhang, L., Yan, R., Chen, X.: Polarity reversal electrochemical process for water softening. Separation and Purification Technology (2018)
14. Emamjomeh, M.M., Sivakumar, M.: Fluoride removal by a continuous flow electrocoagulation reactor. J. Environ. Manag. **90**, 1204–1212 (2009)
15. Donachie, J., Titanium, J.: A Technical Guide, 2nd edn. ASM International, Cleveland (2000)
16. Hien, C.T.T., Ha, V.T.T.: Electrochemical durability of anodic electrodes modified by metal oxides on titanium substrate. J. Sci. Technol. **54**, 118–124 (2016)
17. Kalash, K.R., Ghazi, I.N., Abdul-Majeed, M.A.: Hardness removal from drinking water using electrochemical cell. Eng. Tech. J. **33**, 78–89 (2015)
18. Ministry of Construction: Ventilating, air-conditioning and cooling system - manufacture, installation and acceptance. Vietnam Standards, vol. 232 (1999)

A Compact Set of Smart Phone – Laryngoscope Helps Approaching Intubation Technique with Video Laryngoscope in Mekong Delta

Hong-Phuc Vo-Nguyen[1(✉)], Dao Huynh Tran[2], Quoc-Khanh Huynh[3], Linh Vu Le[1], and Kim-Long Van Vu[1]

[1] School of Medicine, Can Tho University of Medicine and Pharmacy, Can Tho City, Vietnam
{vnhphuc, lvlinh, vvklong}@ctump.edu.vn
[2] Can Tho Central General Hospital, Can Tho City, Vietnam
huynhdaobsgmhs@yahoo.com
[3] College of Engineering Technology, Can Tho University, Can Tho City, Vietnam
hqkhanh@ctu.edu.vn

Abstract. Video intubation was a necessary skill for not only anesthesiologists but also doctors in the emergency department and intensive care unit. Good training of using video laryngoscope in clinical practice would help intubation easier and safer. However, it was not easy to set the video laryngoscope available for everyday training, especially in Mekong Delta, due to the expensive price. This research aimed to develop a new intubation laryngoscope which is more suitable for clinical training conditions in Mekong Delta. The idea was using smartphones as display and process components. A borescope camera was used to capture glottis space and attached on the blade at the end of a 3D printed handle. This compact set of a smartphone-laryngoscope has been designed to be appropriated to training conditions in this area. Test of concept product was done under permission of the Board of medical ethics council of Cantho Central General Hospital. 60 patients who needed operation were chosen randomly to be intubated with the smart phone – laryngoscope set for anesthesia. This device provided succeed intubation at the first access in 82.67% cases; 17,33% needed the second access, none of them cannot be intubated. It took 14.57 ± 3.99 s on average to finish the intubation technique with the smartphone–laryngoscope set. There was no harm to the patients' larynx was found after the technique. The results showed that the new equipment worked safely, helped intubation easier, and on-time.

Keywords: Video laryngoscope · Smart phone · Intubation

B. T. Long et al. (Eds.): MMMS 2020, LNME, pp. 680–684, 2021.
https://doi.org/10.1007/978-3-030-69610-8_90

1 Introduction

Video intubation has been becoming a necessary skill in practical medicine, especially in difficult airway management [1, 2]. S. Dey [3] proved that the use of video laryngoscope both improved the glottis visualization and increased the rate of successful first-pass intubation. S.M. Ayyan, Z. Ali used a Macintosh laryngoscope and a waterproof USB camera connected to a laptop to intubate patients in emergency department within six months. 95.4% cases were successful in the first attempt with the video equipment [4]. R. Maassen claimed that video laryngoscope helped reduce larynx trauma due to difficult airway in obese patients [5].

In brief, studies showed that video laryngoscope was necessary for practical intubation, helped reduce larynx trauma, especially in difficult airway. Besides, the training of intubation needs more time and practices. However, there is not enough space for both trainers and trainees to look at the larynx directly with the laryngoscope. Therefore, video laryngoscope is a good choice in clinical training. Actually, the expensive price of the equipment makes it not available to hospitals and medical training in Mekong Delta. The research aimed to develop a new intubation video laryngoscope which is more suitable for clinical training conditions in this area.

2 Design of the Compact Set of Smartphone-Laryngoscope

The detailed view of the proposed device was described in Fig. 1. This set included the body which was connected with the Macintosh blade (number 3) below and the phone holder at the top.

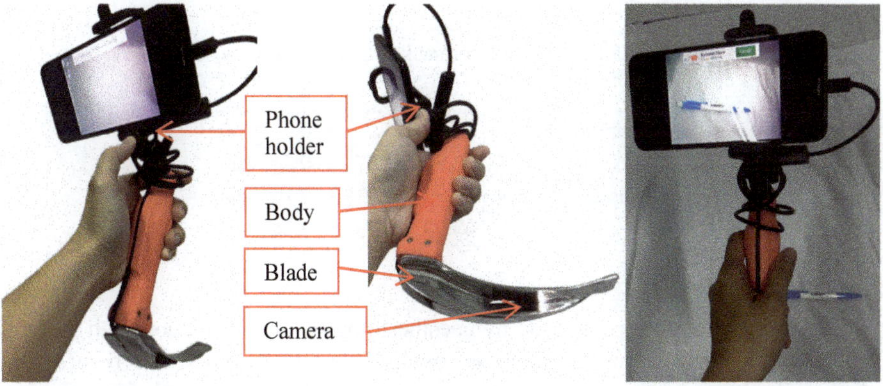

a) General view b) Side-view c) Capturing video

Fig. 1. The views of smartphone-video laryngoscope

A groove was designed on the blade in order to mount the Ø5.5 mm borescope type. In addition, it was also designed with hooks to keep the wire. This blade was made of SUS316L stainless steel due to the direct contact with the biological structure of humans to help ensure the cleanliness.

Figure 1a, b describes one gripping method as well as the parts of this set. The operator holds the plastic body in the middle, not to touch the blade. The smartphone is clamped on the top, and the blade is at the bottom with the borescope camera fixed in the correct position. Figure 1c explains the recording status. The object in form of a pen was selected to capture. Video displaying on the smartphone screen helps the operator to observe the object and perform the endotracheal intubation accurately and quickly.

The body part was made of ABS with 3D printing method. A recessed groove was designed to fix wires, that avoided the external impacts rotating the camera. This part had a curved design that led to more convenient to hold. Initial shape and sizes were chosen to fabricate. Based on the user's evaluations, they were adjusted until appropriate dimensions were found. The phone holder was considered to fit into different phone sizes and could be rotated to adjust the viewing angle of the screen.

3 Test Results and Discussions

3.1 Test Results

60 patients who needed operation were chosen randomly to be intubated with the smart phone – laryngoscope set for anesthesia. Average age was 48 ± 18.4 (years); BMI was 22.46 ± 2.52. Mallampati score and Cormack – Lehane score were used to predict the airway difficulty (Table 1).

Table 1. Mallampati and Cormack – Lehane score evaluation

Score	1	2	3	4	N
Mallampati	20	37	3	0	60
(%)	33.33	61.67	5	0	100
Cormack – Lehane	26	29	3	2	60
(%)	43.33	48.33	5	3.34	100

The intubations were performed by anesthesiologists after induction with fentanyl (2 mcg/kg), propofol (3 mg/kg) and rocuronium (0.6 mg/kg). 90 s after rocuronium, anesthesiologists used the video set to find epiglottis, then vocal cord. The tube was intubating through the vocal cord into the trachea; then fixed the tube, checked, and ventilated.

After-used blade with camera was sterile by cidezyme in 5 min, then rinsed under running water, and routinely storage.

This device provided succeed intubation at the first access in 82.67% cases; 17.33% needed the second access with bougie, none of them cannot be intubated.

Table 2. Time of intubation

	Mean	SD
Time of intubation (second)	14.57 (10–25)	3.99

It took 14.57 ± 3.99 s on average to finish the intubation technique with the smartphone–laryngoscope set. The longest time was 25 s, the shortest time was 10 s (Table 2).

There was no harm to the patients' larynx was found after the technique. 100% of user were satisfied by the set.

3.2 Discussions

Within 60 cases, there were 5% predicted difficult airway due to Mallampati 3; Cormack – Lehane 3 (5%) and Cormack – Lehane 4 (3.34%) were the cases that could not observe the vocal cord or larynx within the laryngoscope. 82.67% cases were successful at the first attempt, the rest needed second try with bougie. The result showed that the set was helpful in intubation, evenly in difficult cases.

The mean time of intubation by the set is 14.57 ± 3.99 s. It was acceptable and safe since the allowed intubation time is within 2 min with 2 attempts.

The handle sizes were adjusted by trial and error method. Although it was not the best one, it was a suitable and quick method that led to results in the context of the incomplete anthropometric data in Vietnam as well as in the Mekong Delta.

The surface of conventional blades should be treated to reduce reflections. However, this device could completely avoid this disadvantage. Therefore, the blades could be in high surface smoothness, thus increasing the ability to ensure hygiene.

Except for the camera that was purchased on the market at a reasonable price, the remaining components were locally manufactured at an acceptable cost. This device was completely feasible in deploying clinical intubation training. The ability to widely deploy in health facilities depends on many factors and needs to be assessed detailed in further studies.

This device was the basis for research in the future. The smartphone's robust connectivity has allowed us to do more from the camera's data. The manipulation of clinical instructions on the model could be sent online, live stream, shown on TV screens, projectors, helping to increase visibility for learners. This is a trend in the time of IoTs.

Another development from this device was to build an artificial intelligence application in order to identify anatomical structures inside the larynx area. This would help prompt suggestions, accurate positioning, avoid confusion, and reduce the intubation time.

The wired camera was selected due to the stable data transmission, and the low cost. Although some structure was additionally designed to support wire positioning, the advantages of using this type of camera were clear and acceptable. The extending study to blade sizes 1, 2, and 4 should be done for a more comprehensive assessment.

4 Conclusion

A compact set of the smartphone-laryngoscope has been designed and fabricated to meet the practical training needs for fast and widespread deployment at an affordable cost. Test results showed that the device was operated safely, useful support the intubation training and had the applicability at a reasonable cost, especially in Mekong Delta. A smartphone application that identifies anatomical structures within the larynx should be developed to expand the capabilities of this device.

Acknowledgements. Thanks to Cantho Central General Hospital for supporting this research, and special thanks to Ass. Prof. Robert Gertler and doctor Craig Nghiem for their useful advises inspiring this research.

This work was supported by the Domestic Master/PhD Scholarship Programme of The Vingroup Innovation Foundation and VINBIGDATA Institute.

References

1. Bjurström, M.F., Bodelsson, M., Sturesson, L.W.: The difficult airway trolley: a narrative review and practical guide. Anesthesiol. Res. Pract. (2019). https://doi.org/10.1155/2019/6780254. Article ID 6780254, 12 pages
2. Gupta, N., et al.: Tracheal intubation using the Airtraq video laryngoscope vs. Macintosh laryngoscope in patients with anticipated difficult intubation (2020). https://doi.org/10.1016/j.mjafi.2020.02.005
3. Dey, S., et al.: Intubation in the intensive care unit: C-MAC video laryngoscope versus Macintosh laryngoscope (2020). https://doi.org/10.1016/j.medine.2019.10.005
4. Ayyan, S.M., Ali, Z.: New assembled video laryngoscope: a study on efficacy and cost-effectiveness. Crit. Care **19**(Suppl 1), P202 (2015). https://doi.org/10.1186/cc14282
5. Maassen, R., Lee, R., Zundert, A.V., Cooper, R.: The videolaryngoscope is less traumatic than the classic laryngoscope for a difficult airway in an obese patient. J. Anesth. **23**, 445–448 (2009). https://doi.org/10.1007/s00540-009-0780-1

Redesign Approach for Increasing Energy Efficiency of Production Machine

Thi Ai Lanh Nguyen[1(✉)], Le Nhat Hoang Tran[1], Henri Paris[2], and Mathieu Museau[2]

[1] The University of Danang – University of Technology and Education, 48 Cao Thang, Danang, Vietnam
ntalanh@ute.udn.vn

[2] G-SCOP Laboratory, Grenoble INP, 46 Félix Viallet, 38031 Grenoble, France

Abstract. Reducing energy consumption is an important target in manufacturing industries. Saving energy not only allows to reduce manufacturing cost but also allows to reduce environmental impact. Production machine is one of the basis energy consumed devices in a manufacturing system. Enhancing energy efficiency of production machine is therefore become an interesting research domain. Taking into account energy efficiency in product manufacturing phase and in machine production, (re)design phase is an important key to decrease total energy consumption of a production machine [1]. Energy efficiency (EE) evaluation is the first step towards energy saving of production machine [2]. This previous research finding give a clear energy consumption mapping of production machine with information on the potential energy gains of the different organs and energy saving opportunities of production machine. Information achieved from this first step allows to propose different redesign scenarios to improve energy efficiency of production machine. By taking into account energy efficiency constraint, in this work, the main objective is to propose a redesign approach based on different guidelines allowing to improve EE of production machine.

Keywords: Energy consumption · Production machine · Energy efficiency improving · Energy interaction matrix · Energy redesign approach

1 Introduction

Nowadays, sustainable development is an important challenge for industrial world. The efficient use of resources has become a priority for industries. In fact, the world's energy consumption is expected to increase considerably due to industrial activity in both developed and developing countries. According to IEA 2019, energy consumption in the world in 2018 have increased at nearly twice the average rate of growth since 2010 [3]. Reducing energy consumption is therefore now considered like one of the most important challenges to achieve a cleaner and more efficient manufacturing.

Currently, energy consumption of production machine attracted a lot of researcher attention [4]. Improving energy efficiency is a source of interest for researchers. There are different ways to improve energy efficiency:

B. T. Long et al. (Eds.): MMMS 2020, LNME, pp. 685–691, 2021.
https://doi.org/10.1007/978-3-030-69610-8_91

- Optimizing process parameters: Different studies [5, 6] shows that the energy gain obtained by optimizing process parameters remains low because the process parameters are already optimized to meet different constraints such as the productivity or the quality of the piece surface. The higher the productivity of a production machine is, the more energy efficiency is better.
- Using innovative alternative processes: Innovative techniques currently enable to produce the mechanical parts by adding material. However, the question on environmental impacts and energy efficiency of this new technology is currently difficult to answer [7].

In summary, there is a little research focus on redesign production machine itself to reduce energy consumption of production machine. This paper aims at proposing a redesign approach based on different guidelines allowing to decrease energy consumption of production machine. In the paper, first of all, a design for energy efficiency method is presented. Then, an energy interaction matrix is proposed. This matrix allows to analyze energy flux within the production machine. Identifying the input energy and output energy of each organ in the production machine allows not only to know how the organ uses energy to perform an associated function but also to determine the energy interaction between different organs of the production machine. From that, different redesign actions allowing to increase energy efficiency of production machine are proposed. In this redesign approach, redesign action proposed to reduce energy consumption is based on 4 major guidelines: Adjusting to "just needed" energy usage, replacing by more efficient technology, optimizing current technology using and energy recovering.

2 Design for Energy Efficiency

Design for energy efficiency has developed significantly in recent years. The aim is to take energy efficiency constraint in the design phase of production machine in order to decrease its energy consumption. In our research, we are basing on design for EE in the redesign phase by keeping the manufacturing process unchanged.

A manufacturing process consists of many different steps through which raw materials are transformed into the final desired product. It creates then a product evolution. The production evolution term is used here to indicate the transformation of product from an initial state to a desired state.

The consummation of energy by a production machine in the manufacturing phase is not used entirely for the product manufacturing. The actual consummation of energy by a production machine is much higher than that required to achieve only the product evolution. The idea of design for energy efficiency is to reduce the part of energy lost during manufacturing phase to move towards ideal target. The ideal target is the part of energy strictly required by production machine to achieve product evolution [2].

The redesign for energy efficiency proposed in our work is based on two phases: a phase of energy analysis and a phase of proposal for alternative redesigns. The energy efficiency analysis method of the production machine and its components was presented in our first research finding [2]. The interest of the proposed analysis method is to obtain a sufficiently accurate energy consumed mapping of production machine with the strictly necessary information. From that, the second step on redesign action proposing to enhance energy efficiency of production machine is realized. This second step will be presented in the following paragraphs of this paper.

3 Redesign Action Proposal for Improving Energy Efficiency

3.1 Energy Interaction Matrix

An organ can transform or transmit material, energy or information. A production machine consists of different interconnected organs that interact with each other to ensure the main function. The interaction between organs can be expressed in different forms such as exchange of material, energy or information.

In this work, the authors are interested in the flux of energy within the production machine. Identifying the input and output energy flux of each organ allows to know how the organ uses energy to perform an associated function but also to determine the energy interaction between different organs of production machine. In fact, the output energy of an organ can become the input energy of another organ of production machine. It creates an energy interaction between these organs. In case of a machine tool, there is an energy interaction between the electric motors (spindle motor and axis motor), the electrical cabinet and the cooling system. Indeed, in term of energy, electric motors in the machine-tool convert electrical energy into mechanical energy to achieve the movement of the cutting tool (spindle motor) and the movement of the table (axis motor) and some of the energy is lost in the form thermal energy (heat). The cooling system is used to evacuate this thermal energy. The electrical energy consumed E_electrical 3 by the cooling system depends therefore the value of thermal energy to be evacuated from the electric motors E_thermal 1, and cabinet electric E_thermal 2:

$$\text{E\_electrical3} = \text{f(E\_thermal 1, E\_thermal2)} \tag{1}$$

This creates an energy interaction between these organs. Indeed, the higher the value of thermal energy (E_thermal 1 and E_thermal 2) must be evacuated is larger, the more E_electrical 3 consumed by the cooling system is great.

If it exists an energy interaction between the organs, the energy consumed by an organ depends not only on the choice of technology itself but also on the technology of other organs that have energy interaction with it. The redesign actions to decrease total energy consumption of an organ can therefore have negative or positive impacts on other organs interacted energy with this organ. Redesign actions to save energy consumption of the production machine cannot be proposed independently. It is also necessary to take into consideration energy interactions in enhancing the energy efficiency of the production machine.

The energy interactions between the organs of the production machine are represented in the form of a matrix. It is a square matrix n x n element whose columns and lines are the organs of the production machine (Fig. 1).

In this matrix: O_i: Organ i; E_i: Energy consumed by organ i; $x_{ij}E_i (0 < x_{ij} < 1)$: Part of energy consumed by organ i that characterizes interaction between organ i and organ j: $x_{ij}E_i = f(E_i)$. This part of energy has therefore an impact on the choice of technology for organ j. This impact is expressed in formed the following equation: $E_j = f(x_{ij}E_i)$.

The symbol ↑: The positive impact sense. That means the greater the value of $x_{ij}E_i$, the greater the energy consumed by organ j. Otherwise, we use the symbol ↓ to show the negative impact sense.

Organ	O_1	O_2	…	O_j	O_i
O_1	E_1	$x_{21}E_2$			
O_2		E_2			
…			…		
O_j				E_j	$x_{ij}E_i$
O_i					E_i

Energy consumed by an organ depends on (→)

Energy consumed by an organ has an impact (↑)

Fig. 1. Energy interaction matrix

The interpretation of the matrix from left to right allows to answer the question: "The energy consumed by an organ depends on which organ?"

The interpretation of the matrix from bottom to top allows to answer the question: "The energy consumed by an organ has an impact on which organ?"

In case of a machine tool, a part of energy interaction matrix is represented in Table 1:

Table 1. A part of energy interaction matrix in case of a machine tool

Organ \ Organ	Spindle motor (1)	Axis motor (2)	Electrical cabinet system (3)	Cooling system (4)	..
Spindle (1)	190 Wh				
Axis (2)		60 Wh			
Electrical cabinet system (3)			80Wh		
Cooling system (4)	19 Wh ↑	6Wh ↑	50Wh ↑	225Wh	
…..					

3.2 Approach for Proposing Redesign Action to Reduce Energy Consumption

By using information about potential energy gain of production machine and its different organs, the energy consumption mapping achieved in the energy analysis step [2] allows us to identify the parts of the energy interaction matrix that we can work on to improve energy efficiency. In our work, we begin to improve the energy efficiency of organs with the greatest potential energy gain.

The parts of the energy interaction matrix we are working on may involve one organ or several inter-linked organs in term of energy. Improving the EE of production machine is therefore not the sum of energy efficiency improvement results from individual organs. Where there is an energy interaction between different organs, redesign actions are proposed by looking at not only the energy usage of an organ but also must ensure that the actions redesign proposed do not create negative impacts on other energy-related organs.

In a machine tool, there is an interaction between electric motors, electrical cabinet system and cooling system as illustrated in Table 1. In order to decrease energy consumption of cooling system, we can work either on the amount of heat to evacuate from electric motor either on the cooling system itself to improve its efficiency.

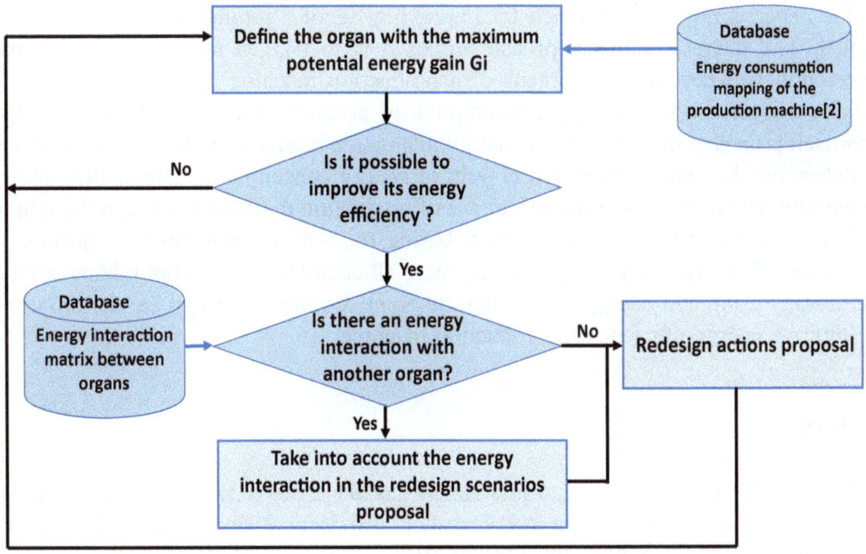

Fig. 2. Redesign actions proposal approach

By using energy interaction matrix, 'key organs' on that redesign actions can be applied on to improve total energy efficiency of production machine are defined. From that, these key organs can be tackled in 4 different ways:

- Replacing with more efficient technology: the parts of energy do not add value to product manufacturing of the production machine may come from components or sub-systems that are not efficient. The idea is to replace these components or subsystems with others that are more efficient.
- Optimizing current technology: the idea is to optimize ‘key organs’ of production machine to improve their energy efficiency. For example: reducing friction, the mass of the parts movement (axis) by changing the material, the structure…
- Adjusting to “just needed” energy usage: the unvalued energy of the production machine can sometimes come from the way how the organs are used. It is therefore possible to improve energy efficiency by using them in more efficient conditions, it means to put the organs into operation just when necessary.
- Energy reusing: to avoid wasting energy, unvalued energy recovery is necessary. The idea is to recover unvalued energy and reuse or transfer it to another more useful form.

The approach for proposing redesign actions to decrease energy consumption of production machine is illustrated in Fig. 2.

4 Conclusion

In the paper, a redesign approach for improving EE of a production machine is presented. The interest of this approach proposed is to take in to account the energy interaction between different organs of a production machine in proposing redesign actions to decrease total energy consumption of production machine. In the redesign approach proposed in our work, an energy interaction matrix is proposed that allowing to determine the organs interacted in term of energy consumption. This matrix allows to identify which organ or group of organs of production machine must be rethought to reduce energy consumption. From that, basing on 4 major guidelines: Adjusting to “just needed” energy usage, replacing by more efficient technology, optimizing current technology using and energy recovering, different scenarios redesign can be proposed to improve energy efficiency of production machine.

References

1. Nguyen, T.A.L., Museau, M., Paris, H.: Méthodologie de (re)conception de l’outil de production pour minimiser sa consommation d’énergie. 21ème Congrès Français de Mécanique CFM 2013, Bordeaux, France (2013)
2. Lanh, N.T.A., Hoang, T.L.N., Henri, P., Museau, M.: Evaluating energy efficiency of production machine. In: Proceedings of the First International Conference on Material, Machines, and Methods for Sustainable Development, MMMS 2018, Danang, Viet Nam, vol. 2, pp. 562–569 (2018)
3. Information on (2020). https://www.iea.org/reports/world-energy-outlook-2019. Accessed 15 July 2020

4. Sihag, N., Sangwan, K.S., Pundir, S.: Development of a structured algorithm to identify the status of a machine tool to improve energy and time efficiencies. In: Proceedings of the 25th CIRP Life Cycle Engineering (LCE) Conference, Copenhagen, Denmark (2018)
5. Diaz, N., Redelsheimer, E., Dornfeld, D.: Energy consumption characterization and reduction strategies for milling machine tool use. In: Proceedings of the 18th CIRP International Conference on Life Cycle Engineering, Brunswick, Germany, pp. 263–267 (2011)
6. Kara, S., Li, W.: Unit process energy consumption models for material removal processes. CIRP Ann. Manuf. Technol. **60**(1), 37–40 (2011)
7. Vayre, B., Vignat, F., Villeneuve, F.: Metallic additive manufacturing: state-of-the-art review and prospects. Mech. Ind. **13**(2), 89–96 (2012)

Gearshift Strategy of a Two-Speed Dual-Clutch Transmission for Energy Efficacy of Electric Vehicles

Van Nhu Tran[1(✉)], Dinh Nghiem Le[2], and Huu Hai Ho[3]

[1] University of Transport and Communications,
No 3, Cau Giay Street, Lang Thuong, Ba Dinh, Hanoi, Vietnam
vannhu.tran@utc.edu.vn
[2] Vietnam Maritime University,
No-484, Lach Tray, Kenh Duong, Le Chan, Haiphong, Vietnam
[3] Hanoi University of Science and Technology,
No 1, Dai Co Viet, Hai Ba Trung, Hanoi, Vietnam

Abstract. In order to increase the energy efficiency of electric vehicles, the research related to controlling traction electric motors and controlling transmission system have been gathering more interest recently. The traditional electric vehicles use a one-speed gearbox. This transmission structure is simple but cannot achieve the best performance because most of the time the electric motor operates at the points where the efficiency is not the highest. Most electric vehicles today use gearboxes with multiple speeds. The transmission systems on traditional cars using internal combustion engines can be used for electric vehicles. In this paper, the author presents the research result on the gearshift strategy of a two-speed dual-clutch gearbox applied on electric vehicles for energy efficiency. The gearshift strategy takes into account the vehicle speed and drag forces.

Keywords: Electric vehicle · Energy efficiency · Gearshift strategy · Electric vehicle transmission · Two-speed gearbox

1 Introduction

Energy efficiency is an important performance criterion for vehicles in general, especially for electric vehicles (EV). EVs are limited in battery capacity and therefore require the optimal use of energy. Energy regeneration during braking, downhill, and optimal energy management are the problems that EVs face. In literature, there has been some research to improve the efficiency of electric motor [2, 7], to improve energy efficiency by optimal design of the transmission ratios [3–5] and design the gearshift strategies for energy efficiency of an electric vehicle. An EV with the multi-gear transmission can enhance both dynamic and economic performances [4]. The EV with a four-gear transmission can save up to 30% energy. [3] shows that only by designing the optimal transmission ratios, energy can be saved up to 15%. In [6], the authors present a graphical method for the development of dynamic shift schedules. Accordingly, with the drag force of vehicle, the point of transition is determined. In this paper, the authors

B. T. Long et al. (Eds.): MMMS 2020, LNME, pp. 692–697, 2021.
https://doi.org/10.1007/978-3-030-69610-8_92

develop a new graphical method for an economic gearshift strategy of electric vehicles. The gearshift strategy takes into account the vehicle speed and drag forces.

2 Gearshift Strategies Modeling

2.1 Calculating the Gear Ratios

A vehicle is designed to tolerate different driving conditions on the road. That means the vehicle can move on the road if the traction generated on the active wheels matches the total resistance acting on the vehicle at that time. The traction force generated at the wheels is supplied by the electric traction motor through the transmission. According to the design point of view, the gear ratios are calculated to meet the vehicle's operating conditions, that means it is necessary to determine the driving forces acting on the vehicle and the required speed that the vehicle can achieve. Therefore, the necessary parameters for the vehicle need to be provided in full to serve the gear ratio calculation. In this study, the vehicle's input parameters are provided as follows: vehicle mass, $m = 1400$ kg; wheel radius, $r_d = 0.26\,\text{m}$; maximum on-road drag coefficient,$\psi_{\max} = 0.4$; frontal area, $F = 1.510\,\text{m}^2$; air resistance coefficient, $K = 0.18\,\text{Ns/m}^4$; transmission efficiency, $\eta = 0.95$.

To perform this study, the authors chose an electric motor model that can meet common road conditions on passenger vehicles with a small weight. The selected electric motor can achieve a maximum torque of 200 N.m and achieve a speed of 8000 rpm as shown in Fig. 1.

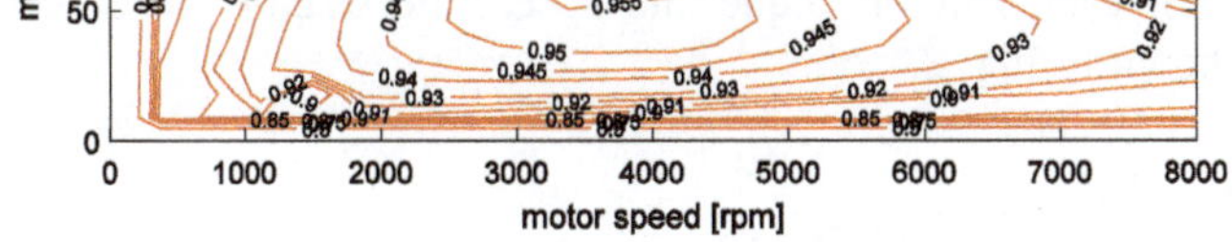

Fig. 1. Efficiency map of electric motor

The powertrain can increase the initial torque of the drive motor, and adjust the vehicle speed through the transmission ratio. The transmission ratio divides into two parts: the gearbox ratio and the final drive ratio. For a two-speed gearbox, the first gear ratio can be calculated as:

$$i_{tr1} = r_d m g \psi_{\max} / (T_{m\max} \eta) \tag{1}$$

and the second gear ratio equation is as follows:

$$i_{tr2} = 3.6\pi n_{m\max} r_d / (30 v_{\max}) \tag{2}$$

Where: g is the gravity acceleration, m/s^2; $T_{m\max}$ is the maximum torque of electric motor, $N.m$; $n_{n\max}$ is the maximum speed of electric motor, rpm; $v_{\max}$ is the maximum speed of vehicle, km/h.

For vehicles using a two-speed gearbox, each gear has a role. The first gear assures that the vehicle can move on the road with the required slope, and the vehicle can achieve maximum speed in second gear. Therefore, the electric motor is limited to a speed of $6000\,rpm$, in the high-efficiency region. The gear ratios are calculated as: the fist gear ratio is $i_{tr1} = 8.2$, and the second gear ratio is $i_{tr2} = 3.3$. For vehicles using a one-speed gearbox, in order for the vehicle to reach the required speed and slope, the engine speed used at its maximum speed is $8000\,rpm$, the gear ratio of the transmission is calculated as $i_{tr} = 4.48$.

2.2 Designing Gearshift Strategies

Assuming that the energy loss in the powertrain is constant, $\eta = 0.95$, and the vehicle moves without acceleration. Thus, the torque and speed of the drive motor required, T_{er}, n_{er} to equal the vehicle drag force equivalent at the time of consideration are calculated as follows:

$$T_{er} = i_{tr}\eta(mgf + KFv_t^2)/r_d \tag{3}$$

$$n_{er} = \pi v_t r_d / (30 i_{tr}) \tag{4}$$

Where v_t is the vehicle speed at the survey time. From Eqs. (3) and (4), it is clear that at the time of the survey, the speed of the drive wheel and the total drag force acting on the vehicle are specific figures.

Corresponding to each gear level selected, there are the corresponding torque and speed pairs. For each pair of torque and speed, the electric motor performance is different as shown in Fig. 2.

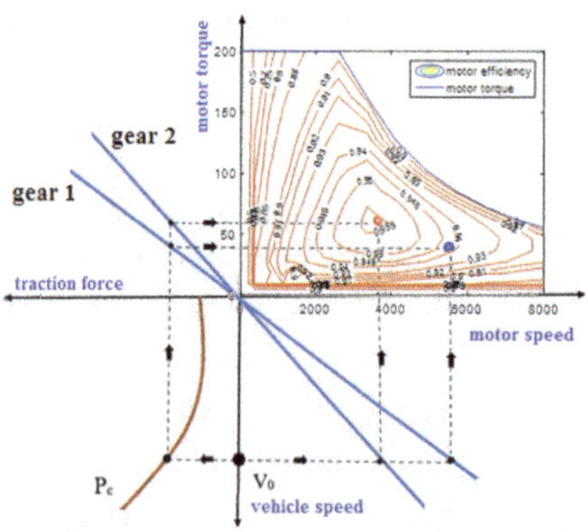

Fig. 2. Operating point of each gear level

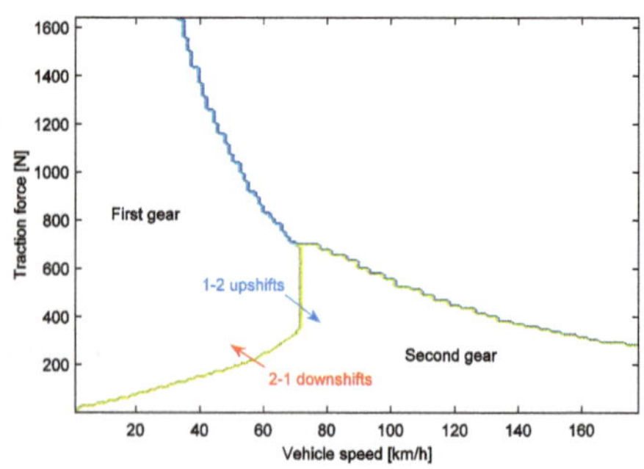

Fig. 3. Optimal operating region of each gear level of two-speed gearbox

According to the above arguments, directly comparing the performance of electric motors at the survey time with each gear level. From then, the controller can make a decision to choose the most relevant number. This is quite easy for current computers.

With the basic characteristics of the electric motor used for the survey, the top tractive torque on all of the active wheels can be reached at 1640 Nm, and the maximum speed of the vehicle is 178 km/h according to the input requirements. Thus, it is possible to identify the operating conditions of the engine, the conditions to succeed the drag on the road throughout the powertrain as shown in Fig. 3. In Fig. 3, the vehicle speed – traction force graph devices in two regions: first gear and second gear region. Once the working conditions of the vehicle are determined, the gearshift strategy can determine the best working region for each gear. In the first gear region, the engine will work with high efficiency when using first gear, the remaining region is optimal for the second gear.

3 Simulation Results and Discussion

The gearshift strategy design is tested and evaluated through a driving test cycle to determine the fuel consumption for an internal combustion engine or the energy required for an electric motor. In this study, the authors use the WLTP test cycle as input data for gearshift simulation (Fig. 4a).

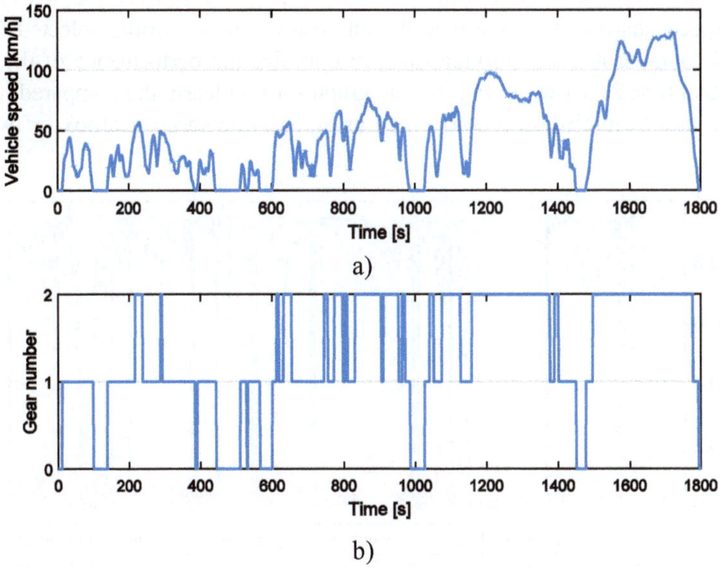

Fig. 4. Gearshift strategy for two-speed gearbox

According to the WLTP test cycle, the vehicle is tested on a flat road. The gear number is selected depending on the vehicle speed and the resistance force conditions. In this simulation case, the gearshift strategy is shown in Fig. 4b. The simulation results in Fig. 5 show that the electric motor efficiency is improved by using a two-speed gearbox. The results showed that the average performance is 90% when using single-speed gearbox and 93% when using two-speed gearbox, improved by 3%.

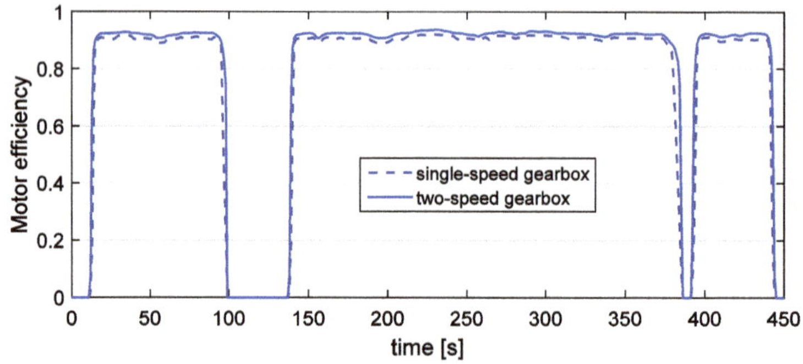

Fig. 5. The motor efficiency with a relevant transmission

For vehicles operating under harsher conditions in the WLTP test cycle, corresponding to each speed at the survey time, there will be an appropriate resistance force. When the vehicle operates in harsher conditions, drag force varies over a wide range, the two-speed gearbox has improved the efficiency. The controller selected the most reasonable number at each working time, to optimize the performance of the traction motor at that time as shown in Fig. 6. The graph shows clearly the compared results of the efficiency of the electric motors when using a single-speed gearbox or two-speed

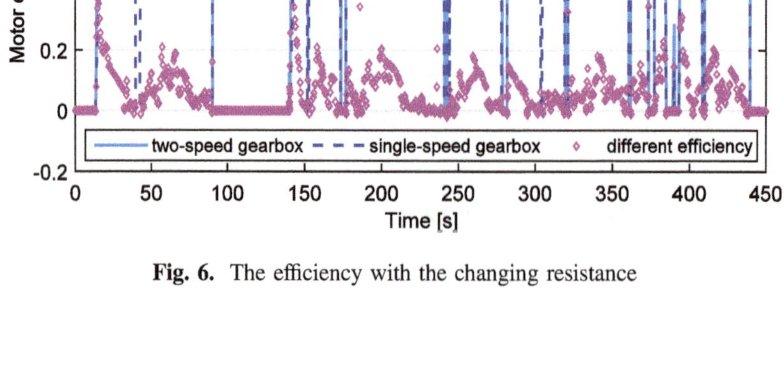

Fig. 6. The efficiency with the changing resistance

gearbox, with a two-speed gearbox much more efficient than a single-speed gearbox. The results show over the entire survey process, in this case, the average performance is 67% with a single-speed gearbox, and 76% with a two-speed gearbox, improved by 13%. There are certain areas where the performance can improve by over 50%.

4 Conclusion

The work aims to demonstrate the effectiveness of the two-speed powertrain used on electric vehicles by using the optimal control strategy. The authors calculated the gear ratios for the powertrain using the single-speed gearbox and the two-speed gearbox. The authors develop a gearshift strategy for the two-speed gearbox transmission system. A gearshift map is developed. The simulation results show that the average efficiency of the drive motor increased significantly compared to the case of a single-speed powertrain. When simulating on a harsher road, the motor efficiency can be improved by 13%. There are certain areas where the motor efficiency rose to 50%. However, when simulating on a good road, with certain resistance conditions, shows that motor efficiency is only increased by 3%.

Acknowledgements. This research is funded by University of Transport and Communication (UTC) under grant number T2020-CK-012TĐ.

References

1. Minh, T.C., Hori, Y.: A sensorless efficiency-optimization control of induction motor suitable for electric vehicles. In: IEE of Japan Technical Meeting Record, Tsuna, Japan, November 2000, pp. 95–100 (2000)
2. Kavitha, C., et al.: Automatic gear shifting mechanism for a sequential transmission system for an electric vehicle. Mater. Today: Proc. **5**(5), 11442–11451 (2018). https://doi.org/10.1016/j.matpr.2018.02.112
3. Wang, Y., Lü, E., Lu, H., Zhang, N., Zhou, X.: Comprehensive design and optimization of an electric vehicle powertrain equipped with a two-speed dual-clutch transmission. Adv. Mech. Eng. **9**(1), 168781401668314 (2017). https://doi.org/10.1177/1687814016683144
4. Ahssan, M.R., Ektesabi, M.M., Gorji, S.A.: Electric vehicle with multi-speed transmission: a review on performances and complexities. SAE Int. J. Alt. Power. **7**(2), 169–181 (2018). https://doi.org/10.4271/08-07-02-0011
5. Gao, B., Liang, Q., Xiang, Y., Guo, L., Chen, H.: Gear ratio optimization and shift control of 2-speed I-AMT in electric vehicle. Mech. Syst. Signal Process. **50–51**, 615–631 (2015). https://doi.org/10.1016/j.ymssp.2014.05.045
6. Zhu, B., et al.: Gear shift schedule design for multi-speed pure electric vehicles. Proc. Inst. Mech. Eng. Part D: J. Automob. Eng. **229**(1), 70–82 (2015). https://doi.org/10.1177/0954407014521395
7. Nguyen, D.-D., Ta, M.C.: New direct torque control scheme for BLDC motor drives suitable for EV applications. In: Presented at the 2014 IEEE Vehicle Power and Propulsion Conference, Coimbra, Portugal, October 2014 (2014)

Effect of Nanosecond Repetitively Pulsed Frequency on Laminar Flame Propagation of Lean n-Butane/Air Mixture with *Le* >> 1

Minh Tien Nguyen(✉) and Minh Tung Phung

Department of Mechanical Engineering, University of Technology and Education, The University of DaNang, Da Nang, Vietnam
{nmtien, pmtung}@ute.udn.vn

Abstract. This study aims to present the effect of pulse repetitive frequency (PRF) from nanosecond repetitively pulsed discharge (NRPD) on laminar flame propagation of lean premixed n-butane/air at an equivalence ratio ϕ of 0.7. The lean n-butane/air with Lewis number $Le \approx 2.1 >> 1$ is ignited in the constant pressure combustion chamber at the atmospheric condition and 0.8-mm spark gap (d_{gap}), where *Le* is determined as the ratio of thermal diffusivity and mass diffusivity. A fixed train of 11 pulses varying PRFs from 5 to 60 kHz, but the same total ignition energy E_{tot} = 23 mJ is applied. The result indicates that the embryonic kernel is quenched at lower PRF = 5 kHz while it becomes the successful flame propagation at the other higher PRFs (10–60 kHz). The fastest combustion process is at PRF = 20 kHz and being detrimental with other PRFs. The non-monotonic effect of PRFs on the combustion process of lean n-butane/air mixture could be attributed to the synergistic effect that strongly depends on the competition between PRF and recirculation frequency of inward reactant flow near the electrode tips. Although NRPD could enhance the combustion process within a specific PRF range, the laminar flame propagation rate after self-sustained propagating is almost constant and independent of PRF values. The unstretched laminar flame speed approximates 25–26 cm/s. The NRPD results are also compared with that of conventional single spark discharge (CSSD). The schlieren flame images are presented to understand the flame structure under the effect of discharge methods.

Keywords: Lean premixed ignition · Nanosecond repetitively pulsed discharge · Conventional single spark discharge · Pulse repetitive frequency · Laminar flame speed

1 Introduction

Lean combustion technology is a possible strategy for increasing the thermal efficiency and reducing the emission due to the increase of the mixture specific heat ratio and the decrease of the net heat loss to the engine cylinder [1]. However, after discharging, the embryonic kernel development is expected to be substantially slower, resulting in unstable flame propagation and incomplete burn as the in-cylinder mixture is leaned out. Furthermore, the misfire may occur because the spark discharge and the kernel

B. T. Long et al. (Eds.): MMMS 2020, LNME, pp. 698–704, 2021.
https://doi.org/10.1007/978-3-030-69610-8_93

development take place at extreme conditions, i.e., high pressure (p), high temperature (T), and high turbulent intensity (u'). One effective way to mitigate the misfire and the slow burn, the most challenging to lean operation, is to take advantage of short duration, non-equilibrium plasma discharge for enhancing ignition probability and combustion process by using nanosecond repetitively pulsed discharge (NRPD).

In NRPD, the non-equilibrium plasma consists of extremely hot electrons with a temperature above 16000 K (1 eV) and relatively cold neutral particles and ions [2]. Consequently, most of the spark ignition energy contributes to the formation of excited radicals such as O atoms [3] that are beneficial for combustion enhancement. In contrast, the electrons, neutral particles, and ions share the same high temperature (above 6000 K for spark discharge) in the equilibrium plasma, resulting in much energy losses due to overheating electrodes and mixture surrounding the spark channel.

The synergy between pulses from NRPD plays a vital role in flame ignitability and the combustion process. The synergy means that the next pulse is enhanced by the previous pulse based on the accumulation of temperature and excited radicals [3]. The synergistic effect depends on the competition between the pulse repetitive frequency (PRF) and the recirculation flow frequency (F_{RC}). When PRF is too low, the first-pulsed kernel will be cooled down by the recirculation flow before the next pulse is created, weakening the synergistic effect. Similarly, the synergistic effect is weak (or negligible) when PRF increases above F_{RC}. This is because the reactant between the spark gap could be dissociated entirely by a first few pulses but no longer refreshed [3], generating fewer excited radicals. The other pulses add more energies to the ignition process but not beneficial for combustion enhancement. The synergistic effect is most profound when PRF is approximately synchronization with F_{RC}- by which the reactant inside the spark gap is refreshed between each pulse, and each pulse dissociates nearly all O_2 molecules presented in the gap [3–5]. The generated O radicals are then ejected into the regions surrounding the electrodes, promoting an ignition kernel, so does the ignitability and the combustion process of lean mixtures.

Recently, NPRD has attracted many researchers for investigating the combustion enhancement by using a pair of pin-to-pin electrodes in flowing methane/air mixtures [6], in 3-D DNS quiescent lean methane/air mixture [5], in quiescent lean propane/air mixture [3, 7, 8]. These studies mostly study the effect of pulse frequency on the combustion process without considering differential diffusion (Le >> 1), where Lewis number (Le) is determined as the ratio of thermal diffusivity and mass diffusivity. With the consideration of Le number effect, Nguyen et al. [4] have studied the ignition of lean n-butane/air mixture (ϕ = 0.7) in both laminar and turbulent conditions using NRPD and conventional single spark discharge (CSSD). However, Nguyen et al. mostly focus on the ignition probability, and the turbulence facilitated ignition (*TFI*) phenomena rather than the laminar flame propagation of lean n-butane/air mixture. Hencce the present work aims to make a clear-cut view of the effect of such synergy or PRFs on the laminar flame propagation of lean n-butane/air mixture with the consideration of Le number.

2 Experimental Setup

Ignition experiments of lean n-butane/air mixture ($\phi = 0.7$, $Le \approx 2.1 >> 1$) were conducted in a large dual-chamber, constant temperature/pressure, and fan-stirred explosion facility using the same 2-mm stainless steel pin-to-pin electrodes with sharp ends at a fixed $d_{gap} = 0.8$ mm and same ignition energy $E_{ig} \approx 23$ mJ. The reader is directed to Ref. [4, 9, 10] (and references therein) for detailed information. An appropriate mole fraction of n-butane and air is injected into the vacuumed chamber by partial pressure method. Two counter-rotating fans at a frequency of 30 Hz are turned on to mix the lean n-butane/air mixture for 3 min. After 10 s of fan-off, the mixture is ignited by either NRPD or CSSD method. In NPRD, a high voltage pulse generator (FID GmbH FPG 20-100NK) is controlled by digital delay GW INSTEK AFG-2225 to generate a train of 11 pulses with PRF from 5 to 60 kHz. The pulse generator has 3–5 ns of full-width haft maximum (FWHM) and can deliver up to 30 kV at up to 100-kHz PRF. In CSSD, the high voltage pulse generator (HV-M25K) is used at pulse duration $\tau = 500$ μs, similar to previous works [10–12].

The applied voltage and current are measured by a high voltage probe (Tektronix P6015A) and a Pearson coil (model 6585), respectively, and recorded by 500 MHz oscilloscopes (Tektronix MD03054). The ignition energy (E_{ig}) is then calculated by the integration of the product of the discharge voltage V(t) and current I(t) over the pulse duration. Figure 1 shows a typical voltage and current waveforms of a single NRP pulse directly measured across the spark gap, $d_{gap} = 0.8$ mm. Independent of the PRF settings, each pulse's energy (E_{pulse}) is about 2.2 mJ except the first initial pulse ($E_{int.pulse} \sim$ 1.0 mJ). The total ignition energy of 11 pulses in NRP, $E_{tot} \approx 23$ mJ, is the same as that of CSSD.

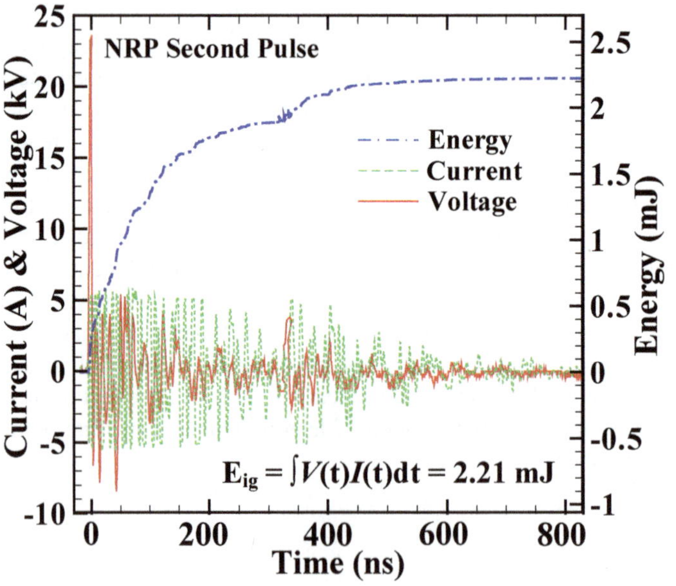

Fig. 1. Measured voltage, current, and energy for a single NRP pulse at $d_{gap} = 0.8$ mm.

3 Results and Discussion

Figure 2 shows schlieren images of the embryonic kernel in quiescence for five different cases, including (a–d) four NRP cases at PRF = 5 kHz,10 kHz, 20 kHz, and 60 kHz, respectively, (e) a CSSD at τ = 500 µs. For NRPD, the embryonic kernel development mostly depends on the deposited energy by subsequent pulses, and the inter-pulse time should be less than 100 µs (PRF > 10 kHz) [13] in order to create synergistic effects [3, 7] before the kernel cools down. At PRF = 5 kHz, the synergy between two consecutive pulses is too weak or negligible due to the long inter-pulse time of 200 µs (5 kHz), resulting in a failure ignition or "*No Go*" case (0% ignitability). Increasing PRF strengthens the synergistic effect, leading to self-sustained flame propagation, as indicated in Figs. 2(b–d). To approach the flame radius R = 20 mm, the time evolution is 23.4 ms, 20.0 ms, and 26.8 ms, corresponding to PRF = 10 kHz, 20 kHz, and 60 kHz. This result reveals the non-monotonic synergistic effect with increasing PRF. The strongest synergy is at PRF = 20 kHz, and detrimental to other PRFs (10 kHz, 60 kHz). Furthermore, at PRF = 60 kHz, the heat loss to electrodes may be increased, causing an elliptic flame kernel different from the spherical flame at other PRFs (10 kHz and 20 kHz). For comparison, the embryonic kernel development by CSSD is presented in Fig. 2(e). The time evolution in CSSD is close to that of PRF = 10 kHz in NRPD. Moreover, a large amount of heat loss to electrodes results in a torus-like (elliptic) kernel.

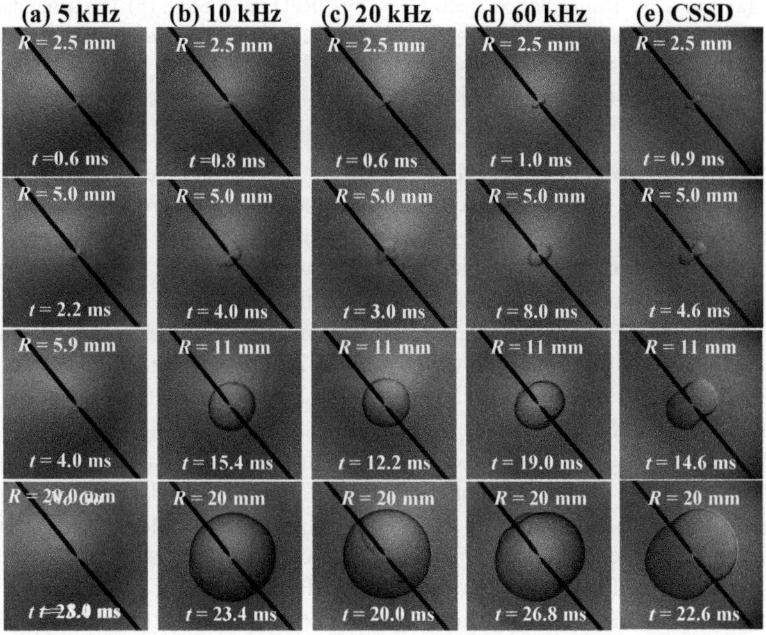

Fig. 2. The embryonic kernel development by NRPD with a fixed train of 11 pulses at four PRFs = 5, 10, 20, 60 kHz (Figs. 2a–d, respectively) and CSSD at τ = 500 µs (Fig. 2e), using the same $E_{tot} \approx E_{ig} \approx 23$ mJ for lean n-butane/air mixture with $Le >> 1$. Field of view 66 × 66 mm.

Figure 3(a) presents the raw data of $R(t)$ vs. t for flame kernel development in Figs. 2(b–e). After reaching a critical radius R_c = 11 mm, the kernel development in both NRPD and CSSD has a constant slope or flame propagation rate but are just translated in time. This is because the synergy from NRPD enhanced the combustion process in the early stage of the flame kernel in which the flame radius (R) is smaller than R_c. It notes that the flame kernel of $Le >> 1$ mixtures need to attain R_c in order to achieve self-sustained propagation [14, 15].

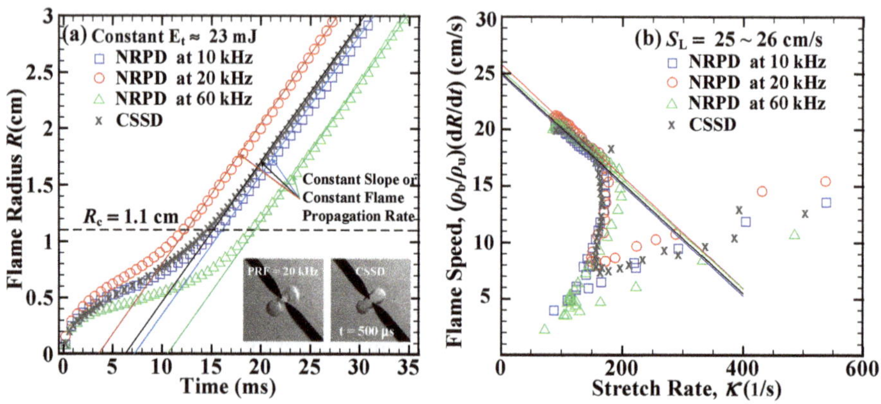

Fig. 3. (a) The flame radius plotted against time by NRPD and CSSD. (b) The same data with (a) are replotted as a function of flame propagation speed and stretch rate to determine laminar flame speed (S_L) at $\kappa = 0$, where S_L = 25–26 cm/s. The inset images reveal different kernel structure between NPRD at PRF = 20 kHz and CSSD due to the synergistic effect.

Since the synergy is most significant at PRF = 20 kHz and the total pulse duration is 500 μs, the image of PRF = 20 kHz is thus selected to compare with that of CSSD as the inset schlieren images in Fig. 3(a). The synergistic effect by NPRD results in the wrinkling structure of the embryonic kernel. The wrinkled flame kernel includes both locally positive and negative curvatures in which the latter could enhance the burning intensity of the flame segments of a lean n-butane mixture with $Le >> 1$ [16], resulting in the combustion enhancement. In contrast, there is no wrinkling inside the torus-like kernel by CSSD. The positive curvatures weaken the reaction rate of flame segments [16] of $Le >> 1$ mixture.

Figure 3(b) represents the same data in Fig. 3(a) in term of flame speed and stretch rate (κ) to determine the laminar flame speed (S_L) at a zero-stretch rate ($\kappa = 0$), where $S_L = (\rho_b/\rho_u)(dR/dt)_{\kappa=0}$ [17]. Using the linear fitting for whole data with $R \geq 11$ mm, the laminar flame speed is approximate to 25–26 cm/s and almost independent of the ignition method (NRPD *vs.* CSSD) or different PRFs. The laminar flame speed is about 12% higher than that of [18], where $S_L \approx 23$ cm/s, probably due to the different experimental configuration.

4 Conclusion

In this study, we applied both NRPD at different PRFs and CSSD to ignite the lean premixed n-butane/air (ϕ = 0.7) with *Le* >> 1 at d_{gap} = 0.8 mm using 2-mm pin-to-pin electrodes. By applying a fixed train of 11 pulses with the same $E_{ig} \approx 23$ mJ, the result shows that increasing PRF could enhance the combustion process. At lower PRF (5 kHz), the embryonic kernel is quenched while it becomes the successful flame propagation as the higher PRF (10–60 kHz). The fastest combustion process is at PRF = 20 kHz and detrimental to other PRFs (10 kHz, 60 kHz). The laminar flame speed, $S_L \approx$ 25–26 cm/s, determined in a range of $R \geq R_c$ = 11 mm, is almost independent of ignition methods (NRPD *vs.* CSSD) or different PRFs.

Acknowledgment. This article was originated from the author's Ph.D. thesis under the supervision of Prof. Shenqyang (Steven) Shy at Department of Mechanical Engineering, National Central University, Taiwan.

References

1. Jung, D., Iida, N.: An investigation of multiple spark discharge using multi-coil ignition system for improving thermal efficiency of lean SI engine operation. Appl. Energy **212**, 322–332 (2018)
2. Tropina, A.A., Shneider, M.N., Miles, R.B.: Ignition by short duration, non-equilibrium plasma: Basic concepts and applications in internal combustion engines. Combust. Sci. Technol. **188**, 831–852 (2016)
3. Lovascio, S., Ombrello, T., Hayashi, J., Stepanyan, S., Xu, D., Stancu, G.D., Laux, C.O.: Effects of pulsation frequency and energy deposition on ignition using nanosecond repetitively pulsed discharges. Proc. Combust. Inst. **36**, 4079–4086 (2017)
4. Nguyen, M.T., Shy, S.S., Chen, Y.R., Lin, B.L., Huang, S.Y., Liu, C.C.: Conventional spark versus nanosecond repetitively pulsed discharge for a turbulence facilitated ignition phenomenon. Proceedings of the Combustion Institute (2020, in press). https://doi.org/10.1016/j.proci.2020.06.020
5. Castela, M., Stepanyan, S., Fiorina, B., Coussement, A., Gicquel, O., Darabiha, N., Laux, C. O.: A 3-D DNS and experimental study of the effect of the recirculating flow pattern inside a reactive kernel produced by nanosecond plasma discharges in a methane-air mixture. Proc. Combust. Inst. **36**, 4095–4103 (2017)
6. Lefkowitz, J.K., Ombrello, T.: An exploration of inter-pulse coupling in nanosecond pulsed high frequency discharge ignition. Combust. Flame **180**, 136–147 (2017)
7. Pancheshnyi, S.V., Lacoste, D.A., Bourdon, A., Laux, C.O.: Ignition of propane–air mixtures by a repetitively pulsed nanosecond discharge. IEEE Trans. Plasma Sci. **34**, 2478–2487 (2006)
8. Lovascio, S., Hayashi, J., Stepanyan, S., Stancu, G.D., Laux, C.O.: Cumulative effect of successive nanosecond repetitively pulsed discharges on the ignition of lean mixtures. Proc. Combust. Inst. **37**, 5553–5560 (2019)
9. Nguyen, M.T., Yu, D.W., Shy, S.S.: General correlations of high pressure turbulent burning velocities with the consideration of Lewis number effect. Proc. Combust. Inst. **37**, 2391–2398 (2019)

10. Shy, S.S., Nguyen, M.T., Huang, S.Y.: Effects of electrode spark gap, differential diffusion, and turbulent dissipation on two distinct phenomena: turbulent facilitated ignition versus minimum ignition energy transition. Combust. Flame **205**, 371–377 (2019)
11. Shy, S.S., Nguyen, M.T., Huang, S.-Y., Liu, C.-C.: Is turbulent facilitated ignition through differential diffusion independent of spark gap? Combust. Flame **185**, 1–3 (2017)
12. Jiang, L.J., Shy, S., Nguyen, M.T., Huang, S.Y., Yu, D.W.: Spark ignition probability and minimum ignition energy transition of the lean iso-octane/air mixture in premixed turbulent combustion. Combust. Flame **187**, 87–95 (2018)
13. Xu, D.A., Shneider, M.N., Lacoste, D.A., Laux, C.O.: Thermal and hydrodynamic effects of nanosecond discharges in atmospheric pressure air. J. Phys. D: Appl. Phys. **47**, 235202–235214 (2014)
14. Kelley, A.P., Jomaas, G., Law, C.K.: Critical radius for sustained propagation of spark-ignited spherical flames. Combust. Flame **156**, 1006–1013 (2009)
15. Chen, Z., Burke, M.P., Ju, Y.: On the critical flame radius and minimum ignition energy for spherical flame initiation. Proc. Combust. Inst. **33**, 1219–1226 (2011)
16. Wu, F., Saha, A., Chaudhuri, S., Law, C.K.: Facilitated ignition in turbulence through differential diffusion. Phys. Rev. Lett. **113**, 024503 (2014)
17. Lipatnikov, A.N., Shy, S.S., Li, W.: Experimental assessment of various methods of determination of laminar flame speed in experiments with expanding spherical flames with positive Markstein lengths. Combust. Flame **162**, 2840–2854 (2015)
18. Davis, S.G., Law, C.K.: Determination of and fuel structure effects on laminar flame speeds of C1 to C8 hydrocarbons. Combust. Sci. Technol. **140**, 427–449 (1998)

Investigation of the Output Parameters of a Cantilever PEG with Two Piezoelectric Elements at Vibration Excitation by Rotating Drive

A. N. Soloviev[1,2], I. A. Parinov[1], A. V. Cherpakov[1,2(✉)], and V. A. Chebanenko[3]

[1] I. I. Vorovich Mathematics, Mechanics and Computer Science Institute, Southern Federal University, Milchakov Str., 8a, Rostov-on-Don, Russia
[2] Don State Technical University, Gagarin Sq., 1, Rostov-on-Don, Russia
[3] Federal Research Center "Southern Scientific Center of the Russian Academy of Sciences", Chekhov Str., 41, Rostov-on-Don, Russia

Abstract. A full-scale simulation of a cantilever piezoelectric generator (PEG) with a proof mass and an active base is considered. The vibration excitation of this piezoelectric generator is carried out by a rotating drive with magnetic devices. A feature of this generator is the presence of two types of elements: plate-type elements undergoing bending deformations and cylindrical-type elements undergoing compression deformations. The drive installation is described. The experimental results for the output characteristics of the generator at low-frequency loading are presented.

Keywords: Energy harvesting · Cantilever piezoelectric generator · Bimorph and cylindrical piezoelectric elements · Proof mass · Active base · Magnet · Rotating drive

1 Introduction

The development and modernization of energy harvesting devices is one of priority tasks of world science and techniques. One of the advanced types of materials that are currently developing are piezoelectrics. Technical goods made from these materials in combination with additional mechanical devices can be the basis for numerous energy harvesting applications. By using piezoelectric materials with purpose of energy harvesting, free mechanical energy of environment is converted into electrical energy and transmitted to feeding the corresponding devices. Some overviews of corresponding results are given in works [1–3].

Devices using energy converters in the form of piezoelectric elements (PEs) are called piezoelectric generators (PEGs). Basic information about PEG, as well as the problems arising at the R&D of energy harvesting devices, were given in papers [4–7], as well as in fundamental monographs [8, 9].

B. T. Long et al. (Eds.): MMMS 2020, LNME, pp. 705–709, 2021.
https://doi.org/10.1007/978-3-030-69610-8_94

It is well-known PEG of cantilever type, taken as a prototype, which contains a cantilever beam made of elastic material, on which PEs are glued either on one side (unimorph), or on both sides (bimorph). One end of the cantilever is fixed at the base, and a proof mass is located at the free end [10–14].

The purpose of the work is to describe a set-up for studying the output parameters of a cantilever PEG with two types of PEs when vibration is excited by a rotating drive.

2 Description of Model Parameters

Piezoelectric converter of cantilever type of mechanical energy into electrical one contains a cantilever, which is a cantilever beam made of elastic material, on which PEs are glued from both sides (bimorph). One end of the cantilever beam is fixed to the base. A proof mass is attached to the free end. Four PEs are additionally installed at the base, two on top and two on the bottom relative to the beam, having opposite (in the upper position) and the same directions (in the lower position) polarization vectors [12–14]. The planes of their electrodes are pressed with the help of base elements to the conductive layers of thin elastic pads metallized on one side (see Fig. 1).

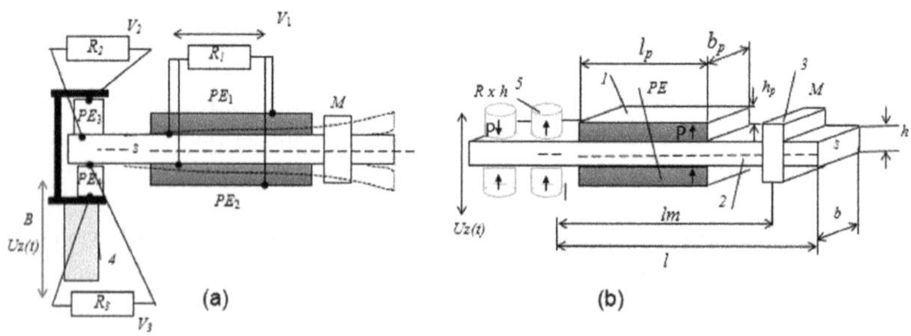

Fig. 1. (a) Electric scheme of compound PEG under active load and (b) structure scheme of PEG with proof mass: 1 – piezoelectric element; 2 – substrate; 3 – proof mass; 4 – place of PEG fixing (B - movable base); 5 – piezoelectric cylindrical elements

Thin symmetrical PEs are polarized in thickness and glued to the base console. The geometrical dimensions of the PEG are shown in Fig. 1b. The characteristics and dimensions of PEG elements are presented in [12–14]. A description of the material parameters is given in [15]. The electric scheme of the PEG with resistive load is shown in Fig. 1a. In the experiment, proof mass M = 17.6 g was used; PEs had sizes: $l_p \times b_p \times h_p$ = 50 × 10 × 0.45 mm^3; piezoelectric cylinders had sizes: $R \times H$ = 10 10 mm^2; Substrate $l \times b \times h$ = 135 × 13.2 × 1.5 mm^3, l_m = 127 mm. Material of the base and substrate was duralumin and proof mass was steel.

The set-up consists of a drive wheel (7) with a magnet of the C25 type (4) fixed on it (see Fig. 2) with a diameter of the magnet equal to 25 mm. The drive is mounted on the base bar structure (8) on a bearing support. The PEG is fixed at a retainer (1). The laser displacement gauge records the oscillations of the PEG cantilever in the vicinity of the end point of fixing the bimorph. The diameter of drive wheel (made of 12 mm plywood) is equal to 67 cm. Inductance sensors located on both sides of the end of the PEG allow recording the transit time of the magnet. The angle between the two sensors relative to the center of the drive wheel is 0.3795 rad. The distance between the magnet and the proof mass was chosen equal to 7 mm.

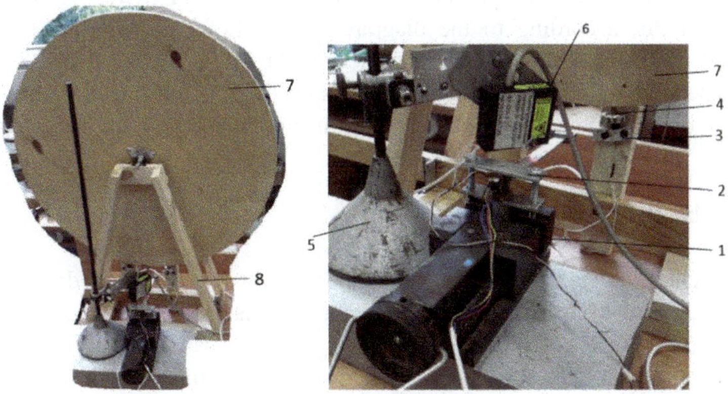

Fig. 2. Vibration set-up for PEG excitation. 1 – retainer; 2 – PEG; 3 – induction sensor; 4 – magnet; 5 – fixing elements of the laser sensor RF603 (6); 6 – laser triangulation displacement meter RF603; 7 – rotating drive; 8 – elements of drive base (7)

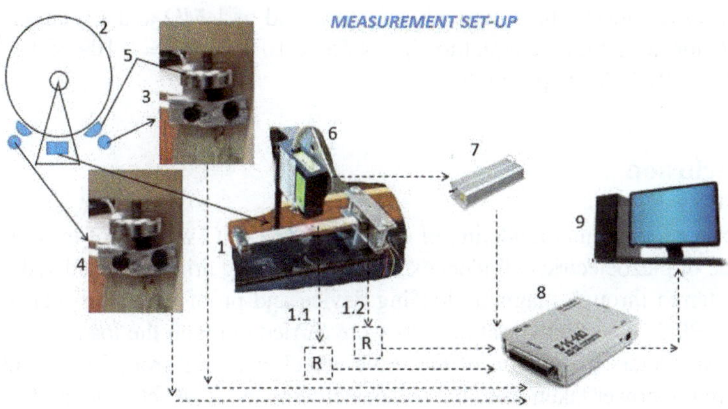

Fig. 3. Measurement set-up: 1 – PEG; 1.1, 1.2 – transmission path from bimorph and PE of cylindrical shape; 2 – drive wheel; 3, 4 – induction sensors before and after PEG; 5 – magnet; 6 – optical sensor of linear displacement RF603; 7 – controller of the optical sensor 6; 8 – external ADC/DAC module L-Card 14-440; 9 – computer

The set-up was operated as follows (see Fig. 3). During the rotation of the drive wheel (2), the oscillations and output characteristics of the PEG (1) were studied. On the drive wheel (2) there was a fixed magnet (5). When the wheel (2) rotated, the magnet (5) excited an electric potential when passing by two induction sensors (3,4) located before and after the attached mass of PEG. Thus, the excitation time of the first and second induction sensors was recorded and the average rotation speed was calculated. When the magnet passed over the attached PEG mass (1), the metal mass was attracted to the magnet (5). After passing through the magnet (5), and thereby removing the load in the form of attraction, PEG (1) continued to oscillate with a certain attenuation. The output excitation voltage of the induction sensors and the PEG output voltages taken from the bimorph and PE at the base of the PEG were recorded (1). In this case, all PE, according to the diagram in Fig. 1, were loaded with an active resistance R = 1 MΩ. Using matching devices (1.1), (1.2), (7), the ADC (8) received signals generated by PEG (1), induction sensors (3), (4), and a laser displacement sensor (6), and recording was performed with using the software on the computer (9).

3 Results and Discussion

The output characteristics of the PEG were obtained for the rotation of drive wheel at a speed from 2.98 to 4.92 rad/s, which corresponds to the speed of rotation of the wheel rim 0.98–1.62 m/s.

The amplitudes of the cantilever oscillation increase linearly with the peripheral speed. For PEs bimorph and cylindrical shapes, the output voltage increases linearly at these speeds and an active load equal to 1 MΩ. At a rotation speed of 4.92 rad/s, $U_1 = 3.095$ V on the bimorph electrodes, $U_2 = 0.47$ V and $U_3 = 0.04$ V on the PEs with cylindrical shape. The total power was measured within 5 s from the moment of excitation of oscillations in the PEG. Analysis of these data shows that the maximum values of power during this time with an active load of 1 MΩ and a rotation speed of 4.92 rad/s for three PEs are equal to $P_1 = 4.55 \times 10^{-6}$ W, $P_2 = 0.109 \times 10^{-6}$ W and $P_3 = 0.01 \times 10^{-6}$ W, respectively.

4 Conclusion

A full-scale experimental modeling of cantilever type PEG with proof mass and active base made of piezoelectric cylinders excited by a rotating drive is considered. The load was transferred through magnetic loading device and proof mass was attached at the end of the PEG. The PEG vibration tests were carried out with the loading time for the possible parameters of rotation of the drive wheel changing from 2.98 to 4.92 rad/s. The maximum power taken from the piezoelectric plates at an active load of 1 MΩ and an output voltage of $U_1 = 3.095$ V was 4.55 μW for a bimorph located on PEG. The peak power value for the top piezocylinders was reached at a load of 1 MΩ with an output voltage of 0.47 V and was equal to 0.109 μW. For the lower piezocylinders, the peak value at a load of 1 MΩ and an output voltage of 0.04 V it was 0.01 μW. This loading did not take place at resonance modes of PEG oscillations. In the future, it makes sense to investigate the resonance modes of excitation by a rotating PEG drive.

Acknowledgements. The study was performed under partial support by the Russian Foundation for Basic Research (No. 19-08-00365) and the Government of the Russian Federation (contract No. 075-15-2019-1928).

References

1. Gaudenzi, P.: Smart Structures: Physical Behavior, Mathematical Modeling and Applications. Wiley, New York (2009)
2. Chopra, I.: Review of state of art of smart structures and integrated systems. AIAA J. **40**(11), 2145–2187 (2002)
3. Chebanenko, V.A., Akopyan, V.A., Parinov, I.A.: Piezoelectric generators and energy harvesters: modern state of the art. In: Parinov, I.A. (ed.) Piezoelectrics and Nanomaterials: Fundamentals, Developments and Applications, pp. 243–277. Nova Science Publishers, New York (2015)
4. Anton, S.R.: Multifunctional piezoelectric energy harvesting concepts. PhD thesis, Virginia Polytechnic Institute and State University. Blackburn, Virginia, USA 1–190 (2011)
5. Badel, A., Guyomar, D., Lefeuvre, E., Richard, C.: Efficiency enhancement of a piezoelectric energy harvesting device in pulsed operation by synchronous charge inversion. J. Intell. Mater. Syst. Struct. **16**(10), 889–901 (2005)
6. Cherpakov, A.V., Chebanenko, V.A., Parinov, I.A., Chang, S.-H., Jani, M.A.: Finite Element Modeling of the piezoelectric cantilever generator type. In: Proceedings of International Symposium on Physics of Lead- free Piezoactive and Relative Materials (Analysis of current state and prospects of development), LFPM-2016, vol. 2, pp. 265–269 (2016)
7. Guyomar, D., Lallart, M.: Recent progress in piezoelectric conversion and energy harvesting using nonlinear electronic interfaces and issues in small scale implementation. Micromachines **2**(2), 274–294 (2011)
8. Erturk A, Inman D J 2011 Piezoelectric Energy Harvesting (John Wiley & Sons)
9. Shevtsov, S., Soloviev, A.N., Parinov, I.A., Cherpakov, A.V., Chebanenko, V.A.: Piezoelectric Actuators and Generators for Energy Harvesting - Research and Development. Innovation and Discovery in Russian Science and Engineering. Springer, Cham (2018)
10. Chebanenko, V.A., Zhilyaev, I.V., Soloviev, A.N., Cherpakov, A.V., Parinov, I.A.: Numerical optimization of the piezoelectric generators. J. Adv. Dielectr. **10**(1–2), 2060016 (2020)
11. Soloviev, A.N., Chebanenko, V.A., Zhilyaev, I.V., Cherpakov, A.V., Parinov, I.A.: Numerical optimization of the cantilever piezoelectric generator. Mater. Phys. Mech. **44**(1), 94–102 (2020)
12. Cherpakov, A.V., Kokareva, Y.A.: Modal analysis of the cantilever type piezo-electric generator characteristics with active based on numerical simulation. Mater. Sci. Eng. **698**(6), 066020 (2019)
13. Cherpakov, A.V., Parinov, I.A., Soloviev, A.N., Rozhkov, E.V.: Experimental studies of cantilever type PEG with proof mass and active clamping. Springer Proc. Phys. **224**, 593–601 (2019)
14. Soloviev, A.N., Parinov, I.A., Cherpakov, A.V., Chebanenko, V.A., Rozhkov, E.V.: Analyzing the output characteristics of a double-console peg based on numerical simulation. Mater. Phys. Mech. **37**(2), 168–175 (2018)
15. Panich, A.A., Marakhovskii, M.A., Motin, D.V.: Crystal and ceramic piezoelectric. Eng. J. Don **1**, 12 (2011). http://www.ivdon.ru/en/magazine/archive/n1y2011/325

Study on the Production Formula Establishment of Multi-component Pellets from Sawdust and Food Plastic Waste

Bui Trung Thanh[1](✉), Le Anh Duc[2], and Nguyen Phi Truong An[3]

[1] Faculty of Heat and Refrigeration Engineering, Industrial University of Ho Chi Minh City, Ho Chi Minh City, Vietnam
buitrungthanh@iuh.edu.vn

[2] Faculty of Mechanical Technology, Nong Lam University of Ho Chi Minh city, Ho Chi Minh City, Vietnam

[3] Ho Chi Minh City University of Technology and Education, Ho Chi Minh City, Vietnam

Abstract. This paper presents the results of an empirical study on the production formula establishment of multi-component pellets (MCP) from the PP plastic powders of food nylon wastes (PPPPW) and sawdust of wood processing plants in Vietnam. The MCP production efficiency and the calculated low heat value (LHV) of the finished-good pellets are two objective functions of the production formulas, while the ratio of the composition of the mixture of PPPPW and sawdust, the size of the materials, and the moisture content of the mixture are the influencing factors. Experimental results have determined that the ratio of the mixed ingredients of 15% PPPPW and 85% sawdust with its moisture content of 13%, the finished product efficiency of 64%, and the calculated LHV of MCP of 5631 kcal/kg could be obtained. The empirical results also showed the new feasibility of combining daily-life PP plastic bag waste treatment with sawdust from wood processing plants to not only create a useful energy source for production but also help solve the current problems of plastic waste treatment for society.

Keywords: Multi-component pellets · Nylon waste · Sawdust · Efficiency of pellet product · Low heat value of pellets

1 Introduction

The production technology of pellets has the remarkable advantage of intensive utilization of wastes from agricultural/forestry production and solid wastes from daily activities to generate useful thermal energy which could be used to resupply to production activities. According to [1], Vietnam plastic waste accounts for roughly 5–10% of municipal waste. That means, on average, there are 1.15–2 million tons of plastic waste among about 23–25 million tons of municipal wastes that were generated per year. Meanwhile, in the US, the yearly average plastic waste generation is approximately 39.33 million tons, in which 6.8% is recycled, 0.7% is used as alternate fuel in cement manufacturing, and 9.9% is mixed with other non-recyclable wastes and is

B. T. Long et al. (Eds.): MMMS 2020, LNME, pp. 710–720, 2021.
https://doi.org/10.1007/978-3-030-69610-8_95

transformed into electricity and steam in facilities that convert waste into energy [2]. Non-recycled plastics (NRP) is currently considered as one of the fascinating wastes, as the calorific value of NRPs is much higher than organic wastes. Most of the NRPs are currently being researched by scientists around the world, to transform them into thermal energy and later convert into electrical energy [3]. Thus, the establishment of NRP treatment facilities would help greatly reduce the number of wastes transferred to landfills, and transform more wastes into usable energy instead. Research from EEC (Earth Engineering Center) in the US has confirmed that the average low heat value (LHV) of NRP exceeds those of common fuels, which are: 19% higher than that of petroleum cokes, 37% higher than that of American coals, and 87% higher than that of woods [2]. Meanwhile, solid wastes generated such as sawdust and nylon bags for covering finished wood products and daily food packaging have become a practical issue of being sources of contamination to wood manufacturing in wood processing plants in Vietnam. These solid wastes could be sources of useful energy with appropriate methods of treatment; but in fact, most of the time they are collected and gathered in heaps to be later dumped in landfills. In order to help solve this fore mentioned solid wastes issue, the researchers of Industry University of Ho-Chi-Minh City (IUH) have proposed the simple treated solution [4]. The scope of this paper presents results of empirical researches on the production formula establishment of multi-component pellets from the PP plastic powders of food nylon waste (PPPW) and sawdust to aim for the effective and economical utilization solution of energy source from solid wastes in wood processing plants in Vietnam [4].

1.1 Paper Review on Fuel Pellets Production

Widodo Wahyu Purwanto and colleagues (2009) [5] created multi-component pellets in which straws, rubberwood, and camphrier wood are the ingredients. The fuel pellet has a diameter of 10 ÷ 20 mm, moisture that varies from (4–20%). The result showed that the multi-component pellets made from rubber woods and camphrier wood, which have a diameter of 10 mm and moisture of 4%, showed the highest calorific value and the best flammability. INyoman Sukarta and colleagues (2018) [6] have conducted research on the formulation of biomass pellets from organic wastes and sawdust. The research applied the moisture of mixed ingredients at 10.07%, and has created fuel pellets from the various mixtures of sawdust and organic wastes which have mixing ratios of (90:10), (50:50), and (80:20). The mixing ratio 90:10% has been defined to produce pellets that have a low heat value of 3822 kcal/kg. Jamradloedluk and colleagues (2017) [7] have conducted research on biomass pellets made from tree bark and fruit skins by using various adhesives. The result showed that, using cashew nutshell liquid as adhesive resulted in pellets that are short and inconsistent in length while using aloe vera latex as adhesive caused cracks on the pellet surface. Tsuchiya and Yoshida (2017) [8] did research on the technology of multi-component pellets formation from brown coal and dust coal. By mixing dust coal with brown coal, the calorific value of brown coal was intensified and its safety during transportation was also improved. The multi-component pellets with the mixing ratio of 50% brown coal and 50% dust coal had the calorific value reached up to 15 MJ/kg, production efficiency rate was achieved at 98%. Zhijia Liu and colleagues (2013) [9] succeeded in producing multi- component pellets from bamboo powder and rice straw with the mixing ratio of ($\leq$3:2). Pellets type has had the calorific value of over 4180 kcal/kg. Nguyen Thi Bich Thuan (2011) [10] has studied the effect of material moisture, the effect of small material size and the effect of contents of supplements on the quality of pellets from rice husks. Tapioca starch content of 2 ÷ 3% added as a supplement of the mixed ingredient to produce pellet. The finished pellet has moisture content from 8,5 ÷ 9,5%, LHV is 3973 kcal/kg, ash content is 13,24%, pellet diameter (d) is 6 mm and length (L) is 25 ÷ 40 mm, bulk density (ρ_b) is 500 ÷ 600 kg/m^3 and density (ρ_p) is 1050 ÷ 1180 kg/m^3.

1.2 Production Process of the Fuel Pellets

Typically, the fuel pellets are produced by the following process as in Fig. 1 [11].

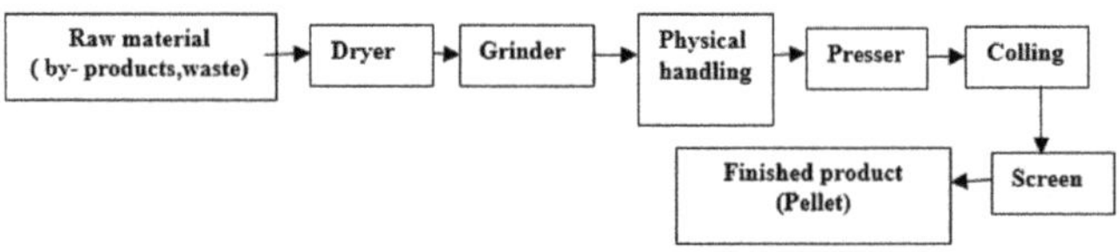

Fig. 1. Process of production technology of fuel pellets [11].

The production process consists of three main operations including ingredient preparing, pellet pressing and pellet forming. The particular operations from Fig. 1 are described as the following:

a) **Ingredient preparing operation**

Drying process: the ingredients are dried around 15%.

Grinding process: for some specific ingredients such as straw, rice stubble, sawdust, after being chopped to reduce the size down to 1 ÷ 3 cm, they would be put into the grinding machine. The ingredient size after being grinded is about 1, 2 ÷ 1,8 mm.

Physical processing process: To enhance the adhesiveness, the pellet durability, and reduce dust during pressing, the grinded ingredients need to be supplemented by several additives such as waste oil, molasses, etc. [11].

b) **Pellet pressing and forming operation**

Pellet pressing process: after the ingredients achieving appropriate ingredients dimension and moisture, the pellet pressing process is initiated. The ingredients are put into the feed hopper of the pellet pressing machine with a determined mass over time at regular intervals using a quantitative feeder. The ingredients in the pressing machine will be compressed into pellet form and then taken out. Since the pellet pressing is the most essential process in the pellet forming operation, choosing the appropriate type of pressing equipment for each type of material with proper conditions and processes will enable calculable productivity and high-quality products.

Finished-product pellet cooling process: since the pellets have a relatively high temperature after being taken out of the pressing machine, they need to be taken into the air-cooling system to reduce to ambient air temperature afterward. The cool finished-product pellets will be taken to packaging and be ready to sell in the markets.

Sieving stage: the pellet fragments would be screened and sorted then returned to the pressing machine. This process is very crucial to pellet pressing, as it helps determine the proportion of finished products after the pressing process in order to have suitable adjustments on affecting factors which result in the best finished-product pellets [11]

2 Materials and Methods

Since the fuel pellet product is applied in providing thermal energy to gasifier furnaces for the drying process, the diameter of the pellet is required to be in the 6 mm to 8 mm range. In this particular empirical experiment presented in this paper, a pressing mold which has the diameter of 6 mm is chosen to form the multi-component pellets [3].

2.1 Production Process of Multi-component Pellets Made from Powdered Nylon Food-Packaging Waste and Wood Processing Sawdust

The pelletizing material consists of 02 primary ingredients: PP powder from FPPW and sawdust. The nylon waste was collected from daily activities in wood processing plants, and the sawdust was collected from a common wood group in furniture factories.

2.2 Production Technical Variables and Objective Functions

Technical variables: (i) - Moisture of each ingredient, %; (ii) - Dimension of each ingredient, mm; (iii) - Mixed moisture, %; (iv) - Ingredient mixing ratio or ingredient proportion,%.

Objective functions: (i) - Efficiency of pellet pressing process, %; (ii) – low heat value of pellet (LHV) kJ/kg (Fig. 2).

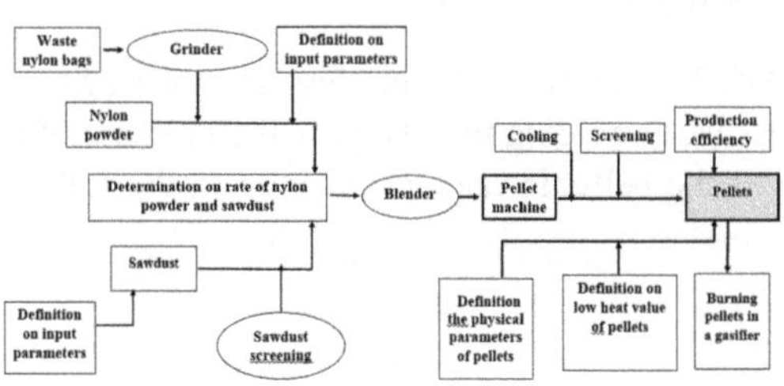

Fig. 2. The empirical production process of multi-component pellets [4].

2.3 Method of Determining Basic Specifications of Production Materials and Multi-component Pellet

a) **Determining the diameter of the ingredients**

The diameters of the nylon powder and sawdust ingredients are determined by using a set of test sieves following the order of opening dimensions from 1 ÷ 4 mm. Starting with screening either sawdust or nylon powder with the 1 mm-opening sieve, then separated material below the sieve is the ≤ 1 mm material. Continuing with using the 1,5 mm-opening sieve, then the 2 mm, and the last is 4 mm.

b) **Determining moisture content of materials, mixed ingredients, and pellets**

The moisture of each ingredient and the pellet production mixed moisture is determined by using Eq. 1:

$$M_{wb} = \frac{G_w}{G} 100\% \tag{1}$$

G - total mass of the material, kg; G_w - water mass within the moist material, kg; M_{wb}- moisture content on a wet basis. The Sartorius Model MA45 sintered moisture meter was used for the measurement. The experiment was conducted with 7 measurements.

c) **Determining the ingredient density following the ingredient dimension**

The bulk density of the nylon powder and sawdust ingredients is calculated by using Eq. 2:

$$\rho_{bnl} = \frac{m_{nl}}{V_{nl}} \tag{2}$$

In which: ρ_{bnl} - bulk density of nylon power, kg/m^3; m_{nl} - mass of nylon powder, kg; V_{nl}- volume of nylon in the test tube, m^3.The bulk density of the nylon powder and sawdust is determined by firstly weighing 20 g of nylon powder with the predetermined diameters, then by pouring this amount of it into a gauge cylinder with ml divisions, the displacement amount in the cylinder defined - which could be now inputted into Eq. 2 to determine.

d) **Determining the density of the pellet**

The density of the MCP depends on the mixing ratio of sawdust and nylon powder, plus the pressing pressure. Providing that the pressing pressure of the pressing machine is fixed, the density of the pellet is calculated by using the following Eq. 3:

$$\rho_P = \frac{m_p}{v_p} \tag{3}$$

In which: ρ_P – density of the pellet, kg/m^3; m_p - mass of a pellet, kg; v_p – specific volume of a pellet, m^3. The density of pellet is determined by firstly weighing a single pellet to define the mass (g), then, with the diameter of the single pellet is measured by using a caliper, the volume of the single pellet (mm^3) is calculated; next, using the Eq. (3), the density could be figured out and converted into kg/m^3.

e) **Determining the bulk density of the pellet**

The bulk density of the MCP depends on the mixing ratio of sawdust and nylon powder as well as the pressing pressure. Thus, the bulk density of MCP is calculated by Eq. 4

$$\rho_b = \frac{M_p}{V_p} \tag{4}$$

In which: ρ_b – bulk density of the pellet, kg/m^3; M_p - total mass of the multiple pellets inside a predetermined volume gauge cylinder, kg; V_b – total volume of the multiple pellet inside the gauge cylinder that needed to be determined, m^3. The density of the pellet is determined by firstly weighing 10 pellets to define the total mass, then dropping the 10 pellets inside the gauge cylinder with ml divisions. The total volume of these 10 pellets is the total volume of replacement in the gauge cylinder. Using Eq. (4) to determine the pellet density.

2.4 Method of Determining Objective Functions of the Production of MCP

a) **Determining the low heat value of the pellet**

The LHV of the pellet is determined by calculating the mass of each type of ingredient inside a single pellet then multiplying with the LHV of the corresponding ingredient in the pellet as in the following Eq. 5

$$Qv = m_{sd}.C_{sd} + m_{nl}.C_{PP} \tag{5}$$

In which: Qv – LHV of the finished-product pellet, MJ/kg; m_{sd} – total mass of sawdust in a single pellet, kg; m_{nl} – total mass of nylon powder in a single pellet, kg; C_{sd} – LHV of the raw sawdust with the value = 20.52 MJ/kg [12]. C_{PP} – LHV of the PP nylon with the value = 40.88 MJ/kg [2].

b) **Determining the production efficiency of pellet**

The efficiency of pellet production is affected by multiple input parameters, which are calculated in Eq. 6.

$$\eta = \frac{M_{mm-M_d}}{M_{nmm}} 100\% \tag{6}$$

In which: η - Efficiency of pellet production; M_{mm} - mixed material mass, kg; M_d- mass of the fragment dust that goes through the sieve, kg.

2.5 Experimental Equipment and Experimental Arrangement

a) Experimental equipment: The mass of the material ingredients used in pellet production was measured by an electronic scale from the South Korean CAS brand, tolerance of ±0,003 g, and an analytical balance by OHAUS PA214 from the US, tolerance of 0.0001 g. The pellet dimension was measured by the mechanical caliper of China, its tolerance is ±.02 mm. The moisture content of materials, mixed ingredients, and pellets were measured by the Sartorius MA45 electronic moisture from Germany with its tolerance of ±0.01%.

b) Experimental arrangement

- Ingredients preparation

+ *Sawdust ingredient*: the collected sawdust from wood processing plants was screened and sorted to retrieve fine particles with consistent dimension, then the moisture is analyzed by the moisture meter instrument.

+ *Nylon ingredient*: the chosen nylon from PP plastic, which is then shredded into small pieces using a shredder, sieved, and sorted into dimension groups of ≤ 1 mm; ≤ 2,5 mm and ≤ 4 mm.

+ *Mixed component formula*: the sawdust and nylon ingredient are weighed up to sufficient amount to form the pellet following the ratio ranges of (90:10); (85:15); (80:20); (75:25); (70:30); (65:35). For each of the different ingredient mixing ratio batches, after achieving the sufficient mass for each ingredient, the two ingredients are mixed in a mixer. The ingredient samples are adjusted in moisture with the moisture range of 12–16%. The values/variables to be determined are averaged over 07 measurements.

- **Pellet pressing machine**

It is a flat mold, power of 2.2 kW. The pressing capacity is 15 kg/h, the mold thickness is 24 mm, and holes diameter is 6 mm.

- **The process of pelleting**

Check the condition of the devices before starting. The pellet pressing machine by tightening the adjusting screw with the rear lock to adjust the pressure between the roller and rotating disc, and then the roller is tested to see if it is in a natural motion. Then start the pressing machine and put the ingredients into the feeding hopper, while tightening the adjustment screw to change the pressure so that the pellets have a uniform shape along with reasonable and stable pressing mode. The pellets after being taken out of the press are cooled with ambient air, then the moisture content of the pellet is measured and the LHV of the pellet is calculated according to Eq. 5. [2], [12] (Table 1).

Table 1. Experimental results to determine the effect of mixing ratio of sawdust and nylon powder ingredients to the mixture density, specific weight, density, calorific value of pellets and tablet production efficiency [4].

No.	The ratio of sawdust and nylon powder ingredients (%)	Moisture (%)	Mass of mixed volume (kg/m^3)	Density of MCP (kg/m^3)	Bulk density of MCP (kg/m^3)	Production efficiency (%)	Calculated LHV (kcal/kg)
1	90/10	12.9	223.85	1107.1	455.82	68.53	5388.4
2	85/15	12.9	207.1	1067.87	445.78	64.33	5631.6
3	80/20	12.9	203.74	924.22	429.01	55	5874.8
4	75/25	12.9	199.85	850.9	415.5	29.7	6118.0
5	70/30	12.9	187.35	708.5	370.89	20.8	6361.2
6	65/35	12.9	185.15	693.67	357.78	18.53	6604.0

3 Results and Discussion

3.1 Determining Pelletizing Ingredient Parameters

Parameters of the material ingredients for pellet production are studied and determined through two criteria, ingredient size and bulk density. The experimental results have been identified and shown in the graph in Fig. 3:

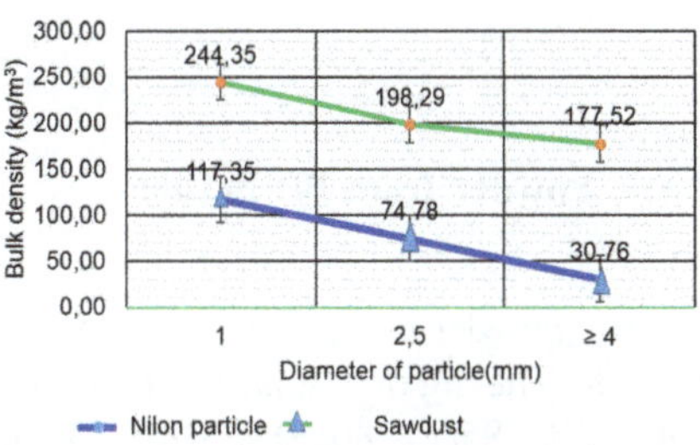

Fig. 3. The graph of the relation between the diameter of sawdust and nylon powder ingredients with bulk density.

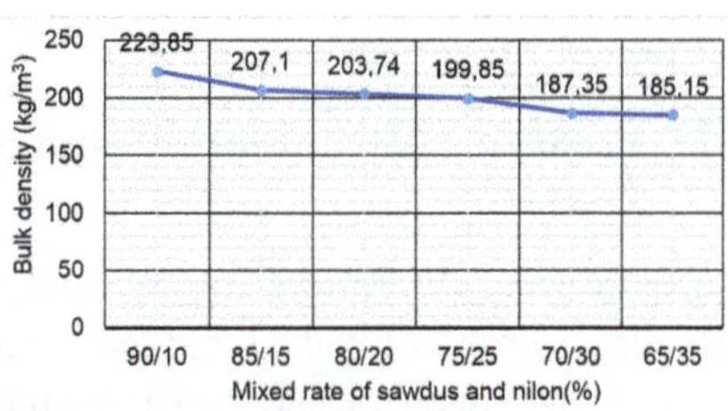

Fig. 4. The graph of the relation of the bulk density of the ingredient mixture

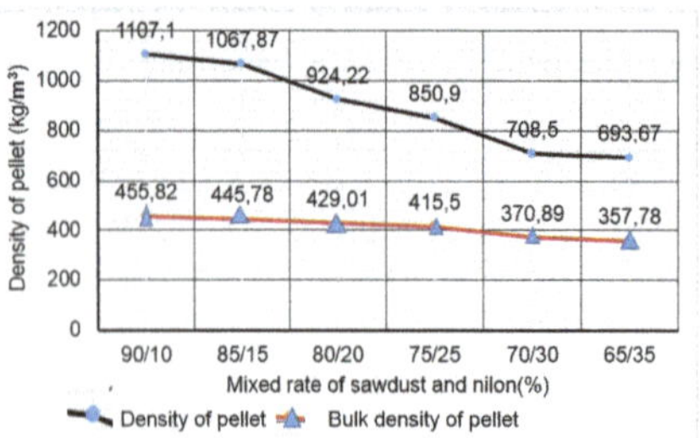

Fig. 5. Graph of the relation of bulk density and density of MCP with the mixing ratio of sawdust and plastic powder

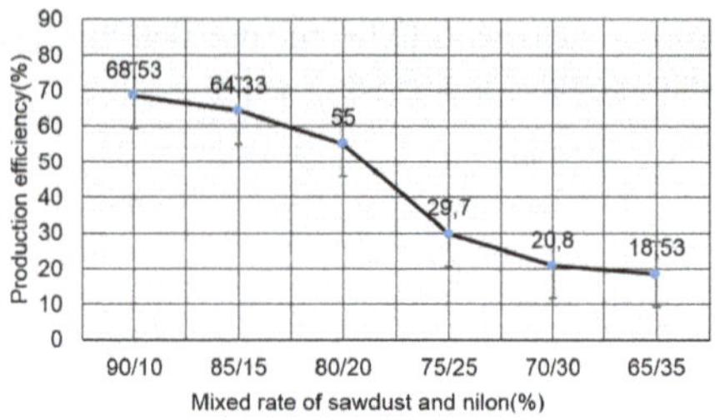

Fig. 6. Graph of the relation of the MCP production efficiency with the mixing ratio.

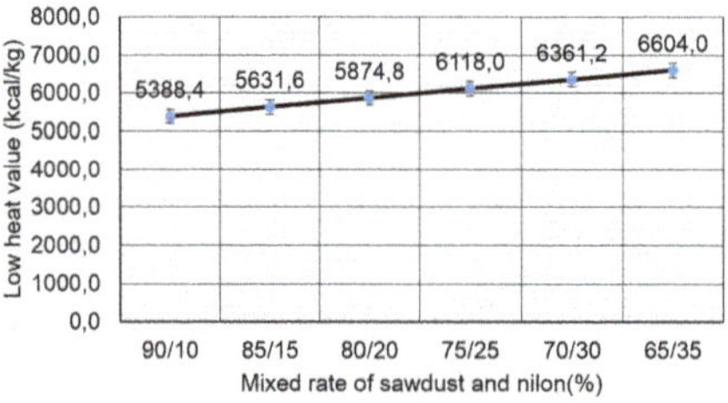

Fig. 7. Graph of the relation of the low heat value of MCP with the mixing ratio of sawdust and nilon powder.

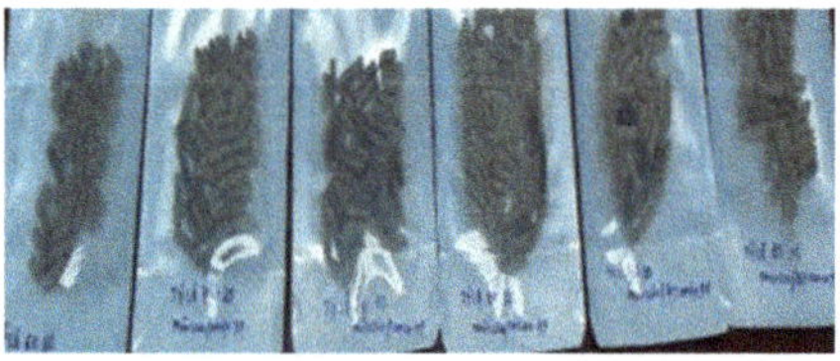

Fig. 8. Samples of finished-product of MCP in this empirical study.

3.2 Producing MCP from Sawdust and Plastic Powder from PP Plastic Powder

In experiments, the chosen size of sawdust and plastic powder is ≤ 1 mm, the average moisture content of the ingredient mixture is 13%. The mixing gradient ratios of sawdust and nylon plastic powders consisted of 90/10; 85/15; 80/20; 75/25; 70/30; 65/35. The experimental results obtained and described following in the Fig. 3, the Fig. 4, the Fig. 5, Fig. 6 and The Fig. 7 (Fig. 8).

Remark figure 3: The smaller the ingredient particle diameter, the higher the bulk density of the material. The sawdust has a density of 224.35 kg/m^3, while the density of plastic powder from nylon is 117.35 kg/m^3.

Remark figure 4: The bulk density of the mixture ingredient decreases along with an increasing proportion of plastic powder into the mixture. At the ratio of sawdust/plastic powder 90:10, the bulk density of the mixture ingredient is the highest and has a value of 223.85 kg/m^3 and the lowest is 185.2 kg/m^3 at the mixing ratio of 65/35.

Remark figure 5: The bulk density of the MCP gradually decreases along with the decreasing proportion of sawdust in the mixture. The highest density is achieved at the ratio of sawdust/plastic powder 90:10, whereas the ratio 65:35 gives the lowest density.

At the mixed ingredient ratio of sawdust/nylon powder of 90:10, the density of MCP has the highest value of 1107 kg/m^3.

Remark figure 6: Proportion of the nylon powder does have an effect on the adhesion between nylon powder and sawdust thus affecting on the compression process. At the mixed ratio of 90% sawdust and 10% nylon powder, the highest pellet production efficiency is achieved value of 68,53%

Remark figure 7:Based on the results of the experiments on pellet pressing and calculation of pellet LHV according to [2] and [12], the research has built the graph relation of the LHV of pellet with the mixing ratio of sawdust and nylon powder. The higher the LHV of MCP due to the higher the nylon powder rate. The highest LHV is found on the MCP with the mixing ratio of 35% nylon and 65% sawdust, and the lowest LHV with the mixing ratio of 90% sawdust and 10% nylon powder.

4 Conclusion and Discussion

This research has fundamentally established the formulas for producing pellets from mixing the sawdust and nylon powders processed from waste to form the multi-component pellets. The experimental results have validated that the input ingredient mixture of 85% sawdust and 15% plastic powder is optimal to create MCP with an average size of 6.2 mm, length of 23 mm, bulk density of 445.8 kg/m^3 and density of 1068 kg/m^3, calculated low heat value is 5631 kcal/kg and production efficiency is 64.3%. The production of MCP is also beneficial for the environment considering its high utilization of plastic waste. Therefore, it is advisable to select the production formula that uses as much plastic powder proportion as position, as long as it satisfies the combined criteria of calorific value and pellet quality. Up to this point, however, the formula to produce MCP which is based on the mixing ratio of sawdust and PP plastic powder established by the results of the empirical study in this paper has these following constraints: the processable material size range is limited to only around one mm with the moisture content of mixed ingredient is 13%, as well the low heat value of the pellet has just solely defined by calculation.

References

1. Van Thuc, H.: Some solutions to prevent plastic waste in Vietnam. J. Environ. ISSN **2615–9597**(6), 2019 (2019)
2. Tsiamis D.A., Castaldi, M.J.: Determining accurate heating values of non-recycled plastics (NRP), Report of Earth Engineering Center, City College City University of New York 23 March 2016

3. Siudyga, R.W.: Energy recovery from waste plastics, Science technique journal, Institute for Chemical Processing of Coal (IChPW), Zabrze, Poland; Tomasz - Silesian University of Technology, Gliwice, Poland (2013)
4. Thanh, B.T.: Applying Research of gasification technology treats solid wastes in the industrial production –foodstuff processing to create heat energy for drying and agricultural products and foodstuffs preservation, Research project of the Ministry of Industry in 2019 (2019)
5. Purwanto, W.W., Supramono, D., Nugroho, Y.S., Lestari, D.E.: Characteristics of biomass pellet as fuel, In: Proceeding of international Seminar on Sustainable Biomass Production and Utilization Challenges and Opportunities (ISOMASS) 3–4 Aug (2009)
6. Sukarta, I., Sastrawidana, I.K., Ayuni, N.P.S.: Proximate analysis and calorific value of pellets in biosolid combined with wood waste biomass. J. Ecolo. Eng. **19**(3), 185–190. https://doi.org/10.12911/22998993/86153 (2018)
7. Jamradloedluk, J., Lertsatitthanakomb, C.: Influences of mixing ratios and binder types on properties of biomass pellets. In: International Conference on Alternative Energy in Developing Countries, Energy Procedía – Elsevier (2017)
8. Tsuchiya, Y., Yoshida, T.: Pelletization of brown coal and dust coal in indonesia: characteristics of the mixture pellets including safety during transportation. Fuel Process. Technol. **156**(2017), 68–71 (2017)
9. Zhijia, L., et al.: The properties of pellets from mixing bamboo and rice straw. Renew. Energy **55** (2013), 1–5 (2013)
10. Thuan, N.T.B.: Process studying of the pellets production from by-products (rice husks) of rice mill. The Master thesis of Hanoi University of Science and Technology. p. 65 (2011)
11. Garcia-Maraver, A., Carpio, M.: Biomass Pelletization process, WIT transaction on state of the art in science and engineering, Chap. 4, Vol. 85, ISSN 1755–8336 (2015)
12. Leory, D.D.: Particulate cleanup of low energy gas produced in biomass fluidized bed gasifier. Ph.D. thesis, Texas A and M University, p. 52 (1983)

A Review on Pulsed Fluidized Bed Dryer

Pham Quang Phu and Bui Trung Thanh(✉)

Industrial University of Ho Chi Minh City, Ho Chi Minh City, Vietnam
{phamquangphu, buitrungthanh}@iuh.edu.vn

Abstract. Pulsed fluidized bed (PFB) dryer is another form of conventional fluidized bed (CFB) dryer, in which the hot air is periodically supplied to the particle bulk in the drying chamber at each part of supporting grid instead of providing to whole surface of grid as in a CFB dryer. Also, each zone of drying materials is fluidized, in turn, by the hot air supplied periodically from a gas distributor. For areas where air has not been supplied, materials are in a static state whereas in the other areas the drying materials are fluidized with liquid-like property. Due to the difference in the way of air supplied to each part of the drying chamber, the PFB dryer has more advantages such as (1) being able to accept more diverse and larger size of materials, (2) can operate with sliced vegetables from 20–30 mm in size and 1.5–3.5 mm in thickness. The energy consumption of the PFB dryers are 30–50% lower than that of CFB dryers as the minimum fluidization velocity is lower 8–25% and pressure drop of the material layer decreases by 7–12%. This paper presents the latest technologies of PFB dryer including types of dryers, structure of dryers, structure of pulsed gas distributor and published results of latest applications of PFB dryer for various materials. Additionally, the paper also presents the selection of the most suitable PFB model for application in drying of refined standard sugar at sugar mills in Vietnam.

Keywords: Drying · Fluidized bed · Drying materials · Pulsed gas distributor · Pulsed fluidized bed dryers · Pulsed flow · Fluidization

1 Introduction

The fluidization of particles is the result of a dynamic balance between gravitational force and force due to the flow of gas through a discrete particles bed. During the fluidization process, the bubbles exist inside a fluidized bed in large size under certain air velocity conditions. This mode is called bubbling fluidized bed. The larger the size of bubbles, the less efficient heat and mass transfer. In PFB technology, the inlet gas flow changes over time, therefore, the gas velocity at the bubbling stage is also changed which leads to the change of the bubbling fluidized bed, also, the stability of the particle layer and affects the hydrodynamic properties [1, 2]. The pulsed flow in gas-solid fluidization was demonstrated effectively in controlling chaos of particles bed and the ability to reduce the bubble size compared to continuous fluidized bed [3, 4].

The fluidized bed dryer (FBD) is one of the most commonly convection dryers for drying types of particulate solids and powder materials. However, FBDs have the main disadvantages that limit their application that The relatively high-pressure loss and gas

B. T. Long et al. (Eds.): MMMS 2020, LNME, pp. 721–730, 2021.
https://doi.org/10.1007/978-3-030-69610-8_96

velocity across the bed that requires the high pressure and flow fans, consequently, increasing investment costs and electricity costs in operation [5]. To solve the problem of energy costs for fans, some authors have studied a new fluidized bed drying method - the PFB drying. The literature showed the positive results in terms of energy saving. The basic principle of rotary PFB dryer is described in the follows: The main difference between a rotary PFB dryer and a FBD is the gas distribution system. The gas distributor is a disk installed under the fixed supporting grid in the gas distributing chamber. The disk has one or several 60° angular opening sectors, allowing the orientation of gas flow through it. As this disk rotates, gas passes through the fixed supporting grid and enters the drying chamber and fluidizes the bed of particles sequentially, causing the pulsed effect. The fixed supporting grid has both a particle support function and a uniform gas distribution function. The rotary disk is driven by a variable speed motor with controllable rotational speed 6–450 RPM [6, 7].

2 Types of Pulsed Fluidized Bed Dryers

There are many ways to classify PFB dryers, but classification by gas supply method is the most widely used. According to Kudra and Mujumdar [5], there are 03 types of PFBD including PFBD with a dual butterfly valve, PFBD with rotary disk and PFBD with relocated gas stream. However, according to Gawrzynski and Rychlik [8], the PFB is divided into two main groups such as PFB with pulsation in a whole the chamber and PFB with zonal pulsation, which consists of 05 basic sub-types of dryers: PFBD with rotating sheet; PFBD with relocated gas stream; Spouted PFBD; FBD with local pulsed jet creation of the fluidization; Rotating jet spouted bed dryer. Meanwhile, Todor [6] proposed a completely different linear PFB dryer, combined with rotation-pulsed fluidization (RPF), called by the common name of jet-pulsed fluidized bed (JPFB). In the range of PFB drying technology, a simple classification is mentioned as Fig. 1. The dryer is divided into two types: (1) Continuous drying and (2) batch drying. The classification according to the gas distributor is also proposed here.

The method of intermittent gas flow using a solenoid valve, a surge tank and a timer to supply gas to the drying chamber has been proposed by Levey since 1965 [9]. This method considered to be the simplest, however, it requires a surge tank which will make the model take up a lot of space. The method of intermittent gas flow with a double butterfly valve was presented by Bühler-Miag GmbH (Germany) where the frequency of pulsed gas changes by adjusting the rotation speed of the butterfly valve [5]. The advantage of this type is that the hydrodynamic characteristics in the drying process are improved but there are still many limitations such as high-pressure loss when the valve is in the closed position and sometimes a vibration device is needed for powdery materials with high moisture content.

As the C-PFB dryer has rectangular section of drying chamber and, the air supply method with rotary disk is not suitable, so Gawrzynski et al. [10] proposed gas supply method by change gas flow position. A gas distribution chamber consists of several small chambers separated by partitions installed below the fixed supporting grid, the gas into the rotary distributor will be changed in turn to each different position in order from the input of wet materials to the output of product. The advantage of this method is that it is applicable to continuous dryers with large capacity, large pulse frequency adjustment range and stable operation without the need for a vibrating device. The drawback of this method is the technology of manufacturing gas distributors, complex piping systems and high investment costs.

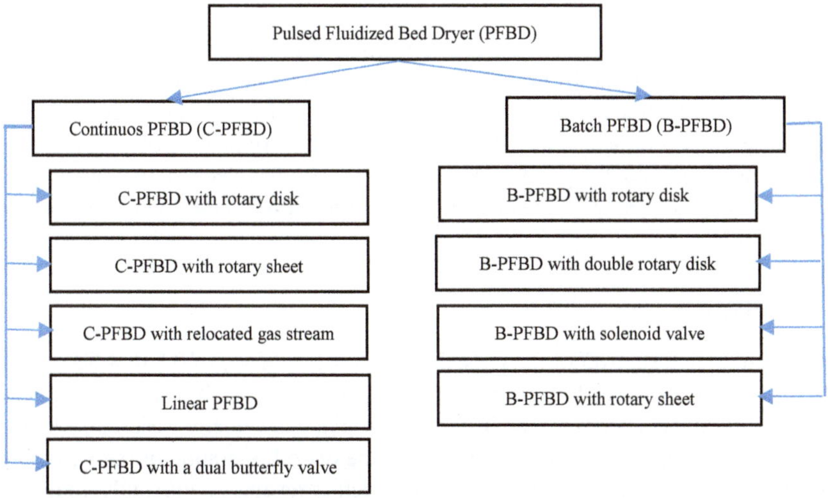

Fig. 1. Classification of PFB dryers

3 Gas Distributor in PFBD

3.1 Solenoid Valve

Solenoid valve was proposed by Levey in 1965 in patent No. 3164440 [9]. With the author's data, the method has improved significantly increase of the fluidizing gas velocity in a wide range of average particle sizes. Wong and Baird [11] conducted a glass beads drying experiment on the PFBD model using a solenoid valve air supply model with a pulse frequency range of 1 to 10 Hz. The results show that, in a bubbling bed, pulsation increases the retention time of the gas by as much as 51%.

3.2 Rotary Disk

Rotary disk was first proposed by Todor [7], and subsequently used by many authors to conduct empirical research. Godoi et al. [12, 13] conducted experiments with this distributor and reported that this equipment can be used to dry such a cohesive particulate material, resulting in the uniform distribution of the gas inside the bed and providing uniform drying of the solid. Similar equipment was employed by Ambrosio-Ugri and Taranto [14].

Another type of rotary disk is patented by Kudra et al. [15] with a 90° semicircular opening that allows the gas to pass through it. As the disks rotate synchronously, the gas is distributed alternately between the four parts of the drying chamber. Experimental research by Camargo et al. [16] conducted with brewer's spent grain on tray dryer and PFB dryers showed that the drying process in PFB was faster than in tray dryer.

3.3 Rotary Sheet

A C-PFBD, of 20-tonnes of paddy/h was mentioned in the research of Somkiat et al. [17]. The air is transported through the main pipe, containing two metal plates that rotate continuously at a specific angular velocity. The pulsed air is then passed through a specific part of the paddy bed (along the side left or right side of chamber drying process), thus fluidizing the paddy. The results showed that electrical energy consumed by the PFB was at least 30% lower than conventional FBD, and drying efficiency was higher [17].

3.4 Butterfly Valve

Simple start up and shut-off of the pulsed gas with a frequency of 0.5 to 4 Hz is used in some dryers in most cases [5], gas pulsation is generated by a butterfly valve installed in the inlet gas duct. Therefore, the instant superficial gas velocity varies across the bed area from 0 to the maximum. However, due to the on–off operation of the valve causes an inverse pressure on the gas supply fan. Such a design significantly reduces the negative effects of a single pulsator when the gas shut off phase is alternated between the two longitudinal sections of the bed.

3.5 Relocated Gas Stream Distributor

The idea of the PFB dryer with relocated gas stream was introduced by Gawrzynski et al. [4] in the 1990s with crosswise pulsation and later versions with lengthwise pulsation [18]. Details of the design of this gas distributor are presented in the literature with a rotating disc and a static disc providing a frequency range of 4 to 12 Hz [8]. Subsequent studies by the authors themselves show that with a bed depth of 0.1 m and a pulse frequency of 8 Hz, the minimum fluidization velocity was 0.64 m/s for continuous flow, whereas this value was 0.28 m/s for crosswise pulsation and 0.43 m/s for lengthwise pulsation [8]. This type of gas distributor is suitable for designing a continuous PFB dryer.

4 Application of Pulsed Fluidized Bed

The study of the hydrodynamic and kinetic characteristics of PFB drying focused on building the relationship between gas velocity and pressure loss at different pulsed frequencies and proved the affects of pulse frequency on pressure loss through particle layer [18–21]. According to [10, 13, 14, 19], the appropriate pulse frequency range for batch drying is from 5–15 Hz, but according to [22, 23], pulse frequency should be lower than 1 Hz. Therefore, for each different material, it is necessary to determine the proper operating range of pulse frequency.

Kudra et al. [19] observed that the pulsed gas flow through particle layer, creating an intense mixing of gas-solid and increasing the surface area of heat exchanger, which helps to improve drying. The results show that for pulp and paper sludge, homogeneous fluidizing phenomenon takes place at gas pulse frequency of 10 Hz. Grzegorz Rogula [21] has also established correlations between pressure loss and Reynolds number based on changes in particle height, particle diameter and pulse frequency. The results showed that if the particle layer was too high, the fluidizing process could not take place despite increasing the pulse rate. In a study of fluidized bed drying kinetics, Sobrino et al. [20] demonstrated that the size of gas orifice and rotational speed of gas distribution disk greatly influence the pressure loss across the particle layer.

Gawrzynski Z. and Bartosz Pieczaba [18] studied the hydrodynamic characteristics of PFB dryer in with lengthwise pulsation – a continuous PFB dryer with pulsed gas supplied in horizontal. This study showed that compared to conventional PFB dryers, the mixing of particles is better and product moisture is more uniform. In addition, the working gas velocity is 2 times higher than the minimum fluidized velocity, so the drying cost and investment cost will be lower than other PFB dryers.

Experimental studies to determine the effect of technological parameters on the PFB drying have been conducted on an experimental scale based on the fluidized bed model published by Gawrzynski et al. [10]. Studies have determined the effects of temperature, gas velocity and pulse frequency on pressure loss, drying time [10, 24–26]. From these studies, several empirical equations to determine the heat transfer and moisture diffusion coefficients have been identified [10, 25, 26].

Gawrzynski et al. [10] established regression equations that determined minimum fluidizing velocity, heat transfer coefficient and volumetric heat transfer coefficient for polypropylene powders on gas-fired fluidized bed dryers with drying temperature changes at levels of 50, 70, 90 °C, pulse frequency 5, 10, 15 Hz and drying gas velocity from 0.178 to 0.25 m/s. Also empirical research in the form of drying model of Gawrzynski, Nitz and Taranto [24] has built a drying curve according to temperature and gas flow, and has evaluated the effects of gas temperature, gas flow, speed of rotating disk to drying process. The authors also found that the pressure loss across the particle layer increased with particle height, gas velocity and pulse frequency. This study is also consistent with the results published by Reyes et al. [25, 26], Godoi et al. [13].

Ambrosio and Taranto [14] studied experimental drying of 2-hydroxybenzoic acid material of Group A (according to Geldart's classification [27]). As a result, the batch PFB dryer is suitable for bulk materials with high moisture content while ensuring the quality of product. According to the authors, the advantage of the PFB dryer is to decrease the gas flow necessary for fluidizing the particle layer. However, the results of this paper only stop at a few simple experiments and have not yet assessed the influence of other factors in the PFB drying.

Godoi et al.[13] analyzed the process of bio-polymer drying in a PFB dryer and determined that the optimal drying mode for this type of product is the rotating frequency of 7 Hz, drying temperature of 90 °C and gas velocity of 0.55 m/s. It also shows that the PFB dryer is suitable for granular materials, which has the characteristics of high cohesive. However, according to Dening Jia et al. [28] for biomass drying, the appropriate pulse frequency is from 0.75 to 1.5 Hz, at frequencies higher than 3 Hz the drying process is similar to common convection drying. Simultaneously, the author has also demonstrated the influence of technological parameters of the drying process on the drying time and product moisture, in which the efficiency of the drying process depends on the pulse frequency (Increasing pulse frequency will increase drying efficiency). The study also showed that the use of a vibrator to support the drying process is no longer necessary, so that the ability to apply PFB drying to replace vibrating fluidized bed drying technique.

Jia et al. [29] by empirical research determined the moisture diffusion coefficient during drying process. The results showed that the moisture diffusion coefficient was proportional to the temperature and gas velocity, consistent with published by Gawrzynski et al. [10]. However, according to Dening Jia et al. the moisture diffusion coefficient achieved high values when drying at a pulse frequency range of 0.75 to 1.5 Hz while Gawrzynski et al. showed that see gas pulse frequency range is from 5 to 15 Hz.

The energy cost of fluidized bed drying is generally determined by two types of costs: electrical cost and thermal cost per kg of evaporated moisture [5]. According to Gawrzynski and Bartosz [18], the fluidizing velocity for PFB drying is lower than that for CFB dryers, so the electrical cost is about 10–30% lower and Somkiat Prachaya-warakorn [17] also reported that with PFB dryer, the electrical cost of fan is 25% lower than that of CFB dryer, the thermal cost of PFB dryer is from 6.3 to 7.8 MJ/kg moisture compared to CFB dryers which are 14 MJ/kg moisture (about 50% thermal energy reduction). Dening Jia et al. [28] conducted an experiment proving that the gas velocity in the PFB is from $0.8U_{mf}$ to $1.3U_{mf}$ (U_{mf} - minimum fluidization velocity). While for the CFB dryer, the gas velocity must reach from $2.0U_{mf}$ to $3.0U_{mf}$ [5]. This result shows that with the pulsed flow, the material can be dried under the lower minimum fluidization velocity while ensuring heat and mass transfer, thus reducing the energy cost [20, 28–30].

In the research of Taranto et al. [31], a PFB dryer for sodium acetate crystals with a cross-sectional area of 0.18 m^2 was used with an effort to reduce energy consumption and increase productivity. The drying rate is higher under pulsed fluidization than conventional fluidization. CFB drying consumes 2.5 times more energy at 80 °C than PFB drying. Based on these studies, the PFB drying method is considered as being able to save energy and decrease the drying time while ensuring the quality of product.

Recently, based on the observation by high speed cameras (>1000 fps), the study authors have provided images of the fluidizing phenomenon that occurs when pulse flow is supplied at very short intervals, thereby showing the influence of air pulse frequency on the drying process. Hamed and Hassan [32, 33] showed that the pulsed delivery reduces the minimum fluidizing velocity and improves the fluidizing mode for fine-grained fine particles, which means that the drying cost is reduced. Similarly, Dong Liang et al. [34] has built an equation to determine the bubble velocity and determine the size of the bubbles in the fluidizing process for magnetite powder. Kaiqiao Wu [35] by empirical research combined with simulation has shown that the phenomena that occur in the process of gas supply at different times and prove the effects of gas velocity and pulse frequency to fluidizing process by PFB method.

The advantage of this empirical research is to identify and evaluate the effect of technological parameters on the pulsed fluid bed drying, thereby finding reasonable parameters for each different material. However, the limitation of these claims is not comparable with theoretical research, so the persuasion is not high.

According to literature review, the advantages of PDB dryers in comparison with CFB dryer are identified as followings:

(1) Easy fluidizing for materials with high cohesive
(2) The particle size range that can be fluidized is larger.
(3) Drying energy consumption is 30 to 50% lower.
(4) Improve the homogeneous fluidizing regime (reduce the channels between the layers of particles).
(5) Reducing the particle breakage.

When compared with other types of convection dryers, PFB dryers have many advantages such as:

(1) High coefficient of heat and mass transfer.
(2) Dryer size is smaller and drying time is lower.
(3) Improve the quality of dried products.
(4) The ability of mixing between materials and drying gas is improved.
(5) Higher thermal efficiency.

5 Refined Standard Sugar Drying by Pulsed Fluidized Bed Dryer

Refined standard sugar (RS sugar) is the material formed by crystallization, the moisture content after centrifugation is usually from 1% – 2% [36]. Unless this moisture is removed, the wet sugar will rapidly deteriorate and be attacked by bacteria. Normally, the moisture content of raw sugar required for preservation must not be greater than 0.2% (TCVN 6961: 2001) and not more than 0.05% (TCVN 6958: 2001) for refined sugar. It is necessary to dry sugar before storage. In addition, the sensory criteria and the physical properties of sugar must be in accordance with national standard.

Therefore, the post-centrifugal drying is necessary for long-term preservation and standard humidity. Through empirical research results, the authors have demonstrated that batch PFB drying is suitable for drying materials with high moisture and cohesive [12–14, 32, 33]. Gawrzynski et al. mentioned one of the industrial PFB coolers for sugar with a throughput of 20.8 tons/h, installed at the Svlelin sugar factory in the early 1980s and the SFC-20 dryer-cooler - designed by the authors themselves. The coolers for cooling sugar from ranging from 54 °C to 75 °C in temperature and 0.02% to 0.066% in moisture content to the final temperature and moisture content of sugar ranged from 34 °C to 38 °C (mean temperature 33.5 °C) and 0.02% to 0.064% (mean moisture content 004%), respectively.

Somkiat et al. [17] showed that the continuous PFB dryer could applied at industrial scale for paddy with capacity of 20 tones/h with a 50% thermal energy reduction at 70–80% of recycled air. Ambrosio-Ugri et al. concluded that the rotating-pulsed fluidized bed is an alternative for the processing of cohesive solids that preserves the final quality of the dry solids [14].

From the above research results, the applicability of PFB dryer for sugar is appropriate, with the hot pulsed gas caused fluidization of the bed, providing an the intense mixing of the solids and the development of an interfacial area that improved the drying process [4, 14]. Figure 2 show that a model of continuous PFB dryer with a capacity of 20 kg/h installed at the Drying laboratory, Industrial University of Ho Chi Minh city. The results of initial tests showed that the ability to dry is suitable for RS sugar at 60–80 °C and air pulse frequency of 0.25–0.75 Hz.

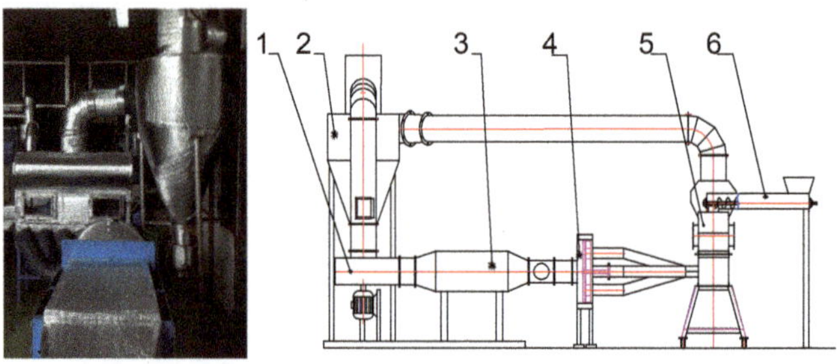

1- Fan; 2- Cyclone; 3-Heater; 4- Pulsator; 5- Drying chamber; 6- Feeder

Fig. 2. The PFB dryer for RS sugar and its diagram

6 Conclusions

The pulsed fluidized bed dryer has been proposed since the 1960s and their ability to save energy has been proven. However, recent researches have not fully concluded the actual energy saving and the development of PFB dryer on an industrial scale for crystalline-originating materials such as RS sugar and sea-salt. Therefore, the study of the PFB drying technique is found to be a new direction that needs to be conducted early in the situation where the world is heading towards saving energy and sustainable development.

References

1. Coppens, M.O., van Ommen, J.R.: Structuring chaotic fluidized beds. Chem. Eng. J. **96**(1–3), 117–124 (2003)
2. Hadi, B., Van Ommen, J.R., Coppens, M.O.: Enhanced particle mixing in pulsed fluidized beds and the effect of internals. Ind. Eng. Chem. Res. **51**(4), 1713–1720 (2012)
3. Pence, D.V., Beasley, D.E.: Chaos suppression in gas-solid fluidization. Chaos **8**(2), 514–519 (1998)
4. Gawrzyński, Z., Glaser, R.: Drying in a pulsed-fluid bed with relocated gas stream. Dry. Technol. **14**(5), 1121–1172 (1996)

5. Kudra, T., Mujumdar, A.S.: Advanced Drying Technologies. CRC Press, Boca Raton (2009)
6. Elenkov, V.R., Djurkov, T.G.: A fluidized bed with jet-pulsed agitation of fluidization. In: Proceedings of the First South East European Symposium–Fluidized Bed in Energy Production, pp. 105–116 (2001)
7. Djurkov, T.G.: Rotation-pulsed fluidized bed apparatus for sesame roasting. In: 2nd Symposium of South-East European Countries (SEEC) on Fluidized Beds in Energy Production, Chemical (1999)
8. Gawrzyński, Z., Rychlik, M.: Zonal gas distribution in pulsed fluidized bed dryers. Chem. Process Eng. Inz. Chem. i Proces. **31**(1), 53–66 (2010)
9. Levey, R.P.: Gas-Solid contacting method. 3,164,440 (1965)
10. Gawrzynski, Z., Glaser, R., Kudra, T.: Drying of powdery materials in a pulsed fluid bed dryer. Dry. Technol. **17**(7–8), 1523–1532 (1999)
11. Wong, W.H., Baird, M.H.I.: Fluidisation in a pulsed gas flow. Chem. Eng. J. **2**, 104–113 (1970)
12. Godoi, F.C., Boin, E.A.S.F., Pereira, N.R., Rocha, S.C.S.: Fluid dynamics and drying of cohesive particles of a biodegradable polymer (poly-hydroxybutyrate) in a rotating pulsed fluidized bed. Dry. Technol. **28**(11), 1297–1306 (2010)
13. Godoi, F.C., Pereira, N.R., Rocha, S.C.S.: Analysis of the drying process of a biopolymer (poly-hydroxybutyrate) in rotating-pulsed fluidized bed. Chem. Eng. Process. Process Intensif. **50**(7), 623–629 (2011)
14. Ambrosio-Ugri, M.C.B., Taranto, O.P.: Drying in the rotating-pulsed fluidized bed. Braz. J. Chem. Eng. **24**(1), 95–100 (2007)
15. Kudra, G.: Glaser, "United States Patent," 5,918,569 (1999)
16. De Camargo, L.L.D., De Souza, L.F.G., Nitz, M.: Study of tray and pulsed fluidized bed drying of brewer's spent grain. Chem. Ind. Chem. Eng. Q. **25**(3), 229–237 (2019)
17. Prachayawarakorn, S., Tia, W., Poopaiboon, K., Soponronnarit, S.: Comparison of performances of pulsed and conventional fluidised-bed dryers. J. Stored Prod. Res. **41**(5), 479–497 (2005)
18. Gawrzyński, Z., Pieczaba, B.: Hydrodynamics of the pulsed fluid bed with cyclically relocated gas stream with lengthwise pulsation. Dry Technol. **24**(9), 1133–1137 (2006)
19. Kudra, T., Gawrzynski, Z., Glaser, R., Stanislawski, J., Poirier, M.: Drying of pulp and paper sludge in a pulsed fluid bed dryer. Dry. Technol. **20**(4–5), 917–933 (2002)
20. Sobrino, C., Sanchez-Delgado, S., Almendros-Ibanez, J.A., de Vega, M., Ruiz-Rivas, U., Santana, D.: Hydrodynamic characteristics of a fluidized bed bed with rotating distributor. In: 2007 ECI Conference on the 12th International Conference on Fluidization - New Horizons in Fluidization Engineering, p. 10 (2007)
21. Rogula, G.: Some research on aerodynamics of a new pulsed fluidized bed dryer. Chem. Process. Eng. **2009**(30), 653–663 (2009)
22. Li, Z., Kobayashi, N., Deguchi, S., Hasatani, M.: Investigation on the drying kinetics in a pulsed fluidized bed. J. Chem. Eng. Japan **37**(9), 1179–1182 (2004)
23. Ali, S.S., Asif, M.: Fluidization of nano-powders: effect of flow pulsation. Powder Technol. **225**, 86–92 (2012)
24. Nitz, M., Souza, L.F.G., Moraes, S.R., Taranto, O.P.: Drying of magnesium sulfate in a pulsed fluid bed dryer, April 2015
25. Reyes, A., Moyano, P., Paz, J.: Drying of potato slices in a pulsed fluidized bed. Dry. Technol. **25**(4), 581–590 (2007)
26. Reyes, A., Vega, R., Garcia, G.: Drying sawdust in a pulsed fluidized bed. Dry. Technol. **26** (4), 476–486 (2008)
27. Geldart, D.: The effect of particle size and size distribution on the behaviour of gas-fluidised beds. Powder Technol. **6**(4), 201–215 (1972)

28. Jia, D., et al.: Fluidization and drying of biomass particles in a vibrating fluidized bed with pulsed gas flow. Fuel Process. Technol. **138**, 471–482 (2015)
29. Jia, D., Bi, X., Lim, C.J., Sokhansanj, S., Tsutsumi, A.: Biomass drying in a pulsed fluidized bed without inert bed particles. Fuel **186**, 270–284 (2016)
30. Jia, D., Bi, X., Lim, C.J., Sokhansanj, S., Tsutsumi, A.: Gas-solid mixing and mass transfer in a tapered fluidized bed of biomass with pulsed gas flow. Powder Technol. **316**, 373–387 (2017)
31. Nitz, M., Taranto, O.P., Mauá, P., Caetano, S., Albert, C.A.: Drying of Beans in a Pulsed-Fluid Bed Dryer – Fluid-Dynamics and the Influence of Temperature, Airflow Rate and Frequency of Pulsation on the Drying Rate, pp. 836–843 (2004)
32. Khosravi Bizhaem, H., Basirat Tabrizi, H.: Experimental study on hydrodynamic characteristics of gas-solid pulsed fluidized bed. Powder Technol. **237**, 14–23 (2013)
33. Khosravi Bizhaem, H., Basirat Tabrizi, H.: Investigating effect of pulsed flow on hydrodynamics of gas-solid fluidized bed using two-fluid model simulation and experiment. Powder Technol. **311**, 328–340 (2017)
34. Dong, L., et al.: Kinetic characteristics of the particles in a dense-phase pulsed fluidized bed for dry beneficiation. Can. J. Chem. Eng. **95**(6), 1133–1140 (2017)
35. Wu, K., de Martín, L., Coppens, M.O.: Pattern formation in pulsed gas-solid fluidized beds – the role of granular solid mechanics. Chem. Eng. J. **329**, 4–14 (2017)
36. Baikow, V.E.: Manufacture and Refining of Raw Cane Sugar. Elsevier, Amsterdam (2013)

A Study on Designing a Machine for Testing the Resistance to Wet Bacterial Penetration of Medical Protective Clothing Based on the British Standard EN 14126:2003

Trong-Thanh Nguyen, Duc-Thang Tran, and Tuan-Anh Bui(✉)

School of Mechanical Engineering, Hanoi University of Science and Technology, No 1 Dai Co Viet Road, Hanoi, Vietnam
anh.buituan@hust.edu.vn

Abstract. The Covid-19 pandemic is happening complicatedly all over the world, affecting the health and daily life of the people and negatively affecting the socio-economy of many countries. Use of masks and other protective clothing is one of the recommended measures in countries around the world to prevent the spread of disease. However, it is difficult to evaluate the effectiveness of those protective clothing. To contribute to the fight against the pandemic, a study on designing and building a machine for testing the resistance to wet bacterial penetration of medical protective clothing has been done. The results show that the machine has been operating stably with the design parameters, under laboratory conditions. In addition, the results of investigating the liquid penetration into the medical mask under an applied force of (0–8) N on a plate containing the test piece rotating at a speed of about 60 rpm, corresponding to the test parameters for the resistance to bacterial penetration according to BS EN 14126: 2003, is also presented. The data will be listed, analyzed, and processed to determine the ability of the sample to prevent bacteria from entering.

Keywords: Wet bacterial penetration · Medical protective clothing · Testing machine

1 Introduction

Vietnam is a country affected by many diseases caused by bacteria and viruses. Notably, the respiratory infections caused by the new virus type SARS-CoV-2. Covid-19 can exist in liquid droplets that are sprayed from infected carriers, diffused in the air under a state of suspension [1]. To prevent epidemics, it is necessary to equip doctors and people with protective equipment, including masks and protective clothing. In order to check the protection of the aforementioned protective clothing according to the published medical standards, and at the same time to serve scientists with equipment and machines for anti-epidemic research, we have studied and build a protective device against bacteria penetration.

B. T. Long et al. (Eds.): MMMS 2020, LNME, pp. 731–737, 2021.
https://doi.org/10.1007/978-3-030-69610-8_97

The equipment for testing the resistance to penetration of solution containing bacteria is used to work in microbiology laboratories, biotechnology products. Hence, we can detect it early and find ways to stop bacterial invaders. This device helps to check the protective gear of the workers to avoid contact or work with the environment such as factory wastewater, medical waste, and bacterial culture,... tested for safety before use. On the other hand, bacteria have many different anomalies in a bacterial set such as shape, size, habitat, bacteria density, spread, viability. The size of a group of bacteria can range from 30 nm for poliovirus to 5 μm and 10 μm for bacteria and even larger. The classification of bacteria is presented in the European Directive 2000/54/EEC [2].

The examination and evaluation of medical protective clothing against the penetration of bacteria or viruses has been of much interest to many scientists. In 2008, W.-Q. Wang et al. presented a study on the performance monitoring tests for the resistance to wet bacterial penetration for several types of medical clothing basing on ISO 22610 [3]. In 2015, Joel Amilcar Ramirez presented his research on the evaluation of the penetration of particles and the breathing resistance of a respirator with N95 filter and an uncertified dust mask. A study presented in the doctoral thesis has shown the performance of respirators in the working place for the particle penetration and breathing resistance. The respirators with N95 filter were used under simulated atmospheric conditions to determine the maximum particle penetration of non-certified dust mask against sodium chloride [4]. R. M. Laing indicated the potential hazards to both workers and patients in typical hospital and operating theatre. The author presented a study on the protection against some of these when using different forms of clothing. Besides, an investigation of the main properties of fabrics and garments used in the working place to suggest further studies was also presented [5].

In the face of complicated developments in the Covid-19 epidemic, the demand for protective gear has increased, many people trade in fake and poor-quality protective gear. Realizing the urgency of assessing the quality of protective gear, it is necessary to design and manufacture to serve the examination of protective requirements for medical masks or protective clothing. The machine is designed and manufactured based on the standard BS EN 14126: 2003 [6]. The purpose is to test and classify protective clothing for the user in a variety of working environments such as microbiological laboratories, doctors, or medical personnel regularly in contact with infected patients or medical waste disposal. Due to the heterogeneity of bacteria, it is not possible to define a criterion or a unified method of classifying its virulence or transmission. Therefore, the experimental methods are used to determine the resistance of bacteria that are carried by a liquid when in mechanical contact with the fluid carrying the bacteria.

2 Design of a Machine for Testing the Resistance to Wet Bacterial Penetration of Medical Protective Clothing

Derived from the need to design and manufacture a device to assess its resistance to the penetration of bacterial fluids through medical protective clothing based on BS EN 14126: 2003, the dynamic structure of machine is described as Fig. 1. Accordingly, the system includes a drive motor (1), mechanical transmissions (2,3) with pulley diameters of d_1, d_2, d_3, d_4, and a turntable system (4) to carry the testing sample. A suitable force F generated by a pendulum arm mechanism will apply on the sample during the experiments of wet bacteria resistance.

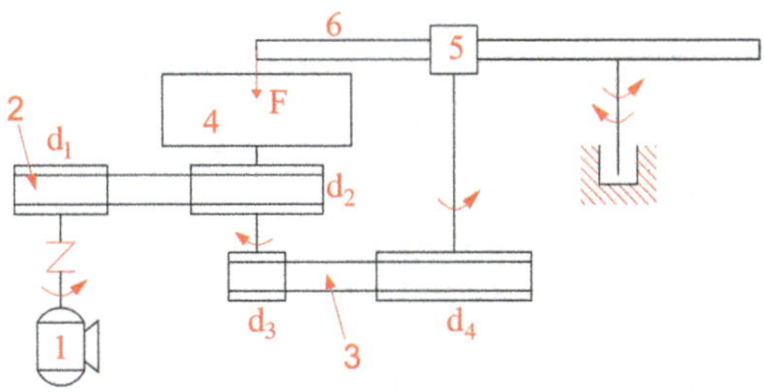

Fig. 1. Dynamic structure diagram of the machine: (1) – Motor (P = 15 W, n = 100 rpm), (2,3) – Toothed belt transmissions, (4) – Turntable, (5) – Cam, (6) – Balancing arm with finger

Table 1. Classification of resistance to penetration by infective agents for mechanical contact with substances containing contaminated liquids [6]

Class	**Breakthrough time, t (*min*)**
6	t > 75
5	60 < t ≤ 75
4	45 < t ≤ 60
3	30 < t ≤ 45
2	15 < t ≤ 30
1	≤ 15 min

The main parts of the machine are designed including a turntable, a camshaft, and a shaking lever. The turntable is attached to the spindle below the Petri dish (with a diameter of 14 cm) with a rotating speed of 60 rpm. Meanwhile, the camshaft is designed with a rotation speed of 10 rpm to push the jib to move back and forth with a force of about 1 N. Above the pendulum arm is fitted with a counterweight at one end to adjust and create a contact force between the pointer (mounted at the other end) and the test piece of about 3 N. In fact, sliding counterweight is to adjust the pressing force exerted by the stainless-steel finger onto the sample in a range from 1 N to 3 N. The structure of the designed machine is shown in Fig. 2.

Furthermore, the machine has been designed to test pieces with parameters based on EN 14126. It is an important criterion in evaluating the quality and performance of protective clothing. Accordingly, protective clothing that meets the standard EN 14126 is a type of clothing that is resistant to chemicals Ultitec 1800, against biological hazards and infectious agents. By using the test method, the standard EN 14126 is used to assess its ability to protect against and against the penetration of bacteria in solution.

Table 1 shows a classification of resistance to penetration by infective agents because of the mechanical contact with substances containing contaminated liquids.

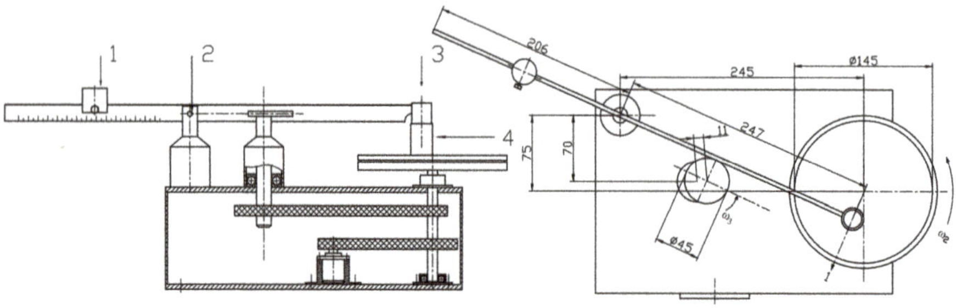

Fig. 2. Structure of the designed machine: (1) – Counterweight; (2) – bearing; (3) – load cell; (4) – stainless steel finger

For the required parameters of the system, an AC motor with rotation speed of about 100 rpm and a power of 15 W (Single Phase AC Motor 3IK15GN-C/3GN15K) can be used [7]. Meanwhile, a toothed belt transmission is designed with a ratio of approximately 1.67 to ensure the rotation speed of the disc around 60 rpm. In addition, the camshafts are also driven via a toothed belt transmission, ensuring a rotating speed of about 10 rpm as designed.

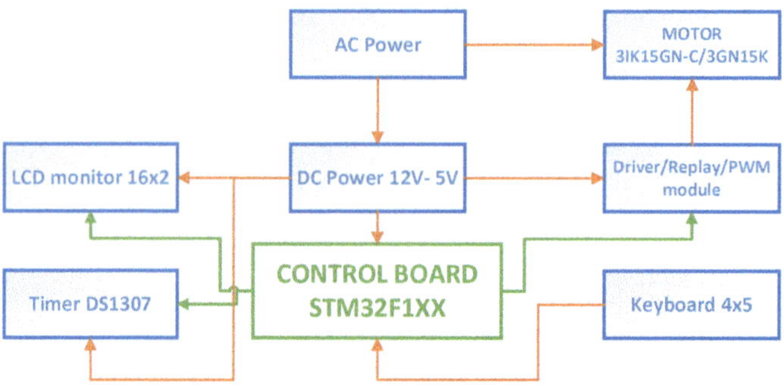

Fig. 3. Structure diagram of the machine control system

The control system of the machine is designed to control actuators and to display necessary information. The system includes a 12 V–5 V power supply, a central processing board STM32F1XX, a LCD monitor 16 × 2, a timer DS1307, a relay module, a power control circuit and a key-press 4 × 5. When testing is performed on the device, the on/off time is adjusted by the keyboard, then the CPU transmits the received signal to the relay and power control circuit. The motor control unit controls the operation of the motor, through which the actuators will operate according to the parameters desired by the user. The structure diagram of the machine control system is descibed in (Fig. 3).

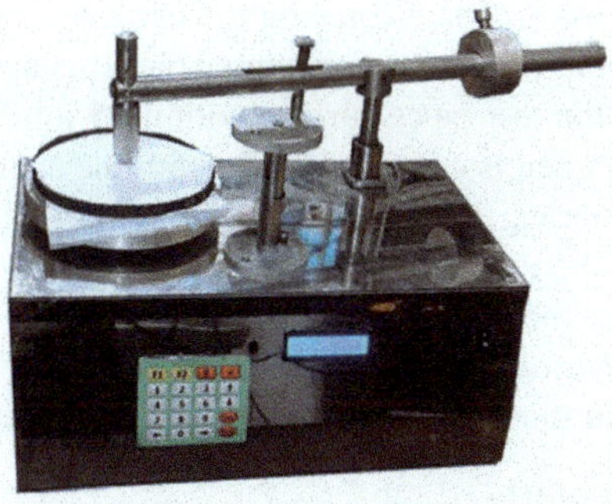

Fig. 4. Experiment setup for the fabricated machine

With the necessary working parameters of the machine for assessing resistance to the ingress of wet bacteria through medical protective clothing, the design parameters have been devised and the device has been successfully fabricated. Figure 4 shows the instrumentation system that has been successfully manufactured and is being prepared for the test. The working parameters were checked after the machine was built and assembled, showing that the machine operation was stable with the design parameters. The process of experimental design and test sample preparation for the machine is performed and described in the following sections.

3 Experiment Process of the Machine

3.1 Sample Preparation

Preparation of test samples consisting of five pieces of 25 cm x 25 cm or a diameter of 25 cm will be cut under aseptic conditions. Test specimens are selected from different positions on a medical protective suit, blouse or fabric material sample. Clothing, protective blouse to take samples including 3 positions: the front of the chest, the back and the area without seams. The fabric sample is taken from a minimum of 3 areas: top, bottom and center of the fabric. Before doing the test, the test samples are packaged and sterilized using the same sterilization methods as the recommendation of the manufacturers.

3.2 Experiment Process

Adjust the counterweight on the lever so that the force from the pointer to the agar plate is about (3 ± 0.02) N. Place the plate on the turntable. Clamp the test piece, carrier material and HD PE film in turn to the double steel frame (using a 90 x 40 mm cylindrical object to provide deflection for the test piece).

The test piece is placed on an uncapped agar plate on a turntable. Above the test piece lay a piece of carrier material and covered with a piece of polyethylene HD film with a thickness of 10 μm corresponding to the test piece size. All are fixed by double steel rings.

The metal pointer exerts a certain force on the carrier material and the test piece so that they meet the agar plate. The shaft lever moves in a translational motion that drives the cursor to move over the test piece surface within 15 min. Notice that, only a very small area of the test specimen is contact directly with the agar surface at a time. Under the combined action of rubbing and movement of the liquid, bacteria can penetrate the material carried through the test specimen.

After every 15 min of the test, the agar plate is replaced, and the test is repeated. During all five 15-min periods, the test is carried out with the same pair of carrier material and test piece. In that way, the test allows to estimate penetration over time.

– ***Calibration of test equipment***

During the test, the test specimen is in contact with the agar surface only at one point at a given time. To ensure the above conditions, perform the test with an upside-down agar plate, place a white sheet of paper on the bottom of the agar plate, charcoal paper and cover with HD PE film. Follow the sequence described above and determine that the pointer moves over the entire surface of the agar plate with uniform contact force (creating a uniform streak on the surface of the white paper).

– ***Experiment procedure***

Place the steel frame with the test piece clamped on the surface of the agar plate on the turntable. Place the cursor on the test piece and adjust the spindle so that the test piece will contact the surface of the agar plate underneath with a force of 3 N for 15 min. After 15 min, lift the steel frame and replace the agar plate below. Repeat the above steps 4 more times using the same test piece and carrier material. At the 6th time remove the carrier material, invert the test piece, cover with a polyethylene HD film and do the experiment with a sixth agar plate for 15 min.

Count the colonies on each plate. Ignore the count in the area of 15 mm radius around the center of the agar plate.

Estimating the expected plate penetration (EPP) as follow:

$$EPP = 6 - (C_1 + C_2 + C_3 + C_4 + C_5)$$

In which:

$$C_1 = X_1/T$$
$$C_2 = (X_2 + X_1)/T$$
$$C_3 = (X_3 + X_2 + X_1)/T$$
$$C_4 = (X_4 + X_3 + X_2 + X_1)/T$$
$$C_5 = (X_5 + X_4 + X_3 + X_2 + X_1)/T$$
$$T = Z + X_5 + X_4 + X_3 + X_2 + X_1$$

Where X_1, X_2, X_3, X_4 and X_5 is the number of stray bacteria per five plates from one of the five specimens; Z is the number of stray bacteria counted from the inverted test sample. The results of the tests will be classified according to Table 1.

Table 2 is an illustrative example of the results of testing the resistance of wet bacteria through several test samples on medical protective clothing that were performed in the laboratory of the Institute for food control. Accordingly, the average EPP value obtained is 5.88, which is equivalent to level 6 according to the classification in Table 1.

Table 2. The expected plate penetration

X_1	X_2	X_3	X_4	X_5	Z	T	EPP	EPP_mean
4	0	1	4	0	200	209	5.85	5.88
3	1	2	5	6	221	238	5.83	
0	1	2	4	5	235	247	5.91	
2	1	1	4	0	248	256	5.90	
0	0	3	5	4	259	271	5.92	

4 Conclusion

With the need to support disease prevention with protective measures, a device to test and evaluate the resistance of wet bacteria to medical protective clothing has been designed and manufactured. The fabricated machine can create a force changing in a range of (0–8) N on a plate containing the test piece rotating at a speed of about 60 rpm, corresponding to the test parameters for the resistance to bacterial penetration based on the British Standard EN 14126:2003.

The machine works stably in the test research laboratory, especially for microbiologists to use anti-epidemic research in the current situation. Initially, several samples of medical clothing have been tested, meeting the requirements for scientists to check, evaluate and classify the quality of medical protective clothing. This contributes to ensuring safe working conditions for people in environments such as microbiology laboratories, waste treatment or in hospitals.

References

1. Asadi, S., Bouvier, N., Wexler, A.S., Ristenpart, W.D.: Aerosol Sci. Technol. **54**, 635–638 (2020)
2. Directive 2000/54/EC of the European Parliament and of the Council of 18 September 2000 on the protection of workers from risks related to exposure to biological agents at work (seventh individual directive within the meaning of Article 16(1) of Directive 89/391/EEC) (2000)
3. Wang, W.-Q., Wu, P., Hou, L., Wang, X., Huang, J.-C.: Chin. J. Med. Instrum. **32**, 290–292 (2008)
4. Ramirez, J.A.: Evaluation of particle penetration and breathing resistance of N95 filtering face-piece respirators and uncertified dust masks, Evaluation of particle penetration and breathing resistance of N95 filtering face-piece respirators and uncertified dust masks, University of Iowa (2015)
5. Laing, R.M.: Int. J. Occup. Saf. Ergon. **14**, 107–115 (2008)
6. BS EN 14126:2003. Protective clothing — Performance requirements and tests methods for protective clothing against infective agents (2004)
7. http://zdmotor.en.hisupplier.com/product-197413-220v-ac-geared-motor.html

Modelling and Simulating Hydraulic System of a Testing Equipment for Bridge Bearings with a Capacity of 6400 Tons

Dinh-Nam Pham[1], Ha-Phi Thai[2], Ngoc-Hai Nguyen[3], and Xuan-Bo Tran[3(✉)]

[1] Institute of Transport Science and Technology, Hanoi, Vietnam
[2] University of Transport and Communications, Hanoi, Vietnam
[3] Hanoi University of Science and Technology, Hanoi, Vietnam
bo.tranxuan@hust.edu.vn

Abstract. To ensure the working ability and longevity of bridge project, bridge bearings need to be strictly tested in a standard laboratory on a specialized testing equipment before being installed and used. For the bearing testing equipment, compressing force acting on the bearing and bearing deformation are two factors that need to be strictly monitored during the test. Hydraulic transmission system is used in the equipment to generate the compression force. Precise simulation of the hydraulic system of the bearing testing equipment can help us to investigate the factors affecting the test results, to optimize the system parameters, and thereby to enhance the measuring accuracy of the testing equipment. This research focuses on developing a suitable mathematical model of the hydraulic system of a testing equipment for bridge bearings with a capacity of 6400 tons. Both numerical simulation and experimental studies are carried out under different types of the bridge bearings to evaluate the accuracy of the system model.

Keywords: Testing equipment for bridge bearing · Hydraulic system · Modelling and simulation

1 Introduction

To ensure the working ability and longevity of bridge project, bridge bearings (Fig. 1) need to be strictly quality checked in a standard laboratory on a specialized testing equipment before being installed and used.

Currently in Vietnam, the bridge bearing testing equipment with a capacity of 6400 tons is the largest and most modern bearing testing equipment [1] (Fig. 2). This is a specialized device to apply compressive force and maintain the force on the test piece. The equipment is designed for testing pot bridge bearings with designed load up to 4000 tons [1]. For the bearing test equipment, measuring accuracy is the most important factor. According to reference [2], the total load applied to the test piece shall not differ by more than 1% from the required load. According to reference [3], the bearing deformation measured after vertical compression must not exceed 5% of the

B. T. Long et al. (Eds.): MMMS 2020, LNME, pp. 738–745, 2021.
https://doi.org/10.1007/978-3-030-69610-8_98

Fig. 1. Bridge bearing

Fig. 2. Photo of the bearing testing equipment at laboratory

bearing thickness. Therefore, the compressive force and the bearing deformation are two factors that need to be strictly controlled during bearing testing.

The hydraulic system is the main driving source to generate the compressive force of the bridge bearing testing equipment [1]. The equipment can generate very large compressive forces (up to 6400 tons) and the force can be maintained for a relatively long time. Currently, the compressive force is measured indirectly through the pressure sensor that is installed on the oil inlet of the hydraulic jacks (Fig. 1b) and the values of the compressive force and the bearing deformation is measured and displayed in steady-states. The pressure (i.e. the compressive force applied to the bearing) as well as the bearing displacement can be fluctuated and greatly influenced by the operating conditions as well as the hydraulic system characteristics. If these two operating characteristics is predicted well by simulation, it is possible to predict the effects of the testing equipment's operating parameters on the system performances. Therefore, it can offer optimal operating conditions of the equipment to improve measuring quality. This paper initially focuses on building a mathematical model for the hydraulic system of the bridge bearing testing equipment with a capacity of 6400 tons. Mathematical model is firstly developed and then is verified by both simulation and experiment.

2 Research Method

2.1 Experimental Hydraulic System

Figure 3 shows the experimental hydraulic circuit of the testing equipment for bridge bearings with a capacity of 6400 tons. In this system, eight large hydraulic jacks (16) with the same dimensions were used to compress the bridge bearing. Jacks' motion was controlled by a directional control valve (10). Oil to the cylinders was supplied by using a small pump (3) and a large pump (4). In the neutral position of the directional control valve, the oil is flowed directly from the pumps to the oil filter tank (17). To create a compression force on the bridge bearing, the directional valve

(10) will operate at the left-side position. At this position, oil is flowed through a one-way throttle valve (12) and a check valve (15) to the below chambers of the jacks (16) causing the piston to go up. Oil from the upper chambers of the jacks will passes through the directional valve and the filter (17) to the tank. In contrast, when the directional control valve (10) operates at the right-side position, oil is fed from the pump through the valve and to the upper chamber of the jacks causing the piston to go down. Oil from the lower chamber of the jacks will flow through the check valve, directional valve, filter and to the oil tank.

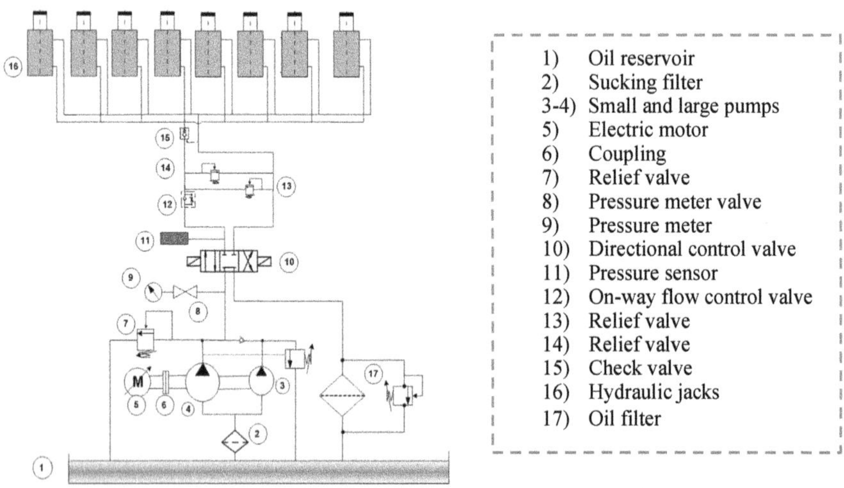

Fig. 3. Experimental hydraulic system

The compressive force acting on the bridge bearing was calculated through the pressure at the lower chambers of the jacks and this pressure was measured using a pressure sensor (11). In the case of the bearing compression, the pressure in the upper chambers of the jacks is connected to the oil tank and therefore the force exerted by the pressure in the upper chambers was considered negligible. To measure the deformation of the bridge bearing, four encoders were used. These encoders are installed at the four corners of the bridge bearing. The average deformation measured by the four encoders is considered as the bearing deformation. The signals from pressure sensors and the encoders were sent to a microprocessor via an analog to digital converter. The microprocessor computed the compressive force from the measured pressure value and the compressive force and bearing deflection values were acquired and displayed by the computer.

2.2 System Modelling

To develop a system model of the testing equipment for bridge bearings with a capacity of 6400 tons as shown in Fig. 3, the following assumptions are considered: variations of flow rate and pressure in the pipe are ignored; the eight jacks are completely identical in structure; the flow rates supplied to 8 jacks are the same; the total reference mass includes the test bearing mass and the intermediate plates is equally divided among the jacks.

The spool displacement of the directional control valve x_v is determined from the input voltage signal u of the directional control valve as follows [4]:

$$\ddot{x}_v + 2\zeta_v\omega_v\dot{x}_v + \omega_v^2 x_v = k_v\omega_v^2 u \tag{1}$$

with ς_v, ω_v, and k_v are the damping coefficient, the natural angular frequency, and the spool position coefficient of the valve, respectively. The valve's input voltage u is given by three values: 24 VDC when the valve is operated in the left position, −24 VDC when the valve is operated in the right position, and 0 VDC when the valve is operated in the center position.

Considering Q_1 and Q_2 are the flow rates supplied to or returned from the lower chamber and upper chamber of the jack, respectively. In addition, the directional control valve is considered with a critically lapped spool. Q_1 and Q_2 are, therefore, determined as follows [5]:

$$Q_1 = C_d\sqrt{\frac{2}{\rho}}wx_v\sqrt{\frac{P_s}{2} + \frac{x_v}{|x_v|}\left(\frac{P_s}{2} - P_1\right)} \tag{2}$$

$$Q_2 = C_d\sqrt{\frac{2}{\rho}}wx_v\sqrt{\frac{P_s}{2} + \frac{x_v}{|x_v|}\left(P_2 - \frac{P_s}{2}\right)} \tag{3}$$

in which C_d, w, ρ, P_s, P_1 and P_2 are the flow coefficient of the valve, the clearance width of the valve, the oil density, the oil source pressure, the pressure of the lower and upper chambers of the jack, respectively.

Applying continuity equation, the relationship between pressure and flow in two-jack chambers can be obtained as follows:

$$\begin{aligned} \frac{V_1}{\beta_h}\dot{P}_1 &= Q_1 - A_1\dot{x} \\ \frac{V_2}{\beta_h}\dot{P}_2 &= A_2\dot{x} - Q_2 \end{aligned} \tag{4}$$

where x, β_h, A_i, and V_i (i = 1–2) are the piston position or the deformation of the bridge, the elastic modulus of hydraulic oil, the effective areas of the piston, and the oil chamber volumes of the cylinder. V_i (i = 1–2) are calculated as

$$\begin{aligned} V_1 &= V_{1d} + A_1 x \\ V_2 &= V_{2d} + A_2(s - x) \end{aligned} \tag{5}$$

with V_{id} (i = 1–2) and s are the dead volumes of the two-cylinder chambers and the piston stroke, respectively. The forces exerted on the bridge bearing are described in Fig. 4. The motion equation of the bridge bearing together with the piston of the hydraulic cylinder is described by the Newton's Law II as follows:

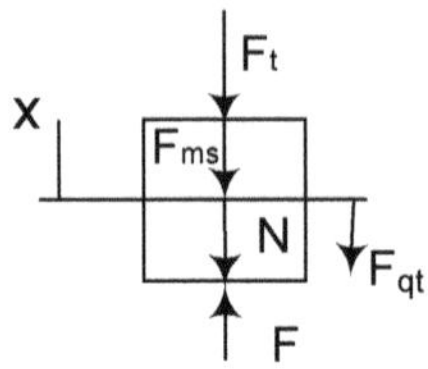

Fig. 4. Forces acting on the bridge bearing

$$M\ddot{x} = P_1\mathrm{A}_1 - P_2\mathrm{A}_2 - F_\mathrm{r} - Mg - F_t \tag{6}$$

with M is the mass of the bearing part acting on one piston. F_t is the feedback force from the upper base of the test exerted by the compressed ball bearing, and it depends on the deforming type of each bridge bearings. F is the force generated by the cylinder pressures and it is calculated by $F = P_1A_1 - P_2A_2$. F_r is the friction force in the hydraulic cylinder, and it is described by a dynamic friction model developed by Tran et al. [6] as follows:

$$F_r = \sigma_0 + \sigma_1\frac{dz}{dt} + \sigma_2\left(v + T\frac{dv}{dt}\right) \tag{7}$$

$$\frac{dz}{dt} = v - \frac{\sigma_0 z}{g(v)}|v| \tag{8}$$

$$g(v) = F_c + [(1-h)F_s - F_c]e^{-(v/v_s)^n} \tag{9}$$

$$\frac{dh}{dt} = \frac{1}{\tau_h}(h_{ss} - h) \tag{10}$$

$$\tau_h = \begin{cases} \tau_{hp} & (v \neq 0,\ h \leq h_{ss}) \\ \tau_{hn} & (v \neq 0,\ h > h_{ss}) \\ \tau_{h0} & (v = 0) \end{cases} \tag{11}$$

$$h_{ss} = \begin{cases} K_f|v|^{2/3} & (|v| \leq |v_b|) \\ K_f|v_b|^{2/3} & (|v| > |v_b|) \end{cases} \tag{12}$$

$$K_f = (1 - F_c/F_s)|v_b|^{-2/3} \tag{13}$$

in which σ_0, σ_1, σ_2, z, and T are the bristle stiffness, the micro-viscosity friction coefficient, the viscous friction coefficient, the average bristle displacement, and the constant the number of times for viscous friction dynamics, respectively. $g(v)$, F_c, h, F_s, v_s, and n are the Tribeck function, the Coulomb friction force, the lubricating film thickness, the static friction force, the Stribeck velocity, and the Tribeck exponent, respectively. h_{ss}, τ_{hp}, τ_{hn}, τ_{h0}, K_f, and v_b are the lubrication film thickness at the steady state, the time constant for the acceleration, the time constant for deceleration and the time constant for rest periods, the proportional constant, the velocity at which the lubricating film thickness is changed, respectively.

3 Results and Discussion

The mathematical model developed for the hydraulic system is evaluated in this section by comparing simulation results with experimental ones on the compressive force - bearing displacement characteristics under three different types of the bridge bearing. Simulation was performed by MATLAB/Simulink software. Dormand-Prince numerical solution method was used in simulation to solve differential and integral equations. The parameters of the system used in the simulation were given as follows: $C_d = 0.009$, $W = 8.10^{-6}, k_v = 0.0096$ m/V, $V_{10} = V_{20} = 0.008$ m^3, $A_1 = 0.125$ m^3, $A_2 = 0.0456$ m^3, and $\rho = 862$ kg/m^3. Static parameters of the friction model were given by: $F_s = 850$ N, $F_c = 180$ N, $v_s = 0.0125$ m/s, $n = 0.6$, $\sigma_2 = 320$ Ns/m for positive velocity case and $F_s = 435$ N, $F_c = 105$ N, $v_s = 0.015$ m/s, $n = 0.9$, $\sigma_2 = 350$ Ns/m for negative velocity case. Dynamic parameters of the friction model were given by: $\sigma_0 = 5.3 \times 10^6$ N/m, $\sigma_1 = 0.2$ N/m, $\tau_{hp} = 0.2$ s, $\tau_{\mathrm{hn}} = 1.5$ s, $\tau_{h0} = 40$, and T = 0.25.

Figure 5 presents the compressive force-displacement characteristics measured and ones simulated by the system model in steady-states for three following types of pot bearings: KAWAKIN-B6 in Fig. 5a, OVM GPZ (2009) 2SX-PBD in Fig. 5b, and OVM GPZ (2009) 4SX-HTB in Fig. 5c under the same operating conditions. In simulation characteristics, the deformations of the three bridge bearings were considered as the elastic deformation and the elastic forces of the bridge bearings were quadratic ratio functions to the bearing deformation, i.e. $F_t = kx^2$. Values of k were given by 1×10^{12}, 5.2×10^{11} and 2.5×10^{11} N/m for the KAWAKIN-B6, OVM GPZ (2009) 2SX-PBD and OVM GPZ (2009) 4SX-HTB bridge bearings, respectively. Figure 5 shows that the characteristics obtained by simulation match well to the experimental ones for all the three bridge bearings. For the KAWAKIN-B6 bearing in Fig. 5a, the average error between the simulation and experimental results is 2.53% and the maximum error is 6.3% at the minimum compression force point of 100 kN. For the OVM GPZ (2009) 2SX-PBD bearing, the mean error is 2.94% and the maximum error is 9.4% at the maximum force point of 187.5 kN. In Fig. 5c, the average error is 3.33% and the maximum error is 6.7% at the maximum compression force point of 375 kN. Comparison for the testing equipment for bridge bearings with a capacity of 6400 tons is relatively accurate.

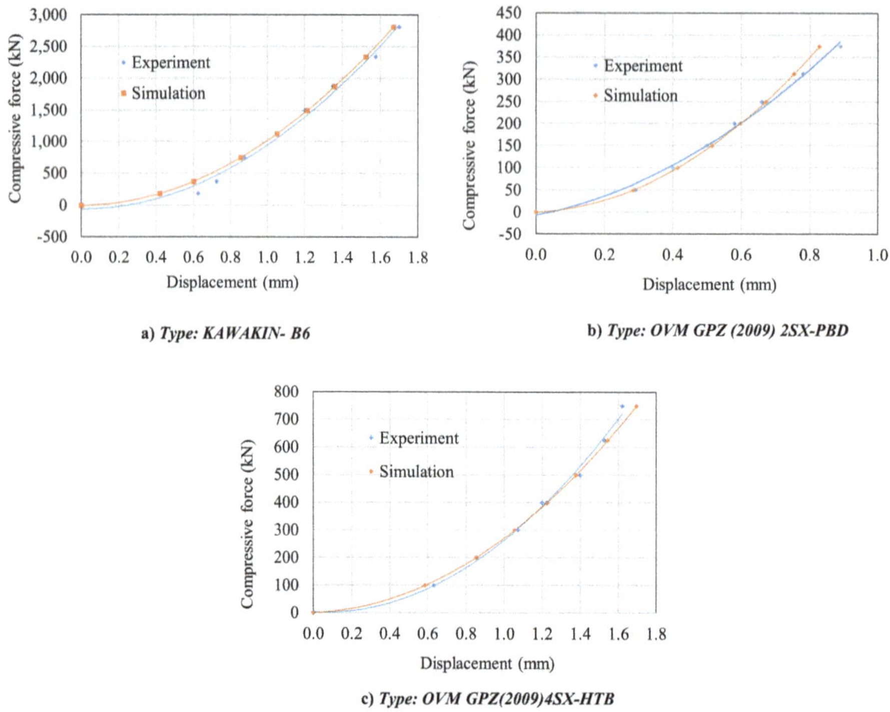

Fig. 5. Comparison between simulation results and experimental ones on compressive force - deformation characteristics for different types of bridge bearing

4 Conclusion

The paper proposed a mathematical model for the hydraulic system of a testing equipment for bridge bearings with a capacity of 6400 tons in Vietnam. The model was built basing on the flow equation, the fluid continuity equation, the Newton's II law, and on the deformation type of the bridge bearing. The system model was evaluated by comparing the compressive force-displacement characteristics measured with ones simulated by the system model in steady-states for different types of the bridge bearing. The results showed that the simulation characteristics matched well with the experimental ones with the maximum average error of 3.33%. This proves that the developed system model is suitable. The development of a suitable mathematical model for the hydraulic system of this testing equipment is an important step to study the optimal operating conditions of the hydraulic system and so that it will be helpful to enhance the measuring accuracy of the testing equipment. Effects of the operating parameters of the hydraulic system on the measuring accuracy of the testing equipment will be carried out in future work.

Funding. This research was funded by the Hanoi University of Science and Technology (HUST) under project number T2020-PC-014.

References

1. Pham, D.N., Nguyen, V.T., Thai, H.P.: Testing equipment for bridge bearings with a capacity of 6400 tons designed domestically and applied in practice for the transportation industry. J. Transp. **8**, 87–90 (2016)
2. Pillow bridge bearings – testing methods, Ministry of Science and Technology (2014). Vietnam Standard 10269:2014
3. Pillow bridge bearings – technical requirements, Ministry of Science and Technology (2014). Vietnam Standard 10268:2014
4. Watton, J.: Fundamentals of Fluid Power Control. Cambridge University Press, Cambridge (2009)
5. Merrit, H.E.: Hydraulic Control Systems. Wiley, New York (1967)
6. Tran, X.B., Khaing, W.H., Endo, H., Yanada, H.: Effect of friction model on simulation of hydraulic actuator. Proc. Inst. Mech. Eng. Part I. J. Syst. Control Eng. **228**, 1–9 (2014)

Optimizing Harmonic Filters in Static Var Compensator of Electric Arc Furnace to Enhance Power System Quality

Le Nhat Hoang Tran(✉) and Thi Ai Lành Nguyen

The University of Danang, University of Technology and Education,
48 Cao Thang, Danang, Vietnam
tlnhoang@ute.udn.vn

Abstract. Electric arc furnaces (EAF) are widely used in the steel production industry. They are unbalanced, nonlinear and time varying loads. These loads cause voltage flickers harmonics and perturbations of the grid, which decrease energy efficiency and power system quality in the system. In order to reduce these impacts, a Static Var Compensator (SVC) is use and connected in parallel with EAF. However, passive elements in SVC occupy a large volume and have a high cost. Therefore, in the paper, the author optimizes passive elements to compensate reactive power and to decrease the harmonics in SVC. Firstly, a studied model is built to study the impact of the arc furnace on the power system. It is implemented in Matlab/Simulink. The EAF arc model based on a linearization of the arc voltage-current characteristic. Secondly, an SVC is added in parallel with EAF in studied model to compensation of the reactive power and to filter some harmonics. Finally, the filters in SVC are designed by optimization to reduce their volume, to filter dominant harmonics (5th and 7th), to improve energy efficiency of EAF and to enhance power system quality.

Keywords: Electric arc furnace · Improving energy efficiency of EAF · Harmonic filters · Compensating reactive power · Sizing filter by optimization

1 Introduction

Since the early twentieth century, electric arc furnace had been widely used in industry for steel production. It uses heat from the electric arc to achieve a temperature large enough to melt the metal [1]. With its flexibility, the electric arc furnace allows to simplify the steel production process. Moreover, the electric arc furnace is one of the most polluting devices in the system because of the power variation that is caused by its operation. The alternating current (AC) arc furnace can be considered as a source of harmonic currents due to its very strongly nonlinear operation. It can generate even, odd, and even non-integer harmonics [2, 3]. In general, the odd orders of harmonic currents are predominant in this type of power supply. The harmonics are only significant for the orders between 1 and 15 and their amplitude can reach 10% of the fundamental. The harmonics generated by the furnace can strongly distort the system voltage. In addition, with the AC arc furnace, the reactive power is subject to greater variations. Therefore, the implementation of a reactive power compensation device is necessary.

B. T. Long et al. (Eds.): MMMS 2020, LNME, pp. 746–752, 2021.
https://doi.org/10.1007/978-3-030-69610-8_99

To study the properties of arc furnaces, the authors need to build a three-phase arc furnace model. Many studies have been carried out on single-phase circuit model [5], this leads to the deviation in the harmonic composition of the phases due to the metal melting process being unbalanced operation [2]. The electric arc is nonlinear and variable over time. So it is easier to simulate its properties in the time domain than in the frequency domain [4]. The method in time domain based on voltage-current characteristics (VIC). Although, this method requires a lot of computational time [2]. With the development of today's computer science, this drawback has been gradually overcome. In order to compensate reactive power and to filter harmonics generated by EAF, SVC is used and connected in parallel with the load. SVC is an electrical component that allows to quickly provide reactive power to the grid where it is connected. It is one of the components of the FACTS (Flexible AC transmission system) family. Normally, an SVC includes one or more capacitors and reactors, fixed or switched. An SVC can also include harmonic filters. However, the main difficulty is the installation cost of SVC. So, the optimization for reducing volume of the harmonic filters in SVC can decrease these costs.

2 Electric Arc Furnace Model

2.1 Arc Model

In order to simulate arc model, the authors use VIC and it is shown in Fig. 1 [2, 6]. Here, V_{ig} is the arc ignition voltage corresponding to the current i_1, and v_{eg} is the arc extinction voltage corresponding to the current i_2. These voltages are calculated by the arc length during the operation of arc furnace.

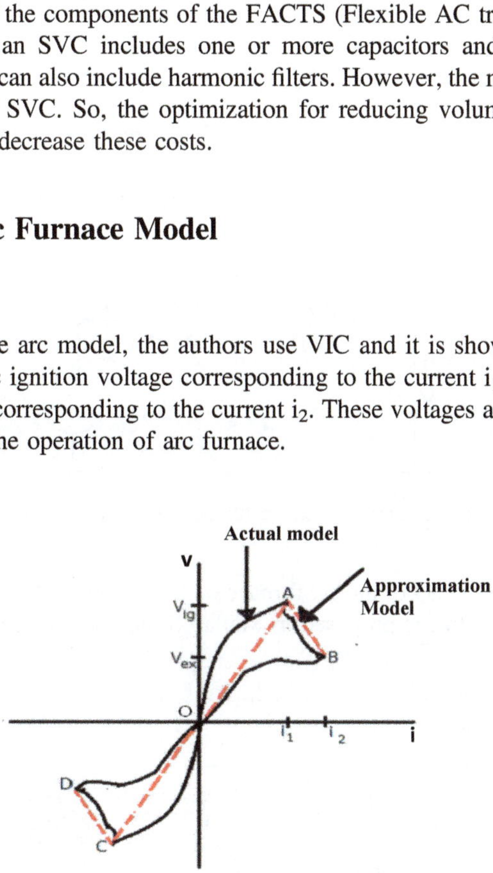

Fig. 1. Actual VIC and approximation model [2]

The actual arc model can be assumed to linearize by two following linear equations corresponding to the segments AB and AC. R_1 and R_2 correspond to the slope of AC and AB. The CD segment is symmetrical with AB through O.

$$v = \begin{cases} R_1 i & \text{with } -i_1 \le i < i_1 \\ R_2 i + v_{ig}\left(1 - \frac{R_2}{R_1}\right) & \text{with } i_1 \le i \le i_2 \\ R_2 i - v_{ig}\left(1 - \frac{R_2}{R_1}\right) & \text{with } -i_2 \le i < -i_1 \end{cases} \tag{1}$$

With

$$i_1 = \frac{v_{ig}}{R_1} \text{ and } i_2 = \frac{v_{ex}}{R_2} - v_{ig}\left(\frac{1}{R_2} - \frac{1}{R_1}\right) \tag{2}$$

2.2 EAF Model

A model of EAF three-phase is shown in Fig. 2, where Zs is the line impedance, PCC bus bar is a common connection point, AF bus bar is the low voltage side of the transformer represented by impedance Zc. These parameters are introduced in [6].

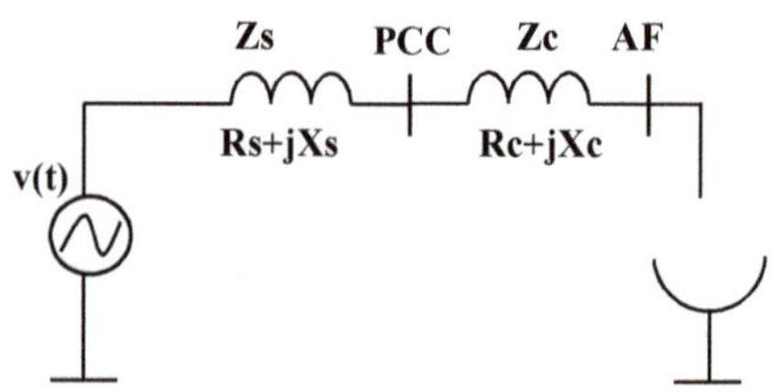

Fig. 2. Three-phase arc furnace model studied [6]

For studying the impact of the arc furnace on the power grid, a model without SVC (Static Var Compensator) is proposed in Matlab Simulink (Fig. 3).

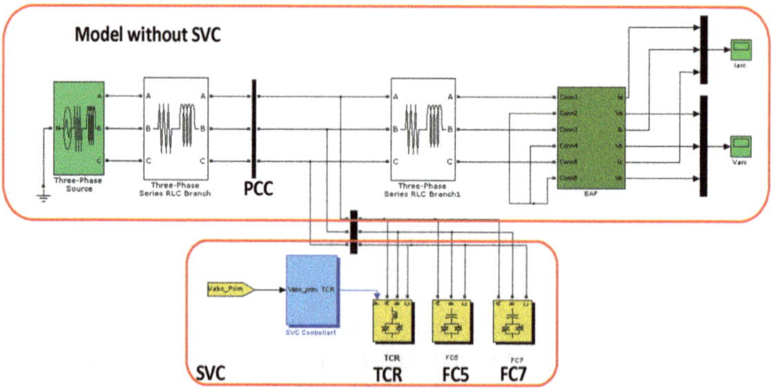

Fig. 3. Studied model in Matlab Simulink

Table 1. THD, 5th rang and 7th rang harmonic of current and of voltage at PCC

Harmonics of current				Harmonics of voltage				Power factor
THD (%)	5th (%)	7th (%)	11th (%)	THD (%)	5th (%)	7th (%)	11th (%)	
6.7	5.89	2.52	0.93	4.54	3.46	2.62	1.39	0.78

By using the model in Matlab Simulink, the total harmonic distortion (THD), 5th rang harmonic (THD5) and 7th rang harmonic (THD7) of current and of voltage of EAF, and power factor at PCC are determined and given by **Table** 1. From these results, the authors constate that, the THD, 5th rang harmonic of both current and voltage, and also power factor at PCC do not satisfy the standard (maximum 5% for THD, 3% for harmonic component, and minimum 0.85% for power factor) [7]. So, it is necessary to filter these harmonics and to compensate the reactive power to enhance the power factor. An SVC is connected at PCC to reduce the disturbances of EAF on grid.

2.3 SVC

SVC is used to control compensated reactive power and to improve the stability of the electrical system. It also helps to regulate the voltage to the point where it is connected. When the voltage decreases, the SVC emit reactive power to the system and vice versa. In this paper, the author studies an SVC connected in parallel with EAF (Fig. 3). It consists a TCR which modulates the reactive compensation power around the fixed value and two passive filters FC5 and FC7 in order to filter 5th harmonic and 7th harmonic respectively. Two filters FC5 and FC7 are the series combination of a capacitor and a reactance according to the chosen harmonic frequency (5th or 7th). These filters have two main functions: compensate reactive power and attenuate the most disturbing harmonics. Taking into account that the cost of passive filters is very high, and these filters can provide reactive power at the fundamental frequency, most applications generally use passive filters. So, it is necessary to size these filters by optimization to reduce installation cost and to improve energy efficiency of EAF.

3 Sizing Harmonic Filters

The objective function based on the total volume of the passive elements which constitute the harmonic filters (the series of a capacitor and a reactance). This gives an idea of decreasing the size and the cost of installing and maintaining these filters.

3.1 Calculation of the Volume of the Reactive Elements

To calculate the volume of the capacitors, the authors use the following equations:

$$C = \frac{\varepsilon_0 \varepsilon_r S}{d} \; and \; d = \frac{V}{E} \quad (3)$$

The volume of the capacitor is determined by:

$$VolC = S.d = \frac{V^2}{\varepsilon_0 \varepsilon_r E^2} C \tag{4}$$

With: S is the armature area (m^2); d is the dielectric thickness (m); V is the voltage across the capacitor (Volt); ε_r is the relative permittivity of the dielectric; E is the dielectric strength (Volt/m).

For harmonic filters in SVC, an air coil is often used. The self-inductance can be calculated from its characteristics.

$$L = \frac{\mu_0 N^2 A}{l} = \mu_0 n^2 A l \left(\text{since} = \frac{N}{l} \right) \tag{5}$$

The volume of the inductance can be approximated by:

$$VolL = Al = \frac{L}{\mu_0 n^2} \tag{6}$$

With: L is the self-inductance of the coil in Henrys (H), μ_0 is the permeability of the vacuum (T.m/A), N is the number of turns, A is the solenoid section (m^2), l is the solenoid length (m), n is the number of turn per unit length.

3.2 Optimization

The objective function is to minimize *Vtotal* (the total volume of two filters in Eq. 7). The limit area of the constraints is given in **Table** 2.

$$Vtotal = V_{filter5} + V_{filter7} = \frac{L_5}{\mu_0 n^2} + \frac{V^2}{\varepsilon_0 \varepsilon_r E^2} C_5 + \frac{L_7}{\mu_0 n^2} + \frac{V^2}{\varepsilon_0 \varepsilon_r E^2} C_7 \tag{7}$$

Table 2. The limit area of constraints

Constraint	Limit area	Constraint	Limit area
THD	[0; 0.05]	THD7	[0; 0.03]
THD5	[0; 0.03]	Power factor	[0.85; 1]

The author uses the Cades tool (developed by G2Elab, France) to size this application by optimization. However, the model, which is dynamic, is built under the Matlab/Simulink software. It is therefore necessary to control the Matlab/Simulink software which contains the simulation model of this application in Matlab/Simulink (Appli.mdl in Fig. 4) to interface it with an Icar component, via a Java function [8, 9]. The whole process of optimization is presented in Fig. 4.

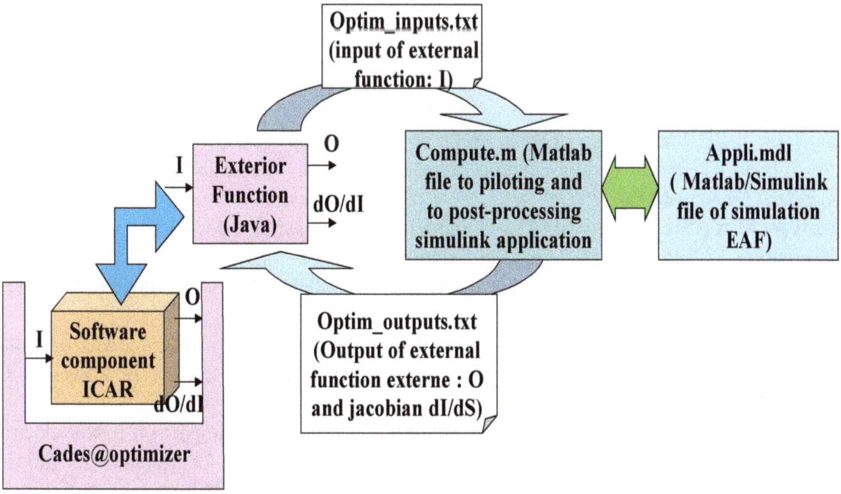

Fig. 4. Piloting Matlab by optimization software to optimize the total volume

The optimization results are given in **Table** 3. The constraints (THD, THD5, THD7, power factor) satisfy the standard. Comparing with the analytical calculation with no sizing, total volume obtained by optimization decreases more than 25%. Furthermore, the power factor increases significantly. This improve energy efficiency of EAF.

Table 3. Sizing results of optimization

Parameter	Value	Parameter	Value	Parameter	Value
C5 (F)	0.049	THD	0.049	V_filtre7 (m^3)	0.206
C7 (F)	0.01	THD5	0.019	V_total (m^3)	1.215
L5 (H)	7.56e-06	THD7	0.015	Power factor	0.865
L7 (H)	1e-06	V_filtre5 (m^3)	1.009		

4 Conclusion

The authors have modeled an electric arc furnace by taking into account the harmonic and power factor problems in Matlab/Simulink. An installation is designed by optimization to solve the problems posed by the pollution generated by this furnace. The optimization result, which is very positive, can use to reduce significantly the volume of the harmonic filters and to improve energy efficiency of EAF. However, the EAF is very dynamic. So, it is necessary to verifies that the optimization remains robust if the operating point of the furnace change, or during the transition phases.

References

1. Four à arc électrique', Wikipédia. Feb. 22, 2018 (2018). Accessed 18 Jul 2018
2. Alonso, M.A.P., Donsion, M.P.: An improved time domain arc furnace model for harmonic analysis. IEEE Trans. Power Deliv. **19**(1), 367–373 (2004)
3. Postiglione, G.: Nouvelle alimentation pour les fours à arc à courant alternatif, Ph.D. thesis, G2Elab, France (2006)
4. Golkar, M.A., Meschi, S.: MATLAB modeling of arc furnace for flicker study. In: IEEE International Conference on Industrial Technology, pp. 1–6 (2008)
5. Zheng, T., Makram, E.B.: An adaptive arc furnace model. IEEE Trans. Power Deliv. **15**(3), 931–939 (2000)
6. Zheng, T., Makram, E.B., Girgis, A.A.: Effect of different arc furnace models on voltage distortion. In: 8th International Conference on Harmonics and Quality of Power, vol. 2, pp. 1079–1085 (1998)
7. Key, T.S., Lai, J.-S.: IEEE and international harmonic standards impact on power electronic equipment design. In: Proceedings of the 23rd International Conference on Industrial Electronics, Control, and Instrumentation, vol. 2, pp. 430–436 (1997)
8. Tran, L.N.H., Gerbaud, L., Retière, N., Nguyen Huu, H.: Use of SQP optimization algorithm to size a multiphysical system: application to an aircraft electrical power channel. In: COMPEL – Int. J. Comput. Math. Electr. Electron. Eng. **37**(2), 661–680 (2018)
9. Gerbaud, L.: Tutorial: ICAr driver of matlab and matlab/simulink applications, using text files to exchange data, Version 1.0 (2011)

Sustainable Machine Design: Metratronics, CAD/CAM/CAE, Maritime Engineering

Machine Vision System for Automated Nuts Classification and Inspection

Hong-Hai Hoang, Thanh-Hung Nguyen(✉), Van-Huong Tran, Tien-Hoang Bui, and Dinh-Manh Nguyen

School of Mechanical Engineering, Hanoi University of Science and Technology, Hanoi, Vietnam
hung.nguyenthanh@hust.edu.vn

Abstract. This paper presents an automated nuts classification and inspection system. The nuts are automated feeding by a mechanical system and their images are captured by a 2-D imaging system each time a nut passing through. The 2-D vision algorithm is developed to classify the nuts and detect the defect on their surfaces. The feasibility and effectiveness of the developed system are verified experimentally. The results indicate that the system can accurately classify the nuts and inspect the defect on them. In the experiments, the accuracy of the classification exceeds 99% and the accuracy of inspection exceeds 92%.

Keywords: Nuts classification and inspection · Deep learning · Machine vision system

1 Introduction

Object recognition is a key problem in image processing. The identification can be performed by a variety of methods such as colors and shapes detection or feature detection and matching. However, today's image processing systems employ the artificial intelligence algorithms for object recognition to achieve human-like accuracy. In the paper, a deep learning technique [1] is applied to classify three kind of nuts.

In the nut products, surface defects are unavoidable. Detecting surface defects promptly plays an important role in order to timely review the technological process. Therefore, all the products must be checked after machining process. The inspection process can achieve the high accuracy by employing the image processing tools. The common defect types on the metal surface are crack, pinhole, spot, corner, edge, and scratch [2]. These defects appear with different shapes, sizes, and colors. From the experiment, the main defects on the surface of nuts are cracks and pinholes.

The methods [3] that can be employed to detect the defect on the nut's surface include the Artificial Neural Networks, Support Vector Machine (SVM), Gabor filter, Classification of principal Component analysis and K-nearest neighbor (PCA, KNN), Fuzzy Cluster Mean (FCM), and Shearlet conversion. Due to the shape of the nuts are complex and contain many edges, the reasonable method can be applied to detect the defect on the nut is artificial neural network [1].

B. T. Long et al. (Eds.): MMMS 2020, LNME, pp. 755–759, 2021.
https://doi.org/10.1007/978-3-030-69610-8_100

The paper is organized as follows. Section 2 presents the algorithm for nuts classification and inspection. The experiments are described in Sect. 3. Finally, the conclusions are given in the Sect. 4.

2 Nuts Classification and Inspection

This section describes in detail the method for nuts classification and inspection. The completed process pipeline is shown in Fig. 1. The process consisted of the following three stages: First step is the object classification. Second is the nuts inspection. These two steps use a deep learning technique to identify the nuts and locate the defects on the surface of each object. Final step is the dimension inspection. This step checks the size of each nut and compare it with the pre-defined values.

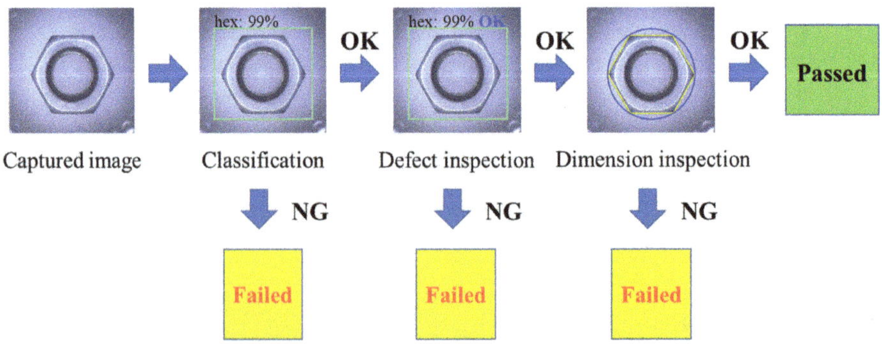

Fig. 1. The vision inspection pipeline.

2.1 Nuts Classification

The nuts are classified based on the YOLOv3 detector [1]. The steps needed are gathering data, labeling, training, and applying the pre-trained model to identify the nuts.

2.1.1 Gathering Data

To generate the dataset for training step, the images are captured with the resolution of 640 × 480 pixels. From the experiment, the identification is stable and accurate if the training dataset is larger than 1,000 images.

2.1.2 Labeling

After collecting the images of different nuts, it is necessary to generate the input parameters that represent the name and position of each nut in an image. These parameters are stored in a text file. Each row in the text file presents a single bounding box of the nut. This can be done with custom script which is written in Python to create .txt files for the collected images in YOLO format.

2.1.3 Training

The annotated images are randomly split into train and test sets in the ratio of 80:20. During the images training, weights of the convolutional neural networks are updated iteratively. By default, the detector's weight is saved for every 100 iterations. Once the training is complete, the generated weights can be used to perform detection. In the experiment, the training can be stopped after the loss has reached below 1.5.

2.1.4 Nuts Classification with Pre-trained Model

After testing the accuracy of the detector with pre-trained model, we can apply it to real time nuts classification. The results of this process are shown in Fig. 2.

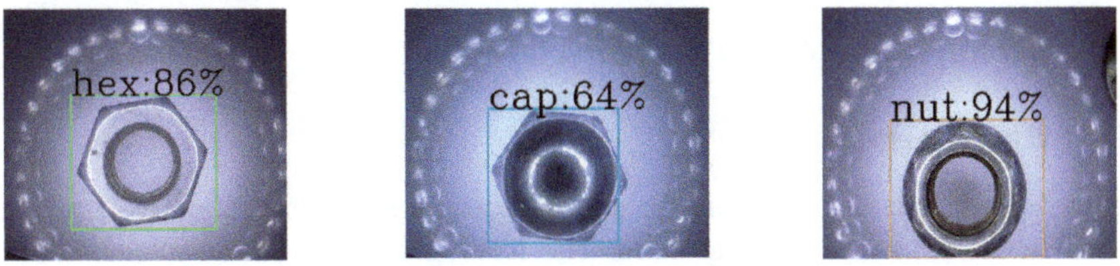

Fig. 2. The results of nuts classification using pre-trained model.

2.2 Nut Inspection

To detect the defects on the surface of a nut, the same procedure as nuts classification is applied. Due to the defects having different shapes, sizes, and colors, the dataset for training must be relatively large. The dataset has collected from both internet and real experiment system. The results of the algorithm after 5250 iterations and the loss of 1.5 are shown in the Fig. 3.

Fig. 3. Defect detection on the nuts' surface.

2.3 Dimension Inspection

After nuts classification process, the ROI region that contains a nut is extracted. Then, the Canny detector [4] is applied to detect the edges on this image. The contours on the image can be found by employing the algorithm [5]. Due to lighting condition, the contour that represents the boundary of the nut is frequently broken. Therefore, the Convex Hull algorithm [6] is applied to joint all outer points of the contours. The boundary of the nut is the largest closed contour. To check the size of the nut, the radius of the minimal enclosing circle of the boundary is compared with a pre-defined value. The minimal circle is estimated using an iterative algorithm [7]. The pipeline of the dimension inspection is shown in the Fig. 4.

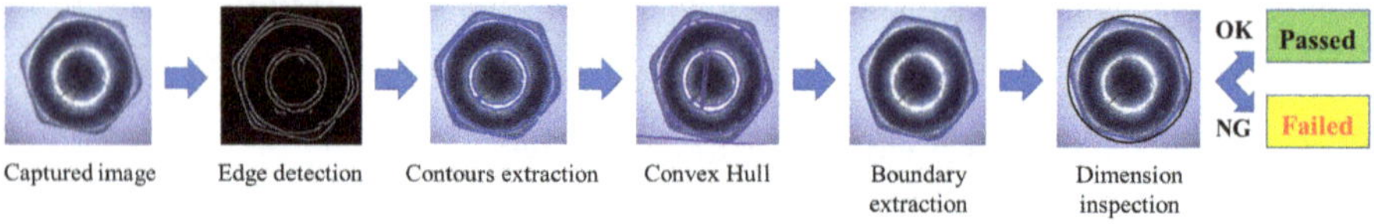

Fig. 4. Dimension inspection pipeline.

3 Experimental Results and Discussions

3.1 System Overview

Figure 5 shows the vision system for nuts classification and inspection, which is used to implement the developed algorithm. The feeding system selects, guides and positions the parts to the rotary indexing table. The vision system includes a ring light and a camera with a resolution of 1280 x 720 pixels. The rotary indexing table moves the objects to the one of the four fixed positions corresponding to the hex, cap, flange, and NG bins. The developed classification algorithm processes the image captured by camera to classify the parts into three categories: hex, cap and flange nuts. After that, the inspection algorithm is applied to detect the defect on the surface of each part.

Fig. 5. The machine vision system for nut classification and inspection: (a) The CAD model; (b) The real system used in the experiment; and (c) The interface of the software.

3.2 Experimental Results

In the experiments, three kind of nuts are used to implement the designed algorithm. The results of the algorithm are shown in Fig. 6. The algorithm has been tested for 150 samples. It can classify the nuts with the accuracy of 100% and detect the defect on the surface of the nuts with the accuracy of 92%.

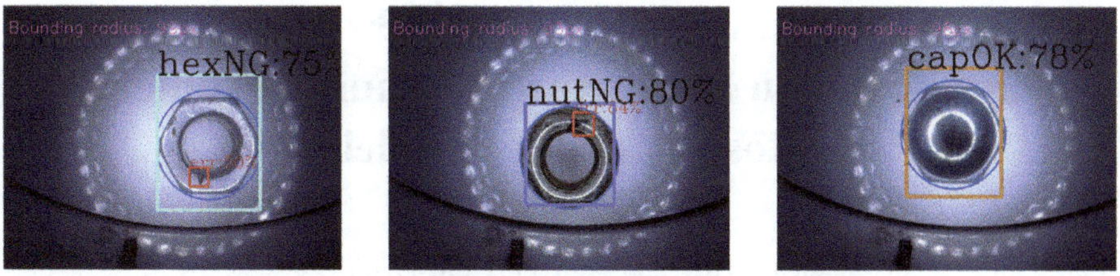

Fig. 6. The results of the nuts classification and inspection.

4 Conclusion

In this paper, we proposed a machine vision system for nut classification and inspection. The algorithm was developed based on the traditional image processing and deep learning techniques. The nut can be identified and the defect on its surface can be detected by applying the YOLO v3. The dimension of the nut is checked by using the traditional image processing methods. The experimental results indicate that the proposed method can accurately classify the nuts. The limitation of the proposed system is that the inspection process still needs to improve the accuracy before applying to the real system. This can be done by optimizing the lighting condition and collecting more data for training process.

References

1. Redmon, J., Farhadi, A.: YOLOv3: An incremental improvement, arXiv:1804.02767
2. Tikhe, C., Chitode, J.S.: Metal surface inspection for defect detection and classification using Gabor filter. IJIRSET **3**, 13702–13709 (2014)
3. Pishyar, E., Emadi, M.: Investigation of different algorithms for surface defects of steel sheet for quality. Int. J. Comput. Appl. **149**, 33–37 (2016)
4. Canny, J.: A computational approach to edge detection. IEEE Trans. Pattern Anal. Mach. Intell. **6**, 679–698 (1986)
5. Suzuki, S., Be, K.A.: Topological structural analysis of digitized binary images by border following. Comput. Vis. Graph. Image Process. **30**, 32–46 (1985)
6. Sklansky, J.: Finding the convex hull of a simple polygon. Pattern Recogn. Lett. **1**, 79–83 (1982)
7. Bradski, G.: The OpenCV Library, Dr. Dobb's J. Softw. Tools **4**, 120–125 (2000)

Stabilization of Constrain Equations in Simulation of Forward Dynamics Problem for a Rostock Delta Parallel Robot

Vu Le Huy and Nguyen Dinh Dzung(✉)

Faculty of Mechanical Engineering and Mechatronics,
Phenikaa University, Hanoi, Vietnam
{huy.vule,dung.nguyendinh}@phenikaa-uni.edu.vn

Abstract. The popular problems in robot are the inverse dynamics and forward dynamics, where the forward dynamics is necessary for simulation in the robot control methods, but it is quite complex for parallel robots. In this paper, the forward dynamics problem of the Rostock Delta robot with specific parameters is simulated numerically by using the system of ordinary differential equations (ODEs). ODEs with redundant coordinates is developed from the system of the differential-algebraic equations (DAEs), which is usually used to formulate the motion of the parallel robots. The simulation is performed with and without of using the Baumgarte stabilization method. The errors of the constrained equations in the two cases are then compared to see the effect of the stabilization method in the forward dynamics problem of the Rostock Delta robot.

Keywords: Forward dynamics · Parallel robot · Rostock delta robot · Differential-algebraic equations · Baumgarte stabilization

1 Introduction

Up to now, parallel robots have been being used widely in industrial applications owing to their light compact structure, high stiffness, and accuracy. Delta robot is a kind of parallel robots invented by Reymond Clavel in the early '80 s [1], which consists of three arms connected to universal joints at the base. Delta robot is one of the most successful parallel robots with thousands of versions created and improved around the world, where their applications are mainly in the food, pharmaceutical factories, and medical field. A kind of Delta robot was improved in 2012 by Johann Rocholl [2] and named as Rostock Delta robot, where the three rotational joints of Delta robot are replaced with the three translational joints. Rostock Delta robot is now used mainly in 3D printing technology.

In the viewpoint of mechanism, parallel robots are the closed-loop multibody systems. Solving the dynamic problem is necessary to design and improve the control quality of parallel robot. There are a lot of literatures on the theory and calculation method for dynamic problems of serial robot [3–7]. In contrast, for the parallel robots, inverse dynamics problems are usually concerned and deeply studied in many literatures such as the literatures [8, 9] using Lagrange-Euler equation and the literatures [5, 10, 11]

B. T. Long et al. (Eds.): MMMS 2020, LNME, pp. 760–768, 2021.
https://doi.org/10.1007/978-3-030-69610-8_101

using virtual work principle. Staicu, *et al.* [12, 13] proposed recursive matrix method to calculate the dynamics problem for parallel robot. Abdellatif, *et al.* [14] used Lagrange equation in multiplier form to solve the inverse dynamics problems for a specific parallel robot. In the recent study of Corves, *et al.* [15] solved numerically the inverse dynamics problem of the Delta robot by using the methods of Lagrange equations with multipliers, virtual work principle, and Newton-Euler equations, where the obtained results of the three methods were similar. The method using Lagrange equations with multipliers gave the calculation time to be smallest.

In the numerical simulation diagram of the robot control method, forward dynamics problem usually need to be solved to calculate the real position and velocity of the robot [16]. However, there are rarely studies mentioning this problem. The equations describing the movement of the parallel robot usually are the differential-algebraic equations (DAEs). The methods to solve DAEs are divided into two groups [17]: The first group is that the methods of transforming the systems to ordinary differential equations, and then using the numerical algorithm to solve them. The second group is that using the numerical methods to solve directly the algebraic differential systems. A method of the first group is used in this paper to calculate the forward dynamic problem of the Rostock Delta parallel robot.

2 Forward Dynamics Problem of Parallel Robot

The requirement of the forward dynamics problem of parallel robot is to find the law of manipulator motion $x = x(t) \in \mathbb{R}^m$ when the equations of motion of the robot as well as the driving moments or forces $\tau_a \in \mathbb{R}^{n_a}$ are known. The movement equations of the parallel is a kind of the motion of multibody system with holonomic constraints of hold, stop and ideal. The differential-algebraic system of equations obtained by using the Lagrange equations with multipliers [17] to describe the movement equations of the parallel robot is written as

$$\boldsymbol{M}(\boldsymbol{s})\ddot{\boldsymbol{s}} + \boldsymbol{C}(\boldsymbol{s},\dot{\boldsymbol{s}})\dot{\boldsymbol{s}} + \boldsymbol{g}(\boldsymbol{s}) = \tau(t) - \boldsymbol{\Phi}_s^T(\boldsymbol{s})\lambda \tag{1}$$

$$\boldsymbol{f}(\boldsymbol{s},t) = \boldsymbol{0} \tag{2}$$

where $\boldsymbol{s} = \left[s_1, s_2, \ldots, s_n\right]^T$ is the vector of redundant generalized coordinates, $\boldsymbol{M}\left(\boldsymbol{s}\right)$ is the generalized mass matrix, $\boldsymbol{\tau}\left(t\right)$ the vector of the generalized forces corresponding to the generalized non-conservative external forces, $\boldsymbol{\lambda} = \left[\lambda_1, \lambda_2, \ldots, \lambda_r\right]^T$ is the vector of Lagrangian multipliers, $\boldsymbol{f} = \left[f_1, f_2, \ldots, f_r\right]^T = 0$ is the constrain conditions, $\boldsymbol{\Phi}_s$ is the Jacobian matrix of the vector $\boldsymbol{f}$ with the size $r \times f$. Matrix $\boldsymbol{C}(\boldsymbol{s},\dot{\boldsymbol{s}})$ is the Coriolis/Centripetal matrix determined by the following formula [17]:

$$\boldsymbol{C}\left(\boldsymbol{s},\dot{\boldsymbol{s}}\right) = \frac{\partial \boldsymbol{M}}{\partial \boldsymbol{s}}\left(\boldsymbol{E}_n \otimes \dot{\boldsymbol{s}}\right) - \frac{1}{2}\left(\frac{\partial \boldsymbol{M}}{\partial \boldsymbol{s}}\left(\dot{\boldsymbol{s}} \otimes \boldsymbol{E}_n\right)\right)^T \tag{3}$$

Vector $\boldsymbol{g}(\boldsymbol{s})$ is determined by the potential energy of the system

$$\boldsymbol{g}(\boldsymbol{s}) = \left(\frac{\partial \Pi}{\partial \boldsymbol{s}}\right)^T \tag{4}$$

The following symbol is used to write conveniently as

$$\boldsymbol{p}_1(\boldsymbol{s}, \dot{\boldsymbol{s}}, t) = \tau(t) - \boldsymbol{C}(\boldsymbol{s}, \dot{\boldsymbol{s}})\dot{\boldsymbol{s}} - \boldsymbol{g}(\boldsymbol{s}) \tag{5}$$

Then Eqs. (1) and (2) are written as

$$\boldsymbol{M}(\boldsymbol{s})\ddot{\boldsymbol{s}} + \boldsymbol{\Phi}_s^T(\boldsymbol{s})\lambda = \boldsymbol{p}_1(\boldsymbol{s}, \dot{\boldsymbol{s}}, t) \tag{6}$$

$$\boldsymbol{f}(\boldsymbol{s}, t) = \boldsymbol{0} \tag{7}$$

The constrained equations are differentiated two times as used in the multibody dynamic documents and obtained as

$$\dot{\boldsymbol{f}} = \frac{\partial \boldsymbol{f}}{\partial \boldsymbol{s}}\dot{\boldsymbol{s}} + \frac{\partial \boldsymbol{f}}{\partial t} = \Phi_s\dot{\boldsymbol{s}} + \boldsymbol{f}_t = \boldsymbol{0} \tag{8}$$

$$\ddot{\boldsymbol{f}} = \Phi_s\ddot{\boldsymbol{s}} + \dot{\Phi}_s\dot{\boldsymbol{s}} + \dot{\boldsymbol{f}}_t = \boldsymbol{0} \Rightarrow \Phi_s\ddot{\boldsymbol{s}} = -(\dot{\Phi}_s\dot{\boldsymbol{s}} + \dot{\boldsymbol{f}}_t) = \boldsymbol{p}_2 \tag{9}$$

Equations (6) and (9) can be rewritten in the form of matrix as

$$\begin{bmatrix} \boldsymbol{M} & \Phi_s^T \\ \Phi_s & \boldsymbol{0} \end{bmatrix} \begin{bmatrix} \ddot{\boldsymbol{s}} \\ \lambda \end{bmatrix} = \begin{bmatrix} \boldsymbol{p}_1 \\ \boldsymbol{p}_2 \end{bmatrix} \tag{10}$$

The equivalence of Eqs. (6) and (7) to the equation system (10) was proved in the literature [17]. This paper uses an algorithm to transform the differential-algebraic system to the system of ordinary differential equations, where the number of the unknowns is equal to the number of the equations, by removing Lagrangian multipliers. Because the generalized mass matrix $\boldsymbol{M}(\boldsymbol{s}, t)$ is the symmetric and positive definite matrix, there is existence of the inverse matrix $\boldsymbol{M}^{-1}(\boldsymbol{s}, t)$. It is inferred from Eq. (6) that

$$\ddot{\boldsymbol{s}} = \boldsymbol{M}^{-1} \;\; \boldsymbol{s}, \mathrm{t} \left[\boldsymbol{p}_1 \;\; \boldsymbol{s}, \dot{\boldsymbol{s}}, t \; - \boldsymbol{\Phi}_s^T \;\; \boldsymbol{s}, t \;\; \boldsymbol{\lambda}\right] \tag{11}$$

By substituting $\ddot{\mathbf{s}}$ from Eq. (11) into Eq. (9), it is obtained as

$$\boldsymbol{\Phi} \;\; \boldsymbol{s}, \mathrm{t} \;\; \boldsymbol{M}^{-1} \;\; \boldsymbol{s}, \mathrm{t} \left[\boldsymbol{p}_1 \;\; \boldsymbol{s}, \dot{\boldsymbol{s}}, t \; - \boldsymbol{\Phi}_s^T \;\; \boldsymbol{s}, t \;\; \boldsymbol{\lambda}\right] = \boldsymbol{p}_2 \;\; \boldsymbol{s}, \dot{\boldsymbol{s}}, t \tag{12}$$

Equation (12) is rewritten as

$$\left[\boldsymbol{\Phi}\ \boldsymbol{s},\mathrm{t}\ \boldsymbol{M}^{-1}\ \boldsymbol{s},\mathrm{t}\ \boldsymbol{\Phi}_s^T\ \boldsymbol{s},t\right]\boldsymbol{\lambda}=\boldsymbol{\Phi}\ \boldsymbol{s},\mathrm{t}\ \boldsymbol{M}^{-1}\ \boldsymbol{s},\mathrm{t}\ \boldsymbol{p}_1\ \boldsymbol{s},\dot{\boldsymbol{s}},t\ -\boldsymbol{p}_2\ \boldsymbol{s},\dot{\boldsymbol{s}},t \tag{13}$$

Equation to calculate λ is inferred from Eq. (13) as

$$\boldsymbol{\lambda}=\left[\boldsymbol{\Phi}\ \boldsymbol{s},\mathrm{t}\ \boldsymbol{M}^{-1}\ \boldsymbol{s},\mathrm{t}\ \boldsymbol{\Phi}_s^T\ \boldsymbol{s},t\right]^{-1}\left[\boldsymbol{\Phi}\ \boldsymbol{s},\mathrm{t}\ \boldsymbol{M}^{-1}\ \boldsymbol{s},\mathrm{t}\ \boldsymbol{p}_1\ \boldsymbol{s},\dot{\boldsymbol{s}},t\ -\boldsymbol{p}_2\ \boldsymbol{s},\dot{\boldsymbol{s}},t\right] \tag{14}$$

By substituting λ from Eq. (14) into Eq. (11), a differential equation system is obtained as

$$\begin{aligned}\ddot{s}=\boldsymbol{M}^{-1}(s,t)\Big\{\mathrm{E}-\Phi_s^T(s,t)\left[\Phi(s,t)\boldsymbol{M}^{-1}(s,t)\Phi_s^T(s,t)\right]^{-1}\Phi(s,t)\Big\}\boldsymbol{p}_1(s,\dot{s},t)\\ -\boldsymbol{M}^{-1}(s,t)\Phi_s^T(s,t)\left[\Phi(s,t)\boldsymbol{M}^{-1}(s,t)\Phi_s^T(s,t)\right]^{-1}\boldsymbol{p}_2(s,\dot{s},t)\end{aligned} \tag{15}$$

where $\boldsymbol{E}$ is a identity matrix with the size of n. Equation (15) is an ordinary differential system with the unknown to be $\boldsymbol{s}$. By using numerical analysis diagram, $\boldsymbol{s}$ and $\dot{\boldsymbol{s}}$ should be obtained. However, it is noted that the numerical analysis method applied with Eq. (15) should usually meet the integral error. After each step of integration, because of the calculation error, the values of $s_i, \dot{s}_i$ do not satisfy the position and velocity constrain equations, which lead to the difference

$$\boldsymbol{f}(\boldsymbol{s}_k,t_k)\neq\boldsymbol{0},k=1,2,\ldots,\ \dot{\boldsymbol{f}}(\boldsymbol{s}_k,\dot{\boldsymbol{s}}_k,t_k)\neq\boldsymbol{0},k=1,2,\ldots \tag{16}$$

In the Baumgarte method [18], instead of solving the equation $\ddot{\boldsymbol{f}}(s,\dot{s},t)=\boldsymbol{0}$, the following equation is considered

$$\ddot{\boldsymbol{f}}(s,\dot{s},\mathrm{t})+2\alpha\dot{\boldsymbol{f}}(s,\dot{s},t)+\beta^2\boldsymbol{f}(s,t)=\boldsymbol{0} \tag{17}$$

in which the factors are selected to satisfy the condition $0<\alpha\leq\beta$. The components $2\alpha\dot{\boldsymbol{f}}$ and $\beta^2\boldsymbol{f}$ play the role of the control terms. By using Eq. (18), the cumulative error should be gradually or completely eliminated in the integration process. Therefore, instead of solving the system

$$\begin{aligned}\boldsymbol{M}(s,\mathrm{t})\ddot{s}+\Phi_s^T(s,t)\lambda-\boldsymbol{p}_1(s,\dot{s},t)=\boldsymbol{0},\\ \ddot{\boldsymbol{f}}(s,\dot{s},t)=\boldsymbol{0}\end{aligned} \tag{18}$$

the system of the following equations is going to be solved

$$\begin{aligned}\boldsymbol{M}(s,\mathrm{t})\ddot{s}+\Phi_s^T(s,t)\lambda-\boldsymbol{p}_1(s,\dot{s},t)=\boldsymbol{0}\\ \ddot{\boldsymbol{f}}(s,\dot{s},t)+2\alpha\dot{\boldsymbol{f}}(s,\dot{s},t)+\beta^2\boldsymbol{f}(s,t)=\boldsymbol{0}\end{aligned} \tag{19}$$

The above Eqs. (21) can be rewritten in matrix form as

$$\begin{bmatrix} \boldsymbol{M}(\boldsymbol{s},\mathrm{t}) & \boldsymbol{\Phi}^T(\boldsymbol{s},t) \\ \boldsymbol{\Phi}(\boldsymbol{s},t) & \boldsymbol{0}_{r\times r} \end{bmatrix} \begin{bmatrix} \ddot{\boldsymbol{s}} \\ \boldsymbol{\lambda} \end{bmatrix} = \begin{bmatrix} \boldsymbol{p}_1(\boldsymbol{s},\dot{\boldsymbol{s}},t) \\ \hat{\boldsymbol{p}}_2(\boldsymbol{s},\dot{\boldsymbol{s}},t) \end{bmatrix} \quad (20)$$

where

$$\begin{aligned} \hat{\boldsymbol{p}}_2(\boldsymbol{s},\dot{\boldsymbol{s}},t) &= \boldsymbol{p}_2(\boldsymbol{s},\dot{\boldsymbol{s}},t) - 2\alpha[\boldsymbol{\Phi}(\boldsymbol{s},t)\dot{\boldsymbol{s}} + \boldsymbol{f}_t] - \beta^2 \boldsymbol{f}(\boldsymbol{s},t) \\ &= -\left[\frac{\partial \boldsymbol{\Phi}}{\partial \boldsymbol{s}}(\boldsymbol{E}_n \otimes \dot{\boldsymbol{s}}) + \frac{\partial \boldsymbol{\Phi}}{\partial t} + 2\alpha \boldsymbol{\Phi}(\boldsymbol{s},t)\right]\dot{\boldsymbol{s}} - 2\alpha \boldsymbol{f}_t - \boldsymbol{f}_{tt} \end{aligned} \quad (21)$$

3 Simulation of Dynamics Problem for the Rostock Delta Parallel Robot

Figure 1 shows the 3D model of a Rostock Delta robot, where the three active links corresponding to the three translational drive joints (Prismatic – P). These joints are connected to the mobile platform by the parallelogram links, and therefore the mo-bile platform is always horizontal. The two ends of each parallelogram link are con-sidered as connecting to the active joint and the mobile platform by Cardan joint (or Universal - U) or Spherical joint (S). Therefore, this robot is also notated to be Delta 3PUS.

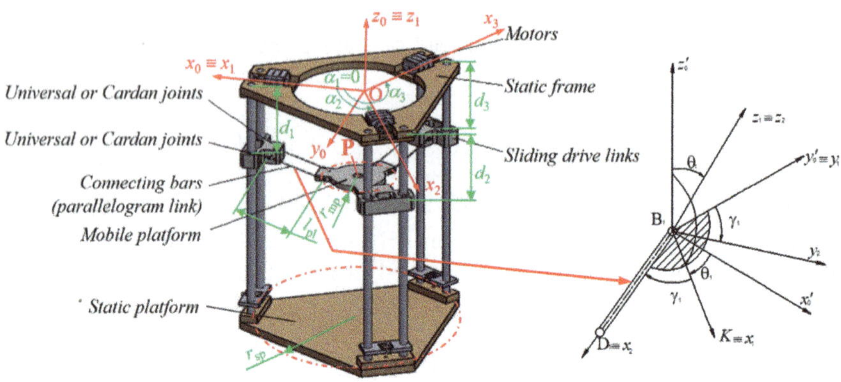

Fig. 1. The model of Rostock Delta robot

The position of the drive links of the robot is determined by the parameters d_1, d_2, d_3, the position of the links B_iD_i (i = 1, 2, 3) is determined by the angles θ_i, γ_i, and the position of the mobile platform is determined by of its center point indicated with the coordinates x_P, y_P, z_P. Therefore, redundant generalized coordinates used to establish the movement equations of the robot are selected as

$$s = \begin{bmatrix} d_1 & d_2 & d_3 & \theta_1 & \theta_2 & \theta_3 & \gamma_1 & \gamma_2 & \gamma_3 & x_P & z_P & z_P \end{bmatrix}^T \quad (22)$$

By applying the matrix form of Lagrangian multiplier equation [4, 6], the movement equations of the robot are established, where their explicit form is presented in the [19]. The established movement equations have the matrix form as presented by Eqs. (1) (2) with 12 differential and 9 algebraic equations. In those equations, $\mathbf{M}(s) \in \mathbb{R}^{12\times12}$ is the generalized mass matrix, $\mathbf{C}(\mathbf{s}, \dot{s}) \in \mathbb{R}^{12\times12}$ is the centrifugal/Coriolis matrix, $\mathbf{g}(\mathbf{s}) \in \mathbb{R}^{12}$ is the gravitational force vector, $\Phi_s(\mathbf{s}) \in \mathbb{R}^{9\times12}$ is the Jacobian matrix, $\lambda \in \mathbb{R}^9$ is the Lagrange multiplier vector, $\tau \in \mathbb{R}^{12}$ is the vector of generalized forces, $\boldsymbol{f}(\boldsymbol{s}) \in \mathbb{R}^9$ is the vector of constraint equations.

Applying the calculation method for the forward dynamics problem presented above, the simulation of that problem was performed here for a specific robot with its parameters as

$$L = 0.242;\ R = 0.16;\ r = 0.03\,(m);\ \alpha_1 = 0;\alpha_2 = 2\pi/3;\alpha_3 = 4\pi/3; m_1 = 0.12;$$
$$m_2 = 2 \times 0.15;\ m_P = 0.2(kg);\ \boldsymbol{I}_1 = diag(0,\ m_1\mathrm{L}^2/12, m_1\mathrm{L}^2/12);$$

Input of the simulation was given by the driven active forces of the three motors as shown in Fig. 2, where the abscissa and the ordinate present the time and the active fores, respectively. The driven active forces were selected to be cyclic in 1 s of period. Then the trajectory of the active joint coordinates d_1, d_2 and d_3, which shows the position of the sliding drive links, was calculated and plotted in Fig. 3, where the ordinate shows the values of d_1, d_2 and d_3. It is obviously seen that since the driven active forces are cyclic, the active joint coordinates also change cyclically. The movement of the center point P of the mobile platform was then calculated with the position presented by x_p, y_p and z_p shown in Fig. 4. The plot shows that the mobile platform move in the horizontal plane at z of −0.45 (m).

In the simulation, the forward dynamics problem was solved with and without using stabilization technique. Figure 5 presents the errors of the constrained equations without using stabilization technique. It shows that the absolute value of the error of the constrained equation f_3 tends to increase up to 3.4×10^{-3} (m) after 1 (s). This error will be accumulated by the time in the movement process, and therefore it is not suitable to control the robot. By using Baumgarte stabilization technique, the errors of the constrained equations were well obtained as shown in Fig. 6. The error of the constrained equation f_3 was fluctuant with the maximum value to be 7×10^{-3} (m), which is much smaller than the case of without using stabilization technique. It means that the error will not be accumulated by the time when using Baumgarte stabilization technique, and therefore the calculation method mentioned above with the Baumgarte stabilization for the forward dynamics problem technique is suitable to control Rostock Delta robot.

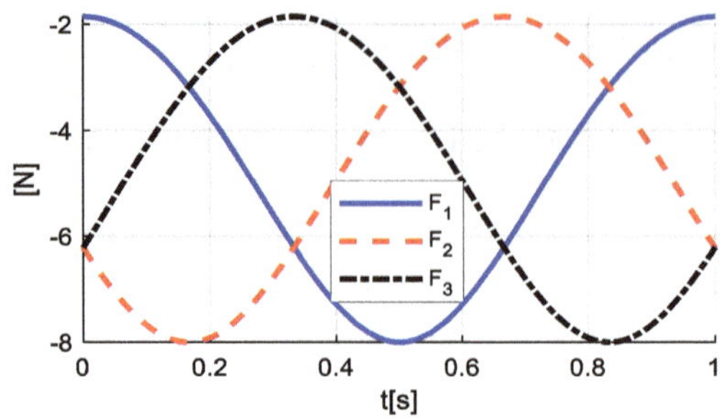

Fig. 2. Driven active forces

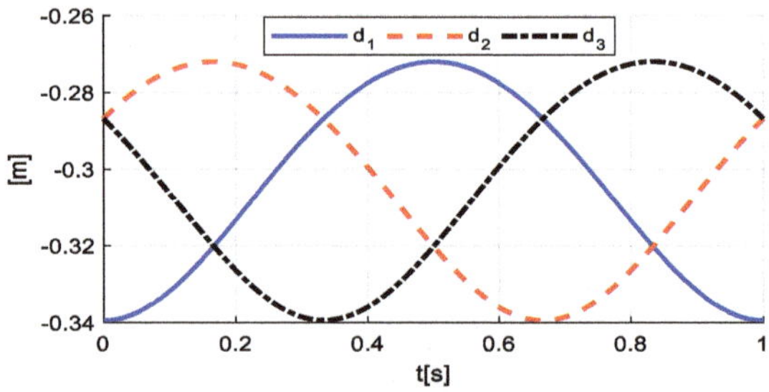

Fig. 3. Trajectory of the active joint coordinates d_1, d_2, d_3

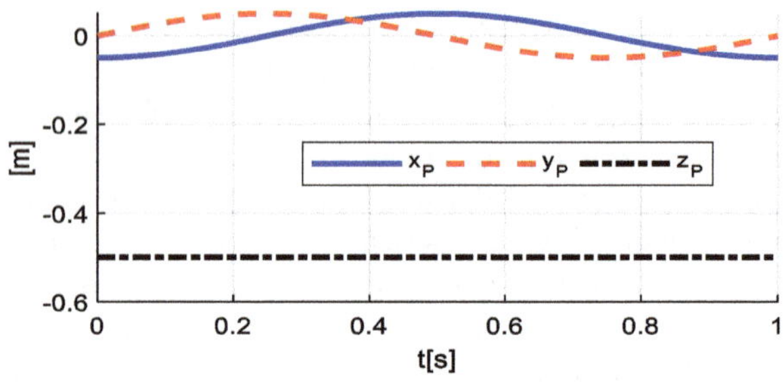

Fig. 4. Movement trajectory of the center point P (x_P, y_P, z_P) of the mobile platform.

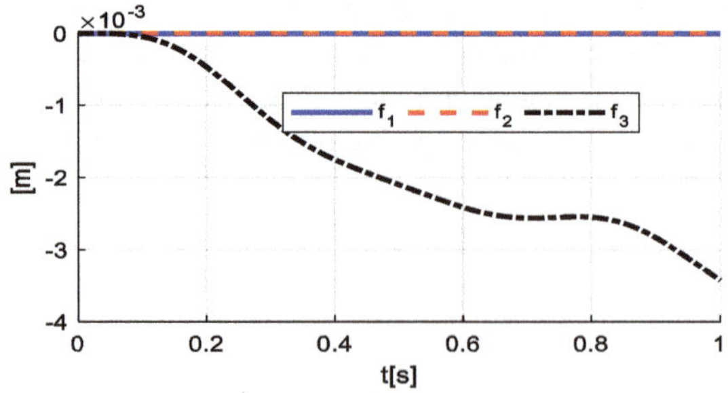

Fig. 5. Error of constrained equations f_1, f_2, f_3 without using stabilization technique

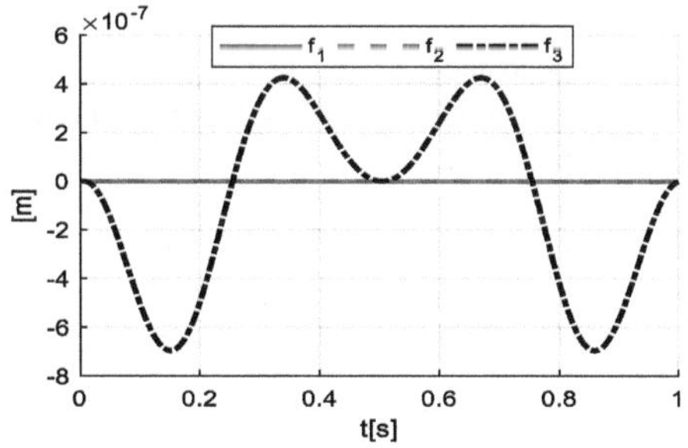

Fig. 6. Error of equations f_1, f_2, f_3 when using Baumgarte stabilization technique

4 Conclusion

This paper used the method of transforming the system of differential – algebraic equations into systems of ordinary differential equations. The method of Baumgarte to stabilize the equations after each step of calculation for the system of differential equations was also applied. The Baumgarte stabilization technique was seen to be efficient in reduce the error of the constrained equations. The forward dynamics problem was then simulated numerically with a specific case of the Rostock Delta robot. The results showed that the method implemented with the Baumgarte stabilization technique in this study gave the maximum error about 7×10^{-7} (m). This is an important problem in simulation of the control methods for parallel robots.

References

1. Merlet, J.P.: Parallel Robots. Springer, Heidelberg (2006)
2. Bell, C.: 3D Printing with Delta Printers. Apress, New York (2015)
3. Wittenburg J.: Dynamics of Multibody Systems 2. edn., Springer, Heidelberg (2008)
4. Shabana, A.A.: Dynamics of Multibody Systems 3 edn., Cambridge University Press, New York (2005)
5. Tsai, L.W.: Robot Analysis the Mechanics of Serial and Parallel Manipulators. Wiley, Hoboken (1999)
6. Jalon, J.Gd., Bayo, E.: Kinematic and Dynamic Simulation of Multibody Systems. Springer, New York (1994)
7. Song, M.W., Vidyasagar, M.: Dynamic and Control of Robot Manipulators. Wiley, Hoboken (1989)
8. Lebret, G., Liu, G.K., Lewis, F.L.: Dynamic analysis and control of a Stewart platform manipulator. J. Robot. Syst. **10**, 629–655 (1993)
9. Liu, M..J, Li, C.X., Li, C.N.: editors. Dynamics analysis of the Gough-Stewart platform manipulator. In: Proceedings of the IEEE International Conference on Robotics and Automation, Cincinnati, USA (1990)
10. Codourey, A., Burdet, E., (eds).: A body oriented method for finding a linear form of the dynamic equations of fully parallel robot. In: IEEE Conference on Robotics and Automation, Albuquerque, New Mexico, U.S (1997)

11. Tsai, L.W.: Solving the inverse dynamics of a Stewart-Gough manipulator by the principle of virtual work. Journal of mechanical design **122,** 3–9 (2000)
12. Staicu, S.: Recursive modelling in dynamics of agile wrist spherical parallel robot. Robotics and Computer-Intergrated Manufacturing 25, 409–416 (2009)
13. Staicu, S., Zhang, D., Rugescu, R.: Dynamic modelling of a 3-DOF parallel manipulator using recursive matrix relations. Robotica **4**, 125–30 (2000)
14. Abdellatif, H., Heimann, B.: Computational efficient inverse dynamics of 6-DOF fully parallel manipulators by using the Lagrangian formalism. Mechanism and Machine Theory **44**, 192–207 (2009)
15. Brinker, J., Corvers, B., Wahle, M., eds.: A comparative study of inverse dynamics based on Clavel's Delta robot The 14th IFToMM World Congress, Taipei, Taiwan (2015)
16. Taghirad, H.D.: Parallel Robots: Mechanics and Control. CRC Press, Boca Raton (2013)
17. Khang, N.V.: Dynamics of Multibody Systems. Scientific and Technical Publishing, (in Vietnamese) (2017)
18. Marques, F., Souto, A., Flores, P.: On the constraints violation in forward dynamics of multibody systems. Multibody Syst. Dyn. **39**, 385–419 (2017)
19. Khang, N.V., Hoang, N.Q., Dung, N.D., Quyen, N.V.: Model-based control of a 3-PRS spatial parallel robot in the space of redundant coordinates. J. Sci. Technol. Tech. Univ. **112**, 49–53 (2016)

Optimizing Magnification Ratio for the Flexible Hinge Displacement Amplifier Mechanism Design

Ngoc-Thai Huynh[1], Tien V. T. Nguyen[2,3], Nguyen Thanh Tam[1], and Quoc-Manh Nguyen[4](✉)

[1] Faculty of Automotive Technology, Industrial University of Ho Chi Minh City, No. 12 Nguyen Van Bao Street, Go Vap District, Ho Chi Minh City 70000, Vietnam
[2] Faculty of Mechanical Technology, Industrial University of Ho Chi Minh City, No. 12 Nguyen Van Bao Street, Go Vap District, Ho Chi Minh City 70000, Vietnam
[3] National Kaohsiung University of Science and Technology, No. 415, Jiangong Road, Sanmin District, Kaohsiung 80778, Taiwan
[4] Faculty of Mechanical Engineering, Hung Yen University of Technology and Education, Khoai Chau, Hung Yen Province, Vietnam
nguyenquocmanh@utehy.edu.vn

Abstract. This investigation analyzed the influence of design variables (*L, y, t,* and *x)* of a new flexible hinge displacement amplification mechanism on (i) magnification ratio, (ii) maximum principal stress, and (iii) the first modal shape frequency. Besides, we proposed a grey relational analysis applied to the three parameters of the mechanism. To achieve the high working performance of the compliant mechanism, the motion scope works long term with high frequency and subjects to force tension with the torque. The finite element analysis (FEA) outcomes indicate that design variables have significantly affected the magnification ratio of the mechanism. The optimal outcomes of displacement, maximum principal stress, and the first modal shape frequency obtain at 0.6727 mm, 32.123 MPa, and 85.383 Hz, respectively while the predicted outcomes of displacement, maximum principal stress, and the first modal shape frequency achieve 0.7095 mm, 31.2077 MPa, and 82.6448 Hz. These results are in good agreement with the error of displacement, stress, and frequency with 5.19%, 1.92%, and 3.11%, respectively. The magnification ratio of the proposed mechanism obtained 67.27 times.

Keywords: Magnification ratio · Flexible hinge · Displacement amplifier mechanism · Grey relational analysis · Computational analysis

1 Introduction

For several recent decades, numerous sorts of flexure-hinge [1–6] have been studied and enhanced for supplanting the conventional systems. Qi et al [7] utilized the kinematics accompanying the elastic beam literature to decide displacement-amplification-ratio (DAR) for the bridge type compliant mechanism flexure hinge (BTCMFH). Liu and Yan [8] proposed an explanatory technique utilizing EBT to

B. T. Long et al. (Eds.): MMMS 2020, LNME, pp. 769–778, 2021.
https://doi.org/10.1007/978-3-030-69610-8_102

decide the impacts of outer power on the DAR and confirmed by FEM [9–11] in ANSYS [12]. Choi et al [13] proposed a completely BTCMFH with input focus and dispersion power to determine the DAR [14–19] which was checked by analysis and past exploration. Lai et al. [20] utilized two L-shape switch type systems and one BTCMFH to wipe out the moments and reactions. The method of stiffness matrix was applied to figure the DAR and checked by FEM along with experiments. To guarantee high unbending nature, huge amplification, high-exactness following, and high-precision situating, Chen et al. [21] presented an exotic relocation intensifier system utilizing FH subject to stacked pressure and bending rather than stacked pressure and bending when redirected. The kinetostatic model [22–24] was applied to determine DAR and the amplification ratio value obtained 40. The results of the model also were confirmed by both FEM and experiment. The rest of this article remains the main body of this investigation as follows. Section 2 introduces the structural design and numerical model. The optimization process will be demonstrated in more detail in Sect. 3. Section 4 illustrates the results and discussion for the study. Section 5 will close the article with several main conclusions and further studies.

2 Modeling of the Mechanisms and Finite Element Model

2.1 Modeling of the Mechanisms

The structure of this designed amplifier mechanism was constructed from two shape lever of L-type and the bridge-type. The total dimension of the compliant mechanism is designed by 140 mm × 132 mm × 10 mm. Figure 1 describes the design variables and the other dimension of this mechanism.

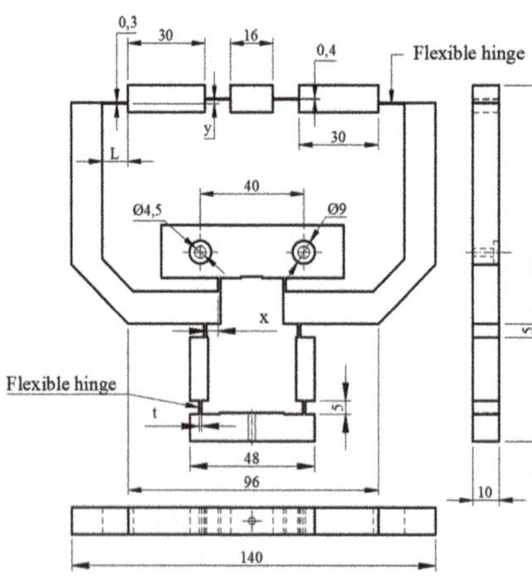

Fig. 1. 2D sketch of the proposed flexible hinge

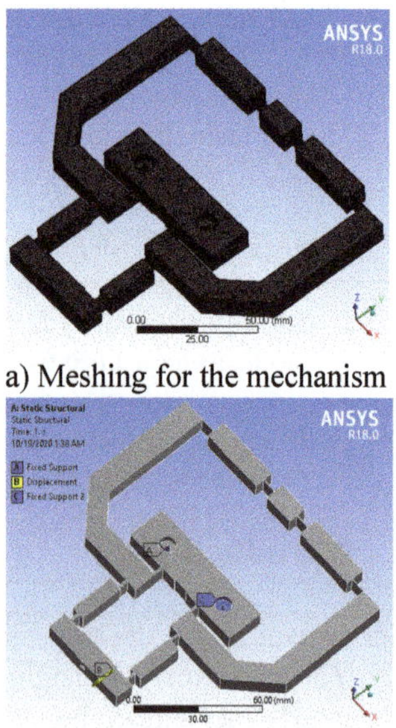

a) Meshing for the mechanism

b) Boundary condition

Fig. 2. Numerical model

2.2 Numerical Methodology

We selected material for the flexible hinge structure model as presented in Table 1. First of all, the structural compliant model was designed via Autodesk Inventor. Next, its geometry could be imported into the Static Structural Module which is one of the tools of the ANSYS commercial software. After that, we determined the mechanical properties for the compliant flexure model such as AL-7075- aluminum. The structural model mesh was demonstrated in Fig. 2(a) having its nodes of 130524 and its elements of 73352. The fixed support was applied by using the two holes in the center of the mechanism and as presented in Fig. 2(b). The A surface with the input displacement of 10^{-2} mm at the input body and the B surface according to y-axis direction are shown in Fig. 2b.

Table 1. Mechanical properties for the compliant flexure

Material	Young's modulus (GPa)	Poisson's ratio	Yield (MPa)
AL-7075	72	0.33	503

3 Grey Relational Analysis

The grey relational analysis is a method for comparison of the changes in mechanisms undergoing analysis to estimate the vitality of the design variables. The GRA is an approach which applied for discretizing sequences. GRA was carried out as follows. The method was presented in [2, 3, 25–28]

Normalization: Re-define each sequence from 0 to 1 as below.
Larger the better - the mathematical formula:

$$D_i^* = \frac{D_i^{(0)}(k) - \min D_i^0(k)}{\max D_i^{(0)}(k) - \min D_i^{(0)}(k)} \tag{1}$$

Smaller the better:

$$D_i^* = \frac{\max D_i^{(0)}(k) - D_i^0(k)}{\max D_i^{(0)}(k) - \min D_i^{(0)}(k)} \tag{2}$$

GRC (grey relational coefficient) (γ) is a quantity approach in the grey relational space. GRC is requested initially before determining the grey-relational-grade, GRG. The deviation is determined as below:

$$\Delta_{0i} = \left\| D_0^*(k) - D_i^*(k) \right\| \tag{3}$$

$$\Delta_{\min} = \max_{\forall j \in i} \min_{\forall k} \left\| D_0^*(k) - D_j^*(k) \right\| \tag{4}$$

$$\Delta_{\max} = \max_{\forall j \in i} \max_{\forall k} \left\| D_0^*(k) - D_j^*(k) \right\| \tag{5}$$

$$\text{Compute grey relational coefficient (GRC)}: \gamma_i(k) = \frac{\Delta_{\min} + \xi \Delta_{\max}}{\Delta_{0i} + \xi.\Delta_{\max}} \tag{6}$$

Here, $\xi \in [0,1]$ stands for the coefficient of distinguishing. We could option the value at 0.5.

Weight decision:

$$\omega_e(x) = x.e^{(1-x)} + (1-x)e^x - 1 \tag{7}$$

where $\omega_e(x)$ stands for the map-function in the entropy measurement and the function achieves max-value while x = 0.5, $e^{0.5} - 1 = 648.7 \times 10^{-3}$ and map-value belonging on the range of [0,1] obtain:

$$w \equiv \frac{1}{(e^{0.5} - 1)} \sum_{i=1}^{m} \omega_e(x), \in = \{\gamma_i(1), \ \gamma_i(2), \ \ldots, \ \gamma_i(n)\}. \text{Note that,} \ i = 1, 2, \ldots, n \tag{8}$$

Decide the entire grey relational coefficient

$$D_j \equiv \sum_{i=1}^{m} \gamma_i(j), j = 1, 2, \ldots, n \tag{9}$$

Estimate the normalized coefficient

$$k = \frac{1}{(e^{0.5} - 1) \times m} = \frac{1}{0.6487 \times m} \tag{10}$$

Define entropy value

$$e_j = k \sum_{i=1}^{m} \omega_e\left(\frac{\gamma_i(j)}{D_j}\right), j = 1, 2, \ldots, n, \ \text{here} \ \omega_e(x) \tag{11}$$

uses Eq. (10)

$$\text{Compute the sum of entropy}: E = \sum_{j=1}^{n} e_j \tag{12}$$

Determine the weight

$$\omega_j = \frac{1}{n - E} \cdot \frac{[1 - e_j]}{\sum_{j=1}^{n} \frac{1}{n-E} \cdot [1 - e_j]}, \ here, j = 1, 2, \ldots, n. \quad (13)$$

GRG (ψ_i) is the average value of GRC and is written as follows:

$\psi_i = \sum_{k=1}^{n} \omega_k \gamma_i(k)$, where n is the quantity of experiment.

4 Results and Discussion

4.1 Simulation Design

In this investigation, four length-dimensions were selected as the four design variables with changed dimensions and as presented in Table 2. It can be seen that the variable *L* has two levels of 8 mm and 10 mm. The variable *y* has three levels (1.2, 2.4, and 3.6 mm) and the variable *t* has the levels (including 1, 1.5, and 2 mm). Besides, the variable *x* has three levels (4.5, 5.5, and 6.5 mm). The orthogonal arrays and FEM outcomes were obtained from Minitab 18.0 and ANSYS as shown in Table 3. The displacement, maximum principal stress, and the first modal shape frequency outcomes were utilized to select the one combination parameter which has maximum displacement, minimum stress, and maximum frequency by employing the grey relational analysis. This method is one of the multi-objective optimization methods.

Table 2. Variables and design levels

Parameters		[Unit]	Level(s)		
			1	2	3
Flexible hinge length	L	mm	8	10	12
Vertical dimension between 2 flexible hinges	y	mm	1.2	2.4	3.6
The flexible hinge thickness	t	mm	1	1.5	2
Horizontal dimension between 2 flexible hinges	x	mm	4.5	5.0	5.5

Table 3. Orthogonal arrays, FEM results

Ord.	*L*	*y*	*t*	*x*	DI (mm)	ST (MPa)	Fre (Hz)
1	8	1.2	1	4.5	0.6104	34.013	47.422
2	8	1.2	1.5	5	0.6932	34.183	50.604
3	8	1.2	2	5.5	0.6583	32.625	52.93
4	8	2.4	1	5	0.8249	39.56	56.798
5	8	2.4	1.5	5.5	0.7681	36.847	63.887
6	8	2.4	2	4.5	0.7355	35.321	64.263
7	8	3.6	1	5.5	0.7779	36.824	70.865
8	8	3.6	1.5	4.5	0.7749	36.403	71.620
9	8	3.6	2	5	0.6483	32.602	81.521
10	10	1.2	1	4.5	0.6825	30.395	46.72
11	10	1.2	1.5	5	0.775	32.639	50.326
12	10	1.2	2	5.5	0.7243	30.745	52.978
13	10	2.4	1	5	0.8869	36.439	57.417
14	10	2.4	1.5	5.5	0.8048	33.279	65.125
15	10	2.4	2	4.5	0.7694	32.479	65.537
16	10	3.6	1	5.5	0.7977	32.152	72.590
17	10	3.6	1.5	4.5	0.7933	32.174	73.397
18	10	3.6	2	5	0.6527	35.101	83.591
19	12	1.2	1	4.5	0.7823	34.262	46.941
20	12	1.2	1.5	5	0.833	31.651	50.969
21	12	1.2	2	5.5	0.769	31.466	53.869
22	12	2.4	1	5	0.9226	33.193	58.756
23	12	2.4	1.5	5.5	0.8199	29.633	66.853
24	12	2.4	2	4.5	0.7829	31.094	67.298
25	12	3.6	1	5.5	0.7991	28.275	74.516
26	12	3.6	1.5	4.5	0.7935	31.267	75.351
27	12	3.6	2	5	0.6727	32.123	85.294

4.2 Results of Grey Relational Analysis

It is clear to see that, in Table 4, the values $D_i^*(1)$, $D_i^*(2)$, $D_i^*(3)$ were obtained by substitute displacement, frequency, and stress as presented in Table 3 into Eq. (3.1) and Eq. (3.2), respectively. The values of $\Delta_{oi}(1)$, $\Delta_{oi}(2)$, $\Delta_{oi}(3)$ were obtained by Eq. (3.3). Therein $D_0^*(k) = 1$, $D_i^*(k)$ *is* $D_i^*(1)$, $D_i^*(2)$, $D_i^*(3)$ for *DI, ST,* and *Fre,* respectively. The $\gamma_i(1)$, $\gamma_i(2)$ $\gamma_i(3)$ values in Table 4 were determined by Eq. (3.6). The GRG values (ψ_i) were calculated by Eq. (3.14). The graph of GRG was painted in Fig. 3 with the GRG values as listed in Table 4. The maximum GRG value achieved at 0.7107 and the minimum value of GRG obtained at 0.3688. The results of the comparison of the predicted and optimal values (see Fig. 4, 5 and 6) of the optimization methods are in good agreement together. Table 5 indicates that the predicted values of GRG, DI, ST, and Fre are 0.655, 0.7095 mm, 31.5077 MPa, and 82.6448 Hz. The optimal values of

Table 4. The greater and the lower is the better of DI and ST, the greater is the better of frequency and the deviation error of DI, ST and Fre, GRC and GRG

Ord.	$D_i^*(1)$	$D_i^*(2)$	$D_i^*(3)$	Δoi(1)	Δoi(2)	Δoi(3)	$\gamma_i(1)$	$\gamma_i(2)$	$\gamma_i(3)$	ψ_i
1	0.000	0.492	0.018	1.000	0.509	0.982	0.333	0.496	0.337	0.377
2	0.265	0.474	0.101	0.735	0.524	0.899	0.405	0.489	0.357	0.404
3	0.153	0.615	0.161	0.846	0.386	0.839	0.371	0.565	0.373	0.423
4	0.687	0.000	0.261	0.313	1.000	0.739	0.615	0.333	0.404	0.437
5	0.505	0.240	0.445	0.495	0.759	0.555	0.503	0.397	0.474	0.444
6	0.410	0.376	0.455	0.599	0.624	0.545	0.455	0.445	0.478	0.445
7	0.537	0.242	0.626	0.463	0.758	0.374	0.519	0.398	0.572	0.481
8	0.52	0.280	0.646	0.473	0.720	0.355	0.514	0.410	0.585	0.488
9	0.121	0.617	0.902	0.879	0.383	0.098	0.363	0.566	0.836	0.571
10	0.340	0.812	0.000	0.660	0.188	1.000	0.431	0.727	0.333	0.486
11	0.527	0.613	0.093	0.473	0.387	0.906	0.514	0.564	0.355	0.463
12	0.365	0.781	0.162	0.635	0.219	0.838	0.441	0.696	0.374	0.488
13	0.886	0.277	0.277	0.114	0.723	0.723	0.814	0.409	0.409	0.527
14	0.623	0.556	0.477	0.377	0.443	0.523	0.570	0.53	0.489	0.513
15	0.509	0.627	0.488	0.491	0.373	0.512	0.505	0.573	0.494	0.508
16	0.600	0.656	0.671	0.400	0.344	0.329	0.555	0.593	0.603	0.566
17	0.586	0.654	0.692	0.414	0.346	0.308	0.547	0.591	0.619	0.568
18	0.135	0.395	0.956	0.865	0.605	0.044	0.366	0.453	0.919	0.562
19	0.551	0.469	0.006	0.449	0.531	0.994	0.527	0.485	0.335	0.435
20	0.713	0.710	0.110	0.287	0.299	0.889	0.635	0.626	0.360	0.523
21	0.508	0.717	0.185	0.492	0.283	0.815	0.504	0.639	0.380	0.492
22	1.000	0.564	0.312	0.000	0.436	0.688	1.000	0.534	0.421	0.631
23	0.671	0.880	0.522	0.329	0.120	0.478	0.603	0.806	0.511	0.620
24	0.553	0.750	0.533	0.448	0.250	0.467	0.528	0.667	0.517	0.553
25	0.604	1.000	0.721	0.396	0.000	0.279	0.558	1.000	0.642	0.711
26	0.587	0.735	0.742	0.414	0.265	0.258	0.547	0.654	0.660	0.601
27	0.194	0.659	1.000	0.806	0.341	0.000	0.383	0.595	1.000	0.639

Table 5. Comparison between the predicted and optimal values

	Combination optimization	GRG	DI (mm)	ST (MPa)	Fre (Hz)
Predicted	L3y3t3x2	0.655	0.7095	31.5077	82.6448
Optimal		0.640	0.6727	32.123	85.383
Error %		2.29	5.19	1.92	3.11

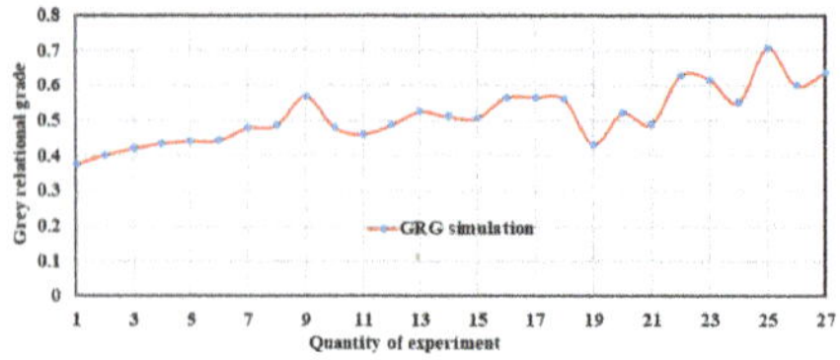

Fig. 3. Plot of GRG

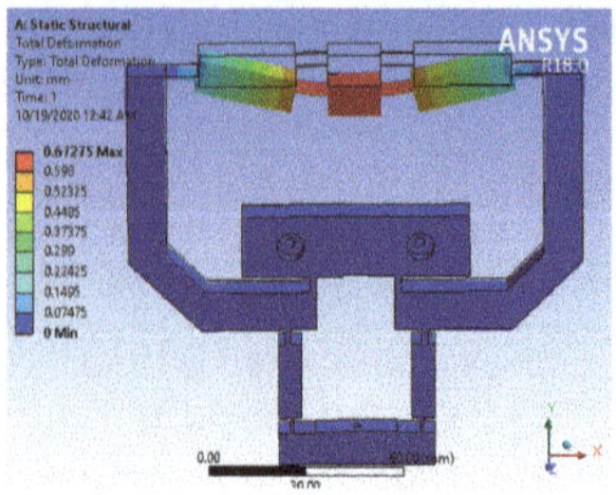

Fig. 4. Optimal DI spectrum

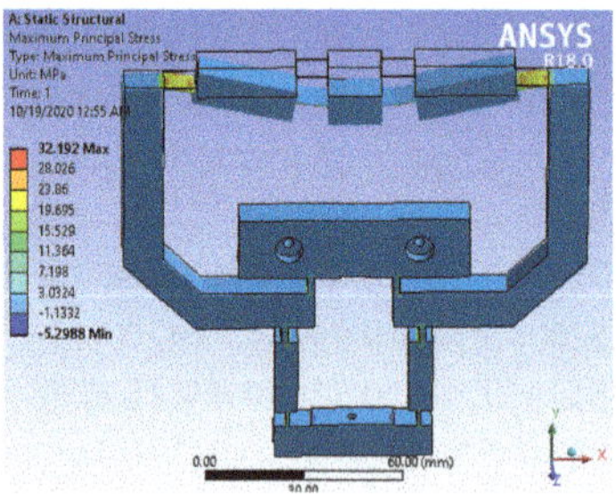

Fig. 5. Optimal results Stress spectrum

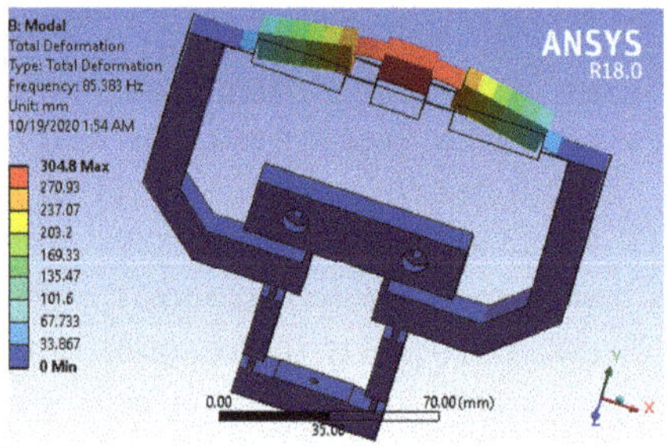

Fig. 6. The optimal Frequency spectrum

GRG, DI, ST, and Fre are 0.64, 0.6727 mm, 32.123 MPa, and 85.383 Hz, respectively. The errors of GRG, DI, ST, and Fre are 2.29%, 5.19%, 1.92%, and 3.2%.

5 Main Conclusions

The effects of the four variables (*L, y, t, x*) on displacement, maximum principal stress, and frequency of the first modal shape were analyzed using FEA. We found that the outcomes identified that these variables have strongly influenced three outputs. The simulation, optimization, prediction results are in good agreement with and better than that of the previous studies. The optimal approach was permitted to utilize optimization analysis variables of the compliant mechanism because the outcomes' errors of this investigation are no larger than 6%.

References

1. Min, K.S., Choi, W.C., Song, S.H., Hwang, E.J.: Static and dynamic analysis of a nanopositioning flexure-hinge stage with a flexible lever mechanism. Proc. Inst. Mech. Eng. Part B-J. Eng. Manuf. **219**, 447–454 (2005)

2. Vu, N.C., Huynh, N.T., Huang, S.C.: Optimization the first frequency modal shape of a tensural displacement amplifier employing flexure hinge by using Taguchi Method. J. Phys: Conf. Ser. **1303**, 012016 (2019)
3. Wang, C.N., Truong, K.P., Huynh, N.T., Hoang, L.Q.: Nhat Optimization effects of design parameter on the first frequency modal of a Bridge-type compliant mechanism flexure hinge by using the Taguchi method. J. Phys. Conf. Ser. **1303**, 012063 (2019)
4. Zhu, J.X., Hao, G.B.: Design and test of a compact compliant gripper using the Scott-Russell mechanism. Arch. Civ. Mech. Eng. **20**, 27 (2020)
5. Tran, N.T., Chau, N.L., Dao, T.P.: A hybrid computational method of desirability, fuzzy logic, ANFIS, and LAPO algorithm for multiobjective optimization design of scott russell compliant mechanism. Math. Prob. Eng. **2020** (2020)
6. Yan, B., Liang, L.: A novel fiber bragg grating accelerometer based on parallel double flexible hinges. IEEE Sensors J. **20**, 4713–4718 (2020)
7. Qi, K.-Q., Xiang, Y., Fang, C., Zhang, Y., Yu, C.-S.: Analysis of the displacement amplification ratio of bridge-type mechanism. Mech. Mach. Theory **87**, 45–56 (2015)
8. Liu, P., Yan, P.: A new model analysis approach for bridge-type amplifiers supporting nano-stage design. Mech. Mach. Theory **99**, 176–188 (2016)
9. Chau, N.L., Dao, T.P., Nguyen, V.T.T.: An efficient hybrid approach of finite element method, artificial neural network-based multiobjective genetic algorithm for computational optimization of a linear compliant mechanism of nanoindentation tester. Math. Prob. Eng. **2018** (2018)
10. Ovchinnikov, M.Y., Tkachev, S.S., Roldugin, D.S., Nuralieva, A.B, Mashtakov, Y.V.: Angular motion equations for a satellite with hinged flexible solar panel. Acta Astronautica, **128**, 534–539 (2016)
11. Petkovic, D., Shamshirband, S., Iqbal, J., Anuar, N.B., Pavlovic, N.D., Kiah, L.: Adaptive neuro-fuzzy prediction of grasping object weight for passively compliant gripper. Appl. Soft Comput. **22**, 424–431 (2014)
12. Chau, N.L., Dao, T.P., Nguyen, V.T.T.: Optimal design of a dragonfly-inspired compliant joint for camera positioning system of nanoindentation tester based on a hybrid integration of Jaya-ANFIS. Math. Prob. Eng. **2018** (2018)
13. Choi, K.-B., Lee, J.J., Kim, G.H., Lim, H.J., Kwon, S.G.: Amplification ratio analysis of a bridge-type mechanical amplification mechanism based on a fully compliant model. Mech. Mach. Theory **121**, 355–372 (2018)
14. Huang, Y.C., Lee, C.Y., Chen, C.R.: Theoretical and experimental design of a new flexible hinged positioning stage. J. Chinese Soc. Mech. Eng. **33**, 59–66 (2012)
15. Khan, S.A., Equbal, M., Islam, T.: A comprehensive comparative study of DGA based transformer fault diagnosis using fuzzy logic and ANFIS models. IEEE Trans. Dielectric. Electric. Insulation **22**, 590–596 (2015)
16. Rodriguez-Barroso, A., Saltaren, R., Portilla, G.A., Cely, J.S., Yakrangi, O.: Potential energy distribution of redundant cable-driven robot applied to compliant grippers: method and computational analysis. Sensors, **19** (2019)
17. Salmasnia, A., Kazemzadeh, R.B., Tabrizi, M.M.: A novel approach for optimization of correlated multiple responses based on desirability function and fuzzy logics. Neurocomputing **91**, 56–66 (2012)
18. Seidemann, V., Butefisch, S., Buttgenbach, S.: Fabrication and investigation of in-plane compliant SU8 structures for MEMS and their application to micro valves and micro grippers. Sens. Actuators A-Phys. **97–8**, 457–461 (2002)

19. Toyoda, T., Tin, O.F., Ito, K., Fujiwara, T., Kumasaka, T., Yamamoto, M., et al.: Crystal structure combined with genetic analysis of the Thermus thermophilus ribosome recycling factor shows that a flexible hinge may act as a functional switch. RNA **6**, 1432–1444 (2000)
20. Lai, L.-J., Zhu, Z.-N.: Design, modeling and testing of a novel flexure-based displacement amplification mechanism. Sens. Actuators A: Phys. **266**, 122–129 (2017)
21. Chen, G., Ma, Y., Li, J.: A tensural displacement amplifier employing elliptic-arc flexure hinges. Sens. Actuators A: Phys. **247**, 307–315 (2016)
22. Ngo, T.H, Tran, H.V, Nguyen, T.A., Dao, T.P., Wang, D.A.: Design and kinetostatic modeling of a compliant gripper for grasp and autonomous release of objects. Adv. Robot. **32**, 717–735 (2018)
23. Hao, G.B., Li, H., Nayak, A., Caro, S.: Design of a compliant gripper with multimode jaws. J. Mech. Robot. Trans. ASME, **10** (2018)
24. Hao, G.B., Zhu, J.X.: Design of a monolithic double-slider based compliant gripper with large displacement and anti-buckling ability. Micromachines, **10** (2019)
25. Huynh, N.-T., Huang S.-C., Dao, T.-P.: Optimal displacement amplification ratio of bridge-type compliant mechanism flexure hinge using the Taguchi method with grey relational analysis. Microsystem Technol. (2018)
26. Huynh, N.-T., Huang, S.-C., Dao, T.-P.: Design variables optimization effects on acceleration and contact force of the double sliders-crank mechanism having multiple revolute clearance joints by use of the Taguchi method based on a grey relational analysis. Sādhanā, **45** (2020)
27. Wang, C.N., Truong, K.P., Huynh, N.T., Nguyen, H.: Optimization on effects of design parameter on displacement amplification ratio of 2 DOF working platform employing Bridge-type compliant mechanism flexure hinge using Taguchi method. J. Phys: Conf. Ser. **1303**, 012053 (2019)
28. Huynh, N.-T., Hoang, V.H, Nguyen, H., Huang, S.C.: Analysis and optimal design a new flexible hinge displacement amplifier mechanism by using Finite element analysis based on Taguchi method. In: 2019 IEEE Eurasia Conference on IOT, Communication and Engineering (2020)

Design of Pressure Testing System for Small Autonomous Underwater Vehicles

Ton Thien Phuong[1(✉)], Huynh Manh Dien[1], Huynh Tan Dat[2], Tran Thien Phuc[1], and Tran Ngoc Huy[2]

[1] Faculty of Mechanical Engineering, Ho Chi Minh City University of Technology, Ho Chi Minh, Vietnam
tonphuong@hcmut.edu.vn
[2] Faculty of Electrical & Electronics Engineering, Ho Chi Minh City University of Technology, Ho Chi Minh, Vietnam

Abstract. Unmanned underwater vehicles are replacing manned underwater vehicles to carry out ocean surveys in dangerous and arduous cases. During the development processes of the underwater vehicles, the testing of operational functions is significantly challenging. The operational functions of autonomous underwater vehicles (AUVs) such as seals, operational propellers and operational rudders will be verified. The equipment should be submerged into the sea until the required depth to inspect whether it has sufficient durability and functions normally as calculated. This testing process is costly because when submerging an underwater device at a great depth, it is necessary to install a sensor system to keep track of the movement of the device or divers may be hired to conduct the same task in some cases. In addition, a multitude of potential risks including accidents, injuries and even the loss of the equipment must be taken into account. Before testing unmanned underwater equipment at a depth of 50-100 m below ocean surface, the equipment was tested at a pressure testing system. This paper aims to present the mechanical and electrical design of the pressure testing system for the small autonomous underwater vehicles. In this research, NX software was used to design the structure of the pressure vessel and its simulation features were deployed to test the durability of system. The other sections describe the controller of the pressure testing system using the pressure drop method. The article should prove that the system can be controlled to maintain, increase, or decrease the inner pressure corresponding to the device movement in the water. The PID algorithm outputs the signal to regulate the speed of a high-pressure pump motor to minimize pressure errors and save energy. The comparison of the experimental tests and simulations is also presented by MATLAB.

Keywords: Pressure vessel · Pressure testing device · Autonomous underwater vehicles · AUVs

B. T. Long et al. (Eds.): MMMS 2020, LNME, pp. 779–786, 2021.
https://doi.org/10.1007/978-3-030-69610-8_103

1 Introduction

Autonomous Underwater Vehicles (AUV) is rapidly developing and playing a more and more important role in the maritime field. It not only applied ability in marine economy and science, but also applied in maritime security and navy. Thanks to the ability to survey and explore under deep water levels, this effective tool can monitor the coastal environment, perform working in hazardous environments such as search rescue and demining. Research underwater robots are challenging task due to the complex in the problems of design and manufacture. Moreover, experimenting to test AUV as water-tight, deformation of the shell under water pressure that the device moves and operational functions is very hard. Testing must be done at the depth of the water level as required by the design. This will cause restrictions in monitoring and inspection, and many other risks can occur such as the loss of the equipment. In this project, the paper presents a pressure testing system, which can flexibly simulate the change of pressure acting on a submersible, this is similar to when the AUV is diving at different depths of the sea. This test system is designed for small AUV objects whose length does not exceed 2.5 m, the largest outer diameter of 0.4 m, and the deepest diving ability of 50 m.

On a global scale, the researches of pressure inspection systems for AUV are still scarce, only a few articles mention this issue as Establish a simplified calculation system to determine the size of the submarine hull and domes under design pressure, an experimental and finite element method analysis were carried out to compare with the derived design values [1]. Give some properties and the layout of various pressure tolerant subsystems due to experiences gained from work, and present an overview of the development of pressure tolerant systems and how it works in a test vehicle [2]. Present some review of conventional testing methods for applying external hydrostatic pressure to buckling-critical shells, show new volume-control pressure testing method to aimed at improving control of specimen deformation near the critical load and the author also gave recommendations for choosing an appropriate pressure testing system for a particular problem [3].

In the article, also has references to some of articles on design and control methods for pressure vessels such as: describes the design pressure vessel that compiles to ASME standards and create various ambient temperature by varying the shell thickness of the vessel to perform thermal analysis [4]. Present the design of hyperbaric chamber for pressure testing according to the ASME Boiler and Pressure Vessel code and structural analysis of the hyperbaric chamber performed by the ANSYS Static Structural software to determine the maximum principal stresses and the corresponding critical regions [5]. Examines the dynamics of valve controlled pressure applications and introduce a set of effective and easy-to-implement PID gain expressions, preceded by an outline about the literature on pressure control [6].

In this study, the presentation of mechanical design that comply with the permitted criteria when manufacturing pressure vessels is shown clearly, including such parts as the design and simulation of vessel shell durability. A hydraulic circuit based on pressure drop method circulates the liquid around the whole system. The article also presents the PID algorithm that output signals to control the speed of a high-pressure pump according the feedback pressure error inside the vessel compared to a predefined pressure reference that describes the ambient characteristic when the AUV is submerged. The ultimate objective of our system is to evaluate shell durability and some functional operation of the AUV. Experimental results are presented at the end for comparison and discussion.

2 Design of Pressure Testing System

2.1 Design of the Pressure Vessel

High pressure hydraulic pump is used to bring water from the tank into the pressure vessel. The pump will fill the pressure vessel with water and continue pumping to increase pressure in the vessel up to the required pressure. The sensors then read the pressure values in the vessel and send signals to the computer. The computer takes the signal and sends it back to the controller to control the pump speed of the motor. The change in motor speed will create pressure, which helps keep the pressure, increase or decrease the pressure according to the preset virtual pressure line. At the same time, when checking AUV, the system also installs sensors to detect water leakage so that it can reduce pressure in time to ensure safety for the device.

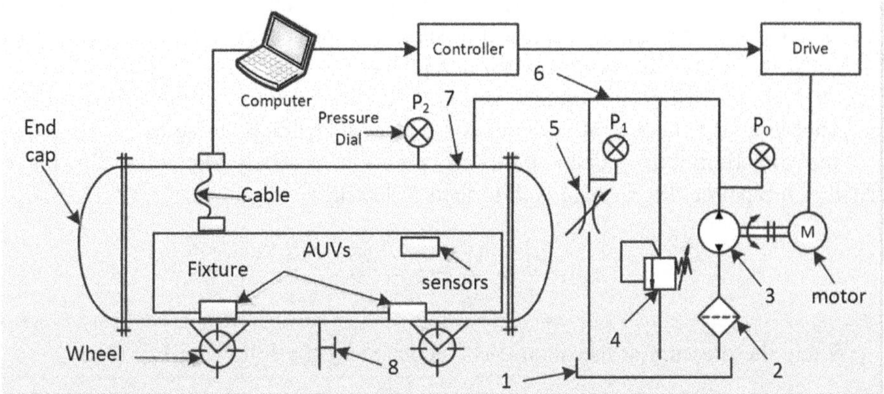

Fig. 1. Diagram of the hydraulic system with 1 tank, 2 filter, 3 piston pumps, 4 relief valve, 5 adjustable flow control, 6 pipeline, 7 pressure vessel, and 8 valve.

With the goal of designing a pressure testing system that can perform the inspections for small AUV, which have 1.5 - 2 m length and external diameter of 0.3 m. The geometry of the vessel has the design characteristics such as 2.5 m length, 0.6 m outer diameter, made of steel material SS400 (tensile strength 400–500 MPa) with rust-proof paint coated externally, work at a maximum pressure of 10bar and using for water at normal temperature. With the thickness of the shell (ts = 8 mm) and the head (th = 16 mm) of the tank that has been calculated, the Simulation of Siemens NX12 module was used to check the durability of the pressure vessel. The results are shown as the Fig. 2:

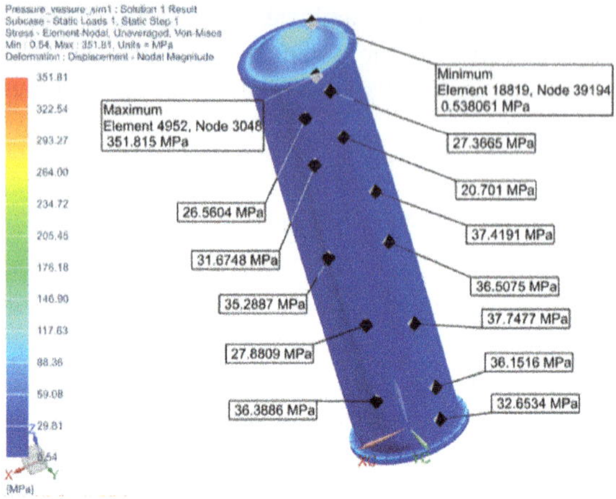

Fig. 2. Von-Mises stress plot on pressure vessel due to pressure loading

2.2 Design of Hydraulic Systems

In order to increase pressure for a pressure vessel, several methods are in place such as compressing gas, compressing water, or using both water and gas. However, the method of using the gas to increase pressure may present an explosion due to the release of the energy of the gas when compressed at high pressure. Furthermore, in order to create a virtual environment similar to the environment of the ship in motion, the group uses the method of compressing water for hydraulic systems.

Applying Bernoulli's equation for two points in a steady streamline flowing of density on pipeline, the Eq. (1) is described following

$$z_1 + \frac{p_1}{\rho} + \frac{V_1^2}{2g} = z_2 + \frac{p_2}{\rho} + \frac{V_2^2}{2g} \tag{1}$$

When, the pressure at the point 2 is calculated by the following Eq. (2):

$$\Delta p = p_2 - p_1 = \frac{\rho v_2^2}{2g} \tag{2}$$

$$\Rightarrow p_2 = p_1 - \frac{\rho Q^2}{2gA^2} \tag{3}$$

Where Q is flow rate; A is cross-sectional area; ρ is density of fluid; V1, p1, and z1 refer to the pressure, speed and height of the fluid at point 1 (on the pipeline), respectively; V2, p2, and z2 refer to the pressure, speed, and height of the fluid at point 2 (in the vessel).

To adjust the pressure in the vessel, the pressure drop method is used to create a difference in pressure on the pipe compared to the pressure in the vessel. To increase and decrease the pressure in the pressure vessel, it is necessary to change the parameter

P1 and Q. Specifically here is increasing (decreasing) P1 by increasing (decreasing) the number of revolutions on the engine and increasing (decreasing) Q flow by the throttle valve. From here the hydraulic diagram is built as shown in Fig. 1.

2.3 Design of Pressure Control System

Pressure controls are available in both manual and electronic variations. Mechanical pressure regulators are more preferable to the applications which requires unchanged long-term pressure set points and minor fluctuations, e.g., pressure switch or throttle valve. However, our objective which is to inspect a wide range of depth requires different pressure values. Therefore, the most effective and economical method is using variable speed drive (VSD) to control the motor speed of high pressure pumps. In industry, PID controller is the most recommended in fan and pump applications for they are popular, simple, easy to implement, and deliver rather good performance. For those advantages, industrial automation manufacturers integrate this feature into their product lines as a fundamental function.

The tank serves as the platform for simulating equivalent underwater ambient pressure. The control system regulate the inner pressure around the reference value by using the analog feedback current from the relative pressure sensor. PLC S7–1200 from SIEMENS utilizes the PID block to calculate the demanded motor speed. The analog output module transmit control signal in the form of 0-10V voltage to the 3-phase industrial inverter which drives the inductive motor in order to minimize the pressure error and reduce the negative impact on the equipment life expectancy. The PC user interface collects information and determines the reference pressure as well as sends command to the whole system. The control framework is presented in Fig. 3.

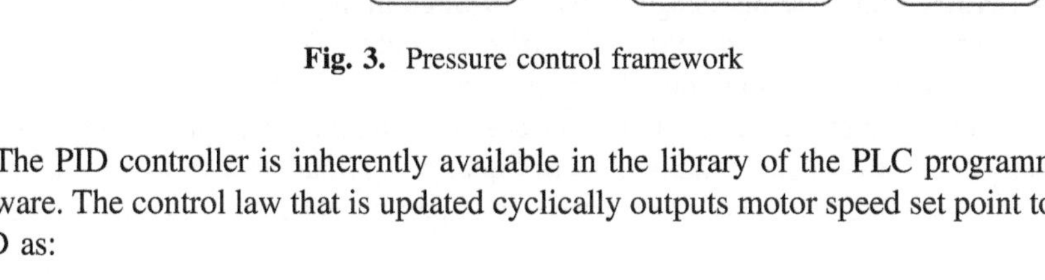

Fig. 3. Pressure control framework

The PID controller is inherently available in the library of the PLC programming software. The control law that is updated cyclically outputs motor speed set point to the VSD as:

$$s*(\%) = K_P\left[(k_1p_d - p) + \frac{1}{T_Is}(p_d - p) + \frac{T_Ds}{k_2T_Ds + 1}(k_3p_d - p)\right] \tag{4}$$

In which, Kp is the proportional gain, TI is the integral time, TD is the derivative time, k1, k2, k3, are the weights of P, I, D terms respectively. The actual pressure is presented by p while the pd denote the desired pressure. Those coefficients can be determined using the fine-tuning feature or manually imported by engineering experience to optimize the response.

The testing process simulates variable pressure conditions to examine the external factors that influence on the AUV. The vehicle's 6-DOF movement is simplified, however, because we only focus on the depth as a function of velocity and altitude. The transformation equation from body-fixed frame to local coordinate frame north-east-down is expressed as follow:

$$\dot{z} = -u\sin(\theta) + v\cos(\theta)\sin(\phi) + w\cos(\theta)\sin(\phi) \tag{5}$$

In the vertical plane, we can ignore the yaw and roll movement. Concurrently, as the vehicle moves forward with a constant velocity, the sway and heave movement can also be considered insignificant. The depth derivative equation can be simplified into:

$$\dot{z} = h_{rate} = -u\sin(\theta) \tag{6}$$

At a certain depth underwater, rigid bodies experience environmental pressure from every direction:

$$p_{fluid} = \rho g h \tag{7}$$

Where ρ is the density of the liquid, whose value in case of sea water is roughly 1030 kg/s3, g is the gravitational acceleration, h is the height of the water column above the object, which means the distance from the object to the surface of the liquid. Finally, we can infer the proportional relationship between the pressure and the depth where the object is, which lead to Eq. (6) and (7):

$$p_{\text{sim}}(t) = \rho g(h_0 + h_{rate}t) \tag{8}$$

2.3.1 Results and Discussion

Figure 4 Shows the simulation process of the pressure acting on the AUV in the sea water environment. The first stage consists of three diving phases in which the AUV moves with variable forward velocities and pitch angles. The stable state is for testing the ability to endure high pressure of the AUV in an interval of time, while the last state simulate the rising of the vehicle to the surface. The pressure error in the whole process is small enough to conclude that the control quality is remarkable as well as the motor speed doesn't fluctuate abruptly, ensures the performance of the pressure testing system.

The operation of AUV is performed inside the vessel to observe the component behaviors such as main thruster, water leak detector, pressure sensor, strain gauge, etc. A microprocessor system collects and analyzes those values in terms of data records. The final results combined from the practical behaviors and collected data can help estimate the performance of the AUV in underwater environment as shown in Fig. 5.

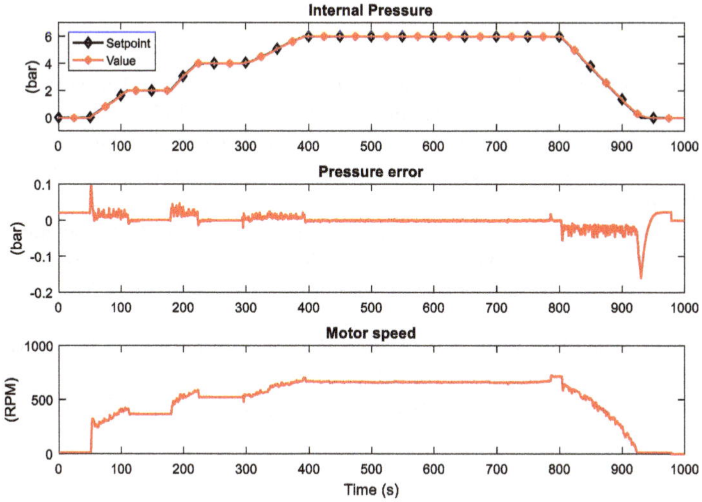

Fig. 4. Underwater pressure simulation testing result

Fig. 5. Pressure testing system during experimental operation

3 Conclusion

This paper demonstrated the detail design of pressure testing system, which can be applied to small autonomous underwater vehicles. The process started with choosing the parameters of the tank meeting the requirements to survey the distortion of the AUV hull under the environmental pressure. Some of the theoretical basis and control diagram are introduced to clarify the operating principles of the testing system. Once the designed platform is fully implemented, the pressure testing experiment is carried out with a suitable AUV model. Finally, the simulation and experimental results in previous sections provide evidences that the quality of the system satisfied the earlier.

References

1. Meschini, A., Ridolfi, A., Gelli, J., Pagliai, M., Rindi, A.: Pressure hull design methods for unmanned underwater vehicles. J. Marine Sci. Eng. **7**(11), 382 (2019)
2. Thiede, C., Buscher, M., Luck, M., Lehr, H., Korner, G., Martin, J., Schlichting, M., Krueger, S., Huth, H.: An overall pressure tolerant underwater vehicle: DNS Pegel. In: OCEANS 2009-EUROPE, pp. 1–6 (2009)
3. MacKay, J.R., van Keulen, F.: A review of external pressure testing techniques for shells including a novel volume-control method. Exp. Mech. **50**(6), 753–772 (2010)
4. Prashanth, D., Hassan, S.: Design and analysis of pressure vessel. Int. J. Mech. Prod. **3**, 1–13 (2019)
5. bin Mohamad Sazali, S.F., Ovinis, M.: Design of a Hyperbaric Chamber for Pressure Testing. In: MATEC Web of Conferences, vol. 13, p. 03004 (2014)
6. Liermann, M.: Pid tuning rule for pressure control applications. Int. J. Fluid Power **14**(1), 7–15 (2013)

Study on Heating Process for Drawing Mold Based on Numerical Method

Thi-Thao Ngo, Luyen The Thanh, and Van-The Than(✉)

Faculty of Mechanical Engineering, Hung Yen University of Technology and Education, Hung Yen, Vietnam
thanthe.ck@gmail.com

Abstract. Heating for drawing mold is widely applied in sheet metal forming process, especially deep drawing technology. Heating process will help to enhance uniform of product thickness and reduce errors during drawing such as fracture, wrinkling and earing. However, using empirical for studying temperature distribution on blank and mold often takes a lot of time and cost due to repeated experiments. Some molds with complex shapes and structures are difficult to set and measure temperature on the blank as well as die. In this study, temperature distribution on the mold is investigated based on numerical method. The heat transfer process on the mold is constructed and simulated by using a finite element software. Temperature results for different heat sources magnitude as well as location are implemented. Results herein will provide useful information to optimize heat source as well as heating time for drawing mold.

Keywords: Drawing mold · Temperature · Comsol multiphysics · Numerical simulation

1 Introduction

A lot of industry brank has been applied warm forming because it will help to significantly enhance formability of material. Therefore, it is necessary to study the heating process in the mold to reach an appropriate blank temperature. R. Kolleck et al. [1] presented a heating systems by using induction coil around boron alloyed steels sheet. In his research, a conveying system is utilized to transport the blank to heating coil region; and blank was not contacted to inductive coil system. Yeong Wei Lean and M. Azuddin [2] studied an effect of elevated dies temperature by conducting different experiments to consider temperature at 4 points. Techniques on heating the die included heating both upper and lower dies, heating only the upper dies, and heating only the lower one. Robert SMUSZ et al. [3] investigated effect of heating conditions on temperature distribution in AMS5604 stainless steel sheet. Termoelements and a FLIR P 640 thermovisual camera were used to measure temperature on the sheet. In their research, the blank was heated outside tool that may affected final temperature distribution on the sheet. M.A.M. Basril et al. [4] applied Taguchi approach (L9 orthogonal array) to find the effects of heating temperature and the most efficient heating position to deep drawing operation of square metal cup. With the same methodology, the effect of the heating temperature on the draw ability of a circular

B. T. Long et al. (Eds.): MMMS 2020, LNME, pp. 787–792, 2021.
https://doi.org/10.1007/978-3-030-69610-8_104

metal cup was carried out in [5]. In references [6–8], authors have been investigated effects of preheat on deep drawing process. They indicated that preheat has significantly affected on quality of products made by deep drawing. From above literature review, it indicates that the heated blank will improve the forming ability of sheet metal. In addition, almost researchers have investigated heating process and its effects by using experiments which usually spend time and increase product price. In this study, the heating process for deep drawing of cup is constructed and simulated to predict temperature distribution not only on blank but also on whole die. In addition, different values as well as locations of heat source are considered.

2 Three-Dimensional Heating Model

2.1 Drawing Mold Structure

In this study, the deep drawing mold is used to make carbon steel (SPCC) cup which is a material grade and designation defined in JIS G 3141 standard. Therefore, 3D model of the mold including 8 main parts is designed by Inventor as in Fig. 1 and then imported to Comsol software. Thermal properties of all parts of the mold are listed in Table 1.

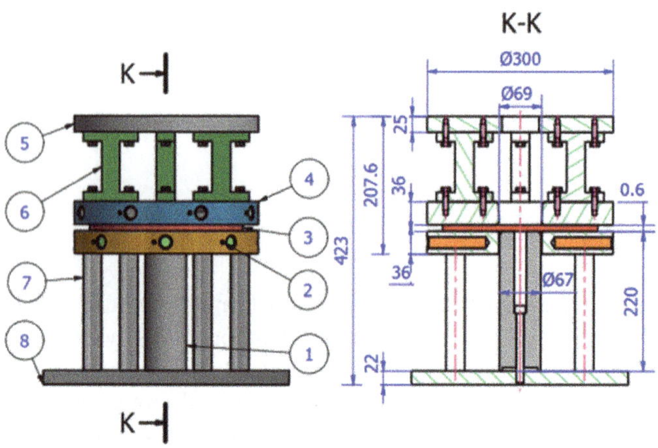

1- punch *3- blank* *5-sole of die* *7- struts*
2- blank holder *4 - die* *6- struts for die* *8- platform*

Fig. 1. Deep drawing mold with blank

Table 1. Material properties

Properties	SKD11 (1,2, 4–8)	SPCC (3)
Density (kg/m3)	7850	7870
Thermal conductivity W/(m.K)	29.3	60
Specific heat (J/(kg.K))	460	460

2.2 Heating Model

A 3D FE thermal model of the heating process for deep drawing mold is established. According to Fourier's law of conduction and the first law of thermodynamics on energy conservation, the transient heat transfer in three-dimentional is expressed by:

$$\frac{\partial}{\partial x}\left(k(T)\frac{\partial T(x,y,z,t)}{\partial x}\right)+\frac{\partial}{\partial y}\left(k(T)\frac{\partial T(x,y,z,t)}{\partial y}\right)+\frac{\partial}{\partial z}\left(k(T)\frac{\partial T(x,y,z,t)}{\partial z}\right)+q_i = \rho C_p(T)\frac{\partial T(x,y,z,t)}{\partial t} \tag{1}$$

with initial and boundary conditions as

$$T(x,y,z,0) = T_\infty \tag{2}$$

$$\partial T(x,y,z,t) = 0 \text{ on insulated surfaces} \tag{3}$$

$$q_{out} = h(T(x,y,z,t) - T_\infty) \text{ on the convective surfaces} \tag{4}$$

Here, $T(x,y,z,t)$ and T_∞ stand for the parts' temperature and the surrounding temperature. $C_p(T)$, $k(T)$ and ρ are the specific heat, thermal conductivity and density, respectively. q_i gives the heat generation at the ith heating rod. Q_{out} and h is the dissipated heat and convection coefficient. Based on Comsol Multiphisics Software, thermal model is established with completely boundary conditions as shown in Fig. 2.

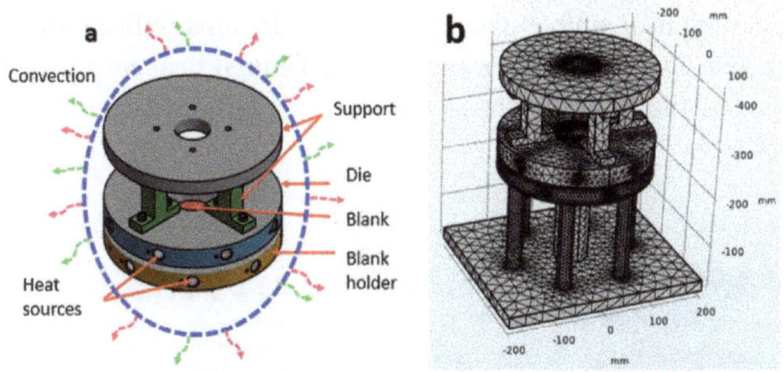

Fig. 2. 3D finite element thermal model; **a**-thermal model; **b**-meshed

3 Results and Discussions

In this paper, different magnitudes and locations of heat sources will be considered. Three cases including heat source placed on die, blank holder and both of them are analyzed. For all situations, inward heat flux is maintained for 1200 s and then it is stopped.

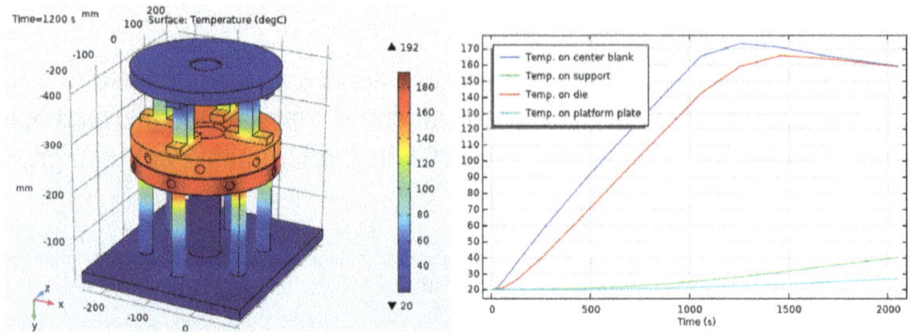

Fig. 3. Temperature results with heat sources (2500 W) located on blank holder

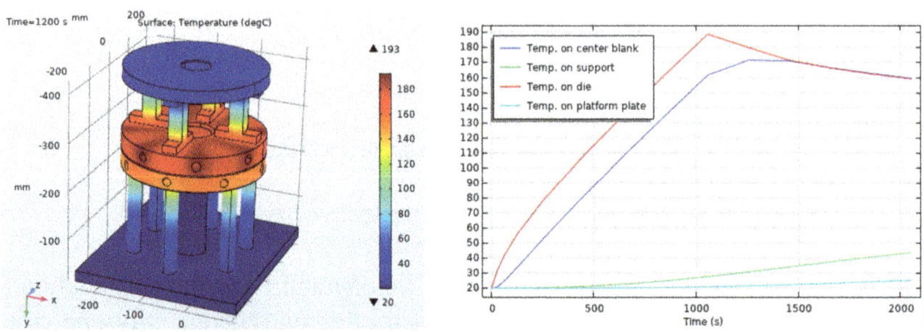

Fig. 4. Temperature results with heat sources (2500 W) located on die

A heat source of 2500 W is utilized for simulating temperature. Figure 3 shows temperature when heat source are assigned on blank holder. Similarly, Fig. 4 presents simulated results with heat source placed on the die. Results indicate that temperature at blank center reaches about 170 °C for both cases. Thus, if heat source is only put on the die or blank holder, the temperature at blank center can't achieved requirement temperature (250 °C).

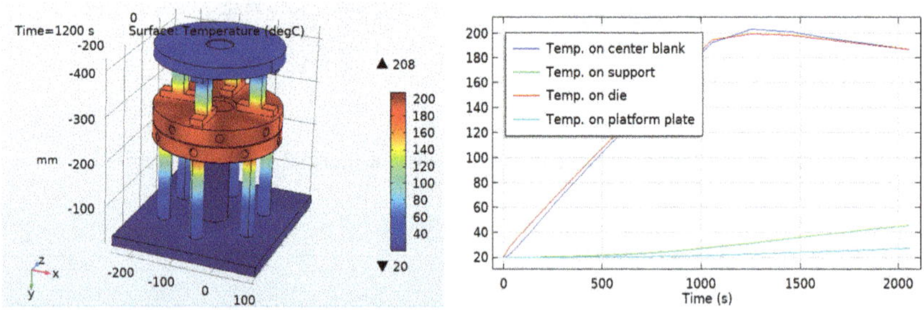

Fig. 5. 1500 W of heat sources on both of die and blank holder

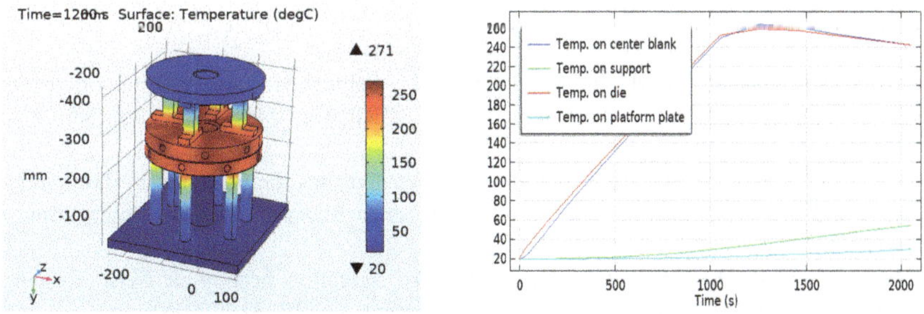

Fig. 6. 2000 W of heat sources on both of die and blank holder

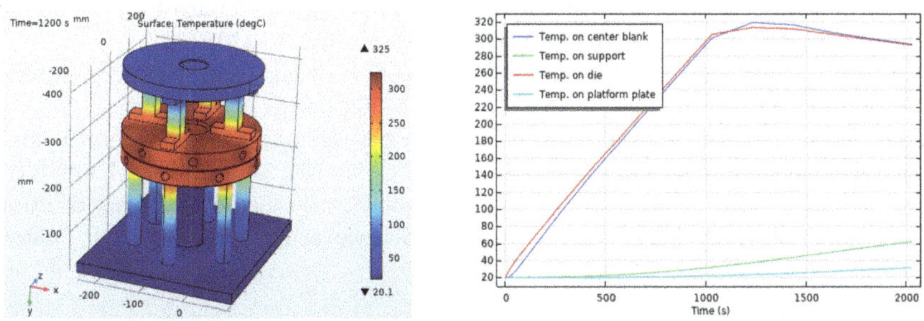

Fig. 7. 2500 W of heat sources on both of die and blank holder

The heat sources with three level of 1500 W, 2000 W and 2500 W which are set on both die and blank holder are continuously simulated to predict temperature distribution on all parts (refer to Figs. 5, 6 and 7). From these figures, it can be seem that temperature at blank center of three cases are subsequently 200 °C, 260 °C and 320 °C. With 2500 W of heat flux, the temperature at blank center is 70 °C higher than required temperature. For case of 1500 W, the temperature at blank center is not matched temperature target. Thus, case of using heating rod with 2000 W is appropriate to warm up present drawing mold reaching about 250 °C. In addition, simulation results can also predict time for heating.

4 Conclusion

This study has successfully applied numerical simulation method to investigate heating process for drawing mold. A three-dimensional thermal model has been established based on Comsol Multiphisics software. The temperature for different situations have been acquired. Results indicate that with heat source of 2000 W can warm up present mold to acquire temperature after 1200 s. One can say that the numerical simulation method is a powerful tool to predict temperature as well as heating time for drawing mold. From numerical results, we will find an appropriate heat source and time to heat any mold with complex structure; it will reduce testing costs as well as save time.

References

1. Kolleck, R., Veit, R., Merklein, M., Lechler, J., Geiger, M.: Investigation on induction heating for hot stamping of boron alloyed steels. CIRP Ann. Manuf. Technol. **58**, 275–278 (2009)
2. Lean, Y.W., Azuddin, M.: Study the effect of elevated dies temperature on aluminium and steel round deep drawing. In: IOP Conference Series Materials Science Engineering, vol. 114 (2016)
3. Robert, S., et al.: Experimental investigations of induction heating in warm forming of stainless steel sheets. Tehnički vjesnik **25**(Supplement 2), 312–318 (2018)
4. Basril, M.A.M., et al.: The effect of heating temperature and methods towards the formability of deep drawn square metal cup. In: IOP Conference Series. Materials Science and Engineering (Online), pp. 210–211 (2017)
5. Basril, M.A.M., Azuddin, M., Choudhury, I.A.: The effect of elevated temperature on the drawability of a circular deep drawn metal cup. Metals **9**, 1303 (2019)
6. Kayhan, E., Kaftanoglu, B.: Experimental investigation of non-isothermal deep drawing of DP600 steel. Int. J. Adv. Manuf. Technol. **99**, 695–706 (2018)
7. Rahiminejad, D., Compston, P.: The effect of pre-heat temperature on the formability of a glass-fibre/polypropylene and steel-based fibre–metal laminate. Int. J. Mater. Form. (2020)
8. Joghan, H.D., et al. Influence of the preheating strategy on the deep drawing of extruded magnesium alloy. In: ME20 sheets. Materials Science and Engineering. IOP Publishing (2019)

Research on the Reverse Dynamic of the Main Structure of the CNC Optical Grinding MB-250

Nguyen Quoc Manh and Vu Xuan Truong(✉)

Faculty of Mechanical Engineering, Hung Yen University of Technology and Education, Hung Yen, Vietnam
nguyenquocmanh@utehy.edu.vn, truongvxktcs@gmail.com

Abstract. Reverse dynamics is crucial in machine structure analysis which determines the relationship between the operating point and the torque excitation on the active link of the machine. The main structure of the MB-250 CNC optical grinding machine is composed of an independent rotated grinding disc that is attached to the planar four-bar mechanism. First, the motion equations of a mechanism will be established using the matrix form of Lagrangian equations using multipliers for constrained multibody systems. Then, a reverse dynamic algorithm was used to determine the torque acting on the active link of the machine. The study was conducted and verified by combining experiments with numerical simulations. The research results are used to actively control the grinding trajectory, velocity, and acceleration of the operating point to enhance the surface quality of the products.

Keywords: Reverse dynamic · Lagrangian equations · The CNC optical grinding

1 Introduction

The problem of studying the effects of moment acting on the mechanism when grinding optical parts, to improve machining accuracy, is a problem posed to technologists. This purpose will be solved when studying the dynamics of the grinding mill structure. After studying the effects of the mechanical factors of the machine structure on the machining accuracy, the improved surface shaping accuracy of optical parts when grinding in the direction of studying the effects of dynamic factors structural force of the machine is one of the effective measures that need to be studied [3, 6].

This paper proposes the model and solving the problem of reverse dynamics of the CNC optical grinding MB-250, the upper stitching mechanism has the form of four-bar linkages [6]. The machine is currently being used at the Laboratory of Precision Mechanics, HUST. By using the matrix form of the Lagrangian equations [o establish a system of motion differential equations of a machine mechanism, determine the dynamic torque applied to active link, thereby allowing control of the grinding process for improving the surface roughness and precision of the machining products [1, 2].

B. T. Long et al. (Eds.): MMMS 2020, LNME, pp. 793–798, 2021.
https://doi.org/10.1007/978-3-030-69610-8_105

2 Modeling and Equations of Motion

2.1 Experiment Data

On the basis of experiments on real machines, by experimental measurements, the parameters of the mechanism are determined (Fig. 1). The topstitching of the machine is essentially a traditional four-bar linkage mechanism. The length and weight of the ith link are presented in Table 1. These data are determined by ruler and electronic scale with high accuracy.

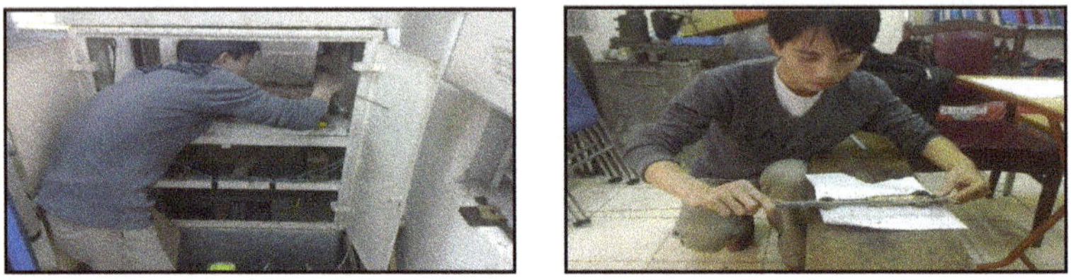

Fig. 1. Experimental measurement of CNC machine MB-250

Table 1. Machine experiment data

Link	Length (mm)	Mass (kg)
0	310	–
1	29.5	2.42
2	158.5	0.56
3	285	8.84

Another important issue to determine from the experimental model before setting up the math model is the determination of the mass moment of inertia and the center of mass of the link. This problem is solved simply by building 3D models of the links and computation on Autodesk Inventor 2020 software (IV 2020). Figure 2 and Table 2 show the results of calculating the mass moment of inertia and center of mass, which is important data for the inverse dynamic problem in this study.

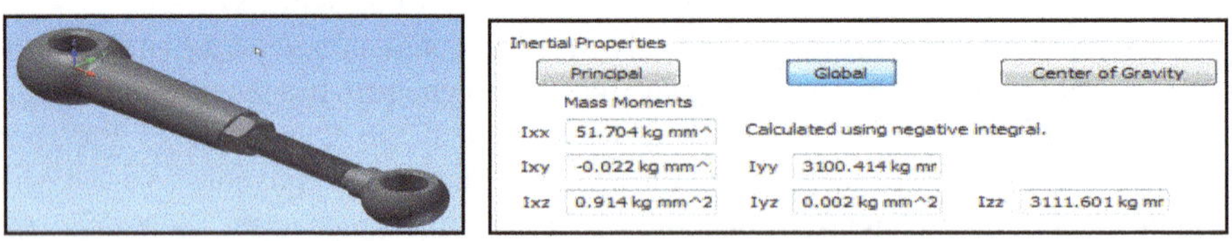

Fig. 2. 3D model and calculation results on IV 2020

Table 2. The calculation results of the links on IV 2020

Link	Mass inertia (kg.mm^2)	Position of center mass (mm)
0	–	–
1	4112.803	(0, –2.624, –1.133)
2	3111.601	(52.633, 0.001, −0.015)
3	159558.986	(65.306, –1.751, 26.874)

3 The Dynamic Model of the Mechanism

Figure 3 describes the real and computational model of the mechanism. The length and mass of the ith link is denoted l_i, m_i. The excitation moment on the active link of the mechanism is τ_1 (Nm). C_i is the symbol for the ith center mass. The absolute coordinate system is Oxy while $\eta_i\xi_i$ is the relative coordinate system attached to the link. The generalized coordinates is q_1, q_2, q_2 which is angle rotation of the ith link. The mechanism moves in a plane that is parallel to the ground.

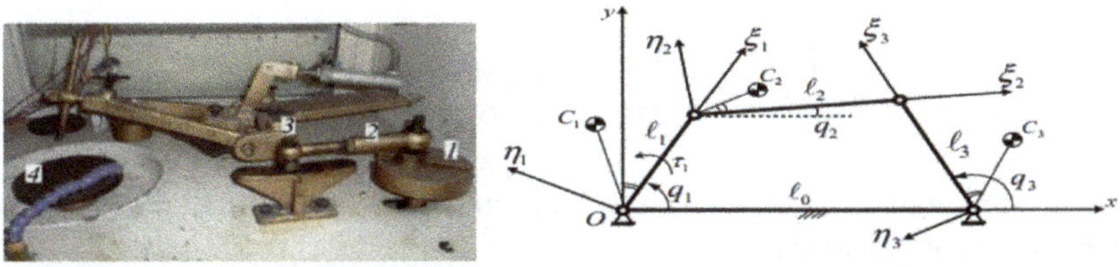

Fig. 3. A real and computational model of the mechanism

3.1 The Equations of Motion

In this study, the survey mechanism has closed-form so we apply the matrix form of Lagrangian equations to establish the equations of motion of the system [1].

$$\mathbf{M}(\mathbf{q})\ddot{\mathbf{q}} + \mathbf{C}(\mathbf{q},\dot{\mathbf{q}})\dot{\mathbf{q}} + \mathbf{F}_q^T\lambda = \mathbf{f}(\mathbf{q}) \tag{1}$$

where the displacements vector is $\mathbf{q} = [q_1\ q_2\ q_3]^T$. $\boldsymbol{\lambda} = [\lambda_1\ \lambda_2]^T$ are the Lagrangian multipliers and $\mathbf{f(q)} = [\tau_1\ 0\ 0]^T$ are the excitation moment vector.

Based on the proposed formulas, a numerical program is developed. The results are presented in the formulas (2–7).

The 3 × 3 inertia matrix $\mathbf{M}(\mathbf{q})$ has the form as

$$\begin{aligned} &m_{11} = m_1\left(\xi_{c1}^2 + \eta_{c1}^2\right) + m_2 l_1^2 + I_1;\ m_{22} = m_2\left(\xi_{c2}^2 + \eta_{c2}^2\right) + I_2; \\ &m_{33} = m_3\left(\xi_{c3}^2 + \eta_{c3}^2\right) + I_3; m_{13} = m_{31} = 0; m_{23} = m_{32} = 0 \\ &m_{12} = m_{21} = m_2 l_1 \xi_{c2}\cos(q_2 - q_1) - m_2 l_1 \eta_{c2}\sin(q_2 - q_1) \end{aligned} \tag{2}$$

The Coriolis/Centripetal vector is determined by [1, 2]

$$\mathbf{C}(\mathbf{q},\dot{\mathbf{q}}) - \frac{\partial \mathbf{M}(\mathbf{q})}{\partial \mathbf{q}}(\mathbf{I}_n \otimes \dot{\mathbf{q}}) - \frac{1}{2}\left(\frac{\partial \mathbf{M}(\mathbf{q})}{\partial \mathbf{q}}(\dot{\mathbf{q}} \otimes \mathbf{I}_n)\right)^T \tag{3}$$

In is a unit matrix which has 3 × 3 size. Therefore, matrix **C** can be rewritten as

$$\begin{aligned}
c_{11} &= -\tfrac{1}{2}m_2 l_1 \dot{q}_2 \left[\xi_{C_2}\sin(q_2-q_1)+\eta_{C_2}\cos(q_2-q_1)\right]\\
c_{12} &= \tfrac{1}{2}m_2 l_1 (\dot{q}_1 - 2\dot{q}_2)\left[\xi_{C_2}\sin(q_2-q_1)+\eta_{C_2}\cos(q_2-q_1)\right]\\
c_{21} &= \tfrac{1}{2}m_2 l_1 (2\dot{q}_1 - \dot{q}_2)\left[\xi_{C_2}\sin(q_2-q_1)+\eta_{C_2}\cos(q_2-q_1)\right]\\
c_{22} &= \tfrac{1}{2}m_2 l_1 \dot{q}_1 \left[\xi_{C_2}\sin(q_2-q_1)+\eta_{C_2}\cos(q_2-q_1)\right]\\
c_{13} &= c_{23} = c_{31} = c_{32} = c_{33} = 0
\end{aligned} \tag{4}$$

Note that according to

$$\mathbf{f} = \begin{bmatrix} f_1 \\ f_2 \end{bmatrix} = \begin{bmatrix} l_1\cos q_1 + l_2\cos q_2 - l_3\cos q_3 - l_0 \\ l_1\sin q_1 + l_2\sin q_2 - l_3\sin q_3 \end{bmatrix} \tag{5}$$

Therefore, the vector of constraint forces can be rewritten as

$$\mathbf{F}_q = \frac{\partial \mathbf{f}}{\partial \mathbf{q}} = \begin{bmatrix} -l_1\sin q_1 & -l_2\sin q_2 & l_3\sin q_3 \\ l_1\cos q_1 & l_2\cos q_2 & -l_3\cos q_3 \end{bmatrix} \tag{6}$$

By substituting Eqs. (2–6) into Eq. (1) algebraic-differential equations of motion of the mechanism are obtained as:

$$\begin{aligned}
&(J_1 + m_2 l_1^2)\ddot{q}_1 + m_2 l_1[\xi_{c2}\cos(q_2-q_1) - \eta_{c2}\sin(q_2-q_1)]\ddot{q}_2\\
&-m_2 l_1[\xi_{c2}\sin(q_2-q_1) + \eta_{c2}\cos(q_2-q_1)]\dot{q}_2^2 = \tau_1 + l_1(\lambda_1\sin q_1 - \lambda_2\cos q_1)
\end{aligned} \tag{7}$$

$$\begin{aligned}
&J_2\ddot{q}_2 + m_2 l_1[\xi_{c2}\cos(q_2-q_1) - \eta_{c2}\sin(q_2-q_1)]\ddot{q}_1\\
&+m_2 l_1[\xi_{c2}\sin(q_2-q_1) + \eta_{c2}\cos(q_2-q_1)]\dot{q}_1^2 = l_2(\lambda_1\sin q_2 - \lambda_2\cos q_2)
\end{aligned} \tag{8}$$

$$J_3\ddot{q}_3 = l_3(-\lambda_1\sin q_3 + \lambda_2\cos q_3) \tag{9}$$

3.2 Analysis of Reverse Dynamics

The purpose of this section is to determine the excitation torque from which the desired drive motor can be controlled for improving the surface roughness and machining accuracy of the machining product. The rotation angle of the active link is usually known (q_1 = ωt), due to the known speed of the drive motor (ω, rpm).

Derived from the constrained equations we obtained

$$\begin{aligned} f_1 &= l_1 \cos q_1 + l_2 \cos q_2 - l_3 \cos q_3 - l_0 = 0 \\ f_2 &= l_1 \sin q_1 + l_2 \sin q_2 - l_3 \sin q_3 = 0 \end{aligned} \tag{10}$$

Equation (10) is solved by Newton-Raphson algorithm on MAPLE software, therefore $q_2(t)$ and $q_1(t)$ are completely determined.

The time derivative of the Eq. (10) is as follows:

$$\begin{aligned} -l_1\ddot{q}_1 \sin q_1 - l_2\ddot{q}_2 \sin q_2 + l_3\ddot{q}_3 \sin q_3 - l_1\dot{q}_1^2 \cos q_1 - l_2\dot{q}_2^2 \cos q_2 + l_3\dot{q}_3^2 \cos q_3 = 0 \\ l_1\ddot{q}_1 \cos q_1 + l_2\ddot{q}_2 \cos q_2 - l_3\ddot{q}_3 \cos q_3 - l_1\dot{q}_1^2 \sin q_1 - l_2\dot{q}_2^2 \sin q_2 + l_3\dot{q}_3^2 \sin q_3 = 0 \end{aligned} \tag{11}$$

The quantities q_1, q_2, q_3, and its derivatives are completely determined by solving the system of Eqs. (10), (11). By substituting the above parameters into Eqs. (7), (8), (9), the excitation torque can be defined as

$$\begin{aligned} \tau_1 = (J_1 + m_2 l_1^2)\ddot{q}_1 + m_2 l_1[\xi_{c2}\cos(q_2 - q_1) - \eta_{c2}\sin(q_2 - q_1)]\ddot{q}_2 \\ -m_2 l_1[\xi_{c2}\sin(q_2 - q_1) + \eta_{c2}\cos(q_2 - q_1)]\dot{q}_2^2 - l_1(\lambda_1 \sin q_1 - \lambda_2 \cos q_1) \end{aligned} \tag{12}$$

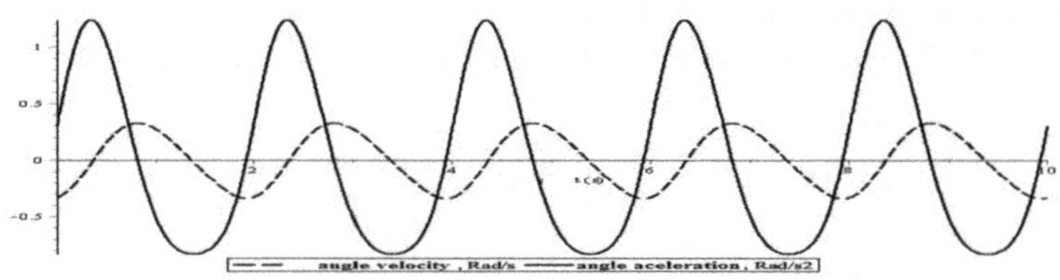

Fig. 4. Calculated dynamic parameters of the 3th link.

From the results obtained in Eq. (12), we can completely control and control the grinding process of the workpiece to achieve stability, improve the surface roughness and accuracy of the machining. Figures 4 and 5 show the kinetic quantities of the actuator and the excitation moment τ_1.

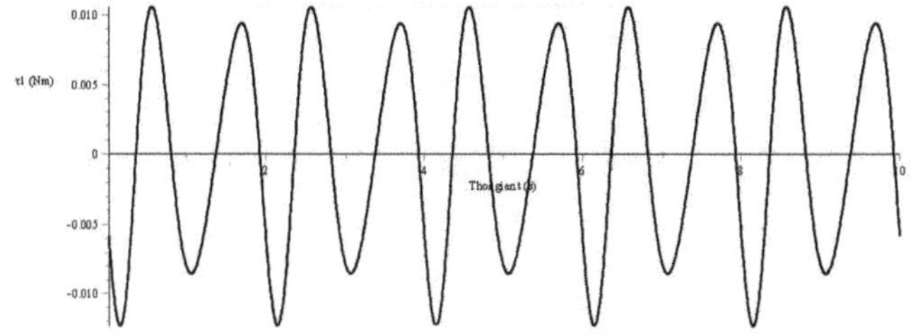

Fig. 5. The driving moment on the active link

4 Conclusions

In this study, an analytical dynamic model of the CNC optical grinding machine MB 250 is presented. A method for analyzing reverse dynamics of an optical grinding machine mechanism (CNC MB-250) is proposed. The system of differential equations for the mechanisms is built using Lagrange's equations method. From the proposed method, the driving moment can be determined. The proposed method also enables determination of the equilibrium conditions for the joints between the contrivance and grinding disc.

References

1. Van, K.-N.: Kronecker product and new matrix form of Lagrangian equations with multipliers for constrained multibody systems. Mech. Res. Commun. **37**(8), 294–299 (2014)
2. Van, K.-N.: Multibody of dynamics system. Science and Technics Publishing House, Hanoi (2007)
3. Trong, H.-N., Van, K.-N., Xuan, T.-V.: Dynamics research of optical parts grinding machine structure CNC MB-250 . J. Sci. Technol. Vietnam Acad. Sci. Technol. **52**(3), 299–309 (2014)
4. Pickarda, J.K, Carretero, J.A, Merletc, J.P.: Appropriate analysis of the four-bar linkage. Mech. Mach. Theory **139**, 237–250 (2019)
5. Salah-M, A.-J., Tatlicioglu, B.: Robust backstepping control for a four-bar linkage mechanism driven by a DC Motor. J. Intell. Rob. Syst. **94**, 327–338 (2019)
6. Truong, V.X.: Technical Master Thesis, Hanoi University of Science and Technology (2014)

Surface Accessibility Detection in CAD Environment for Additive Manufacturing

Thanh Hai Tuan Tran[1], Mouhamadou Mansour Mbow[2], Frédéric Vignat[2], and Dinh Son Nguyen[1(✉)]

[1] The University of Danang, University of Science and Technology, Danang, Viet Nam
{tthtuan, ndson}@dut.udn.vn
[2] University Grenoble Alpes, CNRS, Grenoble INP, G-SCOP Laboratory, Grenoble, France
{mouhamadou-mansour.mbow, frederic.vignat}@grenoble-inp.fr

Abstract. The appearance of Additive Manufacturing (AM) technologies has brought many benefits for industry as well as other areas of life: architecture, medical, etc. In order to create a part by AM technologies especially with EBM technology, it has to go through many steps: designing parts in CAD environment, applying CAM operations, printing the parts and post-print process. Analyzing parts to decide if it is necessary to add the support structures play an important role because of their effects on the quality of the finishing parts. To assess the printability of part, feature recognition is a technique that helps make sure all the features of a part will not exhibit defects before the printing process. There are different studies on how to recognize features after design. Several research groups have developed systems to analyze the characteristics of parts but there are very few systems to analyze parts for the AM process. Therefore, it is necessary to propose a tool to help to perform the feature recognition process in AM technologies. A simple interface presented in the paper has been developed in C++ programming language based on the Open Cascade Technology software to help users automatically recognizing some features of a model. It is applied to a recurrent problem related to support removability. A new type of feature called surface with potential support difficult to remove is defined.

Keywords: Additive Manufacturing · Feature recognition · Support structures · CAD

1 Introduction

Nowadays, in comparison with traditional manufacturing methods in the mechanical engineering industry, additive manufacturing technologies (AM) have many outstanding advantages than the old ones. AM has become a very popular term to describe a manufacturing process by adding layers by layers of material. Instead of being held back by manufacturing constraints, we can design the part we need. Therefore, AM technologies are widely applied in the different fields in industry [1, 2].

B. T. Long et al. (Eds.): MMMS 2020, LNME, pp. 799–807, 2021.
https://doi.org/10.1007/978-3-030-69610-8_106

Although the AM technologies help to make the geometric design space more open than the conventional ones, however they have their inherent manufacturing restrictions [3]. One of them is the removability of support structures. In fact, support structures are additional entities built with the parts for several purposes to ensure their manufacturability. Supports are most of the time build on overhanging surfaces [4] to prevent them from collapsing during the fabrication. In powder-bed fusion technologies, the placement of support structures also improves the building process and anchors [5] the part to avoid different geometrical defects related to shrinkage and residual stresses. Therefore, support structures are necessary to keep these printed materials in the aimed position [6]. However, when designing supports, one must have in mind their removability after the manufacturing, as the surfaces in question might be inaccessible. In fact, for metal AM, for example, different tools are used to cut down the built-in support. Even sometimes, machining operations are led to eliminating the marks left by supports completely. The need to use the support structures depends a lot on the orientation of the manufactured part. Thus many works have been performed on the optimization of part orientation in AM [7–10].

From the benefits and challenges of support structures mentioned above, their placements in a specific position of parts to improve the building process are well appreciated. For simple parts, the identification of removable support structures can be achieved manually, but for the complex parts, this is not trivial. Hence, it is necessary to have a support tool to help identify zones of a part that would potentially have supports difficult to remove. To date, not much effort has been concentrated on it.

In order to solve the problem, a method will be presented in this paper to help us to recognize the positions which have gaps between surfaces that have difficult support structures to remove. From the method, an interface to recognize automatically surfaces where support structures would be difficult to remove in a CAD file a given component is developed in C++ programming language using Open Cascade Technology (OCCT) - open-source software for CAD design [11].

2 Methodology

This paper introduces a method of feature recognition using available libraries of OCCT software based on Boundary Representation (Brep). For the CAD part, there are many kinds of features. In this paper, we just consider the surfaces of part because these are the features related to the support structures during the AM process. The common method to illustrate gaps between surfaces that have difficult support structures to remove in part includes four stages as given in Fig. 1.

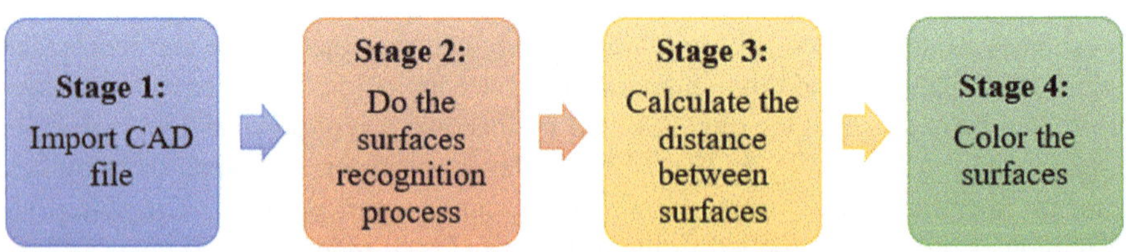

Fig. 1. Main stages to perform the recognition process

2.1 Importation of CAD File

After designing in CAD software, a digital file of parts is exported to different types of file formats depend on the purpose of manufacturing. For Additive Manufacturing, the three most popular types of file formats are OBJ, STEP and STL. The latter two types of files are used in this paper. Therefore, two file readers are developed basing on OCCT libraries.

2.2 The Surfaces Recognition Process

After the CAD file is imported into the interface, all the faces are triangulated. After this step, all faces are made up of a set of triangles. Then, to know the exact triangular face to use, the indexes of all the faces in the part need to be identified. So, the map of faces is the table that contains the indexes of faces.

Among the faces in the map, an initial face is chosen and then an adaptor surface belonging to the face is created because in OCCT, the face is a topological entity and surface is a geometrical entity. From the surface, calculate the vectors created from the vertices of triangular surfaces then, the normal vector and the magnitudes of vectors. Next, find the center of the incircle G of triangular surfaces by Eq. 1.

$$\begin{cases} x_G = \dfrac{\left|\overrightarrow{AB}\right| \times x_A + \left|\overrightarrow{AC}\right| \times x_B + \left|\overrightarrow{BC}\right| \times x_C}{\left|\overrightarrow{AB}\right| + \left|\overrightarrow{AC}\right| + \left|\overrightarrow{BC}\right|} \\ y_G = \dfrac{\left|\overrightarrow{AB}\right| \times y_A + \left|\overrightarrow{AC}\right| \times y_B + \left|\overrightarrow{BC}\right| \times y_C}{\left|\overrightarrow{AB}\right| + \left|\overrightarrow{AC}\right| + \left|\overrightarrow{BC}\right|} \\ z_G = \dfrac{\left|\overrightarrow{AB}\right| \times z_A + \left|\overrightarrow{AC}\right| \times z_B + \left|\overrightarrow{BC}\right| \times z_C}{\left|\overrightarrow{AB}\right| + \left|\overrightarrow{AC}\right| + \left|\overrightarrow{BC}\right|} \end{cases} \quad (1)$$

Where A, B, C are vertices of an initial triangular surface, and G is the center of the inscribed circle of the triangle ABC (see in Fig. 2a).

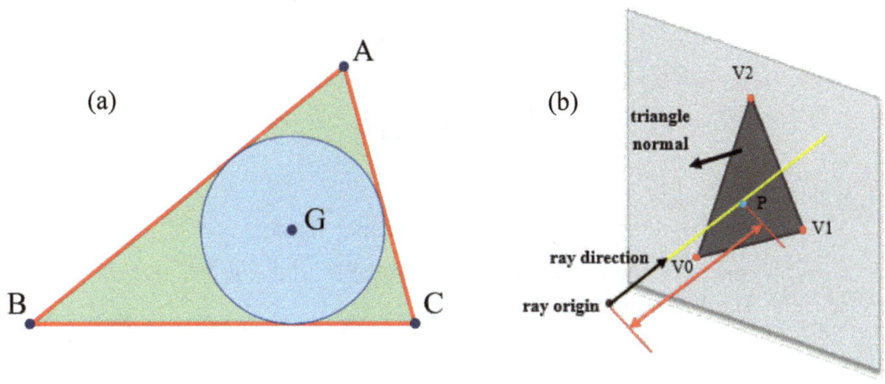

Fig. 2. Vertices A, B, C and center G of the inscribed circle of the triangle ABC.

The incircle of the surface is the reference point from which the distance between two surfaces is calculated. Then, a ray centered in G is traced to check out if it meets any other surface of the map. This ray is used to calculate the width of the gap between the considered surface and an intersected one. The way to find this width will be presented in Fig. 2b.

The purpose of this solution is to find the point P, where the ray from the initial surface intersects the second surface. The ray parametric equation is Eq. 2.

$$\mathrm{P} = \mathrm{G} + \mathrm{tR} \tag{2}$$

Where t is the distance between point G and point P.

In the second surface, we can calculate the plane equation, which contains this surface, as shown in Eq. 3.

$$\mathrm{M'x} + \mathrm{N'y} + \mathrm{P'z} + \mathrm{D} = 0 \tag{3}$$

Where $\mathrm{M'}, \mathrm{N'}$ and $\mathrm{P'}$ are the coordinates of the normal vector $\overrightarrow{n}$ of the second surface and D is the distance between the origin and the plane that contains the second surface. It is calculated by Eq. 4:

$$D = \frac{M' \times x_{V0} + N' \times y_{V0} + P' \times z_{V0}}{\left|\overrightarrow{n}\right|} \tag{4}$$

Where x_{v0}, y_{v0}, z_{v0} are the coordinates of vertex v_0 (see in Fig. 2b) and $\left|\overrightarrow{n}\right|$ is the magnitude of $\overrightarrow{n}$.

Then, the parameter t can be solved by Eq. 5.

$$t = \frac{D + M' \times x_G + N' \times x_G + P' \times x_G}{M' \times M + N' \times N + P' \times P} \tag{5}$$

Where M, N, P are the coordinates of the normal vector of the initial surface and they are also the coordinates of direction R.

Next, we will check if the point P lies on the second surface by condition:

$$\begin{cases} u = (\overrightarrow{PA'}, \overrightarrow{PB'}) \\ v = (\overrightarrow{PA'}, \overrightarrow{PC'}) \\ w = (\overrightarrow{PB'}, \overrightarrow{PC'}) \end{cases}$$

Where:

$\mathrm{A'}, \mathrm{B'}, \mathrm{C'}$ are the vertices of the second triangular surface.

u, v, w are the angles between $\overrightarrow{PA'}$ *and* $\overrightarrow{PB'}$, $\overrightarrow{PA'}$ *and* $\overrightarrow{PC'}$ *and* $\overrightarrow{PB'}$ *and* $\overrightarrow{PC'}$ respectively. The illustration is shown in Fig. 3.

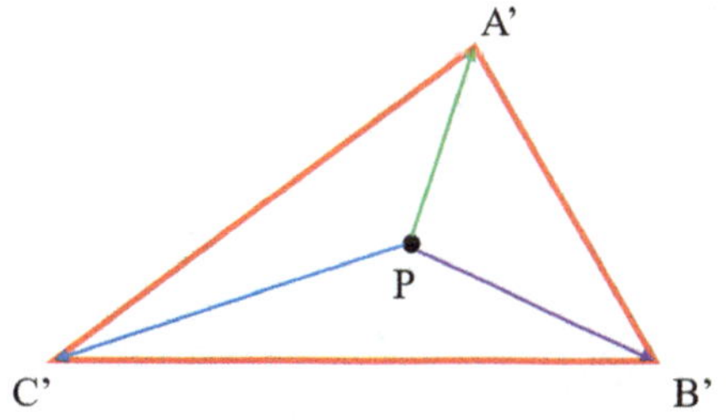

Fig. 3. Relation between vectors in triangle A′ B′ C′

The recognition process above can be done easily with two opposite and parallel surfaces. In order to improve feature recognition accuracy for the two arbitrary surfaces, we can create two more lines that are rotated an angle with the line created from the normal vector. Next, we still do the recognition process for three lines. If there is any line touching an opposite surface, the length or the distance between two surfaces will be calculated and then compared in order to choose the shortest length. This value is the value that will be used to check whether support would be difficult to remove from the considered surface.

2.3 Calculation the Distance Between Surfaces

The distance between two triangular surfaces or the width of the gap between them is the value of t given in Eq. 5. This calculation is the reference for the map of color in the next stage.

2.4 Coloration the Surfaces

The value of the width of gaps between surfaces will be compared with the desired values about width. In fact, support is assumed as removable when this distance is greater than the size of the tool used for it. As the distance is not constant, ranges having different colors are defined. For example, red color for values less than 5 mm; orange color for the values from 5 to less than 10 mm; yellow color for the values from 10 to less than 15 mm and green color for the values greater than 15 mm. These values are just chosen randomly to check the conditions. They can be modified depending on the demand of the user, i.e. his tools dimensions.

3 Result and Discussion

The interface developed in OCCT is shown in Fig. 4. This interface helps users import the CAD file of part, check the features in part and highlight the surfaces having gaps between.

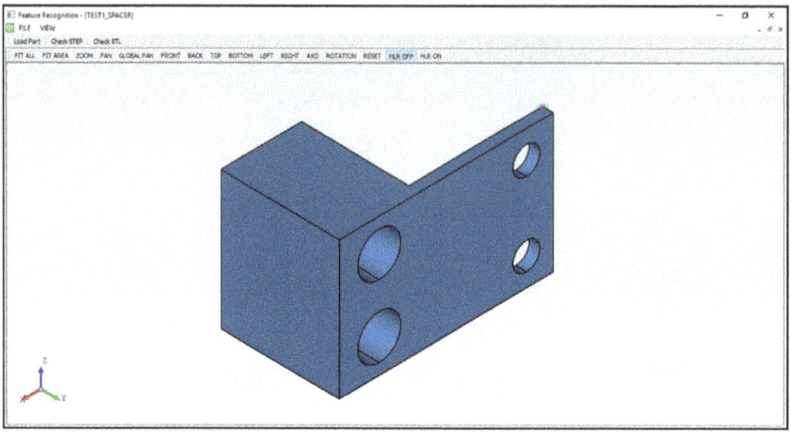

Fig. 4. Interface developed in OCCT

The results of the feature recognition process for some samples of two types of the most popular file format (STEP and STL) for AM are shown in Table 1.

As an illustration, we set up the values of the width of gaps between surfaces changing from less than 5 mm to greater than 15 mm. For the map of color, red corresponds to the tolerances less than 5 mm, orange corresponds to the values from 5 to less than 10 mm, yellow is for the values from 10 to less than 15 mm and green is consistent with the tolerances greater than 15 mm. The smallest width between surfaces represents the most difficult support removal scenario.

Sample 1 includes four slots made up of pairs of planar surfaces with different dimensions.
Sample 2 includes three through all cylindrical holes with different radius arranged in three planes of the cubic base.
Sample 3 is designed by topological optimization with B-Spline surfaces.

Beside the result of the recognition process, the processing time between the two types of file formats is another result of concern. Table 2 will compare time-requiring to finish the recognition process of two types of CAD file format.

The data on processing time in Table 2 shows that the processing time to recognize surfaces in parts depends too much on the type of file format. From the chart above, the time-consuming to recognize surfaces in parts will increase exponentially if the complexity of part or the number of surfaces in part goes up. This complexity is reflected in the size of the CAD file. The recognition time increases as the part contains various features.

There is a big difference between the size of the CAD file in two file format STL and STEP, shown in Table 3. The size of the same part in STL format is bigger than in STEP format. The reason for this difference comes from the characteristic of each file format. STL file format divides surfaces of part into small triangular surfaces while STEP file format illustrates types of surfaces through the parameters such as points, curves, etc. Therefore, recognition of some types of basic surfaces in the STEP file format is easier and more accurate than in the STL file format.

Table 1. Samples with highlighted surfaces in two types of file format

STEP file format	STL file format
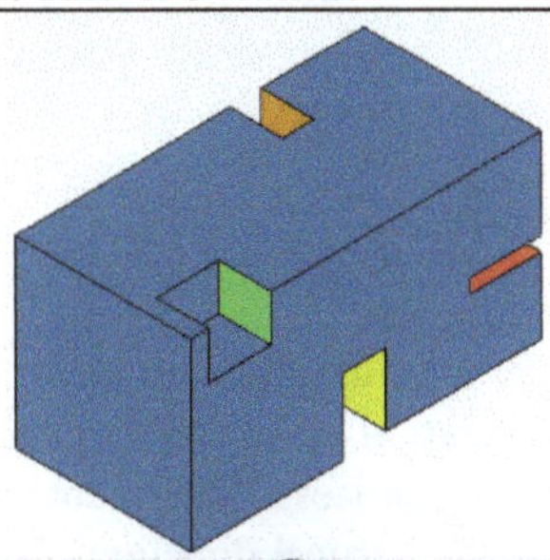	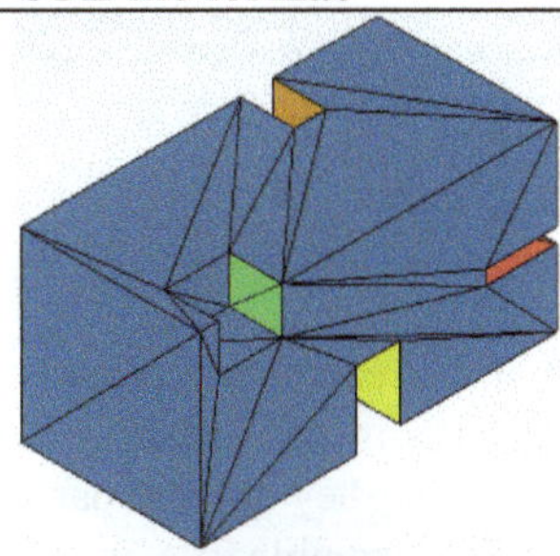
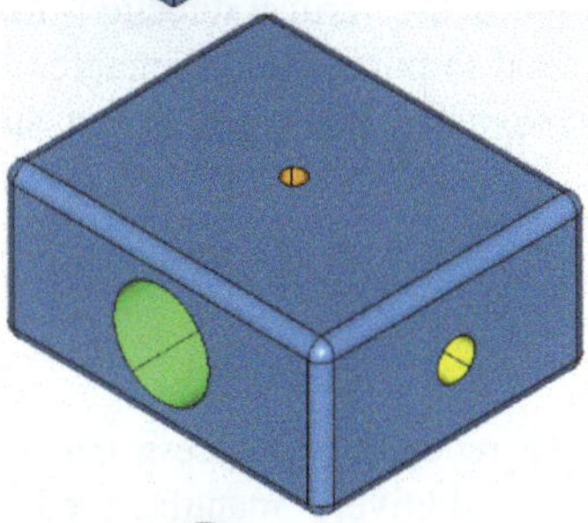	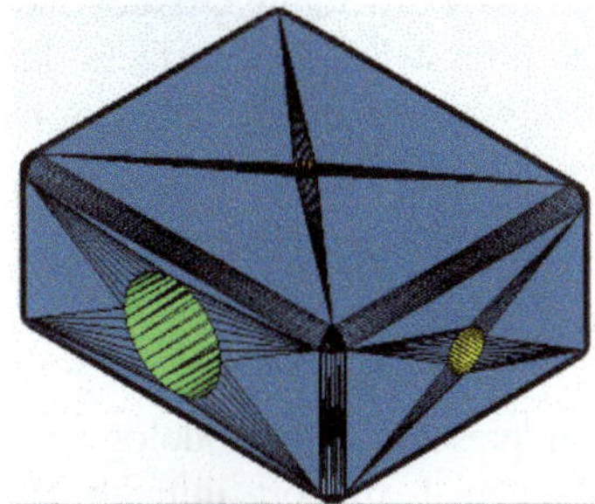
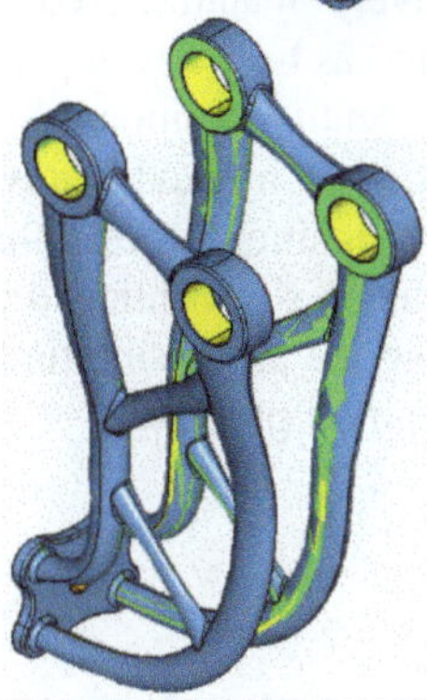	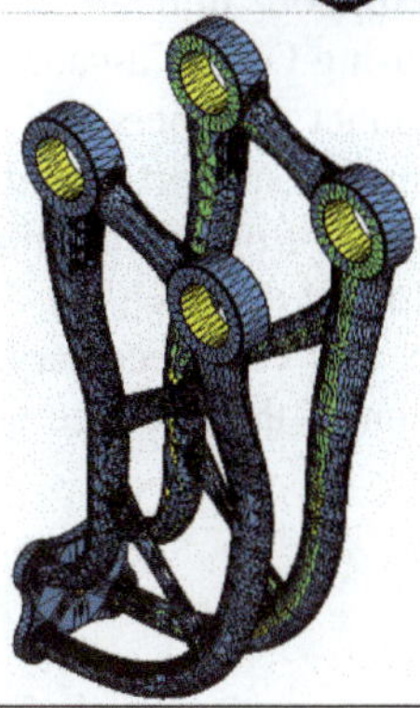

Table 2. Time-consuming to finish the recognition process of STEP and STL file

Name of sample	Time-consuming (s)	
	STEP file	STL file
1	0.05	0.07
2	202.08	73.02
3	3568.08	19260.38

Table 3. Memory size of some sample parts in STEP and STL file

Name of sample	Memory size (KB)	
	STEP file	STL file
1	51	21
2	117	783
3	1830	12694

The result of the recognition process is just a reference for designers to identify the positions of features in part. Making decisions to put support structures depends on them and on the way, they arrange parts during the printing process. The orientation is very important. For example, if a cylinder hole is put in the vertical direction- the best position to print cylindrical surfaces, there is no need to put support structures inside although this is one of the surfaces had been recognized that need to put support structures.

4 Conclusion

This paper presents a methodology to perform the recognition process intended to identify potential support difficult to remove on additively manufactured parts. A simple interface using Open Cascade Technology software has been developed. The interface allows product designers and CAM operators to identify positions that can contain difficult removal support structures when printing a part by Additive Manufacturing. The analysis of processing time to perform the recognition process for two types of CAD file format and memory-consuming to store the files is mentioned in the paper. In the future, it is very necessary to focus on reducing the processing time and improve the precision of the interface to perform the recognition process.

References

1. Comotti, C., Regazzoni, D., Rizzi, C., Vitali, A.: Additive manufacturing to advance functional design: an application in the medical field. J. Comput. Inf. Sci. Eng. **17**, 031006.1–9 (2017)
2. Zhang, F., Campbell, R.I., Graham, I.J.: Application of additive manufacturing to the digital restoration of archaeological artefacts. Int. J. Rapid Manuf. **6**, 75–94 (2016)
3. Shi, Y., Zhang, Y., Baek, S., De Backer, W., Harik, R.: Manufacturability analysis for additive manufacturing using a novel feature recognition technique. Comput. Aided Des. Appl. **15**, 941–952 (2018)
4. Calignano, F., Manfredi, D., Ambrosio, E.P., Biamino, S., Lombardi, M., Atzeni, E., Salmi, A., Minetola, P., Iuliano, L., Fino, P.: Overview on additive manufacturing technologies. In: Proceedings of the IEEE, pp. 1–20 (2017)
5. Hussein, A., Hao, L., Yan, C., Everson, R., Young, P.: Advanced lattice support structures for metal additive manufacturing. J. Mater. Process. Technol. **213**, 1019–1026 (2013)

6. Jiang, J., Xu, X., Stringer, J.: Support structures for additive manufacturing: a review. J. Manuf. Mater. Process. **2** (2018). jmmp2040064, 23 pages
7. Zhang, Y., Bernard, A., Harik, R., Karunakaran, K.P.: Build orientation optimization for multi-part production in additive manufacturing. J. Intell. Manuf. **28**, 1393–1407 (2017)
8. Mbow, M.M., Vignat, F., Marin, P., Perry, N. : Calcul d'Orientation en Fabrication Additive Métallique sur Lit de Poudre - Une Nouvelle Approche Quantitative par Fonction de Désirabilité, 16e Colloque National S-mart (2019). 6 pages
9. Grandvallet, C., Mbow, M.M., Mainwaring, T., Pourroy, F., Vignat, F., Marin, P.: Eight action rules for the orientation of additive manufacturing parts in powder bed fusion: an industry practice. Int. J. Interact. Des. Manuf. 1–12 (2020)
10. Nelaturi, S., Kim, W., Kurtoglu, T.: Manufacturability feedback and model correction for additive manufacturing. J. Manuf. Sci. Eng. **137** (2015). 021015 (9 pages)
11. Open Cascade Homepage. https://www.opencascade.com/

An Integrated Simulation Approach to Study the Propulsion Mechanism of a Biomimetic Fishtail

Anh Tuan Nguyen(✉), Anh Vang Tran, Quoc Nam Tang, Lac Hong Nguyen, Minh Duc Ha, and The Tuyen Nguyen

Le Quy Don Technical University, Hanoi, Vietnam
atnguyen@lqdtu.edu.vn

Abstract. This paper presents an integrated simulation approach to study the propulsion mechanism of a biomimetic silicone fishtail while operating in water. The bending motion of the tail is generated by a cyclic hydraulic water flow that provides a harmonic variation in the inner pressure. A nonlinear finite-element method is employed to obtain the properties of the soft structure; and based on these data, a simplified bar-spring system is created to model the tail. The fluid-structure interaction is handled in the environment of multibody dynamics simulation software which couples the bar-spring system model and an unsteady panel method. Here, the unsteady panel method is used to solve the hydrodynamic problems arising from the complex motion of the fishtail in water. The validations of the bar-spring system model and the unsteady panel method are investigated and confirmed in this paper. The simulation runs for various cases with different values of the hydraulic water pressure and the motion frequency. Based on the propulsion results, it is shown that the present fishtail design is valid for a complete biomimetic fish robot that will be developed soon by the research group.

Keywords: Biomimetic fish robot · Soft robot · Finite–element method · Fluid–structure interaction

1 Introduction

Fish robots belong to a class of biomimetic aquatic robots, which mimic the swimming mechanisms of biological fishes in nature [1]. In terms of geometry, streamlined shapes are normally applied to reduce the fluid drag in an effort to enhance the swimming performance. On the other hand, the relative locations of the center of mass, the geometric center and the center of pressure play an important role in stabilizing the robot while swimming under water. While these abovementioned factors can be easily tuned and rectified in the initial stage of the designing process, the more complex problem is to design the propulsive mechanism of the robot. For both biological and biomimetic fishes, the tail is the main part to generate the propulsive force. While actual fishes can manipulate many joints along their spines to create various motion patterns of their tails so as to generate the propulsive force efficiently [2], robotic fishes have their own mechanical limitations in terms of force generation. In fact, fishes in

B. T. Long et al. (Eds.): MMMS 2020, LNME, pp. 808–814, 2021.
https://doi.org/10.1007/978-3-030-69610-8_107

specific and aquatic animals in general have evolved for millions of years to achieve their current optimal swimming mechanisms. Hence, learning from biological counterparts is a shortcut for robotic fishes to rapidly improve their swimming capabilities.

As noted above, there are always limitations for robotic fishes in copying the propulsive mechanisms of actual aquatic animals. A solution, which has been devised to overcome this trouble, is the use of soft mechanisms [3]. Compared to conventional mechanical joints, soft mechanisms are more flexible and lightweight. However, it should be noted that using soft mechanisms may give more troubles to the development of the control algorithms. In the present study, after compromising many aspects including the efficiency, cost and the availability of the technology, our research group has decided to use the soft propulsive mechanism for the current robot design. This paper shows an integrated simulation approach that is developed to investigate the complex mechanical characteristics of the soft fishtail. The tail is modelled with a simple bar-spring system, and the fluid-structure interaction phenomenon can be studied in detail. Compared to the conventional approach based on the finite-element method, the use of the bar-spring system can reduce the computational cost significantly, which is of great importance in the designing process. The hydrodynamic force is computed by an unsteady panel method [4]. The simulation is performed for a tail model undergoing a harmonic bending motion generated by a cyclic pump at a frequency of 1.0 Hz.

2 Materials and Methods

2.1 The Design of the Fishtail Model

The fishtail comprises the fin layers, the neutral layer and the elastic parts (Fig. 1). The fore part of a fin layer is directly attached to the neutral layer and the elastic parts of the tail. For fishes, fins play a significant role in the generation of the propulsive force. At the trailing edge of the fin, water is shed backward according to the Kutta condition [4] to provide thrust to the model. The neutral layer is the middle layer of the tail, and this component keeps neither compressed nor extended during the pressurization process. The elastic parts, which have the most complex structures, are made of Silicon RTV 25. The centered channel is created with the inlet and outlet are placed at the two ends of an elastic part. Water is carried through this channel to the cavities, which are distributed uniformly along the tail. Inside these hollow chambers, water can be either pressurized or depressurized in order to bend the tail left- or rightward. A cyclic hydraulic pump is connected to the inlets of the tail, and works at a frequency of 1.0 Hz.

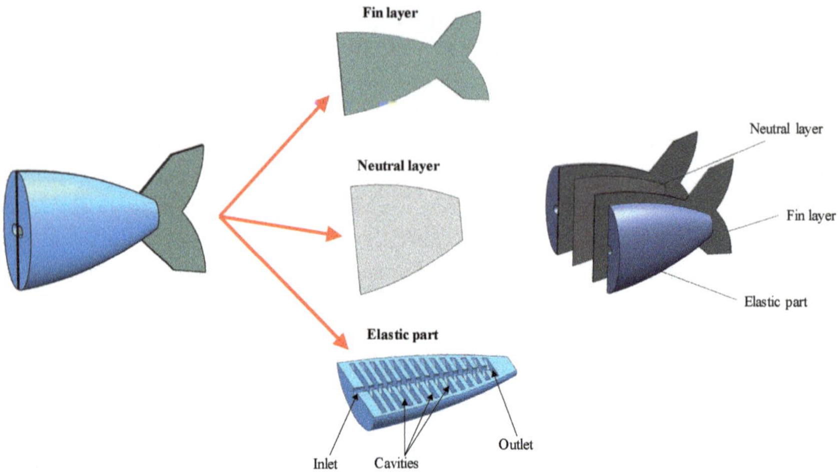

Fig. 1. Fishtail components

2.2 Structural Model

The structure of the fishtail is firstly modelled in finite-element software, in which the solid elements are applied to the elastic part, and the fin and neutral layers are modelled with the shell elements. The front end of the tail is clamped, and water at different pressure levels is pumped to the right half of the tail. Figure 2 shows the deformed shape of the tail when the pressure varies from 0 to 60 kPa.

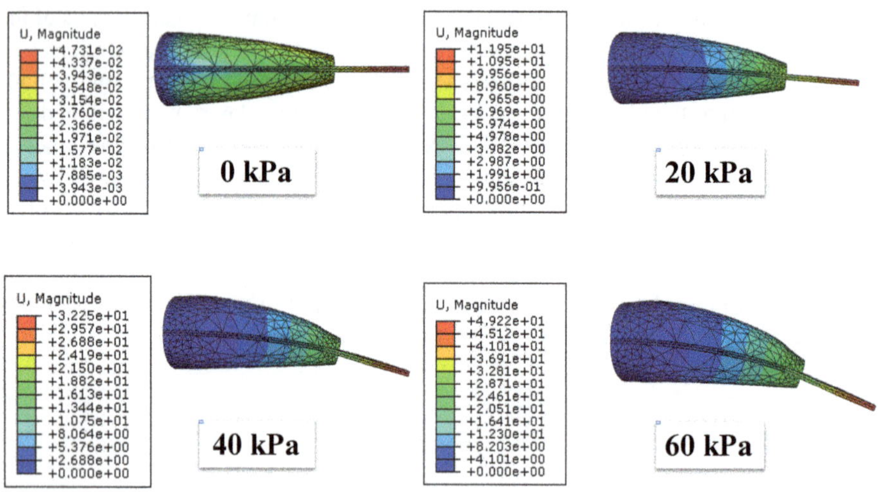

Fig. 2. The deformed shape of the fishtail at various pressure values

Next, an equivalent bar-spring system is developed based on data from finite-element simulation (Fig. 3). Firstly, the stiffness of the torsional springs is obtained based on the bending angles at the cylindrical joints due to a nodal force applied at the end of the tail. In this study, a system of four bars and four springs is used. For the next step, the external torque at each joint of the system is determined as a function of the hydraulic pressure. To validate the present equivalent system, the bending angle θ, which is defined in Fig. 4, is calculated at various pressure values by a finite-element method and a multibody-dynamics solver for the mass-spring system.

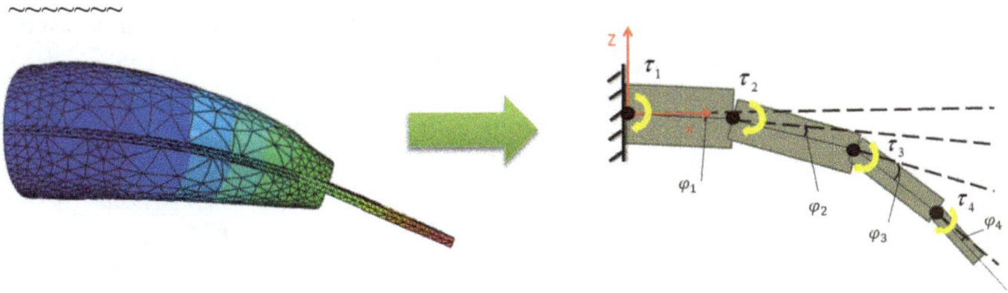

Fig. 3. The equivalent mass-spring system

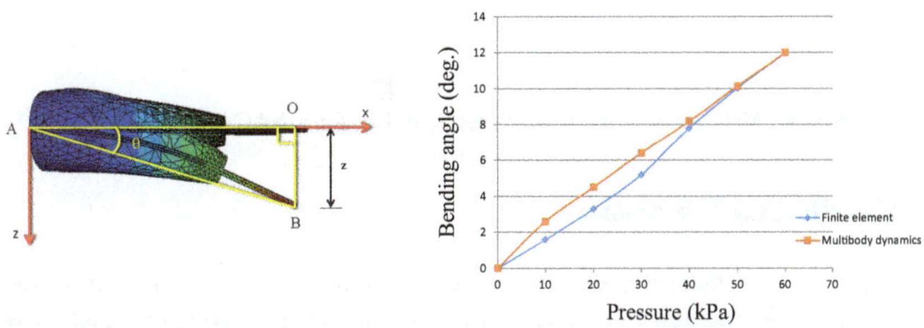

Fig. 4. Validation of the present equivalent mass-spring system

3 Hydrodynamic Model

The hydrodynamic model is based on the unsteady panel method, which is developed to solve the Laplace equation for potential flows [4]. It should be noted that to avoid any abnormalities in the solution of the hydrodynamic solver, the fore part of fish body is included. The geometry and the panel mesh of the fish are shown in Fig. 5. According to the panel method, on each panel we place a source and a doublet, whose strengths are determined by taking the boundary conditions into account. Here, we apply three boundary conditions, which are the far-field, the Dirichlet and the Neumann conditions. The far-field boundary condition guarantees zero perturbations at a large distance. The source strength is obtained by the use of the Dirichlet boundary condition, which prevents the flow from penetrating into the body. The Neumann condition stated that the velocity potential inside the body is zero, or rather no flow appears in this region.

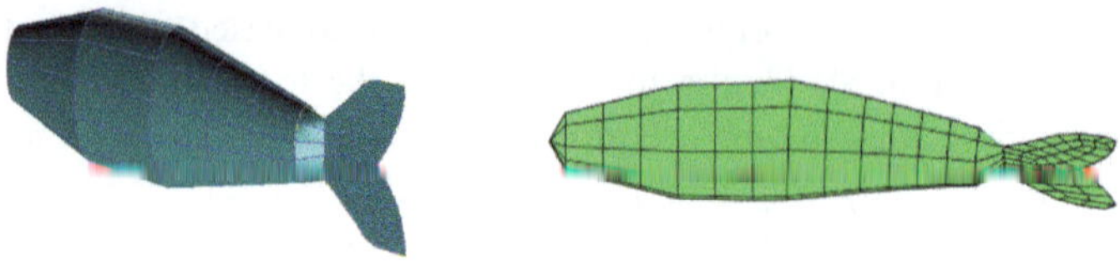

Fig. 5. The geometry and hydrodynamic panel mesh of the robotic fish model

To validate the hydrodynamic model, the lateral force and pitching moment of a pitching/heaving NACA0012 hydrofoil are computed and compared with those from an experiment [5]. Good agreement is observed in Fig. 6; thus, it is relevant to ascertain the validity of the present hydrodynamic model.

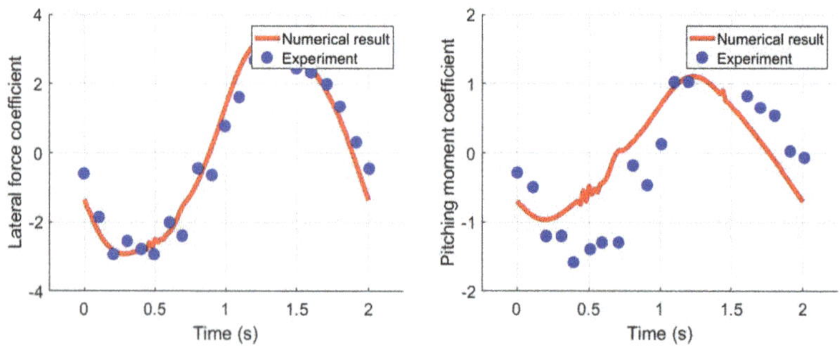

Fig. 6. Validation of the hydrodynamic model for a NACA0012 hydrofoil

4 Results and Discussion

The approach presented in Sect. 2 is applied to the robotic fish model with respect to several hydraulic pressure values. The thrust and side force are computed and shown in Fig. 7.

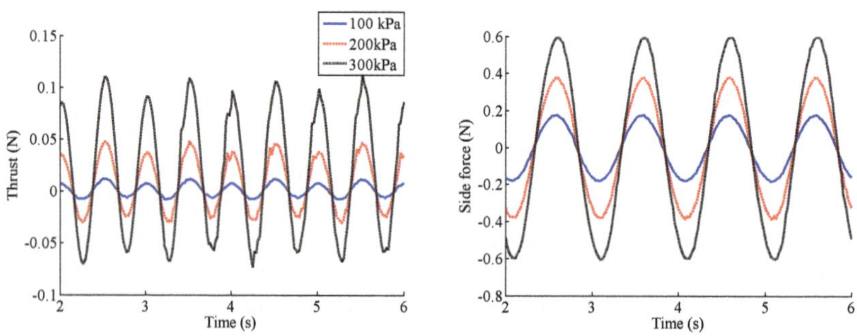

Fig. 7. Thrust and side force at different hydraulic pressure values

It is seen that the forces oscillate at the frequency of the cyclic pump (1.0 Hz). The amplitudes of the forces increase with the hydraulic pressure, and the effect of the pressure on the thrust is greater than that on the side force. The mean side force is almost zero, which indicates the symmetric nature of the bending motion. The mean thrust grows with the pressure as shown in Fig. 8. It appears that the mean thrust increases with a parabolic rule against the hydraulic pressure.

The fluid wake generated by the motion of the tail is shown in Fig. 9. It is seen that at larger pressure, the bending of the tail becomes more significant; hence, the fluid behind the tail is pushed backward more rapidly. According to the momentum conservation principle, a larger propulsive thrust is generated.

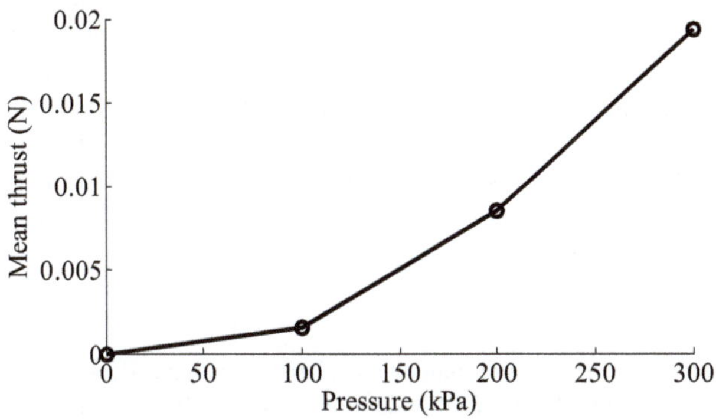

Fig. 8. Mean thrust against the hydraulic pressure

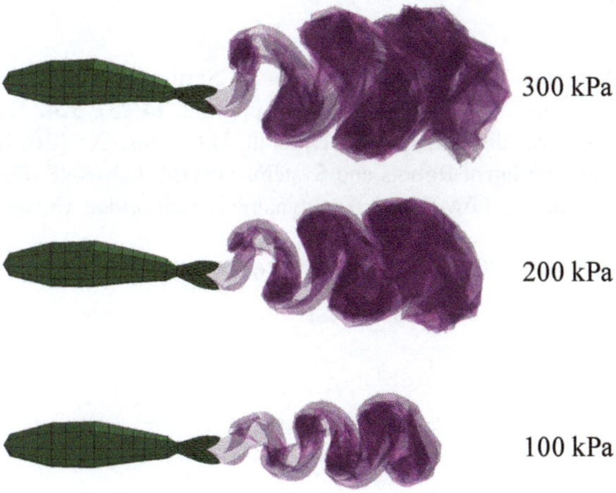

Fig. 9. Wake structure at different hydraulic pressure values

5 Conclusions

The present paper has explored the propulsive mechanism of a biomimetic soft fishtail through a simulation approach that integrates an equivalent mass-spring system and the unsteady panel method. The mass-spring system is built based on data from the finite-element method using the solid and shell elements. The multibody-dynamics solver is used to obtain solutions of the mass-spring system. The unsteady panel method, which is validated in this paper, is employed to compute the hydrodynamic force acting on the model. A cyclic pump is designed to connect to the tail and provide hydraulic pressure to bend it harmonically at a frequency of 1.0 Hz. Compared to the conventional approach using the finite-element method, the present one can reduce the computational cost by decreasing the number of degrees of freedom. The numerical results have shown that the mean thrust increases with the hydraulic pressure provided from the pump by a parabolic rule. The variation of the side force is also derived against the pressure. Moreover, the simulation program can produce the structure of the wake shed from the tail; and according to the momentum conservation principle, the backward motion of this wake accounts for the generation of the thrust when the tail bends along its vertical axis. The results from this paper can form a basis for the further design of a biomimetic fish robot as well as the development of nonlinear control algorithms applied under water.

Acknowledgement. This research is funded by Vietnam National Foundation for Science and Technology Development (NAFOSTED) under grant number 107.01–2018.05. The authors also thank for the supports of the Robotic Research and Development Group at Le Quy Don Technical University.

References

1. Lauder, G.V., James, L.T.: Robot Fish, vol. 25. Springer, Heidelberg (2015)
2. Kiam, B.Y., Loong, W.W., Te, K.T.K.: J. Appl. Sci. **14**(23), 3242 (2014)
3. Katzschmann, R.K., de Maille, A.D., Dorhout, D.L., Rus, D.: 2016 IEEE/RSJ International Conference on Intelligent Robots and Systems (IROS), vol. 3048. IEEE (2016)
4. Katz, J., Plotkin, A.: Low-Speed Aerodynamics. Cambridge University Press, Cambridge (2001)
5. Xue, G., et al.: Appl. Bionics Biomech. **2016**, 6909745 (2016)

Application of GMR Based Probe – for Eddy Current in Flaw Testing System

Xuan Thang Trinh[1(✉)], Jen-Tzong Jeng[2(✉)], Trung Thanh Bui[1], Ngoc Thai Tran[1], and Quoc Tuan Hoang[1]

[1] Faculty of Mechanical Engineering, Hung Yen University of Technology and Education, Hungyen, Vietnam
xuanthangcdt@gmail.com
[2] Department of Mechanical Engineering, National Kaohsiung University of Science and Technology, Kaohsiung, Taiwan
jtjeng@nkust.edu.tw

Abstract. In this research, we have investigated an application of eddy current probe-based on the GMR spin-valve for the flaw testing system. The eddy current probe consists of a GMR spin-valve sensor and a square modulation coil. The sensing direction is performed perpendicular to the modulation field generated by a modulation coil. The GMR spin-valve used in this work is a model GF708 from Sensitec Inc with a bare-die version. The scanning length is 80 mm for six cracks with a different depth on the surface of a metal specimen. The gap between the cracks to each other is about 14.5 mm. The experimental results are obtained by scanning the probe over six cracks of a specimen. The crack positions on the surface of a specimen are determined by the peak or valley values of in-phase and quadrature-phase differential defect field. Similarly, the depth of the cracks is estimated by the phase of the differential defect field.

Keywords: Magnetic sensors · GMR sensor · Eddy current

1 Introduction

Nowadays, the applications of magnetic sensor is an important role in production, research, or medical and so on. The applications must be mentioned such as in magnetic tracking [1, 2], medical [3, 4], navigation, compass application [5, 6], flaw detection, current sensing technique [7, 8]. Among of them, the applicaton of magnetic sensors such as Hall sensor, AMR sensor (Anisotropic magnetoresistance), GMR sensor (Giant magnetoresistance) in flaw detection application has been researched [8–11]. For AMR and GMR sensors, they are the small size sensors and having high sensitivity, low noise at the room temperature. However, the directional property of GMR sensor is better than AMR [12]. In this paper, a testing probe is fabricated to use in flaw testing application. The probe operation is based on Eddy current sensor, it consists of a GMR spin-valve and a square coil. Based on the applications of the GMR sensor in some works mentioned above, we proposed an application of the GMR spin-valve sensor in making a probe based on Eddy current for the flaws testing system with

B. T. Long et al. (Eds.): MMMS 2020, LNME, pp. 815–820, 2021.
https://doi.org/10.1007/978-3-030-69610-8_108

the metal specimen. To reduce the size of probe, the used GMR spin-valve sensor is the bare-die version. The design and test results of this probe are demonstrated and discussed as follows.

2 Experimental Details

2.1 Experimental Setup

The flaw detection system with the probe of GMR sensor-based eddy current include a GMR spin-valve sensor and a square modulation coil was described in [7, 8, 12]. The used GMR spin-valve sensor is commercial model GF708 from Sensitec Inc., with bare die version, [13]. The GMR sensor is formed in the Wheatstone bridge, as shown in Fig. 2. In which, the two GMR cells was shielded to play a role of reference resistor. The GMR sensor is mounted on the PCB so that the sensing direction is perpendicular to the length of PCB. The aluminum wire bonding method is used to connect between the contact pads of the GMR sensor and the PCB. A metal specimen, which has the cracks with the different depths, was used for the scanning detection process. The sensing direction of GMR was fixed parallel to the surface of the scanning direction. The experimental setup diagram of the flaw testing system is shown in Fig. 1.

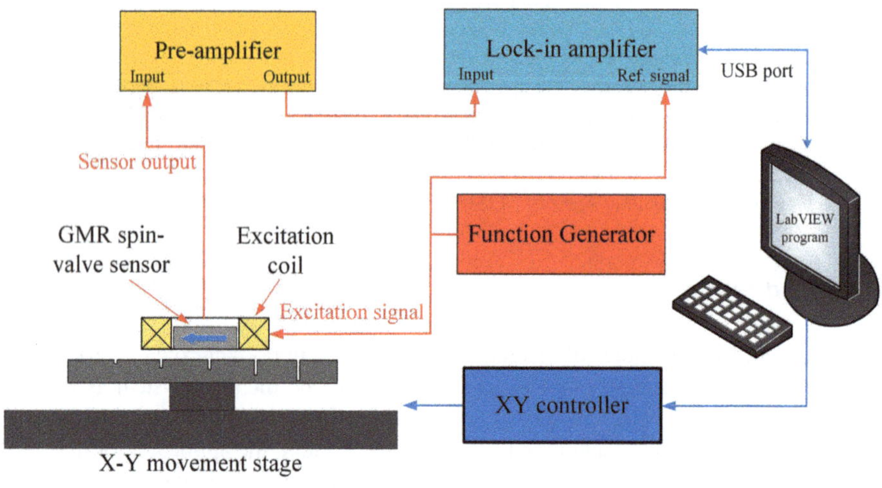

Fig. 1. Block diagram of flaw testing system

The excitation signal used is the sine wave, provided for the excitation coil to generate the vertical component excitation magnetic field. The excitation frequency is 30 kHz. To implement the scanning process, the specimen is moved by an X-Y translation stage while the probe is fixed. The probe is scanned along the length of specimen with the scanning length of 80 cm, covering six flaws on the surface of the specimen. The distance between the cracks to each other is about 14.5 mm. The probe is fixed with distance to the surface of specimen adjusted about 0.5 mm.

The movement stage system is driven by the controller model 08TMC-2 from Unice E-O Services Inc. The controller diver is connected to the PC by USB port. The scan area and velocity can be set by the computer program. The working principle of the probe is based on the perturbation of the eddy-current flow path at the defect location of the scanning surface. At the flaw locations, the sensor output signal is generated by the disturbance of the eddy current flow path only. At the location of the defect-free specimen, these will produce no effect on the sensor output because of the circular symmetry of the induced eddy currents. The sensor output is amplified by the low-noise pre-amplifier with gain of 2. After that, the amplified sensor output is analyzed using the lock-in detection technique, the lock-in amplifier used is model SR865A from Stanford Research System. The sensitivity of lock-in amplifier is set to be 20 mV, corresponding to the gain of 500 with the time constant of 10 ms. Finally, the output data is directly read from lock-in amplifier, the lock-in amplifier is communicated with the PC via a USB connection port. A power supply device is used to provide a 5V dc bias voltage. The program to drive flaw detection system and record the data is implemented by the program coded in LabVIEW software. The output voltage of flaw detection recorded with in-phase and quadrature phase as function of *x*-placement are plotted on the program interface.

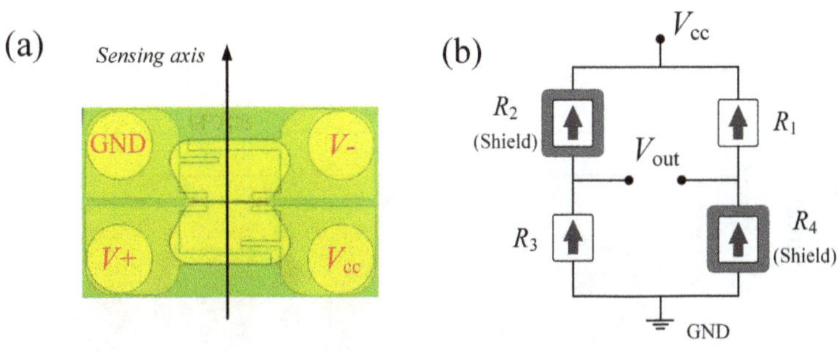

Fig. 2. The used GMR sensor spin-valve. (a) Topview of sensor, (b) Wheatstone bridge configuration of GMR spin-valve.

2.2 Results and Discussion

The application of the eddy-current probe used GMR spin-valve was demonstrated for the flaw detection on the specimen. The probe scanned over the 80 mm length of the specimen. The probe output voltage is recorded and shown in Fig. 3(a). The output of probe is recorded with the in-phase and quadrature phase output voltage. In Fig. 3(a) shows that the amplitude of the defect output voltage is increased when increasing the depth of the flaws. The defect field can be converted from the defect output voltage with the same phase. As the probe sweeps over the crack positions, the resulting obtained defect voltage of in-phase and perpendicular phase is asymmetric across the defect positions, as shown in Fig. 3(a). The asymmetric defect signal, however, is usually accompanied by an offset voltage, possibly due to the excitation coil tilting against the scanning surface. Figure 3(a) also shows that the amplitude of the output voltage of the eddy current probe is proportional to the depth of the flaws.

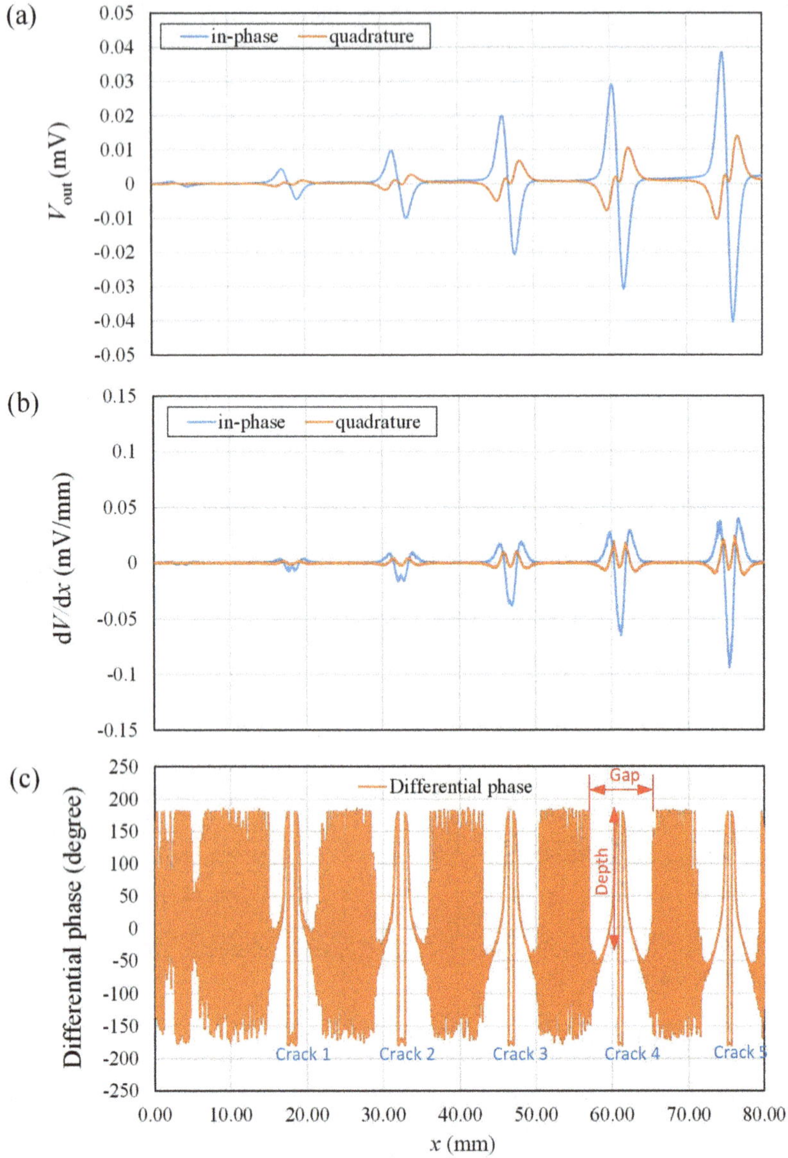

Fig. 3. The variation of the in-phase and perpendicular-phase component in scanning process over the crack of a specimen and scanning length of 80 mm. (a) the defect field in output voltage, (b) differential defect field, and (c) the spatial phase of the differential defect.

Figure 3(b) shows the differential defect output voltage of in-phase and perpendicular-phase when the eddy current probe scan over the various flaw depths on the specimen. Due to the fact that spatial derivative of defect signal get the peak value (maximum or minimum value) at the inflection point. Figure 3(b) indicate that the flaw positions on the specimen is shown by the valley or peak of the in-phase and the perpendicular-phase differential defect fields. In Fig. 3(c), the phase of differential defect field is presented with the various flaw depths. It shows that the phase of the differential defect field near the crack positions is quite similar and uniform. Moreover, we can see that the phase around the position of cracks changes uniformly while the depth of the flaws increases in the scanning process. The behavior of the phase of differential defect field is quite similar to the linear function of the crack depth, presented in [14], it indicates the dependence of the depth to the defect field phase. The formula to calculate the phase of differential defect field was given in [15]. The phase of differential defect field ϕ can be calculated by the in-phase and perpendicular-phase differential output voltage following the equation:

$$\phi \approx \operatorname{atan}\left[\left(\mathrm{d}V_{90}^{\mathrm{o}}/\mathrm{d}x\right)\left(\mathrm{d}V_{0}^{\mathrm{o}}/\mathrm{d}x\right)\right] \quad (1)$$

in which $V_{0^\circ}(x)$ and $V_{90^\circ}(x)$ are the defect output voltage of in-phase and perpendicular-phase. At the flaw positions, the phase of $\mathrm{d}V/\mathrm{d}x$ is equal to the phase of $V(x)$. Although the phase of the differential defect field should be close to the phase of the defect field at the position of cracks, provide that there is no offset in defect field. For the phase of differential defect field, we should consider the signal at the flaw locations and ignore the signal at the other locations where the defect fields are negligibly small. The phase change of differential defect field shows that it is beneficial for the evaluation of the defect detection process. The position of flaws could be easily determined with the curves change of the phase of the differential defect signal, which is indicated in Fig. 3 (c). The peak value of the curve shown in Fig. 3(c) is similar for the crack position while the bottom value of the W shape corresponds to the depth of the crack since these values are equivalent to the phases of the defect field, which are the linear functions of the depth of the crack.

3 Conclusion

The evaluation of flaw depth using the eddy current probe based on GMR spin-valve was also investigated. The phase and magnitude of the differential defect field are correlated to the defect field, directly. By taking the spatial derivative, the unwanted offset field was removed. The positions of flaws on a specimen are determined by the peak or valley values of the in-phase and perpendicular-phase differential defect field while the depth of the flaws can be indicated by the differential phase defect field. The proposed application is promising in non-destructive testing, flaw detection for metal products.

Acknowledgments. The authors acknowledge and thank Micro Magnetic Technologies Lab, National Kaohsiung University of Science and Technology for their support this work.

References

1. Dai, H., Yang, W., Xia, X., Su, S., Ma, K.: A three-axis magnetic sensor array system for permanent magnet tracking. In: IEEE International Conference on Multisensor Fusion and Integration for Intelligent Systems (MFI), pp. 476–480 (2016)
2. Wang, M., Shi, Q., Song, S., Meng, M.Q.H.: A Novel magnetic tracking approach for intrabody objects. IEEE Sensors J. **9**, 4976–4984 (2020)
3. Song, S., Ren, H.: Magnetic tracking in medical robotics. Electromagnetic Actuation and Sensing in Medical Robotics, pp. 141–162 (2018)
4. Pham, D.M., Aziz, S.M.: A real-time localization system for an endoscopic capsule using magnetic sensors. Sensors **14**, 20910–20929 (2014)
5. Chiang, C.-Y., Jeng, J.-T., Lai, B.-L., Luong, V.S., Lu, C.-C.: Tri-axis magnetometer with in-plane giant magnetoresistance sensors for compass application. J. Appl. Phys. **117**, 17A321 (2015)
6. Trinh, X.-T., Jeng, J.-T., Lu, C.-C., Lan, M.-J., Chen, B.-C., Hsu, J.-H., Luong, V.-S., Hsu, H.-Y.: Miniature tri-axis magnetometer with in-plane GMR sensors. IEEE Trans. Magn. **53**, 1–4 (2017)
7. Reig, C., Cubells-Beltrán, M.-D., Muñoz, D.M.: Magnetic field sensors based on giant magnetoresistance (GMR) technology: applications in electrical current sensing. Sensors **9**, 7919–7942 (2009)
8. Ziegler, S., Woodward, R.C., Iu, H.H.-C., Borle, L.J.: Current sensing techniques: a review. IEEE Sens. J. **9**, 354–376 (2009)
9. Tajima, H., Yusa, N., Hashizume, H.: Application of low-frequency eddy current testing to the inspection of a double-walled tank in a reprocessing plant. Nondestructive Test. Eval. **2**, 189–197 (2018)
10. Saari, M.M., Nadzri, N.A., Halil, A.M., Ishak, M., Sakai, K., Kiwa, T., Tsukada, K.: Design of eddy current testing probe for surface defect evaluation. Int. J. Autom. Mech. Eng. **1**, 6357–6367 (2019)
11. Sophian, A., Guiyun, T., Mengbao, F.: Pulsed eddy current non-destructive testing and evaluation: a review. Chin. J. Mech. Eng. **3**, 500–514 (2017)
12. Dogaru, T., Smith, S.T.: Giant magnetoresistance-based eddy-current sensor. IEEE Trans. Magn. **37**, 3831–3838 (2001)
13. Sensitec Inc. GF708 GMR Sensor. https://www.sensitec.com/. Accessed 26 May 2020
14. Jeng, J., Yang, S.-Y., Horng, H.-E., Yang, H.-C.: Detection of deep flaws by using a HTS-SQUID in unshielded environment. IEEE Trans. Appl. Supercond. **11**, 1295–1298 (2001)
15. Horng, H.-E., Jeng, J., Yang, H.-C., Chen, J.: Evaluation of the flaw depth using high-Tc SQUID. Physica C **367**, 303–307 (2002)

An Application of Kinect-Based 3D Scanning in Mechanical Engineering

Van-Bien Bui[1](✉), Tien-Long Banh[2], Duc An Pham[2], and Duc-Toan Nguyen[2]

[1] Hai Phong University, Hai Phong city, Vietnam
bienbv80@dhhp.edu.vn
[2] Hanoi University of Technology and Science, Ha Noi city, Vietnam

Abstract. During last decades, many techniques for CAD/CAM have been applied to industries. For stage of the product design process one can choose the suitable tools for making this process easier and faster. Nonetheless, most of them, both for hardware-based and software-based solutions, are expensive. This paper approaches the study on the use of a low-cost Kinect v2 device for 3D scanning of mechanical parts. Firstly, the influence of scanning factors on the Kinect-based system is considered. Secondly, in the mechanical part modeling process using the Kinect v2 device, the captured depth maps is improved. Finally, an experimental methodology for 3D scanning of mechanical parts is report.

Keywords: Mechanical engineering · Application · Kinect v2 · 3D scanning

1 Introduction

Manufacturing is a major activity in promoting economic growth and development of any country. Recently, reverse engineering (RE) has become an essential sector of manufacturing industries owing to its distinct applications. As reported by the Society of Manufacturing Engineering, RE can be defined as the procedure of working with a exist part and reconstructing to a suitable design format which new parts or molds can be manufactured. With the development in technology, contemporary RE techniques can be realized a design methodology where an existing part is first scanned or measured to obtain data and then a computer aided design (CAD) model is reconstructed as in Fig. 1.

A huge efficient benefit is that parts manufactured by RE engineering possess a low risk failure and the production process spends less time and lower costs than the conventional engineering. As mentioned in Fig. 1, the typical RE process begins with the acquisition of 3D data describing the shape and dimensions of the physical part, specially the free-form surfaces. This is generally achieved thanks to a 3D acquisition system, such as the 3D scanner and CMM; other methods [2, 3] apply a set of orthographic views or 2D images to reconstruct the 3D data of the physical part in a different condition. However, the price of the commercial devices is too high for small businesses and academic tasks.

B. T. Long et al. (Eds.): MMMS 2020, LNME, pp. 821–828, 2021.
https://doi.org/10.1007/978-3-030-69610-8_109

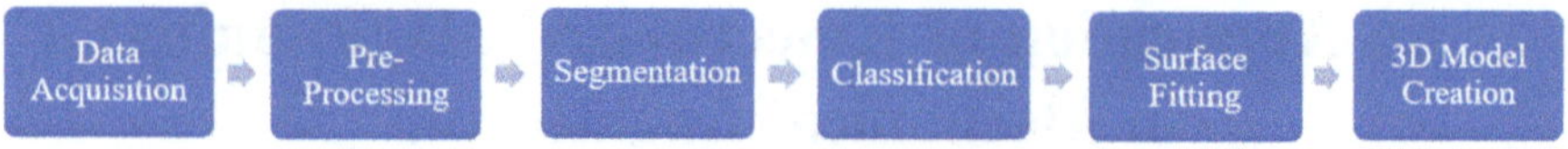

Fig. 1. The RE framework [1]

Being launched in 2014, the Kinect v2 used the Time-of-Flight technology to capture the 3D information. From the start, the purpose is to replace gaming commands and allow users to apply body motions to choose media content. This is to replace through a natural user interface. Kinect v2 features an RGB-B camera, depth sensor which can sense distance, and skeletal tracking. With its powerful technology, considerably low price, and effective functions, the Kinect v2 device is very affordable, which makes it an interesting tool for 3D modelling tasks not only in business but also in academics.

Due to the attractiveness and imaging capacities of the Kinect v2 device, during the last decade, large numbers of works have been completed. Applications for face tracking [4] or cyclical deformations [5] analysis appear in the literature. More recently, the direct reconstruction of 3D meshes is introduced in [6]. Besides, using the Kinect v2 with the Matlab toolbox as the 3D sensor has described good results for the real-time 3D reconstruction [7].

A comprehensive evaluation of Kinect v2 for 3D reconstruction is performed by Lachat et al. [8]. Several experiments have provided to recommending the errors appearing from the properties of the captured object and environment, as well as the errors associated with the device itself. Through error source assessments, all of calibration processes, both geometric and depth, are performed to reduce the effect of the error on the instrument's measuring accuracy. Eventually a 3D surface of an archaeological piece is reconstructed, thereby demonstrating the applicability and equipment accuracy of Kinect v2 in the reconstruction of 3D models of objects.

Although there are lots of research for the ability of Kinect v2 for 3D scanning, many problems and challenges have remained. In this paper, the reconstruction of CAD models of the mechanical part using Kinect v2 is proposed. Beginning from 3D measured data, some problems of this process is also solved. From the view of manufacturing industry applications, this RE process using Kinect v2 reconstruct the CAD model that is as close as desirable to the initial design of the mechanical parts.

2 Methodology

2.1 A Depth Enhancement Strategy

Because of noise inherent to the ToF sensor and its technology, individual frame acquisitions are less accurate. Averaging successive frames is a possible improvement step to overcome this phenomenon. In this section, the effect of frames averaging on the digitization process using Kinect v2 is tested. A statistical analysis with 200 samples is performed. The digitization process with a mechanical part surface is repeated from a single shooting. The point cloud observed with 10, 20, 30, 40, 50, 60, 70, 80, 90 100, 150 and 200 successive depth maps is analyzed.

Using the aluminum part with 120 mm × 120 mm × 14 mm, with the point cloud results obtained at the observation positions, the number of points, its depth values and its standard deviation is calculated and represented in Fig. 2. Specifically, the number of points obtained change from 481 points when the number of successive frames is 10 to 484 points when the number of successive frames is 30 and remains until the end of the experiment. The depth value of the points in the point cloud change a little and oscillate around the reference value. The standard deviation of the depth value of the points in the point cloud decreases as the number of depth maps increases. This shows that the accuracy of this device increases when the distance measurement is taken with the average depth maps.

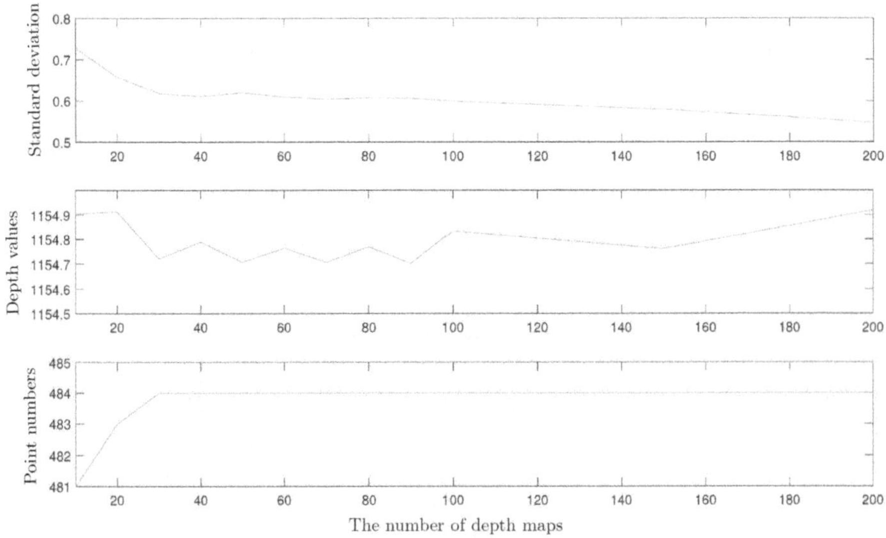

Fig. 2. The results in testing the influence of frames averaging.

Assume that Kinect v2 obtains N numbers of frames, in which the resolution of each frame is U × V pixels. Because of the noise the intrinsic to the camera, at the each pixel position (u, v), all of N data samples may also contain outliers. These outliers should be removed. Based on the median absolute deviation (MAD) method, the each frame acquired are classified and then the outliers are removed [1].

For a normal distribution, the MAD is the median of absolute deviation from the median, and defined by

$$MAD = b \cdot M(|x_i - M(X)|) \tag{1}$$

where the median of a given distribution is abbreviated M, X represents the set encompassing the N samples of all frames acquired, x_i is each particular sample in the data set X. Presuming the depth magnitudes to be a normal distribution, b = 1.4826 is chose [1], neglecting the a irregularities produced by the outliers in the data.

Based on the value of MAD, the benchmark for recognizing outliers is up to a threshold value set. The inequality is shown as follow

$$|x_i - M(X)| \leqslant \sigma \cdot MAD \tag{2}$$

In the case, a given sample x_i of data series X satisfies the inequality 2, this sample appertains to the data series. In the case, more than 50% of the data receive the same value, the MAD value equals zero. Then this method does not implement. According to [9], there are three thresholds for outliers detection: $\sigma = 3$, $\sigma = 2,5$ and $\sigma = 2$.

2.2 Steps of RE Implementation

Principally, the RE process using Kinect v2, shown in Fig. 3, can be consider as a development chain that is composed of three main operations as follows:

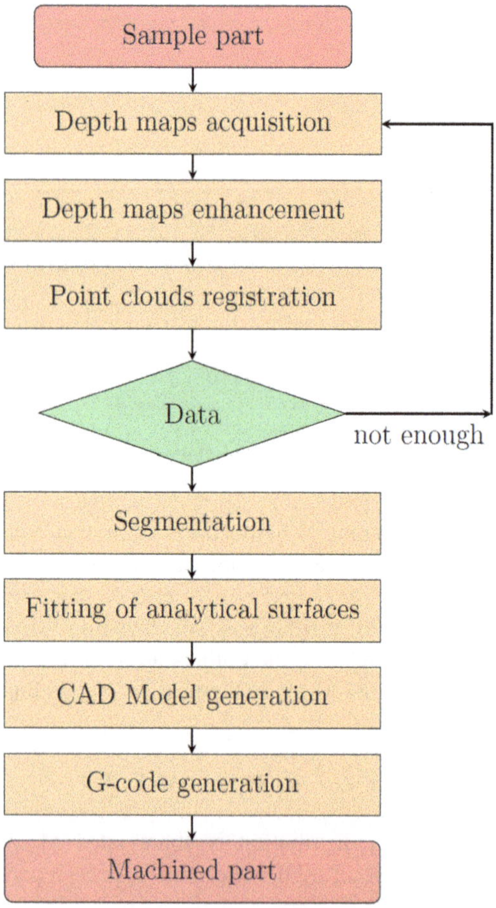

Fig. 3. Main steps in RE implementation.

- Digitization of the mechanical part: The point cloud of product is acquired by Kinect v2 and Matlab program.
- Creation of a CAD model: a CAD model of the mechanical parts must be built to represent all relevant data in Geomagic Design X.
- Manufacturing a complete product: a real part is manufactured by CNC machine.

2.3 A Case Study

The scanning system consists of a Kinect v2, turntable, Matlab and Geomagic design X software. A motorcycle helmets is an object to scan.

Before a scan is attempted, the anti-reflective coating is applied. After finishing from covering, the next step is calibration of Kinect v2 that ensures the position of the turntable and device to record accurately so the 3D scanner can produce the best possible scans.

A sequence consisted of depth maps acquisitions from twelve different standpoints is implemented around the motorcycle helmets under study. An overlap of about 85% on the mechanical part is presented between two successive depth maps. To reduce the noise inherent to the sensor, the depth enhancement process is realized. Based on acquired measurement data, the averaging of the depth maps for each standpoint and the computation of single point clouds from these depth maps are completed. Then, a point clouds registration is manually executed by calibrating with the checkerboard. The complete point cloud of the motorcycle helmets is presented in Fig. 4. Noise persists in the achieved point cloud and then also is hard to eliminate without neglecting geometric information. Therefore, the generation of the CAD model is time-consuming. Undoubtedly, many parameters can be experimentally established during the CAD model generation, which gain a significant influence on the deriving CAD model.

The application of Geomagic Design X will segment the point clouds of the motorcycle helmets into different colored feature region. It has a feature that helps to repair and draw the surfaces of the motorcycle helmets. This region is used to create drawings that define the various features of the initial object. Then sketches drafted and extruded, it is the body of the motorcycle helmets. Finally, the surfaces of the body of the motorcycle helmets are filleted. The CAD model is shown in Fig. 5.

CAD model used after design in CAM system to generate G-code by special software like Solidcam, but it will print in three dimensions so to check the scale of the sample part then manufacture on CNC Manford MCB-850 machine. The product is found in Fig. 6.

For the Kinect v2 device, depth maps quality is a particularly important factor that greatly involves the accuracy of the 3D reconstruction. Based on the median absolute deviation, the noise deduction on depth maps is reported and tested. Experiment shows that noise on the depth maps has decreased significantly. Furthermore, the combination of noise reduction on depth maps and the average depth maps before converting to the point cloud gives good results.

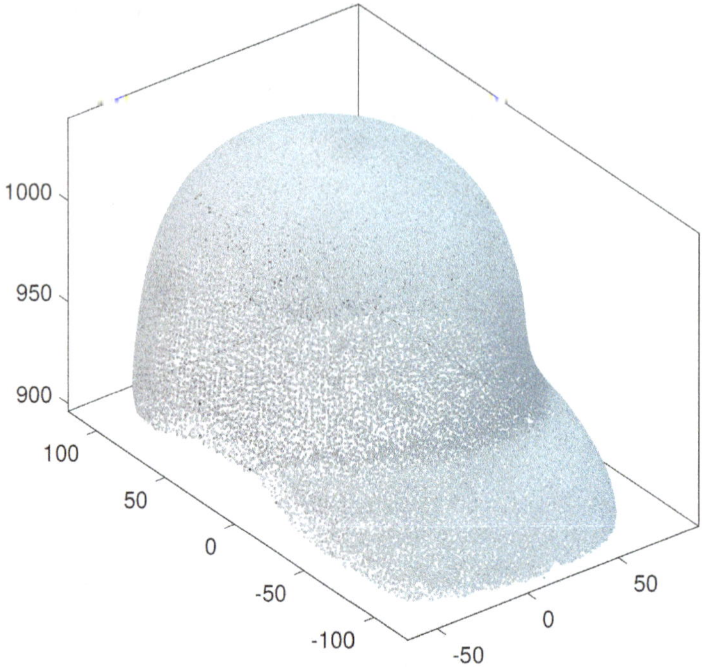

Fig. 4. The complete point cloud of the motorcycle helmets.

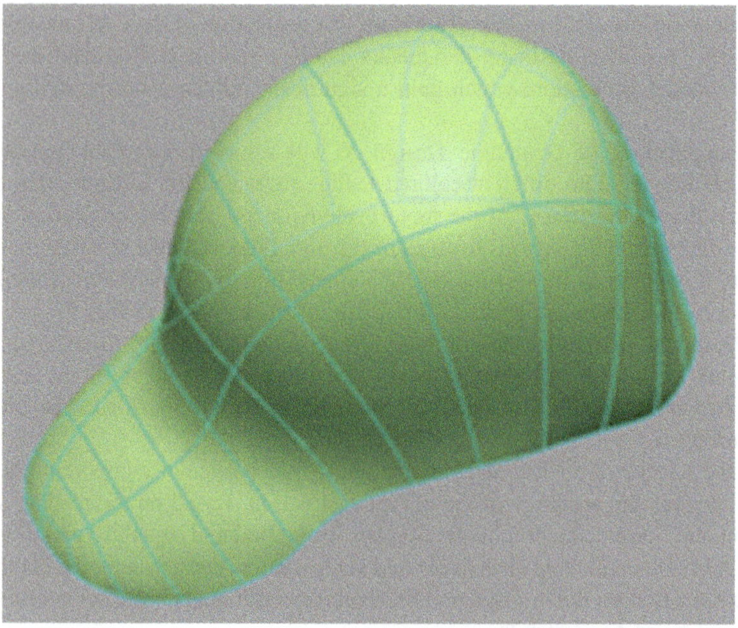

Fig. 5. The CAD model of the motorcycle helmets.

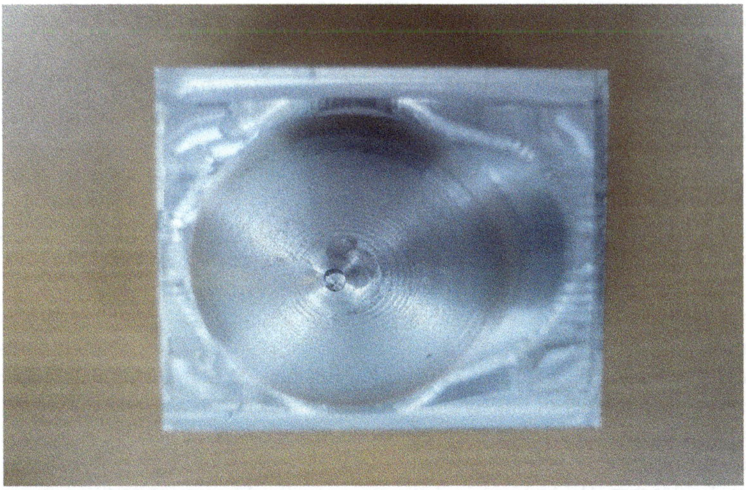

Fig. 6. The machined surface of the motorcycle helmets.

The purpose of this experiment was to validate the capacity of the Kinect v2 device, a low-cost camera, and to study if it could be an alternative for RE process of mechanical parts. Based on the results, it could be achieved that the CAD model of the sample part is created with a suitable accuracy.

3 Conclusion

This paper introduces the main steps of getting a CAD model to showing a simple method using a Kinect-based scanner to generate models accessible for CNC machines. The outputs of Kinect v2 are depth maps and they normally use sophisticated computational procedure to develop them accessible for a digital production. The fundamental procedure for creating the CAD model include depth maps enhancement, point cloud registration, and surface reconstruction. By means of Kinect v2 and our method, the CAD model of an existing object can be built robustly. This model is accepted by CNC machines. This work is also helpful to extend the applications of 3D printers.

In the next study, the CAD model generated by this research is not exactly same as the given mechanical part. Especially, a mechanical part with a detail feature surface may not success to be obtained by this approach. According to this limitation, we are starting to develop an appropriate filter to fix this problem.

References

1. Bui, V.-B., Banh, T.-L., Nguyen, D.-T.: Improving the depth accuracy and assessment of microsoft kinect v2 towards a usage for mechanical part modeling. J. Korean Soc. Precis. Eng. **36**, 691–697 (2019)
2. Bi, Z., Wang, L.: Advances in 3D data acquisition and processing for industrial applications. Robot. Comput. Integr. Manuf. **26**, 403–413 (2010)
3. Burston, M., Sabatini, R., Gardi, A., Clothier, R.: Reverse engineering of a fixed wing unmanned aircraft 6-DoF model based on laser scanner measurements. In: 2014 IEEE Metrology for Aerospace (MetroAeroSpace), pp. 144–149 (2014)
4. Hsieh, C.T.: A new kinect-based scanning system and its application, Applied Mechanics and Materials, pp. 1375–1379 (2015)
5. Lahamy, H., Lichti, D., El-Badry, M., Qi, X., Detchev, I., Steward, J. and Moravvej, M.: Evaluating the capability of time-of-flight cameras for accurately imaging a cyclically loaded beam. In: Videometrics, Range Imaging, and Applications XIII, vol. 9528, pp. 95280V-1–95280V-11 (2015)
6. Lachat, E., Macher, H., Landes, T., Grussenmeyer, P.: Assessment of the accuracy of 3D models obtained with DSLR camera and kinect V2. In: Videometrics, Range Imaging, and Applications XIII, vol. 9528, pp. 95280G-1–95280V-14 (2015)
7. Terven, J.R., Córdova-Esparza, D.M.: Kin2. a kinect 2 toolbox for MATLAB. Sci. Comput. Programm. **130**, 97–106 (2016)
8. Lachat, E., Macher, H., Landes, T., Grussenmeyer, P.: Assessment and calibration of a RGB-D camera (Kinect v2 Sensor) towards a potential use for close-range 3D modeling. Remote Sensing **7**, 13070–13097 (2015)
9. Miller, J., Report, S.: Reaction time analysis with outlier exclusion: bias varies with sample size. Q. J. Exp. Psychol. **43**, 907–912 (1991)

One More Evidence Supporting Damage Growth Inside Silicon MEMS Structures from Comparison of Strength Affected by Cyclic Compressive Stress

Vu Le Huy[1(✉)] and Shoji Kamiya[2]

[1] Faculty of Mechanical Engineering and Mechatronics, Phenikaa University, Ha Dong District, Hanoi, Vietnam
huy.vule@phenikaa-uni.edu.vn

[2] Department of Mechanical Engineering, Nagoya Institute of Technology, Gokiso-cho, Showa-ku, Nagoya, Japan

Abstract. Recently, dislocation in single-crystal silicon has been confirmed to be induced by fatigue. It was found that the fatigue lifetime of silicon is likely to be dominated by accumulation of crystal defects, i.e., dislocations. In previous studies, crystal defects occurring due to cyclic compressive stress were observed. However, there is rare evidence of those defects inducing fatigue damage under cyclic stress. This paper presents a result of strength comparison of silicon structures with and without cyclic compressive stress applying before static strength tests. The reduce of static strength induced by compressive fatigue shows another evidence supporting damage growth inside silicon MEMS structures.

Keywords: Silicon · MEMS · Fatigue · Strength · Compressive stress · Damage

1 Introduction

Silicon is the popular material in Micro electro-mechanical systems (MEMS), and it is increasingly used in MEMS nowadays. MEMS structures are commonly fabricated out of single-crystal silicon wafers and they are frequently applied by cyclic stress. Up to now, fatigue mechanism of silicon MEMS structures is still a debated problem.

The previous model of fatigue process [1, 2] has been assigned to the thickened surface oxide layer induced by cyclic stress and corrosion cracking, which named as the reaction layer model. However, the phenomenon of the thickened surface oxide layer was observed only under tensile cyclic loading, which can be explained that there is no room for oxygen infiltrating into silicon crystal under compression. Surface oxide thickening was not observed on poly-crystal silicon (polysilicon) MEMS devices after cyclic loading [3]. In contrast, there were indirect evidences [4, 5] suggested another possibility of fatigue mechanism based on dislocation mobility. Electron beam induced current (EBIC) [6] images were visualized damage evolution under cyclic compressive stress. They proved that the damage is induced by the growth of defects inside the silicon crystal [7, 8]. Besides, Paris' law with two unknown parameters, which is well-known for fatigue crack extension process in metallic materials, was used to formulate the fatigue lifetime of polysilcion films [9, 10] presented an indirect evidence of the dislocation in silicon.

B. T. Long et al. (Eds.): MMMS 2020, LNME, pp. 829–835, 2021.
https://doi.org/10.1007/978-3-030-69610-8_110

This paper is going to obtain another evidence supporting damage growth inside silicon MEMS structures. The tensile strength of the two groups of silicon structures are tested and compared statistically, where one group is applied by cyclic compressive stress before static strength tests. The reduce of static strength by compressive fatigue will reveal the effect of damage induced by dislocation.

2 Specimen Design

The shape of the specimens in this study is shown in Fig. 1(a), where the total length and the width is 50 mm and 6 mm, respectively. The specimens are fabricated by dicing pattern from (100) silicon wafers with the thickness of 0.4 mm as shown in Fig. 2, where the dicing was performed by using the dicing machine Disco DAD322. By distributing the specimens on the wafer as in Fig. 1(b), the length of the specimens lies along <110> orientation of the crystal.

Each specimen consists of a clamping zone to hold the specimen on the jig and a testing part with the length of 19 mm. Load is applied perpendicularly on the top surface of the specimens at the position with the distance of 19 mm from the clamping zone. There is a notch on the top surface of the specimen and distributing for 8 mm from the clamping zone. The notch was diced by using the dicing saw with the thickness of 30 μm. The depth of the notch is designed as 90 μm. The notch makes the stress concentration at its bottom and therefore the maximum stress in there is the object need to be obtained from the test as the strength of a specimen. If the applied load or displacement at the loading position is known, the maximum stress will be evaluated by the finite element method (FEM). Figure 2 presents the stress distribution on the specimen in the bending tests with the case of applying load to obtain (a) tensile stress and (b) compressive stress at the notch bottom. From here, the stress mentioned below is the stress appearing at the notch bottom.

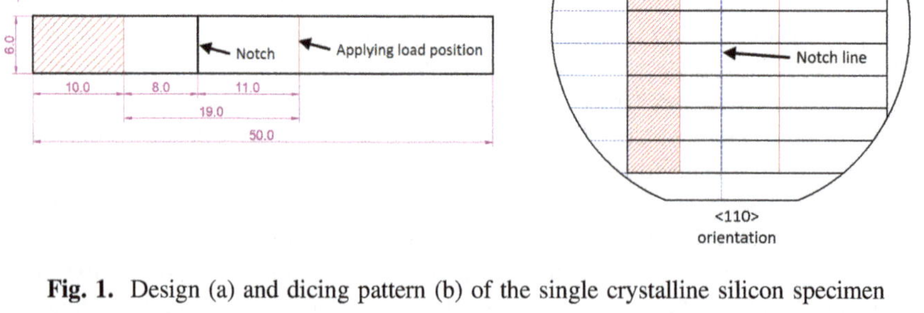

Fig. 1. Design (a) and dicing pattern (b) of the single crystalline silicon specimen

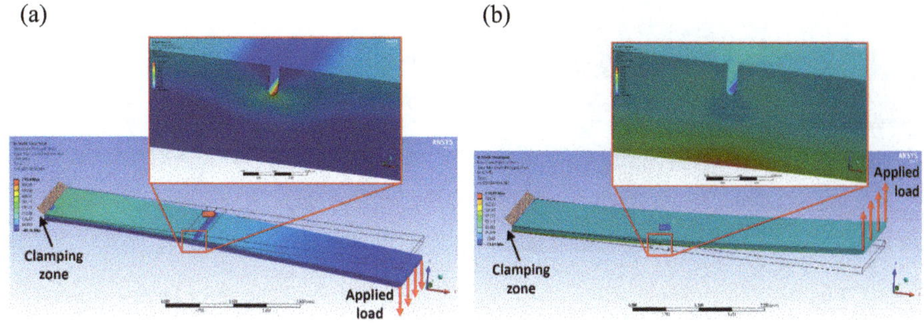

Fig. 2. Stress distribution on the specimen in the bending tests with (a) tensile and (b) compressive stress at the notch

3 Experiment

Experimental setup used in this study is shown in Fig. 3, where the setup is set inside the chamber ETAC TH441HA to control the testing environment. Specimen is kept horizontally at the clamping area on the holder and it is loaded by the loading rod. The specimen holder is mounted on the positioning stage, which is adjusted by the micrometers. They help to set the loading rod contacting to the specimen at the applying load position. The loading rod is fixed on the slider which slides on the rail. The movement of the slider is made by the crank and slider mechanism which consists of the motor, the eccentric, the crank, and the slider. The movement amplitude of the loading rod is adjusted by the eccentric and measured by the dial indicator. Frequency of the movement is set by the inverter. By using this setup, the fatigue test was performed by fixing the stage and controlling the motor, then the loading rod applies the cyclic load on the specimen. In contrast, the static test was performed by fixing the loading rod and adjusting the stage with the micrometer to bend the specimen.

The above specimens were separated into two groups for the strength tests. The first group is tested by the static strength test which was performed under monotonically increasing load until the specimens broken. This group is here called as "without fatigue". The second group was tested by the static strength test after performing the fatigue test, which is named as "with fatigue". All the static tests were performed with the tensile stress on the notch side. In contrast, the fatigue tests were performed by applying the compressive stress on the notch side with the scheme shown in Fig. 4 and the stress ratio is negative infinity, i.e., the maximum stress is zero. The frequency of cyclic loading used for the fatigue test is 10 Hz. The static tests were performed in the lab-air conditions, while the fatigue tests were performed at the temperature and the relative humidity (RH) of 80 °C and 80%, respectively. It is known that humidity in the atmosphere induces the fatigue lifetime of silicon to be shortened significantly. Therefore, this environment conditions is hoped to reduce much strength of the specimen.

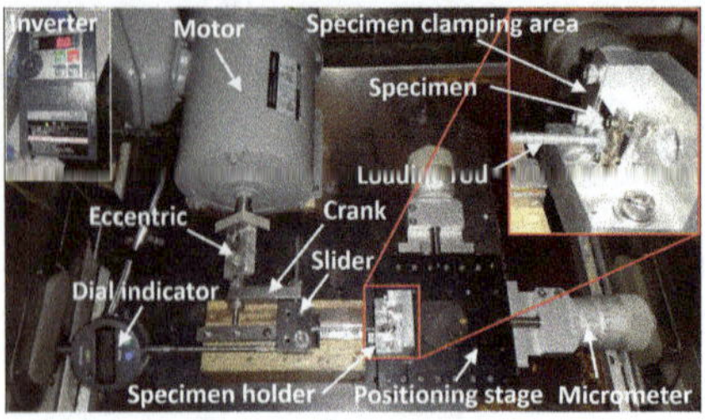

Fig. 3. Experimental setup

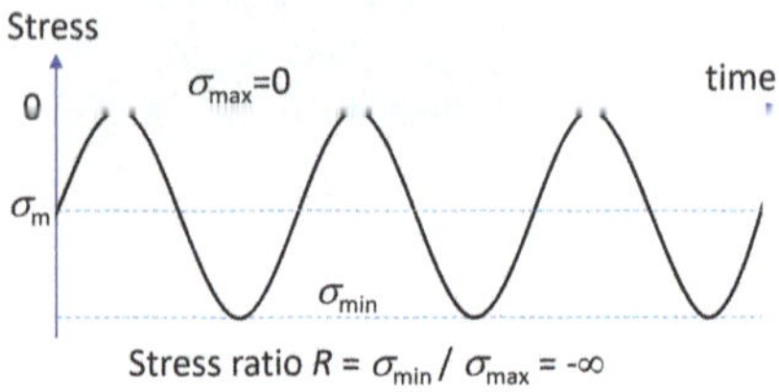

Fig. 4. Scheme of fatigue stress

4 Results and Discussion

The strength tests were performed with the above setup for the two groups of the specimens with and without the fatigue, where the experimental data are shown in Table 1. The data in Table 1 were ranked from the weakest to the strongest specimen based on the fracture stress in each group. The static strength data are shown as the symbols in Fig. 5, where the abscissa presents the fracture stress σ and the ordinate is the cumulative fracture probability F. Using the mean rank method, the value of F for the experimental data was calculated by the formula

$$F = \frac{i}{I + 1} \tag{1}$$

In Eq. (1), the notations i and I are the order number of a specimen ranked by the fracture stress and the total number of tested specimens in the group.

On the other hand, since the strength scatter of brittle materials such as silicon is significant, the Weibull distribution [10, 11] is appropriate to describe the strength distribution for the silicon specimens in this study. The cumulative fracture probability F of the specimens is here defined by using the two-parameter Weibull distribution as

Table 1. Experimental data

Group	Specimen no.	Fatigue test before static		Static test	Optimum values	
		Stress amplitude, σ [GPa]	Number of cycles, N	Fracture stress, σ [GPa]	m	σ_0 [GPa]
Without fatigue	1	–	–	0.525	16.67	0.589
	2	–	–	0.564		
	3	–	–	0.567		
	4	–	–	0.576		
	5	–	–	0.632		
With fatigue	1	−0.561	508800	0.509	17.82	0.556
	2	−0.534	507900	0.509		
	3	−0.567	502800	0.546		
	4	−0.567	502800	0.548		
	5	−0.463	1029600	0.592		

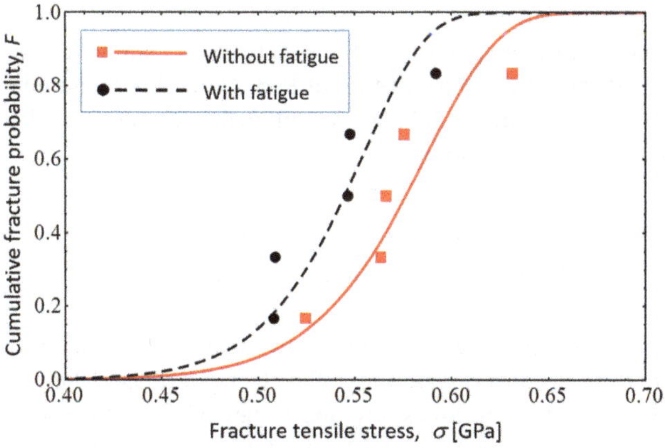

Fig. 5. Static strength comparison

$$F = 1 - \exp\left[-\left(\frac{\sigma}{\sigma_0}\right)^m\right], \tag{2}$$

where m is the Weibull modulus, σ_0 is the scale parameter, and σ is the fracture stress. The scale parameter σ_0 also presents the average strength. The parameters σ_0 and m were estimated by fitting Eq. (2) to the experimental data, where the fitting method used here is the maximum likelihood method (MLM) [10, 12]. To use the MLM, it requires a likelihood function L_s, which is the product of the probability density functions (PDFs) of all the observations. The likelihood function L_s is here calculated from the PDFs notated by $f_s(\sigma|\sigma_0,m)$ with the set of observations $\{\sigma\}$. The function L_s is formed in the natural logarithm as.

$$\ln L_s = \sum_{i=1}^{I} \ln f_s(\sigma_i|\sigma_0, m) = \sum_{i=1}^{I} \left\{ \ln\left[\frac{m}{\sigma_0}\left(\frac{\sigma_i}{\sigma_0}\right)^{m-1}\right] - \left(\frac{\sigma_i}{\sigma_0}\right)^m \right\}, \tag{3}$$

The parameters σ_0 and m are estimated by finding their values maximizing L_s or $\ln L_s$, i.e., solving the follow equation system.

$$\begin{cases} \partial \ln L_s / \partial \sigma_0 = 0 \\ \partial \ln L_s / \partial m = 0 \end{cases}, \tag{4}$$

Thus, the values of σ_0 and m were evaluated by solving Eq. (4) and are listed in Table 1. The results of fitting are shown in Fig. 5 as the curvilinear lines. Figure 5 clearly shows the different levels of static strength of the two groups depending on their state of damage. The average strengths σ_0 of the group with fatigue test is smaller than that of the group without fatigue test about 6%.

To see whether the reduction of the strength corresponding to effect of the fatigue test, the equivalent crack length was estimated from the strength reduction and the fatigue test data. Starting from the equation of Paris' law [9, 10] and integrating it with the crack length to be from the initial crack length to the current equivalent crack length a after N cycles of load, a is estimated as

$$a = \left[\left(\frac{K_{\mathrm{Ic}}}{\beta\sigma_0\sqrt{\pi}}\right)^{2-n} + \frac{C(n-2)}{2}\left(\frac{\beta\sigma\sqrt{\pi}}{K_{\mathrm{Ic}}}\right)^{n} N\right]^{2/(2-n)}, \quad (5)$$

where K_{Ic} is the fracture toughness C and n are two material parameters of Paris' law, σ is the amplitude of the applied cyclic stress. The factor β reflects the geometry of both the crack and the specimen. In this study, the values of C, n, K_{Ic} and β were used as 1.83×10^{-13} m/cycle, 25.8, 1.1 MPam$^{1/2}$ and 1.12, respectively [9, 10]. The crack length is estimated by Eq. (5) and plotted in Fig. 6 as the mesh, where the average strengths σ_0 is 0.589 GPa from the group without fatigue test, which corresponds to the initial strength as well as the initial damage. The equivalent crack length induced by the fatigue test was calculated from the stress amplitude σ and number of cycles N in Table 1 and is shown in Fig. 6 by the symbols. The consistence of the equivalent crack length in obtained from the experiment data and the estimation suggested the strength reduction of the specimens induced by the compressive fatigue test. It is considered here that Eq. (5) based on Paris' law, which is well-known for fatigue crack extension process induced by the dislocation in metallic materials. Besides, in the compressive fatigue, there is no room for oxygen immersing inside silicon crystal to happening oxidization. Therefore, it is suggested that damage growth inside the silicon specimens by the dislocation but not the oxidization.

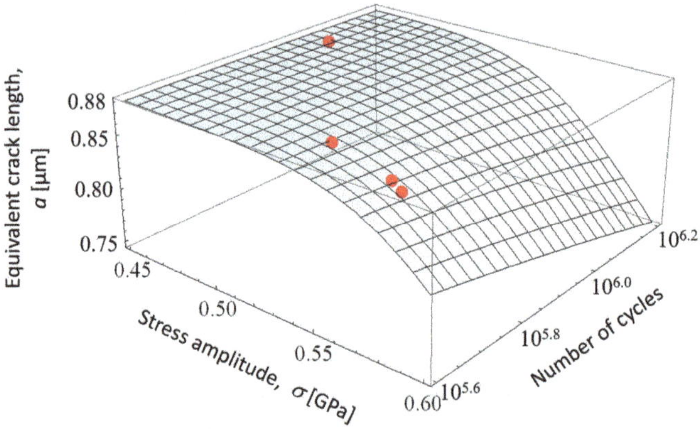

Fig. 6. Comparison of the fatigue data to the equivalent crack length

5 Conclusion

The two groups of the single crystalline silicon specimens were tested with and without applying cyclic compressive stress before measuring the tensile strength at the notch. The experimental data obtained from the two groups were compared to each other. The reduction in the strength of the group tested with fatigue was seen corresponding to effect of applied cyclic compressive stress by comparing the equivalent crack length. Under compression, there is no opportunity for oxygen immersing inside silicon crystal, and therefore the strength reduction was not induced by oxidization but induced by the damage growth inside silicon such as dislocation.

References

1. Muhlstein, C.L., Brown, S.B., Ritchie, R.O.: High-cycle fatigue of single-crystal silicon thin films. J. Micrelectromech. Syst. **10**, 593–600 (2001)
2. Muhlstein, C.L., Stach, E.A., Ritchie, R.O.: A reaction-layer mechanism for the delayed failure of micron-scale polycrystalline silicon structural films subjected to high-cycle fatigue loading. Acta Mater. **50**, 3579–3595 (2002)
3. Kahn, H., Avishai, A., Ballarini, R., Heuer, A.H.: Surface oxide effects on failure of polysilicon MEMS after cyclic and monotonic loading. Scripta Mater. **59**, 912–915 (2008)
4. Kato, N.I., Nishikawa, A., Saka, H.: Dislocations in Si generated by fatigue at room temperature. Mater. Sci. Semicond. Process. **4**, 113–115 (2001)
5. Nakao S., Ando T., Arai S., Saito N., Sato K.: Variation in dislocation pattern observed in scs films fractured by tensile test: effects of film thickness and testing temperature. In: Proceedings of the MRS 2008, 1052, DD03-21 (2008)
6. Parish, C., Batchelor, D., Progl, C., Russell, P.: Tutorial: electron beam-induced current in the scanning electron microscope. Microsc. Anal. **21**, 11–13 (2007)
7. Huy, V.L., Kamiya, S.: A direct evidence of fatigue damage growth inside silicon MEMS structures obtained with EBIC technique. Vietnam J. Mech. **36**(2), 109–118 (2014)
8. Kamiya, S., Kongo, A., Sugiyama, H., Izumi, H.: Electronic imaging of subcritical defect accumulation in single crystal silicon under fatigue loading. Sens. Actuat. A **279**, 705–711 (2018)
9. Huy, V.L., Gaspar, J., Paul, O., Kamiya, S.: Estimation of the parameters determining strength and fatigue behaviors of arbitrarily-shaped polysilicon thin films. In: International Electronic Packaging Technical Conference and Exhibition, vol. 44618, pp. 409–415 (2012)
10. Huy, V.L., Kamiya, S., Gaspar, J., Paul, O.: Fatigue lifetime prediction of arbitrarily-shaped MEMS structures. Microsyst. Technol. **25**(7), 2713–2726 (2019)
11. Weibull, W.: A statistical distribution function of wide applicability. J. Appl. Mech. **18**, 293–297 (1951)
12. Cousineau, D., Brown, S., Heathcote, A.: Fitting distributions using maximum likelihood: methods and Packages. Behav. Res. Methods Instrum. Comput. **36**, 277–290 (2004)

Design Multi-active Devices for High-Rise Building Under Earthquake Excitation

Tien-Hung Pham[1], Xuan-Thuan Nguyen[1(✉)], Nanako Miura[2], Van-Truong Nguyen[1], and Thanh-Lam Bui[1]

[1] Faculty of Mechanical Engineering, Ha Noi University of Industry, 298 Cau Dien, Bac Tu Liem, Hanoi, Vietnam
nxthuan7289@gmail.com

[2] Faculty of Mechanical Engineering, Kyoto Institute of Technology, Goshokaido-cho, Matsugasaki, Sakyo-ku, Kyoto 606-8585, Japan

Abstract. The research analyzed the vibration of a sixty stories structure under earthquake excitation. The amplitudes displacement on the top floor of the building could be very high when the earthquake happened. To reduce the vibration of the building, a passive device Tuned Mass Damper (TMD) is installed at the highest floor of the structure. This paper proposed a control method used for multi-active devices by applying Linear–Quadratic Regulator method (LQR). The location of the multi-active devices is also investigated to find the best place for these devices. Besides, the number of active devices is considered to verify the achievement of the proposed system in the vibration reduction of the building. To examine the effectiveness of the system, the study analyzes the vibration of the high-rise building under different earthquake cases. The numerical simulation results indicate the good performance of the proposed method to restrain the vibration of the building.

Keywords: Building · Vibration · Earthquake · Control · Multi-active devices

1 Introduction

Recently, urbanization is rising rapidly in the world led to increasing numbers of high-rise buildings. For the buildings, when it has external forces such as wind loads, earthquake excitation, they will vibrate, especially for the top stories of the building. These vibrations can cause severe damage to human life in the building as that building. In recent years, there are some big earthquakes magnitude such as in Sichuan (2008), Chile (2010), Ecuador (2016), and Lombok island (2018). These earthquakes have caused severe damage to the civil structure, especial are high-rise buildings. Wherever in the world, designing and constructing the building need to be considered the important problem that the building can withstand vibrations with a maximum magnitude of earthquakes. Therefore, the research is necessary that how to reduce vibration for the high-rise building under unexpected effects for instance wind load and earthquake excitation.

A lot of previous researches proposed methods to reduce vibration for building. TMD is the most popular one due to its simple principle and several successful

B. T. Long et al. (Eds.): MMMS 2020, LNME, pp. 836–843, 2021.
https://doi.org/10.1007/978-3-030-69610-8_111

applications in real-life practice. Den Hartog [1] optimized TMD parameter values applying for the building by using the fixed-point method. C.-L. Lee et al. [2] proposed a method with an efficient procedure to find the optimal parameters of TMD. A. Kaveh et al. [3] used a charged system search to optimizing parameters of TMD of multi-story building systems under seismic excitations. Sadegh Etedali et al. [4] investigated the problem of optimum selection of TMD using a multi-objective cuckoo search. The simulation results of this study confirmed that it was a good solution, a more practical and economic selection of TMD parameters with a choice of smaller mass, stiffness, and damping. Yoshiki Ikeda et al. All of the above researches analyzed the effectiveness of a single TMD on the vibration reduction of the building under external disturbance. However, TMD is a passive device which only has efficient in the specific case of external force.

There are some studies that use multiple tuned mass dampers (MTMD) to restrain the vibration of the structure. Genda Chen et al. [5] introduced and optimized placement of a system MTMD for a high-rise building to reduce the response amplitude under environment excitation. Hamid Radmard Rahmani et al. [6] optimized the properties and locations of multiple tuned mass dampers (TMDs) in high-rise structures. Luciara Silva Vellar et al. [7] used the Search Group Algorithm (SGA) to optimize the parameters and position of MTMD for a 10-story building. Although, MTMD devices have more effective than TMD on the restraint vibration of the building. However, these systems are also passive devices that cannot apply for all cases of earthquakes.

Some studies used active mass damper (AMD) to reduce vibration of the building. Yoshiki Ikeda et al. [8] showed an active system that was placed in a high building in Tokyo to restrain the vibration under environmental disturbances. Chien-Liang Lee et al. [9] analyzed an electric-type AMD system to control the vibration of the structure. Sadegh Etedali et al. [10] proposed a method to optimize the proportional–derivative (PD) and proportional–integral–derivative (PID) controllers to reduce the vibration of a high building under earthquake excitation. Rahmi Guclu and Hakan Yazici [11] proposed a structural system with AMD is modeled and earthquake ground motion is used as input to this building structure. Wonsuk Park et al. [12] presented a new direct pole placement design method for active control systems based on the H∞ method. Satyam Paul et al. [13] proposed a control method using a fuzzy and sliding mode technique to restrain the vibration of the structure caused by bidirectional earthquakes. G. P. Cimellaro et al. [14] addressed the integrated design of civil engineering structures with control systems. The above researches have much energy to restrain the vibration of the structures. Moreover, these studies have not considered the location and the number of AMD.

Unlike previous researches, this study proposes a new method that combines TMD and multi-active devices to reduce the vibration of the building. The control method base on LQR and it is applied for multi-active devices. The location and the number of active devices are also considered in this research. The numerical results indicate the performance of the proposed system in the vibration reduction of the building under earthquake excitation.

2 Motion Equations of the Structure Systems

The high-rise building model has n stories which are constructed with assuming that only horizontal vibrations, ignore vertical vibrations. The linear model of the building is described in Fig. 1.

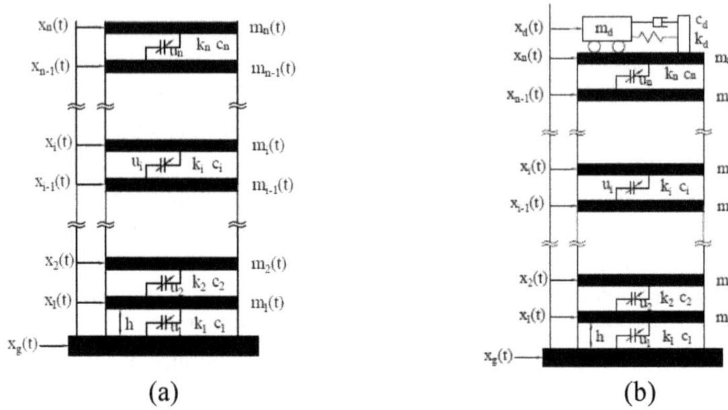

Fig. 1. The high-rise building model (a) with multi-active devices, (b) with TMD and active devices.

The equation of motion of the building is written as follows:

$$M(\ddot{x}(t) + \ddot{x}_g(t)) + C\dot{x}(t) + Kx(t) = fu(t) \tag{1}$$

where:

$$f = \begin{bmatrix} 1 & -1 & 0 & \cdots & 0 \\ 0 & 1 & -1 & \cdots & 0 \\ & & \ddots & \ddots & \\ & & & \ddots & -1 \\ & & & & 1 \\ 0 & 0 & \cdots & 0 & 0 \end{bmatrix}, u(t) = [u_1(t) \quad \cdots \quad u_i(t) \quad \cdots \quad u_n(t)]^T \tag{2}$$

M, C, K are mass matrix, damping matrix, and stiffness matrix of the building, respectively. $x(t)$ is the displacement vector of the floors to the ground. $\ddot{x}_g(t)$ is the absolute acceleration vector of the ground. The active device is a system which is located between two floors. The force control $u_i(t)$ is installed on the ith floor.

3 Design Controller for Multi-active Devices

The control law of multi-active devices is proposed by using Linear Quadratic Regulator method. The equation of motion in case multi-active devices combined with TMD can be written as follows:

$$\dot{X}(t) - G\ddot{x}_g(t) = AX(t) + Bu(t) \tag{3}$$

$$X(t) = \begin{bmatrix} x(t) \\ \dot{x}(t) \end{bmatrix}, A = \begin{bmatrix} O_{(n+1)\times(n+1)} & I_{(n+1)\times(n+1)} \\ -M^{-1}K & -M^{-1}C \end{bmatrix}, B = \begin{bmatrix} O_{(n+1)\times n} \\ M^{-1}f \end{bmatrix}, G = \begin{bmatrix} O_{(n+1)\times(n+1)} \\ -I_{(n+1)\times(n+1)} \end{bmatrix} \tag{4}$$

Matrix I is the identity matrix and O is a zero matrix.

The absolute acceleration of the building is calculated as follows:

$$y_a(t) = \ddot{x}(t) + \{1\}\ddot{x}_g(t) = \begin{bmatrix} -M^{-1}K & -M^{-1}C \end{bmatrix} X(t) + M^{-1}fu(t) = C_{ss}X(t) + D_{ss}u(t) \tag{5}$$

$$C_{ss} = \begin{bmatrix} -M^{-1}K & -M^{-1}C \end{bmatrix}, D_{ss} = M^{-1}f \tag{6}$$

The control object is the acceleration response of the building. Thus, the quadratic cost function is written as follows:

$$J = \int_0^{\infty} \left[X^T QX(t) + 2X^T(t)Su(t) + u^T(t)Ru(t)\right] dt \tag{7}$$

$$Q = C_{ss}^T C_{ss} + Q_x, S = C_{ss}^T D_{ss}, R = Ir + D_{ss}^T D_{ss}, Q_x = \begin{bmatrix} 0 & & & & & & 0 \\ & \ddots & & & & & \\ & & 0 & & & & \\ & & & q_x & & & \\ & & & & 0 & & \\ & & & & & \ddots & \\ 0 & & & & & & 0 \end{bmatrix} \tag{8}$$

where Q_x is a $2(N+1) \times 2(N+1)$ matrix and $Q_{x(N+1)\times(N+1)} = q_x$.

The control force.

$$u(t) = -FX(t), F = R^{-1}(S^T + B^T P) \tag{9}$$

where P is found by solving the continuous-time Riccati differential equation.

$$(A - BR^{-1}S^T)^T P + P(A - BR^{-1}S^T) - PBR^{-1}B^T P - SR^{-1}S^T = O \tag{10}$$

4 Numerical Simulation

Parameters of the high-rise building are referred of the paper of Thuan Xuan Nguyen et al. [15] and are showed in Table 1. Parameters of the TMD system are found based on the Den Hartog method [1]. The parameters of the TMD are presented in Table 2.

The sway of the building is investigated under some earthquake cases. Figure 4 showed the acceleration data of three earthquakes including the GTohoku-Konohana earthquake (Konohana 2011 – a long period earthquake), NChuetsuToka-machi earthquake (Tokamachi 2004 – a short period earthquake), and Toka-chiokiTomakomai earthquake (Osaka 2003 – a long period earthquake).

Table 1. High-rise building parameters for the numerical simulation

Parameter	Symbol	Value
Story height	H	4,0 m
Fundamental natural period	T	5,8 s
Number of stories	n	60
Damping factor	ζ	0,01
Mass of each story	m_i	$2{,}0 \times 10^6$ kg

Table 2. Parameters of TMD

Parameter	Symbol	Value
Mass	md	$6{,}0 \times 10^6$ kg
Damping	cd	$1{,}6544 \times 10^5$ Ns/m
Stiffness	kd	$6{,}3867 \times 10^6$ N/m

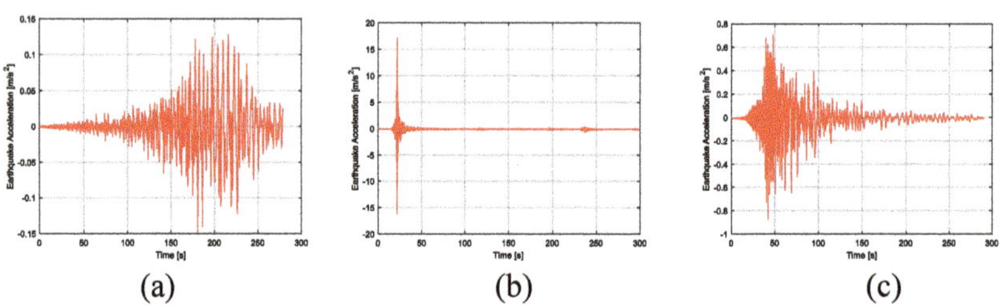

Fig. 2. Acceleration of earthquakes; (a) GTohokuKonohana earthquake; (b) NChuetsuToka-machi earthquake; (c) TokachiokiTomakomai earthquake

The simulation cases are divided into three main groups: without control, with multi-active devices, with both TMD and multi-active devices. Especially, the location and the numbers of the multi-active devices are shown in Table 3. Hence, Fig. 2 , 3 and 4 show the absolute acceleration responses of the building's 60th story, the maximum amplitude displacement of each story, and the maximum control force of multi-active devices are described for the case "20 active devices" highlighted in Table 3. Besides, Table 4 performs values of maximum absolute acceleration of the building's 60th story in all cases.

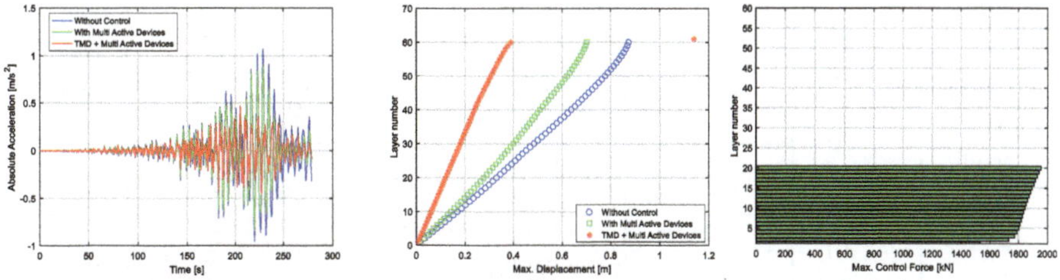

(a) (b) (c)

Fig. 3. (a) Acceleration of the building's top story; (b) Maximum Displacement of the building's stories; (c) Maximum control force of multi-active devices. (Lower case - GTohokuKonohana earthquake)

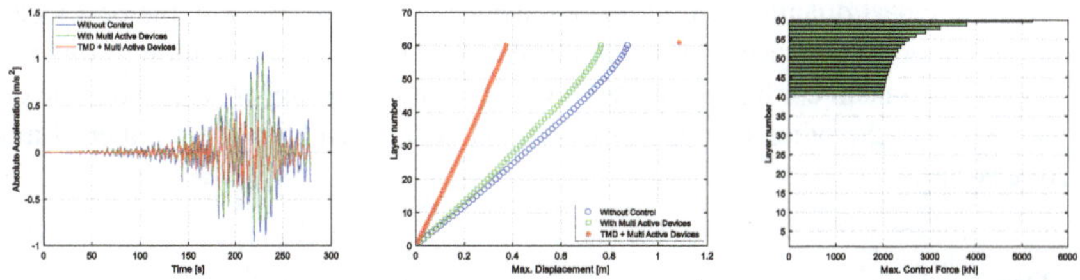

(a) (b) (c)

Fig. 4. (a) Acceleration of the building's top story; (b) Maximum Displacement of the building's stories; (c) Maximum control force of multi-active devices. (Upper case - GTohokuKonohana earthquake)

Table 3. The number and location of active devices for simulation

	Number of devices	Location of devices
Lower	10	1,2,3…10
	20	1,2,3…20
	30	1,2,3…30
Upper	10	51,52,53…60
	20	41,42,43…60
	30	31,32,33...60
Uniform	10	1,7,13… 55
	20	1,4,7…58
	30	1,3,5…59

Table 4. The maximum absolute acceleration of the 60th story (m/s^2)

			GTohok uKon- ohana	NChuetsuTo kam-achi	Tokachiok iTo- makomai
Without Control			1.0761	1.3240	2.1303
With TMD + Multi active devices	10 devices	Lower	0.4833	0.8771	1.4670
		Upper	0.4664	1.3772	1.3038
		Uniform	0.4814	0.7000	1.4163
	20 devices	Lower	0.8914	0.6772	1.4187
		Upper	0.9871	1.3824	1.2268
		Uniform	0.4602	0.6779	1.3210
	30 devices	Lower	0.4600	0.6450	1.3637
		Upper	0.4261	1.2999	1.1936
		Uniform	0.4405	0.7134	1.2560

Table 4 shows the maximum absolute acceleration value of the 60th floor during the earthquake excitations. It can be seen that the reduction of the vibration of the 60th floor by using the proposed method is better than using the multi-active device method. Moreover, the best quantity of multi-active devices is 30. In this case, the best location of the multi-active devices to reducing the vibration of the building under the NChuetsuTokamachi earthquake is the lower case and is the upper case under the other earthquakes. On the other side, the best location also depends on the number of multi-active devices.

5 Discussions

This paper has investigated the reduction of vibration of a 60-story building under the effect of earthquakes using a control method which combines TMD + multi-active devices. The controller for multi-active devices is based on the LQR method. This research also analyzes the location and number of multi-active devices to find the best solution to reduce vibration for the building.

It can be seen that the best location and the number of multi-active devices for reducing the vibration of the building as follows:

- NChuetsuTokamachi earthquake: Lower case and 30 devices.
- GTohokuKonohana earthquake, TokachiokiTomakomai earthquake: Upper case and 30 devices.

The results of the paper will be of great help to engineers in designing and constructing reduced vibration for buildings under earthquake excitations. However, LQR is a local stabilization method, so the performance of the method is not much high. Future research will be considered improving the quality of the control method and do experiments.

Acknowledgment. This research was funded by Vietnam National Foundation for Science and Technology Development (NAFOSTED) under Grant No. "107.01–2019.311".

References

1. Den Hartog, J.: Mechanical Vibration. McGraw-Hill, London (1947)
2. Lee, C.-L., Chen, Y.-T., Chung, L.-L., Wang, Y.-P.: Optimal design theories and applications of tuned mass dampers. Eng. Struct. **28**, 43–53 (2006)
3. Kaveh, A., Mohammadi, S., Khadem Hosseini, O., Keyhani, A., Kalatjari, V.R.: Optimum parameters of tuned mass dampers for seismic applications using charged system search. Trans. Civ. Eng. **39**, 21–40 (2015)
4. Etedali, S., Rakhshani, H.: Optimum design of tuned mass dampers using multi-objective cuckoo search for buildings under seismic excitations. Alexandria Eng. J. **57**, 3205–3218 (2018)
5. Chen, G., Jingning, W.: Optimal placement of multiple tuned mass dampers for seismic structures. J. Struct. Eng. **127**, 1054–1062 (2001)
6. Rahmani, H.R., Konke, C.: Seismic control of tall buildings using distributed multiple tuned mass dampers. Adv. Civ. Eng. **7**, 1–20 (2019)
7. Vellar, L.S., Ontiveros-Perez, S.P., Miguel, L.F.F.: Robust optimum design of multiple tuned mass dampers for vibration control in buildings subjected to seismic excitation. Shock Vib. **7**, 1–9 (2019)
8. Ikeda, Y., Sasaki, K., Sakamoto, M., Kobori, T.: Active mass driver system as the first application of active structural control. Earthquake Eng. Struct. Dynam. **30**, 1575–1595 (2001)
9. Lee, C.-L., Wang, Y.-P.: Seismic structural control using an electric servomotor active mass driver system. Earthquake Eng. Struct. Dynam. **33**, 737–754 (2004)
10. Etedali, S., Tavakoli, S.: PD/PID controller design for seismic control of high-rise buildings using multi-objective optimization: a comparative study with LQR controller. J. Earthquake Tsunami **11**, 1–23 (2017)
11. Guclu, R., Yazici, H.: Vibration control of a structure with ATMD against earthquake using fuzzy logic controllers. J. Sound Vib. **318**, 36–49 (2008)
12. Park, W., Park, K.-S., Koh, H.-M.: Active control of large structures using a bilinear pole-shifting transform with H∞ control method. Eng. Struct. **30**, 3336–3344 (2008)
13. Paul, S., Yu, W., Li, X.: Discrete-time sliding mode for building structure bidirectional active vibration control. Trans. Inst. Meas. Control **41**, 433–446 (2018)
14. Cimellaro, G.P., Soong, T.T., Reinhom, A.M.: Integrated design of controlled linear structural systems. J. Struct. Eng. **135**, 853–852 (2019)
15. Nguyen, T.X., Miura, N., Sone, A.: Analysis and control of vibration of ropes in a high-rise elevator under earthquake excitation. Earthquake Eng. Eng. Vib. **18**, 447–460 (2019)

Predicting and Validating Welding Distortion of Butt Joint

Thi-Thao Ngo, Nguyen Hong Phong, and Van-The Than(✉)

Faculty of Mechanical Engineering,
Hung Yen University of Technology and Education, Hung Yen, Vietnam
thanthe.ck@gmail.com

Abstract. Applying simulation method to predict welding distortion is studied in this paper. Butt joint welding model of low carbon steel is built and simulated based on Comsol Multiphysics software. The welding temperature and distortion under different welding parameters are performed. Experimental distortion are then obtained and compared to predicted results. Results show that simulation results are consistent with measurement results. This proves that current numerical simulation method can accuracy estimate distortion of butt joint by using Gas Metal Arc Welding (GMAW) process.

Keywords: Butt joint · Welding distortion · Comsol multiphysics · Temperature

1 Introduction

Welding deformation is a phenomenon of changing shape, size and position of a weld joint due to effect of welding heat cycle; besides, welding deformation can cause repair costs, decrease quality and aesthetics of products as well as increase construction time. If a welded structure is deformed, it can become unstable when working and leading to structural damage. Therefore, in order to determine the welding distortion and its effect on the working ability of the structure is very necessary.

Recently, many researches have studied about welding distortion. Than Van The et al. [1] presented an application of ANSYS software to simulate temperature, distortion and stress for butt weld and pipe seams by GMAW and Gas Tungsten Arc Welding (GTAW) methods, then evaluating influence of welding conditions on welding temperature, stress and deformation. Stress and deformation state of the composite material layer in T-joint is also studied through finite element method (FEM) [2]. There are also many researchers who have applied FEM to analyze experimental models and propose experimental formulas to calculate the welding distortion. Roger William O'Brien [3] has synthesized a lot of models to calculate longitudinal shrinkage, transverse shrinkage and angular distortion. Then, he compared them with calculated results with actual measurement. Besides, A.S. Gangwar et al. [4] studied the effect of thickness on deformation for low carbon steel plate welds by experimental method. Influence of welding parameters on deformation is also presented by Filipe D.S. Cordeiro [5]. At the same time, welding deformation affecting economy in welding product is presented in detail by N.R. Mandal and C.V.N. Sundar [6]

B. T. Long et al. (Eds.): MMMS 2020, LNME, pp. 844–850, 2021.
https://doi.org/10.1007/978-3-030-69610-8_112

introduced theory of the distortion due to transverse shrinkage in a butt weld joint. M.S. Sulaiman et al. [7] have simulated and experimented on the deformation of butt and T-welded joint using finite element analysis software. In addition, Ngo Thi Thao et al. also studied simulation and calculation of deformation for aluminum alloy butt joint [8] and low carbon steel fillet [9] by GMAW process.

In this paper, COMSOL software is selected to simulate temperature and deformation fields of welded joint. To validating reliability of the simulation method, experimental welding are conducted under different welding conditions and welding distortion is measured and compared with simulated result.

2 Heat Transfer Model

2.1 Thermal Model of GMAW Process

A three–dimensional transient heat transfer analysis is used to simulate GMAW process with below governing and boundary equations [9]

$$\frac{\partial}{\partial x}\left(k(T)\frac{\partial T}{\partial x}\right)+\frac{\partial}{\partial y}\left(k(T)\frac{\partial T}{\partial y}\right)+\frac{\partial}{\partial z}\left(k(T)\frac{\partial T}{\partial z}\right)+q=\rho(T)C_p(T)\frac{\partial T}{\partial t} \quad (1)$$

$$k_n(T)\frac{\partial T}{\partial x}+h(T-T_0)+\sigma\varepsilon\left(T^4-T_0^4\right)=0 \quad (2)$$

where ρ, k and C_p are density, thermal conductivity and specific heat of workpiece material, respectively; t and T refer to time variable and temperature. n, k_n, h, ε, σ and T_O stand for normal direction to surface, thermal conductivity, heat transfer coefficient, emissivity, Stefan-Boltzmann constant, and the ambient temperature, respectively. In Eq. (1), q accounts for the arc heat input of a volumetric heat source given as Eq. (3).

$$q=\frac{6\sqrt{3}f_iP\eta}{\pi\sqrt{\pi}a_ibc}\exp\left(-\frac{3x^2}{a_i^2}-\frac{3y^2}{b^2}-\frac{3z^2}{c^2}\right) \quad (3)$$

in which $P = \alpha U_h I_h$ is the arc power; α, I_h and U_h are efficiency coefficient, current and voltage. η is process efficiency. α and η are assigned at 0.75 and 0.8, respectively [10]. The subscript i indicates 1 and 2 corresponding to the front and rear heat sources.

2.2 3D Modeling on COMSOL Software

Comsol Multiphysics software was used to predict temperature and deformation. A low carbon steel butt weld joint is prepared as shown in Fig. 1. Boundary conditions including convection, radiation and initial condition as well as welding heat source are set as shown in Fig. 1. Theodoe L. Bergman et al. [11] gave following formula to calculate convection coefficient:

For convection of the upper surface:

$$h_t = \frac{2k}{w} 0.15 Ra_L^{1/3} \quad (4)$$

For convection of the lower surface:

$$h_b = \frac{2k}{w} 0.52 Ra_L^{1/5} \quad (5)$$

and

$$Ra_L = \frac{g\beta(T_s - T_\infty)L^3}{\nu\alpha} \quad (6)$$

in which, k is heat conductivity coefficient of the air, w is width of the plate; $g = 9.8\ \text{m/s}^2$; $\beta = 0.0033K^{-1}$; T_s is temperature on the surface; T_∞ is ambient temperature; $L = \frac{A_S}{P_1}$; A_s and P_1 are the surface area and circumference, respectively; The whole process is summarized in Fig. 2.

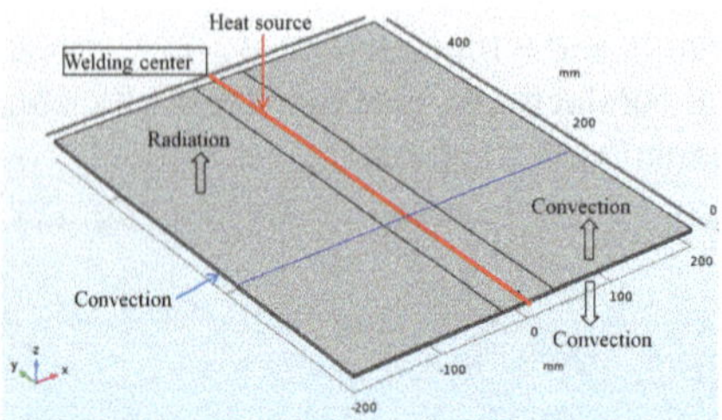

Fig. 1. Modeling of the welding process on Comsol

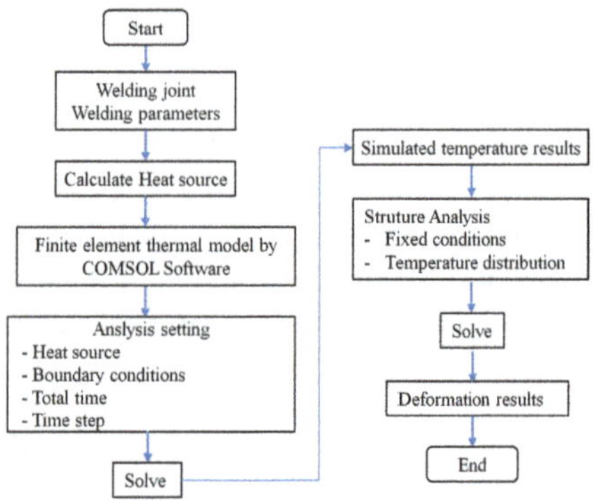

Fig. 2. Diagram of simulation process

3 Experimental Process

Daihen AII-V6 OTC Robot series is equipped with AX21 controller, used for TIG/MIG/MAG/CO2 welding applications and Plasma cutting applications used to perform experimental welding in this study (Fig. 3). After selecting base material, suitable welded joint, welding material, and welding conditions given in Table 1, experiment welding sample is carried out. After welding, actual deformation of these samples are measured. Measured equipment used in this article is also shown in Fig. 3.

Table 1. Welding conditions [12]

Welding condition	Welding sample (600 × 160 × 5.6 mm)	Current I_h (A)	Voltage U_h (V)	Speed V_h (cm/ph)
The first condition	M1	190	23	30
The second condition	M2	185	23	30
The third condition	M3	195	23	30

Fig. 3. Experiment process

4 Results and Discussions

Figure 4 shows temperature field and isothermal distributions on the welding specimen M1 as well as total deformation using the 1st welding condition. As seen, the maximum temperature is about 2170 °C (Fig. 4a). Figure 4b displays direction of heat transfer and the isothermal distribution was extended toward welded part because of different heat transfer rate in the welding samples. While Fig. 4c depicts total deformation of the workpiece with maximum magnitude at 11.8 mm. It can be seen that length of the workpiece is quite large (about 500 mm), so deformation due to the longitudinal shrinkage is relatively large, and prepared weld joint has a chamfer of 60°, thus deformation due to weld transverse shrinkage is also significant. Figure 5a and Fig. 5b show the deformation of welding samples using the second and third welding conditions, respectively. Compared to above case, value of distortion is smaller (Fig. 5a), which can be explained by M2 sample is welded with a smaller heat source (due to smaller I_h), so that thermal expansion phenomenon which causes welding deformation is also smaller. M3 is welded using higher welding current, obtained deformation has a value greater than that of previous two cases. Maximum value of deformation is achieved at 17.2 mm (refer to Fig. 5b).

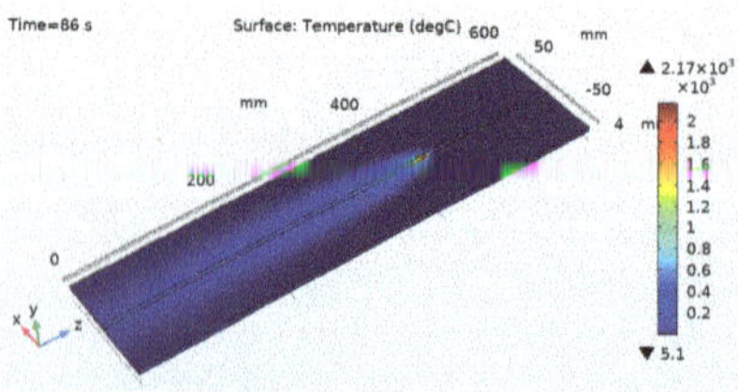

a. Temperature field

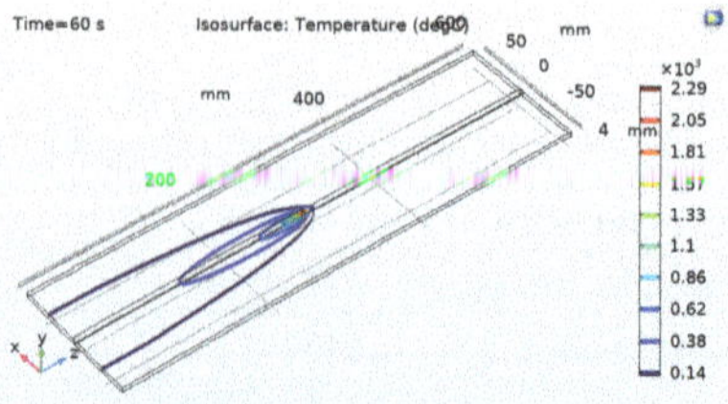

b. Isothermal distribution

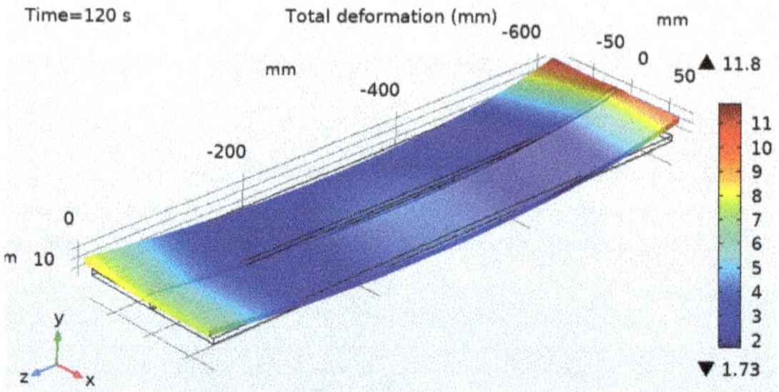

c. Total deformation

Fig. 4. Simulation result using the first condition

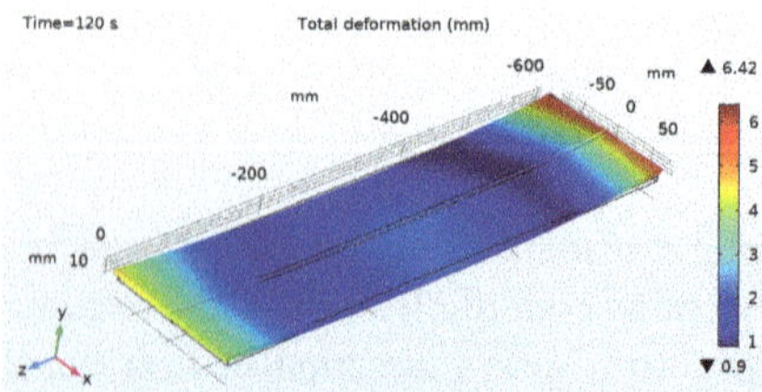

a. The second condition

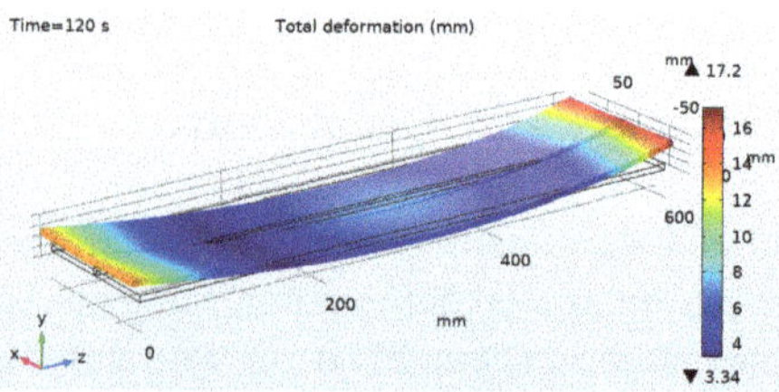

b. The third condition

Fig. 5. Simulation result using different conditions

Table 2. Comparison of simulated and measured deformation

Welding sample	Simulated deformation (mm)	Measured deformation (mm)	Sai lệch (%)
M1 (I_h = 190 A)	11.8	12.2	3.39
M2 (I_h = 185 A)	6.42	6.05	5.76
M3 (I_h = 195 A)	17.2	16.6	3.49

Results of measurement deformation of all 3 samples are summarized in Table 2. Results show that the actual measured deformation changes depending on the welding conditions, when I_h increases the deformation increases and vice versa. Sample M1 is welded with I_h = 190 A, measured value is 12.2 mm, welded sample M2 with the lower current 185 A, measured result is about 6.05 mm and welded sample M3 with the higher current 195 A, actual measured deformation is about 16.2 mm. Moreover, Table 2 also shows a comparison of measured and simulated deformations of welding samples. Error of measured and simulated results is tiny; thus, a relatively high conformity is obtained. The actual measurement results have proved reliability of the simulation method using Comsol software. From results it can be said that simulation model and conditions (heat source, boundary condition, initial condition…) are consistent with experimental model and conditions. This demonstrates that the simulation method based on Comsol software using in GMAW welding is reliable. In the future, it can be completely applied in other welding methods to solve the problem of heat transfer, stress and deformation.

5 Conclusions

Distortion is often unavoidable during welding processed leading to location errors of structures created by welding assembly. However, investigating distortion by experimental methods is often time-consuming and costly. However, in this paper, Comsol software has been explored and applied, so predicting temperature and deformation generated in the welding process is much simpler. To verify the reliability of the simulation method, experiments on the welding sample are conducted and then deformation of the samples after welding are measured and compared with simulation results. With obtained results, it is hoped that this method can be more widely used in the future to predict heat source, temperature distribution in the weld zone and deformation as well as stress for different welding process, different joint type or positions.

References

1. Thế, T.V., Thành, T.N., Thảo, N.T.: Ứng dụng phần mềm Ansys dự đoán nhiệt độ, biến dạng và ứng suất khi hàn liên kết giáp mối, Tạp chí Khoa học và công nghệ, Số 15 - tháng 9, trang 1–8 (2017)
2. Hoa, H.P., Văn Mỹ, N., Hải, Đ.V.: Nghiên cứu trạng thái ứng suất và biến dạng của các lớp vật liệu composite trong liên kết hàn chữ T dưới tác dụng của tải trọng tĩnh bằng phương pháp phần tử hữu hạn, Tạp chí khoa học và công nghệ, Tập 39, Số 4, trang 127–133 (2010)
3. O'Brien, R.W.: Predicting weld distortion in the design of automotive components, trường Durham University (2007)
4. Gangwar, A.S., Srivastava, A., và Rajan, N.G.: Experimental Study of Distortion in Butt Welds of Mild Steel Plates and En-31 Plates Having Different Thickness and Weld Cross Sections, Tạp chí IOSR-JMCE, Tập 14, Số 3 phiên bản IV, trang 51–80 (2017)
5. Cordeiro, F.D.S.: A Critical Analysis on Weld's Distortion, trường Đại học Coimbra, Julho (2015)
6. Mandal và, N.R., Sundar, C.V.N.: Analysis of Welding Shrinkage, Tạp chí Welding Research Supplement, Số tháng 6, trang 233–238 (1997)
7. Sulaiman, M.S., Manurung, Y.H., Haruman, E., Rahim, M.R.A., Redza, M.R., Lidam, R.N.A., Abas, S.K., Tham và, G., Chau, C.Y.: Simulation and experimental study on distortion of butt and T-joints using WELD PLANNER. J. Mech. Sci. Technol. **10**(25), 2641–2646 (2011)

8. Ngô Thị Thảo, Vũ Văn Linh, Thân Văn Thế, “Mô phỏng và tính toán biến dạng của liên kết hàn giáp mối hợp kim nhôm”, UTEHY Journal of Science and Technology, Số 24, trang 7–15,2019.
9. Thảo, N.T., Thế, T.V.. “Mô phỏng và kiểm tra nhiệt độ trong liên kết hàn góc bằng phương pháp hàn GMAW”, Tạp chí cơ khí – Số đặc biệt, trang 134–139, March 2020
10. Pepe, N., Egerland, S., Colegrove, P.A., Yapp, D., Leonhartsberger, A., Scotti, A.: Measuring the process efficiency of controlled gas metal arc welding processes. Sci. Technol. Weld. Join. **16**(5), 412–417 (2011)
11. Bergman, T.L.: Fundamentals of Heat and Mass Transfe, 7th edn. Wiley, Hoboken (2011)
12. The Lincoln Electric Company, Gas Metal Arc Welding Guidelines (2014). www.lincolnelectric.com

Full-Scale Self-propulsion Computations Using Body Force Propeller Method for Series Cargo Ship 12500DWT

Tran Quoc Chuan, Nguyen Kim Phuong, Nguyen Thi Hai Ha, Vu Minh Ngoc, and Tran Ngoc Tu(✉)

Vietnam Maritime University, Hai Phong, Vietnam
tutn.dt@vimaru.edu.vn

Abstract. The paper presents a results of study the full-scale self-propulsion characteristic for series cargo ship 12500DWT built in Vietnam based on the unsteady RANSE method with Body Force Propeller method. Two case studies were analyzed: propeller open-water, and self-propulsion simulations. To predict propeller open water characteristics, the Rotating Reference Frame method has been applied. The dynamic trim and sinkage of the ship were taken into account in the self-propulsion simulation. The free surface effects were considered by applying the volume of fluid method for multi-phase flows. The self-propulsion point was achieved by conducting two different runs at a constant ship speed with different propeller revolutions. The numerical results were compared with sea trial data. The solver is used in this article is the package Star-CCM+.

Keywords: Propeller · Self-propulsion · Body Force Propeller method · Full-scale · RANSE

1 Introduction

Accurate evaluation of self-propulsion characteristic is one of the key factors for the correct design of the ship propulsion system. Traditionally, this self-propulsion characteristic prediction for full-scale vessel has been made by extrapolating the model test results in towing tank. However, due to discrepancy in Reynolds numbers between model-scale and full-scale ship that leads to the flow around the ship at model-scale and full-scale is not the same, hence complete hydrodynamic similarity cannot be attended. Recently, with the fast development of computational resources, many ship hydrodynamics problems can be solved with Computational Fluid Dynamic (CFD) method. The main advantage of CFD methods is that they provide relatively accurate results, savings in time and money compared to scale model test. In addition, CFD approaches are capable to compute the ship simulation problem in both model-scale and full-scale. The post-CFD treatment also gives the designer a lot of detailed information about the flow characteristics around the ship more readily, hence providing a basis for the designers to determine an optimal design plan for the ship from a hydrodynamic

B. T. Long et al. (Eds.): MMMS 2020, LNME, pp. 851–859, 2021.
https://doi.org/10.1007/978-3-030-69610-8_113

perspective. Therefore, direct prediction of ship hydrodynamic in full-scale using CFD method is a new approach.

Among CFD methods used to solve hydrodynamics problems, the most popular approach is the Reynolds Averaged Navier-Stokes equation (RANSE) method, as it provides sufficient accuracy for engineering purposes at the reasonable computational time [1]. Thus, this paper employs the RANSE method for predicting self-propulsion point.

For self-propulsion performance prediction using the RANSE method, there are two approaches, including discretized rotating propeller method and Body Force Propeller method. These approaches have been used by many researchers in determining hull-propeller interaction coefficients Carrica, P.M. et al. [2], Tu T.N. et al. [3], Shen, Z. [4], Win, Y.N., Gokce [5], M.K. et al. [6]. The results of that study have shown that using a discretized rotating propeller method provides efficient and reliable results compared to experimental results. Besides, applied this method could provide the complete detailed description flow field around the propeller. While using Body Force Propeller method allowed the authors to avoid modeling the propeller geometry, complexities associated with moving meshes, and to achieve, as a result, shorter run times. However, the applied method could not provide a detailed flow field around the propeller. The above-mentioned literature has played an important role in further researches in the performance of a propeller working behind the ship hull. This paper presents a numerical prediction the self-propulsion performance in full-scale using the unsteady RANSE method with the Body Force Propeller method. The case study used to verify and validate the computational model was the series cargo ship 12500DWT built in Vietnam. Of which, the geometry and the sea trial data for validation are available on [7].

2 Numerical Simulations

2.1 Reference Vessel

A 12500DWT cargo ship with appendage and superstructure was utilized in this paper. Its geometry is shown in Fig. 1 and Table 1. The main reason for using this hull is that series cargo ship 12500DWT was built with a large number in Vietnam, and their sea trial data at different Froude numbers in ballast condition were published in [7, 8]. The sea trial is conducted at different Froude numbers in ballast condition were executed out by RINA Register, Ship owner and Dong Bac SHIN followed the requirement ISO 15016:2015 [9]. All see trials were carried out at ballast draught with various ship speeds.

Table 1. Main characteristics of cargo ship 12500DWT and its propeller

Main characteristics of cargo ship 12500DWT			
Descriptions		Unit	Value
Length of waterline	L_{WL}	m	130.51
Breadth	B	m	20.20
Depth	D	m	11.30
Design draft	T	m	8.20
volume	Δ	m^3	16050
Block coefficient	CB	–	0.742
Propeller principal particulars			
Diameter of propeller	D_P	m	3.60
Blade area ration	A_E/A_0	–	0.622
Hub ration	D_h/D_P	–	0.175
Number of blades	Z	–	4
Pitch ration	$P_{0.7}/D$	–	0.618
Direction of rotation	–	–	Right-handed

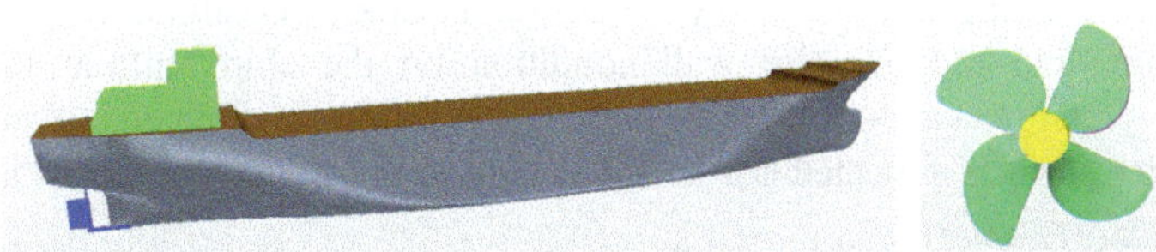

Fig. 1. The geometry of cargo ship 12500DWT and its propeller

2.2 Numerical Setup

2.2.1 Test Case Studies

In order to achieve the self-propulsion point using Body Force Propeller Method, it requires two case studies: computation of propeller open-water performance and self-propulsion computation.

For the case study (1) "propeller open-water performance simulation", the simulation was conducted for different advance coefficients J which varied from 0.4 to 0.6 (to reflect the typical propeller operation range) with the step equal to 0.05. The propeller revolutions were kept constant and equal to n = 215 rpm, and J was changed by changing the advance velocity. The roughness allowance of the propeller equals 30 μm.

For the case study (2) "self-propulsion", the computations were carried out for the ballast draught T = 3.76 m (aft draught T_A = 5.36 m and fore draught T_F = 2.16 m), corresponding to the hull displacement volume Δ = 6615 m^3, and for two speeds 11.05 and 12.75 knots. The setting was performed corresponding to the sea trial conditions [7] as follows: Calm water condition; The vessel is free to trim and sink; The ship with and superstructure. The roughness allowance of hull surface is 150 μm.

2.2.2 Computational Domain and Boundary Conditions

One of the main factors affecting the obtained numerical results is the size of computational domain size and boundary conditions.

For open-water propeller simulation, the computation domain size was set with respect to the recommendations of the International Towing Tank Conference (ITTC) [10] as follows: The inlet and outlet boundaries were located at a distance of 4D and 6D from propeller plane, respectively. The outlet boundary was located at a distance 8D from the propeller axis. The boundary conditions are set as follows: Inlet boundary - velocity inlet condition with advance velocity imposed; Outlet boundary - Pressure outlet condition; Outer boundary - Symmetry plane condition; No-slip wall condition is applied on the propeller.

For Self-propulsion simulation using the body force propeller method, the computational domain consists of only one region that is a stationary region, in which there are a ship hull and virtual disk. The size of the computational domain around the ship hull was set following the recommendations of ITTC [11], as follows: the inlet boundary is located at 1.5Lpp from FP, the outlet boundary is located at 2.5Lpp after aft of AP. Lateral boundaries are located at 2.5Lpp away from the midship plane. The bottom and top boundaries are located at 2.5Lpp and 1.25Lpp away from the free surface, respectively. The boundary conditions were set up on the domain boundaries as well as on the ship hull as follows: a constant velocity condition was used on the inlet, bottom, and top; No-slip wall condition on the ship hull; at outflow, the hydrostatic pressure was specified; Symmetry condition at symmetry plane and side wall. The free surface is located at z = 0. The ship stern (aft perpendicular) is located at x = 0.

2.2.3 Mesh Generation

The mesh used for two case studies was trimmed hexahedral grids. The grid generation process is driven by specifying base mesh size, relative to which all spacing is defined. For propeller open water simulation, the grids were refined around the propeller. For self-propulsion simulation, the grid near the free surface was refined to correctly capture the Kelvin wave pattern generated by ship hull. In order to avoid using a fine grid in which it is not important, local volume refinements were created to the bow and stern regions. The mesh in the near-virtual disk field is further refined in order to capture the hull wake and to properly resolve the mutual interactions between the hull and propeller. To capture the exact flow behavior near the walls, prism layers were used. The result of grid generation is illustrated in Fig. 2 and Fig. 3.

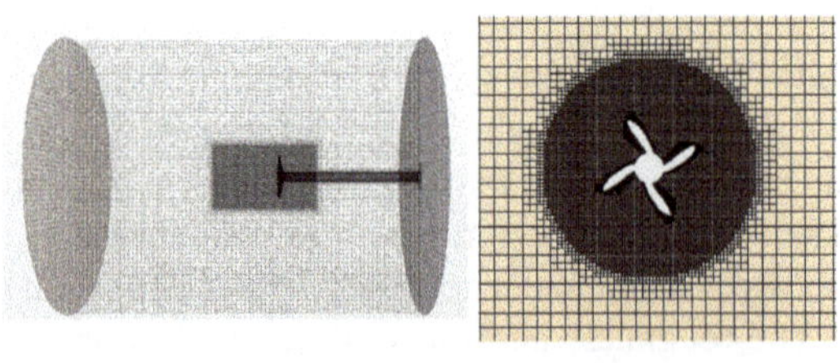

Fig. 2. Result of grid generation for open water propeller simulation

2.2.4 Physical Model

The computation was performed using incompressible viscous RANSE. The turbulence model applied in the calculations was SST K-ω two equation-model. This model has shown to be capable to predict ship hydrodynamics accurately [12]. For self-propulsion computations, the 2-DOF motion and VOF multiphase model are employed to handle heave and pitch motions and also the free surface wave flow around ship hull. The motion of the hull is captured during the computation by using DFBI Equilibrium option, i.e. dynamic fluid-body interaction motion solver optimized for fast heading towards steady-state solution. Wave damping in the region distanced about 1.25 L away from the hull is applied to reduce the resistance force fluctuation due to wave reflections within the domain.

The parameters of the virtual disk of Body Force Propeller method were set following the recommendations of [13], as follows: The outer radius of virtual disk is set as propeller radius, Inner radius is setup as hub radius of propeller, the thickness of the disk is setup as propeller thickness. Inflow plane radius, which is set as 1.1 R (R is the radius of propeller) and inflow plane offset.

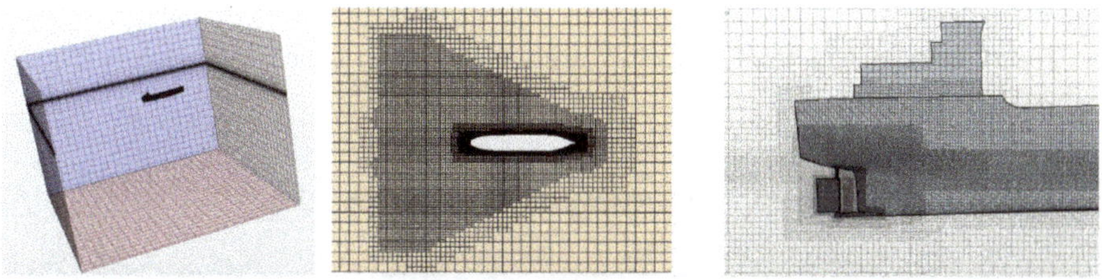

Fig. 3. Result of grid generation for self-propulsion simulation

3 Results and Discussion

The first step of research is to determine grid density at which the difference of numerical obtained results reaches sufficiently low value while maintaining the reasonable computational effort.

Convergence ratio is determined as follows [14]:

$$R_G = \frac{\varepsilon_{i,21}}{\varepsilon_{i,32}} \tag{1}$$

Where: $\varepsilon_{21} = S_2 - S_1$ is the difference between solution achieved using medium (S_2) and fine (S_1); $\varepsilon_{32} = S_3 - S_2$ is the difference between solution achieved using coarse (S_3) and medium (S_2) input parameter.

There are three kinds of possible convergence conditions: divergence ($R_G > 1$), oscillatory convergence ($R_G < 0$), and monotonic convergence ($0 < R_G < 1$).

Table 2. Results of mesh independency study at advance coefficient J = 0.5

Parameter		Mesh density			ε_{32}	ε_{12}	R_G
		Coarse mesh (mesh #3)	Medium mesh (mesh #2)	Fine mesh (mesh #1)			
Thrust coefficient	K_T	0.099	0.0952	0.0936	0.0038	0.0016	0.421
Torque coefficient	K_Q	0.117	0.1154	0.1141	0.0016	0.0013	0.813
Open-water propeller efficiency	η_0	0.673	0.65648	0.640	0.0169	0.0163	0.965

For open water simulation, the grid sensitivity study was conducted at J = 0.5 by using three mesh with the refinement ration r_i equal to $\sqrt{2}$, so that the coarse, medium, and fine mesh corresponded to the cell number of 0.712, 1.405 and 2.948 million cells, respectively. The results of the mesh dependency study for propeller open-water simulations are illustrated in Table 2. As can be seen from Table 2, monotonic convergence was observed, so the fine mesh was used in further studies of propeller open-water simulations.

The numerical results of propeller open-water performance using fine mesh are showed in Table 3.

Table 3. Open-water propeller simulation results

J	K_T	K_Q	η_0
0.40	0.1374	0.1556	0.562
0.45	0.115	0.135	0.610
0.55	0.0936	0.1141	0.653
0.60	0.071	0.0933	0.666

The self-propulsion calculation was carried out with the cell number of 3.8 million cells at ship self-propulsion point, where the rate of propeller revolutions n was adjusted to obtain force equilibrium. In practice, it is difficult to obtain this condition in one run, and a common practice is to perform at least two constant speed runs with different propeller revolutions, hence the discrepancy between the total resistance RT and the thrust of propeller working behind the ship T equals zero. Figure 4 presents the results and procedure to find the self-propulsion point.

The self-propulsion point is searched using linear interpolation, and the obtained self-propulsion point for case study one and two were n = 172.8 rpm and 184.2 rpm respectively (Table 4).

Table 4. Numerical results of two rpm cases for two ship speeds

Propeller revolution n [rpm]	Total resistance RT [kN]	Thrust of propeller [kN]
Case 1: at ship speed V = 11.05 knots		
171	160.80	157.50
175	162.40	166.50
Case 2: at ship speed V = 12.75 knots		
195	192.10	220.05
180	186.02	175.16

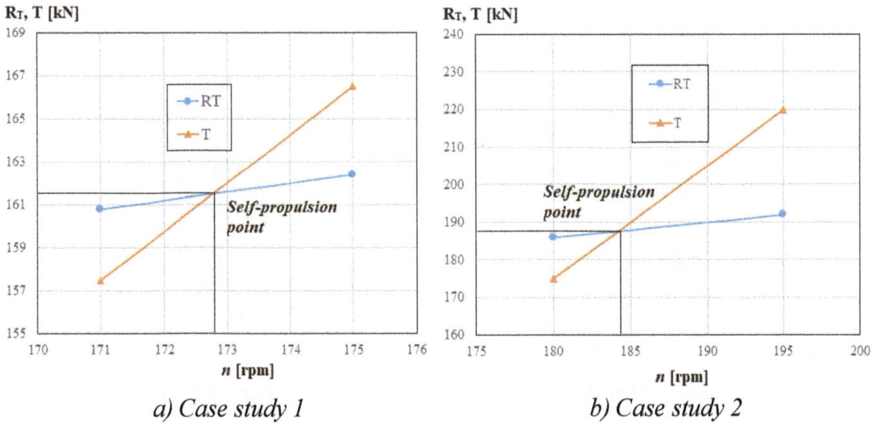

Fig. 4. Procedure to find the self-propulsion point

Table 5. Comparing calculated self-propulsion point results with sea trial data

Parameter	EED [7]	CFD	E%D
Case study 1	171.0	172.8	1.05%
Case study 2	195.0	184.2	5.53%

Table 5 shows the comparison between the self-propulsion point obtained from the computation (CFD) and sea trial data (EED). It can be observed from Table 5 that the predicted self-prolusion point for both case studies agree well with sea trial data with deviation is less than 6.0%.

The details of flow characteristics around ship like wave patterns on free surface and axial velocity distributions in virtual disk's plane in self-propulsion simulation are showed in Fig. 5 and 6 respectively.

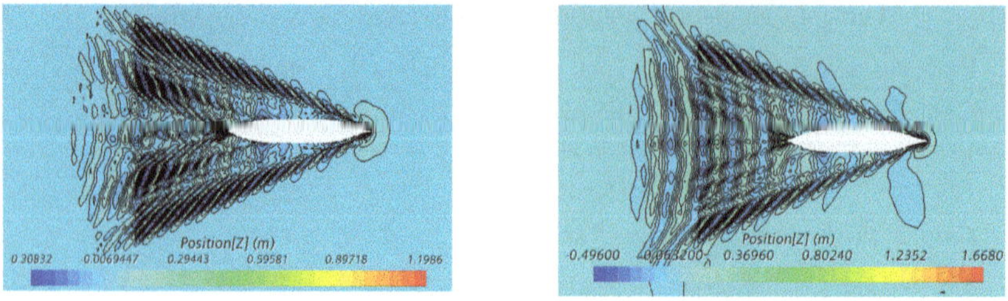

a) V=11.05knots *b) V=12.75knots*

Fig. 5. Wave patterns on free surface

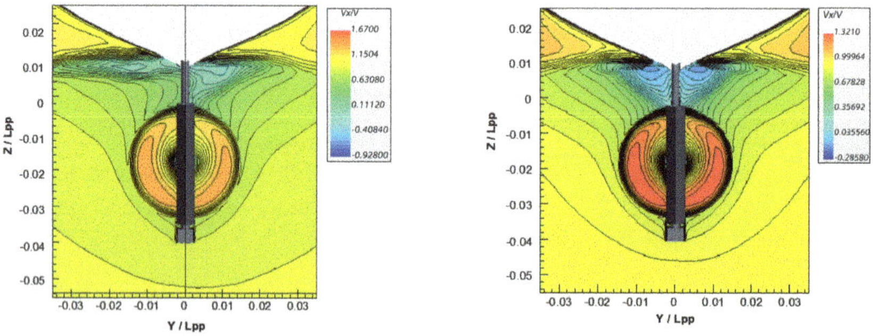

a) V=11.05knots *b) V=12.75knots*

Fig. 6. Axial velocity distributions in virtual disk's plane

4 Conclusions

In this study, the RANSE method has been utilized to simulate self-propulsion for Series Cargo Ship 12500DWT. Two case studies were analyzed: propeller open-water, and self-propulsion simulations. In numerical predictions of propeller open-water performance, the Moving Reference Frame method has been applied. In the self-propulsion simulation, the Body Force Propeller method has been used.

The factors affecting the level of accuracy of the numerical obtained results, such as computational domain, time step, y + value, mesh generation technique, and mesh dependency, were considered in this study.

There is good agreement between CFD and sea trial data. The discrepancies between computed and sea trial data were less than 6.0%. This shows the capability of CFD in the application of RANSE with the Body Force Propeller method in full-scale self-propulsion simulations.

Acknowledgement. The authors are grateful to the Vietnam Maritime University and Dongbac Shipbuilding Industry Join Stock Company for providing necessary research facilities during carrying out this research work.

References

1. Tu, T.N., et al.: Numerical study on the influence of trim on ship resistance in trim optimization process. Naval Eng. J. **130**(4), 133–142 (2018)
2. Carrica, P.M., Castro, A.M., Stern, F.: Self-propulsion computations using a speed controller and a discretized propeller with dynamic overset grids. J. Mar. Sci. Technol. **15**(4), 316–330 (2010)
3. Tu, T.N., et al.: Numerical prediction of propeller-hull interaction characteristics using RANS method. Polish Maritime Research (2019)
4. Shen, Z., Wan, D., Carrica, P.M.: Dynamic overset grids in OpenFOAM with application to KCS self-propulsion and maneuvering. Ocean Eng. **108**, 287–306 (2015)
5. Win, Y.N., et al.: Computation of propeller-hull interaction using simple body-force distribution model around Series 60 CB=0.6. J. Jpn. Soc. Naval Architects Ocean Eng. **18**, 17–27 (2013)
6. Gokce, M.K., Kinaci, O.K., Alkan, A.D.: Self-propulsion estimations for a bulk carrier. Ships Offshore Struct. **14**(7), 656–663 (2019)
7. Result of sea trial "Truong Minh Ocean"_12500. Dongbac Shipbuilding Insdustry Joint Stock Company
8. Ship documents of cargo ship 12500DWT. Dongbac Shipbuilding Insdustry Joint Stock Company
9. ISO. Ship and Marine Technology—Guidelines for the Assessment of Speed and Power Performance by Analysis of Speed Trial Data; ISO: Geneva, Switzerland (2015)
10. ITTC – Recommended Procedures and Guidelines. Practical Guidelines for Ship Self-Propulsion CFD (2014)
11. ITTC 2014. Recommended procedures and guidelines 7.5-03-02-04. https://ittc.info/media/4198/75-03-02-04.pdf
12. Yong, Z., et al.: Turbulence model investigations on the boundary layer flow with adverse pressure gradients **14**(2), 170–174 (2015)
13. Siemens. STAR-CCM+ User Guide (2020)
14. ITTC Specialist Committee, 2017. Recommended procedures and guidelines - uncertainty analysis in CFD verification and validation methodology and procedures. https://www.ittc.info/media/8153/75-03-01-01.pdf

Design and Testing of a New Shear-Mode Magneto–Rheological Damper with Self-power Component for Front–Loaded Washing Machines

Quoc–Duy Bui[1,2], Long–Vuong Hoang[2], Duc–Dai Mai[3], and Quoc Hung Nguyen[2,4](✉)

[1] Faculty of Civil Engineering, HCMC University of Technology and Education, Ho Chi Minh City, Vietnam
duybq.ncs@hcmute.edu.vn, buiquocduy@iuh.edu.vn
[2] Faculty of Mechanical Engineering, Industrial University of Ho Chi Minh City, Ho Chi Minh City, Vietnam
hoanglongvuong@iuh.edu.vn, hung.nq@vgu.edu.vn
[3] Faculty of Mechanical Engineering, HCMC University of Technology and Education, Ho Chi Minh City, Vietnam
daimd@hcmute.edu.vn
[4] Faculty of Engineering, Vietnamese German University, Thu Dau Mot City, Binh Duong, Vietnam

Abstract. This study work focuses on development of a self–powered magneto–rheological (MR) damper to control effectively vibrations of front–loaded washing machines. The proposed damper consists of a shear–mode MR damping part and a power generator featuring permanent magnets and induction coils. The two parts are integrated into a single damper in order to replace traditional passive dampers of washing machines. In the damping part, the gap between the shearing shaft and housing fill up with MR fluid to generate controllable damping force when subjected to a magnetic field. Magnetic coils are placed on the grooves of the housing to produce a closed magnetic loop as electrically excited. The power generator comprises a housing slotted for induction coils and a shaft end carrying permanent magnets. Vibration from the washing tube causes a motion of the damper shaft, which in turn generates induced currents in the induction coils. The induced currents then directly power the magnetic coils of the dampers, creating an adaptive damping force to suppress the vibration.

Keywords: MR fluid · Self–powered · MR damper · Washing machine vibration

1 Introduction

Washer vibration is a challenging problem for investigation. The washing machine vibration is caused by the imbalanced mass of cloths put in the washing drum. During the spin–drying process, the fundamental resonance (rigid body mode vibration) often

B. T. Long et al. (Eds.): MMMS 2020, LNME, pp. 860–866, 2021.
https://doi.org/10.1007/978-3-030-69610-8_114

appears at quite low spindle speed, about 100–200 rpm [1], while flexional resonance due to the rear and side panels vibration usually occurs at much higher drum speed, 1000 rpm or greater. Passive dampers used in conventional suspension systems have a trade-off between the vibration reducibility at low resonant frequency and force transmission preventability at high frequencies since the damping coefficient cannot be controlled. Therefore, smart dampers such as MR fluid based dampers (or MR dampers) were recently implemented. There have been several studies on development of MR dampers for washing machines [2–4]. Carlson [2] proposed a sponge MR damper with low cost and low damping force particularly for suspension system of washing machines. Nguyen et al. [3] optimally designed a small flow–mode MR damper for a well installation and replacement of passive ones in front–loaded washing machines. However, the off–state damping force was relatively high, about 25 N, which causes a severe vibration at high frequencies. In addition, a considerable large volume of MR fluid, resulting in high cost of the damper, is a critical issue for flow–mode MR dampers. To remedy these shortcomings, Nguyen et al. [4] developed a MR damper in shear–mode configuration for front–loaded washing machines. By conducting optimal design, it was shown that the shear–mode damper could provide a damping as high as 120 N while the zero–field damping force was kept as small as 10 N. More recently, Bui et al. [5] investigated a MR damper in shear–mode with self–powered ability for washing machines featuring permanent magnets and induction coils. However, many induction coils (7 coils) are employed which cause difficulties in manufacturing, maintenance and wiring. Consequently, this research deals with development of a new self–powered MR damper for front–loaded washing machines with 4 induction coils. In addition, the whole damper is optimally designed and a prototype damper is fabricated and experimentally tested. The proposed damper is then installed in a prototype washing machine for evaluation of vibration suppression.

2 Configuration and Working Principle

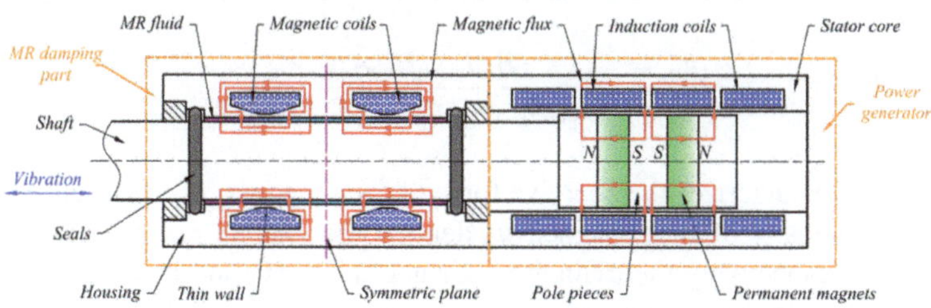

Fig. 1. The proposed shear-mode MR damper with self-power part.

Figure 1 shows configuration of the proposed shear-mode MR damper integrated with a self-power part. The damper consists of a MR damping part and a power generator. In the damping part, the gap between the shearing shaft and housing fill up with MR fluid to generate controllable damping force when subjected to a magnetic field. Magnetic coils are placed directly on the groove features of the housing to produce a closed magnetic loop as electrically excited. The power generator comprises a housing slotted for induction coils and a shaft end carrying permanent magnets. In this study, the ring–shaped axial permanent magnets are used. Vibration from the washing tube causes induced currents in the induction coils, which then directly power the excitation coils of the damper.

3 Modeling and Optimization of the Self–powered MR Damper

The power generator is schematized in Fig. 2. Commercial C45 steel is used to manufacture the stator core and pole pieces. The permanent magnets are made of NdFeB grade N35 and assembled into an aluminum shaft end. Dimensions of the magnets are given as follows: outer diameter of 25 mm, inner diameter of 5 mm and thickness of 5 mm.

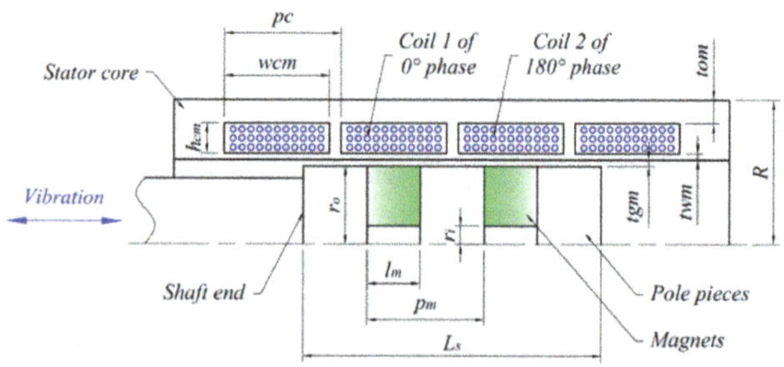

Fig. 2. Schematic of the power generator

Given the magnetic efficiency λ, the magnetic flux across the annular air gap between the stator core and permanent magnets, Φ_g, can be expressed by [6]

$$\Phi_g = \lambda \frac{B_{re} l_m \mu_0 H_{co} A_{gm}}{2 t_{gm} B_{re} + l_m \mu_0 H_{co} (A_{gm}/A_m)} \tag{1}$$

where H_{co} characterizes the coercive force of each magnet, B_{re} represents remanent magnetic flux density of each magnet, μ_0 denotes the relative magnetic permeability, t_{gm} and l_m are respectively the cylindrical air gap thickness and magnet length, while A_m and A_{gm} express the magnet cross–section area and the cylindrical air gap surface area, which are in turn determined as follows

$$A_m = \pi\left(r_o^2 - r_i^2\right) \tag{2}$$

$$A_{gm} = 2\pi\left(r_o + \frac{t_{gm}}{2}\right)\left(\frac{p_m - l_m}{2}\right) \tag{3}$$

in which r_o and r_i are respectively the magnet outer and inner radii and p_m represents the pole pair pitch. In each induction coil, the induced voltage E then can be calculated by

$$E = -N\Phi_g \frac{\pi}{p_m} \sin\left(\frac{\pi}{p_m} x + \varphi_0\right) \frac{dx}{dt} \tag{4}$$

where φ_0 and N are the phase angle and winding turn number of the induction coil, respectively, and x is the displacement of the shaft.

In this study, the coil slot pitch p_{cm} is designed to equal to the magnet pole pair pitch p_m. By this way, the phase angle between two adjacent coils is π and their generated induced voltages are determined by [7]

$$E_1 = -N\Phi_g \frac{\pi}{p_m} \sin\left(\frac{\pi}{p_m} x\right) \frac{dx}{dt} \tag{5}$$

$$E_2 = -E_1 \tag{6}$$

Since the resonance often occurs at low spindle speeds, the coil can be neglected in terms of inductance and assumed to operate as a resistor in that case. Then the generated power is obtained to be the sum of the powers of two adjacent active coils P_1 and P_2 as follows

$$P = P_1 + P_2 = \frac{E_1^2}{R_{cm}} + \frac{E_2^2}{R_{cm}} = \frac{2}{R_{cm}} \left[N\Phi_g \frac{\pi}{p_m} \sin\left(\frac{\pi}{p_m} x\right) \frac{dx}{dt}\right]^2 \tag{7}$$

where R_{cm} represents the resistance of each coil. From Eq. (7), it is observed that the generated power is significantly influenced by geometric characteristics of the power generator. Therefore, a design optimization should be conducted to improve performance of the power generator. Design variables of the optimization are the housing thickness t_{om}, the coil height h_{cm}, the coil width w_{cm}, the magnet pole pair pitch p_m and the coil pitch p_c. By employing gradient steepest optimization algorithm, optimal results of the proposed power generator are obtained and presented in Table 1.

Table 1. Optimal solutions of the self-power part

Parameter (mm)	Optimal value	Parameter (mm)	Optimal value
Coil width w_{cm}	11.52	Air gap thickness t_{gm}	0.8
Coil height h_{cm}	3.91	Thin wall thickness t_{wm}	0.8
Coil pitch p_c	13.84	Housing thickness t_{om}	2
Magnet length l_m	5	Outer radius R	20
Pole pair pitch p_m	13.84	Generated power P (W)	19.3

Figure 3 shows the schematic diagram of the damping part. The modeling and optimal design of damping part are the same as that in the previous research [5] and the results are summarized in Table 2.

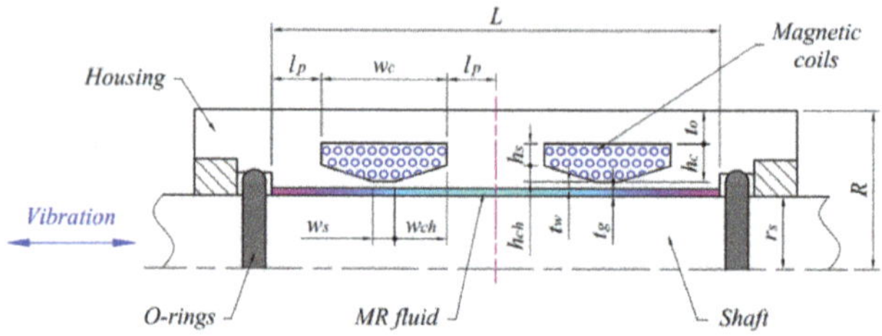

Fig. 3. Schematic diagram of the MRF damping part

Table 2. Optimal solutions of the MRF damping part

Parameter (mm)	Optimal value	Parameter (mm)	Optimal value
Coil width w_c	17.19	Thin wall thickness t_w	0.8
Coil height h_c	6.98	Housing thickness t_o	3.17
Coil width chamfer w_{ch}	7.23	Shaft radius r_s	8.25
Coil height chamfer h_{ch}	3.02	Outer radius R	20
Pole length l_p	5.41	Max. current intensity (A)	0.38
MR fluid gap length L	56.03	Max. damping force (N)	80
MR fluid gap width t_g	0.8	Off–state force (N)	18.5

4 Experiment Test

Figure 4 shows the prototype damper and experimental setup to test performance of the damper.

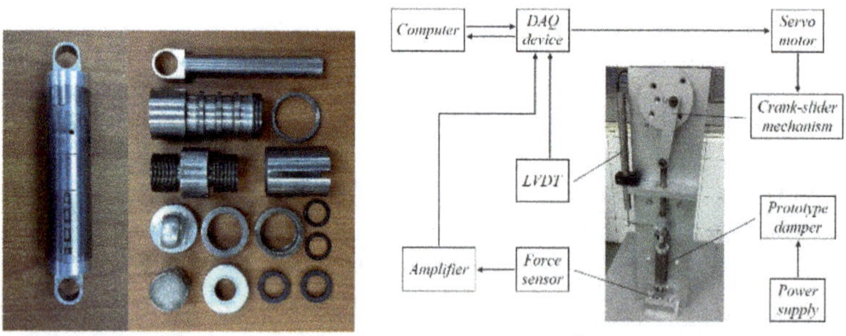

(a) (b)

Fig. 4. (a) Prototype damper and (b) experimental setup to test performance

The measured damping forces of the proposed damper (4 coils) and the previous one (7 coils) [5] are depicted in Fig. 5. From the figure, it is observed that the damping performance of the proposed damper is almost the same as that in [5], although the power generator of the proposed damper possesses less induction coils. T proposed damper can generate a damping force up to 75.5 N, which is nearly similar to that in [5] (75.1 N).

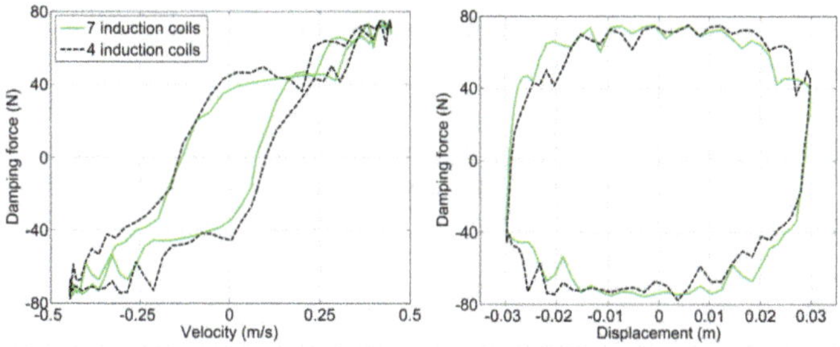

a) force vs. velocity b) force vs. displacement

Fig. 5. Experimental responses of the proposed damper with the power generator

The proposed self–powered MR dampers are installed in a commercial washing machine (WF8690NGW, Samsung). In the experiment, a laundry mass of 7 kg is used and an 3D accelerometer is mounted to the frame of the washing machine for measurement of the frame acceleration. The data is collected during spinning process, in which the drum speed is continuously increased from 0 to 1000 rpm. From the results shown in Fig. 6, it is observed that the self–powered MR dampers are better for vibration suppression of the washing machine than the conventional dampers do at both low speed (resonant) and high speed. It is also noted that the cost for manufacture of a damper is about 5 USD which is applicable for commercial washing machines.

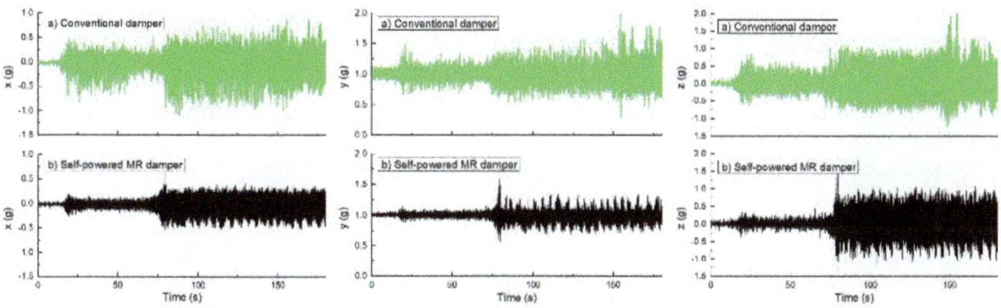

(1) x–direction (2) y–direction (3) z–direction

Fig. 6. Acceleration indices of the washing machine frame

5 Conclusions

A new self–powered MR damper for front–loaded washing machines was designed, manufactured and experimentally tested. It was shown that the proposed self–powered damper with 4 induction coils could produce the same maximum damping force as that with 7 induction coils in [5]. This improves cost in manufacturing, maintenance and increase feasibility in commercialization. It was also shown that the vibration in all directions of the washing machine employing the self–powered MR dampers can be well mitigated as compared to that of the washing machine using conventional passive dampers.

Acknowledgements. This work was supported by the Vietnam National Foundation for Science and Technology Development (NAFOSTED) under grant no. 107.01–2018.335.

References

1. Lim, H.T., et al.: Dynamic modeling and analysis of drum–type washing machine. Int. J. Precis. Eng. Manuf. **11**, 407–417 (2010)
2. Carlson, J.D.: Low–cost MR fluid sponge devices. J. Intell. Mater. Syst. Struct. **10**, 589–594 (1999)
3. Nguyen, Q.H., et al.: Optimal design and performance evaluation of a flow–mode MR damper for front–loaded washing machines. Asia Pac. J. Comput. Eng. **1** (2014)
4. Nguyen, Q.H., et al.: Optimal design of magnetorheological fluid–based dampers for front–loaded washing machines. Proc. Inst. Mech. Eng. C-J. Mec. Eng. Sci. **228**, 294–306 (2014)
5. Bui, Q.D., et al.: Development of a magnetorheological damper with self–powered ability for washing machines. Appl. Sci. **10** (2020)
6. Chen, C., Liao, W.H.: A self–sensing magnetorheological damper with power generation. Smart Mater. Struct. **21**, 025014 (2012)
7. Rhinefrank, K., et al.: Novel ocean energy permanent magnet linear generator buoy. Renew. Energy **31**, 1279–1298 (2006)

Design and Testing of a Novel Magneto-Rheological Clutch with Tooth-Shaped Rotor

Diep Bao Tri[1,2], Nguyen Ngoc Diep[2], Nguyen Xuan Bien[2], Vu Van Bo[2], Huynh Minh Huy[3], and Nguyen Quoc Hung[2,4](✉)

[1] Faculty of Civil Engineering, HCMC University of Technology and Education, Ho Chi Ming City, Vietnam
tridb.ncs@hcmute.edu.vn

[2] Faculty of Mechanical Engineering, Industrial University of Ho Chi Minh City, Ho Chi Ming City, Vietnam
{nguyenngocdiep,nguyenxuanbien,vuvanbo}@iuh.edu.vn, hung.nq@vgu.edu.vn

[3] Faculty of Engineering, Tien Giang University, Tien Giang, Vietnam
huynhminhhuy@tgu.edu.vn

[4] Faculty of Engineering, Vietnamese-German University, Binh Duong, Vietnam

Abstract. This research focuses on development of a new configuration of magneto-rheological clutch (MRC) featuring a tooth-shaped disc with multiple teeth acting as multiple magnetic poles of the clutch. The tooth shaped disc is placed in a clutch housing composed of the left housing and the right housing. The inner face the housing also has tooth shaped features mating with the teeth of the disc via MRF layer. Excitation coils are placed directly on stationary winding cores placed on both side of the clutch housing. A air gap of 0.25 mm is left between the housing and the winding cores to ensure the housing can freely rotate against the winding cores. By using a tooth shaped disc, more active interfacing area between the disc and the housing can be achieved and a high transmitted torque of the clutch is expected. After the introduction part, configuration of the proposed MRC is introduced and the transmitted torque of the MRC is derived. Optimization of the MRC considering mass and transmission torque is then conducted and the optimized MRC is evaluated by both simulation experimental works.

Keywords: MR fluid · MR clutch · Speed control · Tooth-shaped rotor

1 Introduction

Magnetorheological fluid (MRF) is a kind of smart fluid, which is composed of tiny magnetic particles (micro to nanometer scale) dispersed in a carrier fluid. When no magnetic field is applied, the particles are dispersed randomly and MRF rheologically behaves as Newtonian fluid like carrier fluid. In the presence of magnetic field, the particles are magnetized, which consequently experienced attractive forces, forming chain-like structure. At high field intensity, the MRF is almost solidified. This change is

B. T. Long et al. (Eds.): MMMS 2020, LNME, pp. 867–872, 2021.
https://doi.org/10.1007/978-3-030-69610-8_115

very fast (few milliseconds) and completely reversible (when the magnetic field is removed, the MRF behaves like the carries fluid again). With this special effect, the MRF holds great potential in many applications such as clutches, brakes, valves, dampers and engine mount [1, 2]. There have been several researches on clutch featuring MRF (called MR clutch or MRC for short in this paper) [3–6]. In previous researches, the clutch discs have cylindrical shape and the torque is transmitted from the driving shaft to the driven shaft through the MRF in the end-face and cylindrical ducts. Therefore, in order to archive large transmitted torque, the disc should have large size, which results in large size of the MRC and high inertia effect.

The main contributions of this research is to develop a more compact size MRC featuring a tooth-shaped disc. With toothed shaped disc, more contact area between the disc and the MRF can be archived and a large transmitted torque can be archived while size of the MRC is still kept compact.

2 The Proposed MR Clutch with Tooth-Shaped Rotor

Figure 1 shows a configuration of the proposed MRC with tooth-shaped disc. As shown in the figure, a tooth-shaped disc made of magnetic steel is fastened to the driving shaft made of nonmagnetic steel. The disc is embedded inside the clutch housing. The inner face of the clutch housing also has teeth, mating with the teeth of the disc via MRF medium. One side of the clutch housing is connected to the driven shaft made of nonmagnetic steel. In order to prevent the leaking of MRF, a lip- seal is used. Two fixed winding cores, on which the coils are wound, are placed on both side of the clutch housing. To ensure a free rotation of the clutch housing relatively to the

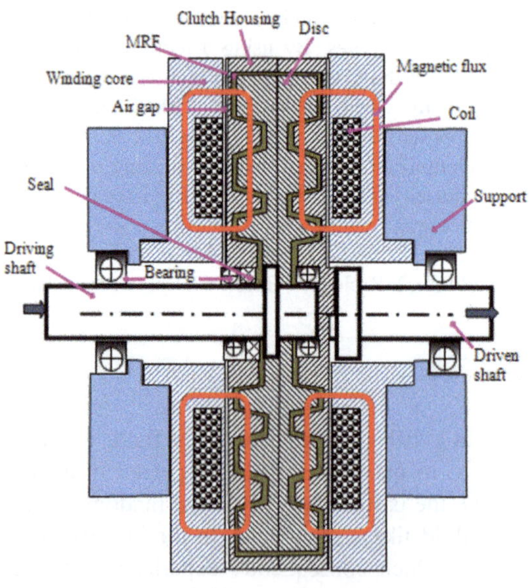

Fig. 1. Configuration of the proposed tooth-shaped MR clutch

winding cores, a small air gap (0.25 mm) is left between the housing and the winding cores. As the power of the coils is turned on, a magnetic field is generated causing MRF in the gap between the disc and the housing becomes solid-like. This results in a controllable torque transferred from the disc (driving) to the housing (driven).

In order to evaluate the transmitted torque of the clutch, it is assumed that the MRF follows Bingham plastic rheological model and shear stress due to yield stress is significantly greater than that due to post-yield viscosity. In addition, friction due to the sealing and bearings are ignored. In this case, the induced transmitted torque can be determined by

$$T_b = 2(\sum_{i=0,2,4,6,8,10} T_{Ei} + \sum_{j=1,3,5,7,9} T_{Ij}) + T_c \tag{1}$$

where

$$T_{Ei} = \frac{2\pi\tau_{yEi}}{3}(R_{i+1}^3 - R_i^3), i = 0, 2, 4, 6, 8, 10 \tag{2}$$

$$T_{Ii} = 2\pi(R_i^2 l + R_i l^2 \sin\phi + \frac{1}{3} l^3 \sin^2\phi)\tau_{yIi};\ i = 1, 3, 5, 7, 9 \tag{3}$$

$$T_c = 2\pi R_{11}^2 (b + 2h)\tau_{yc} + \mu_c \frac{\Omega R_{11}}{d} \tag{4}$$

In the above, T_{Ei} is the friction torque caused by MRF in the E_i gap, T_{Ii} is the friction torque caused by MRF in the I_i gap, T_c is the friction torque caused by MRF in the C gap, R_i is the radius of the point i in the disc profile shown in Fig. 2, l is the length of the inclined gap, ϕ is the inclined angle, h is the height of the tooth, τ_{Ei}, τ_{Ii} and τ_c are respectively the yield stress of the MRF in the E_i, I_i and C gap.

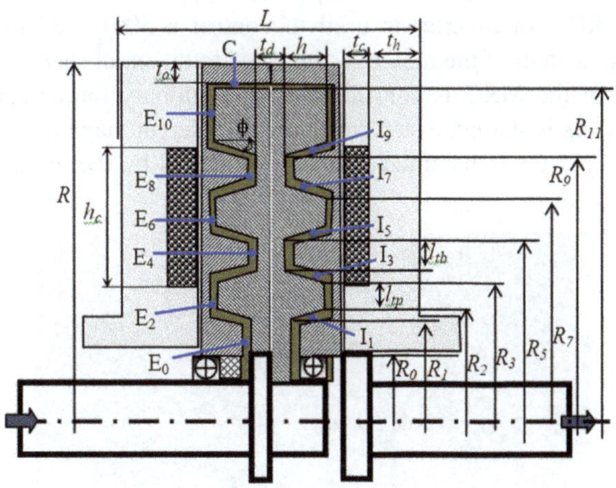

Fig. 2. Illustration of MRF ducts for evaluating transmitted torque

3 Optimal Design of the Clutch

The optimization design problem of the MRC is stated as follows: Find optimal value of significant dimensions of the MR clutch structure so that its maximum transmitted torque determined by Eq. (1) is greater than a required torque and the objective of the optimization is to minimize the overall volume of the MRC determined by

$$V_{MRC} = \pi R^2 L \tag{5}$$

Where R is the overall radius of the MRC and L is the overall length of the MRC. The design variables are: the disc thickness t_d, the tooth-height h, the tooth angle ϕ, tooth-peak length l_{tp}, tooth bottom length l_{tb}, the inner disc radius R_{di} (=R_O), radius of first tooth R_1, the outer disc radius R_{do}(=R_{11}), the clutch outer housing thickness t_o, the winding core thickness t_h, the coil width t_c, the coil height h_c. The MRF duct size (*d*) is empirically set by 1 mm. The thickness of the wall of the housing is set by 1.25 considering manufacturing issue. In order to obtain the optimal solution, a FEA code integrated with an optimization tool is employed. In this study, the first order method with steepest descent algorithm of ANSYS optimization tool is used. It is assumed that the commercial silicon steel is used for magnetic components of the MRC such as the side housing, the envelope and the disc. The coil wires diameter is 0.511 mm, so a maximum current of 2.5 A is applied to the coils, the commercial MR fluid: MRF132-DG is used. The optimal results in case of the transmitted torque is constrained to be greater than 10 Nm are as followings (*mm*): t_c = *7.0*, h_c, = 24.0 (no. of turns: 490), R_O = 32.5, R_1 = 38.0, R_{11} = 63.5, t_d = 4.0, h = 2.0, t_o = 3.0 (R = 67.5), t_h = 5.0 (L = 40.0).

4 Experimental Validations

In this part, a prototype of the optimized MRC is manufactured and its performance is experimentally tested. Figure 3 shows experiment apparatus to evaluate transmitting torque of the MRC. An inverter is used to control a 500W-AC motor rotating at 300 rpm, the output shaft of the motor is connected to the input shaft of the MRC while the output shaft of the MRC is connected to an stationary torque sensor. When the experimental process is started, a step current with different magnitude is applied to the coil of the MRC and the transmitted torque is measured by the torque sensor.

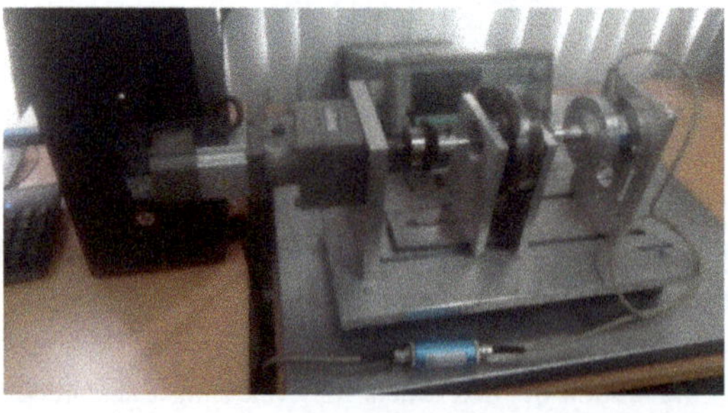

Fig. 3. Experiment apparatus to test the proposed MRC

Figure 4 show the results obtained from experiments. From Fig. 4a, it is observed that the measured transmitted torque at the applied current of 2.5 A is around 10.4 Nm, which is a bit greater than that of the simulation (9.9 Nm). The main reason is that, in the simulated one, the friction torque of the sealing and bearings and the transmitted torque due to post-yield viscosity are ignored. It is also observed that at different applied currents, different values of output torque can be archived and very closed to the simulated ones; the error is within 5%. It is also found the torque response time is around 0.25 s, which is very good time response for application in industry.

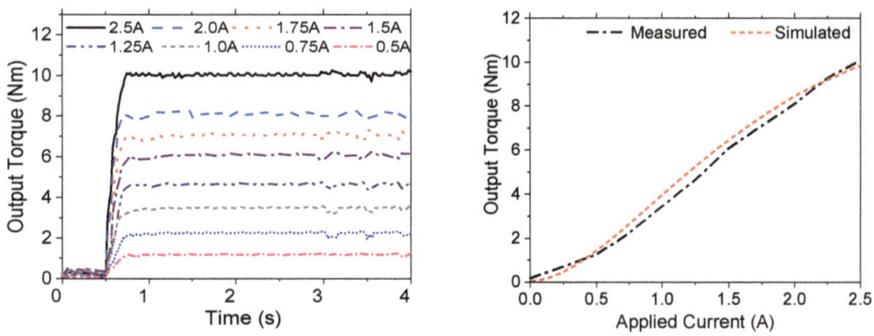

(a) measured torque (b) simulated vs. measured torque

Fig. 4. Experiment results of step response of the MRC

5 Conclusions

In this study, a new configuration of magneto-rheological clutch (MRC) was proposed. In this new configuration, a tooth shaped disc is placed in a clutch housing composed of the left housing and the right housing. The inner face of the housing also has tooth shaped features mating with the teeth of the disc via MRF layer. Excitation coils are placed directly on stationary winding cores placed on both side of the clutch housing. An air gap of 0.25 mm is left between the housing and the winding cores to ensure the housing can freely rotate against the winding cores. Optimization of the proposed MRC considering mass and transmission torque was studied and the optimized MRC was manufactured for testing. Experimental results showed a good agreement with simulated one.

References

1. Wang, J., Meng, G.: Magneto-rheological fluid devices: principles, characteristics and applications in mechanical engineering. J. Mater. Des. Appl. **215**, 165–174 (2001)
2. Muhammad, A., et al.: Review of magnetorheological (MR) fluids and its applications in vibration control. J. Mar. Sci. Appl. **5**, 17–29 (2006)
3. Lee, U., Kim, D., Hur, N., Jeon, D.: Design analysis and experimental evaluation of an MR fluid clutch. J. Intell. Mater. Syst. Struct. **10**, 701–707 (1999)

4. Kikuchi, T., Ikeda, K., Otsuki, K., Kakehashi, T., Furusho: Compact MR fluid clutch device for human-friendly actuator. J. Phys. Conf. Ser. 149 (2009)
5. Nguyen, Q.H., et al.: Development of a new clutch featuring MR fluid with two separated mutual coils. In: Lecture Notes in Electrical Engineering, vol. 371, pp. 835–844 (2016)
6. Nguyen, Q.H., et al.: Development of magneto-rheologial fluid (MRF) based clutch for output torque control of AC motors. In: Proceedings of SPIE, vol. 0602, Denver, Colorado, USA (2018)

Design and Experimental Evaluation of a Novel Damper for Front–Loaded Washing Machines Featuring Shape Memory Alloy Actuator and Wedge Mechanism

Quoc–Duy Bui[1,2], Quy–Duyen Do[3], Long–Vuong Hoang[2], Duc–Dai Mai[4], and Quoc Hung Nguyen[3(✉)]

[1] Faculty of Civil Engineering, Ho Chi Minh City University of Technology and Education, Ho Chi Minh City, Vietnam
duybq.ncs@hcmute.edu.vn, buiquocduy@iuh.edu.vn
[2] Faculty of Mechanical Engineering, Industrial University of Ho Chi Minh City, Ho Chi Minh City, Vietnam
hoanglongvuong@iuh.edu.vn
[3] Faculty of Engineering, Vietnamese–German University, Thu Dau Mot City, Binh Duong Province, Vietnam
{duyen.dq, hung.nq}@vgu.edu.vn
[4] Faculty of Mechanical Engineering, Ho Chi Minh City University of Technology and Education, Ho Chi Minh City, Vietnam
daimd@hcmute.edu.vn

Abstract. In this study, a new shape memory alloy (SMA) damper is developed for suspension system of front–loaded washing machines. The damper is actuated by a spring SMA actuator and its actuating force is amplified by a wedge mechanism and exerted on the inner cylindrical face of the damper housing to create friction against the housing. The friction between the wedges and housing produces damping force of the damper. After an introduction to SMA actuators and the suspension system of front–loaded washing machines, configuration of the proposed SMA damper is presented. Modelling of the proposed SMA damper is then performed considering required damping force, installation space and cost. Based on the modelling, a prototype SMA damper is then designed, manufactured and assessed through experimental works with discussions.

Keywords: Shape memory alloy · SMA actuator · SMA spring · Washing machine

1 Introduction

It is well–known that vibration of a washing machine is one of challenging issues that should be under investigation. The vibration is mainly due to the unbalanced mass of clothes distributed in the washing drum which occurs most frequently in spinning–drying stage. The vibration is transferred to the floor causing noises, uncomfortable feeling for users and gradual failure of the machine. Generally, there are two

B. T. Long et al. (Eds.): MMMS 2020, LNME, pp. 873–878, 2021.
https://doi.org/10.1007/978-3-030-69610-8_116

approaches to the suppression of washing machine vibration. The first approach is based on the control of the tub balance mass to eliminate the source of vibration [1]. In the second one, the vibration is suppressed by suspension system [2]. It was shown that washing machines usually experience the first resonance at quite low frequency, about 100–200 rpm [1–3]. The resonance comes from the rigid vibration mode of the washing drum. When the drum speed exceeds 1000 rpm, the side and rear panels may experience resonances causing noises and walk of washing machines.

There have been many researches on implementation of magnetorheological fluid (MRF) based MR dampers (MR dampers for short) in vibration suppression of washing machines. Carlson [4] developed a MR sponge damper for washing machines. It was shown by Spelta et al. [5] that the MR sponge damper can well eliminate the vibration of washing machines at low frequency. However, stability and lastingness of the sponge are considerable issues. Nguyen et al. [6] researched on design and application of a flow–mode MR damper to control vibration of washing machines. However, high cost and large off–state force of the damper are challenging issues. To overcome these drawbacks, Nguyen et al. [7] and Bui et al. [8] later proposed shear–mode MR dampers for front-loaded washing machines. However, the sealing problems and high cost are still considerable challenges. In order to solve these challenges, in this study a new damper for washing machines featuring shape–memory alloy (SMA) is studied. SMA is a type of smart materials that can remember its original shape and return to that shape when heated due to its superelastic properties, generating a useful force. Inspired by these concepts, a damper employing SMA as an actuator can be archived. It is noted that no sealing is required for SMA actuator and the SMA is very cheap compared with MRF.

2 Configuration of SMA Damper

Figure 1 shows configuration of the proposed SMA damper for washing machines. As shown in the figure, a sleeve shaft slides along a cylindrical housing. Inside the sleeve shaft, a wedge mechanism including an actuating head and four wedges is used to generate friction force on the inner face of the housing.

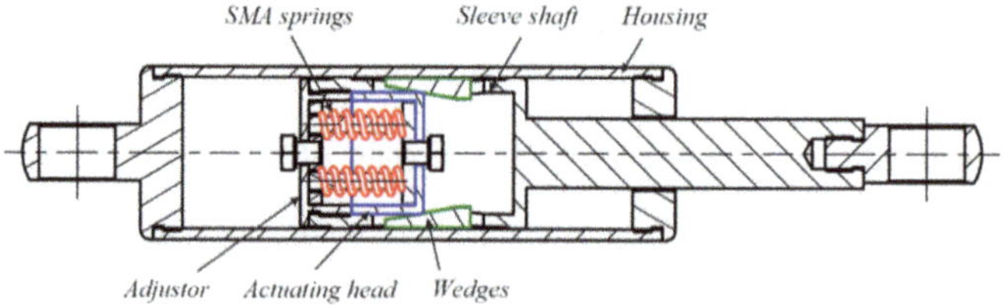

Fig. 1. Configuration of the proposed SMA damper

When the SMA springs are heated, they extend and push the four wedges move outward in contact with the inner face of the housing. The friction between the wedges and the housing produces a damping force. By using a wedge mechanism, actuating force from the SMA springs is magnified to yield a required damping force. At the initial position, the SMA springs are stretched a little to create a restoring force. As the SMA springs are cool down, they are contracted to their original shapes and the restoring force brings the actuating head back to its initial position.

3 Modeling of SMA Damper

In this section, the SMA damper is designed based on the equilibrium of forces acting on the damper components when the SMA springs are active, as shown in Fig. 2.

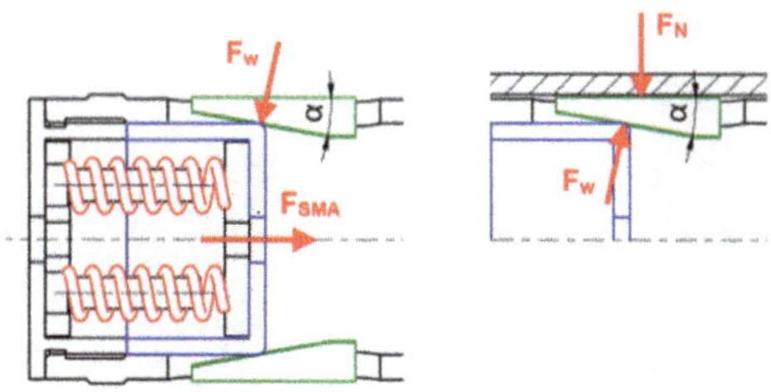

Fig. 2. Equilibrium of forces acting on the actuating head (left) and wedges (right)

As the SMA springs are heated, the actuating force F_{SMA} is generated making the actuating head slide to the right. The sleeve shaft is made of copper and lubricated on the inner cylindrical face to minimize the friction against the actuating head. By neglecting this friction force, the axial equilibrium equation of the actuating head is given as following

$$F_{SMA} - F_W \sin\alpha = 0 \tag{1}$$

where F_W is the total force caused by the four wedges and α is the cone angle of each wedge. As the four wedges are pushed outward, the reaction force F_N at the contact between the wedges and housing is produced. The friction between the four wedges and the sleeve shaft is very small and thus could be neglected. The radial equilibrium equation of the four wedges is then obtained by

$$F_N - F_W \cos\alpha = 0 \tag{2}$$

Damping force of the damper is then determined by

$$F_d = \mu F_{SMA} \mathrm{ctg}\alpha \tag{3}$$

where μ is friction coefficient.

In this design, the cone angle of each wedge is 12°. The SMA springs are made by SAES® Getters Group (SmartFlex ® SMA spring). The mechanical and functional characteristics of the SMA springs are given as follows: mean diameter 10 mm, free length 25 mm, wire diameter 1.5 mm, actuating force 14 N, actuating stroke 8 mm.

From Eq. (3), the damping force produced by one SMA spring is calculated by 42.8 N. Accordingly, to achieve the desired damping force of 80 N [8, 9], two SMA springs should be employed in the configuration of the damper.

4 Experiment Test

In this section, a prototype of the proposed SMA damper is manufactured and its performance is experimentally evaluated. Figure 3 shows the photos of the prototype SMA damper and experimental setup to test performance of the damper.

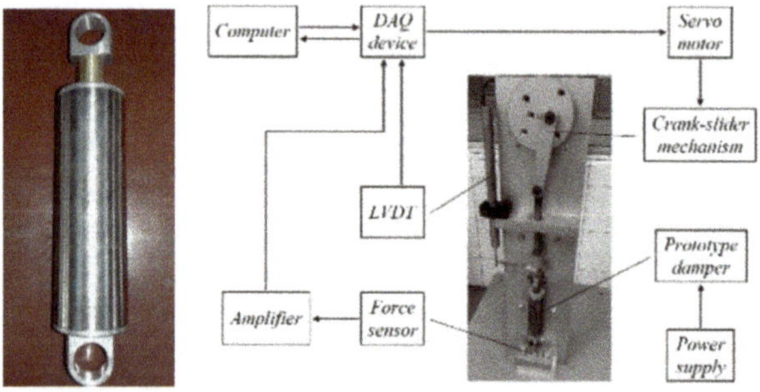

(a) (b)

Fig. 3. (a) Prototype of the SMA damper and (b) experimental setup to test performance

In the figure, a crank–slider mechanism is employed to convert the rotary motion of the motor into the reciprocal motion of the damper shaft. A servo motor controlled by a computer is used to rotate the crank shaft at a constant angular velocity of 4π rad/s (120 rpm). This is also the frequency that the first resonance often occurs. The damping force is measured by a force sensor and the displacement is recorded by a linear variable differential transformer (LVDT). The output signals from the force sensor and LVDT are then sent to the computer via a data acquisition (DAQ). Once the experiment process is stated, a current of 4 A is applied to the SMA springs.

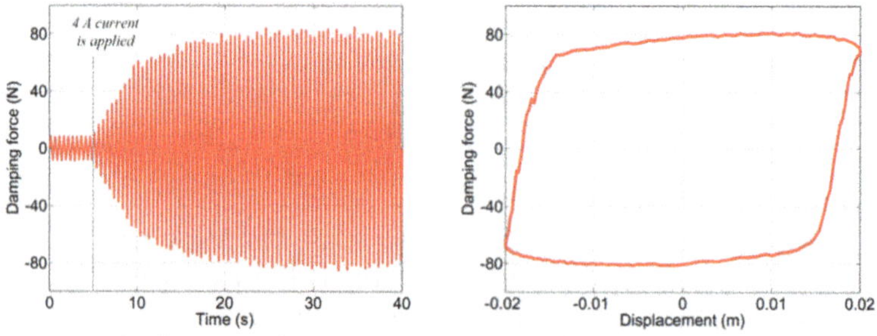

(a) force vs. time (b) force vs. displacement

Fig. 4. Experimental step responses of the SMA damper

Figure 4a shows the responses of the prototype SMA damper in damping force–time relation. From the figure, it is observed that when the SMA damper is not heated (no current is applied to the SMA springs), the off–state force is about 8.2 N. This is mainly the friction force between the housing and the shoulders of the sleeve shaft. When a current of 4 A is applied to the SMA springs, at the steady state, the maximum damping force can reach up to 80.8 N, which is about 94% of the theoretical value (85.6 N). Therefore, this force meets the above desired one of 80 N. The figure also indicates that the rising time to change from the off–state force to the steady value of the damping force is around 20 s. This value is somewhat high and thus is only suitable for open–loop control systems. In order to clearly describe the damping characteristics of the SMA damper, the relation between the damping force and the displacement of the damper shaft is obtained, as presented in Fig. 4b. It is noted that there is a sudden change of damping force when the shaft motion changes its direction. This is obvious because the damping force of the proposed SMA damper is generated by the friction between the wedges and inner face of the housing which is almost Coulomb friction.

5 Conclusions

This research work dealt with design and performance evaluation of a SMA damper for front–loaded washing machines. After an overview of vibration of washing machines due to the unbalanced laundry mass and state of the art of dampers for washing machines, the configuration and principle of the SMA damper were presented. A prototype SMA damper was then designed, manufactured and experimentally investigated. The results showed that the damper behaves almost like a Coulomb friction damper with very small off–state force (about 8.2 N) and high enough maximum damping force (up to 80.8 N) to suppress the rigid mode vibration of most washing machines. However, the rising time to change from the off–state force to the steady value of the damping force is about 20 s, which makes the SMA damper only suitable for on–off control.

Acknowledgements. This work was supported by the Vietnam National Foundation for Science and Technology Development (NAFOSTED) under grant no. 107.01–2018.335.

References

1. Papadopoulos, E., Papadimitriou, I.: Modeling, design and control of a portable washing machine during the spinning cycle. In: Proceedings of the 2001 IEEE/ASME International Conference on Advanced Intelligent Mechatronics, pp. 899–904 (2001)
2. Lim, H.T., et al.: Dynamic modeling and analysis of drum–type washing machine. Int. J. Precis. Eng. Manuf. **11**, 407–417 (2010)
3. Koizumi, T., et al.: Noise prediction of a washing machine considering panel vibration. In: IMAC–XXIV Conference and Exposition on Structural Dynamics (2008)
4. Carlson, J.D.: Low–cost MR fluid sponge devices. J. Intell. Mater. Syst. Struct. **10**, 589–594 (1999)

5. Spelta, C., et al.: Control of magnetorheological dampers for vibration reduction in a washing machine. Mechatronics **19**, 410–421 (2009)
6. Nguyen, Q.H., et al.: Optimal design and performance evaluation of a flow–mode MR damper for front loaded washing machines. Asia Pac. J. Comput. Eng. 1 (2014)
7. Nguyen, Q.H., et al.: Optimal design of magnetorheological fluid–based dampers for front–loaded washing machines. Proc. Inst. Mech. Eng. C -J. Mech. Eng. Sci. **228**, 294–306 (2014)
8. Bui, D.Q., et al.: Design and evaluation of a shear-mode MR damper for suspension system of front-loading washing machines. In: Proceedings of the International Conference on Advances in Computational Mechanics, pp. 1061–1072 (2017)
9. Bui, Q.D., et al.: Development of a magnetorheological damper with self–powered ability for washing machines. Appl. Sci. **10**, 4099 (2020)

An Application of 3D Driving Model for Driver Behavior Training

Duong Thanh Tin[2] and Nguyen Phu Thuong Luu[1(✉)]

[1] Department of Automotive Engineering, Ho Chi Minh City University of Technology (HUTECH), Ho Chi Minh City, Vietnam
npt.luu@hutech.edu.vn

[2] Department of Automotive Engineering, Ho Chi Minh City University of Technical and Education, Ho Chi Minh City, Vietnam

Abstract. In the past few years, many traffic accidents have been occurred directly related to the quality of training. That reasons reflected in the weakness of the driver when handling, identifying situations, through the attitude of disregarding life of passengers and traffic participants. The training process of driving in establishments has many shortcomings, even violating regulations in training activities, reducing practice programs…so Decree 138/2018/ND-CP of government has born to ensure good training of drivers, in this Decree, it is required that driver training centers must equip with more equipment to simulate 3D driver training to lift Highly effective in reviewing and practicing students' driving skills, however, the cost for each driving simulator is high. The article introduces readers to the simulation device application besides training domestic and foreign drivers, thereby proceeding to design and manufacture 3D driving training simulator for digging process. This model was made with reasonable prices to satisfy the driver training and economic situation in Vietnam.

Keywords: 3D model · Simulation · Driving · Training · Control system

1 Introduction

Simulation can be thought of as a method of reproducing a number of situations that are close to reality, but they are still controllable. To achieve the same level of realism, we need to create an environment with identical script content to a real-world situation. This activity turns a simulation program into a flexible scientific research tool with the aim of supporting experiments that can only be performed in a laboratory that could be costly, dangerous, or irreversible in the real world. In the driving school, simulation is performed with driving simulation. These create situations that require feedback like actual driving - but without risk while on the road. The basis for evaluating any simulation program is its applicability. Simulation models must be designed and equipped according to their intended use. Driving simulations have been developed a lot over the years. There are many different simulation models from simple to complex like the model from Daimler-Benz (Fig. 1). By using driving simulations, a lot of research can be done regarding: drivers, cars and routes [1].

B. T. Long et al. (Eds.): MMMS 2020, LNME, pp. 879–885, 2021.
https://doi.org/10.1007/978-3-030-69610-8_117

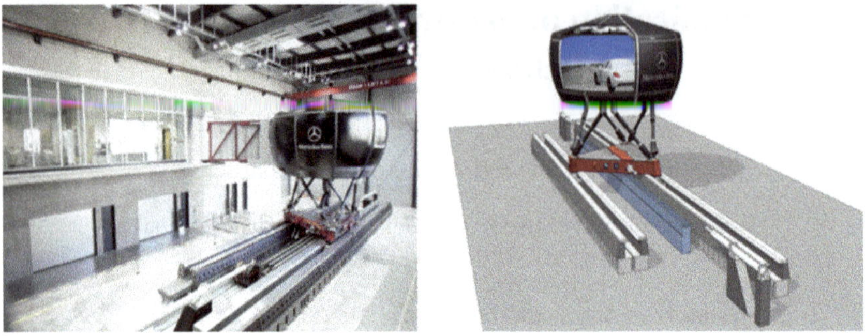

Fig. 1. Daimler-Benz driving simulator

Sometimes simulation activities are useful to classify driving simulation studies according to experimental subjects instead of research objectives. These objects are suitable: drivers, cars, environment and routes. In Table 1 a few specific types of research on simulated driving are listed. The advantages of using an emulator as a research tool can be briefly summarized as follows: situations can be easily repeated, test objects can be easily modified and stored; The effects of fatigue on drivers can be safely studied, given fixed or random situations so that trainee abilities can be analyzed safely. From there calculate and build designs that are suitable for each subject and each different situation.

Table 1. Some studies are done with simulation content depending on the selected subjects

Research content		
About the driver	About the car	Roadmap and environment
Driver's age, health, character, other medical conditions, etc.	Vehicle type, response to driving in the program, vehicle information (form, content and position), visibility of that vehicle. External environmental factors, passengers, etc.	Road geometry, road markings, surface properties, road structure elements, image texture, colors, directions, vehicle speed, traffic regulation, light, weather, pedestrians, and other vehicles, road obstacles, terrain type handling, etc.

With the remarkable development in technology, driving simulation applications are considered a good way for the learner to be aware of the risks and to familiarize the simulation participant with other driving situations. Each other that one cannot predict. Some traffic situations on the road [2] (Figs. 2a, 2b). A perfect simulation program will be able to help the practitioner develop some basic psychological or behavioral problems (how to drive and travel speed on each type of road) in complex traffic conditions. More importantly, good control in situations during the simulation can help trainees with the cognitive abilities needed to handle complex road types and difficult traffic conditions, including receiving. Dangerous situational awareness then makes decisions to handle and implement safe driving techniques to protect yourself and those around you while driving [3]. Dangerous situational awareness then makes decisions to handle and implement safe driving techniques to protect yourself and those around you while driving. The driving simulation program can provide accurate analysis of the practitioner as it will allow the practitioner to collect data with less risk and more control over the practitioner.

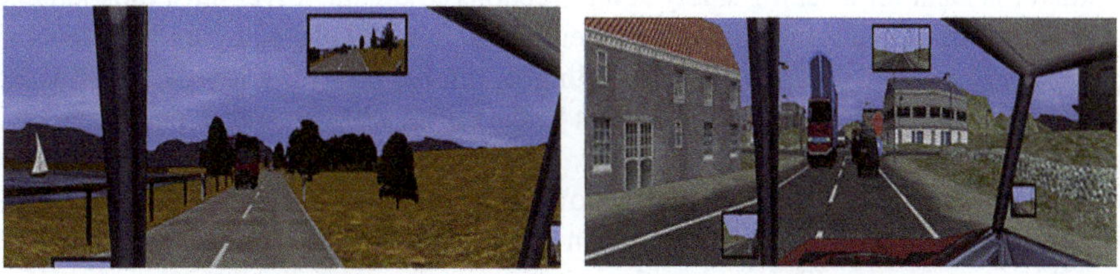

2a. Rural roads **2b.** Urban road

Fig. 2. **a.** Rural roads **b**. Urban road

There are impressive driving simulations, high fidelity, quite realistic environment simulations, such as NADS-I (Fig. 3), Toyota driving simulation (Fig. 4), chapter Daimler AG driving simulation (Fig. 1)… Most of these simulation programs require high investment costs, with many modern technologies. However, there are studies that do not require complicated results, a program simulation can simply be a computer game with a steering wheel and pedals available for the novice to get used to the maneuvers. Depending on quality and price… Although the simulation program is very well-built, the trainees who practice on the device know that they are sitting in an emulator. All you can do is try to create a simulation program that is as realistic as possible, in which the practitioner's actions will perform similar to those in an actual driving situation. An important goal of the current research project is to develop a driving simulation program so that the practitioner can practice and get used to driving like on a real car at a low cost. In this article will describe how to perform and build a driving simulation program like a real car with the lowest cost so that the practitioner can get acquainted with the driving maneuvers.

Currently, in Vietnam, many vocational training centers are studying models for training such as [6] designing and manufacturing models of electric systems in cars (AAES) controlled by smartphones. The system includes security, comfort and safety systems, lighting…. This study focuses on lighting system (front, rear, cockpit and signal lights) and safety system and comfort (windshield wiper and window cleaning…). Luu in [7] presents the design and production of a smartphone-controlled model of the air conditioning system (ACS), the components in the model are installed into the model of the body frame, which is connected with the Arduino board then an Android smartphone device is used to control the ACS model.

2 Simulator Hardware

Use 01 computer with Windows 7 operating system and 2.4 GHz Dual core CPU with 2 GB DDRAM, Graphics Card with 256 MB of GeForce 7600 GT series memory or higher. Or can be used with the recommended configuration with Windows 7 operating system with 3.0 GHz dual-core CPU DDRAM 4 GB processor, graphics card with 1024 MB of GeForce GTS 450 series memory or higher. The wider viewing range allows for a more accurate speed perception [4] The 180° viewing angle in the driving simulator is ideal for teaching safety at intersections where the driver must look left and right before making a turn. To see the potential hazards and check the traffic density at intersections to provide appropriate handling solutions, so with a low-budget driving simulation program using 03 LCD screens Links create 1800 perspectives for practice. Arrange the screens according to the 3 × 1 layout and combine their inputs with the Matrox TripleHead2Go device. This allows the image to appear on the Windows screen driver in the form of a large screen combined into one large screen instead of three separate screens. This makes the driving experience really more engaging and realistic in this configuration because of the larger screen size. The practitioner movement model and control system is a frame system with a seat with steering and pedal parts (clutch, brake, throttle) with simulated programmed motion. The controller uses an Arduino microcontroller which is connected to the simulation program via USB. The actuators for the motion controller have been firmly attached to the trainee's seat on the model. While driving the sense of motion is generated by the 12 v motor connected to the driver seat. However, the drawback of this two-level structure for fast-moving situations like racing will not react promptly and become clear.

Fig. 3. NADS-I driving simulator of Iowa University

Fig. 4. Toyota driving simulator

3 Simulator Software

Using the program Euro Truck Simulator 2 (ETS2) (Fig. 5) is a car driving simulation game of SCS Software for Microsoft Windows, Linux and macOS to build and design an emulator. There are several significant advantages of running a commercial game emulator. With a relatively low cost (about $20) for the game license, with a low price but the game software has given users many tools to control and master the in-game environment, which This causes the simulation environment to change in accordance with the content of the simulation scenario (vehicle type or traffic type). Use ETS2 to study graphic content design with full situations (urban, night, ramps, rain…), graphic content design with exercises and exams in the section of the screen as (content, scoring method and time of exams).

The control program for two degrees of freedom (2DOF) drives is called SimTools (Fig. 6). SimTools can control the movement of actuators which are electric motors. SimTools program runs in conjunction with the ETS2 program on the computer and it receives motion data from the ETS2 game engine. This data is converted into motion on the simulation model. Due to the limited running speed and moderate traffic rules, the motion platform is based on values from SimTools.

Fig. 5. Simulation software game version

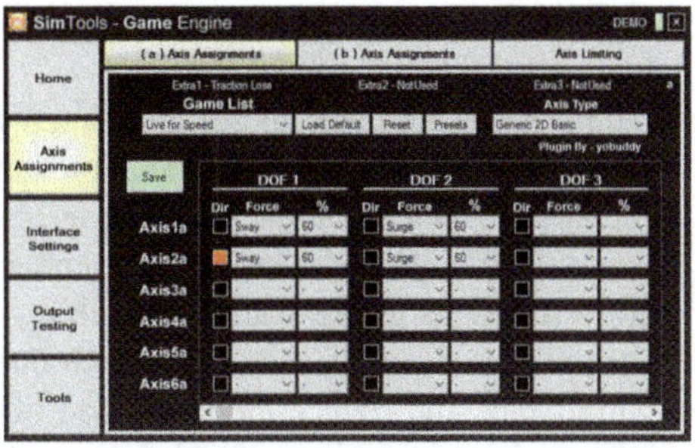

Fig. 6. Software for creating 2° motion effects (2DOF)

4 Design Motion in the 3d Model

To simulate the proposed article movements to a solution (Fig. 7), refer to the website [5] so that people can build their own model when they want to make model for new driver. The above circuit diagram is installed in the charge under the model's seat (Fig. 8) to create movement when driving. And Electrical circuitry is complete (Fig. 9).

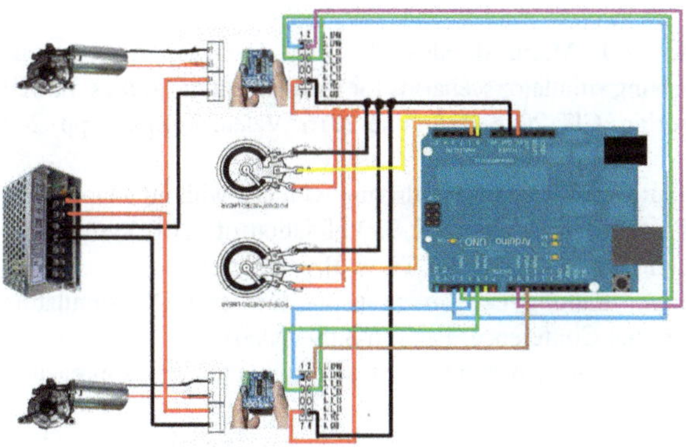

Fig. 7. Circuit diagram of a 2 degrees of freedom motion effect (2DOF)

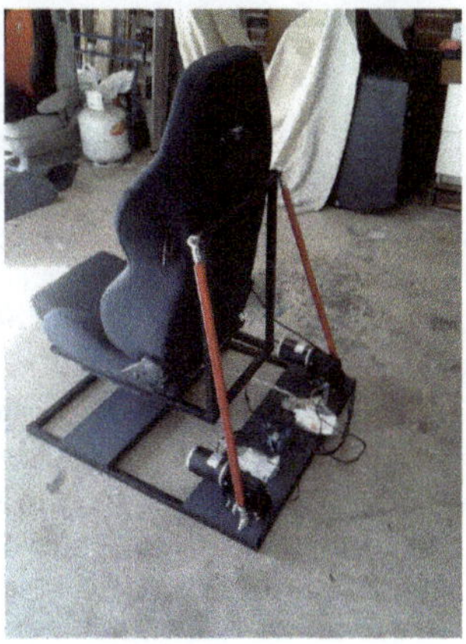

Fig. 8. Motion chair model

Fig. 9. The electrical circuit of a 2 degrees of freedom motion effect (2DOF)

5 Conclusions

In addition to being able to get acquainted with the maneuvers before driving on a real car, trainees on the driving simulation model can also form safe driving behaviors through simulated situations in the program and will help teachers assess the ability of the practitioner. From there, make comments and content of the next episode of the learner. There by giving learners the ability to analyze the important factors in driving. Over time these capabilities will help trainers to drive more successful models in real environments and reduce the likelihood of driving accidents.

The price of the simulation is expected to be about 5000 USD (about 120 million VND), most of the investment in hardware devices such as models, display screens, computers running simulation software. However, with such a relatively low cost of investment to build a model, there are also some limitations such as the novice who can only get used to the basic operations before driving (driving, entering gear…).

However, with a low investment cost, you still have the necessary parts in the real car and can create a driving experience (which may not be the same as driving a real car like crossing the road). (rugged, over ramps) partly helps people who do not know anything about driving a car to understand the basic operations required to control the car on the road so that it can form driving skills for beginners to run a car.

References

1. Dolsa, J.F., Molinaa, J., Camachob, F.J., Marín-Moralesc, J., Pérez-Zuriagab, A.M., Garcia, A.: Design and development of driving simulator scenarios for road validation studies. In: XII Conference on Transport Engineering, CIT 2016, 7–9 June 2016, Valencia, Spain, pp. 2–3 (2016)
2. Verwey, W.B.: The Control of Input Devices by people driving with and without adaptations. DRIVE II PROJECT V2032 TELAID Deliverable No. 9. Validation of ATT Systems for Drivers with Special Needs through Driving Simulator Tests. Brussels (1995)
3. Allen, W., et al.: A novice driver training experiment using low cost PC simulation technology. In: DSC2000 International Conference, Paris, 6-8.09 (2000)
4. Andersen, G.J.: Perception of self-motion. psychophysical and computational approaches. Psychol. Bull. **99**(1), 52–65 (1986)

5. www.xsimulator.net/community/threads/smc3-arduino-3dof-motor-driver-and-windows-utilities.4957. SOFTWARE: - SimTools software - https://www.xsimulator.net/community/ - Euro Truck Simulation – SCS Software – Czech
6. Luu, N.P.T.: Design and manufacture a vehicle auxiliary system model controlled by using smartphone. In: Kahraman, C., Cevik Onar, S., Oztaysi, B., Sari, I., Cebi, S., Tolga, A. (eds.) Intelligent and Fuzzy Techniques: Smart and Innovative Solutions. INFUS 2020. Advances in Intelligent Systems and Computing, vol. 1197. Springer, Cham (2020). https://doi.org/10.1007/978-3-030-51156-2_187
7. Luu, N.P.T.: A study on vehicle air condition system model controlled by using smartphone. In: Kahraman, C., Cevik Onar, S., Oztaysi, B., Sari, I., Cebi, S., Tolga, A. (eds.) Intelligent and Fuzzy Techniques: Smart and Innovative Solutions. INFUS 2020. Advances in Intelligent Systems and Computing, vol. 1197. Springer, Cham (2020). https://doi.org/10.1007/978-3-030-51156-2_188

A Study on Car-To-Truck Rear Underride with and Without Protection Device

Tran The Anh[2] and Nguyen Phu Thuong Luu[1(✉)]

[1] Department of Automotive Engineering, University of Technology (HUTECH), Ho Chi Minh City, Vietnam
npt.luu@hutech.edu.vn

[2] Department of Automotive Engineering, University of Technical and Education, Ho Chi Minh City, Vietnam
trantheanhxl@gmail.com

Abstract. Most of the traffic accidents behind a truck cause heavy damage to the vehicle and safe for people in the under-ride car. This paper focuses on the investigation of truck accidents of over 10 tons with sedans in the rear side; especially underride crashworthiness. Survey the world research on front and rear bumper systems of passenger cars. Using catia software to design rear bumper, Altair Hyper meshing tool and simulate using LS-Dyna software. The data from this analysis we can analyze to improve the performance in rear crash for the truck to ensure safety in case a collision occurs.

Keywords: Rear-end collisions · Structure · Automobile Accidents · Underride

1 Introduction

Many researchers are studying on how to improve the rear bumper structure by software simulation. The results of the test on the computer are re-evaluated the effectiveness of the actual bumper compared to the safety standards of the Insurance Institute of Highway Safety. The results of the collision analysis show that the rear vehicle structure damaged in the first part, affecting driver safety, based on simulation results to improve the vehicle's structure to ensure safety [1–6]. However, most of these researches studied on the passenger car or bus structure there did not focus on the trucks. That the reason for this paper investigates about the car to truck under-ride accidents.

A rear accident between trucks over 10 tons and small cars, most of them are dangerous for the driver if the big truck does not have a rear bumper or the rear bumper does not guarantee safety. Small cars that collide with the rear of the truck absorb the most energy and focus on the front bumper. However, the sizes of cars are different at the ground clearance, so the rear bumper also has different heights. The purpose of the research is to improve the rear bumper for trucks with trucks over 10 tons. It ensures the height of the truck and ensures the collision with small cars in the front bumper position. The rear bumper can increase the bump surface area to ensure that the collision is a direct collision with the front bumper of small cars.

B. T. Long et al. (Eds.): MMMS 2020, LNME, pp. 886–893, 2021.
https://doi.org/10.1007/978-3-030-69610-8_118

Using LS-DYNA software and then perform finite element simulation to analyze and compare the dissipation efficiency in LS-DYNA. In computer simulation, the vehicle collides behind the truck to ensure compliance with national regulations QCVN 09: 2015. The condition of a collision speed of 54 km/h and the form of offset collision added to the test. Virtual test for further testing data from analysis is the result of collision.

2 An Overview of Vietnam Traffic

According to a report of the Vietnam Automobile Manufacturers Association (VAMA) [7] Vietnam market consumed 288,683 types of cars in 2018, according to statistics from the Vietnam Association of Motorcycle Manufacturers. Vietnam (VAMM) [8] in 2018 motorcycle manufacturers sold 3,386,097 units. According to the Adjustment Report of Transport Development Planning [9] road to 2020 and orientation to 2030, in terms of technical standards: roads with high technical standards (expressways, grades I and II) account for very low, only 7.51%. The percentage of roads meeting grade III and IV technical standards accounts for 77.73%; the remaining roads have low technical standards (grade V, grade VI) accounting for 14.77% (Fig. 1).

Fig. 1. The current traffic situation in Vietnam

According to the analysis of the collision data in Table 1 we can see that the frontal collision has the highest rate, the rear collision is relatively low [10].

Table 1. Percentage of all types of collisions in Vietnam in 2001

Type of collision	Number of accidents	Ratio %
Frontal collision	741	53,6%
Rear impact	108	7,8%
Side impact	443	32,1%
Puffer and roar	38	2,7%
Other collisions	52	3,8%
Total	1382	100%

In those cases, traffic accidents between motorbikes and trucks, cars and trucks are leading to death [11–15]. Typically, some photos of traffic accidents from behind heavy trucks without rear bumper protection as shown in Fig. 2.

Fig. 2. The scene of a car traffic accident under the truck

According to the actual survey in the National Highway 1A area in District 12 near the Ga crossroad bridge, for every 100 trucks of over 10 tons, there are about 70 vehicles as shown in Fig. 3 without rear bumper or improper rear bumper. in accordance with the regulations of Vietnam Register.

Fig. 3. Large trucks do not have bumper or rear bumper but do not guarantee traffic safety on Highway 1A.

The causes of large trucks without rear bumper or rear bumper are not safe. Due to the design of the rear bumper is not guaranteed to use for a period of time damaged or due to the large long-tail trucks, low-mounted rear bumper affects the exit angle after the vehicle over the high slopes to be impacted.

3 Crashworthiness Standard

3.1 Asian Standards for Heavy Truck Rear Bumpers

36th Asian Professional Conference on Rear Protection (R58) August 28, 2013. Regulations on rear bumper standards.

The width of the rear protective device shall not exceed the width of the rear axle measured at the outermost points of the wheel, excluding the bulging of the tire near the ground, and the bumpers shall not be shorter than 100 mm at two sides.

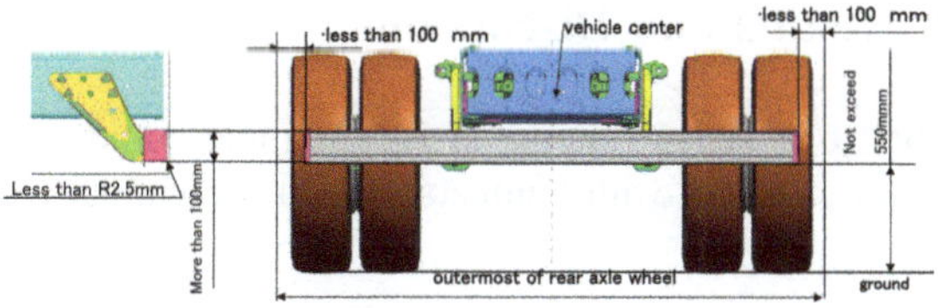

Fig. 4. Asian standard

The rear bumper must be fitted so that the horizontal distance between the rear bumper and the rear of the vehicle does not exceed 400 mm in Fig. 4.

3.2 Relationship of Movement of Two Vehicles During the Collision

Assume for two vehicles to move on the OX coordinate axis.

v1 is vehicle speed 1; v2 is vehicle speed 2.

p is the relative speed of the two vehicles before the collision.

p'is the relative velocity of the two vehicles following the collision.

When two moving vehicles occur in a collision, the following cases are shown in Fig. 5.

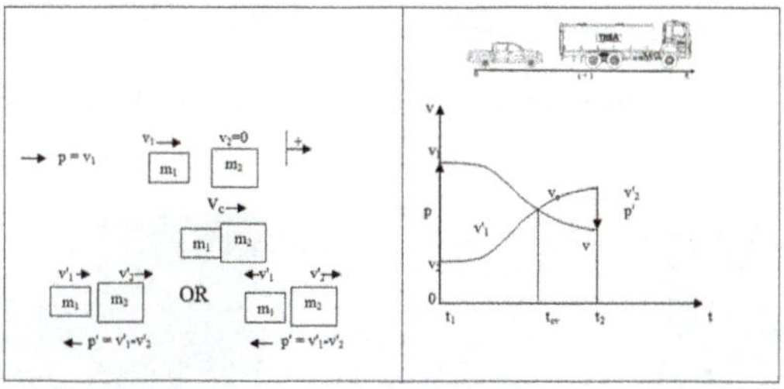

Fig. 5. Collisions between 2 vehicles (standing truck, moving small vehicle)

Call: t is the time of the two vehicles starts colliding.

t′ is the end of deformation of the material and the skeleton of two vehicles.

According to II Newton's law, the force acting on vehicle 1 is equal to the force acting on vehicle 2, F is the force acting on vehicle 1 and car 2 in the period from t to t'. The impulse will then be generated and determined as follows:

$$I = \int_{t}^{t'} Fdt$$

F - is the force of impact

I - is linear impulse

4 Simulation Results and Discussions

Perform a simulation of a sedan running at 56 km/h and collides with the tail of the truck, with two cases of trucks with bumpers and trucks without rear bumper.

Fig. 6. Simulate a sedan with a rear-end collision

As shown in Fig. 6, if a truck is obstructed, the car shall not be intruded into the driver's compartment, the truck without protection must be intruded into the driver's space (Fig. 7).

Fig. 7. The results of the car to rear truck collision without protection device

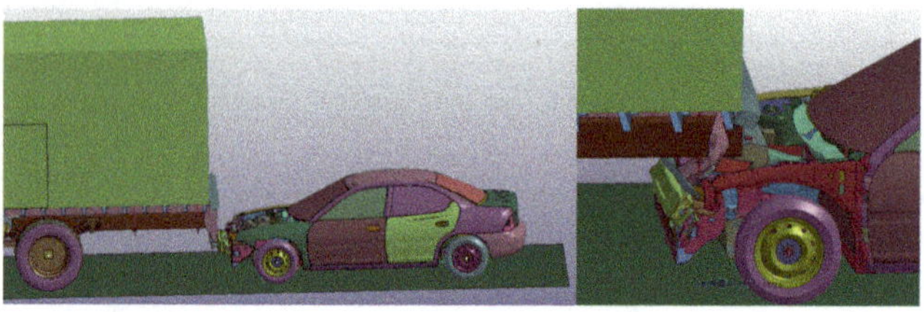

Fig. 8. The results of the car to rear truck collision with protection device

The result of Fig. 8 shows that the rear bumper of the truck and the front bumper are in contact, but up to 0.05 s, the rear bumper cannot reach the front bumper.

The vehicle has the rear bumper to limit the penetration into the driver's compartment, ensuring the safety of the driver.

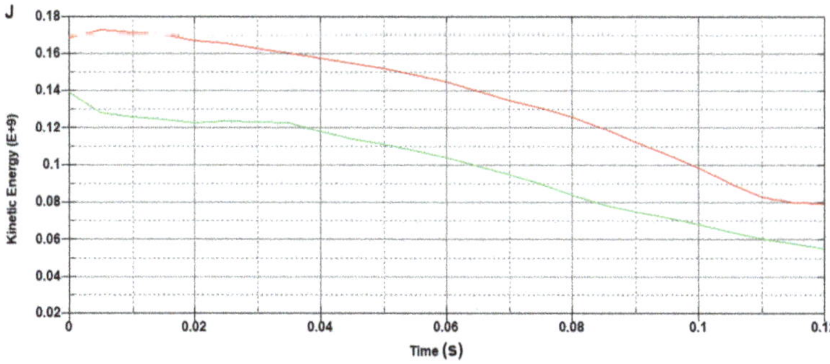

Fig. 9. The comparison of the car kinetic energy in case the truck with and without protection device

Through the energy kinetic chart, we can see that the vehicle with the rear bumper absorbs more energy.

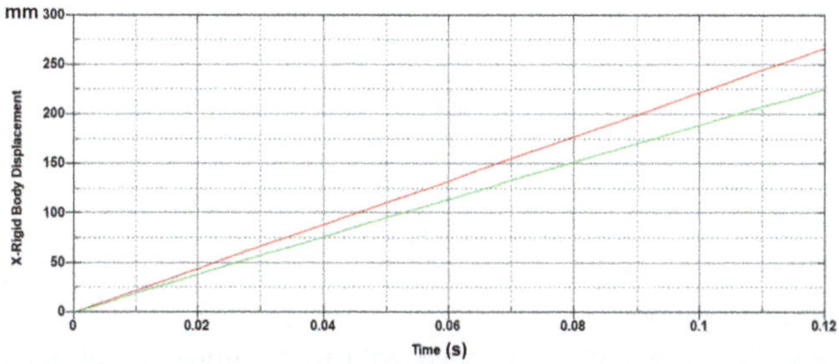

Fig. 10. The comparison of the car displacement in case the truck with and without protection device

Through the chart, we can see the displacement between the vehicle without the rear bumper and the vehicle with the rear guard. Vehicles with a rear bumper will have a shorter travel time compared to vehicles without a rear bumper (Figs. 11 and 12).

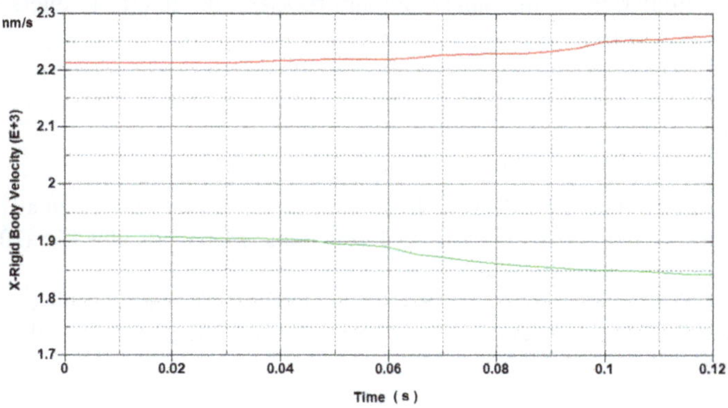

Fig. 11. The comparison of the car velocity in case the truck with and without protection device

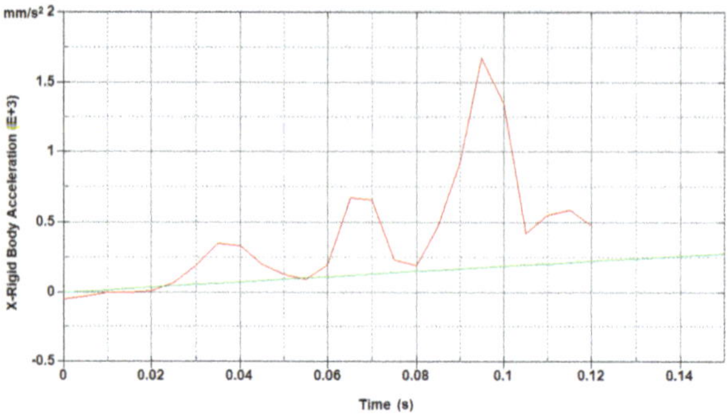

Fig. 12. The comparison of the car acceleration in case the truck with and without protection device

Based on the chart, we can see that the acceleration and speed of a vehicle with the rear bumper is higher than that of a vehicle without the rear bumper.

5 Conclusions

Simulation results show that the truck is equipped with the rear bumper to limit the penetration of the driver's compartment, ensuring the driver's safety.

Vehicle structure with rear bumper breaks smaller vehicle structure at impact than vehicle without rear bumper.

Stopping distance of vehicles equipped with rear bumper is shorter than vehicles without rear bumper.

Based on vehicle simulation results we can improve passenger car structure for safety in rear crash. To decrease the injured or damage for small car, we can improve rear bumper with shock absorbers when absorbing impact force, limiting the breakdown of small vehicle structure when a collision occurs. Limit the intrusion of the driver's compartment to ensure safety for the driver.

Vehicles equipped with rear bumper in Vietnam are not guaranteed to be safe, through this simulation results, analyzing and reassessing the effectiveness of the rear bumper, it needs to be improved.

References

1. Nguyen, P.T.L., Lee, J.Y., Yim, H.J., et al.: Analysis of vehicle structural performance during small-overlap frontal impact. Int. J. Automot. Technol. **16,** 799–805 (2015). https://doi.org/10.1007/s12239-015-0081-7
2. Nguyen, P.T.L., Lee, J.Y., Yim, H.J., et al.: Optimal design of vehicle structure for improving small-overlap rating. Int. J Automot. Technol. **16,** 959–965 (2015). https://doi.org/10.1007/s12239-015-0098-y

3. Nguyen, L.P.T.: An optimisation approach to choose thickness of three members to improve IIHS small-overlap structural rating. Int. J. Crashworthiness **22**(5), 518–526 (2017). https://doi.org/10.1080/13588265.2017.1281203
4. Thuong, L.N.P.: Vehicle frontal impact to pole barrier simulation using computer finite element model. In: 2018 4th International Conference on Green Technology and Sustainable Development (GTSD), Ho Chi Minh City, pp. 273–277. IEEE (2018). https://doi.org/10.1109/GTSD.2018.8595702
5. Thuong Luu, N.P.: Analysis of bus structural performance during full frontal impact. In: 2019 International Conference on System Science and Engineering (ICSSE), Dong Hoi, Vietnam, pp. 635–638, IEEE (2019). https://doi.org/10.1109/ICSSE.2019.8823416.
6. Nguyen, P.T.L., Nguyen, V.D.: Analysis of vehicle body with small - overlap frontal impact on various barriers. In: National Conference on Mechanical and Transportation Engineering, Ho Chi Minh City, pp. 363-368 (2017)
7. Sales Report, source from: https://vama.org.vn
8. Vamm Announces Sales in 1st Quarter 2019 in Vietnam, source from: https://vamm.org.vn
9. Summary of Vietnam Traffic Safety Conference in 2016 Episode 1 Managing Traffic Safety and National Traffic Safety Committee
10. Nguyen, Q.: Study of Dynamics and durability of car body frame in frontal collision, doctoral thesis in engineering
11. Report on Results of Assessing Order and Safety of Traffic In 2018 And Important Tasks in 2019, National Traffic Safety Committee, p. 17, 4 January 2019
12. Motorcyclists rushed into the truck. Source from: https://tintaynguyen.com/tai-nan-chet-nguoi-o-da-lat-nguoi-di-xe-may-lao-vao-gam-xe-tai/880595/.
13. Motorbike got under the truck, 1 person was in danger. Source from: https://baomoi.com/xe-may-chui-vao-gam-xe-tai-1-nguoi-nguy-kich/c/26366889.epi
14. The continuous accident on Thanh Tri Bridge (Hanoi) afternoon 18–1. Source from: https://autopro.com.vn/tai-nan-kinh-hai-xe-taxi-nat-buom-trong-gam-xe-tai-20170118205657765.chn
15. Car accident under the truck. Source from: https://xe.baogiaothong.vn/vu-toyota-camry-chui-gam-xe-tai-vi-sao-tui-khi-khong-bung-d210503.htm

On the Nominal Stiffness of the Foil Structure in a First-Generation Bump-Foil Air Bearing

Minh-Quan Nguyen(✉), Minh-Hai Pham, and Van-Phong Dinh

Hanoi University of Science and Technology,
1 Dai Co Viet Road, Hanoi, Vietnam
quan.nguyenminh@hust.edu.vn

Abstract. With compliant foil structure, foil air bearings have demonstrated strong development, especially in high-speed machines. Deflection of the foil structure plays a significant role in preventing the seizing phenomenon. It is the characteristic of the foil structure including the stiffness that makes a big contribution to the bearing performance although it has shown nonlinear property. In this paper, a bump foil model was proposed to estimate the stiffness of a bump. Based on that, model of a rotor-bearing system was recommended to assess the nominal stiffness of the foil structure. The solution made a good agreement with other works and was confirmed by experiment. The data can be used to investigate bearing performance, while the theoretical calculation can be employed for preliminary design of a first-generation bump-foil bearing in real cases.

Keywords: Foil air bearing · Nominal stiffness · Bearing performance

1 Introduction

A foil-air bearing is an exceptional bearing that has made significant development for about 50 years. Despite the same working principle of hydrodynamic lubrication, what differentiates this kind of bearing from conventional bearings is a flexible foil structure affixed circumferentially onto the inner surface of bearing sleeve. The foil structure acts as an elastic foundation, which means that there is a "soft contact" between rotor and the bearing rather than "hard contact". This structure is considered to be especially meaningful in case of high-speed rotor. In this situation, without compliant foil structure, seizing can appear in the bearing due to centrifugal force and shaft growth. Therefore, this type of bearing has been applied and widely used in turbomachinery such as: turboexpander, turbopump, turbocharger, … [1].

Working capacity of bump-foil bearings have drawn much consideration of scientists in order to enhance their performance in machines. For that purpose, a rotor-bearing system has been analyzed into three subsystems (rotor, air thin film and foil structure) to investigate state variables (rotor displacement, hydrodynamic pressure and foil structure) [2]. Many parameters have been included in differential equations, and one of the most important parameters is the stiffness of the foil structure. Some theoretical models and experiments were proposed and carried out to evaluate the stiffness.

B. T. Long et al. (Eds.): MMMS 2020, LNME, pp. 894–900, 2021.
https://doi.org/10.1007/978-3-030-69610-8_119

Heshmat [3, 4] studied this characteristic of the foil by modelling a foil strip and doing experiment. This demonstrated some good results but the object was only a single horizontal foil strip, not a foil structure assembled in a bearing. In another way, the bump foil was considered to be linear springs linked together by Arghir [5]. The stiffness matrix was calculated through Castigliano's theorem but this model neglected the nonlinearity of the foil structure. Therefore, in this paper, a model of foil structure was proposed to evaluate the stiffness of a single bump with bending moment and radius of curvature considered. From that, another model of rotor-bearing system was suggested to calculate the nominal stiffness of the foil structure. The results were associated with other works and confirmed by experiments.

2 Theoretical Model

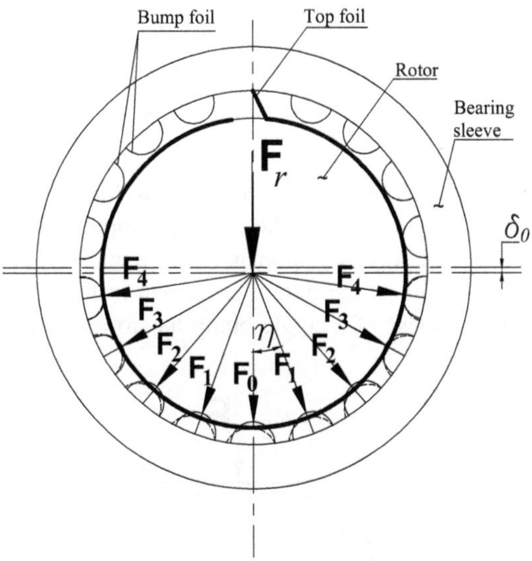

Fig. 1. A static-load model of a rotor-bearing system

A static-load model of first-generation bump-foil bearing is shown in Fig. 1. The foil structures consist of two parts: a top foil (bended sheet metal) embraces a rotor and a bump foil (corrugated shape) is affixed between the top foil and bearing sleeve. In this model, an irrotational rotor supported by a bump-foil bearing is loaded by an external static force F_r. Under the action of this vertical force, the elastic bump-foil structure is supposed to be deflected, causing a vertical displacement δ_o of the rotor. Then the nominal stiffness of the foil structure can be estimated by:

$$K_{nom} = \frac{F_r}{\delta_o} \tag{1}$$

From Eq. (1), assuming that the vertical displacement δ_o was given, the external force F_r needs to be defined for calculating the nominal stiffness. It can be inferred that the external force F_r can be divided into force components that act on foil structure. These force components lead to the deflection of the lower half of the bump foil as shown in Fig. 1. It is noteworthy that some assumptions of contact and distributing conditions should be preset, including:

(i) Under the effect of F_r, the top foil is considered to make point contacts with all of the bumps in the lower half of the bump foil (active bumps).
(ii) The active bumps are regarded to be circumferentially equi-spaced.
(iii) There is no interaction between two consecutive bumps.

With force analysis on Fig. 1, an equilibrium equation can be established:

$$F_r = F_0 + F_1 \cos\theta + \ldots + F_n \cos(n\theta) \tag{2}$$

Where F_i is the force component acting on i^{th} bump.

n is the number of bumps bear the force components.
θ is angle between lines from bearing center to tops of two consecutive bumps.

As mentioned above, it can be inferred that external force components are equal to reaction forces. Therefore, from Eq. (2), these force components need to be calculated to find out F_r. Moreover, the force components can be considered to be elastic forces that cause elastic deformation on the bumps as shown in Fig. 2. As a result, if radial deflection is given, the stiffness of a single bump has to be evaluated.

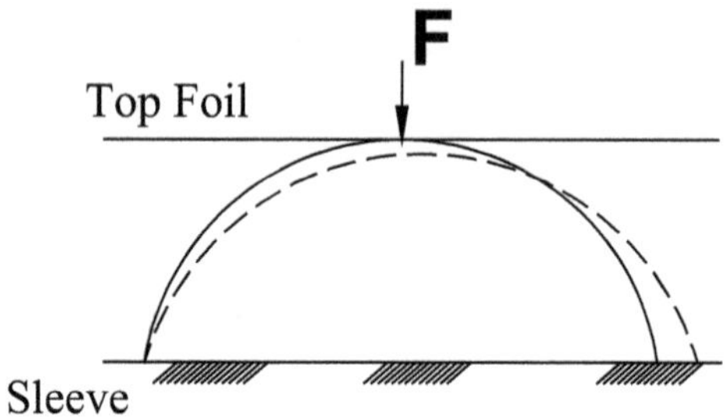

Fig. 2. Bump foil deflection under load

Figure 2 illustrates shape of one bump in the foil-air bearing, with some geometrical constraints: bump height h_O (mm), bump half-length l_O (mm). It is noteworthy that shearing and axial forces can be omitted here because the bump thickness is small compared to bump length, and deflection by bending component M is much greater than by tension or compression.

It can be seen from Fig. 2 that the radius of curvature changed under action of the force component. This transformation can be formulated by an equation [6]:

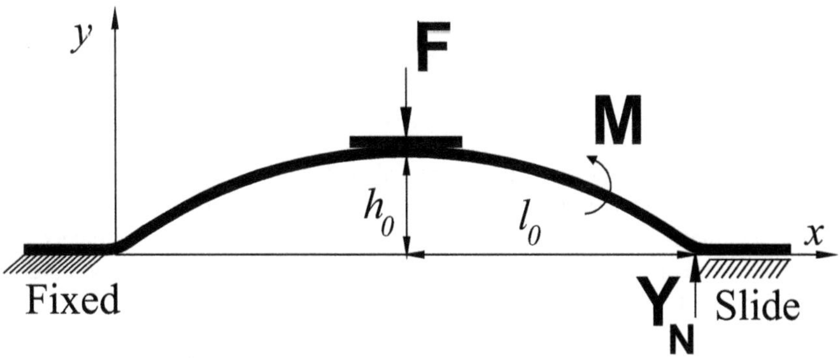

Fig. 3. Force diagram of a bump

$$M_{tot} = EI\left(\frac{1}{\rho_0} - \frac{1}{\rho}\right) \tag{3}$$

Where ρ_o, ρ are the radii of curvature before and after bending, E is the elastic modulus of bump material, I is the moment inertia of cross section and M_{tot} is the total bending moment.

Dividing the bump into lelf half and right half, two equations can be inferred from Eq. (3):

$$\frac{Y_N}{EI}(x_N - x) + \frac{M}{EI} = \frac{1}{\rho_0} - \frac{1}{\rho} \text{ for } x \geq x_F \tag{4a}$$

$$\frac{Y_N}{EI}(x_N - x) + \frac{M}{EI} - \frac{F}{EI}(x_F - x) = \frac{1}{\rho_0} - \frac{1}{\rho} \text{ for x} < \text{x}_\text{F} \tag{4b}$$

Where

Y_N: vertical reaction force at the end point N of the bump.
x_N, x_F: x ordinate of point N and point F.
x: x ordinate of considered point.
M: internal bending moment occurs due to change in radius of curve

Boundary condition:

In this assumption, one end of the bump is fixed and the other end is free to slide as described in Fig. 3. Moreover, an internal bending moment M is included in the model.

Solving the Eq. 4a, 4b with the boundary conditions, the radial deflection at acting point of the force component F can be calculated:

$$\upsilon_F = y_{Fo} - y_F \tag{5}$$

And the stiffness of the bump is computed:

$$K_b = \frac{F}{\upsilon_F} \tag{6}$$

From that, the external force F_r can be calculated from Eq. (2), and the nominal stiffness of the foil structure can be defined from Eq. (1). Using this model with geometric parameters l_0 = 1.778 mm and h_0 = 0.598 mm and other parameters similar to Kim's experiment, the nominal stiffness is $1.181 \times 1\,0^7$ (N/m), in good agreement with the stiffness 1.069×10^7 (N/m) obtained by the author [7].

3 Experimental Work

In the present work, an experiment was established to verify the theoretical model and to study the structural characteristics of the bump foil later. Vertical displacements of the bearing and values of external static force were measured through sensors connected to transducer and monitoring system. The static load-versus-displacement relationship was demonstrated in graph and the nominal stiffness was calculated (Fig. 4).

3.1 Experiment Setup

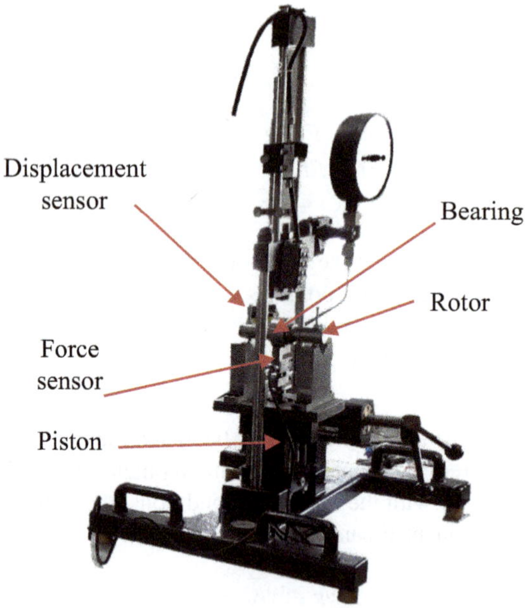

Fig. 4. Rotor-bearing system on WWP-300

The bump foil, made from 0.1-mm-thick SUS304 steel sheet, was shaped by hydrostatic forming to have a corrugated shape in Fig. 1. The number of bump was 22, and some constraints in Fig. 2 were defined: l_0 = 1.5 mm; h_0 = 0.8 mm. The bump foil was then attached onto the sleeve at one end as shown in Fig. 1. The top foil was manufactured and assembled in a similar way.

In the experiment, the bearing was chosen to be loaded rather than the rotor in theoretical model. This means that rotor needed to be fixed and two V-shaped jigs were designed for that purposed. A sensor was attached above the bearing to measure its displacement from the stationary rotor. Moreover, another type of sensor was mounted under the bearing as shown in figure to monitor the external force acting on the bearing. This sensor was also connected to a piston in a cylinder of WP-300 Universal Material Tester. The piston was controlled to move up gradually, generating external force onto the bearing.

3.2 Experiment Procedure

Handle was used to move up the piston in order to create external force onto the bearing. Two sensors recorded values of force and vertical displacement, then transferred the data to a computer through a transducer system. The relationship of force and displacement was illustrated by graph and the nominal stiffness was calculated. The results were then compared to predictions from theoretical model for validation.

3.3 Result and Discussion

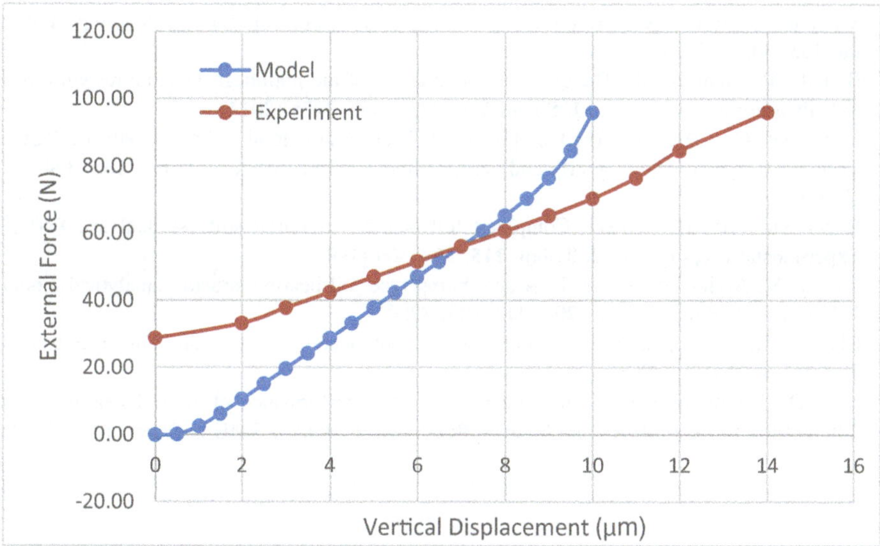

Fig. 5. Load-versus-displacement relationship

The relationship between external force and vertical displacement of the bearing was illustrated in Fig. 5. In general, there is a considerable agreement between prediction from the model and experimental data. However, the slight differences appear at beginning point and some last points. It can be seen that in theoretical model, the vertical displacement seems to directly proportional to the force, and bigger values of force (from 0) causes longer displacement at every point. Meanwhile, in experiment, the range of force from 0 N to about 28 N does not cause any vertical displacement. This can be reasoned that small forces cannot overcome the static state of the bearing structure, which is attributed to static friction among surfaces. Moreover, with the same big values of about more than 60 N, the real displacements are greater than predictions. This can be reasoned that to a certain extent of deflection, there appear gaps between the top foil and bump foil, which means that the number of real active bumps are smaller than predictions, causing the displacement to be bigger.

4 Conclusion

In this paper, a theoretical model of foil structure was proposed in order to evaluate the nominal stiffness of the first-generation bump-foil bearing. In this model, bending moment was included and changes in radius of curvature were considered. The results were compared to foreign researcher and verified by experiment. The data show a considerable agreement although some real phenomena should be included to improve the model.

References

1. Samanta, P., Murmu, N.C., Khonsari, M.M.: The evolution of foil bearing technology. Tribol. Int. **135**, 305–323 (2019)
2. Bonello, P., Pham, H.M.: The efficient computation of the nonlinear dynamic response of a foil–air bearing rotor system. J. Sound Vib. **333**, 3459–3478 (2014)
3. Roger Ku, C.P., Héhmat, H.: Compliant foil bearing structural stiffness analysis: Part I-Theoretical model including strip and variable bump foil geometry. J. Tribol. **135**, 394–400 (1992)
4. Roger Ku, C.P., H. Héhmat: Compliant foil bearing structural stiffness analysis: Part II-Experimental investigation. J. Tribol. **115**, 364–369 (1993)
5. Le Lez, S., Arghir, M., Frene, J.: A new bump-type foil bearing structure analytical model. J. Eng. Gas Turbines Power **129**, 1047–1057 (2007)
6. Chavan, A.P., Zhou, H.: Analysis and simulation of slender curved beams. Int. J. Eng. Res. Technol. **5**, 214–221 (2016)
7. Kim, D.: Parametric studies on static and dynamic performance of air foil bearings with different top foil geometries and bump stiffness distributions. ASME J. Tribol. **129** (2007)

Effects of Computational Geometry to the the Laminar Viscous Flow Around the Vertical Circular Cylinder

Tu Dinh Tran[1(✉)], Jhian-Horng Chen[2], Nguyen-Anh-Vu Le[1], and Le-Hong-Thai Huynh[1]

[1] Nha Trang University, Nha Trang, Vietnam
tutd@ntu.edu.vn
[2] National Taiwan Ocean University, Keelung, Taiwan

Abstract. We numerically investigated the effects of computational domain to the unsteady laminar viscous flow around the vertical confined circular cylinder. The flow field and lift coefficient was computed numerically by finite volume method and laminar model. The results of study, we revealed mainly sinusoidal lift coefficient and flow structure as Reynolds number (*Re*) from 110 to 150, corresponding to block ratio of computational domain (AR - ratio of breadth to the cylinder diameter), 3 and 5. Due to effects of blockage ratio and geometry bottom behind the cylinder, the vortex shedding street development of flow was delayed.

Keywords: Finite volume method (FVM) · Vortex shedding · Unsteady viscous flow

1 Introduction

Flow around cylinder is classical problems in fluid dynamics which have huge application in practical. Particularly, in ocean engineering, this problems has been extensively studied due to its importance in many practical applications, such as heat exchangers, chimneys, hydrodynamic loading on ocean marine piles and offshore platform risers and support legs (Tutar et al. 2001). Therefore, in spite of extensive experimental and numerical studies almost over a century, flow around a circular cylinder still remains a challenging problem in fluid mechanics, where intensive investigations are continued even today to understand the complex unsteady dynamics of the cylinder wake flow. The recent studies which were focused on discover the vortex dynamics behavior should be mentioned such as: (Zhao et al. 2013); (Kanaris et al. 2011); (Rehimi et al. 2008); (Sahin et al. 2004). A comprehensive overview of the recent advances in vortex dynamics in the cylinder wake can be found in the review paper by (Williamson 1996) and (Zdravkovich 1996). Some researchers conducted the flow around the cylinder of low aspect ratio and laminar regime that is (Ribeiro et al. 2012) and (Mittal 2001) which is used to validate current study. Many experimental results shows that, for Reynolds number less than 40 the steady state of the flow was archived within 15 s and the wake exits two symmetric vortices (Braza et al. 1986).

B. T. Long et al. (Eds.): MMMS 2020, LNME, pp. 901–908, 2021.
https://doi.org/10.1007/978-3-030-69610-8_120

The aim of the computations is consequently to investigate the effect of computational geometry; blockage ratio (AR) to structure and lift coefficient of periodic flow past a cylinder.

2 Physical Model and Computational Method

2.1 Physical Model

The flow field around the cylinder is modelled in three dimensions with the axes of the cylinder normal to the direction of flow. The cylinder which has the diameter D = 1 is symmetrically located inside flow domain. In this investigation, we conducted three main cases with different blockage ratio and domain shape. The first one (Case 1) is that blockage ratio BR of the flow field domain which was computed with is 5. The second case (Case 2) was similar with first case, however, the flow field domain was generated blockage ratio BR to 3 domains is described in Fig. 1. In such case, the wall which was located on the bottom was name as flat-bottom. The last case (Case 3) the purpose of this experiment that has been investigated the effect of sudden expand which is located adjacently to the vertical cylinder to on cylinder wake. Therefore, the physical model for such case was set similarly with first case (the BR to 3 and aspect ratio AR to 5), however, at the bottom wall, the domain will have expanded by step with step high to 0.5 D. The computational domain consists of an upstream 5 times the cylinder diameter d to downstream 15 times the diameter. In addition, the parabolic velocity was normal with inlet face of domain and go from left to right.

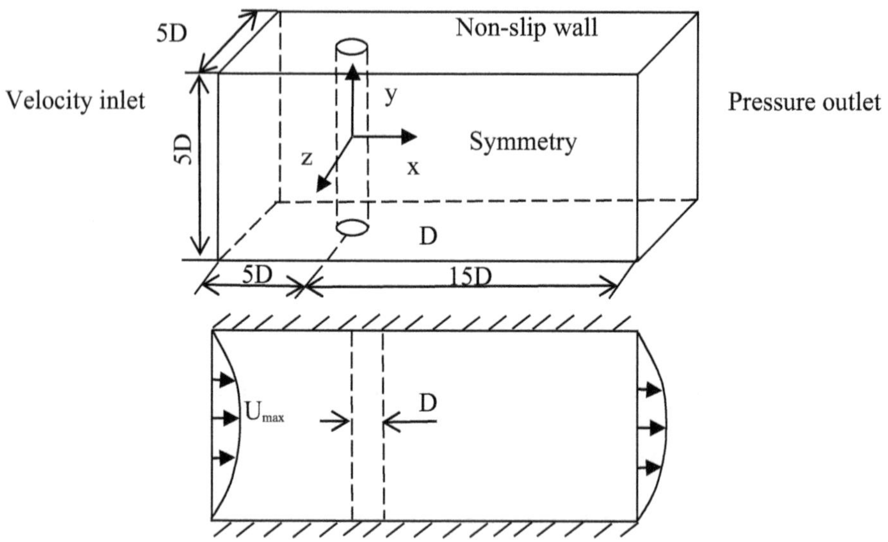

Fig. 1. Domain size; boundary conditions for computation

2.2 Governing Equations

The governing equations of incompressible viscous fluid are the classical continuity and Navier-Stokes equations. They are written under non-dimensional form below:

Continuity equation:

$$\nabla . \overrightarrow{u} = 0 \tag{1}$$

Momentum equations:

$$\frac{\partial(\vec{u})}{\partial t} + \vec{u}\nabla . (\vec{u}) = -\nabla p + \frac{1}{\text{Re}}\nabla^2\vec{u} \tag{2}$$

where u and p and are the velocity and pressure fields, t denotes the time, and ∇ is the Laplacian operator; Re is Reynolds number of the flow around a circular cylinder is defined by,

$$Re = \frac{\rho U_{\max} D}{\mu} \tag{3}$$

Where D is the diameter of the cylinder, U_{max} is the maximum velocity at inlet domain, ρ is a density of fluid and μ is the dynamic viscosity of fluid.

2.3 Compuational Solutions

Mesh Generation and Boundary Conditions

In this computation, the structural mesh is employed for spatial discretization. The detail information is shown in Table 1.

Table 1. Mesh size

Mesh information	Case 1	Case 2	Case 3
Number of nodes	1438710	1144200	1522858
Number of elements	1406608	1102108	1489552
Number of blocks	6	6	8

The wall boundary conditions employed in this study are non-slip condition, i.e., $u = 0$, $v = 0$, $w = 0$ at wall (top, bottom and cylinder surface). In spite of that, an

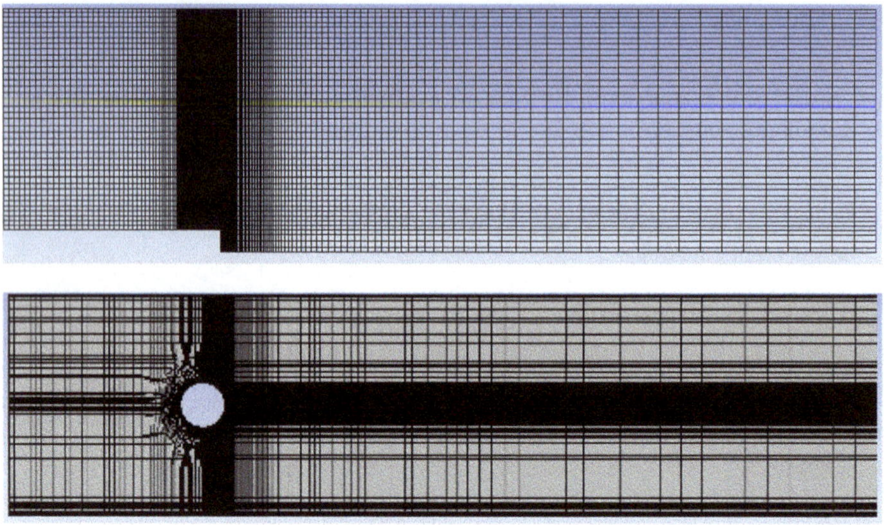

Fig. 2. Mesh distribution for computation

external rectangle boundary enlarges far from the cylinder (Fig. 2). Profile parabolic distribution of velocity condition at inlet face upstream of can be prescribed as:

$$u(y) = \left[1.5 - 2.5\left(\frac{y}{h}\right)^2\right] \tag{4}$$

The symmetry condition was applied at the lateral boundaries. Pressure outlet applies at the face downstream.

Solution Method

A finite volume method was employed to discretize all differential equations in to algebraic equation system. In addition, solution algorithms for pressure-velocity coupling - PISO (Pressure Implicit with Splitting of Operators) was adopted for nonlinear iterations of pressure and velocity fields. For transient regime computations, the implicit time-matching scheme was used. Asymmetry initial condition was input with horizontal velocity component that perpendicular to cylinder axis was set 0.2 m/s everywhere elements of computational domain. The computations for Case 1 and Case 2 with range of the Reynolds number 150, 140, 130, 120 for determine the critical Reynolds number.

3 Results and Discusions

3.1 Results

A series of computations have been conducted to study the 3D flow structure on cylinder wake as well as effect of blockage ratio to development of flow field. The first results show the effect of step of computation geometry to the flow structure at Re = 150. Figure 3 shows the flow structure and lift coefficient for cylinder in flat-bottom (Case 1) while Fig. 4. Illustrates those for case with step bottom (Case 3). It is clear that the existence of step behind the cylinder causes the reduction of lift amplitude and period of flow. In detail, for case 1, the amplitude and period of flow are 0.015 and 50 s, respecively. By contrast, those value are 0.08 and 20 s for case 2.

Due to the effect of blockage ratio (AR) to the flow during computation. Thus, the development of the flow field did not adopt from review of (Williamson 1996). From the paper, the flow regime around smooth, circular cylinder in range Reynolds number from 40 to 200 will be unstable lead to laminar vortex street. However, in this study, the Reynolds numbers was selected at 100. After long time computation, the flow has been being steady. Therefore, it is necessary that increasing the Re from 100 to 150 to determine the critical value of Re, the value determine the flow begin unstable. The lift coefficient CL for five values of Re illustrates in Fig. 5.

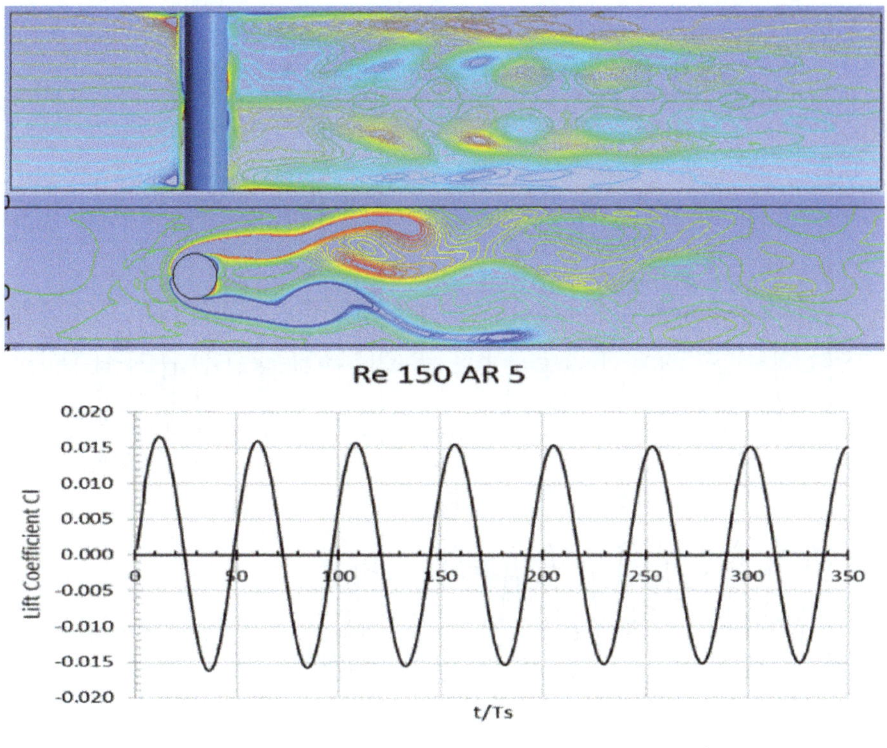

Fig. 3. Flow structure and lift coefficeint for Case 1

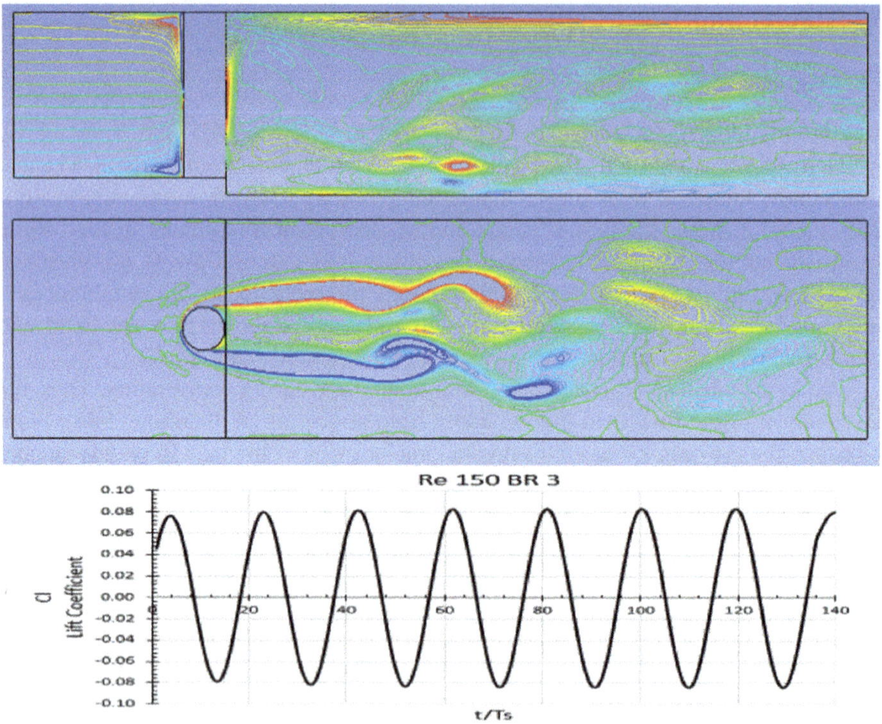

Fig. 4. Flow structure (a) and lift coefficient (b) for Case 3

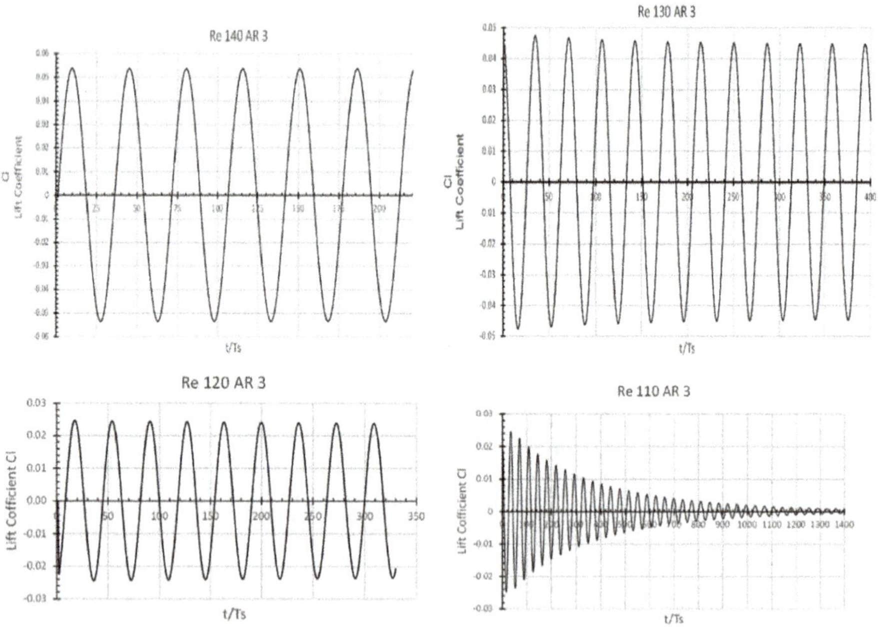

Fig. 5. The Lift coefficient of flow flied at Re = 140, 130, 120,110

By linear approximation function, we found the new critical Reynolds (Re) number of flow field is 110.6 for case AR = 3 while this value is 124.1 for case AR = 5 (Figs. 6 and 7).

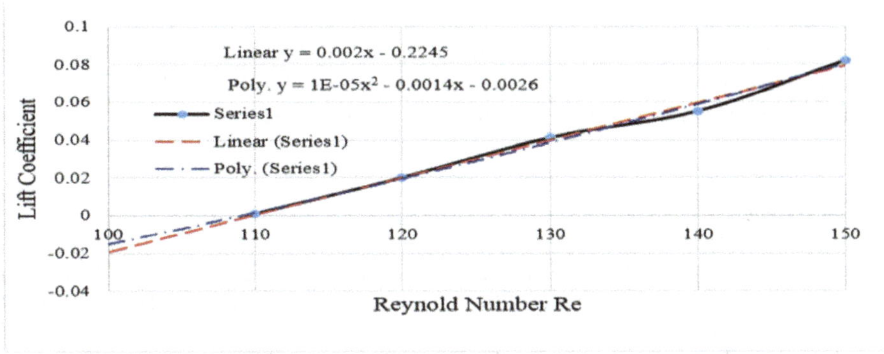

Fig. 6. The relationship between *Re* and amplitude of C_L for AR = 3

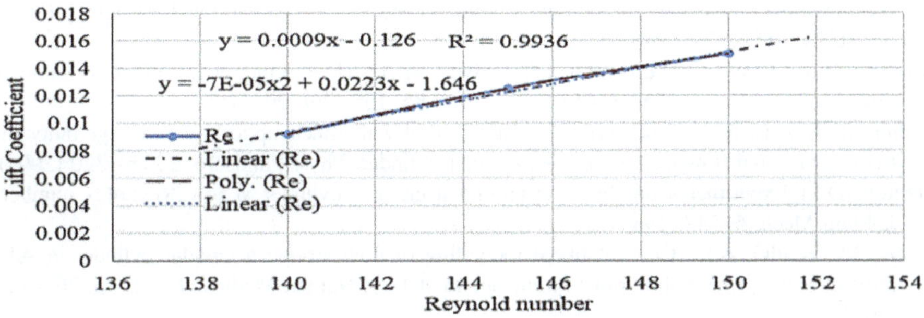

Fig. 7. The relationship between *Re* and amplitude of $C_L$of AR = 5

3.2 Discussion

In the present study, we found that for a sinusoidal lift coefficient of oscillating flow, the two phrases of a Karman vortex street were archived. However, due to the effect of geometry and blockage ratio of the computational domain so that the vortex shedding of flow was delayed comparison with (Tritton 1959). Furthermore, the critical Re of the current study was found at 124.1. This value is larger, comparison 110.6, which was found by (Tritton 1959). Physically, the blockage ratio of the computational domain leads to the delay a development of flow past the cylinder.

4 Conclusion

Results have been presented for three-dimensional unsteady flows past cylinders of low aspect ratios in various computational domain. The structure of vortices in three main investigated case were visualized by CFD-Post tool. The periodic fluctuation flow that was named vortex shedding street has been study at Re from 100 to 150. The effect of aspect ration (AR) causes the delayed of unsteady separation of flow. At AR = 3, the critical Reynolds number is 110.6 while those value is 124.1 for AR = 5. The figuration of complex flow structure at cylinder wake was visualized clearly.

References

Braza, M., Chassaing, P., Ha Minh, H.: Numerical study and physical analysis of the pressure and velocity fields in the near wake of a circular cylinder. J. Fluid Mech. **165**, 52 (1986)

Kanaris, N., Grigoriadis, D., Kassinos, S.: Three dimensional flow around a circular cylinder confined in a plane channel. Phys. Fluids **23** (2011)

Mittal, S.: Computation of three-dimensional flows past circular cylinder of low aspect ratio. Phys. Fluids **13**, 177–191 (2001)

Rehimi, F., Aloui, F., Nasrallah, S.B., Doubliez, L., Legrand, J.: Experimental investigation of a confined flow downstream of a circular cylinder centred between two parallel walls. J. Fluids Struct. **24**, 855–882 (2008)

Ribeiro, V.M., Coelho, P.M., Pinho, F.T., Alves, M.A.: Three-dimensional effects in laminar flow past a confined cylinder. Chem. Eng. Sci. **84**, 155–169 (2012)

Sahin, M., Owens, R.G.: A numerical investigation of wall effects up to high blockage ratios on two-dimensional flow past a confined circular cylinder. Phys. Fluids **16**, 1305–1320 (2004)

Tritton, D.J.: Experiments on the flow around a circular cylinder at low Reynolds number. J. Fluid Mech. **6**, 547 (1959)

Tutar, M., HoldØ, A.E.: Computational modelling of flow around a circular cylinder in sub-critical flow regime with various turbulence models. J. Numer. Methods Fluids **35**, 763–784 (2001)

Williamson, C.H.K.: Vortex dynamics in the cylinder wake. Annu. Rev. Fluid Mech. **28**, 477–539 (1996)

Zdravkovich, M.M.: Different modes of vortex shedding: an overview. J. Fluids Struct. **10**, 427–437 (1996)

Zhao, M., Thapa, J., Cheng, L., Zhou, T.: Three-dimensional transition of vortex shedding flow around a circular cylinder at right and oblique attacks. Phys. Fluids (1994-present) 25 (2013)

Study on Static Obstacles Avoidance Using Advanced Set-Based Guidance with Velocity Obstacle for Unmanned Surface Vehicles

Minh-Hung Vu[1], Ngoc-Huy Tran[2(✉)], and Minh-Tam Phan[2]

[1] Petro Vietnam University, PVN, Bà Rịa, Vietnam
hungvm@pvu.edu.vn
[2] Ho Chi Minh City University of Technology, VNUHCM, Ho Chi Minh City, Vietnam
tnhuy@hcmut.edu.vn

Abstract. In hazardous conditions, the use of an autonomous unmanned surface vessel (USV) may be a solution to an obstacle avoidance problem with minimal human intervention but maximal efficiency. This article describes a method of utilizing Velocity Obstacle algorithm to enhance collision-free paths of Set-Based Guidance (SBG) for USV equipped with Light Detection and Ranging (LiDAR) for detecting obstacles. Additionally, we also propose a novel trapezium path with three waypoints instead of a single waypoint as conventional SBG. Moreover, we build a hardware architecture for USV to realize the obstacle avoidance algorithm in pratice. To verify safety, efficacy, and pragmatics, simulations are conducted in situations of static obstacles. Finally, experiments with a real static obstacle are carried out and experimental results prove that the advanced SBG is able to guide USV to avoid collision autimatically.

Keywords: Unmanned surface vehicle · Obstacle detection · Obstacle avoidance · Velocity obstacle · Set-based guidance

1 Introduction

Today, unmanned surface vehicle (USV) attracts wide attention due to its flexibility, portability, stability and increasingly high intelligence. It is used as a means for environmental monitoring, shallow water surveying, military reconnaissance, homeland security and has become a hot research topic for intelligent ships around the world [1]. Typically, with a path-following task, the guidance system consists of laws for heading, surge velocity to ensure convergence to the desired path. Then the control system calculates the thruster force to track the reference states by the guidance system. A common approach for path following is the Line of Sight (LOS) method, which allows USV to follow straight paths [2, 3] and curved paths [4, 5]. However, in some situations, there are many unknown obstacles which hinders path-following algorithms to be carried out successfully.

Collision avoidance is an important research aspect in the development of USV, with the aim to plan a collision-free path in unknown environments [6]. The local collision-free path can be generated by two approaches, including path searching-based

B. T. Long et al. (Eds.): MMMS 2020, LNME, pp. 909–919, 2021.
https://doi.org/10.1007/978-3-030-69610-8_121

local path planning method, which plans the paths or target waypoints tracked by USV, and the behaviour-based reactive obstacle avoidance method, which generates guidance velocity and guidance angle based on obstacle deeds [7]. Kuwata [8] has proposed a behaviour-based reactive approach, which adopt Velocity Obstacle (VO) to define a cone space of velocity. Kim [9] has implemented genetic algorithm to optimize collision-free path with three objective functions of travel time, environmental loads, and avoiding obstacles. Zhao [10] has introduced a novel approach of real-time automatic avoiding collisions of multiple vessels in compliance with COLREGs rules. Moe and Pettersen [11] have proposed set-based method with path following task and obstacle avoidance task. Myre [12] has modified set-based method to be more appropriate for underactuated systems like USV by using an additional obstacle avoidance waypoint. In this paper, we propose a more effective way of setting new waypoints, which uses three waypoints corresponding three vertices of a trapezium instead of single one. These waypoints are calculated online at the time as avoiding obstacle by adopting velocity algortihm (VO), which allow USV avoiding static obstacles.

This paper is organized as follows: The suggested set-based guidance is presented for collision avoidance in Sect. 2. In Sect. 3, the guidance law ILOS is implemented for path following task. Finally, the results are given in Sect. 4 and conclusions in Sect. 5.

2 Advanced Set-Based Guidance

2.1 Concept of Velocity Obstacle

Velocity Obstacle (VO) approach for generating a collision-free path has been proposed by Kuwata [8]. The basis idea of this algorithm is to turn a general single obstacle avoidance problem to a static one. To simplify the problem, a shape of the obstacle is temporily ignored and replaced by its centroid and a safety circle with radius of Rs (Fig. 1). It can be inferred from Fig. 1 that USV will collide with the target ship when $\vec{v}_{uo}$ belongs to the cone space $T^{-}UT^{+}$ which is formed by two tangents of the safety circle. As a result, the collision between USV and the target ship will not take place when a following condition is satisfied:

$$\arg(\vec{v}_{uo}) \notin \text{ConeSpace}(\vec{R}_{uo}, R_s) = \left(\arg(\vec{R}_{uo}) - \delta; \arg(\vec{R}_{uo}) + \delta\right) \tag{1}$$

where

arg(.): is an argument of vector (.); $\delta = \text{asin}(R_s/\sigma)$:is a haft of cone space angle; R_s: is a radius of the safety circle; $\vec{R}_{uo} = O - U$: is a vector from the obstacle (O) to USV (U); $\sigma = \left\|\vec{R}_{uo}\right\|$: is distance between USV and the obstacle; $\|.\|$ operator is Euclide norm of (.).

Once it is anticipated for a collision between USV and the obstacle according to Eq. (1), USV has to change its course to a desired heading ψ_{ud} such that a new relative velocity $\vec{v}_{uod}$ is out of the cone space. Nevertheless, changing the course angle will cost USV amount of energy as much as its inertia, for the aim of minimizing course change,

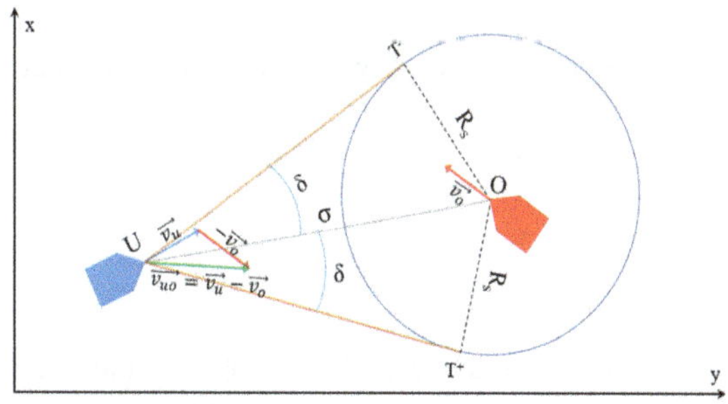

Fig. 1. Illustration of parameters used for assessing and avoding collision according to Velocity Obstacle

$\arg(\vec{v}_{uod})$ is equal to $\vec{R}_{UT^-}$ or $\vec{R}_{UT^+}$ respect to collision scenarios that are regulated by COLREGs rules.

In summary, it has been shown from this section that the concept of VO is used in collision risk assessment and determining the desired heading angle which guide USV to avoid collision even in the worst case. The results in this section will be adopted to plan obstacle avoidance waypoints in the following section.

2.2 Obstacle Avoidance Waypoints

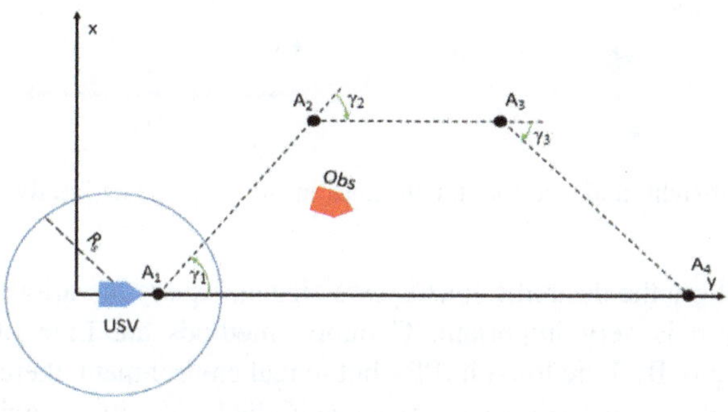

Fig. 2. The trapezium path and its parameters

In an ideal condition, when the heading angle can reach immediately a value of the dersied heading angle ψ_{ud} and the surge velocity is still unchanged, USV is able to avoid obstacle according to the concept of VO algorithm and obstacle avoidance path is a straight line with a slope of $\tan(\psi_{ud})$. Apparently, this ideal condition will never be satisfied by a practical USV model with motion constraints, but USV has capacity of avoiding obstacle when following the ideal obstacle avoidance path. A complete obstacle avoidance algorithm usually includes two tasks of avoiding collision and path following. Along with avoiding hazardous, USV has to keep distance as close as possible with the global path to facilitate the path following task. In this paper, we proposed a trapezium collision-free path (Fig. 2) with parameters determined as follow:

$$\begin{cases} A_i = U(t_i), i - 1 \div 3 \\ \gamma_1 = \psi_{ud}(t_i) \\ \gamma_2 = \alpha_k \\ \gamma_3 = \alpha_k - \text{sign}(\gamma_1 - \alpha_k)\mu \end{cases} \tag{2}$$

where α_k is a slope angle of the global path, ψ_{ud} is the desired heading angle calculated according to VO, μ = pi/4 for head-on and μ = pi/6 for overtaking and crossing, t_1 is time to turn path following task into obstacle avoidance task when σ is less than a given threshold, t_2, t_3 is the moment the relative velocity $\vec{v}_{uo}$ satisfying Eq. (1) but with a assumption that the heading angle ψ_u is equal to γ_2, γ_3, respectively.

3 Guidance System

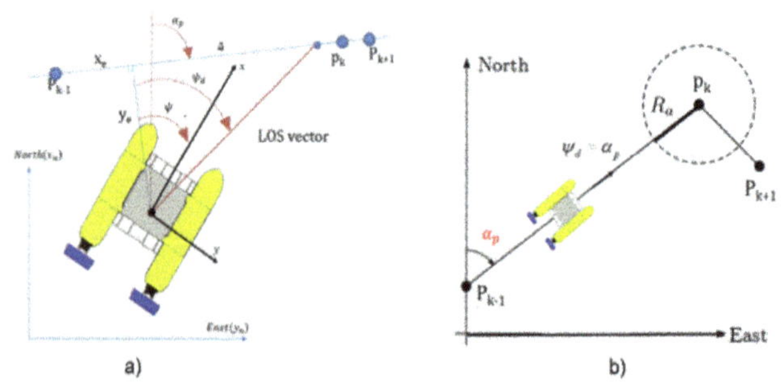

Fig. 3. a) Line-of-sight guidance for straight lines in the xy plane b) Steady state in the path.

Vehicles that follow the desired trajectory are zigzag, square in various missions, so a navigation system is very important. Common methods are Line of Sight (LOS), Constant Bearing (CB), Pure Pursuit (PP), but in real environment, there is disturbance: wave, wind and flow effect to vehicle. As a result, the USV will be deviated by a small angular component called sideslip β during the move. An approach that aims to reduce tilt effects is the Integral line-of-sight (ILOS) method. Therefore, the integral action inserted in the LOS guidance law is used to counteract the presence of current and correctly set the heading of the vehicle. To achieve the goal, in the case of decoupled horizontal plane path-following, the desired heading and surge velocity set-points from the traditional LOS guidance (Fig. 3) are given as:

$$\psi_d = \alpha_p + \arctan(\frac{-y_e}{\Delta}), u_d = u_0 \tag{3}$$

where: $\alpha_p = \text{atan2}(y_{k+1} - y_k, x_{k+1} - x_k)$ is the path-tangential angle,

$y_e = -(x - x_k)\sin\alpha_p + (y - y_k)\cos\alpha_p$ is the cross-track error (the shortest distance to the path) and $\Delta > 0$ is the predefined lookahead distance, selected to fit in the experiment. $P_k(x_k, y_k)$ is the position of the k-th waypoint and u_0 is a predefined desired surge speed.

Finally, the goal is making the cross track converge to zero:

$$\begin{aligned} &\lim_{t\to\infty} y_e(t) = 0, \lim_{t\to\infty} u(t) = u_0 \\ &\lim_{t\to\infty} \psi(t) = \psi_d, \psi_d = \alpha_p \end{aligned} \tag{4}$$

However, convergence is not guaranteed if there are current acting on USV since side-slip β once it has converged to path $\mathcal{P}$ (Fig. 4):

$$\lim_{t\to\infty} \psi(t) = \psi_d, \quad \text{where:} \quad \psi_\mathrm{d} = \alpha_\mathrm{P} + \beta, \beta \in \left(\frac{-\pi}{2}; \frac{\pi}{2}\right) \tag{5}$$

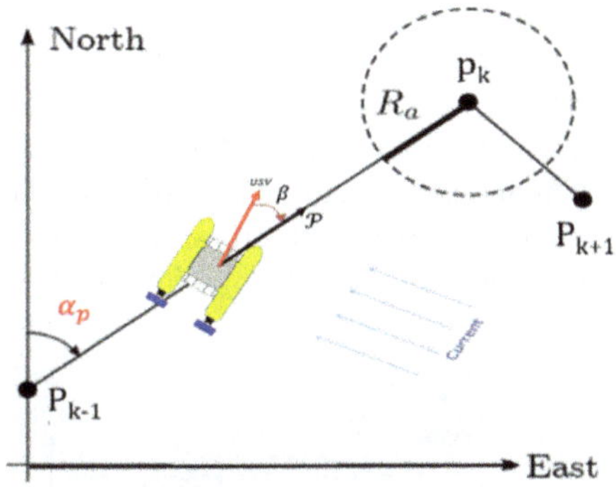

Fig. 4. Steady state: the side-slips β with a constant angle and follows the path

To solve the problem, integral action is introduced in the heading reference. Hence, the integral LOS guidance law is defined as (see Fig. 5b):

$$\begin{aligned} \psi_d &= \alpha_P - \arctan\left(\frac{y_e + \sigma y_{\text{int}}}{\Delta}\right), \\ \dot{y}_{\text{int}} &= \frac{\Delta y_e}{(y_e + \sigma y_{\text{int}})^2 + \Delta^2} \end{aligned} \tag{6}$$

The integral effects give nonzero angle (Fig. 5b), allowing the vehicle to side-slip while staying on the desired path and can counteract the effect of the current. The gain – $\sigma > 0$ is design parameters and the gain σ satisfies:

$$0 < \sigma < u_d - v_{c\ \max} \tag{7}$$

The objective is to make the vehicel follow the path $\mathcal{P}$ with a desired speed u_d(m/s). The maximum intensity of the unknown current is $v_{c\ \max}$(m/s).

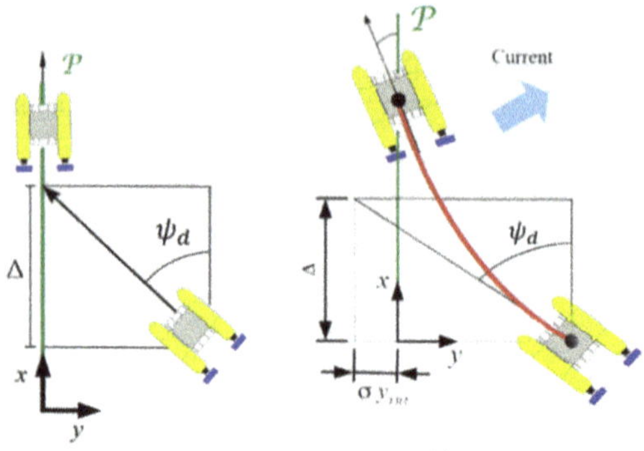

Fig. 5. a) Line of sight guidance, b) Integral line of sight guidance for USV.

4 Results and Discussion

In the previous sections, we presented the dynamic model of USV, obstacle avoidance algorithm and hardware structure to realize the algorithm. To test the algorithm's ability to navigate safely, several simulations were performed with static obstacles collision situation with three different shapes and three collision situations with dynamic obstacles including: head-on, overtaking and crossing. Then, an experiment is carried out to evaluate the obstacle avoidance algorithm with a real static obstacle.

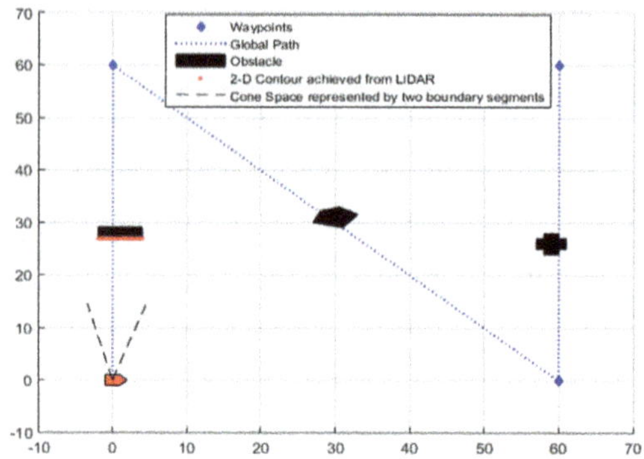

Fig. 6. Notations for the algorithm

4.1 Avoiding Static Obstacles Simulation

In this simulation, USV is located at position (0, 0), heading angle is pi/2 and begins to follow the global path (blue dot) created by waypoints (blue diamond) as Fig. 6. Three black static obstacles with the shape of a rectangle, pentagon and cross are preset on the global path. From the profile of the scanning obstacle used LiDAR, a cone space is defined by VO and is represented by two boundary dash segments.

The results in Fig. 7 pinpoint out that this algorithm will be effective and pragmatic when it is implemented to avoid obstacles. Convergence to the global path after detecting and avoiding obstacles reassures that the obstacle avoidance algorithm did not affect control and following path of USV. Moreover, the simulations verify safe navigation of the algorithm when USV keeps distance to obstacle over time of overcoming and optimization in choosing side to pass such that the course change is as minor as possible.

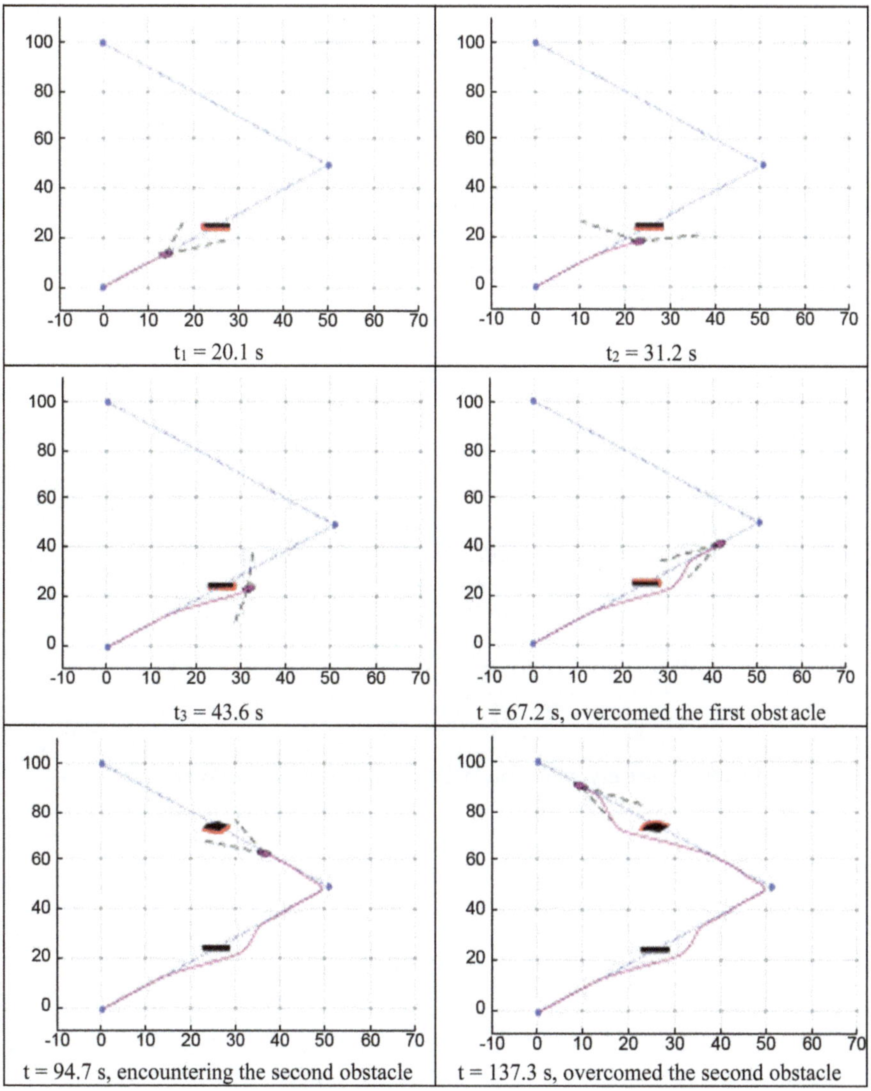

Fig. 7. Simulation results for static obstacles scenario

4.2 Avoiding a Static Obstacle Experimental Result

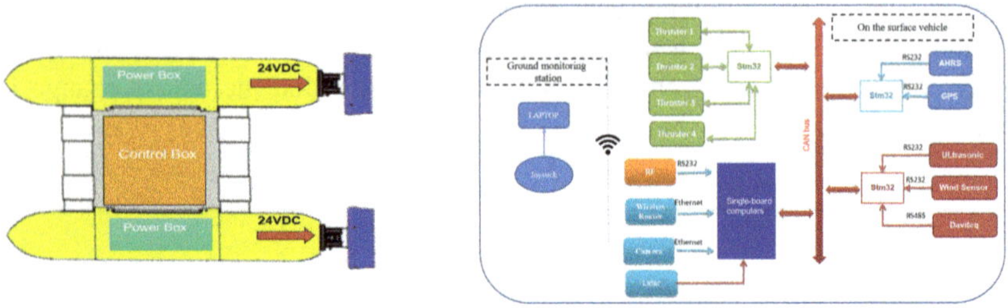

Fig. 8. Structure and control system of the VIAM-USV1000

Figure 8 shows the structure and control system of the VIAM-USV1000 [13] using in experiments. Beside, the LiDAR sensor has an important role in collision avoidance. Also, the vehicle is eqquiped a video camera, using record the entire motion, images and video in different tasks.

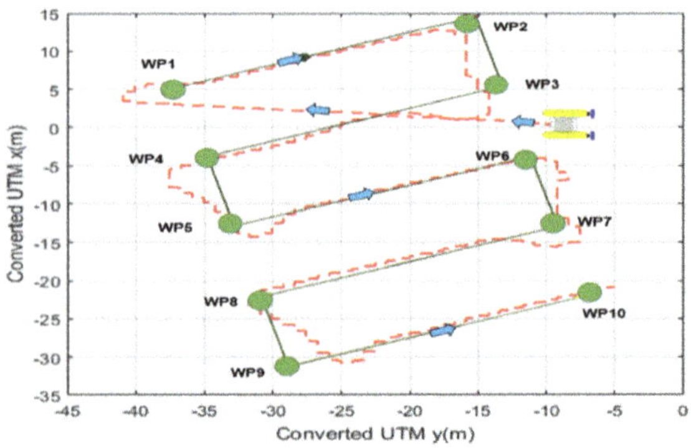

Fig. 9. Trajectory while way point tracking with Square-Wave test

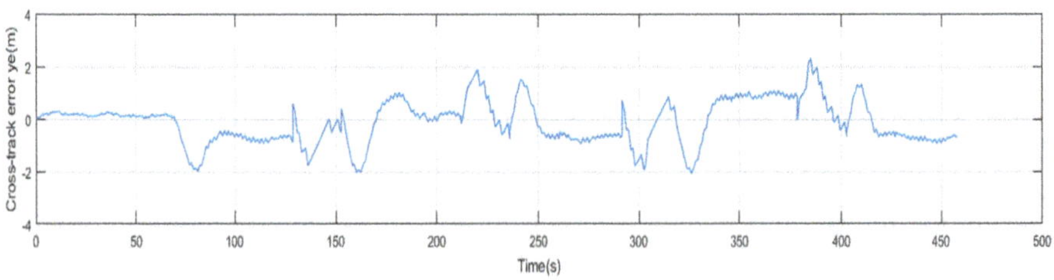

Fig. 10. Cross-track error ye with Square-Wave test

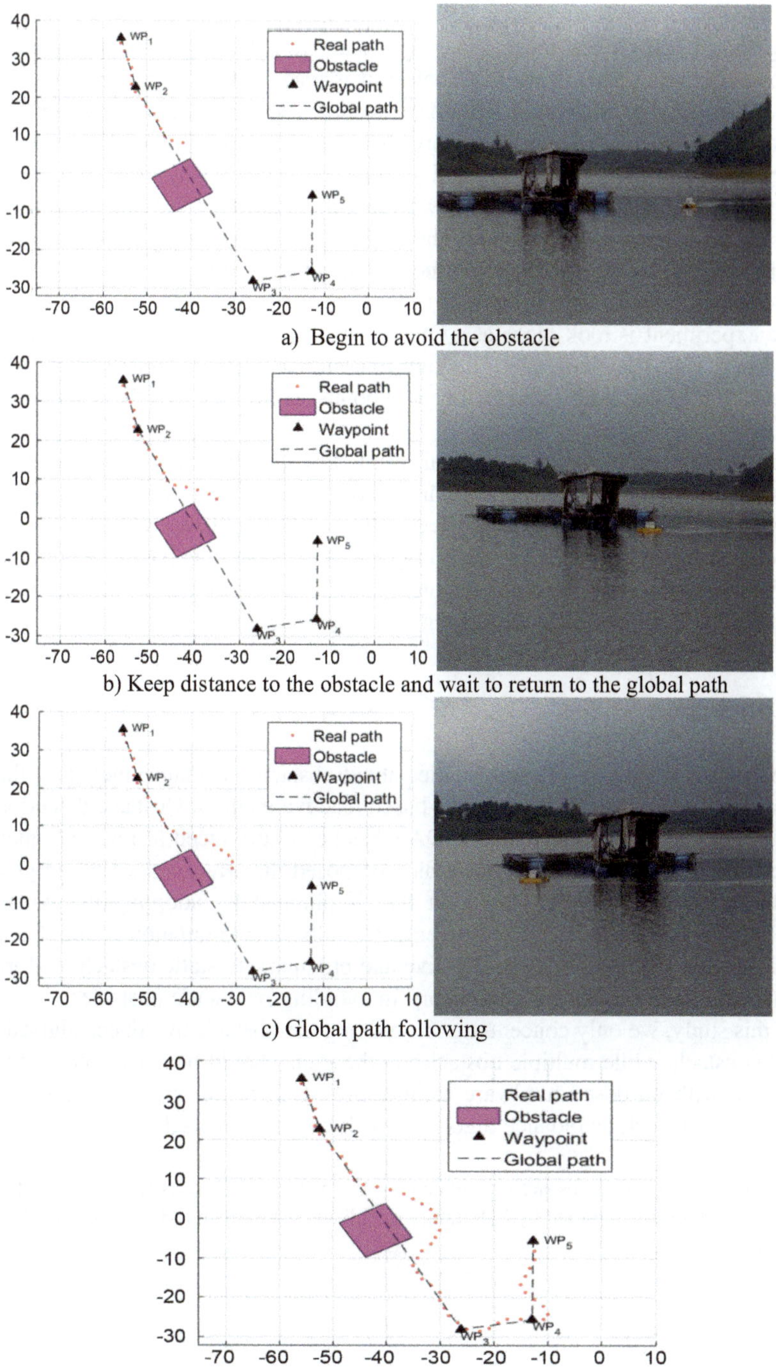

a) Begin to avoid the obstacle

b) Keep distance to the obstacle and wait to return to the global path

c) Global path following

d) Observed path over entire obstacle avoidance process

Fig. 11. Avoiding the static obstacle experimental result.

Waypoint Following Tests: Since USVs utilize LOS to following the obstacle avoidance path of ASBG, we must make sure that USV have good performance with waypoint following tests using LOS before it is ready for obstacle avoidance experiments. The vehicle was preprogrammed with a series of target waypoints. ILOS algorithm and heading control strategy described in Sect. 3. For every time step throughout the mission, the bearing and distance to the target waypoint from the current position was calculated. Figure 9 shows the USV trajectory for one such trial. The results of this experiment were good. The USV tracked its planning path and no wobble heading on its movement. Moreover, the experimental results, as in Fig. 10, show the Cross-track error ye is in the range of 0.7 m.

The experiment is took place at a Quang Thanh lake belonged to Chau Thanh, Ba Ria – Vung Tau, VietNam, and a floating house on the lake is chosen to be the static obstacle. The global path include four waypoints as Fig. 11 and the safety radius is equal to 4 m.

It can be seen from the experimental results that the USV was able to switch from path following mode to obstacle avoidance mode when there is a risk of collision with an obstacle. USV has been guided effectively to avoid collision while keeping a safe distance from an obstacle despite the complicated shape and the path following process after overcoming the obstacle has remained stable, USV can avoid the other obstacles with different shapes without disturbing other operations.

5 Conclusion

This paper has reviewed on some studies about obstacle avoidance including the path searching-based method and behaviour-based reactive method. Obstacle detection with high accuracy and reliability using LiDAR has been introduced to improve quality of the obstacle avoidance task. We have also proposed the advanced set-based guidance with the trapezium collision-free path that is planned by adopting the concept of Velocity Obstacle and LiDAR measurement results. The simulation results have confirmed that this algorithm is safe, effective and pragmatic to static obstacles. Moreover, we have adopted successfully this theory in avoiding real static obstacles.

In this study, we only concentrate on studying the obstacle avoidance algorithm for a single obstacle while multiple obstacles at the same should be concerned in practice. Moreover, with proposed hardware architecture of USV, we are looking forward to realizing the obstacle avoidance algorithm with dynamic obstacles.

Acknowledgement. This research is funded by PetroVietnam University (PVU), under grant number GV2005 and supported by Laboratory of Advance Design and Manufacturing Processes, HCMUT.

References

1. Campbell, S., Naeem, W., Irwin, G.W.: A review on improving the autonomy of unmanned surface vehicles through intelligent collision avoidance manoeuvres. Annu. Rev. Control **36**, 267–283 (2012)
2. Fossen, T.I., Breivik, M., Skjetne, R.: Line-of-sight path following of underactuated marine craft. In: Proceedings of 6th IFAC Conference on Manoeuvering and Control of Marine Craft, pp. 244–249 (2003)
3. Oh, S., Sun, J.: Path following of underactuated marine surface vessels using line-of-sight based model predictive control. Ocean Eng. **37**(2–3), 289–295 (2010)
4. Skjetne, R., Jorgensen, U., Teel, A.R.: Line-of-sight path-following along regularly parametrized curves solved as a generic maneuvering problem. In: Proceedings of IEEE Conference on Decision and Control and European Control Conference, pp. 2467–2474 (2011)
5. Borhaug, E., Pettersen, K.Y.: LOS path following for underactuated underwater vehicle. In: Proceedings of 7th IFAC Conference on Manoeuvring and Control of Marine Craft (2006)
6. Liu, Y., Bucknall, R.: Path planning algorithm for unmanned surface vehicle formations in a practical maritime environment. Ocean Eng. **97**, 126–144 (2015)
7. Wang, Y., Yu, X., Liang, X., Li, B.,:A COLREGs-based obstacle avoidance approach for unmanned surface vehicles. Ocean Eng. pp. 110–124 (2018)
8. Kuwata, Y., Wolf, M.T., Zarzhitsky, D., Huntsberger, T.L.: Safe maritime navigation with COLREGS using velocity obstacles. In: 2011 IEEE/RSJ International Conference on Intelligent Robots and Systems (IROS), pp. 4728–4734. IEEE (2011)
9. Kim, H., Kim, S.H., Jeon, M., Kim, J.H., Song, S., Paik, K.J.: A study on path optimization method of an unmanned surface vehicle under environmental loadsusing genetic algorithm, Ocean Eng. 616–624 (2018)
10. Zhao, Y., Li, W., Shi, P.: A real-time collision avoidance learning system for Unmanned Surface Vessels. Neurocomputing **182**, 255–266 (2016)
11. Moe, S., Pettersen, K.Y.: Set-based line-of-sight (LOS) path following with collision avoidance for underactuated unmanned surface vessel. In: 24th Mediterrian Conference Control Automation MED 2016, Athens, Greece, pp. 402–409 (2016)
12. Myre, H.: Collision Avoidance for Autonomous Surface Vehicles Using Velocity Obstacle and Set-Based Guidance (Master's thesis). NTNU (2016)
13. Tran, N.-H., et al.: The design of an VIAM-USVI000 unmanned surface vehicle for environmental monitoring applications. In: 4th International Conference on Green Technology and Sustainable Development (GTSD), Ho Chi Minh city, Viet Nam (2018)

A Study on an Oil-Air Mixed Lubrication Monitoring System for Spindle Unit of CNC Milling Machine

Tuan-Anh Bui(✉) and Quang-Tu Vu

School of Mechanical Engineering, Hanoi University of Science and Technology, No. 1 Dai Co Viet Road, Hanoi, Vietnam
anh.buituan@hust.edu.vn

Abstract. The stiffness and the load capacity of a spindle unit are important parameters that decide machining ability and accuracy of the machine. Furthermore, the working features of the milling machine spindle depend on ability and method of lubrication and cooling. This manuscript describes a design of an oil-air mixed lubrication monitoring and controlling system of a spindle unit. This is the basis for determining the properties of an oil-air lubrication and cooling system in specific machining and working conditions. The results showed that the system was working properly with the design parameters in the laboratory condition. The lubrication and cooling method by oil – air mixture gives a more outstanding efficient than that of the original water-cooled and grease-lubricated method of the spindle unit. In addition, the system was equipped by sensors that monitor the temperature of the bearings, thereby adjusting the air - oil flow accordingly in a range from 120 to 180 mm^3/h corresponding to the spindle speed change and reach a maximum value of 11000 rpm. This monitoring and controlling process focuses on improving the efficiency of using lubricating oil and power consumption of system.

Keywords: Oil-air mixed lubrication · Monitoring system · Spindle unit · CNC milling machine

1 Introduction

Modern machining increasingly requires high-speed machining to produce more sophisticated products. However, increasing the rotation speed will generate more heat which affects the stiffness and deformation of the spindle of machine tools. In fact, many methods have been introduced to improve the efficiency of lubrication and cooling, including the gas-oil cooling method. In this case, it is possible to lubricate and cool the bearings and save oil efficiently. However, in practical use, this method still exists problem of inaccurate monitoring control with the actual operating status of the ball bearing. This study will focus on improving the efficiency of the air-oil monitoring control system for cooling the spindle unit of machine tools.

As known, increasing the spindle's rotation speed will result in increasing the heat generated in the system bearings. Lubrication methods may be used including the use of grease, oil splash lubrication, or oil mist lubrication. The published studies show that

B. T. Long et al. (Eds.): MMMS 2020, LNME, pp. 920–928, 2021.
https://doi.org/10.1007/978-3-030-69610-8_122

liquid droplets are effective in convection and heat metabolism [1]. When the liquid is broken down into smaller particles, it will easily creep into the heat generating locations, thereby improving lubrication and cooling. To do this, the scientists create a mixture of oil mist or oil-air and put it into the area that needs lubrication and cooling. An US patent has used the oil-gas-cooled lubrication method with high efficiency was presented [2]. This method simultaneously provides the effective cooling and lubrication, saves oil, and it is more optimal than the separate lubrication and cooling methods. It is therefore preferred for use in the modern machining industry. Table 1 lists a comparison between the lubrication methods for the spindle unit of machine tools.

Table 1. A comparison between lubrication methods [3]

Methods	Advantages	Disadvantages
Grease	+ Relatively high speeds up to 1×10^6 dmN + High load capacity; no maintenance; simple design	+ Finite life; moderate power + Speed limitations
Oil Bath	+ High loads + Simple Design	+ Low speeds 0.15×10^6 dmN + Finite life; Moderate oil consumption + Maintenance required
Oil Mist	+ High speeds 1×10^6 dmN + Moderate temperatures; simple design + Infinite lube life	+ Moderate power loss + Environmental hazards + Collection equipment commended + Loss lube moderate oil consumption + Maintenance required
Oil Air	+ High speed up to 1.5×10^6 dmN + Low temperature + Small oil consumption relative to oil mist 10% + Infinite lube life	+ Design complexity + Loss lube collection equipment may be required

From the table comparing the lubrication methods used for the machine tool spindle, each method has its advantages and disadvantages. However, each method is suitable for specific working conditions. In the case of working with high loads, high rotational speeds, or even a simple system design, a suitable lubrication option can be chosen. On the other hand, it can be seen that, with high-speed processing, the lubrication scheme using oil and gas mixture has shown many outstanding advantages, especially the ability to save lubricating oil and does not need regular maintenance. However, precise control of the amount of lubrication and cooling system states is necessary and the efficiency of using the correct lubricating oil is improved, avoiding waste, and causing environmental pollution [4]. This is also a purpose of this study.

2 Building an Experimental Model

To investigate the efficiency of the oil-air cooling lubrication system, the author uses a spindle ER16-80SK 24 k with a power capacity of 1.5 kW. The layout of the oil-air nozzle and temperature sensor as shown in Fig. 1 and Fig. 3. The spindle is direct drive, using three-phase asynchronous motor. It is controlled via an inverter, which receives and responds with operating parameters to the central control cabinet. The inverter will monitor and control the spindle unit, monitor current, voltage, and torque parameters and issue alarm signals about the controller when a problem occurs [5]. The temperature sensors are attached to the ball bearings to monitor the temperature in real time. Temperature signals will be transmitted and monitored by the controller. To create high-pressure air-oil mixture, oil from the tank will be pumped into the mixer (as shown in Fig. 2) [6], where high pressure gas will be shredded by the compressed air pressure into small particles, forming the oil pressure high force, easy to penetrate into high temperature areas. To collect the results, the author uses a computer connected to the central control cabinet to record the temperature of the bearings.

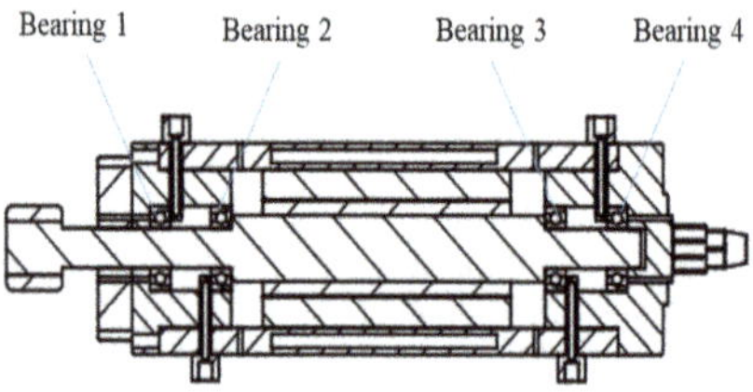

Fig. 1. Cross-section of spindle unit

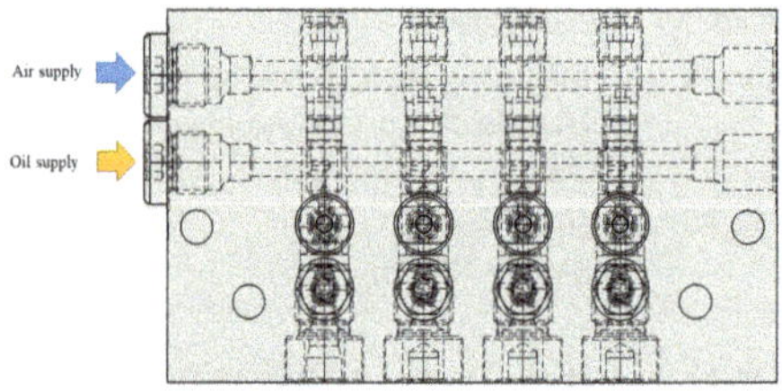

Fig. 2. Oil-air mixture system [6]

Shown in Fig. 1, the spindle uses 4 bearings with the type of 7002. According to the document [4], to make a stable operation for the system, the compressed air pressure needs to reach 5 bar, the amount of oil sprayed into the bearing should be in the range of 120 to 180 mm^3/h.

Based on this parameter, we adjust the on-off cycle of the oil pump by an expression as below:

$$n = \frac{V_{oil}}{d} \quad (1)$$

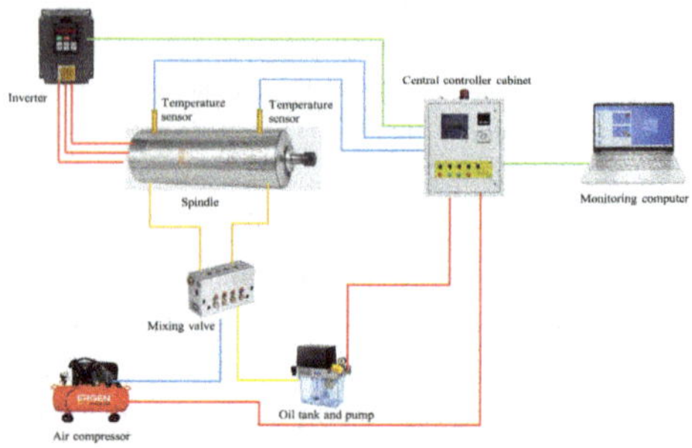

Fig. 3. Principle diagram of an oil-air cooling and lubrication system

Where V_{oil} = 180 mm^3/h; d is the amount of oil to be pumped each time, approximate 10 mm^3/h.

So, the number of on-off cycle of the oil pump should be 18 times an hour.

Pump cycle time:

T = 3.33 (min)

$T_{Pumping}$ = 20% T = 0.67 (min)

$T_{stopping}$ = 80% T = 2.66 (min)

From the above parameters, an experimental model of oil-air cooling and lubrication system was proposed as shown in Fig. 3.

According to the manufacturer's documentation [6], when using a 3-liter oil tank, it takes time to fully pump that amount if the system operates continuously is estimated as:

$$t = \frac{V}{Q} = \frac{3}{1.8 \times 10^{-4}} = 16666.67\,(\mathrm{h}) \approx 1.9\,(\mathrm{year})$$

Thus, it can be seen that the spindle unit of machine tool is continuously lubricated and cooled during the working process. The lubrication status is continuously monitored and ensured. Hence, the system temperature readings are closely and continuously monitored to assess the lubrication status of the spindle bearings. The system monitors some parameters including the temperature of bearings, oil level, air pressure and consumption current value of the spindle. The temperature of bearings is monitored by a sensor which is installed in outer rings of the spindle unit, the oil level in tank is controlled by a float switch. The air pressure needs to be guaranteed at 5 bar for shredding oil flow, optimizing lubrication, and cooling process. Besides, the consumption current of spindle is measured by an inverter and the data is sent to the controller. This will help to prevent the system error in the case the system does not provide the required amount of lubricating oil as designed. The temperature of bearing increases leading to an increase of the consumption current, the system will shut down and trigger alarm. If the temperature of the bearings reaches the preset value and all the above conditions are satisfied, the controller will allow the oil pump and the air valve to operate. Based on the instantaneous temperature, the uptime of the oil pump and air valve will be controlled via the software to optimize the lubrication. Therefore, it will ensure stable operation for a long time without periodic maintenance.

3 Building a Simulation Model of the System

To simulate the heat transfer of the system, we designed a model of the original spindle unit which was grease-lubricated and water-cooled. After examining the initial properties of the spindle unit, the air and oil injectors were placed on the bearings as shown in Fig. 1. The ball bearing used is the 7002 type, which is commonly used for CNC milling with small and medium size. Considering the ball bearing with a ball diameter of "*d*", rotating with speed of V, within a diameter of rotation is "*D*". The ball spins around it with the speed of "*v*". The main factors generating heat in a bearing are the torque caused by load and the torque of the ball [7].

The total heat generated can be calculated by the formula:

$$H_f = \omega_b M_s \tag{2}$$

Where ω_b is the angular velocity of the ball, M_s is the torque generated in the bearing including the torque due to slip and gyroscope of the ball [7]

$$M_s = (M_{ds} + M_{gr})K \tag{3}$$

Where K is the ratio coefficient depending on the accuracy class of the spindle unit.

Sliding torque is calculated by the formula [7]

$$M_{ds} = \frac{PD_0 z}{4d_b}\left(1 - \frac{d_b^2}{D_0^2}\cos^2\alpha\right)(r_0 B_0 - r_i B_i)f \tag{4}$$

Where f: friction coefficient;

$B_0 = B_i = 1$: coefficient depends on the geometry of the inner and outer rim of the bearing;

r_0, r_i: deformation of the inner and outer rim surfaces;

α: contact angle;

The gyroscope torque of the ball is calculated by the formula:

$$M_{gr} = \frac{\pi}{60} d_b^2 \frac{\gamma}{g} \omega_b z \sin\alpha \tag{5}$$

By replacing it with formula (3), we get:

$$M_s = \frac{PD_0 z}{4d_b}\left(1 - \frac{d_b^2}{D_0^2}\cos^2\alpha\right)d_b f + \frac{\pi}{60} d_b^2 \frac{\gamma}{g} \omega_b z \sin\alpha \tag{6}$$

Replacing with the 7002 bearing parameters, the total heat generated is calculated as follows

$$H_f = \omega_b(0.07396 fP + 4.8 \times 10^{-6}\omega_b) \tag{7}$$

From the obtained formula, we calculate and assume the boundary conditions of the machine simulation model. Load on the bearings is calculated based on the unit cutting force and cross-section of the chip [8]:

$$P = p.q \tag{8}$$

Where q is the cross-sectional area of the chip, p is the unit cutting force depending on the machined material. The cutting tool material is high-speed steel, the workpiece material is carbon steel C45, so p = 1 kg/mm^2. The depth of cut chosen is about 1 mm, the end mill tool has a diameter of 16 mm, so q = 16 mm^2. Hence, we get P = 16 N.

In addition, assuming the coefficient of friction of grease and oil-air are about 0.15 and 0.08, respectively. The engine efficiency is said to be 95%, so the heat generated by the stator and rotor are 30 W and 25 W, respectively. From the formula, the heat generated by the bearing is as shown in Table 2.

Table 2. Heat generated by bearings with different lubrication methods

Speed (rpm)	Heat (W)	
	Grease lubrication (bearings 1–4 and 2–3)	Air-oil lubrication (bearings 1–4 and 2–3)
7000	12.34	7.78
	9.08	6.05
9000	16.81	10.95
	12.63	8.72
11000	21.70	14.54
	16.59	11.82

The cooling water flow rate for the spindle unit is about 2 mm^3/s. The ambient temperature is assumed as 25 °C with a pressure of 0.1 MPa, the gravitational acceleration is 9.8 m/s^2. The lubricating oil used is type Tona 68 with the kinematic viscosity 68 mm^2/s at 40 °C. The material of spindle is carbon steel C45. Thus, the heat transfer simulation model was designed on the software with 355237 elements, the element sizes depending on the locations on the model are automatically calculated as 3 mm, 6 mm and 12 mm, respectively as shown in Fig. 4.

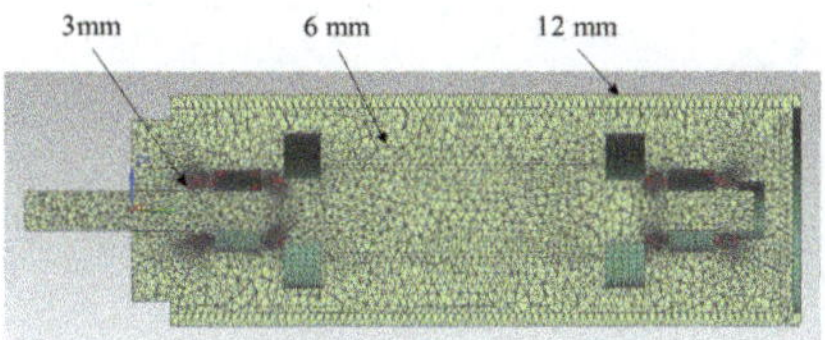

Fig. 4. The element mesh size of the simulation model

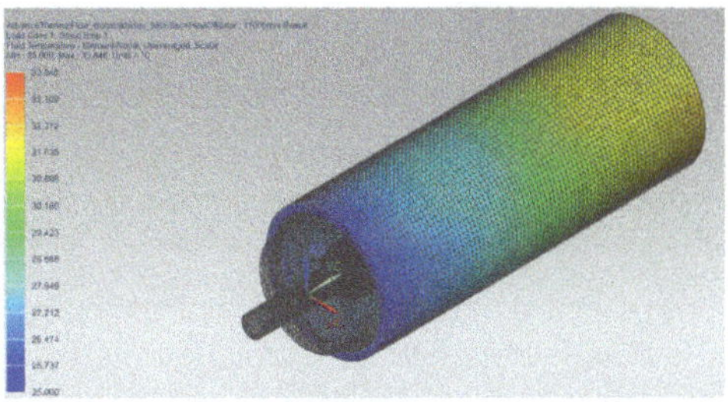

Fig. 5. Heat distribution of the spindle unit

4 Result and Discussion

By setting the spindle unit parameters and working conditions on the simulation model in accordance with the design parameters, the simulation results of the temperature field on the entire spindle of the machine tool and the bearings are depicted in Figs. 5 and 6. Indeed, Fig. 5 depicts the temperature distribution over the entire spindle when operated at a speed of 11000 rpm after a period of working time of 1 h. The highest temperature found at the spindle end is about 33.6 °C. The temperature of other areas is quite low that ensures the working conditions of the spindle unit according to design requirements.

In addition, the temperature distribution on the spindle bearings is also depicted in Fig. 6. Accordingly, a maximum temperature of about 45.3 °C appeared at a single roller bearing after a period of working time approximately 1 h at 11000 rpm. However, this temperature does not affect the machining process, especially the surface quality of the part to be machined on the machine. But it is also a factor that the user needs to consider and adjust the lubrication mode in case of need.

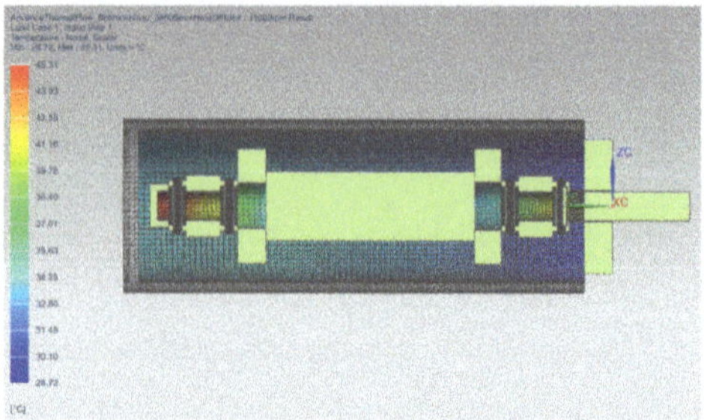

Fig. 6. Heat distribution of the spindle bearings lubricated and cooled by the oil-air mixture after 3600 s at a speed of 11000 rpm

The results of heat transfer and heat distribution on the simulation model help the user observe the heat distributed on the bearings. Accordingly, at the No. 1 and No. 4 bearing position, the temperature is much higher than the bearings No. 2 and No. 3. This is caused by the different rolling friction on the bearings. At No. 4 bearing position, the highest possible temperature is due to the fact that the bearing is in an area where the lubricating mixture is harder to reach than at any other location, and the ability to reduce heat is limited.

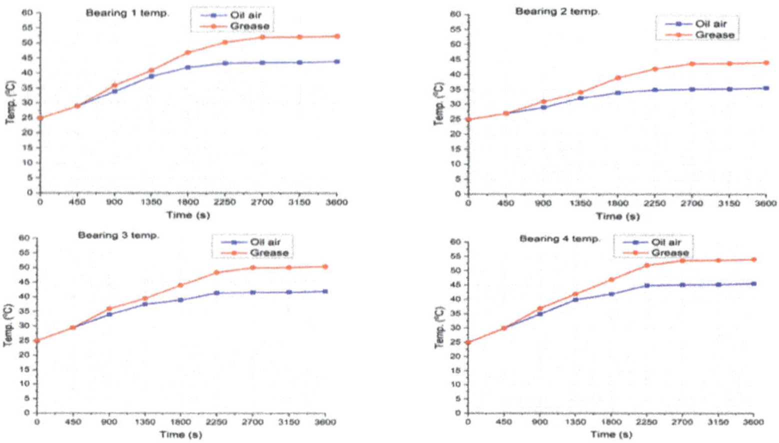

Fig. 7. Temperature changes at 4 bearings corresponding to 2 lubricating and cooling methods

Figure 7 depicts the temperature changes at bearings with different lubrication and cooling methods. Indeed, the bearing temperatures were the same in both methods when started working. After a period of working, the temperature in the grease-lubricated bearings increases faster than the air-oil-lubricating method. The temperature in the bearings is reduced by 8–10 °C. The bearing temperature increases rapidly during the period from 0 to 2000s and then gradually moves into the saturation phase.

Both methods show that the temperature will remain stable after about 2250 s from the start of working. This is also an advantage of the air-oil mixture lubrication and cooling method when compared to the grease lubrication and water-cooling method.

Figure 8 shows the temperature increase at No. 4 bearing at different spindle rotation speeds. At higher cutting speeds, the more heat is generated. However, the rate of temperature increase at different speeds is also different. Accordingly, the highest heating speed at 11000 rpm and lowest at 7000 rpm in different simulation speeds. From this result, the spindle speed will be selected in accordance with the requirements of machining productivity as well as ensuring the most suitable machining conditions when considering the influence of temperature at the bearing on the machining process.

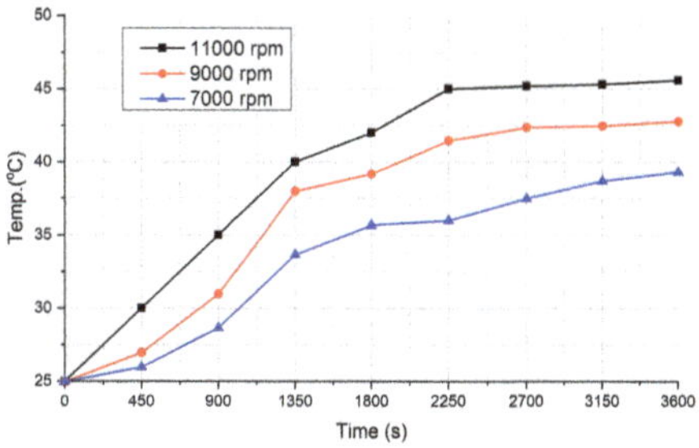

Fig. 8. Temperature rise in bearing No. 4 for 3600 s at different speeds when using oil-air mixture cooling and lubrication method

5 Conclusion

A lubrication and cooling monitoring system design for the spindle of machine tools was presented. In which, the layout diagram of the oil-air cooling and lubrication system, calculations and simulations of the system's operation were presented. In addition, a simulation model of the heat transfer process in the spindle unit was made. The basic parameters of the spindle unit including the diameter, the maximum rotation speed and the power of the spindle are 80 mm, 11000 rpm and 1.5 kW, respectively. With the system layout, calculations for the selection of the system's operating parameters have been performed. From the results obtained during simulation analysis, the oil-air cooling and lubrication method has been more effective than the grease lubrication and water-cooled method, reducing the cumbersome of the water-cooling system. Also, from the simulation results, the main heat-generating areas on outer rim of the bearing have been discovered, thus, it is a basis to arrange the installation of temperature sensors in the actual system.

References

1. Handawi, M., Elshwain, A.E., Mohd Yusof, N., Redzuan, N., Kurniawan, D.: Comparison between nitrogen- oil- mist and air-oil- mist condition when turning of hardened tool stainless steel. Appl. Mech. Mater. **660** (2015)
2. Wilson, F.H., Fails, C., Tire, T.G., Company, R.: United States Patent TO 1C6, pp. 1–6 (1977)

3. Yui, A.: Development of Oil-Air Mist Lubrication System in Ultra-High-Speed Grinding Spindle, pp. 95–101 (2002)
4. Sparham, M., Sarhan, A.A.D., Mardi, N.A., Hamdi, M.: Designing and manufacturing an automated lubrication control system in CNC machine tool guideways for more precise machining and less oil consumption. Int. J. Adv. Manuf. Technol, pp. 1081–1090 (2014)
5. Moon, J.H., Lee, H.D., Kim, S.I.: Lubrication characteristics analysis of an air oil lubrication system using an experimental design method. Int. J. Precision Eng. Manuf. **14**, 289–297 (2013)
6. Jauhari, K.: Oil lubrication on high-speed spindle bearing system: a review Proceeding of Tribology Industry (2015)
7. Stolarski, T.A.: Tribology in mashine design. Eff. Br. mindfulness Interv. acute pain Exp. An Exam. Individ. Differ, pp. 1689–1699 (2015)
8. Maeda, O., Cao, Y., Altintas, Y.: Expert spindle design system. Int. J. Mach. Tools Manuf, **56**, 537–548 (2005)

Time Varying Disturbance Observer Based on Sliding Mode Control for Active Magnetic Bearing System

Van-Nam Giap[1], Shyh-Chour Huang[1], Quang Dich Nguyen[2], and Xuan Thang Trinh[3(✉)]

[1] Department of Mechanical Engineering, National Kaohsiung University of Science and Technology, Kaohsiung 80778, Taiwan R.O.C.
[2] Institute for Control Engineering and Automation, Hanoi University of Science and Technology, No. 1, Dai Co Viet, Hai Ba Trung, Hanoi 100000, Vietnam
[3] Faculty of Mechanical Engineering, Hung Yen University of Technology and Education, Hungyen 160000, Vietnam
xttrinh@utehy.edu.vn

Abstract. This paper proposes a time varying disturbance observer for the active magnetic bearing system. The first derivative disturbance value was considered different to zero to show the effectiveness and originality of the proposed methodology. A novel disturbance observer form was constructed based on the basic nonlinear disturbance observer, which was combined with a proportional integral derivative sliding mode control for designing a control system. The disturbance estimation error was converged exponentially to the zero when the time goes to infinite. The output distance of the active magnetic bearing was closely converged to the desired trajectory. The Lyapunov law was used to analyses the convergent condition of the control system. Simulation results were used to prove the effectiveness of the proposed theory.

Keywords: Active magnetic bearing system · Sliding mode control · Time varying disturbance observer

1 Introduction

This paper reused the active magnetic bearing model of the published paper [1, 2]. Our paper [1] used the nonlinear disturbance observer to reconstruct the system states by considering the existence of the time varying disturbance value and the uncertainty value of the system with a low variation speed. The proposed method was good at tracking the constant disturbance value, where the first derivate disturbance value is equaled to zero. Chen and Lin [2] proposed the recurrent neural network to estimate the disturbance and uncertainty without effects of the time varying disturbance. According to the best author knowledge about the disturbance observer, there is no paper dealt with the time varying disturbance observer by straight consideration the first varying speed of the disturbance. Where the first derivative disturbance is not equaled to zero. Since this weak point, this paper proposed the new disturbance observer named time

B. T. Long et al. (Eds.): MMMS 2020, LNME, pp. 929–935, 2021.
https://doi.org/10.1007/978-3-030-69610-8_123

varying disturbance observer for active magnetic bearing system. Mathematical model of an active magnetic bearing system is rewritten as follows:

$$\ddot{x}(t) = [-\frac{c}{m} + a_1(t)]\dot{x}(t) + [\frac{k_{ap}}{m} + a_2(t)]x(t) + [\frac{k_{ai}}{m} + a_3(t)]i(t) - \frac{1}{m}f_d(t) \quad (1)$$

where m is rotor' mass; k_{ap}, k_{ai} are the amplified values of the position and current, respectively. $x(t)$, $\dot{x}(t)$, and $\ddot{x}(t)$ are the distance, velocity, and acceleration of the rotor respect to the electromagnet poles, respectively. The a_1(t), a_2(t) and a_3(t) are uncertainty coefficients, d(t) is the output disturbance value. The form of the system (1) can be modified as following equation.

$$\begin{bmatrix} \dot{x}(t) \\ \ddot{x}(t) \end{bmatrix} = \begin{bmatrix} 0 & 1 \\ -\frac{c}{m} & \frac{k_{ap}}{m} \end{bmatrix} \begin{bmatrix} x(t) \\ \dot{x}(t) \end{bmatrix} + \begin{bmatrix} 0 \\ \frac{k_{ai}}{m} \end{bmatrix} i(t) + \begin{bmatrix} 0 \\ -\frac{1}{m} \end{bmatrix} f_d(t) \quad (2)$$

$$\text{where } d(t) = [-ma_1(t)x(t) - ma_2(t)\dot{x}(t) - ma_3(t)i(t) + f_d(t)] \quad (3)$$

is called the lumped disturbance and uncertainty value. Basically, these values are well known somehow bounded practically.

Refer the equations below:

$$g_1(x) = \begin{bmatrix} 0 & 1 \\ -\frac{c}{m} & \frac{k_{ap}}{m} \end{bmatrix} \begin{bmatrix} x(t) \\ \dot{x}(t) \end{bmatrix}, g_2(x)u = \begin{bmatrix} 0 \\ \frac{k_{ai}}{m} \end{bmatrix} i(t), \quad (4)$$

$$\text{and } \begin{bmatrix} 0 \\ -\frac{1}{m} \end{bmatrix} d(t) = g_3(x)d. \quad (5)$$

Assuming the lumped uncertainty and disturbance L(t) <ε. The upper boundary value is positively defined. This assumption is reality in the practical system and only for a sliding mode control designation. The sliding mode surface was chosen as following equation.

$$s(t) = \dot{e}(t) + k_1 e(t) + k_2 \int_0^t e(\tau)d(\tau) \quad (6)$$

where $e = x_d - x_m$ with the x_d is the desired distance trajectory output, and x_m is measured distance trajectory. The k_1 and k_2 are positive and satisfied $k_1^2 < 4k_2$. The proposed control algorithm is shown in the following section.

2 Proposed Approach

The control value of an active magnetic bearing system was archived by solving the Eq. (6) with the switching control value is $u_{sw} = -\chi sat(s/\Delta)$. The control input value is represented as following equation.

$$i(t) = -\frac{m}{k_{ai}}[-\frac{c}{m}\dot{x}(t) + \frac{k_{ap}}{m}x(t) - \frac{1}{m}L(t) - \ddot{x}_d + k_1\dot{e}(t) + k_2 e(t)] - \chi sat(\frac{s}{\Delta}) \tag{7}$$

where $\chi > L$ is the switching control gain. The Δ is the layer thickness of switching control.

This paper used the concept of nonlinear disturbance observer in [3] to archive the goals. The basic nonlinear disturbance observer is represented as following equation:

$$\begin{cases} \dot{p} = -L_d(x)g_3(x)p - L_d(x)[g_1(x) + g_2(x)u + g_3(x)q(x)] \\ \hat{d} = p + q(x) \\ q(x) = L_d(x)x \end{cases} \tag{8}$$

The derivative of disturbance error is described as

$$\dot{d} - \hat{d} = \dot{d} - p - q(x) \tag{9}$$

Substituting the Eq. (8) into (9) leads to

$$\dot{e}_d = \dot{d} - L_d g_3 e_d \tag{10}$$

where the $\hat{d}$ is the estimated disturbance of the basic nonlinear disturbance value. According to author knowledge, many published papers assumed that the first derivative disturbance $\dot{d}(t) = 0$. In fact, the first derivative of disturbance value is different to zero, and unknown how these value varying. Since this weak point, a new time varying disturbance observer was constructed as following form:

$$\dot{\hat{d}}_F = (\alpha + 1)[\dot{p} + \dot{q}(x)] + \beta[\ddot{p} + \ddot{q}(x)] \tag{11}$$

where $\hat{d}_F$ is the new estimated disturbance, which is used to compensate the complicated disturbance. By applying the first derivative disturbance to subtract the both sides of Eq. (11) yields

$$\dot{d} - \dot{\hat{d}}_F = \dot{d} - (\alpha + 1)[L_d g_3(d - p - q(x))] - \beta[\ddot{p} + \ddot{q}(x)] \tag{12}$$

or

$$\dot{e}_d = \dot{d} - (\alpha + 1)[L_d g_3 e_d] - \beta[\ddot{p} + \ddot{q}(x)] \tag{13}$$

The second order of disturbance error value is represented as following equation:

$$\ddot{e}_d = -\frac{\alpha}{\beta}\dot{e}_d \tag{14}$$

Solving the Eq. (11) with the zero initial condition leads to

$$e_d(t) = e^{-\frac{\alpha}{\beta}t} \tag{15}$$

Stability analysis: The Lyapunov condition was selected as following:

$$V(t) = \frac{1}{2}s^T(t)s(t) \tag{16}$$

Taking derivative Eq. (16) yields

$$\dot{V}(t) = s(t)\dot{s}(t) \tag{17}$$

or

$$\begin{aligned}\dot{V}(t) =& s(t)[\ddot{e}(t) + k_1\dot{e}(t) + k_2e(t)]\\ =& s(t)[\ddot{x}_m(t) - \ddot{x}_d(t) + k_1\dot{e}(t) + k_2e(t)]\\ =& s(t)[-\frac{c}{m}\dot{x}(t) + \frac{k_{ap}}{m}x(t) + \frac{k_{ai}}{m}i(t) - \frac{1}{m}[d(t) - d(t)] + -\ddot{x}_d(t) + k_1\dot{e}(t) + k_2e(t)]\\ =& s(t)[-\frac{c}{m}\dot{x}(t) + \frac{k_{ap}}{m}x(t) + \frac{k_{ai}}{m}\frac{m}{k_{ai}}[\frac{c}{m}\dot{x}(t) - \frac{k_{ap}}{m}x(t) + \ddot{x}_d - k_1\dot{e}(t) - k_2e(t)]\\ &-\chi sat(\frac{s}{\Delta}) - \ddot{x}_d + k_1\dot{e}(t) + k_2e(t) - \frac{1}{m}[d(t) - d(t)]\end{aligned} \tag{18}$$

The Lyapunov condition in Eq. (18) can be fulfilled as following equation:

$$\dot{V}(t) \leq -|s(t)|\frac{\chi}{\Delta} - \frac{1}{m}\left|d(t) - \hat{d}(t)\right| \tag{19}$$

$$\text{or } \dot{V}(t) \leq -|s(t)|\frac{\chi}{\Delta} - \frac{1}{m}\left|e^{-\frac{\alpha}{\beta}t}\right| \tag{20}$$

$\dot{V}(t) < 0$ when the time goes to infinity.

The designed control method is represented as following Fig. 1. The system parameter are as following; rotor' mass m = 2.565 kg, magnet current stiffness k_{ai} = 40 N/A, position stiffness k_{ap} = 25200 N/m, x (0) = 0.001 mm, damping constant N. s/m, and thrust disk mass m = 0.38 kg. The working condition is a_1(t) = 0.125cos (0.125πt), a_2(t) = 0.1sin(0.125πt), a_3(t) = 0, and the outside disturbance is d(t) = 1.5 sin (0.15πt) + 2.5 cos (0.125πt). The sliding mode control parameter are k_1 = 5, k_2 = 30, Δ = 0.01, and χ = 0.1. The basic nonlinear disturbance observer gain was selected L_d = [−5, −10], the observer gain is negative to fulfil $L_d g_3 > 0$. The time varying disturbance observer gain are α = 50, and β = 0.1.

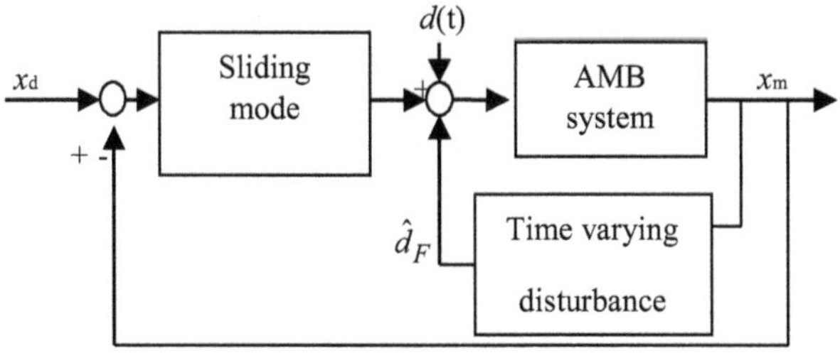

Fig. 1. Time varying disturbance observer based on sliding mode control for active magnetic bearing system.

3 Illustrative Example

This section presents three cases of study. The disturbance tested on three controller to do a self-comparison.

Case 1: Proportional-integral-derivative sliding mode control for active magnetic bearing system.

The sliding mode control with the control input was shown in the Eq. (7). In this proposed method, the disturbance and uncertainty effects were ignored clearly. The controller is simply designed by using the proportional-integral-derivative sliding mode surface to obtain the control input signal. The outcome of this proposed method is given in the below Fig. 2.

Case 2: Basic nonlinear disturbance observer based on sliding mode for active magnetic bearing system.

The sliding mode control and the feedback of basic nonlinear disturbance observer with the control input is as following equation.

$$\begin{aligned} i(t) &= -\frac{m}{k_{ai}}[-\frac{c}{m}\dot{x}(t) + \frac{k_{ap}}{m}x(t) - \frac{1}{m}[d(t) - \hat{d}(t)] \\ &= -\ddot{x}_d + k_1\dot{e}(t) + k_2 e(t)] - \chi sat(\frac{s}{\Delta}) \end{aligned} \tag{21}$$

The control input signal are included the control current in the first case of study and feedback of estimated disturbance value. In this section the basic estimated disturbance value was obtain with the form of Eq. (8), where the first derivative disturbance was considered goes to zero when time goes to infinity. In fact, the first derivative disturbance and uncertainty values are different to zero. In order to get better response for active magnetic bearing system, the time varying disturbance observer can be shown as following section.

Case 3: The time varying disturbance observer based on sliding mode control for active magnetic bearing system.

This section presents the time varying disturbance observer with the form of Eq. (11). The control input signal is as following:

$$\begin{aligned} i(t) &= -\frac{m}{k_{ai}}[-\frac{c}{m}\dot{x}(t) + \frac{k_{ap}}{m}x(t) - \frac{1}{m}[d(t) - \hat{d}_F(t)] \\ &= -\ddot{x}_d + k_1\dot{e}(t) + k_2 e(t)] - \chi sat(\frac{s}{\Delta}) \end{aligned} \tag{22}$$

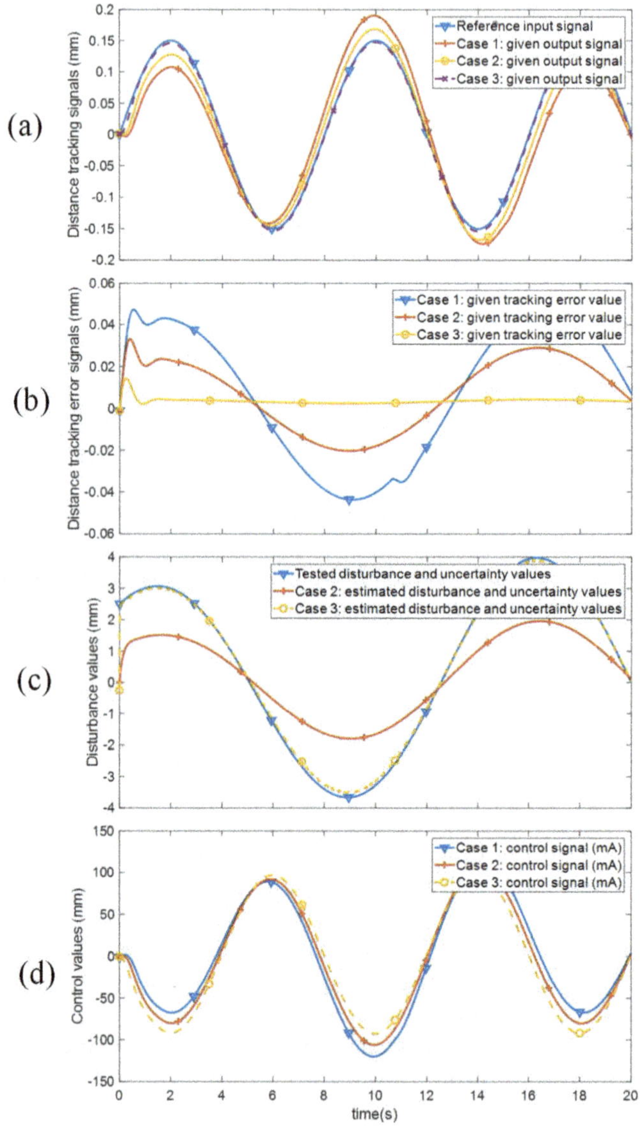

Fig. 2. Given output signals of the small disturbance case. (a) Distance tracking response signals, (b) distance tracking error signals, (c) disturbance response signals, and (d) control input signals.

The parameters of controller in three cases are same, which means the first case was subjected without any disturbance feedback, second case was subjected with basic nonlinear disturbance feedback, and final case used the time varying disturbance observer feedback to replace the basic nonlinear disturbance observer. The given results are displays in the following Fig. 2.

This paper concentrated to reduce the effects of the time varying disturbance value. The given results of three cases was shown in the Fig. 2.

The case 1: Distance tracking error value in steady state are ranged in [−43.720, 54.505] (μm)

The case 2: Distance tracking error value in steady state are ranged in [−20.312, 29.012] (μm). The disturbance was not reached to the tested disturbance.

The case 3: Distance tracking error value in steady state are ranged in [2.149, 4.315] (μm). The estimated disturbance approached to the tested disturbance at 0.04 s.

In fact, the second case can fulfil the tested disturbance with smaller disturbance error value. In comparison of the second case and third case based on the same observer gains, the proposed method named time varying disturbance observer was perfectly archive the disturbance and uncertainty rejection.

4 Conclusion

This paper revealed a time varying disturbance observer based on the sliding mode control for active magnetic bearing system. The disturbance and uncertainty values are rejected completely with a highly exponential convergence speed. The proposed method does not need any information of disturbance varying speed. The condition of stability was proved by using the Lyapunov law. The overshoot of the given output tracking is small, steady state was ranged in narrow and small scale.

References

1. Giap, N., Huang, C.: Effectiveness of fuzzy sliding mode control boundary layer based on uncertainty and disturbance compensator on suspension active magnetic bearing system. Measure. Control **53**, 934–942 (2020)
2. Chen, Y., Lin, J.: Robust nonsingular terminal sliding-mode control for nonlinear magnetic bearing system. IEEE Trans. Control Syst. Tech. **19**, 636–643 (2010)
3. Chen, H.: Disturbance observer based control for nonlinear systems. IEEE/ASME Trans. Mech. **9**, 706–710 (2004)

Non-contact 3D Measurement of Freeform Reflective Surface

Thi Kim Cuc Nguyen, Phuong-Thao Nguyen, Duc An Pham, Toan Thang Vu, and Xuan Binh Cao(✉)

School of Mechanical Engineering, Hanoi University of Science and Technology, Hanoi, Vietnam
binh.caoxuan@hust.edu.vn

Abstract. Non-contact 3D shape measurement has been widely studied and applied to many fields of science and engineering owing to high speed, high accuracy, and simple implementation. However, the measurement system still poses improper acquisition of optical signal when surfaces with high reflectivity are scanned. This paper proposes a new and effective approach to solve the problem of shiny mechanical surface measurement. The appropriate exposure time for the surface measurement is indicated through image intensity histogram analysis. The appropriate exposure times in each region of the histogram are determined. Then, point clouds are merged at appropriate exposure times to obtain a good quality point cloud, avoiding the saturation region on the CCD. Profile parts of two different aluminum surfaces are measured fast and accurately using single exposure time and proposed solution. Experimental results prove that the proposed technique can inspect the 3D surface of the mechanical parts with high surface reflectivity.

Keywords: 3D shape measurement · Fringe projection · Shiny surface · High dynamic range

1 Introduction

Currently, three-dimensional (3D) measurement systems with structured light have been being studied, developed, and widely applied. In industry, this measurement method is applied to measure mechanical parts thanks to its advantages of full-filed inspection, high speed, high resolution, and easy implementation [1–3]. However, measurement system still had difficulty measuring the shiny objects or objects with a large range of reflectivity variation across the surface [4]. It is difficult to achieve high quality measurement across the whole surface of the CNC machining parts which have smooth and high specular surface [5].

Researchers have studied several methods to decrease the influence of highly specular surfaces. Most studies used methods such as high dynamic range scanning, adjusting the intensity of light with adjustable input gray, and using polarizing filters. However, these methods cannot measure highly reflective surfaces. In addition, the brightest unsaturated intensities at each selected pixel, the signal to noise ratio (SNR) is small for low reflectivity regions, is high for the surface with a large range of

B. T. Long et al. (Eds.): MMMS 2020, LNME, pp. 936–942, 2021.
https://doi.org/10.1007/978-3-030-69610-8_124

reflectivity variation. Moreover, the quality of measurement is hardly reliable. During measurement of the object position, it is very hard to ensure features like the table of calibration; as a result, the coordinates after mapping matrix may be inaccurate. Other method [6] presents a new study of structural light to measure shiny objects, where fringe boundaries instead of the raw image intensity are encoded in the projected image. However, if the saturation area in a photo is too large to cover even a bright pattern, information data in these regions cannot be determined.

Therefore, in this manuscript, a solution reduces the influence of the specular surface in non-contact 3D shape measurement is proposed. It utilizes merged point clouds acquired at the appropriate exposure times to achieve the complete image point cloud with conserved surface information. The method does not require the change in exposure time. Based on the total pixel value in image intensity histogram, it is possible to determine the appropriate exposure time. The initial exposure time in the histogram is determined by the maximum percentage of pixels with the intensities of gray level ranging from 50 to 200. To verify the performance of this method, a 3D shape measurement system that uses a phase-shifting combined Gray code (PSCG) algorithm is developed based on previous literature. The method allows a structured light system to successfully measure the 3D surface of mechanical objects with unknown reflectivity.

2 Method of Reducing the Influence of the Shiny Surface

To ensure that the pattern image is projected on the surface of the measuring object, the pixel intensity value $I_c(\mathrm{u},\mathrm{v})$:

$$I_c(u,v) = \mathrm{t}.\xi.R.I_p(u,v) \tag{1}$$

Where t is the exposure time of camera, ξ is the sensitivity of the camera, $I_p(u,v)$ is the intensity of light from the projector and R is surface reflectivity.

According to the previous literatures, the upper threshold of camera is 255 if the camera is quantized with 8bit. Thus, the appropriate image intensity should be $I_c^0(u,v) = 254 < 255$ which is not saturated in the CCD. The optimal exposure time can be achieved when each element in Eq. (1) is determined:

$$\mathrm{t}^{opt} = \frac{I_c^0(u,v)}{I_0}.t_0 \tag{2}$$

Equation (2) shows that each pixel corresponds to an appropriate exposure time t^{opt} and it may be obtained when I_0 and t_0 be determined. The values I_0 and t_0 are obtained through analysis of image intensity distribution on the CCD that is presented in the following. The exposure time range of a specific camera is written as $t_c = [t_{cmin}, t_{cmax}]$. The change in surface reflectivity can be determined by varying the intensity obtained from the CCD. When the distribution of intensity is obtained from CCD through the histogram, this can be used to determine the surface reflectivity and predict the appropriate exposure time for specific measurement surface.

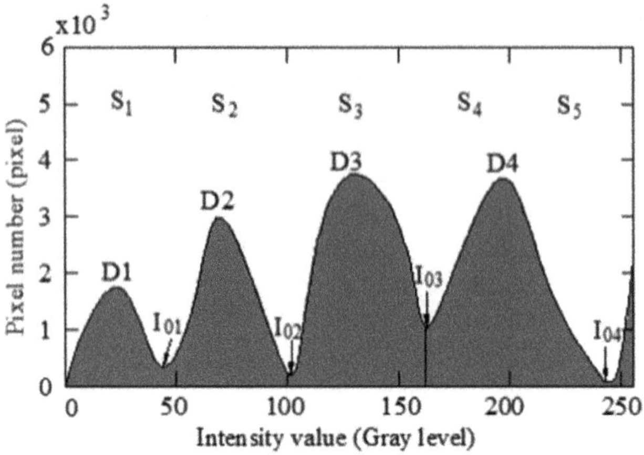

Fig. 1. Histogram of intensity

Figure 1 indicates the intensity distribution of image on CCD according to Gray levels from 0 to 255. The vertical axis shows the number of image points which have same Gray levels. Gray level axis is divided into 5 distinct regions S_i ***(i = 1, 2, 3, 4, 5)*** as S_1 = [0, 50], S_2 = [50, 100], S_3 = [100, 150], S_4 = [150, 200], and S_5 = [200, 255] respectively. The function $p(S_i)$ is a ratio of the total number of image points with Gray level I in region S_i as the following:

$$p(S_i) = \frac{n_I}{n}.100\% \tag{3}$$

For a surface with high contrast, avoiding effect of surface reflectance with appropriate exposure times, the top of histogram lies on the regions S_2, S_3 and S_4, meaning:

$$\sum p(S_{2,3,4}) = (p(S_2) + p(S_3) + p(S_4)) \rightarrow max \tag{4}$$

For specific measurement surface, the reference images are taken with different exposure times of camera in the range $t_c = [t_{cmin}, t_{cmax}]$. The exposure time of which the histogram of image satisfies the condition (4) is chosen as t_o in Eq. (3). The optimal exposure time t^{opt} of each region in the histogram is determine based on the Eq. (2). These charts have been smoothed by removing high frequency noise using low pass filter algorithm. The image points are constructed corresponding to each optimal exposure time and then combined to create a complete image.

3 Setup the Experiment

The feasibility of the proposed method with a specific experimental system is shown in Fig. 2. The experimental system includes a digital camera (DFK 41BU02) with 1280 × 960 resolution, a video projector (InFocus N104) with 1024 × 768 pixels. These devices are arranged as a measurement head connected with a computer equipped by Ram 8G, Core i5–4460, speed 3.20 GH, and separated VGA card. The measurement head is fixed on the rotary and movable table. The table guarantees the scanning area on the measurement object maximized. The number of Gray code patterns in horizontal and vertical directions are integers. Thus, the number of recognizable image points in a stripe period in horizontal and vertical directions are chosen as $T_u = 32$ and $T_v = 32$ and 5 binary Gray code patterns are selected. By overlapping the focal planes of the camera and projector, the whole scanning area is achieved to be 250 × 180 mm in projection distance L = 500 mm and the distance from the projector to the camera is determined to be 130 mm. The camera has exposure time range t_c = [5 ms, 250 ms]. The measurement region is put in a black box to minimize the impact of ambient light as well as reflection from surrounding surfaces.

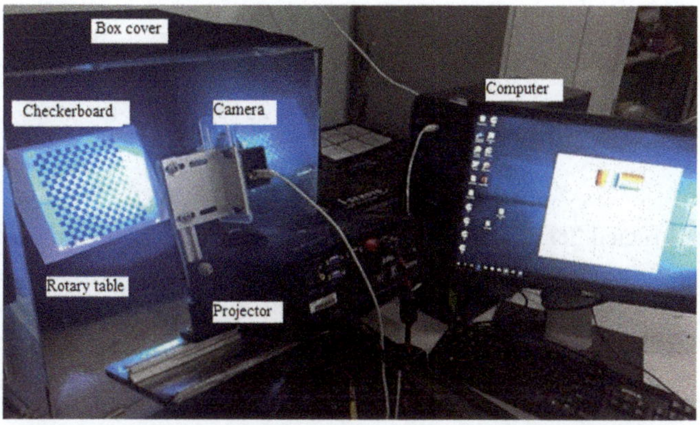

Fig. 2. Setup the experimental system.

4 Experiment Result and Discussion

Aluminum is one of the materials with a surface reflection coefficient of almost 1 which is higher than steel. This material is commonly used in CNC processing. To evaluate the feasibility of technique, 3D measurement experiment is executed on two aluminum workpieces. One aluminum workpiece has complex profile, and another aluminum workpiece has stepped surface profile.

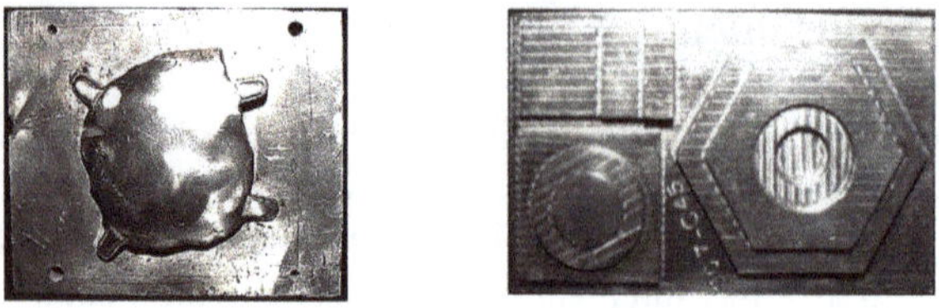

(a) (b)

Fig. 3. Images of tested samples: (a) an aluminum workpiece 1, (b) an aluminum workpiece 2

In the first experiment, with an aluminum workpiece 1 in Fig. 3 (a), the Gray code 2° with light intensity $I_p(255)$ is projected by the projector mapping onto object. The images are then obtained by the camera with different preliminary exposure times t_c. The histogram is constructed with each exposure time and the sum $\sum p(S_{2,3,4})$ is computed. In Table 1, with exposure time $t_c = 16$ ms, the value $\sum p(S_{2,3,4})$ is the largest value.

Table 1. Preliminary exposure timetable

Exposure time	t_{sb}(ms)				
	50	25	16	12.5	10
$\sum p(S_{2,3,4})$	10.66	29.52	71.26	56.29	19.45

The reconstructed 3D results after single exposure time $t_0 = 16$ ms show in Fig. 4 (a) with 6062 points. The point cloud has large areas of holes due to the saturation of pixel in CCD camera. The point cloud obtained after the point clouds with three appropriate exposure times are merged in Fig. 4(b) shows that the pixels show high density of points. The total number of pixels representing the 3D surface is 13135 points.

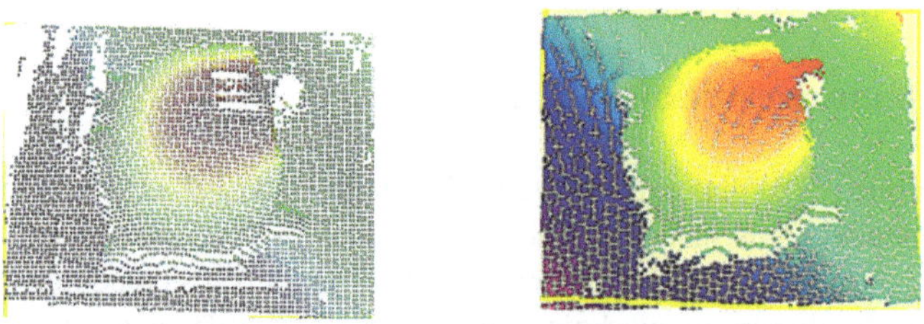

(a) (b)

Fig. 4. 3D point cloud of an aluminum workpiece 1 in single exposure (a), in merged point clouds at 3 appropriate exposures (b).

In the second experiment, an aluminum workpiece 2 with step height (Fig. 3b). The Gray code pattern 2° with light intensity $I_p(255)$ is projected by the projector onto objects. Then images are obtained by the camera with different exposure time t_c: 50 ms, 25 ms, 12,5 ms, 10 ms, 5 ms. The histogram is constructed with each exposure time and calculates the value $\sum p(S_{2,3,4})$ according to Eq. (1). In Table 2, with exposure time $t_c = 12.5$ ms, the value $\sum p(S_{2,3,4})$ is largest. So $t_c = 12.5$ ms is selected is initial exposure time t_o for calculating histogram of I_0.

Table 2. Preliminary exposure time calculated table

Exposure time	t_c(ms)				
	50	25	12.5	10	5
$\sum p(S_{2,3,4})$	9.35	39.8	69.71	65.96	39.32

The experiment for evaluating the effectiveness of method through obtained integrated image point cloud and precision examination based on designated data demonstrate the precision and feasibility of the proposed 3D measurement technique. Using a histogram to determine the appropriate exposure time is a simple but highly effective method. It helps to avoid randomly selection of the initial exposure time.

5 Conclusion

In this paper, the measurement method using phase-shift combining Gray code and process surface measuring high reflectivity have been developed without knowing in advance the characteristics and shape of the object surface measurement. Through the histogram of the raw image, the measurement can determine how much exposure time is appropriate for the image to have the full range of grayscale from 0 to 255. The complete point cloud is obtained in merged point clouds with a sharp surface and conserved information. 3D surfaces of target objects with different morphologies have different reflectivity.

The surfaces of the components are different in shape, the surface reflectivity is different. Experiment results show that the surfaces which have high reflectivity should chose small the exposure time. The presented result demonstrates efficiency of proposed technique for inspection of full-field reflectance surfaces without auxiliary equipment. Surface images with different roughness are constructed exactly by the proposed method which does not requires additional optical systems and hardware controls. A plethora of new techniques utilized for non-contact 3D shape measurement on the basis of the proposed functionality with high precision, multi reflection adaptation, and simplicity can be created as an ideal way to diagnose the 3D topology of unreachable and low reflectance target objects. Future research is concentrating on enhancing resolution and accuracy of point cloud in reducing reflection.

Acknowledgment. This research is funded by Hanoi University of Science and Technology (HUST) under grant number T2020-SAHEP-014.

References

1. Chen, F., Brown, G., Song, M.: Overview of three-dimensional shape measurement using optical methods. Opical Eng. **39**(1), 10–22 (2000)
2. Gaoxu, W., Yanxue, W., Li, L., Liu, F.: High-resolution few-pattern method for 3D optical measurement. Opt. Lett. **44**, 3602–3605 (2019)
3. Lv, S., Sun, Q., Zhang, Y., Jiang, Y., Yang, J., Liu, J., Wang, J.: Projector distortion correction in 3D shape measurement using a structured-light system by deep neural networks. Opt. Lett. **45**, 204–207 (2020)
4. Imran, R., Rogers, T.L.: Resolving Reflection and Resolution in 3D Imaging of Fresh Bone. J. Forensic Sci. **65**(1), 200–208 (2020)
5. Jiang, C., Bell, T., Zhang, S.: High dynamic range real-time 3D shape measurement. Opt. Express **24**(7), 7337–7346 (2016)
6. Song, Z., Chung, R., Zhang, X.T.: An accurate and robust strip-edge-based structured light means for shiny surface micromeasurement in 3-D. IEEE Trans. Ind. Electron. **60**(3), 1023–1032 (2013)

Analyzing Dynamic Characteristics of Working Rotor Pair in Vacuum Pumps

Nguyen Thanh Tung[1,2], Van-The Tran[1(✉)], and Do Anh Tuan[1]

[1] Department of Mechanical Engineering, Hung Yen University of Technology and Education, 39 Road, Hung Yen City, Vietnam
vanct4.hut@gmail.com
[2] Ha Noi University of Mining and Geology, Ha Noi City, Vietnam

Abstract. The vacuum pumps are usually applied in many different life and industrial fields, such as food, medicine and biotechnology. They are used to convey various materials, from low viscosity fluids such as water to very high viscosity fluids such as oil, and even solids. The dynamic characteristics of a meshing rotor pair are greatly affect the working process of the vacuum pumps. In this paper, a method is presented to determine the dynamic characteristics of a meshing rotor pair in vacuum pump working process. The mathematical model and 3D CAD model of rotor pair are also constructed to perform analysis dynamic characteristics of rotor pair. The dynamic load of rotor is defined by the variable forces and variable stiffness of the system by using Adams software. The influence of rotational rotor speed on the dynamic load has been analyzed. Subsequently, the numerical results are presented to verify the merit of constructed model.

Keywords: Dynamic characteristic · Rotor pair · Vacuum pump

1 Introduction

The roots pumps are widely used in the life and industrial applications to transport with various materials including low and high viscosity fluids. The rotor's tooth profile is an important factor for improving performance of the vacuum pump. In this paper, a new tooth profile is designed based on the combination of the different traditional curves such as circular arc, extended epicycloid, involute curve.

The researches for improving rotor's tooth profile of vacuum pump have published in recently years. A tooth profile of a rotor is designed by using a single circular arc [1, 2]. A new tooth profile for rotor is proposed by combining circular arcs and a conjugated epicycloid [3]. A tooth profiles of rotor by comprising four circular arcs that can improve area efficiency of a vacuum pump patented by Fang [4]. Subsequently, Fong and Wang [5] have developed the tooth profile of the rotor that is combined by five circular arcs. For increasing pump efficiency, Niimura et al. [6] patented a tooth profile consisted a circular arc and an involute. For improving pump performance, Hwang and Hsieh [7, 8] have designed a tooth profiles using an extended cycloid curve with a variable trochoid ratio. And Kang et al. [9, 10] developed a new lobe pump rotor

B. T. Long et al. (Eds.): MMMS 2020, LNME, pp. 943–951, 2021.
https://doi.org/10.1007/978-3-030-69610-8_125

profile used circular and epicycloid. A profile design of rotor consisting of a hypocycloid and epicycloid is presented by Chiu [11].

In this paper, based on the previous study [12], the tooth profile of CIC Roots rotors is composed of three curves which comprise orderly one circular arc on the rotor tooth tip, an involute and one conjugated circular arc in the root of rotor tooth. The tooth profile of EE Roots rotors is constructed by an extended epicycloid. 3D rotor models for both EE Roots rotors and CIC Roots rotors are constructed. Numeral example is presented to illustrate and verify that the CIC Roots rotors have more advances than the EE Roots rotors.

2 Mathematical Models for Generating Tooth Profiles of the Roots Rotors

2.1 CIC Tooth Profile

The CIC tooth profile is comprised by a circular arc (addendum portion A_1B_1), an involute curve (pitch circle portion B_1C_1) and a conjugated circular arc (dedendum portion C_1D_1), as shown in Fig. 1. The position locus tooth profiles of Roots rotor are represented in the coordinate system $S_1(x_1, y_1, z_1)$ as follows:

(1) The circular arc A_1B_1:

$$\mathbf{r}_{III_1}^{(A_1B_1)}(\ell) = \begin{cases} x_{III_1}^{(A_1B_1)}(\ell) = r_a \cos\ell \\ y_{III_1}^{(A_1B_1)}(\ell) = r_a \sin\ell \end{cases}, \quad (0 \le \ell \le \ell_1) \tag{1}$$

(2) The involute curve B_1C_1:

$$\mathbf{r}_{II_1}^{(B_1C_1)}(\ell) = \begin{cases} x_{II_1}^{(B_1C_1)}(\ell) = \left(r_p + \ell\cos\alpha_p\right)\cos\beta - \ell\cos\alpha_p\cot\alpha_p\sin\beta \\ y_{II_1}^{(B_1C_1)}(\ell) = \ell\cos\left(\beta - \alpha_p\right)\cot\alpha_p + r_p\sin\beta \end{cases}, \quad (-u \le \ell \le u) \tag{2}$$

with $\beta = \left(\frac{\pi}{2N} - \frac{\ell}{r_p \sin\alpha_p}\right)$

(3) The dedendum circular arc C_1D_1 can be obtained by rotating E_1F_1 (In Fig. 2) an angle of $\vartheta = \frac{(N-1)\pi}{N}$ counterclockwise about z axis:

$$\mathbf{r}_{II_1}^{(C_1D_1)}(\ell) = \begin{cases} x_{II_1}^{(C_1D_1)}(\ell) = 2r_p\cos(\vartheta - \ell) + r_a\cos(\vartheta + 3\ell) \\ y_{II_1}^{(C_1D_1)}(\ell) = 2r_p\sin(\vartheta - \ell) + r_a\sin(\vartheta + 3\ell) \end{cases}. \tag{3}$$

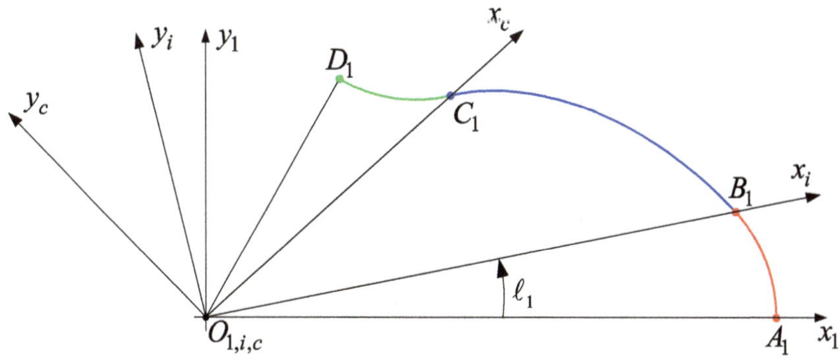

Fig. 1. The tooth profile of the CIC Roots rotors.

In Fig. 2(a), the meshing coordinate system for two Roots rotor is presented, the coordinate systems $S_1(x_1, y_1, z_1)$, $S_2(x_2, y_2, z_2)$, and $S_f(x_f, y_f, z_f)$ are rigidly connected to the driving rotor, the driven rotor, and the frame, respectively. By applying the homogeneous coordinate transformation matrix equation from $S_1(x_1, y_1, z_1)$ to $S_2(x_2, y_2, z_2)$, the locus for CIC tooth profile of the Rotor in the coordinate system $S_2(x_2, y_2, z_2)$ is the same as in Eqs. (4) and (5) and equation of meshing between the driving rotor and driven rotor is presented as in Eq. (6) as follows:

$$\mathbf{r}_{I_2}(\tau, \phi) = \mathbf{M}_{21}(\phi)\mathbf{r}_{I_1}(\tau), \tag{4}$$

$$\text{and } \mathbf{n}_{I_2}(\tau, \phi) = \mathbf{L}_{21}(\phi)\mathbf{n}_{I_1}(\tau), \tag{5}$$

where $\mathbf{L}_{21}(\phi)$ is the upper-left (2×2) sub-matrix of the (3×3) homogeneous coordinate transformation matrix $\mathbf{M}_{21}(\phi)$.

The equation of meshing between the two rotors can be obtained by using Wilson equation

$$(r_p \cos\phi - x_{I_1}(\tau, \phi))x'_{I_1}(\tau, \phi) + (r_p \sin\phi - y_{I_1}(\tau, \phi))y'_{I_1}(\tau, \phi) = 0 \tag{6}$$

The locus for CIC tooth profile of Roots rotor is determined by using Eqs. (1), (2), (3), (4) and (5), simultaneously. The 3D-model of the CIC Roots rotors and 3D-model of two CIC Roots rotor meshing pair are presented in Fig. 2 (b).

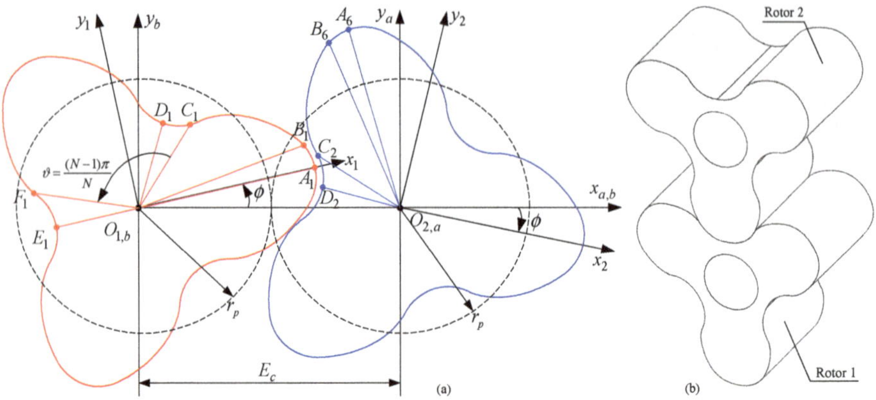

Fig. 2. CIC rotors: a) conjugated coordinate systems and b) 3D meshing rotor pair.

2.2 EE Tooth Profile

The EE tooth profile of rotor is comprised by the epicycloid (A_3B_3) and hypocycloid curves, as shown in Fig. 3. According to Fig. 3(a), the locus tooth profiles of Roots rotor are represented in the coordinate system $S_1(x_1, y_1, z_1)$ as follows:

(1) The extended epicycloid curve A_3B_3:

$$\mathbf{r}_{II_1}^{(A_3B_3)}(\gamma) = \begin{cases} x_{II_1}^{(A_3B_3)}(\gamma) = (r_p + r_e)\cos\gamma + r_e \sin\left(\dfrac{(r_p + r_e)\gamma}{r_e}\right) \\ y_{II_1}^{(A_3B_3)}(\gamma) = (r_p + r_e)\sin\gamma + r_e \cos\left(\dfrac{(r_p + r_e)\gamma}{r_e}\right) \end{cases}, \quad (0 \le \gamma \le \gamma_1) \tag{7}$$

(2) The hypocycloid enveloped by the epicycloid B_3C_3:

$$\mathbf{r}_{II_1}^{(B_3C_3)}(\gamma) = \begin{cases} x_{II_1}^{(B_3C_3)}(\gamma) = (r_p - r_e)\cos(\dfrac{\pi}{N} - \gamma - \gamma_1) - r_e \cos(\dfrac{\pi}{N} - \gamma + \dfrac{r_p}{r_e}\tau - \gamma_1) \\ y_{II_1}^{(B_3C_3)}(\gamma) = (r_p - r_e)\sin(\dfrac{\pi}{N} - \gamma - \gamma_1) - r_e \sin(\dfrac{\pi}{N} - \gamma + \dfrac{r_p}{r_e}\tau - \gamma_1) \end{cases}, \quad (\gamma_1 \le \gamma \le \gamma_2) \tag{8}$$

By applying the homogeneous coordinate transformation matrix equation from $S_1(x_1, y_1, z_1)$ to $S_2(x_2, y_2, z_2)$, shown in Fig. 3(a). The locus for EE tooth profile of the Rotor in the coordinate system $S_2(x_2, y_2, z_2)$ is the same as in Eqs. (4) and (5) and equation of meshing between the driving rotor and driven rotor is presented as in Eq. (6). The 3D-model of the EE Roots rotors and 3D-model of two EE Roots rotor meshing pair are presented in Fig. 3(b).

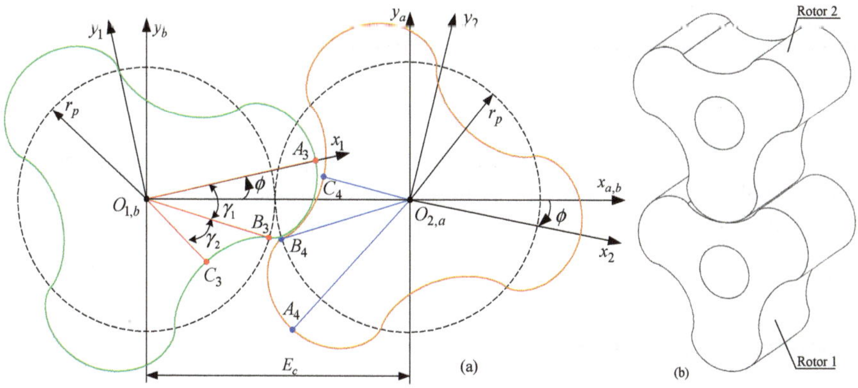

Fig. 3. EE rotors: a) conjugated coordinate systems and b) 3D meshing rotor pair.

3 Contact Force Model

Theoretically, there is no need to distinguish rotor 1 from rotor 2. However, the distinction is often made for numerical convenience, as shown in Fig. 4. The contact condition is the rotor 2 no penetrate into the rotor 1. The contact condition is considered between a node on profile of rotor 2 and a segment on profile of rotor 1. The gap between profile of rotor 1 and profile of rotor 2:

$$g_{ai} = (\mathbf{r}_{p_i} - \mathbf{r}_{q_j}) \cdot \mathbf{n}_i \geq 0 \tag{9}$$

where $\mathbf{r}_{p_i}, \mathbf{r}_{q_j}$ are position vector of rotor profile 1 and rotor profile 2 at points p_i and q_j. If the gap $g_{ai} > 0$, two rotors have no contact and if the gap $g_{ai} < 0$, two rotors have contact.

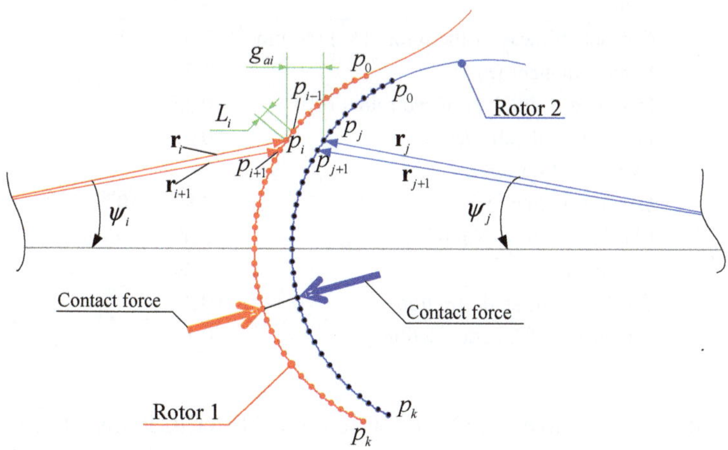

Fig. 4. Gap and contact force between two meshing rotors.

For those contacting rotor pairs, penetration needs to be corrected by applying a contact force and more penetration needs more contact force:

$$F_C = K_n\langle -g_{ai}\rangle \quad (10)$$

where penalty parameter (K_n) is contact stiffness. It depends on material stiffness and allows a small penetration ($g_{ai} < 0$). A bigger contact stiffness K_n => the less allowed penetration. The proper contact stiffness can be determined from allowed penetration by experience.

4 Numerical Example

In order to determine the contact force between two rotors, the rotors are assembled at a practical center distance of A_r = 66 mm. The basic data of the Roots rotor and tooth design parameters are given in Table 1, the contact settings and material properties are given in Table 2 and The input power of rotor is 3 kW and the rotational speed of rotor 3000 rpm. The contact forces are determined for two rotor pairs.

Table 1. Basic data of the root rotor.

Roots rotor		Roots rotor	
Number of rotor lobes (N)	3	Normal pressure angle (α_p)	36.43°
Normal module (m)	22	Center distance (E_c)	66 mm
Pitch radius (r_p)	33.0 mm	Involute parameter (u)	9.00 mm
Outer radius (r_a)	44.88 mm	Top tooth arc angle (θ_1)	4.0°

Table 2. Contact settings and material properties.

Contact settings	
Item	Value
Contact stiffness of the rotors (K_n) (N.mm^{-1})	2. 0 × 10^5
Force exponent (e)	2.2
Damping coefficient of the rotors (C_n)	0.25
Penetration depth (d_s) (mm)	0.01
Material Properties	
Density (kg/mm^3)	7.80 × 10^{-6}
Elastic modulus (N.mm^{-2})	2.07 × 10^5
Poisson's ratio	0.29
Coefficient of static friction	0.2
Coefficient of kinetic friction	0.1

The contact force between two EE Roots rotors and two CIC Roots rotors are shown in Figs. 5, 6, 7 and 8. On time domain, the average contact force between two EE Roots rotors, 1047.1 N, is much higher than that of the contact force between two CIC Roots rotors, 102.4 N. Besides, the maximum contact force between two EE Roots rotors, 5661.4 N, is much higher than that of the contact force between two CIC Roots rotors, 1973.3 N, as shown in Figs. 5 and 6. On frequency domain, the contact force between two EE Roots rotors has much changed extremely abrupt than that of the contact force between two CIC Roots rotors, as shown in Figs. 7 and 8. Therefore, the contact force of CIC Roots rotors have more advantages than that of the EE Roots rotors.

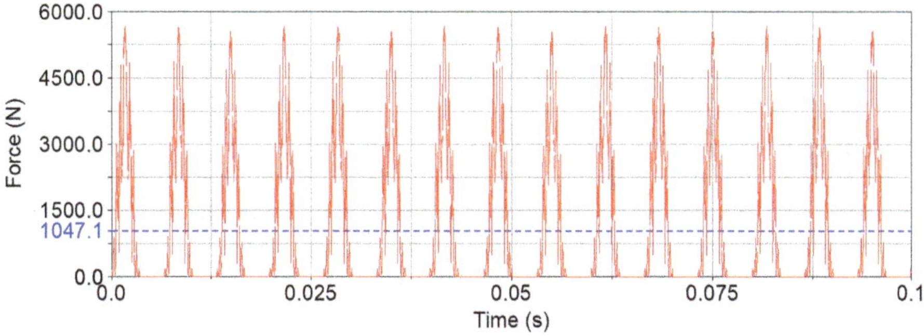

Fig. 5. The contact force between two EE Roots rotors on time domain.

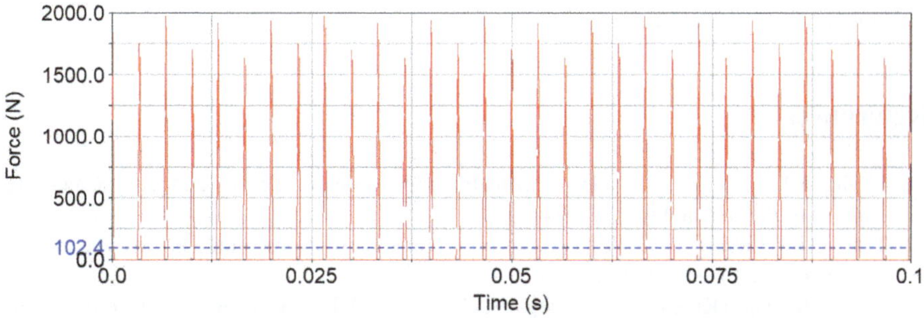

Fig. 6. The contact force between two CIC rotors on time domain.

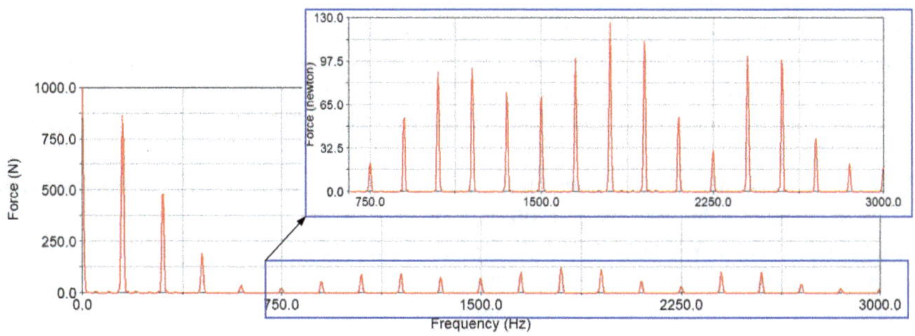

Fig. 7. The contact force between two EE Roots rotors on frequency domain.

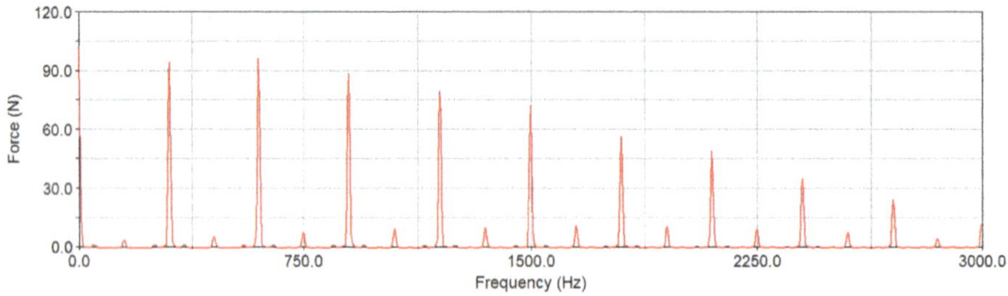

Fig. 8. The contact force between two CIC Roots rotors on frequency domain.

5 Conclusion

In this paper, 3D rotor models for both EE Roots rotors and CIC Roots rotors are constructed. The contact forces of a pair intermeshing rotors are analyzed under combining the torque on the rotor 1 and rotor 2 with the rotation angle of rotor 1. The simulated results show that the contact forces of CIC Roots rotors have more advantages than that of the EE Roots rotors.

Acknowledgments. The authors would like to thank for the subsidy of the Outlay B2019-SKH-03 given by the Ministry of Education and Training, Vietnam, to assist us in successfully finishing this special research.

References

1. Litvin, F.L., Fuentes, A.: Gear Geometry and Applied Theory, 2nd edn, Cambridge University Press, Cambridge (2004)
2. Tsay, C.B.: Solving the conjugated curves of Roots blower lobe profile by analytic method. J. Machine **13**(1), 129–133 (1987)
3. Vecchiato, D., Demenego, A., Argyris, J., Litvin, F.L.: Geometry of cycloidal pump, computer methods and applied. Mech. Eng. **190**, 2309–2330 (2001)
4. Fang, H.S.: Rotor profile for a Roots vacuum pump, U.S. Patent 5,152,684 (1992)
5. Wang, P.Y., Fong, Z.H., Fang, H.S.: Design constraints of five-arc Roots vacuum pumps. J. Mech. Eng. Sci. Part C **216**(2), 225–234 (2002)
6. Niimura, Y., Kikuta, R., Usui, K.: Two-Shaft type rotary machine having a tip circle diameter to shaft diameter within a certain range, US Patent No. 4943214 A (1990)
7. Hwang, Y.W., Hsieh, C.F.: Study on high volumetric efficiency of the Roots rotor profile with variable trochoid ratio. Proc. Inst. Mech. Eng., Part C-J. Mech. Eng. Sci. **220**(C9), 1375–1384 (2006)
8. Hsieh, C.F., Hwang, Y.W.: Tooth profile of a Roots rotor with a variable trochoid ratio. Math. Comput. Model. **48**, 19–33 (2008)
9. Kang, Y.H., Vu, H.H., Hsu, C.H.: Factors impacting on performance of lobe pumps: a numerical evaluation. J. Mech. **28**(2), 229–238 (2012)
10. Kang, Y.H., Vu, H.H.: A newly developed rotor profile for lobe pumps: generation and numerical performance assessment. J. Mech. Sci. Technol. **28**(3), 915–926 (2014)

11. Chiu, H.C.: The mathematical model and computer aided manufacturing of Roots blower gerotor profile". J. Technol. **9**(1), 13–19 (1994)
12. Tran, V.-T., Thanh, B.T., Long, B.T., Tuan, H.Q., Toan, N.D.: Study on the effects of tooth profile design parameters of rotor to performance of vacuum pump. Int. J. Modern Phys. B, 2040141 (2020). https://doi.org/10.1142/s0217979220401414

Mean-Variance Mapping Optimization for Auto-tuning Parameters of Classical Motion Cueing Algorithm

Duc-An Pham(✉)

Hanoi University of Science and Technology, Hanoi, Vietnam
an.phamduc@hust.edu.vn

Abstract. Motion cueing algorithms (MCAs) are commonly used to combine the tilt gravity and translational acceleration components in the restricted workspace of driving simulators. The realistic simulations significantly depend on the MCAs parameters that need typically extensive tuning to generate necessary motion cues (good feeling of motion) and eliminate the false cues (bad feeling of motion). The tuning process is time-consuming for trial and error and requires experts in the loop. Thus, several authors proposed auto-tuning methods based on the fuzzy control theory for several algorithms, such as classical algorithms. Recently, the generic method was used to obtain optimal values for parameters of classical and optimal algorithms. The optimal parameters minimizing the cost function include the sensed errors, the correlation between simulated and target motion sensation, the limited movement, and the fluctuation of sensed errors. This paper introduces an auto-tuning with a mean-variance mapping optimization method that defines the penalty functions to constrain the simulated signals in the allowable range. The methods were applied to the classical MCA to find the maximum amplitude of this algorithm's input signals.

Keywords: Motion cueing algorithms · Auto-tuning · Classical algorithm · Fidelity criteria · Objective assessment

1 Introduction

Motion simulators are applying for both training drivers and researching human motion perception. The simulated motion fidelity depends on the precise reproduction of benefit motion cues that is impossible due to the motion platform's limited workspace. Motion cueing algorithms have inputs, including the target translational and rotational motion of the vehicle and the outputs are the simulated motion of the motion platforms. The algorithms aim to preserve the simulation's perceptual realism by using tilt coordination to emulate sustained translational acceleration motion for the low-frequency part of acceleration via the gravity vector perceived by the human vestibular system. The first kind of the MCA named classical washout filter (CL) was introduced by Conrad and Schmidt [1]. The algorithm is still widely used in commercial simulators because it is simple to implement in a real drive simulator.

The classical algorithm (Fig. 1) was initially developed for a flight simulator and later applied for general drive simulators. The linear acceleration and angular velocity

B. T. Long et al. (Eds.): MMMS 2020, LNME, pp. 952–957, 2021.
https://doi.org/10.1007/978-3-030-69610-8_126

are scaled and filtered by second-order high-pass filters (Translation Motion block) to keep the simulated motion in the workspace and returning the motion base to its neutral position [1, 2]. Moreover, translational accelerations are scaled and filtered by low-pass filters. Afterward, the sustained parts of translational acceleration (low-frequency – Sustained Motion block) are replaced by the appropriated roll and pitch tilt-angles due to the tilt-coordination technique (Tilt Coordination block), respectively. However, the MCA parameters have severe effects on the simulated motion that can lead to motion sickness if the motion and visual cues are not consistent. Thus, for particular motion simulation, tuning the best parameters of an MCA remains a challenging task.

In the literature, the first tuning is to find the appropriate parameters with which the MCAs generate the simulated quantities in the defined ranges. For a specific drive task, the tuning processes were fulfilled mostly by trial and error. Thus, it is very time-consuming and requires drivers' experience in the simulation field [4]. On the other hand, several auto-tuning methods were proposed to search the optimizing parameters for the classical algorithm [5–7]. The methods used fuzzy control theory to constrain angular velocity false cues and the driving simulator's position in the motion platform's workspace. Besides, the fidelity criteria for objective assessment of the simulated quantities were inferred from the subjective assessment. The criteria can save time-consuming for trial and error and be investigated in some research related to flight simulation [9]. Therefore, the parameters' tuning is an open issue because the parameters are not intuitive for inexperienced simulator users. However, automated tuning would be desirable to ease the user to implement the time-consuming task.

This work's main objective is to introduce the auto-tuning method (AuT) for the classical algorithm to obtain high objective motion fidelity, which reduces false cues and increases the scaled signals' amplitude.

2 Auto-tuned Parameters for CLRN with Mean-Variance Mapping Optimization (MVMO)

Mean-variance mapping optimization developed by Erlich et al. (2010) [10] is a novel optimization algorithm mapping method that is a member of the family of the so-called population-based stochastic optimization technique. The technique combinates the excellent performance of a specific number of best individuals to minimize a penalty cost function, which is defined for the motion cueing problem as:

$$F_{MVMO} = \sum w_k J_k \quad \text{with } i \in \{P, fsh, \omega, \dot{\omega}, sc, tr, Wo\}.$$

Here, the penalty functions are defined in the Table 1 with perceptive parameters: δ_O is the otolith threshold value; $\omega, \dot{\omega}$ are the angular velocity and acceleration, respectively; δ_S are the threshold values of the angular velocity perceived by the semicircular organ;

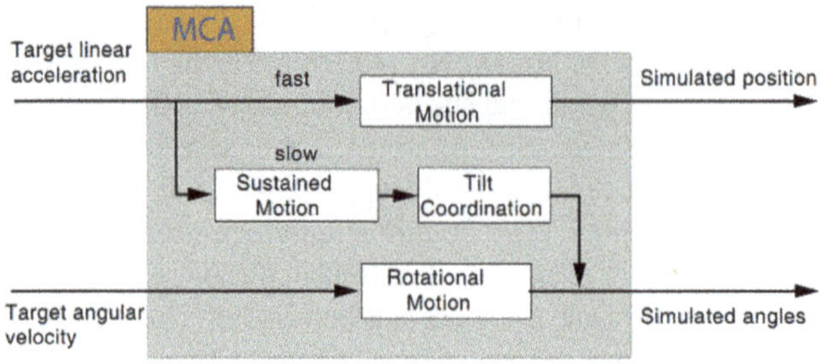

Fig. 1. Classcial motion cueing algorithm (CL)

Table 1. Penalty functions

Penalty	Definition
J_P	$\sum_{j\in\{1\cdots 6\}} \left\{e^{\langle z_j^L - z_i^{min}\rangle^2} + e^{\langle z_j^{max} - z_i^U\rangle^2} - 2\right\}$
J_{fsh}	$e^{\langle \lvert e_{f.sh}\rvert^{max} - k_O\delta_O\rangle^2} - 1$
J_ω	$e^{\langle \lvert\omega\rvert^{max} - \omega_{th}\rangle^2} - 1$
$J_{\dot\omega}$	$e^{\langle \lvert\dot\omega\rvert^{max} - \dot\omega_{th}\rangle^2} - 1$
J_{sc}	$k_{J_{sc}}\left[e^{\langle k_{min} - k_S\rangle^2} + e^{\langle k_S - k_{max}\rangle^2} - 1\right] - 1 + \langle k_{max} - k_S\rangle^0\langle k_S - k_{min}\rangle^0\left[e^{\langle k_{max} - k_S\rangle^2} + k_{J_{sc}}\right]$
J_{W_0}	$e^{g^2\lvert\varphi_S(t_f)\rvert^2} - 1$
J_{tr}	$e^{\langle k_{a,min} - k_a\rangle^2} - 1$

Remarks:

1) Denotes the Heaviside function

$$\langle f(x)^n\rangle = \begin{cases} 0 & for f(x) < 0 \\ f(x)^n & for f(x) \geq 0 \end{cases}$$

2) For J_P, $\underline{z} = [x_E, y_E, z_E, \varphi_E, \theta_E, \psi_E]$ denotes the global coordinates of the end offector, $\underline{z}^L$ and $\underline{z}^U$ denote the lower and upper limits of the feasible motion box within the workspace, and $\underline{z}^{min}$ and $\underline{z}^{max}$ denote the minimal and maximal values attained through the task trajectory, respectively. Note that, if there is no violation of the lower or upper physical limits, $J_P = 0$
3) $|\cdots|^{max}$ denotes the maximum of the absolute value of the argument
4) $k_O \in [0, 1]$ is a shaping parameter for the shape error penalty function. A small value means that the shape error is "punished" more strongly

5) $[k_{S,min}, k_{S,max}]$ is the range of desired scale factor, and $k_{J_{sc}}$ is a shaping parameter that controls the amplitude of the penalty function when k_S is within the desired range.
6) $k_{a,min}$ denotes the minimum desired ratio between simulated and target translational acceleration.

3 Result and Discussion

For the **AuT** case, the scale factor k is in range $[k_{min}, k_{max}] = [0.4, 1]$, the tuning process searches tuned parameters in the defined ranges, and the use of weighting parameters is shown in the Table 2. The weighting parameters are selected to set the most significant priority for the position penalty function, and the next significant level is set for the penalty function related to the restricting potential false cues existing in simulated signals.

Table 2. The setting and optimal parameters

Selected limit-values							
$k_O(-)$	$\delta_O(m/s^2)$	$\delta_S(rad/s)$	$\dot{\omega}_{th}(rad/s^2)$	k_{min}	k_{max}	$k_{a,min}$	
1	0.17	0.1	0.2	0.4	1	0.6	
Searching ranges of the tuned parameters							
	$k_S(-)$	$\omega_{ny}(rad/s)$	$\xi_y(-)$	$\omega_{by}(rad/s)$	ω_{LP_y}		
AuT	[0.4 , 1]	[0 , 1]	[0 , 1]	[0 , 1]	[0 , 3]		
Optimized parameters							
	$k^*(-)$	$\omega_{ny}{}^*(rad/s)$	$\xi_y{}^*(-)$	$\omega_{by}{}^*(rad/s)$	$\omega_{LP_y}{}^*$		
AuT	0.54	0.1563	0.0166	0.3437	1.6583		
Weighting parameters							
	ωJ_P	ωJ_ω	$\omega J_{\dot{\omega}}$	ωJ_{sc}	ωJ_{tr}	ωJ_{fsh}	ωJ_{W_O}
AuT	2^8	2^3	2^3	2^2	1	2^3	0

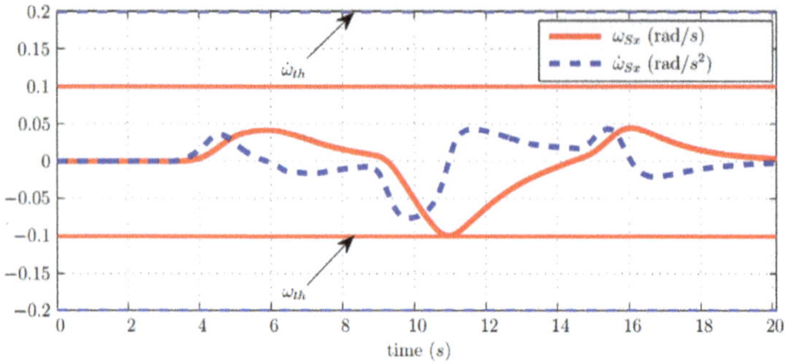

Fig. 2. Simulated angular velocity and acceleration

In order to find the largest k_S, the weighting values of the the scale penalty function were set as those selected for angular acceleration penalty function. Note that the weighting parameter of the washout penalty function wJ_{Wo} is zero because the high-pass filter of the CL has done the task.

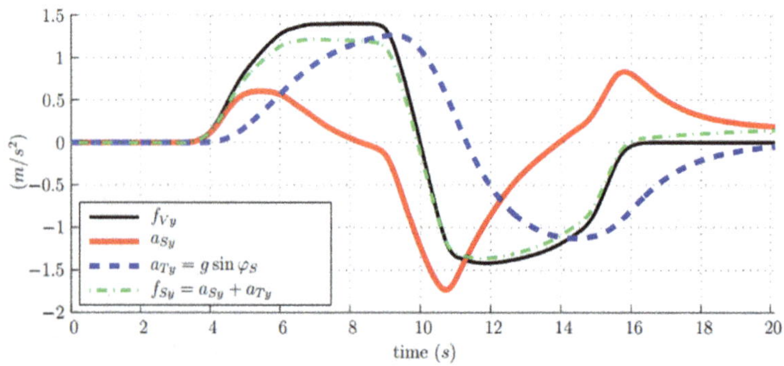

Fig. 3. Simulated specific forces and acceleration and tilt acceleration

Therefore, it is possible to remove the effect of the penalty function. After being implemented in MATLAB R software with the iterative number N = 300, the auto-tuning process generated the optimized parameters (in the Table 2). The symbols with $()^*$ represent the optimized parameters obtained from the auto-tuning.

The simulated signals generated by the CLRN with the optimized parameters shows there are no rotational false cues because all angular velocity and acceleration are not over their threshold values, as shown in Fig. 2. Moreover, in Fig. 3, the specific force error could not disturb the vestibular system. Note that the simulated angle, ϕ_S, goes to zero, although the washout penalty function does not consider.

4 Conclusion

In summary, tuning parameters for MCAs play a crucial role in producing suitable motion cues. Typically, the tuning process focuses on reducing false cues and reproducing benefit motion cues. This paper introduces auto-tuning methods applying in classical MCA with the optimal method (MVMO) to constrain the output signals' physical quantities. The results in Fig. 2 and Fig. 3 show advantages compared to manual tuning methods. Firstly, the method is flexible and saves time-consuming for the tuning process in finding the maximum scale factor while constraining the false cues. A designer can adjust the weighting parameters for tuning purposes. Secondly, the tuning method considers two main quantities: the position, angular velocity, and various physical quantities related to the human perception. Finally, while this has not been mentioned in this paper, the auto-tuning method can be applied to various types of MCAs and different types of drive-tracks in offline mode.

Acknowledgement. This research is funded by the Hanoi University of Science and Technology (HUST) under project number T2020-SAHEP-013

References

1. Conrad, B., Schmidt, S.: Motion drive signals for piloted flight simulators (1970)
2. Stewart, D.: A platform with six degrees of freedom. Proc. Inst. Mech. Eng. **180**(1), 371–386 (1965)
3. Reid, L., Nahon, M.A.: Flight simulation motion-base drive algorithms: part 1. developing and testing equations. University of Toronto (1985)
4. Grant, P.R., Reid, L.D.: Motion washout filter tuning: rules and requirements. J. Aircraft **34** (2), 145–151 (1997)
5. Song, J.-B., Jung, U.-J., Ko, H.-D.: Washout algorithm with fuzzy-based tuning for a motion simulator. KSME Int. J. **17**(2), 221–229 (2003)
6. Hwang, T.-S., Yeh, S.-K., Lin, J.-R., Su, W.-P.: Adaptive motion washout filter design by using self-tuning fuzzy control. In: IEEE/ASME International Conference on Advanced Intelligent Mechatronics, 2009, AIM 2009, pp. 811–815. IEEE (2009)
7. Chen, S.-H., Fu, L.-C.: An optimal washout filter design with fuzzy compensation for a motion platform. IFAC Proc. Vol. **44**(1), 8433–8438 (2011)
8. Reid, L., Nahon, M.A.: Flight Simulation Motion-Base Drive Algorithms.: Part 2, Selecting The System Parameters, UTIAS report, no. N307 (1986)
9. Pham, D.A., Roettgermannand, S., Kecskeméthy, A.: Optimal motion cueing algorithm selection and parameter tuning for sickness-free robocoaster ride simulations. In: Proceedings of the Third MeTrApp Mechanisms, Transmissions and Applications Conference 2015, 06–08. Mai 2015, Aachen (2015)
10. Erlich, I., Venayagamoorthy, G.K., Norawat, N.: A mean-variance optimization algorithm in WCCI. IEEE World Congress Comput. Intell. **2010**, 1–6 (2010)

Numerical Comparison of Offline Motion-Cueing Algorithms for the Ride Simulator RoboCoaster and Auto-tuning Parameters

Duc-An Pham(✉)

Hanoi University of Science and Technology, Hanoi, Vietnam
and.phamduc@hust.edu.vn

Abstract. Motion cueing algorithms (MCAs) are the cheating technique to drive physical simulators along offline-generated trajectories such as to produce proper motion cues to the passenger by a combination of tilt gravity and translational acceleration components (so-called linear specific forces or "g-forces"). In this paper, the numerical perception quality of the aforementioned families of MCAs is compared for a virtual test ride along a simple S-shaped planar curve featuring solely lateral accelerations, using two previously published objective measures. To this end, the MCAs were implemented using originally published MCA parameter values and optimally tuned values concerning the test trajectory. The comparison shows large discrepancies between the different MCAs, both in terms of simulated motion perception and fulfillment of robot workspace constraints. The best-fit tuning of MCA parameters for a given trajectory may significantly improve the objective measures. In this study, a new auto-tuning method based on a generic optimal method called mean variance mapping optimization to minimize the performance indicator was introduced to find the off-line optimal algorithm's tuned parameters. Three testbeds using the auto-tuning process show different tuned parameter sets, however, the off-line optimal MCA has a similar response with all set of tuned parameters.

Keywords: Motion cueing algorithm · Auto-tuning parameters · Objective critera

1 Introduction and Review of Motion Cueing Algorithms

Driving simulation systems are useful and affordable system applying on both training and researching human driving behavior. The systems have first used the parallel robot named Stewart platforms to generate the supporting motion. The classical Stewart platforms. Recently, the serial-robot ride simulators as the KUKA RoboCoaster, have been used in the entertainment industry as well as for research purposes [3, 7]. Common to all motion platforms, including serial robots, is that exact target motions cannot always be precisely reproduced due to the limited workspace. In these cases, the motion cueing algorithms (MCAs) play an essential role in mincing the benefit motion sensations that can be reproduced by motion platforms in the restricted available workspace. MCAs divide the target acceleration into the high-and low-frequency parts. While the high frequency (HF) parts are reproduced by the motion platform's transient

B. T. Long et al. (Eds.): MMMS 2020, LNME, pp. 958–964, 2021.
https://doi.org/10.1007/978-3-030-69610-8_127

linear acceleration, the low-frequency (LF) part of acceleration is simulated by the tilt coordination technique. Recently, various MCAs have still been developed for different motion platforms; thus, selecting the best MCA and tuning its parameters are still the open problems for a specific simulation task.

At the origin of MCAs, classical MCA (or named washout filter) was proposed by Conrad and Schmidt [4] consisting of a low-pass filter (LP) cross-coupled with a high-pass filter (HP), whose parameters were manually adjusted off-line. For better exploitation of the serial robot workspace, such as the KUKA RoboCoaster, the classical MCA was modified into "cylindrical classical algorithm" by Giordano et al. [7]. The authors implemented the washout filters in a defined cylindrical coordinate by which the end-effector can move along both the linear and circular motion. To adjust parameters of washout filters in real-time, the "adaptive washout filter" was introduced [9, 12, 13]. The adaptive method's cost function uses the steepest descent techniques to minimize motion cue errors and restrict linear movements of motion platform deviation of the filtered acceleration signal from the target signal. The adaptive MCAs decreases the false cues compared with the classical MCA; however, the stability of adaptive MCAs significantly depends on its parameters [8] (Table 1).

Table 1. State of the art motion cueing algorithm overview

No.	Name	Description	F. Year[(a)]	R. Year[(b)]	Crd[(c)]
1	CL	Classical algorithm [3]	1970	2010	Cy[(d)]
2	AD	Adaptive algorithm [13]	1974	1985	Ca[(e)]
3	AD-NR	Adaptive algorithm [9]	2011	2011	Cy
4	OpS	Optimal washout filters [14]	1982	1982	Ca
5	OpNR	Optimal washout filters [12]	1985	1985	Ca
6	Tel	Optimal washout filters [15]	2005	2005	Ca
7	Tel-YoMe	Tel with Young and Meiry vestibular system [15]	2005	2005	Ca
8	ZyRo-1	Optimal tracking algorithm [16]	2003	2003	Ca
9	ZyRo-2	Optimal tracking algorithm	2009	2009	Ca
10	MPC-1	Implicit model predictive control [1]	2009	2009	Ca
11	MPC-2	Implicit model predictive control (LMR model)	2012	2012	Ca
12	exMPC	Explicit model predictive control algorithm [6]	2012	2012	Ca

(a): Publication year of original papers; (b): Reference used here for implementation (c): Coordinate system; (d): Cylindrical Coordinates; (e): Cartesian Coordinates

Further development has been to design MCAs that minimize a cost function involving the assumed perception error at the vestibular system and energy terms over the complete track. This family of methods, firstly introduced by Sivan et al. [14], solves the stochastic optimization problem to find higher-order washout filters for 4 couple motions. The Optimal MCAs were later improved by Nahon and Reid [12] and

Telban and Cardullo [15], who use different mathematical models of vestibular systems. Zywiol and Romano [16] considered the simulated and target signal tracking problem to propose a time backward optimal MCA. The algorithm solves the tracking problem with linear model of tilt-coordination in off-line mode to find the best exploitation of the available workspace for both translational and rotational motion. More recently, novel model predictive control (MPC) was applied to motion cueing. The algorithm takes the various restriction into account, such as: workspace limits, the motion perception thresholds (angular velocity, perception of linear acceleration), etc… [1, 2, 6]. These algorithms show substantial improvements in motion sensation fidelity and can fill better the available workspace, but still have some limitations in computational efficiency and avoidance of false cues.

The objective of the present work is to give an overview of existing MCAs for the families of classical, adaptive, optimal control, and model-predictive control algorithms for off-line ride simulators and to numerically compare their performance for a simple virtual test ride along a planar S-shaped curve involving only lateral acceleration, using two existing objective measures for perception evaluation, the "performance indicator" of Pouliot, Gosselin, and Nahon [11]. To this end, the aforementioned MCA families were implemented in MATLAB. Their output was verified in a virtual robot testing environment mimicking a real RoboCoaster ride simulator testbed, confirming the target motions' feasibility in all cases. As it turns out, there are large differences in the behavior and objective measure of MCA responses, and proper MCA parameter tuning for the given trajectory may improve the objective measure significantly along the ride. In this setting, a new approach is proposed for auto-tuning the filter parameters for a given trajectory.

2 Numerical Comparison of MCA Families for a Planar S-Shaped Test Trajectory

To compare the different MCA families' results, a simple virtual ride at constant speed of 6 m/s along a planar left-right S-shaped trajectory with smoothed transitions, hence featuring pure lateral acceleration, was chosen. Figure 1 shows the target acceleration and target roll angular velocity as thick solid lines and the resulting lateral specific forces and the angular velocities for the tested MCAs, respectively. One such objective measure is the "performance indicator" introduced by Pouliot, Gosselin and Nahon [11], which consists of two indicators λ_1 and λ_2, both split into a translational part ("f") for the specific force, and a rotational part ("ω") for the angular velocity, normalized by the maximal acceleration a_{max} and the maximal angular velocity ω_{max} of the simulator, respectively.

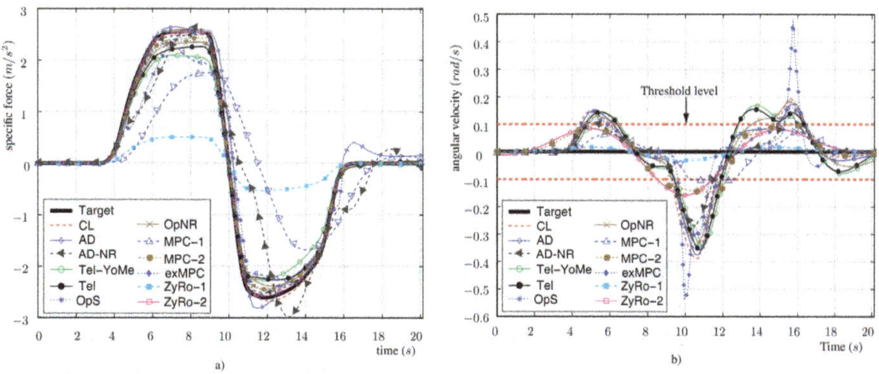

Fig. 1. Results of MCA families for the test track: a) lateral specific forces; b) roll angular velocities

Here, λ_1 describes the normalized average of the square error between the target and simulator motion signal, while λ_2 describing the normalized average square error of the corresponding change rates, respectively. A value of $\lambda_{i,\{f,\omega\}} = 0$ suggests a perfect expected simulator sensation, while the greater $\lambda_{i,\{f,\omega\}}$, the worse the simulation sensation is expected to be. The resulting performance indicators are expressed in percentage values as (Fig. 2).

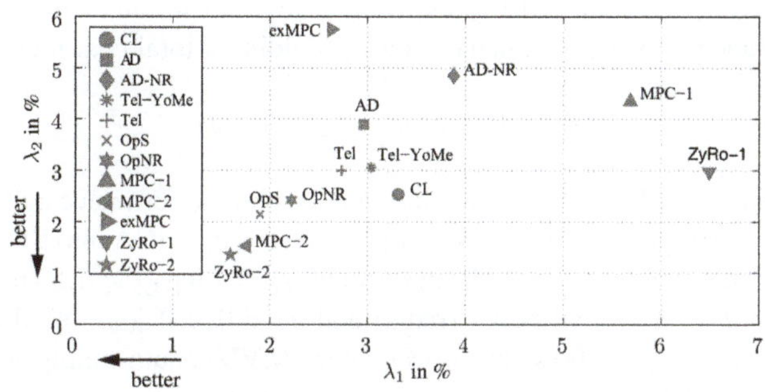

Fig. 2. Comparison of performance indicators

3 Tuning Parameters via Mean Variance Mapping Optimization

As was seen above, a proper selection of the numerical values for the optimal control MCA's filter parameter can have a significant effect on its performance. In order to automatically tune the MCA parameters, an optimization can be run using the objective measure as a cost function and the workspace limits and angular velocity cues as unilateral constraints. This is described in the following for the optimal control MCA case, but can be applied also to other MCA approaches.

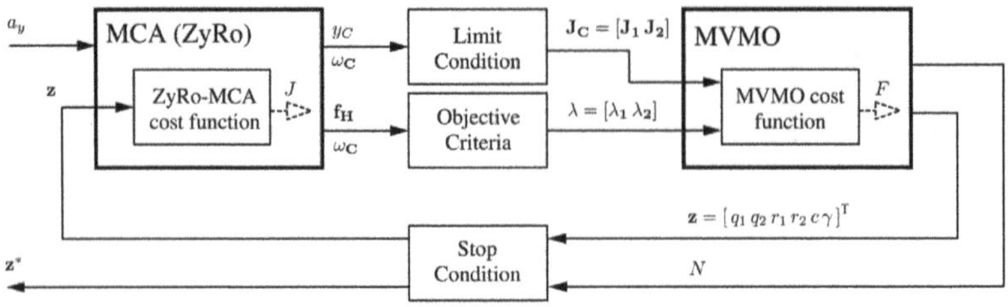

Fig. 3. Auto tuning diagram with MVMO method

Let the free parameters of the Optimal Control MCA of Zywiol and Romano [16] be the diagonal elements q_1, q_2, r_1, r_2 of the weighting matrices and the break frequencies c and γ. These parameters can be understood as components of a design vector $z = [q_1, q_2, r_1, r_2, c, \gamma]$ for which one seeks the special value z^* at which the performance indicators according to [11] are minimized. Hence an alternative approach is proposed using the mean variance mapping optimization (MVMO) method developed by Erlich et al. [5]. The method is a member of the family of so-called population-based stochastic optimization techniques, which incorporates information of the performance of the specific number of best individuals to minimize a particular cost function. In the present case, besides optimizing the performance indicators according to equation…, the following constraints are formulated to achieve a feasible MCA; J_1: The lateral motion y_C of the cabin must remain in the workspace, here assumed to be the lateral interval [−1.5 m, 1.5 m]; J_2: The angular velocity ω_C of the cabin must remain within the threshold interval [−0.1 rad/s, 0.1 rad/s] [14]. These constraints can be added to the actual performance indicators as penalty terms yielding a total cost function

$$F(z) = w_1\lambda_1 + w_2\lambda_2 + J_1 + J_2,$$

with weight factors w_1 distributing the importance within performance indicators, as well as between these and penalty functions. In order to force the MCA into a feasible motion, the following penalty functions have been selected: $J_1 = 1$ if $|y_C| \geq 1.5$ (m) and $J_1 = 0$ if $|y_C| < 1.5$ (m); $J_2 = 1$ if $|\omega_C| > 0.1$ (rad/s) and $J_2 = 0$ 0 if $|\omega_C| \leq 0.1$ (rad/s). The sequence diagram in Fig. 3 shows the process of the MVMO auto-tuning process. At each iteration, the target lateral acceleration ay and the initial values z_0 of parameters are fed into the MCA (ZyRo) block. The corresponding simulated signals ω_C for cabin angular velocity, f_H for passenger's head specific force, and y_C for cabin lateral sway are generated. Upon termination of the MVMO loop, the optimized set z^* of parameters for enhanced motion perception for the objective measure is obtained.

To verify the MVMO auto-tuning method's behavior, we implemented three test-runs with different searching range for the tuned parameters (Table 2). While in Case 1, only four weighting parameters $[q_1, q_2, r_1, r_2]$ were automatically tuned, in Case 2 and Case 3, the tuning process was carried out simultaneously for all four weighting parameters as well as the two break frequencies c and γ. The difference between the Cases 2 and 3 lies on the one side in the different search ranges for c and γ, and on the other side in the different value of the weight factor w_1 of the cost function, which was chosen in case 3, 100 times greater than w_2. As will be seen, these variations have almost no effect on the final result (Fig. 4).

Table 2. Optimal paramters

	$1/c(s)$	$1/\gamma(s)$	q_1	q_2	r_1	r_2
Case 1	0.01	0.05	0.0889	284.3274	72.3126	0.2273
Case 2	2.7567	0.0248	0.8147	183.2356	170.3465	0.6060
Case 3	0.0215	0.0128	0.3001	171.4193	241.3706	0.9829
Case 4	0.01	0.05	0.01	199	199	0.4

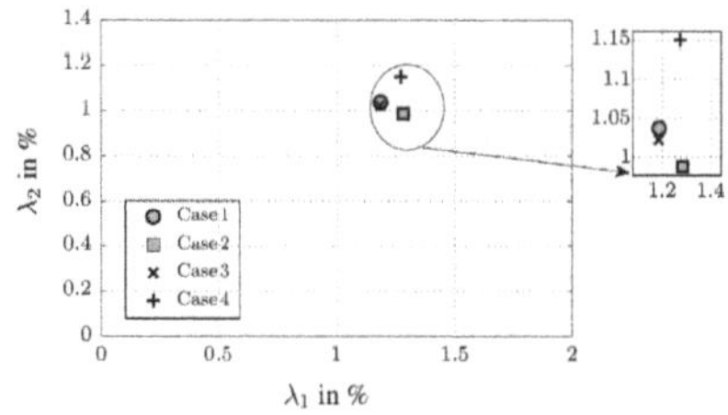

Fig.4. Comparison of λ_1 and λ_2

4 Conclusion

The present paper reviews the families of off-line motion cueing algorithms from the literature, including classical washout, adaptive washout, optimal washout, optimal control, and model-predictive control MCAs, and compares their off-line response. The response is achieved from the implementations and virtual rides over a planar S-shaped curve with lateral acceleration. The performance evaluation is carried out using objective measures from the literature termed as the "performance indicator". The comparison shows that the Optimal Control MCA approach and the best fit parameter tuning with respect to a given trajectory yields the best results compared to MCAs in terms of computational efficiency and fidelity for prescribed specific force and avoidance of false cues. The paper also presents a method for automatic tuning of the Optimal Control MCA parameters based on the mean variance mapping optimization (MVMO) algorithm. The method allows one to find the best parameters for this type of MCA for a given off-line track, extending to other MCA methods.

Acknowledgement. This research is funded by the Hanoi University of Science and Technology (HUST) under project number T2020-SAHEP-013.

References

1. Augusto, B., Loureiro, R.: Motion cueing in the chalmers driving simulator: a model predictive control approach. Master's thesis, Chalmers University of Technology (2009)
2. Baseggio, M., Beghi, A., Bruschetta, M., Maran, F., Minen, D.: An MPC approach to the design of motion cueing algorithms for driving simulators. In: 2011 14th International IEEE Conference on Intelligent Transportation Systems (ITSC), pp. 692–697. IEEE (2011)
3. Bellmann, T., Heindl, J., Hellerer, M., Kuchar, R., Sharma, K., Hirzinger, G.: The DLR robot motion simulator. Part I: design and setup. In: 2011 IEEE International Conference on Robotics and Automation (ICRA), pp. 4694–4701 (2011)
4. Borah, J., Young, L.R., Curry, R.E.: Optimal estimator model for human spatial orientation. Ann. N. Y. Acad. Sci. **545**(1), 51–73 (1988)
5. Erlich, I., Venayagamoorthy, G.K., Worawat, N.: A mean-variance optimization algorithm. In: IEEE World Congress on Computational Intelligence, WCCI 2010, pp. 1–6 (2010)
6. Fang, Z., Kemeny, A.: Motion cueing algorithms for a real-time automobile driving simulator. In: Driving Simulation Conference, pp. 159–174 (2012)

7. Giordano, P.R., Masone, C., Tesch, J., Breidt, M., Pollini, L., Bülthoff, H.H.: A novel framework for closed-loop robotic motion simulation. Part II: motion cueing design and experimental validation. In: 2010 IEEE International Conference on Robotics and Automation (ICRA), pp. 3896–3903. IEEE (2010)
8. Nahon, M.A., Reid, L.D.: Simulator motion-drive algorithms—a designer's perspective. J. Guid. Control Dyn. **13**(2), 356–362 (1990)
9. Parrish, R.V., Dieudonne, J.E., Martin Jr., D.J.: Coordinated adaptive washout for motion simulators. J. Aircr. **12**(1), 44–50 (1975)
10. Pekar, J., Havlena, V.: Design and analysis of model predictive control using MPT toolbox (2004). https://dsp.vscht.cz/konferencematlab/matlab04/pekar.pdf
11. Pouliot, N.A., Gosselin, C.M., Nahon, M.A.: Motion simulation capabilities of three-degree-of-freedom flight simulators. J. Aircr. **35**(1), 9–17 (1998)
12. Reid, L.D., Nahon, M.A.: Flight simulation motion-base drive algorithms. Part 1: developing and testing the equations. Tech. rep. (1985)
13. Schweig, S., Kammers, H.: Bewegungssteuerung eines RoboCoaster Kuka Roboters zur Ride Simulation mit Hilfe von Washout Filtern. University of Duisburg Essen, Duisburg, Germany, Project Report (2011)
14. Sivan, R., Ish-Shalom, J., Huang, J.K.: An optimal control approach to the design of moving flight simulators. IEEE Trans. Syst. Man Cybern. **12**(6), 818–827 (1982)
15. Telban, R.J., Cardullo, F.M., Houck, J.A.: Motion cueing algorithm development: human-centered linear and nonlinear approaches. NASA/CR2005-213747, Tech. rep. (2005)
16. Zywiol, H.J., Romano, R.: Motion drive algorithms and simulator design to study motion effects on infantry soldiers. In: Driving Simulation Conference, North America 2003 (DSC-NA 2003) (2003)

Experimental Investigation and Manufacture of a Multifunction Electric Wheelbarrow

Van Nhut Phan[1] and Tien V. T. Nguyen[2(✉)]

[1] Faculty of Automobile Technology, Industrial University of Ho Chi Minh City, Ho Chi Minh City, Viet Nam
[2] Faculty of Mechanical Technology, Industrial University of Ho Chi Minh City, Ho Chi Minh City, Viet Nam
nguyenvanthanhtien@iuh.edu.vn

Abstract. In this study, we propose a device that is applied to production to minimize human labor in daily work. With some small volume goods and working in a narrow range, workers are mainly porters. Therefore, we have researched and created a multi-function electric trolley operated by electric motors DC with power from a lithium-ion battery and controlled by a Hall sensor. Moreover, containers and tripods can be changed to suit the purpose of the demands. The results show that the vehicle can contain up to 120 kg weight, the maximum slope angle achieves to 23, the operation time is up to 4 h depending on the amount of battery installed. The wheelbarrow can work on many different terrains and durable. Based on the results, the research has met the actual needs of workers, better than other products, and environmentally friendly.

Keywords: Hall sensor · Electric motors DC · Lithium-ion battery · Trolley · Electric-assist · Support tools · Wheelbarrow

1 Introduction

Devices and machines that use battery power are becoming popular. In the future, they will replace the internal combustion engine. The transport of goods, the movement of heavy objects in small spaces, the use of large motor vehicles, and the use of human power are directly mandatory. From these issues, we study the topic of multi-purpose trolleys with electric power to minimize labor consumption and save working time [1–4]. Unskilled, low-skilled workers often have little interest in the tools and working environment, so our research's main purpose is to contribute to solving that problem and improve the quality of the working environment for workers. Based on the mechanical design, calculation of the chassis, and knowledge of electricity – electronics, the authors select the control components, DC motors, and layout of electrical systems on the product. To create a product that ensures technical factors and actual operating conditions [5, 6]. The cargo support device has a mass between 100 and 120 kg, making the work simple, saving working time, and improving the working environment for workers at the construction site, workshops, farms… Improving the working environment for human beings

B. T. Long et al. (Eds.): MMMS 2020, LNME, pp. 965–974, 2021.
https://doi.org/10.1007/978-3-030-69610-8_128

and reducing air pollution in the work area are significantly important. The product has the following requirements:

(i) Calculate the specifications of the car and have 2D, 3D CAD
(ii) Ability to transport goods with a volume between 100 and 120 kg [7, 8].
(iii) Removable cars to replace containers, legs…
(iv) Vehicles carrying construction materials and large goods

For operating in narrow spaces, uneven terrain, construction sites, factories, farms, and households and working in night conditions and waterproof [9, 10], the wheel cart could be adopted. The historical backdrop of handcart can follow back to "hand cart" in China, and before the advanced methods for correspondence generally use, it could be considered as a transporter that is working without any problem. The unique handcart is more convenient than the type of riding. Its containerized shipping is a dynamic, delight, working out, yet exceptionally enormous trouble of riding. Pushcart establishes a reversible pendulum framework which is a traditional multivariate, non-direct period subordinate. Controlling its strolling positively can create extraordinary trouble for humans. The current control hypothesis has had the ready technique to control hand-stand and the development of altered pendulum. Transformed pendulum control method and technology has wide application possibilities in quartz conductor and accuracy instrument handling, robotics, rocket block framework, and air. Therefore, rearranged pendulum control innovation is applied to dandy pony by Segway organization of the U.S.A where has created Segway electrodynamic balance vehicle. This innovation is also applied to single wheel cycles by the Solo wheel organization. They have created an electric single-wheel balance truck. It is self-adjusting previously, then after the fact albeit solo-wheel electric single-wheel balance truck can understand and ride the unique pushcart straightforward better than humans do. However, the left-right party needs a bicyclist's manual control. This controlling ability needs the activity of extensive stretch to get itself balance. In the wake of embracing strategy plot, the utility model contrasted and earlier workmanship have the accompanying focal points: the utility model depending on the self-adjusting electronic handcart of gyroscopic adjustment gadget, reversed pendulum control innovation, gyroscopic adjustment gadget and pushcart naturally mixed, taking care of the dependability issue of all bearings all-around of a handcart, improving security and straightforwardness for utilizing that the handcart rides. It extraordinarily can extend to adapt to a range of climates. It can be considered as a compact climate amicable and green running vehicle, and additionally can be used as a sort of diversion lifting weighing device. Nowadays, natural contamination increases day by day. Therefore, the current utility model all-inclusive can mitigate asset utilization and environmental concerns, for example, elevated tainting [11].

The arrangement of the paper is organized as below: Sect. 2 will present the model and method, show the composition of the components, and the charts have been analyzed. Testing the product under different conditions will be introduced in Sect. 3. Section 4 demonstrates the final result after finishing the product and discussion related. Section 5 will give some conclusions.

2 Structural Model Design and Methods

2.1 Design of Chassis

The chassis is designed through CATIA 3D V5 software [7]. The material to make the chassis is steel; the chassis size is (L × W × H) 1500 mm × 640 mm × 702 mm and the piping thickness is 1.5 mm. The container is made of iron. The thickness is 0.5 mm and its size (L × W × H) is 820 mm × 540 mm × 200 mm.

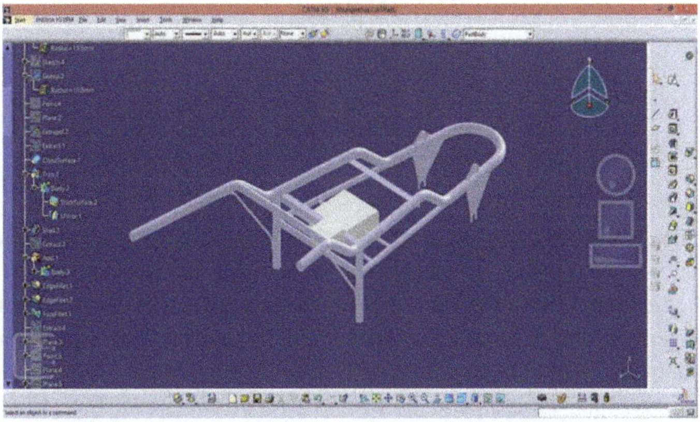

Fig. 1. 3-Dimension model of the designed wheelbarrow

Figure 1 illustrates the 3D model of the chassis of the car is designed on Catia V5 software. The chassis is designed to fit the actual size to calculate the forces and torque [16–18]. The model after design will be analyzed to make sure that its frame has enough strength under initial working conditions. CATIA software was used to employ the model in this study.

2.2 Calculation of the Chassis

After simulating, the force exerted on the chassis on the CATIA V5 software screen could display the bending stress at the positions of the chassis.

$$\sigma_F = \frac{M}{W} \leq [\sigma_F] \tag{1}$$

where M is the moment bending (N.m) and W is standing for modular part value.

2.3 Design of Handle Control

Linear Hall sensors: Linear Hall sensors have analog outputs including voltage regulators (Regulators), hall components (Hall Element), and amplifiers (High Gain

Amplifier). When changing the pole distance S of the magnet. This type of sensor is suitable for the device. The output value of a linear Hall sensor is from 0.5 V to 4.5 V proportional to the strength of the magnetic field acting on the Hall sensor [11, 12].

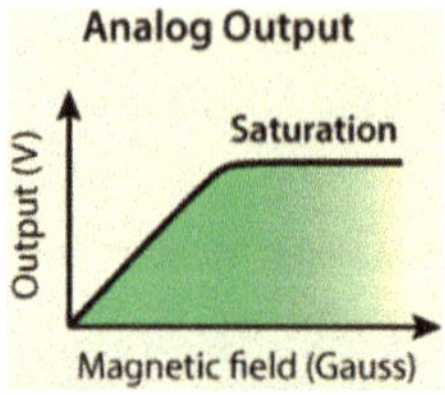

Fig. 2. Hall sensor characteristic chart [3]

Figure 2 demonstrates the relationship between the magnetic force acting on the Hall sensor and the output voltage. When the user exerts a force, the magnet on the handgrip moves, the closer the S pole of the magnet is to the linear Hall sensor, which means that the magnetic field applied to the Hall sensor increases gradually to the maximum value when the S pole of the magnet in contact with the sensor. The signal sensor signal about the controller will change within 0.5 ÷ 4.5 V [3, 15].

Figure 3 presents components of handgrip control Shows the structure of the handgrip control. Part 9 is prismatic within 12 mm and motor torque is according to handgrip position.

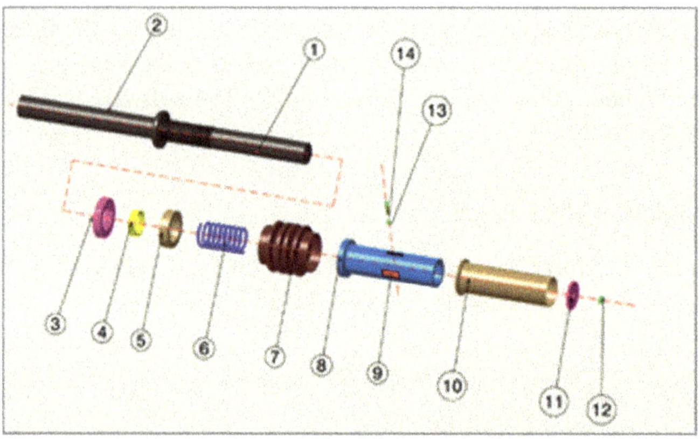

Fig. 3. Components of handgrip control including (1) Hall sensor, (2) Cylindrical pipe, (3) Adjusting nut, (4) Adjusting ring, (5) Spring-ring holder, (6) Spring, (7) Dust shield, (8) Slide tube, (9) Magnet, (10) Handgrip, (11) washer, (12, 13) Nut, (14) Bolt positioning

Moment of the motor is created by magnetic force calculated as

$$M = \frac{\mathrm{N.B_m}.l.\mathrm{I_m}}{2a} \cdot \frac{D}{2}(\mathrm{N.m}) \tag{2}$$

where B_m is the magnetic flux of electro-magnet, l is the conductor length, I_m is the intensity of electric current rotor, D is the conductor diameter, N is the number of conductors in the rotor, a is the number of parallel wires in the rotor. The intensity of the motor current depends on the voltage of the Hall sensor (V_h).

$$V_h = \frac{B_h.I_h}{q.\rho.d}(\mathrm{V}) \tag{3}$$

where ρ is the density of electrons, q is the electron particle charge, d is the thickness of the semiconductor plate, B_m is the magnetic flux of electromagnet. The voltage of the Hall sensor (V_h = 0,5–4,5). The potential value of the hall sensor will vary according to the magnet field at the handgrip position. Changing the magnetic field B_h, the voltage V_h changes proportionally, the change of the magnetic field will cause V_h to change. V_h is the voltage of the Hall sensor (V). The Hall sensor changes the voltage according to the sliding stroke on the magnet bar with magnetic flux B changes with the magnet length l and is directly proportional to the magnetic flux B of the magnet when changing the handgrip position (l = 0 ÷ 12,6 mm) [7].

3 Results and Discussion

3.1 Analysis of the Force and Moments

Figure 4 illustrates the bending stress analysis of the chassis. Safe positions are shown in green, large bearing positions are shown in yellow and red on the chassis.

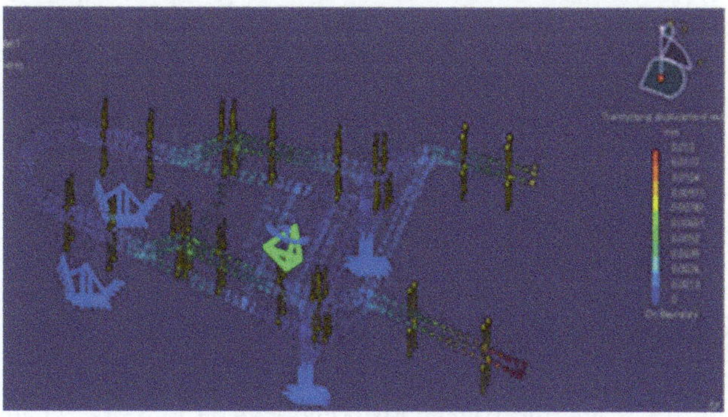

Fig. 4. Bending stress analysis of the chassis

Figure 5 shows the deformed chassis analysis when overloading, Display grid on the frame when in full load state. The load placed on the chassis to calculate is 1200N

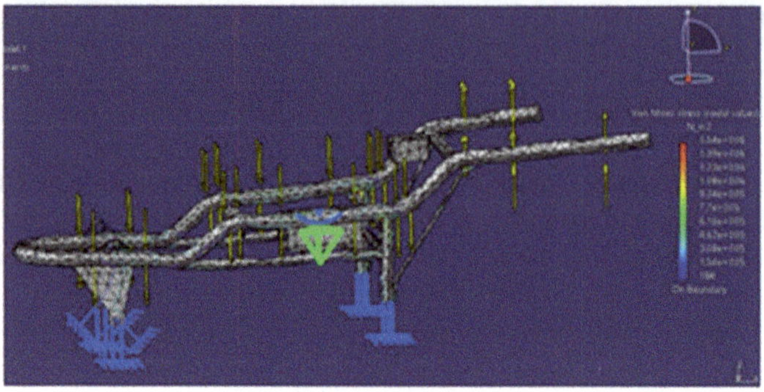

Fig. 5. Deformed chassis analysis when overloading

when the permissible overload is set, the chassis will be deformed [13, 14]. The most bearing position is shown in the two reinforcements of the chassis (Shown in red). Figure 6 shows motor shaft torque according to handgrip position. Diagram of the relationship according to the magnetic force applied to the Hall sensor and the output.

3.2 Electrical Circuit

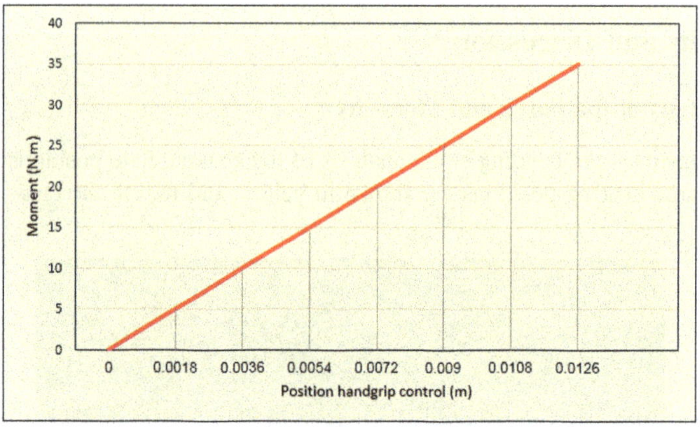

Fig. 6. Motor shaft torque according to handgrip position

3.2.1 Principle of Operation

Source Circuit: ECU and the voltage lowering device (36 V–12 V) are permanently powered by 36 VDC from the vehicle's lithium-ion battery. When the power switch is in the ON position (+) 12 VDC from voltage lowering device → Power switch → Power Relay → Relay closes the contact. We supply (+) 36 to controller Operation controller. Battery parameters are also displayed on the watch on the dashboard.

Motor Control Circuit: When the power switch is in the ON position, supply 5 V to the Hall sensor to E2 (−) and Vc (+). When pushing the handgrip forward, this time the magnet slides on the Hall sensor. The signal voltage was sent to ECU through TH (0.5–4.5 V) E_{CU} was processed and supply the motor current through 1, 2, 3. Hall sensor sends signals to 4, 5, 6, 7, and 8 of the ECU to know the position of the coils in the motor, to control the motor to operate stably.

Brake Control Circuit: When the brake switch is pressed, (+) 12 VDC from the voltage lowering device → Power switch → brake relay (−) 12 VDC → relay closes the contact. We supply (−) 36 VDC from the battery through P_t and P_m. ECM will interrupt the control signal from the Hall sensor to the ECU so that the motor stops operating through the Pt signal, and break the motor through the Pm signal.

Signal Light Circuit: When the power switch is in the ON position. Light: The light switch is in the ON position, (+) 12 VDC from the voltage regulator → Power switch → Light switch (−) 12 VDC Light activity. Hazard Light: Light switch is ON, (+) 12 VDC from voltage lowering device Power switch → hazard switch → light R, L → (−) 12 VDC hazard activity. Figure 7 presents an electrical control circuit that supplies power to control the engine and vehicle safety systems (hazard lights - brake lights). Table 1 demonstrates the specifications of the multifunctional electric wheelbarrow.

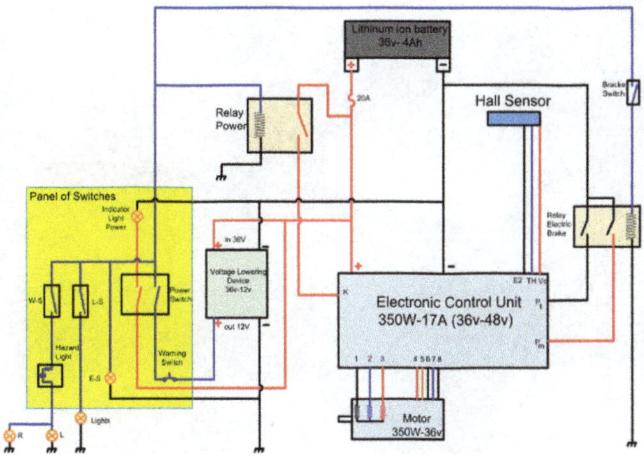

Fig. 7. The electrical circuit of the system

Figure 8 presents the testing products with different cargo weights and terrains tested with 50 kg, 100 kg, 120 kg weight, and testing the product in sloping conditions. Test the product is under good load conditions [14].

Table 1. Design and manufacture multifunctional electric wheelbarrow

Specifications	Values
Size (D × R × C) mm	1500 × 640 × 702
No-load volume	38 kg
The volume of goods	120 kg
Engine type	133 M
Moment	35 N.m
Power	350 W
Rated voltage	36 V
Battery capacity	4000 mAh
Charging time	2 h
Speed	5 km/h
The maximum slope angle	11°
Waterproof	–

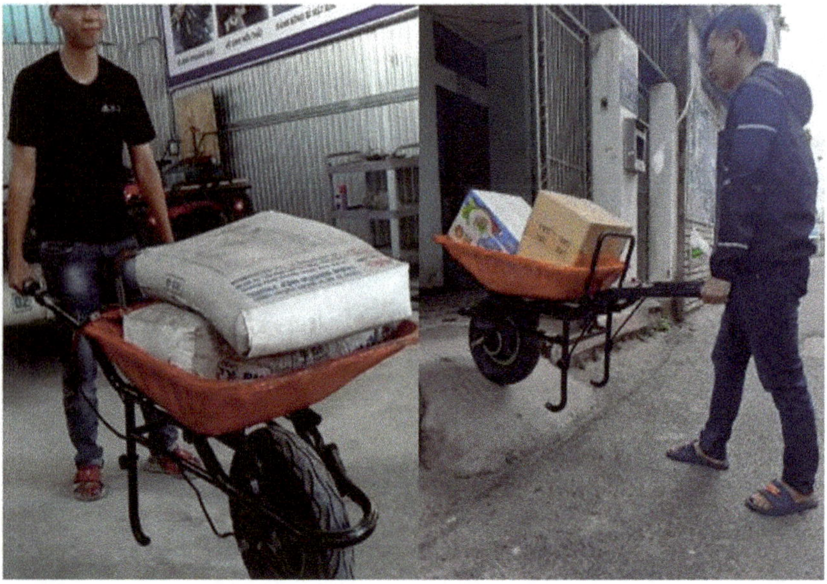

Fig. 8. Testing product after manufacturing

4 Key Conclusions

In this study, authors have done a set of studies including (i) calculating chassis parameters and forces, moments applying for the practical and adapt the model to suitable for the environment of the industrial workshop, (ii) designing and testing models through CatiaV5 software, after simulating, we have checked and then modified the model shape, (iii) fabricating and assembly of actual models, and (iv) testing the product in workshops and constructions. Multi-purpose cargo truck with electric power steering has been researched and developed. The vehicle has a battery design including 2 rows of batteries in parallel, each row of 10 cells, each cell has a voltage of 3.6–3.7 (V), a battery capacity of 4000 mAh. The charger will fully charge the vehicle with a charging time of about 2 h each 3-phase brushless DC electric motor with a capacity of 350 W. Front-wheel diameter 44 cm, rear-wheel diameter 12 cm. Vehicles can operate on the roads with a slope smaller than 20% with a maximum speed of 5 km/h [8], the time of operation and working of vehicles on-site (tested) is about 4h ÷ 5h. The car has a hand brake and emergency brake assist the user in case of emergency or downhill. Versatile cargo vehicles with electric power, with a flip-over mechanism for easy operation, were manufactured and analyzed. Also, the car has hazard lights and lights to work at night. Experiments showed that the stability and safety of the vehicle are ensured according to the calculations set out earlier.

References

1. Loc, N.G.: Mechanical Part Design. Ho Chi Minh City National University Publishing (2015)
2. Dung, D.V.: Engine Electrical and Engine Control, Ho Chi Minh City University of Technical Education (2013)
3. Entler, S., Duran, I., Sladek, P., Vayakis, G., Kočan, M.: Signal conditioning, and processing for metallic Hall sensors. Fusion Eng. Des. **123**, 783–786 (2017)
4. Sumega, M., Zoššák, Š, Varecha, P., Rafajdus, P.: Sources of torque ripple and their influence in BLDC motor drives. Transp. Res. Procedia **40**, 519–526 (2019)
5. Gao, Y., Zhuang, Y., Dong, F., Peng, F., Zhang, P., Yang, L., Ni, Y.: Experimental study on the effect of trolley case on unidirectional pedestrian flow. Phys. A **544**, 122670 (2020)
6. Hung, N., Lim, O.: The effects of operating conditions and structural parameters on the dynamic, electric consumption and power generation characteristics of an electric-assisted bicycle. Appl. Energy **247**, 285–296 (2019)
7. Default system, First steps in CATIA General Structural Analysis, CATIA V5-V6 Tutorials
8. Lin, Y.-H., Shih-Yi, L.: Effects of ground and load on the upper trapezius, biceps brachii muscle and hand forces in one- and two-wheeled wheelbarrow pushing. Appl. Ergon. **88**, 103151 (2020)
9. Jung, M.-C., Haight, J.M., Freivalds, A.: Pushing and pulling carts and two-wheeled hand trucks. Int. J. Ind. Ergon. **35**(1), 79–89 (2005)
10. de la Guerra, A., Alvarez–Icaza, L., Torres, L.: Brushless DC motor control with unknown and variable torque load, IFAC-Papers Online (2018)
11. 曹巧君, 杨定德, 杨国木, Electronic wheelbarrow of self -balancing based on gyroscope stabilizer. Patents: CN205098368U (2015)
12. Girgin, A., Bilmes, M., Amin, H.Y., Karalar, T.C.: A silicon hall sensor SoC for current sensors. Microelectron. J. **90**, 12–18 (2019)
13. Nguyen, V.T.T., Dang, V.A., Tran, N.T., Hoang, N.C., Vo, D.H., Nguyen, D.K., Nguyen, N.L., Nguyen, Q.L., Tieu, T.L., Bui, T.N., Tran, T.B.: An investigation on design innovation, fabrication and experiment of a soybean peeling machine-scale. Int. J. Eng. Technol. (UAE) **7**(4), 2704–2709 (2018)

14. Raghtate, A.S., Handa, D.C.: Design and fabrication of groundnut sheller machine. Int. J. Innov. Res. Sci. Technol. **1**(7) 8 (2014)
15. Ravi, T., GobiGanesh, D., Gokulakannan, R., Kandeeswaran, M., Kesavan, V.: Design and fabrication of solar seed sprayer machine. Int. J. Innov. Res. Multidiscip. Field **2**, (2016)
16. Nguyen, V.T.T., Tran, T.B., Nguyen, T.G., Huynh, H.H., To, V.H., Doan, D.M., Nguyen, T. Q., Van Thanh, D.: An investigation of designing and manufacturing the hard-shell peanut peeling machine with a small scale-size. Int. J. Sci. Technol. Res. **8**, 9 (2019)
17. Foskolos, G.: Measurement-based current-harmonics modeling of aggregated electric-vehicle loads using power-exponential functions. World Electr. Veh. J. **11**, 51 (2020)
18. Yamada, E., Mashiba, T.: Development of technical regulations for fuel cell motorcycles in Japan—hydrogen safety. World Electr. Veh. J. **10**, 48 (2019)

Effect of Lubrication on Deforming the Aluminum Sheet with Two Points Incremental Forming Technology

Ma Van Viet[1], Nguyen Truong Thinh[1], and Le Van Sy[2](✉)

[1] Department of Mechanical Engineering, HCMC University of Technology and Education, Ho Chi Minh City, Vietnam
{vietmv.ncs, thinhnt}@hcmute.edu.vn

[2] PetroVietnam University, Ba Ria, Vietnam
sylv@pvu.edu.vn

Abstract. Facing a complex competition on a global manufacturing market, the companies always change the sharp and size of the products and reducing the designing and development of the products. The manufacturers have been looking for new technologies which are able to response the best customer exigency. These technologies must flexible, simple, inexpensive process, specially, not much time for product design and development. The incremental sheet forming technology has emerged as the best choice for those requirements. Particularly, Two Point Incremental Forming process (TPIF) is method of incremental sheet forming, which was shown the higher geometric accuracy and deformation. This technology is flexible, low cost, economic efficiency which uses the movement of CNC machine tool to form sheet material by layer to layer to the final product shape. The sheet material is clamped firmly on the jig/fixture which can move down along guide bars. The jig/fixture is fixed on the CNC milling machine table. During TPIF processing, the ball end of forming tool always contacts with the metal sheet to deform it following the CNC tool path. The product surface quality will be affected by this contact between metal sheet and tool surface through deformed process. It is also the main cause of wearing the top of forming tool which results on the inaccuracy of a geometric product shape and poor surface quality. This paper focuses on using the kind of lubricants and the lubricating way for TPIF process by investigating the four lubricants such as solid graphite powder, lubrication oil (Gear VG 150 EP), lithium grease, and mixed lubrication. The mixed lubrication is included such as solid graphite powder, lubrication oil, lithium grease with different percentages which helps reducing the contacting friction and improving the surface quality. The influents of machining parameters and their mechanism on the surface quality are analyzed in this study through investigations with TPIF technology for aluminum sheet at room temperature.

Keywords: Lubrication · Two Point Incremental Forming · Roughness surface · deformed profile

B. T. Long et al. (Eds.): MMMS 2020, LNME, pp. 975–982, 2021.
https://doi.org/10.1007/978-3-030-69610-8_129

1 Introduction

Beside forming parameters, machine, toolpath strategy and geometry influence final surface quality of product. Suitable lubricant is also an important factor. It helps improve surface quality of product, reduces forming forces and prevents wear of the end forming tool. Thus, we can overview some studies following.

Diabb et al. [1] study vegetable oil nano lubricants and wear in SPIF process. Sunflower and cornoils, added with 0.0125, 0.025, 0.05 and 0.1 wt% of SiO_2 nanoparticles was used to lubricate 6061 aluminum sheet alloys in SPIF process. The 6061 aluminum sheet surface wears asignificant and reduces when 0.025 wt% of SiO_2 nanoparticles areadded into the vegetable oils. Sornsuwit et al. [2] study lubricants and material properties on formability in SPIF process. Surface roughness of SUS 304 and SUS 316L is low with air blowing as lubricant. Surface roughness of $TiGr_2$ is low with MoS_2 lubricant. Jawale et al. [3] study lubrication study for SPIF of copper. This study result, the mineral oil is used to lubricate. It is an economical and optimal choice for incremental forming of copper and having a significant and positive effect on surface finish. It can also be concluded that the lubrication conditions, in the case of copper, do not affect the formability. Shisode et al. [4] study mixed lubrication for forming of sheet metal. A mixed lubrication was used to determine friction coefficient. It based on local contact conditions and resulting tribological system. The boundary friction direct affects to the measured tool and workpiece surface. The Reynolds equation was used to account the roughness effects. On the other hand, FE software was used to simulate the coupled friction model (boundary friction and the hydrodynamic friction model). Simulation results is clearly showed.

Generally, all studies focused on lubricants suitable for SPIF. In this study, solid graphite powder, lubrication oil (Gear VG 150 EP), lithium grease, and mixed lubrication were studied with lubricating way forming tool submergedinthe, effect of lubrication on deforming the aluminum sheet with two points incremental forming techonology. We will chose lubricants suitable from four lubrications for aluminum sheet with two points incremental forming techonology to obtain the best product surface quality through investigations a frustum cone shaped with TPIF technology for aluminum sheet at room temperature.

2 Experimental Equipment

Experimental equipments incluse a 3 axis CNC milling machine, a Jig and fixture system for TPISF proccss (Fig. 2), a forming tool (Fig. 1) a ball ended forming tool with 12 mm of diameter and etc. A sheet material is a Aluminum sheet A1050 H14 with 400 × 400 mm^2 dimension, 1.5 mm thickness. A sheet material is clamped between movable plate and clamp plate by bolts and move along guide bars. The Jig and fixture system is clamped on the CNC milling machine table (Fig. 3).

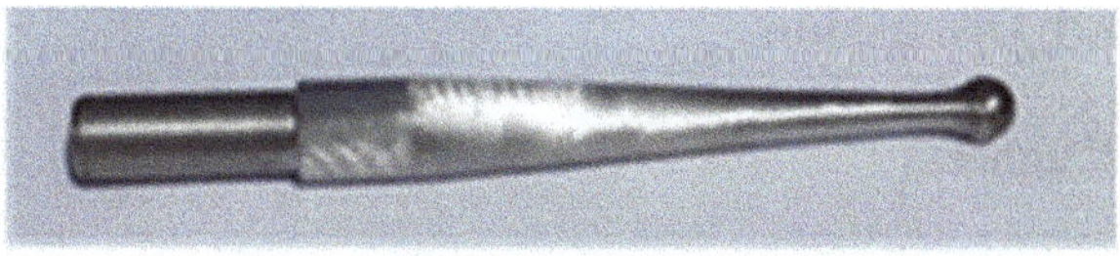

Fig.1. Forming tool

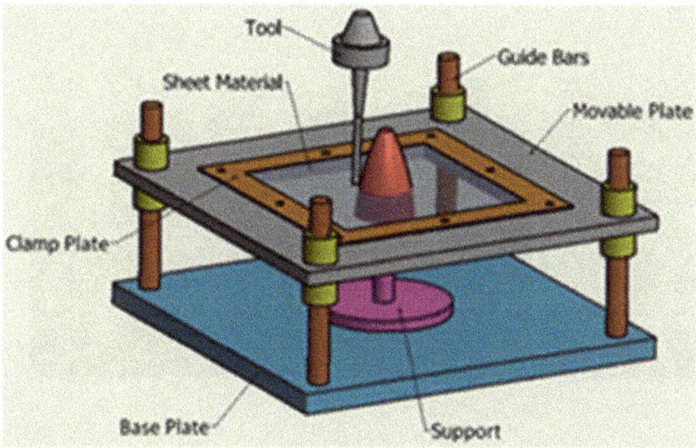

Fig. 2. CAD model of Jig and fixture system for TPISF process

Fig. 3. Practical model of Jig and fixture system for TPISF process

3 Experimental Design

A frustum cone shaped (Fig. 4) is built for experiments. Formed products are investigated about surface roughness and profile through experiments to determine the influence of lubrication on surface roughness and profile with the TPIF for aluminum sheet at room temperature. Experimental parameters, lubricant and lubricating way were chosen as Table 1 (Fig. 5).

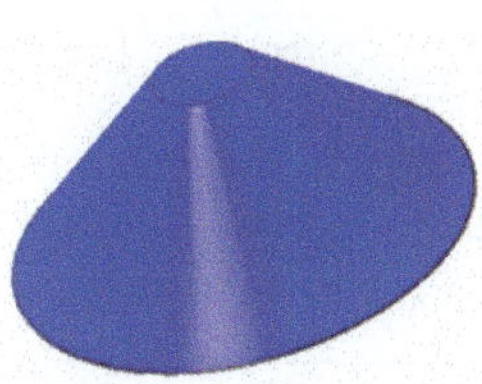

Fig. 4. Mode CAD of a cone shaped

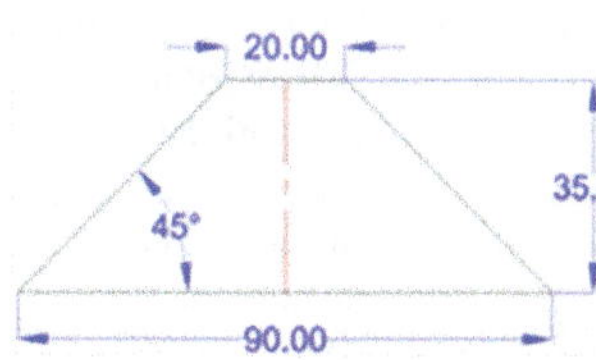

Fig. 5. Profile of frustum cone shape

4 Results and Discussion

4.1 Surface Roughness

Surface quality is characterized by surface roughness. From the measured results, the roughness of the surface exposed to the forming tool is smaller than the other surface, due to many different reasons such as contact conditions, lubrication and machine parameters. Therefore, we can control the best surface quality (Figs. 6, 7 and 8).

a b c d

Fig. 6. Lubrication for TPIF by oil (a), mixed lubrication (b), lithium grease (c), and solid graphite powder solid graphite powder (d)

Table 1. Machining parameters for experimental design, lubricant, and application method.

No	Lubrication		Experimental parameters				Outer surface roughness (μm)
	Lubricant	Application method	Depth step (Δz) mm	Feed rate (Vxy) mm/minute	Tool diameter (D) mm	Spindle speed (n) rpm	
1	Solid graphite powder	Tool submerged in the lubricant	0.2	800	12	1000	3.82
2	Lubrication oil (Gear VG 150 EP)	Tool submerged in the lubricant	0.2	800	12	1000	0.95
3	Lithium grease	Tool submerged in the lubricant	0.2	800	12	1000	1.94
4	Mixed lubrication	Tool submerged in the lubricant	0.2	800	12	1000	0.94

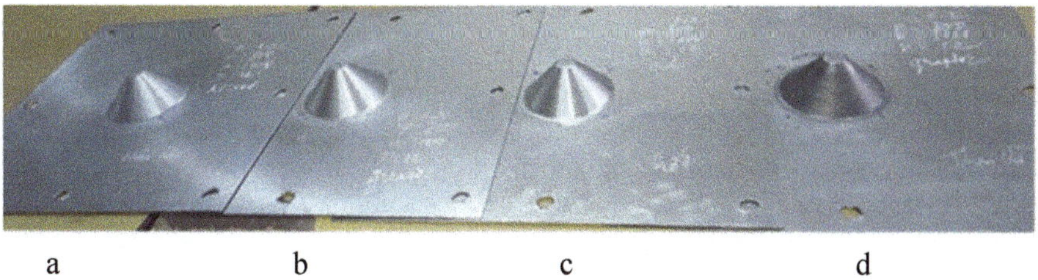

Fig. 7. Products by TPIF. (a: mixed lubrication, b: lithium grease, c: oil lubrication, d: solid graphite powder solid graphite powder)

Fig. 8. Measuring outer surface roughness of a frustum cone product by surface roughness measuring machine

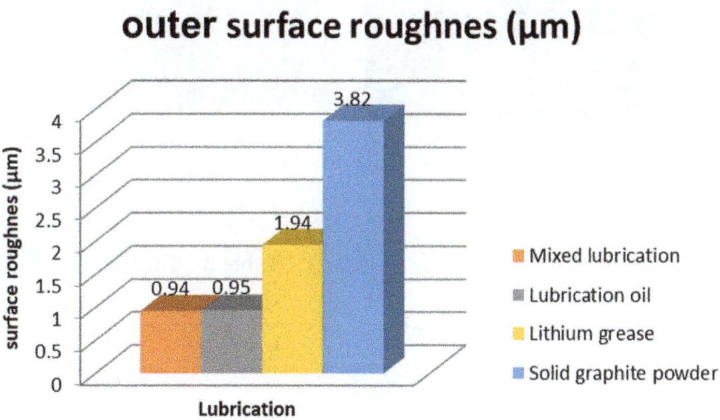

Fig. 9. Outer surface roughnes (µm) with four lubrications

Table 1 shows the product surface roughness of four lubricats with the same lubricating way and experimental parameters. Lubrication help reduces surface roughness. However, product surface roughness is different, it is up to very lubrication. The graph (Fig. 9) shows that lubrication with mixed lubrication is the lowest surface roughness conditions and lubrication with solid graphite powder is the highest surface roughness. Oil lubrication and mixed lubrication are the same the product surface roughness. On the other hand, heating exchange of oil lubrication is very good due to the liquid, so product surface is very smooth and bright (Fig. 10b). Solid graphite powder lubrication, the first, product surface roughness is smooth; after that product surface roughness increases roughened up due to friction and heat increases (Fig. 10 a). Solid graphite powder lubrication is solid condition, so heating exchange of it is very bad.

a b

Fig. 10. a. Solid graphite powder lubrication product, b. oil lubrication product.

4.2 Profile deviation

The products are measured by coordinate measuring machine (CMM) with an accuracy of 2 μm and a probe diameter of 3 mm (Fig. 11). A conical cone profile are rebuilt by the profile measured data. The profiles is presented on the same graph (Figs. 12 and 13).

Fig. 11. Coordinate measuring machine

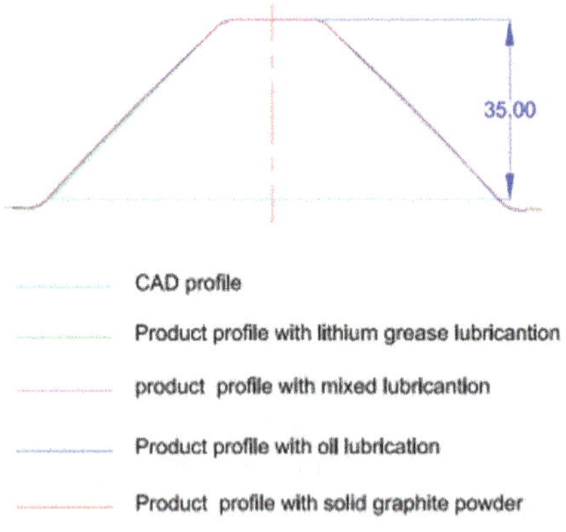

Fig. 12. Compare CAD and product profile.

Table 2. The thickness distribution

No	Thickness (mm)
1	0.512
2	0.505
3	0.51
4	0.511
5	0.507
Average	**0.509**

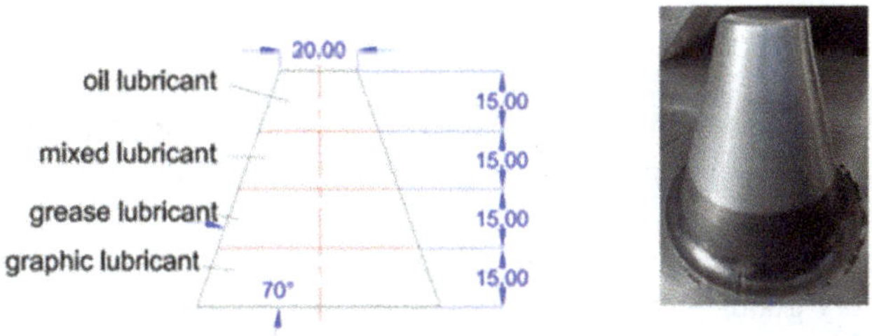

Fig. 13. Integrated lubricants on a frustum cone product

According to graph (Fig. 12). It shows that CAD profile and the experimental profiles obtained from several lubrications is near the same. We said that lubricats is near no significant effect on product profiles.

4.3 Thickness distribution.

A frustum cone mode was chosen with tilt angle 70° (Fig. 14 b) to investigate thickness distribution. The thickness distribution is measured by micrometer with accuracy ± 0.001 mm (Fig. 14 c). Measurement times are five times with average value as (Table 2).

The wall thickness prediction is calculated by the sine law [5].

$t_1 = t_o\mathrm{Sin}(90-\alpha) = t_o\mathrm{Cos}\alpha = t_o\mathrm{Cos}70^o = 1.5*0.342602 = 0.51$ mm, $t_1 = 0.51$ mm.

Percent of error between the wall thickness prediction and the thickness distribution is Δt.

$$\Delta t = (t_0 - t_1) * 100/t_0 = (0.51 - 0.509) * 100/0.51 = 0.196\%$$

According t_o the calculation and measerment results. The wall thickness prediction and the thickness distribution agrees very well with the sine law.

a b c

Fig. 14. a. Frustum cone CAD mode with tilt angle 70°. b. Integrated lubricants on a frustum cone product. c. Measuring thickness of a frustum cone product by micrometer.

5 Conclusion

In order to have good results, the Jig and fixture system is important to investigate the influences of processing parameters on formability. It is designed very stable to creat some results as follows.

The product surface quality and geometric accuracy are formed by The TPIF process, are very good.

The wall thickness prediction and the thickness distribution agrees very well with the sine law.

Lubication is a important factor to achive products with hight quality of surface. Thus, We can chose suitable lubricants and lubricating way. Mixed lubrication is the best lubricant in four lubricants such as solid graphite powder, lubrication oil, lithium grease, and mixed lubrication.

Lubricats is near no significant effect on product profiles.

Oil lubrication and mixed lubrication are the same the product surface roughness.

Oil lubrication is very good heat exchange due to the liquid. Thus, product surface is very smooth and bright. Solid graphite powder lubrication is solid so it is very bad heat exchange, product surface is bad.

Acknowledgments. This research is supported by Key Laboratory of Digital Control and System Engineering (DCSELAB), HCMUT, VNU-HCM under grant number TX2020-20b-01.

References

1. Diabb, J., Rodríguez, C.A., Mamidi, N., Sandoval, J.A., Taha-Tijerina, J., Martínez-Romero, O., Elías-Zúñiga, A.: Study of lubrication and wear in single point incremental sheet forming (SPIF) process using vegetable oil nanolubricants. Wear **376–377 Part A**, 777–785 (2017)
2. Sornsuwit, N., Sittisakuljaroen, S.: The effect of lubricants and material properties in surface roughness and formability for single point incremental forming process. Adv. Mater. Res. **979**, 359–362 (2014)
3. Jawale, K., Duarte, J.F., Reis, A., Silva, M.B.: Lubrication study for single point incremental forming of copper. J. Phys. Conf. Ser. **734**(3), 032038 (2016). IOP Publishing
4. Shisode, M.P., Hazrati, J., Mishra, T., de Rooij, M., van den Boogaard, T.: Modeling mixed lubrication friction for sheet metal forming applications. Procedia Manuf. **47**, 586–590 (2020). Elsevier
5. Li, Y., Chen, X., Liu, Z., Sun, J., Li, F., Li, J., Zhao, G.: A review on the recent development of incremental sheet-forming process. Int. J. Adv. Manuf. Technol. **92**(5), 2439–2462 (2017). Springer, London

A Generalization Approach for the Kinematic Modelling of the Rotational Axes of Five-Axis CNC Centers

Chu Anh My[1(✉)], Chi Hieu Le[2], Xuan Bien Duong[1], Erik L. J. Bohez[3], Minh Duc Vu[1], Tien Anh Nguyen[1], Van Cong Nguyen[1], and Michael Parkianather[4]

[1] Le Quy Don Technical University, Hanoi, Vietnam
myca@lqdtu.edu.vn
[2] University of Greenwich, London, UK
[3] Asian Institute of Technology, Bangkok, Thailand
[4] Cardiff University, Cardiff C24 3AA, UK

Abstract. In practice, the inverse kinematic modelling and analysis of a five-axis CNC center is of importance when transferring and postprocessing a cutter location (CL) data file for machining part surfaces of complex geometry. Decades ago, though there has been a number of investigations focusing on the kinematic modelling and postprocessor development for five-axis CNC centers, a little attention has been paid to the generalization of the inverse kinematic modelling for the rotational axes of five-axis CNC centers. In this paper, a new and general analytical approach for deriving effectively the inverse kinematic equation of two rotational axes of any five-axis CNC center is proposed. It has shown that the proposed technique is very useful for the postprocessor development of any five-axis CNC center.

Keywords: Five-axis CNC center · Inverse kinematic modelling · Postprocessor development

1 Introduction

In theory, when a five-axis CNC center cuts a workpiece, the mechanism of a five-axis CNC center can be viewed as a closed loop 5 DOFs mechanism. This type of mechanism has two kinematic chains which are closed at the cutter contact point (CC point). The first kinematic chain carries the tool and the second one carries the workpiece. During the machining process, the motion of the five joints of both the kinematic chains are constrained all together. In this perspective, the kinematic modelling and analysis of a five-axis CNC center can be investigated based on the methods that were developed for formulating the kinematics model of a robot having closed loop kinematic chains.

In general, a five-axis CNC mechanism is constructed with three translational axes, namely X, Y and Z, and two rotational axes, namely AB, AC or BC. Therefore, the

B. T. Long et al. (Eds.): MMMS 2020, LNME, pp. 983–990, 2021.
https://doi.org/10.1007/978-3-030-69610-8_130

five-axis CNC centers XYZAB, XYZAC and XYZBC are the ones having been widely used in the manufacturing industries.

In the literature, there have been a number of papers related to the kinematic modelling and the inverse kinematics of five-axis CNC centers published in the last decades. Also, a number of geometrical – kinematical techniques were studied and applied to establish the kinematic equation of individual five-axis centers.

The kinematic equation of individual five-axis CNC centers was studied in [1–14]. In the work [1] the kinematic equation was derived for three main types of the five-axis CNC centers (the machines having two rotational axes on the machine table, the machines having two rotational axes on the machine spindle, and the machines having one rotational axis on the machine table and one rotational axis on the machine spindle). In [2] the kinematic equation was derived for of the five-axis CNC center XYZAC with inclined cutter axis. The kinematic equation for some specific five-axis CNC centers was established in [3–9]. For example, the kinematic equation for the five-axis CNC center Spinner U5–620, Maho 600e and DMU 50e was studied in our previous works [5–11]. Note that the papers [10–14] presented the kinematic modelling for the five-axis CNC centers, of which the two rotational axes are not orthogonal. Related to the generalization of the kinematic modelling and analysis for the five-axis CNC centers, there existed some publications [15–18]. The research [15, 16] proposed a kinematic modelling technique for a general five-axis CNC center with seven DOFs since two rotational axes were added on the machine table and other two rotational axes on the machine spindle. A general transformation matrix describing the kinematic relationship between coordinate systems placed in a five-axis CNC kinematic chain was addressed in [17]. By using Screw theory, the authors of [18] generalized the kinematic modelling all five-axis CNC centers. However, it is clearly seen that, the kinematic modelling approaches presented in [15–18] are complex which suffers from a computational burden. This is because, the kinematic equation obtained in [15, 16] was derived with seven generalized coordinates; the kinematic equation in [18] was established with several exponential transformations.

It can be seen that though there has been a number of investigations focusing on the inverse kinematics and postprocessor development for five-axis CNC centers, a generalization of the inverse kinematic modelling for the rotational axes of all five-axis CNC centers was overlooked. Hence, in this paper, we present a new analytical method for deriving effectively the inverse kinematic equation for the rotational axes of any five-axis CNC center. It has shown that the method proposed in this studied is compact and easy-to-use for development of the post-processor for any five-axis CNC center.

2 Kinematic Modelling of a General Five-Axis CNC Center

A generalized five-axis CNC mechanism can be defined as a 5 DOFs mechanism with three prismatic joints and two revolute joints, as shown in Fig. 1 [22]. The three translational joints X, Y and Z characterize the three translational motions of the linear axes. The two rotational joints AB, AC or BC represent the rotational displacements of the two rotational axes of a center.

In the shown mechanism, there are two collaborative kinematic chains. The kinematic chain in the left hand side carries the workpiece on which the workpiece coordinate system $O_w x_w y_w z_w$ is defined. In the right hand side, it is a tool carrying chain, where the cutter coordinate system $O_t x_t y_t z_t$ is defined at the tip of the cutter.

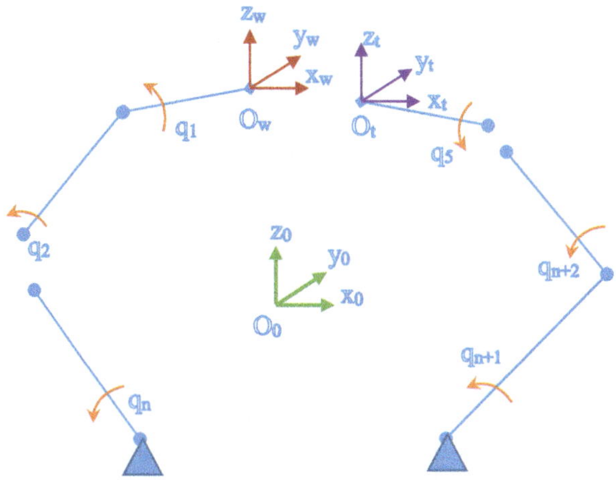

Fig. 1. Mechanism of five-axis CNC centers

We denote $\boldsymbol{q} = [q_1 \quad q_2 \quad q_3 \quad q_4 \quad q_5]^T$ is the vector of the five generalized coordinates corresponding to the five joints of a center. Suppose that on the workpiece carrying chain there exist n (n < 5) generalized coordinates, and the remained (5-n) generalized coordinates are on the tool carrying chain. In this manner, $q_1 \div q_n$ are the notations of the generalized coordinates in the workpiece carrying chain, and $q_{n+1} \div q_5$ are in the tool carrying chain.

The reference frame of the whole mechanism is $O_0 x_0 y_0 z_0$.

Note that three coordinate systems $O_0 x_0 y_0 z_0$, $O_w x_w y_w z_w$ and $O_t x_t y_t z_t$ are parallel all together. For an individual real five-axis CNC center, it is very easy to define these parallel coordinate systems.

In the following equation we defined a transformation matrix which describes the kinematic of a joint i of the generalized mechanism in Fig. 1.

$$\begin{aligned} &\text{For } a \text{ translational axis } i,\ \mathbf{A}_i(q_i) = \begin{bmatrix} \boldsymbol{E} & \boldsymbol{t}_i \\ 0 & 1 \end{bmatrix} \\ &\text{For } a \text{ rotational axis } i,\ \mathbf{A}_i(q_i) = \begin{bmatrix} \boldsymbol{S}_i & \boldsymbol{\tau}_i \\ 0 & 1 \end{bmatrix} \end{aligned} \quad (1)$$

In Eq. (1), $\boldsymbol{E}$ denotes an identity matrix, and $\boldsymbol{t}_i$ denotes the displacement vector of a prismatic joint i. For $q_i = X$, $\boldsymbol{t}_i = [X + X_0 \quad 0 \quad 0]^T$. For $q_i = Y$, $\boldsymbol{t}_i = [0 \quad Y + Y_0 \quad 0]^T$. For $q_i = Z$, $\boldsymbol{t}_i = [0 \quad 0 \quad Z + Z_0]^T$. Note that X_0, Y_0 and Z_0 are the initial distance from O_t to O_0; $\boldsymbol{\tau}_i$ are the offset between the centerline of a rotational axis q_i to O_0x_0, O_0y_0 and O_0z_0, correspondingly.

We denote $\boldsymbol{S}_i$ as a rotation matrix describing the kinematics of a rotational axis i:

$$\text{For } q_i = A,\ \boldsymbol{S}_i = \begin{bmatrix} 1 & 0 & 0 \\ 0 & \cos(A) & -\sin(A) \\ 0 & \sin(A) & \cos(A) \end{bmatrix} \quad (2)$$

$$\text{For } q_i = B,\ \boldsymbol{S}_i = \begin{bmatrix} \cos(B) & 0 & \sin(B) \\ 0 & 1 & 0 \\ -\sin(B) & 0 & \cos(B) \end{bmatrix} \quad (3)$$

$$\text{For} q_i = C,\ \boldsymbol{S}_i = \begin{bmatrix} \cos(C) & -\sin(C) & 0 \\ \sin(C) & \cos(C) & 0 \\ 0 & 0 & 1 \end{bmatrix} \quad (4)$$

Note that if a rotational axis of a five-axis CNC center is inclined at an angle θ, and this inclination is described by a transformation matrix $\boldsymbol{S}(\theta)$, the transformation matrix $\boldsymbol{S}_i(q_i)$ is multiplied by $\boldsymbol{S}(\theta)$ and $\boldsymbol{S}(-\theta)$ in both sides of $\boldsymbol{S}_i(q_i)$, respectively.

The transformation matrices which describe the kinematics of the two kinematic chains can be calculated with the following equations, respectively:

$$\boldsymbol{A}_{0w} = \boldsymbol{A}_n(q_n)\boldsymbol{A}_{n-1}(q_{n-1})\ldots\boldsymbol{A}_1(q_1) \quad (5)$$

$$\boldsymbol{A}_{0t} = \boldsymbol{A}_{n+1}(q_{n+1})\boldsymbol{A}_{n+2}(q_{n+2})\ldots\boldsymbol{A}_5(q_5) \quad (6)$$

The matrix $\boldsymbol{A}_{wt}$ that describes the kinematical relationship between the tool and the workpiece can be calculated as follows:

$$\boldsymbol{A}_{wt} = \boldsymbol{\Theta}_1(q_1)\boldsymbol{\Theta}_2(q_2)\boldsymbol{\Theta}_3(q_3)\boldsymbol{\Theta}_4(q_4)\boldsymbol{\Theta}_5(q_5) \quad (7)$$

In Eq. (7), $\boldsymbol{\Theta}_i(q_i)$ is calculated as follows:

$$\boldsymbol{\Theta}_i(q_i) = \left| \begin{array}{l} \begin{bmatrix} \boldsymbol{E} & \boldsymbol{T}_i \\ 0 & 1 \end{bmatrix}, \text{for } a \text{ translational axis } i \\ \begin{bmatrix} \boldsymbol{R}_i & \boldsymbol{\tau}_i \\ 0 & 1 \end{bmatrix}, \text{for } a \text{ rotational axis } i \end{array} \right. \quad (8)$$

For the joints of the tool carrying chain, we define

$$\begin{array}{l} \boldsymbol{T}_i = \boldsymbol{t}_i, \text{ for } a \text{ translational axis } i \\ \boldsymbol{R}_i = \boldsymbol{S}_i, \text{ for } a \textit{ rotational axis i} \end{array} \quad (9)$$

For the joints of the workpiece carrying chain, we have

$$\begin{aligned} \boldsymbol{T}_i &= -\boldsymbol{t}_i, \text{ for } a \text{ translational axis } i \\ \boldsymbol{R}_i &= \boldsymbol{S}_i^T, \text{ for } a \text{ rotational axis } i \end{aligned} \tag{10}$$

Hence, we can write

$$\boldsymbol{A}_{wt} = \begin{bmatrix} \boldsymbol{R}_{wt} & \boldsymbol{p}_T \\ 0 & 1 \end{bmatrix} \tag{11}$$

In Eq. (11), $\boldsymbol{p}_T = [x, y, z]^T$ denotes the tooltip in $O_w x_w y_w z_w$. $\boldsymbol{R}_{wt}$ is the rotation matrix, in which the third column are the directissson cosines of the tool vector (*i*,*j*,*k*).

Finally, the kinematic equation is obtained as follows:

$$\boldsymbol{p} = \boldsymbol{f}(\boldsymbol{q}) \tag{12}$$

where $\boldsymbol{p} = [x \quad y \quad z \quad i \quad j \quad k]^T$ denotes the tool position.

3 Kinematic Equation for Rotational Axes of Five-Axis Centers

When developing a postprocessor for an individual five-axis CNC center, the derivation of the forward and inverse kinematic equations for the two rotational axes requires complex mathematical transformations. To support practitioners, an effective method is presented in this section to derive easily the kinematic equation for the two rotational axes of any five-axis CNC center. In particular, we provide with a data table presenting the inverse kinematic equations for the two rotational axes of all five-axis CNC configurations.

Based on the Property 1 of the kinematics model has shown in our previous publication [22], the kinematics model of the rotational axes can be formulated based on the rotation matrices $\boldsymbol{R}_u$ and $\boldsymbol{R}_v$ which describe the kinematic displacement of the two rotational axes. When taking into account a given five-axis CNC center, the two basic rotation matrices $\mathbf{S}_u$ and $\mathbf{S}_v$ in the form of Eqs. (2–4) are easily identified. The matrices $\boldsymbol{R}_u$ and $\boldsymbol{R}_v$ are then calculated with Eqs. (9, 10). The forward kinematic equations $i = h_1(q_u, q_v), j = h(q_u, q_v)$ and $k = h_3(q_u, q_v)$ (the last column of $\boldsymbol{R}_{wt}$) can be calculated with Eq. (11). Based on the obtained equations, the inverse kinematic equations for the rotational axes can be achieved accordingly.

Table 1 shows the forward and inverse kinematic solutions obtained with 15 possible configurations of five-axis CNC centers.

In Table 1, (A_t, B_t, C_t) and (A_w, B_w, C_w) are denoted as the rotational axes on the tool chain and the workpiece chain of a five-axis CNC center, respectively. Note that if the both rotational axes are on the tool chain, there exist 3! = 6 combinations of the rotational axes A_tB_t,A_tC_t,B_tA_t,B_tC_t,C_tA_t and C_tB_t. Due to the symmetry, there also exist six combinations of the rotational axes A_wB_w,A_wC_w,B_wA_w, B_wC_w,C_wA_w and C_wB_w. In addition, combining one rotational axis on the tool chain and one rotational axis on the workpiece chain there exist additionally six combinations A_tB_w,A_tC_w,B_tA_w, B_tC_w,C_tA_w

Table 1. The kinematics for the two rotational axes of the five-axis CNC centers

		Forward kinematic solution	Inverse kinematic solution
1	A_tB_t	$i = \sin B; j = -\sin A \cos B; k = \cos A \cos B$	$B = \sin^{-1}(i); A = \tan^{-1}(-j/k)$
2	B_tA_t	$i = \sin B \cos A;\ j = -\sin A;\ k = \cos A \cos B$	$A = \sin^{-1}(-j); B = \tan^{-1}(i/k)$
3	C_tA_t	$i = \sin C \sin A;\ j = -\cos C \sin A;\ k = \cos A$	$A = \cos^{-1}(k); C = \tan^{-1}(-i/j)$
4	C_tB_t	$i = \cos C \sin B;\ j = \sin C \sin B;\ k = \cos B$	$B = \cos^{-1}(k); C = \tan^{-1}(j/i)$
5	A_wB_w	$i = -\cos A \sin B;\ j = \sin A;\ k = \cos B \cos A$	$A = \sin^{-1}(j); B = \tan^{-1}(-i/k)$
6	A_wC_w	$i = \sin C \sin A;\ j = \cos C \sin A;\ k = \cos A$	$A = \cos^{-1}(k); C = \tan^{-1}(i/j)$
7	B_wA_w	$i = -\sin B;\ j = \sin A \cos B;\ k = \cos A \cos B$	$B = \sin^{-1}(-i); A = \tan^{-1}(j/k)$
8	B_wC_w	$i = -\cos C \sin B;\ j = \sin C \sin B;\ k = \cos B$	$B = \cos^{-1}(k); C = \tan^{-1}(-j/i)$
9	A_tB_w	$i = -\sin B \cos A;\ j = -\sin A;\ k = \cos A \cos B$	$A = \sin^{-1}(-j); B = \tan^{-1}(-i/k)$
10	A_tC_w	$i = -\sin C \sin A;\ j = -\cos C \sin A;\ k = \cos A$	$A = \cos^{-1}(k); C = \tan^{-1}(i/j)$
11	B_tA_w	$i = \sin B;\ j = \sin A \cos B; k = \cos A \cos B$	$B = \sin^{-1}(i); A = \tan^{-1}(j/k)$
12	B_tC_w	$i = \cos C \sin B;\ j = \sin C \sin B;\ k = \cos B$	$B = \cos^{-1}(k); C = \tan^{-1}(j/i)$
13	$B_wC_w(45^\circ)$	$i = -\frac{\cos C \sin B}{\sqrt{2}} + \frac{\sin C \cos B - \sin C}{2}$ $j = \frac{\sin C \sin B}{\sqrt{2}} + \frac{\cos C \cos B - \cos C}{2}$ $k = \frac{1}{2} + \frac{1}{2}\cos B$	$B = \cos^{-1}(2k-1) C = \tan^{-1}\left(\frac{(k-1)i + j\sqrt{2(k-k^2)}}{(k-1)j - i\sqrt{2(k-k^2)}}\right)$
14	C_tB_t (45°)	$i = -\frac{\cos C \sin B}{\sqrt{2}} + \frac{\sin C \cos B - \sin C}{2}$ $j = \frac{\sin C \sin B}{\sqrt{2}} - \frac{\cos C \cos B - \cos C}{2}$ $k = \frac{1}{2} + \frac{1}{2}\cos B$	$B = \cos^{-1}(2k-1) C = \tan^{-1}\left(-\frac{(k-1)i + j\sqrt{2(k-k^2)}}{(k-1)j - i\sqrt{2(k-k^2)}}\right)$
15	B_tC_w (45°)	$i = \frac{\cos C \sin B}{\sqrt{2}} - \frac{\sin C \cos B - \sin C}{2}$ $j = -\frac{\sin C \sin B}{\sqrt{2}} - \frac{\cos C \cos B - \cos C}{2}$ $k = \frac{1}{2} + \frac{1}{2}\cos B$	$B = \cos^{-1}(2k-1) C = \tan^{-1}\left(\frac{(k-1)i + j\sqrt{2(k-k^2)}}{(k-1)j - i\sqrt{2(k-k^2)}}\right)$

and C_tB_w. It is noticeable that a five-axis center cannot be constructed with the combinations A_tC_t,B_tC_t,C_tA_w,C_tB_w, C_wA_w and C_wB_w because the displacement of C_t and C_w coincides with the spindle rotation. Therefore, with respect to all feasible combinations of the rotational axes, the family of five-axis CNC centers can be classified into 12 groups shown in Table 1.

Note that the last three rows of Table 1 are also for the configurations B_wC_w, C_tB_t and B_tC_w, however the centreline of the B-axis is inclined at 45^0.

4 Conclusion

Based on the kinematic modelling approach that are applied for modelling the collaborative manipulators, a new method for kinematic modelling of a generalized five-axis CNC mechanism was effectively developed. By using the proposed method, the kinematic equation for the rotational axes of the five-axis CNC centers are formulated in a generalized manner that has never addressed before. The obtained equations in Table 1 can be used for development of the postprocessor for any five-axis CNC center.

Acknowledgement. This research was funded by the UK Department for Business, Energy and Industrial Strategy and delivered by the British Council, under a Newton Fund Project ID 528085858.

References

1. Lee, R.S., She, C.H.: Developing a postprocessor for three types of five-axis machine tools. The International Journal of Advanced Manufacturing Technology **13**, 658–665 (1997)
2. Xu, H. Y., Hu, L. A., Hon-Yuen, Shi, K., Xu, L. A., A novel kinematic model for five-axis machine tools and its CNC applications, The International Journal of Advanced Manufacturing Technology 67 (2013) 1297–307.
3. Farouki, R.T., Han, C.Y., Li, S.: Inverse kinematics for optimal tool orientation control in five-axis CNC machining. Computer Aided Geometric Design **31**, 13–26 (2014)
4. Boz, Y., Lazoglu, I.: A postprocessor for table-tilting type five-axis machine tool based on generalized kinematics with variable feedrate implementation. Int. J. Adv. Manuf. Technol. **66**, 1285–1293 (2013)
5. My, C. A., Bohez, E. L., Multi-criteria optimization approach for five-axis CNC tool path planning: modeling methodology. In Proceedings of the 3rd Asian Conference on Industrial Automation and Robotics (2003) 5–11
6. My, C.A., Bohez, E.L., Makhanov, S.S., Munlinb, M., Phien, H.N., Tabucanon, M.T.: On five-axis freeform surface machining optimization: Vector field clustering approach. International Journal of CAD/CAM **5**, 1–1 (2009)
7. My, C. A., Bohez, E. L. J., & Makhanov, S. S., Critical point analysis of 3D vector field for five-axis tool path optimization. In Proceedings of the 4th Asian Conference on Industrial Automation and Robotics, ACIAR (2005) 11–13
8. Bohez, E.L.J.: Five-axis milling machine tool kinematic chain design and analysis. Int. J. Mach. Tools Manuf **42**, 505–520 (2002)
9. My, C. A., Bien, D. X., Tung, B. H., Van Cong, N., New feed rate optimization formulation in a parametric domain for five-axis milling robots, In International Conference on Computer Science, Applied Mathematics and Applications (2019) 403–411
10. My, C. A., Integration of CAM systems into multi-axes computerized numerical control machines. In 2010 IEEE Second International Conference on Knowledge and Systems Engineering (2010) 119–124.
11. My, C.A., Bohez, E.L.: New algorithm to minimise kinematic tool path errors around five-axis machining singular points. Int. J. Prod. Res. **54**, 5965–5975 (2016)
12. My, C.A., Hai, V.X.: Generalized pseudo inverse kinematics at singularities for developing five-axes CNC machine tool postprocessor. Vietnam Journal of Mechanics **35**, 147–155 (2013)
13. Sørby, K.: Inverse kinematics of five-axis machines near singular configurations. Int. J. Mach. Tools Manuf **47**, 299–306 (2007)
14. She, C.H., Huang, Z.T.: Postprocessor development of a five-axis machine tool with nutating head and table configuration. The International Journal of Advanced Manufacturing Technology **38**, 728–740 (2008)
15. Tutunea-Fatan, Feng H. Y., Configuration analysis of five-axis machine tools using a generic kinematic model, International Journal of Machine Tools and Manufacture 44 (2004) 1235–43
16. Yang, J., Altintas, Y.: Generalized kinematics of five-axis serial machines with non-singular tool path generation. Int. J. Mach. Tools Manuf **75**, 119–132 (2013)

17. She, C.H., Chang, C.C.: Design of a generic five-axis postprocessor based on generalized kinematics model of machine tool. Int. J. Mach. Tools Manuf **47**, 537–545 (2007)
18. My, C.A., Bohez, E.L.: A novel differential kinematics model to compare the kinematic performances of five-axis CNC centers. Int. J. Mech. Sci. **163**, 105117 (2019)

Real Time Inverse Kinematics of Five-Axis CNC Centers

Chu Anh My[1(✉)], Chi Hieu Le[2], Xuan Bien Duong[1], Erik L. J. Bohez[3], Minh Duc Vu[1], Tien Anh Nguyen[1], Van Cong Nguyen[1], Michael Parkianather[4], and Chu Duc Hoang[5]

[1] Le Quy Don Technical University, Hanoi, Vietnam
myca@lqdtu.edu.vn
[2] University of Greenwich, London, UK
[3] Asian Institute of Technology, Bangkok, Thailand
[4] Cardiff University, Cardiff CF24 3AA, UK
[5] Ministry of Science and Technology, Hanoi, Vietnam

Abstract. In most of the previous investigations, the kinematics model of five-axis computer-numerical control (CNC) centers was formulated just at the position level, and the differential kinematic relationships (velocity, acceleration and jerk of the five joints of a five-axis center) that are necessary for several purposes, especially for investigating the relationship between the limits of a machine's drives and the feed rate maximization (the productivity maximization), have been overlooked. Therefore, this paper addresses the differential kinematic modelling and analysis for the five-axis CNC centers. In particular, the differential kinematic equations are formulated in a parametric domain so that they are useful for investigating the kinematic behaviors of the five-axis centers in real time.

Keywords: 5-axis CNC machining · Real time kinematics · Differential kinematics · Inverse kinematics

1 Introduction

In the last decades, the high speed five-axis CNC centers have been widely used to machine complex parts. A five-axis CNC center can be regarded as two cooperative robot arms, one arm carries the cutter and one arm holds the machining part.

In practice, in order to increase the productivity when a five-axis CNC center produces mechanical parts, it is necessary to rise the feed rate which is usually given as a constant in a machining program (G-codes file). When the feed rate of a five-axis CNC center is increased, the velocity and acceleration of the five drives varies rapidly, especially for that case that a surface having high curvature is machined. Hence, real time analysis and monitoring of velocity and acceleration of the five drives of a five-axis CNC center plays an important in optimizing the machining feed rate.

In the literature, several research topics related to 5-axis CNC machining have been taken into account, such as the postprocessor development, the feed rate adjustment, the geometric error compensation, the inverse kinematics, etc. Though there have been

B. T. Long et al. (Eds.): MMMS 2020, LNME, pp. 991–999, 2021.
https://doi.org/10.1007/978-3-030-69610-8_131

numerous research works published, few investigations take into account the real time kinematic modelling and analysis for the five-axis CNC centers.

Years ago, a number of papers have been published which focus on the inverse kinematics problem of the five-axis CNC centers [1–18]. The inverse kinematics model of a general five-axis CNC center has been studied in [1–4]. The authors of the works [1–3] generalize the kinematic formulation for a general five-axis CNC center by adding two more extra axes to a common five-axis mechanism, while the authors of [4] used the Screw Theory to formulate the kinematics model of a five-axis CNC machine. The kinematic modelling and analysis for a class of five-axis CNC center of which the rotational axes are not orthogonal was well documented [5–9]. The works [5, 6] studied the inverse kinematic model of the five-axis CNC center DMU 50e. Other types of five-axis CNC centers were investigated in [7–9]. In the same context, the kinematic model of the class of five-axis CNC center with two orthogonal rotational axes was formulated. Three main types of this machine class were investigated in [10, 11]. The kinematic solution of some individual five-axis CNC centers like Maho 600e and Spinner U5-620 was presented in [12–18].

Obviously, though there have been several research works related to the inverse kinematics of the five-axis CNC centers, none of them takes into account the real time inverse kinematic modelling and analysis for a better kinematic performance analysis of the five-axis CNC centers. In this paper, a novel real time kinematic modelling and analysis of the five-axis CNC centers is presented. It has shown that the method proposed in this paper is useful for analysis and monitoring the kinematic performances of the five drives of any five-axis CNC center.

2 Real Time Kinematic Modelling of a General Five-Axis CNC Center

Based on the kinematic modelling of a general five-axis CNC center presented in our previous work [14], the inverse kinematic equations at differential levels can be formulated in a time domain, and the derivation of these equations are presented in this section. As shown in [14], the forward kinematic equation of a general five-axis CNC center was given as

$$\boldsymbol{X} = \boldsymbol{f}(\boldsymbol{q}) \tag{1}$$

where $\boldsymbol{X} = [x \quad y \quad z \quad \phi \quad \varphi]^T$ is the pose of the cutter, $\boldsymbol{q} = [q_1 \quad q_2 \quad q_3 \quad q_4 \quad q_5]^T$ are the five axes of a five-axis CNC center.

Suppose that $f(t) = \dot{s}(t)$ is the velocity of the cutter along the tool path $\boldsymbol{X}(u) = [x(u) \quad y(u) \quad z(u) \quad \phi(u) \quad \varphi(u)]^T \in \Re^5, u \in [0, 1]$, in which s is the length of a cutting path, and t is time. The inverse kinematic solution can be written as follows:

$$\boldsymbol{q} = \boldsymbol{f}^{-1}(\boldsymbol{X}) \tag{2}$$

From Eq. (1), the following derivative can be yields

$$\dot{X} = J\dot{q} \tag{3}$$

Note that $J = \frac{\partial f}{\partial q}$ is denoted for the Jacobian matrix. Based on Eq. (3), the velocity of the joints can be calculated as:

$$\dot{q} = J^{-1}\dot{X} = J^{-1}\frac{dX}{ds}\frac{ds}{dt} = J^{-1}X_s'\dot{s} \tag{4}$$

Note that

$$ds \simeq \left|\frac{dX}{du}\right| du, \tag{5}$$

$$X_s' = \frac{dX}{ds} = \frac{dX/du}{ds/du} = \frac{dX/du}{|dX/du|} = \frac{X_u'}{|X_u'|} \tag{6}$$

Thus we obtain the velocity as follows:

$$\dot{q} = J^{-1}\frac{X_u'}{|X_u'|}f \tag{7}$$

Equation (7) implies that at any value of time, the velocities of the five axes of a given five-axis CNC center can be computed in time domain with respect to the geometric properties of the prescribed cutting curve $\mathbf{X}_u'$, and the Jacobian $\mathbf{J}$ of a five-axis CNC center. From Eq. (3) we can yield

$$\ddot{X} = J\ddot{q} + \dot{J}\dot{q} \tag{8}$$

Hence

$$\ddot{q} = J^{-1}(\ddot{X} - \dot{J}\dot{q}) \tag{9}$$

Note that

$$\dot{X} = X_s'\dot{s},$$

$$\ddot{X} = \frac{dX_s'}{ds}\dot{s}^2 + X_s'\ddot{s} = X_s''\dot{s}^2 + X_s'\ddot{s} \tag{10}$$

Based on Eq. (5), Eq. (11) is obtained

$$s_u' = |X_{u'}'| \tag{11}$$

Equation (6) now becomes

$$\boldsymbol{X}_s' = \frac{\boldsymbol{X}_u'}{s_u'} \tag{12}$$

Equation (12) can be obtained as follows:

$$\boldsymbol{X}_s'' = \frac{(d\boldsymbol{X}_u'/ds)s_u' + \boldsymbol{X}_u'(ds_u'/ds)}{|\boldsymbol{X}_u'|^2} \tag{13}$$

It is also noted that $\frac{d\boldsymbol{X}_u'}{ds} = \frac{\boldsymbol{X}_u''}{s_u'} = \frac{\boldsymbol{X}_u''}{|\boldsymbol{X}_u'|}; \frac{ds_u'}{ds} = \frac{s_u''}{s_u'} = \frac{(\boldsymbol{X}_u')^T\boldsymbol{X}_u''}{|\boldsymbol{X}_u'|^2}$

Hence

$$\boldsymbol{X}_s'' = \frac{\boldsymbol{X}_u''}{|\boldsymbol{X}_u'|^2} + \frac{\boldsymbol{X}_u'\left((\boldsymbol{X}_u')^T\boldsymbol{X}_u''\right)}{|\boldsymbol{X}_u'|^4} \tag{14}$$

Now Eq. (9) becomes

$$\ddot{\boldsymbol{q}} = \boldsymbol{J}^{-1}\left(\left(\frac{\boldsymbol{X}_u''}{|\boldsymbol{X}_u'|^2} + \frac{\boldsymbol{X}_u'\left((\boldsymbol{X}_u')^T\boldsymbol{X}_u''\right)}{|\boldsymbol{X}_u'|^4}\right)f^2 + \frac{\boldsymbol{X}_u'}{|\boldsymbol{X}_u'|}\dot{f} - \dot{\boldsymbol{J}}\dot{\boldsymbol{q}}\right) \tag{14}$$

Equation (14) is the joint accelerations of a five-axis CNC center. In order to calculate the jerk of the joints, taking time derivative of Eq. (1) yields

$$\dddot{\boldsymbol{X}} = \boldsymbol{J}\dddot{\boldsymbol{q}} + 2\dot{\boldsymbol{J}}\ddot{\boldsymbol{q}} + \ddot{\boldsymbol{J}}\dot{\boldsymbol{q}} \tag{15}$$

Hence

$$\dddot{\boldsymbol{q}} = \boldsymbol{J}^{-1}(\dddot{\boldsymbol{X}} - 2\dot{\boldsymbol{J}}\ddot{\boldsymbol{q}} - \ddot{\boldsymbol{J}}\dot{\boldsymbol{q}}) \tag{16}$$

We calculate

$$\dddot{\boldsymbol{X}} = \frac{d\ddot{\boldsymbol{X}}}{ds}\dot{s} = \frac{d}{ds}\left(\boldsymbol{X}_s''\dot{s}^2 + \boldsymbol{X}_s'\ddot{s}\right)\dot{s} = \left(\boldsymbol{X}_s'''\dot{s}^2 + \boldsymbol{X}_s''\ddot{s}\right)\dot{s} \tag{17}$$

$$\begin{aligned}\boldsymbol{X}_s''' &= \frac{d\boldsymbol{X}_s''}{du}\frac{1}{|\boldsymbol{X}_u'|} \\ &= \frac{\boldsymbol{X}_u'''}{|\boldsymbol{X}_u'|^3} + \frac{\boldsymbol{X}_u''\left[(\boldsymbol{X}_u')^T\boldsymbol{X}_u''\right] + \boldsymbol{X}_u'\left[(\boldsymbol{X}_u'')^T\boldsymbol{X}_u'' + (\boldsymbol{X}_u')^T\boldsymbol{X}_u'''\right]}{|\boldsymbol{X}_u'|^5}\end{aligned} \tag{18}$$

Finally, Eq. (16) becomes

$$\dddot{\boldsymbol{q}} = \boldsymbol{J}^{-1}\left\{ \begin{array}{c} \frac{X_u'''}{|X_u'|^3} f^3 + \frac{X_u''\left[(X_u')^T X_u''\right] + X_u'\left[(X_u'')^T X_u'' + (X_u')^T X_u'''\right]}{|X_u'|^5} f^3 + \\ \left[\frac{X_u''}{|X_u'|^2} + \frac{X_u'\left((X_u')^T X_u''\right)}{|X_u'|^4}\right]\dot{f}f - 2\dot{\boldsymbol{J}}\ddot{\boldsymbol{q}} - \ddot{\boldsymbol{J}}\dot{\boldsymbol{q}} \end{array} \right\} \tag{19}$$

Note that $\dot{\boldsymbol{J}}$ and $\ddot{\boldsymbol{J}}$ are formulated as follows: $\dot{\boldsymbol{J}} = \left[\frac{\partial J_{ij}}{\partial \boldsymbol{q}}\right] \otimes \dot{\boldsymbol{q}}$ and

$$\ddot{\boldsymbol{J}} = \left[\frac{\partial}{\partial \boldsymbol{q}}\left(\left[\frac{\partial J_{ij}}{\partial \boldsymbol{q}}\right] \otimes \dot{\boldsymbol{q}}\right)\right] \otimes \begin{bmatrix} \dot{\boldsymbol{q}} \\ \cdots \\ \dot{\boldsymbol{q}} \end{bmatrix}_{1\times 5} + \left[\frac{\partial J_{ij}}{\partial \boldsymbol{q}}\right] \otimes \ddot{\boldsymbol{q}}$$

Equations (7), (14), (19) are the vectors of velocities, accelerations and jerks of the five drives of a five-axis CNC center that are calculated in time domain. Given an individual five-axis CNC center, the value of this vector can be computed with respect to the velocity of cutter along a given tool path.

3 Numerical Analysis of Real Time Inverse Kinematics for Some Five-Axis CNC Centers

Example 1. The velocities and accelerations of the five drives of the five-axis CNC center DMU 50e

The programmed tool path and the velocity of the cutter are given in Fig. 2.

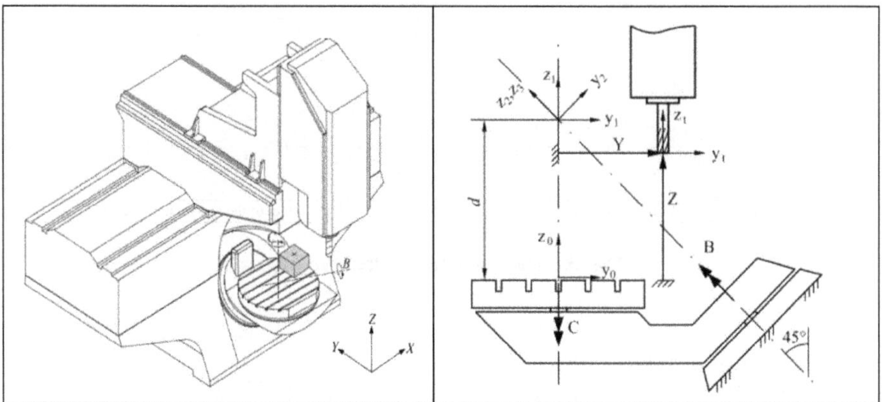

a) The 3D model of the center DMU 50e | b) The kinematic chain of the center DMU 50e

Fig. 1. Five-axis CNC center DMU 50e

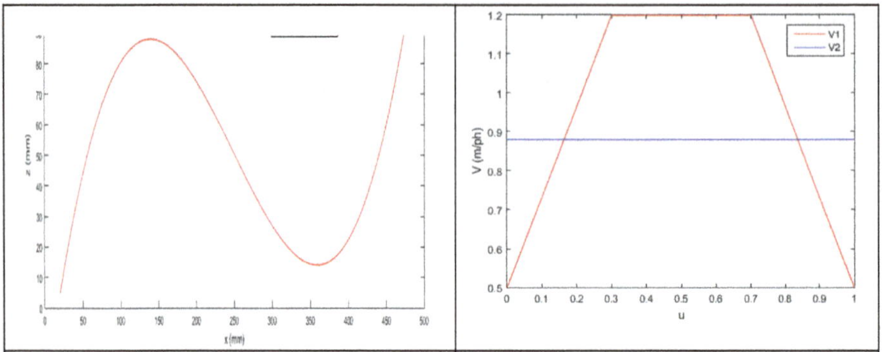

a) b)

Fig. 2. a) A given tool path and b) the prescribed velocity profile of the cutter

Let's consider the model and the kinematic chain of the five-axis CNC center DMU 50e as shown in Fig. 1. Using Eqs. (7), (14), (19) to calculated the real time inverse kinematic solution for the five-axis CNC center DMU 50e, the velocities and accelerations of the machine axes are obtained and presented in the following Figs. 3 and 4.

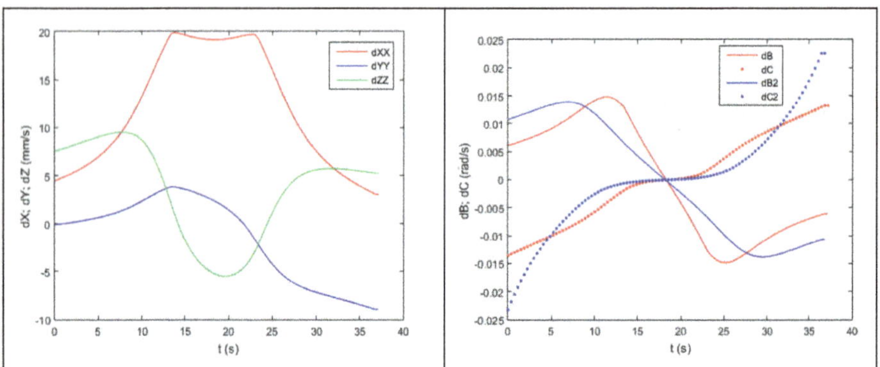

a) Velocity of the three translational axes

b) Velocity of the two rotational axes

Fig. 3. The velocities of the five axes

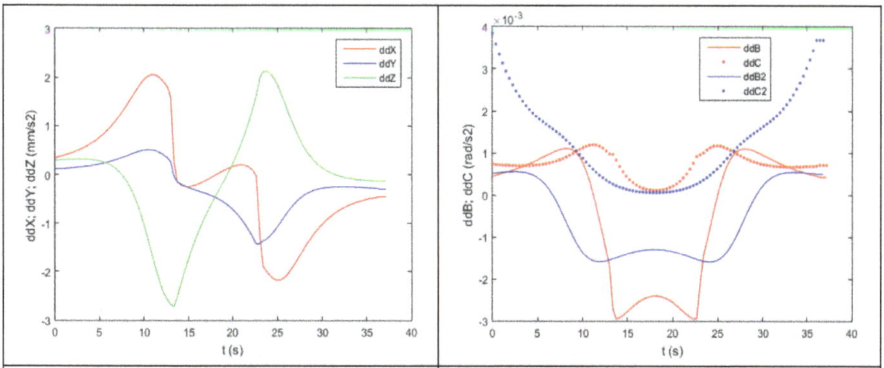

a) Acceleration of the three translational axes | b) Acceleration of the two rotational axes

Fig. 4. The accelerations of the five axes

Example 2. The velocities and accelerations of the five drives of the five-axis CNC center Spinner U5-620

Figure 5 shows the 3D model and the kinematic chain of the five-axis CNC center Spinner U5-620. The velocities and accelerations of the exes of the machine are computed and shown in Figs. 6 and 7, respectively.

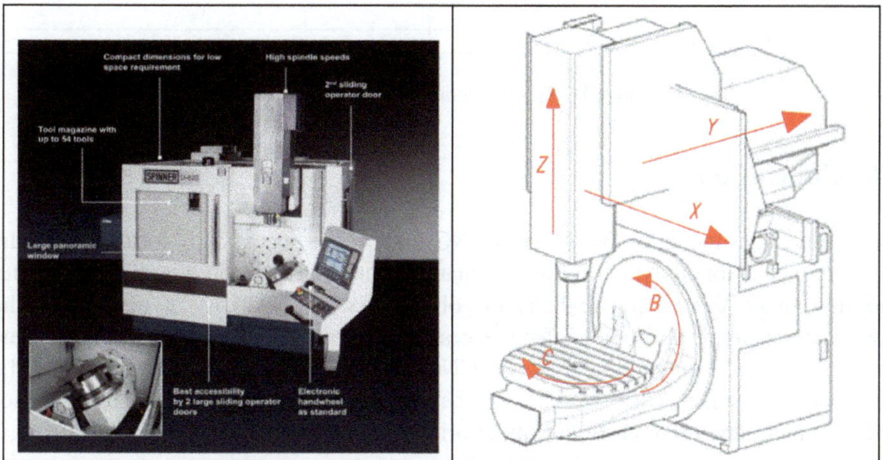

a) The model of Spinner U5-620 | b) Kinematic chain of Spinner U5-620

Fig. 5. The five-axis CNC center DMU 50e

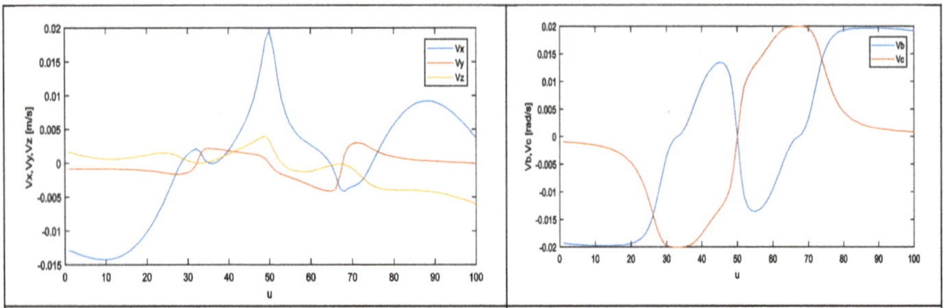

a) Velocity of the three translational axes | b) Velocity of the two rotational axes

Fig. 6. The velocities of the five axes

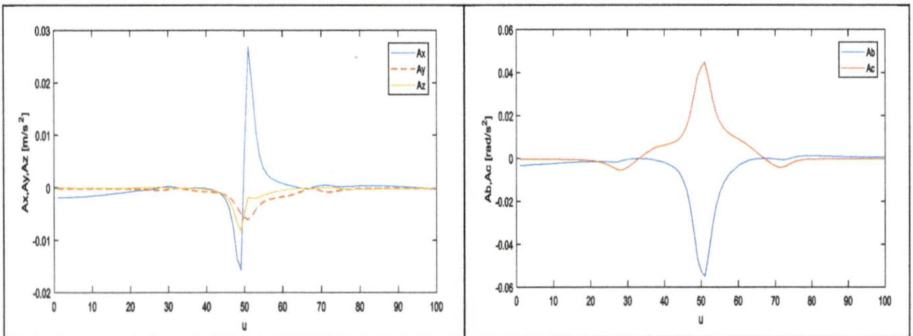

a) Acceleration of three linear axes | b) Acceleration of two rotational axes

Fig. 7. The accelerations of the five axes

4 Conclusion

The kinematics model of the five-axis CNC centers was successfully formulated in the time domain. Note that the inverse kinematic equations yielded in this study are useful to investigate the kinematic behavior of the machine's drives in real time. The numerical example shows the effectiveness of the proposed formulation. The real time kinematic modelling of the five-axis CNC centers presented in this paper can be extended for some future works, e.g. the feed rate optimization, the postprocessor development, the drive capability monitoring, etc.

Acknowledgement. This research was funded by the UK Department for Business, Energy and Industrial Strategy and delivered by the British Council, under a Newton Fund Project ID 528085858.

References

1. Tutunea-Fatan, F.H.Y.: Configuration analysis of five-axis machine tools using a generic kinematic model. Int. J. Mach. Tools Manuf. **44**, 1235–1243 (2004)
2. Yang, J., Altintas, Y.: Generalized kinematics of five-axis serial machines with non-singular tool path generation. Int. J. Mach. Tools Manuf **75**, 119–132 (2013)
3. She, C.H., Chang, C.C.: Design of a generic five-axis postprocessor based on generalized kinematics model of machine tool. Int. J. Mach. Tools Manuf **47**, 537–545 (2007)
4. My, C.A., Bohez, E.L.: A novel differential kinematics model to compare the kinematic performances of five-axis CNC centers. Int. J. Mech. Sci. **163**, 105117 (2019)
5. My, C.A.: Integration of CAM systems into multi-axes computerized numerical control machines. In: 2010 IEEE Second International Conference on Knowledge and Systems Engineering, 119–124 (2010)
6. My, C.A., Bohez, E.L.: New algorithm to minimise kinematic tool path errors around 5-axis machining singular points. Int. J. Prod. Res. **54**, 5965–5975 (2016)
7. My, C.A., Hai, V.X.: Generalized pseudo inverse kinematics at singularities for developing five-axes CNC machine tool postprocessor. Vietnam J. Mech. **35**, 147–155 (2013)
8. Sørby, K.: Inverse kinematics of five-axis machines near singular configurations. Int. J. Mach. Tools Manuf **47**, 299–306 (2007)
9. She, C.H., Huang, Z.T.: Postprocessor development of a five-axis machine tool with nutating head and table configuration. Int. J. Adv. Manuf. Technol. **38**, 728–740 (2008)
10. Lee, R.S., She, C.H.: Developing a postprocessor for three types of five-axis machine tools. Int. J. Adv. Manuf. Technol. **13**, 658–665 (1997)
11. Xu, H.Y., Hu, L.A., Hon-Yuen, S.K., Xu, L.A.: A novel kinematic model for five-axis machine tools and its CNC applications. Int. J. Adv. Manuf. Technol. **67**, 1297–307 (2013)
12. Farouki, R.T., Han, C.Y., Li, S.: Inverse kinematics for optimal tool orientation control in 5-axis CNC machining. Comput. Aided Geometric Des. **31**, 13–26 (2014)
13. Boz, Y., Lazoglu, I.: A postprocessor for table-tilting type five-axis machine tool based on generalized kinematics with variable feedrate implementation. Int. J. Adv. Manuf. Technol. **66**, 1285–1293 (2013)
14. My, C.A., Bohez, E.L.: Multi-criteria optimization approach for 5-axis CNC tool path planning: modeling methodology. In Proceedings of the 3rd Asian Conference on Industrial Automation and Robotics, pp. 5–11 (2003)
15. My, C.A., Bohez, E.L., Makhanov, S.S., Munlinb, M., Phien, H.N., Tabucanon, M.T.: On 5-axis freeform surface machining optimization: vector field clustering approach. Int. J. CAD/CAM **5**, 1–11 (2009)
16. My, C.A., Bohez, E. L. J., Makhanov, S.S.: Critical point analysis of 3D vector field for 5-axis tool path optimization. In: Proceedings of the 4th Asian Conference on Industrial Automation and Robotics, ACIAR, pp. 11–13 (2005)
17. Bohez, E.L.J.: Five-axis milling machine tool kinematic chain design and analysis. Int. J. Mach. Tools Manuf **42**, 505–520 (2002)
18. My, C.A., Bien, D.X., Tung, B.H., Van Cong, N.: New feed rate optimization formulation in a parametric domain for 5-axis milling robots. In: International Conference on Computer Science, Applied Mathematics and Applications, pp. 403–411 (2019)

Comparison of the Efficiency of Numerical Algorithms for Solving the Inverse Kinematics Problem of Redundant Serial-Link Robots

Xuan Bien Duong[1(✉)], Anh My Chu[1], Chi Hieu Le[2], Tien Anh Nguyen[1], Minh Duc Vu[1], and Anh Le[1]

[1] Le Quy Don Technical University, Hanoi, Vietnam
duongxuanbien@lqdtu.edu.vn
[2] Faculty of Science and Engineering, University of Greenwich, London, UK

Abstract. This paper presents the results of the comparison of the use efficiency between two numerical algorithms which are the Adjust of Generalized Vector (AGV) and the Closed-Loop Inverse kinematics (CLIK) method when solving the inverse kinematics problem (IK) for the redundant serial-link robots. AGV method is developed based on the Taylor expansion method but adjusted to improve the accuracy of the IK solution. CLIK is a numerical method that is built as a closed-loop kinematics control system with the position error of the end-effector point in the workspace. The industrial SCARA robots with 4 degrees of freedom (DOF) including rotational and translational joints and PA10-7C industrial robots with 7-DOF are used to illustrate the effectiveness of the algorithms. The comparison results are expressed through the obtained joint variables values, the position error of the end-effector point in the workspace, and the deviation values of joint variables between the two methods. The results of this study allow evaluating the effectiveness of these methods for solving the IK and serve as the basis for choosing a suitable method for each specific robot configuration. However, this article has not considered the effectiveness of the methods mentioned above for other types of robots.

Keywords: Inverse kinematics · Effective algorithm · Redundant robots · Closed-loop feedback

1 Introduction

The characteristics of the redundant robot create the distinctive of the IK. It means that will exist countless options for the configuration of the robot in response to a task in the workspace, or in other words, the system of inverse kinematics equations has countless solutions. This is the advantage of this type of robot because we can choose an optimal result among a multitude of results on the basis of applying quality criteria such as avoidance of limited joints, joint velocity, avoid obstacles, the configuration degradation and is quite useful for minimizing energy consumption [1–6]. However, due to the multitude of solutions to choose from, choosing a viable option that fits the robot's physical configuration is a huge challenge. Research on effective solutions to solve the

B. T. Long et al. (Eds.): MMMS 2020, LNME, pp. 1000–1008, 2021.
https://doi.org/10.1007/978-3-030-69610-8_132

IK problem for redundant robots is the top concern because it is the foundation to analyze the dynamics problem and design the control system.

There are many numerical algorithms which are developed to solve the IK such as Jacobian Transpose [7], Pseudoinverse [8], DLS [9], Quasi-Newton and Conjugate gradient [10, 11], Closed-loop inverse kinematics (CLIK) [12–15]. The Quick IK algorithm [7] to improve the performance solving IK from Jacobian Transpose method due to a great deal of loops. The IK was solved by using CLIK method with the velocity and acceleration constraints [13, 14]. CLIK method is used in [15] for welding robot 6-DOFs combining with a positioner to track a complex 3D curve. The parallel genetic algorithm is used to solve the IK of Puma 500 robot in [16]. The elimination technique is presented in [17] to reduce the complexity of the inverse kinematics formulation based on analytical method. The new solution method is introduced in [18] to avoid joint limitation, avoid singularities and avoid obstacles for redundant robots. The AGV method has been presented and developed in [19, 20, 21]. This numerical method has been applied quite effectively in solving the IK for the redundant robots with 4DOF and 5DOF [19], 6 DOF [20, 21] with small position error in the joint-space and the workspace. It is noted that all of the joints in these studies are rotational joints.

This paper focuses on comparing the efficiency of solving the IK for the redundant serial-link robots between the AGV and CLIK algorithm because of their outstanding compared to the other numerical methods. The highlight of this study is shown through the simulation results of the IK for SCARA robot with 4DOF combining a translational joint and robot PA10-7C with 7DOF [22]. The kinematics equations of these robots are established based on the homogeneous matrix transformation D-H method. The IK solving algorithms are performed on MATLAB/SIMULINK software with kinematics parameters of robots in reality.

2 Materials and Methods

2.1 Kinematics Modeling

Based on the D-H method, a local D-H matrix $\mathbf{H}_i$ for the link i was established as follows [23, 24]. For a robot consisting of n serial of links, the position of the end-effector point on link n relative to the fixed coordinate system is determined as

$$\mathbf{D}_n = \mathbf{H}_1\mathbf{H}_2..\mathbf{H}_n \tag{1}$$

Define the vector $\mathbf{x}$ which includes the first three elements of 4th column of $\mathbf{D}_n$ matrix describes position of the end-effector point (E point) relative to the fixed coordinate system. In essence, the vector $\mathbf{x}$ represents the relationship between the end-effector point position and the joint variable values. Therefore, the forward kinematics equations are determined as $\mathbf{x} = f(\mathbf{q})$ and the end-effecter point velocity is inferred as $\dot{\mathbf{x}} = \mathbf{J}(\mathbf{q})\dot{\mathbf{q}}$. The Jacobian matrix is $\mathbf{J}(\mathbf{q})$ with size $3 \times n$. The inverse kinematics equations are determined as $\mathbf{q}(t) = f^{-1}(\mathbf{x}(t))$. Once the values of $\mathbf{q}$ have been determined from Eq. (4), the joints velocity are calculated as $\dot{\mathbf{q}} = \mathbf{J}^+(\mathbf{q})\dot{\mathbf{x}}$ with $\mathbf{J}^+(\mathbf{q})$ is the pseudo-inverse matrix of $\mathbf{J}(\mathbf{q})$ and defined as [23, 24]. The joints acceleration is inferred as $\ddot{\mathbf{q}} = \mathbf{J}^{+}(\mathbf{q})(\ddot{\mathbf{x}} - \dot{\mathbf{J}}\dot{\mathbf{q}})$. After calculating the value of vector $\mathbf{q}$, the value of vectors $\dot{\mathbf{q}}, \ddot{\mathbf{q}}$ can be determined.

2.2 The Algorithms for the IK Solving

2.2.1 Algorithm for Adjusted of Generalized Vector

Assume that the robot works in the period from $t = 0$ to $t = t_f$. Divide the time interval $[0 \div t_f]$ into N equal segments. The value of each segment is $\Delta t = t_f / N$. The next time step can be calculated as $t_{k+1} = t_k + \Delta t; k = 0 \div (N - 1)$. Operate Taylor's expansion of $\mathbf{q}(t_{k+1})$ near the value of $\mathbf{q}(t_k)$, the values of generalized joints position vector are given as [19, 20].

$$\mathbf{q}(t_{k+1}) = \mathbf{q}(t_k + \Delta t) = \mathbf{q}(t_k) + \dot{\mathbf{q}}(t_k)\Delta t + \frac{1}{2}\ddot{\mathbf{q}}(t_k)(\Delta t)^2 + \ldots \tag{2}$$

The result given (2) is quite rough at $t = t_{k+1}$. Determine a better approximation according to the formula $\mathbf{q}(t_{k+1}) = \tilde{\mathbf{q}}(t_{k+1}) + \Delta\mathbf{q}(t_{k+1})$ by calculating the value of $\Delta(t_{k+1}) = \mathbf{J}^{+}(\tilde{\mathbf{q}}(t_{k+1}))[\mathbf{x}(t_{k+1}) - f(\tilde{\mathbf{q}}(t_{k+1}))]$, $\tilde{\mathbf{q}}(t_{k+1}) = \tilde{\mathbf{q}}(t_{k+1}) + \Delta\mathbf{q}(t_{k+1})$. Operate the loop until $\|\Delta(t_{k+1})\| \leq \varepsilon$ with ε is the allowable error vector then take $\mathbf{q}(t_{k+1}) = \tilde{\mathbf{q}}(t_{k+1})$. So, the value of generalized position joints vector $\mathbf{q}$ has been found. Specific contents of this algorithm were clearly described in [19, 20].

2.2.2 CLIK Algorithm

The general IK solution can be described as [7].

$$\dot{\mathbf{q}} = \mathbf{J}^{+}(\mathbf{q})\dot{\mathbf{x}} + (\mathbf{I} - \mathbf{J}^{+}(\mathbf{q})\mathbf{J}(\mathbf{q}))\dot{\mathbf{q}}_0 \tag{3}$$

where, $\mathbf{I}$ is the unit matrix with size $n \times n$ and $\mathbf{q}_0$ is the initial value. The closed-loop algorithm is used based on the path error $\mathbf{e} = \mathbf{x}_d - \mathbf{x}$ in workspace between the desired and actual path ($\mathbf{x}_d$ and $\mathbf{x}$). The generalized CLIK algorithm can be expressed by [7].

$$\dot{\mathbf{q}} = \mathbf{J}^{+}(\mathbf{q})(\dot{\mathbf{x}}_d + \mathbf{K}_p(\mathbf{x}_d - \mathbf{x})) \tag{4}$$

The IK solution of redundant robot based on closed-loop algorithm is given as [7, 15].

$$\dot{\mathbf{q}} = \mathbf{J}^{+}(\mathbf{q})(\dot{\mathbf{x}}_d + \mathbf{K}_p(\mathbf{x}_d - \mathbf{x})) + (\mathbf{I} - \mathbf{J}^{+}(\mathbf{q})\mathbf{J}(\mathbf{q}))\dot{\mathbf{q}}_0 \tag{5}$$

Where, $\mathbf{K}_p$ is a symmetric positive definite matrix. This algorithm was clearly described in [7, 15].

2.3 Numerical Simulation Results and Discussion

Consider the kinematics model of SCARA robot with 4-DOF as shown in Fig. 1 and PA10-7C robot with 7-DOF as shown in Fig. 2.

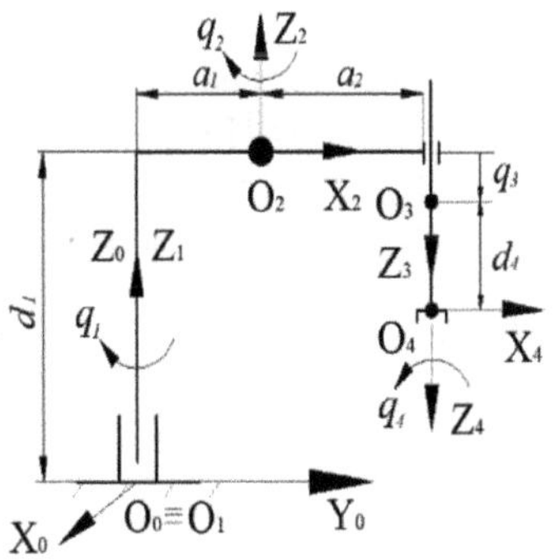

Fig. 1. Kinematics model of SCARA robot $\mathbf{q} = [q_1 \quad q_2 \quad q_3 \quad q_4]^T$, q_1, q_2, q_4 in radian unit and q_3 in meter unit.

Table 1. D-H parameters SCARA

Links	D-H parameters			
	θ_i	d_i	a_i	α_i
1	q_1	d_1	a_1	0
2	q_2	0	a_2	π
3	0	q_3	0	0
4	q_4	d_4	0	0

$\mathbf{q}_{\max} = [2.5 \quad 2.5 \quad 0.4 \quad 6.2]^T$,
$\dot{\mathbf{q}}_{\max} = [6.5 \quad 6.5 \quad 0.3 \quad 31.4]^T$,
$d_1 = 0.42; a_1 = 0.22; a_2 = 0.4(m)$,
$a_2 = 0.4; d_4 = 0.09(m)$

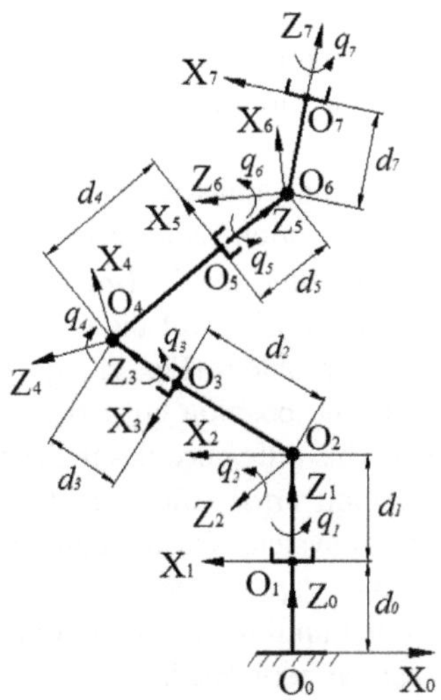

Fig. 2. Kinematics model of PA10-7C robot $\mathbf{q} = [q_1 \quad q_2 \quad q_3 \quad q_4 \quad q_5 \quad q_6 \quad q_7]^T$

Table 2. D-H parameters PA10-7C

Links	D-H parameters			
	θ_i	d_i	a_i	α_i
1	q_1	$d_0 + d_1$	0	$-\pi/2$
2	q_2	d_2	0	$\pi/2$
3	q_3	d_3	0	$-\pi/2$
4	q_4	d_4	0	$\pi/2$
5	q_5	d_5	0	$-\pi/2$
6	q_6	0	0	$\pi/2$
7	q_7	d_7	0	0

Where, $d_0 = 0.21; d_1 = 0.1; d_2 = 0.21; d_3 = 0.13(m)$, $d_4 = 0.28; d_5 = 0.21; d_7 = 0.07(m)$
$\dot{\mathbf{q}}_{\max} = [1 \quad 1 \quad 2 \quad 2 \quad 6.3 \quad 6.3 \quad 6.3]^T$ (rad/s)

The fixed coordinate system is $(OXYZ)_0$ and $(OXYZ)_i$, ($i = 1 \div 4$ with SCARA and $i = 1 \div 7$ with PA10-7C) are the local coordinate systems attached link i. Table 1 and Table 2 describe the kinematic parameters according to the D-H rule [23, 24] of two robots, respectively. Accordingly, the transformation $\mathbf{H}_i$ homogeneous matrices are determined. A given trajectory in the workspace for SCARA robot is described as

$$x_E = 0.3 + 0.08\cos(0.2t)(m); y_E = 0.3 + 0.08\sin(0.2t)(m); z_E = 0.1\cos(0.2t)(m)$$

A given trajectory in the workspace for PA10-7C robot is presented as

$$x_E = 0.2 + 0.4\cos(0.2t)(m); y_E = 0.2 + 0.4\cos(0.2t)(m); z_E = 0.65(m)$$

The given position error value in AGV method is $\varepsilon = 10^{-3}(m)$. The values of coefficient in CLIK method for the SCARA robot and the PA10-PC robot are $k_p = 50$ and $k_p = 100$. The numerical simulation results of the SCARA are shown in Fig. 3, 4, 5, 6, 7, 8, 9, 10, 11 and Fig. 12 and of the PA10-7C robot are described in Fig. 13, 14, 15, 16, 17, 18, 19 and Fig. 25 after applying the AGV and CLIK algorithm to solve the IK problem. Figure 12 and Fig. 25 show the models of both robots in 3D by using the values of joint variables that are obtained from the inverse kinematics problem. These models confirm the reliability of the algorithms.

With SCARA robot, the position of joint 1, joint 2 and translational joint 3 are presented in Fig. 3, Fig. 4 and Fig. 5, respectively. The deviation values between AGV and CLIK is shown in Fig. 6. The maximum deviation value is $10^{-2}(m)$ in joint 2. However, when using the joint variable values to find the position of the end-effector point in the workspace through the forward kinematics problem (Fig. 7, Fig. 8, and Fig. 9), there appears to be quite a difference large in the position error of the end-effector point (AGV with Fig. 10, CLIK with Fig. 11). This difference lies in the nature of the algorithms. With the CLIK method, the position error values continuously changed according to the closed-loop trajectory control system. On the contrary, the position error is calculated specifically in each time step in the AGV method. According to the AGV algorithm, the step i is performed until the position error value smaller than the preset error value, then step $i+1$ can be performed.

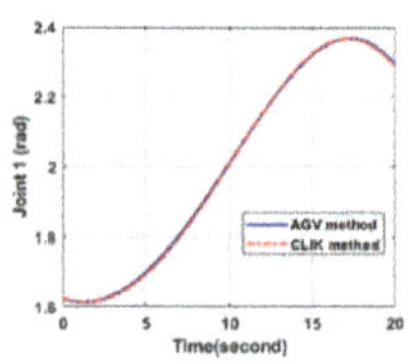

Fig. 3. Joint 1

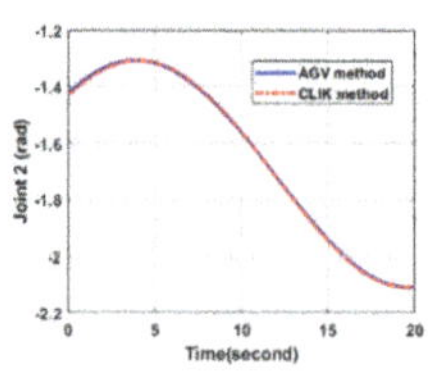

Fig. 4. Joint 2

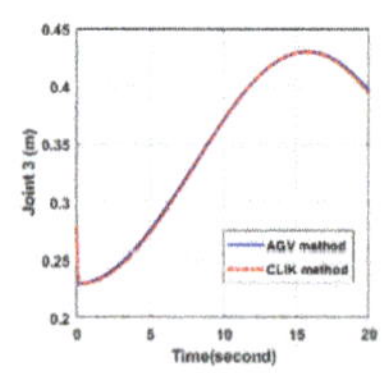

Fig. 5. Joint 3

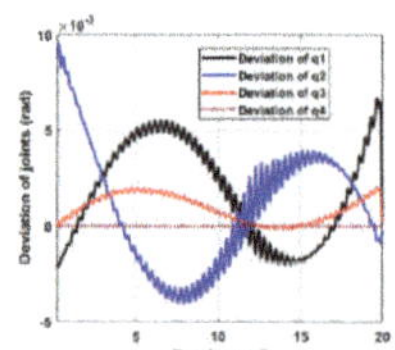

Fig. 6. Deviation

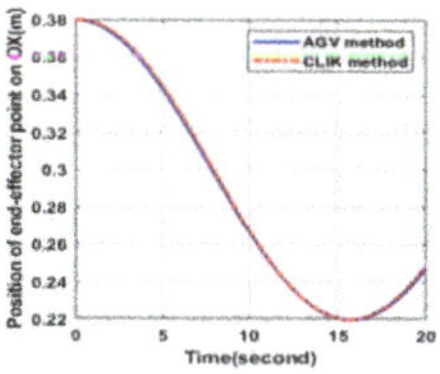

Fig. 7. E point in OX

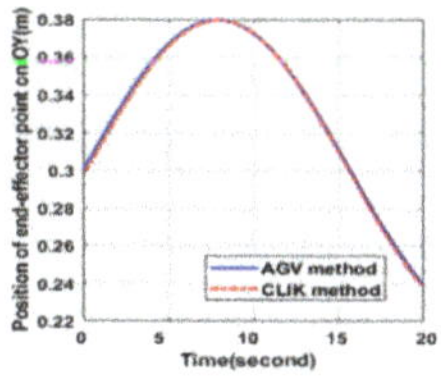

Fig. 8. E point in OY

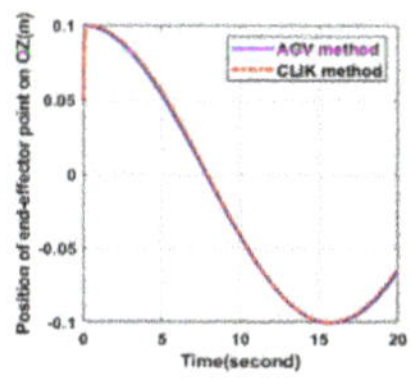

Fig. 9. E point in OZ

For the PA10-7C robot, the values of the joints are depicted in Fig. 13, 14, 15, 16, 17 and Fig. 18. The large differences between the two robots are the translational joint of the SCARA robot and the number of DOF. The number of kinematic equations is also different. This explains the difference in the results of the IK between the AGV and CLIK methods.

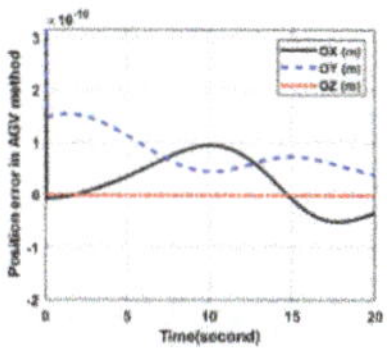

Fig. 10. Error position of E point with AGV

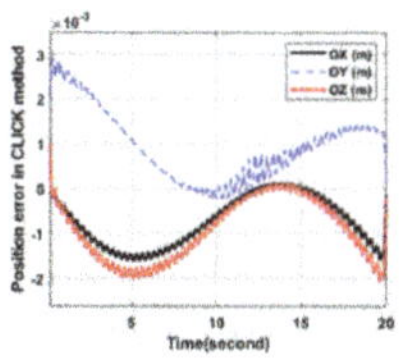

Fig. 11. Error position of E point with CLIK

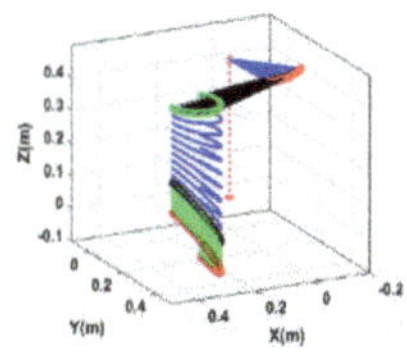

Fig. 12. SCARA model in MATLAB

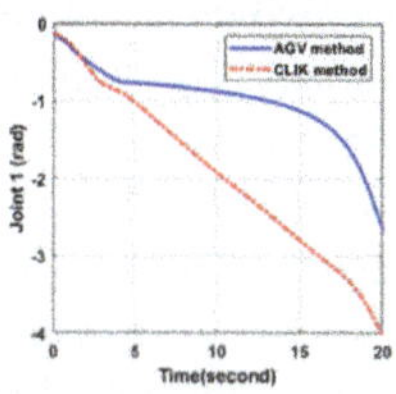

Fig. 13. Joint 1 position

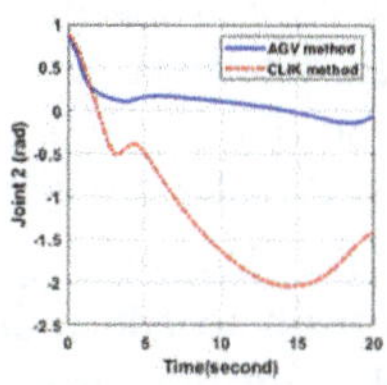

Fig. 14. Joint 2 position

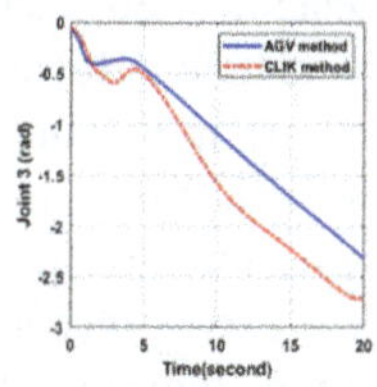

Fig. 15. Joint 3 position

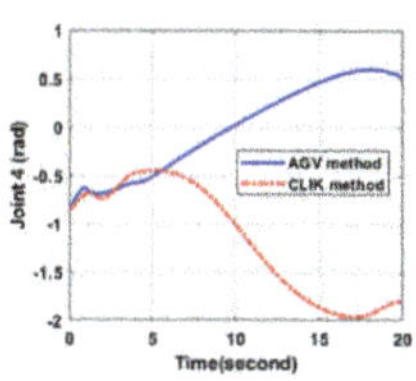

Fig. 16. Joint 4 position

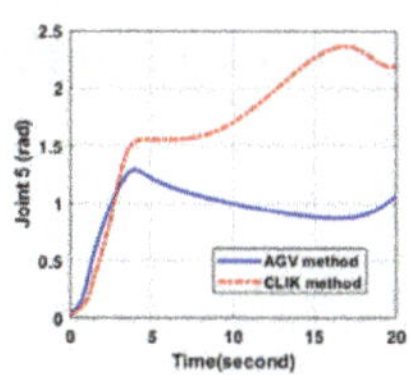

Fig. 17. Joint 5 position

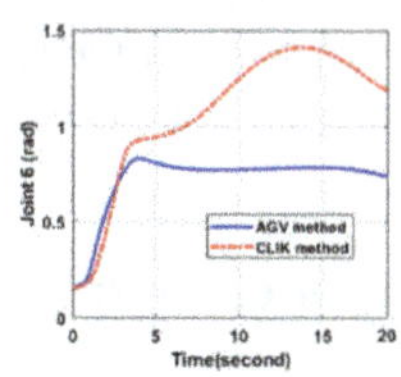

Fig. 18. Joint 6 position

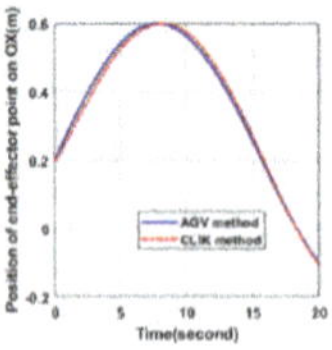

Fig. 19. E point in OX

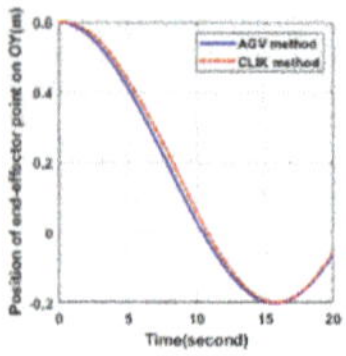

Fig. 20. E point in OY

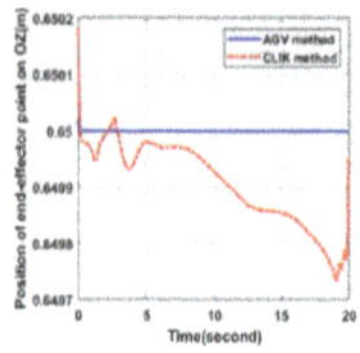

Fig. 21. E point in OZ

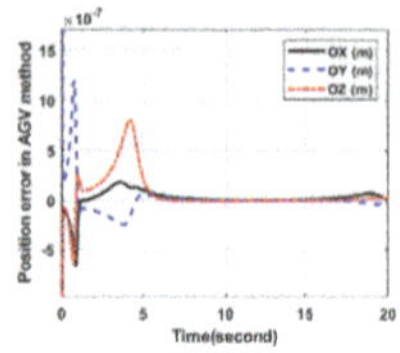

Fig. 22. Error position with AGV

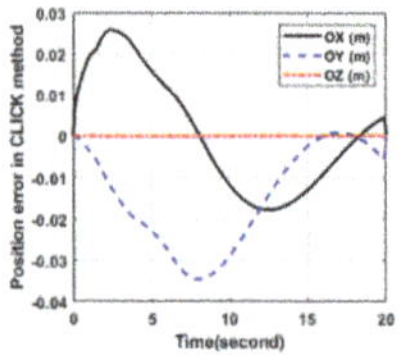

Fig. 23. Error position of E point with CLIK

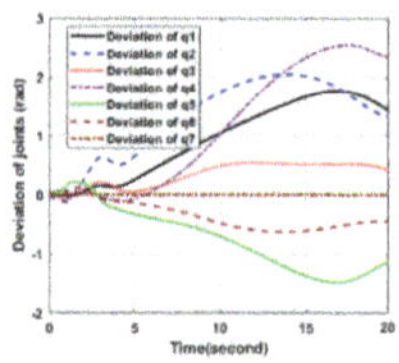

Fig. 24. Joints deviation (AGV and CLIK)

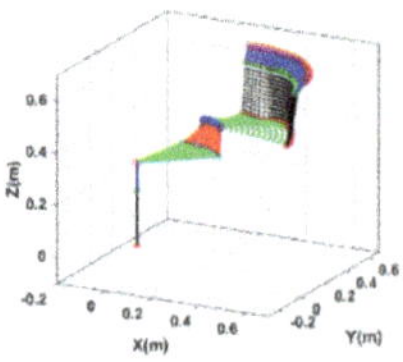

Fig. 25. PA10-7C model in MATLAB

The deviation value of the joint variables is shown in Fig. 24. The end-effector point position is shown in Fig. 19, Fig. 20, and Fig. 21. The position errors of the end-effector point of the two methods are described in Fig. 22 and Fig. 23. The position error of the CLIK method is higher than that of the AGV method. The cause has been considered similar to the case of the SCARA robot.

In general, both techniques have the same level of complexity when calculating Jacobian matrices. However, the CLIK algorithm needs to find a suitable K_p coefficient to ensure the minimum position error. To find this coefficient, it can be to use the automatic adjustment module of MATLAB/SIMULINK software. This technique is easier in designing a position controller. However, given the trajectory in the workspace is a curve, this method is difficult to apply, it is necessary to use a support optimization algorithm to ensure the position error is within the allowable limit. The AGV method takes more computation time than the CLIK method because this algorithm performs for each specific time step. Based on the above evaluation, it can be said that the AGV method is more effective in solving the IK for the redundant serial-link robots than the CLIK method with the given path of the end-effector point in the workspace is a curve.

3 Conclusion

Through building the kinematics equations of the redundant serial-link robots including the SCARA robot with 4DOF combining a translational joint and the PA10-7C robot with 7DOF, two methods AGV and CLIK are applied to solve the inverse kinematics problem of these redundant robots. The efficiency of these methods was compared through numerical simulation results, including the position and deviation of the joint variables, the position, and position error of the end-effector point in the workspace. The results of this study show that both methods have successfully applied to solve the IK of redundant serial-link robots. However, the AGV method is more efficient than the CLIK method for this robot type because the simulation results show that the AGV method gives smaller position errors, higher reliability through each calculation step, and does not depend on finding the coefficient K_p suitable in the CLIK method. This study has not covered IK for other types of robots such as parallel robots, mobile robots.

Acknowledgement. This work was supported by a Research Environment Links, ID 528085858, under the Newton Fund partnership. The grant is funded by the UK Department for Business, Energy and Industrial Strategy and delivered by the British Council.

References

1. Locke, R.C.O, Patel, R.V.: Optimal remote center-of-motion location for robotics-assisted minimally-invasive surgery. In: 2007 IEEE International Conference on Robotics and Automation, pp. 1990–1905 (2007)
2. Chiaverini, S., Oriolo, G., Walker, I.D.: Kinematically Redundant Manipulators. Springer Handbook of Robotics, pp. 245–268. Springer, Heidelberg (2007)
3. Wang, J., Hu, X., Zhang, B.: Real-time motion planning of kinematically redundant manipulators using recurrent neural networks. In: Yu, W. (ed.) Recent Advances in Intelligent Control Systems, pp. 169–193. Springer, London (2009)
4. Maciejewski, A.A., Klein, C.A.: Obstacle avoidance for kinematically redundant manipulators in dynamically varying environments. Int. J. Robot. Res. **4**, 109–116 (1985)
5. Klanke, S., Lebedev, D., Haschke, R., Steil, J., Ritter, H.: Dynamic path planning for a 7-DOF robot arm. In: Proceedings of the 2006 IEEE/RSJ, pp. 3879–3884 (2006)
6. Llama, M.A., Wism, A.Z., Castañon, W.Z., Hernandez, R.G.: Force and position fuzzy control: a case study in a Mitsubishi PA10-7CE robot arm. Advanced Topics on Computer Vision, Control and Robotics in Mechatronics, pp. 165–194. Springer, Cham (2018)
7. Lian, S., Han, Y., Wang, Y., Bao, Y., Xiao, H., Li, X., Sun, N.: Accelerating inverse kinematics for high-DOF robots. In: Proceedings of the 54th Annual Design Automation Conference 2017, Austin, USA (2017)
8. Yoshikawa, T.: Dynamic manipulability of robot manipulators. J. Robot. Syst. **2**, 113–124 (1985)
9. Wampler, C.W.: Manipulator inverse kinematic solutions based on vector formulations and damped least squares methods. IEEE Trans. Syst. Man Cybern. **16**, 93–101 (1986)
10. Wang, L.C.T., Chen, C.C.A.: Combined optimization method for solving the inverse kinematics problem of mechanical manipulators. IEEE Trans. Robot. Automation **7**, 489–499 (1991)
11. Zhao, J., Badler, N.I.: Inverse kinematics positioning using nonlinear programming for highly articulated figures. ACM Trans. Graph. **13**, 313–336 (1994)
12. Sciavicco, L., Siciliano, B.: A solution algorithm to the inverse kinematic problem for redundant manipulators. J. Robot. Autom. **4**, 403–410 (1988)

13. Antonelli, G., Chiaverini, S., Fusco, G.: Kinematic control of redundant manipulators with online end-effector path tracking capability under velocity and acceleration constraints. In: IFAC Robot Control, Austria, pp. 183–188 (2000)
14. Wang, J., Li, Y., Zhao, X.: Inverse kinematics and control of a 7 dof redundant manipulator based on the closed loop algorithm. Int. J. Adv. Robot. Syst. **7**, 1–10 (2010)
15. My, C.A., Bien, D.X., Tung, H.B., Hieu, L.C., Cong, N.V., Hieu, T.V.: Inverse kinematic control algorithm for a welding robot-positioner system to trace a 3D complex curve. In: International Conference on Advanced Technologies for Communications (ATC), pp. 319–323 (2019)
16. Aguilar, O.A., Huegel, J.C.: Inverse kinematics solution for robotic manipulators using a CUDA-based parallel genetic algorithms. In: Mexican International Conference on Artificial Intelligence 2011, Part 1, pp. 490–503 (2011)
17. Husty, M.L., Pfurner, M., Schrocker, H.P.: A new and efficient algorithm for the inverse kinematics of a general serial 6R manipulator. Mech. Machine Theory **42**, 66–81 (2007)
18. Kelemen, M., Virgala, I., Liptak, T., Mikova, L., Filakovsky, F., Bulej, V.: A novel approach for an inverse kinematics solution of a redundant manipulator. Appl. Sci. **8**, 2–20 (2018)
19. Khang, N.V., Dien, N.P., Vinh, N.V., Nam, T.H.: Inverse kinematic and dynamic analysis of redundant measuring manipulator BKHN-MCX-04. Vietnam J. Mech. VAST **32**, 15–26 (2010)
20. Bien, D.X., My, C.A., Hieu, L.C., Tung, B.H.: Inverse Kinematics Analysis of Welding Robot IRB 1520ID Using Algorithm for Adjusting the Increments of Generalized Vector, the 5th RICE2020, Springer. ISSN: 2194-5358 (2020)
21. Bien, D.X., Tuan, P.A., Nhat, D.D., Hiep, D.X., Nghia, T.K., Anh, M.N.: Optimize the feed rate and determine the joints torque for industrial welding robot TA 1400 based on kinematics and dynamics modeling. Int. J. Mech. Eng. Robot. Res. **9**, 1335–1340 (2020)
22. http://www.solutions4u-asia.com/. Accessed Sept 2020
23. Spong, M., Hutchinson, W., Vidyasagar, S.M.: Robot Modeling and Control, First edition. Springer, New York (2001)
24. Lewis, F., Dawnson, L., Abdallah, D.: Robot Manipulator Control Theory and Practice - Second edition, Marcel Dekker INC, New York, USA (2004)

Kinematics and Dynamics Analysis of Geostationary Satellite Antenna System

Quoc Hoang Pham, Xuan Bien Duong(✉), Hoang Cuong Phan, Duc Manh Dinh, Van Duc Nguyen, Van Nam Nguyen, and Van Tuan Pham

Le Quy Don Technical University, Hanoi, Vietnam
duongxuanbien@lqdtu.edu.vn

Abstract. This paper presents kinematics and dynamics analysis of antenna systems for geostationary satellite communications and monitoring. Each component of the antenna system is assumed a rigid body. The mathematical model is build based on multi-bodies kinematics and dynamics theory. The DENAVIT-HARTENBERG (D-H) homogeneous matrix method and Lagrange type II are used to construct the kinematics and dynamics equations. The inverse dynamics problem is analyzed to determine the value of driven torques at the joints under effect of the wind payload, to obtain the given desired path in the workspace. The numerical simulation results kinematics and dynamics were successfully applied in practice, especially for selection of motors and transmission systems for design and development of geostationary satellite antenna systems.

Keywords: Geostationary satellite · Antenna system · Kinematics · Dynamics

1 Introduction

Communication via satellite is the greatly important research result from the combination of the two fields of communication and space science [1]. The biggest advantage of this invention is its range of information transmission and low cost. An indispensable part of this satellite communication technology is the satellite antenna receiver and broadcasting system. Up to the present time, space technology is still a specific science that requires a very high level of scientific development. Therefore, in the world, only a few developed countries are able to develop this field such as Russia, USA, China, Japan [1–7].

Mathematical modelling, kinematics, and dynamics analysis are the important steps in design and development of the antenna system for communications and monitoring of geostationary satellites. The motion characteristics and workspace of the antenna system are determined in details by solving the kinematics problem. The calculation and selection of the transmission system and evaluation of the dynamic behavior of the antenna system under influence of the external forces are mainly based on the analysis results of the dynamics problems.

A few works related to solving the dynamics and control problems of geostationary satellite antenna systems were documented in [8–12]. The basic construction of a satellite antenna was described by Bindi [8]. The mathematical model describing the

B. T. Long et al. (Eds.): MMMS 2020, LNME, pp. 1009–1017, 2021.
https://doi.org/10.1007/978-3-030-69610-8_133

look angle of the geostationary satellite antenna mentioned by Ogundele [9] is based on the use of two control station models. The adjustment model for the satellite antenna viewing angle was proposed by [10]. The woring principles and classifications of the satellite surveillance antenna can be found in [11]. The geostationary satellite antenna system, called Horn Antenna, was designed and simulated by [12]. Researches related to the detailed design and manufacture of geostationary satellite antenna systems have not been well-documented, due to the security issues and design copyrights.

This paper presents the kinematics and dynamics analysis of geostationary satellite antenna systems. The system of dynamic equations of the antenna system is developed, for calculation of drive torque for the joints, under the influence of the maximum wind load. This is the fundamental for selecting the drive motor capacity for the joints of the antenna system. The homogeneous transformation matrix D-H [13] and the Lagrange equations type 2 are used to build the kinematics and dynamics equations of the antenna system. The simulation results are obtained based on the numerical calculation methods.

2 Materials and Methods

2.1 Kinematics Analyzing

Consider the preliminary designed geostationary satellite monitoring antenna model as shown in Fig. 1 and Fig. 2.

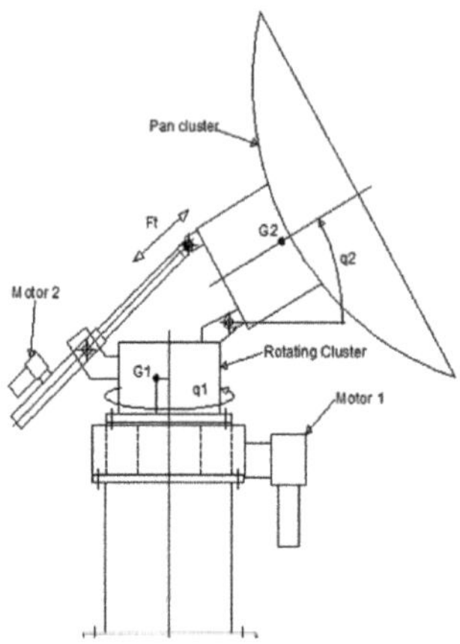

Fig. 1. Preliminary mechanical system

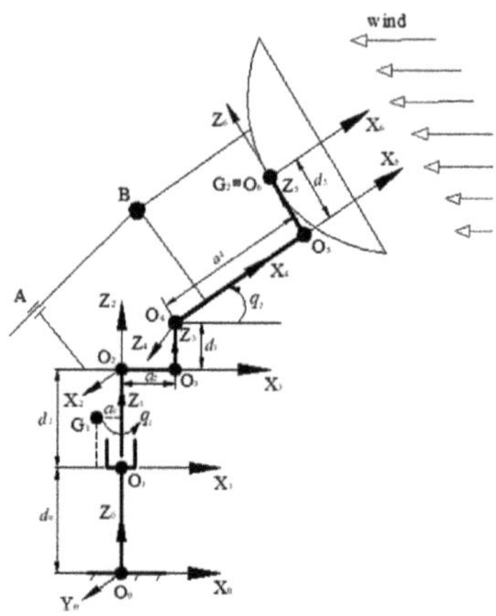

Fig. 2. Mathematical model

Accordingly, the mechanical system consists of two main parts which are the rotating cluster (directional cluster) and the satellite pan cluster (Fig. 1). The movement of the rotating cluster is done by the rotating joint q_1 which is driven by motor 1. The satellite pan cluster height is performed by rotating joint q_2. This joint is driven by motor 2 through the translational joint A and rotational joint B. However, these joints are only responsible for driving the joint q_2, so it is not considered in the kinematic problem. The center of the rotating cluster is G_1, the satellite pan assembly center is G_2. The kinematics model of the mechanical system can be constructed as shown in Fig. 2. Select a fixed coordinate system $(OXYZ)_0$ attached to the ground. The $(OXYZ)_i$ coordinate systems are respectively mounted at the positions shown in Fig. 2. The above fixed and local coordinate systems are attached for the purpose of accurately determining the position of any point on the system. In particular, taking point G_2 is the end-effector point representing the satellite pan cluster and it is necessary to determine the position of this point according to the fixed coordinate system. The D-H parameters tables can be described as shown in Table 1.

Table 1. D-H parameters

Parameters	θ	d	a	α
O_0O_1	0	d_0	0	0
O_1O_2	q_1	d_1	0	0
O_2O_3	0	0	a_2	0
O_3O_4	0	d_3	0	$\pi/2$
O_4O_5	q_2	0	a_4	$-\pi/2$
O_5O_6	0	d_5	0	0

Use the local homogeneous matrix $\mathbf{H}_i$ as in [14, 15]. From local $\mathbf{H}_i$ matrices, the position of the $(OXYZ)_i$ coordinate system compared to the fixed coordinate system $(OXYZ)_0$ through matrix transformation [16] can be determined as follows

$$\mathbf{D}_i = \mathbf{H}_1\mathbf{H}_2..\mathbf{H}_i, i = 1 \div 6 \tag{1}$$

In which, matrix $\mathbf{D}_i$ represents the position and direction of local coordinate systems relative to the fixed coordinate system. Specifically, the position of the end-effector point (G_2) following the fixed coordinate systems is as follows

$$\mathbf{D}_6 = \begin{bmatrix} \cos q_1 \cos q_2 & -\sin q_1 & -\cos q_1 \sin q_2 & \cos q_1(a_2 + a_4 \cos q_2 - d_5 \sin q_2) \\ \sin q_1 \cos q_2 & \cos q_1 & -\sin q_1 \sin q_2 & \sin q_1(a_2 + a_4 \cos q_2 - d_5 \sin q_2) \\ \sin q_2 & 0 & \cos q_2 & d_0 + d_1 + d_3 + a_4 \sin q_2 + d_5 \cos q_2 \\ 0 & 0 & 0 & 1 \end{bmatrix} \tag{2}$$

The position of G_2 point is determined according to the fixed coordinate system as column 4th of $\mathbf{D}_6$ matrix.

2.2 Dynamics Analyzing

The dynamic problem includes the forward and inverse dynamics. The inverse dynamics allows determining the value of torques or driving forces required to ensure that the motion of the system according to the given path that is $\mathbf{x} = \begin{bmatrix} x_{G2} & y_{G2} & z_{G2} \end{bmatrix}^T$ in the workspace or $\mathbf{q}(t) = \begin{bmatrix} q_1 & q_2 \end{bmatrix}^T$ in the joint space. Solving the inverse dynamic problem in the workspace is truly complicated because it is necessary to solve before the already complicated inverse kinematics problem. In this study, the inverse dynamics problem in the workspace will be solved.

The position of the center mass G_1 of the rotating cluster ($\mathbf{R}_{C1}$) and the center mass G_2 of the pan cluster ($\mathbf{R}_{C2}$) in the fixed coordinate system are shown as follows

$$\mathbf{R}_{C1} = \begin{bmatrix} -a_0 \cos q_1 & -a_0 \sin q_1 & d_0 + \frac{d_1}{2} \end{bmatrix}^T, \mathbf{R}_{C2} = \begin{bmatrix} x_{G2} & y_{G2} & z_{G2} \end{bmatrix}^T \tag{3}$$

The kinetic energy of any rigid body $T_i; i = 1, 2$ consists of the kinetic energy resulting from the translational motion T_{Ti} and the kinetic energy resulting from the rotation T_{Qi}. Consider $m_i, \mathbf{I}_i, \mathbf{v}_{Ci}$ and $\boldsymbol{\omega}_i$ that are the mass, moment of inertial, translational and angular velocity of cluster i, respectively. The kinetic energy of cluster i is determined as follows [14]

$$T_i = T_{Ti} + T_{Qi} = \frac{1}{2} m_i \mathbf{v}_{Ci}^T \mathbf{v}_{Ci} + \frac{1}{2} \boldsymbol{\omega}_i^T \mathbf{I}_i \boldsymbol{\omega}_i \tag{4}$$

The total kinetic and potential energy of the antenna system can be determined as [14].

$$T = \sum_{i=1}^{2} T_i = \frac{1}{2} \dot{\mathbf{q}}^T \mathbf{M}(\mathbf{q}) \dot{\mathbf{q}}; \Pi = -\sum_{i=1}^{2} m_i \mathbf{g}_o^T \mathbf{R}_{Ci} \tag{5}$$

where, $\mathbf{M}(\mathbf{q})$ is the mass matrix, $\mathbf{g}_0$ is the vector of gravity acceleration. The Lagrange equations type 2 is determined as [14, 15]

$$\frac{d}{dt}\left(\frac{\partial T}{\partial \dot{\mathbf{q}}}\right)^T - \left(\frac{\partial T}{\partial \mathbf{q}}\right)^T = -\left(\frac{\partial \Pi}{\partial \mathbf{q}}\right)^T + \tau + \mathbf{f}^* \tag{6}$$

where, $\tau = \begin{bmatrix} \tau_1 & \tau_2 \end{bmatrix}^T$ is the driven torque vector at the joints q_1, q_2 and $\mathbf{f}^* = \begin{bmatrix} Q_1^*, Q_2^* \ldots, Q_n^* \end{bmatrix}^T$ is the external force vector. In this study, the external force acting on the antenna system is the wind load. It is assumed that the influence of wind on the rotational cluster is ignored, only considering the effect of wind load on the pan cluster. The value of force caused by the wind on the pan cluster is calculated specifically in [18]. In this case, in order to reduce the calculating complexity, assumed that the wind force is always parallel and opposite to the axial direction OX. The distributed wind load is assumed to be concentrated force acting at the center of mass G_2 (Fig. 2). The maximum value of the wind load on OX axis is $F_{x\max} = 22174(N)$ and on OY axis is

$F_{y\max} = 30670(N)$ in the case $q_2 = 0$ with velocity of the wind is 25 (m/s) [18]. The dynamic equations of the antenna system are presented as

$$\mathbf{M}(\mathbf{q})\ddot{\mathbf{q}} + \mathbf{C}(\mathbf{q}, \dot{\mathbf{q}})\dot{\mathbf{q}} + \mathbf{g}(\mathbf{q}) = \tau + \mathbf{f}^* \tag{7}$$

Where, $\mathbf{C}(\mathbf{q}, \dot{\mathbf{q}})$ is the Coriolis matrix which is determined in [14, 15] and $\mathbf{g}(\mathbf{q})$ is the gravity potential energy vector. The driven torque vector can be found as follows

$$\tau(t) = \mathbf{M}(\mathbf{q})\ddot{\mathbf{q}} + \mathbf{C}(\mathbf{q}, \dot{\mathbf{q}})\dot{\mathbf{q}} + \mathbf{g}(\mathbf{q}) - \mathbf{f}^* \tag{8}$$

2.3 Numerical Simulation Results

2.3.1 Kinematics Problem

The basic parameters of the expected antenna system to be designed and manufactured are as follows

$a_0 = 0.33\,(\mathrm{m}), a_2 = 0.39\,(\mathrm{m}), d_0 = 3.62\,(\mathrm{m}), d_1 = 0.87\,(\mathrm{m}), d_3 = 0.2\,(\mathrm{m}), a_4 = 1.15\,(\mathrm{m}), d_5 = 0.21\,(\mathrm{m}), m_1 = 1050\,(\mathrm{kg}), m_2 = 1350\,(\mathrm{kg})$. The limited values of joints are $0 \le q_1 \le 2\pi; 0 \le q_2 \le \frac{\pi}{2}$. The workspace of G_2 point can be calculated based on the limited joints and the forward kinematics Eqs. (9). Figure 3 shows the workspace of G_2 point.

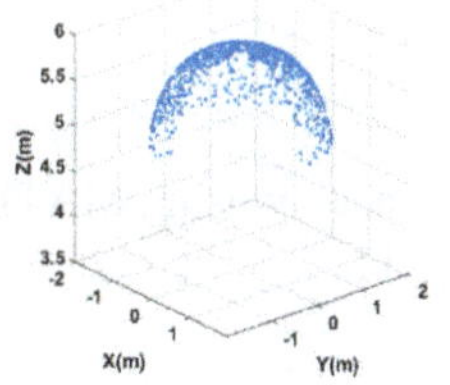

Fig. 3. The workspace of G_2 point

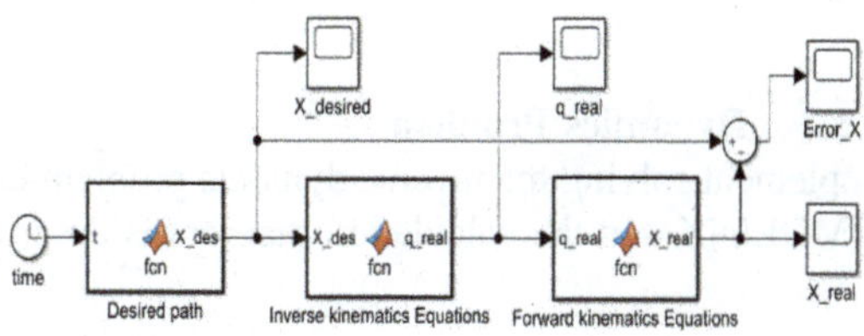

Fig. 4. Inverse kinematics problem calculation diagram

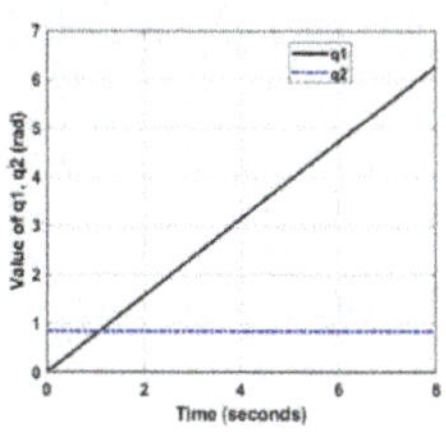

Fig. 5. Joints variables

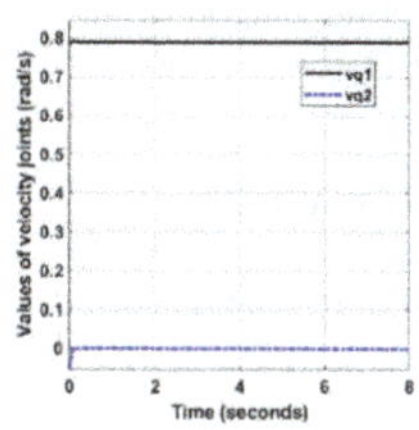

Fig. 6. Joints velocities

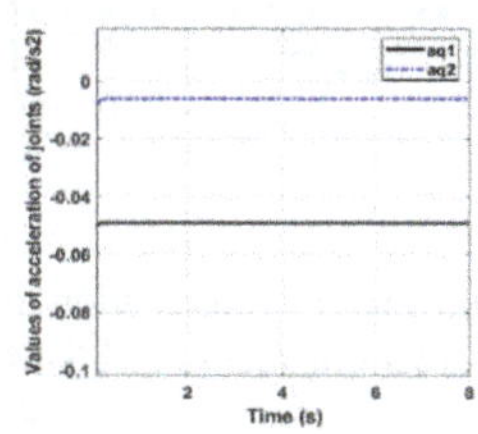

Fig. 7. Joint accelerations

Solve the inverse kinematics problem by using algorithm which is clearly described in [16, 17] with input the desired motion trajectory of point G_2 as follows

$$x_{G2} = \cos\left(\frac{\pi}{4}t\right)\,(\mathrm{m}), y_{G2} = \sin\left(\frac{\pi}{4}t\right)\,(\mathrm{m}), z_{G2} = 5.68\,(\mathrm{m}) \tag{9}$$

The results of the inverse kinematics problem are the values of joint variables $\mathbf{q}_{real} = \begin{bmatrix} q_{1real} & q_{2real} \end{bmatrix}^T$ and the position error of the G_2 point in the workspace $ErrorX = X_{des} - X_{real}$. Figure 4 shows the performing diagram in MATLAB/SIMULINK. These results are described from Fig. 5, 6, 7, 8, 9 and Fig. 10.

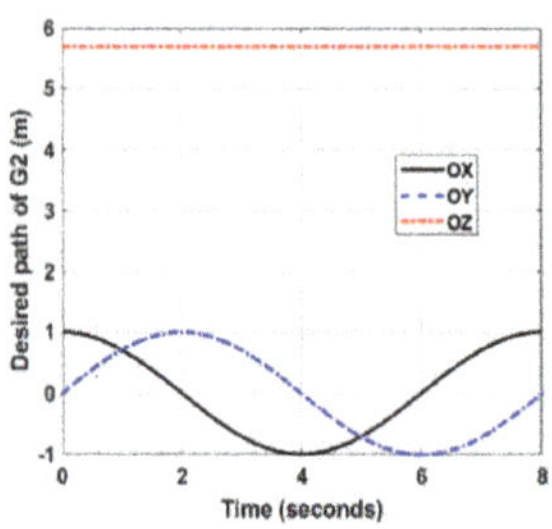

Fig. 8. Position of the G2 point

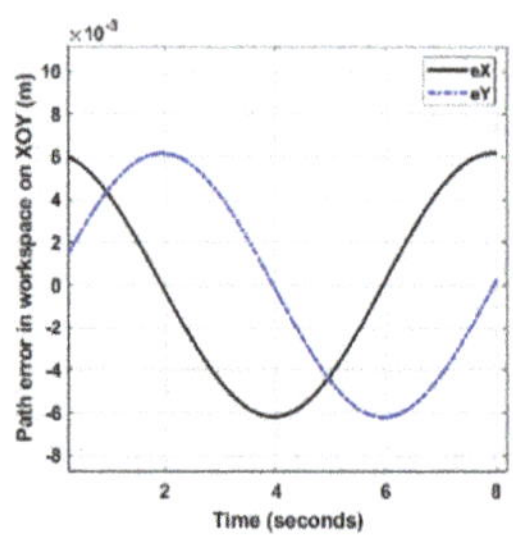

Fig. 9. Position error of the G2 point

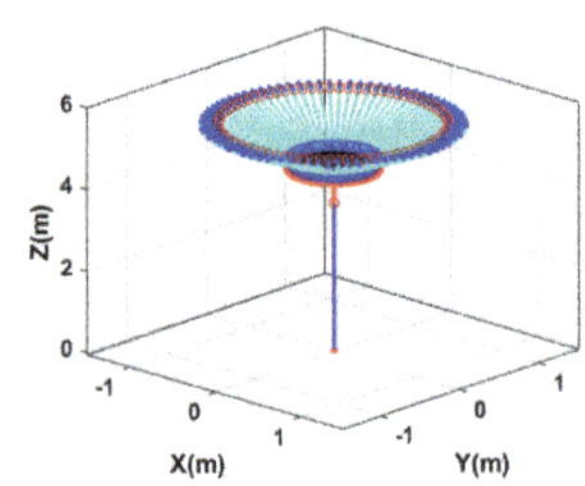

Fig. 10. Actual path of G2 in workspace

2.3.2 Dynamics Problem

Implement solving the inverse dynamic problem in the workspace by using MATLAB/SIMULINK and the calculation diagram is described in Fig. 11.

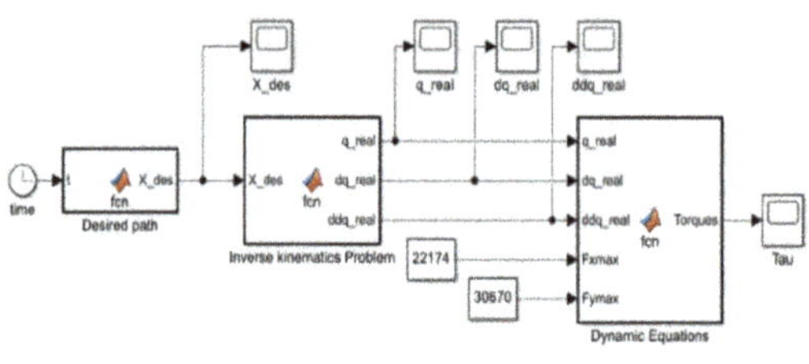

Fig. 11. The calculation diagram for the inverse dynamic problem in MATLAB/SIMULINK

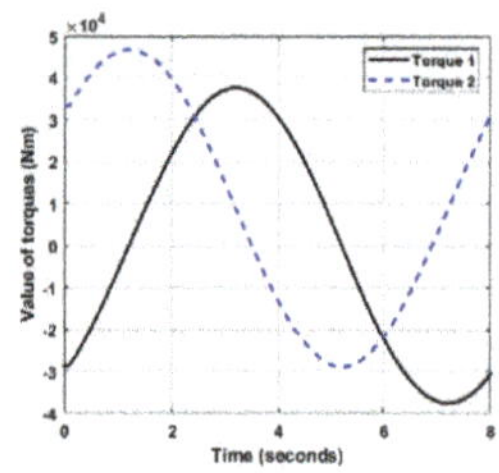

Fig. 12. The driven torques

Determining the values of driven torques at the joints ensuring the given path of the G_2 point in workspace is the goal of this problem. These values can be calculated by using the results of solving the inverse kinematics problem and Eq. (8). The values of the driven torques are shown in Fig. 12.

3 Discussions and Conclusions

The possible working space of G_2 point with the given limit of joints is depicted in Fig. 11. Operate the inverse kinematics problem in MATLAB software, the results of this problem are described from Fig. 5, 6, 7, 8, 9 and Fig. 10. The values of variables joints, velocities and acceleration are shown in Fig. 5, Fig. 6 and Fig. 7, respectively. The position of end-effector point G_2 can be re-calculated through the forward kinematics equations with received $\mathbf{q}(t)$ is shown in Fig. 8 with small position error $(6.2 \times 10^{-3}(\mathrm{m}))$ which is presented in Fig. 9. Following Fig. 12, the maximum values of driven torque at joint 1 is $3.76 \times 10^4(\mathrm{Nm})$ and at joint 2 is $4.68 \times 10^4(\mathrm{Nm})$. It should be noted that, although this study has determined the required drive torque values at the joints q_1,q_2 to ensure the system moves according to the preset trajectory under the influence of wind load, but the exact capacity of the drive motor for joint 2 still has not been determined yet because it is also related to translational joint A and rotational joint B in Fig. 2. This issue will be presented in detail in the next study. Based on the results of the kinematics and dynamics analyzing under the influence of wind loads above, the problems of calculating and designing the whole system has been solved. Figure 13 shows the entire antenna system fabricated in reality and is in the process of installation and experimental implementation.

Fig. 13. Geostationary satellite surveillance antenna was fabricated

In conclusion, this paper presents the kinematics and dynamics of antenna systems for geostationary satellite communications and monitoring, with the focus on the design analysis and optimizations of the mechanical elements and functional units of the antenna systems, taking into account the effects and wind payload.

Each component of the antenna system is assumed a rigid body. The mathematical model is build based on multi-bodies kinematics and dynamics theory. The DENAVIT-HARTENBERG (D-H) homogeneous matrix method and Lagrange type II are used to construct the kinematics and dynamics equations. The inverse dynamics problem is analyzed to determine the value of driven torques at the joints under effect of the wind payload, as well as the necessary behavior of the drive joints, in order to obtain the given desired path in the workspace, to ensure that the antenna system is optimally controlled.

The numerical simulation results kinematics and dynamics were successfully applied in practice, especially for selection of motors and transmission systems for design and development of the antenna system for communications and monitoring of geostationary satellites.

References

1. Maral, G.: Satellite Communications Systems - Systems, Techniques and Technology, 5th edn. Wiley, Hoboken (2009)
2. Satcom and Antenna technologies division. www.cpii.com. Accessed Oct 2020
3. Vincor™ Product Data and Specification Archives. www.catalog.vincor.com. Accessed Oct 2020
4. Hubble Space Telescope. www.nasa.gov. Accessed Oct 2020
5. Featured SATCOM Products. www.satcom-services.com. Accessed Oct 2020
6. Antenna Systems. www.viasat.com/products/antenna-systems. Accessed Oct 2020
7. Satellite Communications equipment. www.digisat.org. Accessed Oct 2020
8. Bindi, Y., Dong, L., Yang, Z., Hao, T., Huibo, Z., Xiangjie, Y., Xiaolei, W.: Dynamics analysis and control of a spacecraft mechanism with joint clearance and thermal effect. Precision Motion Systems: Modeling, Control, and Applications, pp. 163–215. Elsevier Inc (2019)
9. Ogundele, D.A., Akoma, H.E.C.A., Adediran, Y.A.: Mathematical modelling of antenna look angles of geostationary communications satellite using two models of control stations. In: 3rd International Conference on Advanced Computer Theory and Engineering (ICACTE), pp. 236–240 (2010)
10. Ogundele, D.A., Adediran, Y.A., Sikiru, E.O.Y., Oyedeji, E.O., Oseni, O.F.: Model validation and analysis of antenna look angles of a geostationary satellite. In: International Conference on Computer Science and Automation Engineering (CSAE), pp. 509–513. IEEE (2012)
11. Lida, T.: Satellite communications antenna concepts and engineering. In: Pelton, J., Madry, S., Camacho-Lara, S. (eds.) Handbook of Satellite Applications. Springer, New York (2015)
12. Shankar, S.G., Reddy, K.V.: Design and simulation of horn antenna in x-Ku band for satellite communications. Int. J. Res. Sci. Technol. (IJRST) **1**(10), 6 (2014)
13. Spong, M.W., Hutchinson, S., Vidyasagar, M.: Robot Modeling and Control, 1st edn. Wiley, New York (2001)
14. Khang, N.V.: Dynamics of Multi-Bodies. Hanoi Science and Technology Publishing House (2007)
15. Khang, N.V., Dien, N.P., Vinh, N.V., Nam, T.H.: Inverse kinematic and dynamic analysis of redundant measuring manipulator BKHN-MCX-04. Vietnam J. Mech. VAST **32**, 15–26 (2010)

16. My, C.A., Bien, D.X., Hieu, L.C.: Inverse kinematics analysis of welding robot IRB 1520ID using algorithm for adjusting the increments of generalized vector. In: The 5th International Conference on Research in Intelligent and Computing in Engineering. Springer, Singapore (2020) DOI: https://doi.org/10.1007/978-981-15-7527-3
17. Bien, D.X., Tuan, P.A., Nhat, D.D., Hiep, D.X., Nghia, T.K., Anh, M.N.: Optimize the feed rate and determine the joints torque for industrial welding robot TA 1400 based on kinematics and dynamics modeling. Int. J. Mech. Eng. Robot. Res. (IJMERR) **9**(9), 1335–1340 (2020)
18. Hoang, P.Q, Tung, D.M, Toan, B.H.: Study to determine the wind load acting on the antenna serving the calculation and design of the mechanical system controlling the 7.6-diameter geostationary satellite monitoring antenna. In: National Scientific and Technology Conference on Mechanical Engineering-Dynamics, pp. 307–312 (2017)
19. Bien, D.X, My, C.A, Hieu, L.C, Tung, B.H.: Inverse kinematics analysis of welding robot IRB 1520ID using algorithm for adjusting the increments of generalized vector. In: The 5th International Conference on Research in Intelligent Computing in Engineering RICE2020. Proceedings published in Book Series: Advances in Intelligent Systems and Computing. Springer (2020). ISSN: 2194-5358

A Conceptual Digital Twin for Cost-Effective Development of a Welding Robotic System for Smart Manufacturing

Minh Duc Vu[1], Chu Anh My[2(✉)], The Nguyen Nguyen[2], Anh Le[2], Minh Thanh Ta[3], Xuan Bien Duong[4], Tien Anh Nguyen[5], Chi Hieu Le[6], and Duc Hoang Chu[7]

[1] Department of Aerospace Engineering, Le Quy Don Technical University, Hanoi, Vietnam
[2] Institute of Simulation Technology, Le Quy Don Technical University, Hanoi, Vietnam
myca@lqdtu.edu.vn
[3] Deparment of Information Technology, Le Quy Don Technical University, Hanoi, Vietnam
[4] Center of Advanced Technology, Le Quy Don Technical University, Hanoi, Vietnam
[5] Department of Physics, Le Quy Don Technical University, Hanoi, Vietnam
[6] Faculty of Engineering and Science, University of Greenwich, London, UK
[7] Ministry of Science and Technology, Hanoi, Vietnam

Abstract. With rapid technology advancements and impacts of digital transformation technologies, there have been a lot of efforts in recent years at the global and national scope to develop technology eco-systems moving toward Industry 4.0. There has also been a growing number of studies about optimization of process parameters based on AI, IoT and Bigdata analytics, including CNC machining, additive manufacturing, as well as industrial welding with robots, for development of Smart Manufacturing (SM), which is one of the key elements of Industry 4.0. However, there are challenges to fully develop and apply in industrial practices the ideal SM models in the next 5 to 10 years. It is necessary to develop the cost-effective SM models that are easy to understand with a high level of applicability in practice and adoptability for SMEs. In this paper , a conceptual digital twin is proposed, with the focus on cost-effective design and development of a welding robotic system for smart manufacturing.

Keywords: Welding robot · Digital twin · Smart manufacturing

1 Introduction

Manufacturing is becoming smarter and smarter at all levels based on these core advancements, especially digital transformation technologies, AI and bigdata analytics, and abilities to learn, configure and execute with cognitive intelligence, leading to the concept of Smart Manufacturing (SM) and Smart Factory (SF) [1].

There has been a growing number of studies about optimization of process parameters based on AI, IoT and Bigdata analytics, including CNC machining, additive

B. T. Long et al. (Eds.): MMMS 2020, LNME, pp. 1018–1025, 2021.
https://doi.org/10.1007/978-3-030-69610-8_134

manufacturing, as well as industrial welding with robots, for development of SM, which is one of the key elements of Industry 4.0. Development of cost-effective digital twin models for optimization of smart manufacturing shop floors has also obtained growing attentions.

In the welding process with robots, a digital twin model includes not only a digital kinematic model of a robot arm, but also other signal processing layers which constructed with sensors, actuators, data acquisition systems integrated on a robotic system. Specifically, for a smart manufacturing system with welding robots, the signals in the signal processing layers include the following ones: welding voltage, current, weld tip position, seam profile, weld pool, bead dimension.

There are solutions that can be utilized to measure, analyse, monitor and control an integrated welding robotic system. In [2], a sonar sensor was used to estimate the quality of weld bead, together with voltage sensor, current sensor which gathered by a DAQ card and sent to a computer. Force/torque sensor on tool tip was employed in [3] to determine the point on unknow workpiece surface. For recognizing the weld path, a pair of sound microphone set up at two ends of path in [4]. Recently, due to the increasing ability of camera systems with high speed processor and new image recognition algorithm, several parameters of a welding robotic system can be measured without particular sensor. In [5, 6], a camera capturing 2D welding pool shape was integrated with a computation module for calculating the pool shape dimensions to give feedback signals to the controller. 3D images were processed in [7, 9]. A common hardware structure including a robot, a welding machine, sensors, cameras, a control software on PC was demonstrated in [4–6].

There have also been some research works focusing on digital twin model for a robotic system [8–11]. A Kuka robot and the Microsoft Kinetic to integrate a HIRIT platform was investigated in [8]. The work [10] takes into account a 3D digital robot, camera images, ROS to study a digital twin model. Based on the PLM platform, the authors in [11] attempted to a robotic digital twin with a Hybrid 6 DOFs cutting system connected to Labview and Matlab softwares.

The applications of SM in practice are still in the early stage, and there have been a lot of proposed SF conceptualisations with different degree of automation and absence of human workforce [1, 12]. Therefore, there are challenges to fully develop and apply in industrial practices the ideal SF models.

There are challenges to fully develop and apply in industrial practices the ideal SM models in the next 5 to 10 years. It is necessary to develop the cost-effective SM models that are easy to under-stand with a high level of applicability in practice and adoptability for SMEs.

This paper proposes a framework and a case study for cost-effective design and development of a welding robotic system for smart manufacturing. A proposed case study of a digital twin model includes a robot arm and smart elements, for cost-effective development of a smart welding robotic system, with the focus on digital twin model for optimization of welding parameters and robot control, to increase the welding productivity and the welding accuracy.

The rest of the paper is presented as follows. Section 2 presents the materials and methods of a study, with the details about development of the conceptual digital twin for a smart welding robotic system for smart manufacturing. Section 3 presents discussions and conclusions.

2 Materials and Methods

2.1 Hardware and Software for a Smart Welding Robot System

A physical-digital structure of a smart welding robotic system, in which measurement functions, control features and data communication channels, were developed based on the integration of the following functional elements: A robot arm Mitsubishi RV-12SD, welding hardware devices, sensors, data acquisition cards. The following tools and software are used for 3D design and analysis: SolidWorks, LabVIEW and MatLab/SIMULINK.

2.1.1 Industrial Robot Arm

The robot arm Mitsubishi RV-12SD is shown in Fig. 1. The robot arm has 6 Degrees of Freedom (DOF), dedicated by Mitsubishi servo technology plus a high-performance controller CR3D-700 working for high speed (0.66 s cycle time) and accuracy. The robot controller is compatible with not only Mitsubishi's family of automation products, but also other standard devices from other suppliers. This makes the robot arm flexible and durable for industrial applications. The tool head can carry a maximum 12 kg load, moving in 1086 mm radius reach at a top speed of 9600 m/s with 0.05 mm position accuracy.

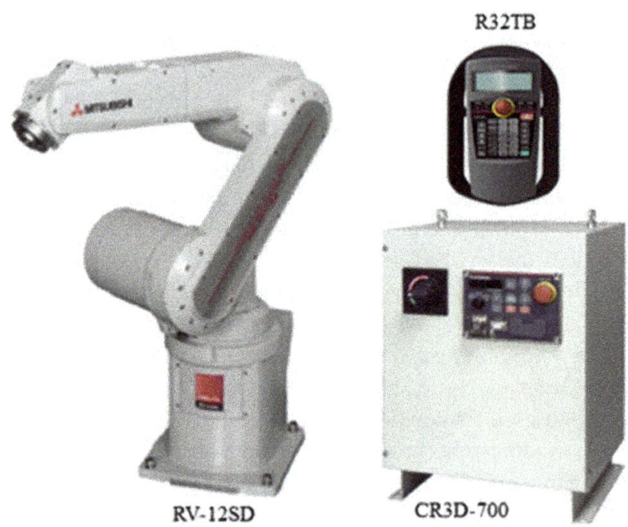

Fig. 1. Standard Robot hardware components for a smart welding system.

The robot controller, CR3D controller, supports the PTP and CP path control methods with the MELFA-BASIC language, which instructions could be sent from the computer. Remote command devices connect to the CR3D controller via either RS 232C or Ethernet ports. Additional teaching pedant R32TB is attached to Improved display performance and operability.

The 3D CAD models of the robot arm Mitsubishi RV-12SD was built in SOLIDWORKS for analysis and simulations.

2.1.2 The Welding Components

The high duty cycle Jasic Mig 250 welding hardware for the integration is used. The supply welding current is from 25 to 250 A at voltage range 11–29 V and wire feeding adjustable 2–15 m/min. The compact accessory set including a welding gun and a pressure air connector are provided to be attached on the robot arm.

2.1.3 Sensors

The voltage sensor module B25, current sensor module Qi-300-I (QEED), encoder Autonics E50S8-100-3-V-5 and air pressure sensor Autonics TPS30-G4KAR2-00 are selected and integrated to measure actual values of the welding voltage, current and wire feed speed respectively.

2.1.4 Data Acquisition Module

The NI USB 6001 module is used which provides functions to measure analog/digital signal from sensors with 8 AI (14-Bit, 20 kS/s), 2 AO (5 kS/s/ch), and 13 DIO. The amplified signal then be given into Labview environment for processing. Simulink model is designed to receive processed data to calculate and simulate the states of the welding robot. The Simulink module also sends separated data back to the robot via Labview/USB 6001.

2.1.5 Camera

In order to determine the dimensions of seam gap, weld bead, torch, etc., the robot is equipped a CCD camera with light filter attached beside the welding tip. The images from camera are transferred to a computer through USB port. In particular cases, it also used to estimate the welding point or path.

2.1.6 Melfa-Basic Remote Command Set

In order to control the robot, the users need programming. For this robot, the commands via RS232C port are accepted. A command usually contains a frame of characters. The movement commands run in either PTP and CP types. The commands could be edited with various software (C#, Matlab, Visual Basic, etc.) to execute the desired tasks. In addition, it helps to set up angle, position or load during operation together with giving the dynamic parameters to remote controller. Remote control ability could change the strategy of movement online, suitable for adaptive welding movement.

Example of commands:

CNTLON set the Robot controller to On for operation.
SRVON to set all the servo a On ready to run all joints.
EXECP1 = (232.00,3.00,400.00,0.00,45.00,0.00)(6,0) set the position 6 DOF.

Figure 2 presents an integration of smart elements for a mart welding system, including the robot arm and controller, welding tool, camera, sensors, and data acquisition module.

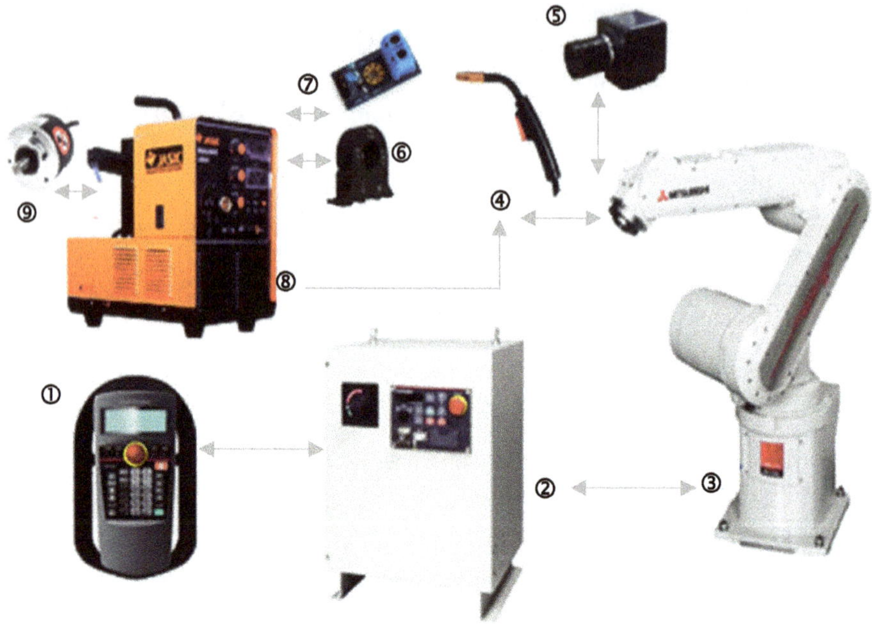

Fig. 2. An integration of smart elements for a mart welding system. (1–3): Robot arm and controller. (4): Welding tool. (5): Camera. (6–9): Sensors, Data acquisition module.

2.2 Integration of a Digital and Communication Model

Figure 3 presents the link between the CAD software and Matlab platform. The robot arm RV-12SD was digitalized with respect to the definition of joints and interfaces, the dimensions of arm bodies, the limit position of the tool head, the range of speed and acceleration of each axis and home position. The 3D CAD model of the robot was built in Solidworks and converted into the Xml format, which is then embedded into the Matlab/Simulink work space with the kinematic features of the physical model.

Figure 4 present the Simulink block diagram of the RV-12SD robot model, a block diagram presents the kinematic features of a robot, in which J1, J2, J3, J4, J5 and J6 are respectively related to the Base, Shoulder, Upper, Elbow, Fore and Wrist of the robot.

The Labview program was built on a PC to read data from NI USB 6001, processes images from the camera to calculate the welding parameters and transfer all the data to Matlab/Simulink work space. In this manner, the states of the physical and digital models can be synchronized. The designed Simulink module also makes tasks and sends Melfa basic commands back to physical robot to make the bidirectional communication. This is an important part of the digital twin creation for the real time parameters updating. In addition, the sensor data, images can be stored in a bigdata platform for smart solution such as IoT and Machine learning.

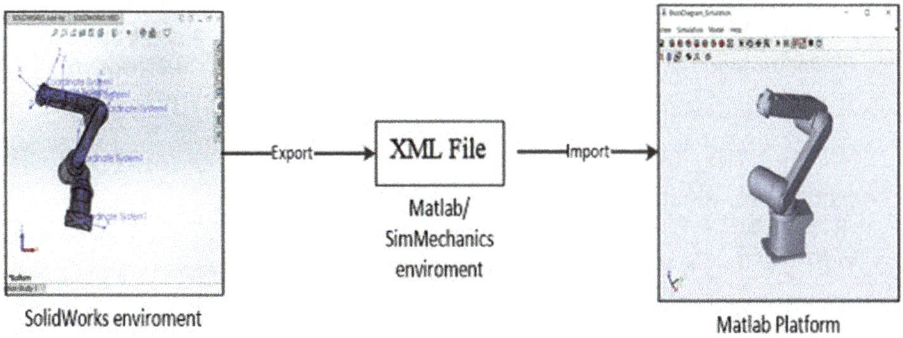

Fig. 3. CAD model to.XML file extension

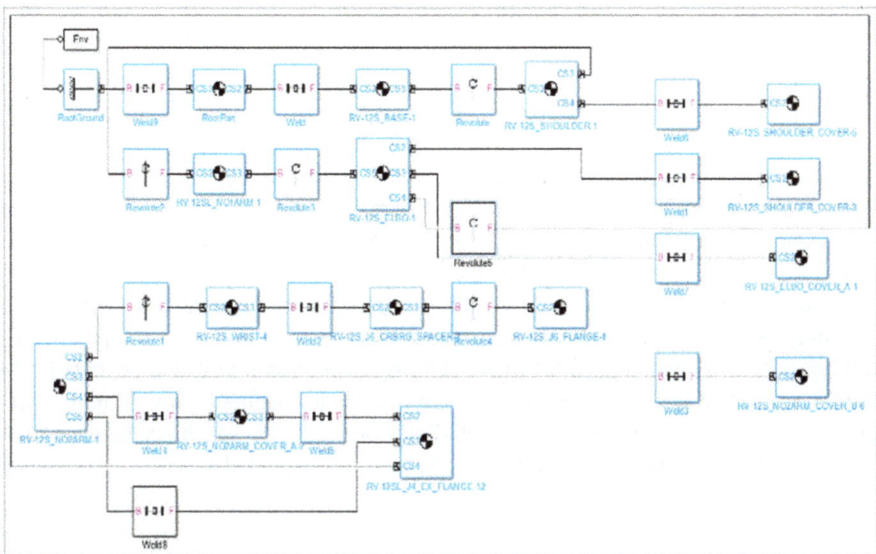

Fig. 4. Simulink block diagram of RV-12SD model

3 Discussions and Conclusions

A conceptual framework for cost-effective development of a smart manufacturing system is presented in Fig. 5, it is aimed to demonstrate the cost-effective digital twin model for a smart manufacturing system, with welding robots.

With the proposed digital twin model, it was successfully demonstrated that the proposed conceptual framework can be extended with additional physical and digital modules, based on the original model, with integration of additional smart elements. The sensor data captured during the operation of the digital-physical system can be stored for related bigdata analytics, from which the machine learning algorithms are applied to optimize the performance of the smart welding robotic system.

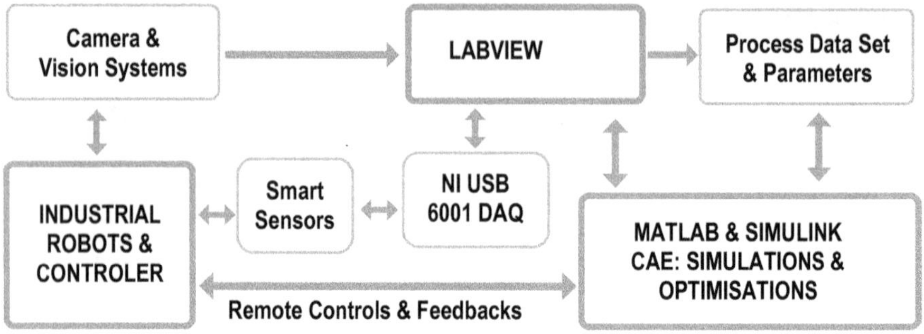

Fig. 5. A conceptual framework for development of a smart manufacturing system with robots.

In conclusion, there have been a lot of efforts in recent years at the global and national scope to develop technology eco-systems moving toward Industry 4.0. However, the manufacturing industry is more vulnerable compared to the service one due to the constraints related to human workforces and manufacturing resources as well as the economic fluctuations and recessions. In addition, applications in practice of SM are still in the early stage. There are challenges to fully develop and apply in industrial practices the ideal SM models in the next 5 to 10 years [1]. In addition, there is also the lack of skilled workforce and financial resources, standardization problems and cybersecurity issues related to development of SM solutions for industrial applications. Therefore, it is necessary to develop the cost-effective SM models that are easy to under-stand with a high level of applicability in practice and adoptability for SMEs. In this study, a conceptual digital twin for cost-effective design and development of a welding robotic system for smart manufacturing is proposed and demonstrated. The study is a foundation for further development of a smart manufacturing system, with a full integration of smart elements, including robots, CNC machines and additive manufacturing systems, collaborative robots, automated guided vehicles (AGVs) and smart sensors.

Acknowledgement. This research is supported by Vingroup Innovation Foundation (VINIF) in project code VINIF.2019.DA08. It is also supported by a Research Environment Links, ID 528085858, under the Newton Fund partnership; the grant is funded by the UK Department for Business, Energy and Industrial Strategy and delivered by the British Council.

References

1. Le, C.H., Le, D.T., et al.: Challenges and conceptual framework to develop heavy-load manipulators for smart factories. Int. J. Mechatron. Appl. Mech. (8) (2020)
2. Zhang, L., Basantes-Defaz, A.C., Ozevin, D., Indacochea, E.: Real-time monitoring of welding process using air-coupled ultrasonics and acoustic emission. Int. J. Adv. Manuf. Technol. **101**, 5–8, 1623–1634(2019). 10/1007/00170-018-3042-2

3. Tian, F., Lin, X., Li, L.: Research on robotic automatic machining for welding groove of complex integral impeller. Int. J. Adv. Manuf. Technol. **105**(12), 5139–5150 (2019). 10/1007/s00170-019-04606-0
4. Na, L.V., et al.: Real-time monitoring of welding path in pulse metal-inert gas robotic welding using a dual-microphone array. Int. J. Adv. Manuf. Technol. **90**(9–12), 2955–2968 (2017). 10/1007/s00170-016-9571-7
5. Xu, Y., Fang, G., Lv, N., Chen, S., Zou, J.J.: Computer vision technology for seam tracking in robotic GTAW and GMAW. Robot. Comput.-Integr. Manuf. **32**, 25–36 (2015). https://doi.org/10.1016/j.rcim.2014.09.002
6. Yu, Z., He, Y., et al.: Vision-based deviation extraction for three-dimensional control in robotic welding with steel sheet. Int. J. Adv. Manuf. Technol. **95**(9–12), 4449–4458 (2018). https://doi.org/10.1007/s00170-017-1546-9
7. Yang, L., Li, E., Long, T., Fan, J., Liang, Z.: A novel 3-D path extraction method for arc welding robot based on stereo structured light sensor. IEEE Sens. J. **19**(2), 763–773 (2019). https://doi.org/10.1109/JSEN.2018.2877976
8. Dröder, K., Bobka, P., Germann, T., Gabriel, F., Dietrich, F.: A machine learning-enhanced digital twin approach for human-robot-collaboration. Procedia Cirp **76**, 187–192 (2018). https://doi.org/10.1016/j.procir.2018.02.010
9. Wu, P., Gan, Y., Dai, X.: The weld extraction algorithm for robotic arc welding based on 3D laser sensor. In: 2019 Chinese Control and Decision Conference, pp. 998–1002 (2019). https://doi.org/10.1109/ccdc.2019.8833358
10. Kousi, N., Gkournelos, C., Aivaliotis, S., Giannoulis, C., Michalos, G., Makris, S.: Digital twin for adaptation of robots' behavior in flexible robotic assembly lines. Procedia Manuf. **28**, 121–126 (2019). https://doi.org/10.1016/j.promfg.2018.12.020
11. López-Estrada, L., et al.: Creation of a micro cutting machine tool digital-twin using a cloud-based model-based PLM Platform: first results. Procedia Manuf. **41**, 137–144. 10.1016.j.promfg.2019.07.039
12. Jeong, Y.W., Min, J.P.: Smart factory adoption in small and medium-sized enterprises: empirical evidence of manufacturing industry in Korea. Technol. Forecast. Soc. Chang. **157**, 120117 (2020)

Real – Time Automatic Laser Focusing System for High – Precision Micromachining by Image Processing

Phuong-Thao Nguyen, Thi Kim Cuc Nguyen, Hong Hai Hoang, Toan Thang Vu, and Xuan Binh Cao(✉)

School of Mechanical Engineering,
Hanoi University of Science and Technology, Ha Noi, Viet Nam
binh.caoxuan@hust.edu.vn

Abstract. In this manuscript, an automatic system for the real-time detection of focal position on a specimen during laser micromachining is presented. The system allows for accurately and rapidly positioning the laser focus on the target surface, wherever it is located along the optical axis. This approach is based on a feedback system that uses a laser source, actuator, focusing lens, specimen and camera to monitor the size of the beam spot. We propose a system to automatically find the focal point of laser by directly observing the shape, size and position of the beam spot on the surface through image processing. Compared with conventional methods, this approach is advantageous because the implementation is simple, easy and inexpensive. The results indicate dramatically high precision. The focal spot analytically maintains its size with a size deviation $\pm 10 \mu m$.

Keywords: Automatic · Detection-focal position · Laser micromachining · Image processing

1 Introduction

Recent enhancements in the laser micromachining capacity have drawn substantial interest from both scientists and engineers. Laser micromachining is very important in the modern world. In laser ablation processing, carefully maintaining the focal position, i.e., the position associated with the minimum waist diameter, on the working surface is a crucial step in the fabrication process. This ensures that the correct laser power density is used to form a groove or a hole on the specimen.

In practice, the quality of laser machined products is highly dependent on machining system parameters such as laser source power and the position of the machining surface (Fig. 1). Just changing a small parameter can affect machining quality. Specifically, laser micromachining of machining quality will be adversely affected when the focal position is not on the surface, especially in high precision micromachining (Fig. 2). When the focal position is not on the surface, the workpieces, especially the drill holes or grooves will be inconsistent and a certain depth cannot be reached, so high accuracy cannot be guaranteed. Conversely, when the focal point is on

B. T. Long et al. (Eds.): MMMS 2020, LNME, pp. 1026–1032, 2021.
https://doi.org/10.1007/978-3-030-69610-8_135

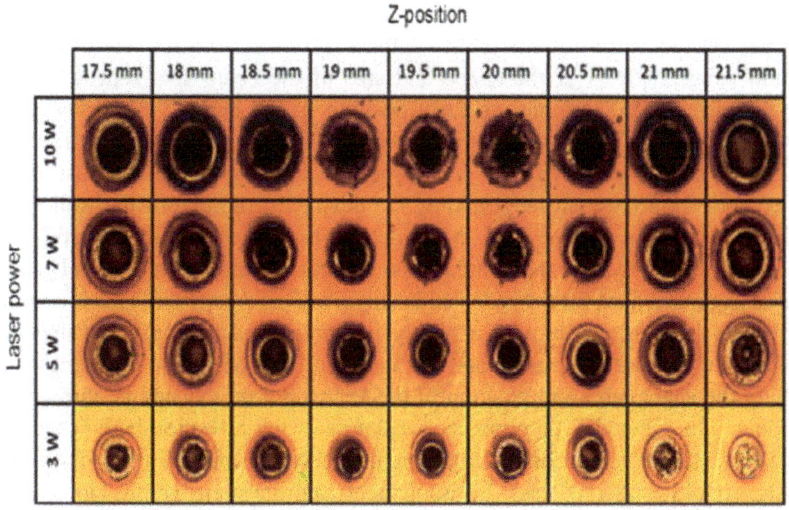

Fig. 1. The percussion drilling results for laser beam propagation measurements.

the machining surface, the workpiece quality is smoother, more homogeneous and has higher precision. This change of the workpieces greatly affects the functionality as well as the aesthetics of the surface. Overcoming this weakness is a challenge for both scientists and engineers.

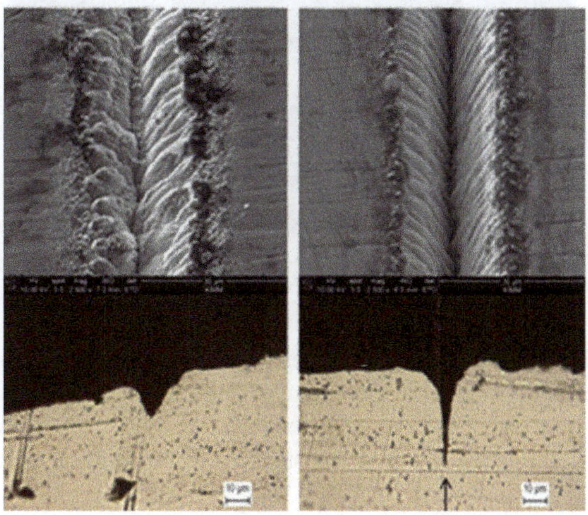

Fig. 2. Laser processing groove pattern when the focal position is deflected (left) and when focal position is right on the surface (right).

Due to the importance of determining the focal position, there are many systematic researches related to this issue. Most of these systems maintain real-time focusing condition using analysis of the shape and size of the beam spot reflected from the material's surface to determine the focal position of the laser [1–6]. However, these concepts of focus positioning cannot be applied to surfaces with low reflectivity. When the surface has low reflectivity or severe warpage, the shape, size and position of the reflected spot from these surfaces greatly are deviated from that of the well-reflective surfaces. This causes a large error in the detection of focal position. In addition, the reflection geometry reduces the compactness of the focus detection system because we have to design the space for the reflected beam as well as the sensor. This is a limitation when processing systems require a highly integrated and compact focusing device. Therefore, we propose a system to automatically find the focal point of laser by directly observing the shape, size and position of the beam spot on the surface through image processing. The method overcomes the weakness of the reflection-based focus detection on low reflectivity surfaces. In addition, this system provides the most accurate focusing conditions. It is easy to implement and inexpensive in industrial laser micromachining. The system is extremely compact and integratable to most of the processing system as well as many other applications of lasers.

2 Theoretical Models

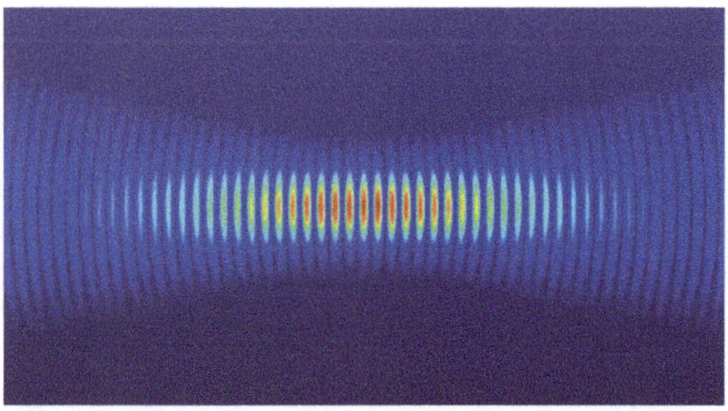

Fig. 3. The energy density distribution of the laser beam

The energy density distribution of the laser beam according to the direction of transmission and the cross-section of the beam with the form of a Gaussian distribution function is shown in Fig. 3. Gaussian beam waist parameters are shown in Fig. 4. The beam waist is w_0, the beam radius at a distance z from the focal point is calculated:

$$w(z) = w_0\sqrt{1 + \left(\frac{z}{z_R}\right)^2} \tag{1}$$

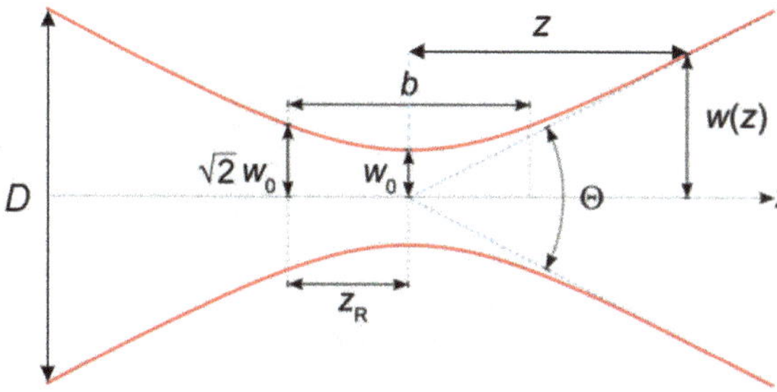

Fig. 4. Diagram of Gaussian beam waist parameters.

Where z_R is Rayleigh length calculated:

$$z_R = \frac{\pi w_0^2}{\lambda} \tag{2}$$

The depth of focus b is the range at which focusing condition is continuously preserved and allowed in the processing. This parameter is calculated as the following:

$$b = 2z_R \tag{3}$$

From theoretical basis, we calculate and give the principle diagram shown in Fig. 5. From the processor, the signal will be transmitted to control the actuator, shifting the focusing lens to change the beam shape on the sample surface. All images are captured by the camera and then returned to the processor.

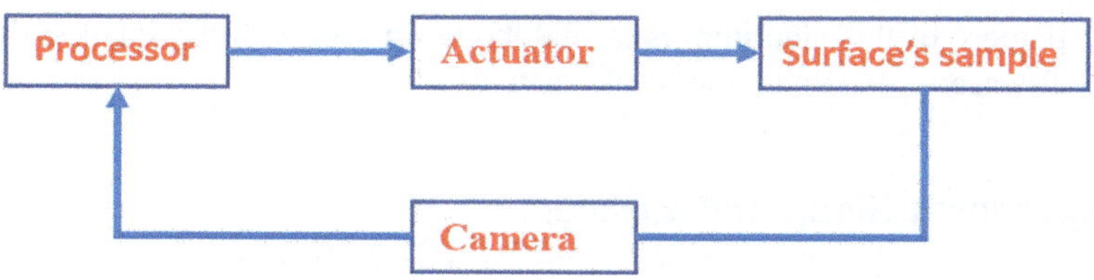

Fig. 5. The principle diagram of the system.

This system comprises laser source, actuator, focusing lens, specimen and camera shown in Fig. 6. The actuator consists of a stepper motor fitted with lead screws and a guide rod. Motor with a minimum displacement of 5 μm is fitted with a lead screw to avoid installation errors and provides straight drive. Additionally, motion is ensured linear by a guide rod. The laser source emits a laser beam passing through the focusing lens to the specimen, the actuator will shift the focusing lens and change the beam spot size on the specimen. The change is recorded and processed continuously by the camera. Subsequently, through the processor, the signal is transmitted to the actuator

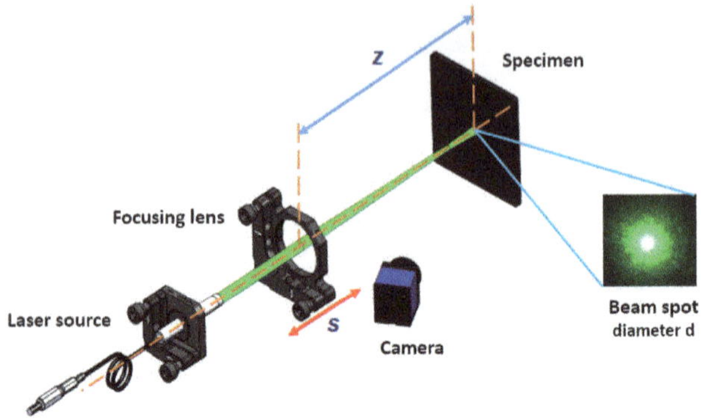

Fig. 6. Operation model of the system.

that moves the laser focus on the specimen. To process recorded images from the camera, we propose the following algorithm:

- Step 1: Move the lens to the z_{max} or “home” position.
- Step 2: Calculate z_{lt} to beam focus on the specimen, depth of focus b with convergence coefficient n, displacement factor k.
- Step 3: Move to position $z_0 = z_{lt} + n.b$, measure the beam diameter on the specimen d_0.
- Step 4: Move with minimum step s to position $z_1 = z_o - k.s$, and determine d_1.
- Step 5: If $d_1 < d_0$ assign $d_0 := d_1$, $z_f := z_1, z_f$ is the distance at which the focal point is detected. If $d_1 > d_0$ then $d_{min} := d_0$ (focal point).
- Step 6: Move to z_f position.

The purpose of this algorithm is to find the beam spot with the smallest diameter and the distance to determine the focal position.

3 Experiment Setup and Results

The experimental setup using a camera for image processing is shown in Fig. 7. We use a laser diode source with wavelength λ = 532 nm to determine the focal position. The focused beam on the specimen is controlled on an actuator attached to a focusing lens with focal length f = 200 mm and minimum displacement s = 0.005 mm. In addition, actuator can move focusing lens with total travel length of 100 mm. The DFK 41 BU02 camera was used to capture the beam spot image while moving the focusing lens.

The experimental results of the system are shown in Fig. 8. This is the experimental result of image processing as well as actuator control of focal position. The focusing lens is adjustable in the range from 0 to 70 mm. Then the image processing software will run the command lines to find the point with the smallest diameter and send the signal to control the actuator to that point. The results show that the focal point is detected at a position with distance of 61.52 mm and a diameter of 20.00 pixels.

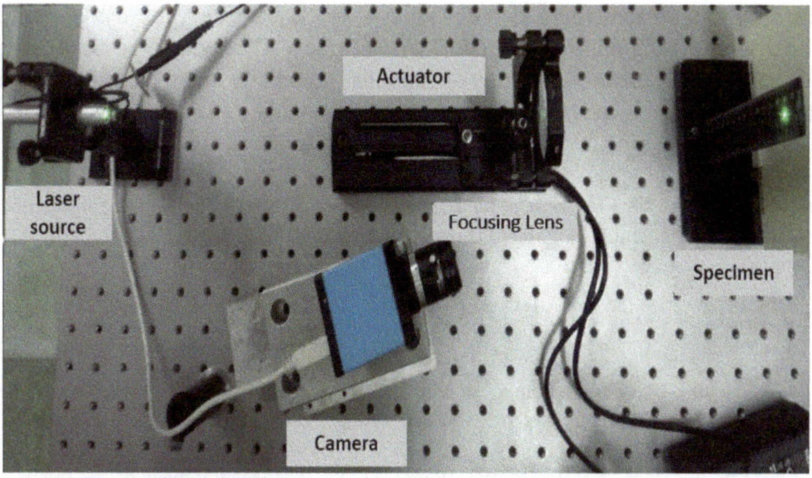

Fig. 7. Setup experiment model of the system.

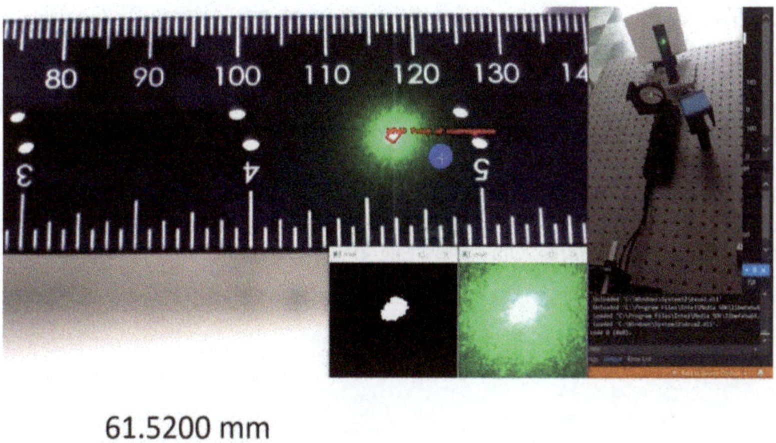

Fig. 8. The experimental results of the system.

4 Conclusion

In this research, an automatic system for the real-time detection of the focal position on a specimen by image processing during laser micromachining is presented. Based on the experimental results and theoretical calculations, the model of the system obtains the depth of focus with deviation of ± 10 μm. Based on the shape and size of the beam spot on specimen, the system is able to maintain the focal position on the specimen during laser micromachining with a deviation of just a few μm. Thus, the workpiece is obtained with great sophistication and high accuracy. The method is more advantageous than existing methods in terms of handling, portability and focusing in real-time. At the same time, this method helps engineers to maintain on-focus state of lasers on low reflectivity specimens, which limits the reflection based focusing techniques.

Acknowledgement. This research is funded by Vietnam National Foundation for Science and Technology Development (NAFOSTED) under grant number 103.03-2020.48.

References

1. Cao, B., Hoang, P., Ahn, S., Kang, H., Kim, J., Noh, J.: High-speed focus inspection system using a position-sensitive detector. Sensors **17**(12), 2842 (2017). https://doi.org/10.3390/s17122842
2. Li, X., Liu, B., Mei, X., Wang, W., Wang, X., Li, X.: Development of an in-situ laser machining system using a three-dimensional galvanometer scanner. Engineering **6**, 68–76 (2020)
3. Cao, B.X., Le Hoang, P., Ahn, S., Kim, J., Noh, J.: High-precision detection of focal position on a curved surface for laser processing. Precis. Eng. **50**, 204–210 (2017). https://doi.org/10.1016/j.precisioneng.2017.05.008
4. Cao, B.X., Le, P.H., Ahn, S., Kang, H., Kim, J., Noh, J.: Automatic real-time focus control system for laser processing using dynamic focusing optical system. Opt. Express **25**(23), 28427 (2017). https://doi.org/10.1364/OE.25.028427
5. Hu, X., Wang, G., Hyun, J.-S., Zhang, Y., Yang, H., Zhang, S.: Autofocusing method for high-resolution three-dimensional profilometry. Opt. Lett. **45**(2), 375–378 (2020)
6. Diaci, J., Bračun, D., Gorkič, A., Možina, J.: Rapid and flexible laser marking and engraving of tilted and curved surfaces. Opt Lasers Eng **49**(2), 195–199 (2011)

A Comparison of Fuzzy-PID Controller, Fuzzy Logic Controller and PID Controller of an Autonomous Robot

Thi Thoa Mac(✉), Nguyen Cao Thanh, Nguyen Cao Cuong, and Nguyen Cong Hoang

School of Mechanical Engineering, Hanoi Universiy of Science and Technology, Hanoi, Vietnam
thoa.macthi@hust.edu.vn

Abstract. Recently, autonomous mobile robots have been increasingly developed for many applications related with surveillance, transportation, and inspection. This paper introduces an autonomous robot motion control based on PID, Fuzzy, and Fuzzy-PID controllers. The experiment is implemented on a mobile robot to compare the performances among three controllers. The comparison results show that Fuzzy-PID controller is the most suitable controller to deal with the rise time, steady sate error and overshoot problems.

Keywords: PID controller · Fuzzy controller · PID-Fuzzy controller · Autonomous navigation

1 Introduction

Autonomous robots are integrated systems with various embedded technologies such as sensors, actuators, controller design, signal processing and electrical component [1, 2]. There are four basic components of autonomous components: *i)* perception, the extraction meaningful information using robot sensor system; *ii)* localization, the determination the robot's location; *iii)* cognition and path planning, the way to reach its goal without collide surrounding obstacles; *iv)* motion control [3]. In which, the motion control is an important part.

Among many control methodologies, PID (Proportional–Integral–Derivative) controller is the most popular in many industrial applications [4]. The three-parts contain a proportional (P), an integral (I), and a derivative (D) term to build a physical system archive a desirable performance in term of small settling time, steady-state error, and overshoot. The three gains can be efficiently tuned by experience or some simple principles, such as Ziegler-Nichols method, FRTool [5]. It is noted that PI or PD controllers are also prefer for many practical applications. However, in general, PID is difficult to solve the time delays, disturbances.

Fuzzy logic control (FLC) is used for developing intelligent system. It uses the If-Then rules. Basically, Fuzzy logic systems is inspired by decision making with uncertain and imprecise information. Beside the conventional FLC, it also has PI-Fuzzy or PD-Fuzzy controllers. To easily implement, the fuzzy control divides into simpler

B. T. Long et al. (Eds.): MMMS 2020, LNME, pp. 1033–1040, 2021.
https://doi.org/10.1007/978-3-030-69610-8_136

behaviors [6–8]. A fuzzy-PID is an integrated controller instead of using sole PID or Fuzzy technique. This integrated controller is used to tune the parameters of PID [10]. Normally, the integrated method provides much better results than individual techniques.

In this study, the performances of PID controller, fuzzy controller and PID-fuzzy controller are compared using a mobile robot. The remainder of this article is organized as follows: Sect. 2 presents the robot system and the model of mobile robot. Three controllers PID, fuzzy controller, Fuzzy-PID are introduced in Sect. 3. The experiment results and discussion are descripted in Sect. 3.4. Section 4 is a conclusion section where the main results of this work is summarized.

2 Robot System Design

In this study, a mobile robot model with differential drive is designed to test three control methodologies. Figure 1 present the mobile robot and kinematics model.

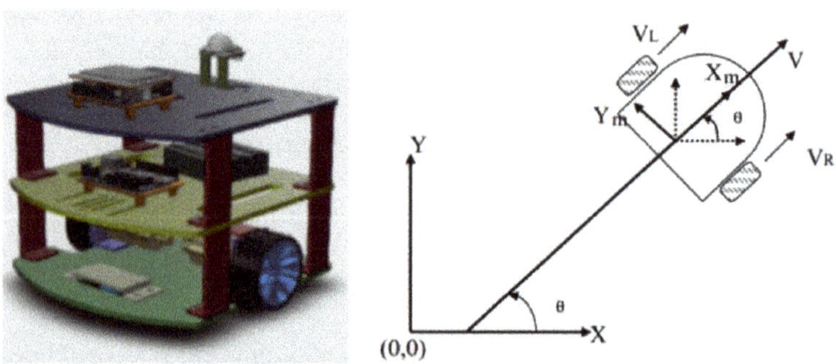

Fig. 1. The mobile robot and Kinematics model

In the internal coordinate frame, The center of the mobile robot is identified by its position and orientation $(x,\ y,\ \theta)$. The movement of the robot in its working space is observed by linear and angular velocities. Therefore, we define the state $\dot{q} = \left(\dot{x}, \dot{y}, \dot{\theta}\right)$. We have

$$\begin{aligned} v &= \frac{R}{2}(v_L + v_R) \\ w &= \frac{R}{L}(v_L + v_R) \end{aligned} \tag{1}$$

Where:

v and ω: the linear velocity and angular velocity of the mobile robot.
v_L, v_R: the left and right wheels velocities.
R: the radius of the wheel.

L: the distance between two wheels.

The tangential velocities according x axis and y axis are:

$$\begin{aligned} \dot{x} &= \frac{dx}{dt} = v\cos\theta \\ \dot{y} &= \frac{dy}{dt} = v\sin\theta \end{aligned} \tag{2}$$

The kinematic model of the mobile robot is presented in following equation:

$$\begin{bmatrix} \dot{x} \\ \dot{y} \\ \dot{\theta} \end{bmatrix} = \begin{bmatrix} \cos\theta & 0 \\ \sin\theta & 0 \\ 0 & 1 \end{bmatrix} \begin{bmatrix} v \\ \omega \end{bmatrix} \tag{3}$$

3 Fuzzy-PID Controller

In this section, PID controller (Subsect. 3.1), Fuzzy controller (Subsect. 3.2), PID-Fuzzy (Subsect. 3.3) controller are presented respectively. Then, the comparison results of three mentioned control methods and the discussion are presented in Subsect. 3.4.

3.1 Conventional PID Controller

The PID controllers is the most popular in many real-life applications. The mathematical form of PID controller is:

$$u(t) = K_p e(t) + K_i \int_0^t e(\tau)d\tau + K_d \frac{de(t)}{dt} \tag{4}$$

Where:

K_p, K_i, K_D: PID' control parameters
e(t): Error between control signal and feedback signal
u(t): Control signal

In this study, PID control is tuned based on Ziegler-Nichols method. Relay-based auto tuning process includes the following steps:

Step 1: Using a random constant input, u_0 to determine the corresponding steady state output y_0.
Step 2: Setting the set point as y_0
Step 3: Running controller in relay mode, which have the input signal as:

$$u = \begin{cases} u_0 + h \; if \; e \geq 0 \\ u_0 - h \; if \; e \leq 0 \end{cases} \tag{5}$$

Where:

u: Control signal
e: Difference between y_0 (set point) and y_t (output signal)
h: Amplitude of input signal.
Step 4: From the signals, define the output amplitude a of reaction signal and ultimate period P_u. The ultimate gain $K_u = \frac{4h}{\pi a}$

After having ultimate gain K_p and ultimate period P_u, applying them with Ziegler-Nichols tuning method to turning PID parameters (Table 1):

Table 1. Ziegler-Nichols tuning parameter

Specification	K_p	K_i	K_d
Original	$0.60 \times K_u$	$2 \times K_p/T_u$	$K_p \times T_u/8$
Little overshoot	$0.33 \times K_u$	$2 \times K_p/T_u$	$K_p \times T_u/3$
No overshoot	$0.2 \times K_u$	$2 \times K_p/T_u$	$K_p \times T_u/8$

3.2 Fuzzy Controller

In this Fuzzy controller, the input signal is the error between controlled speed and setting speed of the motor. The output signal is extra value to add in the PWM control signal. From the feedback signal from the encoder, the Fuzzy controller will modify the PWM signal to get the output to reach the required signal. The input and output membership functions are shown as Fig. 2 and Fig. 3, respectively.

The Fuzzy rule base configurations are:

- If error is negative large then output is add Large negative
- If error is negative medium then output is add Medium negative
- If error is negative small then output is add Small negative
- If error is zero then output is add zero
- If error is positive small then output is add Small positive
- If error is positive medium then output is add Medium positive
- If error is positive large then output is add Large positive.

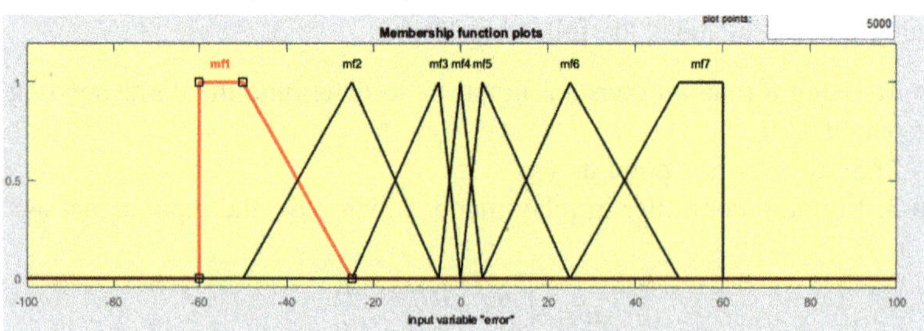

Fig. 2. Fuzzy controller's input membership function

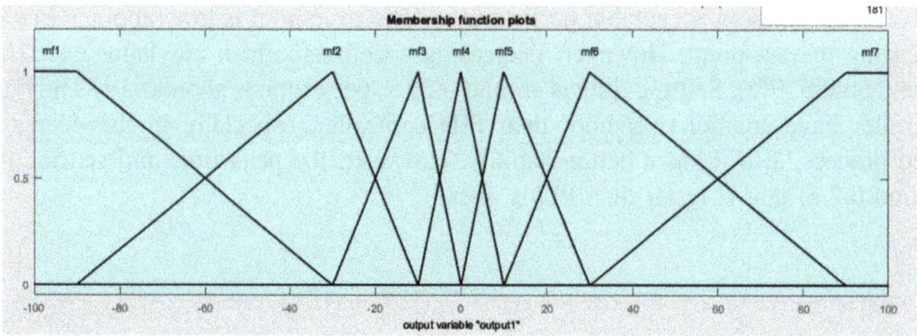

Fig. 3. Fuzzy controller's output membership function

3.3 Fuzzy-PID Controller

In Fuzzy-PID controller, the fuzzy controller rules are applied to obtained the proper PID's parameters as presented in Fig. 4.

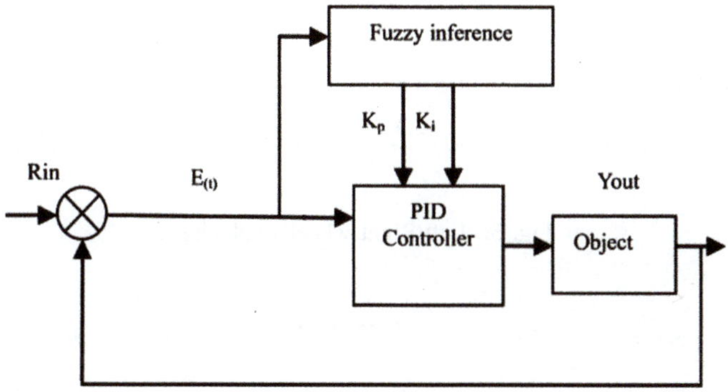

Fig. 4. The self-tuning fuzzy PID controller

In which, the Fuzzy rule base configuration:

- **If** error is *negative large* **then** K_p is *medium* and K*i* is *medium*
- **If** error is *negative medium* **then** K_p is *medium* and K_i is *medium*
- **If** error is *negative small* **then** K_p is *small* and K_i is *small*
- **If** error is *zero* **then** K_p and K_i is *zero*
- **If** error is *positive small* **then** K_p is *small* and K_i is *small*
- **If** error is *positive medium* **then** K_p is *medium* and K_i is *medium*
- **If** error is *positive large* **then** K_p is *large* and K_i is *large.*

3.4 Experiment Results and Discussion

The experiment results of three types of controllers, PID, Fuzzy, PID-Fuzzy are performed and compared on the mobile robot system as shown in Fig. 5 and Fig. 6.

PID controller has an acceptable performance. The overshoot is low (about 3%) when increasing the set point. However, decreasing speed setpoint it has lager over shoot value (about 8–9%). Settling time is around 0.25 s, peek time is about 0.2 s. The Fuzzy controller have smaller overshoot than PID controller, especially in the decreasing speed process, also it has a better stability. However, the peak time and settling time (around 0.2 s) that is faster than PID's ones.

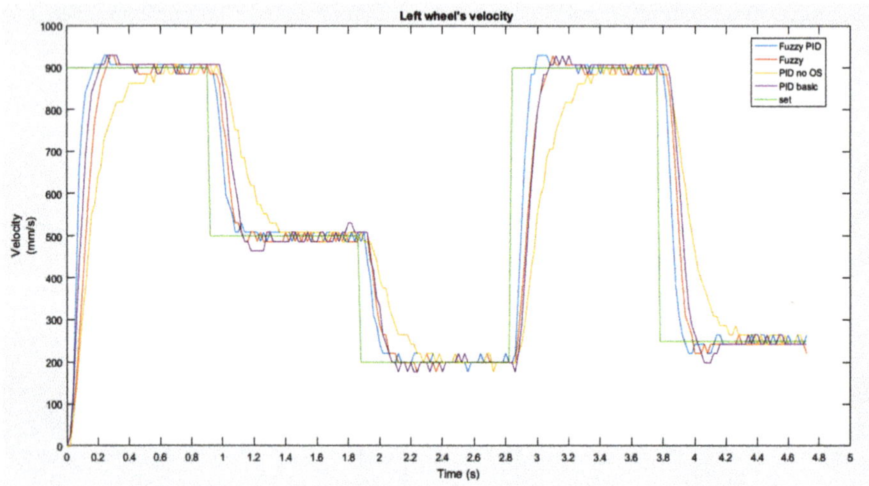

Fig. 5. Left/Right wheel's velocity

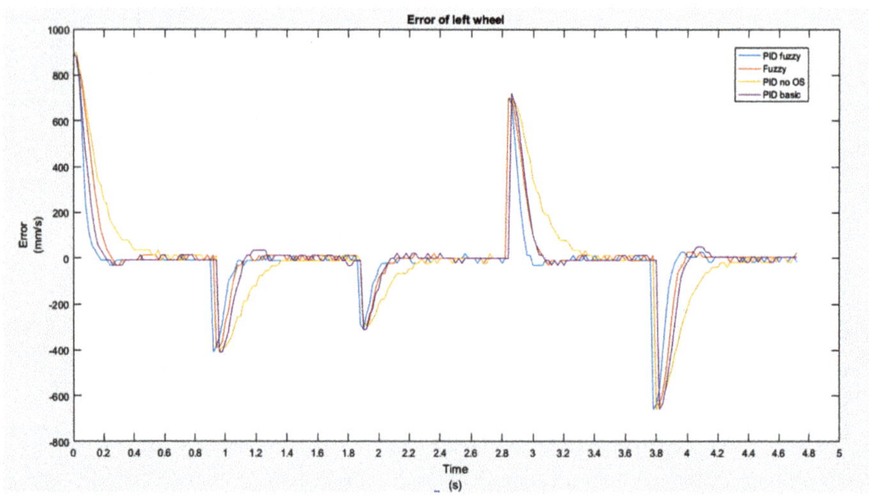

Fig. 6. Left/Right wheel's error

Self-tuning Fuzzy-PID controller is the superior one in the comparison of Fuzzy controller and PID controller. As shown in Fig. 5 and Fig. 6, Fuzzy-PID has the lower peak time and settling time than the others. It has overshoot less than PID, and stability are similar with Fuzzy controller. The wind-up problem is also minimized, and the system working process is smoothly.

4 Conclusion

This paper presents a comparison of Fuzzy-PID controller, Fuzzy logic controller and PID controller performances. Three control methods been implemented in a real robot system. Compare to basic PID controller, Fuzzy controller is more difficult to find the control parameters. The Fuzzy controller have smaller overshoot than PID controller, especially in the decreasing speed process. It also has a better stability. Moreover, its peak-time and settling time are small (around 0.2 s) that is faster than PID's ones. Self-tuning Fuzzy PID Controller is the most powerful controller comparing with pure Fuzzy controller and conventional PID controller in term of overshoot, stability, steady sate error, setting time and rise time. Self-tuning Fuzzy PID is convenient for designers. As a consequence, Fuzzy-PID controller now appropriately increase for autonomous robot applications.

Acknowledgments. This work is granted by the National Foundation for Science and Technology Development of Vietnam NAFOSTED, Vietnam under project number 107.01-2018.331.

References

1. Mac, T.T., Copot, C., Tran, D.T., De Keyser, R.: Heuristic approaches in robot path planning: a survey. Robot. Auton. Syst. **86**, 13–28 (2016)
2. Siegwart, R., Nourbakhsh, I.R.: Introduction to Autonomous Mobile Robot. Massachusetts Institute of Technology press, Cambridge (2004)
3. Mac, T.T., Copot, C., Tran, D.T., De Keyser, R.: A hierarchical global path planning approach for mobile robots based on multi-objective particle swarm optimization. Appl. Soft Comput. **59**, 68–76 (2017)
4. Mac, T.T., Copot, C., Ionescu, C.M.: Design and implementation of a real-time autonomous navigation system applied to lego robots. IFAC-PapersOnLine **51**(4), 340–345
5. Khac, T.D., Lee, K., Im, D.Y.B.: Design of fuzzy-PID controller for path tracking of mobile robot with differential drive. Int. J. Fuzzy Logic Intell. Syst. **18**(3), 220–228 (2018). http://doi.org/10.5391/IJFIS.2018.18.3.220
6. Mac, T.T., Copot, C., Lin, C.Y., Hai, H.H., Ionescu, C.M.: Towards The development of a smart drone police: illustration in traffic speed monitoring. J. Phys. Conf. Ser. **1487**, 012–029 (2020)

7. Mac, T.T., Copot, C., De Keyser, R., Tran, T.D., Vu, T.: MIMO fuzzy control for autonomous mobile robot. J. Autom. Control Eng. **4**(1), 65–70 (2016)
8. Mac, T.T., Copot, C., Verdonck, L., De Keyser, R.: 2016 IEEE International Conference on Systems, Man, and Cybernetics (SMC), pp. 3748–3753 (2016)
9. Mitrovic, S.T., Djurovic, Z.M.: Fuzzy-based controller for differential drive mobile robot obstacle avoidance. IFAC Proc. **43**(16), 67–72 (2010)
10. Chao, C.-T., Sutarna, N., Chiou, J.-S., Wang, C.-J.: An optimal fuzzy PID controller design based on conventional PID control and nonlinear factors. Appl. Sci. **9**, 1224 (2019)

Measurement Range Extension of an Industrial Tomography and Profilometry Using Comb-less Interferometry

Tuan Cong Truong[1(✉)], Thoa Thi Mac[1], Tuan Quoc Banh[1], Tung Thanh Vu[1], Tatsutoshi Shioda[2], and Thang Toan Vu[1]

[1] School of Mechanical Engineering, Hanoi University of Science and Technology, 1st Dai Co Viet Street, Hai Ba Trung District, Hanoi, Vietnam
tuan.truongcong@hust.edu.vn

[2] Graduate School of Science and Engineering, Saitama University, 255 Shimo-Okubo, Sakura-ku, Saitama 338-8570, Japan

Abstract. A measurement extension of a 2D single-shot tomography and profilometry is proposed for the industrial applications using a comb-less broadband optical interferometry. The measurement range can be extended and flexibly varied using a spatial phase modulator (SPM) installed in the interferometer, providing a zooming operation in the axial axis. The measurement range can be controlled from 200 μm to 3.5 mm by controlling the diffraction orders of the SPM. Tomogram and surface profile interference 2D image can be simultaneously acquired by a single - shot with an axial resolution of 2.0 μm (for tomography) and 0.6 μm (for profilometry), respectively.

Keywords: Profilometry · Tomography · Optical zooming

1 Introduction

Recent developed measuring techniques for surface profilometry and tomography often involve of using optical devices specifically designed for the demands of industrial applications. Target development of these techniques are usually the improvements of measurement speed, axial and lateral resolutions, sample non-invasive, and system setup flexibility. Conventional optical techniques for measuring surface profile such as microscopy [1], laser scanning [2], optical interferometry [3, 4] have been developed since the last four decades. These technologies have the advantages of high resolution, non-invasive and useful for the off-line measurements where the number of the objects to be measured are not too large. The recent developed tomography and profilometry mostly employs the low-coherent interferometry in either time-domain [5] or frequency-domain [6]. This method has the advantage of having no blind region due to normal-incidence and can be non-contact optical configuration. It, however, suffers from a time-consuming problem owing to the mechanical scanning [7] in time-domain or the Fourier transform data processing [8] in frequency - domain. The measurement time for these interferometry to achieve an interference image is typically in the range

B. T. Long et al. (Eds.): MMMS 2020, LNME, pp. 1041–1052, 2021.
https://doi.org/10.1007/978-3-030-69610-8_137

of few tenths milliseconds to several minutes. This limits their usages in the industrial factories where real time operation in the in-line productions is a very important factor.

Advanced methods to overcome the time consuming problem of the conventional optical interferometry have been recently introduced. For instance, optical frequency comb light source [9] or supercontinuum light source [10] based interferometer was introduced. However, the efficiency of these systems was not sufficient because of the system complexity and high cost. In the previous research [11–13], we introduced a 2D, long range, single-shot profilometry and tomography using a virtually image phase array (VIPA) [11], and an optical frequency comb interferometry combined with the spatial phase modulator (SPM) and CCD camera [12, 13]. These systems had overcome the time-consuming of the conventional time-domain interferometry. Also, the measurement range could be extended with optical frequency comb technology in a range of several tenths' millimeters. It has the advantage of realizing a comb light source very easily by using an etalon. However, using comb interferometry suffered the loss of the light source power owing to the optical frequency comb generator (etalon) and the comb-interference order must be distinguished in order to calculate the physical dimension of the sample [13]. In the previous our work, the axial zoomable 2D single - shot comb interferometry [14] have been propped. However, the Fabry – Perot etalon, which is used for generating the filtered optical frequency comb, is also the cause of optical power loss of the light source.

In this paper, we propose another technique for the measurement range extension of the single - shot 2D tomography and profilometry without using optical frequency comb so that the discrimination of the interferometer fringe order will be neglected and the optical power loss of the Fabry – Perot etalon will be solved. By a utilization of the diffraction lights of the spatial phase modulator (SPM) installed in the broadband Michelson interferometer, the measurement range can be flexibly changed so that we can extend or shrink the measurement range as a zooming operation. The zooming operation in the axial axis with the maximum magnification of 20x is realized by changing the diffraction lights of the SPM, giving a dynamic measurable range in axial axis can be changed from 200 μm to approximate 3.5 mm. The zooming function is useful for the industrial application since it gives a flexible way for the users to monitor various depths of sample. This technique has an affinity for measuring the samples with smooth surfaces (like semiconductor surface, car body, or continuous surface, etc.). For these samples we do not need such a wide measurement range for single-shot images as provided by the single-shot comb interferometry. Instead of that we need an efficient measurement range that will be easy for detecting the initial position of the sample as well as easy to monitor the sample dimension within a CCD camera screen. Therefore, with the changeable measuring range using a controllable SPM, we can omit the etalon to enhance the optical power, also the setting of the initial sample position will be much easier compared with that of the single-shot comb interferometer.

A broadband optical interferometer in the visible wavelength region is also used for improving the resolution of the tomography and profilometry. 2D interference images are continuously monitored in real time (acquisition time of 10 μs). The sample's information in axial (depth) and horizontal planes can be calculated from the 2D interference images. The tomogram and profile axial resolutions are estimated of approximate 2.0 μm, and 0.6 μm with the light-source center wavelengths of 610 nm and the bandwidth of 150 nm, respectively. With this high resolution, one of the typical application to measure the defect crashes on the sample surfaces can be easily seen. These high-resolution measurements are very important for the industry. The lateral resolution of the system is also improved by processing the scattered lights back from the sample using 4-f configuration lenses. Lateral resolution depends on the depth positions of the sample. It varies from 10 μm to approximate 80 μm (corresponding to the maximum measurable depth of 3.5 mm). In addition, 3D reconstructed image of the sample can be made from the proposed 2D single-shot system by adding an additional scan in lateral axis. Operating principle as well as the experimental results for demonstrate the proposed system will be presented.

2 Measurement Range Expansion Principle

Figure 1 shows the top view of optical setup of the proposed 2D single-shot tomography and profilometry. A visible wavelength region ($\lambda \approx 550$ nm–700 nm) of a meter of the beam is expanded to about 10 mm by a beam expander. The beam diameter is an important factor to make a reflection grating become a spatial phase modulation, which is the key process to make the single-shot operation. The expanded beam then comes to the Michelson interferometer where it is divided by two paths: measurement path and reference path. In the reference path, the beam passes a set of plano-convex (PCX) lenses with 4-f configuration before being reflected by the SPM. The SPM is a reflection grating, it separates the incident beam into many diffraction lights at the diffraction angle αm (m = 1, 2, 3….: diffraction order). The SPM is tilted by the angle αm to make sure that the selected mth order diffraction light will return to the same direction of the incident beam. The returned beam will be merged with the reflected light from the measurement arm of the Michelson interferometer. In Fig. 1, the scattered lights at a position on the SPM (green dash-line) are collected and focused on the CCD camera by the 4-f configuration of the PCX lenses. These lenses are used to optimize the quality of the interference images. On the other measurement path, the beam passes to a cylindrical lens (CYL) and a plano-convex lens (PCX) before reaching to a sample. This lenses configuration creates a line-focused on the sample. The reflection beam from the sample is re-combined with the beam from reference arm. The generated interference images will be seen on a CCD camera. Owing to the line-focused on the sample with 4-f lenses configuration, the lateral resolution of the interference images will be improved. In the top view of the optical setup in Fig. 1, every single point on the SPM (from left to right) means a line from the top to the bottom of the SPM. It means that if the interference resonant condition is satisfied at a point on the SPM in the top view Fig. 1, then a set of points on a line from the top to the bottom of the SPM will also satisfy the interference resonant condition. Therefore, on the CCD camera, line-shape interference fringes will be seen. Different position of the sample will generate different line – shape fringes at different location on CCD camera. This makes single-shot operation become reliable.

The measurement range in axial axis for single-shot operation is estimated from the grating equation:, where d, α_i, α_m, λ are the grating pitch, incident angle, the mth order diffraction light, and the center wavelength of the beam, respectively.

Because the mth order diffraction beam comebacks to the same direction of the incident beam, the condition must be satisfied. Therefore, the diffraction angle α_m can be calculated by

$$\alpha_m = \sin^{-1}\left(\frac{m\lambda}{2d}\right) \tag{1}$$

The measurement range in axial axis (Δz) is calculated by: $\Delta z = D\tan(\alpha_m)$ where D is the diameter of the incident beam (Fig. 1a). From (1), measurement range is derived by

$$\Delta z = D\tan\left[\sin^{-1}\left(\frac{m\lambda}{2d}\right)\right] \tag{2}$$

From Eq. (2), if the diffraction order m is varied, the measurement range Δz will be changed. It means that when the SPM generates more diffraction orders, the dynamic measurement range will change in a longer range. Changing the diffraction order m can be done by rotating the SPM at the step angle of ($\alpha_{m+1} - \alpha_m$), this is similar to the zoom in and zoom out effect in the normal digital camera but in axial axis. Figure 1(a) shows the two difference positions of SPM, the green line corresponding to the diffracted angle α_1 and the violet line corresponding to the diffracted angle α_2. ($\alpha_2 > \alpha_1$). By rotating the SPM from α_1 to α_2, the measurement range can expand from $\Delta z_1 = D\tan(\alpha_1)$ to $\Delta z_2 = D\tan(\alpha_2)$.

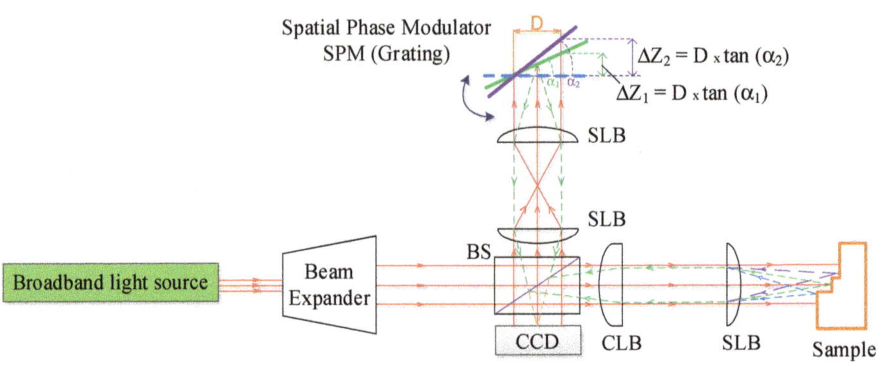

Fig. 1. Schematic of the proposed system

By changing the diffraction orders of the SPM, the measurement range can be flexibly changed. It will be very helpful for the setting the sample position before a measurement, which is typically take time for a conventional whitelight interferometry system. For the conventional white-light interferometry system, the optical path length difference (OPD) between the reference arm and signal arm of the system must be smaller than the coherence length of the light source. Which mean, the setting position of sample has typically limited in few micrometers. And it takes long time to set the initial position of the sample in conventional system. In our proposed system, with the changeable measurement range, the initial setting position of sample can be expanded to few millimeters. So that it will effectively reduce the setting time for the sample measurement. In case if the size of the sample is bigger than the measurement range of initial system setup.

3 Experimental Results

3.1 The Measurement Range Expansion

In this part, we will investigate the relationship of the SPM diffraction orders to the measurement range, giving the zooming operation of our proposed tomography and profilometry. For the demonstration, blaze grating as a SPM, which has 79 LPM and 63^0 blaze angle, was utilized. From Eq. (1) and (2) in the second part, we can simulate the measurement range vs the SPM order. To compare with the simulation, an experiment on the relationship between the SPM order and the measurement range was performed using a reference 100 μm step-height sample. Figure 2 shows a series of the 2D interference fringe image of the step-height sample at different SPM orders. The SPM diffraction order was switched by consequently rotating the SPM to its corresponded diffraction beam that entered to the CCD camera.

The measurement range is defined by the range from the left side to the right side of the 2D images. In Fig. 2, we can see that the measurement range varies by the SPM order. Figure 3 shows the relationship between the SPM orders and the measurement range. The experimental result of the measurement of the dynamic range and that of the simulation, which based on Eq. (2), were very similar. It is clear that the dynamic measurement range can be changed by controlling the SPM order.

In the experiment, total 20th diffraction lights were generated by the SPM compared to the 8th diffraction order of the fine - scale of our previous work [14]. It meant that the measurement range could be changed from $1\times$ to $20\times$, corresponding to the range from 200 μm ($1\times$) to 3.5 mm ($20\times$). Therefore, the proposed system can expand the axial measurement range to several mm without using optical frequency comb. The zooming function is useful for the inspection in industry since it gives a flexible way

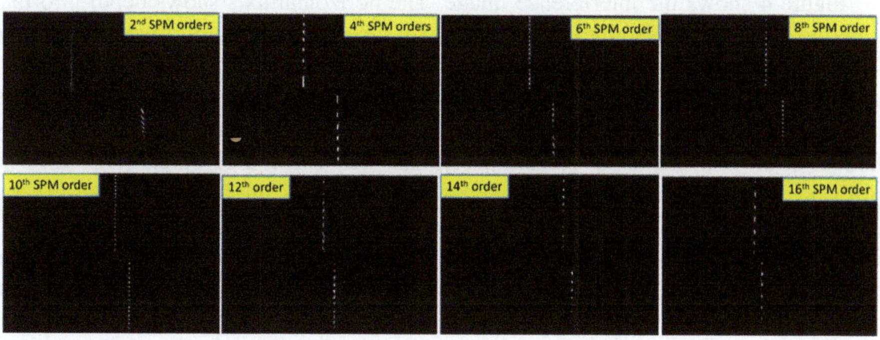

Fig. 2. 2D interference fringes images of 100 μm the step-height sample at different SPM diffraction orders

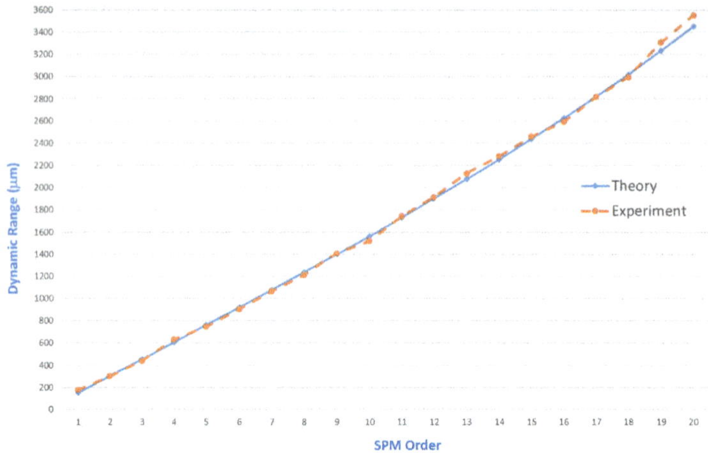

Fig. 3. Relationship between the SPM diffraction orders and the measurement range of the tomography and profilometry

for the users to monitor various depths of sample, especially for the real-time measuring system.

3.2 Calibration for the Zooming Single-Shot System

To calibrate the resolution of the tomography and profilometry of our proposed system, a reference step-height sample, of Mitutoyo of 20 µm ± 0.2 µm, was used. For the demonstrated SPM, the energy is focused on 3rd and 4th diffraction order. Hence, the 3rd or 4th diffraction order of SPM has ability to create the high intensity for 2D interference fringe image. But the 4th order of SPM can provide higher measurement range than one of 3rd order. So that, for easier setting, the SPM was set at 4th order. The incident light is line-focused on the sample (expressed by the red line in Fig. 4). Owing to the two reflection surfaces of the step sample, line-shape interference fringes will be generated and displayed on the CCD camera.

Figure 4 shows the interference image. The horizontal axis shows depth information (the height of the sample), the vertical axis shows the lateral information along the focused beam on the sample. Two lines-shapes interference fringes, corresponding to two reflection surfaces on the sample can be easily seen. The distance between two line-shape interference fringes means the step-height of the sample, which is already known of 20 µm. The CCD camera pixels size was 960 × 687 pixels (H x V). The pixel size of the horizontal axis is then calculated by analyzing the line profiles (1) and (2) in Fig. 4. Figure 5 shows the plot of the lines profile (1) and (2). Full-width at half maximum (FWHM) was then estimated by approximate 2 µm. The calculated axial resolution of the tomography and profilometry from the experimental result shows an agreement with the theoretical estimation (~1.8 µm). The profilometry resolution is estimated by a single pixel size in horizontal axis. It was calculated approximately of 0.6 µm from the line-profile analysis. The lateral resolution of the system was

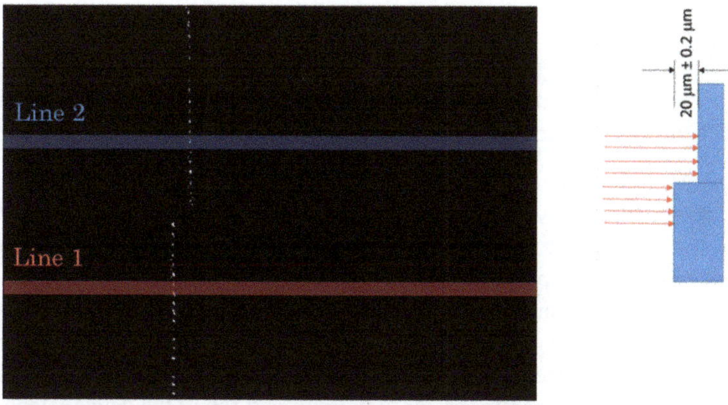

Fig. 4. 2D interference image of 20 μm reference step-height sample at 4th diffraction order of SPM.

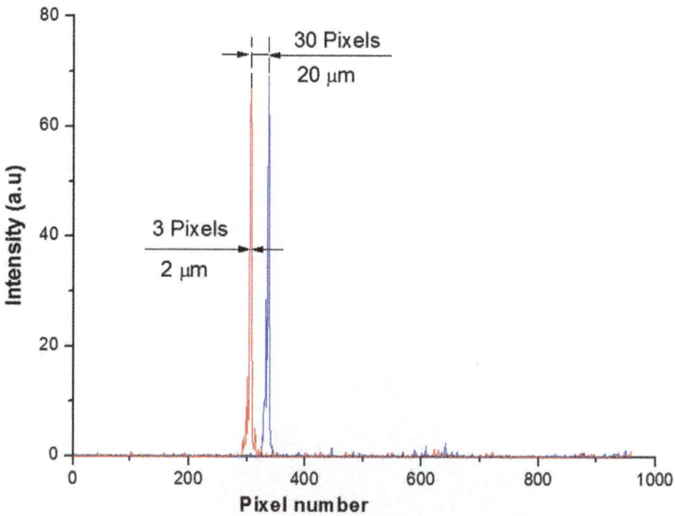

Fig. 5. Line-profiles plot of 20 μm reference step-height sample.

estimated of 17 μm by using a standard test target on which the known dimensional patterns were measured.

Theoretically, the tomogram resolution depends on the bandwidth of the light source and the working distance between CCD to the sample. When the working distance become longer, the spectral losses will be increased due to the limit size of the CCD sensor and the diffraction angle of each diffracted light. Therefore, once the diffraction order is larger, it may cause the axial resolution of the system become lower. This matter will not be a problem because we can use the other bigger size CCD sensor to reduce the spectral loss of the light source.

3.3 Demonstration for Profilometry and Tomography of the Zooming Single-Shot

Several measurements to demonstrate the proposed profilometry were performed after the experiment to calibrate the axial resolution.

3.3.1 Demonstration for Profilometry of a Highly Scattered Sample

The first demonstration was measured the 3D surface profile of a highly scattered sample. A fraction area on the 10 Japanese yen coin was used as a sample. The Coin is representative for the sample, which has very small profile and bumpy surface. A lot of the energy of incident beam will be diffused by the that surface. In this case, there are two requirements for the SPM. First, diffraction order, which has high energy efficiency, is requested. Second, the diffraction order can zoom out the measurement region on sample for the operator easier to monitor and found the defect. Hence, the 3rd order of SPM, which can provide high energy efficiency as well as the higher magnification in comparison with 4th order, was used. Figure 6 shows the coin sample indicating where and how incident beams were focused. A single line corresponded to a 2D interference profile image. Figure 7 shows an example of 4 line-interference fringe images corresponded to 4 positions on the sample (1st, 2nd, 3rd and 4th in Fig. 6, respectively). By moving the sample along the horizontal axis, all 2D images were recorded and used to re-construct a 3D profile image of the coin. Figure 8 shows the 3D surface profile that was reconstructed from a series of 2D by our developed software. The information of a cross section of the coin within the measurement area was estimated and clearly showed in Fig. 9. It was confirmed that by using $4 - f$ configuration with a cylindrical lens for our system, the scattered lights from the sample were effectively eliminated.

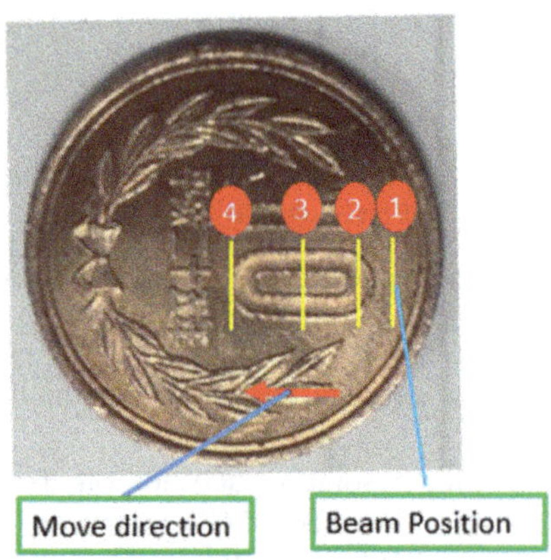

Fig. 6. 10 Japanese yen coin sample

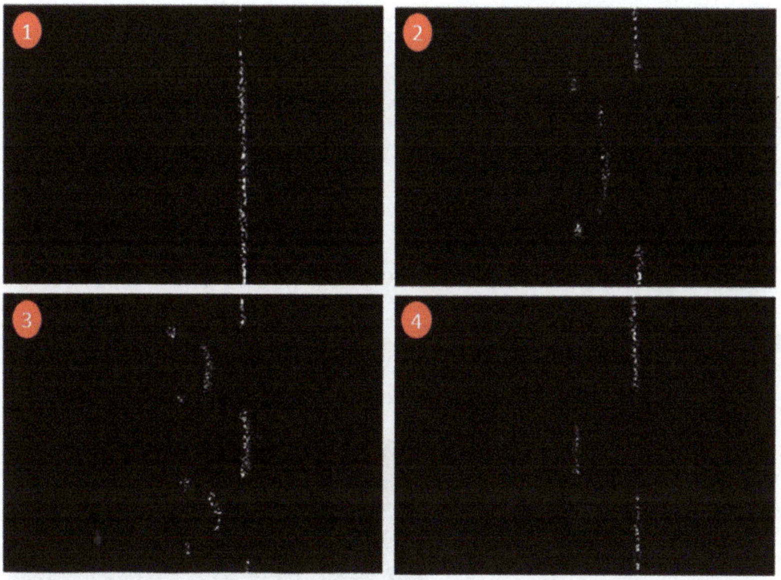

Fig. 7. 2D single-shot interference images at different location of the 10 Japanese yen coin using 3rd order of SPM

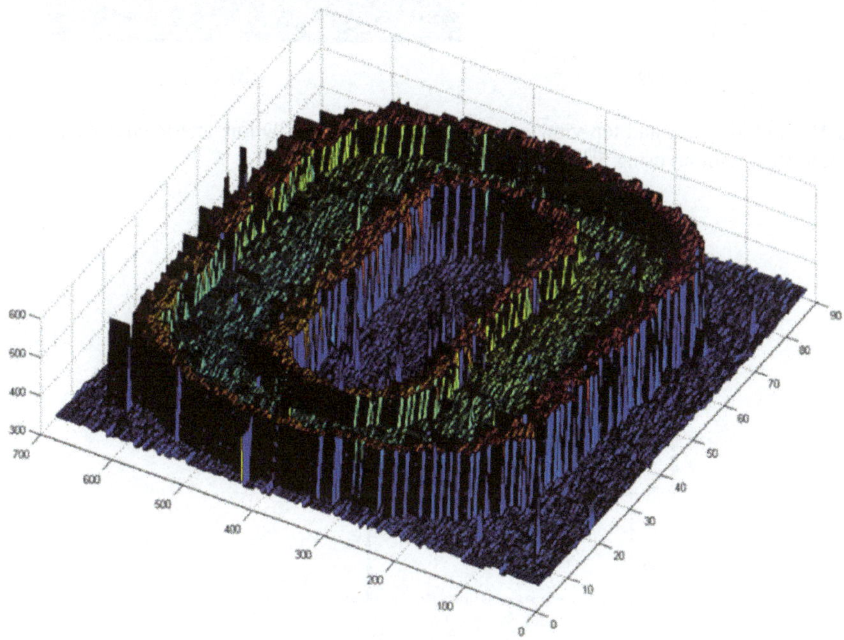

Fig. 8. Reconstructed 3D surface profile of number 0 of the 10 Japanese yen coin

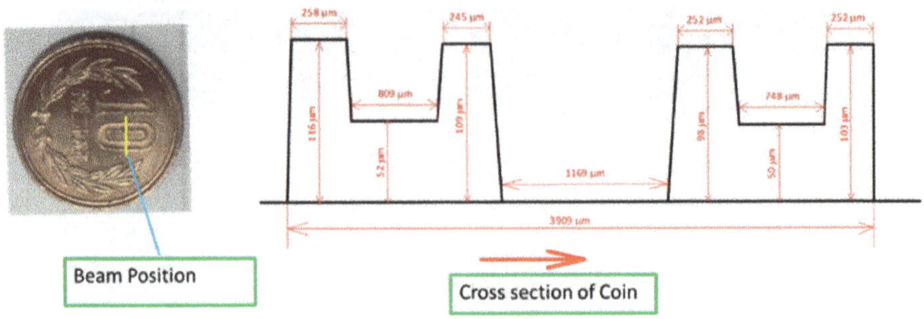

Fig. 9. Cross-section of a surface profile of number 0 of the 10 Japanese yen coin

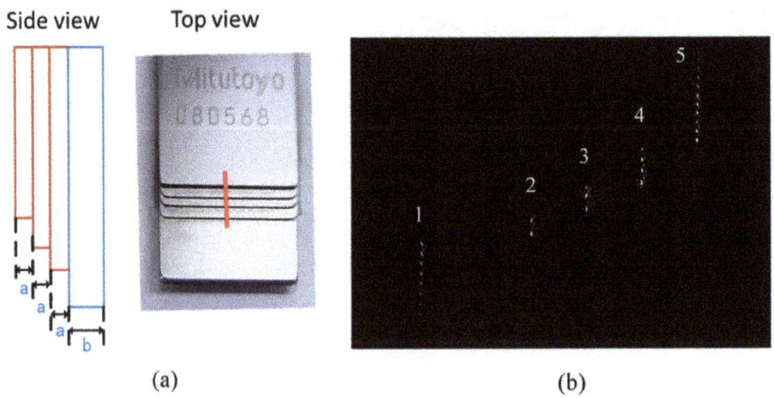

Fig. 10. a) Side-view and top-view of multi-steps sample. b) 2D line-interference fringe image of multi-steps sample using 5th order of SPM

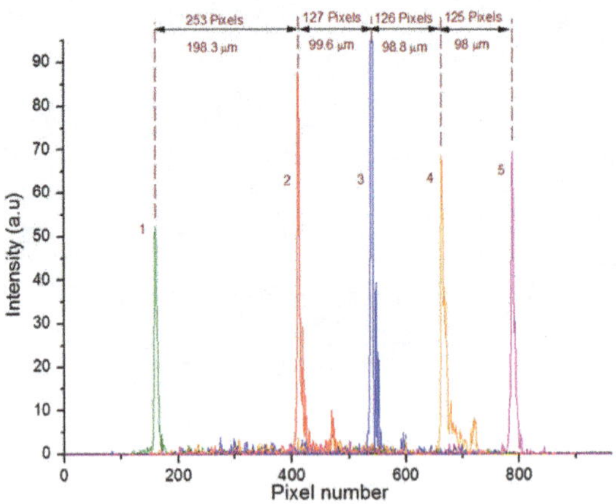

Fig. 11. Line-profile of the multi-steps sample

3.3.2 Demonstration for Profilometry of a Big Size Sample

Next step, the second sample for the profilometry demonstration was a multi-steps sample, which was made by several 100 μm and 300 μm Gauge-blocks. Figure 10(a) shows a photo of the sample in both side-view and top-view. There were three Gauge blocks whose step-height were 100 μm, and one Gauge block whose step-height was 200 μm. The incident beam was line-focused, which crossed 5 reflection surfaces on the sample. The sample has flat surface and over 500 μm size in axial axis (depth direction). When the SPM set at 3rd order, the measurement range of system was around 460 μm in axial axis. It was suitable for coin sample but is not sufficient for the multi-step sample. It's necessary to increase the measurement range of the system. Instead of replace the optical set up such as increase the beam diameter or CCD camera size, the zooming single-shot system can provides the most convenience way to expand the measurement range by just turning the SPM until order 5th, we can easy upgrade the measurement up to 750 μm. Figure 10(b) shows the 2D interference profile image of the multi-steps sample when the order 5th of SPM was used. We can see all five line-interference fringes corresponding to five reflection surfaces of the step sample. The position of the line-interference fringe is derived from analyzing their corresponded line-profiles. Figure 11 shows the calculated step-heights of the sample. The step-height results were 198.3 μm, 99.6 μm, 98.8 μm, and 98 μm, respectively. It showed a good result compare to the real size of the. The demonstration above also confirmed the efficiency of the zooming single-shot system to measure with variable size of difference samples.

4 Conclusion

The 2D single-shot tomography and profilometry using broadband light source and spatial phase modulator were presented. The tomogram resolution (2.0 μm) and profile resolution (0.6 μm) were experimentally achieved. These resolutions are 5 times better than our previous developed system [11]. The 3D reconstructed surface profile of the highly scattered sample was also demonstrated by our developed data processing software that processes a series of 2D interference images acquired by this system. The effectiveness of the diffraction orders of the SPM to the dynamic measurement range was investigated and confirmed by the experiment. The comb-less zooming profilometry has been demonstrated with the magnification of 20x based on the effectiveness of the SPM order was also presented. We have provided a useful technique for the industrial applications where requires fast and continuous data acquisition process.

For the further development related to the measurement range, we will develop this zooming tomography and profilometry by using optical frequency comb technology, the expected dynamic measurement range will be possible to be extended up to 100 mm based on the theoretical estimation that we have done before in the previous reports. This work was partially supported by Ministry of Education and Training under grant number B2020-BKA-16.

References

1. Minsky, M.: Memoir on inventing the confocal scanning microscope. Scanning **10**, 128–138 (1988)
2. Aung, S.C., Ngim, R.C.K., Lee, S.T.: Evaluation of the laser scanner as a surface measuring tool and its accuracy compared with direct facial anthropometric measurements. Br. J. Plast. Surg. **48**, 551–558 (1995)
3. Wyant, J.C.: Advances in interferometric surface measurement. Proc. of SPIE **6024**, 602401–602411 (2005)
4. Farahi, F., Jackson, D.A.: A fibre optic interferometric system for surface profiling. Rev. Sci. Instrum. **61**, 753 (1990)
5. Huang, D., Swanson, E.A., Lin, C.P., Schuman, J.S., Stinson, W.G., Chang, W., Hee, M.R., Flotte, T., Gregory, K., Puliafito, C.A., Fujimoto, J.G.: Optical coherence tomography. Science **254**, 1178–1181 (1991)
6. Leitgeb, R., Wojtkowski, M., Kowalczyk, A., Hitzenberger, C.K., Sticker, M., Fercher, A.F.: Spectral measurement of absorption by spectroscopic frequency-domain optical coherence tomography. Opt. Lett. **25**, 820–822 (2000)
7. Kaori, S., Sumit, S., Takuhiro, Y., Peter, K.K.: Comparison of spectral-domain versus time-domain optical coherence tomography in management of age-related macular degeneration with Ranibizumab. Ophthalmology **116**, 947–955 (2009)
8. Chandan, S.R., Vishal, S.G.: Optical Coherence Tomography Systems and signal processing in SD-OCT, Special issue for International Technological Conference ITechCON-2014, EXTC-04 (2014). ISSN 2319-4847
9. Choi, S., Yamamoto, M., Moteki, D., Shioda, T., Tanaka, Y., Kurokawa, T.: Frequency-comb-based interferometer for profilometry and tomography. Opt. Lett. **31**, 1976–1978 (2006)
10. Choi, S., Tamura, N., Kashiwagi, K., Shioda, T., Tanaka, Y., Kurokawa, T.: Supercontinuum comb generation using optical pulse synthesizer and highly nonlinear dispersion-shifted fiber. Jpn. J. Appl. Phys., **48** (2009). 09LF01
11. Shioda, T., Morisaki, T., Banh, T.Q., Suzuki, K.: two dimensional single-shot tomography using a virtually imaged phased array and a spatial phase modulator. Appl. Opt. **51**, 5224–5230 (2012)
12. Banh, T.Q., Suzuki, K., Shioda, T.: Development of an incoherent optical frequency comb interferometer for longrange and scanless profilometry and tomography. Opt. Commun. **296**, 1–8 (2013)
13. Banh, T.Q., Suzuki, K., Kimura, M., Shioda, T.: Two-dimensional high-speed and long-range tomography and profilometry using liquid-crystal Fabry-Perot resonator. Appl. Opt. **54**, 912 (2015)
14. Truong, T.C., Banh, T.Q., Kim, H., Shioda, T.: Axial zoomable 2D single-shot comb interferometry using spatial phase modulator for profilometry and tomography. Optics Commun. **439**, 276–283 (2019)

The Development of 2D Slam-Based Navigation for an Autonomous Nursing Robot in Global Covid 19 Period

Thi Thoa Mac[1(✉)], Nguyen Trong Doanh[1], Pham Nguyen Trung Hieu[1], and Hoang Van Quy[2]

[1] School of Mechanical Engineering, Hanoi Universiy of Science and Technology, Hanoi, Vietnam
thoa.macthi@hust.edu.vn

[2] Hai Phong University, Haiphong, Vietnam

Abstract. Recently, robotic nurses or nurse robots during Covid 19 period. Many studies show that robots can provide necessary support and remedies in hospital. The nurse robots can assist patients/doctors or perform some common tasks such as taking vital signs or delivering medicine. In this study, the development of 2D-Slam based navigation for an autonomous nursing robot is proposed which includes mapping and localization system to cope with challenging hospital scenarios. Our system is able to maintain a high quality global map as well as local map when the robot faces to obstacles. The simulation results in Robot Operation System (ROS) show that the proposed approach can achieve 2D Slam-based navigation system in hospital working environment.

Keywords: Nursing robot · 2D Slam · Navigation · Autonomous robot · COVID-19

1 Introduction

Today, robotic nurses (nurse robots) are robots that help patients physically move around or perform simple task like taking vital signs or delivering medicine. Some nurse robots serve as interfaces for doctors to use over distances to communicate with patients [1].

In these days, Coronavirus disease (COVID 19) is very dangerous. Corona virus disease caused by a newly discovered coronavirus. It is easily transmitted through droplets of infected person. People can be infected COVID-19 by breathing in the virus or by touching a contaminated surface. There are 5,404,512 confirmed cases have been announced over the world with 343,514 casualties [2]. Therefore, the help of nurse robot to prevent transmit or spread of the virus is very important.

Recently, there are various robotic systems which used to support doctors/nurses/patients in hospitals [3]. For example, the robot can observe patient regularly like a nurse. With the specific devices, robot can take people temperature, blood pressure. The patient's profiles are stored in the memory of the robot. The camera and screen install on the robot can help patients and doctors to interface to each other.

B. T. Long et al. (Eds.): MMMS 2020, LNME, pp. 1053–1060, 2021.
https://doi.org/10.1007/978-3-030-69610-8_138

In [4], a ambulance robot with the wheel system and capability of locating of some devices such as infrared sensor and ultrasonic can help the nurse robot easily to move in the crowed place like hospital.

In [5] the nurse robot has capability of delivery food, water and medicines to the patients is appeared in the hospital for various patients. This robot is controlled through internet connection and be observed by camera. These two robots above are controlled manually through internet connection.

In this study, we propose a complete 2D mapping and localization process for nursing robot. The robot is able to (1) incrementally building 2D map and localizing using 2D point cloud, (2) the robot can autonomous go to the target position and be able to avoid obstacles effectively.

The paper is organized as follows: Sect. 2 presents 2D SLAM-Based Navigation. Section 3 introduces the simulation results in Ros-Gazebo and discussion. Section 4 is a conclusion and future work section with the main outcome of this work.

2 2D SLAM-Based Navigation

2.1 LiDAR-Based SLAM

Robots are able to explore and map while estimating its orientation and position based on internal sensors' information by SLAM (Simultaneous Localization and Mapping) [6]. Encoders, IMU (inertial measurement unit), GPS (global positioning system), cameras are typically used for pose estimation of an autonomous robot. The most common estimation methodologies are Kalman filter, Markov localization, Monte Carlo. Robots use distance sensors such as infrared scanners, light detectors, ultrasonic sensors, laser range finders, radio detectors for mapping purpose. Besides, cameras and liars are more advanced sensors to obtain the surrounding environment of robots.

The localization (pose estimation) is a very important part in robot's navigation and mapping and is being actively studied. In theory, if the localization process is implemented properly then SLAM can be easily executed. However, in real-time process, robots face with many problems such as uncertainty of the sensor observation information, the low sensors' accuracy, noises.

For the nursing robot application in this application, the maximum distance of Rplidar-A1 rangefinder (500 cm) is used for the measurements. To create the robot map, the robot working environment is divided to cells. In this application, the grid cell size is 5 cm and the sensor is located at cell (0, 0). The grays area are the unexplored and the unknown cells, the white area are the obstacle-free cells and the black area are the occupied cells as shown in Fig. 1.

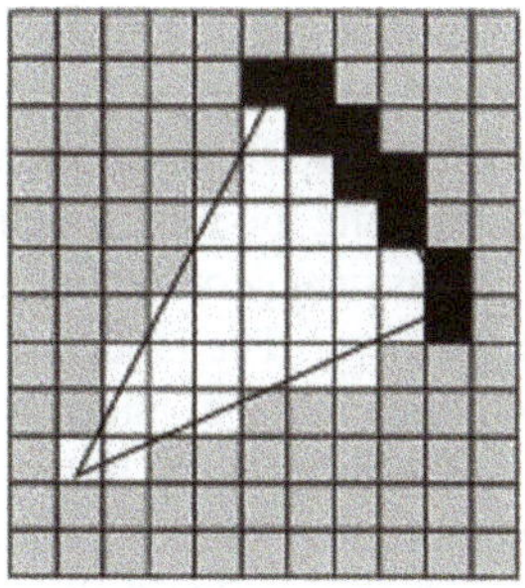

Fig. 1. LIDAR Data point occupancy grid map

3 Simulation Results in Ros-Gazebo and Discussion

3.1 World Description

For our nursing robot, we assume the robot is served in a room of the hospital which has several patients' beds and other furniture. The robot missions are building room map, avoiding obstacles and reaching the designed location. For simulation purpose, we use the ".world" file which will build an environment with walls and obstacles. The robot can autonomously avoids obstacles and goes to the patient's bed to deliver the medicine. For simulating, we used the "corridor.world" file from gazebo's library. Figure 2 presents the hospital's room in Gazebo simulation.

Fig. 2. Simulation on Gazebo

3.2 Building the Differential Drive Mobile Robot URDF

Two basic URDF components are used to define a tree structure that describes a robot model. The link component describes a rigid body by its physical properties (dimensions, the position of its origin, color, and so on). Links are connected together by joint components. Joint components describe the kinematic and dynamic properties of the connection (that is, links connected, types of joint, the axis of rotation, amount of friction and damping, and so on). The URDF description is a set of these link elements and a set of the joint elements connecting the links together. The first component of the robot is a simple chassis box. The physical geometry of each link needs to be defined in the form of visual and collision geometry. The model of the robot includes a cubic box presenting the base footprint, body link and 2 cylindrical shapes presenting two wheels. The robot wheels are attached to the robot body by using continuous joints.

Beside, the robot's inertial parameters includes the mass, center of the mass, and the moment of inertia tensor matrix which are obtained from Inventor. The model of the robot is presented in Fig. 3.

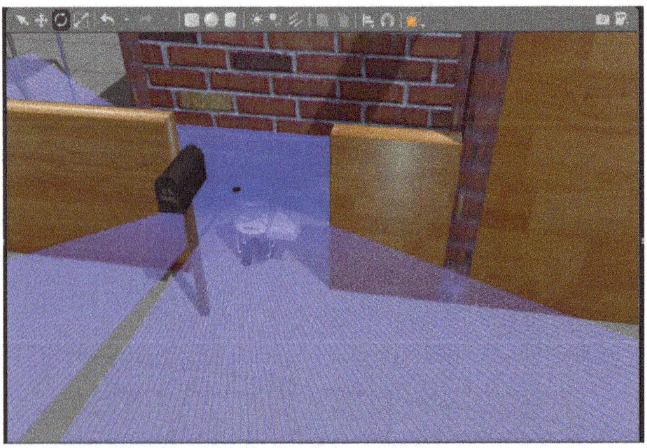

Fig. 3. Simulation model of robot

3.3 Sensor Modelling and ROS Navigation

With many different sensors on the robot, each sensor can be simulated independently as a Gazebo plugin and must be physically attached to the robot model as a link. These plugins are included in the URDF file. These plugins output sensor information in form of standard ROS messages and services. Common parameters, such as error characteristics and transformation frame of each sensor, can be defined as well as other parameters related to each sensor. By default, sensors in Gazebo have no noise and they sense the simulated environment perfectly. To make them more realistic, noise can be added.

We have created a simulated differential drive model that satisfies the necessary pre-requisites for autonomous navigation: the transform tree, odometry, laser scanner, and base controller. We are now making our simulated robot move around on its own by using ROS's navigation Meta-package as shown in Fig. 4. In this part, we are going to create a map to navigate the robot with the map: (*i*) Plug in the ROS navigation Meta-package; (*ii*) Create a map; (*iii*) Navigation with the map.

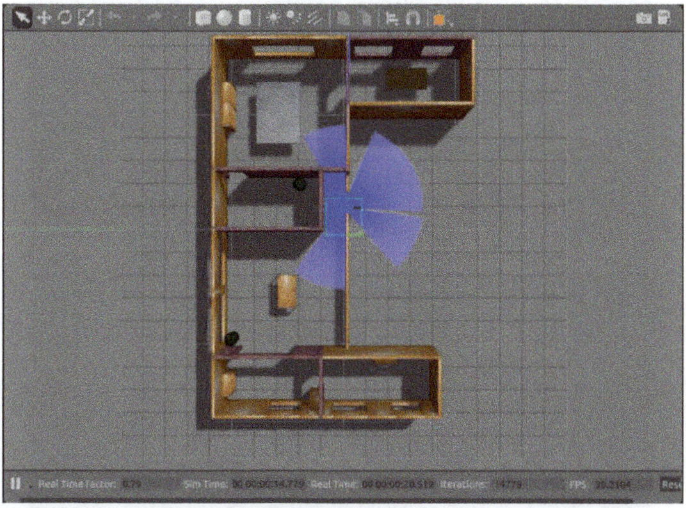

Fig. 4. Simulation model of robot

3.4 Implement the Dijkstra into ROS with Navigation Meta-package

The navigation stack of the Robot Operating System (ROS) open-source middleware incorporates both global and local path planners to support ROS-enabled robot navigation. However, only basic algorithms are defined for the global path planner including Dijkstra, A*, and carrot planners. In this part, we present Dijkstra global path planner into the ROS navigation system. Figure 5 presents the trajectories from current robot position to the destination.

Figure 6 depicts the situation when Liar sensor scans a new obstacles. The left pictures present the global paths with known obstacles and the right pictures presents the updated local paths with sensed new obstacles. The robot is successfully to find the shortest path to avoid that obstacles and go to the goal.

Fig. 5. Testing Dijkstra algorithm with known obstacles

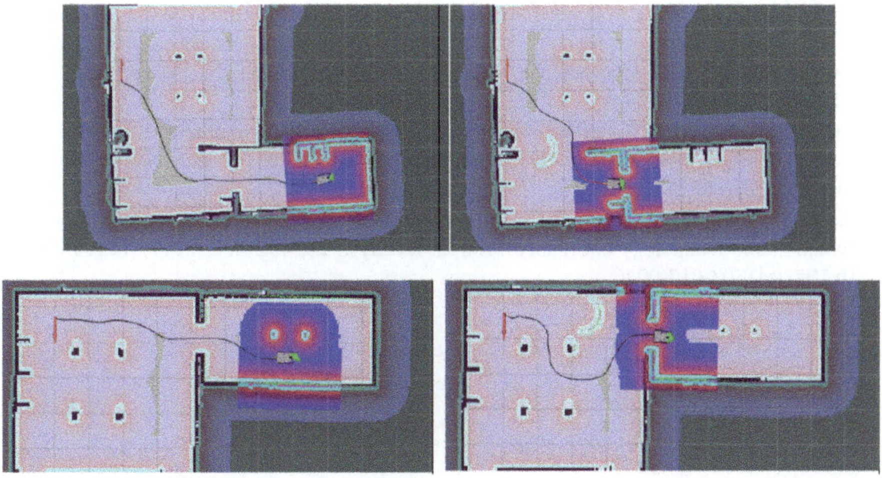

Fig. 6. Testing Dijkstra' algorithm with unknown obstacles

4 Conclusion

This research aims to develop 2D Slam-based navigation for an autonomous nursing robot. The approach includes mapping and localization system to cope with very crowded hospital working environment. The system is capable to maintain a high quality global map as well as an updated local map when the robot near to obstacles. The simulation is performed in Robot Operation System (ROS). The results show that the robot can easily to get 2D information of working space through point cloud. Based on this information, the robot is able to generate 2D map and move easily in hospital working environment. This research is a crucial study for the development of autonomous nursing robot. It shows the robotics potential in medicine and hospital areas with special meaning to the handle of the COVID-19 pandemic. Future work includes an extension of the proposed approach on the real robot system.

Acknowledgments. This work is granted by the National Foundation for Science and Technology Development of Vietnam NAFOSTED, Vietnam under project number 107.01-2018.331.

The authors would like to gratefully thank Nguyen Manh Cuong and Chu Ngoc Son for their assistance in conducting the experiments in which some results were part of their thesis guided at our lab, Hanoi University of Science and Technology, Vietnam.

References

1. Khan, Z.H., Siddique, A., Lee, C.W.: Robotics utilization for healthcare digitization in global COVID-19 Management. Int. J. Environ. Res. Public Health. **17**(11), 3819 (2020)
2. World Health Organization Coronavirus Disease 2019 (COVID-19) Situation Report, 127 (2020). https://www.who.int/docs/default-source/coronaviruse/situation-reports/20200526-covid-19-itrep-127.pdf?sfvrsn=7b6655ab_8. Accessed 27 May 2020

3. Vänni, K.J., Salin, S.E.: A need for service robots among health care professionals in hospitals and housing services. In: Kheddar, A., et al. (eds.) Social Robotics, ICSR 2017. Lecture Notes in Computer Science, vol. 10652. Springer, Cham (2017). https://doi.org/10.1007/978-3-319-70022-9_18
4. Samani, H., Zhu, R.: Robotic automated external defibrillator ambulance for emergency medical service in smart cities. IEEE Access. **4**, 268–283 (2016). https://doi.org/10.1109/ACCESS.2016.2514263
5. Mettler, T., Sprenger, M., Winter, R.: Service robots in hospitals: new perspectives on niche evolution and technology affordances. Eur. J. Inf. Syst. **26**, 451–468 (2017). https://doi.org/10.1057/s41303-017-0046
6. https://wiki.ros.org/.

Author Index

B. T. Long et al. (Eds.): MMMS 2020, LNME, pp. 1061–1065, 2021.
https://doi.org/10.1007/978-3-030-69610-8

Lightning Source UK Ltd.
Milton Keynes UK
UKHW020630310321
381298UK00001B/1